# 3rd International Conference on Frontiers of Materials Synthesis and Processing (FMSP 2019)

AA007109

IOP Conference Series: Materials Science and Engineering
Volume 772

Sanya, China
9-10 November 2019

ISBN: 978-1-7138-4710-6
ISSN: 1757-8981

**Printed from e-media with permission by:**

Curran Associates, Inc.
57 Morehouse Lane
Red Hook, NY 12571

**Some format issues inherent in the e-media version may also appear in this print version.**

This work is licensed under a Creative Commons Attribution 3.0 International Licence.
Licence details: http://creativecommons.org/licenses/by/3.0/.

No changes have been made to the content of these proceedings. There may be changes to pagination and minor adjustments for aesthetics.

Printed with permission by Curran Associates, Inc. (2022)

For permission requests, please contact the Institute of Physics at the address below.

Institute of Physics
Dirac House, Temple Back
Bristol BS1 6BE UK

Phone: 44 1 17 929 7481
Fax: 44 1 17 920 0979

techtracking@iop.org

**Additional copies of this publication are available from:**

Curran Associates, Inc.
57 Morehouse Lane
Red Hook, NY 12571 USA
Phone: 845-758-0400
Fax: 845-758-2633
Email: curran@proceedings.com
Web: www.proceedings.com

# TABLE OF CONTENTS

IOP Conf. Series: Materials Science and Engineering **772** (2020) 011001 doi:10.1088/1757-899X/772/1/011001

# Preface

This proceeding includes original and peer-reviewed research papers from the 3rd International Conference on Frontiers of Materials Synthesis and Processing (FMSP 2019), which held from 9 to 10 November 2019, in Sanya City, China.

FMSP 2019 aims at creating a meeting point for researchers and practitioners that from both China and abroad to exchange ideas in the domain of Materials Synthesis and Processing and related topics with its prospective application in the various interdisciplinary domains of engineering, including not only technological issues but also social issues.

FMSP 2019 received 791 manuscripts, and 120 submissions had been accepted by our reviewers. Articles submitted to the conference should report original, previously unpublished research results, experimental or theoretical and must not be under consideration for publication elsewhere. We firmly believe that ethical conduct is the most essential virtual of any academic. Hence any act of plagiarism is a totally unacceptable academic misconduct and cannot be tolerated. The committees of FMSP 2019 invested great efforts in reviewing the papers submitted to the conference and organizing the sessions to enable the participants to gain maximum benefit.

Hopefully, all participants and other interested readers will benefit scientifically from the proceedings and also find it stimulating in the process.

With our warmest regards,

Lin Liu

Content from this work may be used under the terms of the Creative Commons Attribution 3.0 licence. Any further distribution of this work must maintain attribution to the author(s) and the title of the work, journal citation and DOI.
Published under licence by IOP Publishing Ltd

IOP Conf. Series: Materials Science and Engineering **772** (2020) 011001 doi:10.1088/1757-899X/772/1/011001

Conference Organizing Chair

Wuhan, China

IOP Conf. Series: Materials Science and Engineering **772** (2020) 011002 doi:10.1088/1757-899X/772/1/011002

# Committees

**Conference Chairmen**

Prof. M.H. Cao, Huazhong University of Science and Technology, China
Prof. S.Y. Li, Central South University, China
Prof. L. Liu, Wuhan University of Technology, China

**Scientific Committee**

Prof. W.L. Su, China University of Geosciences, China
Prof. B.W. Lai, Central South University, China
Prof. T.X. Tu, Central South University, China
Prof. Z.H. Zhu, Central South University, China
Prof. X.Q. Shi, Zhejiang University, China
Prof. X.Q. Zheng, The Institute of Physics, Chinese Academy of Sciences, China
Prof. G.D. Sun, The Shanghai Institute of Ceramics of the Chinese Academy of Sciences, China
Prof. H.G. Ye, Wuhan University of Technology, China
Prof. Q.R. Deng, Wuhan University of Technology, China
Prof. L.M. Long, Wuhan University of Technology, China
Prof. H.Y. Zhu, Chongqing University, China
Prof. C.Q. Shi, Wuhan University, China
Prof. Y.L. Tan, Wuhan University, China
Prof. Z.W. Zhu, Hunan University, China
Prof. D.X. Zhou, Hunan University, China
Associate Prof. S.B. Hu, Wuhan University of Technology, China
Dr. Q. Chou, University of Birmingham, UK
Dr. T.H. Wu, Purdue University, USA
Dr. Tse Guan Tan, Universiti Malaysia Kelantan, Malaysia
Dr. Lei Jiang, Nanjing University of Science and Technology, China

**Organizing Committee**

Prof. L. Liu, Wuhan University of Technology, China
Prof. H.G. Ye, Wuhan University of Technology, China
Dr. Z.Y. Su, Central South University, China
Dr. Q.P. Liu, Central South University, China
Dr. J.W. Zhang, Central South University, China
Dr. S.Z. Hu, Huazhong University of Science and Technology, China
Dr. X.L. Wu, Huazhong University of Science and Technology, China
Dr. X.Z. Su, Wuhan University of Technology, China
Dr. Y.M. Feng, Wuhan University of Technology, China
Dr. W.W. Zhang, Wuhan University of Technology, China
Mr. H.T. Ke, Hubei New Wen Sheng Conference Co., Ltd., China
Mr. R.L. Xia, Hubei New Wen Sheng Conference Co., Ltd., China

Content from this work may be used under the terms of the Creative Commons Attribution 3.0 licence. Any further distribution of this work must maintain attribution to the author(s) and the title of the work, journal citation and DOI.
Published under licence by IOP Publishing Ltd

IOP Conf. Series: Materials Science and Engineering **772** (2020) 011002 doi:10.1088/1757-899X/772/1/011002

Mr. X.T. Su, Hubei New Wen Sheng Conference Co., Ltd., China
Mr. D.S. Zhao, Hubei New Wen Sheng Conference Co., Ltd., China
Ms. Y.F. Ma, Hubei New Wen Sheng Conference Co., Ltd., China
Ms. B.M. Zhou, Hubei New Wen Sheng Conference Co., Ltd., China

| No. | First name | Surname | Institution | Email address |
|---|---|---|---|---|
| 1 | Qian | Jia | Grikin Advanced Materials Co. Ltd., Beijing, China | jq@grikin.com |
| 2 | Liwen | Zhang | Xi'an Aeronautical Polytechnic Institute, Xi'an, 710089, China | 375350782@qq.com |
| 3 | Jiping | Song | Chinese Flight Test Establishment, Xi'an, 710089, China | 595513529@qq.com |
| 4 | Jinshun | Xue | Hubei University of Arts and Science. Xiangyang, China | jinshunx@chd.edu.cn |
| 5 | Zhengxun | Yang | Xinjiang Communication Construction Group Co., LTD, Urumqi, China | 16709263@qq.com |
| 6 | Liming | Dong | School of Chemistry and Chemical Engineering, Xuzhou University of Technology, Xuzhou, China | polymerdlm@163.com |
| 7 | Junshan | Guo | State Grid ShanDong Electric Power Research Institute, Jinan, 250003, China | junshan_guo@sina.com |
| 8 | Jianyi | Zheng | School of Aerospace Engineering Xiamen University Xiamen, China | zjy@xmu.edu.cn |
| 9 | Yuting | Zhang | School of Aerospace Engineering Xiamen University Xiamen, China | 35120191151197@stu.xmu.edu.cn |
| 10 | Zhaobin | Su | School of Aerospace Engineering Xiamen University Xiamen, China | 35120171150989@stu.xmu.edu.cn |
| 11 | Shihui | Yao | National Key Laboratory on Ship Vibration and Noise, Wuhan, China | yaoshihuicn@foxmail.com |
| 12 | Linghui | Kong | The Yellow River Institute of Hydraulic Research, the Yellow River Water Conservancy Commission, Zhengzhou, 450003, china | lhkong0416@163.com |
| 13 | Haixing | Liu | Chemistry & Chemical and Environmental Engineering College, Weifang University, Weifang 261061, P.R. China | haixingliu@tom.com |
| 14 | Yan | Li | Shanghai University of Electric Power, Shanghai, China | liyandy2000@sina.com |
| 15 | Xin | Zhou | School of Management, Tianjin University of Technology, Tianjin 300384, China | 1228313436@qq.com |
| 16 | Xiaodong | Zhao | Army Engineering Shijiazhuang Campus, Hebei Shijiazhuang 050003, China | 307794209@qq.com |
| 17 | Xihui | Mu | 32181 Troops of the Chinese People's Liberation Army, Hebei, Shijiazhuang 050003, China | 1572660524@qq.com |
| 18 | Ling | Chen | School of elementary education, Wanjiang University of Technology, Ma'anshan, Anhui, China | imidazol@163.com |
| 19 | Xiaoqi | Chen | Institute of Energy Source, Hebei Academy of Sciences, Shijiazhuang, China | cxq198508@163.com |
| 20 | Lin | Li | Key Laboratory of Functional Materials and Chemistry for Performance and Resource of Guizhou Education Department, Anshun University, A | 404003375@qq.com |
| 21 | Xianglong | Li | School of Jinan, Qilu normal University, Jinan 250000, China | 858199882lxl@gmail.com |
| 22 | Zikun | Quan | School of Weihai, Shandong University, Weihai 264200, China. | 304916550@qq.com |
| 23 | Dongping | Liu | CO., LTD of Shenzhen, SHENZHEN FENGPING A.I. CO.,LTD, Shenzhen 580000, China. | liud@fengping.tech |
| 24 | Rongbin | Ju | School of Mechanical Engineering, Xi'An Shiyou University, Xi'An, China | 806465856@qq.com |
| 25 | Jinwei | Fan | College of Mechanical Engineering and Applied Electronics Technology, Beijing University of Technology, Beijing, China | 710484127@qq.com |
| 26 | Xingming | Xu | Shandong University of Science and Technology, Taian, China | xxm1601@163.com |
| 27 | Lulu | Gao | School of Mechanical Engineering, University of Science and Technology Beijing, Beijing 100083, China | ustb_gll@163.com |
| 28 | Jing | Yang | School of Resources and Materials, Northeastern University at Qinhuangdao, Qinhuangdao 066004, PR China | yjwzb@163.com |
| 29 | Zhenhua | Li | School of ship and electromechanical engineering, zhejiang Ocean University, Zhejiang, China | lizh760905@zjou.edu.cn |
| 30 | Jianan | Xu | College of Mechanical and Electrical Engineering, Harbin Engineering University, Heilongjiang, China | xujianan@hrbeu.edu.cn |
| 31 | Shuhui | Jiang | Chongqing Jianzhu College, Chongqing 400072, China. | jsh0529@126.com |
| 32 | Li | Tang | Chongqing Jianzhu College, Chongqing 400072, China. | 398681892@qq.com |
| 33 | Hui | Ju | Chongqing Jianzhu College, Chongqing 400072, China. | 17720444@qq.com |
| 34 | Lihong | Bao | School of Materials Science and Engineering, Beijing Institute of Fashion Technology, Beijing 100029, China | clyblh@bift.edu.cn |
| 35 | Lijuan | Wan | Nanjing Vocational Institute of Transport Technology, Nanjing, China | bartty_ym@163.com |
| 36 | Wei | Liu | Key Laboratory of Rubber-Plastics, Ministry of Education/Shandong Provincial Key Laboratory of Rubber and Plastics, Qingdao University of Sc | qdlorin@126.com |
| 37 | Xu | Yang | Shenyang Aerospace University, Shenyang 110136, China | 312428849@qq.com |
| 38 | Yajun | Gong | Wuhan Second Ship Design and Research Institute, Wuhan, China | 2863535115@qq.com |
| 39 | Zhengjiang | Wu | Wuhan Second Ship Design and Research Institute, Wuhan, China | yin@auit.org |
| 40 | Hong | Wang | Hubei University of Technology Engineering and Technology College, Wuhan 430068, China | 515971726@qq.com |
| 41 | Jing | Wang | School of Materials Science and Engineering & Liaoning Key Laboratory of Advanced Polymer Matrix Composites, Shenyang Aerospace Univer: | wally9932@126.com |

Content from this work may be used under the terms of the Creative Commons Attribution 3.0 licence. Any further distribution of this work must maintain attribution to the author(s) and the title of the work, journal citation and DOI.
Published under licence by IOP Publishing Ltd

| | | | | |
|---|---|---|---|---|
| 42 | Yuguo | Li | School of Physics and Electronics, Shandong Normal University, Jinan 250014, China | liyuguosd@hotmail.com |
| 43 | Jing | Hu | College of Physics and Electronic Information, Huaibei Normal University, Anhui Huaibei, China | hujing200407696@126.com |
| 44 | Lei | Qin | School of Civil Engineering and Architectural Engineering, University of Jinan, Jinan 250022, China | cea_qinl@ujn.edu.cn |
| 45 | Xiaorong | Liu | School of Management, Tianjin University of Technology, Tianjin, China | 2365861939@qq.com |
| 46 | Xuezhi | Feng | Qingdao University of Science and Technology, Qingdao, China | 2500875646@qq.com |
| 47 | Jian | Zhang | School of Materials Science and Engineering, Southeast University, Nanjing, China | 2364963216@qq.com |
| 48 | Chunxiang | Qian | Key Laboratory of Materials Microbiology, Southeast University, Nanjing, China | seucyq@163.com |
| 49 | Yanqiang | Chen | Jiangsu Key Laboratory for Construction Materials, Southeast University, Nanjing, China | cxqian@seu.edu.cn |
| 50 | Di | Hu | School School of Civil Engineering, Central South University, Changsha, China | hudibridge@csu.edu.com |
| 51 | Yafei | Wang | College of Civil Engineering, Hunan University, Changsha, China | 854634352@qq.com |
| 52 | Cheng | Peng | College of Materials Science & Engineering, Chongqing University of Technology, Chongqing 400054, China | pc123@2017.cqut.edu.cn |
| 53 | Lijuan | Zeng | College of Materials Science & Engineering, Chongqing University of Technology, Chongqing 400054, China | 1422173163@qq.com |
| 54 | Fuxiang | Huang | College of Materials Science & Engineering, Chongqing University of Technology, Chongqing 400054, China | hfx@cqut.edu.cn |
| 55 | Mingjun | Zhong | College of Materials Science & Engineering, Chongqing University of Technology, Chongqing 400054, China | 463064227@qq.com |
| 56 | Shuang | Liang | College of Materials Science & Engineering, Chongqing University of Technology, Chongqing 400054, China | ls@2017.cqut.edu.cn |
| 57 | Xiaojie | Ran | College of Materials Science & Engineering, Chongqing University of Technology, Chongqing 400054, China | 1071602004@qq.com |
| 58 | Lu | Zhou | College of Materials Science & Engineering, Chongqing University of Technology, Chongqing 400054, China | 1137662948@qq.com |
| 59 | Zhongzhi | Han | CNPC Engineering Technology Research Co., LTD, Tianjin, China | hanzz01@cnpc.com.cn.com |
| 60 | Xiaojun | Guo | CNPC Engineering Technology Research Co., LTD, Tianjin, China | guoxj@cnpc.com.cn.com |
| 61 | Guoquan | Tao | School of Aeronautics Science and Engineering, Beihang University (BUAA), Beijing, China | taogq_buaa@163.com |
| 62 | Zhou | Yan | Engineering Training Center of Shenyang Aerospace University, Shenyang, China | sdyanzhou16888@163.com |
| 63 | Pengcheng | Xie | College of Mechanical and Electrical Engineering, Beijing University of Chemical Technology Beijing, China | 641676092@qq.com |
| 64 | Wenli | Wang | AVIC Manufacturing Technology Institute, Beijing 100024, China | bkxie@buaa.edu.cn |
| 65 | Qi | Zou | China Aviation Manufacturing Technology Research Institute, Beijing, China | 13329392517@126.com |
| 66 | Shan | Jiang | School of Electromechanical, Northwestern Polytechnical University University, Xian, China | 1123750226@qq.com |
| 67 | Xiaochen | Wang | National Engineering Research Center of Flat Rolling Equipment, University of Science and Technology Beijing, Beijing 100083, China | wangxiaochen@ustb.edu.cn |
| 68 | Jiamin | Zhang | National Engineering Research Center of Flat Rolling Equipment, University of Science and Technology Beijing, Beijing 100083, China | zhangjm_1993@163.com |
| 69 | Huaqiang | Liu | National Engineering Research Center of Flat Rolling Equipment, University of Science and Technology Beijing, Beijing 100083, China | Lhuaqiang@163.com |
| 70 | Quan | Yang | National Engineering Research Center of Flat Rolling Equipment, University of Science and Technology Beijing, Beijing 100083, China | yangquan@nercar.ustb.edu.cn |
| 71 | Dong | Xu | National Engineering Research Center of Flat Rolling Equipment, University of Science and Technology Beijing, Beijing 100083, China | xudong@ustb.edu.cn |
| 72 | Bijie | Xue | Institute of Telecommunications Satellite, China Academy of Space Technology, Beijing, China | xuebijie@126.com |
| 73 | Zemin | Wu | The Northeast Petroleum University, Daqing 163318, China. | jaywu1980@163.com |
| 74 | Yugang | Liu | Beijing Institute of Spacecraft Environment Engineering, Beijing, 100094, China | liuyugang46497@sina.com |
| 75 | Sang | Xiong | Jiangsuzhongse Composite Materials Co., LTD., Wuxi 214000, China | jxbjxs2008@163.com |
| 76 | Zhengping | Shang | Jiangsuzhongse Composite Materials Co., LTD., Wuxi 214000, China | 641483235@qq.com |
| 77 | Huan | Gao | Jiangsuzhongse Composite Materials Co., LTD., Wuxi 214000, China | 411622074@qq.com |
| 78 | Qijie | Zhai | School of Materials Science and Engineering, Shanghai University, Shanghai 200000, China | 1271800877@ qq.com |
| 79 | Xinhui | Lu | College of Forestry, Sichuan Agricultural University, Sichuan, China | 278102491@qq.com |
| 80 | Xiaobing | Luo | Division of Structural Steels, Central Iron and Steel Research Institute, Beijing, China | luoxiaobing@cisri.com.cn |
| 81 | Jianhong | Gong | School of Electrical, Mechanical and Information Engineering, Shan Dong University, Shandong, China. | gongjh@sdu.edu.cn |
| 82 | Jing | Xia | National Institute of Clean and Low Carbon Energy, Beijing, 102211, China | Jing.xia.e@chnenergy.com.cn |
| 83 | Zhaoxin | Zhou | School of Navigation, Shandong Jiaotong University, Weihai, China | 15065135036@126.com |

| | | | | |
|---|---|---|---|---|
| 84 | Xilong | Guo | School of Mechanical and Electrical Engineering, Qingdao, China | 961616153@qq.com |
| 85 | Zhongfu | Bao | Guangdong Mechanical & Electrical Polytechnic, Guangzhou 510515, China | bzfjixie@163.com |
| 86 | Yu | Wang | School of Qingdao University of science and technology, Qingdao 266061, China. | 4570536@qq.com |
| 87 | Yantao | Li | Institute of Energy Source, Hebei Academy of Sciences, Shijiazhuang, 050000, China | 13903116163@163.com |
| 89 | Xiangjun | Bi | Aerospace Research Institute of Special Material and Processing Technology, Beijing, China | bxiangj@163.com |
| 90 | Guohuai | Liu | State Key Laboratory of Rolling and Automation, Northeastern University, Shenyang 110819, China | liugh@ral.neu.edu.cn |
| 91 | Meng | Song | Key Laboratory of Metallurgical Emission Reduction & Resources Recycling (Anhui University of Technology), Ministry of Education, Maanshan | songmeng@ahut.edu.cn |
| 92 | Yong | Yang | Anhui Province Key Laboratory of Metallurgical Engineering & Resources Recycling, Anhui University of Technology, Maanshan, China | yangyongboxian@163.com |
| 93 | Huijie | Guo | Beijing Institute of Radio Metrology and Measurement, Beijing, China | ghj203@126.com |
| 94 | Weijie | Gou | Beijing Polytechinc, Beijing, China | gvjie@126.com |
| 96 | Yan | Li | School of economics and management, Tianjin University of Science & Technology, Tianjin 300222, China; | 1198221091@qq.com |
| 97 | Yunhai | Jia | Research and development center, Beijing Institute of Electro-Machining, Beijing 100191, China | jyh308401@sina.com.cn |
| 98 | Zhenzhu | Yu | College of Electromechanical Engineering, Qingdao University of Science &Technology, Qingdao, China | yuzhenzhu@qust.edu.cn |
| 99 | Yunlong | Hu | Zhejiang University Kunshan Innovation Institute, Suzhou, China | 3257063992@qq.com |
| 100 | Jin | Hao | Guangxi University of science and technology, Liuzhou, China | 15731069297@163.com |
| 101 | Zhaoping | Tang | School of Information Engineering, East China Jiaotong University, Nanchang Jiangxi, 330013, China. | tzp@ecjtu.edu.cn |
| 102 | Min | Wang | School of Information Engineering, East China Jiaotong University, Nanchang Jiangxi, 330013, China. | 1291966861@qq.com |
| 103 | Yi | Wu | College of Science and Technology, Nanchang University, Nanchang, 330029, China | 1443059097@qq.com |
| 104 | Min | Zhao | CSR Qishuyan Locomotive & Rolling Stock Technology Research Institute Co., Ltd, Changzhou Jiangsu, 213011, China | zhaomin.qs@crrcgc.cc |
| 105 | Xianliang | Zheng | CSR Qishuyan Locomotive & Rolling Stock Technology Research Institute Co., Ltd, Changzhou Jiangsu, 213011, China. | zhengxianl@163.com |
| 106 | Guo | Xu | Jiangsu XCMG Construction Machinery Research Institute Ltd, Xuzhou Construction Machinery Group, Xuzhou, China | 1454014925@qq.com |
| 107 | Yuhao | Zhao | School of Chongqing University, Chongqing, China | zhaoyuhao1992@126.com |
| 108 | Guiling | Ju | Basic Courses Department, Army Academy of Armored Forces, Beijing 100072, China | guilingju@163.com |
| 109 | yifei | Li | School of Wuhan Textile University, Wuhan, China | 18986072242@163.com |
| 110 | Wenxiang | Du | School of petroleum engineering, China University of Petroleum, Qindao, China | wenxiang_du973@163.com |
| 111 | Qin | Yang | School of Science and Technology Liaoning, Liaoning, China | 1535569532@qq.com |
| 112 | Chenghui | Sun | School of Materials Science & Engineering, Beijing Institute of Technology, Beijing, China | sunch@bit.edu.cn |
| 113 | Lijie | Qiao | Huadian Electric Power Research Institute Co., Ltd, Hangzhou, China | qiaolj@126.com |
| 114 | Yanju | Wang | Aviation Engine Corporation of China, Beijing Institute of Aeronautical Materials, Beijing, 100095, China | wyjbiam@163.com |
| 116 | Houdi | Xiao | School of Aeronautic Science and Engineering, Beihang University, Beijing, 100191, P. R. China | xhdbuaa@126.com |
| 117 | Zhongliang | Yang | Jiangsu XCMG Construction Machinery Research Institute Ltd organization, Jiangsu Xuzhou, China | zhengshuping123@126.com |
| 118 | Ziyan | Fan | Irvine High School, 4231 Walnut Ave, Irvine, CA, 62604, USA | 20fanziyan@gmail.com |
| 119 | Di | Sun | Marine Engineering College, Jimei University, Xiamen, China | sundi6329@sina.com |
| 120 | Kai | Li | Shandong University of Science and Technology, College of transportation, Qingdao, China | 825840462@qq.com |
| 121 | Fulong | Jiang | Inner Mongolia University of Technology, Mechanical Engineering, Inner Mongolia, Huhehot, 010000, China | 2803541366@qq.com |
| 122 | Yi | Hu | School of Automotive Engineering, Wuhan University of Technology, Wuhan, China | 68235238@qq.com |
| 123 | He | Zhang | Ministry of Civil Affairs of the People's Republic of China, Beijing, China | mishanxixi@126.com |
| 124 | Yuchen | Chen | School of Laval University, Quebec | carmenchen24@outlook.com |
| 125 | Yongkuo | Wang | Ministry of Civil Affairs of the People's Republic of China, Beijing, China | yongkuow@163.com |
| 126 | Chunmei | Xu | School of Civil engineering, Southwest Forestry University, Kunming, China | 2219139473@qq.com |
| 127 | Shiwei | Zhang | School of Metallurgical Engineering, Xi'an University of Architecture and Technology, Xi'an 710055, China; | 2234789895@qq.com |
| 128 | Zihao | Yang | School of Information and Control Engineering, Xi'an University of Architecture and Technology, Xi'an 71005, China | 1105453436@qq.com |

IOP Conf. Series: Materials Science and Engineering **772** (2020) 011004 doi:10.1088/1757-899X/772/1/011004

# Peer review statement

All papers published in this volume of *IOP Conference Series: Materials Science and Engineering* have been peer reviewed through processes administered by the proceedings Editors. Reviews were conducted by expert referees to the professional and scientific standards expected of a proceedings journal published by IOP Publishing.

Content from this work may be used under the terms of the Creative Commons Attribution 3.0 licence. Any further distribution of this work must maintain attribution to the author(s) and the title of the work, journal citation and DOI.
Published under licence by IOP Publishing Ltd

IOP Conf. Series: Materials Science and Engineering **772** (2020) 012001 doi:10.1088/1757-899X/772/1/012001

# Study on Calculating Method of External Load on Digging Arm of Underground Excavator Truck

**Shuping Zheng, Zengjie Yang and Zhongliang Yang ***

Jiangsu XCMG Construction Machinery Research Institute Ltd organization, Jiangsu Xuzhou, China

*Corresponding author e-mail: zhengshuping123@126.com

**Abstract.** The digging arm of underground excavator truck has many movable joints, complex structure and variable operation properties which results in the complicated loading. So it is difficult to obtain the external working load directly through the test. For this problem, the calculation of the external load on the digging arm of the underground excavator truck is studied by combining test and virtual prototype in this paper. The details include: Firstly, the static test of the digging arm was conducted using pressure sensor and displacement sensor, and the parameters of the virtual prototype built by Motion View software were modified. Then the pressure and displacement curves of each hydraulic oil cylinder are measured during the operation on site. The working process of digging arm is simulated and reconstructed by taking the measured displacement curves of each cylinder as the driving inputs, and outputs the curves of centre of mass coordinate, curves of acceleration and corresponding curves of torque. The rotary torque of external forces and gravity on a rotation axis (rotation axes of movable arm and pitch) should be equal to the sum of torque of the product of each component's mass and the acceleration of the centre of mass on that point according to the law of torque calculation. Thus, the non-homogeneous linear equations of external loads (horizontal, vertical and lateral loads) are derived under the actual conditions. And external loads were solved by substituting the above curve parameters into the equations. Finally, the obtained external load is reloaded into the virtual prototype of the digging arm to simulate the force of actual digging process. The error is less than 5% by comparing the calculated and test values of the oil cylinders force of rod and bucket and the effective strain error of the movable arm is less than 10%. So the correctness of the derivation method of external load and the dynamic model considering inertial force is verified. It provides theoretical and algorithm support for the next acquisition of load spectrum and fatigue analysis of digging arm.

## 1. Introduction

The underground excavator truck [1] is a new type of construction machinery integrating, which integrates the functions of excavation, milling, and bolting. It is one of the key supporting equipment for the mechanized operation of underground excavation tunnels, which is mainly used in the construction of subway tunnels and municipal pipeline projects. The underground excavator truck is a double arm structure. In this paper, the external load calculation method of the left arm (digging arm) is

Content from this work may be used under the terms of the Creative Commons Attribution 3.0 licence. Any further distribution of this work must maintain attribution to the author(s) and the title of the work, journal citation and DOI.
Published under licence by IOP Publishing Ltd

studied. There are few such studies due to the multi-link arm structure of the digging arm (figure 1). Most of the research is about dynamic simulation of excavator arm and finite element static simulation under typical dangerous working conditions, etc. [2-4]. In reference [5], the external load is approximately obtained and the transient dynamic simulation analysis is carried out, which takes the maximum theoretical impact force of the digging arm as the driving force. In reference [6], a load spectrum test method is proposed which can reflect the actual working conditions of excavator working device. In reference [7], the load spectrum of transmission system is studied. In reference [8], the parameters and posture control of hydraulic excavator are studied. In reference [9], the digging force under typical excavation attitude is calculated and analyzed. In reference [10]-[16], the virtual prototype model establishment and dynamic simulation of excavator working device were studied, and the motion and digging force performance of working device were calculated and analyzed. Based on the above analysis, it is found that there is little research on the calculation of external load in the whole working process of the digging arm. In the above literature, the model is greatly simplified to obtain approximate solutions, and the digging force is only calculated for some typical or dangerous working conditions, rather than for all the loads in the work engineering. The biggest difference of this paper is that the method of combining the real and the virtual is adopted. At the same time, the influence of lateral load and inertial force is considered in the calculation of external load, so that the bucket tip load obtained is more accurate, and the external load of each position in the working process of the digging arm is obtained.

The digging arm of the underground excavator truck is mainly composed of movable arm, transition joint, bucket bar, bucket and other parts, which can not only realize forward excavation but also horizontal and lateral excavation in front and vertical excavation on both sides of the core soil. Due to narrow working space and harsh geological environment, it is difficult to measure the external load on bucket end directly. Therefore, in this paper, the force condition of the digging arm under static conditions was first tested with a weighing sensor to modify the Motion View virtual prototype. Then, the data of each cylinder was obtained by testing in the user site which was imported to the Motion View virtual prototype. By substituting the motion parameters of each component output from the virtual prototype and the cylinder force obtained by the test into the derived external load matrix equation, the external load curve of bucket end under actual conditions is obtained. The oil cylinder forces of bucket and bucket rod obtained by experiment and simulation are compared by reloading the external load curve. At the same time, the strain values of the moving arm measuring points are also compared. So the correctness of the calculation method of the external load of the digging arm proposed in this paper is further verified.

**Figure 1.** The structure of underground excavator truck.

IOP Conf. Series: Materials Science and Engineering **772** (2020) 012001 doi:10.1088/1757-899X/772/1/012001

## 2. The verification of calculation method of external load of digging arm under static conditions and model modification

By testing the cylinder force and stroke of the digging arm under no-load condition, the model quality distribution of each component of the digging arm is adjusted, so that each cylinder force of the test and the calculation are consistent, and finally the purpose of model modification is achieved. In order to verify the correctness of the method of deducing the external load of digging arm by the cylinder force, the digging arm is pressed down under the non-working condition (as shown in figure 2) to achieve the force, length of each cylinder and the data of weighing sensor. Each cylinder forces of the digging arm are obtained by loading the specified external loads (as shown in figure 3) on the models with different postures. The forces of each cylinder and external load obtained by test and calculation under different postures are listed in table 1. It can be seen from the table that the cylinder force errors are all within 6%, which shows that the corrected virtual prototype meets the next test requirement.

**Figure 2.** The external loading test of digging arm in position A under static condition.

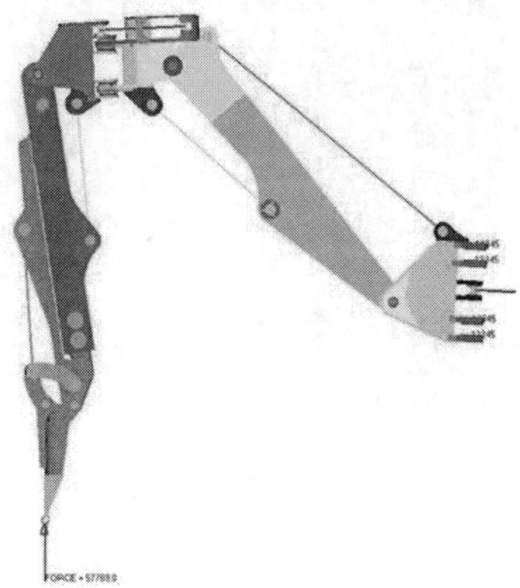

**Figure 3.** The model of digging arm in position A under static condition.

**Table 1.** Comparison of cylinder force, external load test and calculated in different poses.

| | posture | movable arm oil cylinder | pitch oil cylinder | bucket rod oil cylinder | bucket oil cylinder | load |
|---|---|---|---|---|---|---|
| | test values | -204.0 | 218.0 | 71.0 | -78.0 | |
| A | calculated value | -202.0 | 216.0 | 66.9 | -74.3 | 57.9 |
| | error /% | -1.0 | -0.9 | -5.8 | -4.7 | |
| | test values | -204.5 | 215.0 | 73.0 | -9.5 | |
| B | calculated value | -205.0 | 215.0 | 68.8 | -75.8 | 57.0 |
| | error /% | 0.3 | 0.0 | -5.8 | -4.7 | |
| | test values | -210.0 | 218.0 | 74.2 | -84.4 | |
| C | calculated value | -209.5 | 217.7 | 74.2 | -84.4 | 56.9 |
| | error /% | -0.2 | -0.1 | -5.9 | -4.4 | |
| | test values | -237.4 | 217.2 | 135.6 | 46.8 | |
| D | calculated value | -233.0 | 211.0 | 129.0 | 47.2 | 43.2 |
| | error /% | -1.8 | -2.9 | -4.9 | 0.8 | |

## 3. Study on calculation method of external load of digging arm under dynamic working conditions

### *3.1. The data acquisition of digging arm under construction conditions*

Data collection is carried out under the working condition of underground excavator truck, as shown in figure 4. The excavation medium is sand pebble soil, which is loose and adherent. The big cavity pressure, small cavity pressure, displacement and other physical quantities of each cylinder are measured by using oil pressure sensors and pull rope displacement sensors. After processing the test data, the force and displacement curves of each cylinder are obtained respectively. The force curves of the pitching cylinder and the movable arm cylinder are shown in figure 5. Figure 6 shows the displacement variation curves of the above oil cylinders.

**Figure 4.** The load test of digging arm in working condition.

IOP Conf. Series: Materials Science and Engineering **772** (2020) 012001 doi:10.1088/1757-899X/772/1/012001

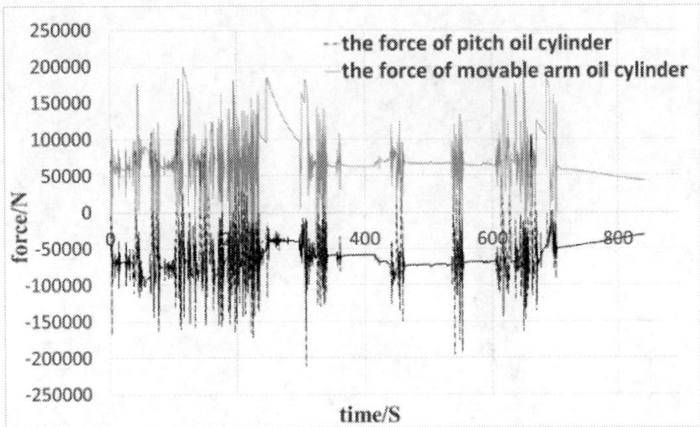


**Figure 5.** The cylinder force curves of pitch and movable arm in working condition.

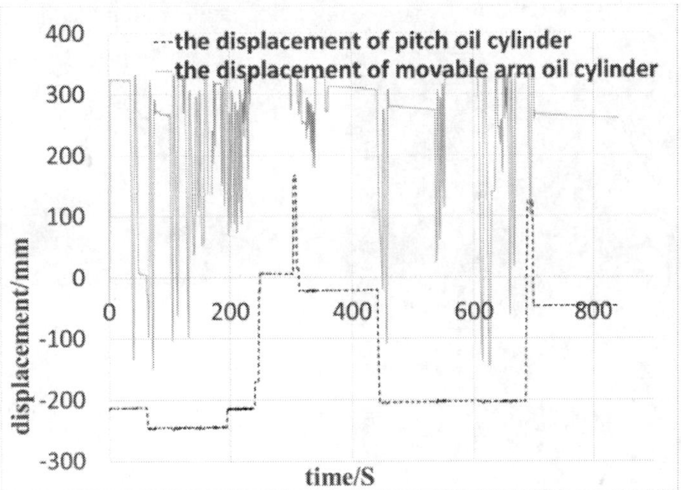


**Figure 6.** The cylinder displacement curves of pitch and movable arm in working condition.

*3.2. The establishment of virtual prototype of digging arm*

The corresponding dynamic simulation model (as shown in Figure 7) was established in the Motion View according to the model modification parameters under the static working condition of the digging arm previously. The model is symmetric about the XY plane, and the revolute pairs, sliding pairs, fixed pairs and spherical hinges are established in the corresponding position according to the actual working state. A local coordinate system is established for each cylinder, and X axis is set along the axis of the cylinder. The other two directions are determined according to the right hand rule. In order to simulate the external load, three perpendicular forces (F1, F2 and F3) are established at the bucket end, where F3 is perpendicular to the outside of the paper. Figure 8 shows the summary of the results after the time synchronization of video and the simulation motion of digging arm. The figure on the right shows the displacement loading curve of the movable arm oil cylinder. The displacement curves of each cylinder obtained in the working conditions were taken as the cylinder motion of the virtual prototype to simulate the actual working process of the digging arm.

**Figure 7.** The dynamic simulation model of digging.

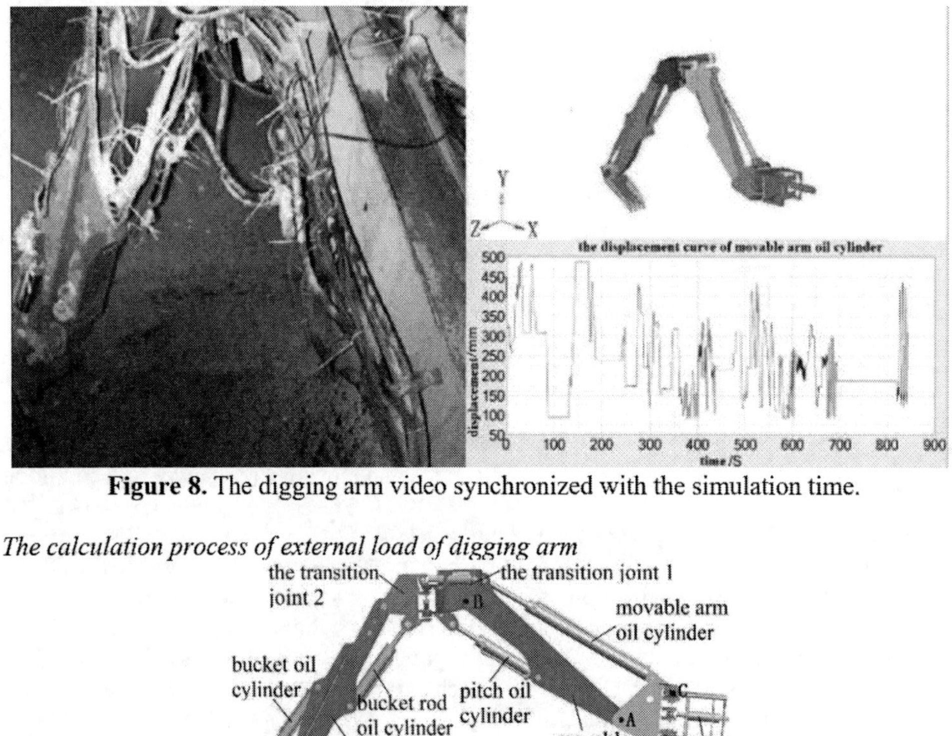


**Figure 8.** The digging arm video synchronized with the simulation time.

### *3.3. The calculation process of external load of digging arm*

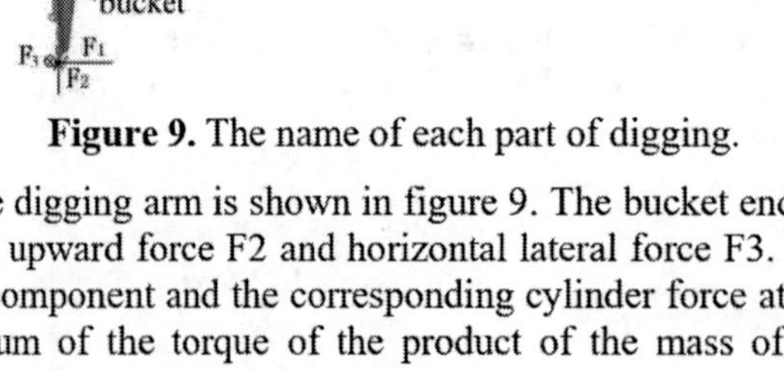


**Figure 9.** The name of each part of digging.

Each component of the digging arm is shown in figure 9. The bucket end is subjected to horizontal forward force F1, vertical upward force F2 and horizontal lateral force F3. The torque of the external load, the gravity of each component and the corresponding cylinder force at the specific rotation point should be equal to the sum of the torque of the product of the mass of each component and the acceleration of the center of mass at this point during the movement of the digging arm, as shown in the following formula:

$$\sum F_i l_i + \sum m_j g\, h_j + F_{\mathrm{oil}} l_{\mathrm{oil}} = \sum m_j a_j n_j \tag{1}$$

The detailed calculation process is as follows:

Firstly, the rotation of bucket, bucket rod, transition joint, each oil cylinder and movable arm around point A at the root of movable arm was analyzed, and the following items were outputted:

Output the vertical projection distance of the center of mass of each component relative to the rotation axis of point A, namely the moment arm $h_{Ai}$ generated by the gravity of each component relative to the rotation axis of point A, and records the weight $m_i$ of each component.

According to the center of mass of each component and the rotation axis of A at the root of the moving arm, the local coordinate system of the center of mass relative to A is established. The origin of the coordinate system is the center of mass, the X-axis points to the rotation axis of point A, the Z-axis lies in the plane composed of the center of mass and the rotation axis of point A, the Y-axis is determined by the right hand rule, and the counterclockwise rotation of the center of mass around the rotation axis of point A is positive. Output acceleration $a_{Ai}$ of the center of mass of each component along the Y axis of its local coordinate system, and output the coordinates of point A relative to each component along the X-axis in the local coordinate system, which is the moment arm $(n_{Ai})$ of the center of mass's acceleration.

A local coordinate system along the direction of F1 and F2 and F3 at the end of the bucket is established. In the local coordinate system, output the moment arms ($l_{A1}$, $l_{A2}$ and $l_{A3}$) of the F1, F2 and F3 relative to the axis of rotation at point A.

Output the vertical distance of point A relative to the axis of movable arm oil cylinder, that is, the moment arm ($l_{Ad}$) of the movable arm cylinder's force relative to point A.

Therefore, the moment equation at point A of the rotating center of the moving arm can be established:

$$F_1 l_{A1} + F_2 l_{A2} + F_3 l_{A3} + \sum m_i g\, h_{Ai} + F_d l_{Ad} = \sum m_i a_{Ai} n_{Ai} \tag{2}$$

Where: $F_d$ is the force of the movable arm cylinder obtained by field test.

Then, take the rotation point B between the transition joint and the moving arm as the reference point, and output the vertical projection distance of the center of mass of bucket, bucket rod, transition joint 1, transition joint 2, bucket cylinder, bucket rod cylinder and other components relative to the rotation axis of point B, that is, the force arm $h_{Bj}$ generated by the gravity of each component relative to the rotation axis of point B.

The local coordinate system is established according to the centroid point of each component and the rotation axis of the point B between the transition joint and the moving arm. The origin of the coordinate system is the center of mass, the X-axis points to the rotation axis of point B, the Z-axis lies in the plane composed of the center of mass and the rotation axis of point B, the Y-axis is determined by the right hand rule, and the counterclockwise rotation of the center of mass around the rotation axis of point B is positive. Output acceleration $a_{Bj}$ of the center of mass of each component along the Y axis of its local coordinate system, and output the coordinates of point B relative to each component along the X-axis in the local coordinate system, which is the moment arm $(n_{Bj})$ of the center of mass's acceleration.

According to the established local coordinate system along the direction of F1, F2 and F3 at the bucket end, the moment arms ($l_{B1}$, $l_{B2}$ and $l_{B3}$) of the load of F1, F2 and F3 relative to the rotation axis of point B is output.

Output the vertical distance of point B relative to the axis of pitch oil cylinder, that is, the moment arm ($l_{Bf}$) of the pitch oil cylinder's force relative to point B.

Therefore, the torque equation at point B of the rotation center can be established:

$$F_1 l_{B1} + F_2 l_{B2} + F_3 l_{B3} + \sum m_j g\, h_{Bj} + F_f l_{Bf} = \sum m_j a_{Bj} n_{Bj} \tag{3}$$

Where: $F_f$ is the force of pitch oil cylinder obtained by field test.

Finally, the local coordinate system is established on the basis of the centroid point of each component and the rotation axis of C point connected between the movable arm seat and the frame. The origin of the coordinate system lies at the center of mass of the component, the X-axis points to the rotation axis of point C, The Z-axis is located in the plane composed of the center of mass and the rotation axis of described above, the Y-axis is determined by the right hand rule, and the counterclockwise rotation of the center of mass around the rotation axis of point C is positive. Output acceleration $a_{Ck}$ of the center of mass of each component along the Y axis of its local coordinate system, and output the coordinates of point B relative to each component along the X-axis in the local coordinate system, which is the moment arm $(n_{Ck})$ of the center of mass's acceleration.

According to the established local coordinate system along the direction of F1, F2 and F3 at the bucket end, the moment arms ($l_{C1}$, $l_{C2}$ and $l_{C3}$) of the load of F1, F2 and F3 relative to the rotation axis of point C is output.

Output the vertical distance of point C relative to the axis of the swinging arm cylinder, that is, the moment arm ($l_{Bf}$) of the swinging arm cylinder's force relative to point C.

Therefore, the torque equation at point C of the rotation center can be established:

$$F_1 l_{C1} + F_2 l_{C2} + F_3 l_{C3} + F_b l_{Cf} = \sum m_k a_{Ck} n_{Ck} \quad (4)$$

Where: $F_b$ is the force of movable arm swinging cylinder obtained by field test.

According to formula (2)-(4), the following non-homogeneous linear equations concerning F1, F2 and F3 are composed:

$$\begin{bmatrix} l_{A1} & l_{A2} & l_{A3} \\ l_{B1} & l_{B2} & l_{B3} \\ l_{C1} & l_{C2} & l_{C3} \end{bmatrix} \begin{bmatrix} F_1 \\ F_2 \\ F_3 \end{bmatrix} = \begin{bmatrix} \sum m_i a_{Ai} n_{Ai} - \sum m_i g\, h_{Ai} - F_d l_{Ad} \\ \sum m_j a_{Bj} n_{Bj} - \sum m_j g\, h_{Bj} - F_f l_{Bf} \\ \sum m_k a_{Ck} n_{Ck} - F_b l_{Cf} \end{bmatrix} \quad (5)$$

Where, I, j and k correspond to the serial number combination of different parts at points A, B and C.

The parameters previously output and the cylinder force curves obtained from test were substituted into formula (5), and the change curves of F1, F2 and F3 were obtained, as shown in figure 10:

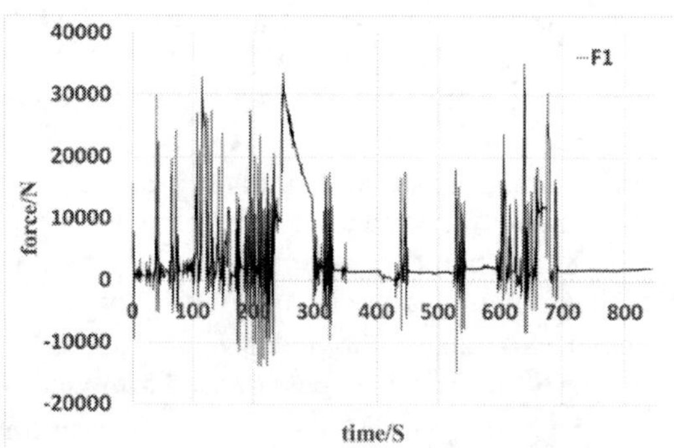


(a). The curve of external loading $F_1$ of digging arm.

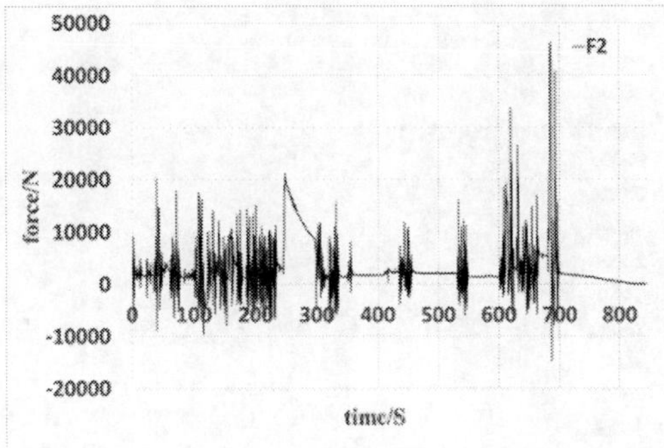


(b). The curve of external loading $F_2$ of digging arm.

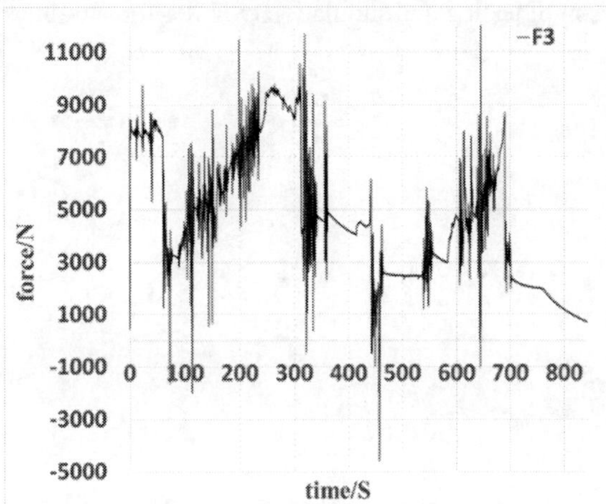


(c). The curve of external loading $F_3$ of digging arm.

**Figure 10.** The curves of external loading $F_1$, $F_2$ and $F_3$ of digging arm in working condition.

## 4. The verification of external loads

### *4.1. The oil cylinder force verification*

The calculated external loads F1, F2 and F3 were loaded into the bucket end, and the simulation analysis of the digging process was carried out performed again. The oil cylinder forces of bucket rod and bucket obtained by the simulation are compared with that obtained by actual test. Due to the large amount of data, only part of the data is displayed for clarity, as shown in figure 11 and figure 12. The data between 107 s and 109 s are intercepted for magnification comparison in figure 11. The data between 176 s and 178s are intercepted for magnification comparison in figure 12. It can be seen from the figure that the simulated cylinder force curves of bucket rod and bucket are in good agreement with the test. Figure 13 and 14 respectively show the relative error curves of cylinder force of bucket rod and bucket of simulation and test, and the error values of both are within 5%, which proves the correctness of the calculation methods of external loads F1, F2 and F3 of the digging arm.

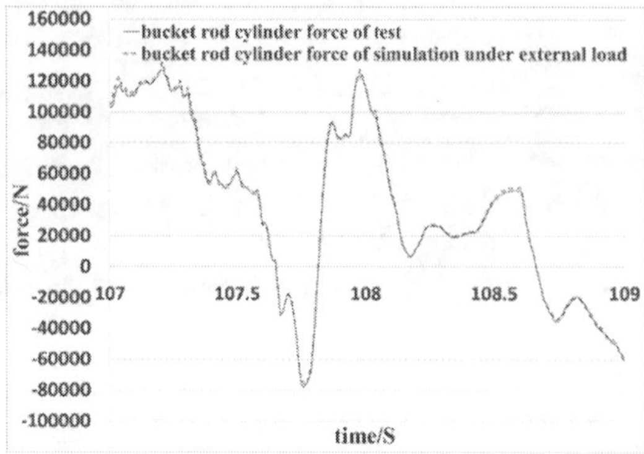


**Figure 11.** The comparison of testing and simulation cylinder force of bucket rod between107 s and 109 s.

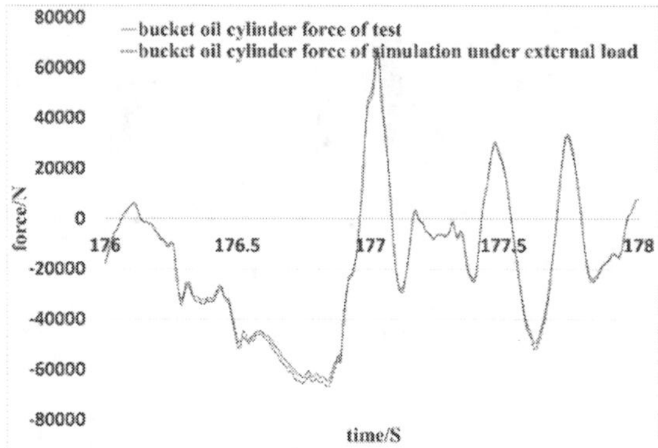


**Figure 12.** The comparison of testing and simulation cylinder force of bucket between 176 s and 178 s

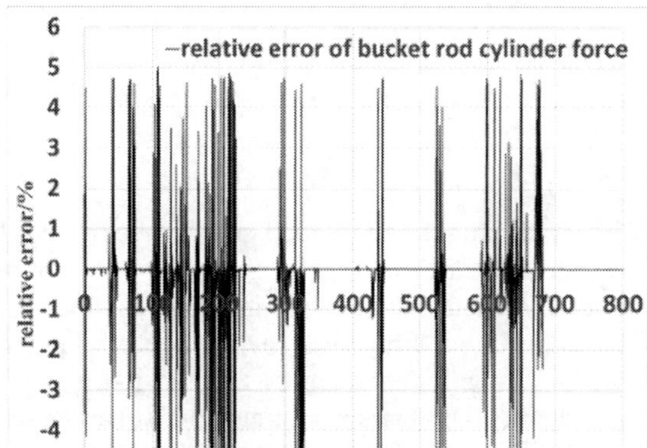


**Figure 13.** The relative error of cylinder force of testing and simulation of bucket rod in working condition.

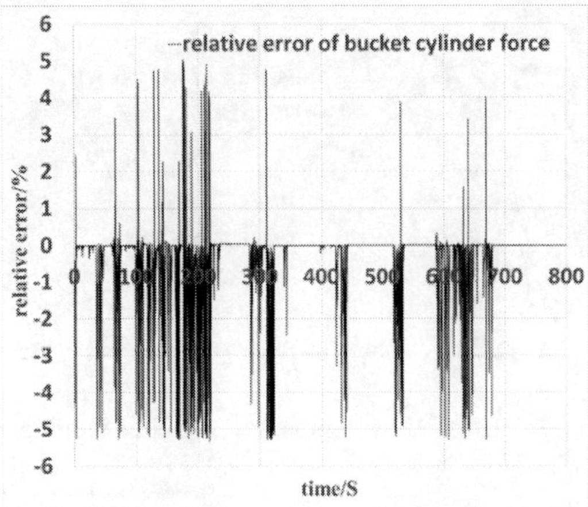


**Figure 14.** The relative error of cylinder force of testing and simulation of bucket in working condition.

### *4.2. The strain validation*

The structure strain of movable arm obtained from the test and simulation are also compared and analyzed except oil cylinder force of bucket rod and bucket, so as to further verify the correctness of the dynamic simulation model and the calculation method of external load. Figure 15 shows the strain test location of movable arm in working condition. Figure 16 is the rigid-flexible coupling simulation result after replacing the movable arm with a flexible body, from which the simulated strain data of the measuring point can be extracted and compared with the test. The strain-time history curves of movable arm between the simulation and test were compared, as shown in figure 17. The data between 190 and 200 s at a certain point of the moving arm are intercepted for comparison for clarity. It can be seen from the figure that the values of both are basically the same. Figure 18 shows the relative error curve of movable arm's strain obtained by the simulation and test. In the figure, the error value at some small micro-strain is very large, which is caused by the large measurement error of the small strain by the strain testing system. Therefore, after considering the test system error and fatigue life requirements, measurement points with macrostrain larger than 80 are selected for relative error comparison (as shown in figure 19). As can be seen from the figure, when the strain of movable arm is large, the error is also large, but the overall error is within 10%.

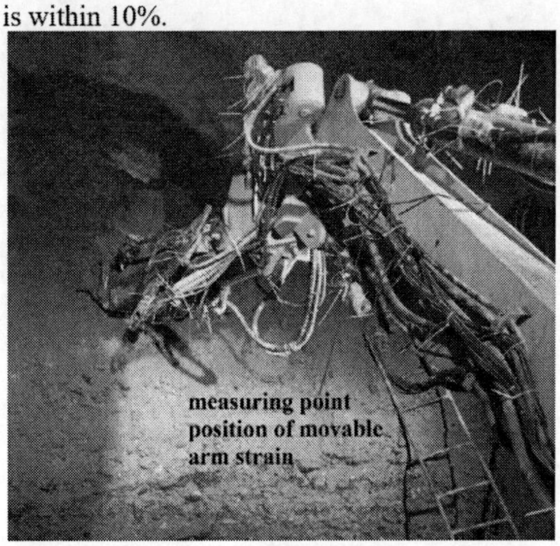


**Figure 15.** The strain test of movable arm in working condition.

IOP Conf. Series: Materials Science and Engineering **772** (2020) 012001 doi:10.1088/1757-899X/772/1/012001

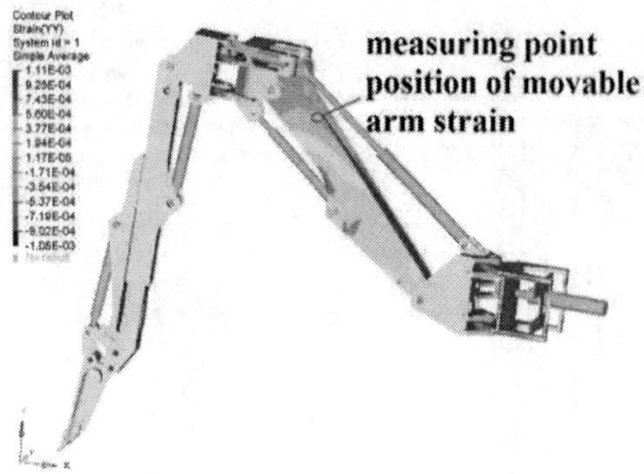


**Figure 16.** The rigid and flexible coupling simulation model of digging arm in working condition.

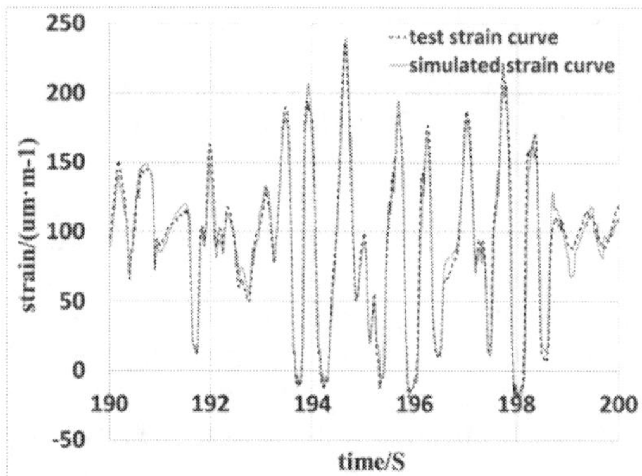


**Figure 17.** The comparison of movable arm's strain between simulation and experiment in working condition.

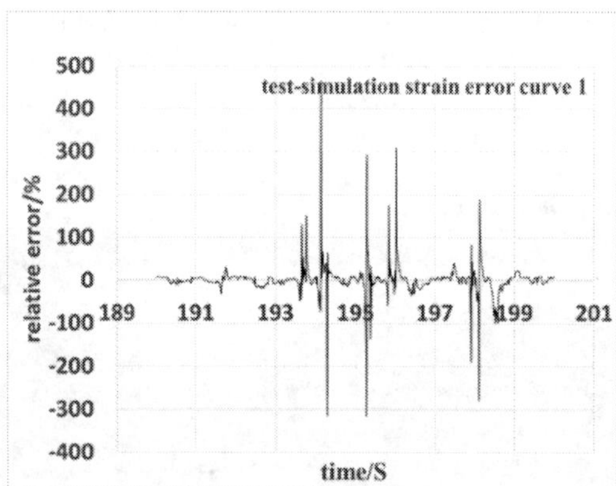


**Figure 18.** The relative error curve of movable arm's strain of simulation and experiment in working condition.

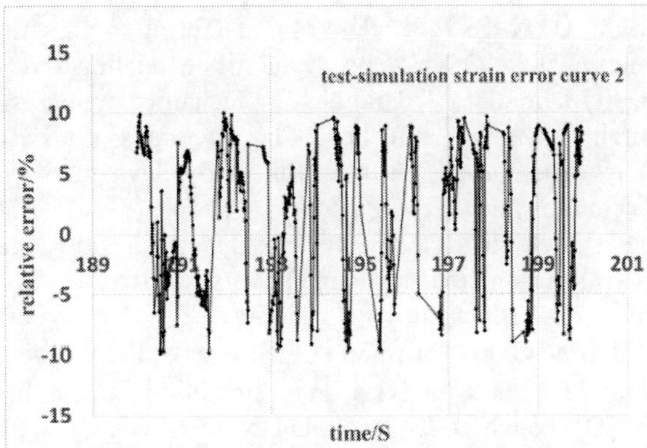


**Figure 19.** The relative error curve of movable arm's strain of simulation and experiment considering system error.

## 5. Conclusion

1) The method to calculate the external load of the digging arm of the underground excavator truck is studied by using the virtual and real combined technology of dynamic simulation and test in this paper. The matrix equation of external load is established by considering the influence of inertial force during the movement of digging arm. And then it was solved to obtain the external load data of digging arm in the actual working process by importing the relevant data obtained from test and simulation. The calculated oil cylinder force error is within 5% and the effective strain error within 10% by verifying the obtained external load data of cylinder force and structural strain. So the correctness of the external load calculation method is proved, which provides theoretical and algorithm support for the next step of load spectrum collection, fatigue analysis and structure design optimization of the digging arm.

2) This method can be further extended to the calculation of external loads of other mechanical structures with many kinematic joints, large acceleration and complex theoretical derivation process.

## References

[1] Liu Gang. New techniques of underground excavation of XCMG promote new course on the construction of rail transit [J].Construction Machinery and Equipment, 2016, (1).

[2] LiI Xiangliang. Simulation of high altitude operational arm's performance based on ADAMSS and Workbench [J]. Journal of Hebei University of Engineering, 2017, 99−108.

[3] Wan Yunlong, XU Hongxia. Dynamic simulation of robot manipulator [J]. Computer Simulation, 2016, 33(12):346−351.

[4] Peng Dan. Dynamic simulation and finite element analysis of loader working device [o] :( Master's thesis). Chang 'an university, 2015.

[5] Zhang Weiguo, QUAN Long, CHENG Heng, et al. Transient dynamic analysis of excavator working device based on real load [J]. Journal of Mechanical Engineering, 2011, 47(12):144−149.

[6] Xiang Qingyi, LV Pengmin, WANG Binhua et al. Load Spectrum Test Method of Hydraulic Excavator Working Device[J]. China J. Highw. Transp, 2017, 30(9):151−158.

[7] CHEN H, SUN Y, WU K. Load Spectrum Testing and Analysis for Transmission System of Closed High-speed Press[J]. Procedia Engineering, 2014, 81:1645−1650.

[8] Lv Pengmin, WU Yuwen, ZHANG Bei. Parametric Stance Control System of Hydraulic Excavator[J].Road Machinery & Construction Mechanization,2017,34(3):84−86.

[9] Zhou Xiaojin. Mechanical Properties Analysis of Key Components of Excavator [D].Xi'an:Chang'an University,2015.

[10] HOUSTON R L, LIU Youwu. The Dynamics of Multibody System: I [M].Tianjin University Press, 1987.

[11] Jin Long, LI Yuqi, ZENG Hao. The Modeling and simulation analysis of a large hydraulic

IOP Conf. Series: Materials Science and Engineering **772** (2020) 012001 doi:10.1088/1757-899X/772/1/012001

excavator based on ADAMS [J]. Modern Manufacturing Engineering, 2017, (11):144−149.
[12] Liu Qiang, GE Weimin, WANG Xiaofeng. Dynamics Modeling and Simulation of Constrained Flexible Load with Manipulator Operation [J]. Mechanical transmission, 2018, 42(3):111−116.
[13] She Yini. A Research on the Dynamic Load Characteristics of a Large Face-shovel Hydraulic Exacavator Attachment based on the Muti-body Dynamics [D]. Hangzhou: Zhejiang University of Technology, 2013.
[14] Wang Xiangbing. Research on Structure System Dynamics and Characteristics of Engineering Mechanical Arm [D]. Hangzhou:Zhejiang University, 2014.
[15] Li Xu. Research on Mechanical Properties of Key Structures on Boom-type Roadheader[D].Beijing:China University of Mining & Technology, 2017.
[16] Sun Keyi. Transient Dynamic Analysis and Simulation Research Of Hydraulic Excavator Working Device [D]. Lanzhou: LANZHOU UNIVERSITY OF TECHNOLOGY, 2014.

# Research on Tolerance Redistribution Method Based on Measurement Data

**Tianzhao Fu[1, 2], Ruiqiang Lyu[1], Mingyang Wang[2], Weimin Yang[2], Mingyang Wang[1], Pengcheng Xie[2, *]**

[1]Aeronautical Key Laboratory for Digital Manufacturing Technology, AVIC Manufacturing Technology Institute, Beijing, China
[2]College of Mechanical and Electrical Engineering, Beijing University of Chemical Technology Beijing, China

*Corresponding author e-mail:641676092@qq.com

**Abstract.** As the performance requirements of aircrafts increase, the mechanical structure of the product is more complicated and the requirements for assembly are getting higher and higher. Due to the accumulation of errors in the processing and assembly process, assembly may not meet the product design requirements. It is necessary to repair some of the processed parts before performing the assembly process. This paper proposes an assembly simulation analysis and a tolerance redistribution approach based on measurement data that is making decisions on which part should be repaired. Firstly, the processed parts and assembly fixture are measured by measuring devices and fitted into the digital model with measurement data. Secondly, the assembled digital simulation is performed to verify whether the processed parts can be assembled successfully. Finally, if the parts cannot be assembled successfully, an improved genetic algorithm is used to optimize the tolerance to form a repair scheme.

**Keywords:** Measurement data; Assembly simulation; Iterative optimization; Assembly repair

## 1. Introduction

Aircraft products are more complicated compared to general mechanical products. The number of parts is more, the assembly execution is more difficult, and the control process of assembly is more complicated. Meanwhile, since the aerodynamic performance and structural strength of the aircraft are related to the error in the manufacture and assembly of the aircraft components and products, the error in manufacturing and assembly may lead to a decrease in the quality of the assembly, thereby loss the performance of the aircraft. Consequently, assembly results may not satisfy the process requirements. Therefore, the analysis and optimization of assembly error during aircraft assembly is an urgent problem to be solved.

With the development of computer technology, Computer Aided Tolerancing (CAT) has been widely used in the current processing and assembly process. It can solve the problems that the occurrence of assembly errors leading to unsatisfactory process requirements to some extent. With the development of CAT technology, many experts and scholars have focused on the relationship between assembly

Content from this work may be used under the terms of the Creative Commons Attribution 3.0 licence. Any further distribution of this work must maintain attribution to the author(s) and the title of the work, journal citation and DOI.
Published under licence by IOP Publishing Ltd

tolerance and assembly performance. Zhang established the relationship between the assembly performance and tolerance based on traditional assembly dimension chain with the consideration of the assembly performance [1]. Qureshi proposed a mathematical formula for tolerance analysis that verifies that the assembly requirements of the product are met within acceptable tolerances and acceptable tolerances using the Monte Carlo method [2]. Ghali employed CAT methods to study and analyze the effects of tolerance values on post-production and assembly early in the design process [3]. Tang disassembled the large-scale nonlinear problem into a series of sub-problems by analyzing the nonlinear relational view of the weighted graph tree model, and the complexity of the overall tolerance distribution solution by solving the sub-problems were reduced [4]. Those scholars analyzed the tolerance distribution of the aircraft model where the theoretical model is used, but in practice many random error may occur, resulting in the tolerance of the actual parts or assembly process may not meet the actual assembly needs.

The theoretical model cannot establish the information transmission mechanism in the design, manufacture and assembly process of the product completely. In order to solve these problems, scholars were no longer satisfied with the analysis on the theoretical model, but paid more attention to simulating the real assembly situation and solving practical problems. Some research worked on the error accumulation in the process of dimensional chain transfer. Based on the minimum region constraint degree of freedom associated with geometric error, Guo combined the three-dimensional convex hull to construct the minimum region. A geometric error evaluation method was proposed based on the minimum region method [5]. Liu studied the error in the manufacturing and assembly process, and dealt with the difficulty to express and error transmission of parts manufacturing error information [6]. Zhang predicted the assembly accuracy by assembly positioning of parts, and proposed an assembly simulation positioning method based on surface constraint matching algorithm [7]. According to multi-color set theory, Li combined the joint surface and error relation contour matrix (JSS) which described the assembly relationship, joint surface type and error transfer characteristics, the error transfer characteristics of the surface of common assembly joints were then analyzed [8]. Zhao established a method for calculating the error bound by establishing a planar single-ring mechanism assembly uncertainty analysis model. In this work, the randomness of the assembly joint gap was solved for the purpose of satisfying the product design requirements [9]. Chen proposed the statistical analysis method of assembly error for geometric error and material error of coupled parts by using first-order perturbation theory and finite element method [10]. Cai and Qiao proposed a rigid-flexible mixed variable model, which applied the transformation relationship between different coordinate systems to represent the spatial relationship between the theory and error of the assembly unit [11]. The research of above mainly focus on mathematical modeling analysis at a certain point in the assembly process, and lack analysis of the overall assembly performance, which is difficult to practically apply in actual product processing and assembly processes.

This paper combines traditional tolerance technology and actual assembly error analysis, using measurement technology and model reconstruction technology to analyze the error of real product assembly. On the basis of obtaining the actual measured point cloud data of the actual machined parts, the part reconstruction technology is used to re-fit the measured point cloud data to form the model of measurement data. The model with measurement data is used to replace the theoretical model for assembly error analysis, in order to find the suitable parts being repaired. Using the model with measurement data analysis, on the one side, the actual assembly situation of the product can be predicted more accurately. On the other side, by analyzing the assembly simulation results of the product, the problems in the assembly can be found as well as in manufacture, process and design phases. Hence, the product design can be then optimized and the process methods can be improved. In the meantime, the information transmission in the whole product development phases could be enhanced, the product development cycle could be shortened, and the development cost is reduced. The overall process of assembly simulation based on the model with measurement data is shown in Fig. 1.

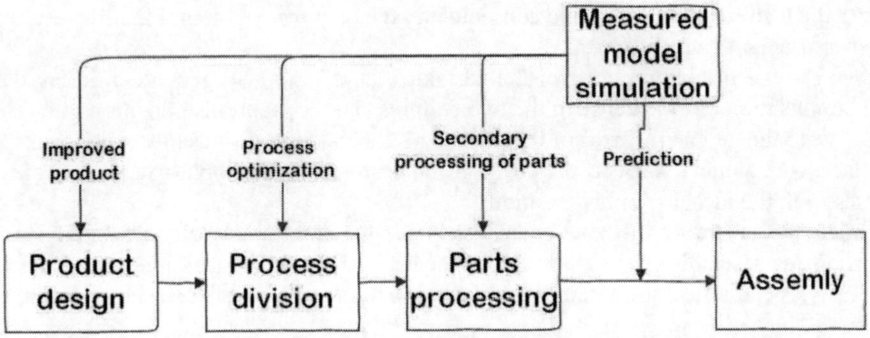


**Fig. 1** Assembly Simulation based on the Model with Measurement Data

## 2. Error analysis during assembly modeling

The error types from the establishment of the part processing model to the assembly model establishment mainly include positioning error, measurement error, machining error, fitting reconstruction error, assembly constraint error between different parts and random error at the assembly site. The measurement error, machining error and fitting error constitute the feature error of the part in the digital space. The part error positioning error, assembly constraint error and assembly site random error together constitute the assembly error of the assembly simulation, as shown in Fig. 2.

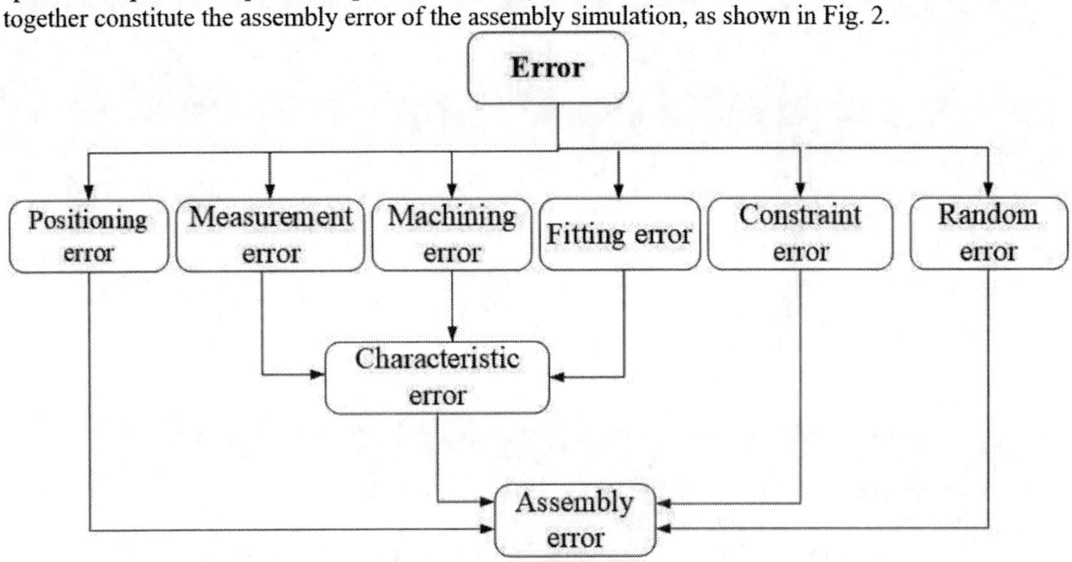


**Fig. 2** Assembly error analysis

In this paper, the model reconstruction is based on the feature unit, and the measured point cloud is re-fitted into the measured feature to form the measured data part model. In order to enhance the accuracy of the assembly simulation, it is necessary to minimize the error generated during the measurement and fitting process in the simulation. Selecting the appropriate measuring equipment and making a reasonable measurement plan during the measurement process can reduce the measurement error.

This paper applies the three-coordinate measuring instrument as the data acquisition device for the actual processed parts, and the measurement data point cloud is generated according to the measurement plan set in the process stage. In the fitting stage, the point cloud obtained by the coordinate measuring instrument is used for noise reduction and data simplified. Then the point cloud is divided into regions according to the features on the part, and consequently the features are fitted. Finally the new part model based on measurement data is formed.

In addition to the machining error of the parts, deviations are also generated during the assembly process. The constraint error is between the two features and the theoretical position of the actual phase matching. Due to the geometric error of the surface of the matching feature, there are peaks and valleys on the surface, which directly affects the position and posture of the two parts when they are assembled, and deviates from the ideal assembly position.

Combining the errors of the above analysis with the traditional tolerance analysis, the actual manufacturing and assembly errors are added on the basis of the theoretical dimension chain, so that the tolerance analysis can reflect the assembly of the product more accurately, as shown in Fig. 3.

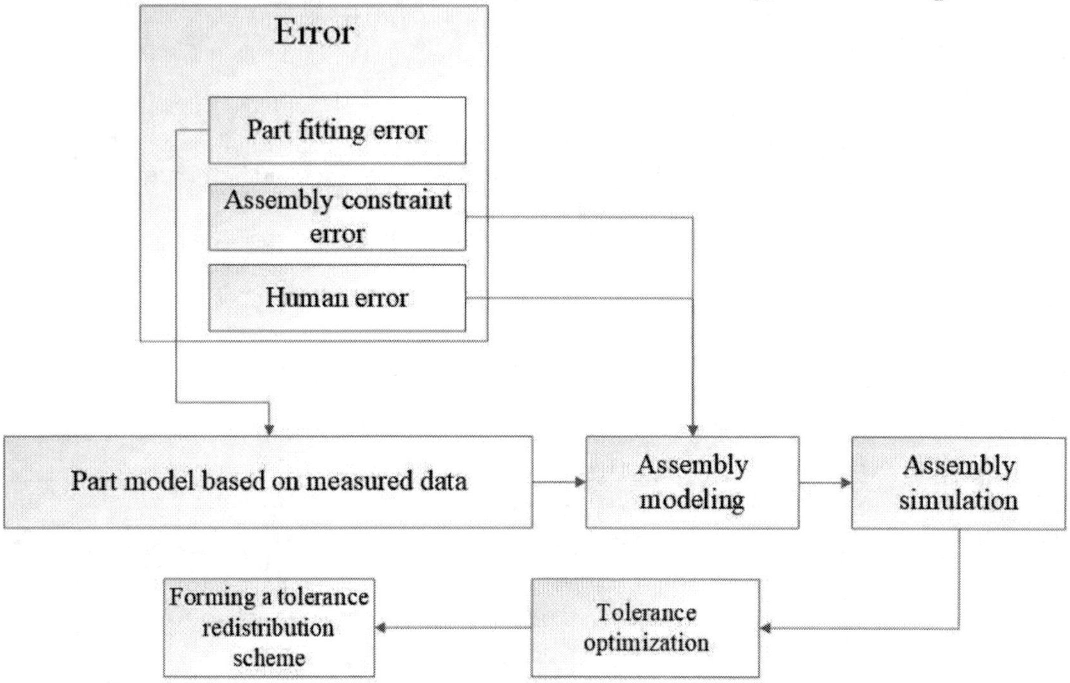


**Fig. 3** Tolerance analysis and optimization based on model of measurement data

Through the analysis of the dimensional chain, the unreasonable part of the dimensional chain is found and the tolerance is re-assigned. The tolerance distribution no longer refers to the amount of error allowed on the theoretical model, but the amount of parts that need to be repaired in the model with measurement data.

## 3. Assembly simulation tolerance analysis based on measurement data

As described above, this paper works on the error analysis based on the traditional tolerance analysis. The theoretical model used for the assembly simulation is replaced by the model with measurement data. The tolerance analysis is then performed under the measurement data model. Tolerance analysis based on measurement data mainly includes tolerance in modeling and assembly simulation calculation, as shown in Fig. 4.

### *3.1. Tolerance modeling based on measurement data*

First, according to the assembly process, the assembly sequence and positioning reference of the product are determined. The assembly relationship between the parts is defined to form the assembly process model of the product. The dimensional chain is searched based on the model to form an assembly dimension chain. The assembly deviations are added, as mentioned in the previous chapter, to the corresponding features to simulate the real assembly environment.

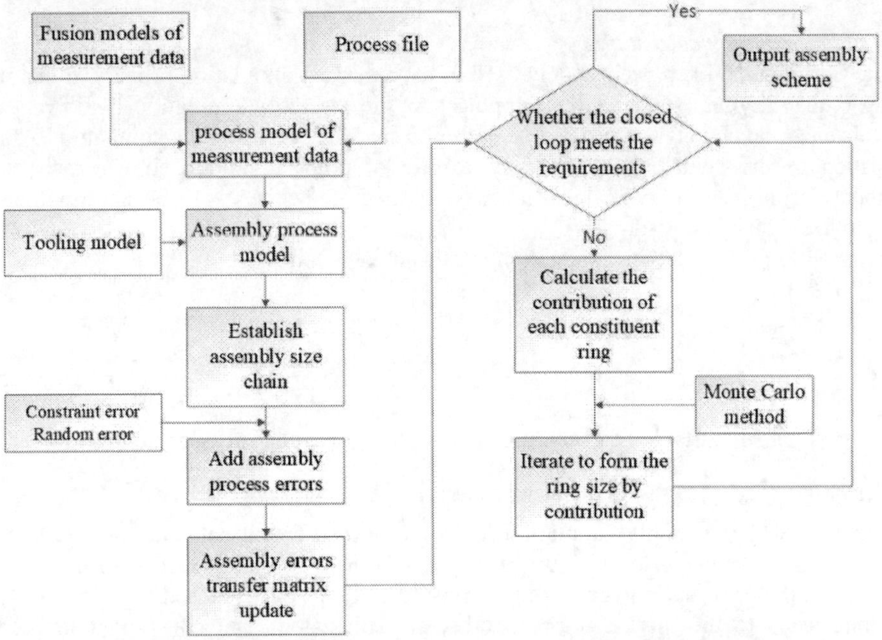


**Fig. 4** Tolerance redistribution based on measured data

Consequently, it is necessary to define a closed loop. The closed loop is a natural part (indirectly obtained) of the product assembly. It is a ring that needs to be guaranteed in the dimensional chain and is also an important factor of verifying the assembly quality of the product. Closed rings are mostly judged by measuring the critical dimensions of the product, such as the gap or step between two part planes. The closed loop is formed by the transfer of individual parts through the dimensional chain. The key factor affecting the size of the closed loop measurement is the accumulation of the deviations in individual constituent loops. The measurement of the closed loop is mainly divided into distance measurement and angle measurement. The distance measurement mainly includes plane gap measurement, plane step difference measurement, and axis position measurement.

The calculation for tolerance analysis in this paper is based on the analysis of 3DCS (Dimensional Control Systems), a commercial size analysis software that examines whether the product assembly meets the design requirements by analyzing the assembly dimension chain. In this paper, 3DCS is used as the computational solver for tolerance distribution. It is verified whether the tolerance value of the model redistribution can make the assembly of the product meet the required satisfaction. Therefore, it is necessary to convert the dimensional chain in the model into a 3DCS identifiable size chain, wherein the closed loop is an important one. The requirement of the closed loop needs to be converted into the measurement in the 3DCS. In order to reduce the complexity of assembly simulation, it is often necessary to convert feature measurements into measurements for discrete points.

For example, for two plane gap measurements, the points belonging to three different lines on the plane can be used instead of the related plane, and the corresponding three points can be found on the other plane that cooperates with it. Thus, the gap measurement between the two planes is converted into the distance measurement between the three pairs of points. Similarly, the axial position measurement is the comparison between the actual axial position of the product and the theoretical position. The theoretical position is mostly calibrated in the fixture, while the actual position is located on the part. By comparing the theoretical position and the actual position of the axis, it can be determined whether the axial position meets the design requirements of the product.

Based on the steps described above, the tolerance analysis model is established. Next, the assembly simulation needs to be performed.

### *3.2. Assembly simulation calculation*

The Monte Carlo method has been widely used in the assembly simulation process. It is a method for simulating probability and statistics by computer. As the assembly simulation is run repeatedly, the statistical data can be obtained in order to acquire the credible results of the simulation. Afterwards, a list consisting of the contribution of every feature or point is calculated for each closed loop measurement. If the measurement does not meet the requirements, it can be appropriately adjusted according to the contribution value.

The contribution of the statistical situation is calculated as follows.

$$contrib_{x_i} = \left(\frac{\left(\frac{\partial U}{\partial x_i} \times \sigma_{x_i}\right)^2}{\sigma_U^2}\right) \times 100\% \tag{1}$$

where $x_i$ is the ith variable representing each feature, U is the measurement object, $\sigma_{x_i}$ is the standard error of $x_i$, and $\sigma_U$ is the standard error of U.

If the results of the assembly simulation using real measurement data cannot meet the design or process requirements, it needs to find out the very part or parts in the dimension chain that affects the assembly. Then, the contribution values of the constituent loops to the closed loops in the dimensional chain, which leads to the unsatisfactory results, are calculated. The constituent loops are sorted according to the contribution values. Finally, according to the difference of contribution in the simulation results, the ring size is adjusted iteratively, in order that the size range which satisfies the assembly requirements can be found.

## 4. Iterative adjustment of tolerance based on measured data

Through the simulation calculation of the above process, the overshoot of all closed loop measurements can be obtained from the simulation results. If the tolerance is within an acceptable range (the proportion of the simulated 2000 assembly simulations that meets the requirements is higher than the specified percentage, eg ±3σ), then it is expected that the currently processed parts can be assembled under the current assembly process plan, while the design and process requirements can be achieved.

Conversely, if the out-of-tolerance rate is within an unacceptable range, then tolerance-based redistribution using the current measurement data is required to be performed. In other words, without changing the assembly process plan, through the analysis of the assembly dimension chain, the component ring that affects the assembly success rate needs to be found and the scope of the repair processing needs to be decided. Furthermore, the repair position and the repair size needs to be determined.

Before the assembly repair scheme performs, the size chain relationship needs to be converted into a relationship that the 3DCS solver can recognize. In 3DCS, the direction and length of the constituent loops and closed loops in the dimensional chain are converted into directions and distances between corresponding points in space. As shown in the Fig. 5, the closed loop represents the measurement point pair of the part where the two part features match in assembly. The constituent loop represents the distance between every two assembly control points, which are formed by the assembly constraint relationship.

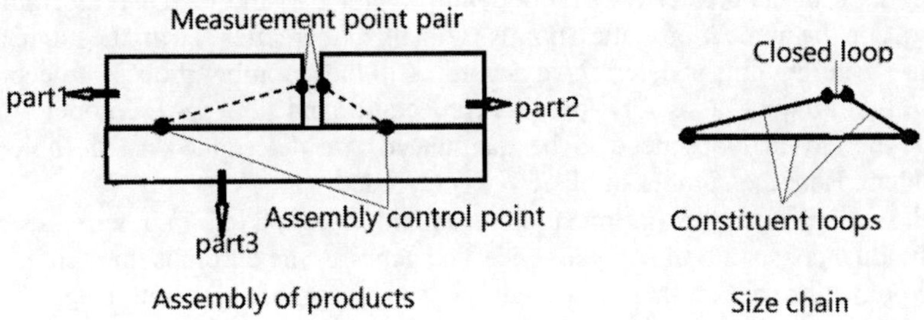


**Fig. 5** Representation of the size chain in the solver

Then, the closed loop which does not meet the assembly requirements needs to be found. The size chain is analyzed where the closed loop is located. The constituent loops which have a large influence on the closed loop is then determined. This process is performed with the contribution analysis as described above. With the experimental practice, it is found that the selection set of features with the contribution value greater than 10% can generally solve the problem of closed loop. These features that may be repaired are used as an alternative feature of tolerance redistribution.

Since the features has been converted into discrete points, i.e., using three points to represent the plane feature. The feature points on the constituent loops can be used as the input variables in the calculation. The measurement of the closed loop also replaces the gap or step of the plane with 3 pairs of points as described above, which converts the distance measurement between the corresponding point pairs and uses the measured value as the output. Here, the difference between the measured distance and the theoretical distance of the point pairs is smaller, the performance is better. In that case, the tolerance allocation problem is transformed into a multivariate and multi-objective optimization problem.

Because there may be conflicts and competition among the various objectives, the outcome of this optimization problem is usually a solution set. The traditional genetic algorithm cannot solve the problem appropriately. Therefore, this paper proposes a combined genetic algorithm to seek for optimal or near optimal solutions.

The factors in combined genetic algorithm GA are represented as the following seven-dimensional vector form.

$$GA = (N_{pop}, N_{gen}, \Omega, fitness, f_{sel}, FITNESS, F_{sel}) \tag{2}$$

where $N_{pop}$ is the population size, $N_{gen}$ is the iterative times, $\Omega$ is the genetic operator (crossover and mutation) and their probability set, $fitness$ is the fitness function of the subpopulation, $f_{sel}$ is the selection rule of the subpopulation, $FITNESS$ is the fitness function of the recombined population, and $F_{sel}$ is the selection rule of the recombined population.

In the combined genetic algorithm, the multi-objective optimization problem is divided into several independent single objective optimization problems before performing the selection operation. According to the number of optimization objectives, all individual chromosomes in the initial population are divided into corresponding number of subpopulation groups, and a selection operation based on the fitness function is performed for each subpopulation group.

Consequently, the selected subpopulation individuals with relatively high fitness value are recombined into new large populations, and the squares of the single objective function are summed up to form a new objective function, which is employed in the selection, crossover, mutation operations for the recombined population. After several iterations, a solution that satisfies all the closed loop measurement objectives can be obtained, shown as Fig. 6. The specific steps of the combined GA approach are described as follows:

The feature points with the greater contribution values than 10% in the 3DCS assembly simulation report are selected, and the number of feature points is obtained thereafter. Then, the range of acceptable repair values of the feature points is determined according to the assembly process procedure.

According to the product assembly process procedure and the CAD model, the assembly requirements that the closed loops need to be guaranteed, are determined as the objectives of the optimization problem. Here the number of objective is recorded.

A form of the solution to the optimization problem is converted into a representation of a chromosome individual consisting of several genes.The genes in the chromosome represent the repair values of control points for each of the constituent loops. The number of iterations and the number of populations are set, and an initial population is generated.

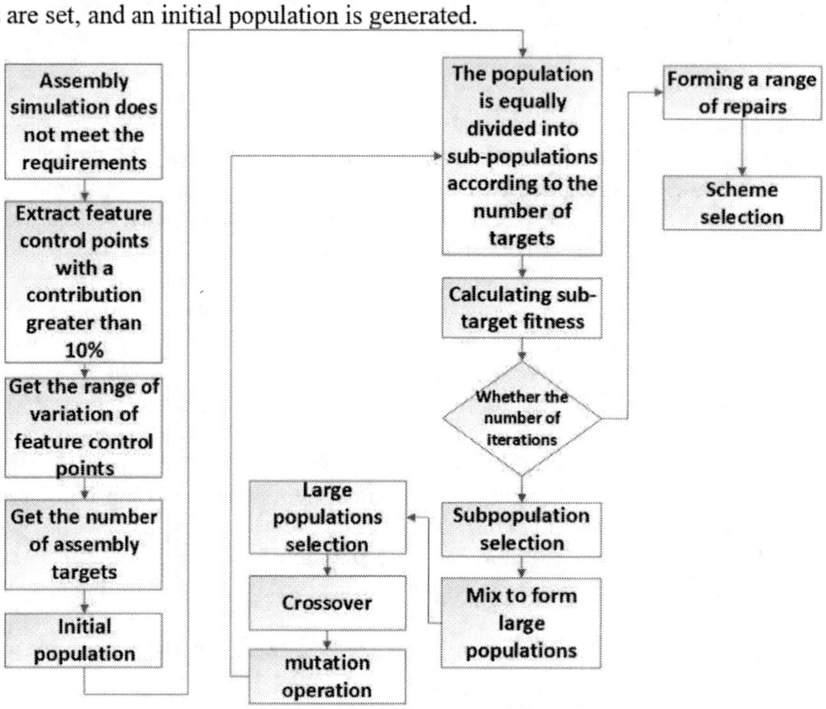


**Fig. 6** Genetic algorithm optimization

The populations are grouped equally according to the number of objectives, and each group corresponds to an optimization objective.

Each individual chromosome, representing a solution consisting of repair values of the control points, is set to be the input of 3DCS solver for simulation and calculation. For each subpopulation, the chromosomes with higher fitness values, corresponding to each respective objective, are selected. Here, the fitness function is calculated as shown in Eq. (3). The greater the function value is, the higher the fitness of the individual is in the corresponding subpopulation, and the easier this chromosome is to be selected. .

$$fitness = \frac{1}{|mean - Normal|} \tag{3}$$

Where fitness is the contribution value of each measurement area, mean is the statistical median value of the corresponding measurement area, and Normal is the theoretical value of the measurement.

The Roulette Wheel selection method is employed as the subpopulation selection rule $f_{sel}$. The possibility of each chromosome being selected as a parent is the proportional to its fitness value.

Parents selected from subpopulations are recombined into a new population group.

In the newly recombined population, the simulation and calculation result of the solver is used again, where the contribution values represented by chromosome fitness functions. The function here is calculated as shown in Eq. (4). The descendant selection rule $F_{sel}$ of large populations is the roulette method, and the individuals with higher fitness are more likely to be selected.

$$FITNESS = \frac{1}{\sum_{1}^{N}(mean_i - Normal_i)^2} \tag{4}$$

where FITNESS is the total fitness value of all measurements; mean is the statistical median of the corresponding measurement area; Normal is the theorctical value of the measurement, n is the number of measurements.

The Roulette Wheel is adopted again and newly formed group of parent chromosomes are then generated.

Parent chromosomes produce offsprings, based on single-cutoff-point crossover. The probability of crossover execution is set to be Ωc. When the crossover probability values of two parents are greater than Ωc, the crossover is executed. Fig. 7 shows an example of the crossover operation.

| | ***Before crossover*** | ***After crossover*** |
|---|---|---|
| $H_1$ | $a_1a_2\cdots a_ia_{i+1}\cdots a_n$ | $a_1a_2\cdots a_ib_{i+1}\cdots b_n$ |
| $H_2$ | $b_1b_2\cdots b_ib_{i+1}\cdots b_n$ | $b_1b_2\cdots b_ia_{i+1}\cdots a_n$ |

**Fig. 7** Crossover operation

In order to increase the diversity of the population, the mutation operation is performed after the crossover. The mutation probability is set to be Ωv. When the mutation probability of a certain gene in a certain chromosome is greater than Ωv, the gene value is changed to complete the mutation operation. For example, if the gene ai is mutated, the new gene becomes (ai + u), where u is a random number in a fixed range such as [-0.2, 0.2].

Determine whether the pre-defined number of iterations is reached. If not, return to step (3).

After a certain iterations finish, or the individuals in the population stop changing, the chromosomes are converged to one or several solutions which are able to satisfy the assembly requirements. The mechanism ends. These optimal or near optimal solutions can be adopted before the assembly execution. Thus which processed parts needs to be repaired and the repair values are then determined.

## 5. Example analysis of typical assembly

In this paper, a typical frame-like component is used as an example for analyzing the approach described above, as shown in Fig. 8. The assembly procedure is performed as follow. First, the skin (part1) is positioned on the basis of the frame. Then the rib (part3) is assembled onto the skin. Finally, the stringer (part2) is assembled onto the skin. There is a gap between the stringer and the rib, and the required gap size is $6 \pm 0.5$ mm.

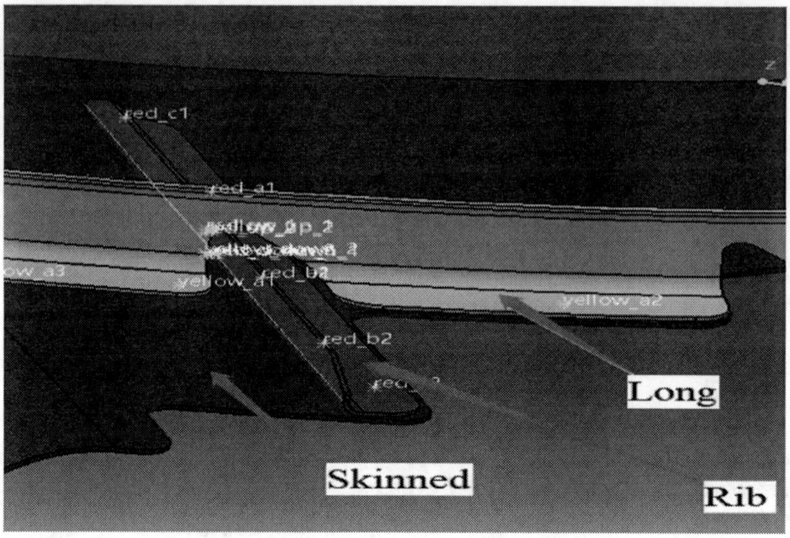


**Fig. 8** Modeling of measured models

*5.1. Assembly simulation based on measured data*

Using the tolerance analysis software 3DCS, the assembly simulation analysis model of the assembly is established according to the actual assembly process plan, and the gap requirements of the assembly are converted into the measurement scheme recognized by 3DCS.

The specific method is taking four measurements of the four vertices on the measurement gap surface of Part 3 and Part 2, respectively named up_1, up_2, down_1, down_2. The gap measurement between Part2 and Part3 is replaced by the measurement between these four pairs of point. Then, the error in the assembly process is added for all the assembly joints. The deviation value is determined according to the assembly process procedure and the error analysis calculation. The measurement and assembly error here is assigned to be 0.1mm. The measurement points are shown in Fig. 9. The inside of the frame is the gap between the stringer and the rib.

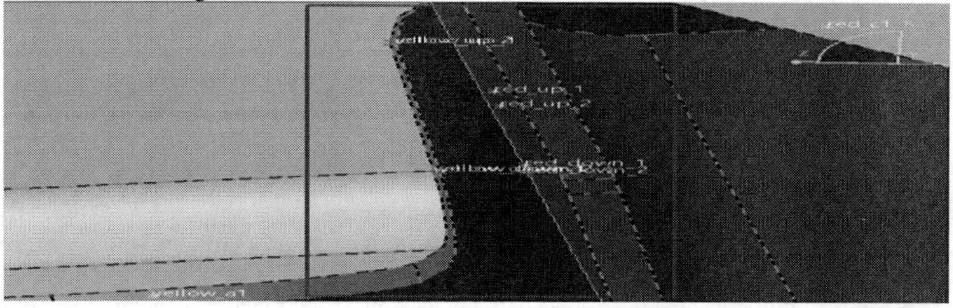

**Fig. 9** Closed ring measurement point and position

*5.2. Simulation analysis based on measured data*

As shown in Fig. 8, the point cloud measured by the three coordinates is simplified. The assembly simulation has been run for 2000 times, and the results are shown in Fig. 10.

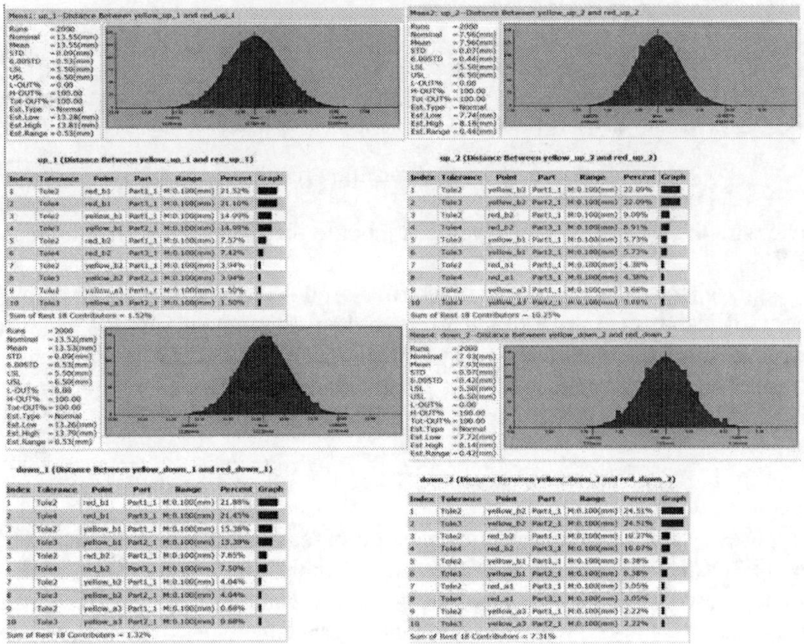

**Fig. 10** Analysis of measured data

It can be found from the simulation assembly results that the points with the maximum contribution values are red_b1 and red_b2 of Part3, yellow_b1 and yellow_b2 of Part2, red_b1 and red_b2, yellow_b1 and yellow_b2 of Part1. Consequently, these points with greater contribution values are mainly located on the connection between Part3 and Part1, as well as on the connection between Part2 and Part1. From the analysis of the measurement data, it can be found that there are certain processing errors on the plane belonging to Part1, where the Part3's corresponding plane is matched during assembling. In that case, we can estimate that the results of the simulation analysis were credible. Moreover, through the analysis of the contribution value for each point, we can also find other solutions to meet the gap assembly requirements, so the 3DCS contribution analysis can find out where the assembly problems.

### *5.3. 3DCS-based tolerance redistribution with measurement data*

By analyzing the computational results of 3DCS simulation, the measurement data points with a contribution greater than 10% are selected. These points will be used as variable inputs for tolerance distribution. Next, the tolerance variable is initialized, and the tolerance variable value is assigned to be the input for the 3DCS.

The Monte Carlo method is used for 2000 assembly simulation, and the results are statistically analyzed. The median of the statistical measurements was compared with the theoretical median, and individuals with a small median difference were retained.

In the GA steps as described in the last chapter, the offspring need to be selected twice. The first selection is for each measurement objective, and the individuals whose statistical median values deviate from the median of the measurement theory are selected thereafter. After chromosomes recombination, the reciprocal of the square sum for all the measured statistical median values and the theoretical median values are selected. The individual chromosome with the larger value is more likely to be selected. Then, after crossover and mutation, a new child chromosome is generated. Its corresponding solution is assigned as the new input to the 3DCS solver for the next iteration. After 500 times iteration, we can find the optimal individual with the smallest square sum. The iteration results are shown in Fig. 11.

| Tole2 | Tole10 | Tole11 | Tole12 | Tole13 | Tole14 | Tole15 | Tole3 | Tole7 | Tole9 | Tole4 | Tole6 | Tole8 | Tole1 | FITNESS |
|---|---|---|---|---|---|---|---|---|---|---|---|---|---|---|
| -5.491317272 | -0.561601639 | 0 | -0.047975097 | -1.417279601 | 0 | -0.033143103 | -1.590319514 | 0 | -2.404797554 | -3.346171379 | -3.277688026 | -1.903622508 | 0 | 4.818182124 |
| -5.491317272 | -0.561601639 | 0 | -0.047975097 | -1.417279601 | 0 | -0.033143103 | -1.590319514 | 0 | -2.404797554 | -3.629810572 | -3.277688026 | -1.903622508 | 0 | 1.884130468 |
| 0 | -0.633198023 | 0 | -2.639545918 | -2.416699648 | 0 | 0 | -2.035645723 | 0 | -0.814661086 | -3.075716496 | -3.277688026 | -5.296304226 | 0 | 0.913060834 |
| 0 | -2.038941622 | 0 | -0.110782191 | -2.158146858 | -4.059389114 | 0 | -1.590319514 | 0 | -2.404797554 | 0 | -1.082735658 | -1.903622508 | -2.092776299 | 1.399350852 |

**Fig. 11** Optimization result

The variable with the largest FITNESS value is input to the solver for verification, and the result is shown in Fig. 12.

As can be seen from the results of the iterative process, all the measurements are within the allowable tolerance. The statistical median value of the errors in the maximum closed loop measurement is 6.27, and the difference from the theoretical measurement value is 0.27. The difference between the measurement value and the theoretical value is small, resulting in assembly requirements satisfaction. The feasibility of using the combined genetic algorithm to iteratively optimize the amount of repair scheme is verified.

## 6. Summary and outlook

This paper proposes a tolerance redistribution method based on measurement data. The dimensional analysis software 3DCS is applied for acquiring the assembly simulation results, while the combined genetic algorithm is used to adjust the size of the processed parts. The tolerance adjustment amount is then calculated to achieve the optimal assembly quality. Since the types of assembly features currently established are relatively simple, in future the approach described in this paper will be applied with enriched assembly feature types and with the consideration of more complex surfaces. Moreover, in order to increase the authenticity of the assembly simulation, finite element analysis could be involved to solve the problems of clamping deformation, loose clip rebound and other force deformation during the assembly process of flexible parts.

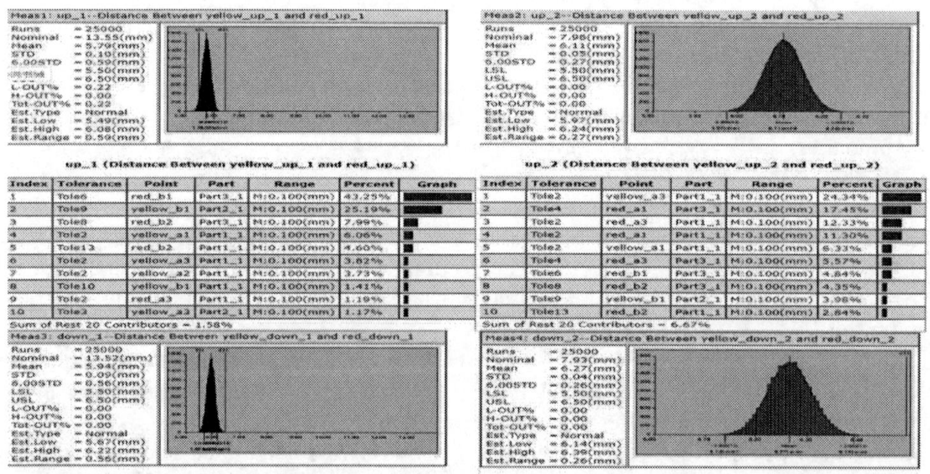

**Fig. 12** Optimization Analysis Report

IOP Conf. Series: Materials Science and Engineering **772** (2020) 012002 doi:10.1088/1757-899X/772/1/012002

**References**

[1] Dissertation: (Zhang K. Research on cumulative analysis and tolerance optimization method for aircraft component assembly error (PhD thesis, Northwestern Polytechnical University, China 2006)).

[2] Qureshi A. J., Dantan J. Y., Sabri V., et al. A statistical tolerance analysis approach for over-constrained mechanism based on optimization and Monte Carlo simulation. Computer Aided Design. Vol. 44 (2012) No. 02, p. 132-142.

[3] Ghali M., Tlija M., Aifaoui N., et al. A CAD method for tolerance allocation considering manufacturing difficulty based on FMECA tool. The International Journal of Advanced Manufacturing Technology. Vol. 91 (2017) No. 05-08, p. 2435-2446, doi: 10.1007/s00170-016-9961-x.

[4] Dissertation: (Tang W. (2015). Research on Aircraft Nonlinear Assembly Error Analysis and Tolerance Cooperative Distribution Method. (PhD thesis, Northwestern Polytechnical University, China 2015.) p. 146.)

[5] Guo C. Geometric Error Assessment and Assembly Accuracy Prediction of Complex Products. Beijing Institute of Technology. (2015) No. 03, p. 171.

[6] Liu J., Jin J. , Shi J. State Space Modeling for 3-D Variation Propagation in Rigid-Body Multistage Assembly Processes. IEEE Transactions on Automation Science and Engineering. Vol. 7 (2010) No. 02, p. 274-290.

[7] Zhang Q., Jin X., Zhamg Z. Assembly simulation positioning method based on surface constraint matching algorithm. Journal of Mechanical Engineering. Vol. 54 (2018) No.11, p. 70-76.

[8] Li C., Liu Z. , Qin H. , et al. Method of Calculating the Assembly Error Transfer Path Based on JSS Matrix. part of verifying the assembly quality of the product. Vol. 51 (2015) No. 20, p.177.

[9] Zhao Q., Guo J. , Hong J. Uncertainty Analysis of Assembly Error of Planar Single-loop Mechanisms Based on the Rotatability Laws of Linkages. Jixie Gongcheng Xuebao/Journal of Mechanical Engineering. Vol. 54 (2018) No. 11, p. 29-38.

[10] Chen H., Tan C. , and Wang Z. Statistical Analysis of Flexible Assembly Errors of Coupling Geometry and Material Error. Journal of Aviation. Vol. 41 (2014) p. 45-64.

[11] Cai N., Qiao L. Rigid-Compliant Hybrid Variation Modeling of Sheet Metal Assembly with 3D Generic Free Surface. Journal of Manufacturing Systems. Vol. 41 (2016) p. 45-64.

# The Modeling Method of Digital Twin Models for Machining Parts

**Qi Zou[1, a], Zhixia Hou[1], Mingyang Wang[1] and Shan Jiang[2, b]**

[1] China Aviation Manufacturing Technology Research Institute, Beijing, China;
[2] School of Electromechanical, Northwestern Polytechnical University University, Xian, China.

[a] 13329392517@126.com, [b] 1123750226@qq.com

**Abstract.** This paper introduces digital twin models for machining parts based on key features and gives the modeling method. In this paper we propose three layers structure of the digital twin model, including geometric layer, data layer and document layer, in order to make machining process optimization and quality prediction. The digital twin model realizes the expression of physical parts in digital space and the information correlation with theoretical processing features. Besides, this paper also analyzes the potential application scenarios and development trends of digital twin models of machined parts.

**Keywords:** Digital twin model; Machining; Key features; Processing quality.

## 1. Introduction

With the rapid development of communication level and information technology, digital representation of physical entities formed at various stages of industrial product manufacturing has become the trend of intelligent manufacturing [1]. At present, the Model of Definition (MBD) model can reflect the design size and design requirements of the design model according to the digital quantity, and on this basis, add the 3D mark and engineering annotation part. Some scholars solve many problems in the machining process by constructing part models of different machining stages.

Tian et al [2] proposed the concept of the process MBD model and its construction method; the process MBD model is divided into the three-level structure of "physical layer-extension layer-application layer", and the annotation information and process attributes of the processing features are extracted. Hu [3] proposed the management method and application of the MBD model of the process based on the process steps in the product manufacturing stage. Liu et al [4] proposed a modeling method for 3D process models with process assisted features to improve the efficiency of process planning. The model technology has been widely used in the development of the product "design-manufacture-test", which greatly improved the manufacturing efficiency and quality of the product. The manufacturing based on the theoretical model alone cannot adapt to the changes brought by the process and environmental uncertainty of the complex product manufacturing process. The new generation of aircraft uses a large number of difficult-to-machine materials and integrated structures. The processing technology is complex, the deformation control is difficult, and the processing process is large. The

Content from this work may be used under the terms of the Creative Commons Attribution 3.0 licence. Any further distribution of this work must maintain attribution to the author(s) and the title of the work, journal citation and DOI.
Published under licence by IOP Publishing Ltd

existence of uncertain factors poses a challenge to the product manufacturing model based on theoretical models.

With the development of sensor technology and various digital devices for different machining processes, processing data and quality inspection data such as the rotational speed, force and wear values of the tool can be obtained in the machining stage of the part. At present, these data have not been fully exploited and utilized, and have not been associated with design models and process models. They cannot effectively support the optimization and continuous improvement of product manufacturing processes, and have not formed a model closed loop of theoretical models and physical manufacturing processes, and processes oriented to manufacturing processes. The iterative optimization model with manufacturing instructions has not been formed, and the model technology needs to be extended to the physical manufacturing process.

Digital Twin refers to the use of digital technology to describe and model a process or method consistent with the physical properties, behavior, and performance of a physical entity, also known as digital twinning technology [5], which was first taught by GRIEVES, a professor at the University of Michigan, USA. Presented in the course of cycle management. In 2015, General Electric Company of the United States built a digital twin model of the engine that can be used to monitor the engine's operating status, check engine performance in real time, and provide maintenance and maintenance recommendations based on its operating status [6]. From the perspective of industrial applications, digital twinning technology has also received a lot of research and breakthroughs. Chen et al [7] proposed the idea of constructing an aircraft digital twin assembly workshop. The workshop can collect various data of the aircraft assembly process in real time, realize the virtual simulation of the aircraft equipment process and control the assembly process of the aircraft according to the collected data. The intelligence of the aircraft assembly process. Yu et al [8] applied the digital twin model to product configuration management, and proposed to use the ontology technology to construct the product digital twin model to express the product configuration information. This scheme can be used to solve the product configuration management problem under the full 3D development model. Production workshop manufacturing IoT [9] and data integration technology are the direction of digital hygiene research. Tao [10] used digital twinning technology to solve the problem of the fusion of massive data in the workshop and the fusion of multi-source heterogeneous data. Huang [11] made a research of the construction of the IoT in discrete workshops. Greyce N. Schroede [12] proposed the use of AutomationML modeling to construct digital twin attributes to realize the integrated CPS (Information Physics System) for future manufacturing systems and product service systems and give industrial application examples. At present, digital hygiene technology is mainly applied to processing equipment, production lines and workshops, and the research on digital twin model for process optimization is in a blank.

This article proposes to construct a digital twin model of machined parts in the physical manufacturing stage to solve the mentioned problems. The product development and quality control model based on the part digital twinning model (Fig.1) has the following characteristics:

1) Modeling and management of various theoretical and measured data of machined parts, fully exploited and utilized to improve design and process quality.

2) The part processing process is closely integrated with the measurement, the manufacturing process is dynamically optimized, and the equipment, tooling, and component deformation are fully measured and controlled.

3) The execution process of the manufacturing instruction based on the model realizes closed-loop control by means of a model, which can be evaluated and predicted.

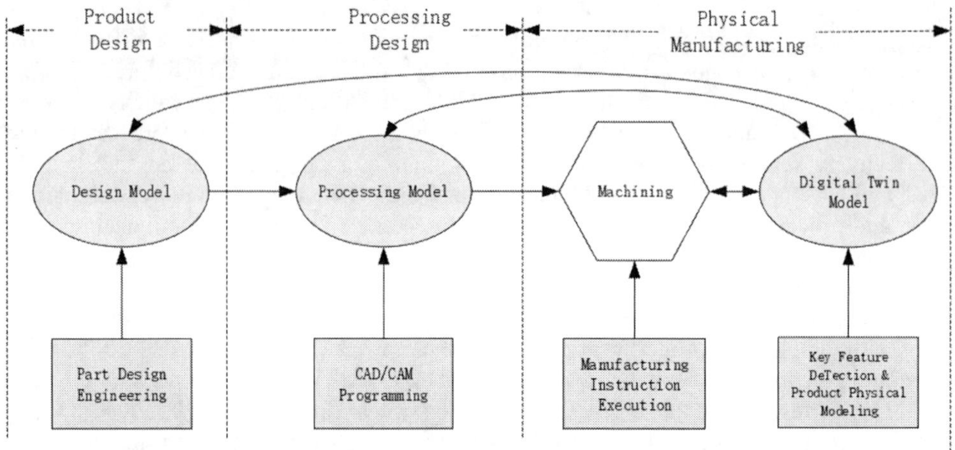


**Fig. 1** Product development and quality control mode under digital twin model

## 2. Method for Constructing Digital Twin Model of Parts

The digital twin model of parts provides a way to digitally reflect the state of the physical entities of the part during machining. In fact, each machining process stage of each part corresponds to a physical entity of the product, and corresponds to a digital twin model. Each digital twin model should contain all the information that can express the real physical state of the part.

For the digital twin model of machined parts, it needs to contain process information, machining information, inspection information, shape and material information, etc., and completely record the data and real geometric model of each feature that expresses the part will greatly reduce modeling efficiency, not only that, the huge amount of engineering data is also a very big challenge for computer hardware systems. To this end, this paper proposes a digital twin model based on key features. The key features refer to the typical structures and locations that are difficult to control in the machining of parts or that have a large impact on product quality and performance. These include the design key features (such as designing the separation surface, bearing axial surface, etc.), manufacturing key features (such as machining datums, positioning holes, etc.), testing key features. It is an effective and reasonable modeling method to sort and organize the information that needs to be expressed according to the key features. The construction of digital twinning model based on key features essentially embodies a modeling concept of demand-oriented digital twin model. This method does not pursue the large and complete model, and appropriately abandons some data that does not require special attention during the machining phase. The formation of a scientific, systematic digital twin model structure with key features as a carrier is effective and reasonable for recording the state of real machined parts.

The demand-oriented digital twin model modeling method closely combines the data content and structure of the model with the application requirements. The process digital twin model for process and instruction optimization not only needs processing information including measured key features, but also needs theory. The model implementation is closely related. The digital twin model proposed in this paper contains two parts: the theoretical model and the actual model for the part entity in this state. The theoretical data expressed in the theoretical model part of the digital twin model must be organized according to the established key features. The geometric modeling of the theoretical model can be performed only for key features. A part has only one digital twinning theoretical model under one process stage, but there are a series of digital twin measured models obtained with the batch variation or detection data changes. Such a theoretical model and a measured model together constitute a digital twin model of an actual machined part (Fig.2).

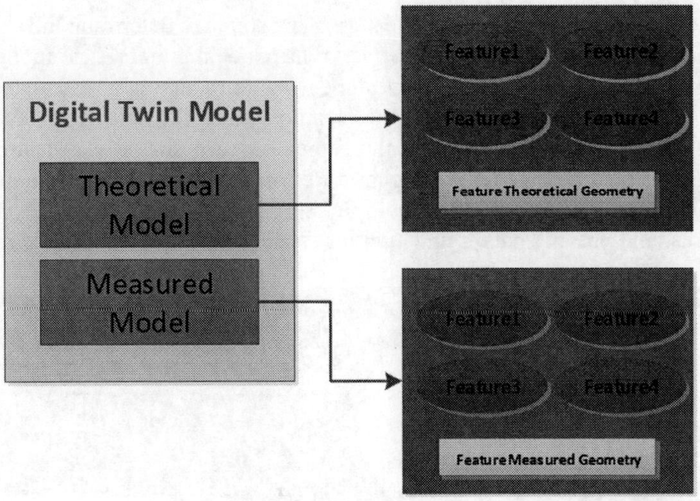


**Fig. 2** The content of the digital twin model of machined parts

*2.1. Three-layer Structure of Digital Twin Model*

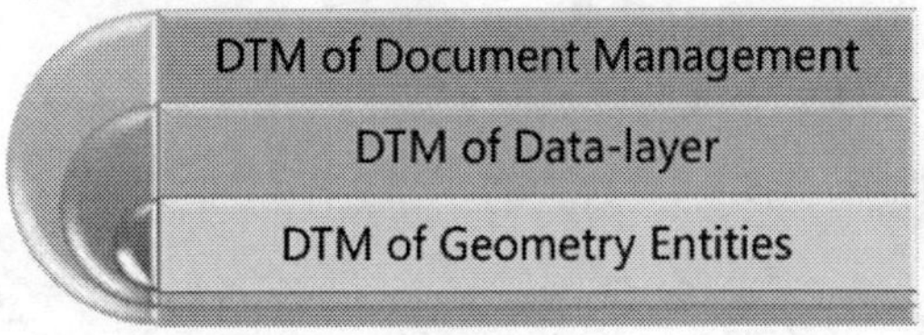


**Fig. 3** Three-layer Structure of DTM

Digital twin modeling of machined parts can be performed at three levels. The resulting model is a model with a three-layer structure (Fig.3). The first layer is a geometric layer, which is mainly used to create and reconstruct geometric entities of key features, and obtains geometric entities with universal formatting such as valuable part models or triangular patches. Its purpose is to provide the corresponding geometric entities based on the characteristics of the machining simulation and process parameters optimization. The second layer is a data layer, which fills in the processing feature attributes for the detection data and the manufacturing processing data, and forms a digital twin measurement model information representation with the feature as a carrier. Its purpose is to track processing parameters online and to analyze the quality of the processing offline. The third layer is the twin model layer, which realizes the association between the theoretical model and the measured model, and stores the digital twin model in terms of parts, processes, batches, etc. It provides the data source and platform for process personnel to analyze big data based on parts or process flows.

In the digital twin model, the relationship between the model and the theoretical geometric entity needs to be established. The geometric entity in the theoretical model is used as the 'virtual' in the digital twin; and the key features of each real product are reconstructed by point cloud fitting or parameterization. The geometric entity is regarded as 'real'; in the key feature attribute, the value in the theoretical entity is regarded as 'virtual'; the value obtained by the actual measurement in each real product is regarded as 'real'. It realized the "virtual and real" mapping of theoretical models and measured models.

*2.2. Modeling Process of DTM*

The process of selecting a machined product to create a digital twin model is divided into four parts. The first step is to define the key features of the product and define the corresponding attributes and feature reconstruction methods for each key feature. This step can determine the content of the model geometry layer and data layer. The second step is to introduce the theoretical model and calibrate the theoretical values of the attributes of the key features. The third step is to calculate and determine the measured values of the attribute values of the key features when the part entity detection data and the online collection data arrive, and reconstruct the geometric parts for the key features that need to be reconstructed, and the obtained model organization Released as a whole to form a digital twin model. The fourth step is to organize and manage multiple digital twin models according to the standards of parts and processes, and provide data foundation for subsequent model calls. The modeling process is shown in Fig.4.

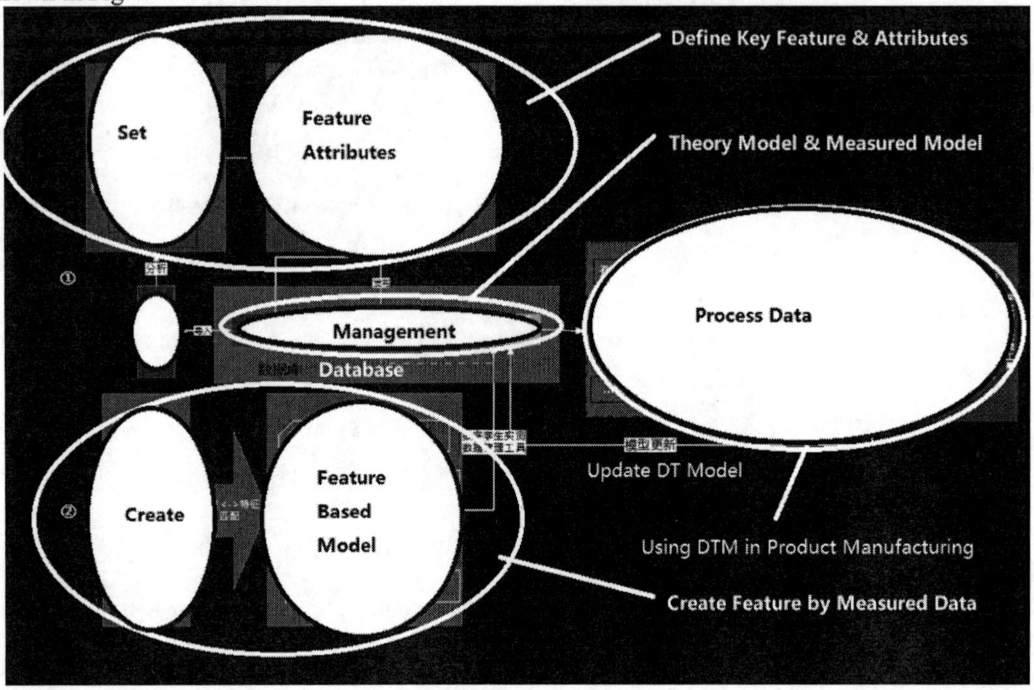


Fig. 4 Modeling data stream of DTM

From a usage perspective, the digital twin modeling tool opens two interfaces to provide two-part staffing. For designers, inspectors and process personnel, provide the ability to modify key feature (attribute) libraries, define key features under part or component CAD models, publish theoretical model files, import measured model files, and task-oriented output. Designers and technicians can perform optimization analysis on geometric or attribute metrics based on theoretical and measured models. For the machining technician, a data entry operation interface is provided, and an import interface is prepared for the automatically picked up data, and the non-structural data is manually entered.

## 3. Key Technologies for DTM

The key features in the digital twin model serve as the carrier of the part entity data, and the types need to be rigorously screened and determined. The data items contained in various key features, also called the attributes of key features, also require strict paradigms and definitions. The screening and definition must be adapted to the current state of the process and the application of the digital twin model. In addition, there are many methods for the construction of feature geometry entities in digital twin models. This chapter discusses the key feature libraries and geometric layer modeling methods of data layer features and feature attributes. Finally, the construction method of digital twin database is expounded.

### *3.1. Key Feature Library*

Through the investigation of complex part processing technology and inspection process, the key features of manufacturing are defined on the basis of geometric features, such as the plane of component fitting, the axis and hole of assembly, and easy-scratched surfaces, etc., are constructed as key features in the digital twin model. Before establishing a digital twin model for each real part, you need to define a key feature library, that is, the type of key features and the attribute information contained under the features. For different machining processes, in-depth research on process methods, process inspection standards and process knowledge maps are carried out in advance, and then the types and information of key features are fully defined.

Among the attributes included in the key features, one is the geometric information of the features, such as the hole diameter, the surface normal, the height of the boss, etc.; the accuracy information such as flatness, roughness, verticality, plane gap; and the parameters that are customized according to process requirements are referred to herein as alternate feature parameters such as hole center point, axis minimum diameter, predefined assembly point position, and feature scan point cloud. According to the knowledge map of a certain combat wing box processing process, this paper abstracts five key features and defines related attributes as shown in Table 1. The other type is non-geometric information about features, including processing conditions, process documentation, and large amounts of data generated by machine tool processing. The author abstracts the processing information into the attributes quantified in Fig. 5 for modeling and storage.

**Table 1.** Key feature attributes

| Feature | Hole | Plane | Surface | Pad | Slot |
|---|---|---|---|---|---|
| Attr. 1 | Type | Type | Type | Height | Width |
| Attr. 2 | Center point | Flatness | Cylindricity | Length | Depth |
| Attr. 3 | Diameter | Parallelism | Roughness | Diameter | Position |
| Attr. 4 | Depth | Roughness | Concentricity | Symmetry | Symmetry |
| ... | ... | ... | ... | ... | ... |

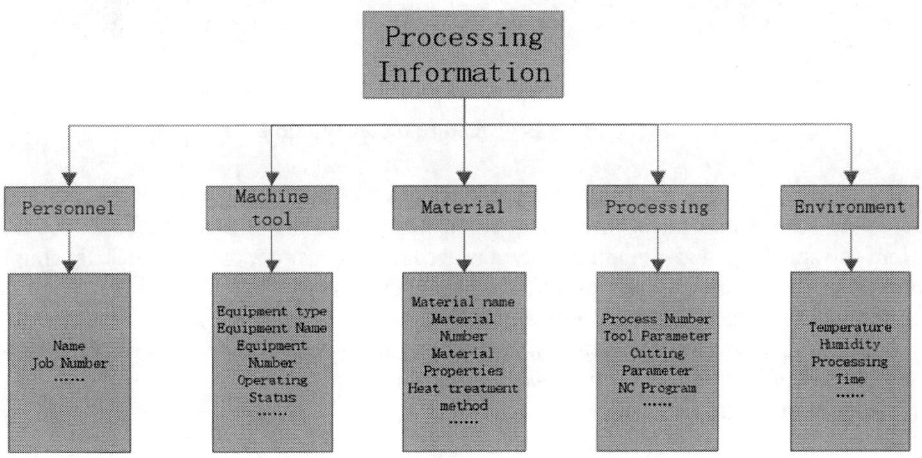


**Fig. 5** Processing information of DTM

### *3.2. The Geometry Entity Construction of Key Features*

The attributes of key features cannot truly represent the shape and exact location of key features. Many simulation optimization tasks must be performed based on the model geometry entities. Therefore, while defining key features in the theoretical model, the extraction and association of feature geometric elements should be performed simultaneously. For example, the hole feature, while defining its precision parameter, geometric parameter, machining parameter and spare feature parameter, needs to extract the geometric elements such as the upper and lower surface of the hole, the inner wall cylindrical surface and the rounded corner on the theoretical three-dimensional geometric entity. In the measured model, the feature needs to be reconstructed according to the mode selected by the user, and then the reconstructed entity is associated with various parameters of the key feature. There are two main ways of feature reconstruction. One is to reconstruct the triangular patches, and the other is to reconstruct the feature entities. The flow of the two reconstruction methods is given in Fig. 6.

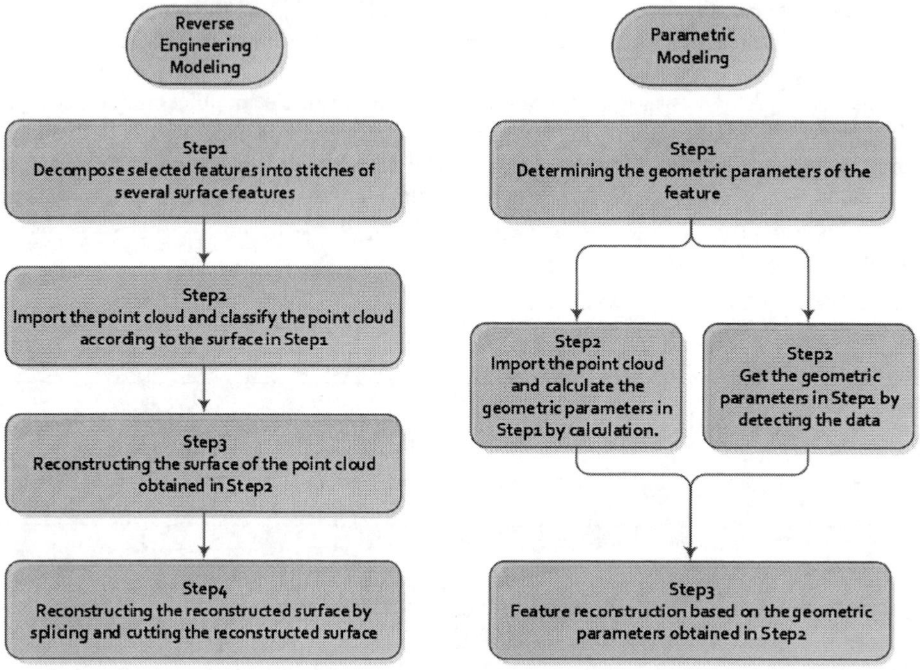


**Fig. 6** Process of feature reconstruction

The method of reverse engineering modeling can preserve the true state of the measured features more accurately. The feature reconstruction method in Fig. 6 transforms the feature reconstruction method into the stitching and cropping after the surface reconstruction. In order to simplify the reconstruction steps of various types of surfaces (including plane, cylindrical, conical and NURBS surfaces, etc.), all the surfaces are transformed into the more common Non-Uniform Rational B-Splines (NURBS) surfaces in 3D computer-aided design modeling software for reconstruction. Use the NURBS surface representation [13] to fit the actual machined surface.

The representation of the NURBS surface is:

$$p(u,v) = \sum_{i=0}^{m}\sum_{j=0}^{n} d_{ij}\frac{\omega_{ij}N_{ik}(u)N_{jl}(v)}{\sum_{r=0}^{m}\sum_{s=0}^{n} N_{rk}(u)N_{sl}(v)}$$

Where $d_{ij}$ is the control vertex; $\omega_{ij}$ is the weight factor; $N_{ik}$,$N_{jl}$ are the B-spline basis functions along u and k times, respectively.

To realize the reconstruction of NURBS surface, the values of control vertices, weight factors and node vectors must be adjusted according to the actual measurement points on the basis of the theoretical model to achieve the purpose of surface reconstruction. In this paper, the control point adjustment method based on constraint optimization is used to complete the surface reconstruction [14]. The specific steps are shown in Fig. 7.

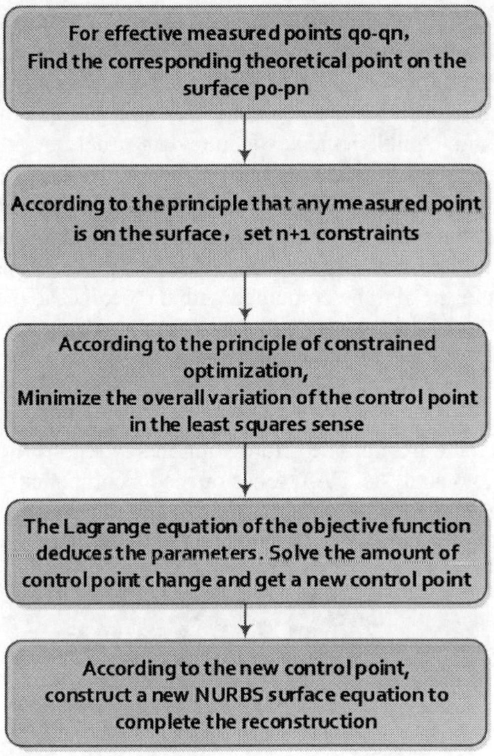


**Fig. 7** Reconstructing surface method according to measurement points

According to the needs of the craftsman to simulate and calculate the machining process of the model, the method of feature reconstruction is selected, and the obtained real geometric model is the geometric layer of the digital twin model. The geometry contained in the geometry model thus obtained is only a key feature inside the part. This layer of work can be done with tools such as Geomagic, Polyworks, and Catia.

### *3.3. Data Management of Digital Twin Model*

Data management for digital twin models is divided into three main categories. The first category is the key feature library, which contains the attribute fields and types corresponding to several key features. The second class stores information (part-feature-attribute) according to the tree structure, so that the part number and batch number can be used. The feature number, the attribute number finds the corresponding value; the third type is the model file information, which contains the time information of the file, the file type, and the like. The first type of information stores the established key feature attribute table, and the reserved field is used by the subsequent model to customize the feature attribute. The second type of tree structure can be designed as a theoretical feature table and an actual feature table. These table contents are defined by the key feature definition personnel through the interactive interface and stored in the database, representing the attributes of each defined feature. Shown. The third type of file information to be stored includes a theoretical model, a measured model, and a measured data file package, wherein the data file is hung in the form of a path under the feature attribute field, and a data table is created for the theoretical model and the measured model.

## 4. Digital Twin Modeling Tool Based on Catia CAA

The Product Digital twin modeling tool provides users with two functions in a CATIA environment, based on key features. The first function is to create key features and select the concerned attributes and the way of model reconstruction according to the key feature library. At the same time of creation, through the information reading of the theoretical model, key feature sets are generated on the CATIA feature structure tree. The theoretical value of the attributes, the way of reconstruction, and the time of feature creation are attached to each feature geometry set. After all the features have been created, you can click the release command to publish the key feature data structure and CATPart file to the database to maintain the data persistence; this process obtains the theoretical model in the digital twin model. The second function is to import the measured data of the part according to the part number, batch number and time number, and reflect the real feature attribute under the feature geometry set, and use the feature point cloud file according to the previously set reconstruction mode. Structural features. A part of the machine plus feature attributes need to be combined with the geometric model for calculation, and then hung under the CATIA tree geometry set; after the creation, the newly generated CATPart file is imported into the database; this process obtains the measured model in the digital twin model.

### *4.1. CATIA CAA*

In order to correlate, extract and manipulate information in key features, CATIA software can be used for display, reconstruction and analysis. CATIA has opened a component interface CAA for secondary development programming, which can be programmed with features such as features, inputs, outputs, interactions, and structures in the component to implement digital twin modeling tools.

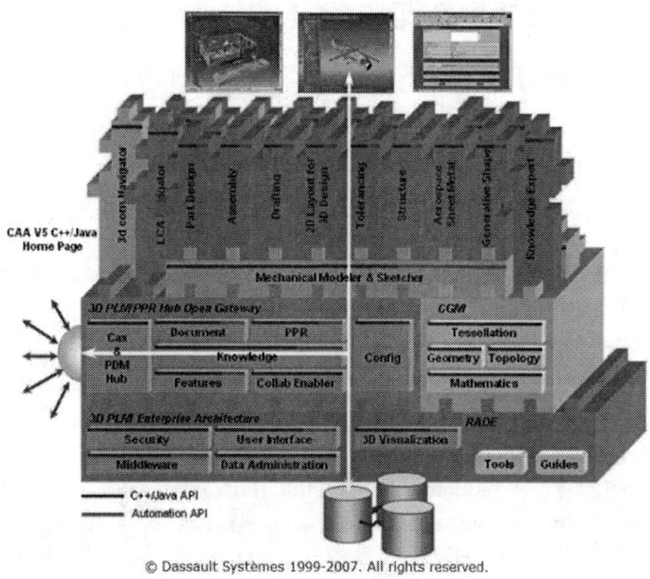


**Fig. 8** CATIA CAA developed tool

The interactive interface operation is designed with the CATDialog module, and an interface for modifying the key feature database, creating features, publishing the theoretical model, and importing the measured model data is designed. Use the CATMechanicalRootFactory module to create and edit geometry sets for key features and their properties; use the CAPIPARAmeterEditor module to create and edit property parameters. The CATICGMFactory module is used to extract and construct the topology based on the theoretical model, and the feature point cloud is reconstructed using the CATIGSMFactory module. File input and output are programmed using the CATDocument module and the standard template class std. Use the MYSQL module to read and write to the database. Features are labeled using the CATIPSDDocument module.

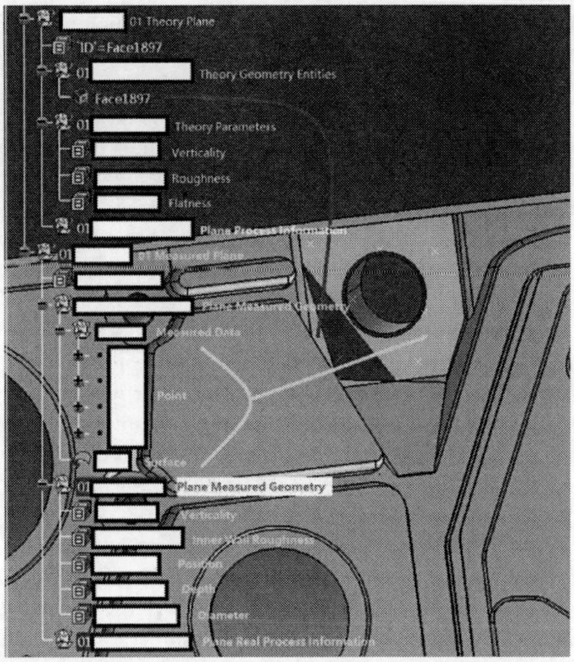


**Fig. 9** Feature reconstruction based on imported point cloud

The theoretical and measured feature data are associated with the feature identification number and are linked under the same Catia structure tree.

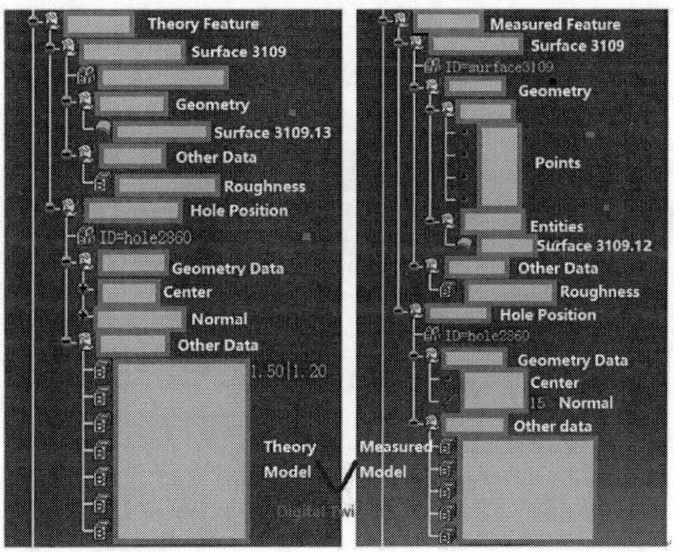


**Fig. 10** Building a digital twin model based on measured data

### *4.2. Database System of Digital Twin Model*

The digital twin model database is built based on MySQL, which provides the technician with an interface for detecting measurement data and stores all the information of the digital twin model.

## 5. Summary

This paper introduces a digital twin model of machining parts, which can effectively apply the digital virtual model to each process by modeling the state of real products in the machining process. In order to streamline manufacturing data, standardize model paradigm and improve modeling efficiency. This paper puts forward the concept of key features and analyses the attributes of key features. It uses key features as the carrier of manufacturing information, and expounds the storage methods of key feature information and its geometric reconstruction methods. This paper elaborates on the three-level structure of the digital twin model and briefly analyses the application method. The digital twin model integrates and integrates the information in the process. It is oriented to the process flow of the part machining, and the theoretical model and the measured model can be used to analyse and optimize the process parameters.

At the end of the article, an example of a digital twin modeling tool using CATIA CAA is introduced, which basically achieves the intended purpose. Its engineering application value remains to be developed. Faced with the environment of intelligent manufacturing, the author still needs to continue to explore the relationship between data and technology in the manufacturing process, enrich the digital twin model, and apply the model to continuous improvement and innovation in production.

## References

[1] Zhou Yang, Yu Yong, Gu Li, et al. Research on Feature-based MBD Model Retrieval Method[J]. Combined Machine Tool & Automatic Processing Technology, 2018, 536(10): 156-160+165.

[2] Tian Fujun, Tian Xitian, Qi Junhao, et al. Modeling and Application of Process Information Based on Model Definition [J]. Computer Integrated Manufacturing Systems, 2012(5): 913-919.

[3] Hu Wenquan, Zhang Xiaoling, Hui Hongchao, et al. Three-dimensional annotation information management method based on Pro/E [J]. Machinery Industry Standardization & Quality, 2018.

[4] Liu Jinfeng, Wang Pengjun, Jing Xuwen, et al. Method for creating process models with auxiliary features [J]. Computer Integrated Manufacturing Systems, 2019(8).

[5] Zhuang Cunbo, Liu Qiaohua, Xiong Hui, et al. The Connotation, Architecture and Development Trend of Product Digital Hybrids [J]. Computer Integrated Manufacturing Systems, 2017(4).

[6] Warwick, Graham. GE advances analytical maintenance with digital twins [N]. Aviation Week & Space Technology, 2015-10-19.

[7] Chen Zhen, Ding Xiao, Tang Jianwei, et al. Exploration of Production Control Model of Aircraft Assembly Workshop Based on Digital Hybrid [J]. Aeronautical Manufacturing Technology, 2018, 61(12): 46-50.

[8] YU Yong, FAN Sheng-ting, PENG Guan-wei, et al. Application of digital twin model in product configuration management [J]. Aviation Manufacturing Technology, 2017, 60(7): 41-45.

[9] Shariatzadeh N, Lundholm T, Lindberg L, et al. Integration of Digital Factory with Smart Factory Based on Internet of Things [J]. Procedia Cirp, 2016, 50:512-517.

[10] Tao Fei, Cheng Ying, Cheng Jiangfeng, et al. Theory and Technology of Information Physics Fusion in Digital Hygiene Workshop [J]. Computer Integrated Manufacturing Systems, 2017(08): 4-12.

[11] Huang Shaohua, Guo Yu, Cha Shanshan, et al. Overview of Research and Application of IoT and Its Key Technologies in Discrete Workshops [J]. Computer Integrated Manufacturing Systems, 2019, 25(2).

[12] Schroeder G N, Steinmetz C, Pereira C E, et al. Digital Twin Data Modeling with AutomationML and a Communication Methodology for Data Exchange [J]. IFAC-PapersOnLine, 2016, 49(30):12-17.

[13] Fazhong Shi. Computer-aided geometric design and non-uniform rational B-spline [M]. Beijing: Higher Education Press, 2013

[14] Hu S M, Li Y F, Ju T, et al. Modifying the shape of NURBS surfaces with geometric constraints [J]. Computer Aided Design, 2001, 33(12): 903-912.

IOP Conf. Series: Materials Science and Engineering **772** (2020) 012004 doi:10.1088/1757-899X/772/1/012004

# Preparation and Force Sensitive Properties of ZnO/Graphene Composite Films

**Jianming Wang, Zhaomin Luo, Feng Zheng, Qiaozhi Wang and Xingming Xu***
Shandong University of Science and Technology, Taian, China

*Corresponding author email: xxm1601@163.com

**Abstract.** In this paper, the preparation and force sensitivity of ErGo-ZnO composite films were studied. In order to improve the application value of graphene composite film in optical devices, sensors, photocatalysis, etc., ZnO is added. In this experiment, the ErGo-ZnO film was successfully prepared by vacuum filtration. At the same time, by adjusting the self-assembled layer of the composite film and the concentration of graphene oxide in the self-assembly growth process, the optimal design parameters are further optimized. The experimental results show that the self-assembled 6 layers have the best performance of the graphene oxide concentration of 0.5mg/ml.The crystallinity of the 0.05wt% graphene oxide ErGo-ZnO composite film is obviously improved. The amount of phase in the film after axial stretching is greatly increased (from 20% before stretching to 92%). In addition, comprehensive characterization and experimental results show that the excellent mechanical properties and specific surface area of graphene oxide and the excellent compatibility with ErGo-ZnO materials are the key factors to improve the interface stress transfer efficiency. This makes the crystallinity and mechanical properties of the composite film greatly improved, and the piezoelectric properties are further enhanced, which greatly improves the application value of the graphene composite film in optical devices, sensors, photocatalysis and the like.

## 1. Introduction

The basic structural unit of graphene is the most stable benzene six-membered ring in organic materials, and it is the most ideal two-dimensional nanomaterial. The ideal graphene structure is planar hexagonal lattice, which can be regarded as a layer of stripped graphite molecules. Each carbon atom is sp2 hybridized and contributed to the remaining electrons in the p orbital to form a large bond. The electrons can move freely, giving the graphene good electrical conductivity. The two-dimensional graphene structure can be seen as the basic building block for forming all sp2 hybrid carbonaceous materials.

The introduction of graphene into inorganic semiconductor materials can increase the surface area increase the charge transfer rate of electrons, compensate for the weak conductivity of metal oxides, and reduce the recombination of photogenera electron pairs by reducing the interaction of graphene-bonds. The photocurrent is increased, and the synergistic effect of graphene and ZnO makes the composite material have great application prospects in the field of optoelectronics. Graphene and ZnO composites have many synthetic routes such as thermal decomposition, electrochemical pathway, ultrasonic spray pyrolysis, electrokinetic atomization and solvent heat treatment.

Content from this work may be used under the terms of the Creative Commons Attribution 3.0 licence. Any further distribution of this work must maintain attribution to the author(s) and the title of the work, journal citation and DOI.
Published under licence by IOP Publishing Ltd

IOP Conf. Series: Materials Science and Engineering **772** (2020) 012004 doi:10.1088/1757-899X/772/1/012004

At present, there are few researchers' studied force-sensitive properties of graphene composite films. Based on the research of the existing literature, the forcesensitive- properties of graphene composites, while adjusting the self-assembled layers of composite films in self-assembly growth process, The concentration of graphene oxide is further optimized to screen out the best design parameters.

## 2. Basic theory and test method of flexible force sensor

### *2.1. Basic theory and test method of flexible strain sensor*

Compared with the traditional strain gauges, the flexible strain sensor not only has the characteristics of the flexible folding of the traditional strain gauge, but also has the stretchable characteristics, which greatly expands its application field. In practical applications, the flexible strain sensor is closely attached or fixed to the object to be tested, and the slight deformation will cause the resistance or capacitance of the sensor to change, and the signal is visualized and recorded in real time by connecting a resistance or capacitance test device. Therefore, in the laboratory test, it is necessary to build a comprehensive test platform, which includes dynamic motion platform and real-time data monitoring and recording platform.The integrated test platform system is composed of a dynamic movable platform system and a real-time data monitoring and recording platform system.For flexible ductile strain sensors, the deformation is mainly divided into transverse tensile strain and longitudinal compression deformation.Therefore, according to different deformation modes, the test principle is slightly different.

### *2.2. Basic theory and test method of flexible piezoelectric sensor*

The device is an electrical component based on the piezoelectric effect of a flexible material, which can convert the mechanical change amount applied to the object to be measured into a visible electrical signal through a certain law, thereby achieving the purpose of detection.The specific principle is shown in the Figure below.

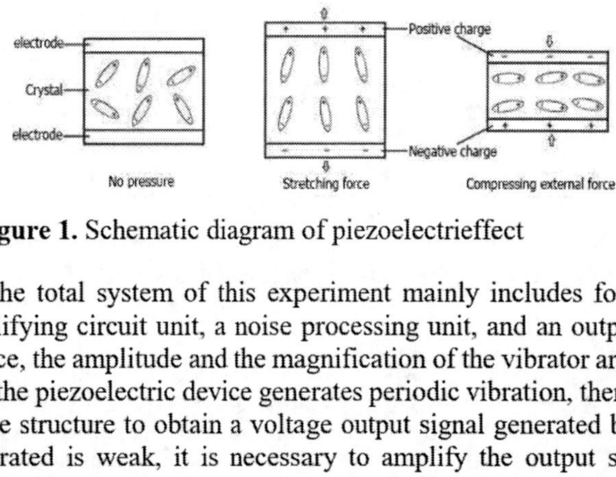


**Figure 1.** Schematic diagram of piezoelectrieffect

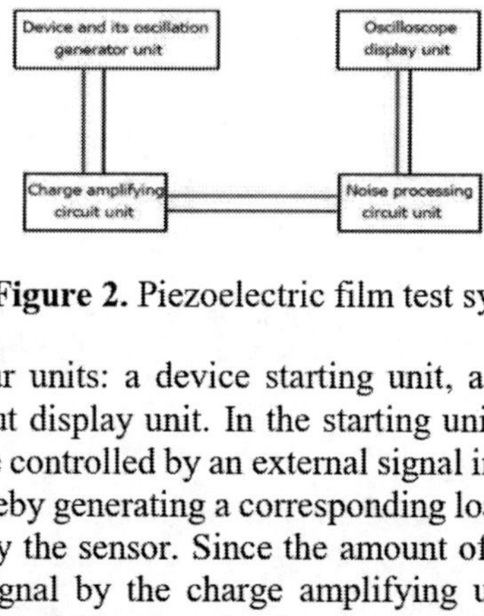


**Figure 2.** Piezoelectric film test system

The total system of this experiment mainly includes four units: a device starting unit, a charge amplifying circuit unit, a noise processing unit, and an output display unit. In the starting unit of the device, the amplitude and the magnification of the vibrator are controlled by an external signal input, so that the piezoelectric device generates periodic vibration, thereby generating a corresponding load form in the structure to obtain a voltage output signal generated by the sensor. Since the amount of charge generated is weak, it is necessary to amplify the output signal by the charge amplifying unit.The amplified signal contains the piezoelectric device output signal and the external noise signal, so the noise processing unit is required for noise reduction processing, and then the output electrical signal is collected by the oscilloscope in the display unit.Finally, the collected data is analyzed and processed.

## 3. Preparation and Force Sensing Properties of 3 ZnO/Graphene Composite Films

### *3.1. Preparation process of ErGo-ZnO nanocomposite film*

In this experiment, the ErGo-ZnO film was successfully prepared by vacuum filtration.First, the EuGo nanosheets were prepared by chemical oxidation method by controlling the amount of strong oxidant added, and the zinc acetate dihydrate was used as raw material.The epoxy groups, hydroxyl groups and carboxyl groups on the Zn nanorod 9-1001ErGo nanosheets were prepared under solvothermal conditions as Celsius. The ZnO nanorods with abundant hydroxyl groups on the surface can pass hydrogen bonding and electrostatic interaction. The effect is attached to the ErGo nanosheet, and the bonding of the layered ErGo nanosheet under vacuum pressure results in the embedding of the ZnO nanorod in the interlayer of the ErGo nanosheet.The unique advantages of ErGo nanosheets and ZnO nanorods will reasonably adjust the surface morphology and channel structure of the ErGo-ZnO film, so that the prepared ErGo-ZnO film can effectively separate the dye solution.

### *3.2. Composite film characteristics and analysis*

Fourier transform infrared spectroscopy (FT-IR) analysis

During the preparation of the composite membrane, FT-IR was used to characterize the changes in functional groups at different stages.Figure 3 is an infrared spectrum of the sample ErGo nanosheet, ZnO nanorod and ZnO film, respectively.A curve from Figure 3 shows that the absorption peaks corresponding to 3350 $cm^{-1}$, 1820 $cm^{-1}$, 1640 $cm^{-1}$, 1420 $cm^{-1}$, and 1030 $cm^{-1}$ are respectively water OH, carboxyl C=, aromatic carbon ring. C=C, CO in the carboxyl group and vibration of CO in the hydroxyl group, which is consistent with the characteristic peaks of the reported ErGo nanosheets, indicating that the ErGo nanosheets were successfully prepared.As can be seen from the b curve in Figure 3, a characteristic peak of ZnO appeared at 524 $cm^{-1}$, indicating that the ZnO nanorods were successfully prepared.As can be seen from the c curve in Figure 3, the absorption peak at 1640 $cm^{-1}$ is significantly weakened due to the electrostatic interaction between the ErGo nanosheet and the ZnO nanorod.In addition, the characteristic peaks in ErGo nanosheets and ZnO nanorods were successfully retained, indicating that the ErGo-ZnO film was successfully prepared.

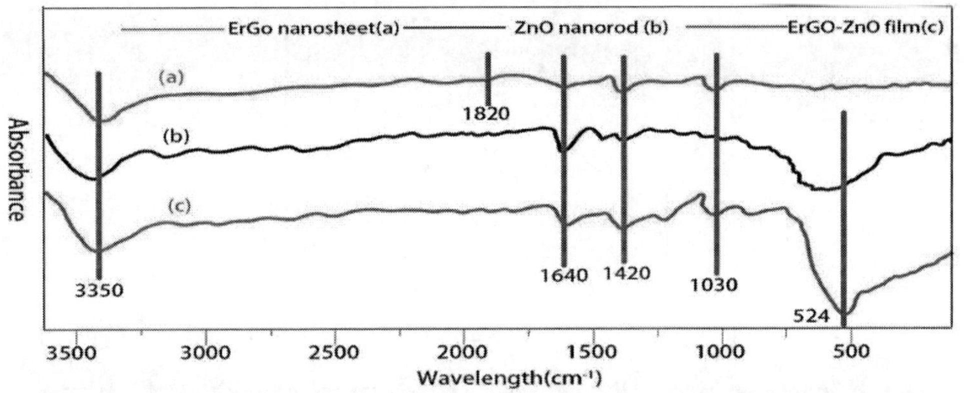


**Figure 3.** Infrared spectrum of sample ErGO nanosheet (a), ZnO nanorod (b) and ErGO-ZnO film (c)

### *3.3. Study on the performance and sensitivity mechanism of flexible strain sensors*

*3.3.1. Study on compressive strain characteristics of flexible strain sensors.* Compared with other composite film preparation methods, the layer-by-layer self-assembly technology constructs a multi-layer structure by alternating adsorption, which has the characteristics of simple operation and controllable film thickness.Therefore, the influence of the self-assembled layer of the composite film and the concentration of graphene oxide on the overall performance of the device must be considered during the experiment.Here we define a single-layer polyethyleneimine film and a single-layer graphene film as a composite film. In this experiment, a 1% w/v PEI solution and a graphene solution with a concentration of 0.5 mg/ml were used for layer-by-layer self-assembly.

Figure. 3 Infrared spectrum of sample ErGO nanosheet (a), ZnO nanorod (b) and ErGO-ZnO film (c)

Four, six, eight and ten layers of ZnO-ErGo composite films were prepared by the process.The flexible composite film was packaged as strain sensor S (4L-GO/0.5mg/ml), strain sensor S (6L-GO/0.5mg/ml), strain sensor S (8L-GO/0.5mg/ml). And strain sensor S (10L-GO/0.5mg/ml).

In order to study the influence of the number of self-assembled layers on the device, we tested the initial resistance of these devices. The experimental results are shown in Figure 4. It can be concluded that as the number of self-assembled layers increases, the initial resistance of the device decreases. small.It shows that with the increase of the number of self-assembled layers, the conductive network in the composite film is gradually improved, and the initial resistance is drastically reduced.

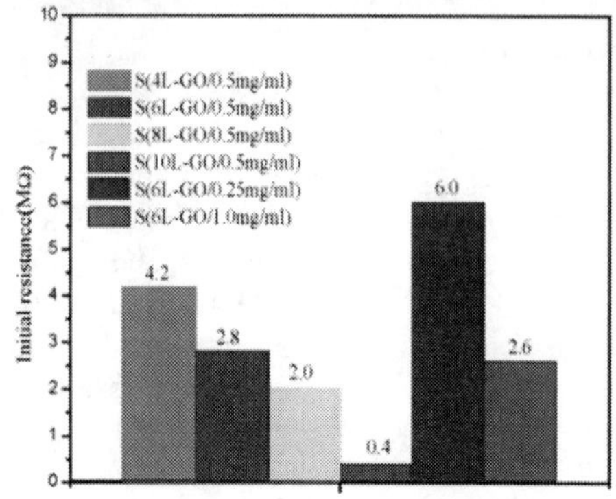


**Figure 4.** Initial resistance of six flexible composite devices

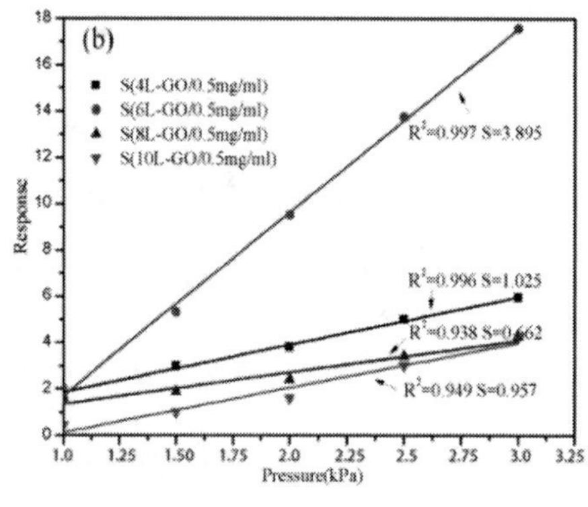


**Figure 5.** Pressure-response curves of four strain devices

In order to further understand the influence of the number of self-assembled layers on the performance of the device, this paper studied the resistance change response of several devices under small pressure (1kPa-5kPa). Several devices have excellent response under small strain. At the same time, it can be found that the device has excellent linearity (R2>0.9) in a small compressive strain range, so the device has excellent performance and broad application prospects.According to the formula defined by the strain coefficient formula (1) and the sensitivity formula (2), the sensitivity coefficients of self-assembled 4-layer, 6-layer, 8-layer and 10-layer strain sensors can be calculated, and a self-assembled 6-layer strain sensor can be obtained. Has the largest response, the best linearity and the maximum sensitivity factor (S = 3.895).Therefore, the following conclusions can be drawn. After comprehensive testing of different layer-count strain devices prepared by the layer self-assembly process, it is found that the self-assembled 6 layers are the optimal self-assembled layer number design parameters.

$$GF = \frac{\Delta R}{R_0} / \frac{L}{L_0} = \frac{R/R_0 - 1}{\varepsilon} \quad (1)$$

$$S = (\Delta R / R_0) / \Delta \varepsilon \quad (2)$$

In order to further study the effect of graphene oxide concentration on the performance of the device in the self-assembly process, this chapter based on the previous optimal layer design parameters (6 layers), using a layer-by-layer self-assembly process to prepare 6 layers of PEI based on 1% w/v. Solution and concentration of 0.25 mg/ml, 0.5 mg/ml and 1.0 mg/ml of graphene solution ZnO-ErGo composite film.The flexible composite film was packaged as strain sensor S (6L-GO/0.25mg/ml), strain sensor S (6L-GO/0.5mg/ml) and strain sensor S (6L-GO/1.0mg/ml). .The test results are analyzed by using a self-made test system. The experimental results are shown in Figure 6. It can be seen that the three strain devices still have excellent response and linearity in a small pressure range.At the same time, it can be found that in three kinds of strain devices based on different graphene oxide concentrations, the self-assembled 6-layer graphene oxide concentration of 0.5mg/ml strain sensor has the maximum response value and sensitivity as well as the optimal linearity.

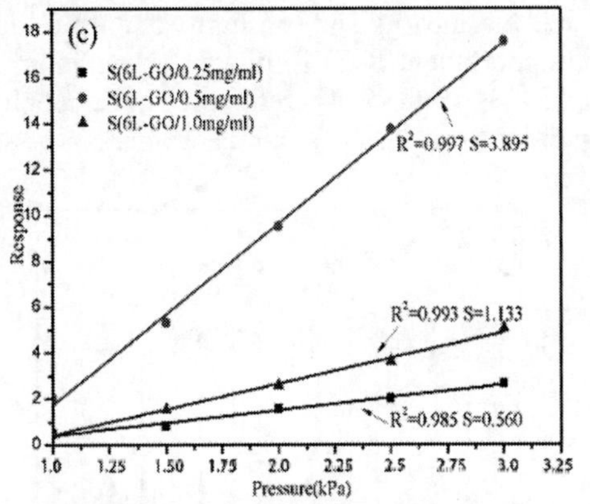


**Figure 6.** Curves of different graphene concentration strain devices

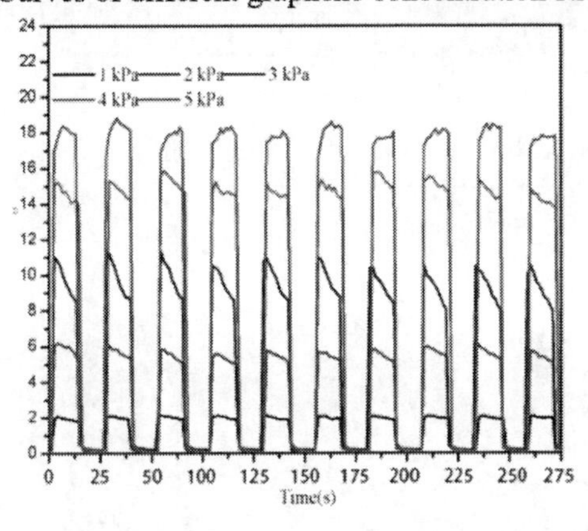


**Figure 7.** Dynamic response of strain sensor S (6L-0.5mgml)

Comprehensive above, this alsoIt is obvious that under the small compression pressure strain, the strain sensor S (6L-GO/0.5mg/ml) prepared by the layer self-assembly process has the best performance.Therefore, it can be concluded that the optimum layer number design parameters of the strain sensor prepared by the layer self-assembly process are 6 layers, and the optimal GO concentration parameter is 0.5 mg/ml.In order to further verify the other properties of the device, the stability test and repeatability test of the optimal design strain sensor S (6L-GO/0.5mg/ml) are shown in this experiment. It is obvious from Figure 7. that the strain sensor is Under the small pressure range (1kPa-5kPa), the relative change of resistance remains stable, and the curve has no obvious delay or drift, which indicates that the device has excellent stability performance.In addition, it can be seen from the repeatability test result curve of the device that the relative change of the resistance of the device at different pressures is substantially maintained without a large increase or decrease, indicating that the device has excellent repeatability.

*3.3.2. Study on Tensile Strain Characteristics of Flexible Strain Sensor.* Similar to the test under compressive strain, firstly, in response to the influence of the number of layers and the concentration in the self-assembly process, several experiments of the devices prepared in the previous section were tested in the small strain range.Aiming at the influence of the number of layers in the self-assembly process, the tensile test system is used to test the relative resistance value of four different layer strain devices at 1%-5% strain.The dynamic test, the experimental test results are shown in Figure 8.It can be seen from the Figure that the resistance fluctuation curves of several devices at 2.5% tensile strain are stable, indicating that the devices have good stability under tensile strain.Combined with the strain response test results of several devices under different strains, the comprehensive analysis can clearly find that 4 The device has excellent linearity and sensitivity and ultra-high response in the small strain range. According to the strain coefficient (GF) formula, the maximum strain coefficient of the four devices (at 5% strain) is 105, 754 ,326 and 79. Therefore, it can be concluded that the optimum number of layers of the strain sensor under compressive strain is designed. The self-assembled 6 layers are the optimal layer design parameters.

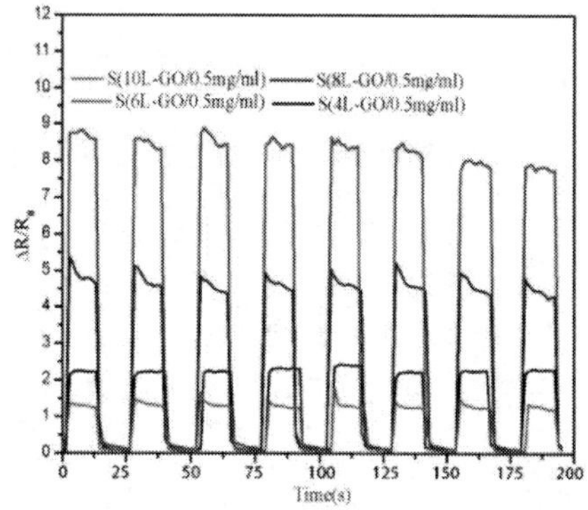


**Figure 8.** Dynamic response curves of different self-assembled layer strain devices

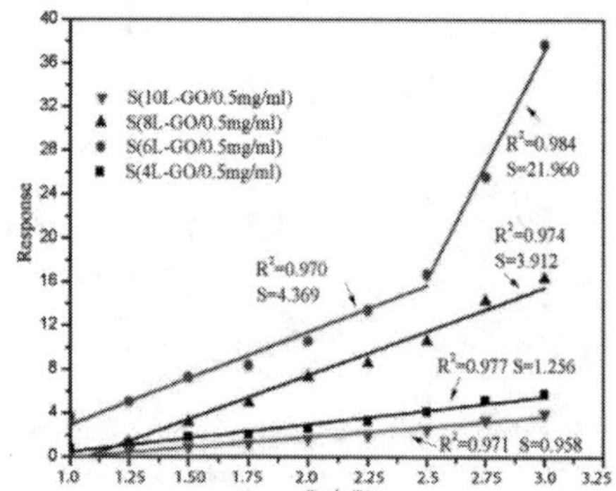


**Figure 9.** Pressure-response curves of four strain devices

After determining the optimal layer number design parameters of the layer self-assembly process, in order to further study the influence of the graphene oxide concentration on the performance of the variable device, the dynamic test of the three strain devices is shown in Figure10.It can be seen from the Figure that the curves of the relative resistance of the three devices under 2.5% tensile strain are undulating and there is no large increase or decrease, which indicates that the device has good stability under tensile strain. performance.Combined with the response curve of different tensile strains (Figure11), it can be seen that the optimized graphene oxide concentration design parameter is 0.5 mg/ml.

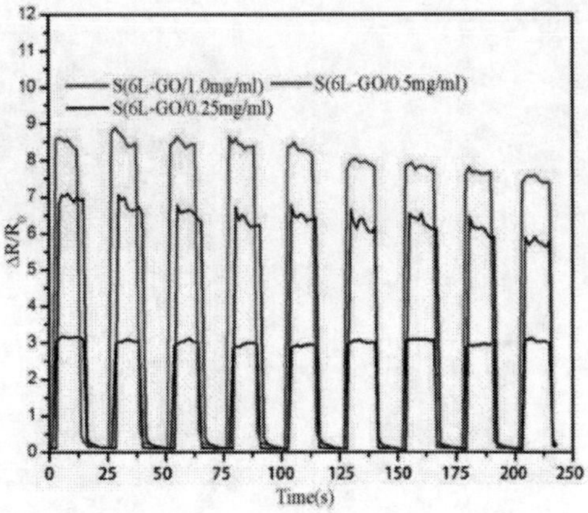


**Figure 10.** Dynamic response curve of different graphene concentration strain devices

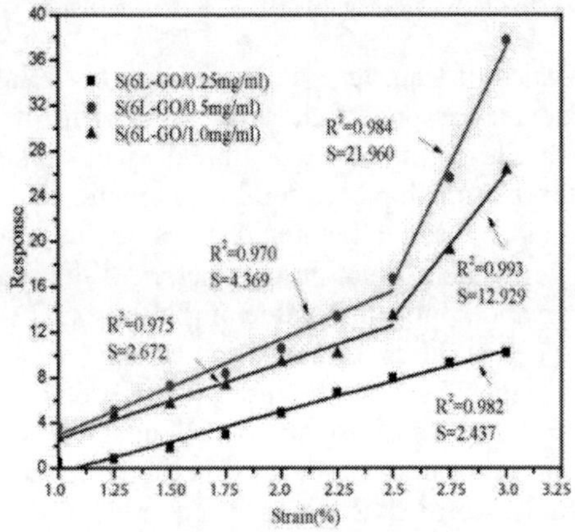


**Figure 11.** Pressure response curves of three kinds of strain devices

In order to further analyze other performance parameters of strain sensor S (6L-GO/0.5mg/ml), the ohmic behavior, repeatability, tensile mechanical properties, minimum detection limit, response time and practical application of the device were analyzed. test.Figure12 shows the real-time resistance value of the strain sensor S (6L-GO/0.5mg/ml) under 1%-5% tensile strain. The device can have excellent dynamic resistance response. At the same time, it has excellent stability under different strains.In order to further study the repeatability of the device, this paper uses the dynamic test platform to repeat the test at 2.5% tensile strain for 500 times. The experimental results are shown in Figure13.The black part of the Figure is the change curve of the relative change value of the resistance of 500 cycles. The two embedded pictures below the Figure are the initial change curve and the change curve of 500 cycles.It can be clearly seen that after 500 cycles of testing, the overall response value change is almost negligible, and there is no significant change in response value and stability before and after the test.A comprehensive analysis can conclude that the flexible strain sensor S (6L-GO/0.5mg/ml) has excellent repeatability.

IOP Conf. Series: Materials Science and Engineering **772** (2020) 012004 doi:10.1088/1757-899X/772/1/012004

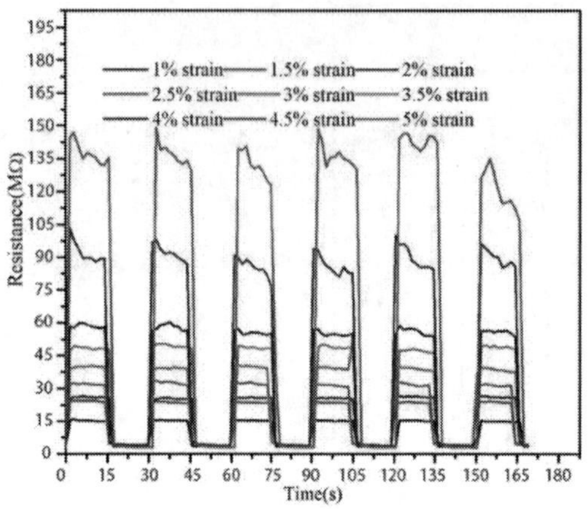


**Figure 12.** Dynamic response at 1%-5% strain

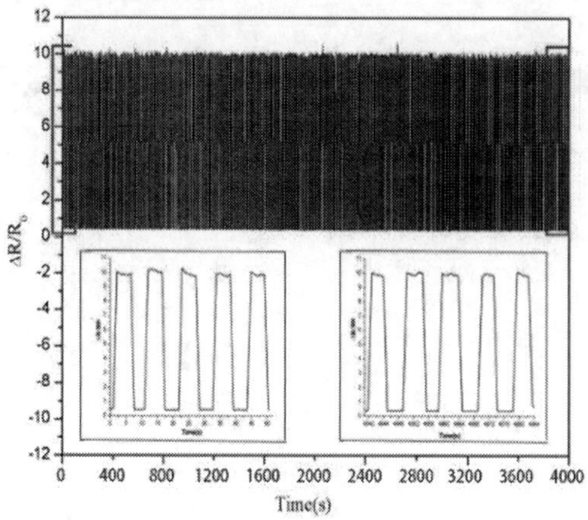


**Figure 13.** 500 cycle stability test

According to the characteristics of transverse tensile strain, the relationship between tensile strain and tensile force of flexible devices is one of the important performance parameters of the device, because it can not only reflect the flexibility of the flexible device, but also reflect the strain and tension of the flexible device. The linear relationship between.In the practical application, the excellent linearity can not only calculate the strain size through the response value, but also calculate the tensile force.As one of the most important performance evaluation parameters of the sensor, the detection limit largely determines the application range and application value of the sensor.The flexible strain sensor detection limit is defined as the ability to detect the minimum strain change of the object under test, significantly different from the background signal.Through the test and analysis of the prepared strain sensor, it is found that the device still has good responsivity and excellent stability under 0.1% minimum strain.In addition, as one of the most important performance evaluation parameters of the sensor, the response time of the strain sensor is usually defined as the time required for the sensor to reach the maximum value after the test amount changes by one step.It can be clearly seen from Figure14 that the response time of the flexible strain sensorLess than 0.6s.This result further indicates that such devices not only have great sensitivity, but also can quickly detect small strains.In order to further verify the practical application value of this kind of device, the flexible strain sensor prepared in this paper is closely attached to the joints of human fingers, and the dynamic response of the device under the minimal strain of the finger is measured. The test results are shown in Figure15.It can be seen from the Figure that during the bending of the finger, since the flexible device is closely attached to the joint of the finger, the device undergoes a slight deformation as the finger joint is bent, resulting in an increase in the resistance value of the device.By analyzing the measured resistance response data, it is found that the resistance response of the device is almost unchanged and has excellent repeatability. These results are a good proof of the composite film strain sensor prepared by the layer-by-layer self-assembly process. Has a unique advantage and use value.

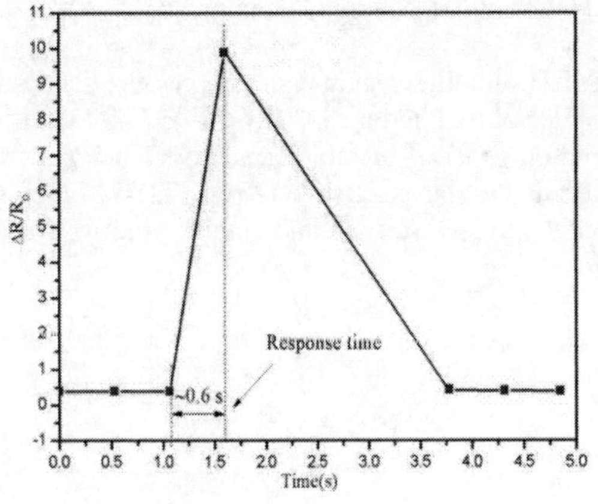


**Figure 14.** Response time curve

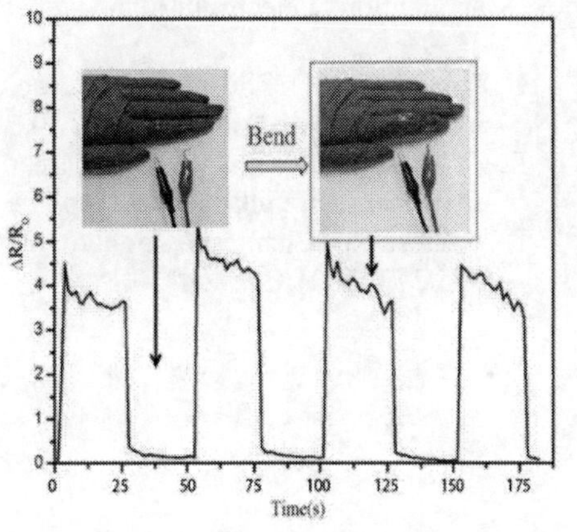


**Figure 15.** Human hand joint bending actual test

## 4. Conclusion

In this paper, five kinds of samples were prepared by adjusting the mass percentage of graphene oxide doping (0wt%, 0.02w%, 0.05wt%, 0.08wt%, 0.1wt%). To show the effect of graphene oxide doping on PDF-HFP film Four kinds of composite films were tested by Fourier transform infrared spectroscopy (FT-IR). The results show that with the increase of the mass percentage of graphene oxide doping, the content of β crystal phase in the composite film increases first. The trend of reduction.

The test results show that as the number of self-assembled layers increases, the conductive network in the region becomes more complete.In order to further study the self-assembled layer number and the optimal design parameters of graphene oxide concentration, four kinds of composite films were packaged to prepare flexible strain sensors.For the different strain formation methods, compressive strain test and tensile strain test were performed on all devices.The comprehensive test results showed that the self-assembled 6 layers and the concentration of graphene oxide were 0.5 mg/ml as the optimal parameters.The device exhibits an excellent linearity in the 1% 5% strain range with a minimum detection limit of 0.1% strain, a response time of less than 0.6 s, and a maximum strain factor (GF) of 754. The device has excellent stability and repeatability. Sex, 500 cycles test still has good stability, so it can be used to detect human finger bending in real time.Based on the comprehensive characterization test results and strain test results, the response mechanism of the graphene-based flexible strain sensor is preliminarily explored.

**Acknowledgments**
This work was financially supported by the overseas program of Shandong University of science and technology.

**References**

[1] Jiang Yadong, Xie Guangzhong. Sensitive materials and sensors.[M] Chengdu University of Electronic Technology Press, 2008.
[2] Guo Yuansheng, Zhu Zuoyun, Dong Hualimin Sensor Market Forecast Analysis Research [Sensor World, 2001, 7(3): 1-4].
[3] Cai Le, Zhou Wei, Jie Sishen, et al. Research on flexible strain sensor of carbon nanotubes [j] Modern Physics, 2014, 10(1): 24-28.
[4] Hung, C. Lee, S. O. Choy. Ubiquitous health monitoring:Integration of wearable sensors, novel sensing techniques, and body sensor networks [M]. Gewerbestrasse:Springer International Publishing, 2015: 319-342.
[5] G. Schwartz, B. C. Tee, Mei, et al Flexible polymer transistors with high pressure sensitivity for

application in electronic skin and health monitoring [J]. Nature Communications, 2013, 4 (5): 1859.

[6] C. Pan L. Dong, G. Zhu, et al High-resolution electroluminescent imaging of pressure distribution using a piezoelectric nanowire LED array [J]. Nature Photonics, 2013, 7 (9): 752-758.

[7] R. Z. Li, A. Hu, T. Zhang, et al. Direct writing on paper of foldable capacitive touch pads with silver nanowire inks [J]. Acs Applied Materials Interfaces, 2014, 6 (23): 21721-21729

[8] B. Cantor.i.t.h .Chang.p. Knight,a..b. Vincent Materials Science and Engineering:A Volumes 375-377, 2004, 213-218

# Design and finite element analysis of thick shell mussel shell opener

**Zhenhua Li *; Erlin Luo, Jinling Tan, Yang Ni, Zuchano Song, Jian Wang**

School of ship and electromechanical engineering, zhejiang Ocean University, Zhejiang, China

*Corresponding author e-mail: lizh760905@zjou.edu.cn

**Abstract.** In order to obtain fresh mussel meat, this paper comes up with a design of mechanical automatic thick-shell mussel open shell machine. According to the physiological parameters and structural characteristics of thick shell mussel, it is determined that the scheme of automatic shell opening and the related parameters of open shell cutter tool. At the same time, Using Solidworks simulation and ANSYS Workbench on the open shell tools and pneumatic cylinder and other major component of finite element analysis, which is ensure that thick-shell mussel open shell machine open shell with stable and efficient. And the relevant part of parameters are confirmed that the thick shell mussel open shell machine avoid some adverse vibration frequency in theory. The purpose of the thick-shell mussel open shell machine is to replace manual opening shell, improve the efficiency of mussel shell opening and provide further reference for the production design of the automatic thick shell mussel opening shell equipment.

## 1. Introduction

Mussel meat has high nutritional value, rich in protein, a variety of essential amino acids, vitamins and trace elements, and enjoys the reputation of "eggs in the sea", which is a rare nutritional and health food [1,2]. Fresh mussels are a favorite of seafood lover, but they're hard to obtain, In order to meet the demand of consumers in the market, the mussel processing industry in China still use the traditional method of manually opening shells to obtain fresh mussels [3,4] (figure1).

However, this method has high labor intensity and low production efficiency, and it is difficult to ensure the integrity and cleanliness of fresh mussels. Therefore, many researchers have carried out a lot of research: He. H [5] and Gipsy [6] et al. proposed a method to obtain fresh mussels by using ultra-high pressure technology. Zhu [7], Wang [8] et al. also used ultrahigh pressure technology to obtain fresh mussels, and confirmed some key parameters of ultrahigh pressure treatment for mussels. This technology can process a large number of mussels and obtain fresh mussel meat with good meat quality, but it is difficult to promote the mussel processing industry in China due to the high cost of equipment and difficult maintenance. Wheaton [9] and Singh [10] et al proposed infrared shell opening method and laser shell opening method respectively. They tested that their respective schemes could obtain relatively perfect fresh shell meat. But these high-tech technologies are not yet ready for opening thick shells mussel now. Hiroaki. S [11] and Hu [12] et al. carried out an experiment to open shell processing of mussels by means of high-pressure water jet. After the experiment, fresh mussel meat are available but

Content from this work may be used under the terms of the Creative Commons Attribution 3.0 licence. Any further distribution of this work must maintain attribution to the author(s) and the title of the work, journal citation and DOI.
Published under licence by IOP Publishing Ltd

badly damaged. Because this method is still immature and many key parameters have not been obtained, far from being applied in practice.

In order to improve the efficiency of obtaining intact fresh shellfish, this paper puts forward a design of thick-shell mussel open shell machine. It is beneficial to improve the open shell efficiency and reduce the wear of the open shell cutter that the size parameters of thick-shell mussel open shell machine are designed in a good adaptability. And the open shell cutter of thick-shell mussel open shell machine is verified by finite element analysis, and the strength failure and mechanical resonance failure of the open shell cutter are avoided [13-17].

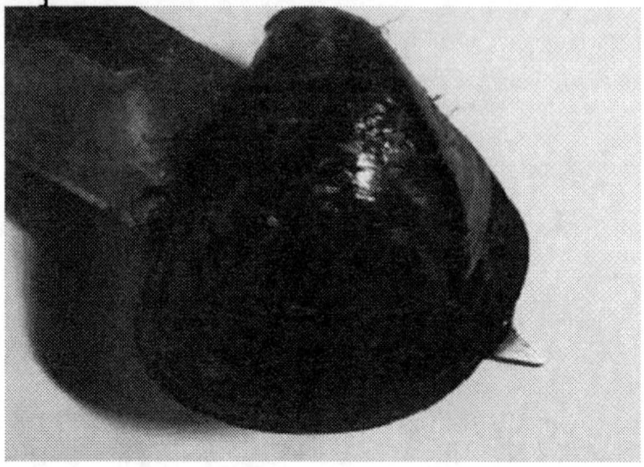

**Fig. 1** Opening thick shell mussel with a knife

## 2. Structure and working principle of the thick-shell mussel open shell machine

### *2.1. The key to opening thick-shell mussel*

The physiological structure of thick shell mussel was analyzed, the adductor muscle is the most important factor to control the closure and opening of the mussel shells (figure2). When the thick shell mussel is in the state of death, the adductor muscle loses its activity and the thick shell mussel is always in the state of open shell. Therefore, the thick shell mussel is to destroy the adductor mussel by cold-processed, so as to achieve the autonomous shell opening of thick shell mussel [18].

**Fig. 2** the adductor muscle was destroyed

### *2.2. The Structure design of thick-shell mussel open shell machine*

The thick-shell mussel open shell machine is mainly composed of four parts which are respectively thick-shell mussel fixation platform, the open shell cutter tool, Angle adjustment device and pneumatic cylinder system (figure3 and figure4). The function of thick-shell mussel fixing platform is to fix the thick-shell mussel to be processed, avoid the movement of the thick-shell mussel in the process of processing. The function of the open shell cutter tool is to cut the adductor muscle connecting the two shells; The Angle adjusting device is to adjust the deviation Angle of the shell cutter, so as to ensure the best deviation Angle when the shell cutter enters the mussel body to cut off the adductor muscle. The pneumatic cylinder system provides the power to cut off the adductor muscle.

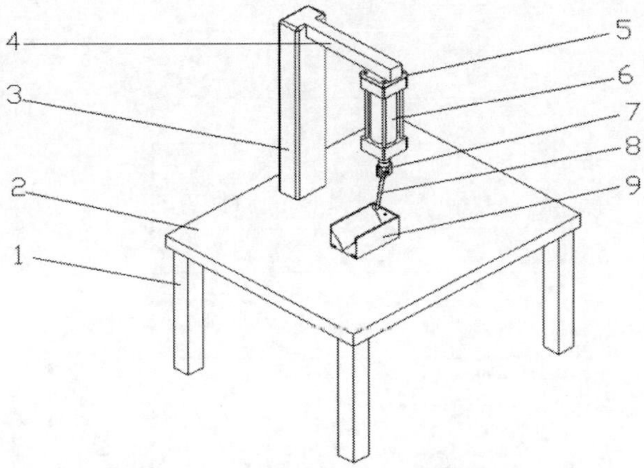


1. Stabilizer blade 2. Table top 3. The bracket 4. Cantilever 5. Land 6. Pneumatic cylinder
7. Angle adjusting connection device 8. Open shell cutter tool 9. The thick-shell mussel fixation platform

**Fig. 3** The Structural sketch of thick-shell mussel open shell machine

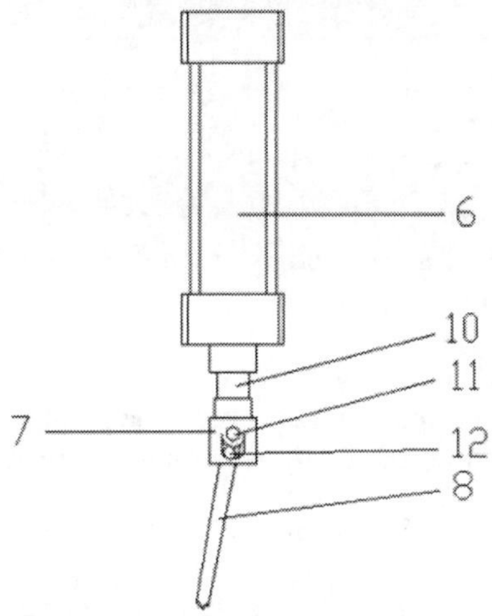


10. Pneumatic cylinder piston rod 11. Bolt 12. Angle adjusting bolt

**Fig. 4** Assembly drawing of cutting tool for thick shell mussel

*2.3. Working principle of the thick-shell mussel open shell machine*

Here's how open shell machine works(figure5): The staff place the thick shell mussel to be processed on the fixed table, and then start PLC control program. Under the instruction of PLC control program, the pneumatic cylinder push the open shell cutter tool to move down to the designated position to cut off the adductor muscle. After cutting off the adductor muscle, the PLC control program continues to run, and the piston in the pneumatic cylinder moves up which drive the open shell cutter tool to move back to the position. The above is the overall workflow.

IOP Conf. Series: Materials Science and Engineering **772** (2020) 012005 doi:10.1088/1757-899X/772/1/012005

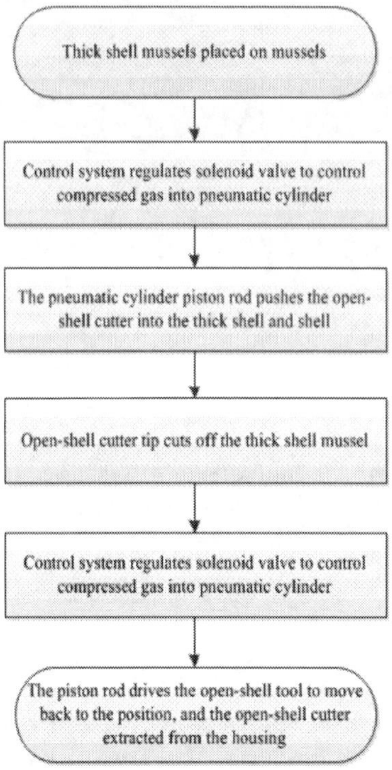


**Fig. 5** Workflow diagram of shelling machine for the thick shell mussel

## 3. The design of open shell cutter tool

The open shell cutter tool is the key component of the thick-shell mussel shell opener. Therefore, the open shell cutter tool is accurately designed and checked. The pointer type digital display tension meter was used to perform the posterior closed shell muscle pulling experiment on 20 active thick-shell mussel samples (figure6 and figure7), and the maximum value of 600N in the experiment was taken as the external load on the handle end of the shell cutter. Because of pointer display tension meter retractor and closed shell after muscle column tensile process belongs to the surface contact, and in the actual tool edge and the adductor muscle belongs to the line contact. In order to avoid producing larger force analysis error, The force measured by the digital display tensiometer needs to be converted into the force of cutting off the adductor muscle by the open shell cutter tool, then the other related parts are design[19].

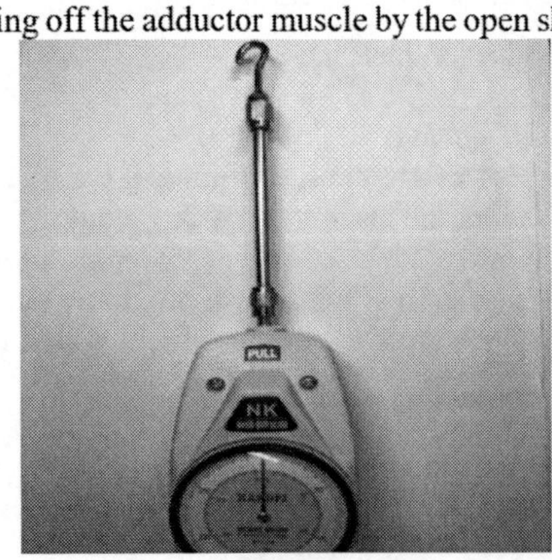


**Fig. 6** Pointer tensiometer

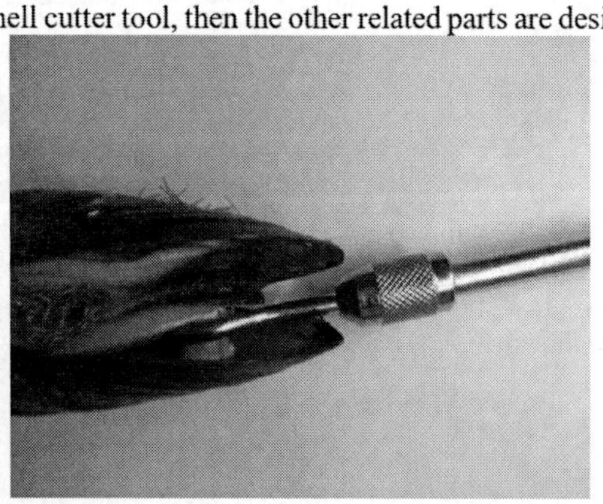

**Fig. 7** Tension experiment of the adductor muscle

IOP Conf. Series: Materials Science and Engineering **772** (2020) 012005 doi:10.1088/1757-899X/772/1/012005

The average diameter of the adductor muscle was determined by measuring the diameter of the adductor muscle trace of 10 thick-shell mussels (figure8), and the contact area between the hook and the adductor muscle was further calculated. The measurement results are shown in table 1.

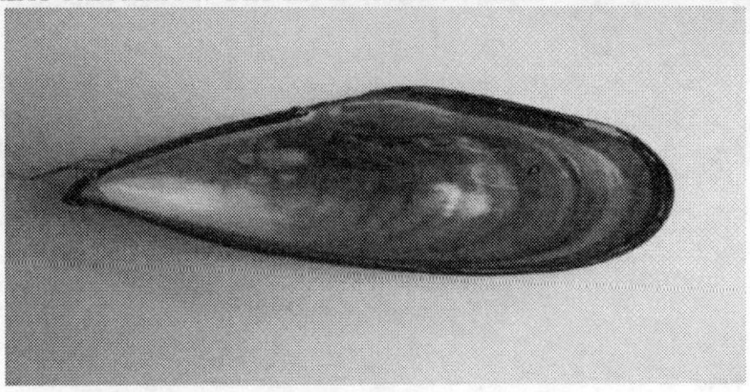

**Fig. 8** the adductor muscle scar diameter

**Table 1.** Measurement of the diameter of adductor muscle scar of 20 thick shell mussels

| No. | 1 | 2 | 3 | 4 | 5 | 6 | 7 | 8 | 9 | 10 |
|---|---|---|---|---|---|---|---|---|---|---|
| Group 1 Diameter (mm) | 9.9 | 5.1 | 9.6 | 8.4 | 10.1 | 7.9 | 10.0 | 6.1 | 9.6 | 10.2 |
| Group 2 Diameter (mm) | 5.0 | 7.7 | 9.8 | 10.0 | 10.9 | 9.9 | 9.8 | 10.0 | 8.9 | 9.2 |

The average values of the 20 data in table 1 were taken as 8.9mm, rounded to 9mm, and used as the average diameter of the adductor muscle of the thick-shell mussel. In order to improve the cutting efficiency of the shell cutter for the adductor muscle, the width of the designed open shell cutter should be larger than the average diameter of the adductor muscle, and the designed width of the shell cutter should be 12mm.

The height of the cylindrical retractor was measured as 5mm, and the area of the retractor and the rear closed shell muscle was calculated as $S_1 = 5 \times 9 = 45mm$.

The contact area between the cutting edge of the adductor muscle of the rear closed shell was estimated to be $S_2 = 1 \times 1 = 45mm$,

According to the pressure formula [20-22]:

$$\frac{F_1}{S_1} = \frac{F_2}{S_2} \qquad (1)$$

F1: pointer type digital display tension meter tension 600N;

F2: strength of the closed shell after cutting off the adductor muscle.

After solving the cutting edge of the open shell cutter, the muscle strength of adductor muscle is 13N. According to this parameter, the applied load of the pneumatic cylinder to the open shell cutter and related models can be further determined.

To avoid the slippage of the open shell machine in the process of the adductor muscle after cutting, the edge of the shell cutter is designed as a symmetrical semi-arc. In order to facilitate the open shell cutter tool entering from the thick shell mussel abdomen closed gap position, the middle position of the cutting edge is designed as a small square tip (figure 9 and 10).

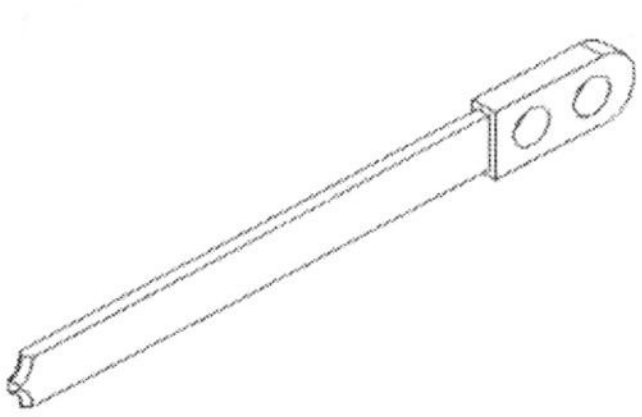

**Fig. 9** Three-dimensional drawing of open shell cutter tool

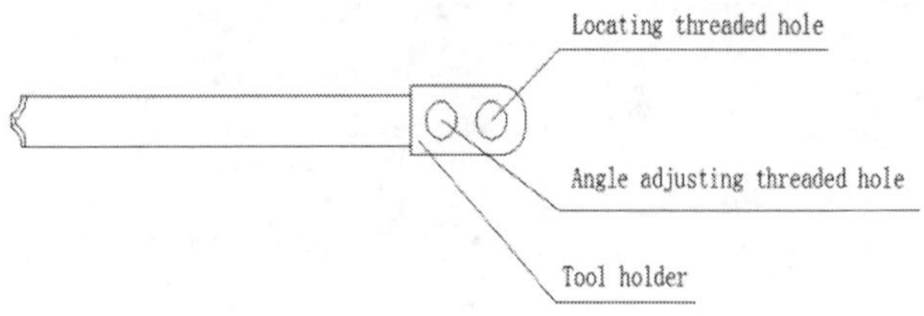


**Fig. 10** The plane graph of open shell cutter tool

At the position where the thick shell mussel develops long filament attachments in its abdomen, there is usually a gap about 2mm wide for the position where the thick shell mussel cannot be completely closed (figure11).The open shell cutter tool can be effectively reduced the resistance when it insert the open shell cutter tool from this position, and facilitate the open shell cutter tool to enter the shell smoothly. Therefore, the thickness of the open shell cutter tool designed is 2mm.

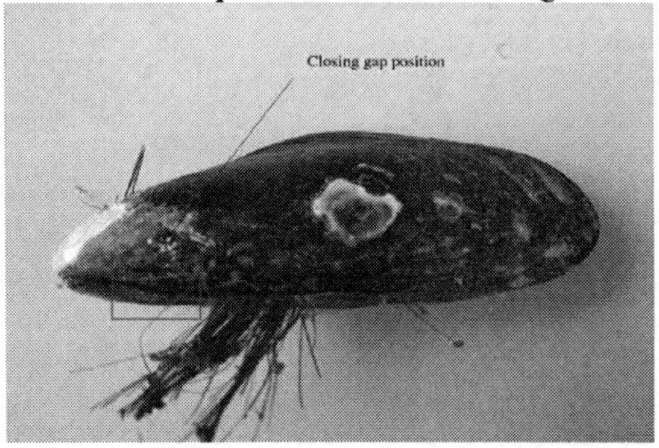


**Fig. 11** Location of Closed Gap in Abdomen of Mussel

By measuring the distance between the closed gap position of 10 thick-shell mussels and the center position of the adductor muscle, the average distance was calculated to be 45mm, In order to ensure that the shell cutter can achieve some cutting for most thick-shell mussels, the design length of the open shell cutter tool is 50mm.

## 4. Finite element analysis of the thick-shell mussel open shell machine

### *4.1. The open shell cutter tool simulation*

Alloy steel was selected as the material of the open shell cutter tool, and Solidworks Simulation was used to simulate the state of the adductor muscle after the cutting of the thick shell mussel. The analysis results are shown in Fig12-13.

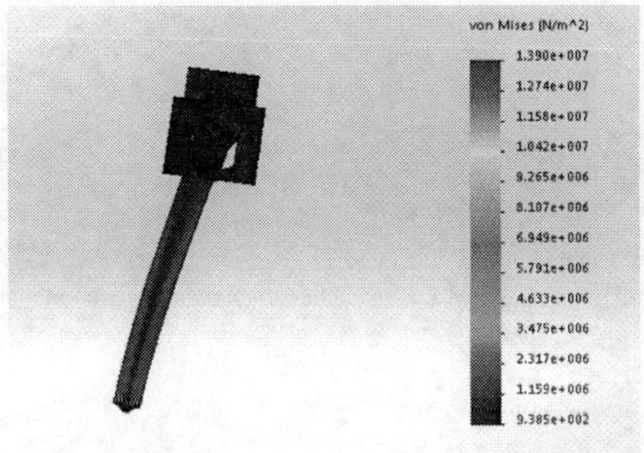


**Fig. 12** Stress analysis of the open shell cutter tool device for thick shell mussel

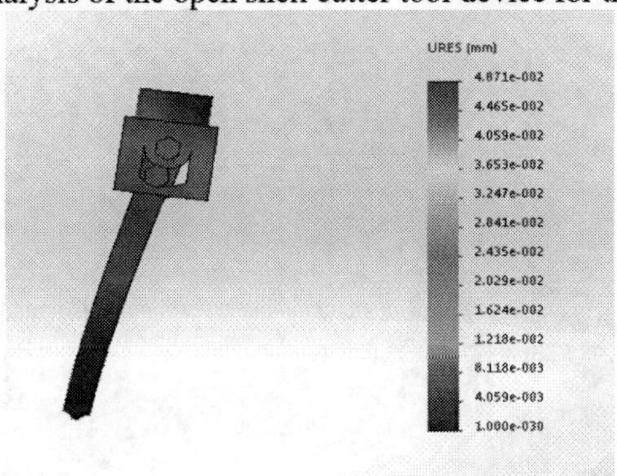


**Fig. 13** Displacement analysis of the open shell cutter tool device

It can be seen from figure 12 that the maximum stress carried by the open cutting tool device is $1.39 \times 10^7$N, Less than the yield limit W of the open cutting tool device material, Therefore the material of open shell cutter tool meets the requirements. It can be seen from figure 13 that the offset displacement of the shell cutter is $4.871 \times 10^{-2}$mm, satisfying the operating requirements.

*4.2. Modal analysis of pneumatic cylinder*

Modal analysis is the basic content of dynamic analysis, which can avoid resonance before the product is put into production. Assuming that the thick shell mussel open shell machine material is ordinary steel, its vibration is free vibration and the damping is ignored, the equation is

$$[M]\{\ddot{u}\} + [M]\{u\} = \{0\} \tag{2}$$

When you have a simple resonance, $u = U\sin(\omega t)$,The equation for

$$([K] - \omega_i^2[M])\{\phi_i\} = 0 \tag{3}$$

The above formula is a classical eigenvalue problem, and the square root of the eigenvalue of the equation is $\omega_i^2$, The square root $\omega_I$ is the natural circular frequency of the structure. The natural frequency is $f_i = {}^{\omega_i}/_{2\pi}$ ,The range of I is 1 degree of freedom and the corresponding vector is $\{u\}_I$ ,It's also called an eigenvector. The eigenvalue $\omega_I$ is

corresponding eigenvector $\{u\}_i$ , which are the natural vibration frequency corresponding mode. ANSYS was used to conduct the six-order modal analysis of the thick-shell mussel open shell machine as a whole. The analysis results are shown in figure 14-19 and table 2.

IOP Conf. Series: Materials Science and Engineering **772** (2020) 012005 doi:10.1088/1757-899X/772/1/012005

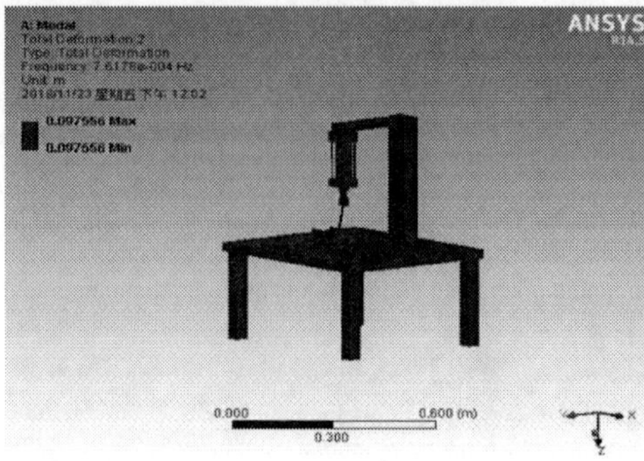


**Fig 14.** The first stage vibration mode

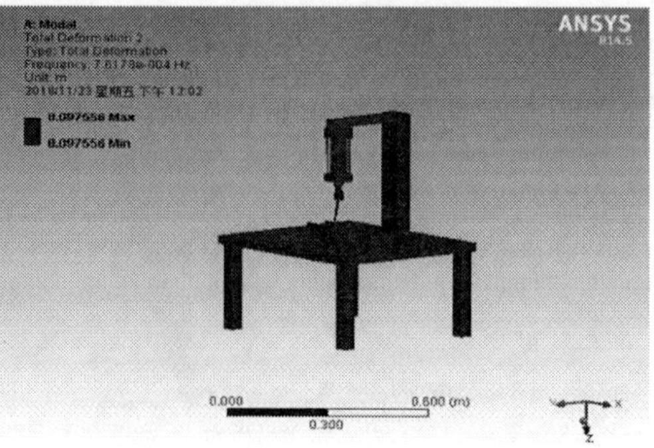


**Fig 15.** The second stage vibration mode

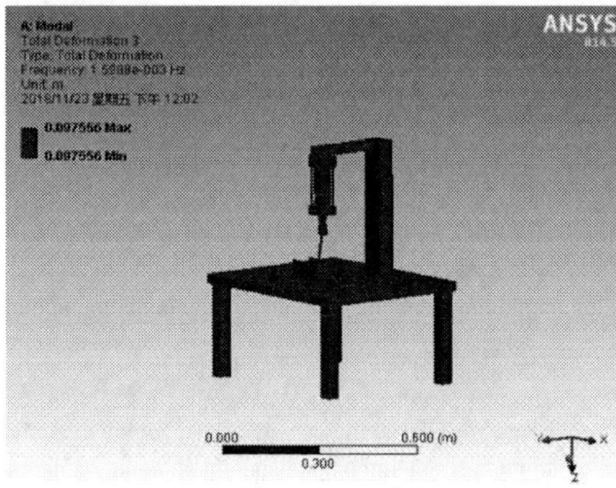


**Fig 16.** The third stage vibration mode

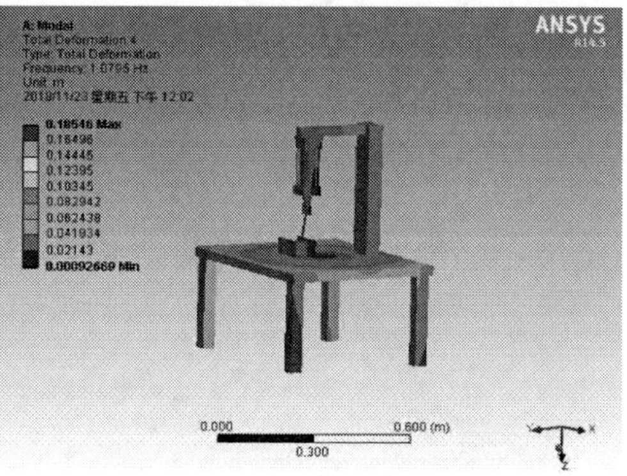


**Fig 17.** The fourth stage vibration mode

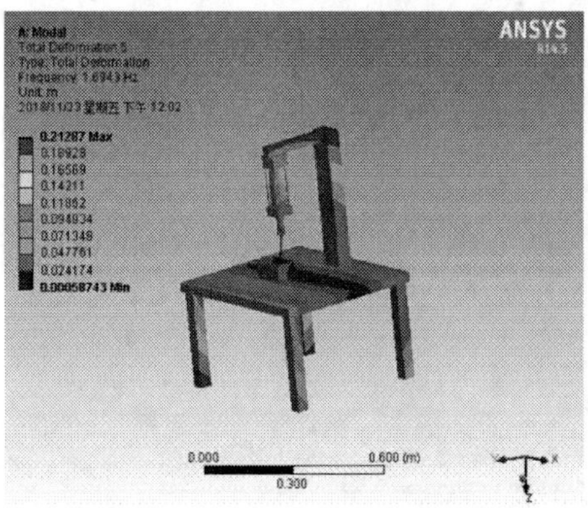


**Fig 18.** The fifth stage vibration mode

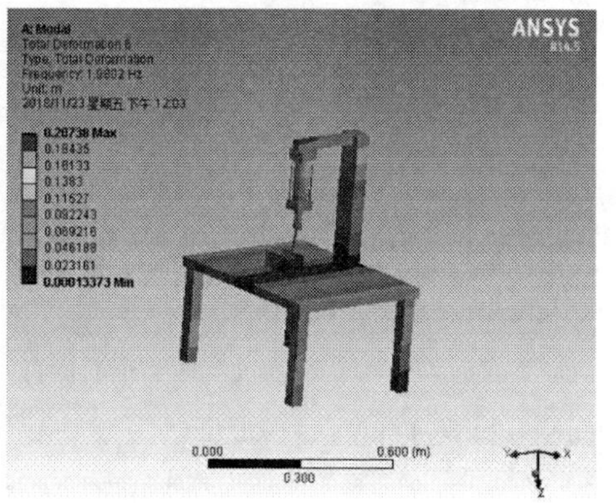


**Fig 19.** The sixth stage vibration mode

**Table 2.** Sixth-order modal analysis table of the thick-shell mussel open shell machine

| Mode | Frequency [Hz] |
|---|---|
| 1 | 0 |
| 2 | $7.6178 \times 10^{-4}$ |
| 3 | $1.5988 \times 10^{-3}$ |
| 4 | 1.0795 |
| 5 | 1.6943 |
| 6 | 1.9802 |

As can be seen from table 2, the reversing frequency of pneumatic cylinder and solenoid valve should mainly avoid 1.0795Hz, 1.6943Hz and 1.9802Hz frequencies to prevent damage caused by resonance with the thick-shell mussel open shell machine.

## 5. Conclusion

In the design process of this thick shell mussel opening shell machine, the following conclusions are drawn,

(1) The length and width of the open-shell cutter are related to physiological parameters and structure of thick shell mussel, which are determined to be 50mm and 12mm respectively.

(2) Finite element analysis is carried out to confirm that the selected alloy steel open shell cutter meets the requirements.

(3) It was concluded that the air pressure cylinder and electromagnetic valve should avoid the three frequencies of 1.0795Hz, 1.6943Hz and1.9802Hz,which prevent the thick-shell mussel open shell machine from resonating with air pressure cylinder and electromagnetic valve.

## Acknowledgements

This work was funded by Zhejiang province public welfare technology application research social development project (No.2014C33086) and Zhoushan Science and Technology Planning Project (No.2018C21006).

## References

[1] Sun C, Niu T T. International competitiveness analysis of Chinese shellfish products [J]. Chinese agricultural science bulletin,2015(20):44-50.

[2] Guo W. Changes of ripe mussel quality at different storage temperatures [D]. Zhejiang university of technology,2018.

[3] Lu X C, Sun X Z, Wang W. Bivalves hulling pretreatment technology present situation and development trend of [J]. Modern agricultural machinery, 2015 (9):74-76.

[4] Hu J Y, Li Z H, Gu P C, et al. Research progress on the method of mussel shucking [J]. Scientific fish farming,2017(05):73-74.

[5] He.H, Aams.R.M, Farkas.D.F, et al. Use of high-pressure processing for oyster shucking and shelf-life extension[J].Journal of Food Science, 2010, 67(2): 640-645.

[6] Gipsy Tabilo-Munizaga, Santiago Aubourg, Mario Pérez-Won. Pressure Effects on Seafoods [J]. High Pressure Processing of Food. 2016(01):625-669.

[7] Zhu S M, Yu Y, Wang M. Methods of mussel shell removal [P]. China, CN105851195A. 2016.08.17.

[8] Wang M. Effects of ultra-high pressure on mussel hulling and quality [D].Zhejiang university,2012.

[9] Wheaton, F.W. Oyster Processing Study[J].National Marine Fisheries Service,1971:57

[10] Singh.G.Laser modernizes oyster shucking [J]. Food Technology,1972(26):60-61.

[11] Sugiyama. H. Shell processing method and shell processing device used in the method [P]. United States Patent: No.09,787425.2004-05-18

[12] Hu J Y, Tan J L, Li Z H. Numerical simulation study on the internal flow field of water jet mussel single-side sheller nozzle [J]. Electromechanical engineering,2018.35(8): 828-832.

[13] Zhu B W, Xue C H. Marine aquatic product processing and food safety [M]. Beijing: science press,2016.

[14] Li N S, Xue C H. Modern processing technology and quality safety of Marine aquatic products in China [M]. Beijing: ocean press,2010.

[15] Xin H Z, Chen L, Wu J, et al. Internal quality traceability system for shengsi mussel processing enterprises [J]. Chinese fishery quality and standards,2013,3(3):88-92.

[16] Duan W W, Luo W, DuanZ H et al. Research progress of mussel processing and utilization [J].

Modernization of fisheries,2013,40(3):51-55.
[17] Song G L. Processing, utilization and development of mussels [J]. Modernization of fisheries,2003 (3):30-31.
[18] Hu J Y. Research on new mussel unilateral shell removal technology [D]. Zhejiang: zhejiang ocean university, 2018.
[19] Zhang F F, Wang J C. Design and experiment of seedling clamping mechanism of vegetable grafting machine [J]. Research on agricultural mechanization, 2018, (12) : 135-138.
[20] Song P W, Liu J. Solidworks and ANSYS finite element comparison of cantilever beam model [J]. Modern manufacturing technology and equipment, 2018, (3) : 71-73.
[21] Pu G Y. ANSYS Workbench basic course and detailed explanation of examples [M]. Beijing: China water conservancy and hydropower press, 2014.3:240-242.
[22] Liu X H, Liu J H, et al. Modal analysis of knife plate operation of circular orchard trenching machine [J]. Research on agricultural mechanization, 2016,(6):102-105.

IOP Conf. Series: Materials Science and Engineering **772** (2020) 012006 doi:10.1088/1757-899X/772/1/012006

# Recently Progress on Polylactide/Nanocellulose Nanocomposites

**Zhuoyuan Xu**
Rice University, USA

**Abstract.** The development of renewable resources is a trend in the future. Polylactide (PLA), as one member of bio-polymers, has obtained a widely focus in industry and academy because of its renewability, biodegradability, biocompatibility and good thermomechanical properties. PLA is used in a wide variety of applications such as biomedical, food packaging, electrical and electronic equipment, and the automotive industry. However, PLA still has some shortcomings, such as lower heat distortion temperature, slower crystallization rate, lower thermal resistance, etc. These shortcomings also limit the further widespread use of PLA. One effective method to expand and improve these properties of PLA is adding nanofillers. This review summarizes and describes the role of nanocellulose as a filler in the modification of PLA matrix and some of the problems faced.

**Key words:** Polylactide/ Nanocellulose, Nanocomposites, Bio-polymers.

## 1. Introduction

Bio-polymers which are obtained from renewable resources are widely considered as promising alternatives to traditional petro-polymers because they fulfill current environmental concerns in terms of environmental pollution [1, 2]. As one of the members of bio-polymer, poly(lactic acid) (PLA) is a leader biopolymer due to its attractive mechanical properties, reproducibility and biodegradability [3], with which a Young modulus of ~3 GPa, a strength of 50~70 MPa, an elongation at break of ~4 %, and an impact strength close to 2.5 kJ/m [2, 4]. These mechanical properties, particularly the Young's modulus, are better than the most of commodity polymers such as polyethylene (PE), polypropylene (PP), polystyrene (PS) and polyethylene terephthalate (PET), leading it as an excellent candidate for exchanging commodity polymers in short-time packaging. However, the low impact strength of PLA limits the widely application. [3] Furthermore, PLA shows poor crystallization ability, which largely limiting its industrial implementation in various applications. [5] One way usually developed to enhance both the crystallization ability and impact strength over the past years is incorporating reinforcements within the polymer matrix, yielding so-called nanocomposite materials [6, 7]. Much of different types of nanofillers have been incorporated into PLA matrix as an enchant agent to modify its thermo- and mechanical properties as well as to provide additional functionalities like conductivity and flame retardancy.

Nanocomposites refer to the fillers with at least one dimension at the nanometer level. Lots of researches have been carried out about polymeric nanocomposites due to their properties substantially improved at very low nanofiller loadings [8]. Environmentally friendly nanocomposites are the key

Content from this work may be used under the terms of the Creative Commons Attribution 3.0 licence. Any further distribution of this work must maintain attribution to the author(s) and the title of the work, journal citation and DOI.
Published under licence by IOP Publishing Ltd

IOP Conf. Series: Materials Science and Engineering **772** (2020) 012006 doi:10.1088/1757-899X/772/1/012006

direction for future nano-material development. Nanocellulose, as a renewable green material extracted from organisms, has received extensive attention in recent years. Cellulose has good mechanical properties and is the main material that constitutes the cell wall of plants, which can improve the resistance to deformation of key parts. Cellulose has a wide range of sources, low prices, and no pollution to the environment, so as to one of the ideal fillers for the formation of polymer composites. Since nanocellulose is mostly fibrous, when it is used as a filler, it can be laterally separated from the entire amorphous portion present on its axis, resulting in defect-free rod-shaped nanoparticles. According to the aggregate structure of the rod-shaped nanoparticles, it can be classified into cellulose nanocrystals (CN) and nanofibrillated cellulose (NFC). Their morphology are shown in Figure 1 and Figure 2, respectively. Among them, nanofibrillated cellulose can also be biosynthesized by microorganisms, and is known as bacterial cellulose (BC). This paper will summarize recent research on PLA-based nanocellulose nanocomposites and look forward to future opportunities.

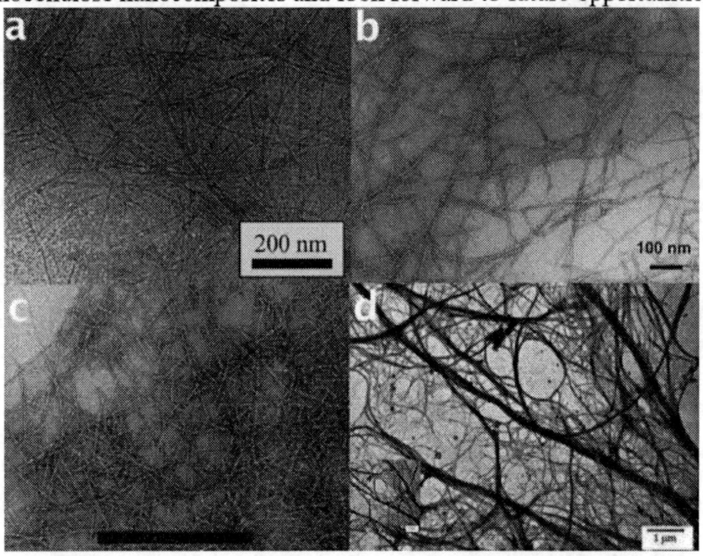


**Figure 1.** TEM micrographs of a dilute suspension of NFCs obtained from wood fibers by mechanical processing that was employed in conjunction with (a) enzymatic treatment [9]; (b) TEMPO-mediated oxidation [10]; (c) carboxylmethylation pretreatment [11]; (d) NFCs extracted from Opuntia ficus-indica [12].

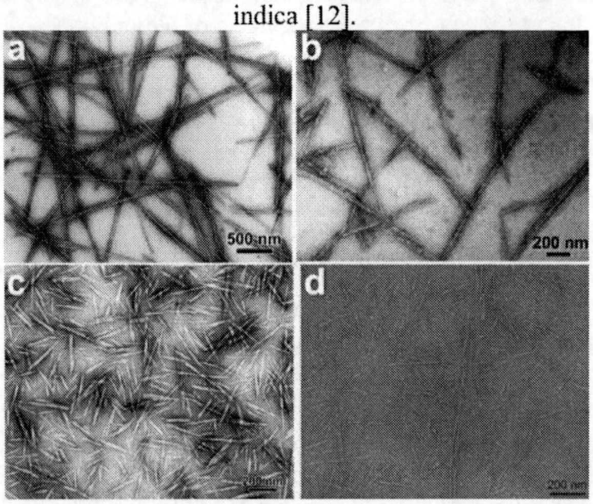


**Figure 2.** TEM images of dried dispersion of cellulose nanocrystals derived from (a) tunicate [13], (b) bacterial [13], (c) ramie [14] and (d) sisal [15].

## 2. PLA/nanocellulose nanocomposites

Nanocellulose substrates have high surface area, low density, high mechanical strength and controllable nanoscale size, which has attracted great interest in the preparation of high performance nanocomposites. A larger number of studies have shown that nanocrystalline cellulose has comparable or even better mechanical properties than other materials. In addition, nanocellulose crystals have some advantages not found in other types of fillers, such as being more readily available, and renewable and biodegradable. Nanocellulose fillers including NFC, CN and BC, have been incorporated into a variety of polymer matrices such as PLA. However, obtaining good dispersion of the filler in the PLA matrix remains one of the major challenges in the preparation of nanocomposites. Nanocellulose has a large number of hydroxyl groups on the surface, so its good hydrophilicity makes them poorly dispersible in hydrophobic matrices such as PLA or common non-polar solvents, so it is very easy to induce nanocellulose particles aggregation phenomenon in the PLA matrix. In addition, due to the hydrogen bond network formed by a large number of hydroxyl groups and hydroxyl groups present on the surface of the nanocellulose, irreversible aggregation of the nanocellulose substrate is also caused upon drying. Therefore, they usually remain dispersed in an aqueous medium or a polar solvent after preparation. At present, different processing methods have been reported, and several chemical strategies have been developed to adjust the interfacial compatibility between PLA and nanocellulose substrates. However, the final results and uses of nanocellulose-based PLA composites have rarely been reported.

### *2.1. PLA/CNs nanocomposites*

The first report on the preparation of PLA-based nanocomposites for nanocellulose as a filler was to inject the aqueous suspension of PLA into the PLA melt during extrusion. However, due to the strong aggregation properties of NFC, the lack of compatibility with PLA matrix, the hydrolytic degradation of PLA induced by the presence of water, and the fact that no thermal or mechanical properties of the final product have been studied in this attempt, it is suspected that the first report is mediocre. [16] Oksman et al. [17] It has recently been reported that a mixture of lithium chloride (LiCl) and N,N-dimethylacetamide (DMAc) is useful to swell and partially separate CN, and finally the suspension is extruded to the PLA melt during the extrusion process. A composite material in which CNs are well dispersed in a PLA matrix can be obtained. However, the disadvantage of this study is that DMAc / LiCl appears to cause degradation of the composite at high processing temperatures. One of the improvement methods of this processing strategy is to use polyvinyl alcohol (PVOH) as a processing aid [18], and mix PVOH with PLA before extrusion or directly into the extruder by mixing PVOH with CN suspension. However, in both cases, the resulting nanocomposites showed poor mechanical properties, mainly due to the phase separation structure of the PLA and PVOH phases, where PLA is the continuous phase and PVOH is the discontinuous phase. The CN is mainly located in the discontinuous phase PVOH, which greatly limits the load transfer between the nanofiller and the PLA matrix. To some degree, this strategy is not conducive to the improvement of the comprehensive mechanical properties. In addition to this, it is also possible to coat the surface of the CNs with an anionic surfactant and then incorporate the surface-modified CNs into the PLA melt in a dry state during the extrusion of the PLA in order to improve the CNs in the PLA matrix. [19] The dispersing ability of the anionic surfactant is usually composed of a phosphate having an alkylphenol tail. However, when the surfactant content is low, the dispersion state of CNs in the PLA matrix is not improved. As the surfactant content increases, the dispersion of CNs in the PLA matrix becomes good, but excess surfactant tends to cause significant thermal degradation of the PLA matrix at processing temperatures. The mechanical properties of the reinforced nanocomposite are improved compared to unreinforced nanocomposites. Further studies have shown that the use of solvent casting instead of extrusion to combine surfactant-modified CN with the same surfactant into PLA results in a better dispersion state. However, the thermomechanical properties of the PLA composites prepared by this method have decreased, although the results of TGA indicate that the obtained nanocomposites can still exist stably at 220 ° C, but the DMA test results (as shown in Figure 3) show that the excess surfactant can reduce the composite materials. The storage modulus is mainly due to the very strong interaction between PLA and surfactant, but the interaction between CN and surfactant is relatively weak, so reinforcement is not achieved to some extent. [20]

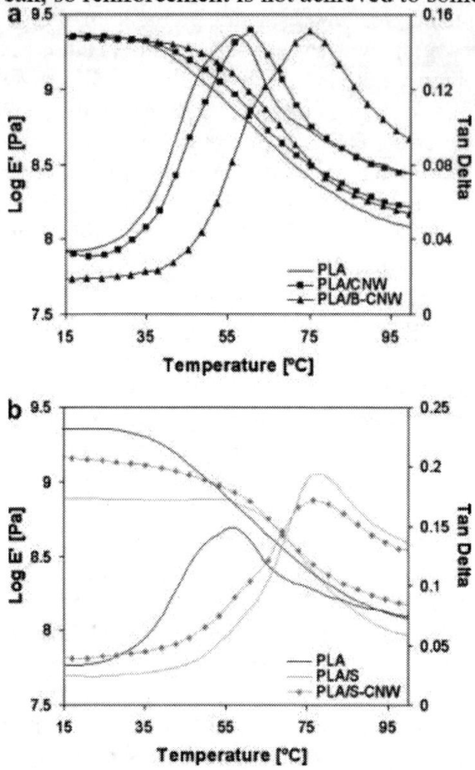


**Figure 3.** Storage modulus curves and tan δ peaks from DMTA analysis. (a) PLA, PLA/CNW and PLA/B-CNW (b) PLA, PLA/S and PLA/S-CNW. [20]

On the other hand, the surfactant-modified CNs significantly enhanced the crystallization nucleation ability of PLA, which to some extent highlighted the influence of crystal and its surface modification on the thermal and mechanical properties of nanocomposites. Studies have shown that although aggregates are still present on very small scales, CN promotes nucleation and subsequent crystallization of PLA more efficiently in composites of amorphous PLA than in composites [21-23] of crystalline PLA. In addition to mechanical properties, the study also found that, when the content of CNs is more than 3 wt%, not only the crystallinity of PLA is significantly improved, but also the water and oxygen permeability of PLA nanocomposites can be significantly improved, which can be increased to 82 % and 90 % [22]. It is worth noting here that the presence of CNs does not impair the hydrolytic degradation of PLA nanocomposites as expected, but instead causes a strong delay. The reason is the physical barrier effect of highly crystalline CN, which inhibits water absorption and thus delays the hydrolytic degradation of the bio-nanocomposite. The results of the study indicate that CNs may control biodegradability and extend the life of PLA by incorporating a small amount of nanocellulose. [24]

*2.2. PLA/NFC nanocomposites*

NFC is also one of the main green fillers for the preparation of reinforced PLA nanocomposites. Solution casting can generally be used to prepare high levels of NFC PLA/NFC masterbatch, followed by melt blending with a PLA matrix by twin screw extrusion using a strategy of diluting the wood. [25] That is, a PLA nanocomposite with a relatively good NFC dispersion state and good mechanical properties can be obtained by a two-step process. Acetone, chloroform, etc. are common organic solvents in industry and laboratories (as shown in Figure 4). Unmodified NFC can be pre-mixed with PLA using a low-boiling organic solvent such as acetone. After stirring, the organic solvent can be removed and dried.

IOP Conf. Series: Materials Science and Engineering **772** (2020) 012006 doi:10.1088/1757-899X/772/1/012006

The NFC uniformly disperses the nanocomposite in the PLA, and finally the sheet is prepared by means of hot pressing or the like. [26] The well-dispersed NFC particles reinforced the PLA matrix, which increased the Young's modulus and tensile strength of the PLA by 40 % and 25 %, respectively, and the yield strain did not decrease when the content of the nanofiller was 10 wt%. Further studies have found that the storage modulus of the composite remains constant above the $T_g$ of the PLA matrix. In addition, the authors also found that the addition of NFC accelerates the crystallization of fully amorphous PLA [27] and has the same ability to promote PLA crystal nucleation as CNs. [28] And the promotion effect exhibited in the amorphous PLA matrix is significantly better than that of the semi-crystalline PLA matrix. Recently, the same group of researchers further found that when NFC is combined with another nucleating agent, zinc phenylphosphonate (PPA-Zn), the combination of the two has a synergistic effect on the crystallization and mechanical properties of the PLA composite. The mechanism of synergy is: The effective nucleation of PPA-Zn accelerates the crystallization of PLA and the enhancement of NFC, thus providing higher strength and hardness for the composite. This synergistic effect greatly increases the dimensional stability of the material, so that the composite is not deformed even when the composite is injection molded at a high temperature (95 ° C) for a holding time of 10 s [29].

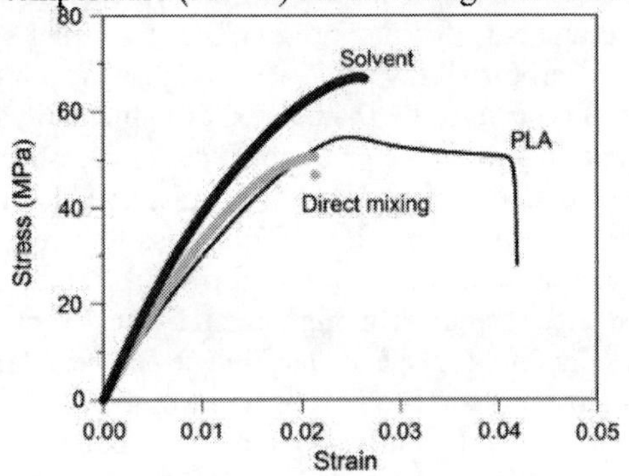


**Figure 4.** Typical stress–strain curves comparing the solvent method, direct mixing method compounding, and neat PLA. [26]

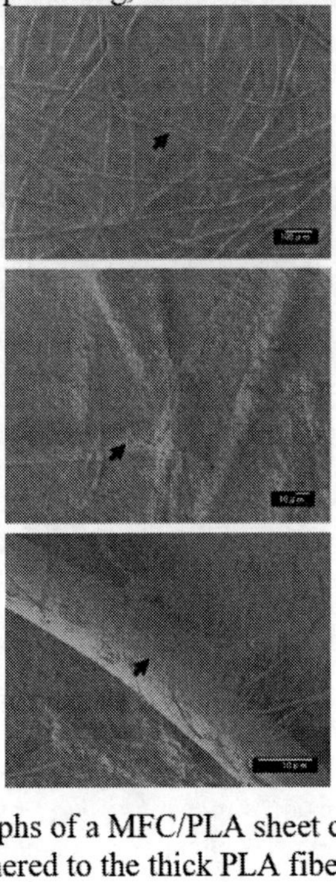

**Figure 5.** Scanning electron micrographs of a MFC/PLA sheet containing 70 wt% MFC. The network of tiny MFC fibrils is uniformly adhered to the thick PLA fibers. Arrows point to PLA fibers. [30]

In addition, Nakagaito et al. [30] developed a new manufacturing process similar to papermaking for the production of NFC-reinforced PLA nanocomposites (as shown in Figure 5). Since this method has a significant advantage of high productivity and rapid dehydration, this method can be industrially mass-produced. The thin-plate composites produced by this method have excellent mechanical properties. Studies have shown that the modulus, strength and strain of the composites at break are linearly increased with the NFCs content. Furthermore, slight modifications in this process are also enabled direct mixing of the different components allows for good NFC dispersion in the PLA in the bio-nanocomposite, improving the mechanical properties of the composite even at high NFC levels. The greater the content of NFC, the stronger the composite will become and the more sensitive it becomes to defects. NFC nanoparticles can also play a good role in bio-based nanocomposites. For example, the formation of a strong NFC network of NFC particles in a PLA matrix can improve the thermal stability of the composite. Since the surface of the NFC has a large amount of hydroxyl groups, the NFC can carry a part of water, and the water carried by the NFC can be used as a natural foaming agent to prepare a foam structure of the PLA-based composite material. [31] Studies have shown that due to the high activity of hydroxyl groups, NFC can be modified by carboxymethylation or acetylation to change the NFC hygroscopicity, and the amount of water present in the starting system can be adjusted under different humidity conditions. Carboxymethylated NFC can provide a higher water content, making it possible to tailor the density of biocomposite foams. For example, 9.6 wt% water and 8.3 wt% can be used to prepare foams with a density of 0.68 $g/cm^3$ and 0.84 $g/cm^3$, respectively. However, studies have found that only water adsorbed on the fibrils can act as an effective blowing agent.

It has been reported that electrospinning is a useful process for producing PLA nanocomposites in which the filler is a nanocellulose matrix in the form of fibers. For example, the PLA/CNs nanocomposite fiber can be electrospun at a high temperature from a solution of PLA in N-N-dimethylformamide (DMF) containing suspended cellulose nanocrystals, thereby obtaining a well-dispersed PLA/CNs nanocomposite fiber. The fibrous nanocomposites prepared by the electrospinning method have a smooth morphology and a uniform nanometer diameter [32] even at a filler content of up to 10 wt%. The increase in the CNs content caused a slight decrease in the fiber diameter, but the elongation at break of the electrospun nonwoven fabric decreased. Fine research has found that the distribution of CN in the fibers obtained by electrospinning is not very uniform, and the amount of CN available on the surface is larger than the amount of CN inside, indicating that there is no adhesion between CN and PLA in the fiber. Fortunately, despite the lack of compatibility, CN can act as a nucleation site during electrospinning, resulting in an increase in crystallinity of the electrospun PLA nanocomposite as the CN content increases. Due to the reduction in the diameter of the electrospun fiber and the increase in crystallinity, the strength of the electrospun nonwoven fabric can be achieved by adding CNs to the PLA. Liu et al. used a similar method to prepare uniform smooth fibers with an average diameter of 300 nm, and the mechanical properties of the composites were greatly improved. [33]

### 2.3. *PLA/BC nanocomposites*

In order to achieve good dispersion of nanocellulose in the PLA matrix and increase the compatibility of the nanofiller with the polymer matrix, the researchers also tried to use covalent chemical functionalization of nanocellulose. Several chemical modifications that have been successfully implemented include esterification, silanization, and polymer grafting etc. Lee et al. [34] have extensively studied the esterification of organic acids with different chain lengths (C2 to C12) with BC nanofibers in order to adjust their hydrophobic equilibrium as shown in Figure 6. When the chain length of the organic acid used for the modification is larger than C2, the tensile modulus and tensile strength of the nanocomposite will be significantly improved after the esterified BC filler and the PLA matrix are combined by the twin-screw extrusion technique. However, when the BC particles are modified by the acetylation of an organic acid of the same chain length (chain length of C2), the ester bond will be degraded during the twin-screw extrusion process. Further release of the acetic acid molecule will catalyze the degradation of the PLA matrix, causing a decrease in the molecular weight of the PLA, further impairing the mechanical properties of the final nanocomposite. Interestingly, when the acetylation reaction occurs under moderate conditions, the degree of acetylation is significantly reduced, and not only good dispersion of BC nanofibers in PLA can be obtained. Moreover, no degradation was observed in the nanocomposite produced by the twin-screw extrusion processing at a high temperature of 190 °C. A significant increase in elastic and Young's modulus and tensile strength (increased by about 100%, 40% and 25%, respectively) at a very low nanofiller content demonstrates a significant improvement in thermal and mechanical properties.

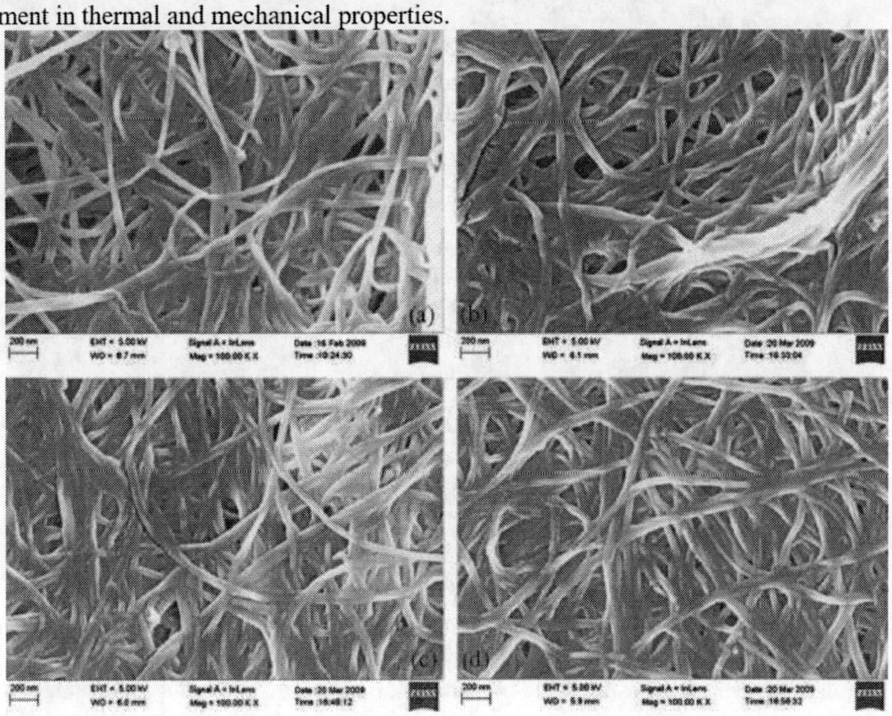

**Figure 6.** Scanning electron micrographs of: (a) neat BC, (b) C2-BC, (c) C6-BC and (d) C12-BC. [34]

Moreover, these nanocomposites exhibit interesting and very meaningful physical characteristics such as humidity insensitivity, high transparency, and resistance to aging. Studies have shown that when acetylated NFC [35] or acetylated CNs [36] is added to the PLA matrix by casting-evaporation techniques at room temperature, the dispersion of nanoparticles is improved due to the good interfacial adhesion between the surface of the nanoparticles and the PLA matrix. The state of dispersion in the polymer matrix ultimately results in the material not only resisting degradation, but also significantly enhancing the mechanical properties of the composite. Another example is the incorporation of maleic anhydride-modified BC nanofibers into a PLA matrix by solution casting, followed by compression to prepare composites. The composites prepared by this method also have excellent mechanical properties [37]. In addition, the partial silanization process of CNs extracted from cotton was studied. [38] The CNs extracted from cotton are first reacted with n-dodecyldimethylchlorosilane. The purpose of this method is to obtain good dispersion of CN in the composite. It was found that when the silanized CNs content is only 1%, the crystallization rate of PLA will be significantly increased as shown in Figure 7. In contrast, unmodified CNs have limited impact on the crystallization ability of PLA. Furthermore, studies have found that the increase in crystallinity can effectively improve the tensile modulus and strength of the composite, but the increase in crystallinity is not conducive to the increase in elongation at break.

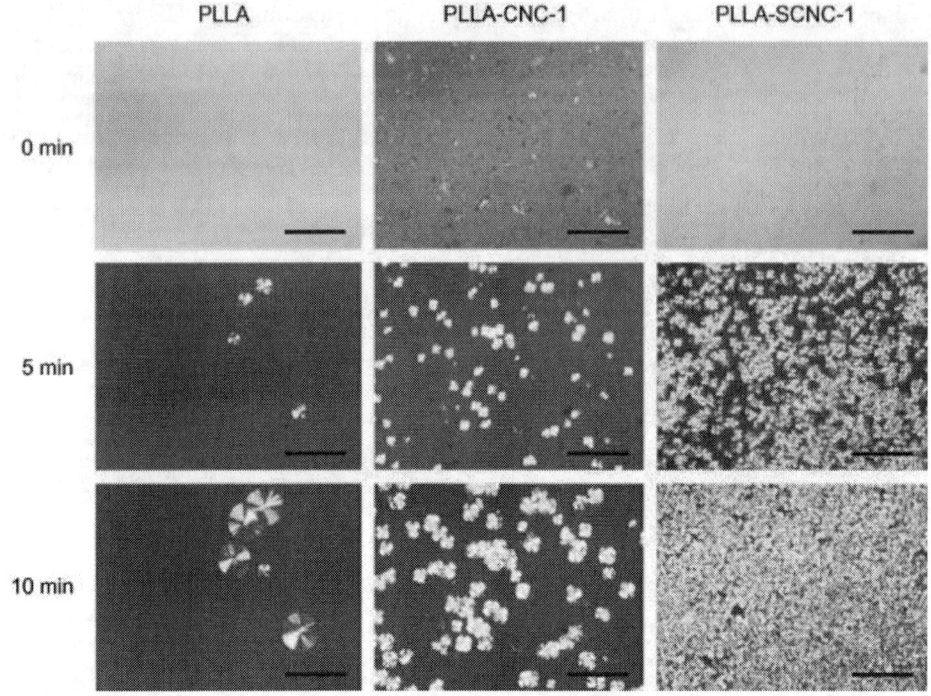


**Figure 7.** Polarized optical microscope images of PLLA, PLLA–CNC-1, and PLLA–SCNC-1 acquired on the 0, 5th and 10th min at 125 °C after quenched from melt at 210 °C. Scale bar, 200 μm. [38]

In addition, the researchers also attempted to adjust the compatibility between nanocellulose and PLA by silanizing the surface of NFC fibers and CN fibers with 3-aminopropyltriethoxysilane. [39] Then the PLA-based nanocomposites were prepared by twin-screw extrusion processing, but the final mechanical properties were only slightly increased. This may be due to the aminolysis reaction of the amine groups during high-temperature processing, which caused the degradation of the PLA molecular chains. Recent studies have found that CNs can be directly combined with silanes having different functional groups (i.e. alkyl, amino, methacryloxy) [40] after extraction. The reaction process is simple and easy to control, and can be efficiently carried out in an aqueous medium at room temperature. After the reaction product is freeze-dried, the obtained silylated CN can be successfully incorporated into the PLA matrix by twin-screw extrusion. It is found that there is no change in nanostructures and a decrease in molecular weight even in the high-temperature twin-screw extrusion process. In addition, Li et al. [41] successfully grafted epoxy-terminated polylactide- glycidyl methacrylate copolymer onto the surface of BC nanofibers by surface grafting. The nanofibers obtained by this method showed good dispersibility in the PLA matrix, but unfortunately the study did not report any mechanical properties. Furthermore, by means of a "grafting" method, a polycaprolactone (PCL) chain can be grown from the surface of the CNs by microwave-assisted ring-opening polymerization, and then the surface-modified CNs fibers are dispersed into the PLA matrix [42], although PCL and PLA are The thermodynamic entropy exhibits incompatible properties, but the resulting PCL-g-CNs show good dispersibility in PLA, thus greatly improving the mechanical properties of the composite. Goffin et al. [43] also used a "grafting" method to graft a PLA molecular chain directly on the surface of the CNs by ring-opening polymerization (as shown in Figure 5), and then redisperse it into the PLA by twin-screw melt extrusion technology. The PLA chain grafted on the surface of the CNs not only enhances the compatibility of the CNs with the PLA matrix, but also co-crystallizes with the PLA matrix to form a co-continuous network. The mechanical properties of the final PLA-based nanocomposite are improved.

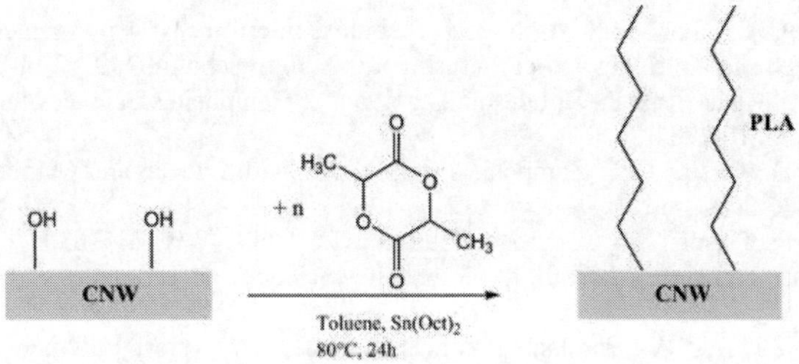


**Figure 8.** Grafting from method based on ROP of LA carried out via hydroxyl functionalities present at the CNs surface. [44]

## 3. Conclusion

The development of bio-based polymers is closely related to consumer and industrial needs. In this regard, PLA, as a member of bio-based polymers, is one of the most popular and promising "green" environmentally friendly materials in the application of food packaging, medical equipment, electrical and electronic equipment, and mechanical and automotive components. At the same time, however, PLA must be enhanced in terms of heat resistance, heat distortion temperature, crystallization rate, and impact resistance. As shown in this review, the preparation of PLA-based nanocomposites by the addition of nanocellulose is one of the effective methods for improving the properties of PLA. Nanocellulose is easily added to the PLA matrix by appropriate pre-dispersion, pre-mixing, masterbatch, surface treatment and grafting methods. For future research, although the dispersion of nanoparticles in PLA matrix is still a challenging problem in the preparation of high performance composites, it is interesting to note from the recent trends that the development of synergistic combinations of multiple fillers will be the basis for the preparation of high performance composites in the future. This will be an effective ways to greatly improve the overall performance of nanocomposites.

## References

[1] Jenck, J. F.; Agterberg, F.; Droescher, M. J. Green Chemistry 2004, 6, (11), 544-556.
[2] Kümmerer, K. Green Chemistry 2007, 9, (8), 899-907.
[3] Auras, R.; Harte, B.; Selke, S. Macromolecular Bioscience 2004, 4, (9), 835-864.
[4] Anderson, K. S.; Schreck, K. M.; Hillmyer, M. A. Polymer Reviews 2008, 48, (1), 85-108.
[5] Rasal, R. M.; Janorkar, A. V.; Hirt, D. E. Progress in Polymer Science 2010, 35, (3), 338-356.
[6] Alexandre, M.; Dubois, P. Materials Science and Engineering: R: Reports 2000, 28, (1), 1-63.
[7] Bordes, P.; Pollet, E.; Avérous, L. Progress in Polymer Science 2009, 34, (2), 125-155.
[8] Kumar, A. P.; Depan, D.; Singh Tomer, N.; Singh, R. P. Progress in Polymer Science 2009, 34, (6), 479-515.
[9] Pääkkö, M.; Ankerfors, M.; Kosonen, H.; Nykänen, A.; Ahola, S.; Österberg, M.; Ruokolainen, J.; Laine, J.; Larsson, P. T.; Ikkala, O.; Lindström, T. Biomacromolecules 2007, 8, (6), 1934-1941.
[10] Saito, T.; Kimura, S.; Nishiyama, Y.; Isogai, A. Biomacromolecules 2007, 8, (8), 2485-2491.
[11] Wågberg, L.; Decher, G.; Norgren, M.; Lindström, T.; Ankerfors, M.; Axnäs, K. Langmuir : the ACS journal of surfaces and colloids 2008, 24, (3), 784-795.
[12] Malainine, M. E.; Mahrouz, M.; Dufresne, A. Composites Science and Technology 2005, 65, (10), 1520-1526.
[13] Elazzouzi-Hafraoui, S.; Nishiyama, Y.; Putaux, J.-L.; Heux, L.; Dubreuil, F.; Rochas, C. Biomacromolecules 2008, 9, (1), 57-65.
[14] Habibi, Y.; Goffin, A.-L.; Schiltz, N.; Duquesne, E.; Dubois, P.; Dufresne, A. Journal of Materials Chemistry 2008, 18, (41), 5002-5010.
[15] Garcia de Rodriguez, N. L.; Thielemans, W.; Dufresne, A. Cellulose 2006, 13, (3), 261-270.

[16] Chakraborty, A.; Sain, M.; Kortschot, M., Cellulose microfibrils: A novel method of preparation using high shear refining and cryocrushing. In Holzforschung, 2005; Vol. 59, p 102.
[17] Oksman, K.; Mathew, A. P.; Bondeson, D.; Kvien, I. Composites Science and Technology 2006, 66, (15), 2776-2784.
[18] Bondeson, D.; Oksman, K. Composites Part A: Applied Science and Manufacturing 2007, 38, (12), 2486-2492.
[19] Bondeson, D.; Oksman, K. Composite Interfaces 2007, 14, (7-9), 617-630.
[20] Petersson, L.; Kvien, I.; Oksman, K. Composites Science and Technology 2007, 67, (11), 2535-2544.
[21] Liu, D. Y.; Yuan, X. W.; Bhattacharyya, D.; Easteal, A. J. Express Polymer Letters 2010, 4, (1), 26-31.
[22] Sanchez-Garcia, M. D.; Lagaron, J. M. Cellulose 2010, 17, (5), 987-1004.
[23] Hossain, K. M. Z.; Ahmed, I.; Parsons, A. J.; Scotchford, C. A.; Walker, G. S.; Thielemans, W.; Rudd, C. D. Journal of Materials Science 2012, 47, (6), 2675-2686.
[24] Luiz de Paula, E.; Mano, V.; Pereira, F. V. Polymer Degradation and Stability 2011, 96, (9), 1631-1638.
[25] Jonoobi, M.; Harun, J.; Mathew, A. P.; Oksman, K. Composites Science and Technology 2010, 70, (12), 1742-1747.
[26] Iwatake, A.; Nogi, M.; Yano, H. Composites Science and Technology 2008, 68, (9), 2103-2106.
[27] Suryanegara, L.; Nakagaito, A. N.; Yano, H. Composites Science and Technology 2009, 69, (7), 1187-1192.
[28] Suryanegara, L.; Nakagaito, A. N.; Yano, H. Cellulose 2010, 17, (4), 771-778.
[29] Suryanegara, L.; Okumura, H.; Nakagaito, A. N.; Yano, H. Cellulose 2011, 18, (3), 689-698.
[30] Nakagaito, A. N.; Fujimura, A.; Sakai, T.; Hama, Y.; Yano, H. Composites Science and Technology 2009, 69, (7), 1293-1297.
[31] Boissard, C. I. R.; Bourban, P.-E.; Tingaut, P.; Zimmermann, T.; Månson, J.-A. E. Journal of Reinforced Plastics and Composites 2011, 30, (8), 709-719.
[32] Xiang, C.; Joo, Y. L.; Frey, M. W. Journal of Biobased Materials and Bioenergy 2009, 3, (2), 147-155.
[33] Liu, D.; Yuan, X.; Bhattacharyya, D. Journal of Materials Science 2012, 47, (7), 3159-3165.
[34] Lee, K.-Y.; Blaker, J. J.; Bismarck, A. Composites Science and Technology 2009, 69, (15), 2724-2733.
[35] Tingaut, P.; Zimmermann, T.; Lopez-Suevos, F. Biomacromolecules 2010, 11, (2), 454-464.
[36] Lin, N.; Huang, J.; Chang, P. R.; Feng, J.; Yu, J. Carbohydrate Polymers 2011, 83, (4), 1834-1842.
[37] Li, Z. Q.; Zhou, X. D.; Pei, C. H. Polymer-Plastics Technology and Engineering 2010, 49, (2), 141-146.
[38] Pei, A.; Zhou, Q.; Berglund, L. A. Composites Science and Technology 2010, 70, (5), 815-821.
[39] Frone, A. N.; Berlioz, S.; Chailan, J.-F.; Panaitescu, D. M.; Donescu, D. Polymer Composites 2011, 32, (6), 976-985.
[40] Raquez, J. M.; Murena, Y.; Goffin, A. L.; Habibi, Y.; Ruelle, B.; DeBuyl, F.; Dubois, P. Composites Science and Technology 2012, 72, (5), 544-549.
[41] Li, Z. Q.; Zhou, X. D.; Pei, C. H. International Journal of Polymeric Materials and Polymeric Biomaterials 2010, 59, (9), 725-737.
[42] Lin, N.; Chen, G.; Huang, J.; Dufresne, A.; Chang, P. R. Journal of Applied Polymer Science 2009, 113, (5), 3417-3425.
[43] Goffin, A.-L.; Raquez, J.-M.; Duquesne, E.; Siqueira, G.; Habibi, Y.; Dufresne, A.; Dubois, P. Biomacromolecules 2011, 12, (7), 2456-2465.
[44] Raquez, J.-M.; Habibi, Y.; Murariu, M.; Dubois, P. Progress in Polymer Science 2013, 38, (10), 1504-1542.

IOP Conf. Series: Materials Science and Engineering **772** (2020) 012007 doi:10.1088/1757-899X/772/1/012007

# Modeling and Simulation of Fuel Engine Rail Pressure System

**Hong Wang***

Hubei University of Technology Engineering and Technology College, Wuhan 430068, China

*Corresponding author e-mail: 515971726@qq.com

**Abstract.** The rail pressure system of fuel engine is a nonlinear dynamic system with strong disturbance and its system modeling is especially important, the engine rail pressure system is taken as the research object in this paper, and the "pump-pipe-nozzle" mathematical model is established, the dynamic relationship among the cam speed of oil pump, the fuel injection quantity of fuel spray nozzle and the fuel pressure in the common rail pipe is explored in theory and simulation.

**Keywords:** fuel engine; rail pressure system modeling; oil pressure control; simulation.

## 1. Introduction

The essence of pressure control problem of the rail pressure system is the transmission problem of fuel pressure waves, from the source high-pressure oil pump, through the high-pressure oil pipe, to the fuel spray nozzle(abbreviated as "pump-pipe-nozzle"); finally enter into the engine combustion cavity, during the transmission process, the pressure, flow and density of fuel are dynamically changing. Therefore, the key to solve the problem is to establish a mathematical model of the pump-pipe-nozzle three parts.

The pressure stability of high-pressure fuel pipe in high-pressure fuel system is taken as the research goal in this paper, by establishing the mathematical model of each "pump-pipe-nozzle", the system model of the rail pressure system is formed, and this paper conducts theoretical research and simulation exploration for the dynamic relationship among the cam speed of oil pump, the fuel injection quantity of fuel spray nozzle and the fuel pressure in the common rail pipe.

## 2. Raise the Problem

The operation process diagram of a fuel engine oil rail pressure system is shown in Fig.1, the fuel enters the high-pressure oil pipe from A through the high-pressure oil pump, and the length of the oil supply time is controlled by the one-way valve switch, and then sprayed by many nozzles B.

The intermittent operation process of fuel into and out of the common rail pope (high pressure oil pipe) causes the pressure of the fuel system of the fuel engine to present nonlinear dynamic fluctuations, which causes the sprayed fuel amount to deviate from the expected value, thus affecting the operating efficiency of engine.

The fuel at the high-pressure oil pipe A is from the plunger cavity outlet of the high-pressure oil pump, and the oil injection is controlled by the needle valve of the fuel spray nozzle. The pressure oil process of the high-pressure oil pump plunger is as follows: when the plunger moves upward, the fuel

Content from this work may be used under the terms of the Creative Commons Attribution 3.0 licence. Any further distribution of this work must maintain attribution to the author(s) and the title of the work, journal citation and DOI.
Published under licence by IOP Publishing Ltd

in the plunger cavity is compressed, when the pressure in the plunger cavity is greater than the pressure in the high pressure-oil pipe, the one-way valve connected to the plunger cavity and the high-pressure oil pipe is opened, and the fuel enters the high-pressure oil pipe.

The spray nozzle structure of fuel injector is shown in Fig.2, when the needle valve lift is 0, the needle valve is closed; when the needle valve lift is greater than 0, the needle valve is opened, the fuel flows to the injection hole sprays through the injection hole.

Under the high-pressure oil pipe size and initial pressure, determine the angular velocity of the cam and the fuel injection amount of the fuel injector are determined, so that the pressure in the high-pressure oil pipe is as stable as possible to a specified value (such as 100 MPa) or so.

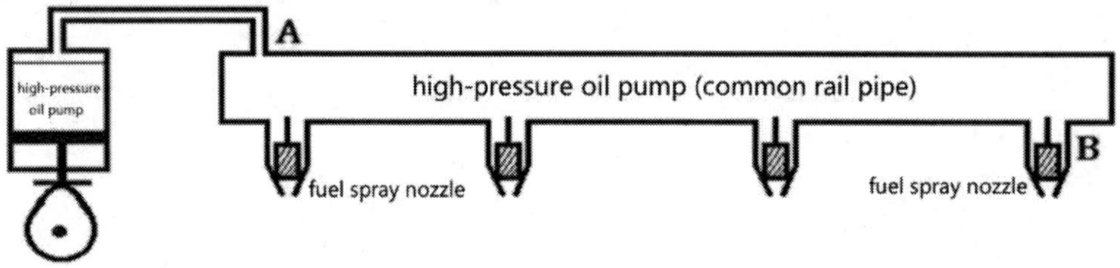


**Fig. 1** diagram of the operation process of rail pressure system of fuel engine

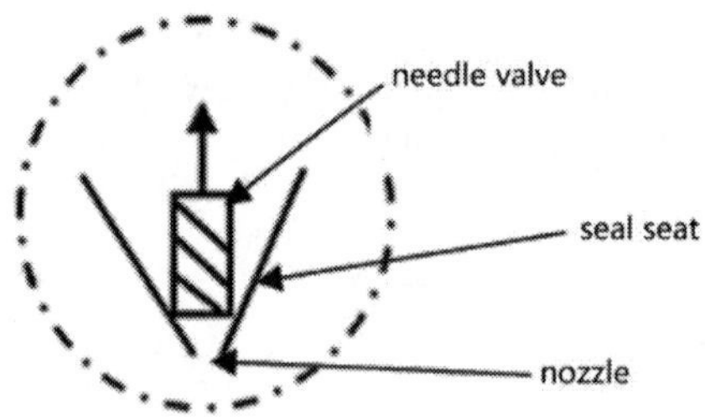


**Fig. 2** diagram of the spray nozzle of fuel injector after enlargement

## 3. Establish the Model

The solution to the pressure stability problem of high-pressure oil pipe is to establish a mathematical model of "pump-pipe-nozzle" three parts, the system model of the rail pressure system is constituted, the dynamic relationship among the cam speed of oil pump, the fuel injection quantity of fuel injector and the fuel pressure in common rail pipe is studied. The following models are established to discuss separately.

### *3.1. Establishment of dynamic equation of fuel pressure in high-pressure oil pipe*

Since the pressure of the fuel in the high-pressure oil pipe is dynamically changing, The fuel is subjected to pressure and its volume changes, this nature is called the compressibility of the fuel, the fuel with $V$ volume, when the change amount of pressure is $\Delta P$, and the absolute change amount is $\Delta V$, the relative change of volume of the fuel under the change of unit pressure, namely the compression coefficient of the fuel is:

$$k = -\frac{1}{V}\frac{\Delta V}{\Delta P} \tag{1}$$

As the pressure increases, the volume of the fuel will decrease, so the right side of formula (1) must be added with a minus sign to make the pressure coefficient positive.

The volume of the high-pressure oil pipe is set to be $V$ : at the $t$ moment, the pressure of fuel in the high-pressure oil pipe is $P_t$, the fuel density is $\rho_t$, the fuel compression coefficient is $k_t$, at the

$t+\Delta t$ moment, the pressure of fuel in the high-pressure oil pipe is $P_{t+\Delta t}$, at the $\Delta t$ moment, the mass of fuel injected into the high-pressure oil pipe through the high-pressure oil pump is $Q_A$, the mass of the fuel injected from the nozzle is $Q_B$, The relationship among mass, density and volume can get $\Delta V=\frac{Q_B}{\rho_t}-\frac{Q_A}{\rho_t}=\frac{Q_B-Q_A}{\rho_t}$, and $\Delta P=P_{t+\Delta t}-P_t$, the above $\Delta V$ and $\Delta P$ are put into formula (1),

$$k_t=-\frac{1}{V}\frac{\frac{Q_B-Q_A}{\rho_t}}{P_{t+\Delta t}-P_t}$$

Namely:

$$P_{t+\Delta t}=P_t+\frac{Q_A-Q_B}{\rho_t k_t V} \tag{2}$$

From $t$ moment to $t+\Delta t$ moment, the fuel mass $Q_A$ entering the high-pressure oil pipe from the one-way oil valve and the fuel mass sprayed from the fuel spray nozzle are compared, if $Q_A=Q_B$, then at the $\Delta t$ moment, the fuel pressure in the high-pressure fuel pipe does not change during the period, if, then the pressure of fuel in the oil pipe does not change, if $Q_A\neq Q_B$, and the pressure in the oil pipe change, pressure in the oil pipe is determined by the formula (2), the formula (2) dynamically describes the dynamic change among the pressure, mass, density and compression coefficient of the fuel in the high-pressure oil pipe, therefore, this article calls formula (2) as dynamic equation of fuel pressure of high pressure oil pipe.

In order to control the pressure in the pipe through the above-mentioned pressure dynamic equation of high-pressure oil pipe, it is necessary to solve several variables involved in dynamic equation(2): through the fuel pressure $Q_A$ of the high-pressure oil pipe pressed from the high-pressure oil pump, the function relationship of fuel amount $Q_B$ sprayed through the fuel spray nozzle, the fuel density $\rho$, fuel compression coefficient $k$ and pressure $P$, the following are solved separately.

### *3.2. Dynamic model establishment of fuel pressure in high-pressure oil pump*

It can be seen from analysis of the pressure oil process in high pressure oil pump plunger that the cam drives the plunger up and down, and the plunger compresses the fuel in the plunger cavity when the plunger moves upwards, so that the fuel pressure in the plunger cavity gradually increases, when the fuel pressure in the plunger cavity is greater than the fuel pressure in the high-pressure oil pipe, the fuel pressure dynamic models are established separately before and after the one-way valve is opened.

The angular velocity of the cam is $\omega$ rad/md, the polar angle and the polar diameter in the rotation is $\theta$ rad and $camp$ mm, respectively, and the diameter of the plunger cavity is $d$ mm, when the plunger moves to top dead center position, the residual volume of the plunger cavity is $V_1$ mm3. When plunger moves to the bottom dead center, the low-pressure fuel will fill the plunger cavity(including residual volume), and the pressure of low-pressure fuel is $P_1$ MPa.

(1) The plunger moves upward from the bottom dead center to the stage before the one-way valve is opened

The plunger moves upward from the bottom dead center to the stage before the one-way valve is opened, the fuel quality in the plunger does not change, the volume is gradually decreasing, so that the pressure of fuel is gradually increased. According to invariable fuel mass in the plunger cavity, the equation can be obtained.

$$\rho_{zs}V_{zs} = \rho_{down}V_{down} \tag{3}$$

The $\rho_{zs}$ and $V_{zs}$ represents the density and volume of fuel in the plunger cavity at the $t$ moment. $\rho_{down}$ is the density of fuel in the plunger at the bottom dead center, namely the corresponding density when the pressure of the low-pressure fuel is $P_1$ MPa, $V_{down}$ is the volume of fuel in the plunger at the bottom dead center, through the analysis of relationship between the polar radius and the polar angle when the cam rotates, when the polar radius take maximum value $camp\rho_{\max}$, the corresponding plunger is at the top dead center, and when the polar radius take minimum value $camp\rho_{\min}$, the corresponding plunger is at the bottom dead center, therefore:

$$V_{down} = \pi(camp\rho_{\max} - camp\rho_{\max})\left(\frac{d}{2}\right)^2 + V_1 \tag{4}$$

Assuming the plunger is at the bottom dead center, $t = 0$, the volume of fuel in the plunger cavity at the $t$ moment.

$$V_{zs} = \pi\left(\frac{d}{2}\right)^2 \cdot camp\rho \cdot \sin\theta \tag{5}$$

The formula (4) and (5) are put into formula (3), the density of fuel in the plunger cavity can be obtained at the $t$ moment.

$$\rho_{zs} = \rho(P_1)\left[\pi(camp\rho_{\max} - camp\rho_{\max})\left(\frac{d}{2}\right)^2 + V_1\right] \Big/ \left[\pi\left(\frac{d}{2}\right)^2 \cdot camp\rho \cdot \sin\theta\right] \tag{6}$$

The function relationship $camp\rho = camp\rho(\theta)$ between polar radius $camp\rho$ and polar angle of the cam $\theta$ can be obtained by fitting the statistical data of the polar radius and polar angle when the cam rotates. Further, according to the function relationship $\rho = \rho(P)$ between fuel density and pressure (discussed in paragraph 2.4 below), the pressure $P_{zs}$ of the fuel in the plunger cavity is available at the $t$ moment.

$$P_{zs} = P(\rho_{zs}) \tag{7}$$

(2) Stage from one-way valve opening to one-way valve closing

In the stage from the of one-way valve opening to the one-way valve closing, since the pressure of fuel in the plunger cavity is greater than the pressure in the high-pressure pipe, which makes the one-way valve open and the fuel in the plunger cavity flows into the high-pressure pipe, at this time, it is necessary to establish the continuous flow equation of fuel in the plunger cavity, namely the dynamic balance equation of fuel flow velocity.

On the one hand, the flow velocity $\frac{\Delta V}{\Delta t}$ of fuel in the plunger cavity can be represented as

$$\frac{\Delta V}{\Delta t} = \frac{1}{V}\frac{\Delta V}{\Delta P} \cdot V \cdot \frac{\Delta P}{\Delta t} = -k \cdot V \cdot \frac{\Delta P}{\Delta t} \tag{8}$$

Among them, $V$ is the volume of the corresponding fuel in the plunger cavity when the one-way valve starts to open, $k$ is the compression coefficient of the fuel in the plunger cavity, $P$ is the pressure of the fuel in the plunger cavity.

IOP Conf. Series: Materials Science and Engineering **772** (2020) 012007 doi:10.1088/1757-899X/772/1/012007

On the other hand, considering the process in which the plunger is pushed upward and the fuel is discharged to the high pressure oil pipe via the one-way valve, the flow velocity $\frac{\Delta V}{\Delta t}$ of the fuel amount in the plunger cavity can be represented as

$$\frac{\Delta V}{\Delta t} = S_{zs} \cdot \frac{\Delta h}{\Delta t} - Q'_A \tag{9}$$

Among them, $S_{zs} = \pi \left(\frac{d}{2}\right)^2$ is the cross-sectional area of the plunger cavity, $\frac{\Delta h}{\Delta t}$ is the change rate of the plunger lift, $Q'_A$ is the amount of fuel that is pressed out through the one way valve per unit time, it is obtained by formula (8) and (9).

$$-kV \cdot \frac{\Delta P}{\Delta t} = S_{zs} \cdot \frac{\Delta h}{\Delta t} - Q'_A$$

The current limit on both sides of the above equation is solved when $\Delta t \to 0$, and the corresponding differential equation is obtained:

$$-kV \cdot \frac{dP}{dt} = S_{zs} \cdot \frac{dh}{dt} - Q'_A \tag{10}$$

Among them, $h = camp \cdot \sin\theta$. Formula (10) is s the continuous flow equation of the fuel in the plunger cavity.

Formula (10) is transposed and obtain

$$Q'_A = kV \cdot \frac{dP}{dt} - S_{zs} \cdot \frac{dh}{dt} \tag{11}$$

The formula (10) is the fuel flow from the high-pressure oil pump to the high-pressure oil pipe, and the fuel mass $Q_A$ from the high-pressure oil pump to the high-pressure oil pipe can be obtained by further integral.

### *3.3. Dynamic model establishment of oil injection amount of fuel spray nozzle*

The oil injection hole is shown in Fig.3, the diameter of the needle valve is known to be $d_1$ mm, the seal seat is a cone with $\alpha$ rad half angle, and the diameter of the lowermost nozzle is $d_2$ mm.

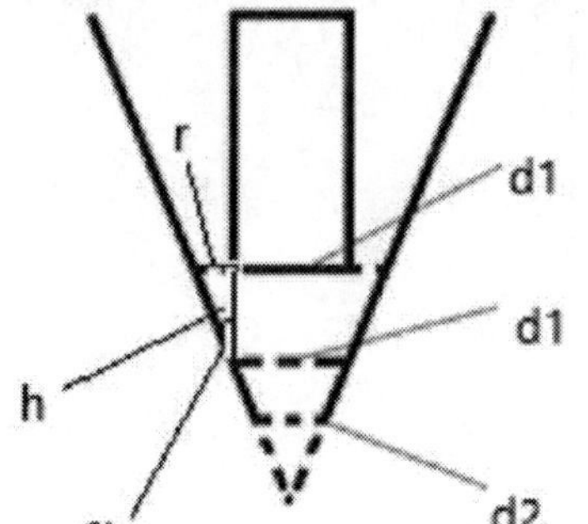


**Fig. 3** diagram of the geometric relationship of the fuel injection hole of the fuel injector

IOP Conf. Series: Materials Science and Engineering **772** (2020) 012007 doi:10.1088/1757-899X/772/1/012007

In a oil injection cycle $T$, the lift of the needle valve is set to $h = h(t)$ at the $t$ moment, the distance from the edge to the seal seat of the needle valve on the same level is $r = r(t)$. As can be seen from the geometric relationship in the figure, when the lift $h > 0$, $\frac{r}{h} = \tan\alpha$, namely $r = h\tan\alpha$, the area $S_1$ of the gap between the needle valve and the seal seat is

$$S_1 = \pi\left(\frac{d_1}{2} + r\right)^2 - \pi\left(\frac{d_1}{2}\right)^2$$
$$= \pi r^2 + \pi r d_1 = \pi h^2 \tan^2\alpha + \pi h d_1 \tan\alpha$$

Namely:

$$S_1 = \pi h^2(t)\tan^2\alpha + \pi h(t) d_1 \tan\alpha \qquad (12)$$

There are three main stages in the movement of the needle valve:

The first stage: when the needle valve is opened, the fuel flows to the nozzle hole and is sprayed through the nozzle hole, as the needle valve lift increase, the area $S_1$ of the gap between the needle valve and the seal seat gradually increases, when $S_1$ is equal to spray hole area $S_2 = \frac{\pi d_2^2}{4}$, the needle valve has the highest lift at the $S_1 = S_2$, this moment is $t_1$, the oil injection rate reaches the maximum. At this stage, namely during $[0, t_1)$ time period, the actual nozzle area of the fuel is $S_1$, and the oil injection rate is gradually increased;

The second stage: then needle valve enters into a steady state, the lift remains unchanged, at this stage, the actual nozzle area of fuel is the nozzle area $S_2$, and the oil injection rate remains unchanged;

The third stage: then the needle valve lift begins to decrease, and the time when the lift begins to decrease is $t_2$: during this period, namely $[t_2, T]$ time period, as the needle valve lift decreases, the nozzle area begins to decrease, the actual nozzle area of fuel is $S_1$, the oil injection rate decreases gradually until the needle valve lift is zero, and the oil injection stops.

The actual nozzle area of the fuel is recorded as $S = S(t)$, it can be known from the above analysis that the actual nozzle area of the fuel is at three different stages in one injection cycle $T$. The actual nozzle area of the fuel is

$$S = S(t) = \begin{cases} \pi h^2(t)\tan^2\alpha + \pi h(t) d_1 \tan\alpha, & t \in [0, t_1), \\ \dfrac{\pi d_2^2}{4}, & t \in [t_1, t_2), \\ \pi h^2(t)\tan^2\alpha + \pi h(t) d_1 \tan\alpha, & t \in [t_2, T]. \end{cases}$$

The above formula replaces the small hole area $A$ in the flow formula of the high-pressure oil pipe $Q = C \cdot A \cdot \sqrt{\frac{\Delta P}{\rho}}$, the oil injection rate is available fuel, namely the amount of oil injection $Q'_B$ through the nozzle per unit time., namely,

$$Q'_B = Q'_B(t) = \begin{cases} C[\pi h^2(t)\tan^2\alpha + \pi h(t)d_1\tan\alpha]\cdot\sqrt{\dfrac{\Delta P}{\rho}}, & t\in[0,t_1), \\ \dfrac{C\pi d_2^{\,2}}{4}\cdot\sqrt{\dfrac{\Delta P}{\rho}}, & t\in[t_1,t_2), \\ C[\pi h^2(t)\tan^2\alpha + \pi h(t)d_1\tan\alpha]\cdot\sqrt{\dfrac{\Delta P}{\rho}}, & t\in[t_2,T]. \end{cases} \quad (13)$$

$\Delta P$ is the pressure difference between the inside and outside of the oil injection nozzle, assuming that the outside of the oil injection nozzle is a low-pressure oil system, which can be ignored compared with the pressure in the oil injection nozzle, $\Delta P$ can be taken as the real-time pressure of the high-pressure oil pipe; $\rho$ is the pressure density at the high-pressure side, namely the real-time fuel density in the high-pressure oil pipe; $C = 0.85$ is the flow coefficient. In addition, the lift of the needle valve $h = h(t)$ at the $t$ moment can be obtained by the statistical data of lift and time, $t_1$ and $t_2$ can be obtained by solving the equation $S_1 = S_2$.

Thus in an oil injection cycle $T$, the amount of oil injection $Q_B$ from the high-pressure oil pipe through the injection nozzle at the $t$ moment is

$$Q_B = Q_B(t) = \int_0^T Q'_B(t)\cdot\rho(t)\cdot dt \quad (14)$$

### *3.4. Establishment of function relationship between fuel density $\rho$ and pressure P*

The change amount of fuel pressure is proportional to the change amount of density, the ratio coefficient is $\dfrac{E}{\rho}$

$$\Delta P = \frac{E}{\rho}\Delta\rho \quad (15)$$

Among them, $\rho$ is density of fuel; $E$ is modulus of elasticity, and E is fuel pressure. The differential equation can be obtained from formula (15)

$$\frac{d\rho}{dP} = \frac{\rho}{E}$$

Solve the differential equation and get

$$\rho = \rho(P) = \rho_0 e^{\int_{P_0}^{P}\frac{dP}{E(P)}} \quad (16)$$

Thus the function relationship between the fuel density $\rho$ and the pressure $P$ and the function relationship between the elastic modulus and the pressure in formula (16) are obtained.

$E = E(P)$ can be obtained by conducting data fitting for fuel elastic modulus and statistical data of pressure.

### *3.5. Function relationship determination of fuel compression coefficient k and pressure P*

The fuel compression factor is used to define formula (1) and the relationship $\Delta V = \dfrac{\Delta M}{\rho}$ among mass, density and volume, and the compression coefficient of the fuel in the high-pressure oil pipe is obtained.

IOP Conf. Series: Materials Science and Engineering **772** (2020) 012007 doi:10.1088/1757-899X/772/1/012007

$$k = -\frac{1}{V}\frac{1}{\Delta P} \cdot \frac{\Delta M}{\rho} \tag{17}$$

$V$ is the internal volume of the high-pressure oil pipe among them. The formula (16), (17) and the pressure and fuel elastic modulus statistical data can be used to obtain a series of data of fuel pressure coefficient in the high-pressure oil pipe at corresponding time, further conduct data fitting for the relationship between fuel compression coefficient and pressure, the relationship formula between the fuel compression coefficient and the pressure can be obtained.

$$k = k(P) \tag{18}$$

## 4. Model Simulation and Parameter Optimization

The mathematical model of the rail pressure system established in MATLAB/Simulink is shown in Fig.4.

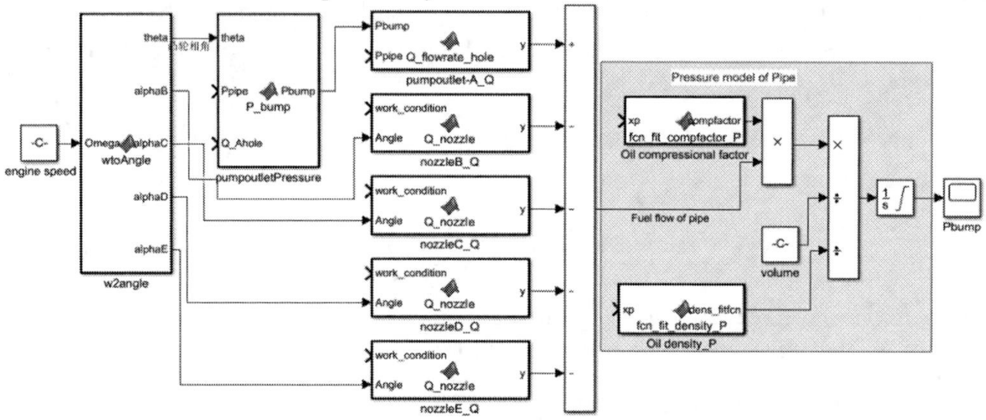


**Fig. 4** simulation model of rail pressure system of engine fuel

The model parameter optimization process is shown in Fig.5, first, the function relationship between the oil pump pressure and the cam angle is established (abbreviated as the pressure corner function), and the fuel compression coefficient and the fuel density pressure function are combined, the fuel mass $Q_A$ entering the high-pressure oil pipe via the one-way oil valve is solved, then according to the fuel spray nozzle needle valve system, the fuel mass $Q_B$ sprayed from the fuel spray nozzle is solved, finally, $Q_A$, $Q_B$, fuel compression coefficient $k_t$, the fuel density pressure function $\rho_t$ and the fuel volume in the high-pressure pipe $V$ (namely internal volume of high pressure pipe) are put into pressure dynamic equation of the high-pressure oil pipe (namely formula (2)) to conduct iterative optimization calculation, under the given working condition and initial pressure, the angular velocity of cam and the oil injection amount of the fuel injector are calculated when the pressure in the high-pressure oil pipe as stable as possible at a specified value (such as 100 MPa). In the specific solution, it is necessary to find the proper angular velocity of the cam and the oil injection amount of the fuel injector, the optimization process is shown in Fig.5:

IOP Conf. Series: Materials Science and Engineering **772** (2020) 012007 doi:10.1088/1757-899X/772/1/012007

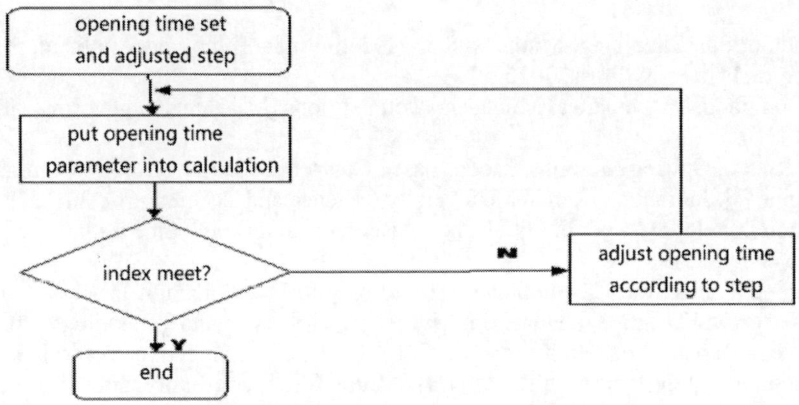


**Fig. 5** flow chart of parameter optimization

The specific data of high-pressure oil pump and fuel spray nozzle are as follows: the pressure oil process of high-pressure oil pump plunger is shown in Fig.1, the cam drives the plunger up and down, and the relationship between the cam edge curve and the angle adopts a sine function. The diameter inside plunger cavity is 5mm, when the plunger moves to the top dead center position; the residual volume of plunger cavity is 20mm3. When the plunger moves to the bottom dead center, the low-pressure fuel will fill the plunger cavity (including residual volume), and the pressure of the low-pressure fuel is 0.5MPa. The nozzle structure of fuel injector is shown in Fig.3, the diameter of the needle valve is 2.5mm, the seal seat is a cone with 9° half angle, and the diameter of the lowermost nozzle is 1.4mm. The fuel injector is set to be operated 10 times per second, the inner cavity length of a high-pressure oil pipe is 500mm, its inner diameter is 10mm, and the initial pressure in the high pressure oil pipe is 100 MPa, the angular velocity of the cam is determined, so that the pressure in the high-pressure oil pipe is as stable as possible at about 100 MPa.

In order to verify the practicability and algorithm effectiveness of the engine fuel rail pressure system model established in this paper, the rotational speed under different working conditions are input into the mathematical model of rail pressure established in this paper, under the specific data conditions given above, the simulation curve of fuel pressure fluctuation in high-pressure fuel pipe is shown in Fig.6.

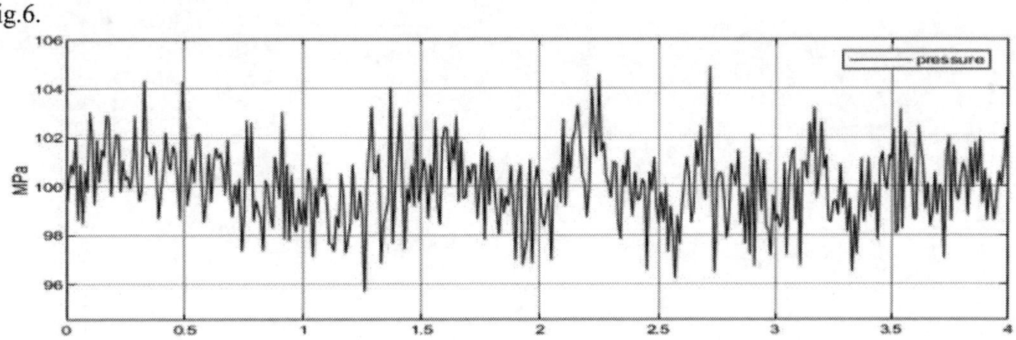


**Fig. 6** simulation curve of fuel pressure fluctuation in high-pressure oil pipe

Through the simulation curve of pressure fluctuation of rail pressure system in the fuel engine under different operation conditions, it can be seen that the rail pressure fluctuation calculated by the simulation of the established rail pressure control method has less error with the target rail pressure, and it meets the requirements of rail pressure control.

## References

[1] Si Shoukui, Sun Zhaoliang, Mathematical Modeling Algorithm and Application [M]. National Defense University Press, 2015.

[2] Zhao Dongfang. Mathematical Model and Competitions [M]. Huazhong Normal University Press, 2014.

[3] Zhang Kexun, Ouyang Minggao. Model-based Controller Framework for a Common Rail Diesel Engine [J]. Journal of Tsinghua University (Science and Technology), 2010(7): 1077-1081.

[4] Wu Yuqi. Development of GDI Fuel System Research and Development Platform [D]. Shandong University, 2019.

[5] Cai Liping. Simulation Calculation of High-Pressure Fuel Injection Process of Diesel Engine Based on MATLAB [D]. Huazhong University of Science and Technology, 2005.

[6] Pei Haing. Research on High-pressure Common Rail Oil Injection Control Strategy and Its Common Rail Optimization Design [D]. Central South University, 2009.

[7] Lu Yaozu. Simulation of High Pressure Fuel Injection System of Diesel Engine [J]. Journal of Xi'an University of Architecture & Technology, 2011, 33(4): 379~382.

# Machining Quality Evaluation and Optimization of Aircraft Structural Parts Based on Multi-factor Analysis of Variance

**Bukun Xie and Wenli Wang***

AVIC Manufacturing Technology Institute, Beijing 100024, China

*Corresponding author e-mail: bkxie@buaa.edu.cn

**Abstract.** Quantitative evaluation of machining quality of aircraft structural parts plays an important role in improving machining quality of parts. The rate of error limit (REL) is proposed as a new evaluation index of the machining quality, which realizes quantitative evaluation of machining quality and is applied to optimization of machining quality. Aiming at the disordering of the measurement data and lack of effective analysis in actual production of engineering, a Machining Quality Evaluation and Optimization software (MQEO) is designed and developed with REL as the evaluation index. And the multi-factor analysis of variance is used to find out the main factor influencing the machining quality, and propose optimization suggestions.

**Keywords:** REL; Quantitative evaluation of machining quality; Multi-factor analysis of variance; MQEO.

## 1. Introduction

Aircraft structural parts are generally obtained by numerical control machining, and the machining quality is directly related to aircraft performance and flight safety. The manufacturing of aircraft structural parts uses a large amount of new lightweight high-strength difficult-to-cut materials, along with complex thin-walled structures and integrated structures. The machining processes is complex, with high requirements of dimensional accuracy and geometric tolerance, also with long machining cycles and high costs. The evaluation and optimization of the machining quality are very important for the guarantee of product quality. The research can effectively improve the machining quality of aircraft structural parts, used to guide the processing process. A large amount of data is generated during machining, including machine data, process data, and measured data. But the value of these data is not being fully utilized. In the actual production process, a lot of data is not collected, or the collection is only simply checked to render certain if it is qualified, the real value of the data is not deeply mined.

Experts and scholars have done in-deep research on the evaluation and optimization analysis of machining quality. Some scholars analysed the influencing factors and optimization method of machining quality from the perspective of errors sources. Liu Qingjie, Li Hai, et al, achieved the machining accuracy evaluation and optimization, from the degree of machine composition and the impact on the parts machining quality [1, 2]. Wang Wei analysed the influence of position ring and speed ring on the milling accuracy of parts, from the perspective of dynamic and static error sources [3]. Hu Xiaochen studied the influence of machine dynamic factors and process parameters on surface roughness [4]. Wan Xiaojin established the model of "parts - fixture – tool - machine" system, analysed

Content from this work may be used under the terms of the Creative Commons Attribution 3.0 licence. Any further distribution of this work must maintain attribution to the author(s) and the title of the work, journal citation and DOI.
Published under licence by IOP Publishing Ltd

the influence of the deviation of machine, fixture and tool on the parts machining error, and predicted the parts error [5].

Some scholars set up mathematical models and put forward optimization strategies from the perspective of error evaluation index. Yao Changfeng, Wang Dong et al studied the influence of milling parameters on surface roughness, and gave the best combination of parameters for high-speed milling by orthogonal test [6, 7]. Wang Xiaoming et al. used multiple linear regression analysis method to study the influence of process parameters on the surface roughness of parts [8]. Some scholars introduced cost and efficiency as evaluation indicators as well to optimize the process planning. Wu Xiaorong, Deng Zhaohui, et al, various methods such as gray level correlation analysis were applied to analyse the selection of machining technology scheme, taking energy consumption and efficiency as the evaluation index [9-12]. Some scholars take machining quality, machining time, machining cost and other objectives as optimization indicators to study the optimization method of machining [13-15].

Scholars have used various methods to study the evaluation and optimization of machining quality. Wu Xiaorong applied fuzzy theory to evaluate and analyse the process parameters to conduct multi-objective optimization analysis for multiple objectives of time, cost and quality, and to find the optimal combination of parameters [12]. Meng zhaoyuxi applied the grey system theory to analyse the machining quality evaluation and prediction of the transmission shaft [16]. Zang Xuebai and Wen Hua applied simulation method to analyse the optimization method of parts machining parameters [17, 18]. Zhao Shuangfeng, Qiao Renjie et al. studied a large amount of data generated in the manufacturing process, and applied data analysis methods such as neural network or R language to realize the quality evaluation and control of the machining [19, 20].

In the previous studies, experts and scholars have analysed the evaluation and optimization methods of machining quality from many perspectives such as error sources and evaluation indexes. They designed experiments, adopted fuzzy mathematics, neural network and other methods to obtain and verify the evaluation and optimization methods of machining quality. However, there is little research about quantitative evaluation of parts machining quality. In this paper, the process planning, machine tool, cutting tool, operator and so on are taken as the error source, and the characteristic quantity of the rate of error limit (REL) is put forward as a new evaluation index of machining quality. So that it can quantitatively evaluate the machining quality of various aspects of parts machining quality more comprehensively. A Machining Quality Evaluation and Optimization software (MQEO) is developed, with the application of REL. The data characteristics of REL are studied, the main error sources of machining quality are analysed with multi-factor analysis of variance, and suggestions for optimization of machining quality were put forward.

## 2. Machinging Quality Error Sources

There are many factors that affect machining accuracy in machining process. According to the source of error, it can be divided into the following aspects: machining principle error, error caused by process planning, error caused by process system, error caused by thermal deformation, error caused by deformation under force and error caused by residual stress, as shown in figure 1.

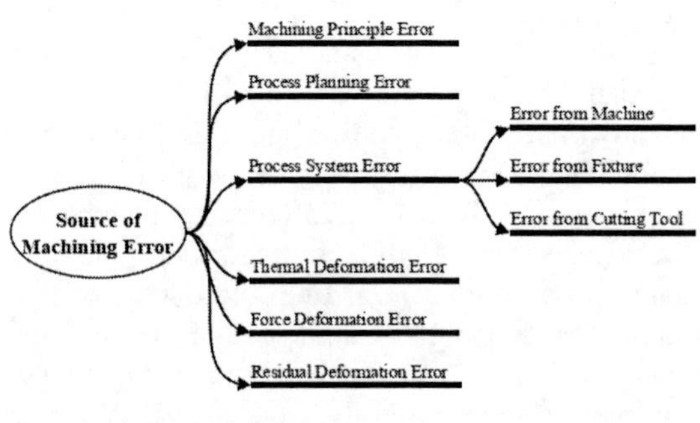


**Fig. 1** Source of machining error

The error of machining principle refers to the difference between the actual shape and the ideal state caused by the approximate forming and cutting movement. The process planning error refers to the error caused by the applicability of process parameters design, such as spindle speed, feed, cutting depth, etc. The different process parameters design has different influence on the quality of the parts. Process system mainly includes machine tools, fixtures, cutting tools, workpiece, etc., each link of the error will affect the workpiece machining accuracy to varying degrees. The manufacturing of machine tools inevitably exists errors, along with the installation, use, maintenance and other processes, resulting in the parts machining error. Particularly, spindle error, guide error and error caused by the transmission chain of machine tools produce greater impact on the accuracy of the parts. Tool manufacturing accuracy and tool wear will have a certain impact on the machining accuracy. Similarly, the manufacturing accuracy of the fixture, the positioning error, the size of clamping force will also cause the machining error. The deformation under force during the machining process will cause errors in the parts machining. The cutting force, cutting vibration, frequent changes in acceleration and other factors will cause the parts to be deformed under force and affect the machining accuracy. The main heat sources such as cutting heat, machine power source heat and frictional heat of moving parts can cause thermal expansion and contraction of the process system, affecting the relative position between the various links of the process system, resulting in machining errors. The residual stress also causes the error of the parts. Before machining, internal stress is generated due to various reasons such as blank processing and heat treatment. The internal stress is redistributed during the cutting process, and the internal structure changes continuously until the internal stress reaches equilibrium. The redistribution of internal stress causes deformation of the part and affects the machining accuracy of the part.

In the actual machining and production of aircraft parts, the error of machining principle is inevitable and little generally. The deformation caused by stress and heat during the machining process, as well as the machining error caused by residual stress after machining are closely related to the process planning. The fixture error is mainly caused by the difference of operators, and it is more convenient to measure it by operators Therefore, this paper mainly analyses the machining quality from four aspects, the process planning, machine error, tool error and operator error.

The parts are processed according to the process parameters, cutting tools and machining paths designed according to the process procedure, and the required parts shape and surface can be obtained. However, the actual state obtained by machining cannot be completely consistent with the ideal state. The degree of conformity between the actual state and the ideal state is called machining accuracy, and the degree of deviation between the actual state and the ideal state is called machining error. Machining error generally includes dimensional error, geometric error and surface quality error. Each item needs to be tested, as shown in figure 2. The surface quality includes the geometric features of the surface layer and the physical and mechanical properties of the surface layer. Physical and mechanical properties are not commonly used to detect the machining quality of parts, and it is difficult to quantify them, limited to the current situation. Therefore, this paper does not include them in the evaluation index, generally using surface roughness to evaluate the machining surface quality of parts. To quantitatively analyse the machining quality of three kinds of evaluation indexes, REL is proposed as the quantitative evaluation index of the machining quality. The measurement results of the machining quality such as dimension error, geometric error and surface quality error are converted into REL value, and the evaluation results of the machining quality are obtained by analysing the data characteristics of REL.

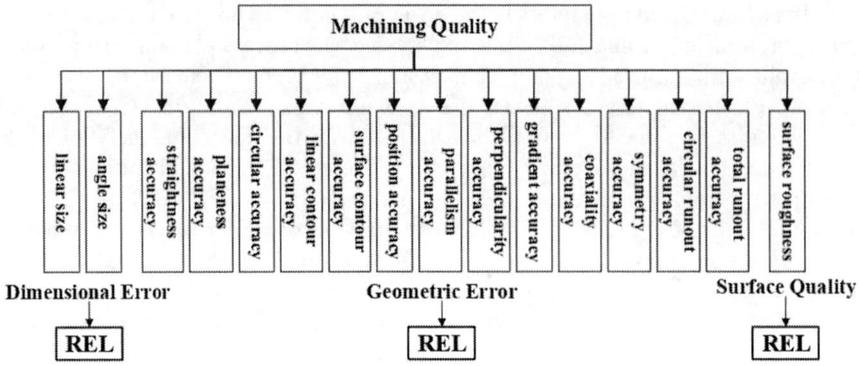


A series of REL values are obtained, and the machining quality evaluation results are obtained by analysing the characteristics of REL data

**Fig. 2** Machining quality details

## 3. Machining Quality Analysis Method

### *3.1. The Evaluation Index REL*

The quality of aircraft structural parts includes dimension error, geometry error and surface roughness. However, the evaluation criteria of the three indicators are not consistent, and the order of magnitude of the test data is also inconsistent. In order to normalize the evaluation indexes, and realize the quantitative evaluation of machining quality, the characteristic quantity REL is put forward as the evaluation indictor of parts machining quality, the computational formula is shown in formula (1).

$$REL = \frac{X_m - X_p}{X_{max} - X_{min}} \tag{1}$$

Where, $X_m$ refers to the measured value of the component test item, $X_p$ refers to the process design value of this test item, and $X_{max} - X_{min}$ is the design tolerance limit of this test item. The calculation methods of the three evaluation indicators are shown in table 1.

**Table 1.** Definition of formula symbols under three evaluation indicators

| Evaluation indicators | | $X_m$ | $X_p$ | $X_{max} - X_{min}$ |
|---|---|---|---|---|
| Dimensional error | | Measured value | Designed value in the NC program | That is the dimension tolerance. The maximum allowable value of the detection item minus the minimum. |
| Geometric tolerance | The shape tolerance | | 0 | That is the maximum allowable error |
| | The position tolerance | | 0 | That is the maximum allowable error |
| Surface roughness | | | 0 | That is the maximum allowable surface roughness |

REL is proposed as the evaluation index of the machining quality. The difference of the measured value and the process design value is used to measure the actual machining state of the part. The tolerance limit is used as the basis of the requirements of parts processing tolerance. The rate between them take the information into account in the evaluation indicators. When REL is bigger than 1, the test item is in state of out-of-tolerance and unqualified. When REL is between 0.5 and 1, or the test size is not within the tolerance limit, the machining quality of the test item can be optimized. When REL is less than 0.5 and the size is within the tolerance limit, the machining quality is relatively ideal.

### *3.2. Multi-factor Variance Analysis Method*

There are many factors that affect the quality of machining. This paper mainly studies the influence of process planning, machine, tool and operators on the quality of machining. First, the machining quality is evaluated according to REL, and then the order of the factors that affect the machining quality is analysed by the multi-factors analysis method of variance, the influence on the machining quality is studied, and the optimization suggestions are put forward if it's needed to be improved.

(1) REL characteristics calculation

Analyse the mean and variance of REL under each factor, and observe the fluctuation of REL. And then calculate range of each factor, the greater the difference of the range, the greater the impact on the machining quality.

(2) The sum of squares of deviations and significances calculation

By calculating and comparing the variance of each factor, the main and significant effects of each factor on machining quality were judged.

Total deviation sum of squares S:

$$S = \sum_{i=1}^{a}(y_i - \bar{y})^2 = \sum_{i=1}^{a} y_i^2 - \frac{1}{a}(\sum_{i=1}^{a} y_i)^2 \tag{2}$$

Where, $y_i$ refers to the rate of error limit of each item, a is the total number of samples, $\bar{y} = \frac{1}{a}\sum_{i=1}^{a} y_i$.

Freedom of the sum of squares of total deviations f:

$$f = a - 1 \tag{3}$$

The sum of deviation squares in column $S_j$:

$$S_j = \frac{a}{b}\sum_{k=1}^{a}(\bar{y}_{jk} - \bar{y})^2 = \frac{b}{a}\sum_{k=1}^{a} y_{jk}^2 - \frac{1}{a}(\sum_{k=1}^{a} y_i)^2 \tag{4}$$

Where, b is the number of factor levels, $\bar{y}_{jk} = \frac{b}{a}\sum_{i=1}^{k} y_i$, $y_{jk} = \sum_{i=1}^{k} y_i$,

Freedom degree of the square sum of column deviations $f_j$:

$$f_j = b - 1 \tag{5}$$

Freedom of the sum of squares of error deviations $f_e$:

$$f_e = f - \sum f_j \tag{6}$$

Calculate the significance of each factor and judge whether each factor has a significant impact on the machining quality, that is, judge whether the factor has a fault. If there is a fault, the influence is significant; if there is no fault, the influence is not significant.

$$F_j = \frac{({}^{S_j}/{}_{\sigma^2})/f_j}{({}^{S_e}/{}_{\sigma^2})/f_e} = \frac{{}^{S_j}/{}_{f_j}}{{}^{S_e}/{}_{f_e}} = \frac{\hat{\sigma}_A^2}{\hat{\sigma}_e^2} \sim F(f_j, f_e) \tag{7}$$

(3) Error source analysis and suggestions

According to the REL and the results of previous analysis, the error source of a certain data is analysed and optimization suggestions are given. There are few quality unqualified contents due to the importance of aircraft parts processing. Therefore, this paper mainly analyses the machining quality evaluation and optimization methods under each qualified test item. As for hole characteristics, if REL is too small, it may be due to tool wear, clamping and positioning problems, etc. If REL is too large, it may be improper tool diameter, improper program allowance, etc. As for shaft features, the reverse turns out to be true.

## 4. MQEO Development

In order to facilitate the application of engineering production, and easy to operate, to realize the quantitative evaluation of parts machining quality, and the improvement of the machining quality in production, MQEO is developed by using C#.NET and SQL database. The composition of each functional module of the system is shown in figure 3. The software is used to evaluate the machining quality, analyse the influencing factors of machining quality, and put forward optimization suggestions for the parts that need to be optimized. The interface of the software system is shown in figure 4.

IOP Conf. Series: Materials Science and Engineering **772** (2020) 012008 doi:10.1088/1757-899X/772/1/012008

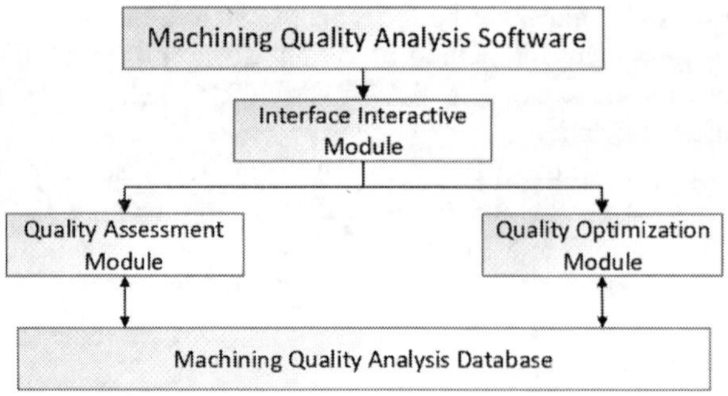


**Fig. 3** Function module of MQEO

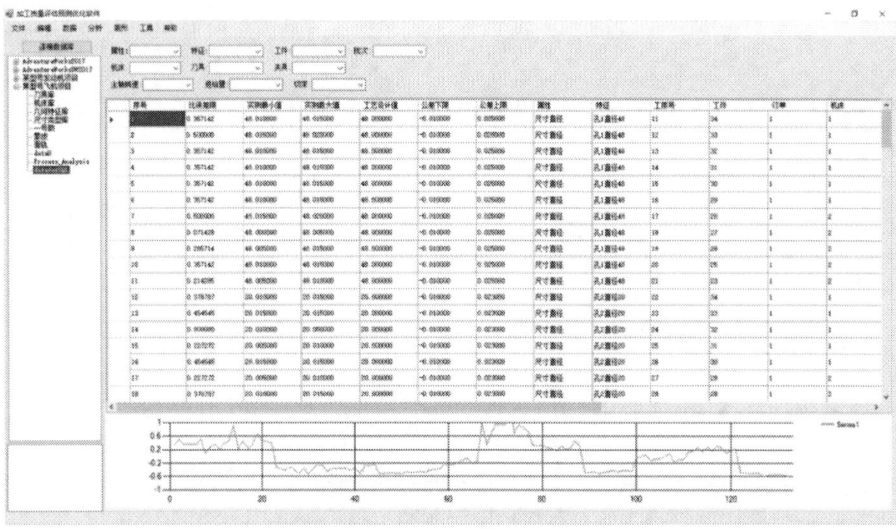

**Fig. 4** Interface of MQEO

## 5. MQEO Application in Aircraft Structural Parts

A key structural part of a certain aircraft is shown in figure 5. This part is an important bearing part in the aircraft structure, with many processing features, complex structure and complex processing procedures. The quantitative evaluation and optimization of the processing quality of the parts can greatly reduce the waste of productivity and cost and improve the processing efficiency. During the process of machining, 10 sorties data of this part were collected in this processing, including features of test items, measurement value of test items, process design value, machine tool, cutter, operator and other information. They were organized and saved into SQL database.

**Fig. 5** A part of a certain aircraft

1) REL characteristics calculation

The machining data includes the measure data, operator, machine, cutting tool and other information of part machining. The REL characteristics of these factors are calculated and shown in table 2-5.

**Table 2.** RELunder operator factor

| Operator | A | B | C | D | E |
|---|---|---|---|---|---|
| Number | 32 | 33 | 32 | 15 | 20 |
| Average | 0.1634 | 0.1143 | 0.1333 | 0.0590 | 0.0377 |
| The Variance | 0.126 | 0.65 | 0.73 | 0.006 | 0.004 |
| Range | $l_o = 0.1634 - 0.0377 = 0.1257$ | | | | |

**Table 3.** REL under machine tool factor

| Machine | I | II | III. |
|---|---|---|---|
| Number | 36 | 48 | 32 |
| Average | 0.1146 | 0.0992 | 0.1129 |
| The Variance | 0.071 | 0.067 | 0.064 |
| Range | $l_m = 0.1146 - 0.0992 = 0.0154$ | | |

**Table 4.** REL under cutting tool factor

| Cutting Tools Number | D0601 | D0602 | D0603 | D0604 | D1001 |
|---|---|---|---|---|---|
| Number | 10 | 12 | 14 | 19 | 17 |
| Average | 0.0312 | 0.1582 | 0.1089 | 0.1101 | 0.1272 |
| The Variance | 0.024 | 0.141 | 0.008 | 0. 044 | 0.096 |
| **Cutting Tools Number** | **D1002** | **D1003** | **D2001** | **D2002** | **D2003** |
| Number | 17 | 15 | 14 | 8 | 6 |
| Average | 0.0941 | 0.1502 | 0.0945 | 0.2153 | 0.0031 |
| The Variance | 0.066 | 0.119 | 0.057 | 0.093 | 0.012 |
| Range | $l_t = 0.2153 - 0.0031 = 0.2122$ | | | | |

**Table 5.** REL under feature factors

| The Features | Bore Diameter | Linear dimension | Axle Diameter |
|---|---|---|---|
| Number | 22 | 44 | 11 |
| Average | 0.1017 | 0.0576 | 0.0198 |
| The Variance | 0.027 | 0.005 | 0.002 |
| **The Features** | **Surface Profile** | **Vertical Accuracy** | **Cylindricity Accuracy** |
| Number | 22 | 11 | 22 |
| Average | -0.1439 | 0.7440 | 0.2425 |
| The Variance | 0.018 | 0.079 | 0.007 |
| Range | $l_f = 0.2425 - (-0.1439) = 0.3864$ | | |

2) Machining quality evaluation

The calculation results of deviation sum of squares and degrees of freedom under each factor are shown in table 6. With the calculations in these tables, the analysis shows that the feature factor has the biggest range around the factors. And from Table 6, only the feature factor's significance test smaller than the limit. They prove that feature factor, as well as the process planning, is the main factor affecting the machining quality. When the parts are processed in different positions, the machining quality will be significantly different, and the machining quality of some parts that are difficult to be processed is worse. But the machine, the cutter, the operator has little influence on the machining quality, namely these several kinds of factors do not have the obvious breakdown questions.

**Table 6.** Calculation results of deviation sum of squares and degrees of freedom

| Check Item | Deviation Square | Degrees of Freedom | Mean Square Error | F | Significant |
|---|---|---|---|---|---|
| **The Operator** | 0.011 | 20 | 0.003 | 0.163 | 0.957 |
| **Machine** | 0.016 | 1 | 0.008 | 0.466 | 0.629 |
| **Cutting Tool** | 0.139 | 4 | 0.015 | 0.874 | 0.551 |
| **Feature** | 6.16 | 2 | 1.232 | 69.872 | 0 |
| **Error** | 1.957 | 111 | 0.18 | - | - |
| **Total** | 10.344 | 132 | - | - | - |

3) Quality optimization suggestions

From the analysis data above, it can be found that under the factors of operator, machine and cutter, the difference of REL is small. That is, these factors have little influence on the machining quality. There is no obvious fault problem in these factors and no need for optimization. But under the feature factor, the difference of REL is obvious. It means that the process planning is the important influence factor of the part machining quality. Optimization suggestions should be given at this time: optimize the parameters in the process planning, especially in the machining parts with poor machining quality.

## 6. Summary

In this paper, REL is proposed as the quantitative evaluation index of processing quality, MQEO software is designed and developed for quantitative evaluation and optimization of processing quality, and processing quality data obtained from the production and processing of key structural parts are collected to analysis and verification. Through analysing REL's data characteristics, the processing quality of parts was quantitatively evaluated, and the main factors affecting the processing quality of parts were analysed with the multi-factor analysis method. It was verified that REL could be used as the evaluation index of the processing quality, and the processing quality could be optimized according to the analysis results of this index.

With the increasingly in-depth research and more and more applications of digital manufacturing and intelligent manufacturing, the machining and production of aircraft parts is increasingly pursuing high efficiency, high quality and high convenience machining methods. Quantitative analysis and scientific analysis of machining quality can provide a theoretical basis for machining and production, and will play an increasingly important role in machining and production, providing a basis for the subsequent prediction and optimization of machining quality.

## References

[1] Qingjie L. Research on machining accuracy evaluation technology of CNC machine tools (Master, Southwest jiaotong university, China, 2012).

[2] Hai L, Yingguang L, Yinghao C, et al. Shape and position error prediction of nc machining based on geometric error model of machine tool[J]. Tool technology. Vol. 53 (2019) No.6, p.74-79.

[3] Wei W, Congzhi Z, Xin Z. Comprehensive prediction and evaluation of machining error of five-axis CNC milling machine based on multi-error source coupling[J]. China mechanical engineering. Vol. 26 (2015) No. 1, p.85-91.

[4] Xiaochen H. Prediction method for surface roughness of nickel-based superalloy micromilling (Master, Dalian university of technology, China, 2017) .

[5] Xiaojin W. Study on error analysis and prediction of work-jig and cutter system (Master, Huazhong university of science and technology, China, 2010).

[6] Changfeng Y, Daoxia W, Qichao J, et al. Study on surface roughness and surface morphology of TB6 titanium alloy high-speed milling[J]. Aeronautical manufacturing technology. (2012) No. 21 p. 90-93.

[7] Dong W. Parameter optimization of surface roughness of AA5083 aluminum alloy (Master, Tianjin University, China, 2014).

[8] Xiaoming W, Jiang H. Study on surface roughness of high speed milling titanium alloy of TC4[J].

Mechanical design and manufacturing. (2019) No. 05 p. 232-236.
[9] Ruixue Y. Energy consumption assessment model in machinery manufacturing process and its application in process planning[J]. Machinery design and manufacturing. (2014) No. 9 p.270-272.
[10] Zhaohui D, Yahui F, Linlin W, et al. Multi-objective optimization of milling process parameters for green and efficient manufacturing[J]. China mechanical engineering. Vol.28 (2017) No.19 p. 2365-2372.
[11] Wei Y, Hua Z, Zhigang J. Multi-objective optimization model of nc machining system for energy saving and efficient demand[J]. China mechanical engineering. Vol. 29 (2018) No.21 p. 2571-2580.
[12] Xiaorong W, Shuyou Z, Lemiao Q, et al. Decision method and application of processing parameters for green manufacturing[J]. Chinese journal of mechanical engineering. Vol. 49 (2013) No.07 p.91-100.
[13] Renbin X, Hongfu Z, Zhenwu T. Study on tolerance design multi-objective model and particle swarm optimization algorithm[J]. Computer integrated manufacturing system. (2006) No.07 p. 976-980.
[14] Sihua Q, Yetai F. Multi-objective optimization of manufacturing quality control based on genetic algorithm[J]. Chinese journal of agricultural machinery. (2006) No.06 p.110-112.
[15] Jianlin C, Jie S, Jianfeng L. multi-objective optimization of titanium alloy milling parameters[J]. China mechanical engineering. Vol. 25 (2014) No.02 p.169-173.
[16] Zhaoyuxi M. Prediction of machining quality of transmission shaft based on grey system theory[J]. Equipment manufacturing technology. (2013) No.10 p.80-82.
[17] Xuebai Z, Hongwei Z, Zhimin P, et al. Vibration drilling virtual integrated environment for laminated materials[J]. Journal of jilin university of technology (engineering edition), (2002) No.02 p.54-58.
[18] Hua W, Lin W, Yuchen Z, et al. Design and simulation of CNC machining fixture for typical aircraft structural parts[J]. Machinery, Vol. 45 (2018) No.06 p.43-46.
[19] Shuangfeng Z. Research on quality control method and system of machining workshop manufacturing process based on MES (Master, Chongqing university, China, 2016).
[20] Renjie Q, Tingxin S. Manufacturing quality control and prediction based on R language[J]. Integrated machine tool and automatic processing technology. (2019) No. 02 p. 127-130.

# Shimmy analysis of steering system with independent suspension based on coupled feedback model

**Lulu Gao[1, *], Fei Ma[1], Chun Jin[1] Yuchao Liu[1] Xinze Li[2]**

[1]School of Mechanical Engineering, University of Science and Technology Beijing, Beijing 100083, China
[2]College of Mechanical and Electrical Engineering, Beijing University of Chemical Technology, Beijing 100029, China

*Corresponding author e-mail: ustb_gll@163.com

**Abstract.** Shimmy of steering system causes the snaking and deteriorates the handling stability and safety of vehicles, which has not been solved effectively for the ambiguity of shimmy characteristic. In order to uncover the mechanism of shimmy, a 5-DOF dynamic model of the steering system with independent suspension is established, and the coupled feedback model of the steering system is further obtained based on the dynamic model. Then the frequency characteristics of feedback are analysed, and the work and damping energy dissipation of each feedback link in a unit period are calculated. Based on the range of natural frequency and unit periodic energy of steering system, the range of wheel shimmy frequency and velocity interval is determined. The analysis results show that the shimmy frequency is the first natural frequency of steering system. The amplitude of wheel shimmy is calculated based on energy conservation, and the energy transfer of steering system is elaborated. The mechanism of shimmy was uncovered based on the energy transfer of coupled feedback system. The theoretical analysis was verified by numerical simulation. The results provide theoretical support for the mechanism research and active control of shimmy.

## 1. Introduction

Shimmy brings about the continuous vibration of wheels about the kingpin, deteriorates the handling stability, ride comfort, and safety [1, 2]. The uncertainty of the shimmy mechanism leads to these problems can not be thoroughly solved in engineering, which affects the safe driving of vehicles seriously, even nose landing gear of aircraft[3]. The shimmy can be avoided during design stage or usage stage by design and active control [4], while the clear mechanism of shimmy is the basis of these efforts. Different methods had been applied to research the mechanism of shimmy and the influence factor.

The shimmy is usually described as a bifurcation or instability of the steering dynamic system, and nonlinear dynamics and bifurcation theory are widely used in the shimmy research [5, 6, 7]. And the influence of sideslip force, clearance of pairs, and dry friction were analysed by numerical simulation during the analysis. The simulation can explicit the influence of different factors, while the mechanism of shimmy still was covered. Guo et al. [8] revealed the mechanism of shimmy by the effect of negative damping, and focused on the influence of sideslip characteristic. The results shown the negative damping of alignment torque and sideslip force in low frequency range might lead to shimmy. Besselink [9]

Content from this work may be used under the terms of the Creative Commons Attribution 3.0 licence. Any further distribution of this work must maintain attribution to the author(s) and the title of the work, journal citation and DOI.
Published under licence by IOP Publishing Ltd

IOP Conf. Series: Materials Science and Engineering **772** (2020) 012009 doi:10.1088/1757-899X/772/1/012009

applied the method of energy flow in the research of shimmy, and the dynamic characteristic of tire was mainly analysed. Then Ran [10, 11] treated the contact area as a 2 DOF motion based on Von Schlippe model, and analysed the effect of velocity on the input energy of steering system. Jiang et al. [12] treated the tires as delayed feedback blocks, and the simulation results shown that the phase difference between alignment torque and shimmy motion was the reason of shimmy. Similar thought emerged in the research of Mi [6] and Yang et al [13].

In this paper, the steering system with independent system is taken as the research object. Firstly, a 5 DOF dynamic model of steering system is established. And the coupling feedback model of steering system is further established based on the dynamic model. Secondly, the work and energy consumption of feedback block are calculated, and the frequency range of shimmy is obtained according to the characteristic of feedback block. Then the natural frequency of steering system is calculated to obtain the velocity range of shimmy. And the amplitude of shimmy is obtained based on the energy conservation, and the relationship of amplitude, frequency, and velocity range is revealed. The mechanism of shimmy is uncovered based on the energy transfer of coupled feedback system.

## 2. 5 DOF model of steering system

### 2.1. *Structure of steering system*

The structure of 5 DOF steering system with independent suspension was shown in Figure 1. The forward direction of vehicle was set as X-axis, the vertically up direction was set as Z-axis, and the Y-axis conform to right-hand coordinate system. The structures of steering system in Y-Z plane and X-Y plane were shown in (a) and (b) of Figure 1. The symbol $\varphi_1$ and $\varphi_2$ indicated the rotary motion of left and right tire around kingpin, respectively, $\theta_1$ and $\theta_2$ indicated the swing motion of left and right tires, respectively. And $\varphi_3$ indicated the swing motion of middle rod. The parameters of steering system was shown in Table 1.

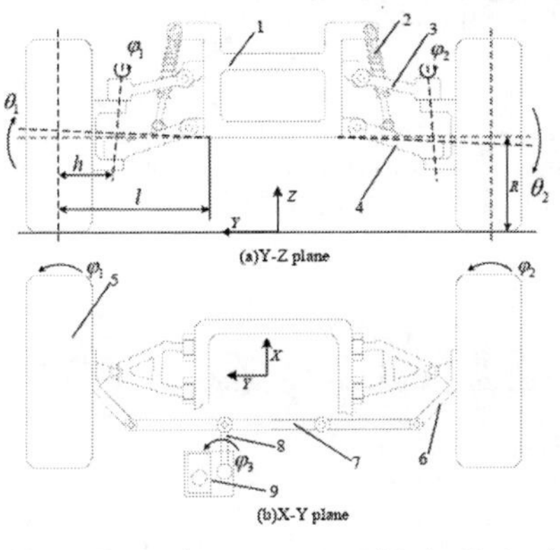


**Figure 1.** Structure of steering system with independent suspension

1-vehicle frame, 2-suspension, 3-upper arm, 4-lower arm, 5-wheel, 6-knuckle arm, 7-tie rod, 8-middle rod, 9-gearbox

### 2.2. *Dynamic model of steering system*

According to the structure in Figure 1, the dynamic model of steering system could be obtained based on Lagrangian equation:

$$\begin{cases} J_1\ddot{\varphi}_1 + c_1(\dot{\varphi}_1 - \dot{\varphi}_3) + k_1(\varphi_1 - \varphi_3) + c_{e1}\dot{\varphi}_1 + J_0\frac{V}{R}\dot{\theta}_1 = -F_{y1}*(e+\gamma R) \\ J_2\ddot{\varphi}_2 + c_2(\dot{\varphi}_2 - \dot{\varphi}_3) + k_2(\varphi_2 - \varphi_3) + c_{e2}\dot{\varphi}_2 + J_0\frac{V}{R}\dot{\theta}_2 = -F_{y2}*(e+\gamma R) \\ J_3\ddot{\varphi}_3 + c_1(\dot{\varphi}_3 - \dot{\varphi}_1) + c_2(\dot{\varphi}_3 - \dot{\varphi}_2) + k_1(\varphi_3 - \varphi_1) + \\ k_2(\varphi_3 - \varphi_2) + c_3\dot{\varphi}_3 + k_3\varphi_3 = 0 \\ J_4\ddot{\theta}_1 - J_0\frac{V}{R}\dot{\varphi}_1 + k_4\theta_1 + c_4\dot{\theta}_1 + (k_z l^2 + k_y R^2)\theta_1 = F_{y1}R \\ J_5\ddot{\theta}_2 - J_0\frac{V}{R}\dot{\varphi}_2 + k_5\theta_2 + c_5\dot{\theta}_2 + (k_z l^2 + k_y R^2)\theta_2 = F_{y2}R \end{cases} \tag{1}$$

Where the sideslip force $F_{yi}$ was nonlinear, and can be expressed as:

$$F_{yi} = S_y + D\sin\{\mathrm{C}\arctan[\mathrm{B}(\alpha_i - \mathrm{S}_x)(1-\mathrm{E}) + \mathrm{E}\arctan\mathrm{B}(\alpha_i - \mathrm{S}_x)]\} \tag{2}$$

Where where $\alpha_i$ is side slip angle of left and right tire, respectively. The parameters, Sx, Sy, B, C, D, and E are constant and obtained from the experimental data. We take Sx=0, Sy=0 and other parameters are calculated by equations with vertical load [13].

And the non holonomic constrained equation between the shimmy angle $\varphi_i$ and $\alpha_i$ can be given by[6]:

$$\begin{cases} \dot{\alpha}_1 + \frac{V}{\sigma}\alpha_1 + \frac{V}{\sigma}\varphi_1 - \frac{a - R\gamma}{\sigma}\dot{\varphi}_1 - \frac{R}{\sigma}\dot{\theta}_1 = 0 \\ \dot{\alpha}_2 + \frac{V}{\sigma}\alpha_2 + \frac{V}{\sigma}\varphi_2 - \frac{a - R\gamma}{\sigma}\dot{\varphi}_2 - \frac{R}{\sigma}\dot{\theta}_2 = 0 \end{cases} \tag{3}$$

**Table 1.** Parameters of Steering System

| Symbol | Parameters | Value |
|---|---|---|
| $J_1$,$J_2$ | Mass moment of inertia of left and right tire w.r.t kingpin | 7 kg·m² |
| $J_3$ | Mass moment of inertia of middle rod w.r.t Z | 5 kg·m² |
| $J_4$,$J_5$ | Mass moment of inertia of left and right wheel w.r.t. X | 6 kg·m² |
| $J_0$ | Mass moment of inertia of wheel w.r.t. Y | 4 kg·m² |
| $k_1$,$k_2$ | Equivalent torsional stiffness of tie rod | 1.9kN·m·rad$^{-1}$ |
| $c_1$,$c_2$ | Equivalent torsional damper of tie rod | 10N·m·s·rad$^{-1}$ |
| $k_3$ | Equivalent torsional stiffness of mid rod | 3.9kN·m·rad$^{-1}$ |
| $c_3$ | Equivalent torsional damper of mid rod | 30 N·m·s·rad$^{-1}$ |
| $k_4$,$k_5$ | Equivalent torsional stiffness of suspension | 15 kN·m·rad$^{-1}$ |
| $c_4$,$c_5$ | Equivalent torsional damping of suspension | 1130N·m·s·rad$^{-1}$ |
| $c_{e1}$,$c_{e2}$ | Torsional damper of left and right wheels rotation around kingpin | 40 N·m·s·rad$^{-1}$ |
| $\gamma$ | Caster angle | 3° |
| $e$ | Pneumatic trail | 0.09 m |
| $l$ | Distance of lower arm hinge point and tire centre | 0.8 m |
| $h$ | Distance of kingpin axis and tire centre | 0.2 m |
| $R$ | Radius of tire | 0.4 m |
| $a$ | Half length of tire | 0.22 m |
| $\sigma$ | Relaxation length of tire | 0.4 m |
| $k_z$ | Vertical stiffness of tire | 2200 kN·m$^{-1}$ |
| $k_y$ | Lateral stiffness of tire | 35 kN·m$^{-1}$ |

### 2.3. *Coupled feedback model of steering system*

In order to establish the coupled feedback model of steering system, it is necessary to linearize the nonlinear block in Equation (1). And the previous research shown that the linearization at the equilibrium point will not bring significant deviation.

The nonlinear sideslip tire force in Equation (2) can be linearized as:

$$F_{yi} = BCD\alpha_i \tag{4}$$

Taking Equation (4) into (1), and making Laplace transform for (1) and (2):

$$\begin{cases} \left(J_1 s^2+\left(c_1 s+k_1+c_{e1} s\right)\right)\varphi_1(s)-\left(c_1 s+k_1\right)\varphi_3(s)+J_0 \frac{V}{R} s\theta_1(s)=-BCD^*(e+\gamma R)\alpha_1(s) \\ \left(J_2 s^2+\left(c_2 s+k_2+c_{e2} s\right)\right)\varphi_2(s)-\left(c_2 s+k_2\right)\varphi_3(s)+J_0 \frac{V}{R} s\theta_2(s)=-BCD^*(e+\gamma R)\alpha_2(s) \\ \left(J_3 s^2+\left(c_1 s+c_2 s+c_3 s+k_1+k_2+k_3\right)\right)\varphi_3(s)-\left(c_1 s+k_1\right)\varphi_1(s)-\left(c_2 s+k_2\right)\varphi_2(s)=0 \\ \left(J_4 s^2+c_4 s+k_4\right)\theta_1(s)-J_0 \frac{V}{R} s\varphi_1(s)+(k_z l^2+k_y R^2)\theta_1(s)=BCD\alpha_1(s)R \\ \left(J_5 s^2+c_5 s+k_5\right)\theta_2(s)-J_0 \frac{V}{R} s\varphi_2(s)+(k_z l^2+k_y R^2)\theta_2(s)=BCD\alpha_2(s)R \\ s\alpha_1(s)+\frac{V}{\sigma}\alpha_1(s)+\frac{V}{\sigma}\varphi_1(s)-\frac{a-R\gamma}{\sigma}s\varphi_1(s)-\frac{R}{\sigma}s\theta_1(s)=0 \\ s\alpha_2(s)+\frac{V}{\sigma}\alpha_2(s)+\frac{V}{\sigma}\varphi_2(s)-\frac{a-R\gamma}{\upsilon}s\varphi_2(s)-\frac{R}{\sigma}s\theta_2(s)=0 \end{cases} \tag{5}$$

The rotation motion of tires around kingpin can be treated as forward block, and the alignment torque and other torques can be treated as feedback block, then the coupled feedback model of steering system can be obtained from Equation (5), which is shown in Figure 2.

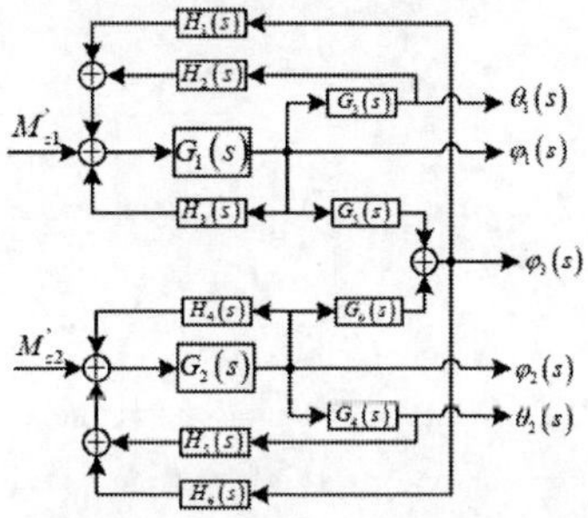


**Figure 2.** Coupled feedback model of steering system

As shown in Figure 2, $M_{z1}'$ and $M_{z2}'$ were the interference torque around Z axis, and the $G_1$ and $G_2$ indicated the forward block of rotation motion of two tires around kingpin, $G_3$ and $G_4$ indicated the forward block of two tires motion in Y axis. Similarly, $G_5$ and $G_6$ indicated the forward block of two tie rods. $H_1$ and $H_6$ represented the feedback block of middle rod, $H_2$ and $H_5$ indicated the feedback block of gyroscopic torque, $H_3$ and $H_4$ indicated the feedback block of sideslip force and alignment torque. And the details of different block in Figure 2 can be given by:

$$\begin{cases} G_1=\frac{1}{J_1 s^2+\left(c_1 s+k_1+c_{e1} s\right)}, G_2=\frac{1}{J_2 s^2+\left(c_2 s+k_2+c_{e2} s\right)} \\ G_3=\frac{J_0 \frac{V}{R} s(\sigma s+V)+(as-V)BCDR}{\left[J_4 s^2+c_4 s+k_4+(k_z l^2+k_y R^2)\right](\sigma s+V)-BCDR^2 s} \\ G_4=\frac{J_0 \frac{V}{R} s(\sigma s+V)+(as-V)BCDR}{\left[J_5 s^2+c_5 s+k_5+(k_z l^2+k_y R^2)\right](\sigma s+V)-BCDR^2 s} \\ G_5=\frac{\left(c_1 s+k_1\right)}{\left(J_3 s^2+\left(c_1 s+c_2 s+c_3 s+k_1+k_2+k_3\right)\right)}, G_6=\frac{\left(c_2 s+k_2\right)}{\left(J_3 s^2+\left(c_1 s+c_2 s+c_3 s+k_1+k_2+k_3\right)\right)} \end{cases} \tag{6}$$

$$\begin{cases} H_1=c_1 s+k_1, H_2=\frac{-BCD(e+\gamma R)Rs-J_0 \frac{V}{R} s(\sigma s+V)}{\sigma s+V}, H_3=\frac{C_1 e(as-V)}{(\sigma s+V)} \\ H_4=\frac{C_2 e(as-V)}{(\sigma s+V)}, H_5=\frac{-BCD(e+\gamma R)Rs-J_0 \frac{V}{R} s(\sigma s+V)}{\sigma s+V}, H_6=c_2 s+k_2 \end{cases} \tag{7}$$

## 3. Energy analysis of steering system

### 3.1. *Frequency character of feedback block*

Due to the symmetry of steering structure, the frequency character of the alignment torque feedback block $H_3$ and $H_4$ can be given by

$$\begin{cases} A_1(\omega)=\dfrac{|BCD|(e+\gamma R)\sqrt{\left(\dfrac{a\omega}{V}\right)^2+1}}{\sqrt{\left(\dfrac{\sigma\omega}{V}\right)^2+1}} \\ \phi_1(\omega)=\pi-\arctan\left(\dfrac{\sigma\omega}{V}\right)-\arctan\left(\dfrac{a\omega}{V}\right) \end{cases} \tag{8}$$

Similarly, the frequency character of the gyroscopic torque and middle rod torque feedback block $G_3H_2$ and $G_5H_1$ can be given by

$$\begin{cases} A_2(\omega)=\dfrac{\sqrt{\left(1+\dfrac{J_0\sigma}{R}\omega^2\right)^2+\left(\left(\dfrac{J_0V^2}{R}+BCDRa\right)\omega\right)^2}}{\sqrt{\left(KV-(c_4\sigma+J_4V)\omega^2\right)^2+\left(\left(K\sigma+c_4V-BCDR^2\right)\omega-J_4\sigma\omega^3\right)^2}} \\ \phi_2(\omega)=-\arctan\left(\dfrac{\left(\left(K\sigma+c_4V-BCDR^2\right)\omega-J_4\sigma\omega^3\right)}{\left(KV-(c_4\sigma+J_4V)\omega^2\right)}\right)-\arctan\left(\dfrac{\left(\dfrac{J_0V^2}{R}+BCDRa\right)\omega}{\left(1+\dfrac{J_0\sigma}{R}\omega^2\right)}\right) \end{cases} \tag{9}$$

$$\begin{cases} A_4(\omega)=\dfrac{k_2^2\sqrt{\left(1-\dfrac{c_2^2}{k_2^2}\omega^2\right)^2+\left(\dfrac{2c_2k_2}{k_2^2}\omega\right)^2}}{(k_1+k_2+k_3)\sqrt{\left(1-\dfrac{J_3}{(k_1+k_2+k_3)}\omega^2\right)^2+\left(\dfrac{(c_1+c_2+c_3)}{(k_1+k_2+k_3)}\omega\right)^2}} \\ \phi_4(\omega)=\arctan\left(\dfrac{\dfrac{(c_1+c_2+c_3)}{(k_1+k_2+k_3)}\omega}{1-\dfrac{J_3}{(k_1+k_2+k_3)}\omega^2}\right)+\arctan\left(\dfrac{\dfrac{2c_2k_2}{k_2^2}\omega}{1-\dfrac{c_2^2}{k_2^2}\omega^2}\right) \end{cases} \tag{10}$$

### *3.2. Work of feedback block per unit period*

According to literature [11], the rotation motion of left and right tire around kingpin can be assumed as:

$$\begin{cases} \varphi_1(t)=A_1\sin\omega t \\ \varphi_2(t)=A_1\sin(\omega t+\phi) \end{cases} \tag{11}$$

Where $A_1$ is the amplitude of motion.

Then the motion around Y and Z axis can be given by:

$$\begin{cases} \theta_1(t)=|G_3|A_1\sin(\omega t+\angle G_3) \\ \theta_2(t)=|G_4|A_1\sin(\omega t+\phi+\angle G_4) \\ \varphi_3(t)=|G_5|T\sin(\omega t+\xi) \end{cases} \tag{12}$$

Where $T=\sqrt{\left[(A_1+B_1\cos\phi)^2+\sin^2\phi\right]}$, $\xi=\arctan\left(\dfrac{A_1+A_1\cos\phi}{\sin\phi}\right)$

According the frequency character in 3.1, the work per unit period of sideslip feedback block can be given by:

$$W_{in1}=\int_0^{\frac{2\pi}{\omega}}M_{z1}\dot{\varphi}_1(t)dt=\int_0^{\frac{2\pi}{\omega}}A_1^2A_1(\omega)\omega\sin(\omega t+\phi_1(\omega))\cos(\omega t)dt=\pi A_1^2A_1(\omega)\sin(\phi_1(\omega)) \tag{13}$$

Considering the phase difference between the rotation motion of left and right tire, the work per unit period of gyroscopic and middle rod can be given by:

$$W_{in2}=\int_0^{\frac{2\pi}{\omega}}M_{z2}\dot{\varphi}_1(t)dt=\pi A_1^2A_2(\omega)A_3(\omega)\sin(\phi_2(\omega)+\phi_3(\omega)) \tag{14}$$

$$W_{in3}=\int_0^{\frac{2\pi}{\omega}}M_{z3}\dot{\varphi}_1(t)dt=\pi A_1^2TA_3(\omega)\sin(\xi+\phi_3(\omega)) \tag{15}$$

$$W_{in4}=\int_0^{\frac{2\pi}{\omega}} M_{z3}\dot{\varphi}_2(t)dt=\pi A_1^2 T A_3(\omega)\sin\left(\xi+\phi_3(\omega)-\phi\right) \tag{16}$$

### 3.3. *Energy consumption of damping per unit period*

As shown in Figure 2, the damping of the steering system included the damp in kingpin, suspension, and the middle rod. According to the assumption (11), the energy consumption of kingpin damping can be given by:

$$W_{out1}=\int_0^{\frac{2\pi}{\omega_s}} c_e\dot{\varphi}_1(t)\dot{\varphi}_1(t)dt=A_1^2 c_e\pi\omega \tag{17}$$

The energy consumption of suspension damping can be given by:

$$W_{out2}=\int_0^{\frac{2\pi}{\omega}} c_4\dot{\theta}_1(t)\dot{\theta}_1(t)dt=c_4\pi\omega\left(A_1A_2(\omega)\right)^2 \tag{18}$$

And the energy consumption of middle rod and tie rod can be given by:

$$W_{out3}=\int_0^{\frac{2\pi}{\omega_z}} c_3\dot{\varphi}_3(t)\dot{\varphi}_3(t)dt=c_3\pi\omega\left(A_4(\omega)T\right)^2 \tag{19}$$

$$W_{out4}=\int_0^{\frac{2\pi}{\omega_z}} c_1\left(\dot{\varphi}_1(t)-\dot{\varphi}_3(t)\right)^2 dt=\pi c_1\omega\left[A_1^2+A_3^2T^2-2A_1A_3T\cos(\xi)\right] \tag{20}$$

Similarly, the amplitude of the energy consumption was directly related to the frequency characteristic of shimmy vibration. And the amount of energy in system determined the state of steering system.

### 3.4. *Energy analysis of steering system*

According to the results of section 3.2 and 3.3, the energy of steering system can be given:

$$W_{total}=W_{in}-W_{out} \tag{21}$$

Where

$$\begin{cases} W_{in}=2W_{in1}+2W_{in2}+W_{in3\_1}+W_{in3\_2} \\ W_{out}=2W_{out1}+2W_{out2}+W_{out3}+2W_{out4} \end{cases} \tag{22}$$

We take the velocity V=60 km/h, $A_1$=0.1 rad, and $\phi=2\pi$ considering the snaking of vehicle when the shimmy occurred. Then we obtained the work of different feedback block per unit period, as shown in Figure 3 (a). Similarly, the energy consumption can also be obtained, as shown in Figure 3(b). And the total energy of system was shown in Figure 3(c), where the $\omega_c$ was the corresponding frequency when $W_{total}$=0.

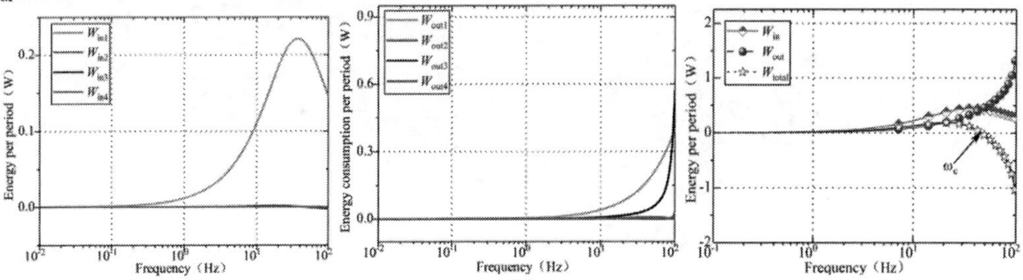


(a) (b) (c)

**Figure 3.** Energy of unit period

As shown in Figure 3(a), the main part of the feedback work is the alignment torque block, and the work increased firstly and then decreased with the frequency. While the energy consumption of damping

IOP Conf. Series: Materials Science and Engineering **772** (2020) 012009 doi:10.1088/1757-899X/772/1/012009

increased with the frequency increased, and the damping in kingpin and middle rod accounted for a large proportion according to the details of Figure 3(b). For the energy of steering system, when the $\omega<\omega_c$, the total energy of system was positive, and when the $\omega>\omega_c$, the total energy of system was negative.

## 4. Mechanism analysis of shimmy

### *4.1. Frequency and velocity interval of shimmy*

According to the results of section 3.4, the total energy of system was related to the frequency of steering system, and the frequency was always same as the natural frequency of steering system.

Then we linearized the Equation(1)~(3), and obtained:

$$\dot{X} = AX + g(X) \tag{23}$$

where $X \in R^{12}$, and $A$ is the Jacobian matrix at equilibrium point, and the g(X) is the high order term. And the eigenvalues of the Jacobian matrix can be calculated with:

$$\det\left|\lambda I - A\right| = 0 \tag{24}$$

Combining the parameters in Table 1, we can obtain the eigenvalues of $A$, $\omega_1$=35.2 rad/s, $\omega_2$= 51.1 rad/s, $\omega_3$=129.8 rad/s, $\omega_4$= 599.7 rad/s, $\omega_5$= 599.7 rad/s, and they were also the natural frequency of steering system. While it was necessary to explicit the relationship of different natural frequencies and velocity range of shimmy. The total energy of steering system at different velocity and frequency can be obtained with the values and Equation (13)~(20). And the result was shown in Figure 4(a), where the red region was the negative region of steering system, and the green region was the positive region, and the corresponding range of frequency and velocity were marked as vermilion. As we can see, only the first natural frequency $\omega_1$ had intersection with the positive energy region. Therefore, we can believe that the frequency of shimmy was the first natural frequency $\omega_1$.

And we taken the value of $\omega_1$ as the frequency in Equation (11), the total energy of steering system was obtained and shown in Figure 4(b). The energy consumption was not affected by velocity according to the results in Figure 4(b), therefore the total energy of steering system increased firstly and then decreased with the increase of velocity. And the total energy was positive during 34 km/h~72 km/h, which can be considered as the velocity interval of shimmy. According to the result in Figure 5, when the steering system occurred shimmy, the amplitude of vibration would increase continuously, while it was not the case in practice. It can be explained by the nonlinear characteristic of sideslip force of tire. The error between the linear model and nonlinear model increased with the increasing of sideslip angle, which also implied that the amplitude of shimmy increased. And the input energy originated from sideslip force was less than the calculated values. Therefore the amplitude of shimmy can be stable in practice. This also shown that the calculation of amplitude should use nonlinear model of tire.

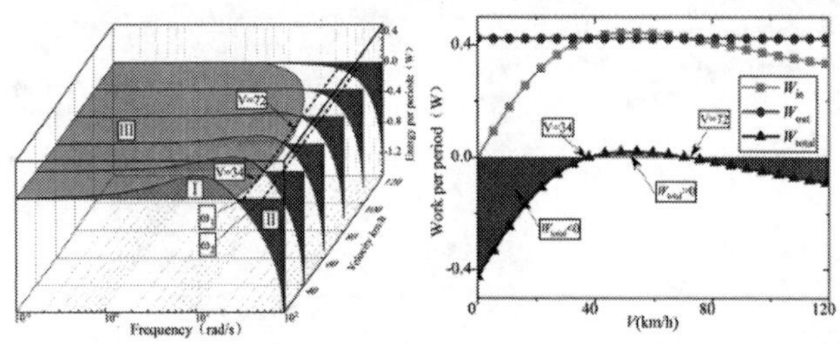


(a) (b)

**Figure 4.** Total energy of steering system at different velocity and frequency

### *4.2. Calculation of shimmy amplitude*

When the shimmy occurred, the state of steering system satisfied

$$W_{total} = W_{in} - W_{out} = 0 \tag{25}$$

Then we calculated the shimmy amplitude with the Equation (1)~(3), (21), and (25) based on the conservation of energy. The velocity interval of shimmy was 34 km/h~72 km/h, the maximum amplitude of shimmy was 5.2°.

When the steering system was disturbed in the static equilibrium position and the velocity was within the range where the shimmy might occur, where $W_{in}>W_{out}$, the energy of the steering system will increase gradually with time, showing that the amplitude of the shimmy will increase, including the tire sideslip angle. And there was a non-linear relationship between the sideslip angle and the sideslip force, which resulted in the reducing of feedback block $W_{in}$ with the increase of the amplitude until the condition $W_{in}=W_{out}$ was met. Then the work brought by the feedback block was equal to the energy consumption by damping in steering system. The continuous reciprocating vibration occurred, i.e., shimmy.

### 4.3. *Numerical simulation verification*

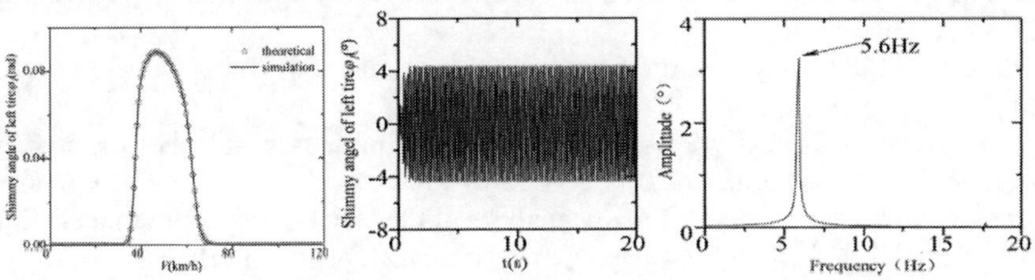


(a) (b) (c)

**Figure 5.** Dynamic response of steering system

Numerical simulation of the equation (1)~(3) was carried out with Runge-Kutta method combining with the parameters in Table 1, assuming that the left wheel encountered perturbation. The comparing results of theoretical and numerical simulation were shown in Figure 5. The velocity interval and the amplitude of shimmy were all consistent according to the Figure 5 (a). The amplitude of shimmy was 4.4° and the frequency was 5.6 Hz according to the results in Figure 5 (a) and (b), which means that the frequency of shimmy was consistent with theoretical results. Therefore the results of theoretical analysis based on the coupled feedback model were verified by the numerical simulation.

## 5. Conclusion

(1) A 5 DOF dynamical model of steering system with independent suspension was established firstly, and then the coupled feedback model of system was obtained based on the dynamical model.

(2) On this basis, the frequency characteristic of feedback block was analyzed. The instability velocity interval, frequency, and amplitude of shimmy were obtained based on conservation of energy. The relationship of them and mechanism of shimmy were uncovered further according to the results.

(3)The results obtained by theoretical analysis were verified by numerical simulation, which can provide theoretical support for the suppression and active control of shimmy in steering system..

**Acknowledgments**

This work was financially supported by the National Key Research and Development Program of China (Grant No. 2016YFC0802900).

## References

[1] Mi T, Stepan G, Takacs D, et al. Shimmy model for electric vehicle with independent suspensions [J]. Proceedings of the Institution of Mechanical Engineers, Part D: Journal of Automobile Engineering, (2017), 24, p. 1-11.

[2] Wei D, Xu K, Jiang Y, et al. Hopf Bifurcation Characteristics of Dual-Front Axle Self-Excited

IOP Conf. Series: Materials Science and Engineering **772** (2020) 012009 doi:10.1088/1757-899X/772/1/012009

Shimmy System for Heavy Truck considering Dry Friction [J]. Shock and Vibration, 2015(2015)1-20.

[3] Beregi S, Takács D, Hős C. Nonlinear analysis of a shimmying wheel with contact-force characteristics featuring higher-order discontinuities [J]. Nonlinear Dynamics, (2017), 90, p.877-888.

[4] Wei D, Wang Y, Jiang T, et al. Chaos vibration of pinion and rack steering trapezoidal mechanism containing two clearances[J]. Mechanical Systems and Signal Processing, 92(2017)146-155.

[5] Lu J, Xin J, Vakakis A F, et al. Influences of system parameters on dynamic behavior of the vehicle shimmy system with clearance in steering linkage[J]. Journal of Vibration and Control, 21(2015), 2, p. 359-370.

[6] Mi T, Chen N, Stepan G, et al. Energy Distribution of a Vehicle Shimmy System with the Delayed Tyre Model [J]. IFAC-PapersOnLine, 51(2018), 14, p. 7-12.

[7] Guo K, Liu Q. Influences of Non-Steady State Tire Cornering Properties on Automobile Shimmy[C]. International Truck and Bus Meeting and Exposition, 1999, Detroit, Michigan, USA.

[8] Besselink I J M. Shimmy of aircraft main landing gears [D]. Delft University of Technology, 2000.

[9] Ran S, Besselink I, Nijmeijer H. Application of nonlinear tyre models to analyse shimmy [J]. Vehicle System Dynamics. (2014), 53, p.387-404.

[10] Ran S, Besselink I, Nijmeijer H. Energy analysis of the Von Schlippe tyre model with application to shimmy [J]. Vehicle System Dynamics. (2015).53, p.1795-1810.

[11] Jiang J, Lu, Yang S. Mechanism analysis of vehicle shimmy with time-delayed displacement feedback and energy transfer [J]. Chinese Journal of Computational Mechanics. 2018(01): 35-43.

[12] Yang S, Lu D, Zhang H, et al. Mechanism Analysis of the front wheel shimmy [J]. Science Technology and Engineering. 2015(32): 221-223.

[13] Hb P. Tire and Vehicle Dynamics [M]. Oxford: Butterworth-Heinemann, 2012.

IOP Conf. Series: Materials Science and Engineering **772** (2020) 012010 doi:10.1088/1757-899X/772/1/012010

# Effect of double quenching process and tempering temperature on the microstructure and mechanical properties of a High Strength Low Alloy Steel

**Fei Zhu, Xiaobing Luo *, Feng Chai, Caifu Yang and Zhengyan Zhang**

Division of Structural Steels, Central Iron and Steel Research Institute, Beijing, China

*Corresponding author e-mail: luoxiaobing@cisri.com.cn

**Abstract.** Effect of double quenching process and tempering temperature on the microstructure and mechanical properties of a High Strength Low Alloy Steel was investigated by OM, SEM, TEM methods and so on. The results demonstrated that double quenching process could refine prior austenite grains (PAG) effectively and the average diameter of PAG was refined by 5.4μm compared with single quenching process. Yield strength and tensile strength of double quenching and tempered specimens (QQT) and single quenching and tempering specimens (QT) exhibited similar variation trend with tempering temperature. Yield strength increased monotonously with increasing tempering temperatures. However, tensile strength descended at first and then recovered with rising tempering temperature. QQT specimens displayed greatly improved impact absorbing energy at -80℃ but a little improved yield strength compared with QT specimens owing to the refined microstructure caused by double quenching process. Considerable austenite formed during holding process of tempering (670℃) in which elements of nickel, manganese and copper aggregated. And some of the aforementioned austenite transformed to hard phases (secondary Martensite and M-A islands) and some are retained as islands of austenite after water cooling. The presence of hard phases is the main reason for recovering tensile strength and deteriorating impact absorbing energy at -80℃.

## 1. Introduction

Low-carbon, copper-strengthened high strength low alloy steel 100 (HSLA100) is broadly applied in hull structure, pipeline and equipment of mining and dredging [1-5]. The low carbon content (⩽ 0.07wt.%) is adopted in HSLA100 to avoid pre-heating process before welding, greatly curtailing the processing cost. The strength loss caused by lower carbon content is offset by strengthening of Cu-rich precipitates during the aging process. Low temperature toughness is also another important mechanical property in this steel owing to the wicked service circumstance and loading conditions. Generally, Niobium (Nb) was added as a kind of micro-alloying element in HSLA100 to refine the microstructure by the pinning effect of Nb(C, N) precipitating during hot-rolling process on grain boundaries of austenite. Quenched-tempered process is normally applied in HSLA100 after hot-rolling and the typical applied microstructure of this steel consists of tempered Martensite (M) and (or) Bainite (B) matrix and Cu-rich precipitates of diameters ranging from 2nm to 25nm precipitating in the matrix [1].

Content from this work may be used under the terms of the Creative Commons Attribution 3.0 licence. Any further distribution of this work must maintain attribution to the author(s) and the title of the work, journal citation and DOI.
Published under licence by IOP Publishing Ltd

Steel with higher strength under the condition of superior low temperature toughness can reduce the manufacturing cost, lose weight of constructions and lower the carbon emission, generating great benefits [6]. It is a well-known fact that refined grains of prior austenite could improve both yield strength and low temperature toughness due to the increased areas of grain boundaries [7, 8]. And process of heat treatments is greatly valued in producing medium and heavy plate because of the advantages for adjusting microstructure without complex operations. Double quenching process, a relatively novel heat treatment, is extensively studied due to the refining effect on the size of prior austenite grains (PAG) and thus refining the size of packet and block of M[9-11]. Consequently, we present here the effect of double quenching process on microstructure and mechanical properties of HSlA100 steel. Furthermore, evolution of microstructure and mechanical properties with increased tempering temperature under heat treatments of double-quenching and tempering was also investigated.

## 2. Materials and methods

The tested steel plates were commercially hot-rolling of 20 mm thickness and the composition was C0.02~0.07, Si0.39, Mn0.5~0.8, (Ni+Cr+Mo)>4.0, Cu1.50, Nb0.03, Al0.026, Fe balance (mass %). As-rolled plates were firstly heated to 900 ℃, holding for 1 h and then water quenched to room temperature (Q). The single quenched plates were again heated to 880 ℃, holding for 10 min and water quenched to room temperature (QQ). Q and QQ specimens were then tempered at 620 ℃~670 ℃ for 2h and then water cooled to room temperature respectively, termed as QT and QQT process.

Mechanical properties of both QT and QQT specimens were tested. The tensile properties were evaluated with rod specimens of 5 mm diameter according to GB/228.1-2010. The axial direction of rod tensile specimens was perpendicular to rolling direction. Charpy impacting tests were conducted using standard transverse impacting specimens of 10*10*55mm at -80 ℃ according to GB/T 229-2007, and the V notch was on the plane of transverse direction and normal direction. Double tensile specimens and three impacting specimens were tested for each tempered condition and the average values were applied.

Grain boundaries of prior austenite were etched using supersaturated solution of picric acid mixed with some detergent and the average intercept of PAG was obtained by intercepting method according to GBT 6394-2002. The macro microstructures were observed with Optical Microscope (OM), Scanning Electron Microscope (SEM). Transmission Electron Microscope (TEM) was applied to observe the refined microstructure and Cu-rich precipitates of tempered specimens.

## 3. Results and discussion

### *3.1. Mechanical properties*

Fig. 1 shows the average values of yield strength (RP0.2), tensile strength (Rm) and impact absorbing energy for QT and QQT specimens. Variations of RP0.2 and Rm of both QT and QQT specimens show similar trend with increasing temperature. Taking as QQT specimens for example, RP0.2 declines with increasing tempering temperature and the lowest and highest RP0.2 were 690 MPa, 870.5 MPa respectively. Variation of Rm with tempering temperature, however, exhibits two stages. During tempering temperature of 620~650 ℃, Rm of QQT specimens show similar variation trend with RP0.2. During tempering temperature of 660~670 ℃, Rm of QQT specimens recover with increasing tempering temperature. QQT specimens all obtain impact absorbing energy more than 200J at -80 ℃ and exhibit more excellent low temperature toughness during the whole range of tempering temperature compared with QT specimens, shown as Fig. 1b. And the best toughness is obtained in QQT specimens tempered at 640 ℃. It is also seen from Fig.1 that QQT specimens show a little improvement on RP0.2 but obvious enhancement for low temperature toughness.

### *3.2. Microstructure*

*3.2.1. Quenched microstructure.* SEM images of Q and QQ specimens are illustrated in Fig. 2ab. Microstructures of both Q and QQ specimens consist of M matrix and minor B, and M packet boundaries are readily visible within prior austenite grain boundaries (PAGB). Vickers hardness of Q and QQ specimens is HV367, HV361 with 5 Kg loading respectively. Fig. 2cd shows PAGB of Q and QQ specimens, and PAG of QQ specimen are evidently refined compared with Q specimen. The average intercept of PAG for Q and QQ specimen is 11.95 μm, 7.64 μm respectively, and the average diameter of PAG for Q and QQ specimens is 13.3 μm, 7.9μm respectively.

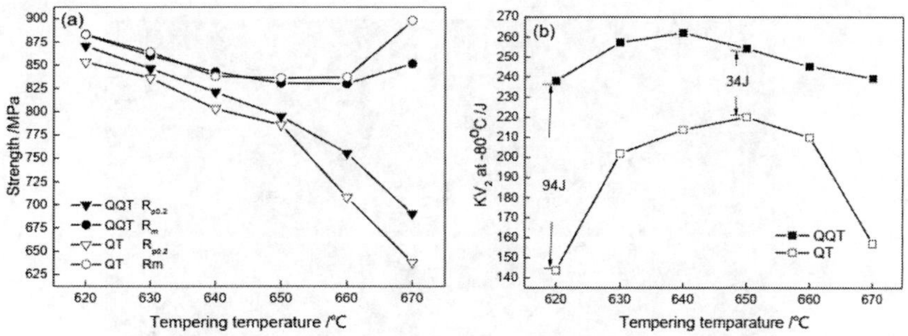


**Figure 1.** Mechanical properties of tested steel as a function of tempering temperature: (a) strength, (b) impact absorbing energy at -80 ℃.

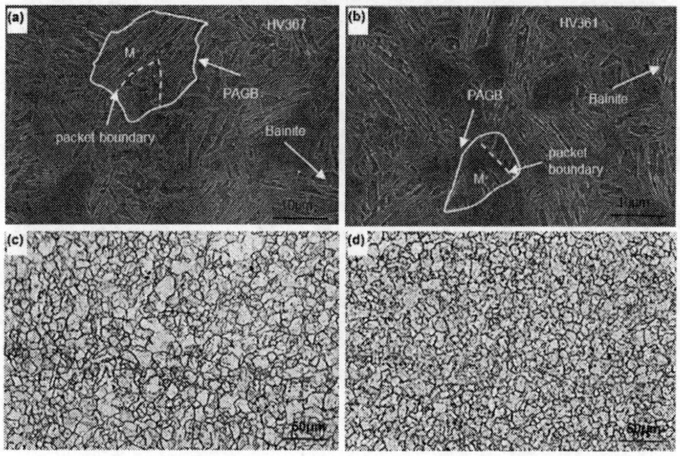


**Figure 2.** Microstructure and prior austenite grain boundaries for Q and QQ specimens: (a) and (c) for Q specimens, (b) and (d) for QQ specimens.

*3.2.2. Tempered microstructural evolution.* Fig. 3 shows SEM images for microstructural evolution with increasing tempering temperatures ranging from 620~670 ℃ for QQT specimens. Certain laths of M seem to be merged and recovery after tempering [12], and white-bright phases with rod and chain-like shape exist within tempered M/B matrix. It is also found that those white-bright phases exhibit parallel arrangement indicated by broken circle (Fig 3d) and evidently coarsen at high tempering temperatures (Fig 3cd). Those white-bright phases may be the mixture of cementite and transformation product of austenite formed in holding process of tempering after water cooling at low tempering temperature (620 ℃). And the above white-bright phases are mainly the transformation product of austenite at high tempering temperature (660, 670 ℃). Because cementite gradually resolves and transformation product of austenite gradually augments with increasing tempering temperature.

Fig. 4 shows the results of EDS analysis of white-bright phase and tempered M matrix labeled as A and B respectively in Fig. 3d. White-bright phases formed at 670 ℃ occur segregation of Ni, Cu and Mn alloy elements compared with tempered M matrix. The corresponding micro-hardness indentation of white-bright phase and tempered M matrix in specimens tempered at 670℃ are displayed in Fig. 5 and the value of micro-hardness is HV414, HV319 with 10g loading respectively. It is reasonable to deem coarsened hard white-bright phases as secondary M or M/A islands transformed from austenite after water cooling. And this need to be further discussed in Section 3.3.1.

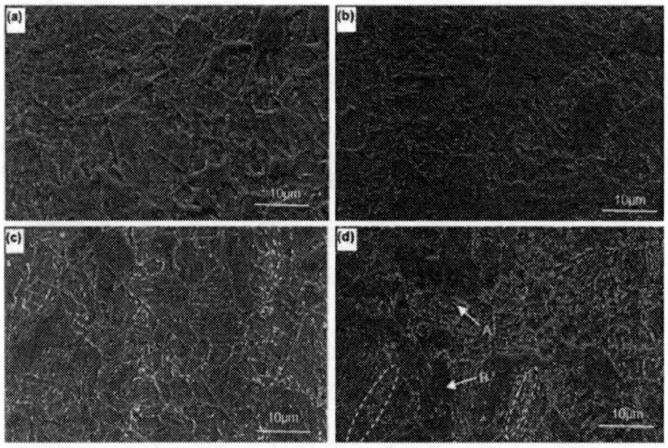


**Figure 3.** Microstructural evolution of double-quenched specimens at different tempering temperatures: (a) 620℃, (b) 640℃, (c) 660℃, (d) 670℃.

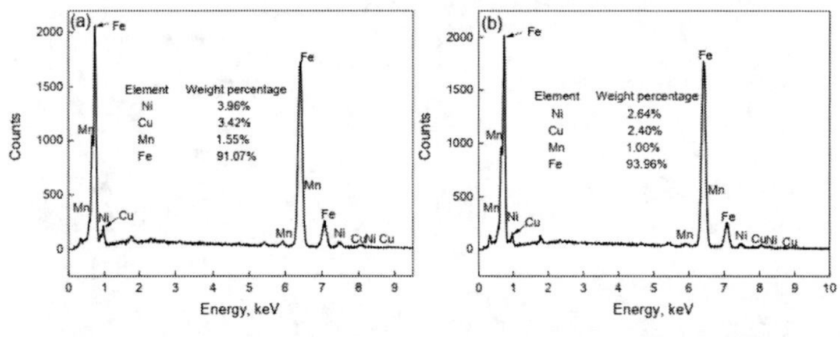


**Figure 4.** Elemental analysis of A and B spots selected from figure3d by EDS.

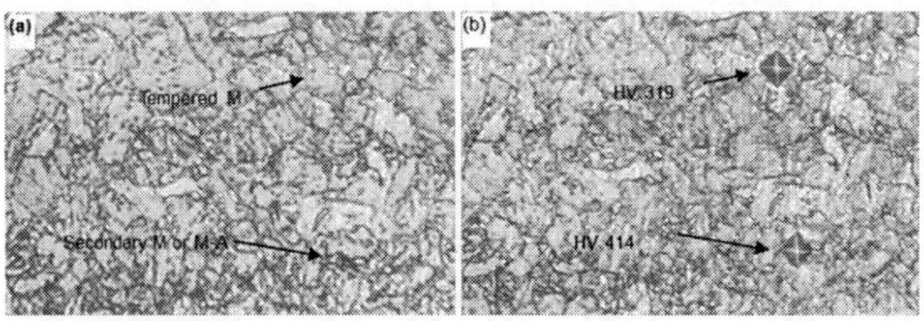


**Figure 5.** Microstructure and micro-hardness indentation of double-quenched specimen tempered at 670℃: (a) original microstructure, (b) micro-hardness indentation.

*3.2.3. TEM analysis.* Fig. 6 displays the refined microstructural evolution of QQT samples with increasing tempering temperature. Tempered LM and elliptical spherical Cu-rich precipitates precipitating in LM are easily observed in Fig.6a, and dislocations with un-fully recovery are readily visible within certain laths. Fig.6bc illustrates the film-like secondary M transformed from austenite between laths after water cooling and the corresponding selected area diffraction pattern. Film-like secondary M with length of about 0.43 μm and width of about 0.12 μm is consistent with the observation of white-bright phase in SEM (Fig. 6). Cell-like sub-grain forms and incises laths of M indicating the occurrence of recrystallization, and coarseness of Cu-rich precipitates within laths is also shown in Fig.6d. Average diameter of Cu-rich precipitates in specimens tempered at 620, 640, 670℃ is 17.05、22.59、28.86 nm respectively. Fig. 6def shows bright image, dark image and selected area diffraction pattern of a film-like M/A island transformed from austenite after water cooling. Retained austenite and M contained in M/A island conform to Kurdjumov-Sachs (K-S) relationship, i.e., (-1-11)γ //(01-1)α , [-110]γ //[1-1-1]α . Fig7abc shows bright image, dark image and selected area diffraction pattern of island shaped austenite locating at crystal corner transformed from austenite due to be of sufficient thermal stability. Fig7d shows bright image, selected area diffraction pattern of island shaped secondary M between laths.

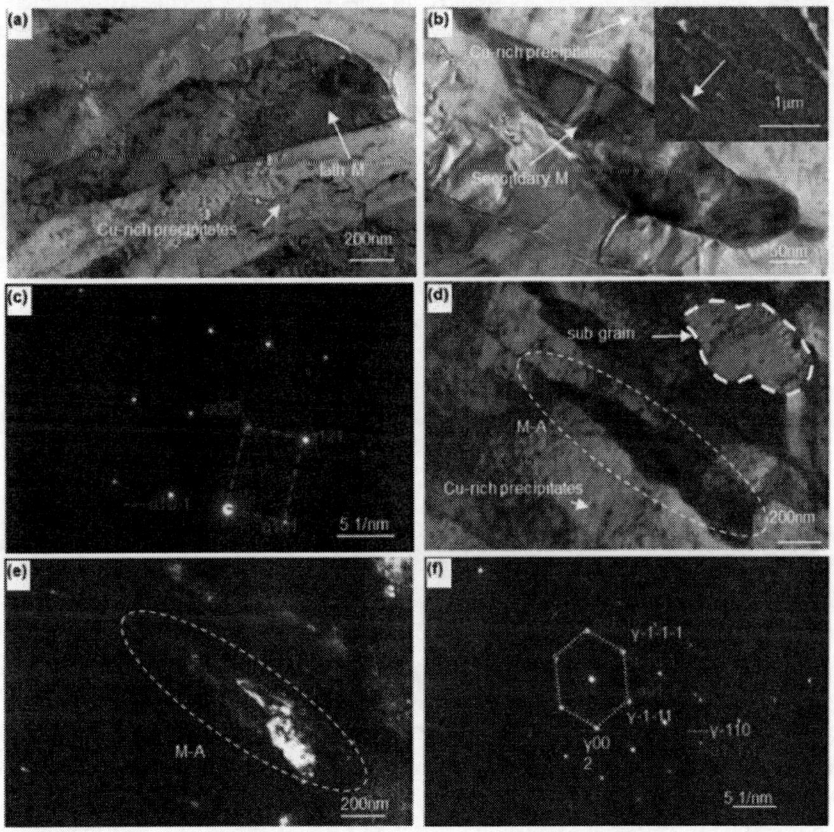


**Figure 6.** TEM image and selected area diffraction pattern of double-quenched specimen tempered at different temperatures: bright images at 620 ℃(a), 640 ℃(b), diffraction pattern of secondary M(c) in (b), bright image of M/A island at 670 ℃(d), dark image of M/A island at 670 ℃(e), diffraction pattern of M/A island (f).

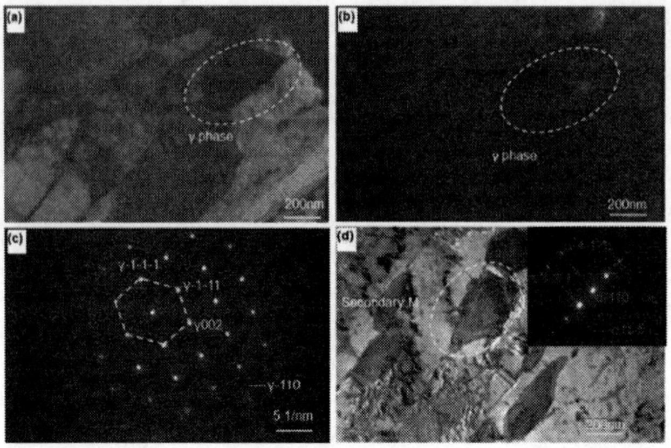


**Figure 7.** TEM image and diffraction pattern of austenite and secondary M of the tested steel tempered at 670℃: (a) bright image of austenite, (b) dark image of austenite, (c) diffraction pattern of austenite, (d) bright image and diffraction pattern of secondary M.

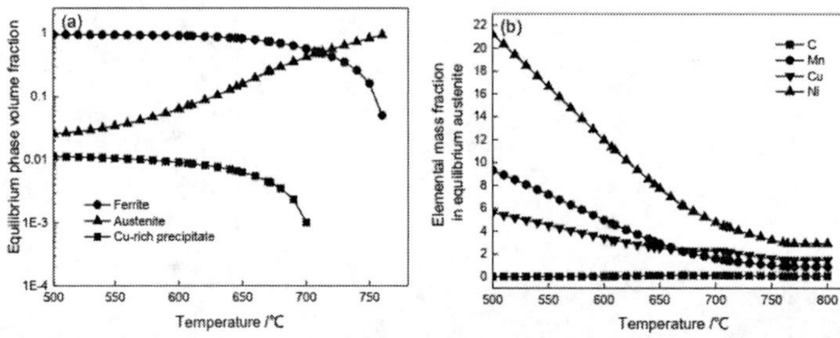


**Figure 8.** Thermal equilibrium calculation of tested steel: (a) volume fractions of phases as functions of temperature, (b) alloy elements contained in austenite as functions of temperature.

### *3.3. Discussion*

*3.3.1. Thermal calculation of equilibrium phases and transformation products of austenite.* Thermal calculation of main equilibrium phases of tested steel was conducted by thermal-calculation software, and the relevant results are shown in Fig. 8. Fig. 8a shows austenite could exist during 500~760 ℃, and the volume fraction increases with increasing temperature. The volume fraction of austenite is 8.8%, 13%, 24.8% when the temperature is 620, 640, 670 ℃ respectively. Volume fraction of Cu-rich precipitates declines with increasing temperature, and the dissolving temperature for Cu-rich precipitates is nearly 700 ℃. The volume fraction of Ferrite is also reduce with increasing temperature during 500-760 ℃. Fig. 8b exhibits variation of main alloying elements (C, Mn, Ni, Cu) contained in austenite with increasing temperature and the enriching extent for the above alloys contained in austenite increases with dropping temperatures. Taking austenite formed at 670 ℃ for instance, mass contents of C, Mn, Ni, and Cu contained in it are 0.136%, 2.27%, 6.27%, 2.32% respectively and are several fold than the contents of relevant alloys shown in nominal composition. The calculated results for enrichment of the above elements contained in austenite formed at 670 ℃ are consistent with the results of EDS analysis of white-bright phases.

Austenite formed during holding process of tempering transforms to different kind of products, which was determined by grain size and the content of alloying elements contained in austenite after water cooling. On the one hand, the higher contents of alloying elements such as C, Mn, Ni, and Cu contained in austenite, the lower the temperature of M starting transformation. And austenite with high contentment of the above alloy elements is inclined to be retained at room temperature after quenching. On the other hand, the smaller the austenite grains, the easier it is supposed to be retained at room temperature after quenching, i.e., size stabilizing effect of austenite. Chen et.al [13] shows that this effect is caused by lacking in effective nucleation positions in extremely fine austenite grains. The combination of observation made by SEM, TEM, micro-hardness and tensile tests could make a conclusion that partial austenite formed at 670 ℃ transformed to hard phase of secondary M and M/A island after water cooling, and partial austenite were still retained at room temperature. It is reasonable to make a hypothesis that austenite formed at temperature lower than 670 ℃ is inclined to be retained at room temperature, taking grain size and enriching extent of alloy elements contained in austenite into consideration. And coarsening of Cu-rich precipitates, recovery of dislocation and soften of M/B matrix also occurs with increasing tempering temperature [14].

*3.3.2. Strengthening and toughening mechanism.* Yielding phenomenon in steel is the outcome of slipping of massive dislocations [15]. In other words, yield strength of steels depends on how easily a large number of dislocations slip. The mechanism for yield strength of tested steel declined with increased tempering temperatures in QQT specimens is as follows. The average diameters of Cu-rich precipitates increased with increasing tempering temperature, however, the volume fraction of Cu-rich precipitates exhibited the opposite trend with increasing tempering temperature. The comprehensive changes of Cu-rich precipitates led to the reduce of precipitating strengthening effect according to the Orowan mechanism expressed by Eq.(1) [16]:

$$\Delta\sigma_{orowan} = \frac{MGb}{\left(\sqrt{\frac{4\pi}{3f}}-2\right)r} \tag{1}$$

Where M is Taylor factor, G is the shear modulus of matrix taken as 80GPa, b is the burgers vector, r is the average radius of Cu-rich precipitates and f is the volume fraction of Cu-rich precipitates. Furthermore, strengthening effect caused by dislocation and solidification also reduce due to dislocation recovery and solidification atoms precipitating from M/B matrix during tempering.

Mechanism for Variation of tensile strength with tempering temperature ranging from 620~650 ℃ is similar to the one of yield strength. However, tensile strength of QQT specimens tempered at 670 ℃ recovers, which is caused by the formation of hard phases of secondary M and M/A island after tempering. Microstructure of QQT specimens tempering at 670 ℃ consists of soft tempered M/B matrix and hard phases. Model of tensile strength of dual-phase steel is expressed by Eq. (2) [17]:

$$R_m = \lambda_\alpha(1-f_\beta)R_m^\alpha + \lambda_\beta f_\beta R_m^\beta \tag{2}$$

Where Rm is the tensile strength of dual-phase steel, $\lambda_\alpha$ and $\lambda_\beta$ is the constant about uniform true strain respectively, $f_\beta$ is the volume fraction of β phase and $R_m^\alpha$, $R_m^\beta$is the tensile strength of α, β phase respectively. It could be seen from Eq. (2) that formation of hard phases is a key reason for recovering of tensile strength.

Fig. 9 shows the impact fracture surfaces for QT and QQT specimens. All fracture surfaces of QQT specimens consist of fibrous area and shear lip zone, and all ones of QT specimens consist of fibrous area, evident cleavage area and shear lip zone. The ratios of cleavage area accounting for the whole fracture surface for QT specimens tempered at 620, 670 ℃ are larger than others, indicating the relatively poor total absorbing energy. Detailed observation for central areas of QT and QQT specimens tempered at 660 ℃ were shown in Fig. 9bc. Central area of QQT specimen is characteristic of numerous dimples connecting (Fig. 9b), indicating ductile fracture mode. However, ones of QT specimens consist of flat cleavage facets connected by tear ridges (Fig. 9c), indicating quasi-cleavage fracture mode. It could make a conclusion that the refinement of PAG caused by double quenching process converts the fracture mode. Beneficial effect of lower P, S brittle elements contained in zones of grain boundaries caused by grain refinement is also attributed to this conversion.

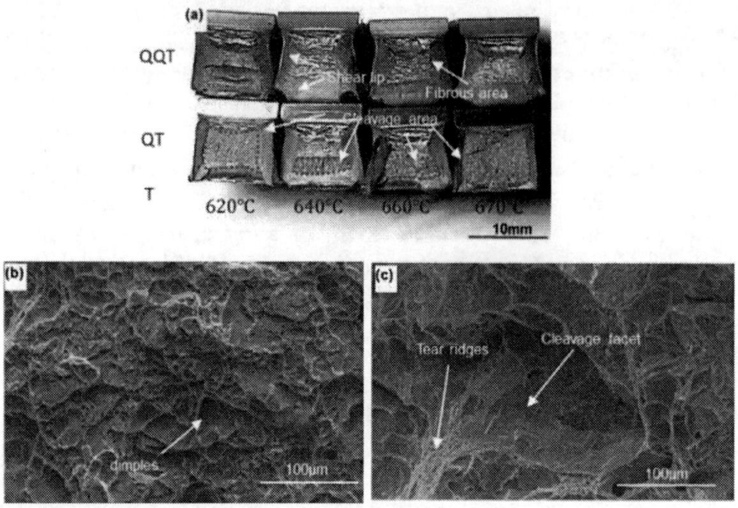


**Figure 9.** Characteristics of impact fracture surfaces for QT and QQT specimens: (a) macro image of QT and QQT specimens tempered at 620, 640, 660, 670 ℃, (b)detailed observation for central area in fracture surface of QQT specimen (b) and QT specimen (c) tempered at 660 ℃.

Variation of low temperature toughness of QQT specimens with increasing temperature show similar trends with QT specimens. Taking QQT specimens for instance, toughness improved with increasing tempering temperature ranging from 620~640 ℃, subsequently declined with increasing tempering temperature. The best toughness (258.3J) was obtained in specimens tempered at 640 ℃. The toughening mechanisms for tempered specimens during tempering temperature of 620~670 ℃ are determined by integrated effects of following factors.

(1) coarsening of Cu-rich precipitates and elevating extent of dislocation recovery in M/B matrix with increasing tempering temperature improved the toughness; (2) austenite formed at holding process of tempering absorbed P and S brittle element from matrix, purifying the matrix, and 'TRIP effect' of island shaped retained austenite in the course of impacting deformation process improved toughness[18]; (3) hard phases of secondary M and M/A island formed after tempering at high temperatures deteriorated toughness, because hard phases are usually the initiation resource of crack nucleation and the interfaces between hard phases and matrix are usually the propagation path during crack propagation.

## 4. Conclusion

(1) Double quenching process effectively refined the prior austenite grains compared with traditional single quenching process, the average diameter of austenite grains for Q and QQ specimens were 13.3 μm, 7.9 μm respectively.

(2) Cu-rich precipitates coarsened with increasing tempering temperature, and the average diameter of Cu-rich precipitates at tempering temperature of 620, 640, 670 ℃ is 17.05, 22.59, 28.86 nm respectively. Film-like and island shaped austenite formed in holding process of tempering (670 ℃) obtained complicated transforming products after water cooling, including secondary M, M/A island and island-shaped austenite depending on grain size and enriching extent of alloying elements.

(3) QQT specimens during the tempering temperature of 620~670 ℃ exhibited greatly improved low temperature toughness and a little higher yield strength caused by grain refinement compared with QT specimens. And the optimal combination of strength and low temperature toughness is $R_{P0.2}$=821MPa, Rm=842.5MPa, and total impact absorbing energy at -80 ℃ is 258.3J. Hard phases

such as secondary M and M/A island formed after tempering (670 ℃) led to recovery of tensile strength and deterioration of toughness for QQT specimens.

**References**

[1] Wilson A D, Hamburg E G, Colvin D J, et al. Properties and microstructure of copper precipitation aged plate steels[C]// Microalloying. 1988:259-275.
[2] Czyryca E J , Link R E , Wong R J , et al. Development and Certification of HSLA -100 Steel for Naval Ship Construction, J. Naval Engineers Journal. 2010, 102(3):63-82.
[3] Dhua S K, Mukerjee D, Sarma D S. Influence of tempering on the microstructure and mechanical properties of HSLA-100 steel plates, J. Metallurgical & Materials Transactions A. 2001, 32(9):2259-2270.
[4] LIU Q D, ZHAO S J. Comparative Study on Austenite Decomposition and Cu precipitation during continuous cooling transformation, J. Metallurgical and Materials Transactions A. 2013, 44(1):163-171.
[5] Isheim D, Kolli R P, Fine M E, et al. An atom-probe tomographic study of the temporal evolution of the nanostructure of Fe–Cu based high-strength low-carbon steels, J. Scripta Materialia. 2006, 55(1):35-40.
[6] Liu Zhen-yu, Chen Jun, Tang Shuai, et al. State of the Art Development of the manufacturing technologies of new generation war ship steels, J. Materials China, 2014.
[7] Zhang Li-li, Zhang Xiao-yong, Gao Hui-lin, et al. Microstructure and Mechanical Properties of a High Nb-microalloyed X80 Pipeline Steel, J. Journal of Materials Engineering. 2009, 23(6):1530-1535.
[8] Gao Kuan, Wang Liu-ding, et al. Refinement of grain and enhancement of impact toughness for low-alloying ultra-high strength bainite steels, J. Acta Metallurgica Sinica. 2007, 43(3):315-320.
[9] Kim H J, Kim Y H, MORRIS J W. Thermal mechanisms of grain and Packet refinement in an lath martensitic steel, J. Transactions of the Iron & Steel Institute of Japan. 1998, 38(11):1277-1285.
[10] Yin Zhi-xin, Yu Wei-cheng, Yao Ge, et al. Effect of cycle quenching on structure and fatigue property of 42CrMo, J iron and steel. 2002,37(10):52-56.
[11] Morris J W, Kim J I, Syn C K. Advances in the heat treatment of steels.[C]// Sagamore Army Materials Research Conference. Springer US, 1978.
[12] Liu D, Cheng B. Fine microstructure and toughness of low carbon copper containing ultra high strength nv-f690 heavy steel plate. Acta Metallurgica Sinica , J. Acta Metallurgica Sinica, 2012, 48(3):334.
[13] Chen I W, CHIAO Y H, TSUZAKI K. Statistics of martensitic nucleation, J. Acta Metallurgica. 1985, 33(10):1847-1859.
[14] Thompson S W. Fine–scale structural features of intercritically aged HSLA-100 plate steel and their influence on yield strength and low-temperature impact toughness, J. Materials Characterization, 2017.
[15] Ma Ming-tu, Wu Bao-rong. Dual phase steel: Physical and mechanical metallurgy[M]. Beijing: Metallurgical Industry press.
[16] JIAO Z B, LUAN J H, ZHANG Z W, et al. High-strength steels hardened mainly by nanoscale NiAl precipitates, J. Scripta Materialia. 2014, 87(4):45-48.
[17] Luo Xiao-bin, Yang Cai-fu, Su Hang, et al. Effect of aging temperature on microstructure and properties of HSLA ship-hull steel, J. Transactions of Materials & Heat Treatment, 2011, 32(6):73-77.
[18] Olson G B. Transformation Plasticity and toughening, J. Journal De Physique IV. 1996, 6(1):407-418.

# Effects of Welding Currents on Microstructure and Properties of 5052 Aluminum Alloy TIG Welded Joint

**Weijie Gou *, Lihong Wang**

Beijing Polytechinc, Beijing, China

*Corresponding author e-mail: gvjie@126.com

**Abstract.** To study the influence of welding current on the welding quality of 5052 aluminum alloy TIG welding, the microstructure, mechanical properties and corrosion resistance of the welded joints obtained under different currents from 100 A to 120 A were discussed. The results showed that when the welding current is 100～120 A, with the increases of welding current, the strength of the welded joint (of 5052 aluminum alloy TIG) increases, while the plasticity decreases.

## 1. Introduction

5052 aluminum alloy belongs to Al-Mg series aluminum alloy. It has good fatigue performance and welding performance, and it is resistant to marine atmospheric corrosion. Therefore, it is widely used as a welded component in ships [1-2], aviation [3-4] and vehicles [5], etc. Among many aluminum alloy welding methods, friction stir welding [6-9] and laser welding [10-13], which have outstanding welding quality, are studied most. However, friction stir welding is limited to the welding of rotary parts, while laser welding has high energy coupling barriers for highly reflective and highly thermally conductive materials (such as aluminum and copper) [14], which has extremely high requirements for the output power of the laser and beam quality while welding. Therefore, based on practicality and cost considerations, it is still necessary to study other welding methods besides the above two welding methods.

During the process of TIG welding, groove form, fixture and pad design, and welding parameters, etc. directly affect weld formation, welding quality and the strength of the welded joint. Among them, the most important thing is to adjust the welding parameters; in the welding parameters, the welding current plays the leading role. Kumar and other people [15, 16] studied the influence of peak current, base current and welding speed on the welding quality of AA 5456 aluminum alloy plate pulsed tungsten argon arc welding, and found that the welding current has great influence on the melt depth: If the current is too large, the workpiece is easily burnt through; if the current is too small, it may cause incomplete soldering. Xu Zheng [17] used the TIG welding method to weld 7075 aluminum plates at the currents of 160 A and 200 A, and found that the tensile strength was only 219 MPa and 188 MPa, and he did not conduct studies on lower-current welding. Wang Xiling [18] studied the TIG welding process with welding currents of 170 A and 180 A; Liao Chuanqing et al [19] studied the TIG welding process with the welding current of 250 A; the TIG welding current studied by Wang Peiji et al [20] is higher, 300 A. The above literature [18-20] only studied the TIG welding with single-current parameters, and did not conduct TIG welding research under different currents; and the welded joints obtained from the 7-

Content from this work may be used under the terms of the Creative Commons Attribution 3.0 licence. Any further distribution of this work must maintain attribution to the author(s) and the title of the work, journal citation and DOI.
Published under licence by IOP Publishing Ltd

series aluminum alloy sheet were lower in strength and only reached 40% to 50% of the base metal. [18-20].

## 2. Experimental materials and methods

In this paper, the influence of welding current on the mechanical properties and corrosion resistance of 3mm-thick 5052 aluminum alloy sheet with TIG welding is studied, when the welding current is changed from 100 to 120 A. The influence of TIG welding current on the microstructure and properties of the alloy is obtained. The law is a supplement to the TIG welding aluminum alloy sheet process in the small current range, and provides reference for engineering practice.

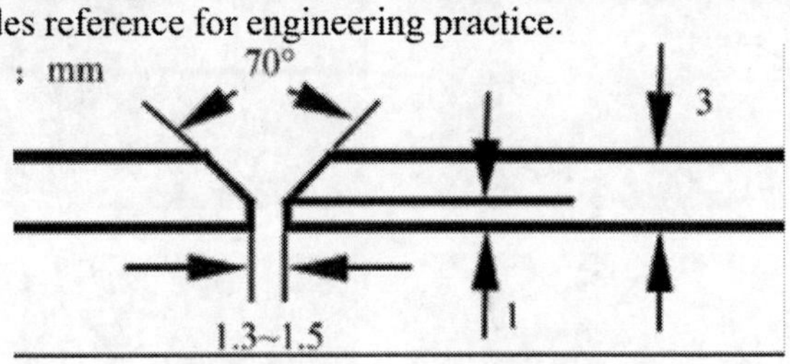


**Fig. 1** Type of the welding joint

**Tab. 1** Chemical compositions of 5052 aluminum alloy and ER 5356 welding wire wt %

| Material | Mg | Mn | Cr | Ti | Fe | Si | Cu | Zn | Be | Al |
|---|---|---|---|---|---|---|---|---|---|---|
| 5052 | 2. 4 | 0. 06 | 0. 17 | —1 | 0. 2 | 0. 12 | 0. 08 | 0. 08 | — | margin |
| ER 5356 | 5. 00 | 0. 12 | 0. 10 | 0. 13 | 0. 30 | 0. 25 | 0. 10 | 0. 10 | 0. 0003 | margin |

Note 1: – indicates that there is no such element.

After the welding, the metallographic samples were taken along the cross section of the weld. After grinding, polishing and corroding with Keller test strip, the microstructure was observed with a DMC 4000 /MC 170 optical microscope, and the microhardness was measured with a HXD-1000 /LCD Vickers hardness tester. The electrochemical corrosion sample was a Φ 14 mm × 3 mm wafer and was ground to 5,000 mesh with water sandpaper. For the PS-286 C electrochemical workstation, the auxiliary electrode is Pt, the reference electrode is saturated calomel electrode, the electrolyte is 3.5% NaCl aqueous solution, the potential measurement range is 1200-400 mV, and the potential scanning speed is 60 mV/min. Salt spray corrosion was carried out according to the "Artificial Atmosphere Corrosion Test Salt Spray Test" (GB / T 10125-2012), using a YFX-150 salt spray corrosion machine. The sample is cuboid of 20 mm × 10 mm × 3 mm, is ground up to 2,000 mesh with sandpaper, and is accurate to 0. 000 1 mg when weighing. When corroding, the temperature of the pressure tank is 47 °C, the experimental temperature is 35 °C, the spray pressure is 0.6 MPa, the corrosion time is 168 hours. During the process, the sample is taken, rinsed, dried and weighed every 24 hours.

The stretching test was carried out on the WDW-1000 electronic universal testing machine according to "Stretching Test Method for Welded Joints" (GB/T 2651-2008), and the moving speed of beam was 3 mm/min. The fracture morphology was observed with Hitachi SU 8020 scanning electron microscope.

## 3. Results and analysis

### *3.1. Microstructure of welded joints*

The microstructure of the welded joint during TIG welding at 100~120 A current is shown in Fig. 2. When the welding current is changed from 100 to 120 A, the microstructure of the 5052 welded joint is mainly composed of weld zone, fusion zone and heat affected zone, but the boundaries between the regions are not obvious. When the welding current is 100 A, 110 A and 115 A respectively, the matrix is equiaxed crystal, the weld zone is dendritic crystal, and the fusion zone and the heat affected zone are between the equiaxed crystal and the dendritic crystal. As the welding current increases from 100A to 115A, the fusion zone and the heat-affected zone become larger. The morphology of these regions is between equiaxed crystal and dendritic crystal, and is biased toward the finer dendritic crystal; when the welding current is 120 A, the heat affected zone is larger and the structure is larger, because the welding current is larger, the welding heat input is also increased, and the heat affected zone is enlarged and the structure is larger.

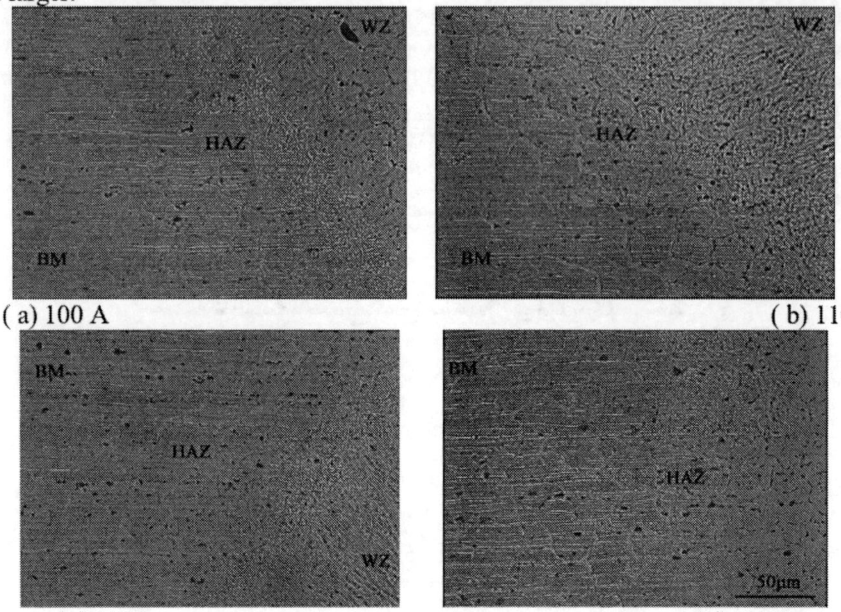


( a) 100 A ( b) 110 A

( c) 115 A ( d) 120 A

**Fig. 2** Microstructure of TIG joints with welding currents of 100～ 120 A

Note 1: WZ represents the weld zone; HAZ represents the heat affected zone; BM represents the Base Metal zone.

### *3.2. Mechanical properties of welded joints*

Table 2 shows the mechanical properties of TIG welded joints at the welding current of 100～120 A. The shrinkage of the joint section decreases as the welding current increases, indicating that the plasticity of the joint decreases as the welding current increases. As the welding current is increased from 100 A to 115 A, the tensile strength of the joint is gradually increased from 160. 4 MPa to 212.1 MPa, reaching 90.8% of the tensile strength of the base material. At 100-110A, the tensile strength of the joint is the most sensitive as the change of the welding current, and the tensile strength is increased from 160. 4 MPa to 201. 9 MPa. When the welding current is increased to 120 A, the tensile strength of the joint is 210. 2 MPa, which is almost unchanged compared with the joint at the welding current of 115 A.

**Tab. 2** Mechanical properties of TIG joints with welding currents of 100～ 120 A

| Welding current /A | Tensile strength /MPa | Percentage of base metal strength /% | Reduction of area /% |
|---|---|---|---|
| Base metal | 233.5 | | 50.5 |
| 100 | 160. 4 | 68.7 | 46.8 |
| 110 | 201.9 | 86.5 | 31.9 |
| 115 | 212.1 | 90.8 | 28.9 |
| 120 | 210.2 | 90.0 | 16.3 |

When the welding current is too small, the temperature of the molten pool is too low, and the solidification speed is increased, thus the solute in the molten pool is too late to be diffused and easy to get segregated, and the strength of the welded joint is reduced. When the welding current is increased, the degree of segregation is reduced and the mechanical properties are also improved. However, the increase of the welding current will make the grain in the weld zone coarse, and the heat affected zone will be "softened" because the heat is higher than the recrystallization temperature, which will also affect the strength of the joint [21-22]. This is also the reason why the tensile strength of the joint no longer increases when the welding current increases from 115 A to 120 A.

The microhardness of the TIG welded joint at 100~120 A welding current is shown in Figure 3. The HV of the fusion zone is the highest, 63 to 68; the HV of the matrix is 60 to 63; the HV of the weld zone is 50 to 55; A "softening" zone appears in the area close to the matrix, with a minimum HV of 51, called heat affected zone. These four regions are consistent with the microstructure of Figure 2: the matrix is equiaxed crystal, the weld zone is dendritic crystal, and the mixed region between equiaxed crystal and dendritic crystal. The hardness is the lowest in the weld zone and rises to the highest in the fusion zone. After the "softening" in the heat-affected zone, the hardness gradually increases toward the hardness of base metal (63 HV) as the distance from the center of the weld increases. The order of microhardness in the weld zone is 115 A>110 A>100 A>120 A; the order of microhardness in the fusion zone is 115 A >100 A>120A >110 A; the hardness of microhardness in the heat affected zone is 115 A>120 A>110A>100 A. In weld zone, fusion zone and heat affected zone of the TIG welded joint, the hardness is the highest when the welding current is 115 A.

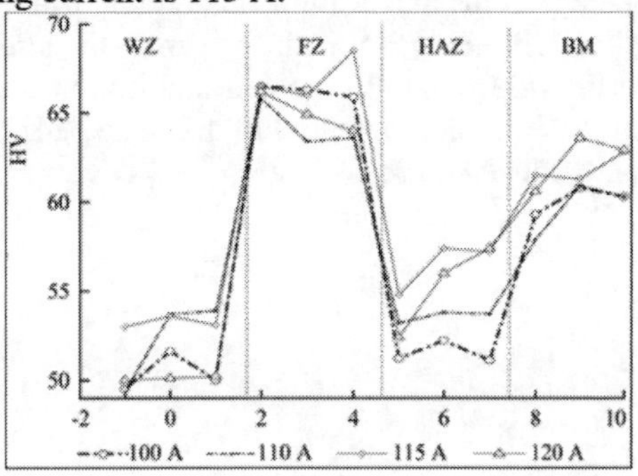


**Fig. 3** Microhardness of welding TIG joints with welding currents of 100~ 120 A (perpendicular to the welding direction)

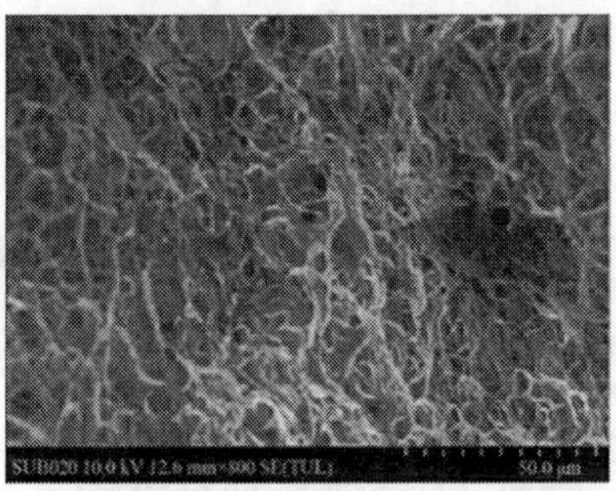

(a) 100 A

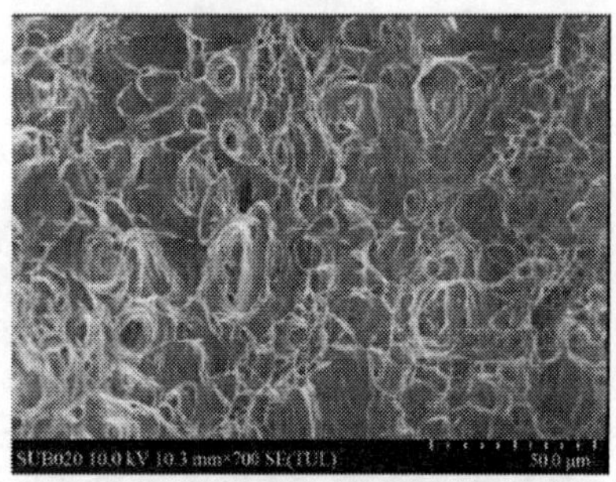

(b) 110 A

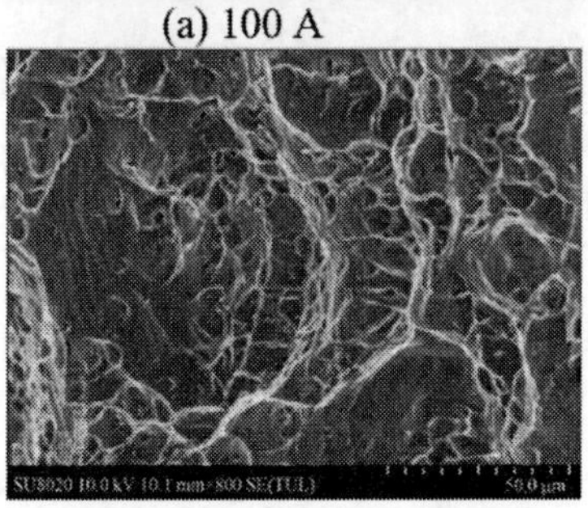

(c) 115 A

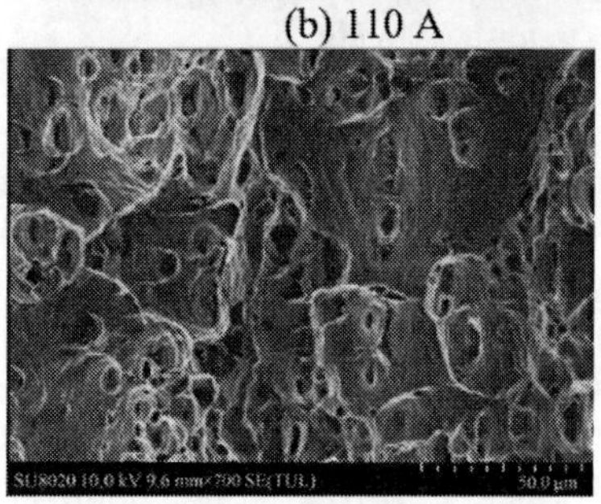

(d) 120 A

**Fig. 4** Tensile fracture morphology of TIG joints with welding currents of 100~ 120 A

Figure 4 shows the tensile fracture morphology of TIG welded joints at 100-120 A welding current. When the welding currents are 100 A and 110 A, it is a ductile fracture, and the dimple is smaller and shallower; but when the welding currents are 115 A and 120 A, it's mainly brittle fracture, and has a small number of dimples.

### *3.3. Corrosion resistance of welded joints*

As shown in Figure 5, the relationship between the weight-loss rate and time of the TIG welded joint in the salt spray box at the welding current of 100～120 A. The corrosion rate of the sample decreases with increasing corrosion time, and the weight loss rate of the sample is the highest during the 24 hours before corroding. Because the sample which is initially placed in the salt spray corrosion chamber is not protected by a dense oxide film, the corrosion rate is very fast; especially when the welding current is 120 A, the corrosion rate is at most 0.86 g / (m2 · h). The corrosion rate of the sample is significantly reduced from 24 to 96 hours, because the aluminum in the sample reacts with the oxygen in the chamber to form a dense oxide film, which is in the magnesium-rich region, while being eroded by the salt spray atmosphere at 35 °C. Magnesium will also react with oxygen in the tank to form a corrosion-resistant magnesium hydroxycarbonate film [23], and also copper and iron in the wire will shift the surface of the 5052 aluminum alloy from the corrosion potential to the right [ 24], improved corrosion resistance of the 5052 aluminum alloy, and reduced corrosion rate.

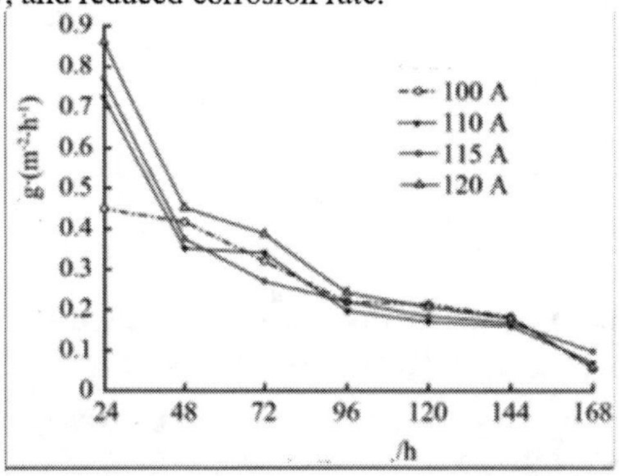


**Fig. 5** kinetics curve of salt spray Corrosion of TIG welded joints with current of 100～ 120 A

Figure 6 shows the morphology of TIG welded joints after being corroded by salt spray for 168 h at 100 to 120 A welding current. The weld zone of welded joints with welding currents of 100 A and 110 A is mainly pitting, and the weld zone of welded joints with welding currents of 115 A and 120 A is mainly pitting and intergranular corrosion. In the environment of neutral salt spray corrosion, as the TIG welding current of the 5052 aluminum alloy increases, the tendency of intergranular corrosion in the weld zone gradually increases. Intergranular corrosion greatly weakens the bond between the grains and reduces the strength of materials.

The Cl- in the small droplets of NaCl has a destructive effect on the passivation film on the surface of the 5052 sample. The oxygen in the salt spray corrosion chamber forms OH- and O2-in the small droplets of NaCl through the reaction formulas (1) and (2). In addition, since the chemical sites of Al3+ and Mg2+ in the NaCl solution are not equal to those of 5052 aluminum alloy, Al3+ and Mg2+ in the 5052 aluminum alloy will be transferred to the solution, through the reaction formulas (3) and (4) Al3+ and Mg2+ are formed in NaCl solution. Finally, corrosion products such as Al2O3, MgO, Al(OH)3, and Mg(OH)2,etc. are formed. Combined with the A-point energy spectrum analysis, see Table 3, it is known that the corrosion product is mainly Al2O3. Because in the 5052 aluminum alloy, there is less alloying element of magnesium, and the spray is neutral, Mg2+ and OH- are produced less, and the formed Mg(OH)2 does not exceed its solubility in brine [24], so The corrosion product observed was mainly Al2O3.

$$O2+ 2H2O + 4e^{-} \rightarrow 4\,OH^{-} \quad (1)$$

$$O2+4e^{-} \rightarrow 2O2^{-} \quad (2)$$

$$Al - 3e^{-} \rightarrow Al3+ \quad (3)$$

$$Mg - 2e^{-} \rightarrow Mg2+ \quad (4)$$

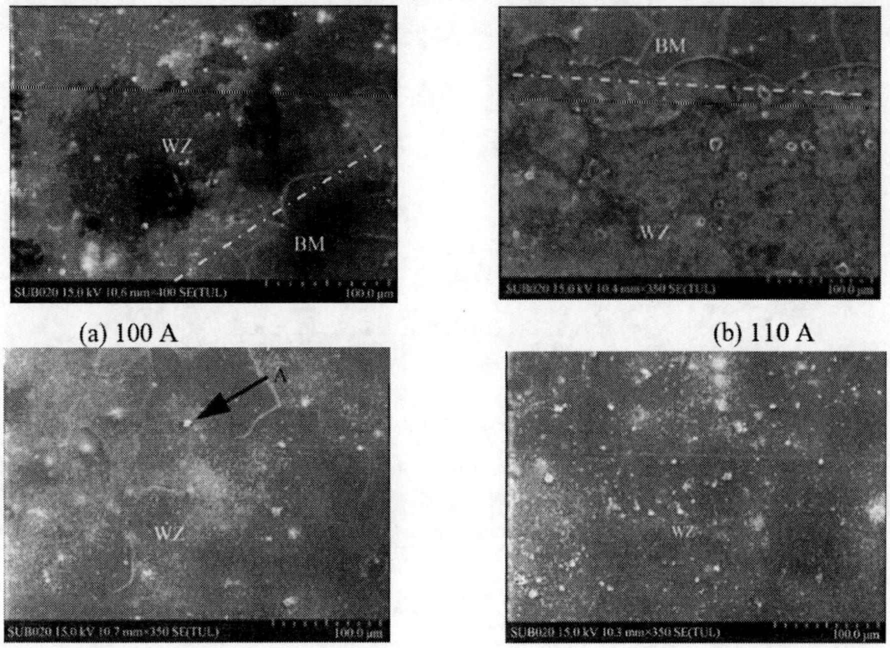


(a) 100 A (b) 110 A

(c) 115 A (d) 120 A

**Fig. 6** Microstructure of TIG joints with welding current of 100~ 120 A after being corroded in Salt spray for 168 h

Figure 6 shows the morphology of TIG welded joints after being corroded by salt spray for 168 h at 100 to 120 A welding current. The weld zone of welded joints with welding currents of 100 A and 110 A is mainly pitting, and the weld zone of welded joints with welding currents of 115 A and 120 A is mainly pitting and intergranular corrosion. In the environment of neutral salt spray corrosion, as the TIG welding current of the 5052 aluminum alloy increases, the tendency of intergranular corrosion in the weld zone gradually increases. Intergranular corrosion greatly weakens the bond between the grains and reduces the strength of materials.

**Tab. 3** A-point Energy spectrum analysis in Fig. 6

| Element | Mass percentage /wt% | Atomic percentage /% |
|---|---|---|
| O | 53.13 | 65.77 |
| Mg | 0.62 | 0.50 |
| Al | 44.94 | 32.99 |
| Cl | 1.32 | 0.74 |

Figure 7 shows the Tafel curve for a TIG welded joint at the welding current of 100 ~ 120 A. Table 4 shows the corrosion current density and corrosion potential of TIG welded joints at the welding current of 100 ~ 120 A. The corrosion potential of the TIG welded joint at 100 to 120 A is between -720 and -680 mV, and the corrosion current density is -5.22 to -4.48 mA/cm2. The corrosion potential and corrosion current density of all samples were slightly shifted to the right compared with the parent metal.

The electrochemical corrosion tendency of all joints and the corrosion rate after corrosion were smaller than that of the parent metal. When the welding current is 115 A, corrosion tendency of the joint is the smallest; when the welding current is 100 A, the corrosion rate after corrosion is the smallest. When the welding current is 110 A, the corrosion current density and the corrosion potential are both negative; compared with the joints under the other welding currents, the corrosion tendency and corrosion rate are both at a high level, and the corrosion resistance is the worst.

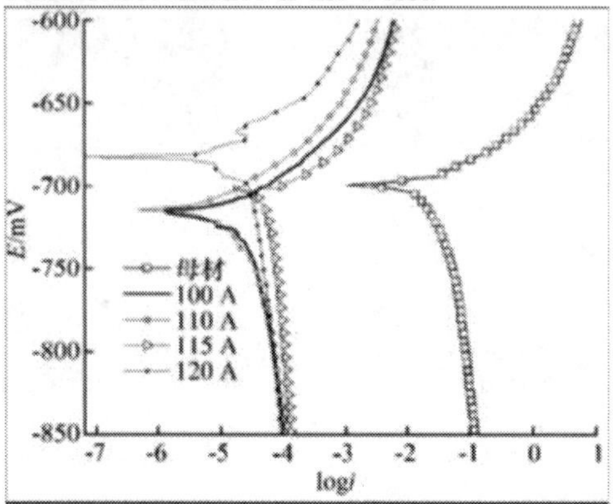


**Fig. 7** Tafel curves of the TIG welded joints at different welding currents

**Tab. 4** Corrosion current densities and the corrosion potentials of TIG welded joints at different welding currents

| welding current /A | Corrosion current density /( mA·cm—2) | Corrosion voltage /mV |
|---|---|---|
| 0(Base metal) | −5.72 | −725 |
| 100 | −4.78 | −720 |
| 110 | −5.12 | −716 |
| 115 | −5.22 | −680 |
| 120 | −5.07 | −700 |

## 4. Conclusion

All manuscripts must be in English, also the table and figure texts, otherwise we cannot publish your paper. Please keep a second copy of your manuscript in your office. When receiving the paper, we assume that the corresponding authors grant us the copyright to use the paper for the book or journal in question. Should authors use tables or figures from other Publications, they must ask the corresponding publishers to grant them the right to publish this material in their paper. As show in Fig. 1 and Table 1, three scheme comparing.

TIG welding of 5052 aluminum alloy at the welding current of 100～120 A: when the welding current is 115 A, the welded joint has better mechanical properties. The tensile strength of the joint is 212.1 MPa, which is 90.8 % of the tensile strength of the base metal, and the reduction ratio of the section is 28.9%, which is nearly half of the reduction of the section shrinkage of 50.3 % of the base metal.

The salt spray corrosion after 168 h showed that when the welding current is 100 A and 110 A, TIG welded joint of the 5052 aluminum alloy is mainly pitting corrosion and the corrosion was average; when the current is 115 A and 120 A, intergranular corrosion Intensified, and converted to pitting and intergranular corrosion. When the welding current is between 100 and 120 A, the change of corrosion current density is not obvious, which is related to the unevenness of the weld structure.

**Acknowledgments**
This work was financially supported by Beijing Polytechinc Research fund.

**References**

[1] PEREIRA A, TREVELYAN P M J, THIELE U, et al. Mechanical and electrochemical characteristics in sea water of 5052－O aluminum alloy for ship [J].Chinese Journal of Nonferrous Metals (English) , 2013, 23( 3) : 636-641.

[2] KIM S J, PARK J C, CHONG S O. Investigation on Corrosion Protection Potential by Electrochemical Experiment in SeaWater of 5052－O Al Alloy for Leisure Ship [M]. Mokpo city, South Korea: Mokpo City Pivision of Morinc System Engineering, 2010.

[3] SINGH D, DASH P K. Role of damage parameters in remaining life assessment of composites in aircraft structures [J]. ProcediaEngineering, 2012, 38: 2702-2712.

[4] MIN H E, FUGUO L I, WANG Z. Forming limit stress diagram prediction of aluminum alloy 5052 based on GTN modelparameters determined by in situ tensile test [J]. Chinese Journal of Aeronautics (English), 2011, 24( 3) : 378-386.

[5] KIM J S, KIM I J, KIM Y G. Optimization of welding current waveform for dissimilar material with DP590 and Al5052 by Delta-spot welding process [J]. Journal of Mechanical Science & Technology, 2016, 30(6) : 2713-2721.

[6] KWON Y J, SHIM S B, PARK D H. Friction stir welding of 5052 aluminum alloy plates [J]. Transactions of Nonferrous Metals Society of China, 2009, 19(s1) : 23-27.

[7] ZHANG Z, YANG X, ZHANG J, et al. Effect of welding parameters on microstructure and mechanical properties of friction stir spot welded 5052 aluminum alloy [J]. Materials & Design, 2011, 32( 8-9) : 4461-4470.

[8] ZHANG H J, WANG M, ZHU Z, et al. Impact of shoulder concavity on non-tool-tilt friction stir welding of 5052 aluminum alloy [J]. International Journal of Advanced Manufacturing Technology, 2018: 96( 1-4) : 1497-1506.

[9] REDDY P J, KAILAS S V, SRIVATSAN T S. Effect of Tool Angle on Friction Stir Welding of Aluminum Alloy 5052:Role of Sheet Thickness [J]. Advanced Materials Research, 2012, 410( 1) : 196-205.

[10] SONG C Y, PARK Y W, KIM H R, et al. The use of Taguchi and approximation methods to optimize the laser hybridwelding of a 5052－H32 aluminium alloy plate [J]. Proceedings of the Institution of Mechanical Engineers Part B Journal of Engineering Manufacture, 2008, 222( 4) : 507-518.

[11] LI Y Q, ZHAO X H, ZHAO H, et al. Study on the effect of welding current during laser beam-resistance seam welding of aluminum alloy 5052 [J]. China Welding, 2008, 17( 2): 42-46.

[12] ZHANG H G, JIN X Z, CHEN G Y, et al. Study on the burning loss of magnesium element in fiber laser welding aluminum alloy 5052 [J]. Laser Technology, 2012, 36( 6) : 713-718.

[13] ZHANG J, ZHANG W, LIU W. Research on the welding process of aluminum alloy based on high power fiber laser [C]/ / Society of Photo-Optical Instrumentation Engineers. Society of Photo-Optical Instrumentation Engineers ( SPIE) Conference

[14] Series. Shanghai, China: Los Anaeles SPIE, 2017: 10339.

[15] Xiao Rongshi, Chen Wei, Zuo Tiezhen. New progress in laser welding of high-strength aluminum alloy [J]. Applied Laser, 2002, 22(2): 206-208.

[16] Huang Wangfu, Huang Gang. Guide for welding aluminum and aluminum alloys [M]. Changsha: Hunan Science and Technology Press, 2004: 125-126.

[17] KUMAR A, SUNDARRAJAN S. Optimization of pulsed TIG welding process parameters on mechanical properties of AA 5456 Aluminum alloy weldments [J]. Materials & Design, 2009, 30( 4) : 1288-1297.

[18] Xu Zheng. Application research of 7005 /7075 medium and high strength aluminum alloy [D]. Nanjing: Nanjing University of Science and Technology, 2007.

[19] Wang Xiling. Research on TIG welding process and performance of 7xxx aluminum alloy sprayed [D]. Zhenjiang: Jiangsu University of Science and Technology, 2011.

[20] Liao Chuanqing, Su Guoyou, Gao Yanfang, et al. Microstructure and mechanical properties of

7075 /5A06 dissimilar aluminum alloy TIG welded joints [J]. Chinese Journal of Nonferrous Metals, 2015, 25(1): 43-48.
[21] Wang Peiji, Chen Kanghua, Jiang Huili, et al. Microstructure and properties of 7003 aluminum alloy TIG welded joints [J]. Powder Metallurgy Materials Science and Engineering, 2016, 21(6) : 832-839

# Experimental study on dielectric properties of SiC material and temperature distribution of rubber materials containing silicon carbide coated metal via microwave heating

**Qingling Li, Ruzhou Ye, Ruizhu Jin, Hailong Chen, Guangzhen Xia and Tao Li**

College of Electromechanical Engineering,Qingdao University of Science and Technology, Qingdao, China

**Abstract.** Microwave heating has not been widely used in the heating process of the rubber products containing metal skeleton, because the metal reflects microwaves that leads to a large amount of heat in the metal-rubber contact surface due to the skin effect and the eddy current effect in the microwave field. This experiment investigated the effects of microwave frequency and temperature on the dielectric parameters of silicon carbide coating materials. The microwave heating process of rubber composites containing metal skeleton coate d with SiC mixture was also investigated experimentally. The results are as follows: As the microwave frequency increases, the dielectric parameters of the SiC coating material increase slowly first, when the frequency reaches a certain point, the dielectric parameters increase sharply to a maximum after that fall rapidly; Based on microwave frequency of 2450MHz, the dielectric constants of three different compositions of SiC coating materials decrease with increasing temperature; The temperature of the rubber composite containing the steel wire coated with the SiC is lower than the rubber composite containing the steel wire naked. The larger the content of SiC, the more uniform the temperature distribution, and the more effective the weakening of the sparks phenomenon caused by the skin effect.

**Key words:** rubber composites; microwave heating; metal skeleton; dielectric parameters; silicon carbide.

## 1. Introduction

Rubber is widely used due to its beneficial proprieties like strength, long lasting, water resistance and heat resistance [1]. In conventional vulcanization process, rubber is heated by hot air or water. Because of the lower conductivity of rubber material, inner temperature of rubber products is frequently less than surface temperature, which leads to the nonuniformity in temperature distribution of rubber material [2]. Finding more efficient ways to heat rubber matierial is essential in research and particular industry.

Microwave heating is one of new advancing techniques for heating rubber material. Its major advantages are rapidly emerging, precise control, fast start-up, and so-called volumetric heating, because the microwave energy that penetrates into the material will be absorbed and converted into internal energy in the material [3]. In recent years, microwave technology has been used increasing rapidly in

Content from this work may be used under the terms of the Creative Commons Attribution 3.0 licence. Any further distribution of this work must maintain attribution to the author(s) and the title of the work, journal citation and DOI.
Published under licence by IOP Publishing Ltd

rubber industry, including drying, vulcanization and preheating, but these rubber products can only be without metal skeleton [4].

The metals can reflect microwave radiation, where delocalized electrons are free to move in relatively broad regions. The kinetic energy of some electrons may increase enabling them to jump out of the material, resulting in the ionization of the surrounding atmosphere. At a macroscopic level, this phenomenon is perceived as sparks or electric arcs formation. The problem of metal sparks must be dealt with when microwave technology is used in heating rubber products containing metal skeleton materials. One way to solve this problem is to coat the metal surface with absorbing materials.

The materials which interact with microwaves to generate heat are called microwave absorbers. The ability to heat material in the presence of a microwave field is defined by its dielectric loss tangent: $\tan\delta = \varepsilon''/\varepsilon'$. The dielectric loss tangent is composed of two parameters, the dielectric constant (or real permittivity), ε', and the dielectric loss factor (or imaginary permittivity), ε"; i.e., $\varepsilon = \varepsilon' - i\varepsilon''$, where ε is the complex permittivity. The dielectric constant (ε') determines the amount of reflection and absorption of incident energy, while the dielectric loss factor (ε") measures the electric energy dissipation in form of heat within the material [5].

Silicon carbide (SiC) is a typical semiconductor material. Due to its low density, good oxidation resistance and chemical stability, excellent mechanical strength and adjustable electrical conductivity, it is an ideal choice for lightweight, high temperature resistant microwave-absorber.[6-12] Hongyu Wang[13]et al. compared the microstructure, mechanical properties, composite dielectric constant and microwave absorption properties of two different types of silicon carbide and epoxy composites of KD-ISiCf and SLF-SiCf. It was found that the absorption rates of the two composites were not high.

Yongshan Wei [14] et al. proposed a new method to improve the microwave absorption properties of SiC fibers. The properties of FeCo/SiCf composites in low temperature and high vacuum environments were investigated by RF magnetron sputtering. Compared with SiC fiber, FeCo/SiCf composites have good microwave absorption properties in the microwave frequency range.

Jae-Hun Choi [15] et al. mainly studied the mechanical properties and electromagnetic properties of SiC/epoxy composites, and measured the electromagnetic properties of epoxy-based composites under different pressures. The microwave absorption properties of single-layer and multi-layer composites were analyzed and found to have excellent microwave absorption properties.

Wei Zhou [16] et al. used nickel nanoparticles as catalysts to carry out catalytic chemical vapor deposition (CCVD) on the surface of carbon fibers to form silicon carbide nanofibers (SiCNFs). And the test found that the microwave absorption characteristics were effectively improved compared with pure carbon fibers.

The aim of the study is to investigate dielectric properties of SiC mixture and it's effects on the temperature distribution of rubber composites containing steel wire during microwave heating.

## 2. Experimental

### *2.1. Materials and apparatus*

Abrasion resistant compound; Micron-sized silicon carbide (Average particle size 10 μ m); Round iron sheet (d=25mm, $\delta$=1mm); Phenolic resin powder; N, N-Dimethylformamide.

The microwave heater(Mls1200, operating frequency is 2450MHz); Fiber optic temperature sensor(Temperature range -50 ℃ ~200 ℃ ). Compared with traditional

temperature sensor, optical fiber temperature sensor has the advantages of high sensitivity, small volume, light weight, easy bending, no electromagnetic interference and good corrosion resistance; Impedance Analyzer (Keysight E4991B Impedance Analyzer, Frequency range 1MHz~3GHz); Scanning electron microscope (JSM-7001F); The SZQ type tetrahedral preparator(Tianjin Kexin Testing Machine Factory); Electric oven(KWS0709J-02H(XP)).

### *2.2. Preparation*

Silicon carbide coating solution: The phenolic resin and the solvent N,N-dimethylformamide (DMF) were mixed uniformly, and micron-sized silicon carbide powder was added to make silicon carbide solutions with 200g/L, 400g/L and 600g/L respectively.

Silicon carbide coating sample: A suitable amount of silicon carbide solution was dropped from a dropper onto a Teflon plate to form a circle with a diameter greater than 12 mm.

Iron sheet coated with silicon carbide: First, stir the pre-configured silicon carbide coating solution evenly with a stirrer, then, the 200g/L, 400g/L, and 600g/L silicon carbide coating solutions are respectively applied to a plurality of iron sheets by the SZQ type applicator. Then heat them in an electric oven. Set 5 groups for each composite due to different coating thickness. The coating solution of 200g/l is set as group A; the coating solution of 400g/l is set as group B; and the coating solution of 600g/l is set as group C. Grouped as shown in the following table:

**Tab. 1** The thickness of silicon carbide coating

| Number | A1 | A2 | A3 | A4 | A5 | B1 | B2 | B3 |
|---|---|---|---|---|---|---|---|---|
| Thickness (mm) | 0.1032 | 0.1553 | 0.2022 | 0.2543 | 0.3019 | 0.1041 | 0.1562 | 0.2013 |
| Number | B4 | B5 | C1 | C2 | C3 | C4 | C5 | |
| Thickness (mm) | 0.2540 | 0.3011 | 0.1029 | 0.1549 | 0.2031 | 0.2549 | 0.3014 | |

### *2.3. Testing of dielectric performance parameters.*

The test process is carried out in two cases, at a normal temperature of 19℃ and temperature rise of 30℃~120℃.

Before the normal temperature test, the RF coaxial extension cable needs to be removed from the heating furnace; The measurement frequency range from 1MHz to 3GHz, the interval is 5MHz, and the data is tested and recorded.

In the heating experiment, the furnace is heated to a specified temperature for a period of time. When the temperature sensor measures the temperature of the sample to reach the test temperature, the scanning measurement starts at the set frequency; after the measurement is completed, continue to heat up, maintain temperature, scan and measure until the last temperature point. After the measurement is completed, save the measurement data. The temperature rise is measured at a temperature interval of 5℃, and the measurement frequency is 2450 MHz. Test and record the data.

### *2.4. Study on Temperature Field of Microwave Heating Rubber Composite*

The rubber sheets are made into sheets of 70 mm in length, 70 mm in width and 6 mm in thickness. Take two pieces of rubber sheet for each test, the iron piece is placed in the center of the two rubber pieces, and two pieces of rubber piece are wrapped around the piece of iron. The temperature measurement point is located at the outer edge of the iron piece in the two rubber sheets. The rubber sheets are placed in the microwave generator, and turn on the fiber temperature online monitoring device, then start measuring and recording the temperature data. The heating power is 100W, and the data is recorded every second. The heating is stopped when the temperature sensor detects that the temperature reaches 100 ℃ . Observing whether there is sparking during the experiment, and study the temperature rise law of the iron carbide rubber composite containing different compositions of silicon carbide.

## 3. Results and discussion

### *3.1. Effect of Microwave Frequency on Dielectric Properties of Silicon Carbide Materials*

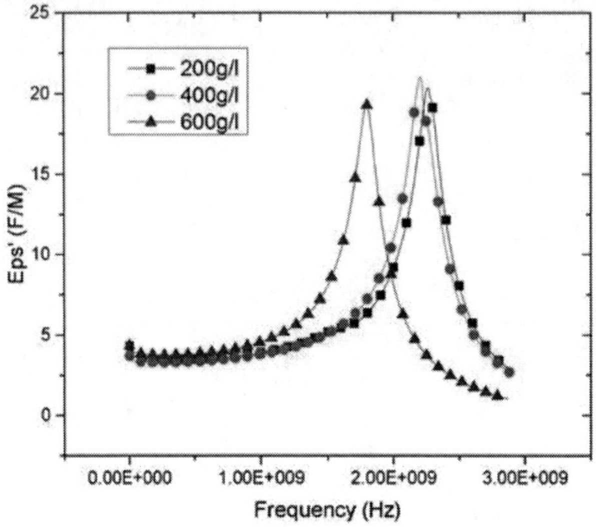


**Fig. 1** The $\varepsilon$ of different content silicon carbide coating material

As can be seen from Figure 1, at low frequency of the microwave, the $\varepsilon'$ is relatively stable; when the frequency increases to a certain value, the $\varepsilon'$ increases sharply and a peak appears. And the higher the silicon carbide content, the earlier the peak appears. As the frequency continues to increase, the value drops sharply and then gradually stabilizes.

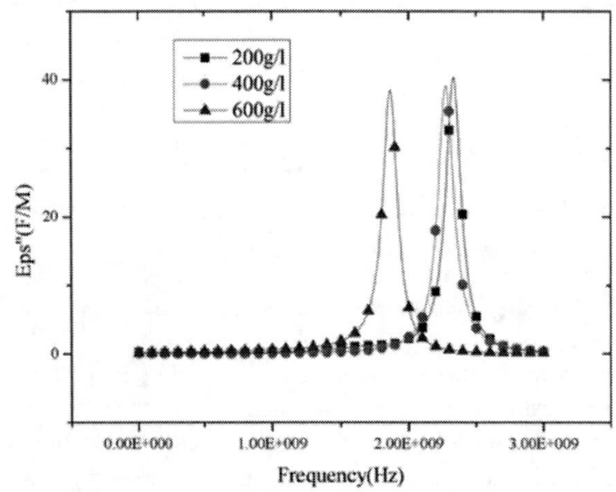


**Fig. 2** The $\varepsilon''$ of different content silicon carbide coating materials

As can be seen from Figure 2, as the microwave frequency increases, three curves are roughly the same. At the low frequency of the microwave, the $\varepsilon''$ rises slowly, and the value of the 600g/L silicon carbide coating material is slightly larger than the other two. When the frequency increases to a specific value, since the polarization gradually cannot keep up with the electric field change, the loss gradually becomes larger, and thus the value increases sharply, and a peak appears. And 600g/L silicon carbide coating material peaks earlier than the other two. As the frequency continues to increase, the value drops abruptly and then slowly returns to a plateau value.

IOP Conf. Series: Materials Science and Engineering **772** (2020) 012012 doi:10.1088/1757-899X/772/1/012012

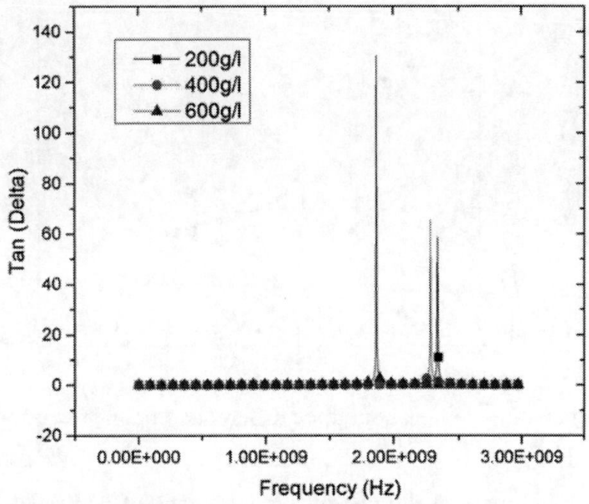


**Fig. 3** The tan$\delta$ of different content silicon carbide coating material

As can be seen from Figure 3, the trend of the three contents of the silicon carbide coating material is consistent. First, it is relatively stable. As the frequency continues to increase, the value of tan$\delta$ increases abruptly and a peak appears, and the higher the silicon carbide content, the earlier the peak appears. This indicates that as the content of silicon carbide in the coating material increases, the tan$\delta$ of the silicon carbide coating material becomes larger.

*3.2. Effect of Temperature on Dielectric Properties of Silicon Carbide Materials*

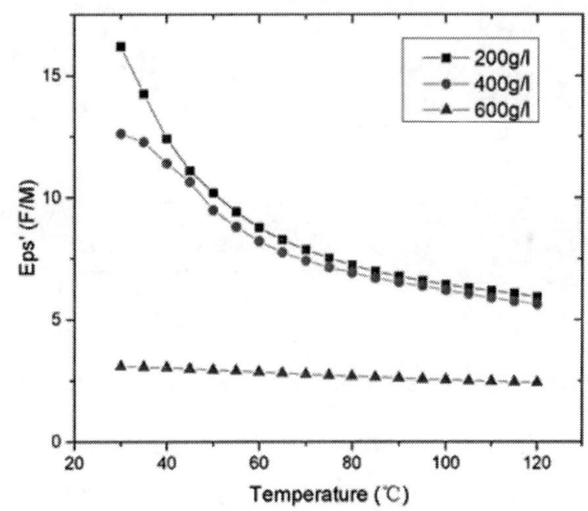


**Fig. 4** The effect of temperature on $\varepsilon$ of three different contents of silicon carbide coating at 2450MHz

As shown in Figure 4, at the microwave frequency of 2450 MHz, the $\varepsilon$ of three different contents of SiC coating materials are decreasing from 30 ℃ to 120℃.

The $\varepsilon$ of 200g/l SIC coating material is higher than the other two.

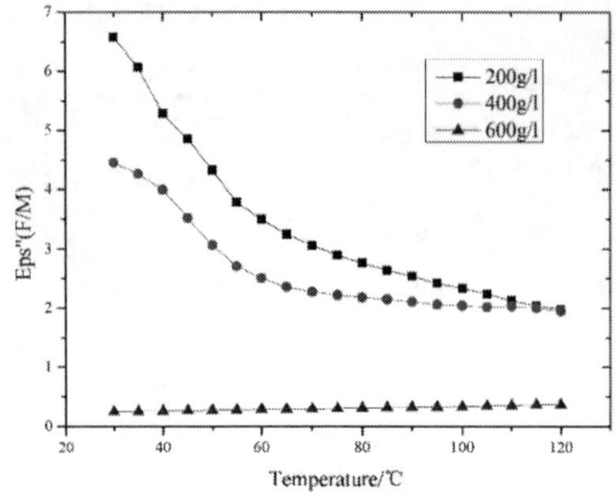


**Fig. 5** The effect of temperature on $\varepsilon''$ about three different components of ' silicon carbide coating at 2450MHz

As can be seen from Figure 5, at the microwave frequency of 2450MHz, the $\varepsilon''$ of three different contents of SiC coating materials decrease with increasing temperature. Among them, 200g/L coating material curve has the largest slope, indicating that the $\varepsilon''$ of the coating material of this composite is sensitive to temperature change at this frequency. The $\varepsilon''$ of the coating material of 200 g/L is greater than the other two, it is indicated that it has a good microwave dissipation capability.

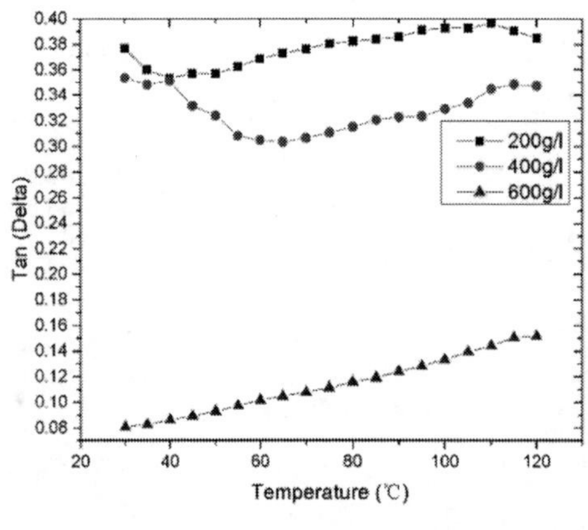


**Fig. 6** The effect of temperature on tan$\delta$ about three different components of silicon carbide coating at 2450MHz

It can be seen from Fig. 6 that when the microwave frequency is 2450MHz, the tan$\delta$ of three different compositions of silicon carbide coating materials increase with increase of temperature. The smaller the content of silicon carbide coating, the larger the value. That is, at the microwave frequency of 2450 MHz, 200 g/L silicon carbide coating material absorbs microwaves more easily, and 600 g/L silicon carbide coating material is more difficult to absorb microwaves.

### *3.3. Temperature field study of rubber composite*

Microwave heating experiment was carried out on the rubber composites containing iron sheets. During the first few seconds of the experiment, there was an arc sound in the cavity of the microwave generator. At the same time, the ventilator of the microwave generator emits a pungent smell of scorching. The fiber optic thermometer detected a sharp up and down fluctuation in temperature. At this time, stop heating and took out the rubber sheet, and it was found that the rubber was obviously scorched. In combination with the arc sound heard and the smell of the burnt rubber smelled before, it can be judged that the rubber composite has a "discharge and spark" phenomenon. The charred rubber in microwave field can be seen in figure 7.

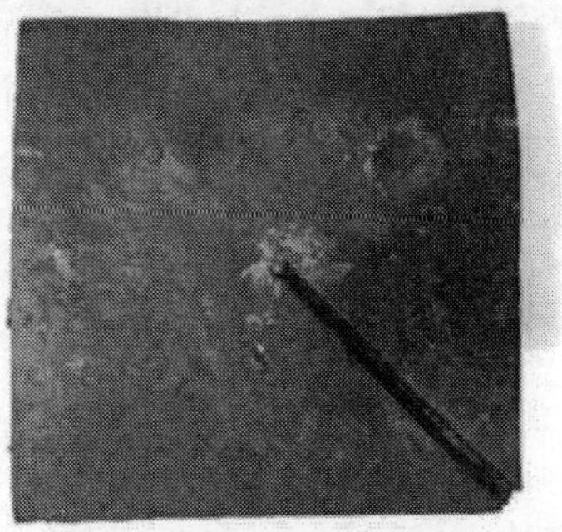

**Fig. 7** The charred rubber in microwave field

The temperature rise law of the rubber composites containing silicon carbide coated iron sheet measured by the experiment was obtained. The results are as follows:

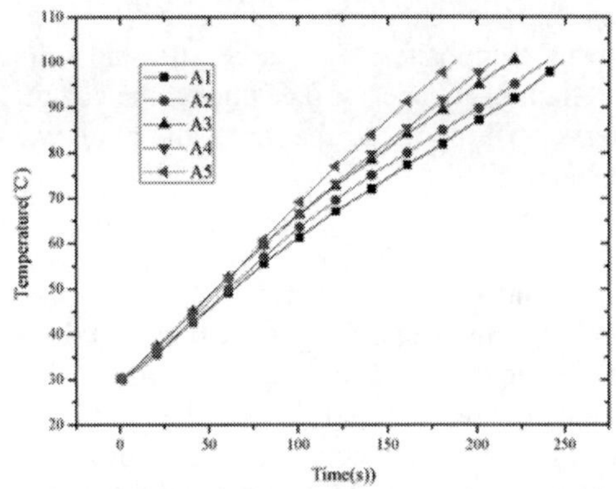


**Fig. 8** Temperature rise curve of rubber composites with different thickness silicon carbide coating (The content of silicon carbide is 200g/L)

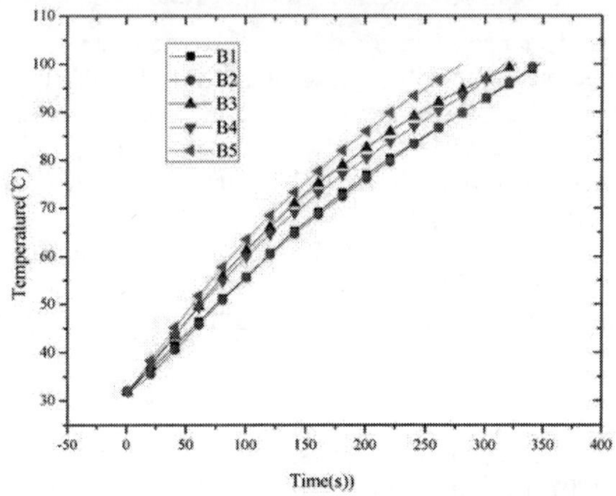


**Fig. 9** Temperature rise curve of rubber composites with different thickness silicon carbide coating (The content of silicon carbide is 400g/L)

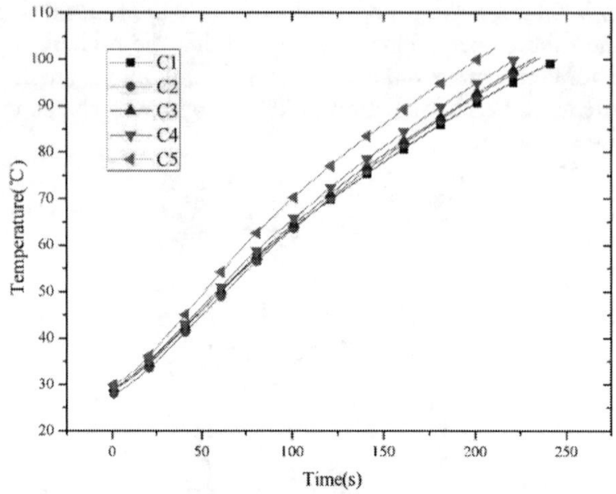


**Fig. 10** Temperature rise curve of rubber composites with different thickness silicon carbide coating (The content of silicon carbide is 600g/L)

It can be seen that there is no "ignition" phenomenon in the rubber composites coated with the silicon carbide coating during microwave heating process. And the thickness of the silicon carbide coating has a significant effect on the temperature of the rubber composites: The thicker the silicon carbide coating, the higher the temperature of the composites. This is because the silicon carbide coating can absorb microwave energy and convert it into heat energy. The thicker the coating, the higher the silicon carbide content, the more the microwave absorption, so the larger the microwave energy to be converted, the higher the temperature of the composites.

## 4. Conclusion

The purpose of this study is to investigate the effect of silicon carbide coating materials on the temperature distribution of rubber composites. The conclusions obtained are as follows.

(1) At room temperature, in the microwave frequency range of 1MHz to 3GHz, the dielectric properties of the silicon carbide coating material do not change much at low frequencies. When the frequency increases to a certain value, the dielectric parameter peaks and then fall sharply, and finally stabilize. And the higher the silicon carbide content, the larger the parameter.

(2) At the microwave frequency of 2450MHz, in the temperature range of 30 ℃to 120 ℃ , the $\varepsilon'$ and the $\varepsilon''$ of three different compositions of silicon carbide coating materials are continuously decreasing. And to a certain extent, the larger the silicon carbide content, the smaller the value. And the tan$\delta$ increases with increasing temperature, indicating that the ability of the material to absorb microwaves increases with increasing temperature, and to a certain extent, the lower the content of silicon carbide coating, the stronger the ability of the coating material to absorb microwaves.

(3) The rubber composites coated with silicon carbide coating material heating by microwave is studied and the conclusions are as follows:

There is no "ignition" phenomenon in the rubber composite coated with the silicon carbide coating during microwave heating. It is indicated that the absorbing material coated with silicon carbide can effectively prevent the phenomenon of "ignition". And the thicker the silicon carbide coating, the higher the temperature of the composite.

**Acknowledgements**

Great support was given by Shandong provincial Natural Science Foundation, China during the research. It is a part of the items (ZR2013EEL009). And Technology development Program of Shandong Province. It is a part of the items (2012GSF11604).

IOP Conf. Series: Materials Science and Engineering **772** (2020) 012012 doi:10.1088/1757-899X/772/1/012012

**References**


[1] Lee S.Y, You J.H, Lee Y.S, Kim H.S, Cheong S.K and Shin K.H 2016 Journal of the Korean Society of Safety 31(1) pp 19–24.

[2] Chen H.L, Li T, Liang Y, Sun B and Li Q.L, Heat and Mass Transfer ,201753(3) pp 1051–1060.

[3] P. Keangin et al. Analysis of temperature profile and electric field in natural rubber glove due to microwave heating: effects of waveguide position[C]. 2018 IOP Conf. Ser.: Mater. Sci. Eng. 297.

[4] J.A. Menéndez*, A. Arenillas, B. Fidalgo, Y. et al.. Microwave heating processes involving carbon materials[J]. Fuel Processing Technology, 2010, 91 (1), 1-8.

[5] Meredith R. Engineers' Handbook of Industrial Microwave Heating. The Institution of Electrical Engineers, London, UK 1998.

[6] Fabiula D.B.de Sousa, Carlos H. Scuracchio. et al. Devulcanization of waste tire rubber by microwaves[J]. Polymer Degradation and Stability,2017,138:169-181.

[7] K. Osouli-Bostanabad, E. Hosseinzade, A. Kianvash, A. Entezami, Modified nano-magnetite coated carbon fibers magnetic and microwave properties[J], Appl. Surf. Sci. 2015, 356: 1086-1095.

[8] Dou Y, Li J, Fang X, Jin H, Cao M, The enhanced polarization relaxation and excellent high-temperature dielectric properties of N-doped SiC, Appl. Phys. Lett.2014,104: 052102-052104.

[9] Peng C.H, Shiu Chen P, Chang C.C, High-temperature microwave bilayer absorber based on lithium aluminum silicate/lithium aluminum silicate-SiC composite, Ceram. Int. 2014,40 :47-55.

[10] Yuan J, Yang H.J, Hou Z.L, et.al. Ni-decorated SiC powders: enhanced high-temperature dielectric properties and microwave absorption performance, Powder Technol. 2013, 237: 309–313.

[11] Han M, Yin X, Duan W, S. et.al. Hierarchical graphene/SiC nanowire networks in polymer-derived ceramics with enhanced electromagnetic wave absorbing capability[J]. Eur. Ceram. Soc, 2016,36: 2695–2703.

[12] Wu R, Zhou K, Yue C.Y, et.al. Recent progress in synthesis, properties and potential applications of SiC nanomaterials, Prog. Mater. Sci, 2015,72: 1–60.

[13] Wang H.Y, Zhu D.M, Wang X.F, Influence of silicon carbide fiber (SiCf) type on the electromagnetic microwave absorbing properties of SiCf/epoxy composites[J]. Composites: Part A, 2017,93: 10-17.

[14] Wei Y.S, Yue J.L, Tang X.Z, Enhanced microwave-absorbing properties of FeCo magnetic filmfunctionalized silicon carbide fibers fabricated by a radio frequency magnetron method[J]. Ceramics International, 2017, 43: 16371-16375.

[15] Jae-Hun Choi, Young-Woo Nam, et.al. Characteristics of silicon carbide fiber-reinforced composite for microwave absorbing structures [J]. Composite Structures, 2018

[16] Zhou W, Long L, Xiao P, et al. Silicon carbide nano-fibers in-situ grown on carbon fibers for enhanced microwave absorption properties [J]. Ceramics International, 2017, 43:5628-5634.

IOP Conf. Series: Materials Science and Engineering **772** (2020) 012013 doi:10.1088/1757-899X/772/1/012013

# Dynamic stability analysis and Experiment of the Mobile Elevating Work Platforms

**Xuhong Gao[1], Di Sun[2, *], Yan Ge[1], Xiantao Yang[1] and Jin Li[1]**

[1]Liaoning Aerospace Linghe Automobile Co., Ltd, Liaoning, China
[2]Marine Engineering College, Jimei University, Xiamen, China

*Corresponding author e-mail: sundi6329@sina.com

**Abstract.** In order to calculate dynamic stability of the Mobile Elevating Work Platforms (MEWP) during driving across obstacles and operating on the slope, the analysis method based on mathematical models and simulation models of the MEWP was put forward, then the MEWP dynamic simulation models where the tires and booms were set as flexible bodies were made by ADAMS. Four kinds of the most dangerous working conditions were analyzed, and then the dynamic curves of tire loads and dynamic stability safety coefficient were obtained. Next, the dynamic stability of the MEWP during operating on the slope was verified by field experiments, and the results show that the minimum dynamic stability safety coefficient is 0.157. By comparison and analysis of dynamic stability safety coefficient between simulation analysis and field experiment, it can be seen that the deviations are all positive and less than 5%, therefore the results show that the dynamic stability of the MEWP during driving across obstacles are satisfied. At the same time, the analysis method combining dynamic simulation and some field experiments of the MEWP is proved reliability, accuracy and economy. In all, the dynamic stability of the MEWP can be calculated accurately by ADAMS, in the meantime, the overload operation test (1.5 times rated load) must be performed on the 5° cross slope before the MEWP are selled, and so long as the minimum dynamic stability safety coefficient is no less than 0.1, the MEWP are proved stability and safety.

## 1. Introduction

The Mobile Elevating Work Platforms (MEWP), also known as one of the aerial devices, are mechanical devices used to provide temporary access for people or equipment to inaccessible areas [1, 2], which are widely used in Aerospace, Shipbuilding, Construction and other industries [3, 4].

In RR961 research report of 2013, at least 82 overturning accidents happened and led to more than 20 people injury seriously even death [5]. Two kinds of typical overturning accidents are as shown in Figure 1.Thus, the most important performance of the MEWP is stability and safety in the process of driving or operating. Now, in the Europe standard EN 280 and the Chinese national standard GB 25849 illustrations can be clearly seen that "In the case of allowance by the manufacturer, the MEWP must be enough stability, at the same time, the MEWP can drive across the kerb or depression with a height of 0.1 m at the maximum permitted travel speed without overturning" [6, 7].

Content from this work may be used under the terms of the Creative Commons Attribution 3.0 licence. Any further distribution of this work must maintain attribution to the author(s) and the title of the work, journal citation and DOI.
Published under licence by IOP Publishing Ltd

(a) Overturning forward

(b) Overturning backward

**Figure 1.** Overturning accidents

In recent years, a lot of research works about dynamic stability analysis of the MEWP have been carried out. The calculation method, such as the subjective and objective evaluation method, the torque balance method and the minimum tire load method were all put forward [8]. Based on the accident statistics, the dynamic stability boundary conditions with different types of the MEWP were analyzed [9, 10]. In the aspect of control strategy, the velocity and acceleration limitation were regarded as the key factor to keep the dynamic stability of the MEWP [11, 12]. With the help of simulation tools, the overturning relation between cross-obstacle collision and dynamic stability of the MEWP were also analyzed [13].

Based on the research results mentioned above, it can be seen that the dynamic stability of the MEWP is related to more factors such as ground conditions, operation velocity and acceleration, obstacles structure etc. However, at present the analysis methods based on dynamic simulation and field experiments of the MEWP haven't been found yet. So in this study, a reliable and economical method combining simulation analysis with some field experiments was put forward and the effect of this method was discussed and verified.

## 2. Mathematic model

The MEWP are composed of chassis, turntable, booms, platform, power assembly, hydraulic actuators (including bumps, motors, hubs, valves and cylinders, etc.) and electric control units (including the controller, the junction boxes, the switches and lamps, etc.). When the MEWP driving or operating on level ground or the slope, the dynamic stability of the MEWP includes forwards stability and backwards stability, and the overturning load and force moment combination is as shown in Figure 2.

The anti-overturning force moment M is calculated as follows.

$$M = \sum_{i=1}^{8} G_i \cdot L_i + \sum_{i=1}^{3} f_i \cdot l_i + 0.1 \cdot \sum_{i=2}^{8} G_i \cdot L_i \tag{1}$$

Where the parameters Gi, fi, Li, and li are mean the gravity of different component assembly, wind loads with manual forces, level distance between the canter of mass of different component assembly and the tipping line and vertical distance from the point of wind loads with manual forces to the ground. Herein, the dynamic coefficient of the forces produced by acceleration or deceleration of structural moving components is 0.1 [6, 7].

If the force moment M is zero or negative, it means that the MEWP have overturned. On the contrary, even the M is positive, which must be greater than a certain value in order to ensure safety at work. Therefore, the tire reaction forces and dynamic stability safety coefficient of the MEWP were calculated as follows.

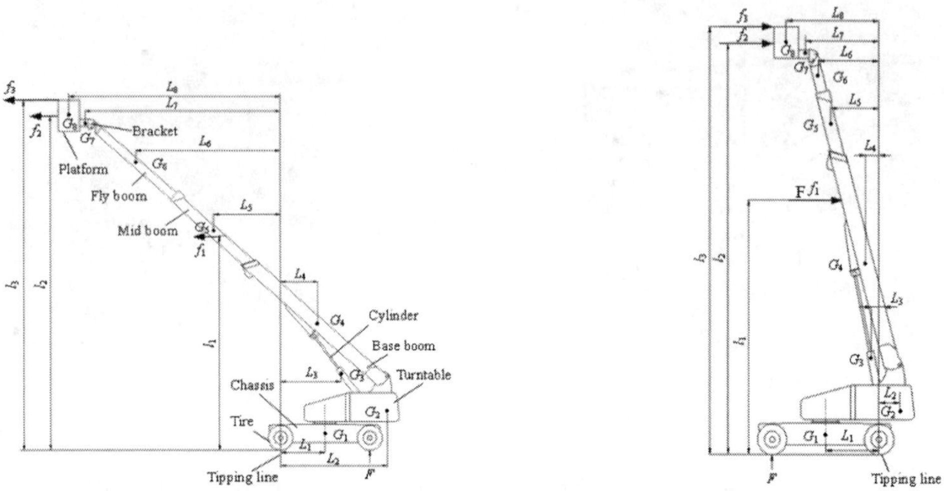


(a) Force diagram of overturning forwards (b) Force diagram of overturning backwards

**Figure 2.** Force diagram of overturning status

$$F = \frac{M}{b} \tag{2}$$

Where the parameters F and b are mean the minimum tire reaction forces (including two tires on the same side) and vertical distance from the point of the tire reaction forces F to the tipping line.

$$\theta = \frac{F}{\sum G_i} \tag{3}$$

Where the parameter θ means the minimum dynamic stability safety coefficient. While the MEWP are driven across obstacles (including the kerb or depression with a height of 0.1 m) by means of front side two tires or rear side two tires simultaneously or operated on the maximum 5° cross slope, the MEWP should be in the most dangerous working conditions. In this case, the dynamic stability and safety of the MEWP can be determined by the minimum θ.

## 3. Modeling and Simulation

### *3.1. Modeling*

In order to improve assembly accuracy of the dynamic simulation analysis models, the 3D models of the MEWP were set up by Pro/E, then were converted *.x_t format into ADAMS [14]. In ADAMS, most of structural components including the chassis, turntable, platform and cylinders were set as rigid bodies, while booms using FEM and tires using Fiala module were set as flexible bodies. According to all of settings above, the elastic deformations of booms extension and tires loads were both taken into consideration, thus the simulation models of the MEWP would be closer to physical structures and the simulation calculation accuracy should be improved. Then the revolute pairs and sliding pairs with friction were set in different joints so that the simulation analysis models of the MEWP were in the same gravitational field [15, 16]. The simulation analysis models of the MEWP are as shown in Figure 3.

### *3.2. Working conditions*

In According to structural calculation method stipulated in EN 280 and GB 25849, four kinds of the most dangerous working conditions were analyzed as shown in Table 1.

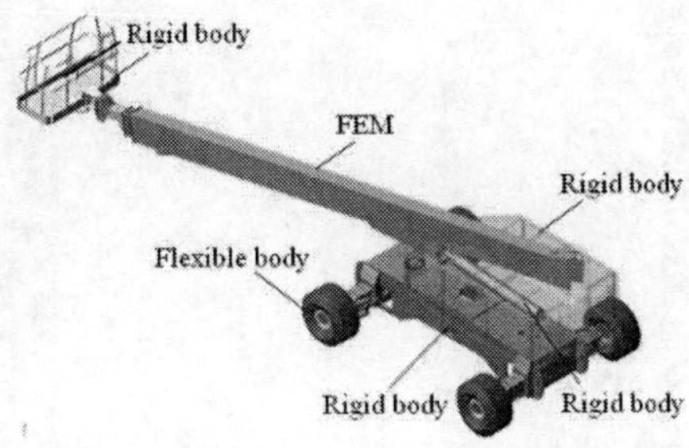


**Figure 3.** Simulation models of the MEWP

**Table 1.** Dangerous working condition

| Working condition | Ground | Platform loads/kg | Boom state | Velocity/km/h |
|---|---|---|---|---|
| 1 | K/D | 345 | EM /53° | 0.5 |
| 2 | K/D | 80 | RM /75° | 0.5 |
| 3 | 5°/S | 345 | EM /53° | 0 |
| 4 | 5°/S | 80 | RM /75° | 0 |

Where the parameters K, D, S, EM and RM are mean the kerb, the depression, the slope, the maximum boom extension and the minimum boom retraction.

In consideration of feasibility and safety of field experiments next, the wind loads, the manual forces and the dynamic structural loads were substituted by means of overload [6, 7] so that the loads of 345 kg (1.5 times rated load of 230 kg) were used in simulation analysis.

### *3.3. Simulation process*

For driving across obstacles simulation analysis, the road spectrums were set up as shown in Figure 4.

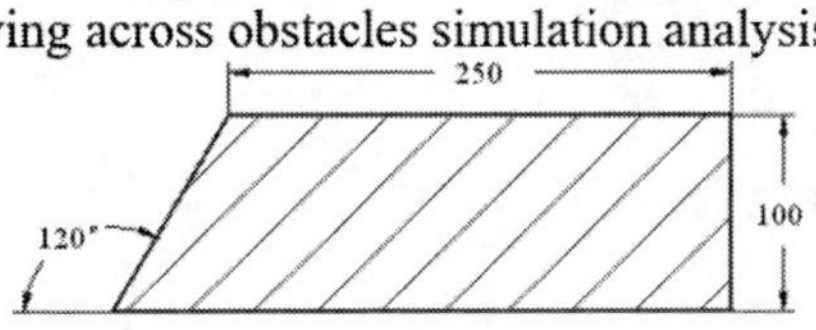


(a) Section of the kerb

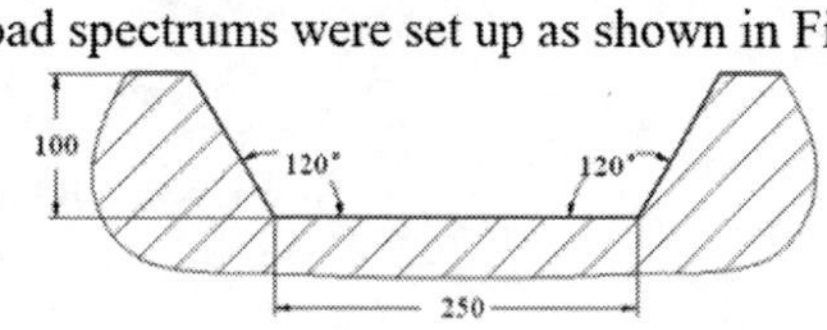


(b) Section of the depression

(c) Combined road spectrums in ADAMS

**Figure 4.** Road spectrums

In ADAMS, the MEWP were driven in turn across the kerb and depression as shown in Figure 5. For operating on the slope simulation analysis, the maximum 5° cross slope was used as shown in Figure 6. In ADAMS, the MEWP were operated on the cross slope (left padding 5°) as shown in Figure 7.

### *3.4. Simulation results and analysis*

After driving across obstacles simulation analysis, the curves of dynamic tire loads were obtained as shown in Figure 8.

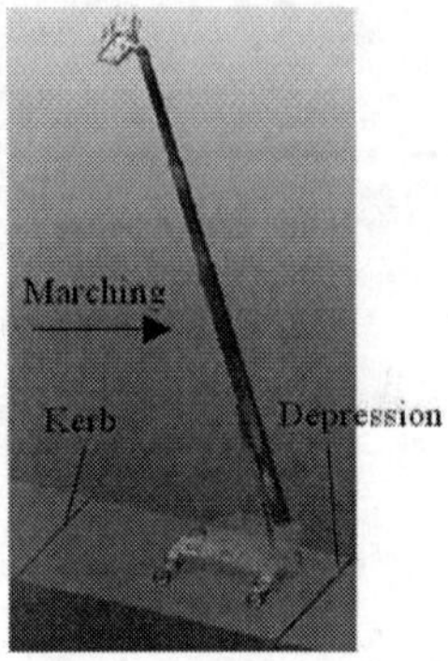


(a) Working conditions 1

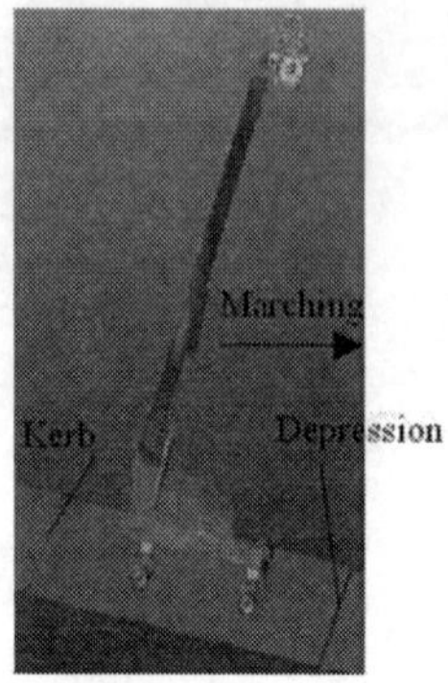


(b) Working conditions 2

**Figure 5.** Driving across obstacles on the road spectrums

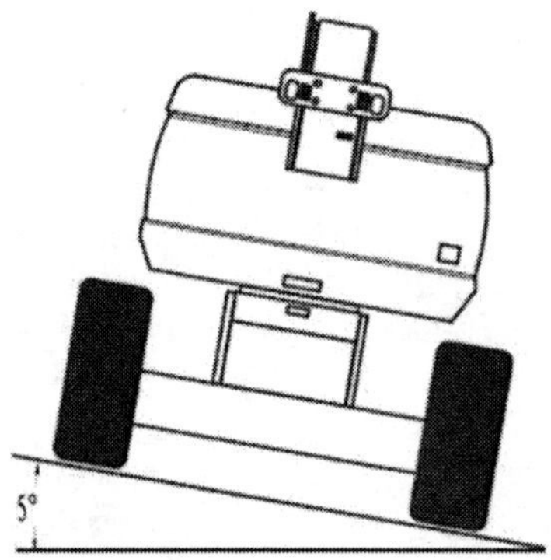


**Figure 6.** The maximum 5° cross slope

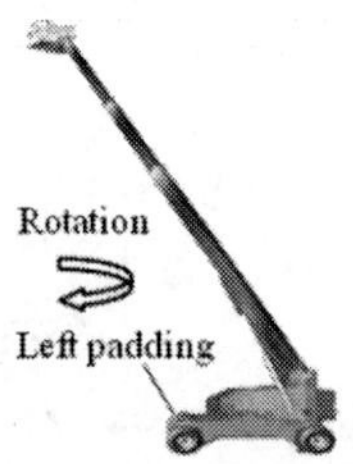


(a) Working conditions 3

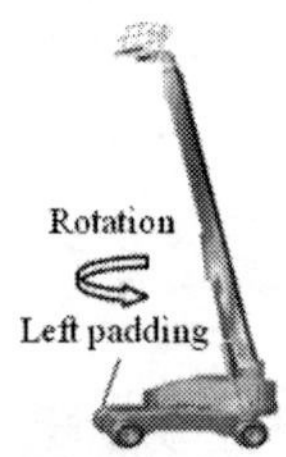


(b) Working conditions 4

**Figure 7.** Operating on the cross slope

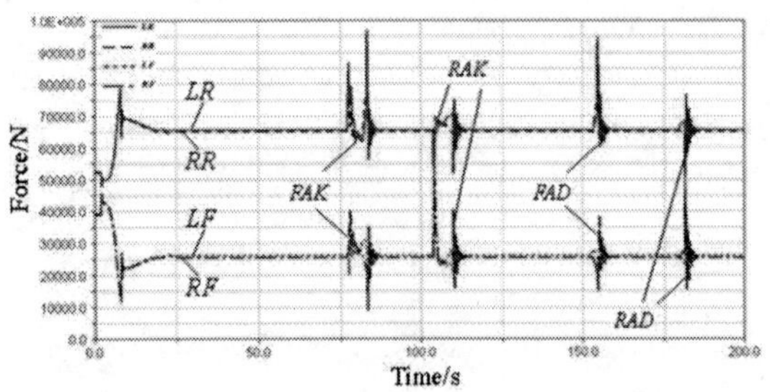


(a) Tire loads of working conditions 1

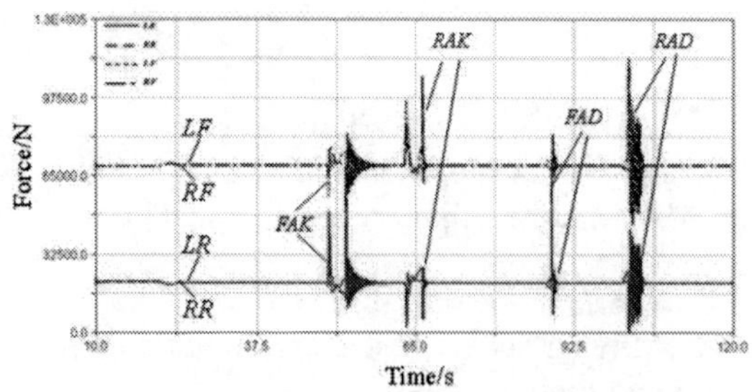


(b) Tire loads of working conditions 2

**Figure 8.** Simulation results of driving across obstacles

Where the parameters *LF*, *LR*, *RF* and *RR* mean left front tire, left rear tire, right front tire and right rear tire respectively, and the annotations *FAK*, *RAK*, *FAD* and *RAD* are mean the process of the front axle climbing over the kerb, the rear axle climbing over the kerb, the front axle climbing over the depression and the rear axle climbing over the depression respectively.

After operating on the cross slope simulation analysis, the curves of the dynamic tire loads were obtained as shown in Figure 9. The simulation results of tire loads were shown in Figure 8 and Figure 9, and were analyzed as follows.

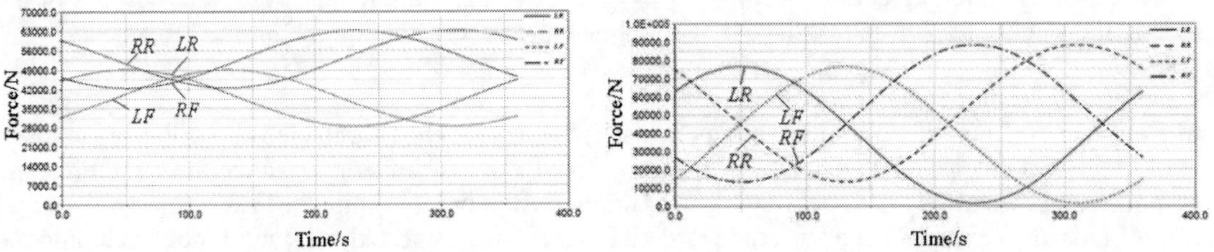


(a) Tire loads of working conditions 3 (b) Tire loads of working conditions 4

**Figure 9.** Simulation results of operating on the cross slope

a) None of tire loads is zero during driving across obstacles and operating on the slope, which shows that the MEWP can keep dynamic stability and safety without overturning even in the most dangerous working conditions.

b) As shown in Figure 8(a), the minimum side tire loads are 20718 N (including *LF*=10736 N and *RF*=9982 N) in climbing over the kerb instantaneously. By contrast, as shown in Figure 8(b), the minimum side tire loads are only 3830 N (including *LR*=2020 N and *RR*=1810 N) in climbing over the depression instantaneously. Thus the minimum dynamic stability safety coefficient θ is 0.108 (20718/(19500*9.8)) and 0.020 (3830/(19500*9.8)) in driving across obstacles on working conditions 1 and 2 respectively (the design weight of the MEWP is 19500 kg and the gravitational acceleration g is 9.8 kg /m$^3$).

c) As shown in Figure 9(a), the minimum side tire loads are 60452 N (including *LR*=30198 N and *LF*=30254 N) when the booms rotate to 270°. By contrast, as shown in Figure 9(b), the minimum side tire loads are 21662 N (including *LR*=10765 N and *LF*=10897 N) when the booms rotate to 270°. Thus the minimum dynamic stability safety coefficient θ is 0.316 (60452/(19500*9.8)) and 0.113 (21662/(19500*9.8)) in operating on the cross slope respectively.

d) Comparing the change amplitude of tire loads in Figure 8(a) and 8(b), Figure 9(a) and 9(b), along with different dynamic stability safety coefficient, all of these show clearly the overturning backwards is more dangerous than the overturning forwards.

e) The curves of tire loads during driving across the kerb and the depression showed fluctuated, as shown in Figure 8(a) and 8(b). The fluctuation was caused by structure moving components (including turntable, booms, platform and sandbags, etc.) waggling during driving across obstacles. Even if the MEWP kept stability and safety when driving across obstacles, the great panic was brought to the operator. Thus, when the MEWP were driven across obstacles, the stability and safety of the MEWP belong to inherent design safety properties once the danger appeared suddenly. In fact, it is not allowed to drive across obstacles in any common operation, so irregular behaviour and habit must be all prohibited strictly.

IOP Conf. Series: Materials Science and Engineering **772** (2020) 012013 doi:10.1088/1757-899X/772/1/012013

## 4. Field experiments and analysis

### *4.1. Test method*

In consideration of safety, feasibility and actual costs of field experiments, only the overload operation experiments were done. Like the same method used in simulation analysis, the wind loads, the manual forces and the dynamic structural loads were substituted by means of overload [6, 7], thus the sandbags of 345 kg in forward overturning test and the sandbags of 80 kg in backward overturning test were placed on the platform respectively. The MEWP were driven to four weighting sensors (the maximum measurement is 10000 kg and the error is ±5 kg), then two tires in left side were raised by use of the pads at the height of 0.3 m so that the MEWP inclined 5° as shown in Figure 10.

### *4.2. Test results and analysis*

In field experiments, all of moving components (including booms, turntable, oil cylinders, bracket, platform and sandbags, etc.) were rotated together from 0° to 360 ° continuously, at the same time, the proportional flow valves were opened to the maximum flow in order to simulate the electric proportional control fails, then every tire loads were recorded. Each test was repeated at least 6 times (including the clockwise and the anticlockwise rotation three times respectively), then all the test data were processed by MATLAB and the curves of tire loads were as shown in Figure 11.

**Figure 10.** Overturning experiments

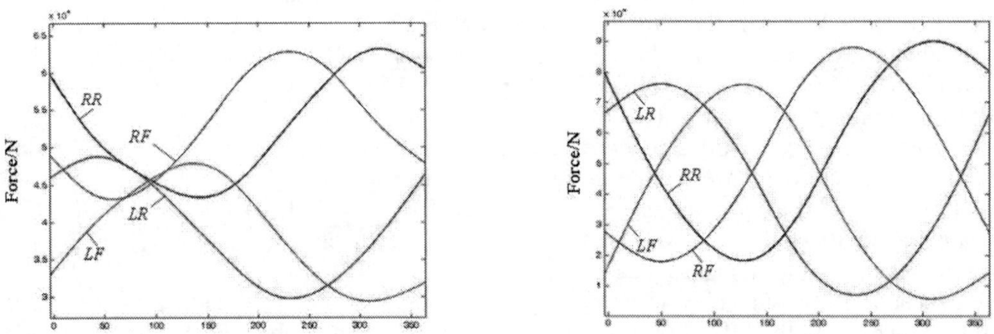


(a) Tire loads of working conditions 3 in the test (b) Tire loads of working conditions 4 in the test

**Figure 11.** Test results

The test results of the tire loads were as shown in Fig. 11 and were analyzed as follows.

a) The change trend of test curves of tire loads is similar with simulation analysis results, which show the simulation models of the MEWP are satisfied and almost the same as the physical structures.

b) As shown in Figure 11(a), the minimum side tire loads are 62898 N (including *LR*=31067 N and *LF*=31831 N) when the booms rotate to 270°. By contrast, as shown in Figure 11(b), the minimum side tire loads are 29616 N (including *LR*=15629 N and *LF*=13987 N) when the booms rotate to 270°. Thus the minimum dynamic stability safety coefficient θ is 0.334 (62898/(19216*9.8)) and 0.157 (29616/(19216*9.8)) during operating on the cross slope respectively (the actual weight of the MEWP is 19216 kg and the gravitational acceleration g is 9.8 kg/m$^3$).

c) The dynamic stability safety coefficient θ errors between test results and simulation results are +1.8% (0.334-0.316) and +4.4% (0.157-0.113), which is proved that simulation analysis can truly reflect the actual working status. Thus it can be seen that the flexible bodies (including booms using FEM and tires using Fiala module) used in simulation analysis are good for improving calculation accuracy.

d) By contrast, even the test of driving across obstacles were not carried out due to security risks and high costs, the simulation analysis results of driving across the kerb or depression and operating on the cross slope by using the same dynamic models, therefore the simulation results during driving across obstacles should be regarded as more accurate and effective evaluation and reference to dynamic stability of the MEWP in actual driving.

## 5. Conclusion

With the dynamic stability analysis of the MEWP above, the main conclusions can be drawn as follows.

a) By simulation analysis, the dynamic curves of tire loads and the minimum dynamic stability safety coefficient $\theta$ of the MEWP were obtained, and the results of which prove that the MEWP have enough stability and safety even in the most dangerous working conditions.

b) In operating field experiments, the dynamic stability safety coefficient errors between simulation results and test results are all positive deviation and less than 5%, which is proved that simulation analysis can truly reflect the actual working status. It also infers that the simulation analysis results of driving across the kerb and depression should be regarded as more effective and accurate evaluation and reference.

c) An accurate, simple and economical field experiments method for dynamic stability of the MEWP was put forward. So long as on the 5° cross slope, the overload test (1.5 times rated load) was carried out and the tested minimum dynamic stability safety coefficient $\theta$ is no less than 0.1, which guarantee stability and safety of the MEWP sales to the customers.

d) Next, one new field experiment method based on small scale models of the MEWP in driving across obstacles would be considered and carried out.

Manuscripts must be in English, also the table and figure texts, otherwise we cannot publish your paper. Please keep a second copy of your manuscript in your office.

**Acknowledgments**

This project was financially supported by the Natural Science Foundation of Fujian Province, China (Grant No. 2018J01498), the Scientific Research Foundation of Jimei University, China and Li Shang Da Scientific Research Foundation of Jimei University, China.

## References

[1] Stewart L, Specialized aerial work platforms stretch makers' market share, J. Construction Equipment. 2 (2007) 56-61.
[2] Zhang L Q, Market forecast of domestic and international aerial work platform, J. Construction Mechanization. 4 (2012) 42-43.
[3] Chen J P, Thinking on the development of China's aerial work platform industry, J. Construction Mechanization, 9 (2010) 35-37.
[4] Euan Y, Moving with the times, J. Access International, 4 (2015) 28-29.
[5] Christine L, David R, Amy J, RR961-Mobile elevated work platform (MEWP) incident analysis 2013, Health and Safety Executive, R. Derbyshire, 2013, pp. 6-8.
[6] EN 280 Mobile elevating work platforms-Design calculation-Stability criteria-Construction-

safety-Examinations and tests, S. 2010.
[7] GB 25849, Mobile elevating work platforms- Design calculations, safety requirements and test method, S. 2010.
[8] Zhao J Y, Miao Z, Chen F, et al, Calculation of the Support Reaction and Planning of Working Radius on Aerial Platform, J, Machine Tool & Hydraulics, 16 (2011) 1-3.
[9] Pan C S, Hoskin A, McCann M, et al, Aerial lift fall injuries: A surveillance and evaluation approach for targeting prevention activities, J, Journal of Safety Research, 6 (2007) 617-625.
[10] Ross B, Mcdonald B, Vijay Saraf S.E., Big blue goes down, The Miller Park crane accident, J, Engineering Failure Analysis, 6 (2007) 942–961.
[11] Ghigliazza R. M, Holmes P, On the dynamics of cranes, or spherical pendula with moving supports, J, International Journal of Non-Linear Mechanics, 7 (2002) 1211-1221.
[12] Ruzzene M, Baz A, Dynamic stability of periodic shellswith moving loads, J, Journal of Sound and Vibration, 4 (2006) 830-844.
[13] Behrooz M, Hanif N, Automatic control of a modified tractor to work on steep side slopes, J, Journal of Terramechanics, 6 (2009) 299-311.
[14] Bosˇnjak S M, Gnjatovic´ N B, Momcˇilovic´ D B, etal, Failure analysis of the mobile elevating work platform, J, Engineering Failure Analysis, 3 (2015) 80-87.
[15] Derlukiewicz D, Przybyłek G. Chosen aspects of FEM strength analysis of telescopic jib mounted on mobile platform, J, Automation in Construction, 3 (2008) 278-283.
[16] Wang S Y, Hu Z C, Peng B L, et al, Simulation of auto-follow row detection mechanism in beet harvester based on ADAMS, J, Transactions of the Chinese for Agricultural Machinery, 12 (2013) 62-67.

# Optimization design of hoisting column for multi-casting system of solid rocket motor based on response surface method

**Fulong Jiang[1,*], Zhaoxia Cui[1], Yang Tao[1], Le Zhou[2]**

[1]Inner Mongolia University of Technology, Mechanical Engineering, Inner Mongolia, Huhehot, 010000, China
[2]Aerospace Sixth Institute, Inner Mongolia, Huhehot, 010000, China

*Corresponding author e-mail: 2803541366@qq.com

**Abstract.** The erection of the solid rocket motor multi-casting system is optimized. The three-dimensional parametric model of the column is established, and the static analysis is carried out to determine the total deformation and the equivalent stress distribution of the hoisting column. Take the important dimensions such as shaft diameter and length of column as design variables, mass, equivalent stress and maximum deformation as target variables. The multi-objective genetic algorithm based on response surface is used to optimize the size of the hoisting column. The results show that after optimization, the overall quality of the column is 28.931kg, which is reduced by 6.16kg, which is 17.7% lower than the original plan.

## 1. Introduction

The production process of solid rocket motor includes steps such as preheating, pouring, vulcanization and other steps. Pouring is an important part in the manufacturing process of the solid rocket motor. Due to the long casting process cycle, single shot casting has low productivity, it can't meet the needs of actual production. Therefore, the multiple pouring system is used to improve the pouring efficiency [1-2]. Its structure is shown in Fig. 1. The hoisting of the column components plays an important role in the entire system. After the pouring engine is filled with the medicine, the whole structure needs to be lifted from the vacuum casting cylinder through the lifting system. At this time, the overall structure weight of the hoisting parts reaches the maximum, and all the weight is borne by three hoisting columns in the hoisting process. At present, the design of hoisting column in China is traditional experience design. Excessive quality of the column will not only increase the manufacturing cost, but also increase the weight of the entire gating system and affect the work efficiency. Therefore, lightweight and good performance are the goals of the hoisting column design.
At present, the design of mechanical structures has begun to change from traditional experience design to modern design. Many domestic scholars began to use modern design methods to optimize the design of mechanical structures. Response surface analysis is widely used in the optimization design of mechanical structures [3]. For example, Chongyang Huang [4] et al. Structural optimization design of bulk carrier based on response surface method, Hanqi Yang [5] et al. Optimization Design of Segmental Structure of Steel Box Bridge Based on Response Surface Model, Peng Wu [6] et al. Optimization design of pipe racking column based on ANSYS and response surface method, the

Content from this work may be used under the terms of the Creative Commons Attribution 3.0 licence. Any further distribution of this work must maintain attribution to the author(s) and the title of the work, journal citation and DOI.
Published under licence by IOP Publishing Ltd

IOP Conf. Series: Materials Science and Engineering **772** (2020) 012014 doi:10.1088/1757-899X/772/1/012014

weight of the optimized scheme is reduced by 3.2%, and the maximum total deformation is reduced by 14.1%, therefore, it is feasible to lighten the column based on the response surface method. However, there is little research on the lightweight design of the hoisting column of the solid rocket motor multi-casting system.

In this paper, the multi-objective variable parameterized model is established for the hoisting column of the multi-engine gating system of solid rocket motor and the static analysis is performed by ANSYS Workbench software. Under the premise of ensuring the strength and deformation of the column, the response surface multi-objective genetic algorithm is used to determine the importance of the design variables through sensitivity, and the design variables are optimized to determine the appropriate lifting column size.

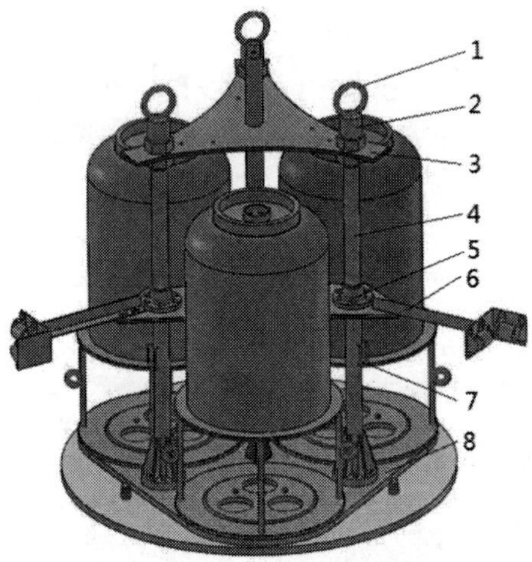


**Figure 1.** Multiple casting system

1.Hoist ring 2. Hoist ring connection 3. Distribution plate 4. Hoisting column 5. Protective flange 6. Guide part 7. Strap ring 8. Hoisting platform

## 2. Finite element analysis of hoisting columns

### *2.1. The establishment of model*

According to figure 1, the size of the hoisting column model is determined according to the solid rocket motor multiple casting system. As shown in Figure 2, the length of the column is 1367 mm, the diameter is 57 mm, and the thickness of the reinforcing plate is 8 mm. The material of the hoisting column is structural steel, the modulus of elasticity is 2.0×105 MPa, the Poisson's ratio is 0.3, and the density is 7890 kg/m3.

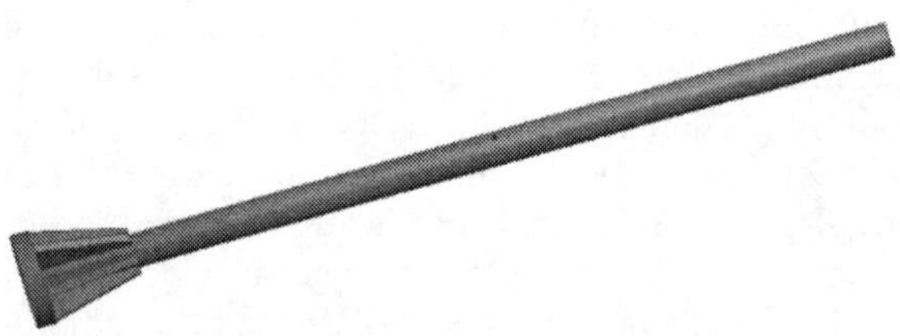

**Figure 2.** Lifting column model

*2.2. Meshing*

The upper end of the hoisting column is connected to the sling ring by a threaded connection, due to the small size of the thread structure, it's even much smaller than the cell size, it will affect the calculation process. So before the meshing, the column was unthreaded [7-8]. The quality of the grid determines the pros and cons of the analysis results, the finite element analysis adopts automatic mesh generation, element size is 3mm, generating 1434749 elements and 2006496 element nodes in total.

*2.3. Loads and constraints*

The lifting column and column connecting flange are connected by welding, and the rib plate is welded to enhance the bearing capacity. The flange surface at the lower end of the lifting column is connected with the lifting platform by high-strength bolts. After each pouring container is filled with slurry, the weight is about 800kg, after the weight of the third engine vessel plus the dead weight of the tooling, the overall hoisting weight is about 3t. Therefore, the maximum weight of each hoisting column allocated by the total weight is about 1t. Figure 3 shows the load and constraint conditions of the hoisting column.

**Figure 3.** Loads and constraints

*2.4. Analysis of calculation results*

After calculation, as shown in Figure 4, the maximum deformation of hoisting column is 0.025351mm. In the actual working condition, the precision of the pouring system is required to be 0.05mm, which can meet the maximum precision requirement of the hoisting column. The yield strength of the column is 235 MPa. Considering the manufacturing accuracy of column and other uncertain factors, the safety factor is 1.8.Therefore, the maximum allowable stress of hoisting column is 235 / 1.8 = 130.5Mpa. As shown in Figure 5, the maximum equivalent stress of the column is 11.102 MPa, which meets its own strength requirement, so the size and quality of the hoisting column can be optimized.

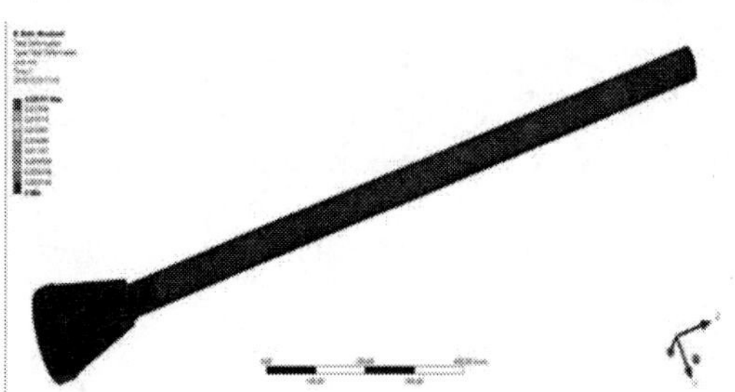

**Figure 4.** Deformation cloud diagram of the lifting column

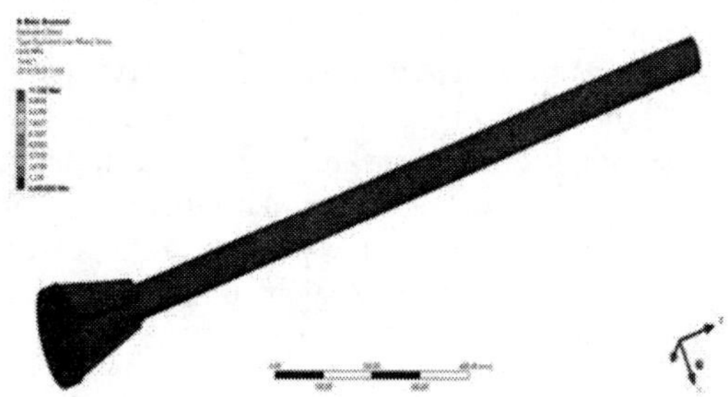

**Figure 5.** Equivalent stress cloud diagram of the lifting column

### *2.5. The establishment of a parametric model*

In this optimization, the column length H, column diameter D and stiffener plate thickness L are selected as design variables, as shown in Table 1. According to the working conditions and installation coordination, the variable size ranges are established. The mass M, the maximum deformation amount Q and the equivalent stress F of the column are taken as target variables. Among them, the weight of the column M is the most important target variable, and the weight of the column can reduce the quality of the whole system, shorten the process time, and reduce the waste of resources. However, it is required that the maximum deformation Q shall not exceed 0.05mm and the equivalent stress F shall not be greater than the allowable stress 130.5Mpa.

**Table 1.** Variation range of design variables

| Parameter | Initial value(mm) | lower limit (mm) | Upper limit (mm) |
|---|---|---|---|
| H(length) | 57 | 50 | 65 |
| D(diameter) | 1367 | 1355 | 1375 |
| L(thickness) | 8 | 5 | 10 |

The established mathematical model of optimal design is shown in formula (1) (2) (3):

Optimization target:

$$\begin{cases} & \min(M) \\ \text{Min } Z(x) & \min(Q) \\ & \min(F) \end{cases} \tag{1}$$

Constraint function:

$$\begin{cases} M \le 40\text{kg} \\ Q \le 0.05\text{mm} \\ F \le 130.5\text{MPa} \end{cases} \tag{2}$$

The range of values of the optimization variables:

$$\begin{cases} 50\text{mm} \le D \le 65\text{mm} \\ 1355\text{mm} \le H \le 1375\text{mm} \\ 5\text{mm} \le L \le 10\text{mm} \end{cases} \tag{3}$$

### *2.6. The multi-objective genetic algorithm*

The lightweight model of hoisting column is a multi-objective optimization problem, its solution is more complex, but compared with the original genetic algorithm, its convergence speed is faster and the calculation result is more accurate. The multi-objective genetic algorithm based on response surface avoids the tedious process of repeated modeling calculation and effectively improves the optimization rate [9]. The calculation process is shown in Fig. 6.

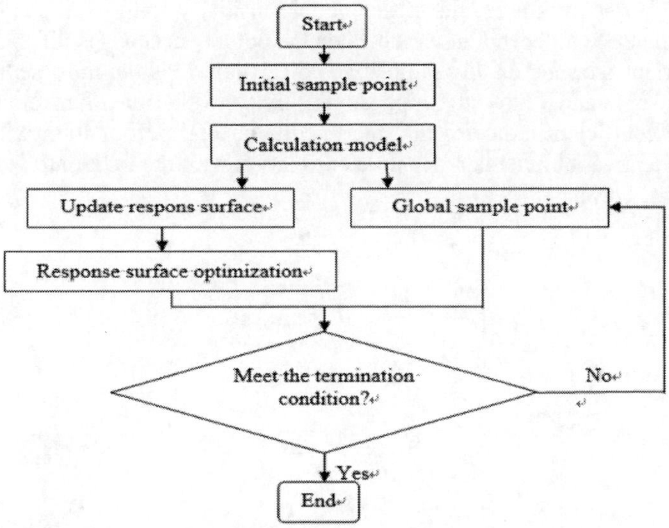


**Figure 6.** Flow chart

### *2.7. The calculation of the design poins Parameter sensitivity and response surface analysis*

The generation of design points is the key to multi-objective optimization design. Fifteen design points are randomly generated by Design of Experiments. The three groups of variables are randomly assigned and combined together, and then the calculation results of fifteen sets of target variables are obtained through analysis and calculation. The fifteen groups of data are shown in Table 2, where H (mm) is the length of the column, D (mm) is the diameter of the column, L (mm) is the thickness of the reinforcing rib, M (kg) is the column mass, Q (mm) is the maximum deformation of the column, and F(Mpa) is the equivalent stress.

**Table 2.** Experimental design points

| Serial number | Column length(mm) | Column diameter(mm) | Stiffener thickness(mm) | Mass (kg) | Maximum deformation(mm) | Equivalent stress(MPa) |
|---|---|---|---|---|---|---|
| 1 | 1365 | 57.5 | 7.5 | 35.293 | 0.024949 | 11.218 |
| 2 | 1365 | 50 | 7.5 | 28.774 | 0.035536 | 14.786 |
| 3 | 1365 | 65 | 7.5 | 42.623 | 0.019763 | 7.8521 |
| 4 | 1355 | 57.5 | 7.5 | 35.089 | 0.024761 | 10.813 |
| 5 | 1375 | 57.5 | 7.5 | 35.497 | 0.025138 | 10.872 |
| 6 | 1365 | 57.5 | 5 | 34.253 | 0.025282 | 10.742 |
| 7 | 1365 | 57.5 | 10 | 36.357 | 0.024707 | 11.049 |
| 8 | 1356.9 | 51.402 | 5.4674 | 28.931 | 0.033464 | 24.316 |
| 9 | 1356.9 | 63.598 | 5.4674 | 40.22 | 0.020682 | 7.9955 |
| 10 | 1373.1 | 51.402 | 5.4674 | 29.196 | 0.033848 | 26.122 |
| 11 | 1373.1 | 63.598 | 5.4674 | 40.625 | 0.020933 | 8.0465 |
| 12 | 1356.9 | 51.402 | 9.5326 | 30.744 | 0.033434 | 19.308 |
| 13 | 1356.9 | 63.598 | 9.5326 | 41.75 | 0.020305 | 8.0621 |
| 14 | 1373.1 | 51.402 | 9.5326 | 31.009 | 0.033819 | 19.079 |
| 15 | 13373.1 | 63.598 | 9.5326 | 42.156 | 0.020556 | 7.9239 |

### *2.8. Parameter sensitivity and response surface analysis*

Through the sensitivity analysis, we can get the degree of influence of the independent variable parameters on the target output, and then optimize the optimization of some independent variables based on this [8]. Fig. 7 shows the results of the global sensitivity analysis, it can be seen from Figure 7: the influence of the three independent variables on the output variables is different. The influence of the independent variables on the quality is positive correlation. D is the most sensitive to the quality, followed by L and H. D and l have the highest sensitivity to the maximum total deformation. The negative sensitivity coefficient indicates that the maximum total deformation will decrease with the increase of independent variable. D is a negative correlation with the maximum equivalent stress, and L and H are positively correlated.

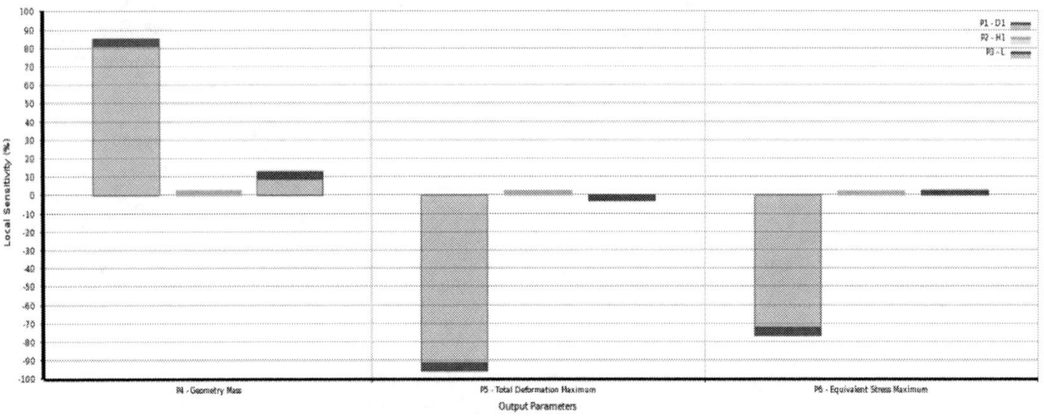


**Figure 7.** The global sensitivity analysis

Through the results of the local sensitivity analysis shown in Fig. 8, the sensitive trend of the respective variable parameters to the maximum mass can be quickly viewed. Each sensitivity curve intersects the response point at which the sensitivity of the respective variables to the maximum mass is consistent.

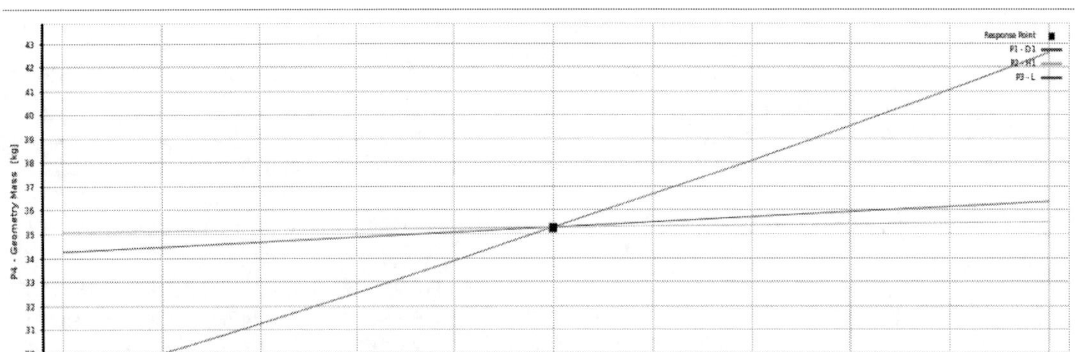


**Figure 8.** The local sensitivity analysis

The Design Explorer module uses a full second-order polynomial to fit the response surface function. Fig. 9 shows the fit degree distribution map.

IOP Conf. Series: Materials Science and Engineering **772** (2020) 012014 doi:10.1088/1757-899X/772/1/012014

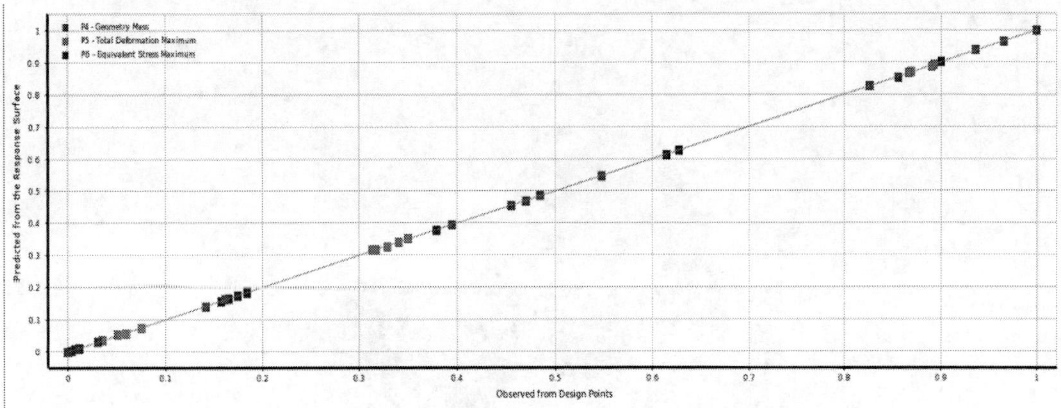


**Figure 9.** The distribution of fit conformity

Combined with the sensitivity analysis results, the focus is on the corresponding surface of the column diameter D and the thickness of the stiffener rib L for the maximum mass, maximum deformation and maximum equivalent stress. Fig. 10 shows the maximum mass response surface. It can be seen from Figure 10 that the effect of D and L on the maximum mass is a linear positive correlation, and increasing D and l will increase the maximum mass. Fig. 11 shows the maximum deformation response surface. It can be seen from Fig. 11 that the influence of D and L on the maximum total deformation is a linear negative correlation, and increasing the values of D and L can effectively reduce the maximum total deformation. The corresponding surface results shown in Fig. 12 show that the influence of D and L on the maximum equivalent stress is nonlinear. As D increases, the maximum equivalent stress first increases and then decreases, and the maximum equivalent stress increases as L increases.

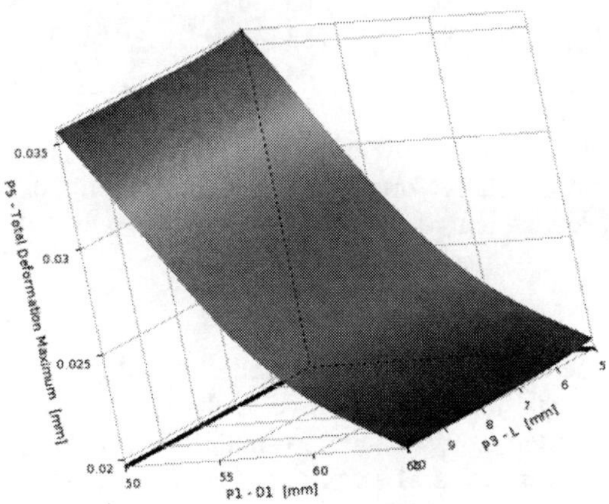


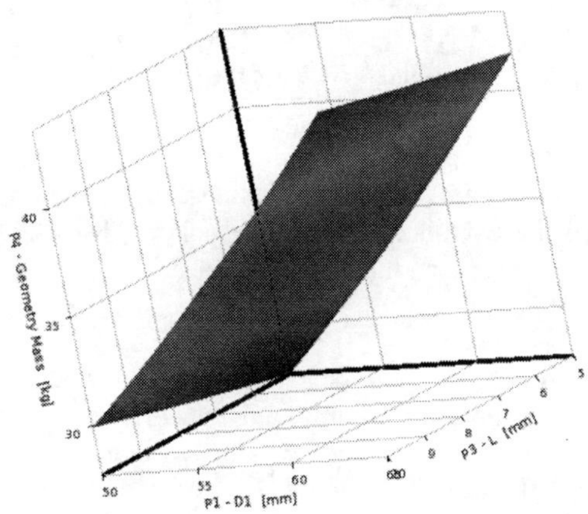


**Figure 10.** Maximum mass response surface **Figure 11.** Maximum total deformation response surface

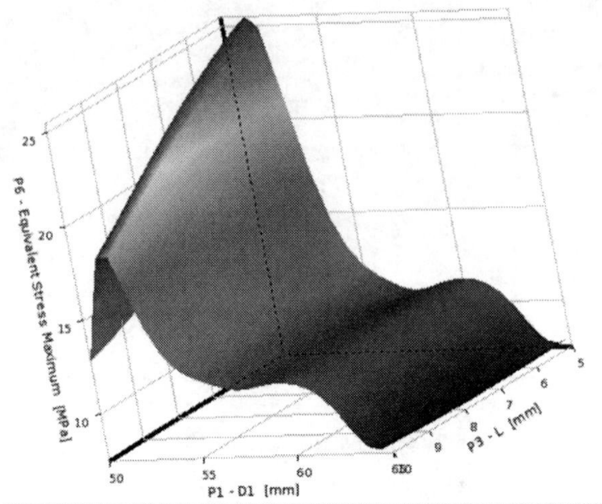


**Figure 12.** Maximum equivalent stress response surface

### *2.9. Optimization design point*

Based on the weight reduction, through the analysis and comparison of fifteen sets of test design points, four design points are selected as alternatives. Preferred design points are shown in Table 3.

**Table 3.** Preferred design points

| Serial number | Column length(mm) | Column diameter(mm) | Stiffener thickness(mm) | Mass (kg) | Maximum deformation(mm) | Equivalent stress(MPa) |
|---|---|---|---|---|---|---|
| 1 | 1365 | 50 | 7.5 | 28.774 | 0.035536 | 14.786 |
| 2 | 1356.9 | 51.402 | 5.4674 | 28.931 | 0.033464 | 24.316 |
| 3 | 1373.1 | 51.402 | 5.4674 | 29.196 | 0.033848 | 26.122 |
| 4 | 1356.9 | 51.402 | 9.5326 | 30.744 | 0.033434 | 19.308 |

Considering the maximum mass and maximum deformation of the hoisting column, the second group of optimization schemes is selected. After rounding, H=1357mm, D=51mm, L=5.5mm. At this time, the overall quality of the column was 28.931 kg, which was reduced by 6.16 kg, which was 17.7% lower than the original plan.

## 3. Conclusion

(1)Through parameter sensitivity analysis and response surface analysis, it is found that the diameter D of the hoisting column and the thickness H of the rib are more sensitive to the maximum total deformation. Increasing the values of D and L can effectively reduce the maximum total deformation. The influence of D and L on the maximum equivalent stress is nonlinear. As D increases, the maximum equivalent stress first increases and then decreases. When L increases, the maximum equivalent stress increases.

(2)The response surface method is used to get the optimization scheme, which reduces the mass by 17.7% compared with the original scheme. Under the premise of satisfying the rigidity and strength of the column, the purpose of reducing the quality of the column is achieved, and the rapid optimization design of the column size is realized, which provides a certain reference value for the actual production and application.

IOP Conf. Series: Materials Science and Engineering **772** (2020) 012014 doi:10.1088/1757-899X/772/1/012014

**References**

[1] Zeping Wu,Donghui Wang,Wenjie Wang,Okolo Patrick N,Weihua Zhang. Solid rocket motor design employing an efficient performance matching approach [J]. Proceedings of the Institution of Mechanical Engineers, 2019, 233(11).

[2] Dan Wu. Development status and trend of large solid rocket engines [A]. China Aerospace Third Major Proceedings of the 38th Technical Exchange Conference and the 2nd Aerospace Power Joint Conference on Information Network - Solid Propulsion Technology [C]

[3] Daiyan Xu. Research on Acoustic Optimization Design of Rail Traffic Box Beam Structure Based on Response Surface Method [D]. East China Jiaotong University, 2016.

[4] Chongyang Huang. Structural optimization design of bulk carrier based on response surface method [D]. Dalian University of Technology, 2011.

[5] Hanqi Yang.Optimization Design of Segmental Structure of Steel Box Bridge Based on Response Surface Model [J]. China High Technology, 2017, 1(07): 57-59.

[6] Peng Wu, Tao Shi, Shigang Zhao, Shuanghui Wei.Optimization design of pipe racking column based on ANSYS and response surface method [J]. Petroleum Machinery, 2019, 47(02): 40-48.

[7] Yue Zuo. A finite element model correction method based on Bayesian probability sampling and sensitivity analysis [D]. Huazhong University of Science and Technology, 2016.

[8] Junpeng Zhao, Dawei Jiang, Zhi Gao, Wei Chen, Guoya Song.Finite Element Analysis of the Sub-frame of Special Transport Vehicle [J]. Journal of Changchun University of Technology, 2015, 36(04): 421-425.

[9] Hailian Yu, Yongquan Wang, Hualing Chen, Huaying Cun.Optimization of machine tool column parameters based on response surface model and multi-objective genetic algorithm [J]. Journal of Xi\'an Jiaotong University, 2012, 46(11): 80-85.

# Optimum Analysis of Microporous Foaming/Mould Decoration Compound Forming Technology for Automobile Door Decoration Panel

**Yi Hu[a], Tieqiang Yuan**

School of Automotive Engineering, Wuhan University of Technology, Wuhan, China

[a]68235238@qq.com

**Abstract**. All Energy conservation and environmental protection are increasingly concerned by the automotive industry. Lightweight vehicles are one of the effective ways to achieve energy conservation and environmental protection. As an important method in plastic processing, the microcellular foam injection molding process can further meet the requirements of lightweight vehicles while meeting the performance requirements of the parts. However, a large number of bubbles exist on the surface of the microcellular foamed injection molded product, and the problems such as large surface roughness, low gloss, and poor apparent quality restrict the application of the process in automotive plastic parts. In order to improve the surface defects of microcellular foamed parts, this paper establishes experimental schemes and mathematical models based on partial factor test design method, response surface design and genetic algorithm, and obtains the optimal micro-foaming/in-mold decorative composite injection molding. Process parameter plan.

## 1. Introduction

Microcellular foam injection moulding plastic parts have shorter processing cycle and relatively stable size compared with traditional injection moulding plastic products, but there are crazes, whirlpool flow marks and bubble marks on the surface of microfoam plastic. Rough surface quality restricts the application prospects of this moulding technology in plastic parts [1].

The micro-foaming/in-mold decorative composite injection molding process has different thermal conductivity on both sides of the mold due to the intervention of the decorative film, which changes the flow characteristics of the gas/polymer mixed melt and the cell formation process, and improves the mechanics of the part. Performance and molding defects, so its process control is more complicated. Chen H Lobtained the following conclusions by studying the relationship between the composite film and the surface quality of microporous foamed products: The surface quality of products decreases with the increase of film thickness [3]. He Yongyuan studied the effect of temperature on warpage, shrinkage and residual stress by simulating and analyzing the process of microcellular foaming. According to rheological theory, the influence mechanism of process parameters was revealed [5] .Hu Guanghong established the relationship among the three technological parameters in the process of Microcellular Foaming Injection moulding, i.e. cell nucleation rate, saturated melt pressure and melt temperature, and improved the mathematical model of bubble growth. The mathematical model was analyzed and validated by numerical calculation [6].

Content from this work may be used under the terms of the Creative Commons Attribution 3.0 licence. Any further distribution of this work must maintain attribution to the author(s) and the title of the work, journal citation and DOI.
Published under licence by IOP Publishing Ltd

IOP Conf. Series: Materials Science and Engineering **772** (2020) 012015 doi:10.1088/1757-899X/772/1/012015

In this paper, based on the partial factor test design method, the experimental scheme is established. The main influence parameters of the response model are melt temperature, mold temperature, V/P transition point and initial gas concentration. Based on the response surface design, the cell radius and warpage are respectively established. The prediction mathematical relationship model of tensile modulus; based on genetic algorithm to establish the process parameter optimization method under the defect minimization, the optimal process parameter scheme of micro-foaming/in-mold decorative composite injection molding is obtained.

## 2. Test Design

### *2.1. Modeling Analysis*

The experiment selects the typical automobile trim-door trim panel as the research object, and builds the mold model diagram shown in Figure 1 according to the mold installation and design criteria. Using MOLDFLOW finite element software to simulate and analyze it, divide the grid and establish the gating system and cooling system. As shown in Figure 2, attach the decorative film to the outer surface of the door trim panel, and mesh and detect it. Establish a corresponding microcellular foaming/in-mold decorative composite injection molding system.

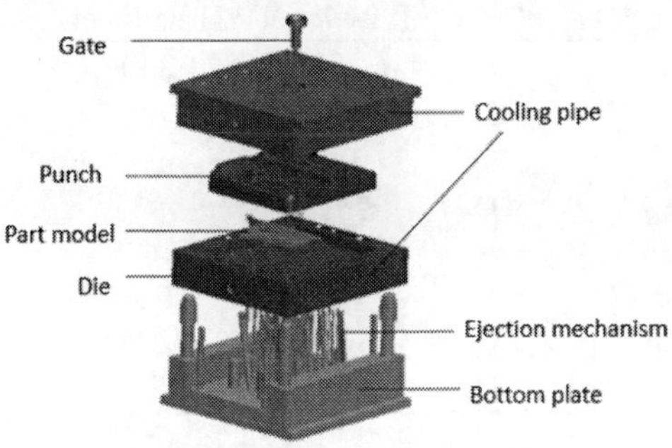


**Figure 1.** Mould model diagram

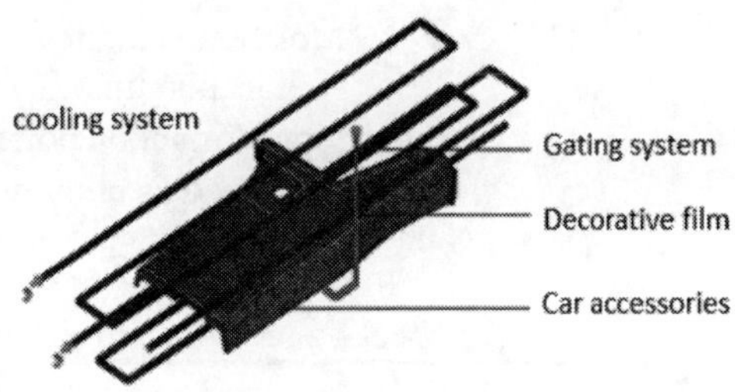


**Figure 2.** Finite element model

### *2.2. Material and Process Conditions*

The experimental materials include polymer, decorative film and gas foaming agent. Polymers are PP composite materials, their viscosity properties and PVT properties are shown in Figure 3. TPU produced by Lubrizol manufacturer is selected for the film. Supercritical nitrogen (N2), critical pressure 3.4MPa, critical temperature - 147.05 C, gas constant R8.3145J/(mol k) and density 1.25kg/m3 were selected as blowing agent. The mould is manufactured by Generic manufacturer and the brand is TOOL STEEL P-20.

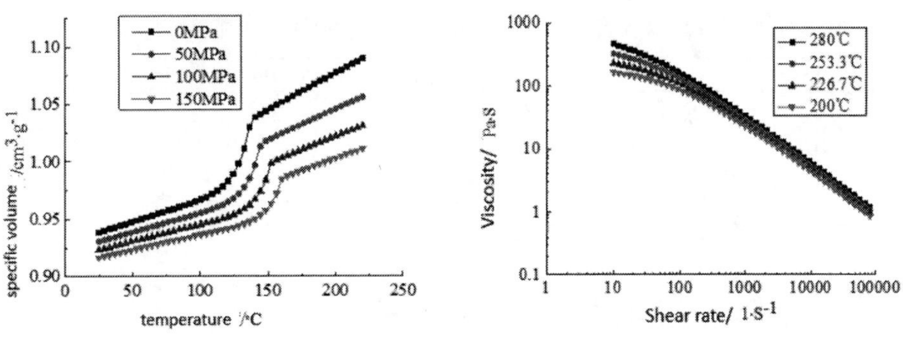


**Figure 3.** Viscosity and PVT properties

In order to study the influence of the coupling effect between injection molding factors and factors on the forming properties, the analysis focuses on melt temperature, mold temperature, injection time, cooling temperature, V/P conversion point, initial gas concentration, number of gas cores per volume, and cooling. The influence of medium Reynolds number and cooling medium temperature on the composite forming process of automobile door trim.

### *2.3. Test*

*2.3.1. Partial Factor Test.* In this section, melt temperature, mould temperature, injection time, cooling temperature, V/P conversion point, initial concentration of gas, number of gas nuclei per volume, Reynolds number of cooling medium and temperature of cooling medium are selected as initial factors. For the convenience of later elaboration, these eight factors are replaced by letters A, B, C, D, E, F, G, H respectively. The specific values of high and low levels are determined, as shown in Table 1. The experimental design of these eight factors is carried out and 16 groups of experimental arrays are obtained.

**Table 1.** Selection of parameters

| number | factor | Code | Low level | Hihg level |
|---|---|---|---|---|
| 1 | Melt Temperature /℃ | A | 200 | 240 |
| 2 | Mold temperature /℃ | B | 40 | 60 |
| 3 | Injection time /s | C | 0.8 | 1.2 |
| 4 | V/Ptransformation point/% | D | 95% | 99% |
| 5 | Initial gas concentration /% | E | 0.3 | 0.7 |
| 6 | Number of gas nuclei per volume /$m^{-3}$ | F | 1.8e+6 | 2.2e+6 |
| 7 | Reynolds number of coolant | G | 8000 | 12000 |
| 8 | Cooling medium temperature /℃ | H | 20 | 30 |

*2.3.2. Response Surface Factor Test.* Response Surface Methodology (RSM) can obtain the functional relationship between response and factor, so it has more advantages in optimum design. According to the results obtained from the experimental design of some factors, the main influencing factors are selected as follows: A (melt temperature), B (mould temperature), D (V/P conversion point), E (initial concentration of gas) as the research object, and the level is selected as shown in Table 2.The four process parameters that have no significant influence are as follows: injection time 0.8s, number of gas nuclei 2E+05 per volume, Reynolds number 10000 of cooling medium, temperature of cooling medium 25℃.After the process parameters except the main effect factor are set up in the same way, the central composite experiment is used to design the factors, and the response results of the central composite factor test array shown in Table 2 and extracted from MOLDFLOW model are obtained.

**Table 2.** Central Compound Test Factor and Its Level

| number | factor | Low level | Medium level | Hihg level |
|---|---|---|---|---|
| A | Melt Temperature /℃ | 200 | 220 | 240 |
| B | Mold temperature /℃ | 40 | 50 | 60 |
| D | V/P transformation point/% | 95 | 97 | 99 |
| E | Initial gas concentration /% | 0.3 | 0.5 | 0.7 |

## 3. Results and Analysis

### *3.1. Analysis of Partial Factor Test*

Three groups of results were obtained by MOLDFLOW simulation: cell radius (um), warpage result (mm) and tensile modulus (MPa), as shown in Table 3.

**Table 3.** Partial Factor Test Array and Response Results

| number | A | B | C | D | E | F | G | H | Bubble radius(um) | Warpage results(mm) | Tensile modulus(MPa) |
|---|---|---|---|---|---|---|---|---|---|---|---|
| 1 | -1 | 1 | -1 | -1 | 1 | -1 | 1 | 1 | 7.36 | 10.03 | 874.76 |
| 2 | 1 | -1 | -1 | -1 | -1 | 1 | 1 | 1 | 5.32 | 9.86 | 950.83 |
| 3 | 1 | 1 | -1 | 1 | -1 | -1 | -1 | 1 | 5.48 | 10.10 | 908.96 |
| 4 | -1 | -1 | 1 | 1 | -1 | -1 | 1 | 1 | 5.45 | 9.78 | 990.71 |
| 5 | 1 | -1 | 1 | -1 | 1 | -1 | -1 | 1 | 8.52 | 9.78 | 905.84 |
| 6 | -1 | -1 | -1 | 1 | 1 | 1 | -1 | 1 | 7.19 | 9.72 | 974.48 |
| 7 | -1 | 1 | 1 | -1 | -1 | 1 | -1 | 1 | 5.19 | 10.31 | 942.72 |
| 8 | -1 | -1 | -1 | -1 | -1 | -1 | -1 | -1 | 5.1 | 9.99 | 947.62 |
| 9 | -1 | 1 | 1 | 1 | 1 | -1 | -1 | -1 | 8.16 | 10.24 | 931.93 |
| 10 | -1 | 1 | -1 | 1 | -1 | 1 | 1 | -1 | 4.58 | 10.37 | 957.53 |
| 11 | -1 | -1 | 1 | -1 | 1 | 1 | 1 | -1 | 7.57 | 9.92 | 906.90 |
| 12 | 1 | -1 | -1 | 1 | 1 | -1 | 1 | -1 | 8.28 | 9.88 | 932.76 |
| 13 | 1 | 1 | 1 | -1 | -1 | -1 | 1 | -1 | 6.37 | 9.79 | 859.61 |
| 14 | 1 | 1 | -1 | -1 | 1 | 1 | -1 | -1 | 7.63 | 9.69 | 784.12 |
| 15 | 1 | -1 | 1 | 1 | -1 | 1 | -1 | -1 | 6.02 | 9.94 | 962.64 |
| 16 | 1 | 1 | 1 | 1 | 1 | 1 | 1 | 1 | 8.03 | 10.03 | 874.30 |

From Figure 4, it can be seen that the critical value for the influence of the radius of the bubble on the surface of the workpiece is 0.546, the critical value for the warpage result is 0.2139, and the critical value for the influence of the tensile modulus is 39.01. As shown by the dotted line in the Pareto diagram, the three factors that have significant influence on the radius of the bubble on the surface of the workpiece are the initial concentration of gas, the melt temperature and the injection time, in turn, which affect the warpage result. The most important factor is the temperature of the die. The notable factors affecting the tensile modulus are the temperature of the die, V/P conversion point, melt temperature and initial gas concentration.

Normal effect maps are used to screen the factors. Among these uncertain factors, the main factors which have obvious influence on the main effect are selected. As shown in Figure. 4, the significant factors are identified by red dots, which are gas initial concentration, melt temperature, injection time, mould temperature and V/P conversion point, while the other process parameters which have little influence on the results show blue.

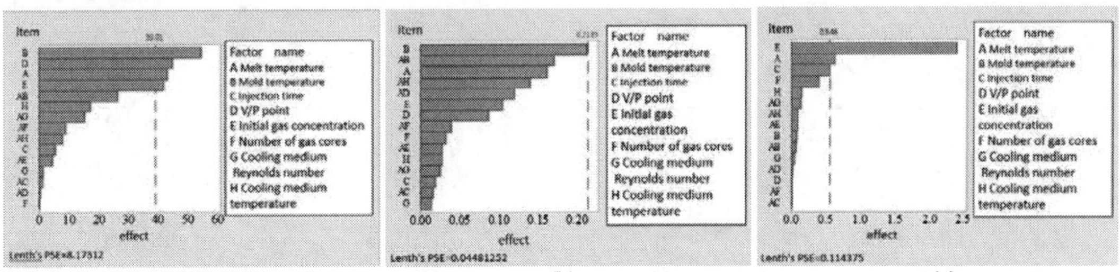


(a) (b) (c)

**Figure 4.** Paretochart (a) Bubble radius (b) Warpage (c) Tensile modulus

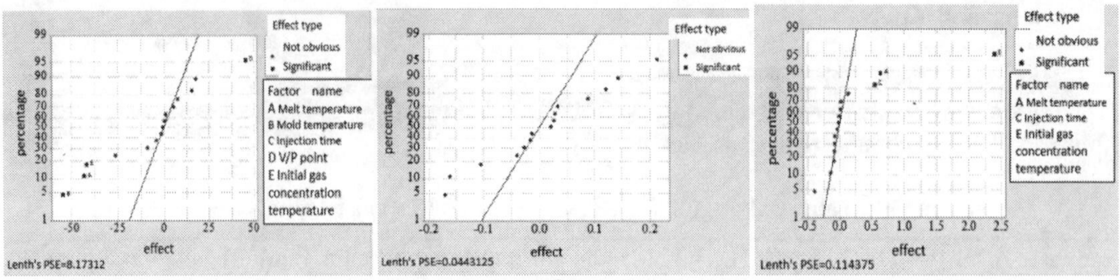


(a) (b) (c)

**Figure 5.** Normal Effect Map (a) Bubble radius (b) Warpage (c) Tensile modulus

The Pareto plot is consistent with the significance of the influence factor screened by the normal effect diagram. According to the results obtained, the initial gas concentration, melt temperature, mold temperature and V/P transition point are selected as the research object.

### 3.2. *Response Surface Analysis*

**Table 4.** Central Composite Test Factor Array and Response Results

| number | A | B | D | E | Bubble radius(um) | Warpage results(mm) | Tensile modulus (MPa) |
|---|---|---|---|---|---|---|---|
| **1** | 0 | 0 | 0 | -1 | 5.53 | 10.12 | 902.83 |
| **2** | 1 | 0 | 0 | 0 | 7.79 | 9.923 | 834.58 |
| **3** | 0 | 0 | 0 | 1 | 7.76 | 9.971 | 853.43 |
| **4** | 1 | -1 | 1 | -1 | 5.82 | 10.16 | 899.06 |
| **5** | -1 | 1 | -1 | -1 | 5.12 | 10.16 | 919.07 |
| **6** | 1 | 1 | 1 | 1 | 8.16 | 9.855 | 818.32 |
| **7** | -1 | 1 | -1 | 1 | 7.86 | 10.03 | 876.71 |
| **8** | -1 | 0 | 0 | 0 | 6.61 | 10.09 | 908.39 |
| **9** | 1 | -1 | -1 | 1 | 8.18 | 9.969 | 818.54 |
| **10** | 0 | -1 | 0 | 0 | 7.4 | 9.883 | 849.61 |
| **11** | -1 | -1 | -1 | -1 | 4.98 | 9.885 | 883.68 |
| **12** | -1 | -1 | 1 | 1 | 7.66 | 9.758 | 854.2 |
| **13** | 0 | 0 | 0 | 0 | 7.06 | 10.04 | 879.21 |
| **14** | -1 | 1 | 1 | -1 | 5.12 | 10.16 | 936.91 |
| **15** | -1 | -1 | 1 | -1 | 4.98 | 9.864 | 903.34 |
| **16** | -1 | -1 | -1 | 1 | 7.66 | 9.783 | 832.4 |
| **17** | 0 | 1 | 0 | 0 | 7.27 | 10.01 | 875.53 |
| **18** | 1 | 1 | 1 | -1 | 5.78 | 9.99 | 878.49 |
| **19** | 0 | 0 | 0 | 0 | 7.06 | 10.04 | 879.21 |
| **20** | 1 | 1 | -1 | -1 | 5.78 | 9.972 | 856.07 |
| **21** | 1 | 1 | -1 | 1 | 8.16 | 9.854 | 796.34 |
| **22** | 0 | 0 | 0 | 0 | 7.06 | 10.04 | 879.21 |
| **23** | 0 | 0 | 1 | 0 | 7.06 | 10.06 | 892.3 |
| **24** | 0 | 0 | -1 | 0 | 7.06 | 10.03 | 867.65 |
| **25** | 0 | 0 | 0 | 0 | 7.06 | 10.04 | 879.21 |
| **26** | 0 | 0 | 0 | 0 | 7.06 | 10.04 | 879.21 |
| **27** | 1 | -1 | 1 | 1 | 8.18 | 10 | 844.54 |
| **28** | -1 | 1 | 1 | 1 | 7.86 | 10.05 | 901.09 |
| **29** | 0 | 0 | 0 | 0 | 7.06 | 10.04 | 879.21 |
| **30** | 0 | 0 | 0 | 0 | 7.06 | 10.04 | 879.21 |
| **31** | 1 | -1 | -1 | -1 | 5.82 | 10.12 | 876.57 |

The central composite experiment was used to design the factors. The experimental array of the central composite experiment factors shown in Table 4 and the response results extracted from the MOLDFLOW model were obtained.

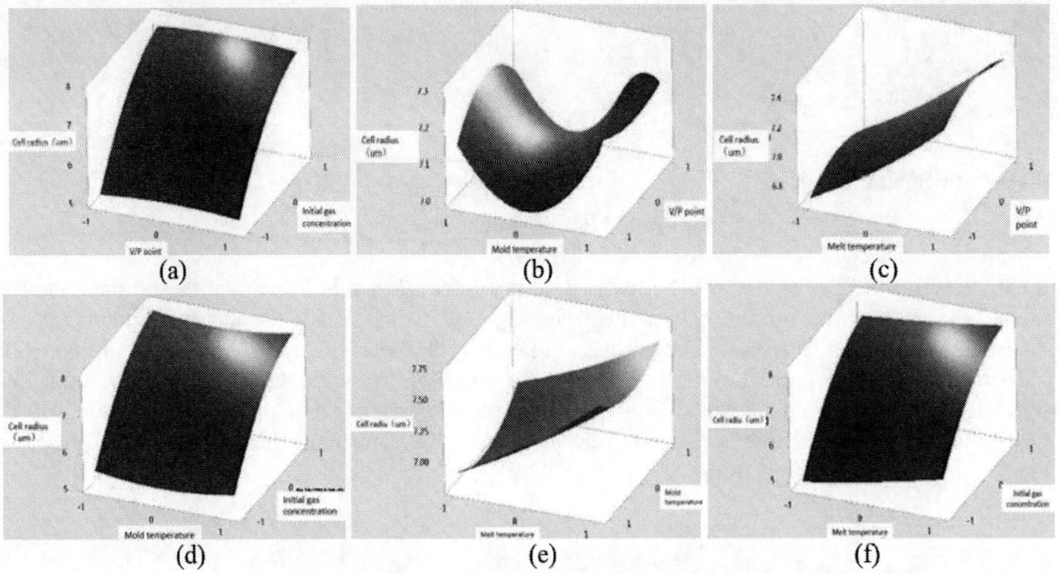


**Figure 6.** Factor Interaction Surface Diagram

Figure 6 shows that the interaction between the V/P transition point and the initial gas concentration is small. When the V/P transition point is at a different level, the bubble radius decreases with the initial concentration of the gas, ie, at V/ A smaller initial gas concentration is used to reduce the radius of the cell when the P-switching point is at different levels.

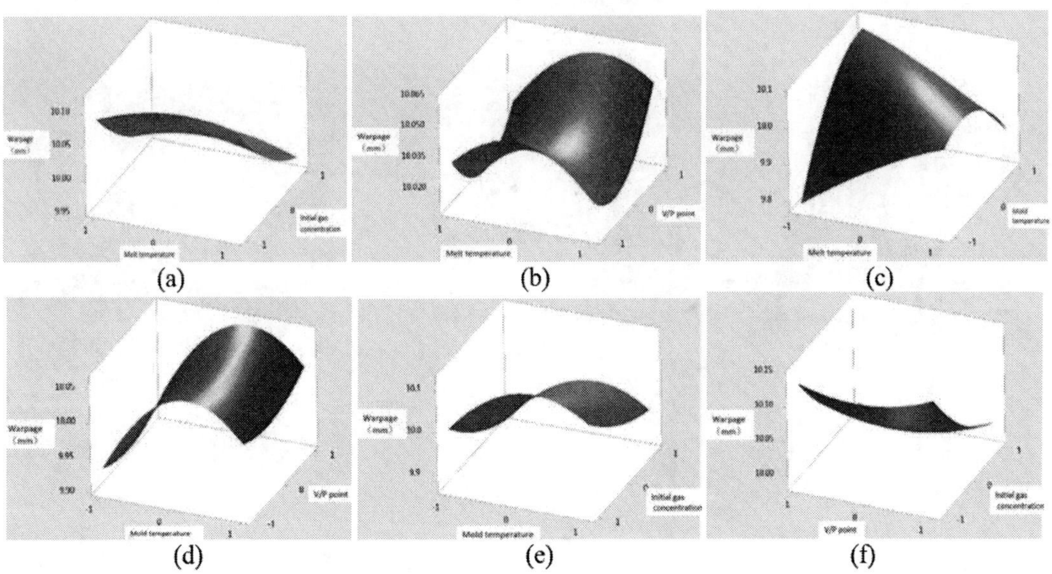


**Figure 7.** Factor Interaction Surface Diagram

### *3.3. Defect Prediction and Process Optimization*

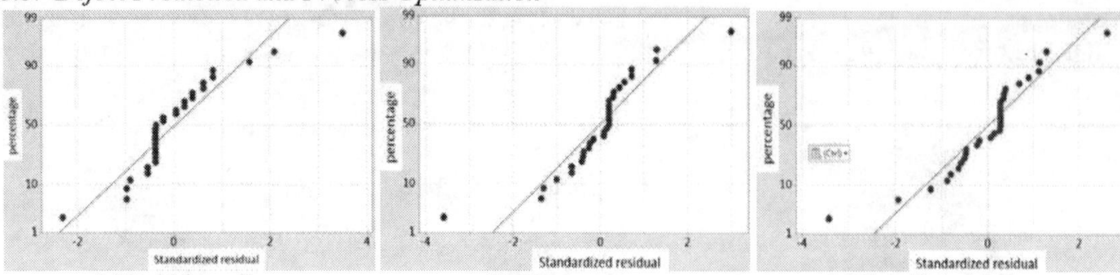


(a)Bubble radius (b)Warpage (c)Tensile modulus

**Figure 8.** Normal probability diagram

It can be seen from the normal probability map of the defect shown in Fig. 8 that the residual between the experimental value of the cell radius, warpage and tensile modulus and the fitted value are distributed near the standard oblique line, indicating that the residual obeys the normal state. Distribution; the defect and the fitted value map are as shown in Fig. 9, and the residual value is kept equal to the variance; as shown in Fig. 10, the residual values are not regularly distributed on the horizontal axis; the cell radius and warpage are explained. The mathematical prediction model of tensile modulus is highly accurate.

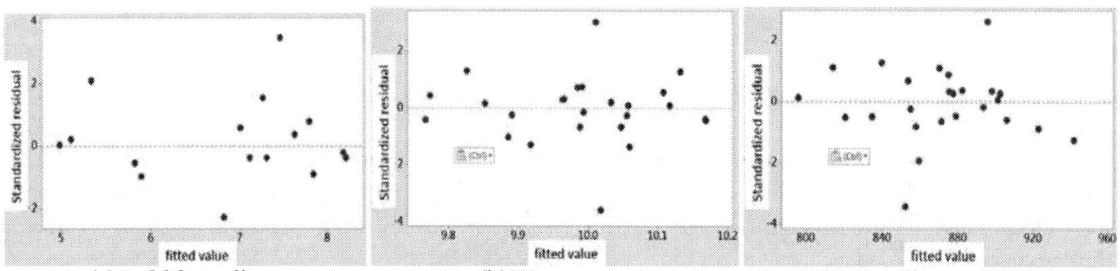


(a)Bubble radius (b)Warpage (c)Tensile modulus

**Figure 9.** And fitting value

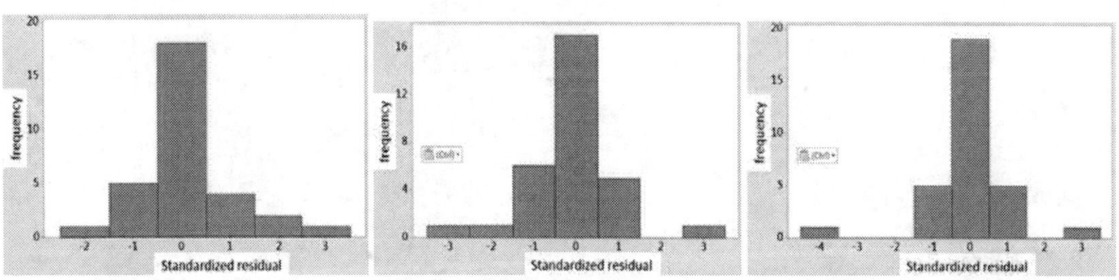


(a)Bubble radius (b)Warpage (c)Tensile modulus

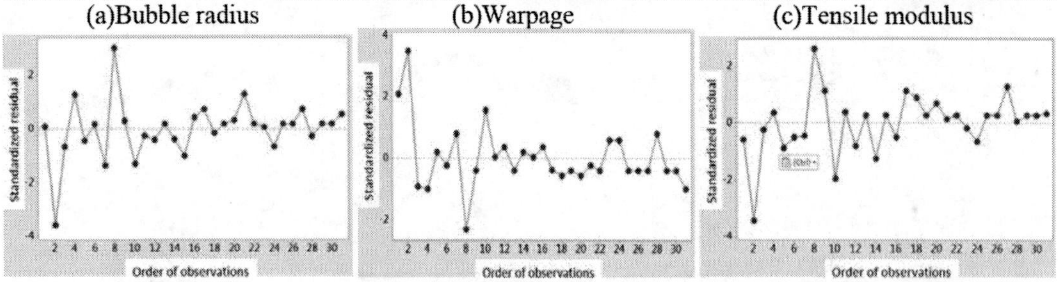


(d)Bubble radius (e)Warpage (f)Tensile modulus

**Figure 10.** And sequence diagram

Therefore, the mathematical regression equation of bubble radius is:

$$Y_1 = 7.1067 + 0.3233A + 0.0239B + 1.2528E + 0.0388A^2 + 0.1738B^2 - 0.1012D^2 - 0.5162E^2 - 0.0500AB - 0.0850AE + 0.0100BE \quad (1)$$

The mathematical regression equation of warpage results is:

$$Y_2 = 10.0333 + 0.00350A + 0.03661B + 0.00522D - 0.06450E - 0.0190A^2 - 0.0790B^2 + 0.0195D^2 + 0.0200E^2 - 0.10550AB + 0.00725AD - 0.00725AE + 0.00088BD + 0.00163BE - 0.00063DE \quad (2)$$

The regression equation of tensile modulus is:

$$Y_3 = 877.43 - 21.85A + 5.37B + 11.18D - 25.58E - 3.86A^2 - 12.78B^2 + 4.63D^2 + 2.78E^2 - 15.60AB + 0.58AD - 3.37AE - 0.21BD + 0.93BE + 0.73DE \quad (3)$$

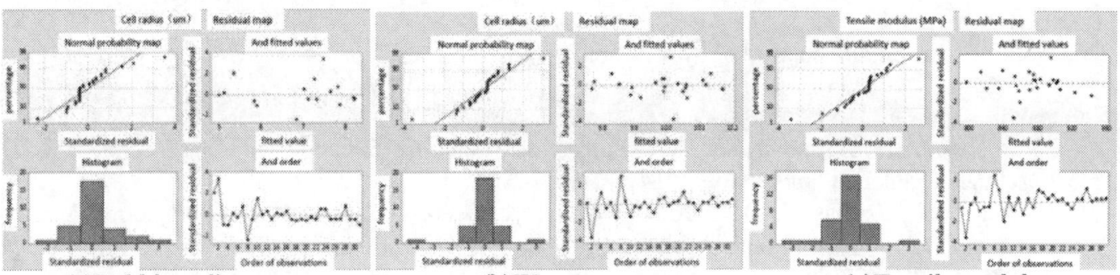

(a)Bubble radius (b)Warpage (c)Tensile modulus

**Figure 11.** Results residual plot

In the target space, the convergence point of the single objective parameter combination is taken as the ideal solution of the multi-objective problem. In the process of microcellular foaming/die decorative composite injection molding, the minimum cell radius and warpage and the maximum tensile modulus are the ultimate goals of the parts.Therefore, the three objective responses are transformed into single objective responses, and the mathematical model of genetic algorithm as shown in formula 5-4 is established by simultaneous (1), (2), (3).

$$F(x) = \frac{Y_1^2 + Y_2^2}{Y_3} \quad (4)$$

S.T. $-1 \le xi \le 1$;

In the formula, $F(x)$ represents the optimal result obtained by fitting.

The GA genetic algorithm toolbox is used to iteratively calculate the function, and the four process parameters A, B, D, and E are optimized. When the optimization target result is 0.134487, the four parameters are combined (-1, -0.221, 1). , -1), the melt temperature after fitting was 200 ° C, the mold temperature was 47.79 ° C, the V/P conversion point was 99%, and the initial gas concentration was 0.3%.

In order to test the optimization results of the genetic algorithm, the optimized process parameters were substituted into Moldflow to obtain a cell radius of 4.7 um, a warpage result of 10.17 mm, and a tensile modulus of 938.09 MPa, which was 0.1338, which met the accuracy requirement. Then, when the melt temperature is 200 ° C, the mold temperature is 47.79 ° C, the injection time is 1 s, the V/P conversion point is 99%, the initial gas concentration is 0.3%, the number of gas cores per volume is 2E+05, the Reynolds number of the cooling medium is 10000, and the temperature of the cooling medium is 25 At °C, it helps to obtain the best part quality.

## 4. Conclusion

(1) According to the partial factor test design, the three factors that have significant influence on the cell surface radius are the initial gas concentration, melt temperature and injection time. The most influential factor on the warpage result is the mold temperature and pull. The significant factors affecting the modulus of elongation are mold temperature, V/P transition point, melt temperature, and initial gas concentration. The accuracy between the model and the experimental values was verified by a central composite test design.

(2) Using genetic algorithm analysis, the optimal combination of process parameters was obtained, that is, when the melt temperature was 200 ° C, the mold temperature was 47.79 ° C, the injection time was 1 s, the V/P conversion point was 99%, the initial gas concentration was 0.3%, and the number of gas cores per volume was 2E. +05, the Reynolds number of the cooling medium is 10000, and the temperature of the cooling medium is 25 °C, which helps to obtain the best quality parts.

**Acknowledgments**
This work was financially supported by xxx fund.

**References**

[1] Rot. Research on Supercritical N2 Microporous Foaming Injection Molding Technology [D]. Fuzhou: Fuzhou University, 2014.
[2] Turng L S, Kharbas H. Development of a Hybrid Solid-Microcellular Co-injection Molding Process [J]. International Polymer Processing Journal of the Polymer Processing Society, 2004, 19(1): 77-86.
[3] Chen H L, Chien R D, Chen S C. Using Thermally Insulated Polymer Film for Mold Temperature Control to Improve Surface Quality of Microcellular Injection Molded Parts [J]. International Communications in Heat & Mass Transfer, 2008, 35(8):991-994.
[4] Brendan J C. Effects of Process Conditions Materials and Manufacturing Processes, 2008, on In-mold Film Decoration [J]. 32(8)64-72.
[5] Zhou Guofa, He Yongyuan. NUMERICAL ANALYSIS OF VOLUME CONTRACTION AND WARPING DEFORMATION OF MICROPOROUS PLASTICS INJECTION MOLDING PRODUCTS [J].CHINESE PLASTICS, 2007(10): 61-65.
[6] Hu Biao, Hu Guanghong. The relationship between surface roughness and surface bubble rupture deformation behavior of microcellular foamed injection moulded products [J]. Journal of Shanghai Jiaotong University 2016, 50(08): 1300-1306.
[7] Huang H X, Wang J K. Equipment Development and Experimental Investigation on the Cellular Structure of Microcellular Injection Molded parts [J]. Polymer Testing, 2008, 27(4): 513-519.
[8] Wang Yue. Study on Improving Surface Quality of Microporous Foaming Injection Molded Products [D]. Shanghai: Shanghai Jiaotong University, 2012.
[9] Li Bei, Mao Huajie, Guo Wei, et al. Study on the relationship between the cell structure and the tensile properties of PP microcellular foamed samples [J]. Application of engineering plastics, 2015, 43(10):52-58

# Fatigue Life of the 316L Stainless Steel Based on Dual-parametric Weibull Distribution

**Ziliang An[1], Ziye Wang[1,*], Yifeng Tang[2], Weixin Wang[1] and Tao Chen[1]**

[1]School of Railway Transportation, Shanghai Institute of Technology, Shanghai, China

[2]School of electromechanical engineering and information, Shanghai Urban Construction Vocational College, Shanghai, China

*Corresponding author e-mail: wzy960413@163.com

**Abstract.** In order to investigate the fatigue strength of the 316L stainless steel (316L-SS) diffusion welded joint, it is necessary to study the fracture time of the 316L-SS. The fracture time test of the 316L-SS has many problems, such as high cost, long time consumption and little data acquisition. Regarding the issue above, the Weibull distribution method is adopted to process the test data of small sample size in this paper. At the same time, the P-S-N curve of the 316L-SS is established to obtain the relationship between maximum stress and fatigue life under different reliability.

## 1. Introduction

Due to the low efficiency and precision of the traditional fusion welding, it is difficult to meet the micro-mechanical connection requirements such as micro-heat exchangers and micro-reactors [1-4]. With the development of the welding technology and the application of special materials, diffusion welding technology has been widely used in aerospace, nuclear energy, stations, petrochemical, electric power, transportation, mechanical manufacturing and other fields, becoming one of the most commonly used material connection technologies [5-8]. As a type of austenitic stainless steel, 316L-SS has many unique advantages, such as excellent mechanical properties, corrosion resistance and high temperature oxidation resistance. Therefore, with operating temperature between 200°C and 650°C, the 316L-SS is often chosen as the base metal in diffusion welding [9]. For the recent decades, the research on the diffusion welding methods has been relatively mature, whereas, there are few literature about the fatigue strength of the diffusion welded joint based on the 316L-SS as the base metal [10, 11]. Therefore, it is necessary to study the rupture time of the 316L-SS.

However, the high cost of the fracture time test and the long experimental time make it difficult to obtain a large amount of the test data analysis [12]. Therefore, for the limited experimental data processing analysis, the correct model and its accurate parameter evaluation are the key to predict the 316L-SS fracture time [13-15]. The Weibull distribution model has great advantages and can well describe the problem to be solved.

The Weibull distribution is one of the most suitable statistical distribution models for describing the mechanical life distribution, which has a strong adaptability to all kinds of test data, especially for small sample sampling tests [16-18]. The Weibull distribution has good data fitting ability and convenient expression processing. It can effectively fit the data of the whole product life stage, and can also be used

Content from this work may be used under the terms of the Creative Commons Attribution 3.0 licence. Any further distribution of this work must maintain attribution to the author(s) and the title of the work, journal citation and DOI.
Published under licence by IOP Publishing Ltd

to simulate the change of failure efficiency. In automobile and other engineering fields, the Weibull distribution is used to analyze the life of the failure parts such as hubs, bearings, gearboxes, etc., so as to accurately predict automobile faults and improve the reliability of automobile products. The Weibull distribution mainly has the three-parameter Weibull distribution ($\gamma$ is the positional parameter, $\eta$ is the size parameter, $\beta$ is the shape parameter) and the dual-parameter Weibull distribution (when $\gamma = 0$, the model is degraded) two forms [19, 20]. The three-parameter Weibull distribution is cumbersome to calculate and suitable for more data. In general, by assuming gamma = 0, the dual-parameter Weibull distribution is often adopted in the engineering. The actual meaning is that the specimen may fail at any time from the beginning of the test. The dual-parameter Weibull distribution will be considered in this study.

## 2. Experiments

The material of the sample used in the experiment is 316L-SS, and the chemical composition of 316L-SS is shown in table 1.

**Table 1.** Chemical composition of the 316L-SS.

| Element | wt % |
|---|---|
| C | 0.01 |
| Fe | 67.10 |
| Ni | 12.43 |
| Si | 0.41 |
| Mo | 2.16 |
| S | 0.01 |
| Cr | 17.84 |
| P | 0.04 |

Since the 316L-SS is easy to work harden, the surface condition of 316L-SS weld is determined to have a roughness of 0.8μm and a flatness of 0.02cm. In order to avoid errors caused by processing and bonding batches, two 316L-SS cylindrically specimens with a gauge length of Φ80mm×50mm are used to make butt diffusion joints. According to the national standards GB/T 228-2002 “metal material - ambient temperature tensile test” and GB/T 4338-2002 “metal material - high temperature tensile test”, the fracture test is carried out on the Shimadzu material testing machine at about 550°C, and the temperature was controlled in high temperature furnace. Five parallel experiments were performed in each group at three groups stress levels ($S_1 = 160$, $S_2 = 140$, $S_3 = 120$), and the fracture time of the joint in each experiment was recorded.

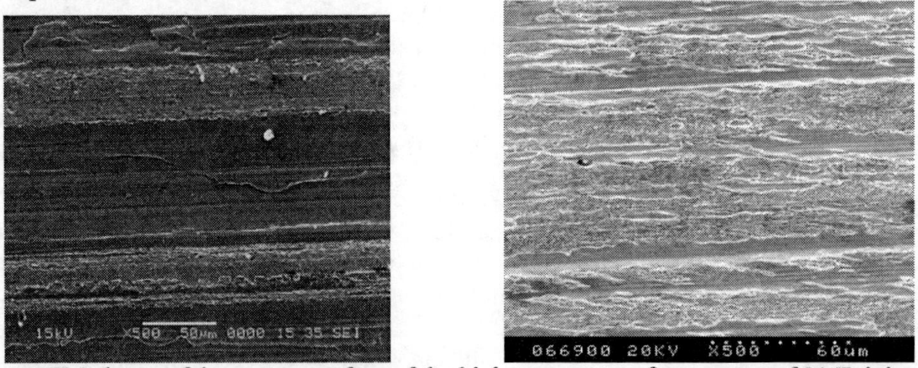

**Figure 1.** SEM photos of the rupture surface of the high-temperature fracture test of 316L joint: (a) the rupture surface of the joint after high temperature tensile, (b) the rupture surface of the joint after testing at 550°C( t= 266.3 hours ).

IOP Conf. Series: Materials Science and Engineering **772** (2020) 012016 doi:10.1088/1757-899X/772/1/012016

## 3. Theoretical model

The Weibull distribution is being used to mitigate the extreme values of the fracture strength and the failure time. Two forms of the method are dual-parameter and three-parameter Weibull distributions. In this case, the distribution function and the probability density function (PDF) can be written as follows, respectively:

$$F(x)=1-\exp\left[-\left(\frac{x-\gamma}{\eta}\right)^{\beta}\right],\quad \gamma\geq 0,\ \eta\geq 0,\ \beta\geq 0 \tag{1}$$

$$f(x)=\frac{\beta}{\eta}\left(\frac{x-\gamma}{\eta}\right)^{\beta-1}\exp\left[-\left(\frac{x-\gamma}{\eta}\right)^{\beta}\right],\quad \gamma\geq 0,\ \eta\geq 0,\ \beta\geq 0 \tag{2}$$

Where $\gamma$ is the positional parameter, when $\gamma=0$, the model is degraded to the dual-parameter Weibull distribution; $\eta$ is the size parameter and $\beta$ is the shape parameter.

As a rule, the three-parameter Weibull distribution is suitable for more data. In this work, the dual-parameter Weibull distribution will be considered. In the circumstances, the distribution function can be expressed as:

$$F(x)=1-\exp\left[-\left(\frac{x}{\eta}\right)^{\beta}\right],\quad \eta\geq 0,\ \beta\geq 0 \tag{3}$$

In the context of this work, F(x) represents the probability that the fatigue strength is equal to or less than x. Using the equation F(x) + R(x) = 1, the reliability R(x), that is, the probability of the fatigue strength is defined as:

$$R(x)=\exp\left[-\left(\frac{x}{\eta}\right)^{\beta}\right],\quad \eta\geq 0,\ \beta\geq 0 \tag{4}$$

Under the certain stress level, if the fatigue life N of the specimen follows the Weibull distribution, then the probability density function is:

$$f(N)=\frac{b}{N_a-N_0}\left(\frac{N-N_0}{N_a-N_0}\right)^{b-1}\exp\left[-\left(\frac{N-N_0}{N_a-N_0}\right)^{b}\right] \tag{5}$$

Where $N_0$ is the minimum life parameter; $N_a$ is the characteristic life parameter; b is the shape parameter. The unreliability function is:

$$F(N)=P(n\leq N)=\int_{N_0}^{N} f(N)dN \tag{6}$$

The distribution function of fatigue life is obtained by substituting equation (5) into equation (6) and integrating it:

$$F(N)=1-\exp\left[-\left(\frac{N-N_0}{N_a-N_0}\right)^{b}\right] \tag{7}$$

The expression of the fatigue life reliability function that conforms to the three-parameter Weibull distribution is:

$$R(N)=\exp\left[-\left(\frac{N-N_0}{N_a-N_0}\right)^{b}\right] \tag{8}$$

The mathematical expectation and variance of the Weibull distribution variables are:

$$E(N) = N_0 + (N_a - N_0)\Gamma\left(1 + \frac{1}{b}\right) \tag{9}$$

$$\mathrm{var}(N) = (N_a - N_0)^2\left[\Gamma\left(1 + \frac{2}{b}\right) - \Gamma^2\left(1 + \frac{1}{b}\right)\right] \tag{10}$$

Let the probability of fatigue life $n \le N$ be:

$$P(n \le N) = p \tag{11}$$

Where p is the reliability, and the reliability corresponding to the fatigue life of p can be expressed as $N_p$.

When the p equals to 0.5 and $N_p$ equals to $N_{50}$, $N_{50}$ corresponds to the median value of the fatigue life samples, substituting $N_p = N_{50}$ into (7):

$$0.5 = 1 - \exp\left[-\left(\frac{N_{50} - N_0}{N_a - N_0}\right)^b\right] \tag{12}$$

Take the logarithm for both sides of formula (12):

$$\ln 0.5 = -\left(\frac{N_{50} - N_0}{N_a - N_0}\right)^b \tag{13}$$

When the fatigue life is affected by the three-parameter Weibull distribution, the reliability function expression of the fatigue specimen is shown in equation (8), and the natural logarithm is taken for both sides of (8).

$$\ln R(N) = -\left(\frac{N - N_0}{N_a - N_0}\right)^b \tag{14}$$

If the reliability R is given, the reliable lifetime $N_R$ is expressed as:

$$N_R = N_0 + (-\ln R)^{1/b}(N_a - N_0) \tag{15}$$

Let the fatigue life N follow the three-parameter Weibull distribution at any stress level S. In reality, each unknown parameter estimated in the probability density function is related to the stress level S. At different stress levels, the distribution parameters of the fatigue life are different, and the three parameters of the Weibull distribution can be expressed as a function of the stress level S. Equation (15) is rewritten as:

$$N_R(S) = N_0(S) + (-\ln R)^{1/b(S)}[N_a(S) - N_0(S)] \tag{16}$$

For $S_{0R}$, aR and $C_R$, the reliability is R, corresponding to the S-N curve equation, as shown in equation (17):

$$(S - S_{0R})^{a_R} N_R = C_R \tag{17}$$

Considering that the fatigue life is a function of stress level, it can be obtained from equation (17):

$$N_R(S) = \frac{C_R}{(S - S_{0R})^{a_R}} \tag{18}$$

Substitute formula (18) into formula (16):

$$\frac{C_R}{(S - S_{0R})^{a_R}} = (N_a(S) - N_0(S))(-\ln R)^{1/b(S)} + N_0(S) \tag{19}$$

IOP Conf. Series: Materials Science and Engineering **772** (2020) 012016 doi:10.1088/1757-899X/772/1/012016

Consequently, on the premise of the given stress level and reliability, the relationship between the three parameters of the Weibull distribution and the three parameters of the curve equation is established by equation (19).

In addition, three stress levels were selected for the fatigue test. Statistical analysis was performed on the results of the three groups of fatigue life tests. The distribution parameters $N_a(S_i)$, $N_0(S_i)$, and $b(S_i)$ ($i=1,2,3$) of the three-parameter Weibull distribution were obtained. Substitute equation (16) into an equation system composed of three equations:

$$\begin{cases} N_{R,S_1} = (N_a(S_1) - N_0(S_1))(-\ln R)^{1/b(S_1)} + N_0(S_1) \\ N_{R,S_2} = (N_a(S_2) - N_0(S_2))(-\ln R)^{1/b(S_2)} + N_0(S_2) \\ N_{R,S_3} = (N_a(S_3) - N_0(S_3))(-\ln R)^{1/b(S_3)} + N_0(S_3) \end{cases} \quad (20)$$

By formula (17):

$$\begin{cases} \ln C_{R,S_1} = \ln N_{R,S_1} + a_R \ln(S_1 - S_{0R}) \\ \ln C_{R,S_2} = \ln N_{R,S_2} + a_R \ln(S_2 - S_{0R}) \\ \ln C_{R,S_3} = \ln N_{R,S_3} + a_R \ln(S_3 - S_{0R}) \end{cases} \quad (21)$$

Separately subtract two by using the formula in (21):

$$\begin{cases} a_R[\ln(S_1 - S_{0R}) - \ln(S_2 - S_{0R})] = \ln N_{R,S_2} - \ln N_{R,S_1} \\ a_R[\ln(S_2 - S_{0R}) - \ln(S_3 - S_{0R})] = \ln N_{R,S_3} - \ln N_{R,S_2} \end{cases} \quad (22)$$

Divide the two formulas in equation (22):

$$\frac{\ln(S_1 - S_{0R}) - \ln(S_2 - S_{0R})}{\ln(S_2 - S_{0R}) - \ln(S_3 - S_{0R})} = \frac{\ln N_{R,S_2} - \ln N_{R,S_1}}{\ln N_{R,S_3} - \ln N_{R,S_2}} \quad (23)$$

Let

$$\frac{\ln N_{R,S_2} - \ln N_{R,S_1}}{\ln N_{R,S_3} - \ln N_{R,S_2}} = m \quad (24)$$

Replace equation (24) with equation (25):

$$\ln\left(\frac{S_1 - S_{0R}}{S_2 - S_{0R}}\right) = \ln\left(\frac{S_2 - S_{0R}}{S_3 - S_{0R}}\right)^m \quad (25)$$

$$(S_2 - S_{0R})^{m+1} - (S_1 - S_{0R})(S_3 - S_{0R})^m = 0 \quad (26)$$

Solving the nonlinear equation of equation (26), and the estimated value $S_{0R'}$ of $S_{0R}$ is obtained. According to any formula in equation (27), the estimated value $a_{R'}$ of $a_R$ is estimated.

$$a_R' = \frac{\ln N_{R,S_2} - \ln N_{R,S_1}}{\left[\ln(S_1 - S_{0R}') - \ln(S_2 - S_{0R}')\right]} \quad (27)$$

The value of $C_{R'}$ and $C_R$ can be estimated from the equation (21). On the basis of equation (17), the P-S-N curve equation of the material can be obtained.

## 4. Results and discussion

Table 2 shows the rupture time under three different stresses. The Weibull probability plot of the experimental results is shown in Figure 1.

**Table 2.** The results of the rupture testing for 316L-SS diffusion bonding joint.

| No. | $S_1$=160MPa | $S_2$=140MPa | $S_3$=120MPa |
|---|---|---|---|
| 1 | 141 | 236 | 402 |
| 2 | 164 | 243.5 | 423 |
| 3 | 175 | 255 | 436 |
| 4 | 192 | 261 | 442 |
| 5 | 209 | 273 | 450 |

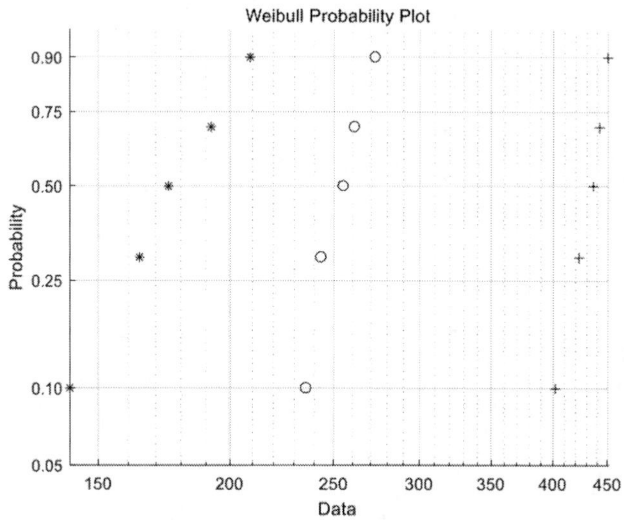


**Figure 2.** The Weibull probability plots of the three groups of experimental results.

The Weibull probability plots of the three groups of data are plotted respectively. It can be seen from the figure 2 that the fracture time in each group is linearly distributed under three different stresses. Therefore, the relevant parameters can be obtained by using the formula described above.

As shown in table 3, based on the formula to establish a nonlinear equation group, the distribution parameters of the three groups of data are found.

**Table 3.** Probability distribution parameters for each group of data.

| Statistical feature value | Calculation results | | |
|---|---|---|---|
| | $S_1$=160MPa | $S_2$=140MPa | $S_3$=120MPa |
| Mean | 176.2 | 253.7 | 430.6 |
| Median | 175 | 255 | 436 |
| Standard deviation | 26.052 | 14.533 | 18.783 |
| Shape parameters | 2.695 | 7.287 | 3.683 |
| Characteristic life | 184.311 | 259.699 | 463.461 |
| Minimum life | 111.088 | 163.911 | 394.546 |

Taking the reliability R equal to 0.5, the P-S-N curve equation can be obtained as:

$$(S-92.11)^{1.0788}N=16565 \tag{28}$$

The P-S-N curve is shown in Figure3.

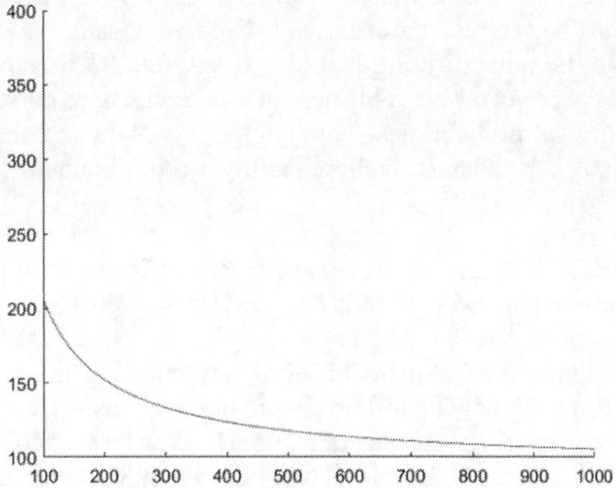


**Figure 3.** P-S-N curve (R=0.5).

Moreover, taking the reliability of R equal to 0.1, 0.3, 0.8, 0.95 respectively, the corresponding P-S-N curve equation can be obtained:

$$(S-111.46)^{0.4746}N=1331.2 \tag{29}$$

$$(S-102.397)^{0.7607}N=4138.6 \tag{30}$$

$$(S-64.495)^{1.9475}N=1.099\times10^{6} \tag{31}$$

$$(S-31.039)^{3.0823}N=4.33\times10^{8} \tag{32}$$

The P-S-N curves are shown in Figure 4.

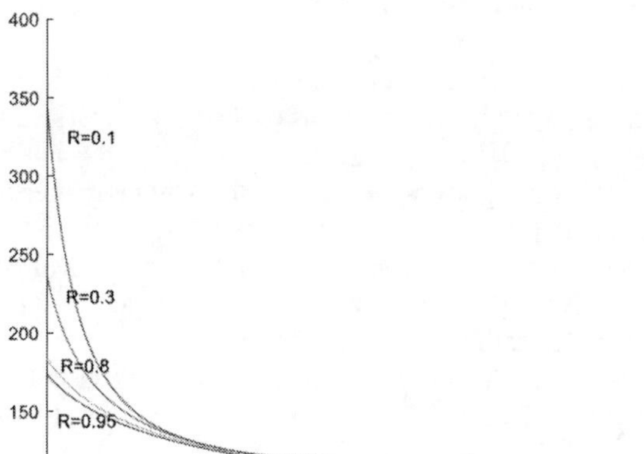


**Figure 4.** P-S-N curve(R=0.1,0.3,0.8,0.95).

## 5. Conclusion

In order to investigate the fatigue strength of the 316L-SS diffusion welded joint, in this work, the dual-parameter Weibull distribution is used to analyze the distribution of the fracture time under the small sample experimental data. The fracture time of the samples at different stress levels is independent and identical.

First, the reliability R equals to 0.5. Under different stress levels, multiple tests are carried out to obtain the fracture time of the sample, and then the S-N curve equation is obtained though regression analysis. In addition, take the reliability R equal to 0.1, 0.3, 0.8, 0.95, respectively. By repeating the above method, the P-S-N curve can be established, that is, each curve corresponds to a group of S-N curves under different survival probabilities. Through the P-S-N curves, the relationship between the maximum stress and fatigue life under different reliability can be obtained.

**References**

[1] Bahiraei M, Rahmani R, Yaghoobi A, et al. Recent research contributions concerning use of nanofluids in heat exchangers: a critical review[J]. Applied Thermal Engineering, 2018, 133: 137-159.
[2] Bahiraei M, Jamshidmofid M, Amani M, et al. Investigating exergy destruction and entropy generation for flow of a new nanofluid containing graphene–silver nanocomposite in a micro heat exchanger considering viscous dissipation[J]. Powder technology, 2018, 336: 298-310.
[3] Tiwari A, Rajesh V M, Yadav S. Biodiesel production in micro-reactors: A review[J]. Energy for sustainable development, 2018, 43: 143-161.
[4] Samih S, Latifi M, Farag S, et al. From complex feedstocks to new processes: The role of the newly developed micro-reactors[J]. Chemical Engineering and Processing-Process Intensification, 2018, 131: 92-105.
[5] Dunford D V, Wisbey A. Diffusion bonding of advanced aerospace metallics[J]. MRS Online Proceedings Library Archive, 1993, 314.
[6] Bhanumurthy K, Joyson D, Jawale S B, et al. Diffusion bonding of nuclear materials[J]. BARC Newsl, 2013, 331: 19-25.
[7] Steinberg M A. Materials for aerospace[J]. Scientific American, 1986, 255(4): 66-73.
[8] Trask R D. Diffusion bonding machine and method: U.S. Patent 9,561,558[P]. 2017-2-7.
[9] Kliauga A M , Travessa D , Ferrante M . Al 2O 3/Ti interlayer/AISI 304 diffusion bonded joint : Microstructural characterization of the two interfaces[J]. Materials Characterization, 2001, 46(1):65-74.
[10] Cai Q, Zhu W, Ma Y, et al. Rational design of composite interlayer for diffusion bonding of tungsten–steel joints[J]. International Journal of Refractory Metals and Hard Materials, 2018, 70: 155-161.
[11] Sharma G, Tiwari L, Dwivedi D K. Impulse pressure assisted diffusion bonding of low carbon steel using silver interlayer[J]. Transactions of the Indian Institute of Metals, 2018, 71(1): 11-21.
[12] Smith T R, Sugar J D, San Marchi C, et al. Strengthening mechanisms in directed energy deposited austenitic stainless steel[J]. Acta Materialia, 2019, 164: 728-740.
[13] Ganesh Kumar J , Vijayanand V D , Nandagopal M , et al. Evaluation of variation of tensile strength across 316LN stainless steel weld joint using automated ball indentation technique[J]. Materials at High Temperatures, 2015, 32(6):619-626.
[14] Riemer A, Leuders S, Thöne M, et al. On the fatigue crack growth behavior in 316L stainless steel manufactured by selective laser melting[J]. Engineering Fracture Mechanics, 2014, 120: 15-25.
[15] Tapia G, Khairallah S, Matthews M, et al. Gaussian process-based surrogate modeling framework for process planning in laser powder-bed fusion additive manufacturing of 316L stainless steel[J]. The International Journal of Advanced Manufacturing Technology, 2018, 94(9-12): 3591-3603.
[16] Bachour G, Bou-Chacra N A, Löbenberg R. Evaluation of the rupture test for stability studies of soft-shell capsules[J]. Dissolution Technologies, 2017, 24(2): 16-19.
[17] Maillet E, Godin N, R'Mili M, et al. Analysis of Acoustic Emission energy release during static fatigue tests at intermediate temperatures on Ceramic Matrix Composites: Towards rupture time prediction[J]. Composites Science and Technology, 2012, 72(9): 1001-1007.

[18] Sharif M N, Islam M N. The Weibull distribution as a general model for forecasting technological change[J]. Technological Forecasting and Social Change, 1980, 18(3): 247-256.
[19] Pham H, Lai C D. On recent generalizations of the Weibull distribution[J]. IEEE transactions on reliability, 2007, 56(3): 454-458.
[20] Cousineau D. Fitting the three-parameter Weibull distribution: Review and evaluation of existing and new methods[J]. IEEE Transactions on Dielectrics and Electrical Insulation, 2009, 16(1): 281-288.

# Fluid-Structure Coupling Vibration of an Elastically Restrained Circular Plate

**Shihui Yao [1, 2, a], Aihua Fu [1, 2], Junhua Xing [1, 2], Jinhua Zhou [1, 2] and Bo Tian [1, 2]**

[1] National Key Laboratory on Ship Vibration and Noise, Wuhan, China
[2] China Ship Development and Design Center, Wuhan, China

a yaoshihuicn@foxmail.com

**Abstract.** In this paper, an analytical method is presented to calculate the fluid-structure coupling vibration of an elastically restrained circular plate. The displacement of the plate is expanded in series of dry mode shapes and the motion of the fluid is described by velocity potential function. Considering the equilibrium differential equation of the plate and the velocity continuous conditions at the interface of fluid and structure, the Galerkin method and expansion of Fourier-Bessel series are used to establish the governing equations of the system. The free vibration characteristics of the circular plate in contact with fluid are studied and the present method is verified by comparison with numerical results. Furthermore, the influence of spring stiffness and fluid depth on the free vibration characteristics of the circular plate is investigated.

**Keywords:** elastic boundary conditions, circular plate, fluid-structure interaction, Galerkin method, Fourier-Bessel series.

## 1. Introduction

Circular plates are widely applied in engineering. In many cases, circular plates are in contact with fluid, such as micro pumps, the nuclear reactor internal components, disk brakes, etc. The existence of fluid decreases the free vibration frequencies of the circular plate, which has a significant influence on the dynamic characteristics of the circular plate. Therefore, it is meaningful to study the fluid-structure coupling vibration of circular plates.

Kwak [1] studied the vibration characteristics of a circular plate floating on a infinite fluid region. By using Nondimensionalized Added Virtual Mass Incremental (NAVMI) to evaluate the influence of the fluid, the coupled natural frequencies of the plate are derived approximately. Kwak [2] applied Hankel transform to calculating the hydrodynamic characteristics of a circular plate clamped in a rigid baffle. Kwak and Han [3] calculated the vibration characteristic of a circular plate floating on a fluid with finite depth. References [1]~[3] ignored the influence of fluid on the mode shapes of the circular plate, thus the results obtained are not accurate. By using Rayleigh-Ritz method to establish the governing equation, Amabili and Kwak [4] gained more accurate results. It was also found that the calculation deviation by using NAVMI method is less for lower order frequencies. Bauer [5] studied the coupling vibration of an elastic cover of a cylindrical container and the fluid inside the container. On basis of [5], Bauer and Chiba [6] considered the influence of fluid viscidity. Amabili [7] investigated the vibration charastics of a circular plate floating on the fluid in a container, and the influence of free surface wave is studied.

Content from this work may be used under the terms of the Creative Commons Attribution 3.0 licence. Any further distribution of this work must maintain attribution to the author(s) and the title of the work, journal citation and DOI.
Published under licence by IOP Publishing Ltd

IOP Conf. Series: Materials Science and Engineering **772** (2020) 012017 doi:10.1088/1757-899X/772/1/012017

Yousefzadeh [8] analysed the vibration characteristics of a functionally graded circular plate floating on fluid. Cheung and Zhou [9] studied the free vibration of a flexible thin plate placed into a circular hole and elastically connected to the rigid bottom slab of a circular cylindrical container. Eftekhari [10] introduced differential quadrature method (DQM) to solve the fluid-structure vibration of the circular plate. Tariverdilo et al. [11] solved the coupled natural frequencies of the plate in contact with fluid by employing Fourier-Bessel series and variational formulation, respectively. But the method was not applicable to axisymmetric modes of the circular plate. Jeong [12] calculated the fluid-structure coupling vibration of two identical circular plates, and the elements of the added mass matrix was adjusted for axisymmetric modes. However, the result error of the out of phase mode was relatively large.

On basis of [11] and [12], this paper investigates the fluid-structure coupling vibration of an elastically restrained circular plate in contact with fluid. The expansion of Fourier Bessel series and Galerkin method are combined to establised the governing equations. Considering the properties of Bessel functions of index 0, the added constraint equation is introduced to calculate the coupled frequencies of axisymmetric modes. Furthermore, the effects of spring stiffness and fluid depth on the vibration characteristics of the plate are studied.

## 2. Theoretical Background

### *2.1. Formulations for Circular Plates*

Figure 1 depicts an elastically restrained circular plate coupled with an incompressible and inviscid fluid. The radius and thickness of the plate are *a* and *h*, respectively. The fluid depth of is *d*. The side walls at $r=a$ are considered to be rigid walls. The distributed stiffness of translational springs and rotational springs are *K* and $K_\phi$, respectively.

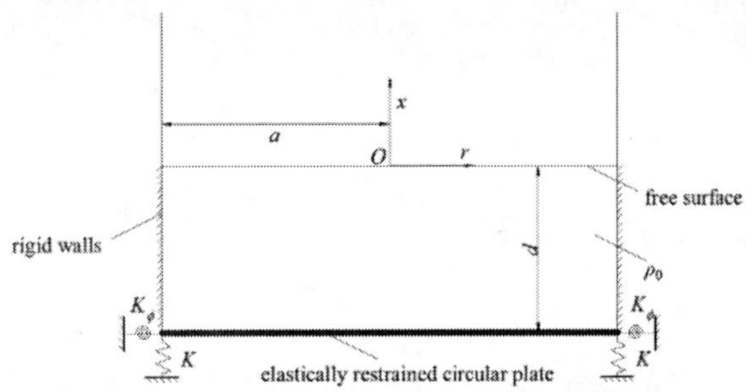


**Figure 1.** Elastically restrained circular plate in contact with fluid

Assuming the circular plate satisfies Kirchhoff's thin plate theory, the dynamic equilibrium equation of the system can be written as

$$D\nabla^4 w + \rho h \frac{\partial^2 w}{\partial t^2} = P \tag{1}$$

where $w$, $D$, $\rho$ are the deflection, bending stiffness and density of the circular plate, respectively. $P$ is the fluid pressure and $\nabla^2$ is the Laplace operator in polar coordinates.

According to modal orthogonality theory, the plate deflection can be expanded in series of mode shapes in vacuum:

$$w(r,\theta,t) = \sum_{n=0}^{\infty}\sum_{m=0}^{\infty} q_{nm} W_{nm}(r,\theta) e^{i\omega t} \tag{2}$$

where $W_{nm}$ is the mode shapes of the circular plate in vacuum, $q_{nm}$ is a generalized coordinate, $n$ and $m$ represent numbers of nodal diameters and nodal circles, respectively.

The mode shape of the circular plate in vacuum can be written as [13]:

$$W_{nm}(r,\theta) = \left[J_n(\lambda_{nm} r) + C_{nm} I_n(\lambda_{nm} r)\right]\cos(n\theta) \tag{3}$$

IOP Conf. Series: Materials Science and Engineering **772** (2020) 012017 doi:10.1088/1757-899X/772/1/012017

where $J_n$ and $I_n$ are the Bessel function and the modified Bessel function of the first kind, $\lambda_{nm}$ is the frequency parameter defined by

$$\lambda_{nm}=\sqrt[4]{\omega_{nm}{}^2\rho h/D} \tag{4}$$

Where $\omega_{nm}$ represents the natural freequency of the circular plate in vacuum. When the size and material of the plate are certain, the parameters $\lambda_{nm}$ and $C_{nm}$ are determined by $K$ and $K_\phi$.

### 2.2. *Velocity potential function of the fluid*

Assuming the movement of the fluid is irrotational, the velocity potential function $\Phi(r, \theta, x, t)$ can be introduced to describe the fluid movement. Considering the fluid is incompressible and inviscid, $\Phi(r, \theta, x, t)$ satisfies the Laplace equation:

$$\nabla^2\Phi=0 \tag{5}$$

Separate the function $\Phi$ into time and space terms:

$$\Phi(r,\theta,x,t)=i\omega\phi(r,\theta,x)e^{i\omega t} \tag{6}$$

where $\phi$ is the spatial velocity potential, which also satisfies the Laplace equation.
Using method of separation of variables, the general solution of the function $\phi$ can be derived:

$$\phi(r,\theta,x)=(a_n x+b_n)\delta_{n0}\cos n\theta+$$
$$\sum_{s=1}^{\infty}J_n(\beta_{ns}r)\left[E_{ns}\sinh(\beta_{ns}x)+F_{ns}\cosh(\beta_{ns}x)\right]\cos n\theta \tag{7}$$

Where $\delta_{n0}$ is Kronecker delta function, $a_n$, $b_n$, $E_{ns}$, $F_{ns}$, $\beta_{ns}$ are undetermined coefficients and $\beta_{ns}>0$.
The boundary condition at the rigid walls is:

$$\left.\frac{\partial\phi(r,\theta,x)}{\partial r}\right|_{r=a}=0 \tag{8}$$

yields

$$J_n'(\beta_{ns}a)=0 \tag{9}$$

The boundary condition at the free surface is

$$P|_{x=0}=\rho_0\left.\frac{\partial\phi}{\partial t}\right|_{x=0}=i\omega\rho_0 e^{j\omega t}\phi\Big|_{x=0}=0 \tag{10}$$

Substitution of Eq. (3) into Eq. (10) gives

$$b0=0 \tag{11}$$
$$Fns=0 \tag{12}$$

Hence the velocity potential function $\phi$ can be reduced to

$$\phi(r,\theta,x)=\cos n\theta\left[a_0 x\delta_{n0}+\sum_{s=1}^{\infty}J_n(\beta_{ns}r)E_{ns}\sinh(\beta_{ns}x)\right] \tag{13}$$

### 2.3. *Velocity Continuity at the Fluid-Structure Interface*

At the fluid-structure interface, the velocity continuity condition should be satisfied:

$$\frac{\partial w}{\partial t}=\left.\frac{\partial\Phi}{\partial x}\right|_{x=-d} \tag{14}$$

yields

$$\sum_{m=0}^{\infty}q_{nm}R_{nm}(r)=a_0\delta_{n0}+\sum_{s=1}^{\infty}E_{ns}\beta_{ns}J_n(\beta_{ns}r)\cosh(\beta_{ns}d) \tag{15}$$

where $R_{nm}(r)=J_n(\lambda_{nm}r)+C_{nm}I_n(\lambda_{nm}r)$ is the radial component of the $W_{nm}$. According to the orthogonality of the eigenfunctions of Bessel equations, multiplying both sides of Eq. (15) by $rJ_n(\beta_{ns}r)$ , and integrating on domain [0, $a$] gives

$$E_{ns}=\frac{\sum_{m=0}^{M} q_{nm}\int_0^a r_{nm}J_n(\beta_{ns}r)R_{nm}(r)\mathrm{d}r}{\beta_{ns}\cosh(\beta_{ns}d)\int_0^a r\left[J_n(\beta_{ns}r)\right]^2\mathrm{d}r} \tag{16}$$

When $n$=0, constant function is also eigenfunctions of the Bessel equation of order 0. Multiplying both sides of Eq. (15) by $r$, and integrating on domain [0, $a$] yields

$$\sum_{m=0}^{\infty} q_{0m}\int_0^a rR_{0m}(r)\mathrm{d}r-\frac{a^2}{2}a_0=0 \tag{17}$$

Substitution of Eq. (16) into Eq. (13), the velocity potential function can be written as

$$\phi(r,\theta,x)=\cos n\theta\left[a_0x\delta_{n0}+\sum_{m=0}^{\infty} q_{nm}\sum_{s=1}^{\infty}\frac{T_{nms}}{\cosh(\beta_{ns}d)}J_n(\beta_{ns}r)\sinh(\beta_{ns}x)\right] \tag{18}$$

where

$$T_{nms}=\frac{\int_0^a rR_{nm}(r)J_n(\beta_{ns}r)\mathrm{d}r}{\beta_{ns}\int_0^a r\left[J_n(\beta_{ns}r)\right]^2\mathrm{d}r} \tag{19}$$

### *2.4. Establishment of the Governing Equations*

Employing linearized Bernoulli's equation, the fluid pressure exerted on the circular plate is

$$P=\rho_0\left.\frac{\partial\Phi}{\partial t}\right|_{x=-d}=\rho_0\omega^2e^{i\omega t}\left.\phi\right|_{x=-d} \tag{20}$$

where $\rho_0$ is the density of fluid. And the equilibrium equation can be expressed as

$$D\sum_{m=0}^{\infty} q_{nm}\nabla^4W_{nm}-\rho h\omega^2\sum_{m=0}^{\infty} q_{nm}W_{nm}-\rho_0\omega^2\left[a_0d\delta_{n0}+\sum_{m=0}^{\infty} q_{nm}\sum_{s=1}^{\infty}\frac{T_{nms}}{\coth(\beta_{ns}d)}J_n(\beta_{ns}r)\right]\cos n\theta=0 \tag{21}$$

Multiplying both sides of Eq. (21) by $W_{nk}(r,\theta)$ and integrating over the circular plate gives

$$\rho h\left(\omega_{nm}^2-\omega^2\right)q_{nm}\int_0^a rR_{nk}^2\mathrm{d}r-\rho_0\omega^2\left[a_0d\delta_{n0}\int_0^a rR_{nk}\mathrm{d}r+\sum_{m=0}^{\infty} q_{nm}\sum_{s=1}^{\infty}\frac{T_{nms}}{\coth(\beta_{ns}d)}\int_0^a rJ_n(\beta_{ns}r)R_{nk}\mathrm{d}r\right]=0 \tag{22}$$

Truncating $m$ and $s$ to $M$ and $S$, Eq. (22) can be expressed in matrix form. When $n>0$, Eq. (22) can be written as

$$\left[\rho h\mathbf{P}_n-\omega^2\left(\rho h\mathbf{Z}_n+\rho_0\mathbf{G}_n\right)\right]\mathbf{q}_n=\mathbf{0},\quad n>0 \tag{23}$$

where $\mathbf{q}_n=\{q_{n0},q_{n1},...,q_{nM}\}^T$ is the coefficient vector and $\mathbf{P}_n$, $\mathbf{Z}_n$, $\mathbf{G}_n$ are $(M+1)\times(M+1)$ matrixes whose elements are given by

$$P_{n,km}=\omega_{km}^2\int_0^a rR_{nk}^2(r)\mathrm{d}r\delta_{km},\quad k,m=0,1,2...,M \tag{24}$$

$$Z_{n,km} = \int_0^a rR_{nk}^2(r)\mathrm{d}r\delta_{km},\ k,m = 0,1,2...,M \tag{25}$$

$$G_{n,km} = \sum_{s=1}^{S} T_{nms}\tanh(\beta_{ns}d)\int_0^a rJ_n(\beta_{ns}r)R_{nk}(r)\mathrm{d}r \tag{26}$$

When n=0, to derive the coefficients $a_0$ and $q_{0m}$, Eq. (17) and Eq. (22) should be solved simultaneously. And the governing equation for $n$=0 can be written as:

$$\begin{bmatrix} \rho h\mathbf{P}_0 - \omega^2(\rho h\mathbf{Z}_0 + \rho_0\mathbf{G}_0) & \rho_0\omega^2\mathbf{c}_0 \\ \mathbf{l}_0 & -a^2/2 \end{bmatrix}\begin{bmatrix} \mathbf{q}_0 \\ a_0 \end{bmatrix} = \mathbf{0},\quad n=0 \tag{27}$$

where $\mathbf{c}_0$ and $\mathbf{l}_0$ are column vector and row vector of dimension $M$+1, respectively. The elements of the vectors are defined by

$$c_{0,k} = \int_0^a rR_{0k}(r)\mathrm{d}r \tag{28}$$

$$l_{0,m} = \int_0^a rR_{0m}(r)\mathrm{d}r \tag{29}$$

Eqs. (23) and (27) give the governing equations in matrix form for any fixed number $n$. By solving the generalized eigenvalue problem, the natural frequencies and the mode shapes of the circular plate in contact with fluid can be derived.

## 3. Numerical Results and Discussion

### 3.1. *Verification of Results*

To check the validity of the present method, the finite element analysis is carried out by using FEM software ANSYS. The values of the parameters used in computation are listed in Table 1.

0Gives the comparison of the coupled natural frequencies obtained in this paper with the FEM results. It can be found that, the results of the present method agree well with the FEM results, and the relative error is less than 1%.

**Table 1.** Values of parameters used in computation

| **Parameter** | **Value** |
|---|---|
| plate radius | $a$=0.12m |
| plate thickness | $h$=0.002m |
| plate density | $\rho$=2700kg/m$^3$ |
| plate Young modulus | $E$=69Gpa |
| plate Poison's ratio | $\nu$=0.3 |
| fluid density | $\rho_0$=1000kg/m$^3$ |
| fluid depth | $d$=0.02m |
| distributed stiffness of translational springs | $K$=10$^3$N/m$^2$ |
| distributed stiffness of rotational springs | $K_\phi$=10$^3$N/rad |

**Table 2.** Comparison of the coupled natural frequencies between the present method and fem

| $n$ | $m$ | **FEM (Hz)** | **Theory (Hz)** | **Discrepancy (%)** |
|---|---|---|---|---|
| 0 | 0 | 4.1 | 4.1 | 0.0 |
| 0 | 1 | 192.6 | 192.2 | -0.2 |
| 0 | 2 | 753.6 | 748.9 | -0.6 |
| 0 | 3 | 1801.8 | 1779.5 | -1.2 |
| 1 | 0 | 34.9 | 34.9 | 0.0 |
| 1 | 1 | 403.3 | 401.9 | -0.3 |
| 1 | 2 | 1192.3 | 1181.6 | -0.9 |
| 1 | 3 | 2516.8 | 2477.2 | -1.6 |
| 2 | 0 | 112.5 | 112.3 | -0.2 |
| 2 | 1 | 684.0 | 680.7 | -0.5 |
| 2 | 2 | 1720.8 | 1702.8 | -1.0 |
| 2 | 3 | 3329.4 | 3273.8 | -1.7 |
| 3 | 0 | 234.5 | 233.7 | -0.3 |
| 3 | 1 | 1039.8 | 1031.8 | -0.8 |
| 3 | 2 | 2347.0 | 2312.9 | -1.5 |
| 3 | 3 | 4262.2 | 4167.2 | -2.2 |

### *3.2. Effect of Spring Stiffness and Fluid Depth*

0lists the coupled natural frequencies of the circular plate with different spring stiffness. The table shows that for range of $K$ and $K_\phi$ between 10 and $10^9$, the natural frequencies of the circular plate increase significantly as the spring stiffness increases. When the spring stiffness range from $10^9$ to $10^{10}$, the coupled natural frequencies remain nearly the same, which means the boundary conditions can be seemed to be clamped boundary conditions when the spring stiffness is higher than $10^9$.

**Table 3.** Coupled natural frequencies of circular plate with different spring stiffness

| $K$ | $K_\phi$ | $(n,m)$ | | | | | |
|---|---|---|---|---|---|---|---|
| | | (0,0) | (0,1) | (0,2) | (1,0) | (1,1) | (1,2) |
| 10 | 10 | 0.4 | 148.9 | 687.6 | 4.9 | 350.6 | 1114.5 |
| $10^3$ | $10^3$ | 4.1 | 192.2 | 748.9 | 34.9 | 401.9 | 1181.6 |
| $10^5$ | $10^5$ | 40.0 | 243.2 | 873.6 | 68.9 | 482.7 | 1348.1 |
| $10^7$ | $10^7$ | 155.6 | 510.6 | 1021.1 | 305.5 | 707.3 | 1451.9 |
| $10^9$ | $10^9$ | 162.8 | 675.7 | 1630.8 | 348.0 | 1071.5 | 2273.7 |
| $10^{10}$ | $10^{10}$ | 162.8 | 675.7 | 1630.8 | 348.4 | 1075.9 | 2296.0 |

To evaluate the fluid effects on the free vibration of the circular plate, the normalized natural frequency, defined as the ratio of the coupled natural frequencies to the corresponding natural frequencies in vacuo, is introduced. As can be seen in 0, the values of normalized natural frequencies are always smaller than 1. It means that, due to the contribution of added mass, the existence of fluid reduces the natural frequencies of the circular plate. As can be also inferred from the figure, with the increase of $n$ or $m$, the normalized frequencies increase. That is to say, for the mode with higher order, the added fluid mass acting on the plate is less. And this conclusion applies to circular plates with different spring stiffness.

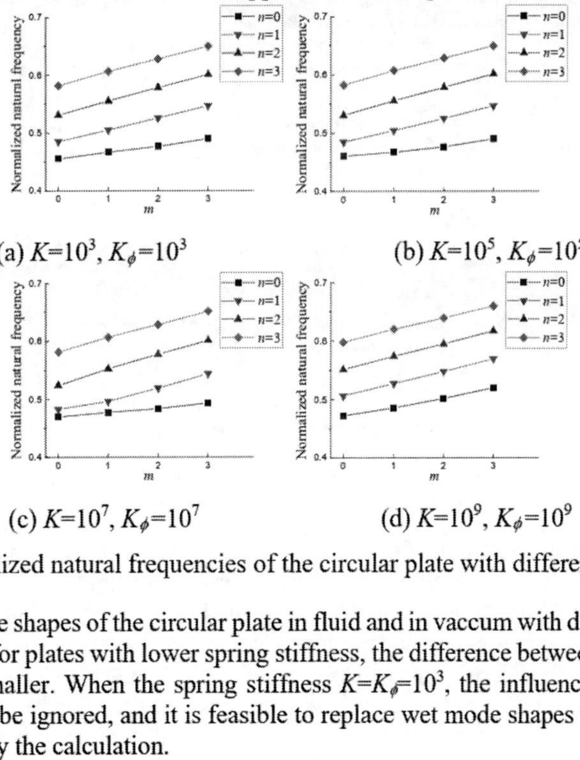


**Figure 2.** Normalized natural frequencies of the circular plate with different spring stiffness

0 compares the mode shapes of the circular plate in fluid and in vaccum with different spring stiffness. It can be inferred that, for plates with lower spring stiffness, the difference between dry mode shapes and wet mode shapes is smaller. When the spring stiffness $K=K_\phi=10^3$, the influence of fluid on the mode shapes of the plate can be ignored, and it is feasible to replace wet mode shapes with corresponding dry mode shapes to simplify the calculation.

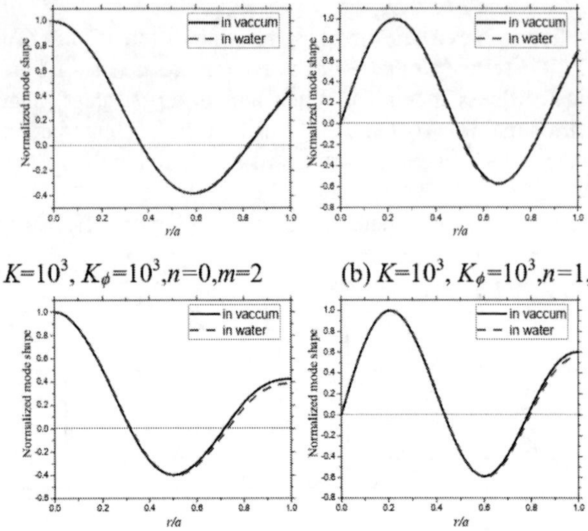


(a) $K$=$10^3$, $K_\phi$=$10^3$,$n$=0,$m$=2 (b) $K$=$10^3$, $K_\phi$=$10^3$,$n$=1,$m$=2

(a) $K$=$10^7$, $K_\phi$=$10^7$,$n$=0,$m$=2 (b) $K$=$10^7$, $K_\phi$=$10^7$,$n$=1,$m$=2

**Figure 3.** Comparison of the mode shapes of the circular plates in vacuum and in fluid

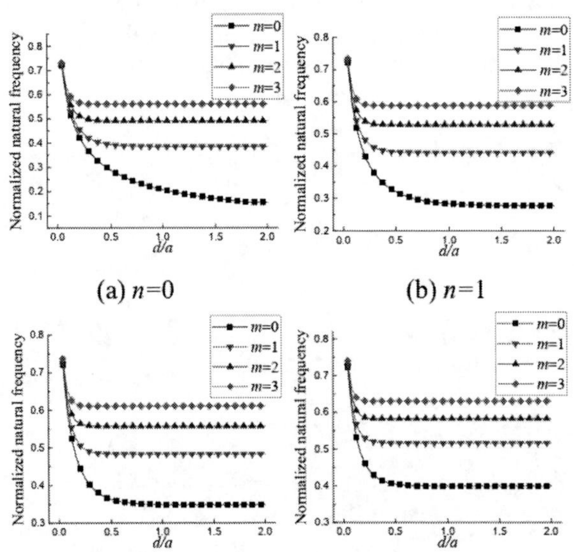


(a) $n$=0 (b) $n$=1

(c) $n$=2 (d) $n$=3

**Figure 4.** Evolution of normalized natural frequency with fluid depth

0 depicts the evolution of normalized natural frequencies with fluid depth. As shown in the figure, when the normalized fluid depth $d/a$ ranges from 0.01 to 1, the normalized natural frequencies reduce significantly, which means increase of added fluid mass.

With increase of the fluid depth, the decrease rate of the frequencies reduces. It can be seen from the figure that for each frequency of the plate there is a threshold fluid depth. And when the fluid depth reachs the threshold, the natural frequency remains unchanged with growth of $d/a$. This threshold is smaller for the mode with bigger $n$ or $m$. When $d/a$ >1.8, the coupled frequency of arbitrary mode of the circular plate remains unchanged with variation of the fluid depth.

## 4. Conclusion

In this paper, the free vibration characteristics of an elastically restrained circular plate in contact with fluid is obtained by using Fourier Bessel series and Galerkin method. The present method is proved to by accurate by comaring with FEM.

Then the vibration characteristics of circular plates with serveral spring stiffenss is studied. The results show that the existence of fluid reduces the natural frequencies of the plate significantly, and it has more influence on the lower order frequencies.

Furthermore, the influence of fluid depth is investigated. Results show that in low depth range, the coupled natural frequencies reduce rapidly with fluid depth. When the fluid depth is greater than 1.8 times the radius of the circular plate, the influence of the increase of fluid depth can be ignored.

## References

[1] Kwak M K. Vibration of circular plates in contact with water[J]. Journal of Applied Mechanics, 1991, 58(2): 480-483.
[2] Kwak M K. Hydroelastic vibration of circular plates[J]. Journal of Sound and Vibration, 1997, 201(3): 293-303.
[3] Kwak M K, Han S B. Effect of fluid depth on the hydroelastic vibration of free-edge circular plate[J]. Journal of Sound and Vibration, 2000, 230(1): 171-185.
[4] Amabili M, Kwak M K. Free vibrations of circular plates coupled with fluids: revising the lamb problem[J]. Journal of Fluids and Structures, 1996, 10(7): 743-761.
[5] Bauer H F. Coupled frequencies of a fluid in a circular cylindrical container with elastic fluid surface cover[J]. Journal of Sound and Vibration, 1995, 180(5): 689-704.
[6] Bauer H F, Chiba M. Hydroelastic viscous oscillations in a circular cylindrical container with an elastic cover[J]. Journal of fluids and structures, 2000, 14(6): 917-936.
[7] Amabili M. Vibrations of circular plates resting on a sloshing fluid: solution of the fully coupled problem[J]. Journal of sound and vibration, 2001, 245(2): 261-283.
[8] Yousefzadeh S, Jafari A A, Mohammadzadeh A. Effect of hydrostatic pressure on vibrating functionally graded circular plate coupled with bounded fluid[J]. Applied Mathematical Modelling, 2017, doi: 10.1016/j.apm.2017.11.009
[9] Cheung Y K, Zhou D. Hydroelastic vibration of a circular container bottom plate using the Galerkin method[J]. Journal of fluids and structures, 2002, 16(4): 561-580.
[10] Eftekhari S A. Pressure-based and potential-based differential quadrature procedures for free vibration of circular plates in contact with fluid[J]. Latin American Journal of Solids and Structures, 2016, 13(4): 610-631.
[11] Tariverdilo S, Shahmardani M, Mirzapour J, et al. Asymmetric free vibration of circular plate in contact with incompressible fluid[J]. Applied Mathematical Modelling, 2013, 37(1): 228-239.
[12] Jeong K H. Free vibration of two identical circular plates coupled with bounded fluid[J]. Journal of Sound and Vibration, 2003, 260(4): 653-670.
[13] Zhang W B. Mathematical and physical equation[M]. Beijing: Tsinghua University Press, 2007.

# Atomic layer deposition of Ti and its deposition method

**Xingyu Zhang**

University of Pennsylvania, USA

**Abstract.** Atomic layer deposition (ALD) is a very popular thin-film technique, and it is considered to be a method with great potential because it can produce conformal thin film with control of the thickness of layers and composition of the films at the nano-scaled level. In fact, there are other two types deposition techniques: Chemical Vapor Deposition (CVD) and Physical Vapor Deposition (PVD), which are also commonly used in surface treatment. However, the treatment conditions are different as CVD requires up to 1000°C while PVD requires lower than 500 °C. However, because of the self-controllability in ALD, this technique is widely applied in the fields including semiconductors, nanotechnology, and catalysts. In our current knowledge, substantial numbers of element could be used for ALD to form substrate layers. Some notable elements are carbide, Nitride, oxide, metal and so on. Among all kinds of possible deposited materials, titanium contained precursor are special and they receive tremendous attention because they could strengthen the base materials or protect the base materials from oxidization so that they improve the mechanical or chemical properties of pristine materials. Specifically, it could improve the properties of base materials such as reducing the work function, which is discussed in the passage. The purpose of this review is to provide an overview of current technology of deposition titanium contained chemicals and point out possible research topics in the future.

**Keywords:** Atomic layer deposition; thin-film; Ti.

## 1. Introduction

Generally, PVD is a process in which the material goes from a condensed state to a vapor phase and then goes back to a thin film state. PVD is commonly used in the manufacture of items such as thin film solar panels [1], aluminized PET film for food packaging and balloons [2], and titanium nitride coated cutting tools for metalworking. CVD is highly used in semiconductor industry to produce thin films. In typical CVD, the substrate is exposed to one or more volatile precursors and the precursors would react and/or decompose on the surface to produce the desired deposit. The application of CVD could be found almost anywhere. Commonly, CVD can be used to deposit conformal films in ways that traditional surface modification techniques are not capable of. Some examples are CVD of metal-organic frameworks, amorphous polysilicon, and membrane coatings [3,4]. Speaking of "Atomic Layer Deposition" (ALD), it is also a thin-film deposition technique which is actually a subclass of chemical vapor deposition. Generally, the ALD reactions has two chemical called precursors. These precursors react with the surface of a material one at a time in a sequential, self-limiting, manner [5]. The thin film will then gradually deposit on the surface of the base material. Semiconductor processing has been one of the main applications. One of main specific examples is that ALD has been included for high

Content from this work may be used under the terms of the Creative Commons Attribution 3.0 licence. Any further distribution of this work must maintain attribution to the author(s) and the title of the work, journal citation and DOI.
Published under licence by IOP Publishing Ltd

dielectric constant gate oxides in the MOSFET Structure and copper diffusion barriers in backend interconnects [6]. All three methods play a very important role in current industrial field. Depending on the requirements of a component, different deposited methods can be applied. However, ALD has the edge of thickness control and uniform distribution. As mentioned above, depending on the number of precursors, ALD could stop by itself.

## 2. Literature review of ALD of Ti

Among several elements, titanium is widely used as thin film deposited on the surface of other materials because its ability to provide outstanding properties on base materials. In the followings, several methods in different parameters of depositing Ti contained thin film are classified. For example, various precursors could make a huge difference and play an important role in determining the properties of thin film [7]. In addition, it is highly used in work function engineering. The work function of a metal gate material could be tuned by manipulating TiC or TiN thin film [8, 9, 10]. Furthermore, it is learned that the TiN thin film could possibly be deposited by plasma assisted atomic layer deposition for copper metallization. Its plasma pulse time, power, and composition will have a huge effect on the electrical properties of TiN [11]. Sometimes, in order to produce highly conformal TiN thin film. The combination of thermal and plasma-enhanced atomic layer deposition can be applied [12, 13]. Coatings are another common usage for ALD of Ti, which acts as an adhesion layer and promotes the bonding [14, 15].

### *2.1. Effect of Precursors on ALD of Ti*

To make titanium thin-film, ALD technique is commonly used. However, different treatment methods will meet different requirements and properties of materials for various usages. Namely, different kinds of precursors could be a factor on changing properties of materials. In Gihee Cho and Shi-Woo Rhee's paper, they used three types of titanium precursors: tetrakis dimethyl amido titanium (TDMAT), tetrakis ethylmethyl amido titanium (TEMAT), and tetrakis diethyl amido titanium (TDEAT) and compared them in growth rate, resistivity, and the composition as function of substrate temperature. According to the result, the TDMAT, TEMAT and TDEAT had different ALD window. TDMAT has a range of 125-175 °C, and TEMAT has range between 175-200 °C while TDEAT is between 175-250 °C.Specific growth rate data is shown below in Figure.1. The plot has shown that the TDMAT presents much faster growth rate at any temperature from 100 -300 °C range, TEMAT is not far behind and TDEAT has the slowest growth rate.

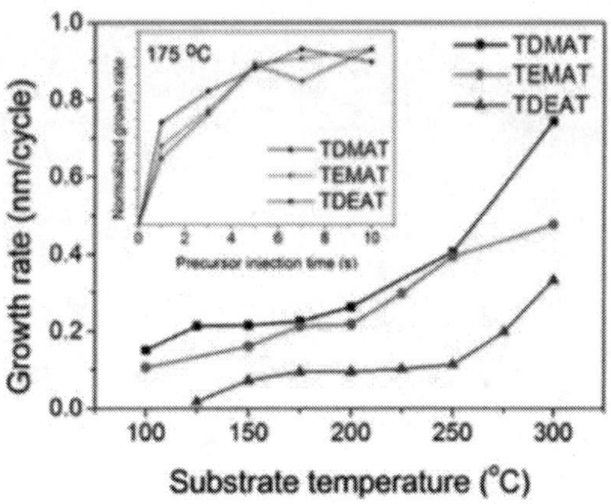


**Figure 1.** The growth rate for TDMAT, TEMAT and TDEAT precursors

IOP Conf. Series: Materials Science and Engineering **772** (2020) 012018 doi:10.1088/1757-899X/772/1/012018

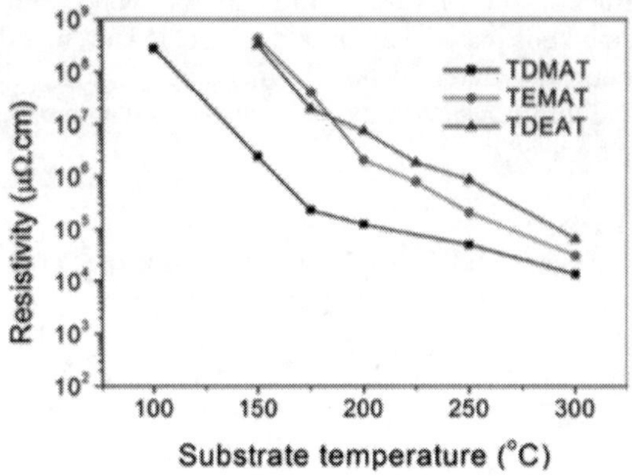


**Figure 2.** Resistivity changes for TDMAT, TEMAT and TDEAT at different temperatures

Figure.2 shows the resistivity changes as function of substrate temperature. The graph shows resistivity decreases with temperature going up with TDMAT having the lowest resistivity across the temperature range and TDEAT presenting the highest resistivity. Therefore, it illustrates the higher thermal stability will increase resistivity because TDEAT is the most stable one among all three according to the ALD temperature windows range.

Finally, the paper reveals composition of film of three precursors as function of temperature. TDMAT and TEMAT both showed decomposition of carbon content as the composition of carbon increases, resulting in the reduction of oxygen and forming in TiC phase filling in the vacancy. However, the temperature in the project is only up to 300 °C. How will things change above that temperature is still unknown.

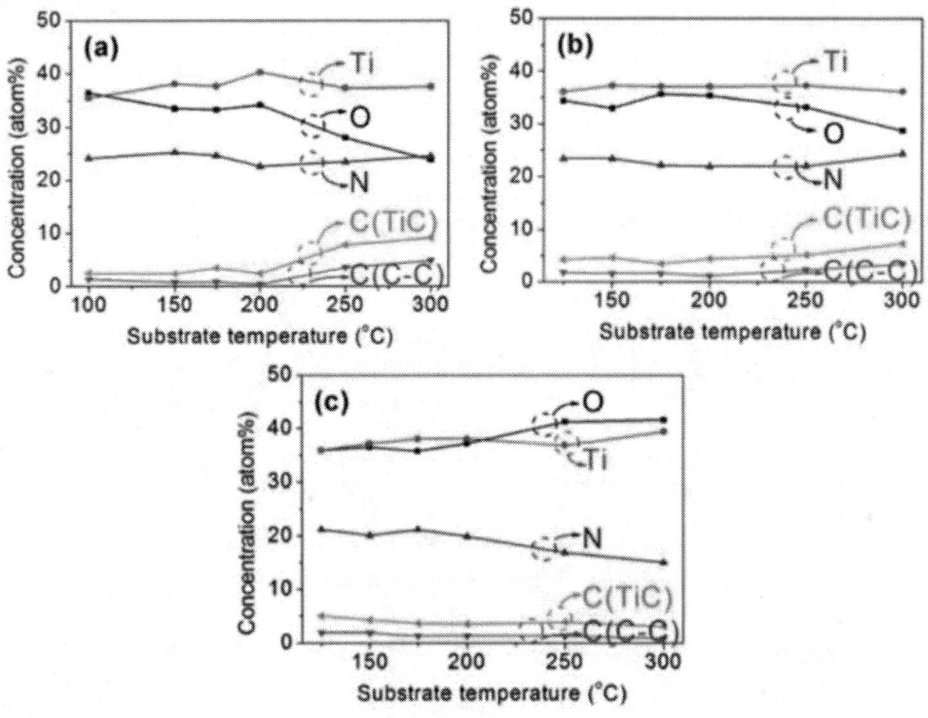


**Figure 3.** composition of thin films by using (a)TDMAT, (b)TEMAT and (c)TDEAT at different temperatures

*2.2. Effect of ALD on work function*

ALD process of titanium also plays a very important role in tuning work function. It is proved that TiN is suitable for metal gate material. However, there is no such a technique to develop it in a conformal film growth, which is required for 3D devices. In Sanghun Jeon and Sungho Park' paper [8], they adjusted deposition temperature, the carbon content could be altered. More specifically, they changed deposition mode from ALD mode to CVD like mode at temperature above 200 °C which increased bonding concentration and lower the work function of metal gate. The range of work function they focused on was 5.0 eV, where it is suitable for a p type FET, to 4.6 eV where it is suitable for a fully depleted silicon-on-insulator FET. Similarly, there is another TiC film which is almost as important as TiN. Choong-Ki Kim and his company used slightly different temperatures which ranged from 250°C to 500°C [9]. It was discovered that work function of TiC was affected by crystal orientation and determined by the average WF of each oriented sub-plane.

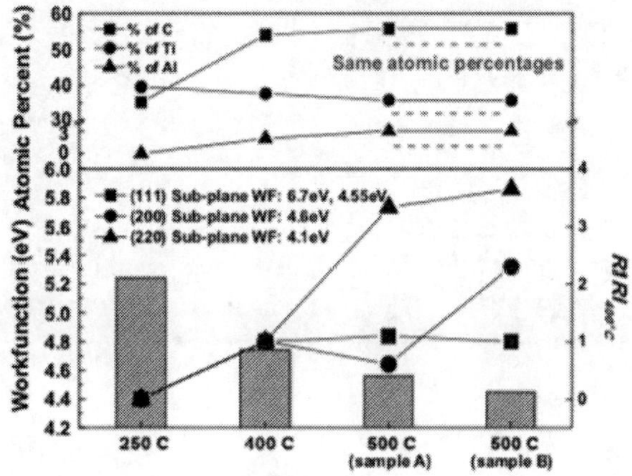


**Figure 4.** Work function and summary of atomic percentages of C, Ti and Al.

The previous two papers are both related to temperature. There is another one published by JinJun Xiang and Tingting Li regarding thickness changes on work function [10].

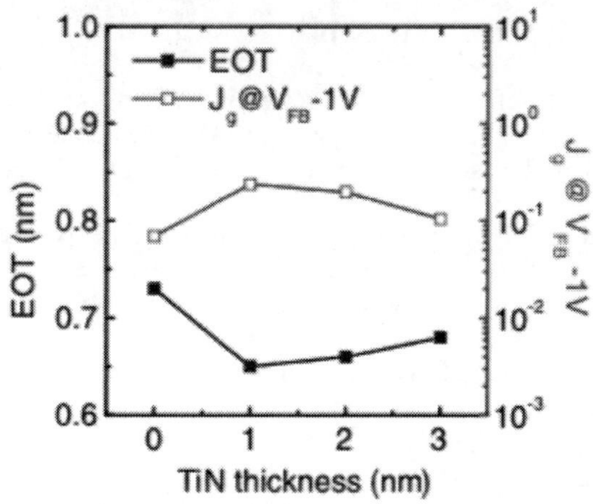


**Figure 5.** The EOT and $J_g$ values of the capacitors with different TiN thickness

The trend shows the thickness would initially increase the current densit J and lower the EOT. With the thickness becoming thicker, The EOT goes up and current density goes down. The result shows that metal-oxide-semiconductor (MOS) without TiN barrier exhibits lower $J_g$ and higher EOT. They argued that was due to the TiN could suppress Al diffusion and boost Ti diffusion into substrate. However, the sample pool was not quite substantial enough as they only tested four points. In order to make more convincing conclusion, more data points should be expected.

### *2.3. Effect of plasma on ALD of Ti*

Sometimes, plasma could be helped to prepare Titanium ALD. Do-Heyeoung Kim, Young Jae Kim and others introduced plasma method because the traditional metalorganic TiN ALD method, high resistivity remains a problem of the sample [11]

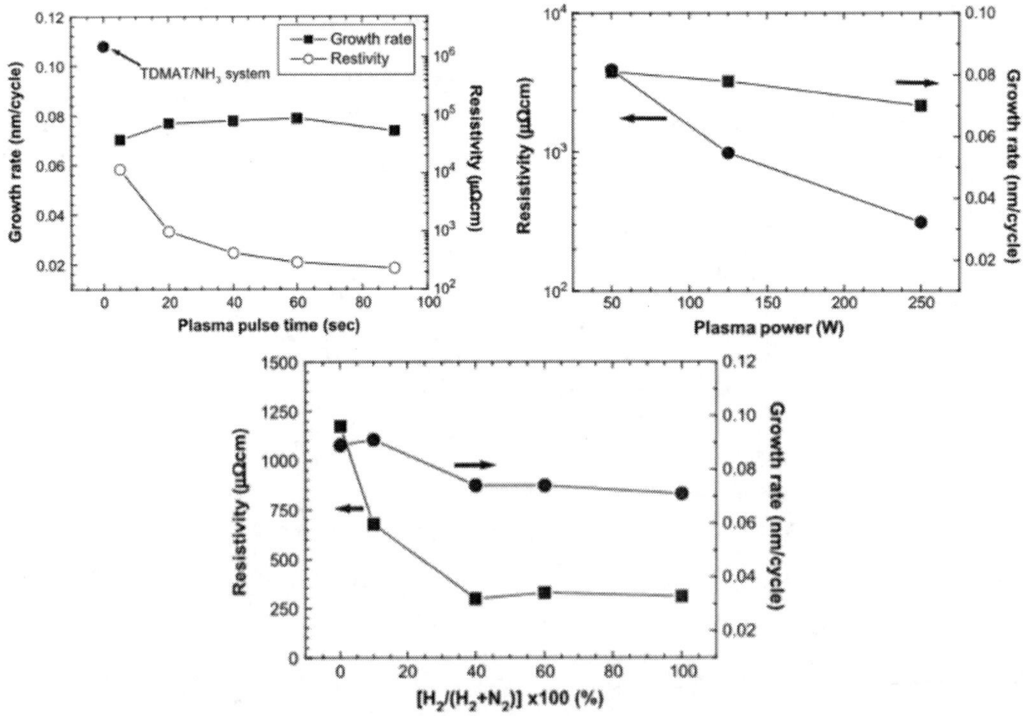


**Figure 6.** Top left: The effect of plasma treatment on film growth rate and resistivity.
Top right: The effect of plasma power on film growth rate and resistivity.
Bottom: The effect of plasma gas composition on film growth rate and resistivity.

From those diagrams, it mainly illustrates plasma treatment time, power and the ratio of Nitrogen over hydrogen could be a factor on the resistivity, growth rate and stability of TDMAT-based PAALD TiN films. Based on the graphs, the resistivity goes down as a longer plasma treatment time while the growth rate doesn't change significantly. Also, as the plasma power increases, the resistivity goes down. The higher hydrogen percentage could lower the resistivity as well. The method really showed a big improvement on electrical properties of thin films.

Some scientists such as Delphine Longrie and his research group, and Loic Assaud tried combination of plasma and thermal ALD methods [12,13]. According to Delphine and his company, they used a rotary reactor to deposit TiN on ZnO powder. The most important takeaway from this article is the resistivity decreases and the conductivity increases with the increasing power of plasma-enhanced ALD shown in Figure.7

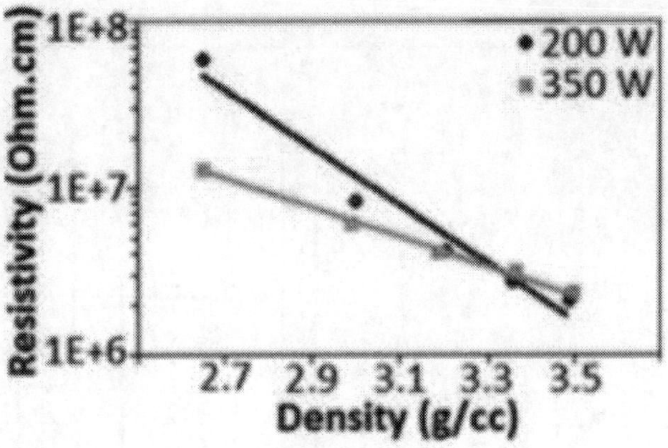


**Figure 7.** Resitivity of the AnO powder coated with 100 cycles of TiN using PE-ALD process with 200W and 350W plasma power

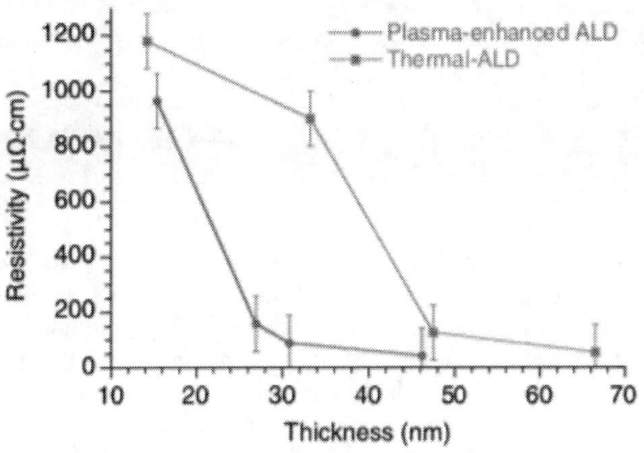


**Figure 8.** Resistivity as function of thickness deposited by Thermal-ALD and plasma-enhanced ALD

Figure.8 also clearly showed that both Thermal and Plasma-enhaced ALD would lower the resistivity value as layers become thicker, but the plasma-enhanced ALD shows a significantly better performance Therefore, it is declared that plasma-enhanced ALD has an edge over Thermal-ALD.

*2.4. Effect of ALD on mechanical properties*

Lastly, there are a couple of articles describing Ti coating to improve the mechanical properties. J.B Zang and J. Lu used ALD method to deposit Titanium on Nanocrystalline which they employed hydrogen gas and $TiCl_4$ binary reaction sequence to react with diamond matrix and form TiC in coating [14,15]. They suggested the formation of TiC could improve the internal bonding of nanocrystal diamond. Tommi O.Kaariainen has used ALD of Ti to improve the adhesion to PMMA substrate. $TiO_2$ and $Al_2O_3$ were used for two deposition processes, and ozone was used as a source of oxygen for both. Then, the deposition of the metal/metal carbide layer was followed by using pulsed DC reactive magnetron sputtering. The fracture strength prior to sputtered Ti is shown in Figure.9:

IOP Conf. Series: Materials Science and Engineering **772** (2020) 012018 doi:10.1088/1757-899X/772/1/012018

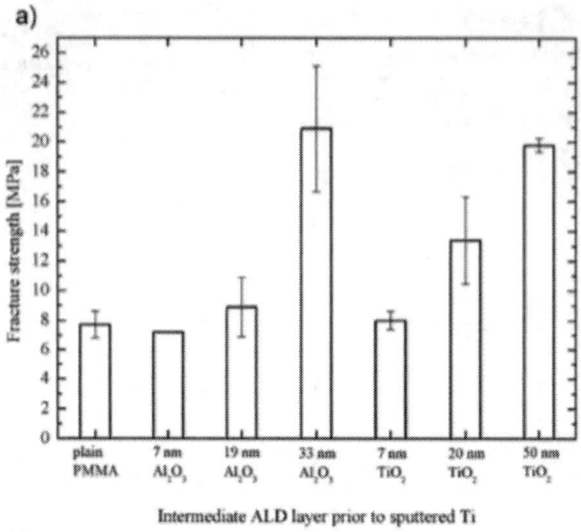


**Figure 9.** The relationship between fracture strength and deposition thickness

Clearly, we can see that the thickness of $Al_2O_3$ reaches 33 nm and $TiO_2$ reaches 50 nm, the fracture strength is significantly better than original PMMA, suggesting a big improvement on mechanical properties and adhesion.

## 3. Conclusion

The purpose of this review is to get an idea on how Ti ALD is currently used in application and method. To summarize, Ti ALD is heavily used in semiconductor devices such as metal gate materials. They could also be prepared by thermal and assisted by plasma pulse. Sometimes, they could be used as a coating method to improve the internal bonding and enhance mechanical properties of base materials. Sometimes, different precursors could also have a completely different performance as well. Therefore, in future, several research topics could be made based on this review. For instance, it could be test other precursors and compare their electrical properties. Some notable materials are Titanium diisopropoxidebis, Titanium isopropoxide. Also, regarding on the papers about work function, all the treatment temperatures were above 200°C. How is work function going to change under 200°C could be an interesting topic. Also, plasma assisted properties is proved to be competent on improving electrical properties. But some topic such as whether there weill be any effect on mechanical properties of the materials and how mechanical properties of materials would change with plasma power and power time as well as the stability of the Ti coating applied by ALD method, such as corrosion resistance. Those can all be problems to be solved.

## References

[1] Selvakumar, N.; Barshilia, Harish C. (2012-03-01). "Review of physical vapor deposited (PVD) spectrally selective coatings for mid- and high-temperature solar thermal applications". Solar Energy Materials and Solar Cells. 98: 1–23. doi:10.1016/j. solmat.2011.10.028.

[2] Hanlon, J. (1992). 1st ed. Handbook of Package Engineering, Lancaster, PA, Technomic Publishing: ISBN 0-87762-924-2. Chapter 4 Coatings and Laminations

[3] Stassen, I; Styles, M; Grenci, G; Van Gorp, H; Vanderlinden, W; De Feyter, S; Falcaro, P; De Vos, D; Vereecken, P; Ameloot, R (2015). "Chemical vapour deposition of zeolitic imidazolate framework thin films". NatureMaterials. 15 (3): 304-10. Bibcode:2016 NatMa.15.. 304S. doi:10.1038 /nm at4509. PMID26657328.

[4] Servi, Amelia T.; Guillen-Burrieza, Elena; Warsinger, David M.; Livernois, William;

Notarangelo, Katie; Kharraz, Jehad; Lienhard V, John H.; Arafat, Hassan A.; Gleason, Karen K. (2017). "The effects of iCVD film thickness and conformality on the permeability and wetting of MD membranes". Journal of Membrane Science. 523: 470–479. doi: 10.1016/ j.memsci. 2016.10.008. hdl: 1721.1/108260. ISSN 0376-7388.

[5] International Technology Roadmap for Semiconductors, 2007 Edition, http://www.itrs.net/.

[6] George, S. M. (2009). Atomic layer deposition: an overview. Chemical reviews, 110(1), 111-131.

[7] Cho, Gihee, and Shi-Woo Rhee. "Effect of the amido Ti precursors on the atomic layer deposition of TiN with NH3." Journal of Vacuum Science & Technology A: Vacuum, Surfaces, and Films 31.1 (2013): 01A117.

[8] Jeon, Sanghun, and Sungho Park. "Tunable work-function engineering of TiC–TiN compound by atomic layer deposition for metal gate applications." Journal of The Electrochemical Society 157.10 (2010): H930-H933.

[9] Kim, Choong-Ki, et al. "Temperature control for the gate workfunction engineering of TiC film by atomic layer deposition." Solid-State Electronics 114 (2015): 90-93.

[10] Xiang, Jinjuan, et al. "Understanding the role of TiN barrier layer on electrical performance of MOS device with ALD-TiN/ALD-TiAlC metal gate stacks." ECS Journal of Solid State Science and Technology 5.6 (2016): P327-P329.

[11] Kim, Do-Heyoung, et al. "Preparation of TiN films by plasma assisted atomic layer deposition for copper metallization." Materials Science and Engineering: C 24.1-2 (2004): 289-291.

[12] Assaud, Loïc, et al. "Highly-conformal TiN thin films grown by thermal and plasma-enhanced atomic layer deposition." ECS Journal of Solid State Science and Technology 3.7 (2014): P253-P258.

[13] Longrie, Delphine, et al. "Thermal and plasma-enhanced atomic layer deposition of TiN using TDMAT and NH3 on Particles agitated in a rotary reactor." ACS applied materials & interfaces 6.10 (2014): 7316-7324.

[14] Zang, Jian Bing, et al. "Ti coating of nanocrystalline diamond by atomic layer deposition." Key Engineering Materials. Vol. 304. Trans Tech Publications, 2006.

[15] Kääriäinen, Tommi O., David C. Cameron, and Mari Tanttari. "Adhesion of Ti and TiC coatings on PMMA subject to plasma treatment: effect of intermediate layers of Al2O3 and TiO2 deposited by atomic layer deposition." Plasma Processes and Polymers 6.10 (2009): 631-641.

IOP Conf. Series: Materials Science and Engineering **772** (2020) 012019 doi:10.1088/1757-899X/772/1/012019

# Experimental Research on Physical Indices and Mechanical Properties of Sandstone under Combined Action of Chemical-Freeze-Thaw Cycles

**Xuezhi Feng*, Yongyan Wang, Qiang Ge and Zhuoqun Yu**

Qingdao University of Science and Technology, Qingdao, China

*Corresponding author e-mail: 2500875646@qq.com

**Abstract.** In order to study the effects of chemical solution erosion and freeze-thaw cycles on the physical parameters and mechanical properties of rocks, sandstone samples from central Shandong Province were taken as the research objects. The freeze-thaw cycle test and uniaxial compression test were carried out for rock samples eroded by 0.01mol/L H2SO4 solution, 0.01mol/L NaCl solution and 0.01mol/L NaOH solution, respectively. The results show that the mass loss and porosity of rock samples eroded by different chemical solutions increase with the increase of freeze-thaw cycles, and the peak strength and elastic modulus decrease exponentially with the increase of freeze-thaw cycles. The damage degree of acidic solution to freeze-thaw cycling sandstone samples is the greatest, that of alkaline solution is the next, and that of neutral solution is the smallest. The conclusions can be used for reference in support and design of underground engineering.

## 1. Introduction

In recent years, with the vigorous construction of a large number of projects in cold regions, freeze-thaw cycles have a great impact on rock properties [1-6]. Scholars domestic and abroad are gradually aware of the practicability of the frozen rock problem, so there are a lot of research on the frozen rock problem. DelRoa [7] carried out 56 freeze-thaw cycles tests on different types of granite in the temperature range of - 12 C-20 C. After ultrasonic testing, it was found that the wave velocity decreases with the increase of freeze-thaw cycles; Tan [8] carried out granite research. After many freeze-thaw cycles, it is found that the compressive strength of rock decreases with the increase of freeze-thaw cycles, but increases with the increase of confining pressure. Hall [9] simulates the freeze-thaw damage process of quartz-mica shale through experiments. Yang Gengshe and Shen Yanjun [10] studied the identification methods and damage mechanics mechanism of freeze-thaw damage under different scales, proposed the development direction of multi-scale damage identification and evaluation mechanism of rock mass freeze-thaw; Zhang Huimei [11] carried out freeze-thaw cycle tests on saturated red sandstone under open system, and recorded the freeze-thaw degradation of red sandstone. Characteristic of process and damage. In the actual natural environment, rocks are not only affected by freeze-thaw cycles, but also by groundwater and acid rain. There are a lot of anions and cations in rocks, showing acid-alkaline properties. However, there are few studies on the physical and mechanical properties of rocks under the coupling of the two factors. Therefore, the study on the physical and mechanical properties of rocks

Content from this work may be used under the terms of the Creative Commons Attribution 3.0 licence. Any further distribution of this work must maintain attribution to the author(s) and the title of the work, journal citation and DOI.
Published under licence by IOP Publishing Ltd

under the combined action of chemical solution erosion and freeze-thaw cycle has certain theoretical guiding significance for the construction and protection of rock engineering.

## 2. Experimental equipment and programme

### *2.1. Experimental samples and equipment*

Sandstone samples are taken from mines in central Shandong Province, mainly consisting of quartz, feldspar and other cements.Standard cylindrical specimens with diameter of 50 mm and height of 100 mm were prepared by coring and grinding. The density of the specimens ranged from 1.97 to 2.04 g/cm3, and the specimens with obvious defects were removed. Some sandstone specimens are shown in Figure 1. Five specimen were grouped into three groups. Each group is subjected to three repeated tests.

**Figure 1.** Part of Standard Cylindrical Sandstone Samples

The main equipment used in Experimental:

RPH-80 controllable constant temperature and humidity test box was used in freeze-thaw cycle test.The test box is controlled by PID mode, which can realize the functions of refrigeration and heating. The adjustable temperature range is - 20 ~150 C, as shown in Figure 2 (a). The uniaxial compression test is carried out by TAW-200 electronic multifunctional material mechanical testing machine developed jointly by Qingdao University of Science and Technology and Changchun Chaoyang Test Machine Factory, with the maximum axial pressure of 200 KN. The confining pressure is 30 MPa, as shown in Figure 2 (b).

(a) Constant temperature and humidity box (b) TAW-200 electronic multifunctional material mechanics testing machine

**Figure 2.** Main experimental equipment

Experimental programme:

Chemical solution preparation: The sandstone specimens were dried, then saturated in vacuum, and the quality of the samples was measured. The specimens which met the test requirements were divided into three groups: 0.01mol/L H2SO4 solution, 0.01mol/L NaCl solution and 0.01mol/L NaOH solution, and the PH values of each solution were determined as PH=3, PH=7 and PH=13, respectively. The samples were immersed for 30 days, and the changes of PH in the solution were monitored every day to ensure the stability of PH by replenishing the solution.

Freeze-thaw cycle test: referring to the Standard for Test Method of Engineering Rock Mass [12], the soaked rock sample is put into a constant temperature and humidity box. The freezing temperature is set to -20 C, the freezing time is 6 h, and then the freezing time is kept at 20 C for 6 h in each solution. That is, the freezing-thawing cycle time is 12 h, and the five freezing-thawing cycles of 0, 15, 30, 45 and 60 are set up. The mass of saturated specimens was measured after freeze-thaw cycles.

Uniaxial compression test: after connecting the testing machine with the control system, the pre-stress of the specimen is controlled by the way of displacement loading. The loading process is controlled by the axial displacement rate. The displacement rate is 50N/s. The stress-strain curve of the specimen is recorded. The loading process is completely controlled by the computer terminal. The test software records and outputs the experimental data directly.

## 3. Analysis of physical indicators

### *3.1. Mass loss rate*

The saturated mass of rock samples after the coupling action of chemical solution erosion and freeze-thaw cycle is weighed, and the average quality of each group is calculated, such as table 1.

**Table 1.** Sample saturated mass under the coupling action of chemical solution erosion and freeze-thaw cycles

| number of times | saturation mass of sandstone in different chemical solutions | | |
|---|---|---|---|
| | 0.01mol/l $H_2SO_4$ Acidic solution | 0.01mol/l NaCl Neutral solution | 0.01mol/l NaOH Alkaline solution |
| 0 | 425.2g | 428.1g | 427.0g |
| 15 | 422.6g | 425.1g | 424.0g |
| 30 | 421.0g | 423.5g | 422.4g |
| 45 | 418.5g | 421.0g | 420.1g |
| 60 | 416.2g | 419.5g | 418.3g |

According to the freeze-thaw damage rate formula provided in the *Standard for Test Method of Engineering Rock Mass*:

$$L_f = \frac{m_s - m_f}{m_s} \times 100 \tag{1}$$

$L_f$—Freeze-thaw mass loss rate, %; $m_s$—Saturated mass of specimen before freeze-thaw test, g; $m_f$—Saturated mass of specimens after freeze-thaw test, g.

In this test,The average mass of rock samples saturated in distilled water was 430.5g. The damage rate of rock samples was calculated by ms=430.5g. The curve of damage rate of rock samples under different chemical solution erosion with the number of freeze-thaw cycles was obtained, as shown in Figure 3.

IOP Conf. Series: Materials Science and Engineering **772** (2020) 012019 doi:10.1088/1757-899X/772/1/012019

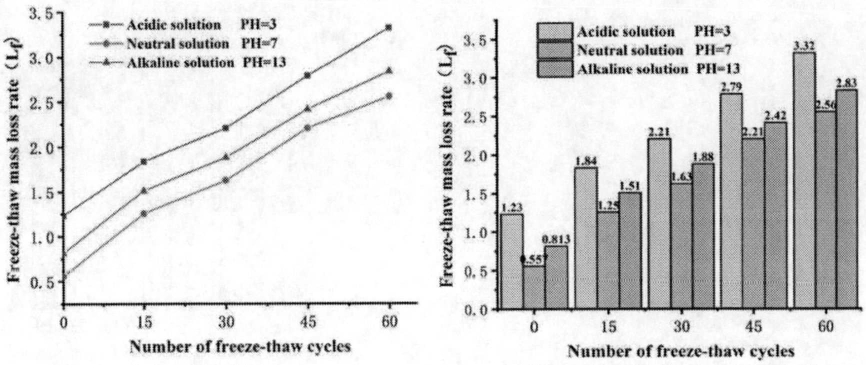


(a) Point-line change graph (b) Column contrast graph

**Figure 3.** Relation curve between freeze-thaw damage rate and freeze-thaw cycles number of times of rock samples eroded by different chemical solutions.

From that, it can be seen that the freeze-thaw damage rate of rock samples eroded by three kinds of chemical solutions increases continuously with the increase of freeze-thaw cycles, and the curve increases approximately linearly as a whole. Under the same freeze-thaw cycles, the freeze-thaw damage rate of rock samples eroded by acid solution is the highest, and that of rock samples eroded by alkaline solution is the next, and that of rock samples damaged by neutral solution is the smallest. The results show that the acidity solution has the most serious erosion on rock samples, and the particles peeling damage is the highest. The alkaline solution is the next, and the rock sample particles have the smallest corrosion degree under the neutral solution.

### *3.2. Porosity*

Formula for calculating porosity according to *Standard for Test Method of Engineering Rock Mass*.

$$n = (1 - \frac{\rho_d}{\rho_s}) \times 100 \tag{2}$$

Where, $n$—rock porosity, %; $\rho_d$—dry density, g/cm$^3$; $\rho_s$—saturation density, g/cm$^3$

Figure 4 is a graph showing the variation of rock porosity with the increase of freeze-thaw cycles under the erosion of different chemical solutions. From Figure 4, it can be seen that the porosity of rock samples eroded by different chemical solutions increases with the increase of freeze-thaw cycles. Under the same freeze-thaw cycles, the porosity of rock samples eroded by acidic solution is the largest, and that of rock samples eroded by alkaline solution is the next, and that of rock samples eroded by neutral solution is the smallest, but the variation of porosity of rock samples eroded by alkaline solution and neutral solution is similar. During the 0-15 freeze-thaw cycles, the porosity of rock samples increases fastest under the erosion of three solutions, which indicates that in the early stage of freeze-thaw, the particles inside the rock shrink sharply, and the original small pore expands rapidly. The initial freeze-thaw changes the spatial structure of the original rock particles, and the freeze-thaw has a greater impact on rock damage.

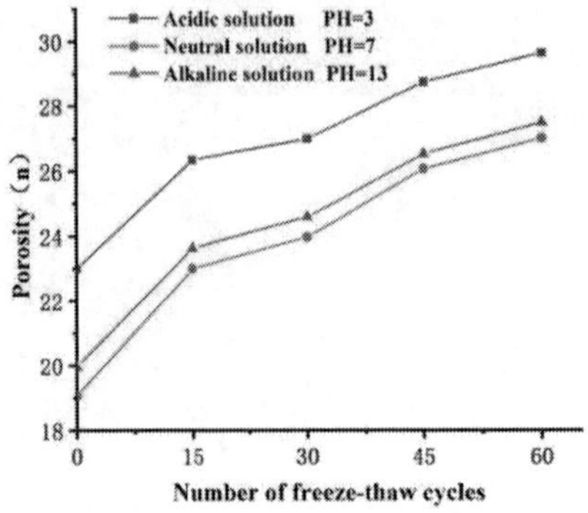


(a) Point-line change graph

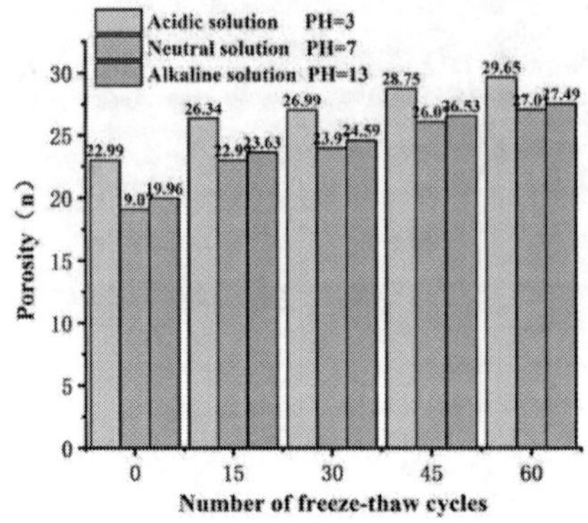


(b) Column contrast graph

**Figure 4.** Relation curve between porosity and freeze-thaw cycles number of times under different chemical solution erosion

## 4. Analysis of uniaxial compression experimental results

According to the sample data, the average values of peak strength, peak strain and modulus of elasticity of each group of rock samples are calculated, and analysis of various mechanical parameters.

### *4.1. Peak Strength*

Figure 5 shows the curve of peak strength and freeze-thaw cycles of rock samples under different chemical solutions. The peak strength of rock samples under different chemical solutions decreases with the increase of freeze-thaw cycles. When the number of freeze-thaw cycles is the same, the peak strength of rock samples under acid solution erosion is the lowest, and that under alkaline solution next time, the peak strength of rock samples under neutral solution is the highest, but under alkaline solution and neutral solution. There is little difference between the two.The peak strength of rock samples eroded by different chemical solutions increases with the increase of freeze-thaw cycles, which indicates that rock damage is not proportional to the number of freeze-thaw cycles, but increases sharply with the deterioration of freeze-thaw cycles.During the freeze-thaw cycle of 0-60 cycles, the peak strength of neutral solution and alkaline solution decreased by 31.69% and 35.02%, respectively, while that of acidic solution decreased clearly by 40.61% . The results show that the acid solution has the most serious erosion and damage to rock samples.

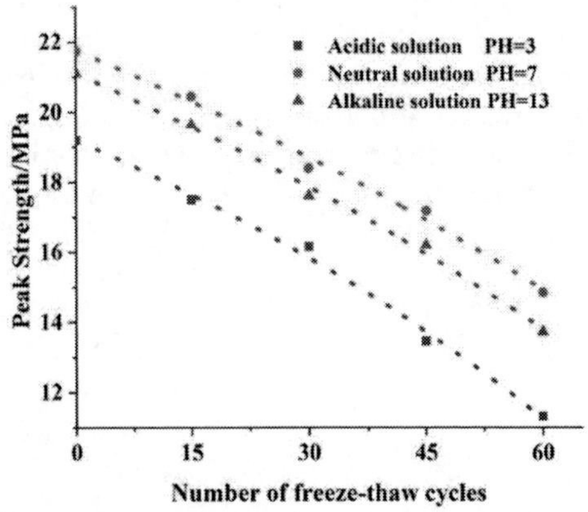


**Figure 5.** Variation curves of peak strength and freeze-thaw cycles of rock samples eroded by different chemical solutions.

The law of change of peak strength of sandstone specimens with freeze-thaw cycles in different chemical solutions is obtained.

Acidic solution:

$$S = 28.124 - 8.943\exp(0.011m), \quad R^2 = 0.989 \tag{3}$$

Neutral solution:

$$S = 34.895 - 13.139\exp(0.007m), \quad R^2 = 0.987 \tag{4}$$

Alkaline solution:

$$S = 33.969 - 12.869\exp(0.075m), \quad R^2 = 0.992 \tag{5}$$

Formula S is the peak strength of rock samples eroded by different chemical solutions through M freeze-thaw cycles.$R^2$ is the correlation coefficient.It can be seen that the peak strength of rock samples eroded by different chemical solutions decreases exponentially with the increase of freeze-thaw cycles.

### *4.2. modulus of elasticity*

According to the Test Method of Engineering Rock Mass of national standard engineering, the slope of straight section in stress-strain curve is usually used to define the elastic modulus (average modulus) of elastic modulus as follows:

$$E = \frac{\sigma_2 - \sigma_1}{\varepsilon_{2a} - \varepsilon_{1a}} \tag{6}$$

where, $E$——Modulus of Elasticity(GPa); $\sigma_1$、$\sigma_2$——Stress at the beginning and end of the linear segment on the stress-strain curve; $\varepsilon_{1a}$、$\varepsilon_{2a}$—— Longitudinal strain at the beginning and end of the linear segment on the stress-strain curve.

Figure 5 shows the curves of elastic modulus and freeze-thaw cycles of rock samples eroded by different chemical solutions.The elastic modulus of rock samples eroded by three chemical solutions decreases continuously with the increase of freeze-thaw cycles, and the overall slope of the curve increases with the increase of freeze-thaw cycles. Under the same freeze-thaw cycles, the elastic modulus of rock samples eroded by acidic solution is the smallest, followed by alkaline solution, and the elastic modulus of rock samples eroded by neutral solution is the largest. After 60 times freeze-thaw cycles, the elastic modulus of rock samples increases with the initial The modulus of elasticity of sandstone samples eroded by three kinds of solutions decreased by 32.13%, 24.31% and 26.50% respectively, indicating that the damage degree of sandstone samples caused by freeze-thaw cycle is the greatest in strong acidic environment, and the damage degree in neutral environment is the smallest, which is not significantly different from that in strong alkaline environment.

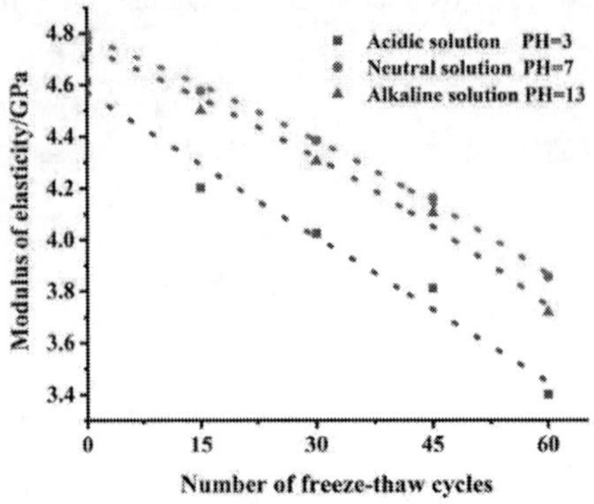


**Figure 6.** Curves of elastic modulus and freeze-thaw cycles of rock samples eroded by different chemical solutions

The curve was fitted exponentially, and the law of change elastic modulus of sandstone specimens with the numbers freeze-thaw cycles under different chemical solution was obtained.

Acidic solution:

$$E = 5.314 - 0.743\exp(0.034m),\quad R^2 = 0.956 \tag{7}$$

Neutral solution:

$$E = 6.171 - 1.385\exp(0.008m),\quad R^2 = 0.995 \tag{8}$$

Alkaline solution:

$$E = 5.961 - 1.216\exp(0.010m),\quad R^2 = 0.980 \tag{9}$$

The elastic modulus of rock samples eroded by three chemical solutions decreases exponentially with the increase of freeze-thaw cycles.

## 5. Conclusion

In this paper, laboratory experiments are carried out on sandstone samples from mines in central Shandong Province. The variation of physical and mechanical properties of sandstone samples under the coupling of chemical and freeze-thaw cycles is analyzed, which provides theoretical basis for engineering protection and construction. Based on the analysis of experimental results, the following conclusions are obtained:

(1) The mass loss and porosity of rock samples eroded by different chemical solutions increase with the increase of freeze-thaw cycles, while the peak strength, modulus of elasticity and coefficient of freeze-thaw decrease with the increase of freeze-thaw cycles.

(2) Under the same number of freeze-thaw cycles, the damage degradation degree of acid solution to freeze-thaw cycled sandstone samples is the greatest, and that of alkaline solution is the next, and that of neutral solution is the smallest, but the damage difference between neutral solution and alkaline solution is not obvious.

(3) The peak strength and modulus of elasticity of sandstone eroded by different chemical solutions decrease exponentially with the increase of freeze-thaw cycles, and the decreasing rate increases gradually.

(4) After 45 times freeze-thaw cycles of sandstone in different chemical solutions, the internal particle damage is aggravated and the compressive strength decreases sharply. The softening coefficient of rock sample in acid solution is the lowest, and the deterioration characteristics are the most obvious.

**Acknowledgments**
This work was supported by the National Natural Science Foundation of China (grant nos.51674149) and the Natural Science Foundation of Shandong Province (ZR2019MEE082).

**References**

[1] Hall. K. A laboratory simulation of rock breakdown due to freeze-thaw in a Maritime Antarctic environment [J]. Earth Surface Processes and Landforms, 1998, 13: 369-382.
[2] Fukuda, M. Rock weathering by freeze-thaw cycles [J]. Low TemP. Sci. Series A. Phys. Sci., 1974, (32): 243-249.
[3] Matsuoka, N. Mechanics of rock breakdown by frost action:an experimental approach [J]. Cold Regions Science and Technology, 1990, 17(3): 253-270.
[4] G.L.Golewski, T.Sadowski. An analysis of shear fracture toughness K and microstructure in concrete containing fly-ash [J]. Construction and Building Materials. 2014, 51: 207-214.
[5] MATSUOKA N. Microgelivation versus macrogelivation: towards bridging the gap between laboratory and field frost weathering [J]. Permafrost and Periglacial Processed, 2001, 12(3): 299-313.
[6] Wu Yang. Strength, damage and creep properties of rock-like materials under freeze-thaw cycles [D]. Qingdao University of Science and Technology, 2018.
[7] Del Roa L M, Lopez F, Esteban F J, et al. Ultrasonic study of alteration processes in granites caused by freezing and thawing [C]. 2005 IEEE Ultrasonics Symposium, 2005, 1:415-418.
[8] Tan X J, Chen W Z, Yang J P, et al. Laboratory investigations on the mechanical properties degradation of granite under freeze-thaw cycles [J]. Cold Regions Science and Technology, 2011, 68: 130-138.
[9] Hall K. Evidence for freeze-thaw events and their implications for rock weathering in northern Canada: II. The temperature at which water freezes in rock [J]. Earth Surface Processes and Landforms, 2007, 32:242-259.
[10] Shen Yanjun, Yang Gengshe, et al. Discussion on the suggested test scheme of rock freeze-thaw cycle [J]. Journal of Geotechnical Engineering, 2016, 38 (10): 1775-1782.
[11] Zhang Huimei, Yang Gengshe. Study on rock damage model under freeze-thaw and load coupling [J]. Journal of Rock Mechanics and Engineering. 2010, 29 (3): 471-476.
[12] Standard for engineering classification of rock mass(GB/T 50266-2013), National standards of People's Republic of China.

# SLP-based Technical Plant Layout Planning and Simulation Analysis

**Xiaorong Xu**

Beijing Jiaotong University, Beijing, 100044, China

**Abstract.** This paper analyzes the current situation and shortcomings of the assembly and the whole material flow in a technical workshop. Through the process analysis, this paper uses the SLP method to optimize, and combines the data analysis to adjust the unreasonable place, and gives the optimized layout plan. Finally, using Flexsim modeling, the whole process of the program is simulated, and the time-consuming reduction of the whole process is verified.

**Keywords:** Layout Planning; Process Optimization; Simulation; SLP.

## 1. Existing problems and research trends

In the production process, materials are stored and transported in 90% to 95% of the time. Therefore, it is of great significance to optimize the production system for logistics [1], and good layout design is the best means to improve the logistics process. The material handling cost is reduced, and the moving distance between personnel and transportation equipment is shortened, thereby improving the overall production efficiency and optimizing the production system. This paper will use the SLP method [2] to optimize the layout of a technical plant to achieve testing, decomposition, storage, assembly, and transshipment of a large product. Initially set up seven functional areas: storage area, test area, decomposition area, transfer area, storage area, assembly area, and outbound area: the test area is tested after the product is put into storage and before the warehouse is released; the decomposition area decomposes the whole machine. The storage area stores large and small sections and other parts and facilities; the assembly area carries out the work of connecting large and small sections. All relevant parts of the operation use the corresponding means of transport. The design purpose is as follows: the product takes the shortest time from the segment state to the whole machine. The overall process consists of two processes. The overall process is schematically shown in Fig 1. Warehousing process: warehousing testing decomposition packaging storage; Outbound process: picking processing assembly loading testing out of the library.

Content from this work may be used under the terms of the Creative Commons Attribution 3.0 licence. Any further distribution of this work must maintain attribution to the author(s) and the title of the work, journal citation and DOI.
Published under licence by IOP Publishing Ltd

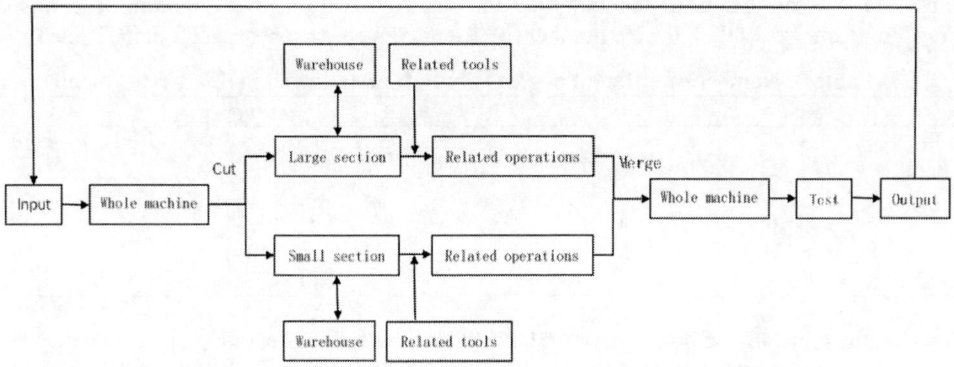


**Figure 1.** Schematic diagram of assembly and decomposition

Based on this, the research content and technical route of this paper are determined as follows:

Layout planning: space layout and facility planning, and promote the U-shaped layout of the overall assembly line to avoid the location and construction difficulties brought by the linear type;

II) Simulation evaluation: simulation verification of the assembly process and layout

## 2. Plant planning and spatial layout

The method and technology of factory layout has always been a problem that is constantly explored in the field of industrial engineering. Among the many layout methods, SLP is the most famous system design [3]. This method can be used not only for plant and production system design, but also for hospital, school, and office building design [4]. This paper also uses the Systematic Arrangement Method (SLP) for the layout design of the plant.

### *2.1. Determination of the Relative Position of the Operating Unit*

The layout of the facility is determined by the closeness of the process between the operating units. There are two kinds of relationship between logistics and non-logistics. The former uses the logistics intensity to indicate the close relationship between the two operating units. The latter cannot be quantitatively expressed and can only be distinguished by qualitative analysis [5]. In the research of this paper, logistics movement is the main part of the whole process, and its turnover efficiency directly affects the completion speed of the whole process. Therefore, quantitative logistics analysis becomes the core work of factory layout design. Logistics analysis should not only consider the route of each component and facility in the factory, but also follow the principle of “two minimum” and “two avoidance” [6]: Two minimum principles: minimum distance and minimum logistics cost; Two avoidance principles: avoiding roundabouts and avoiding crisscross;

*2.1.1. Analysis of the Relationship between the Operating Units.* Here, we mainly analyze the seven functional areas closely related to the whole process: the storage area, the test area, the decomposition area, the transfer area, the storage area, the assembly area, and the delivery area, in which the products are in storage and before delivery. It is necessary to pass the test area for performance and accessory inspection, and the whole machine packing work before leaving the warehouse is arranged in the test area. Before estimating the degree of relationship between the operating units, first agree on the meaning and nature of the different levels of representation, as shown in Table 1:

**Table 1.** Work unit relationship classification

| symbol | A | E | I | O | U | X |
|---|---|---|---|---|---|---|
| significance | Absolutely important | very important | important | general | unimportant | Don't approach |
| Quantitative value | 4 | 3 | 2 | 1 | 0 | -1 |
| Number of lines | 4 | 3 | 2 | 1 | nothing | 1 fold line |
| proportion(%) | 2-5 | 3-10 | 5-15 | 10-25 | 45-80 | base on needs |

IOP Conf. Series: Materials Science and Engineering **772** (2020) 012020 doi:10.1088/1757-899X/772/1/012020

There are 7 areas that need to determine the relative position, and 21 pairs of corresponding relationships. Refer to Table 1 for the number of different levels to appear, and obtain Table 2:

**Table 2.** Number of different degrees of relevance allowed in this arrangement

| Relevance | Number |
|---|---|
| A | ≤1 |
| E | ≤2 |
| I | 1~3 |
| 0 | 2~5 |
| U | 10~17 |
| X | base on needs |

Consider the logistics and non-logistics relationships between the operating units: The transfer speed from the storage area to the assembly area has a great impact on the whole "outbound process", and the relationship between them is closely connected, so the degree of correlation between them is set to "A"; in addition, the outbound process is also Including the end of the assembly into the test area, the end of the test is ready to go out of the library, so the degree of their relevance is set to "E"; in order to reduce congestion when warehousing, the "incoming area" and "decomposition area" should be reasonably placed. The degree of correlation between the two is set to "I"; most of the segments that have been decomposed by the "decomposition zone" can smoothly enter the "transportation zone" and the rational design of the "transportation zone" has a great effect on the smooth storage of most segments, so this will be The correlation between the two groups was also set to "I". The relatively important correlations of the above six groups A, E and I are assigned to the directly connected working unit groups, which means that the non-logistics analysis is based on the logistics related strength. In addition, it is determined that the degree of correlation "O" is 5 and the number of "U" is 10. The "assembly area" and "decomposition area", "inflow area" and "outbound area" are not excluded from joint placement or space. There is an intersection on the upper side and the value is "O". The "U" with the most occurrences indicates that there is no interference or interference between the two operation units. Combined with the above description, the following Fig 2 is obtained:

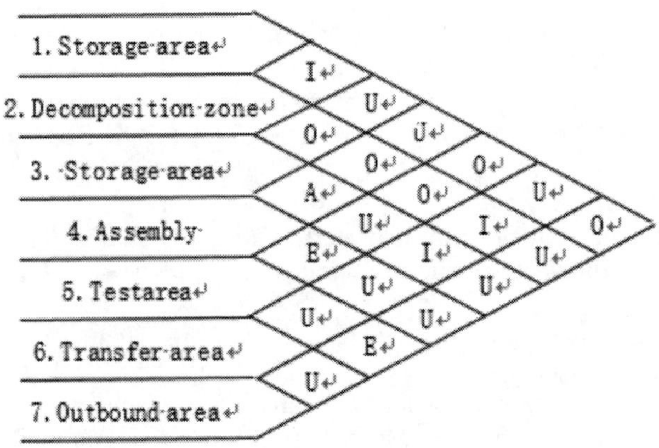


**Figure 2.** Comprehensive relationship diagram of the operating unit

In the SLP, the facility layout does not directly consider the floor space and geometry between the work units, but arranges the relative positions between the units based on the close relationship between the work units. The distance is near and the distance is low. According to the "comprehensive proximity", the unit of work with the same degree of correlation is solved: the comprehensive proximity of an operating unit is equal to the sum of the quantified degree between the unit and all other units. Based on this, the following table as Table 3:

IOP Conf. Series: Materials Science and Engineering **772** (2020) 012020 doi:10.1088/1757-899X/772/1/012020

**Table 3.** Work unit comprehensive proximity ranking table

| Unit code | 1 | 2 | 3 | 4 | 5 | 6 | 7 |
|---|---|---|---|---|---|---|---|
| 1 | - | I | U | U | O | U | O |
| 2 | I | - | O | O | O | I | U |
| 3 | U | O | - | A | U | I | U |
| 4 | U | O | A | - | E | U | U |
| 5 | O | O | U | E | - | U | E |
| 6 | U | I | I | U | U | - | U |
| 7 | O | U | U | U | E | U | - |
| Proximity | 4 | 7 | 7 | 8 | 8 | 4 | 4 |
| Sort | 3 | 2 | 2 | 1 | 1 | 3 | 3 |

The level of the comprehensive proximity value reflects whether the work unit is in the center position or the edge position in the layout drawing. Accordingly, the relative position of each work unit is obtained by an appropriate planar arrangement method.

*2.1.2. Plane Layout.* The commonly used layout methods are the Muther line graph method and the Tompkins relation table technique, and the former is used here. The line graph method uses different numbers of equal line segments to indicate the degree of correlation between the two operating units. It first puts the work units with A and E relationship into the layout drawing. The same level is represented by the same length of the line segment. The A level is the shortest, indicating that the contact is the closest, taking it as one unit; the E level is twice the length of the A level. And so on.I, O, and U-level relationships are arranged according to the same rules. Then adjust the position of each department to meet the limit of multiple degrees of intimacy; finally put the area of each department into the layout diagram to generate a spatial relationship diagram. Combined with the comprehensive approach of Table 3, and referring to the three factors of the degree of correlation between the arranged units, the ranking of comprehensive proximity, and the highest degree of correlation carried by themselves, 7 operating units will be arranged on the layout map in turn. As Fig3:

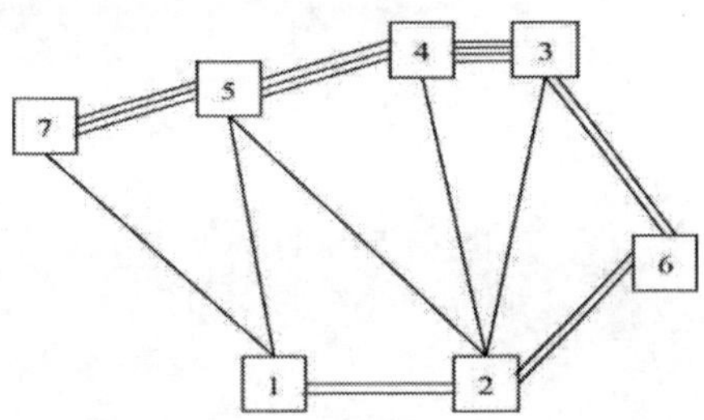


**Figure 3.** Job unit position correlation diagram

The connection is removed, and the arrangement is relatively dense. The overall process follows the U-shaped route [7]. Combined with the needs of the layout method, the floor area of different areas is estimated, and the area factor is added to fine-tune the relative position arrangement. . The main area of the operating unit is mainly from the static land occupation of facilities and materials [8], but because of the large scale of the plant facilities in this paper, the floor space mainly comes from the dynamic operation of the facilities, and the safety around the large facilities. Distance, personnel access and redundancy design to meet the rule layout [9]. Adding the in/out storage area and the non-working unit in the whole site, the layout design of the whole factory is shown in Fig 4:

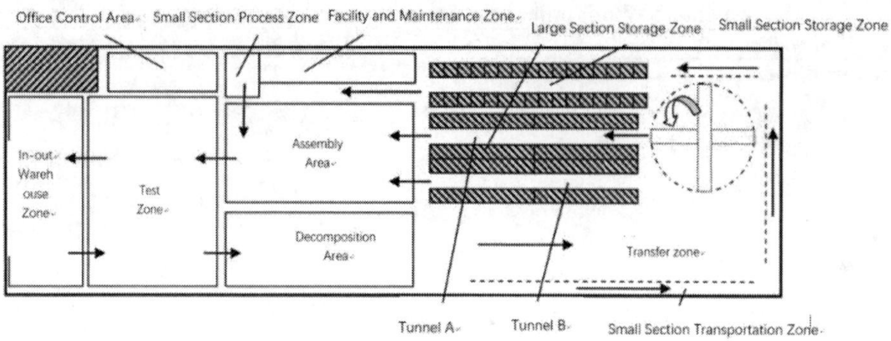


**Figure 4.** The final layout of the plant
Note: The arrows in the figure indicate the flow of material

*2.2. Summary of This Chapter*

This chapter carries out the overall layout of the plant, combined with the assembly line analysis, according to the system layout design method (SLP) to carry out the relative position arrangement of each functional area, and finally gives the design of the overall layout of the plant.

## 3. Simulation verification and optimization

In this paper, Flexsim is used for simulation modeling and experimental process to observe whether the design scheme is scientific. Among them, Fig 5 shows the main facilities layout and relative position layout of the factory building:

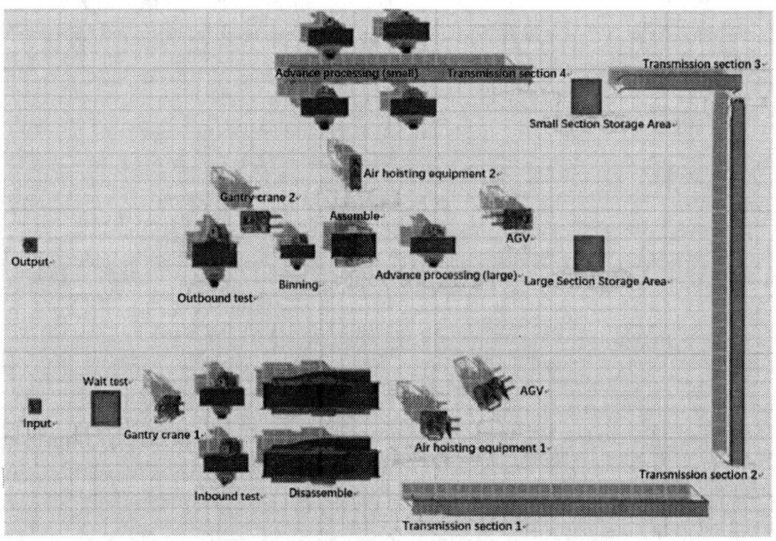


**Figure 5.** Schematic diagram of facility layout

Fig 6 shows the process relationship after the connection, which is used to indicate the whole process of the process, and compares the impact of the number of facilities and relative spacing on the actual process and facility idle rate. The processing facilities with successive relationships use the “A” connection. The “S” connection between the upstream processing facilities and the handling equipment that needs to use the transportation equipment during the connection process, and the “Use transportation tools” in the properties of the processing facilities are selected.

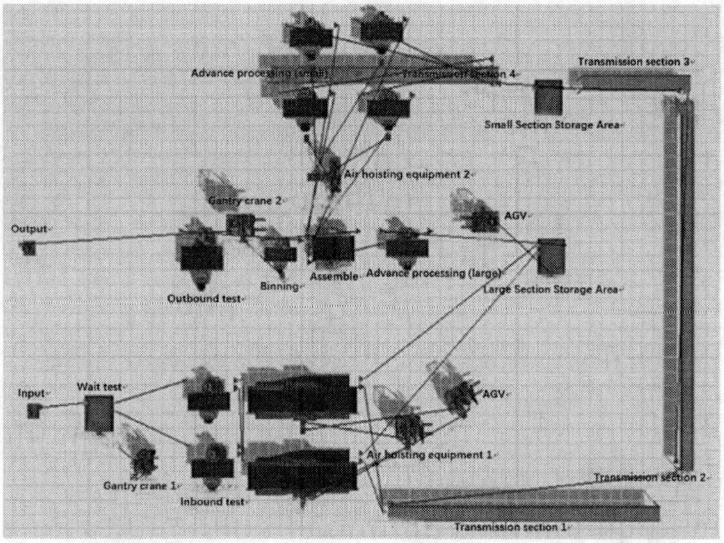


**Figure 6.** Implementation of the connected facility layout model

For the warehousing process, first disconnect the storage area of the large and small sections and the downstream, and simulate the process of putting the product into the warehouse. The time interval between the arrival of the product and the factory is subject to a normal distribution with a mean of 500 and a variance of 10; the "conditional port" setting is adopted for the two parallel decomposition activities, so that the two entities generated by the decomposition flow to the two outputs respectively. Simulate the decomposition-storage process and get the statistical report of Table 4

**Table 4.** Decomposition - Stored Procedure Statistics Report

Flexsim Summary Report-1
Time: 9,984.204

| Object | Class | stats content avg | idle | processing | blocked | waiting for transporter |
|---|---|---|---|---|---|---|
| Wait test | Queue | 2.433 | 0 | 0 | 0 | 1355.090 |
| decompose | Separator | 1.530 | 2152.889 | 7831.315 | 0 | 0 |
| | Queue | 5.927 | 0 | 0 | 0 | 0 |
| Large part | Queue | 3.364 | 0 | 0 | 0 | 0 |
| Input | Source | 0.956 | 0 | 0 | 0 | 0 |
| Gantry crane1 | Transporter | 0.488 | 218.357 | 0 | 0 | 0 |
| AGV | Transporter | 0.224 | 0 | 0 | 0 | 0 |
| Input test | Processor | 0.445 | 5544.471 | 4211.431 | 228.303 | 0 |
| | Processor | 0.481 | 5184.204 | 4800 | 0 | 0 |
| Temporary104 | Queue | 0.263 | 0 | 0 | 0 | 2629.083 |

In the table, "waiting for test" and "temporary storage area 104" wait for a long time for the forklift, indicating that the number of gantry cranes 1 and AGV has an influence on the upstream process, and the "decomposition zone" has a long idle time and also verifies the number and installation of the gantry crane 1. The deficiencies caused by the unloading efficiency. Consider adding these two transportation facilities and collaborating using the task distributor controller. Add the outbound process and simulate the acquisition-assembly-outbound process, as shown in Fig 7:

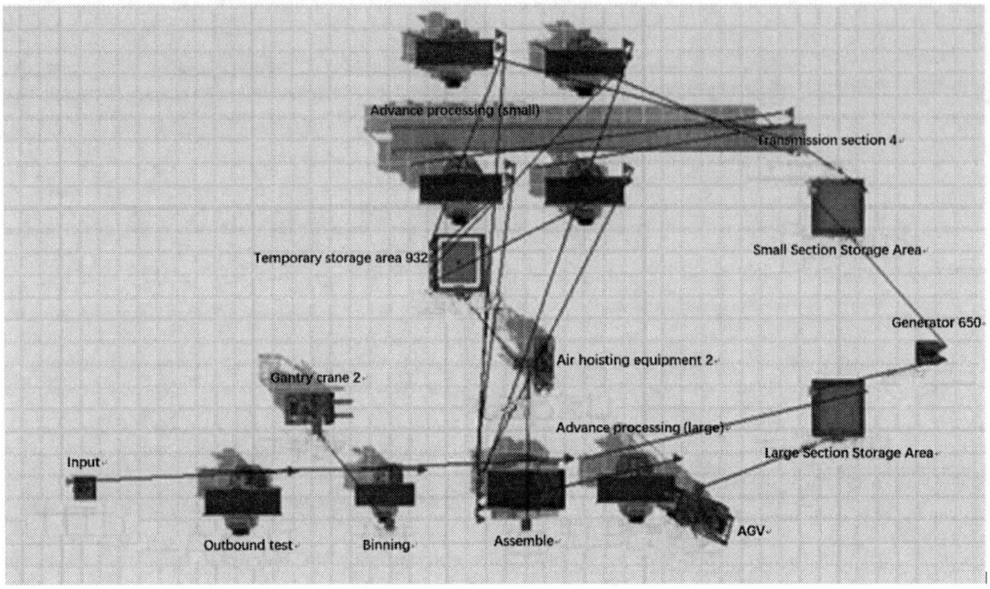


**Figure 7.** Acquisition-assembly-outbound process simulation

The temporary storage area 982 indicates the waiting line area of the section line after the preceding processing is completed. After 100,000s of simulation (32 finished products out of the warehouse), the idle rate of upstream equipment is between 10% and 15%, and the congestion rate is below 10%, which verifies the effectiveness of layout optimization

Based on the second chapter, this chapter carries out the modeling and simulation of the current layout scheme, including the assembly area assembly line and the overall planning of the plant. Combined with the analysis of statistical data, the superiority of the layout scheme is verified..

## 4. Conclusion

Based on the background of the project and the knowledge of industrial engineering, human factors engineering and facility planning, this paper designs a technical workshop with relatively reasonable layout and relatively efficient process, and uses Flexsim to simulate and verify. In conclusion, the layout of the plant designed by using relevant methods and tools has played a corresponding role, achieving efficient storage and assembly of heavy-duty columns, and giving additional design outside the main layout process, taking into account humanization and actual conditions. But limited by time and effort, modeling and simulation analysis part of this article failed to implement a realistic model to demonstrate the actual facilities, only use the same type or the same purpose of the software internal entity module instead, after assigning time and route data, although It will not affect the results of the experimental verification, but it is not visually prominent and needs to be improved.

## References

[1] Yang Baode. Strengthening enterprise logistics research is an important way to improve efficiency - Commemorating the international logistics expert Richard Joseph to teach in ChinaSHA, SLP20 Anniversary [J]. Engineering Construction and Design, 2003 (07): 51-52.

[2] Deng Bing, Lin Guangchun. Optimization of Workshop Equipment Layout with Improved SLP and Genetic Algorithm [J]. Combined Machine Tool & Automatic Processing Technology, 2017(08):148-151+156.

[3] Sadan Kulturel-Konak. A Matheuristic Approach for Solving the Dynamic Facility Layout Problem [J]. Procedia Computer Science, 2017, 108.

[4] Seyedeh Sabereh Hosseini, Seyed Ali Mirzapour, Kuan Yew Wong. Improving Multi-Floor

Facility Layout Problems Using Systematic Layout Planning and Simulation [M]. Springer, 2013, 409:58- 69

[5] Xue Shun, Xu Zhipei, Hong Hao, Cui Ping, Liu Xiaohui. Production facility layout design using system layout design method [J]. Industrial Engineering, 2011, 14(4): 135-138.

[6] Wang Jiashan. Facilities Planning and Design [J]. Industrial Engineering, 1998 (01): 11-14.

[7] Zhang Yu, Zhu Mo, Yin Shizhen, Tang Xiuli, Wang Rui. Overall layout planning of cableway assembly line workshop [J]. Crane Transportation Machinery, 2017(07): 83-88.

[8] Zhang Zhiqiang, Hu Shanying, Hu Xueyao, Guan Yanling, Jia Shengju. Development and application of intelligent factory integrated management information system [J]. Chemical Industry and Engineering Progress, 2016, 35(04): 1000-1006.

[9] Shang Yonggui, Du Lizhen. Analysis and Optimization of Work Environment Based on Human Factors Engineering [J]. Logistics Engineering and Management, 2014(4): 116-118.

# Study on Mechanical Properties of Natural Rubber by Carbon Black/Graphene/White Carbon Black

**Zhenzhu Yu *, Shaofei Zhang, Lu Su and Fei Nan**

College of Electromechanical Engineering, Qingdao University of Science &Technology, Qingdao, China

*Corresponding author e-mail: yuzhenzhu@qust.edu.cn

**Abstract.** Graphene (GE) and white carbon black (SiO2) were added to natural rubber (NR) to partially replace carbon black (CB). The effects of different filling ratios of CB/GE/SiO2 on the mechanical properties of NR were investigated. The data were processed by orthogonal experiment. The results showed that when the mass fraction of CB was 41g, the mass fraction of GE was 4g, and the mass fraction of SiO2 was 1g, the maximum true strain in the mechanical properties of NR was 139.8%. The maximum true stress is 85.695 MPa, and it is indicated that41g CB, 4 g GE, and 1 g SiO2 are added to the NR as the optimum filling ratio.

**Key words:** Graphene; Carbon black; White carbon black; Natural rubber; Mechanical properties; Stress strain.

## 1. Introduction

The mechanical properties of pure natural rubber (NR) are poor, and the mechanical properties of NR are improved by filling the reinforcing filler. Carbon black and white carbon black are two reinforcing fillers in the rubber industry. Both fillers can form aggregates in rubber compounds [1, 2, 3, 4]. Due to their different surface characteristics, the mechanical properties are different, but the overall interaction has a good improvement on the mechanical properties of NR [5, 6]. The filler-filler interaction of carbon black is mainly produced by relatively weak van der Waals forces, which are easily destroyed during the mixing process [7, 8, 9]; In addition to Van der Waals force and other physical interactions, the agglomeration of silica has hydrogen bonding, and the filler-filler interaction is stronger. Carbon black/ white carbon black has poor compatibility between molecules in the rubber. The addition of graphene can enhance the compatibility between molecules to some extent, improve the mechanical properties of rubber. Graphene itself has excellent mechanical properties [10, 11]. For example, the original graphene prepared by vapor phase deposition has an elastic modulus of 1 TPa and an ultimate strength of 130 GPa [12]. All three fillers can enhance the mechanical properties of NR [13, 14, 15], but the ratio of the three of them has not been studied by scholars.

In this paper, some carbon black were replaced by graphene and white carbon black, and three kinds of fillers were simultaneously added into natural rubber. The effects of different filling ratios of carbon black/graphene/white carbon black on the vulcanization characteristics and mechanical properties of NR were studied.

Content from this work may be used under the terms of the Creative Commons Attribution 3.0 licence. Any further distribution of this work must maintain attribution to the author(s) and the title of the work, journal citation and DOI.
Published under licence by IOP Publishing Ltd

## 2. Experimental part

### *2.1. Raw Materials*
Natural rubber (NR), grade 3L, Dongguan Kunhe Plastic Chemical Co., Ltd. products. Carbon black, grade N330, Tianjin Yihuachang New Material Technology Co., Ltd. products. White carbon black, DuPont chemical business products in Jinshui District, Zhengzhou City. Zinc oxide content of 99.7%, accelerator M (MBT), anti-aging agent 4020NA and stearic acid 1801, Shijiazhuang Youlian Chemical Co., Ltd. products. Silane coupling agent, brand SI-69, Dongguan Changping Yuxin Plastics Business Department. Graphene, liquid phase stripping method, product of the production and research base of Chongwen Avenue, High-tech Zone, Jining City, Shandong Province. Sulfur, Jiangsu Huaxing New Material Technology Co., Ltd.

### *2.2. Sample Preparation*
Basic formula: NR100 parts (mass parts, the same below), zinc oxide 5 parts, stearic acid (1801) 2 parts, accelerator M (MBT) 1 part, antioxidant (4020NA) 2 parts, sulfur 2.5 parts, silane 3 parts of coupling agent (Si69), carbon black / graphene / white carbon black are 47/1/1, 47/2/2, 47/3/3, 47/4/4, 44/1/2, respectively. 44/2/1, 44/3/4, 44/4/3, 41/1/3, 41/2/4, 41/3/1, 41/4/2, 38/1/4, 38/ 2/3, 38/3/2, 38/4/1.

Mixing rubber and vulcanizate: Place the NR at room temperature on the machine and roll it evenly. The mixture was prepared in the order of zinc oxide, stearic acid, accelerator M (MBT), anti-aging agent (4020NA), carbon black, the mixture was prepared by adding the graphene, white carbon black and sulfur. The mixture was allowed to stand at room temperature for 18 h, and then the vulcanization curve was measured at 150 °C using a non-rotor vulcanizer to obtain the vulcanization time (Tc90). It was corrected to obtain a vulcanization time of 1.3*Tc90, and the vulcanizate was vulcanized on a flat vulcanizing machine to obtain a carbon black/graphene/white carbon black mass ratio of 47/1/1, 47/2/2, 47/3/3, 47/4/4, 44/1/2, 44/2/1, 44/3/4, 44/4/3, 41/1/3, 41/2/4, 41/3 /1, 41/4/2, 38/1/4, 38/2/3, 38/3/2, 38/4/1,marked as 1#, 2#, 3#, 4#, 5#, 6#, 7#, 8#, 9#, 10#, 11#, 12#, 13#, 14#, 15#, and 16#.

### *2.3. Analysis and Testing*
Adopting X(S)K-160 open-type rubber (plastic) machine produced by Shanghai Double Wing Rubber Machinery Co., Ltd. with the roller speed ratio of 1:1.35; using XLB type flat vulcanizing machine produced by Qingdao Yadong Rubber Machinery Co., Ltd. The AI-7000M tensile testing machine produced by the high-speed rail testing instrument Co., Ltd. was used to measure the tensile properties according to GB/T 529-2008; the M2000-AN type produced by the high-speed rail testing instrument Co., Ltd. The vulcanization characteristic curve was tested by a rotor fluidizer under the test condition of 150 °C × 30 min; the morphology was observed by a SU8010 cold field emission scanning electron microscope (SEM) manufactured by Hitachi, Ltd., Japan.

## 3. Result and Discussion

### *3.1. Vulcanization characteristics*
As can be seen from Table 1, the maximum torque (MH), minimum torque (ML), and MH of the composite rubber increase when the parts of GE/SiO2 increase when the carbon black parts are 47 g and 44 g. The difference in ML (ΔM) generally shows a gradual increase. This may be because the addition of GE/SiO2 has a greater restriction and hindrance to the movement of rubber molecular chains, which is manifested by an increase in MH and ML, and an increase in ΔM is due to the small amount of GE/SiO2 particles. The diameter size makes it play the role of physical cross-linking point when dispersed in the rubber molecular chain, which leads to the improvement of the cross-linking degree of the system, thus improving the mechanical properties of the NR rubber compound.

When the mass fraction of graphene reaches 4 g of the maximum fraction and when the fraction of CB/SiO2 gradually decreases, the vulcanization time and the scorch time tend to increase first and then decrease. This may be related to the specific surface area of carbon black and white carbon black respectively. The smaller the specific surface area is, the faster the vulcanization rate of the rubber compound and the shorter the vulcanization time will be, which in turn affects the mechanical properties of the NR vulcanizate.

**Table 1.** Curing properties of CB/GE/$SiO_2$ for NR mixture.

| CB/GE/$SiO_2$ | MH/ (dN • m ) | ML/ (dN • m ) | (MH- ML)/ (dN • m) | Tc90/mi n | Ts2/min |
|---|---|---|---|---|---|
| 1# | 23.92 | 0.86 | 23.06 | 8.17 | 1.75 |
| 2# | 23.76 | 1.1 | 22.66 | 10 | 2.22 |
| 3# | 30.37 | 1.34 | 29.03 | 11.02 | 2.03 |
| 4# | 32.04 | 1.31 | 30.73 | 9.91 | 1.79 |
| 5# | 11.59 | 0.55 | 11.04 | 9.86 | 2.62 |
| 6# | 12.97 | 0.6 | 12.37 | 8.57 | 2.16 |
| 7# | 13.07 | 0.58 | 12.49 | 10.12 | 2.6 |
| 8# | 13.68 | 0.51 | 13.17 | 10.19 | 2.6 |
| 9# | 11.94 | 0.45 | 11.49 | 9.47 | 2.57 |
| 10# | 13.01 | 0.59 | 12.42 | 9.3 | 2.32 |
| 11# | 12.34 | 0.52 | 11.82 | 7.74 | 2.09 |
| 12# | 12.96 | 0.53 | 12.43 | 9.06 | 2.34 |
| 13# | 11.93 | 0.54 | 11.39 | 9.46 | 2.57 |
| 14# | 10.67 | 0.47 | 10.2 | 9.11 | 2.65 |
| 15# | 12.31 | 0.52 | 11.79 | 8.52 | 2.2 |
| 16# | 11.53 | 0.39 | 11.14 | 8.49 | 2.25 |

CB: carbon black; GE: graphene; SiO2: white carbon black; MH: maximum torque; ML: minimum torque; Tc90: vulcanization time; Ts2: scorch time.

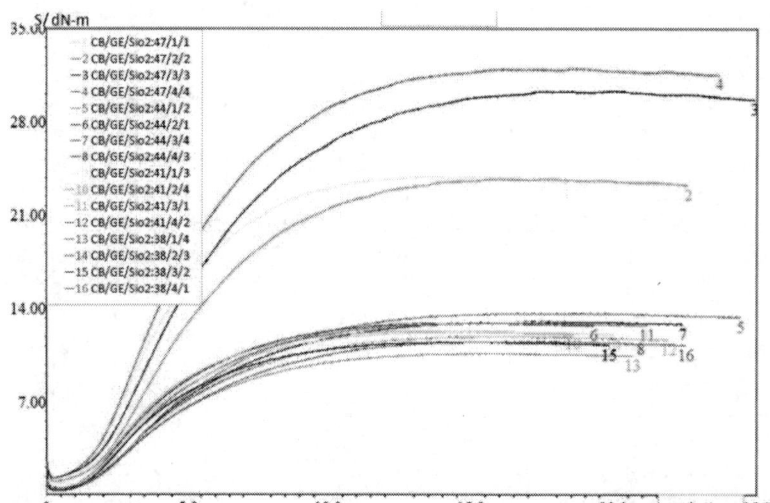


**Figure 1.** The vulcanization characteristic curve of CB/GE/$SiO_2$ for NR mixture.

As can be seen from Figure 1, when the mass fraction of CB is 47g, the maximum torque (MH) is the largest group of all the compounds. While MH is the characterization of the stiffness or shear modulus of the rubber when it is completely vulcanized and minimum torque (ML) is the characterization of the initial modulus of the unvulcanized rubber. After adding GE/SiO2, the ML basically shows an upward trend, indicating that GE/SiO2 has a greater effect on the initial modulus of the rubber mixture than CB. Probably because the ability of CB to adsorb rubber molecular chains is weaker than that of GE/SiO2, which in turn affects the mechanical properties of NR vulcanizate.

### *3.2. Mechanical Properties*

In this paper, orthogonal test design method was used to study the effect of different filling ratios of CB/GE/SiO2 on the mechanical properties of NR. Orthogonal arrangement of experiments can save time and reduce costs. Through a few representative experiments, the influence of each factor on the results is ascertained, and the order of factors is determined according to the size of the influence, and the optimal combination of parameters is found. The comprehensive balance method designed by multi-index orthogonal experiment can find the optimal conditions of single index, and balance these conditions to find out the production conditions that take into account the various indicators.

In this paper, according to the orthogonal test to improve the mechanical properties of rubber, the tensile strength and elongation at break are the test indicators. CB/GE/SiO2 are three factors, which are represented by A, B and C respectively. For the four levels, the number of designed factors is shown in Table 2. The orthogonal test plan and test results are shown in Table 3.

**Table 2.** Factor levels.

| Level | CB (g) A | GE (g) B | $SiO_2$ (g) C |
|---|---|---|---|
| 1 | 47 | 1 | 1 |
| 2 | 44 | 2 | 2 |
| 3 | 41 | 3 | 3 |
| 4 | 38 | 4 | 4 |

**Table 3.** Orthogonal experimental scheme and experimental results.

| Experiment number | CB(A) (g) | GE(B) (g) | $SiO_2$(C) (g) | Tensile strength (MPa) | Elongation at break (%) |
|---|---|---|---|---|---|
| 1 | 1(47) | 1 | 1 | 20.467 | 295.116 |
| 2 | 1 | 2 | 2 | 15.362 | 247.755 |
| 3 | 1 | 3 | 3 | 17.367 | 241.453 |
| 4 | 1 | 4 | 4 | 19.719 | 266.98 |
| 5 | 2(44) | 1 | 2 | 20.33 | 287.859 |
| 6 | 2 | 2 | 1 | 21.537 | 271.945 |
| 7 | 2 | 3 | 4 | 18.877 | 258.322 |
| 8 | 2 | 4 | 3 | 20.499 | 268.444 |
| 9 | 3(41) | 1 | 3 | 18.109 | 253.866 |
| 10 | 3 | 2 | 4 | 20.951 | 291.297 |
| 11 | 3 | 3 | 1 | 21.822 | 306.193 |
| 12 | 3 | 4 | 2 | 20.559 | 303.392 |
| 13 | 4(38) | 1 | 4 | 16.198 | 269.908 |
| 14 | 4 | 2 | 3 | 20.273 | 299.572 |
| 15 | 4 | 3 | 2 | 20.221 | 305.174 |
| 16 | 4 | 4 | 1 | 20.747 | 303.392 |

The results of the 16 tests were visually compared. As shown in Table 3, the good conditions that can be obtained are: the good condition of the tensile strength index is the No. 11 test, the test condition is A3B3C1, and the good condition of the elongation at break index is Test No. 11, the test condition was A3B3C1. As shown in Table 4, by calculating the range, the test conditions were as follows: a good test condition for the tensile strength index was A3B4C1, and a good test condition for the elongation at break index was A4B4C1.

**Table 4.** Range analysis of test results.

| | Factors | A | B | C |
|---|---|---|---|---|
| Tensile strength | K1 | 18.229 | 18.776 | 21.143 |
| | K2 | 20.311 | 19.531 | 19.118 |
| | K3 | 20.36 | 19.572 | 19.062 |
| | K4 | 19.36 | 20.381 | 18.936 |
| | R | 2.131 | 1.605 | 2.207 |
| Elongation at break | K1 | 262.826 | 276.687 | 294.161 |
| | K2 | 271.642 | 277.642 | 286.045 |
| | K3 | 288.687 | 277.785 | 265.834 |
| | K4 | 294.512 | 285.552 | 271.627 |
| | R | 31.686 | 8.865 | 28.327 |

It can be seen from the magnitude of the range R that the order of significance of each factor to each index is: tensile strength C→A→B, and elongation at break is A→C→B.

Factor A is in the first position for the elongation at break index, and is in the second position for the tensile strength index, while factor C is in the first position for the tensile strength index and the second position for the elongation at break index. While considering the contribution of the mass fraction of A and C to the index, the comprehensive factor C should be selected as 1 level. The factor B is not a significant factor for the tensile strength index and the elongation at break index, and considering the contribution of the mass fraction of the factor B to the index, the comprehensive factor B should be selected as 4 level. Factor A is a significant factor for the tensile strength index and the elongation at break index. Considering the third and fourth levels of the A factor, the tensile strength of the 3 level is 4.91% higher than that of the 4 level, and the elongation at break is 1.98%. It is consistent with the shortest results of vulcanization time and scorch time in Table 1, which further shows that the mechanical properties of NR vulcanizate are the best. Therefore, after comprehensive consideration, the factor A should be selected as 3 levels, and the good test condition after comprehensive balance is A3B4C1.

It can be seen from Figure 2 that the values of true stress and true strain of NR vulcanizate are different under different ratios, which indicates that different ratios have significant effects on the mechanical properties of NR, and the values of true stress and true strain in test No. 11 are the largest, and they are 88.58 MPa, 1.402, with the corresponding mass parts of 41g CB, 3g GE, 1g SiO2; The values of true stress and true strain in test No. 12 are 82.81 MPa and 1.394 respectively after test No. 11 and test No. 16.The corresponding mass fractions are 41 g CB, 4 g GE, 2 g SiO2. When the mass fraction (4g) of the factor GE is fixed, the test factor CB/SiO2 mass fraction of the No. 4 test, the No. 8 test, the No. 12 test and the No. 16 test is gradually smaller, the true stress increases from 72.31MPa to 83.64MPa, and the true strain increases from 1.3 to 1.395. The value shows an increasing trend. This may be the result of the specific surface area of GE, further indicating that 4g GE has a greater effect on NR vulcanizate than CB/SiO2. When the mass fraction (3g) of the factor GE is fixed, the true stress and true strain values first increase and then decrease when the mass fraction of factor CB/SiO2 in test No. 3,No. 7,No. 11 and No. 15 changes. The maximum value appeared in the No. 11 test, which was 88.58 MPa and 1.402 respectively, and the corresponding mass parts were 41 g CB, 3 g GE, and 1 g SiO2. This indicates that the CB/SiO2 ratio of 41/1 has a great influence on the mechanical properties of NR vulcanizate, which is consistent with the maximum values of tensile stress and elongation at break shown in Table 3, which indicates that the mechanical properties of NR vulcanizate are the best. After a comprehensive balance, a good test condition is A3B4C1, and the corresponding mass fraction is 41 g CB, 4 g GE, 1 g SiO2.

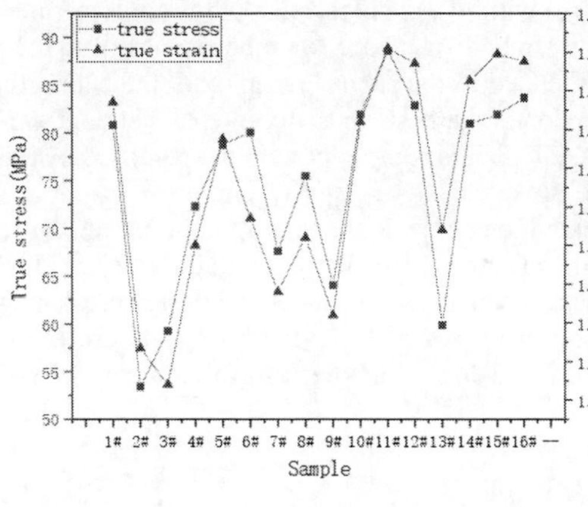


**Figure 2.** The mechanical properties of NR vulcanized rubber were studied under different ratios.

### *3.3. Microstructure Analysis*

In the analysis of vulcanization characteristics and mechanical properties of NR vulcanizate, the optimum ratio of CB/GE/SiO2 was obtained, and the corresponding mass fraction was 41 g CB, 4 g GE, 1 g SiO2. The microscopic analysis of the cross-section structure of NR vulcanizate helps to further explain that under the optimal ratio of CB/GE/SiO2, the dispersion state of the filler has a great influence on the mechanical properties of NR vulcanizate, Therefore it is necessary to carry out cross-sectional microstructure analysis of the proportioned NR vulcanizate.

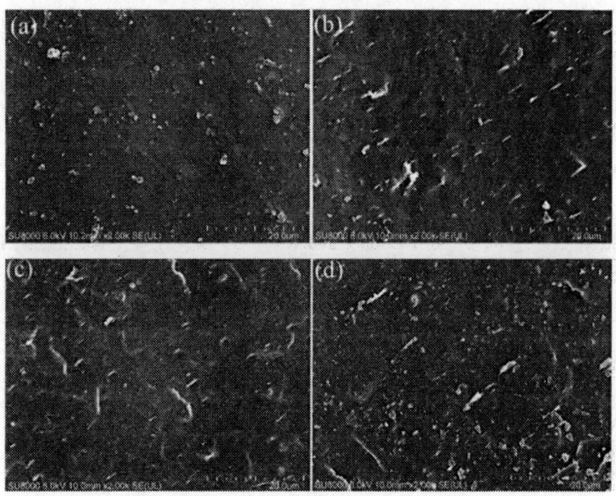


Figure 3. SEM photographs of different proportions of CB/GE/$SiO_2$.
CB/GE/$SiO_2$ (mass ratio): a—50/0/0;b—38/4/1;c—41/3/1;d—41/4/2

In Figure 3(a), the fracture surface of the NR rubber compound is relatively smooth, in which the white particles are attributed to the unreacted ZnO particles as the active agents in the vulcanization system. The cross-section of the NR rubber compound in (b) and (d) is rough, and the inside is dispersed with a plurality of sheets having a width of 2-8 μm and a thickness of 200-500 nm, which may be a surface-attached rubber-bonded graphene; at the same time, there are some irregular-shaped smaller-sized protrusions which may be a fillers such as CB/SiO2/ZnO containing a bonding rubber. After observation, (d) is more uniform than (b), which may be related to the synergistic enhancement of CB/SiO2. When comparing (c) with (d), the black shading is agglomerated carbon black particles. The carbon black particles are uniformly dispersed in the rubber matrix and the interface is the most blurred. Most of the particles are embedded in the rubber compound, indicating that the materials have strong bonding effect, so the dispersion is the best, so that the tensile strength and elongation at break of the NR vulcanizate are the largest, which is consistent with the results shown in Figure 3. Moreover, in (d) there are some small white spherical aggregates which form agglomerates. The presence of these agglomerates leads to the generation of stress concentration points, as the result, the tensile strength and elongation at break of NR vulcanizate are slightly lower than those of (c) NR vulcanizate, which are the most consistent with the values of tensile stress and elongation at break shown in Table 3, which further indicates that the mechanical properties of the NR vulcanizate are the best. After a comprehensive balance, a good test condition is A3B4C1, and the corresponding mass fraction is 41 g CB, 4 g GE, 1 g SiO2.

## 4. Conclusion

a) In the analysis of vulcanization characteristics, when the mass fraction of graphene is 4g, the fraction of CB/SiO2 gradually decreases, and ΔM also decreases, which may be due to the tiny diameter size of the particles with a smaller role of physical cross-linking points in the dispersion of rubber molecular chains, which in turn leads to a decrease in the degree of cross-linking of the system. The vulcanization time and scorch time increase first and then decrease, which may be related to the specific surface area of carbon black and white carbon black. The smaller the specific surface area is, the faster the vulcanization rate of the rubber compound, and the shorter the vulcanization time are, thus the mechanical properties of the NR vulcanization are improved.

b) Design multi-index orthogonal test and calculate the size of the range R by calculation. The order of significance of each factor to each index is: tensile strength is C→A→B, and elongation at break is A→C→B. The good test condition for the tensile strength index is A3B4C1, and the good test condition for the elongation at break index is A4B4C1. Among the calculated values of true stress and true strain, the value under the condition of A3B3C1 is the largest, a good test condition after comprehensive balance is A3B4C1, and the mass fraction of the factors are 41g CB, 4g GE, 1g SiO2, the maximum true strain in the corresponding mechanical properties is 139.8%, and the maximum true stress is 85.695MPa.

c) The results of cross-sectional morphology analysis show that when the mass fraction of CB/GE/SiO2 is 41/3/1, the dispersion of the filler is best in the NR matrix, and the mass fraction of CB/GE/SiO2 is 41/4. The results of /2-degree cross-sectional morphology analysis show that the mechanical properties of NR vulcanizate are very good. The best ratio after final equilibrium is consistent with the results obtained by orthogonal test, which is 41g CB, 4g GE, 1g SiO2.

**Acknowledgments**
This work was financially supported by Shandong Key Research and Development Project (2018GGX102007) fund.

## References

[1] Lei Wang, SuheZhao. Study on the structure - mechanical properties relationship and antistatic characteristics of SSBR composites filled withSiO2/CB [J]. Journal of Applied Polymer Science. 2010 (1): 338-345.
[2] Li Y, Han B Y, Wen S P, et al. Effect of the Temperature on Surface Modification of Silica and Properties of Modified Silica Filled Rubber Composites [J]. Composites. Part A. Applied Science and Manufacturing. 2014, 62 (17): 52-59.
[3] Yang G W, Liao Z F, Yang Z J, et al. Effects of Substitution for Carbon Black with Graphene Oxide or Graphene on the Morphology and Performance of Natural Rubber/Carbon Black Composites [J]. Journal of Applied Polymer Science. 2015, 132 (15): 41832-41840.
[4] Xu T W, Jia Z X, Luo Y F, et al. Interfacial Interaction between the Epoxidized Natural Rubber

and Silica in Natural Rubber/Silica Composites [J]. Applied Surface Science. 2015, 328: 306-313.

[5] Fei Zhao, Xinyan Shi, Ping Zhang. Influence of Sulfur on Cure Kinetics of Natural Rubber/Sulfur/N-t-butylbenzothiazole-2-sulfenamide (TBBS) System Studied by Rubber Processing Analyzer [J].Journal of Macromolecular Science, Part B. 2009 (4): 663-669.

[6] Haiyan Xu, Jingjie Han, Liang Fang, Fei Shen, Chifei Wu.Effect of Grafted Carbon Black on Properties of Vulcanized Natural Rubber [J]. Polymer Bulletin. 2007, 58 (5-6): 951-965.

[7] Fei Zhao, Xinyan Shi, Xiang Chen, Shugao Zhao. Interaction of vulcanization and reinforcement of CB on dynamic property of NR characterized by RPA2000 [J]. Journal of Applied Polymer Science. 2010 (2): 1168-1172.

[8] Johann G.Meier, Manfred Klüppel. Carbon Black Networking in Elastomers Monitored by Dynamic Mechanical and Dielectric Spectroscopy [J]. Macromol. Mater. Eng. 2008 (1): 12-38.

[9] Boripat Sripornsawat, Sitisaiyidah Saiwari, Charoen Nakason. Thermoplastic vulcanizates based on waste truck tire rubber and copolyester blends reinforced with carbon black [J]. Waste Management. 2018, 79: 638-646.

[10] Geim A K, Novoselov K S .The rise of graphene .Naturematerials.2007, 6 (3): 183-191 .

[11] Lee C ., Wei X ., Kysar J .W ., et a1 .Measurement of the elastic properties and intrinsic strength of monolayer graphene .Science. 2008, 321 (5887): 385-388 .

[12] Lu Gan, Songmin Shang, Chun Wah Marcus Yuen, et a1. Facile preparation of graphene nanoribbon filled silicone rubber nanocomposite with improved thermal and mechanical properties [J] . Composites( Part B). 2015 (69): 237-242.

[13] Xiao Liu, Suhe Zhao. Study on structure and properties of SSBR/SiO2 co - coagulated rubber and SSBR filled with nanosilica composites [J]. Journal of Applied Polymer Science. 2008 (6): 3900-3907.

[14] L.Chazeau, J. D.Brown, L. C.Yanyo, S. S.Sternstein. Modulus recovery kinetics and other insights into the payne effect for filled elastomers [J]. Polymer Composite. 2000 (2): 202-222.

[15] Yuanbo Liu, Li Li, Qi Wang. Effect of Carbon Black/Nanoclay Hybrid Filler on The Dynamic Properties of Natural Rubber Vulcanizates [J]. Journal of Applied Polymer Science. 2010, 118 (2): 1111–1120.

IOP Conf. Series: Materials Science and Engineering **772** (2020) 012022 doi:10.1088/1757-899X/772/1/012022

# Modeling and solving the dynamic problem of a porous rotational shell based by modified Boit's theory

**Jin Hao *, Xiang Yu and Lu Jing**

Guangxi University of science and technology, Liuzhou, China

*Corresponding author e-mail: 15731069297@163.com

**Abstract.** Combined the Boit's theory and the classical elastic theory of thin shell, a new model for analyzing the vibrations of a thin fluid-saturated porous rotational shell is proposed and solved by the extended homogeneous capacity precision integration method and the precise element method. The present model is precise and closer to the real materials as it takes the fluid-solid coupling effect into account and doesn't make any hypothesis for the fluid displacement. At the end of the paper, the accuracy of the model had been verified by experiment and FEM methods.

## 1. Introduction

In this article, taking account for the relative action between the fluid and solid phase, we derive the dynamics governing equations of the porous rotational shell based on the classical thin shell theory and Boit's theory. The equation is written as a first-order differential matrix and solved by the precise integrate method. The experiment is given to verify the accuracy of the present method.

Porous materials have been widely used for passive absorption and noise control in many fields, such as the vehicle, aerospace, geophysics and civil engineering. In the last years, many researchers have paid attention to the vibrations and acoustics characteristics of the poroelastic structures. In 1956, Boit established the three-dimensional theory of the wave propagation in fluid-saturated porous elastic solid [1, 2]. With some hypothesis, Boit's theory has been applied to porous plates and shells. Manuel et al derived a mix displacement-pressure formulation for the bending vibration of porous plates by the thin plate theory and Boit's theory [3]. This rigid frame is useful for the material saturated by air. In order to evaluate the energy dissipation by viscous friction, Leclaire and Horoshenkov proposed a analytical description of a fluid-saturated plate by considering the fluid loading terms and relative transverse fluid/solid displacements [4]. However, it is pointed out in the article that the model is only valid in the low frequency since the thickness of the plate is taken to be smaller than any acoustic wavelength and the variation of amplitude of the fluid transverse displacement is small. Ref [5] assumed the in-plane fluid flow relative to the motion of the solid is negligible to the transverse one, and the governing equation written in terms of the three solid displacement and fluid transverse displacement is derived to characteristics the flexural plate dynamics in the frequency domain. Based on the above hypothesis, some works about the porous cylindrical shell were published. Ahmed Shah et al present the frequency equation of axially symmetric vibrations for poroelastic circular cylindrical shells of infinite extent with different boundary conditions [6]. Recently, employing the first-order shear deformation theory and 3D Boit theory, Julien et al derived an analytical model of an infinite sandwich cylindrical shell to demonstrated the influence of the structural damping [7]. Those infinite cylindrical shells are quite far

Content from this work may be used under the terms of the Creative Commons Attribution 3.0 licence. Any further distribution of this work must maintain attribution to the author(s) and the title of the work, journal citation and DOI.
Published under licence by IOP Publishing Ltd

away from real industrial problem. Owe to the difficulty in the analytical modeling, the finite element method is implemented for the calculation of finite porous cylindrical shells [8]. The finite element method is convenient and widely used, but the finite element model of the porous structure is usually large and complex, in conjunction with time-consuming, so this method can only be applied in low frequency. Boily and Charron found the finite element model of a sandwich cylindrical shell [9]. In the model, the poroelastic core is simplified by the equivalent fluid model, and the modal reduction method is used to decrease the size of the system. To the knowledge of the authors, the vibration analyses for the porous rotational porous shell are not available in the literature.

## 2. Governing equations

### *2.1. Constitutive equations of the porous rotational shell*

The stresses-strain relations of a homogeneous, isotropic poroelastic material are

$$\begin{aligned} &\sigma_{ij} = 2Ne_{ij} + \left(Ae + Q\varepsilon\right)\delta_{ij}\left(i,j=1,2,3\right), \\ &-\phi p = Qe + R\varepsilon. \end{aligned} \tag{1}$$

The first formula describes the constitutive relation of the solid phase, and the second formula describes the constitutive relation of the fluid phase. The coefficients A and N correspond to the familiar Lamé coefficients in the theory of elasticity. The coefficient R is a measure of the pressure required on the fluid to force a certain volume of the fluid into the aggregate while the total volume remains constant. The coefficient Q represents the coupling between the volume changes of the solid to that of liquid. $\delta_{ij}$ is the Kronecker's delta. p is the fluid press. The coefficient $\phi$ is the porosity. $\sigma_{ij}$ and $e_{ij}$ are the solid stresses and strains tensor respectively. $e$ and $\varepsilon$ are the dilations of solid and fluid respectively, given by

$$e = e_{kk} = u_{k,k}, \quad \varepsilon = \varepsilon_{kk} = U_{k,k}, \quad k = 1,2,3, \tag{2}$$

where u is the solid displacement tensor, and U is the fluid displacement tensor; the repetition of the indices k imply summation; commas indicate differentiation with respect to space; $\varepsilon_{ij}$ denotes the fluid strain tensor.

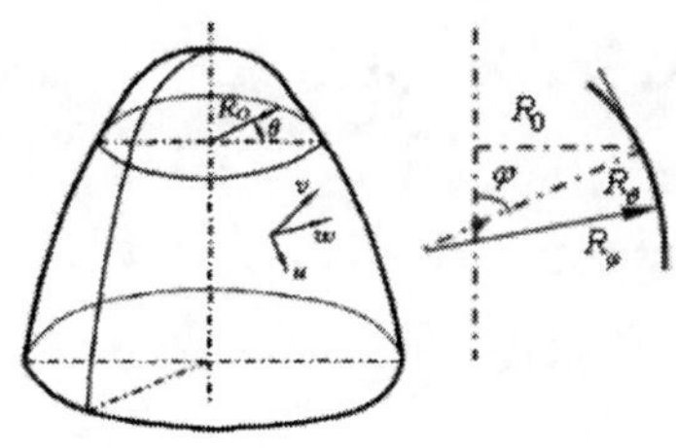

**Fig. 1** The porous rotational shell

Consider a thin porous rotational shell, whose middle plane coincides with the natural coordinates, as shown in Fig.1. In the figure, $\varphi$ and $\theta$ are meridian and circumferential coordinates, whose relevant curvature radius are $R_\varphi$ and $R_\theta$ respectively. The mid-plane displacement in the meridian, circumferential, and normal directions are u, v, w respectively. The mid-plane Lamé coefficients in meridian and circumferential directions are $R_\varphi$ and $R_0$ respectively, which are both the function of $\varphi$.

For a thin shell, the normal line (z direction) of the middle plane is assumed as a straight line, and whose angles between the vertical line of the meridian and circumferential direction are unchanged. Furthermore, the effect of the total normal stress can be neglect. These mean that

$$e_{\varphi z}=0,\ e_{\theta z}=0,\ \sigma_z-\phi p\eta=0. \tag{3}$$

The fluid dilation can be evaluated from the second formula in Eq.(1) as

$$\varepsilon=\frac{-\phi p-Qe}{R}. \tag{4}$$

Substituting Eq.(4) into the first formula in Eq.(1) yields

$$\sigma_\varphi=2Ne_\varphi+\left(A-\frac{Q^2}{R}\right)e-\frac{Q}{R}\phi p,$$

$$\sigma_\theta=2Ne_\theta+\left(A-\frac{Q^2}{R}\right)e-\frac{Q}{R}\phi p,$$

$$\sigma_z=2Ne_z+\left(A-\frac{Q^2}{R}\right)e-\frac{Q}{R}\phi p,$$

$$\tau_{\varphi\theta}=Ne_{\varphi\theta}. \tag{5}$$

By eliminating $e_z$ with the aid of Eq. (3) and Eq. (5), we can obtained the solid dilation from the second formula in Eq.(1)

$$e=\left(1-\bar{B}\right)\left(e_\varphi+e_\theta\right)+\bar{A}\phi p, \tag{6}$$

Where $\bar{A}$ and $\bar{B}$ are constants, with $\bar{A}=\dfrac{Q+R}{2NR+AR-Q^2},\bar{B}=\dfrac{AR-Q^2}{2NR+AR-Q^2}$.

Combing Eq.(5) and Eq.(6), the solid constitutive equations of a porous rotational shell can take the form

$$\sigma_\varphi=\bar{\bar{N}}e_\varphi+\bar{\bar{A}}e_\theta+\bar{\bar{B}}\phi p,$$

$$\sigma_\theta=\bar{\bar{N}}e_\theta+\bar{\bar{A}}e_\varphi+\bar{\bar{B}}\phi p,$$

$$\tau_{\varphi\theta}=Ne_{\varphi\theta}, \tag{7}$$

Where $\bar{\bar{A}}$, $\bar{\bar{B}}$ and $\bar{\bar{N}}$ are constants, with $\bar{\bar{A}}=\left(A-\frac{Q^2}{R}\right)\left(1-\bar{B}\right),\bar{\bar{B}}=\left[\left(A-\frac{Q^2}{R}\right)\bar{A}-\frac{Q}{R}\right],\ \bar{\bar{N}}=\left[2N+\left(A-\frac{Q^2}{R}\right)\left(1-\bar{B}\right)\right]$.

Inserting Eq.(6) into Eq.(4), the fluid constitutive equations of a porous rotational shell can take the form

$$\varepsilon=\bar{\bar{A}}'\phi p+\bar{\bar{B}}'\left(e_\varphi+e_\theta\right), \tag{8}$$

Where $\bar{\bar{A}}'$, and $\bar{\bar{B}}'$ are constants, with $\bar{\bar{A}}'=-\dfrac{\left(1+Q\bar{A}\right)}{R}$, $\bar{\bar{B}}'=-\dfrac{Q}{R}\left(1-\bar{B}\right)$.

*2.2. Governing equations of the solid frame*

According to the Reissner model in the classical thin shell theory, the solid displacements can be written as

$$\hat{u}_s = u_s + z\theta_\varphi,\ \hat{v}_s = v_s + \frac{z}{R_0}\left(v_s \sin\varphi - \frac{\partial w_s}{\partial \theta}\right),\ \hat{w}_s = w_s, \tag{9}$$

where $\hat{u}_s = \hat{u}_1$, $\hat{v}_s = \hat{u}_2$, $\hat{w}_s = \hat{u}_3$ are the solid displacements at any point, and $u_s$, $v_s$, $w_s$ are the corresponding solid displacements in the middle plane, in which the mid-plane rotation $\theta_\varphi$ being defined as

$$\theta_\varphi = \frac{1}{R_\varphi}\left(u_s - \frac{\partial w_s}{\partial \varphi}\right). \tag{10}$$

The solid strains at any point can be represented by the mid-plane solid displacements as

$$\begin{cases} e_\varphi = \left(\frac{1}{R_\varphi}\right)\frac{\partial u_s}{\partial \varphi} + \left(\frac{1}{R_\varphi}\right) w_s + \frac{z}{R_\varphi}\frac{\partial \theta_\varphi}{\partial \varphi} \\ e_\theta = \left(\frac{\cos\varphi}{R_0}\right) u_s + \left(\frac{1}{R_0}\right)\frac{\partial v_s}{\partial \theta} + \left(\frac{\sin\varphi}{R_0}\right) w_s + \frac{z}{R_0}\frac{\partial}{\partial \theta}\left(\frac{\sin\varphi}{R_0} v_s - \frac{1}{R_0}\frac{\partial w_s}{\partial \theta}\right) + \left(\frac{z\cos\varphi}{R_0}\right)\theta_\varphi \\ e_{\varphi\theta} = \frac{1}{R_0}\frac{\partial u_s}{\partial \theta} + \left(\frac{1}{R_\varphi}\right)\frac{\partial v_s}{\partial \varphi} - \left(\frac{\cos\varphi}{R_0}\right) v_s + \frac{2z}{R_0}\frac{\partial \theta_\varphi}{\partial \theta} \end{cases} \tag{11}$$

It is assumed that the dynamic excitation is harmonic in time and hence the shell response is also harmonic. In the other hands, the variables in the rotational shell can be expanded with the Fourier series along the circumferential direction, so that that the function $f(\varphi,\theta,t)$ has the form

$$f(\varphi,\theta,t) = \left(\sum_n \tilde{f}^{(n)}(\varphi) e^{jn\theta}\right) e^{j\omega t}, \tag{12}$$

Where $\tilde{f}^{(n)}$ is the amplitude of $f$, $j = \sqrt{-1}$ and $\omega$ is the circular frequency. In the following, the factors $e^{j\omega t}$, $e^{jn\theta}$, and the superscript $n$ is omitted in the follows for simplicity conveniences.

By combing Eq.(7) and Eq.(11), and implement the Fourier series expansion, the Internal force – displacement relations of the solid frame are derived in frequency domain. Based on the thin shell theory, and taking into the dynamics equations of the solid frame, the solid dynamics equations of the fluid-saturated porous rotational shell can be derived in the frequency domain as

$$R_\varphi \cos\varphi \tilde{N}_\varphi + R_0 \frac{\partial \tilde{N}_\varphi}{\partial \varphi} + R_\varphi \frac{\partial \tilde{N}_{\theta\varphi}}{\partial \theta} - R_\varphi \cos\varphi \tilde{N}_\theta + R_0 \tilde{Q}_\varphi + R_\varphi R_0 \tilde{q}_\varphi - R_\varphi R_0 h\left(\Omega_{11}\tilde{u}_s - \Omega_{12}\tilde{u}_f\right) = 0,$$

$$R_\varphi \frac{\partial \tilde{N}_\theta}{\partial \theta} + 2R_\varphi \cos\varphi \tilde{N}_{\varphi\theta} + R_0 \frac{\partial \tilde{N}_{\varphi\theta}}{\partial \varphi} + R_\varphi R_0 \frac{\tilde{Q}_\theta}{R_\theta} + R_\varphi R_0 \tilde{q}_\theta - R_\varphi R_0 h\left(\Omega_{11}\tilde{v}_s - \Omega_{12}\tilde{v}_f\right) = 0,$$

$$R_\varphi \cos\varphi \tilde{Q}_\varphi + R_0 \frac{\partial \tilde{Q}_\varphi}{\partial \varphi} + R_\varphi \frac{\partial \tilde{Q}_\theta}{\partial \theta} - R_\varphi R_0 \frac{\tilde{N}_\varphi}{R_\varphi} - R_\varphi R_0 \frac{\tilde{N}_\theta}{R_\theta} + R_\varphi R_0 \tilde{q}_z - R_\varphi R_0 h\left(\Omega_{11}\tilde{w}_s - \Omega_{12}\tilde{w}_f\right) = 0,$$

$$R_\varphi \frac{\partial \tilde{M}_\theta}{\partial \theta} + R_\varphi \cos\varphi \tilde{M}_{\varphi\theta} + R_0 \frac{\partial \tilde{M}_{\varphi\theta}}{\partial \varphi} + R_\varphi \cos\varphi \tilde{M}_{\theta\varphi} - R_\varphi R_0 \tilde{Q}_\theta = 0,$$

$$R_\varphi \cos\varphi \tilde{M}_\varphi + R_0 \frac{\partial \tilde{M}_\varphi}{\partial \varphi} + R_\varphi \frac{\partial \tilde{M}_{\beta\varphi}}{\partial \theta} - R_\varphi \cos\varphi \tilde{M}_\theta - R_\varphi R_0 \tilde{Q}_\varphi = 0, \tag{13}$$

Where $\Omega_{11} = b\omega j - \rho_{11}\omega^2$ and $\Omega_{12} = \rho_{12}\omega^2\tilde{u}_f + b\omega j$ are frequency parameters. $\tilde{u}_f = U_1$, $\tilde{v}_f = \tilde{U}_2$ and $\tilde{w}_f = \tilde{U}_3$ are the amplitude of fluid mid-plane displacements. $\tilde{Q}_\varphi$ and $\tilde{Q}_\theta$ are the amplitude of shear forces per unit length along $\varphi$ and $\theta$ direction respectively, which can be transformed to the Kelvin—Kirchhoff equivalent shear force

$$\tilde{S}_\varphi = \tilde{N}_{\varphi\theta} + \frac{\sin\varphi}{R_0}\tilde{M}_{\varphi\theta},$$

$$\tilde{V}_\varphi = \tilde{Q}_\varphi + \left(\frac{nj}{R_0}\right)\tilde{M}_{\varphi\theta}. \tag{14}$$

### 2.3. *Governing equations of the fluid*

Combining the dynamics equations of the fluid of the Boit theory and thin shell theory, we inferred the governing equations of the fluid.

$$\begin{aligned}
&-R_\varphi\left(\Omega_{21}-\Omega_{22}\bar{\bar{B}}'\right)\frac{d\tilde{u}_s}{d\varphi}+\left[\left(\frac{R_\varphi\cos\varphi}{R_0}\right)-\left(\frac{R'_\varphi}{R_\varphi}\right)\right]\frac{dP}{d\varphi}+\frac{d^2P}{d\varphi^2}= \\
&\left(\frac{R_\varphi^2\cos\varphi}{R_0}\right)\left(\Omega_{21}-\Omega_{22}\bar{\bar{B}}'\right)\tilde{u}_s+\left(\frac{R_\varphi^2 nj}{R_0}\right)\left(\Omega_{21}-\Omega_{22}\bar{\bar{B}}'\right)\tilde{v}_s+R_\varphi^2\left(\frac{\sin\varphi}{R_0}+\frac{1}{R_\varphi}\right)\left(\Omega_{21}-\Omega_{22}\bar{\bar{B}}'\right)\tilde{w}_s- \\
&R_\varphi^2\left(\frac{n^2}{R_0^2}+\Omega_{22}\bar{\bar{A}}'\right)\tilde{P}-\left(\frac{\sin\varphi}{R_0}+\frac{1}{R_\varphi}\right)\left(\frac{R_\varphi^2}{h}\right)\phi\left(p_1-p_2\right) \\
&-\left(\frac{1}{R_\varphi^2}\right)\frac{d^2\tilde{M}_P}{d\varphi^2}-\left[\left(\frac{n^2}{R_0^2}\right)+\Omega_{22}\bar{\bar{A}}'\right]M_P+\left(\frac{R'_\varphi}{R_\varphi^3}\right)\frac{d\tilde{M}_P}{d\varphi}+\left(\Omega_{21}-\Omega_{22}\bar{\bar{B}}'\right)\left(\frac{nj\sin\varphi}{R_0^2}\right)\tilde{v}_s- \\
&+\left(\Omega_{21}-\Omega_{22}\bar{\bar{B}}'\right)\left(\frac{n^2}{R_0^2}\right)\tilde{w}_s+\left(\Omega_{21}-\Omega_{22}\bar{\bar{B}}'\right)\left(\frac{1}{R_\varphi}\right)\frac{d\theta_\varphi}{d\varphi}+\left[\Omega_{21}-\Omega_{22}\bar{\bar{B}}'\right]\left(\frac{\cos\varphi}{R_0}\right)\theta_\varphi=0
\end{aligned} \tag{15}$$

Where $\tilde{p}_1$ and $\tilde{p}_2$ are the amplitude of the press on the surfaces of the shell, and $R'_\varphi = \frac{\partial R_\varphi}{\partial\varphi}$.

Eq.(15) are the governing equation of the fluid in the fluid-statured porous rotational shell. It should be emphasized that, in the above derivation, we have not make any assumption for the displacement, and consider the interactions between the fluid and solid phases fully.

### 2.4. *The integrated governing equations of a porous rotational shell saturated in fluids*

Combing the solid governing equations Eq.(13) and the fluid governing equations Eq.(15), we can obtain the integrated governing equations of a porous rotational shell by eliminating some dependant variables. Then we implant the dimensionless procedure, letting

$$\begin{aligned}
&\tilde{u}_s = L\bar{u}_s, \tilde{v}_s = L\bar{v}_s, \tilde{w} = L\bar{w}_s, \tilde{\theta}_\varphi = \bar{\theta}_\varphi, \tilde{N}_\varphi = \left(h\bar{\bar{N}}\right)\bar{N}_\varphi, \tilde{S}_\varphi = \left(h\bar{\bar{N}}\right)\bar{S}_\varphi, \tilde{V}_\varphi = \left(h\bar{\bar{N}}\right)\bar{V}_\varphi, \\
&\tilde{M}_\varphi = \left(\frac{\bar{\bar{N}}h^3}{12L}\right)\bar{M}_\varphi, \tilde{P} = \frac{\bar{\bar{N}}}{\bar{\bar{B}}}\bar{P}, \tilde{M}_p = \left(\frac{\bar{\bar{N}}}{\bar{\bar{B}}L}\right)\bar{M}_p, \tilde{q}_\varphi = \left(\frac{h\bar{\bar{N}}}{L}\right)\bar{q}_\varphi, \tilde{q}_\theta = \left(\frac{h\bar{\bar{N}}}{L}\right)\bar{q}_\theta, \tilde{q}_z = \left(\frac{h\bar{\bar{N}}}{L}\right)\bar{q}_z.
\end{aligned} \tag{16}$$

Where $\tilde{q}_\varphi$, $\tilde{q}_\theta$, $\tilde{q}_z$ are the amplitudes of the load along $\varphi$, $\theta$ and z directions. Finally, the integrated governing equations can be expressed as the first-order differential matrix equation as

$$\boldsymbol{A}\frac{d\boldsymbol{Z}}{d\varphi}=\boldsymbol{B}\boldsymbol{Z}+\boldsymbol{F}, \tag{17}$$

Where $\boldsymbol{Z}=\left\{\bar{u}_s \ \bar{v}_s \ \bar{w}_s \ \bar{\theta}_\varphi \ \bar{N}_\varphi \ \bar{S}_\varphi \ \bar{V}_\varphi \ \bar{M}_\varphi \ \bar{P} \ \frac{d\bar{P}}{d\varphi} \ \bar{M}_p \ \frac{d\bar{M}_p}{d\varphi}\right\}^T$ is the integrated state vector, $\boldsymbol{A}$, $\boldsymbol{B}$ are coefficient matrixes and $\boldsymbol{F}$ is the integrated load vector respectively. Their non-zero elements are listed in Appendix.

## 3. Validation of model correctness

### *3.1. Comparison with the experiment result*

To verify the present model, a porous cylindrical shell is considered which is degenerated by the porous rotational shell. In this case, the geometric parameters $R_\varphi \to \infty, R_0 = R, \varphi = \frac{\pi}{2}$. The coefficients and experimental set-up and are shown in table 1 and figure 2 respectively. A cantilever porous cylindrical shell is excited by a point force $2e^{j\omega t}$N at the bottom of the free end ($\theta = 0$), and the deflection of the bottom point and top point of the free end are both measured and compared with the present method. The results are shown in figure 3 and figure 4 respectively, in which the response function is $\mathrm{FRF} = \log_{10}\left(\left|\bar{w}\right|\right)$.

It is seen from figure 3 and figure 4 that the deflections curves given by the present method and experiment are almost the same trend. Moreover, we can know from figure 3 that, the first order natural frequencies of the numerical and experimental results are both 116Hz, while the numerical and experiment results are 116Hz and 115Hz respectively in figure 4. The above comparisons have indicated the reliability of the present method.

**Table 1.** Coefficients of the porous cylinder shell

| $l$ (mm) | $R$ (mm) | $h$ (mm) | $\rho_s$ (kg/m$^3$) | $\rho_f$ (kg/m$^3$) | $\mu$ | $E$ (GPa) | $\phi$ | $q$ (m$^2$) |
|---|---|---|---|---|---|---|---|---|
| $5\times10^3$ | 100 | 10 | 2700 | 1.213 | 0.3 | 0.38 | 0.7 | $2.7\times10^{-10}$ |

**Fig. 2** Vibration experiment of porous cylinder shell

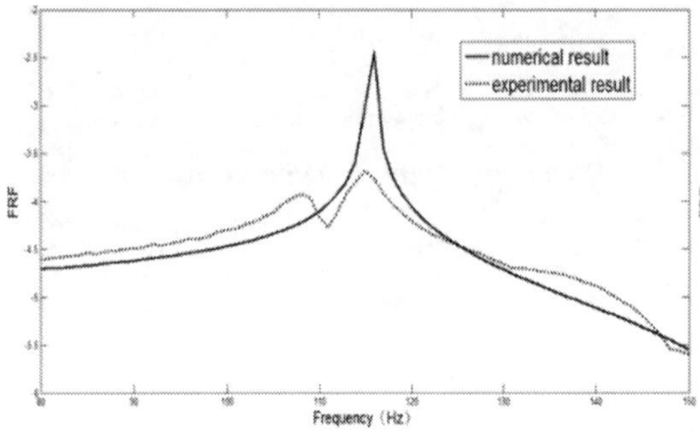


**Fig. 3** Comparison of results on bottom point

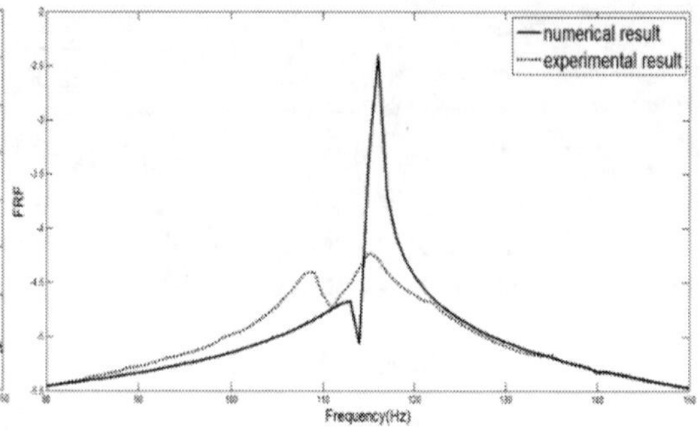


**Fig. 4** Comparison of results on top point

*3.2. Comparison with the equivalent models*

In existing work of modeling porous materials with saturated fluid, the equivalent models are widely used, and the effective medium method and equivalent density fluid method are the most popular. The effective medium method is applied to the porous materials with rigid solid frames. It assumes that the micro elements of porous materials can be seen as an effective elastic element [10]. The equivalent density fluid method is used to the porous materials with soft solid frames. In this model, it is assumed that the frame does not resist to external excitations, the frame of the porous medium is ignored, and the micro elements of porous materials can be seemed as an effective limp element. Both of the two models describe the porous mediums by equivalent mass density and ignore the solid-fluid coupling.

In this section, we further compare the results with the effective medium method and the equivalent density fluid method to show the utility of the present method in the case of vibration of performance. A porous rotational shell with major axis $a = 210mm$, minor axis $b = 105mm$, start angle $\varphi = \frac{\pi}{6}$, end angle $\varphi = \frac{5\pi}{6}$ and layer thicknesses $h = 10mm$ is considered. The material parameters and boundary condition are the same as section 3.1. According to the effective medium method, the equivalent elastic modulus $E_e$ is equal to $0.04E$ and the equivalent density $\rho_e$ is equal to $0.3\rho_s + 0.7\rho_g$ in the finite element method. The results of the natural frequency of the shell are listed in Table 2, and some FEM results are shown in figure 5. In which, m denotes the half wave numbers of the meridian direction and n denotes that of the circumferential direction. The close agreement between the present solution and FEM solution indicates the correction of the proposed method again.

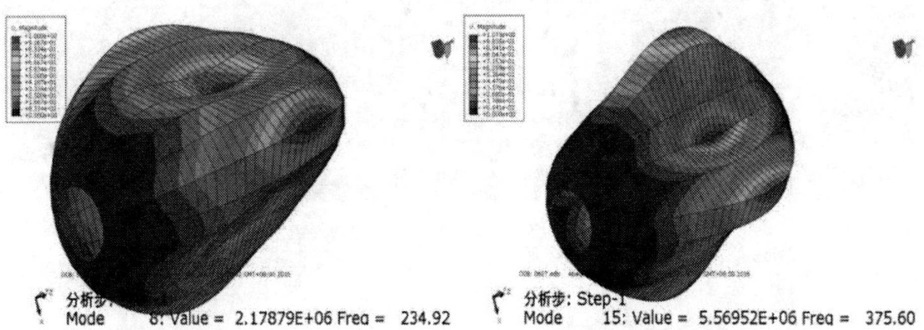


m=1 n=3 Freq=234.92Hz

m=1 n=4 Freq=375.60Hz

**Fig. 5** several typical FEM results of porous rotational shell

**Table 2.** Comparison of natural frequency

| order (m,n) | model ( Hz ) | FEM ( Hz ) |
|---|---|---|
| (1,1) | 76.37 | 74.724 |
| (1,2) | 181.93 | 180.92 |
| (1,3) | 233.78 | 234.92 |
| (1,4) | 369 | 375.60 |
| (1,5) | 610.5 | 621.37 |

## 4. Conclusion

In this paper, a new theoretical model of thin porous rotational shell has been established based on the Biot's theory and the classical elastic theory of thin shell. The accuracy of the present model had been verified by comparing with the results of experiment and finite element method. When comparing with the effective density fluid model, the two methods give almost identical results in the low and medium frequency range, but the resonance amplitudes of the two models have a discrepancy in higher frequency range.

## References

[1] M. A. Biot, Theory of propagation of elastic waves in a fluid-saturated porous solid. Ⅰ. Low-frequency range, Journal of the Acoustical Society of America 28(2) (1956) 168-178.
[2] M. A. Biot, Theory of propagation of elastic waves in a fluid-saturated porous solid. II. High-frequency range. Journal of the Acoustical Society of America 28(1956) 179-191.
[3] E. Manuel, S. Sohbi, B. Bruno, Bending vibration of a rectangular poroelastic plate, Comptes Rendus de I'Académie des Sciences 329(2001) 615-620
[4] P. Leclaire, K.V. Horoshenkov, Transverse vibrations of a thin rectangular porous plate saturated by a fluid, Journal of Sound and Vibration 247(1) (2001) 1-18
[5] D.D. Theodorakopoulos, D.E. Beskos, Flexural vibrations of poroelastic plates, Acta Mechanica 103(1994) 191-203.
[6] S. Ahmed Shah, Axially symmetric vibrations of fluid-filled poroelastic circular cylindrical shells, Journal of Sound and Vibration 318(2008) 389-405
[7] M. Julien, A.H. Mohamed, C. Jean-Daniel et al, A mixed "Boit- Shell" analytical model for the prediction of sound transmission through a sandwich cylinder with a poroelastic core, Journal of Sound and Vibration 360(2016) 203-223
[8] Y.J. Kang, J.S. Bolton, Finite element modeling of isotropic elastic porous materials coupled with acoustical finite elements, Journal of the Acoustical Society of America 98(1) (1995) 635-643
[9] S. Boily, F. Charron, The vibroacoustic response of a cylindrical shell structure with viscoelastic and poroelastic materials, applied Acoustics 58(1999) 131-152
[10] QIN Bo, LIANG Bin, ZHU Zhemin, et al. Effective medium method of slightly compressible elastic media permeated with air-filled bubbles[D]. Acth Acustica. 2007, 3,32(2) :110-115.(in chinese)
[11] R. Panneton. Comments on the limp frame equivalent fluid model for porous media[J]. Acoustical Society of America,2007,122(6).
[12] K. L .Williams, D. R. Jackson, E. I. Thorsos et al. Comparison of sound speed and attenuation measured in a sandy sediment to predictions based on the Biot Theory of porous media[J]. IEEE journal of oceanic engineering, 2002,27(3):413-426.

IOP Conf. Series: Materials Science and Engineering **772** (2020) 012023 doi:10.1088/1757-899X/772/1/012023

# Simulation and Physical Experiment Study on Powder Metallurgy Preparation of Silicon Composite Powder

**Qian Jia[1, 2, *], Zhaochong Ding[1, 2], Xiaona Zhang[1, 2], Xiaomeng Cao[1, 2], Haitao Teng[1, 2] and Yongjun Li[1, 2]**

[1] Grikin Advanced Materials Co. Ltd., Beijing, China
[2] Bejing Technology Research Center for Sputtering Target Material Engineering of High Pure Metals, Beijing, China

*Corresponding author e-mail: jq@grikin.com

**Abstract.** The compaction and sintering processes of multi-component silicon powders, including single-action die compaction, uniaxial hot pressing and bidirectional hot compaction, were numerically performed by FEM (finite element method) modelling. The relationships of initial packing structure, compaction method, pressing pressure and sintering temperature between overall relative densities of multi-component silicon components were systematically discussed and studied. Meanwhile, various macro- and microscopic properties of silicon composite powders were carefully characterized, such as overall and local relative density, local stress distributions and void filling behaviour. The results show that fully compacted silicon composite targets with low internal stress can be obtained by using uniaxial hot pressing or bidirectional hot compaction method. And the physical experiments were carried out to verify the accuracy of the simulation results.

## 1. Introduction

Memory is an important part of integrated circuit semiconductor components, accounting for up to 20% of semiconductor products [1-5].The most mainstream memory in the moment is DRAM, NAND Flash, NOR Flash, which accounts for about 95%. But the three major semiconductor memories themselves also have various shortcomings, such as the data of DRAM is volatile and it has small capacity; NAND Flash long delay; NOR Flash has a small capacity and a slow write and erases speed. Silicon alloy target is used in phase change memory chip manufacturing. Phase change memory has many advantages, such as low latency, read and write time balance, long life, low power consumption, high storage density and anti-irradiation, etc. However, there are still numerous in CM's current research and development, such as that device power consumption and work speed are difficult to balance, thermal crosstalk problem at high density, the material needs to have both a high crystallization temperature and a low melting point, volume change before and after phase change affects device reliability. Many technical difficulties have not yet been completely overcome from materials and processes [6-9].

In this paper, the compaction and sintering processes of multi-component silicon powders, including single-action die compaction, uniaxial hot pressing and bidirectional hot compaction, were numerically performed by FEM (finite element method) modeling. The relationships of initial packing structure, compaction method, pressing pressure and sintering temperature between overall relative densities of

Content from this work may be used under the terms of the Creative Commons Attribution 3.0 licence. Any further distribution of this work must maintain attribution to the author(s) and the title of the work, journal citation and DOI.
Published under licence by IOP Publishing Ltd

multi-component silicon components were systematically discussed and studied. Meanwhile, various macro- and microscopic properties of silicon composite powders were carefully characterized, such as overall and local relative density, local stress distributions and void filling behaviour. The results show that fully compacted silicon composite targets with low internal stress can be obtained by using uniaxial hot pressing or bidirectional hot compaction method. And the physical experiments were carried out to verify the accuracy of the simulation results.

## 2. Simulation method and conditions

FEM method was used to setup Single-action die compaction, uniaxial hot pressing and bidirectional hot compaction models of Si composite powders. In order to realistically simulate the whole PM process, the initial packing structures of multi-component powders with different particle sizes, which conform to a normal distribution, was firstly generated in DEM program. DEM, as a mature and effective dynamic simulation method, has been broadly applied in modeling the particle packing. Therefore, the governing equations and force models in DEM are not detailed and can be found in [10-14]. And then the initial packing densities of various heights were calculated by sectional area method. These data of initial packing densities were imported into FEM model as initial conditions, which was different with previous research. After that, other normal parameters were inputted, as well as materials property, boundary condition setting, contact definition, loading mode and temperature setting etc. Finally, subsequent compacting and sintering were simulated by using FEM method. The flowchart of the whole PM process is shown in Fig. 1(a).

The mesh division is shown in Fig. 1 (b), where each particle contains 5356 nodes and 4480 elements. Fig. 1 (c) gives the initial FEM model in the die before compaction. In the simulation, the die and punches are set to be rigid and Si composite powder is set to be elasto-plastic. The compaction is performed from the upper punch with the lower punch and the die being fixed.

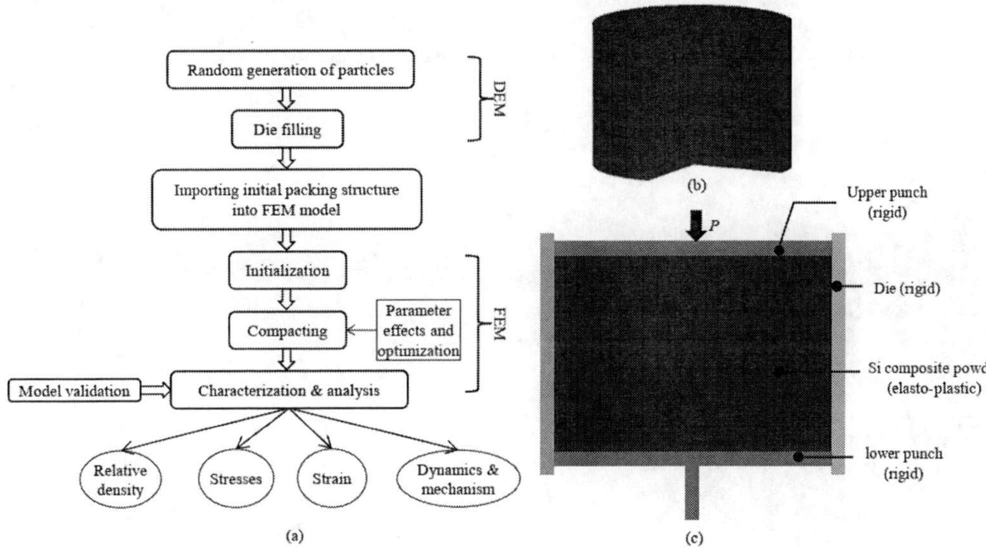


**Figure 1.** (a) Flowchart in the numerical modeling on the PM process of Si composite powders; (b) mesh division for FEM model and (c) initial FEM model in the die before compaction.

In the simulation, the modified Shima-Oyane model [15, 16] is used to define the properties of multi-component Si alloy. In this model the flow yield stress is expressed as:

$$F = \frac{1}{\gamma}\left(\frac{3}{2}\sigma^d \sigma^d + \frac{p^2}{\beta^2}\right)^{\frac{1}{2}} - \sigma_y \tag{1}$$

Where $\sigma_y$ is uniaxial yield stress, $\sigma^d$ stands for deviatoric stress tensor, $p$ represents hydrostatic pressure and γ, β are material parameters. γ and β can be expressed as:

$$\gamma = (q_1 + q_2 \rho^{q_3})^{q^4} \tag{2}$$

$$\beta = (b_1 + b_2 \rho^{b_3})^{b_4} \tag{3}$$

For multi-component Si alloy, $q_1 = 0$, $q_2 = 1$, $q_3 = 3$, $q_4 = 1$, $b_1 = 6$, $b_2 = -6$, $b_3 = 1$, $b_4 = -0.5$. Poisson's ratio v and elastic modulus of multi-component Si alloy can be represented by:

$$\nu = 0.5e^{-12.5(1-\rho)^2} \tag{4}$$

$$E = E_c \rho^{3.4} \tag{5}$$

When defining contact, powders in the model are considered as deformable, while the die and punches are set as rigid bodies. The coefficient of friction between powders is set to be 0.2 using Coulomb friction model, while the contacts between the powders and the rigid bodies are considered as smooth. During compaction, the pressure from the upper punch will increase from 0 MPa to 150 MPa, and then it will be kept constant for a while and released. The temperature of hot pressing is 600 °C.

## 3. Results and discussion

### 3.1. Initial packing density

Fig. 2 (a) gives stable static initial packing structure with different particle sizes by DEM. It can be seen that large pores exist on the die boundary, which caused by the boundary effect of die wall.

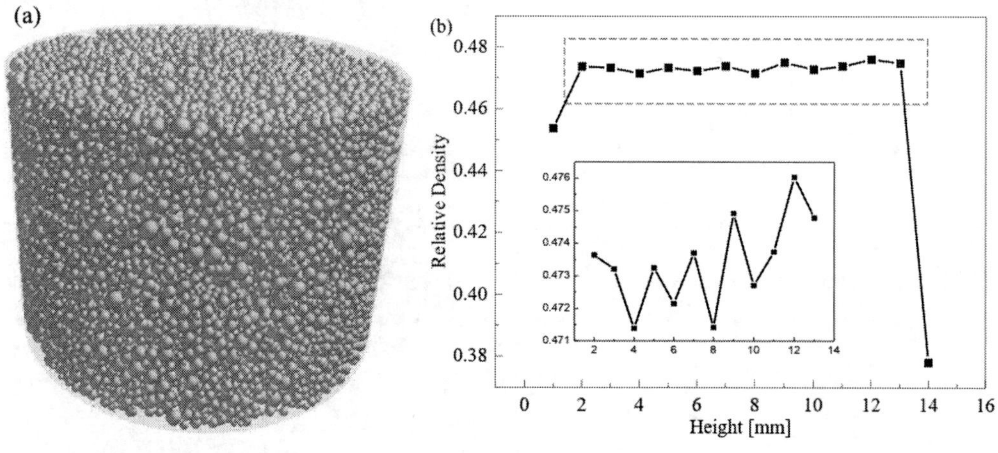


**Figure 2.** (a) Initial packing structures with different particle sizes by DEM; (b) evolution of relative density of initial packing structure with height, where the inset figure shows the local results presented by red dotted lines.

Meanwhile, Fig. 2 (b) shows that the evolution of relative density of initial packing structure with height, where the inset figure shows the local results presented by red dotted lines. It can be found that stratification generates in the distribution of initial packing density, which consistent with Fig. 2 (a), the initial packing density of bottom and end faces is the lowest. In addition, one more case can be seen in

Fig. 2 (b) that the initial packing density of middle part fluctuates with increasing height, which shows that this part is an irregular stacked structure.

### *3.2. Macro characteristics*

Relative density-pressure relationship for Si composite powder mass using different pressing methods was shown in Fig. 3 (a), where the inset figures indicate the distribution of relative density at 150 MPa. As shown in Fig. 3 (a), with the increase of pressure, the initial relative density increase, and the packing structure becomes denser. However, high relative density component can be obtained with hot pressing comparing die compaction at same compaction pressure. This phenomenon can be explained by material softening. Temperature can decrease yield strength of Si Alloys, so high relative density can be obtained at low compaction pressure using hot pressing method. At the same time, we can see that the relative density distribution of hot pressing is more uniform, indicating that hot pressing can gain more uniform and denser green target.

Meanwhile, internal stress distribution is studied. Here, the equivalent von Mises stress is given by [17, 18]:

$$\bar{\sigma} = [(\sigma_1 - \sigma_2)^2 + (\sigma_1 - \sigma_2)^2 + (\sigma_1 - \sigma_2)^2]^{1/2} / \sqrt{2} \tag{6}$$

Where $\sigma_1$, $\sigma_2$, $\sigma_3$ are the principal Cauchy stresses along three main axes. Fig. 3 (b) gives the equivalent von Mises stress of different compaction methods as a function of pressure during compaction. From this figure, it can be found that lower stress can be obtained in hot pressing, so high quality green target is easily gained by using hot pressing.

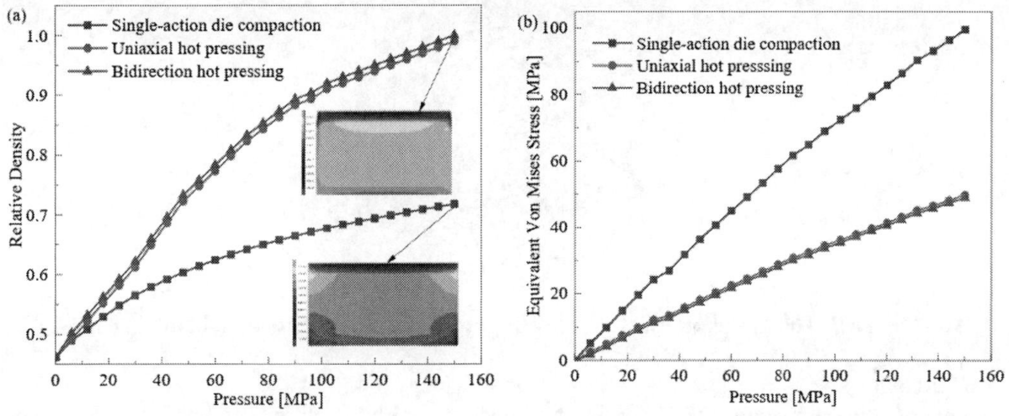


**Figure 3.** (a) Relative density-pressure relationship for Si composite powder mass using different pressing methods, where the inset figures indicate the distribution of relative density at 150 MPa; (b) equivalent von Mises stress of different compaction methods as a function of pressure during compaction.

### *3.3. Micro characteristics*

Axial flow and radial flow of Si composite powders are studied in this part. Fig. 4 (a) shows the evolution of axial displacement with relative height during compaction using different compaction methods, where the inset figures represent overall particulate flow. It can be found that hot-pressed axial flow is slower and smaller than single-action die compaction flow. Meanwhile, Fig. 4 (b) gives the evolution of radical displacement with radical distance during single-action die compaction. From Fig. 4 (b), it can be seen that the particles in the upper part of the molding flow toward the center, and the particles in the lower part flow to both sides. The flow in the middle part is the most intense.

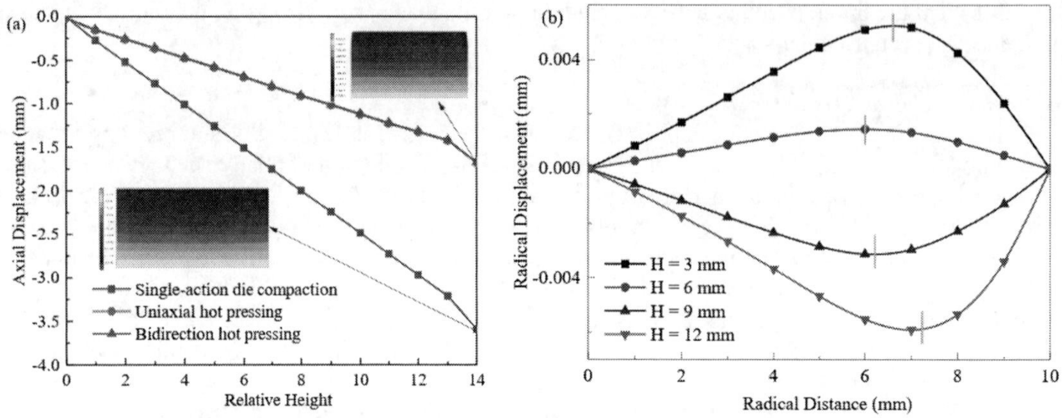


**Figure 4.** (a) Evolution of axial displacement with relative height during compaction using different compaction methods, where the inset figures represent overall particulate flow; (b) evolution of radical displacement with radical distance during single-action die compaction.

### *3.4. Physical experiment*

Fig. 5 gives the morphology of Si alloy target prepared by bidirection hot pressing. As shown in Fig. 5, complete crack-free Si alloy target has been produced by bidirection hot pressing, indicating our simulations are robust and reliable.

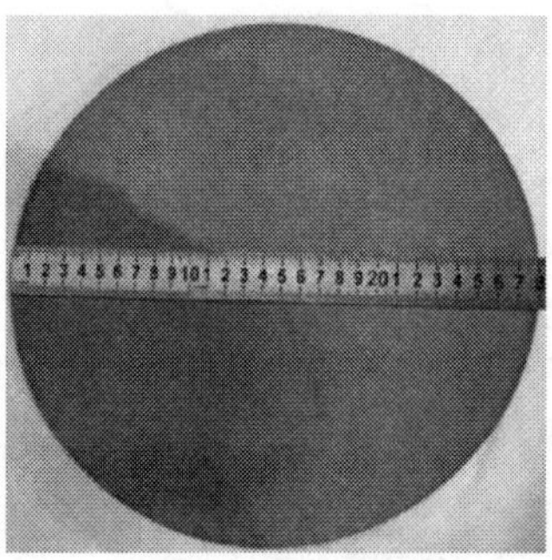

**Figure 5.** The morphology of Si alloy target prepared by bidirectional hot pressing

## 4. Conclusion

Cold uniaxial die compaction, single-action hot pressing and bidirection hot pressing of Si composite powders were numerically reproduced by using FEM method. The impacts of initial packing structure, compaction method, pressing pressure and sintering temperature on overall relative densities of multi-component silicon components were studied. Following conclusions can be drawn: (1) High relative density and low internal stress can be obtained at low compaction pressure using hot pressing method. (2) Hot-pressed axial flow is slower and smaller than single-action die compaction flow.

**Acknowledgments**

The authors are grateful to National Key Development Program of the Ministry of Science and Technology (2017YFB0305603) of China for the financial support of current work.

**References**

[1] K. Gu, X.Z. Jia, H.L. You, The yield estimation of semiconductor products based on truncated samples, Int. J. Metro. Qual. Eng. 4 (2014) 215-220.W. Strunk Jr., E.B. White, The Elements of Style, third ed., Macmillan, New York, 1979.

[2] H.S.P. Wong, S. Raoux, S.B. Kim, Phase Change Memory, Proc. IEEE 98 (2010) 2201-2227.

[3] B.C. Lee, E. Ipek, O. Mutlu, Architecting phase change memory as a scalable dram alternative, Acm Sig. Comp. Ar. News 37 (2009) 2-13.
[4] G.W. Burr, M.J. Breitwisch, M. Franceschini, Phase change memory technology, J. Vac. Sci. Technol. B 28 (2010) 223-262.
[5] J. Hegedüs, S.R. Elliott, Microscopic origin of the fast crystallization ability of Ge-Sb-Te phase-change memory materials, Nat. Mater. 7 (2008) 399-405.
[6] R.E. Simpson, P. Fons, A.V. Kolobov, Interfacial phase-change memory, Nat. Nanotechnology 6 (2011) 501-5.
[7] G. Bruns, P. Merkelbach, C. Schlockermann, Nanosecond switching in GeTe phase change memory cells, App. Phys. Lett. 95 (2009) 824-3.
[8] D. Loke, T.H. Lee, W.J. Wang, Breaking the speed limits of phase-change memory, Science 336 (2012) 1566-1569.
[9] B. Huang, J. Robertson. Bonding origin of optical contrast in phase-change memory materials, Phy. Rev. B 81(2010) 237-245.
[10] X.Z. An, R.Y. Yang, K.J. Dong, R.P. Zou, A.B. Yu, Micromechanical simulation and analysis of one-dimensional vibratory sphere packing, Phys. Rev. Lett. 95 (2005) 205-502.
[11] X.Z. An, A.B. Yu, Analysis of the forces in ordered FCC packings with different orientations, Powder Technol. 248 (2013) 121-130.
[12] X.Z. An, R.Y. Yang, R.P. Zou, Effect of vibration condition and inter-particle frictions on the packing of uniform spheres, Powder Technol. 188 (2008) 102-109.
[13] H.Y. Zhao, X.Z. An, DEM modeling on stress profile and behavior in granular matter, Powder Technol. 323 (2018) 149-154.
[14] A.R. Ragab, C.A.R. Saleh, Evaluation of constitutive models for voided solids, Int. J. Plasticity 15 (1999) 1041-1065.J. Cedergren, N.J. Sorensen, A. Bergmark. Three-dimensional analysis of compaction of metal powder, Mech. Mater. 34 (2002) 43-59.
[15] J. Cedergren, N.J. Sorensen, A. Bergmark. Three-dimensional analysis of compaction of metal powder, Mech. Mater. 34 (2002) 43-59.
[16] P. Han, X.Z. An, Y.X. Zhang, Particulate scale MPFEM modeling on compaction of Fe and Al composite powders, Powder Technol. 314 (2017) 69-77.
[17] F. Huang, X.Z. An, Y.X. Zhang, Multi-particle FEM simulation of 2D compaction on binary Al/SiC composite powders, Powder Technol. 314 (2017) 39-48.
[18] Q. Jia, X.Z. An, H.Y. Zhao, Compaction and solid-state sintering of tungsten powders: MPFEM simulation and experimental verification, J. Alloy. Compd. 750 (2018) 341-349.

# Self-Healing Performance of Protective Coating with Inhibitor Loaded Nanoparticles of Mesoporous Silica on Copper Alloy

**Liwen Zhang[1, *], Jiping Song[2, a]**

[1] Xi'an Aeronautical Polytechnic Institute, Xi'an, 710089, China
[2] Chinese Flight Test Establishment, Xi'an, 710089, China

* Corresponding author email: 375350782@qq.com, [a] 595513529@qq.com

**Abstract.** The self-healing coating on the copper alloy surface was prepared by adding inhibitor loaded composite nanoparticles into alkyd varnish coating with a content of 4 wt.%. The self-healing performance of the scratched coating immersed in 3.5% NaCl solution was evaluated by morphology observation, Fourier transform infrared spectroscopy, electrochemical impedance spectroscopy and scanning Kelvin probe. It has been proved that the scratched coating containing the composite nanoparticles can be self-healed with the formation of the inhibitive film at the coating crevice during the immersion.

**Key words:** A. copper; A. organic coating; B. EIS.

## 1. Introduction

Copper and its alloy are widely used in such industries as electrical engineering, electronic industry, marine and offshore engineering and so on. Although they have good performance in applications, the corrosion resistance of copper and its alloy still needs to be further improved in some hostile environment, such as sea water or coastal atmosphere [1-5].

Coating is an effective means for corrosion protection of metallic structures [6-9]. However, when coating is damaged, the exposed metal surface at the defects will be prone to suffer from corrosion. To solve this problem, smart coatings with self-healing ability have attracted much attention and shown bright prospect of development [2, 10]. One of the routes to achieve self-healing is to add corrosion inhibitor into the coating. If the corrosion inhibitor is directly added to the coating, the protective performance of the coating will be reinforced but with a low efficiency. And tiny pores will be left on the coating once the corrosion inhibitor dissolves, which speeds up coating failure. In addition, the corrosion inhibitor may react with the ingredients of the coating. So adding the inhibitor-loaded nanocontainers into the coating becomes a promising and effective way to realize self-healing and enhance coating performance [10]. When the coating is broken, the corrosion inhibitor will release from the nanocontainers into the defects, making the exposed metallic surface inactivated or forming a layer of protective film to inhibit the corrosion. Since the corrosion inhibitor is stored inside the nanocontainers, its period of validity for corrosion prevention can be prolonged.

With focus on the corrosion protection of copper alloy, we prepared a new type of composite nanoparticles by loading a typical inhibitor 1-hydroxybenzotriazole (HOBT) in mesoporous silica nanocontainers. The structures of the mesoporous silica were characterised; the release kinetics of

Content from this work may be used under the terms of the Creative Commons Attribution 3.0 licence. Any further distribution of this work must maintain attribution to the author(s) and the title of the work, journal citation and DOI.
Published under licence by IOP Publishing Ltd

HOBT from the nanoparticles in 3.5% NaCl solution was investigated, and the corrosion inhibition process of the composite nanoparticles on a Cu-Zn alloy was analysed. In this article, we present the self-healing performance of the coating on the Cu-Zn alloy surface containing the composite nanoparticles in the alkyd varnish, and discuss the mechanism of corrosion protection with self-healing effect.

## 2. Experimental Process

### *2.1. Material*

The Cu-Zn alloy of H68 brass used as substrate was manufactured by Shandong Qingteng Metallic Materials Co., Ltd. with the composition, wt. %: 67.59 Cu, 0.10 Fe, 0.03 Pb, and balance Zn. SiC sandpaper (600#, 800#, 1000#, 1200#) was purchased from Shanghai Yihe Grinding Materials Co., Ltd.. Alkyd varnish was acquired from Lehua Inc. China. Ethanol and sodium chloride (analytical grade) were purchased from Sinopharm Chemical Reagent Co., Ltd.. The water utilised in all of the experiments was produced with the Millipore Milli-Q Plus 185 purification system having a resistivity higher than 18.1 MΩ cm.

### *2.2. Preparation of self-healing coatings*

The Cu-Zn alloy plate of 3 mm thick was used as the coating substrate, which was abraded using the SiC sandpaper in the order from 600 to 1200 mesh, then washed with purified water and ethanol successively, and finally, air-dried at room temperature.

The detailed descriptions on the synthesis and characterisation of the composite nanoparticles were given in the previous paper. The prepared mesoporous silica nanoparticles had an average diameter of about 60 nm and pore size distributed between 2 nm and 8 nm with most pores of around 3 nm. The silica nanocontainer had a high surface area of 951.6 $m^2/g$ and a large pore volume of 1.045 mL/g, loaded with about 40 wt.% of the corrosion inhibitor HOBT.

The composite nanoparticles were added into the alkyd varnish with a content of 4 wt.%. The mixture was stirred vigorously for 20 min to obtain the paint. Then the paint was coated on the surface of the prepared copper alloy substrate by brushing. Finally, the samples were heated at 30 oC for 72h. As a comparison, the alkyd varnish coating without nanoparticles was also prepared on the copper alloy samples using the same procedure. The coatings for the tests had a dry film thickness of 100±5 μm. The scratches of 5mm long were made with a sharp blade through the coating to the substrate.

### *2.3. Characterisation*

The morphologies of the coatings after immersion testing in 3.5% NaCl solution were observed with KH-8700 digital microscope (HIROX).

The measurement of electrochemical impedance spectroscopy (EIS) was conducted with Princeton PARSTAT 2273 electrochemical workstation at the open circuit potential in 3.5% NaCl solution. A three-electrode cell was used in the electrochemical measurement with the coated copper alloy sample served as the working electrode, the saturated calomel electrode (SCE) as the reference electrode and the platinum sheet as the counter electrode. The working electrode was insulated using epoxy resin with an exposed testing area of 1 $cm^2$. The amplitude and frequency range of the applied disturbing signal were 10 mV and $10^{-2} \sim 10^5$ Hz respectively with 49 points recorded. The EIS data simulation based on the equivalent circuit were conducted using the software ZsimpWin.

Volta potentials of the coated samples with scratchs were recorded using VERSA SCAN scanning Kelvin probe (SKP, Princeton). The SKP measurement was carried out in the air atmosphere with about 65% relative humidity. The tungsten probe with a diameter of 500 μm was utilised and lifted up from the surface to a constant distance of 100 μm. The scanning area was a square of 2000μm × 2000μm with the scratches in the middle of the scanning area (at X=1.0 mm). The scanning step both in X axis and Y axis direction was 100 μm. The measurement sensitivity was 50 μV.

The infrared spectra were measured with Thermo Scientific Nicolet iS 10 spectrometer (Thermo Fisher Scientific Inc.) with a FT-IR scanning range of 2000 $cm^{-1}$~ 550 $cm^{-1}$ and a resolution of 2 $cm^{-1}$.

## 3. Results And Discussion

### *3.1. Morphology observation*

Fig. 1 (a) shows the appearance of the blank alkyd varnish coating with a scratch after 150h immersion in the NaCl solution. The coating is transparent, therefore, the condition at the interface between the coating and the substrate can be observed. The solution spread into the interior along the interface between the coating and the substrate from the crevice. The failed area of the blank coating was large after 150h immersion. The copper alloy substrate corroded evidently under the coating around the scratch. While at the scratched region of the self-healing coating with composite nanoparticles (Fig.1(b)), a protective film was formed to retard the development of the coating defect and the corrosion of the substrate at the defect was hindered.

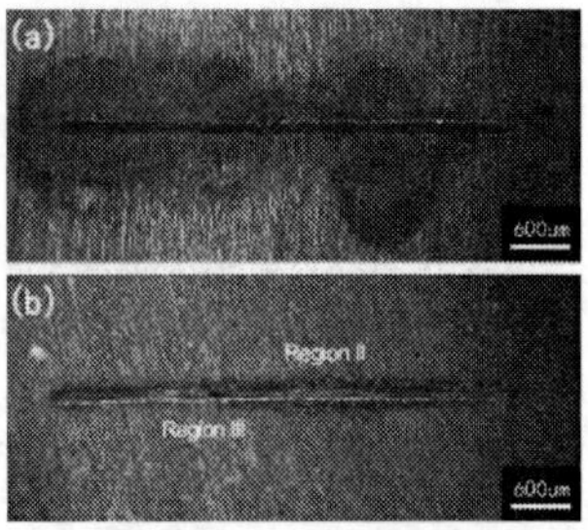


**Figure 1.** The appearance of the blank alkyd varnish coating with a scratch after 150h immersion in the NaCl solution

### *3.2. FT-IR Analysis*

FT-IR analysis was used to detect the release of the inhibitor HOBT from the mesoporous nanoparticles and the chemical changes after self-healing process of the coating. The alkyd varnish coating contains the most functional groups of the inhibitor HOBT, such as benzene ring, hydroxyl, etc., which makes the FT-IR spectrum of the alkyd varnish covering the most characteristic bands of the inhibitor. Fig.2 shows the FT-IR spectra of the inhibitor HOBT (I), the crevice region (II), and the peripheral region (III) of the coating containing composite nanoparticles after 150h immersion in 3.5% NaCl solution. The bands at 668 $cm^{-1}$ and 1450 $cm^{-1}$ correspond to the characteristic vibration frequencies of HOBT aromatic structure and the stretching vibration frequencies of C-N respectively, with an obvious shift to the low wave number because of conjugation effect, which are the unique characteristic bands of the inhibitor HOBT (Fig.2 I) comparing with alkyd varnish coating[24, 25]. At the crevice region, the FT-IR spectrum (Fig.2 II) displayed the unique characteristic bands of the inhibitor HOBT, while there were not these bands in the peripheral region (Fig.2 III), which indicated that the inhibitor was released and the film formed at the crevice.

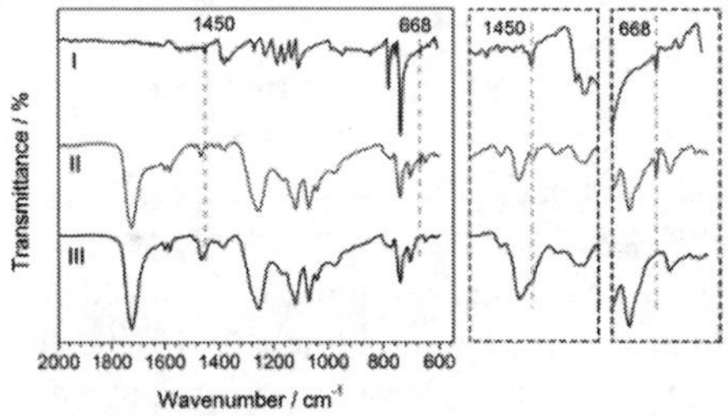


**Figure 2.** The FT-IR spectra of the inhibitor HOBT (I), the crevice region (II), and the peripheral region (III) of the coating containing composite nanoparticles after 150h immersion in 3.5% NaCl solution

### *3.3. EIS Measurements*

EIS measurements can be used to study the self-healing process of protective coating. The prepared self-healing coating containing composite nanoparticles with a scratch of 5 mm long was immersed in 3.5% NaCl solution. The EIS measurements were conducted at the open-circuit potential during the immersion. Fig.3 gives the evolution of the EIS spectra of the self-healing coating having scratch with immersion time.

The impedance modulus (|*Z*|) at the low frequency of 0.01 Hz can be used to evaluate the protective effect of the organic coatings. Fig.4 shows the change of the impedance modulus (*f*=0.01 Hz) of the scratched self-healing coating immersed in the NaCl solution. In the initial 2h of the immersion, the solution penetrated into the crevice and reached the copper alloy substrate, leading to the |*Z*| value (*f*=0.01 Hz) at a low level of 2.37 MΩ cm$^2$. Then the |*Z*| value (*f*=0.01 Hz) was increased sharply to 9.01 MΩ cm$^2$ at 14h and fluctuated within a range of 4.30 MΩ cm$^2$ ~ 6.67MΩ cm$^2$ from 14h to 112h. This is because that the inhibitor HOBT can release from the mesoporous composite nanoparticles exposed in the solution at the crevice of the coating, which can be absorbed on the exposed copper alloy surface and form a film covering the coating crevice with the increase of the immersion time. Because the area of the scratched crevice is small, a small amount of inhibitor released can form a protective film healing the coating defect. On the other hand, the chloride ions in the solution can destroy the healed film· The little fluctuation of the |*Z*| value (*f*=0.01 Hz) at 14h to 112h may be attributed to the alternative growing and damage of the formed protective film in the self-healing process. The self-healing process was repeated until the released inhibitor exhausted after 112h. After that time, the |*Z*| value (*f*=0.01 Hz) of the coating began to decrease from 6.67MΩ cm$^2$ to 1.22MΩ cm$^2$ at the end of testing, which means the failure of protection by the coating.The EIS plots in Fig.3 revealed two time constants for the scratched coating from the beginning of immersion in the NaCl solution, corresponding to the relaxation processes of the coating and the electric double layer on the interface with the solution respectively.

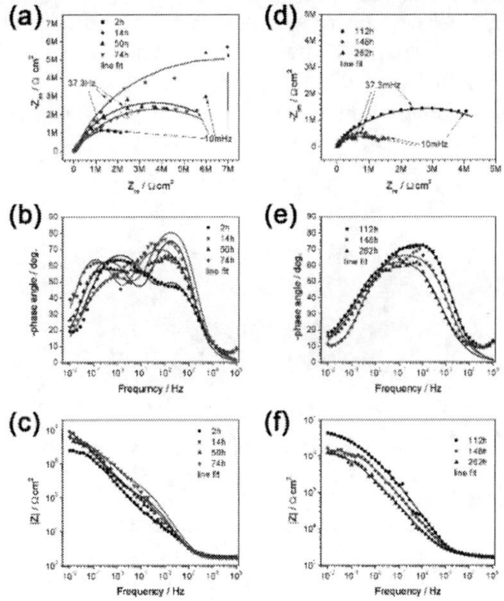


**Figure 3.** Tthe evolution of the EIS spectra of the self-healing coating having scratch with immersion time

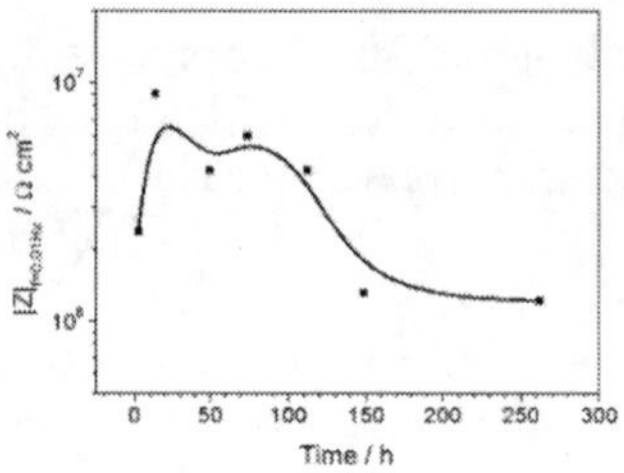


**Figure 4.** The change of the impedance modulus (f=0.01 Hz) of the scratched self-healing coating immersed in the NaCl solution

### *3.4. SKP Measurement*

SKP technique, which can measure the distribution of Volta potential of the local area around the defects of the coating, was employed to analyse the healing process of the scratched alkyd coating with composite nanoparticles. For comparison, an alkyd varnish coating with scratch was prepared as blank contrast.

Fig.5 shows the SKP measurement results of the blank alkyd varnish coating at different immersion time. At the place of X=1.0 mm in the Volta potential maps, there appeared an obvious line of peaks, which was referred to the scratched crevice, and the both sides corresponded to alkyd varnish coatings. The solution penetrated into the crevice with immersion, causing the copper alloy substrate exposed at the crevice corroded. The exposed copper alloy surface with corrosion product had higher Volta potential than that of the around alkyd coating. After immersed for 2h, the Volta potential maps showed obvious peak in the range of Y=1.0 mm ~ 2.0 mm at the crevice, while the peak was not obvious in the range of Y=0.0 mm ~ 1.0 mm (Fig.5a), showing the inconsistence of the prepared scratch. The latter may result from the drawing close of the two side walls of the scratched crevice at this part due to the inherent elasticity of alkyd varnish coating. After 22h of immersion, all of the parts of the crevice were infiltrated by solution and gradually covered by corrosion products, and the Volta potential peaks became more obvious in all the range of the crevice (Fig. 5(b), (c) and (d)).

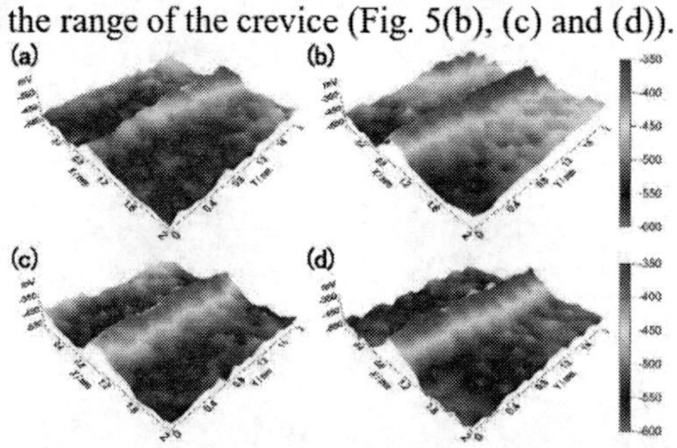


**Figure 5.** The SKP measurement results of the blank alkyd varnish coating at different immersion time

With the immersion time increased, the maximum Volta potential of the alkyd varnish coating went up, to about -350mV. The range of the high Volta potential was slightly enlarged, indicating that the solution penetrated into the bottom of the crevice and began to spread slowly to both sides at the interface of the coating and the metal substrate.

Fig.6 shows the Volta potential maps of the self-healing coating containing composite nanoparticles with a scratched crevice at different immersion time, with the crevice at the X=1.0 mm. Like the blank vanish coating, some obvious Volta potential peaks can be seen at the part of the crevice after immersed for 2h. However, with the immersion time increased, the Volta potential peaks at the crevice became lower and lower. And the peaks were almost invisible after 50h of immersion. This phenomenon can be related to the self-healing effect of the coating. When the solution penetrated into the crevice, the inhibitor could be released quickly from the mesoporous nanoparticles exposed in the coating. The dissolved inhibitor would adsorb on the surface of the exposed copper alloy substrate at the crevice and form a layer of organic film, making the Volta potential shift negatively close to that of the alkyd coating.

## 4. Conclusion

A self-healing coating on the Cu-Zn alloy surface was prepared by adding the composite nanoparticles with a content of 4 wt. % of mesoporous silica loaded inhibitor HOBT into the alkyd varnish. The performance of the scratched coating immersed in 3.5% NaCl solution was investigated using optical microscope, FT-IR spectroscopy, EIS and SKP. Morphology observation indicated that the coating with the composite nanoparticles presented better protective performance for the copper alloy substrate than the blank vanish coating. The FT-IR analysis proved the formation of the inhibitive film at the coating crevice with immersion. The self-healing process of the scratched coating was well characterised by EIS and SKP. In the initial period of immersion, the solution penetrated into the crevice, and the coating demonstrated low impedance and high Volta potential peaks at the scratch. With the increase of immersion time, the coating resistance and the charge transfer resistance presented a trend of fast increasing first, then maintaining for a period of time and finally decreasing, while the coating capacitance and the double layer capacitance showed an opposite trend, which indicated that the scratched crevice was self-healed with inhibitive film formation and then failed after the inhibitor was exhausted. The Volta potential peaks of the scratched alkyd coating containing composite nanoparticles decreased continuously with immersion time increased, and became not obvious, indicating the crevice was well self-healed.

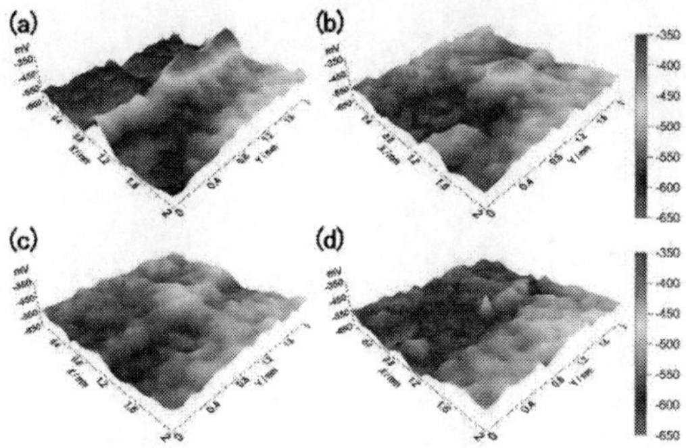


**Figure 6.** The Volta potential maps of the self-healing coating containing composite nanoparticles with a scratched crevice at different immersion time, with the crevice at the X=1.0 mm.

## References

[1] Y. Liu, S. Li, J. Zhang, J. Liu, Z. Han, L. Ren, Corrosion inhibition of biomimetic super-hydrophobic electrodeposition coatings on copper substrate, Corros. Sci., 94 (2015) 190-196.
[2] E. Abdullayev, V. Abbasov, A. Tursunbayeva, V. Portnov, H. Ibrahimov, G. Mukhtarova, Y. Lvov, Self-healing coatings based on halloysite clay polymer composites for protection of copper alloys, ACS Appl. Mater. Interfaces, 5 (2013) 4464-4471.
[3] G. Tansuğ, T. Tüken, E. Giray, G. Fındıkkıran, G. Sığırcık, O. Demirkol, M. Erbil, A new corrosion inhibitor for copper protection, Corros. Sci., 84 (2014) 21-29.
[4] A. Shaban, G. Vastag, A. Pilbáth, I. Kék, L. Nyikos, Electrochemical study of copper corrosion inhibition in acidic environemnt by 5-(4'-isopropylbenzylidene)-2, 4-dioxotetrahydro-1, 3-thiazole, J. Mater. Environ. Sci., 7 (2016) 2572-2582.
[5] M.N. El-Haddad, Chitosan as a green inhibitor for copper corrosion in acidic medium, Int. J. Biol. Macromol., 55 (2013) 142-149.
[6] R. Figueira, C.J. Silva, E. Pereira, Organic–inorganic hybrid sol–gel coatings for metal corrosion protection: a review of recent progress, J. Coat. Technol. Res., 12 (2015) 1-35.
[7] H. Leidheiser, Electrical and electrochemical measurements as predictors of corrosion at the

IOP Conf. Series: Materials Science and Engineering **772** (2020) 012024 doi:10.1088/1757-899X/772/1/012024

metal—organic coating interface, Prog. Org. Coat., 7 (1979) 79-104.
[8] H. Huo, Y. Li, F. Wang, Corrosion of AZ91D magnesium alloy with a chemical conversion coating and electroless nickel layer, Corros. Sci., 46 (2004) 1467-1477.
[9] I. Dehri, M. Erbil, The effect of relative humidity on the atmospheric corrosion of defective organic coating materials: an EIS study with a new approach, Corros. Sci., 42 (2000) 969-978.
[10] X. Guo, B. Hurley, R.G. Buchheit, Encapsulation of NaVO3 as corrosion inhibitor into microparticles and its active corrosion protection for AA2024 based upon inhibitor control release, Corrosion, 71 (2015) 1411-1413.

# Influence of Wood Fiber on the Laboratory Road Performance of Recycled Asphalt Pavement (RAP) Asphalt Mixture

**Fang Chen[1], Jinshun Xue[2,] *, Zhengxun Yang[1], Kaikai Gao[1] and Zhiqiang Huo[3]**

[1]Xinjiang Communications Construction Group Co., LTD, Urumqi, China.
[2]Hubei University of Arts and Science. Xiangyang, China
[3]Sichuan Vocational and Technical College of Communications, Chengdu, China

* Corresponding author e-mail: jinshunx@chd.edu.cn

**Abstract.** The influence of wood fiber content on the road performance of asphalt mixture with high recycled asphalt pavement (RAP) content was studied based on the Marshall test, high temperature rutting test, low temperature crack resistance test and water stability test. And the optimum wood fiber content in RAP asphalt mixture with high RAP content was recommended. The results show that the dynamic stability and low temperature bending strain of RAP asphalt mixture increased and then decreased with the increase of wood fiber content. When the content of wood fiber is 0.3 %, the road performance of RAP asphalt mixture is the best. Compared with RAP asphalt mixture without wood fiber, the Marshall stability and low temperature flexural tensile strain of RAP asphalt mixture with 0.3% wood fiber increase by 7.1% and 4.6%, respectively, and the high temperature rutting resistance and water stability decreases.

## 1. Introduction

In order to solve the problem of recycled asphalt pavement (RAP) obtained during the maintenance of asphalt pavement, road engineers at home and abroad mostly adopt recycling, crushing and screening of waste asphalt mixture, then mixing it with new aggregate, new asphalt and recycling agent in proper proportion to prepare recycled asphalt mixture [1, 2]. Zhou et al investigated the effect of reclaimed asphalt pavement (RAP) on the rheological properties of a polymer modified binder and the fatigue cracking performance of polymer modified asphalt mixture [3]. Zhang et al using the binder blending chart to select a performance grade of the virgin binder may not be applicable for mixtures containing more than 25% reclaimed asphalt pavement (RAP) [4].Most studies on the design and construction of RAP asphalt mixture with RAP materials content less than 25%; thus, information about the performance of recycled asphalt mixture with high RAP content is limited, and it is difficult to meet the pavement performance requirements under special circumstances [5].

In order to improve the performance of asphalt mixture, road engineering researchers have carried out a series of studies on fiber reinforced asphalt mixture. The addition of fibers can effectively improve the low temperature crack resistance and fatigue resistance of asphalt mixture, and improve the durability of asphalt pavement. Wood fibers are mainly used in SMA mixtures, while the application of wood fibers in cold recycled asphalt mixtures is less. The performance of high RAP content mixtures, the influence of wood fiber content on the performance of asphalt mixtures with high RAP content was

Content from this work may be used under the terms of the Creative Commons Attribution 3.0 licence. Any further distribution of this work must maintain attribution to the author(s) and the title of the work, journal citation and DOI.
Published under licence by IOP Publishing Ltd

studied through Marshall test, high temperature rutting test, low temperature crack resistance test and water stability test. The optimum wood fiber content in asphalt mixtures with high RAP content was recommended.

## 2. Materials and methods

### 2.1. Materials

Table 1 shows the the general specifications of AH-90 asphalt which obtained from Karamay, Xinjiang, China. The recycled asphalt pavement (RAP) materials for this study were obtained from the reconstruction project of Wusu to Sailimu Lake first class highway in Urumqi, Xinjiang, China. The virgin aggregate (VAG) was obtained from Urumqi, Xinjiang, China, and it was graded into six types of coarse aggregates, fine aggregate and mineral powder with the following sizes: 23-30 mm, 18-23 mm, 10-18 mm, 5-10 mm, 0-5 mm and mineral powder. Their technical indicators which tested according to the technical requirements of Technical Specification for Construction of Highway Asphalt Pavements (JTG F40-2004) are omitted. The technical indicators of wood fiber are shown in Table 2. The mineral gradation of the AC-25 mixture is listed in Table 3, and the RAP materials content is 45%.

**Table 1.** General index of asphalt

| Tested Item | | Tested results | Specified value |
|---|---|---|---|
| Penetration(25°C,100g,5s) ( 0.1mm) | | 89 | 80~100 |
| Penetration index (PI) | | 0.13 | -1.0 ~ +1.0 |
| Ductility(5cm/min,15°C) (cm) | | 151 | ≥100 |
| Softening point(°C) | | 48 | ≥46 |
| Thin-film heating test(63°C) | Mass loss (%) | -0.24 | -0.8 ~ +0.8 |
| | Residual penetration ratio (%) | 72.5 | ≥61 |
| | Ductility(10°C) (cm) | 31 | ≥6 |

**Table 2.** General index of wood fiber

| Fiber types | Fiber length (mm) | Ash content (%) | Oil absorption rate (%) | Water content (%) |
|---|---|---|---|---|
| Wood fiber | ≤5 | 16 | 7 time fiber mass | ≤4 |

**Table 3.** Mineral aggregate gradations of AC-25 RAP asphalt mixture

| Gradating | Percentage passing(%)for sieve size(mm) | | | | | | | | | | | | |
|---|---|---|---|---|---|---|---|---|---|---|---|---|---|
| | 31.5 | 26.5 | 19 | 16 | 13.2 | 9.5 | 4.75 | 2.36 | 1.18 | 0.6 | 0.3 | 0.15 | 0.075 |
| Study Gradation | 100 | 97.3 | 85.1 | 77.1 | 68.6 | 54.6 | 38.9 | 25.6 | 19.2 | 15.9 | 11.0 | 6.7 | 5.4 |
| Upper limit | 100 | 100 | 90 | 83 | 76 | 65 | 52 | 42 | 33 | 24 | 17 | 13 | 7 |
| Lower limit | 100 | 90 | 75 | 65 | 57 | 45 | 24 | 16 | 12 | 8 | 5 | 4 | 3 |

### 2.2. Test methods

The preparation steps of RAP asphalt mixture specimens are as follows: (1) The RAP materials were placed in an oven for 2 hours with an air temperature of 110 oC; (2) The VAG and asphalt were separated place in two oven for 4 hours with an air temperature of 190 oC and 150 oC, respectively; (3) The throwing order of materials was wood fiber, RAP, aggregates, asphalt and mineral powder. The stirring time was controlled within 90 seconds after the aggregates, asphalt and mineral powder added, respectively. And the total stirring time was controlled within 270 seconds.

To adequately study the effect of the wood fiber content on the road performance of RAP asphalt mixture, five possible content of wood fiber were chosen: 0, 0.2 %, 0.3 %, 0.4 % and 0.5 %, and the corresponding optimum asphalt-aggregate ratio were 4.9 %, 5.3 %, 5.4 %, 5.6 % and 5.7 %.

The performance evaluation indicators in this study include volumetric property, Marshall stability (*MS*), and road performance. Volumetric property indicators include the volume of air void (*VV*), voids filled with asphalt (*VFA*), voids in mineral aggregate (*VMA*). Road performance measures include dynamic stability (DS), which is the indicator used to evaluate the high temperature rutting resistance of RAP asphalt mixtures. The freeze-thaw splitting strength ratio (*TSR*) was used to evaluate the water stability of RAP asphalt mixtures. Flexural tensile strength ($R_B$) and flexural tensile strain ($\varepsilon_B$) were used to evaluate the low temperature performance of RAP asphalt mixture. The *VV*, *VMA*, *VFA*, *MS*, *TSR*, $R_B$, and $\varepsilon_B$ of asphalt mixture were all measured according to the Standard Test Methods of Bitumen and Bituminous Mixtures for Highway Engineering (JTG E20-2011).

## 3. Test results and discussion

### *3.1. The Marshall index test results of RAP asphalt mixtures*

Table 4 shows the Marshall index test results of RAP asphalt mixtures with different wood fiber content.

**Table 4.** The Marshall index test results of RAP asphalt mixtures with different wood fiber content

| Wood fiber content (%) | optimum asphalt-aggregate ratio (%) | VV (%) | VMA (%) | VFA (%) | MS (kN) | FL (mm) |
|---|---|---|---|---|---|---|
| 0 | 4.9 | 4.2 | 13.2 | 67.0 | 9.1 | 3.6 |
| 0.2% | 5.3 | 4.4 | 13.4 | 67.0 | 9.5 | 3.5 |
| 0.3% | 5.4 | 4.5 | 13.7 | 67.2 | 9.8 | 2.9 |
| 0.4% | 5.6 | 4.5 | 14.0 | 69.2 | 9.4 | 3.0 |
| 0.5% | 5.7 | 4.6 | 14.4 | 68.1 | 9.2 | 3.6 |

As shown in Table 4, compared with the RAP asphalt mixture, which wood fiber content is 0, the Marshall index test results were different based on different wood fiber content. With the wood fiber content of mixture increased, the optimum asphalt-aggregate ratio, VV and VMA increased, and the MS increased to peak values then decreased, however the VFA decreased to lower values then increased.

### *3.2. The high temperature rutting resistance of RAP asphalt mixtures*

The high temperature rutting test results of RAP asphalt mixtures with different wood fiber content were shown in the table 5, and the relationship between wood fiber content of RAP asphalt mixtures and deformation difference, DS were shown in the Figure 1.

**Table 5.** The high temperature rutting resistance of RAP asphalt mixtures with different wood fiber content

| Wood fiber content (%) | Optimum asphalt-aggregate ratio (%) | 45min deformation (mm) | 60min deformation (mm) | Deformation difference (mm) | DS (times/mm) |
|---|---|---|---|---|---|
| 0 | 4.9 | 2.895 | 3.187 | 0.292 | 2154 |
| 0.2 | 5.3 | 4.058 | 4.322 | 0.264 | 2387 |
| 0.3 | 5.4 | 3.993 | 4.310 | 0.317 | 1986 |
| 0.4 | 5.6 | 3.714 | 4.078 | 0.364 | 1732 |
| 0.5 | 5.7 | 4.239 | 4.672 | 0.433 | 1456 |

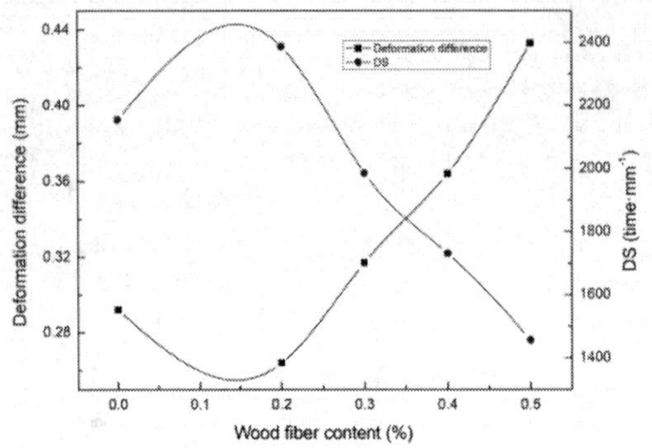


**Fig. 1** Wood fiber content of RAP asphalt mixtures versus deformation difference and DS of RAP asphalt mixtures

As shown in the Table 5 and Fig.1, the 45 and 60 min deformation of RAP asphalt mixtures increased with the increasing of wood fiber content of RAP asphalt mixture.

It is also observed from the Table 5 and Fig.1 that with the wood fiber content of mixture increased, the deformation difference decreased to lower values then increased, and DS of RAP asphalt mixtures increased to peak values then decreased, which is proof that the high temperature rutting resistance of RAP asphalt mixtures increases as the wood fiber content increases to a maximum at the optimum point (0.2 %). These results could be due to the wood fiber is flocculent structure, which cannot be well distributed in the mixture when the content reaches a high value, so the high temperature rutting resistance of RAP asphalt mixtures decreased with the high wood fiber content. And the wood fiber content of RAP asphalt mixtures should be controlled within 0.3 %.

### *3.3. The water stability of RAP asphalt mixtures*

The water stability test results of RAP asphalt mixtures with different wood fiber content were shown in Table 6, and the relationship between wood fiber content of RAP asphalt mixtures and freeze-thaw splitting strength ratio (TSR) was shown in the Fig.2.

**Table 6.** The water stability of RAP asphalt mixtures with different wood fiber content

| Wood fiber content (%) | Optimum asphalt-aggregate ratio (%) | Freeze-thaw splitting strength(MPa) | Splitting strength(MPa) | Freeze-thaw splitting strength ratio (%) |
|---|---|---|---|---|
| 0 | 4.9 | 1.089 | 1.296 | 84.0 |
| 0.2 | 5.3 | 1.049 | 1.352 | 77.6 |
| 0.3 | 5.4 | 1.038 | 1.396 | 74.3 |
| 0.4 | 5.6 | 0.994 | 1.415 | 70.2 |
| 0.5 | 5.7 | 0.938 | 1.359 | 69.0 |

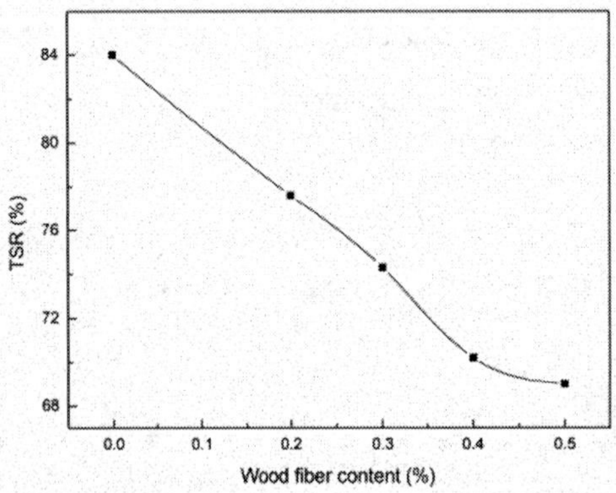


**Fig. 2** Wood fiber content of RAP asphalt mixtures versus TSR of RAP asphalt mixtures

As shown in the Table.6 and Fig.2 that compared with splitting strength of RAP asphalt mixtures specimens without wood fiber, the splitting strength of RAP asphalt mixtures specimens with wood fiber was increased. And compared with splitting strength of RAP asphalt mixtures specimens without wood fiber after freeze-thaw cycle, the splitting strength of RAP asphalt mixtures specimens with wood fiber after freeze-thaw cycle was decreased.

It is also observed from the Table.6 that with the wood fiber content of mixture increased, the freeze-thaw splitting strength ratio decreased, and compared with the freeze-thaw splitting strength ratio of RAP asphalt mixtures without wood fiber, the freeze-thaw splitting strength ratio of RAP asphalt mixtures with 0.2 % wood fiber was decreased about 11.9 %.

### *3.4. The low temperature performance of RAP asphalt mixtures*

The low temperature performance test results of RAP asphalt mixtures with different wood fiber content were shown in the table 7, and the relationship between wood fiber content of RAP asphalt mixtures and flexural tensile strength, flexural tensile strain were shown in the Figure 3.

**Table 7.** The ow temperature performance of RAP asphalt mixtures with different wood fiber content

| Wood fiber content (%) | Optimum asphalt-aggregate ratio (%) | Maximum load (N) | Midspan deflection (mm) | Flexural tensile strength (MPa) | Flexural tensile strain (με) | Stiffness modulus (MPa) |
|---|---|---|---|---|---|---|
| 0 | 3.9 | 924 | 0.553 | 7.63 | 2905 | 2627 |
| 0.2 | 5.3 | 965 | 0.590 | 7.66 | 2984 | 2567 |
| 0.3 | 5.4 | 879 | 0.610 | 7.48 | 3091 | 2420 |
| 0.4 | 5.6 | 932 | 0.611 | 7.35 | 3134 | 2345 |
| 0.5 | 5.7 | 810 | 0.622 | 6.61 | 3227 | 2048 |

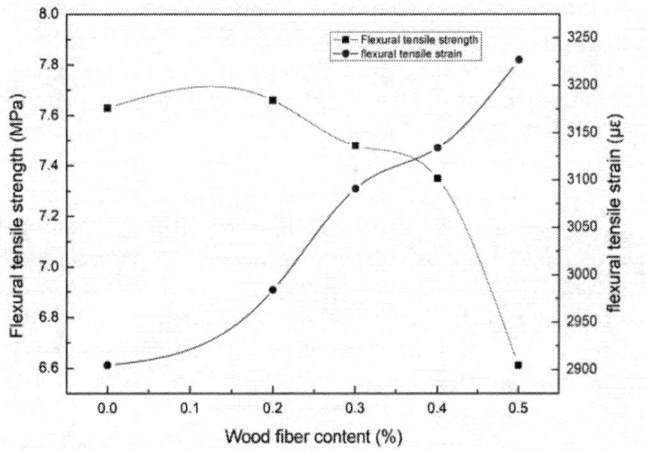


**Fig. 3** Wood fiber content of RAP asphalt mixtures versus flexural tensile strength and flexural tensile strain of RAP asphalt mixtures

It is observed from the Table 7 and Figure 3 that with the wood fiber content of mixture increased, the flexural tensile strength decreased, thus the flexural tensile strain increased. And compared with the flexural tensile strain of RAP asphalt mixtures without wood fiber, the flexural tensile strain of RAP asphalt mixtures with 0.2 % wood fiber was increased a little. However, compared with the flexural tensile strain of RAP asphalt mixtures with 0.2 % wood fiber, the flexural tensile strain of RAP asphalt mixtures with 0.3 % wood fiber was increased 3.6%. That shows that the low temperature performance of RAP asphalt mixture can be improved when the content of wood fiber reaches a certain level, but the effect is not significant. When the content of wood fiber exceeds a certain amount, the tensile property of the RAP asphalt mixture decreases and the strain resistance improves.

## 4. Conclusion

The effect of the wood fiber content on the road performance of RAP asphalt mixture were studied, and the optimum wood fiber content of high RAP content asphalt mixtures was recommended. Based on the results the following conclusions can be drawn:

(1) With the wood fiber content of mixture increased, the optimum asphalt-aggregate ratio, VV and VMA increased, and the MS and FL increased to peak values then decreased, however the VFA decreased to lower values then increased.

(2) The addition of wood fibers can improve the high temperature and low temperature performance of RAP asphalt mixtures; however the improvement of water stability performance of v mixtures is not enough. It is suggested that the RAP asphalt mixture with wood fibers can be used in high temperature areas under more economical and reasonable conditions, and the content of wood fibers should be controlled within 0.3%.

**Acknowledgments**

This work was financially supported by National Key R&D Program of China fund.

**References**

[1] M K Idham,S N A Jeffry, M R Hainin,H Yaacob,N A Hassan,M N M Warid,A Mohamed, N Z M Yunus. Effect of different rejuvenating agents on the mechanical performance of recycled bituminous mixture [J]. IOP Conference Series: Materials Science and Engineering, 2019, 527 (1).

[2] Da Li, Effect analysis on durability of warm mix-recycled asphalt mixture by used material content[J].Journal of Chang'an University(Natural Science Edition), 2018, 38(05): 25-31+48.

[3] Zhou Zhou,Xingyu Gu,Jiwang Jiang,Fujian Ni,Yanxu Jiang. Fatigue cracking performance evaluation of laboratory-produced polymer modified asphalt mixture containing reclaimed asphalt pavement material[J]. Construction and Building Materials, 2019, 216.

[4] Gang Zhou, Peiguo Li, Tianquan Chen, Huoming Wang, Jinsheng Shen. Pavement performance investigation on high percentage of RAP in asphalt mixture by polyester fibres and asphalt [J]. Journal of Chongqing Jiaotong University(Natural Science), 2016, 35(04): 40-46+105.

[5] Kun Zhang,Shihui Shen,Justin Lim,Balasingam Muhunthan. Development of Dynamic Modulus - Based Mixture Blending Chart for Asphalt Mixtures with Reclaimed Asphalt Pavement [J]. Journal of Materials in Civil Engineering, 2019, 31 (2).

[6] Sheng Dai, Xiaoguang Yao, Yan Wang, Zhengqi Zhang, Peilong Li. Design Parameters investigation of mixture ratio for recycled asphalt mixture[J]. Jorunal of Functional Materials, 2018, 49(09): 9141-9147.

[7] Shaowen Du. Influence of chemical additives on mixing procedures and performance properties of asphalt emulsion recycled mixture with reclaimed cement-stabilized macadam [J]. Construction and Building Materials, 2016, 118.

IOP Conf. Series: Materials Science and Engineering **772** (2020) 012026 doi:10.1088/1757-899X/772/1/012026

# Study on the Performance of Road Cement Stabilized Base Using Natural Graded Gravel

**Chenhui Yang[1], Jiaqi Yang[1], Zhiqiang Huo[1] and Zhengxun Yang[2,*]**

[1]Sichuan Vocational and Technical College of Communications, Chengdu, China
[2]Xinjiang Communication Construction Group Co., LTD, Urumqi, China

*Corresponding author e-mail: 16709263@qq.com

**Abstract.** In order to alleviate the increasing supply shortage of building materials, this paper uses the method of comparative test to carry out a stability study of cement stabilized base with green cement bonded natural graded gravel on compressive strength, splitting strength, resilient modulus and other indicators. The research shows that the natural graded gravel, which is a kind of energy-saving and environmental natural building materials, has good road performance as a raw material for cement stabilized base with green cement binding. The green cement bonded base shows higher compressive strength, splitting strength and resilient modulus than those of ordinary Portland cement.

## 1. Introduction

Construction industry is a high energy consumption industry. With the development of infrastructure construction, the demand for raw building materials is increasing day by day, which leads to over-exploitation of natural resources and causes excessive consumption of natural resources, ecological environment damage and other environmental problems. The state has come up corresponding policies to encourage the research and application of green building materials instead of natural minerals. Natural graded gravel is a common raw material in Xinjiang area, which is composed of pebble, sand and soil. The pebbles in the natural gravel have good carrying performance, the sand, soil and other fine aggregates make the structure compact and lead to high carrying capacity and low porosity. The correct and reasonable use of natural grade gravel can not only improve the current situation of shortage of building materials, but also realize the local materials of building materials, reduce the cost of construction and promote the sustainable development of highway engineering construction.

Traditional cement could not bind the natural graded gravel well for the high soil content. In this study, cement stabilized base using natural graded gravel was fabricated by adding a new type of cementitious materials call green cement. This kind of material could bind the fine particles on the soil surface to generate gel material at room temperature to solidify the structure. The main components of this material are CaO, $SiO_2$, $Al_2O_3$, $Fe_2O_3$ and activator. The composition could be adjusted to show different adaptive performance which could be able to bind construction waste, industry waste slag and etc. Compared with traditional cement, the concrete fabricated with this kind of cementitious material have higher strength, lower cost and shorter construction time.

Content from this work may be used under the terms of the Creative Commons Attribution 3.0 licence. Any further distribution of this work must maintain attribution to the author(s) and the title of the work, journal citation and DOI.
Published under licence by IOP Publishing Ltd

## 2. Raw materials

### *2.1. Cement*

Ordinary Portland cement 42.5 and green cement 42.5 were used as binder material. In accordance with the test specifications of the standard consistency, the water consumption, initial coagulation time, stability and strength of two cements were measured. The test data are shown in Table 1:

**Table 1.** Basic properties of cement.

| | Test projects | Green Cement | Ordinary Portland Cement | Specification values |
|---|---|---|---|---|
| | Standard Consistency (%) | 137 | 151 | / |
| | Initial condensation time (min) | 192 | 150 | ≥45 |
| | Final condensation time (min) | 252 | 180 | ≤600 |
| | Stability | Qualified | Qualified | Qualified |
| | Fineness (%) | 6.22 | 9.89 | <10 |
| 3d | Flexural Strength (MPa) | 4.3 | 3.5 | ≥2.5 |
| | Compressive strength (MPa) | 30.5 | 25.2 | ≥10 |
| 28d | Flexural Strength (MPa) | 7.8 | 6.6 | ≥6.5 |
| | Compressive strength (MPa) | 51.9 | 45.9 | ≥42.5 |

### *2.2. Natural grade with gravel*

The natural grade gravel selected in the experiment is the typical natural grade gravel in Hami area. Sieving test was carried out for No. 1 and No. 2 natural graded gravel which were taken from 2 road construction site in Hami. The sieving test was carried out according to Communication Industry standard (JTJ034-2000). Table 2 showed the screening data of 2 type of raw materials.

**Table 2.** Screening data of 2 type of raw materials.

| **Sieve hole size** (mm) Grading Range | 31.5 | 26.5 | 19 | 9.5 | 4.75 | 2.36 | 0.6 | 0.075 |
|---|---|---|---|---|---|---|---|---|
| Specification | 100 | 90~100 | 72~89 | 47~67 | 29~49 | 17~35 | 8~22 | 0~7 |
| 1# | 97.8 | 94.4 | 89.1 | 75.5 | 63.7 | 54.6 | 43.7 | 18.2 |
| 2# | 100 | 100 | 97.0 | 87.9 | 72.5 | 55.7 | 34.1 | 8.2 |

## 3. Experimental program

The optimal water content of natural graded gravel made cement stabilized base was determined by heavy 98 hit method according to standard T0804-1994. To determine the optimal water content and maximum dry density, the cement content for hit test is 4.5% , and water contents is ranged from 2-8% in accordance with the standard. The specimens for compressive strength test were made in accordance with standard T0843-2009. The test specimens with 150mmdiameter and 150mm height were made, then cured in standard condition. The compressive strength, splitting strength and resilient modulus were measured after 7d and 28d curing.

## 4. Analysis of test results

As shown in Table.1, the performance of ordinary cement met the requirements of the specification and the performance of green cement was better than ordinary cement. The standard consistency water consumption of green cement could save about 10% compared with ordinary cement. The initial condensation time of green cement was 28% and 40% higher than that of ordinary cement. The compressive strength of green cement was 17% higher than ordinary cement, and the flexural strength was increased by 20%. The main reason was that green cement particles were smaller so the hydration reaction with sand was more complete, and the consolidation capacity of cement was maximized. The standard consistency water consumption of green cement was less than that of ordinary 42.5# ordinary cement under the same label. The chemical ion exchange reaction occurred to reduce the water absorption effect caused by capillary, powder voids and surface tension of soil powder. And make the treated soil powder from "hydrophilic" into "hydrophobic" which directly reduced the water demand greatly.

Data analysis can be seen in Table 2: 1# natural graded gravel was mixed with some large size pebbles, resulting in passing rate of 31.5mm and 26.5mm size was less than the specification value. The particle size of 2# was smaller than 1#. Both of two kinds of material were relatively smaller than standard specification, which were not recommended to be used as raw materials for cement stabilized base with traditional cement.

### *4.1. Liquid and plastic limit*

**Table 3.** Liquid plastic limit test results.

| Aggregate type | Liquid limit (%) | Specification values | Plastic limit (%) | Specification values | Plastic index |
|---|---|---|---|---|---|
| 1# | 19.8 | < 40 | 14.1 | < 17 | 5.7 |
| 2# | 19.8 | | 12.33 | | 7.5 |

The liquid limit of 1# and 2# is same for the similar composition of less than 5mm particles in two kinds of natural graded gravels, and the plastic limit of 1# is greater than 2# mainly because the soil content of 1# is 18.2%, which is much higher than the soil content of 2# 8.2%.

### *4.2. Transcript test*

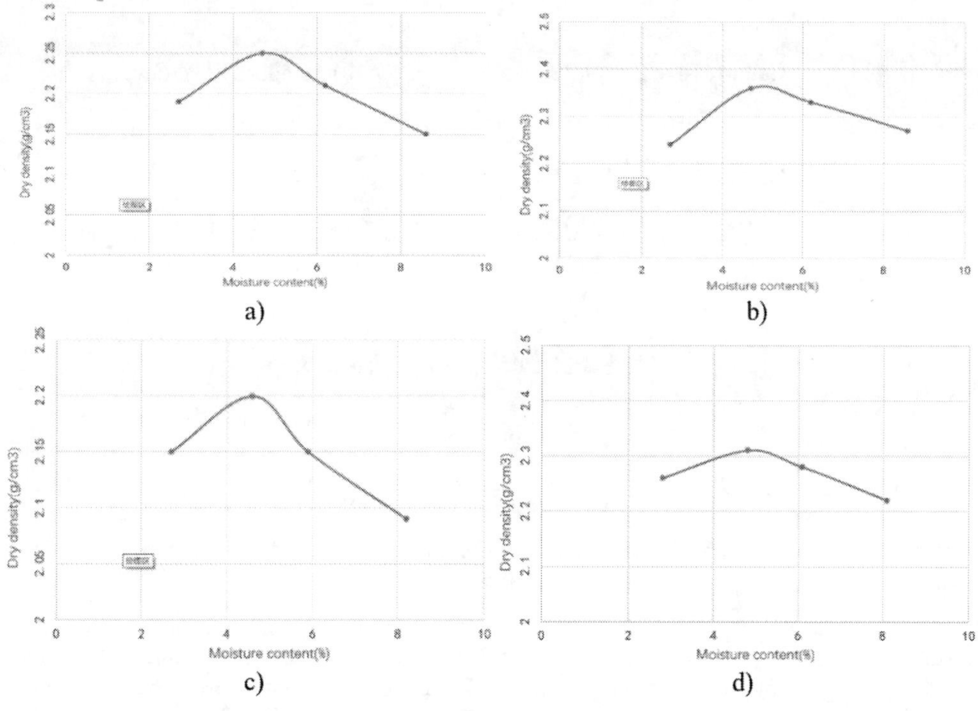


**Figure 1.** The optimal water content and maximum dry density of a) 1# with ordinary cement, b) 1# with green cement, c) 2# with ordinary cement and d) 2# with green cement

**Table 4.** Transcript test results (Cement content 4.5%).

| | 1# Ordinary | 2# Ordinary | 1# Green | 2# Green |
|---|---|---|---|---|
| Optimal water content (%) | 4.7 | 4.5 | 4.8 | 4.7 |
| Maximum dry density (g/cm$^3$) | 2.25 | 2.21 | 2.37 | 2.32 |

IOP Conf. Series: Materials Science and Engineering **772** (2020) 012026 doi:10.1088/1757-899X/772/1/012026

Table 4 indicated the same trend: the use of green cement can significantly increase the maximum dry density of inorganic binding material, mainly because green cement could fill the natural graded gravel compactly, and green cement could react well with fine soil particles, and increase the overall compactness of natural graded gravel, reduce the impact of water damage, and improve the durability of natural graded gravel of cement stabilized base. The fineness of green cement is greater than that of ordinary Portland cement, and the optimal water content is not varied significantly, mainly because the green cement contains high-efficiency superplasticizer as a additive.

### *4.3. Compressive strength test*

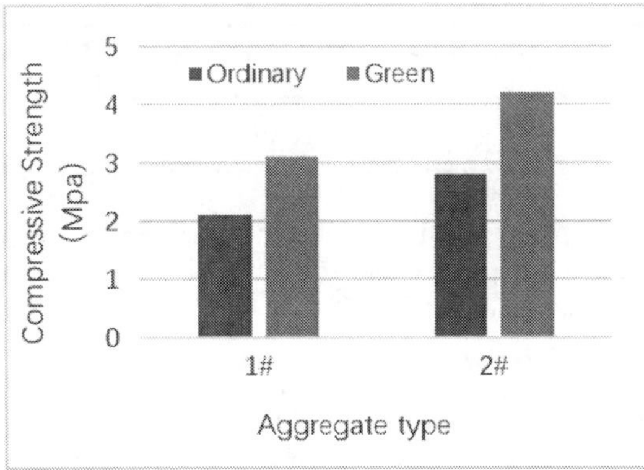


a) Compressive Strength.

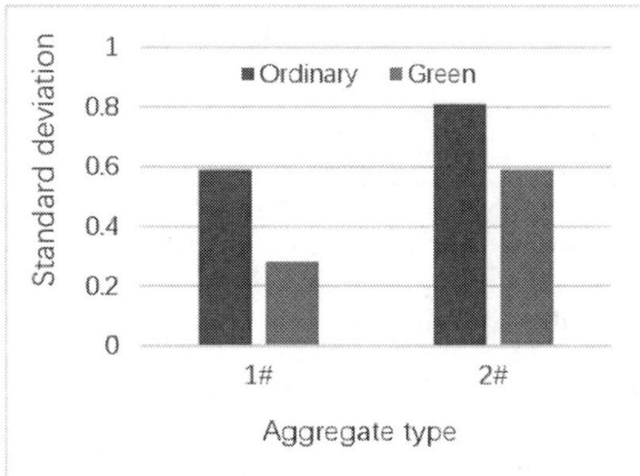


b) Standard deviation.

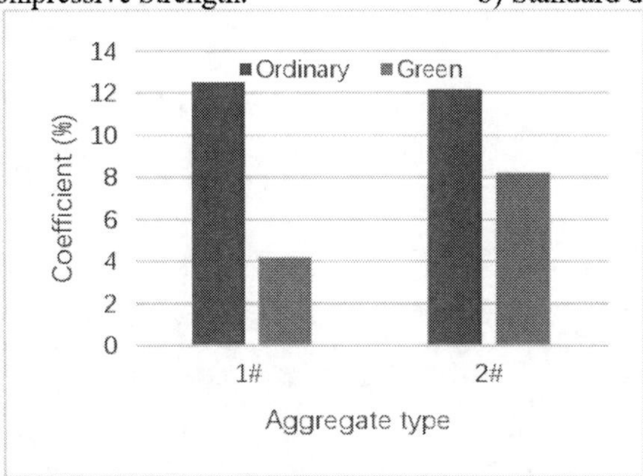


c) Coefficient of variation.

**Figure 2.** The comparison of a) compressive strength, b) standard deviation and coefficient of variation

As can be seen from Figure 2, the compressive strength of both specimens have been significantly improved by using green cement. In the compressive strength test, 15 specimens were tested each group, the standard deviation of green cement bonded specimens is significantly lower than that of ordinary Portland cement. It is proved that the effective strength of green cement bonded base is more stable than that of ordinary Portland cement, and the coefficient of variation of compressive strength by using green cement is much smaller than the specification value, which indicated that the bonding of green cement is strong enough to bind fine soil particles in the natural graded gravel structure, and make up for the defect of the larger soil content of natural graded gravel.

*4.4. Splitting strength test*

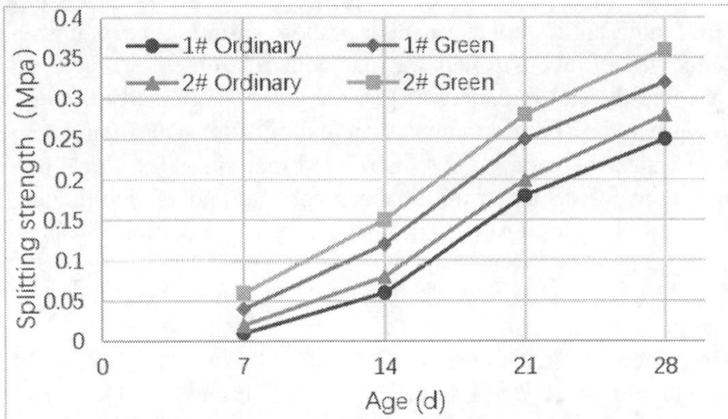


**Figure 3.** Splitting strength comparison of cement stabilized base specimens with 2 type of binders.

In order to explore the anti-pull resistance of the natural grade gravel of green cement curing, the splitting strength test was carried out. As shown in Figure 3, the splitting strength increased with curing period increasing, the splitting strength of 28d of 1# is 0.32 Mpa, and the splitting strength of 28d of the 2# material is 0.36 Mpa when using green cement. The splitting strength of specimens made with green cement is significantly higher than that of ordinary silicate cement, mainly because the green cement can form a three-dimensional spatial mesh slurry, which could wrap the soil particle surface well, then enhance the strength of the bonding transition zone between the aggregates, and improve the overall strength of the specimens.

*4.5. Resilient modulus test*

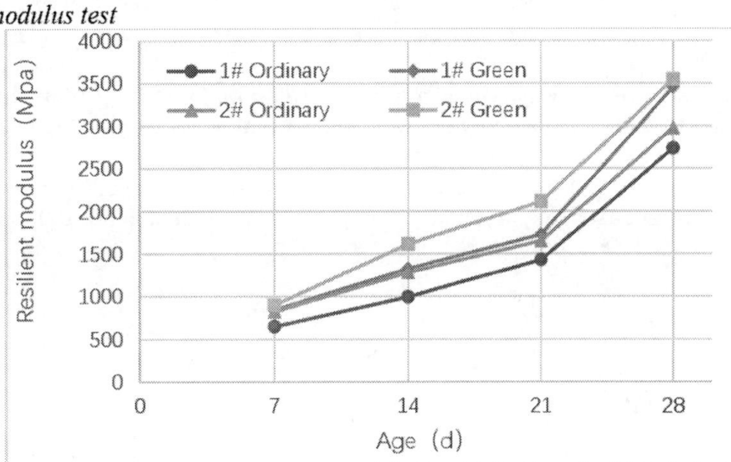


**Figure 4.** Resilient modulus

As can be seen from the Figure 4, the resilient modulus increased with the growth of the aging period. The resilient module of 28d of specimens made of green cement is 3155 Mpa and 3544 Mpa respectively, and has reached the highest value of the various cement stabilized base materials resilient modulus (3000-4200 Mpa) given in JTG D50-2006 Road Asphalt Pavement Design Specifications. It is shown that the natural graded gravel with green cement adding has good stiffness and meets the requirements of the use of the road base material.

## 5. Conclusion

Natural graded gravel has good performance on forming cement stabilized base when using green cement as the binder, which could replace manual graded natural stone. Shows an energy-saving and environmental-friendly way of raw materials for road construction.

Green cement gel could solidified natural graded gravel compactly, shows higher compressive strength, spitting strength and resilient modulus than that of ordinary Portland cement.

Both kinds of aggregates are wide-ranged distributed materials from local construction site. Green cement can make up for the defects that traditional cement could not deal with wide-ranged raw material and high soil content well. This kind of material has a broad application prospect in road construction field.

## References

[1] Talling B., Brandštetr in: A Progress in Cement and Concrete, Vol. 4, Mineral admixtures in cement and concrete, p. 297-339. Editors S. L. Sarkar and S. N.Gosh, ABI Books Private Ltd., New Delhi 1993.

[2] Krivenko P. in: Proc. 1st Intern. Conf. "Alkaline cementsand concretes", Vol. 1, p. 11-130, VIPOL Stock Comp., Kiew 1994.

[3] Škvára F. in: MRS Symp. Proc, Vol. 370, Microstructure of Cement-based systems / Bonding and Interfaces in Cementitious Materials, p. 153-158, editors S. Diamond,S. Mindess, Materials Research Society, Pittsburgh,1994.

[4] JTG D50-2006 "Road Asphalt Pavement Design Specifications"

[5] JTJ 034-2000 "Communication Industry standard"

[6] L. B. Andrade, J. C. Rocha and M. Cheriaf, "Aspects of Moisture Kinetics of Coal Bottom Ash in Concrete", Cement and Concrete Research, Volume 37, Issue 2, February 2007, Pages 231-241

[7] L. B. Andrade, J. C. Rocha and M. Cheriaf, "Evaluation of Concrete Incorporating Bottom Ash as a Natural Aggregates Replacement", Waste Management, Volume 27, Issue 9, 2007, Pages 1190-1199.

[8] Apha Sathonsaowaphak, Prinya Chindaprasirt and Kedsarin Pimraksa, "Workability and Strength of Lignite Bottom Ash Geopolymer Mortar", Journal of Hazardous Materials, Volume 168, Issue 1, 30 August 2009, Pages 44-50.

[9] Haldun Lurama and Mine Kaya, "Usage of Coal Combustion Bottom Ash in Concrete Mixture", Construction and Building Materials, Volume 22, Issue 9, September 2008, Pages 1922-1928.

IOP Conf. Series: Materials Science and Engineering **772** (2020) 012027 doi:10.1088/1757-899X/772/1/012027

# Optimization determination of three-dimensional solubility parameters of fluororubber F2603

**Hongmei Li [1], Xiaoqin Shi [1], Xuesong Feng [2] and Liming Dong [1,*]**

[1]School of Chemistry and Chemical Engineering, Xuzhou University of Technology, Xuzhou, China
[2]Xi'an Modern Chemistry Research Institute, Xi'an, China

*Corresponding author e-mail: polymerdlm@163.com

**Abstract**. Turbidity titration experiment on fluororubber F2603 using a variety of good solvents and poor solvents, the minimum volume closure ellipsoidal fit is performed on the 48 cloud point data generated by KY (Kumar and Yildirim) first-order algorithm, thus determining the ellipsoid The ball center is a three-dimensional solubility parameter: $\delta_d$=16.63(MPa)$^{1/2}$; $\delta_P$=11.25(MPa)$^{1/2}$; $\delta_H$=12.02(MPa)$^{1/2}$, and the solubility parameter is 23.40(MPa)$^{1/2}$.The reliability and accuracy of the fitting results are determined by the group contribution method, The solubility parameters have important guiding significance for selecting solvents with superior performance and determining better formulation design.

## 1. Introduction

Fluoro-polymers have excellent weather resistance, heat resistance, corrosion resistance and so on, which mainly because of the high electro-negativity of fluorine elements and strong fluorocarbon bonds [1]. For example, PTFE, which is referred to as a "plastic king"[2], but the application is limited because it has poor dissolution process-ability and is almost insoluble in all reagents. In order to eliminate this disadvantage, a high molecular elastomeric fluororubber is prepared by copolymerization of a perfluoromonomer and a non-perfluoromonomer [3]. The non-perfluoro structure of the main chain destroys the crystallinity of the polymer, making the solvent molecules easily invade the inside of the chain structure, thereby exhibiting solubility characteristics. For example, the fluororubber-26 can be sufficiently dissolved in ketone, ether, ester, amine, ammonia, chlorosulfonic acid, or phosphoric acid solvent. This soluble property broadens the application of such materials, such as anti-corrosion coatings, automotive gaskets, O-rings, oil seals and connecting tubes [4-6]. The solubility parameter is an important parameter to weigh the compatibility between different substances, and is also an important tool to study the solubility characteristics of polymers[7].The solubility parameter of the fluororubber accurately obtained by the solubility parameter theory can quickly and accurately select the polymer solvent and better determine the formulation, which is of great significance.

Since the polymer cannot directly measure its heat of vaporization, its parameters are often obtained by the indirect method of pure solvent method. Liu W.P. used inverse gas chromatography to determine the solubility parameters of fluororubber, but the workload is too large [8]. Liu J. indirectly measured the solubility parameter of fluororubber by intrinsic viscosity method, but the accuracy of the obtained results was poor [9]. Hansen solubility parameter (HSP) [10] can accurately describe the effects of

Content from this work may be used under the terms of the Creative Commons Attribution 3.0 licence. Any further distribution of this work must maintain attribution to the author(s) and the title of the work, journal citation and DOI.
Published under licence by IOP Publishing Ltd

dispersive force, polar force and hydrogen bond on the solubility parameters of polymers. It is recognized as a scientific method with the good applicability and the selection of solvents by the polymer field [11]. Therefore, it is of great practical value to study the three-dimensional solubility parameters of fluororubber. However, the measurement method of HSP has the disadvantages of requiring large variety of solvents, large workload, long time, high cost, and low accuracy.

In the previous work, a method for accurately fitting the three-dimensional solubility parameter of polymer HSP by turbidity titration has been proposed. The method has the characteristics of simple operation, simple calculation and less solvent[12] .Based on this work, the calculation of the three-dimensional solubility parameters of the polymer was carried out with the commonly used fluororubber F2603 as the object, in order to obtain more accurate three-dimensional solubility parameter results, and provide a theoretical basis for the selection of solvents.

## 2. experimental part

### 2.1. Experimental materials

The F2603 produced by Shanghai Sanaifu New Materials Co., Ltd., which was synthesized by copolymerization of 20% hexafluoropropylene and 80% vinylidene fluoride (mol). Its structure is shown in Figure 1:

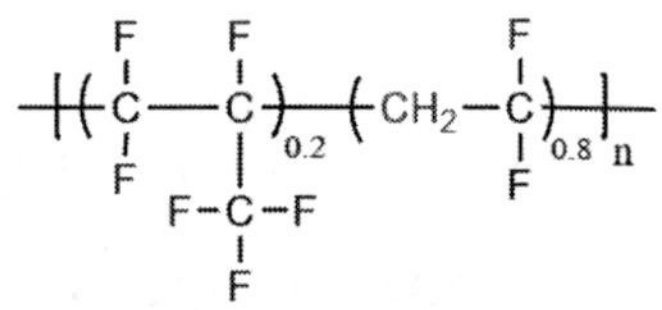

**Figure 1.** Structure of F2603

According to the solubility experiment, it can be determined that the polymer good solvent includes: N,N-dimethylformamide (DMF), N,N-dimethylacetamide (DMAc), Methyl Isobutyl ketone(MIK), Ethyl acetate(EA), acetone, N-methyl-2-pyrrolidone (NMP), methanol, dimethyl sulfoxide (DMSO). Poor solvents include: Benzene, Chlorobenzene(CB), Xylenes, chloroform (CF), tetrachloroethylene (PCE), formamide (FA), 1,1,2,2-tetrachloro Ethane(TCE), water, n-hexane. The solvents were derived from Sinopharm Group, Tianjin Fuchen Chemical Reagent Factory and so on. Analytical purity. All solvents are subjected to water removal and vacuum distillation. The three-dimensional solubility parameters of each solvent were obtained by querying [13].

### 2.2. Experimental process

F2603 samples 0.1g were accurately weighed and dissolved in 100ml good solvent, then the transparent solution was prepared, 5 ml of the polymer solution was titrated from the poor solvent by using an acid burette, and was identified as the end point of the titration when the solution became cloudy and the turbidity no longer disappeared. The volume of bad solvent was recorded, and tested three times at the same conditions. The average value of the results was finally obtained.

## 3. Results and discussion

The three-dimensional solubility parameter of cloud point can be calculated by the experimental results according to the document [12]. The calculated three-dimensional solubility parameters of the cloud point are shown in the following Table 1.The coordinates of 48 cloud points were fitted by (Minimum Volume Enclosing Ellipsoid) of minimum volume closure ellipsoid. The result of the fitting is shown in Figure 2.

The fitted ellipsoid sphere can be considered to be soluble and insoluble interface, and the spherical center coordinate is the three-dimensional solubility parameter value of polymer, then the three-dimensional solubility parameter values of F2603 are: $\delta_d$=16.63(MPa)$^{1/2}$; $\delta_p$=11.25(MPa)$^{1/2}$; δ

IOP Conf. Series: Materials Science and Engineering **772** (2020) 012027 doi:10.1088/1757-899X/772/1/012027

$_H$=12.02(MPa)$^{1/2}$. According to the theory of three-dimensional solubility parameters of Hansen, the solubility parameters can be obtained : $\delta$ = 23.40(MPa)$^{1/2}$.

**Table 1.** The three-dimensional Solubility parameters of Cloud Point (MPa)$^{1/2}$

| Solvent | | Benzene | CB | Xylenes | CF | PCE | FA | TCE | water | n-Hexane |
|---|---|---|---|---|---|---|---|---|---|---|
| DMF | φs/% | 80.35 | 79.59 | 84.49 | 63.44 | 90.75 | 72.10 | 78.19 | 71.32 | / |
| | $\delta_d$ | 17.60 | 17.74 | 17.43 | 17.55 | 17.49 | 17.34 | 17.71 | 16.88 | / |
| | $\delta_p$ | 12.28 | 12.38 | 12.60 | 11.07 | 13.17 | 18.08 | 12.35 | 14.40 | / |
| | $\delta_H$ | 10.17 | 10.12 | 10.46 | 9.64 | 10.76 | 13.88 | 10.29 | 24.58 | / |
| DMAc | Φs/% | 73.75 | 77.2 | 79.49 | 41.03 | 61.32 | 63.81 | 69.4 | 63.89 | / |
| | $\delta_d$ | 17.23 | 17.33 | 16.97 | 17.4 | 17.4 | 16.95 | 71.44 | 16.34 | / |
| | $\delta_p$ | 9.88 | 10.31 | 10.26 | 7.74 | 9.68 | 18.24 | 9.99 | 13.3 | / |
| | $\delta_H$ | 8.82 | 9.01 | 9.2 | 7.86 | 7.99 | 14.04 | 8.99 | 26.69 | / |
| MIK | Φs/% | 78.9 | 83.88 | 81.94 | 49.29 | 82.89 | / | 71.63 | / | 31.03 |
| | $\delta_d$ | 16 | 15.95 | 15.74 | 16.61 | 15.85 | / | 16.36 | / | 15.03 |
| | $\delta_p$ | 5.42 | 5.85 | 5.54 | 4.82 | 6.03 | / | 5.84 | / | 3.4 |
| | $\delta_H$ | 3.76 | 3.84 | 3.94 | 4.98 | 3.73 | / | 4.47 | / | 2.83 |
| EA | Φs/% | 81.65 | 90.88 | 93.29 | 82.48 | 88.46 | 84.52 | 94.78 | / | 42.8 |
| | $\delta_d$ | 16.31 | 16.12 | 15.93 | 16.17 | 16.11 | 16.02 | 15.97 | / | 15.29 |
| | $\delta_p$ | 4.79 | 5.22 | 4.16 | 4.99 | 5.35 | 11.4 | 5.29 | / | 3.47 |
| | $\delta_H$ | 6.56 | 6.89 | 7 | 6.96 | 6.77 | 9.98 | 7.11 | / | 4.47 |
| Acetone | Φs/% | 83.77 | 90.79 | 91.74 | 89.79 | 93.19 | 80.05 | 94.24 | 72.13 | 47.07 |
| | $\delta_d$ | 16.01 | 15.85 | 15.69 | 15.75 | 15.71 | 15.85 | 15.71 | 15.5 | 15.18 |
| | $\delta_p$ | 9.52 | 10 | 9.97 | 9.9 | 10.15 | 14.95 | 10.17 | 12.22 | 7.14 |
| | $\delta_H$ | 6.46 | 6.7 | 6.76 | 6.88 | 6.76 | 10.55 | 6.91 | 23.11 | 4.8 |
| NMP | Φs/% | 52.16 | 60.67 | 59.5 | 42.75 | 75.26 | 63.21 | 77.42 | 75.81 | / |
| | $\delta_d$ | 18.19 | 18.4 | 17.84 | 17.89 | 18.07 | 17.71 | 18.18 | 17.43 | / |
| | $\delta_p$ | 8.88 | 9.95 | 9.51 | 8.38 | 11.04 | 18.66 | 11.09 | 13.29 | / |
| | $\delta_H$ | 5.38 | 5.75 | 5.89 | 6.38 | 6.25 | 12.87 | 6.82 | 21.73 | / |
| Methanol | Φs/% | 76.1 | 79.07 | 87.42 | 58.68 | 78.67 | 47.66 | 91.11 | 68.43 | / |
| | $\delta_d$ | 15.95 | 16 | 15.44 | 16.27 | 15.84 | 16.23 | 15.46 | 15.23 | / |
| | $\delta_p$ | 10.73 | 11.11 | 11.51 | 9.63 | 11.22 | 20.79 | 11.84 | 13.58 | / |
| | $\delta_H$ | 19.48 | 19.85 | 2.88 | 17.47 | 19.78 | 20.64 | 21.3 | 30.09 | / |
| DMSO | Φs/% | 64.98 | 66.72 | 71.49 | 64.53 | / | 72.11 | 86.69 | 26.8 | / |
| | $\delta_d$ | 18.4 | 18.6 | 18.18 | 18.19 | / | 18.07 | 18.45 | 16.33 | / |
| | $\delta_p$ | 13.22 | 13.62 | 13.88 | 13.3 | / | 19.63 | 15.38 | 16.11 | / |
| | $\delta_H$ | 8.31 | 8.41 | 8.78 | 8.87 | / | 13.25 | 9.69 | 36.57 | / |

"/" means that the two solvents are not soluble in each other.

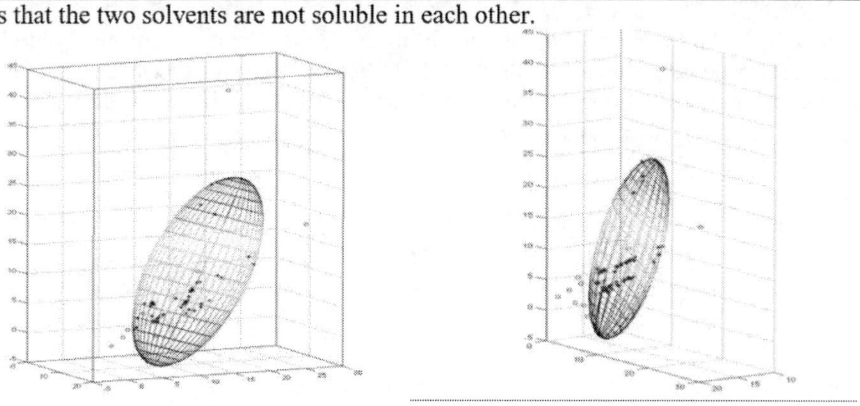

**Figure 2.** The ellipsoid fitting graph of cloud points

**Table 2.** The mathematical expression of ellipsoid

| Vector c | Positive definite matrix | | |
|---|---|---|---|
| $\delta_d$=16.63 | 0.3023 | -0.0505 | 0.0157 |
| $\delta_p$=11.25 | -0.0503 | 0.0280 | -0.0088 |
| $\delta_H$=12.02 | 0.0157 | -0.0088 | 0.0068 |

The distance that from the center of the ball to the coordinates of the solvent participating in the fitting experiment was calculated from small to large as follows: DMAc(1.84), DMF (2.67), NMP (5.12), and acetone(5.22), EA(7.70), TCE(9.36), MIK(9.54), CF(10.38), CB(12.42), PCE (13.34), Xylenes (15.17), FA (16.51), n-hexane (16. 55), benzene (18.96), water (30.67). Theoretically, the closer the distance from the spherical center coordinate to the good solvent, the better the solubility of fluorine rubber samples, and the actual result is consistent with the theory. DMAc value is closest to the spherical center coordinates, and the dissolution effect on the fluororubber sample is also optimal at normal temperature. As can be seen from figure 2, most of the coordinates of cloud points are located in the ellipsoid surface, the coordinates of good solvents are located in the interior of the ellipsoid, then the coordinates of bad solvents are all located in outside the ellipsoid, and the center of the sphere is the point represented by the red circle. According to the above distance calculation results, TCE (9.36) is less than MIK (9.54), however, it was found that MIK is a good solvent in the sphere, TCE is a poor solvent outside the sphere. This undoubtedly proves that the ellipsoid fitting accuracy is better than Hansen's positive sphere.

Small [14] believes that the cohesive energy density of a polymer is equal to the sum of the cohesive energy densities of all atoms and radicals in the molecular chain. According to this principle, the solubility parameter of the polymer can be calculated from the sum of the molar gravitational constants F of all atoms or groups in the macromolecular structural unit. The relationship between the solubility parameter and the molar gravity constant is as follows:

$$\delta = \rho \sum F / M_0 \tag{1}$$

ρ—represents the relative density of the polymer;

$M_0$—represents the relative molecular mass of the structural unit.

According to the query data, the molar gravity constant of group were as follows: $CH_2$=272.0, $CF_2$=307.0, $CF_3$=561.0, F=84.5[$(J/cm^3)^{1/2}$/mol].The relative density of F2603 was 1.81 g/cm$^3$, the relative molecular mass of the structural unit is 122.8 g/mol, then the solubility parameter is calculate ed according to formula (1) is:

$$\delta = \rho \sum F / M_0 = \frac{1.81(272.0 + 307.0 \times 2 + 561.0 + 84.5)}{122.8} = 22.57\left(J/cm^3\right)^{1/2}$$

The solubility parameter obtained by the fitting is 23.40 $(J/cm^3)^{1/2}$, and the solubility parameter calculated by the group contribution method is 22.57 $(J/cm^3)^{1/2}$. The difference between the two is small, indicating that the turbidity titration method is combined with HSP. Theoretical measurement of solubility parameters is reliable and accurate.

## 4. Conclusion

The three-dimensional solubility parameter value of fluororubber F2603 was obtained by spherical matching method based on turbidity titration method and Hansen HSP theory. The result is: $\delta_d$=16.63 $(MPa)^{1/2}$; $\delta_p$=11.25$(MPa)^{1/2}$; $\delta_H$=12.02$(MPa)^{1/2}$; δ =23.40$(MPa)^{1/2}$.The solubility parameter (22.57 $(MPa)^{1/2}$) measured by the group contribution method is similar to the fitting result, which proves the reliability of the fitting result. The turbidity titration fitting HSP three-dimensional solubility parameter method has proved to be an accurate and effective method in this work. it can be extended to the estimation of other polymer solubility parameters.

**Acknowledgments**
This work was financially supported by National Natural Science Foundation (51873232) and Xuzhou University of Technology (XKY2018123).

**References**
[1] Stahl T., Klare H. F. T., Oestreich M. Main-Group Lewis Acids for C–F Bond Activation[J]. ACS Catalysis, 2013, 3(7):1578-1587.
[2] Fujiwara, Masahiro. Water desalination using visible light by disperse red 1 modified PTFE membrane[J]. Desalination, 2017, 404:79-86.
[3] Xie Yizhi, Liu Dengxiang, Zhou Mingwei. Handbook of Rubber Industry[M]. Chemical Industry Press, 1989.598-621.
[4] Zhang Yulong, Zhang Jinsheng. Special rubber and application[M]. Beijing: Chemical Industry Press, 2011.
[5] Cai Shuming, Cai Hongzhi, Liang Xingyu. Application Technology of Fluororubber[M]. Beijing: Chemical Industry Press, 2009.
[6] Lu Gang. Structural characteristics of fluorine rubber and its application and development [J].Chemical Industry, 2014, 32(7):32-38.
[7] Moganty S S, Baltus R E. Regular Solution Theory for Low Pressure Carbon Dioxide Solubility in Room Temperature Ionic Liquids: Ionic Liquid Solubility Parameter from Activation Energy of Viscosity[J]. Industrial & Engineering Chemistry Research, 2010, 49(12):5846-5853.
[8] Liu Weiping, Zhou Jiwen. Determination of Solubility Parameters of Fluororubber by Inverse Gas Chromatography[J]. Science and Technology Bulletin, 1989(01):63-64
[9] Liu Jie. Research on new fluororubber-coated metal gaskets for internal combustion engines[D]. Nanchang University, 2007
[10] Hansen C.M. The three-dimensional solubility parameter-key to paint component affinities: solvents, plasticizers, polymers, and resins. II. Dyes, emulsifiers, mutual solubility and compatibility, and pigments. III. Independent calculation of the parameter components[J]. Journal of paint technology, 1967, 39(511): 505-510, (505):104-117, (511):511-514.
[11] Hansen C.M. 50 Years with solubility parameters—past and future[J]. Progress in organic coatings, 2004, 51(1): 77-84.
[12] Dong Liming, Wang Renli, Chen Hao. Estimation of Three-Dimensional Solubility Parameters of Poly(phthalazinyl) Poly(aryl ether nitrile) by Turbidity Titration[J]. Guangzhou Chemical Industry,2014,42(18):99-102.
[13] Hansen C.M. Hansen solubility parameters: a user's handbook Second Edition[M]. CRC press, 2007.
[14] Small, P. A . Some factors affecting the solubility of polymers[J]. Journal of Applied Chemistry, 2007, 3(2):71-80.

# Diagnostic method and application of low speed sliding bearing wear fault

**Junshan Guo [1,*], Wei Zheng [1], Yue Han [1], Linkai Zhu [1], Yongqiang Che [1], Jin Wang [2], Ming Wang [2]**

[1] State Grid ShanDong Electric Power Research Institute, Jinan, 250003, China
[2] State Grid ShanDong Electric Power Company, Jinan, 250001, China

*Corresponding author e-mail: junshan_guo@sina.com

**Abstract.** The static thermal and dynamic characteristics of the oil film affect the motion stability of the rotor system and directly determine whether the entire equipment can operate safely and stably. In this paper, the characteristics of sliding bearing oil film are studied by theoretical analysis, and the bearing wear fault diagnosis model is proposed. The model is verified by the actual production. The results show that the proposed diagnostic model can quickly locate the cause of the fault and ensure the safe and stable operation of the equipment.

## 1. Introduction

Sliding bearings have the advantages of high bearing capacity, long service life, convenient processing and maintenance, etc., and thus are widely used in rotating machinery, especially in rotating machinery of thermal power plants [1]. In China, the support of medium and low-speed rotating equipment such as circulating water pumps and condensate pumps in active units is based on sliding bearings. The stability of these equipments will be severely tested when the rotating machinery or the bearings themselves deviate from normal operation. Research on the characteristics of sliding bearings and the establishment of fault diagnosis models for sliding bearings are of great significance for the stable operation of these equipments.

The static and dynamic characteristics of the sliding bearing oil film are one of the key factors affecting the bearing capacity and stability of the system [2]. Therefore, the study of the static and dynamic characteristics of the oil film of the sliding bearing is of great significance in the analysis of the characteristics of the bearing rotor system. According to the theory of sliding bearing oil film dynamic pressure, the fluid dynamic pressure bearing is carried out. When analyzing the characteristics of the rotor system, it is found that there are many characteristic parameters that affect its characteristics. Including the bearing structure such as the bearing's aspect ratio, bearing clearance, physical properties of the medium, as well as operating conditions such as external load and speed, etc., will affect the performance of the bearing rotor system. In this regard, many relevant scholars at home and abroad have carried out extensive and in-depth research. In 1982, based on experience, the analytic solution of the oil film pressure of a finite-length sliding bearing was obtained based on the analytic function of the infinite-length and infinite-short bearing oil film pressure. However, since the analytical solution is based on the steady-state solution, it is not suitable for Transient oil film pressure study [3]. In 1986, Capone proposed a method for calculating the oil film force of a symmetrical supported sliding bearing

Content from this work may be used under the terms of the Creative Commons Attribution 3.0 licence. Any further distribution of this work must maintain attribution to the author(s) and the title of the work, journal citation and DOI.
Published under licence by IOP Publishing Ltd

IOP Conf. Series: Materials Science and Engineering **772** (2020) 012028 doi:10.1088/1757-899X/772/1/012028

[4]. In 1991, the nonlinear oil film force model under the assumption of short bearing theory was proposed. Because of the good precision and convergence of the model, it is widely used [5]. In 2002, Kraiko proposed a bearing optimization approximation method for the lubrication of incompressible fluids based on Elrod-Burgdorfer boundary conditions [6]. At present, domestic and foreign scholars' research on fault diagnosis of sliding bearings is mostly limited to theoretical research and simulation and simulation. The established analytical model also requires a large amount of detailed measurement data support. At present, the vibration signal of the low-speed sliding bearing is difficult to accurately measure, and most of the currently obtained diagnostic models relying on detailed measurement data are not available. It is necessary to study the fault diagnosis model of sliding bearing based on the actual operation of the site.

## 2. Sliding bearing lubrication principle

The friction surface of two mutually moving objects, the viscous fluid film generated by the action of the pump oil when the shaft diameter itself rotates completely separates the two friction surfaces, and the pressure generated by the fluid film balances the external load, which is called hydrodynamic lubrication. There are three elements of the fluid-generating dynamic pressure lubricating film: 1. There are two solid surfaces 2 which can form a wedge-shaped gap with each other. The wedge-shaped gap formed is filled with a fluid having a certain viscosity, and the relative relationship between the two solid surfaces. The motion causes the viscous fluid filled therein to move from the large end of the gap to the small end of the gap to generate supporting pressure. The structure of the radial sliding bearing and its working form just make it necessary to form the dynamic pressure lubricating film.

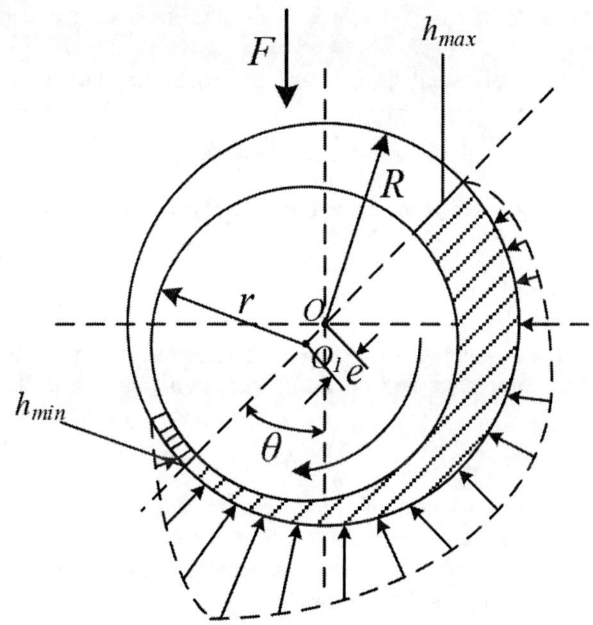


**Figure 1.** Schematic diagram of sliding bearing lubrication principle

Under the action of the external force load $F$, in order to reach the equilibrium state, the shaft center $O_1$ needs to be at an eccentric position with respect to the bearing center $O$. Where e is the eccentricity and $\theta$ is the deviation angle. On the extension line of the continuous heart line $OO_1$, one end is the maximum oil film thickness $h_{max}=R\text{-}r+e$, and the other end is the minimum oil film thickness $h_{min}=Rre$; the shaft diameter is clockwise rotation direction, in the semicircle of $h_{max}$ to $_{hmin}$, the gap From the large to the small convergent wedge gap, this is the most important set of conditions for the lubricating oil film to generate the lifting pressure to withstand the load F. In the semicircle from $h_{min}$ to $h_{max}$, the wedge-shaped gap is enlarged from small to large, so the pressure distribution in the lubricating oil film drops sharply after $h_{min}$; if lubricating oil is used as the lubricant, it is usually after $h_{min}$. At a small distance, the oil film breaks due to the inability to withstand too much negative pressure, forming a void area. This creates a complete pressure distribution in a gap slightly larger than 180°.

## 3. Diagnostic model

When the rotating shaft starts to start, the journal depends on the frictional force and crawls upward along the inner surface of the bearing. After reaching a certain position, the frictional force can not support the weight of the rotor and begins to slip. At this time, it is semi-liquid friction. As the rotational speed continues to increase, the journal brings the viscous lubricating oil into the wedge-shaped gap (oil wedge) between the bearing, because the wedge-shaped gap is convergent, and its inlet section is larger than the outlet section, so A certain oil pressure is generated in the oil wedge, and the journal is pushed by the pressure of the oil to the other side.

If the flow of lubricating oil introduced into the wedge gap is continuous, the oil pressure in the oil will increase, causing the average flow rate at the inlet to decrease and the average flow rate at the outlet to increase. The oil layer accumulated in the gap is called an oil film, and the oil film pressure can lift the rotor journal. When the oil film pressure is balanced with the external load, the journal operates stably without contact with the inner surface of the bearing.

According to the conditions established by the oil film, for low-speed sliding bearings, there are three main problems that may cause oil film damage: 1. The bearing lubrication system is insufficiently supplied with oil. 2. Mechanical vibration exists in the shaft or bearing bush. 3. The bearing load is too large. The manifestations and hazards of the three types of problems are shown in Table 1:

**Table 1.** Analysis of wear failure of sliding bearings.

| Fault type | Manifestations | harm |
|---|---|---|
| Lack of oil supply | Lubricating oil temperature rises rapidly and oil return decreases | Insufficient lubrication damage bearing bush |
| Mechanical vibration | Lubricating oil temperature rises, equipment parameters are abnormal | Worn bushing, damaged equipment |
| Excessive load | The temperature of the lubricating oil rises and returns to normal after the output of the equipment drops. | Long-running wear bearing |

According to Table 1, the bearing wear accidents encountered in the actual production can be analyzed. The production parameters and operating conditions can basically locate the cause of the accident and solve it.

## 4. Case analysis

The circulating water pump of #4 unit of a power plant adopts G48SH-19 horizontal pump with YD1250/780-12/14/1730 motor. After the overhaul, the temperature of the motor bearing bush has been high, and the tungsten gold phenomenon has been worn. After the power plant technicians adjusted and replaced the bearing bush several times, the wear phenomenon still exists.

**Figure 2.** Schematic diagram of bearing wear

Based on the diagnostic model given in Section 3, we made the following diagnosis:

a. In-situ inspection of the wear of the bearing bushings, according to the wear characteristics of the black gold, the wear caused by poor lubrication is judged.

b. Look for the cause of poor lubrication of the bearing bush. Due to the low motor speed (495r/min), the bearing lubrication requirements are low, and it is judged that there is an external force to cause the rotor to rotate irregularly. It is recommended to start the test, observe the actual operation and contact the vibration expert to try to test the vibration.

c. During the trial operation, the temperature of the lubricating oil at the driving end bearing bush is raised from 34 °C to 70 °C within 15 min, and the inspection is stopped. It is found that the phenomenon of black gold wear occurs. During the trial operation, it was found that the oil ring was severely beaten.

d. Through trial operation, it is preliminarily determined that the bearing bush wears due to insufficient oil supply due to severe oil slamming at the drive end bushing. The cause of the oil ring vibration is suspected to be the following: the roundness of the shaft is poor; the shaft sway is too large at the oil ring; the oil ring size does not match and cannot work normally; the motor force is unbalanced, causing abnormal vibration.

Disposal: Check the roundness of the shaft of the motor rotor oil ring; measure the oil slinger and connect the wheel sway (should be no more than 50μm); check that the oil ring specifications and models match the original design; contact the motor manufacturer to try the test The rotor at the oil ring vibrates.

The motor was disassembled by the motor manufacturer, and the following problems were found:

a. The shaft extension is beating 0.14-0.16;

b. the iron core file beats 0.18-0.24;

c. the bearing profiles at both ends are different from each other;

d. the bearing pad babies alloy grinding;

e. The rotor imbalance is distorted and the data deviation is large.

According to the above situation, the technical analysis believes that the motor shaft is deformed and the coaxiality is extremely poor, resulting in poor stability during the operation of the motor, increased temperature, and large vibration, resulting in damage to the bearing bush. The result is consistent with model analysis results.

## 5. Conclusion

In this paper, the characteristics of sliding bearing oil film are studied by theoretical analysis, and the bearing wear fault diagnosis model is proposed, which outlines the manifestations and hazards of the three main types of failures. The model is verified by the actual production. The results show that the proposed diagnostic model can quickly locate the cause of the fault and ensure the safe and stable operation of the equipment.

## References

[1] S. S. Kirn, H. N. Yu, D. G. Lee.The Sliding Friction of Hybrid Composite Journal Bearing under Various Test Conditions [J]. Tribology Lettle, 2009: 21 1-219.

[2] D.S.H.Chan.Nonlinear Analysis of Rotor Dynamic Instability in High—Speed Turbo machinery. ASME JournalofEngineeringforGasTurbines andPower. 1996 (118): 122—129.

[3] H F De Castro, K. L. Cavalca, R. Nordmann. Whirl and whip instabilities in rotor-bearing system considering a nonlinear force model [J]. Journal of Sound and Vibration, 2008, 317(1-2): 273-293.

[4] G. Capone. Orbital motions of rigid symmetric rotor supported on journal bearings. La Meccanica Italiana, 1986, 199 (199): 37-46.

[5] Q. Ding, A. Y. Leung. Numerical and experimental investigations on flexible multi-bearing rotor dynamics. Journal of Vibration and Acoustics, 2005, 127 (4): 408-415.

[6] A. N. Kraiko. The formulation of problems of lubrication theory for journal bearings. Journal of applied mathematics and mechanics, 2002, 66 (5): 757-765..

# Research on Vibration Isolation Platform Based on Vibration Transfer Theory

**Zhaobin Su [a], Yuting Zhang [b], Jianyi Zheng***

School of Aerospace Engineering Xiamen University Xiamen, China

*Corresponding author e-mail: zjy@xmu.edu.cn, [a] 35120171150989@stu.xmu.edu.cn, [b] 35120191151197@stu.xmu.edu.cn

**Abstract.** Aiming at the interference problem of various vibration sources to laboratory precision instruments, combined with the theory of vibration transmission, the basic block design research based on large vibration isolation platform is proposed. Firstly, through the theoretical analysis of the vibration isolation platform, the transfer function of the vibration isolation effect with different damping ratios is obtained. Secondly, the parameter knowledge of the basic block model is analyzed, and the basic block is analyzed based on the analysis of the system resonance. Finite element analysis; finally, the vibration of the foundation and the basic block was measured by high-precision acceleration sensor. By analyzing the data, it was verified that the vibration isolation platform has obvious blocking effect on high-frequency vibration.

**Keywords:** Vibration isolation platform, base block, high frequency vibration, finite element analysis.

## 1. Introduction

The laboratory usually has a variety of sources of interference, including not only indoor short-range sources but also long-range sources of interference [1]. However, a variety of experimental equipment, such as scanning probe microscopes, require a low vibration and noise environment, which is often much more stringent than the environmental requirements of industrial semiconductor plants. For example, in the scanning tunneling microscope/spectrum (STM/STS) experiment [2], it is very important to reduce the transmission of external low-frequency vibrations to the microscope, because the subtle changes in probe sample separation are exponentially increased in the tunneling current [3]. Scanning tunneling microscopes must be able to detect surface undulations of 0.01 nm [4] when operating in constant current scanning mode. Such sophisticated testing requires very stringent requirements [5] for vibration isolation of the surrounding environment.

During the entire vibration isolation process, we need to focus on the overall design of the vibration isolation platform and the overall cooperation of the platform and the air spring. Scientifically and reasonably design the vibration isolation platform so that its center of mass is in an optimal position to ensure that no unnecessary torque is generated during the vibration adjustment process. At the same time, we should also ensure that the platform and the air spring do not resonate during the process of mating, which requires natural frequency measurement and modal analysis of the platform. In this article, we report an effective vibration isolation platform that mates with six air springs and is placed on a thick

Content from this work may be used under the terms of the Creative Commons Attribution 3.0 licence. Any further distribution of this work must maintain attribution to the author(s) and the title of the work, journal citation and DOI.
Published under licence by IOP Publishing Ltd

base plate that is directly connected to the building foundation. In order to compare the vibration level on the vibration isolation platform and the foundation, we evaluated the vibration isolation performance of the vibration isolation platform.

## 2. Vibration Transfer Analysis

The common methods of vibration isolation can be divided into active control and passive control according to the control mode [6]. As our vibration isolation platform is mounted on the air spring, for the convenience of research, we design the whole vibration isolation system into a dynamic model with damping and stiffness only in the z-direction [7], as shown in Figure 1.

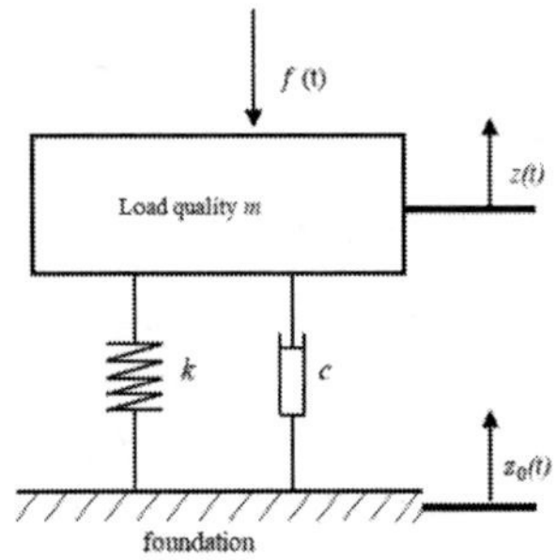


**Figure 1.** Vibration isolation platform simplified model diagram.

In this system, where m is the mass of the vibration isolation platform, and $z(t)$, $z_0(t)$ represent the vibration displacement of the vibration isolation platform and the ground respectively. Due to the air spring is simplified as a linear vibration isolation system with stiffness k and damping c, and $f(t)$ represents the external force which actively applied to the platform. Hence, the dynamic equation of the system can be derived as follows:

$$m\ddot{z}+c\left(\dot{z}(t)-\dot{z}_0(t)\right)+k\left(z(t)-z_0(t)\right)+f(t)=0 \tag{1}$$

Therefore, according to ratio of the vibration displacement between the platform and the ground, we can characterize the vibration suppression effect of the platform, and the transfer function is given by:

$$h(t)=\frac{z(t)}{z_0(t)} \tag{2}$$

$$H(s)=\frac{cs+k}{ms^2+cs+k} \tag{3}$$

$$\omega_0=\sqrt{\frac{k}{m}} \tag{4}$$

$$\xi=\frac{c}{2m\omega_0}=\frac{c}{2\sqrt{mk}} \tag{5}$$

and let $s=j\omega$, therefore the vibration displacement transfer function of the platform and the ground can be described by :

$$\left|H(j\omega)\right|=\left|\frac{j2\xi\omega_0\omega+\omega_0^{\ 2}}{-\omega^2+j2\xi\omega_0\omega+\omega_0^{\ 2}}\right| \tag{6}$$

Then we draw the transfer function diagram under different damping ratios by MATLAB, which relates the vibration isolation effect of the platform, as shown in Figure 2. From the image of transmission rate, we can clearly see that the combination of the air spring and the vibration isolation platform can theoretically isolate the high frequency disturbance in the external vibration. However, the disadvantage is that the disturbance near the natural frequency of the platform will be amplified. If we want to solve this problem, we need to reduce the natural frequency of the platform as much as possible. According to the natural frequency calculation formula $\omega_0 = \sqrt{\frac{k}{m}}$ , we can conclude that the natural frequency of the platform can be effectively reduced by increasing the mass of the isolation platform or reducing the equivalent stiffness of the system.

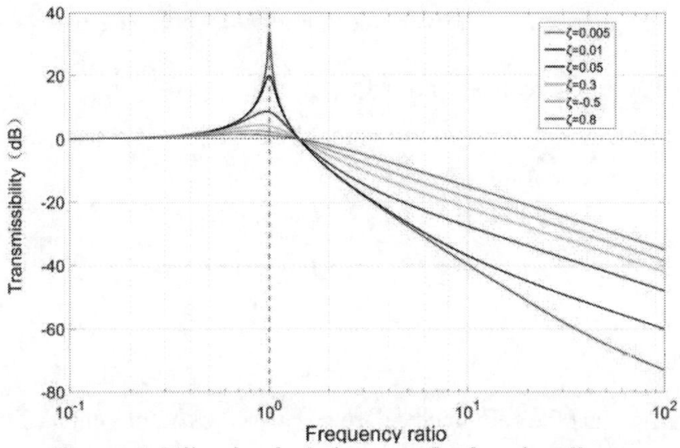


**Figure 2.** Vibration isolation transfer function diagram.

## 3. Vibration Reduction Laboratory

As shown in the figure 3, we can visually see the effect of the air spring and the vibration isolation platform. The building foundation uses a thick floor that is directly connected to the building foundation. The large volume of the thick bottom plate can overcome the disadvantage of unnecessary vibration caused by the high inertia center, but it is required to reduce the vibration noise transmitted from the surrounding environment. Therefore, an important component of the vibration isolation system is the basic block placed on the air spring. Obviously we focus on the size design and natural frequency measurement of the basic block that needs attention.

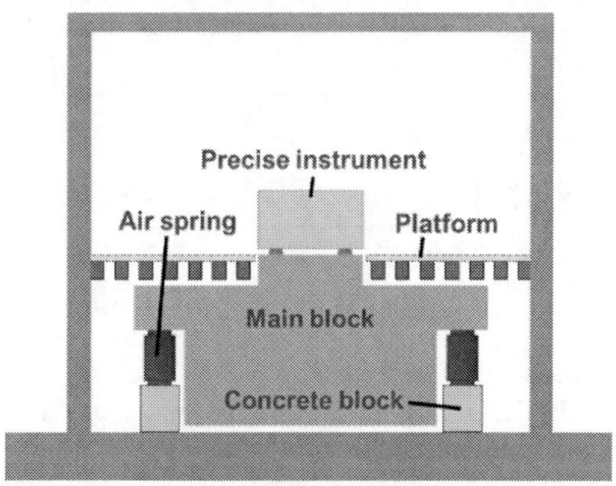


**Figure 3.** The schematic cross-sectional view of the vibration reduction laboratory: (1) The single laboratory covers an area of about 30 square meters. (2) The vibration-damping system based on air springs and air cushions. (3) The vibration isolation platform only carries precision instruments.

### *3.1. Base Block and Air Spring*

For this large-quality basic block, the most important thing we need to pay attention to is the centroid problem of the basic block. The multi-degree-of-freedom vibration isolating device comprises a base block and six air springs arranged on the lower edge of the base block, and the six air springs are evenly arranged in two rows. That is to say, the basic block we designed is that if it is only a rectangular parallelepiped, it is certainly not guaranteed that the center of mass is on the adjustment plane. With this in mind, we design the base block into a shape as shown in Figure 4. When the center of mass of the vibration isolation platform is at the center of the adjustment plane of the air spring, it is extremely important for us to ensure that no torque is generated during the actual adjustment process.

We can assume that the platform is composed of three rectangular parallelepipeds. The three parts are $M_1$, $M_2$, $M_3$ (top-down). The centroid coordinates are $G_1(x_1, y_1, z_1)$, $G_2(x_2, y_2, z_2)$, $G_3(x_3, y_3, z_3)$. The centroid coordinates are $\left(0, 0, c_2 + \frac{c_1}{2}\right)$, $\left(0, 0, \frac{c_2}{2}\right)$, $\left(0, 0, -\frac{c_3}{2}\right)$. The combined centroid are $(x, y, z)$ .

$$x = (M_1 x_1 + M_2 x_2 + M_3 x_3) / (M_1 + M_2 + M_3) \tag{7}$$

$$y = (M_1 y_1 + M_2 y_2 + M_3 y_3) / (M_1 + M_2 + M_3) \tag{8}$$

$$z = (M_1 z_1 + M_2 z_2 + M_3 z_3) / (M_1 + M_2 + M_3) \tag{9}$$

Then we need to ensure that the combined centroid $(x, y, z)$ is at the center of the contact surface of the assembly and the air spring, that is (0, 0, 0) , so that the air spring is always in the process of adjustment. An ideal state, that is, no excess torque is generated.

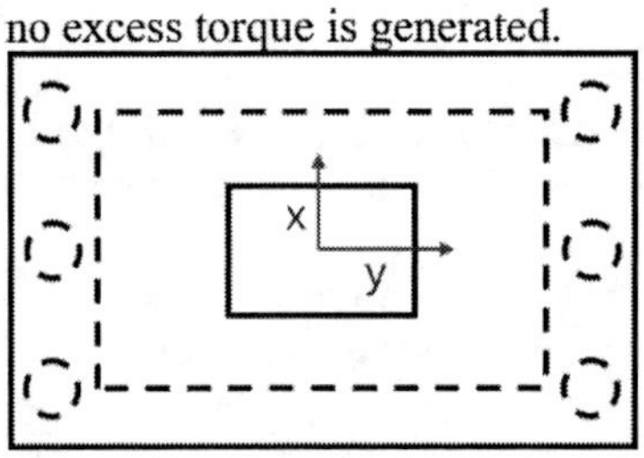


a

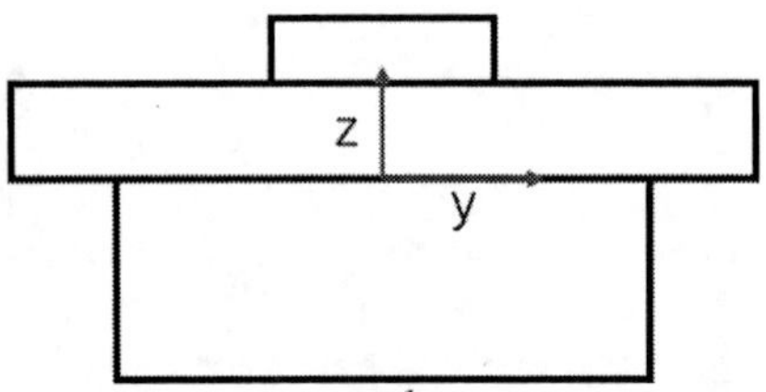


b

**Figure 4.** Base block details: (a) main view of the base block (b) top view of the base block.

### *3.2. Base Block Natural Frequency Measurement*

The most common method for determining the natural frequency of the vibration isolation platform is the simple harmonic excitation method, which causes the system resonance by the simple harmonic force, so as to find the natural frequencies of the system. Hence, the equation of motion is represented as:

$$m\ddot{x} + C\dot{x} + Kx = F_0 \sin \omega_e t \tag{10}$$

The solution of the equation can be written as:

$$X = A\sin\left(\omega_e t - \varphi\right) \tag{11}$$

$$A = \sqrt{A_1^2 + A_2^2} = \frac{q/\omega^2}{\sqrt{\left(1 - \frac{\omega_e^2}{\omega^2}\right)^2 + \frac{4\varepsilon^2\omega_e^2}{\omega^4}}} \tag{12}$$

$$\varphi = arctg\frac{A_2}{A_1} = arctg\left(\frac{2\omega_e\varepsilon}{\omega^2 - \omega_e^2}\right) \tag{13}$$

$$\mu = \frac{\omega_e}{\omega},\ \varepsilon = D\omega \tag{14}$$

$$A = \frac{q/\omega^2}{\sqrt{\left(1-\mu^2\right)^2 + 4\mu^2 D^2}} \tag{15}$$

$$\varphi = arctg\frac{2D\mu}{1-\mu^2} \tag{16}$$

$$q/\omega^2 = \frac{F_0}{m} / \frac{K}{m} = \frac{F_0}{K} = x_{st} \tag{17}$$

Due to the static displacement caused by the peak force of the mechanism, the amplitude can be expressed as:

$$A = \frac{1}{\sqrt{\left(1-\mu^2\right)^2 + 4\mu^2 D^2}} x_{st} = \beta x_{st} \tag{18}$$

Where β is the called power amplification factor:

$$\beta = \frac{1}{\sqrt{\left(1-\mu^2\right)^2 + 4\mu^2 D^2}} \tag{19}$$

Dynamic coefficient of forced vibration, β, the ratio of the dynamic amplitude to the static amplitude, is important for the vibration study of the vibrator and the single-degree-of-freedom system. When $\mu=1$, that is, the forced vibration frequency equals to the natural frequency of the system, and at this moment the dynamic coefficient increases rapidly, which results in the resonance of the system. From the formula: $X = A\sin\left(\omega_e t - \varphi\right)$, It can be seen that both the amplitude and phase of the resonance change significantly. By measuring these two parameters, we can determine whether the system reaches the common vibration point, and thus determine the vibration frequency of each stage of the system.

### *3.3. Basic Block Modal Analysis*

Based on the theoretical analysis of the natural frequency, combined with the actual working conditions of the basic block and the air spring, we perform modal analysis on the basic block. The following table shows the material parameters of the basic block. The total volume is about 18m and the total mass is 43ton. In order to eliminate the interference of the resonance point, we use the Ansys finite element simulation software to perform modal analysis on the basic block and take 6 for the object. The modal mode, the simulation results are shown in the figure 5.

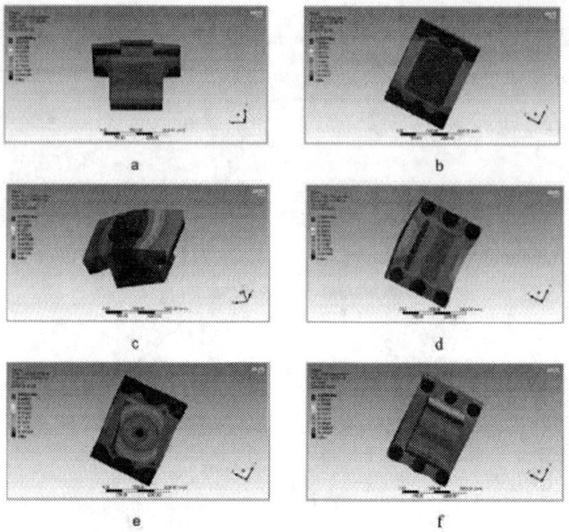

**Figure 5.** Basic block modal analysis diagram

## 4. Vibration Data Analysis

Through the above theoretical analysis and modal analysis, we can clearly understand that our basic block design can meet the initial vibration isolation requirements. Then, it is also necessary to compare whether the effect of the basic block and the air spring can match our original theoretical analysis. Therefore, we have made in-depth measurements of the vibration level of the foundation and the vibration level on the foundation block. The specific results of the measurement are shown in the figure 6. By analyzing the vibration image, it is obvious that the vibration isolation platform has a very direct vibration isolation effect for high frequency vibration. At the same time, due to the natural frequency of the air spring itself, there is a significant vibration amplification phenomenon in the low frequency band, which is also an element that we need to pay attention to.

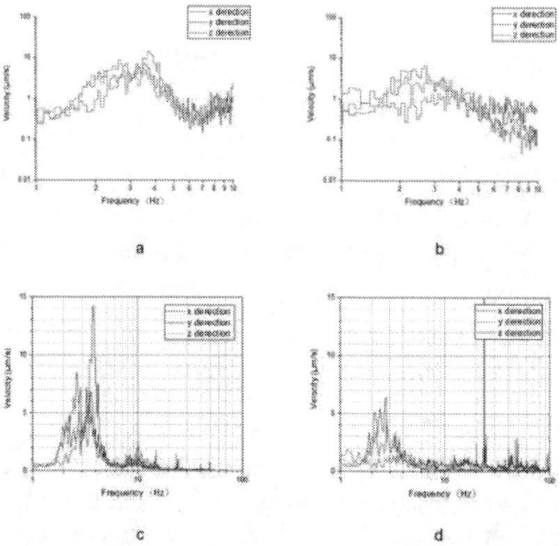

**Figure 6.** Detailed information of vibration (a) 1~10Hz vibration ladder diagram of x-axis direction vibration isolation platform (b) 1~10Hz vibration step diagram on the x-axis direction (c) vibration line diagram on the x-axis direction vibration isolation platform (d) ground vibration line diagram in the x-axis direction

## 5. Conclusion

Based on the theoretical analysis of vibration transmission, the modal analysis of the basic block is performed by ANSYS, which verifies that the basic block design effectively avoids the resonance point. Finally, the high-precision acceleration sensor is used to measure the vibration of the foundation and the vibration isolation platform respectively. It is proved that the vibration isolation platform has obvious effect on blocking some high-frequency vibrations, but the blocking effect on the ultra-low frequency vibration interference is not satisfactory.

## Acknowledgments

This work was financially supported by Sciencc and Technology Planning Project of Fujian Province (2018J01082).

## References

[1] cIwaya K, Shimizu R, Hashizume T , et al. Systematic analyses of vibration noise of a vibration isolation system for high-resolution scanning tunneling microscopes [J]. Review of Scientific Instruments, 2011, 82(8): 3313.

[2] Katsuya Iwaya and Ryota Shimizu , et al. Design of an effective vibration isolation system for measurements sensitive to low-frequency vibrations, 2012, 30(6): 063201-1.

[3] C. J. Chen, Introduction to Scanning Tunneling Microscopy, 2nd ed. (Oxford University Press, New York, 2008).

[4] Emanuel Lörtscher, Widmer D, Gotsmann B. Next-generation nanotechnology laboratories with simultaneous reduction of all relevant disturbances [J]. Nanoscale, 2013, 5.

[5] Y. J. Song, A. F. Otte, V. Shvarts, Z. Zhao, Y. Kuk, A. Blankenship, S. R. Band, F. M. Hess, and J. A. Stroscio, Rev. Sci. Instrum. 81, 121101 (2010).

[6] Song Li. Research on the active isolation system controlled by AEROSTATIC and voice coil motor [D]. Harbin Institute of Technology, 2015.

[7] Iwaya K, Shimizu R, Teramura A, et al. Design of an effective vibration isolation system for measurements sensitive to low-frequency vibrations [J]. Journal of Vacuum Science & Technology A Vacuum Surfaces & Films, 2012, 30(6): 063201-063201-5.

[8] Longqi Cai, Zihao Liu, Bo Zhang, et al. Simulation research on vibration and shock resistance integrated control technology of complex vibration isolation system [J]. Journal of Harbin Engineering University, 2013

IOP Conf. Series: Materials Science and Engineering **772** (2020) 012030 doi:10.1088/1757-899X/772/1/012030

# Research Progress on Typical Fault Monitoring and Diagnosis of Aircraft Engine

**Jiayan Li**

International Department, Nanning No.2 High School, China

**Abstract.** As one of the core components of aircraft, whether engine could be operated normally will directly effect on the safety of aircraft. How to monitor and identity different fault condition about core components of engine plays a vital role in ensuring air safety and reducing maintenance costs as well as has important value in safety and economy. This paper summarizes the typical fault and monitoring means of aircraft engine, which can provide some references for related research.

**Key Words:** aircraft, engine, fault, diagnosis.

## 1. Introduction

Aircraft engine, a highly complex and sophisticated thermal mechanical, provides power needed for flight. With the rapid and high-level development of engine technology, the comprehensive flight performance of aircraft has been constantly improved and perfected. While due to its complex component parts, long-term service in high-speed, high temperature and high load of work environment contribute to corrosion, wear and failure condition easily which are very likely to cause a major safety accident resulting in aircraft destruction and human death[1]. Therefore, how to improve the safety performance of aircraft engines and to ensure their working reliability is becoming a high-profile issue [2].

Healthy status of aircraft engine is the most direct reflection of aircraft performance, analysis and diagnosis to each fault mode is a significant basis for maintenance, changing, formulation and implementation of spare parts strategies. Hence that, analyzing health status and studying its typical ways to monitor fault of aircraft engine are effective solutions to ensure safe flight. Based on this, this paper focuses on the research progress of typical fault monitoring and diagnosis technology of aircraft engine.

## 2. Significance of Aircraft Engine Fault Monitoring

Monitoring for aircraft engine fault is an essential work during normal flight and mission execution. Generally, aircraft engine fault monitoring refers to basing on process status information collected by fault monitoring system, analyzes and processes fault mode of essential parts of aircraft engine and obtain its characteristic parameters of its health state establishing a diagnostic model to diagnose the fault mode of key components and master their health state also fault information, then report the obtained information to the unit or ground crew timely so that they can do relevant maintenance work according to this information.

Mechanical system of aircraft engine is very complicated, results from this, monitoring its working condition effectively can ensure safety and reliability during work process, master the health status of

Content from this work may be used under the terms of the Creative Commons Attribution 3.0 licence. Any further distribution of this work must maintain attribution to the author(s) and the title of the work, journal citation and DOI.
Published under licence by IOP Publishing Ltd

aircraft engine in time, and judge the fault modes accurately then provide related information to pilots and ground maintenance staffs, so that they can take effective measures to deal with potentially dangerous condition quickly and timely ensuring the working reliability of aircraft engine which has important practical value to aircraft flight safety.

## 3. Typical faults analysis and fault monitoring methods of aircraft engine

### *3.1. Typical Fault of Aircraft Engine*

With the development of aircraft engines towards high performance, the increase of its parts and the random coupling of the faults during work progress lead to the increasing of aircraft engine fault modes and complex mechanism. Especially wide envelope of military aircraft , the plane engine operates under a bad environment conditions, high strength, large overload, so-long flight time and other job characteristics, leads to frequent multiple and dangerous fault, which are easy to give a rise to deviation of normal working state of engine and leads to the reduction of engine stability margin and deterioration of the working performance, even major flight safety accidents such as air parking. Therefore, according to the research and incomplete statistics, common engine failures probably can be divided into the following several categories:

*3.1.1. Stability Fault.* Excursion of engine normal working state and reduced stability contribute to aerodynamic performance instability and so that happen great-harm, fast-failure and difficult-recovery faults such as rotating stall, blade flutter, coupled vibration, surge, oscillating combustion, over-temperature, structural damage, air shutdown and so on. Stability failure is a key factor that affects the safety, reliability, and stability of whole aircraft engine in the whole development periods and service life terms.

*3.1.2. Air Passage Fault.* Surge, the main manifestation of airflow passage fault and an abnormal working state of the engine, is caused by the deviation of the air flow and the excessive compressor speed of the design state having a strong impact on the normal operation of the engine, it even can cause flameout in flight of engine in severe cases so that a significant safety accident would happen.

*3.1.3. Vibration Fault.* Vibration is a a vibration failure mode, which results from local resonance, fatigue break, rotating stall and surge, imbalance of high and low pressure rotor, change of heat bending concentricity of rotor, failure of sealing device, wear of blade, damage of gear and bearing, looseness of bearing seat and clearance of supporting structure, failure of elastic support, failure of squeeze oil film damper and other factors, it is the main fault mode of engine and has the major effects on reliability of engine.

*3.1.4. Stable Combustion Fault.* Mechanical kinetic converted by the heat energy in its combustion chamber contributes to the normal operation of the engine. Combustion in aircraft engines is the coupling and interaction of physical and chemical processes. The main factors affect the normal operation of combustion chamber are fuel atomization and evaporation, turbulent transport, finite rate chemical reaction, high temperature radiation and particle motion characteristics, etc. In addition, the combustion chamber is affected by many other factors, such as temperature, height, and airflow distortion and so on.

*3.1.5. Fatigue Fracture Fault.* Fatigue fracture fault usually occurs on engine blades, and the common parts are the maximum steady-state stress point and the highest temperature point of the blade, vibration nozzles, the vulnerable parts, the connecting and contacting parts.

*3.1.6. Control System Fault.* The control system fault of the aircraft engine is manifested in the fault alarm of the software system during the operation of the engine, which leads to the faults of aircraft engine such as starting interruption, flameout in flight and so on.

### *3.2. Main methods of aircraft engine fault monitoring and diagnosis*

*3.2.1. Composition and working principle of aircraft engine fault monitoring system.* The fault monitoring of aircraft engine mainly consists of airborne equipment and ground equipment. Airborne equipment generally includes sensors, signal regulators, data acquisition systems, data transmission and recording devices, data processing systems, alarm and simple print processing devices, etc. Common fault monitoring systems includes Aircraft Integrated Data System (AIDS), Engine Indicating and Crew Alerting System(EICAS) ,etc. Ground equipment mainly consists of transmission equipment, decoding equipment or data processing equipment, ground maintenance station, computing center and corresponding software system for status monitoring and fault diagnosis. Data acquisition system would collect the sensor signal in accordance with the sequence and time and transmit to the airborne recording device, some parameters can do preliminary processing through the airborne computer of data, given real-time fault warning in the cockpit and print various reports engine when it is necessary, also has a development of Space Radio Communication System (Datalink), transmits the flight test data to the ground maintenance computer center directly to conduct a comprehensive and in-depth analysis of data in real time, the analysis results are directly fed back to the crew, and the necessary maintenance work shall be timely notified to the destination port of the aircraft and all preparations shall be made. Thus, these reduce the aircraft standing time on the ground greatly and the flight delay rate. In the future, satellite communication technology will be used to allow aircraft to exchange information with maintenance bases anywhere in the world.

*3.2.2. Key technology of aircraft engine fault monitoring.* At present, with the continuous deepening and progress of research, engine monitoring and fault diagnosis theory have been constantly developed and improved. According to incomplete statistics, the key technologies can be summarized as follows:

The key technologies of signal processing and feature extraction mainly include multi-signal separation technology, noise elimination technology, feature generation and extraction technology, nonlinear dynamic analysis technology, time-domain processing technology, frequency-domain processing technology, spectral analysis technology, non-stationary signal processing technology, and wavelet analysis technology, etc.

Condition monitoring and fault diagnosis of key technology is mainly divided into graphical diagnostic methods (such as the major fault graph, fuzzy fault graph, the retina figure, etc.), integrated information fusion diagnosis methods (such as the use of data filtering algorithm, characteristics, many methods of fusion method of weak classifier fusion algorithm), statistical decision methods (e.g., decision tree, decision well, fisher discriminant method, etc.), artificial intelligence methods (e.g., neural network method, expert system, etc.), evolutionary computation methods (for example, a cloud model evolutionary computation trend monitoring and diagnostic classification method), signal detection method (e.g., singular value detection, wavelet detection, etc.), statistical learning methods (e.g., support vector Machine, all kinds of learning machine), covering learning method (from the perspective of cognitive science research of fault diagnosis of new machine learning theory), dynamic diagnosis method, the dynamic model of fault analysis methods, such as a variety of detection, isolation, identification algorithm etc. Comprehensive monitoring engine technology should make full use of all kinds of information fusion theory in order to improve the level of fault diagnosis.

## 4. Current situation of aircraft engine fault monitoring and diagnosis and research

### *4.1. Current situation of overseas research*

In order to reduce the plane accident probability, ensuring the safety and reliability of the flight, researching the monitoring of aircraft engine health state has always been concerned by domestic and researchers. Fault monitoring of PHM system has been applied to the engineering practice early in abroad, such as the United States Department of Defense, National Aeronautics and Space Administration(NASA) especially attach importance to the study of health management system, their PHM system has long been applied in the f-35, the fourth-generation advanced fighter of the United States, which greatly guarantees the safety and reliability of the fighter and significantly improves the operational efficiency of the fighter. At the same time, because it can quickly locate the fault mode reducing the maintenance investment significantly [8].

Since the beginning of the 21st century, the development of aircraft engine PHM technology in foreign countries has been greatly accelerated, and the research on aircraft engine health monitoring has been gradually deepened. Typical researches among them are: Jaiwon, ey. al in the process of PHM technology research for system the problem of high failure rate of the engine to improve the safety and reliability in the process of aircraft engine flight, in improving the stability of the system at the same time reduce the incidence of engine, its engine fault monitoring were elaborated in details. In the study the technology has been proved in engineering application [9]. Salahshoor studied the health status of engine fuel pump to some extent, PHM system was firstly introduced into engine fuel system, which provided an effective technical solution for fault diagnosis of aircraft engine and had high engineering application value [10, 11].

In recent years, in order to achieve the "enhanced maintenance based on state," the United States all services offered some plans and initiatives and carried out several projects. The new projects introduce some new and improved maintenance technology and the application method into the engine fault maintenance practice. In the engine monitoring management system of JSF project, choosing a set of engine sensor subsystem includes detection system of gas-path debris detection I EDMS DMS and motor loss detection system. In addition, the American Society of Automotive Engineering (SAE) E - 32 aviation gas turbine monitoring committee developed and issued a series of guidelines, which has played a good role in guiding and promoting the development of aero engine condition monitoring and fault diagnosis, including aviation gas turbine engine monitoring system guides, limited monitoring system, the oil system monitoring guide, vibration monitoring system guide, the service life monitoring system and parts management guidelines, etc.

### *4.2. Current Domestic Research Situation*

Domestic engine health monitoring technology started late, have not reached the mature applications condition, yet. The relevant research work on health condition monitoring and fault diagnosis of large aircraft engines carried out by civil aviation companies and colleges in China has achieved initial results. The ground of the real-time diagnosis avionic subsystem ARDMS of Northwestern Polytechnical University, Beijing University of Aeronautics and Astronautics and Avic Qianshan Avionics , send back real-time flight data to the ground terminal through space data-chain, after processing of ground terminal, the data will be distributed to the engine subsystem achieving the effect of closed-loop real-time health monitoring. At the same time, domestic colleges and universities carry out a series of related theory research work, such as a control system based on Kalman Filter Sensor Fault Diagnosis, based on the fault diagnosis expert system for aero engine vibration testing, engine gas path fault diagnosis based on neural network, etc. Compared with foreign countries, domestic aircraft engine health condition monitoring still is at the theoretical level. At present, many research results are mainly focused on fault diagnosis, and the integrated system of online real-time status monitoring is rarely reported.

Despite the rapid development of China's aviation industry in recent years, aircraft engine is a complex systems engineering, the application of engine condition monitoring and diagnosis technology is just getting started. Especially for the newly developed high-performance engine, implement of state monitoring should be listed as important technical and tactical indicators, we will carry out comprehensive research in this area. But in general, the domestic work is not enough, it is urgent to learn from the successful experience of foreign countries in a planned and step-by-step way, to develop and promote our own engine health status monitoring and fault diagnosis technology for adapting to the needs of the development and aircraft. Now serving military aircraft and civil aircraft engines do not have a complete engine status monitoring and diagnosis system, yet. How to develop an effective monitoring and diagnosis device for burst fault based on the existing engines and ensure that the operating state of the engine is evaluated with fewer effective parameters are becoming a difficult problem.

## 5. Conclusion

With the development of aviation technology, the reliability of aircraft engine is increasing. But increasingly complex structure, high strength, large overload, long flight time, strong fault coupling and so on result in many fault modes and complex mechanism of the aircraft engine, which can easily lead to the change of engine working point, the decrease of engine stability margin and the deterioration of performance, and lead to multiple and dangerous faults. Through monitoring engine, we can effectively evaluate and control the use of the engine to determine the engine and its unit body, the state of the residual life and spare parts, detection, evaluation, segregation engine failure and inspection, adjustment of correction; We should track the changing trend of engine state, support the decision of engine life management and logistics protection, improving the reliability, maintainability, guarantee and economy of aircraft engine effectively , and ensuring flight safety.

## Acknowledgements

Thanks for the support of the National Natural Science Foundation of China (Item NO. 91860205).

## References

[1] Shanghai Aviation Measurement and Control Technology Research Institute. Aviation Fault Diagnosis and Health Management Technology [M]. Beijing: Aviation Industry Press, 2013.
[2] Zhang shaoji. Research and Prospect of Aircraft Engine Fuel and Control System [J]. Aircraft Engine, 2003, 29(3): 1-5.
[3] Chang qi, yuan shenfang. Current Situation and Development of Aircraft Integrated Health Management (IVHM) System Technology [J]. Systems Engineering and Electronics Technology, 2009, 31(11): 2652-2657.
[4] Zuo hongfu, Cai jing, Wu hao ,etc. Aviation Maintenance Engineering [M]. Beijing: Science Press, 2011. 194-206.
[5] Cao huiling, Du Peng. Fault Diagnosis of Aircraft Engine Fuel Pump Based on Bayesian Network [J]. Journal of Civil Aviation University of China, 2016,34(4): 27-30.
[6] Wang Lei. Fault Diagnosis of Aircraft Engine Combustion Control System [D]. Shanghai: Shanghai Jiaotong University, 2012.
[7] Wang Kai. Study on Common Faults and Diagnosis Technology of Fuel System of Aviation Piston Engine [J]. Electronic Test, 2016(15): 113-114.
[8] Litt J S,Simon D L, Gary S,et. al. A Survey of Intelligent Control and Health Management Technologies for Aircraft Propulsion Systems [C], NASA/TM, 2005.
[9] Jaiwon S. The NASAAviation Safety Program: Overview [C]. NASA/TM-2000-209810.
[10] Stephen O T Ogaji, Singh Riti. Advanced engine diagnostics using artificial neural networks [C]. 2002 IEEE International Conference on Artificial Intelligence Systems (ICAIS 2002).2002: 236-241.
[11] Karim Salahshoor, Mojtaba Kordestani, Majid S. Khoshro. Fault detection and diagnosis of an industrial steam turbine using fusion of SVM (support vector machine) and ANFIS (adaptive neuron-fuzzy inference system) classifiers [J], Energy, 2010(35), 5472-5482.
[12] Wang zhongsheng, Li Ming. Study on Identification and Monitoring Methods of Aircraft Engine

Burst Fault [J]. Journal of Northwest Polytechnical University, 2013, 31(3): 104-405.
[13] Wei xunkai ,Liu Fang,etc. Diagnostic Prediction Index System for Aircraft Engine Health Users Management [J]. Aircraft Engine, 2012, 38(5): 27-35.
[14] Xu Jiuping, Xu Lei. Health management based on fusion prognostics for avionics systems [J]. Journal of Systems Engineering and Electronics,2011, (3): 428-436.

# Viscosity and rheological behavior of aqueous nanofluids containing $TiO_2$ nanoparticles

**Linghui Kong*, Guixun Li, Yong Yang, Jun Zheng**

The Yellow River Institute of Hydraulic Research, the Yellow River Water Conservancy Commission, Zhengzhou, 450003, china

*Corresponding author e-mail: lhkong0416@163.com

Abstract. Nanofluids containing different concentration of $TiO_2$ nanoparticles were prepared. Their rheological behavior was investigated on LVDV III ultra-programmable Brookfield viscometer at different temperature and particles loading. Results show that rheological behavior of nanofluids is determinated by the concentration of $TiO_2$ in base fluid, nanofluids shows good Newton behavior when nanoparticles loading is low, while shear thinning becomes evident when the volume fraction of particles was continuously increased. Besides, viscosity of nanofluids shows strong nanoparticles loading and temperature dependence, but quite differs from classical viscosity model, as the theory models severely underestimate the value of viscosity, which is attributed to electroviscous effect of nanoparticles in base fluids. The primary electroviscous coefficient is evaluated at different pH values. Modified primary electroviscous coefficient is in good agreement with Booth's theory.

## 1. Introduction

Nanofluids are dispersions of nanomaterials in base fluids. Since the concept of "nanofluids" was coined by Choi in the year 1995, it has been studied in all fields include heat transfer in automobiles, nanolubrication, waste heat recovery and environment engineering. Viscosity is the resistance of fluid to flow, it is one of the important thermophysical parameters of nanofluids, while rheology is the study of how a fluid flows when external force are applied. They are both important factors for the potential application of nanofluids.

More recently, viscosity and rheology of nanofluids has been intensively investigated. These studies reveal that the viscosity of nanofluids is a function of temperature [1], concentration [2], shape and size [3] of nanoparticles. Zhou [4] deeply investigated temperature and shear rate dependencies of viscosity by analyzing alumina polyaphaolefins nanofluids, he found that the relative viscosity of alumina polyaphaolefins nanofluids decreases with an increased temperature. Baratpour [5] studied the dynamic viscosity of single wall carbon nanobubes in ethylene glycol at different temperature and solid volume faction, he pointed out that the dynamic viscosity of SWCNTs/EG nanofluids increases with decreasing temperatures and increasing solid volume fractions.

Despite numerous experimental and theoretical researches have been put into practice, an effective or integrated theory to establish the mechanism of enhanced viscosity of nanofluids still remains ambiguous. The failure of the well-known models to predict the viscosity of nanofluids gives rise to various theoretical hypotheses and empirical formulas. Most scholars attribute it to the addition of

Content from this work may be used under the terms of the Creative Commons Attribution 3.0 licence. Any further distribution of this work must maintain attribution to the author(s) and the title of the work, journal citation and DOI.
Published under licence by IOP Publishing Ltd

dispersant or pH value. However, it is not everyone thinks that was right, According to Rubiohernández [6], the phenomenon of viscosity enhancement is briefly summarized in two factors. Firstly, the flow field of base fluid is distorted with the appearance of solid particles, which would generate an increase in the energy dissipation. Secondly, additional energy dissipation is produced due to the electrical double layer (EDL) that surrounds the particles, which may exert an extra distortion on flow field.

This work aims at investigating the effect of volume fraction, temperature and shear rate on rheological behavior of aqueous nanofluids. It is believed that this research would be beneficial for deeply understanding the rheological behavior of nanofluids, and then perfecting the theoretical foundation for the development of nanofluids research.

## 2. Experimental methods

### *2.1. Preparation of $TiO_2$ nanofluids*

The $TiO_2$ nanoparticles chosen in this experiment are purchased from Sinopharm group Co. Ltd, China, nanoparticles are analytical grade and used directly without any further treatment. Firstly, nanoparticles were dispersed in deionized water at different volume fraction, followed by vigorous stirring for 30mintes, and then the mixture was sonicated for 30 minutes to obtain stable nanofluids. No dispersant is involved during the whole process to avoid the effect of surfactant to viscosity of the nanofluids. A drop of nanofluids dropped on a copper grid coated by carbon and evaporated under ambient atmosphere, and then the morphology was evaluated by JEM 2010 TEM. Size distribution of nanoparticles in nanofluids was measured with Zetasizer ZS from Malvern as well.

### *2.2. Measurement of dynamic viscosity*

Viscosity experiment will be conducted as soon as the nanofluids were prepared to reduce the error caused by aggregation. The dynamic viscosity was firstly measured under continuously increased shear rate from 1 to 1000 $s^{-1}$ at different particles loading. Then the volume fraction dependence of the viscosity was performed at the stationary shear rate and temperature range from 20℃ to 40℃. Relative viscosity was defined as $\eta_{nf}/\eta_{bf}$ and made a comparison with well-known viscosity models.

## 3. Results and discussion

### *3.1. Characterization of as-prepared nanofluids*

According to the results of Malvern Zetasizer ZS, The solid particles distribution of nanofluids containing 1 vol. % $TiO_2$ nanoparticles is presented in Fig. 1. Evidently, the average size of nanoparticles in nanofluids is approximately 30 nm, it is much larger than the primary size of 10nm, possibly due to the absence of dispersant, which arouse agglomeration among nanoparticles.

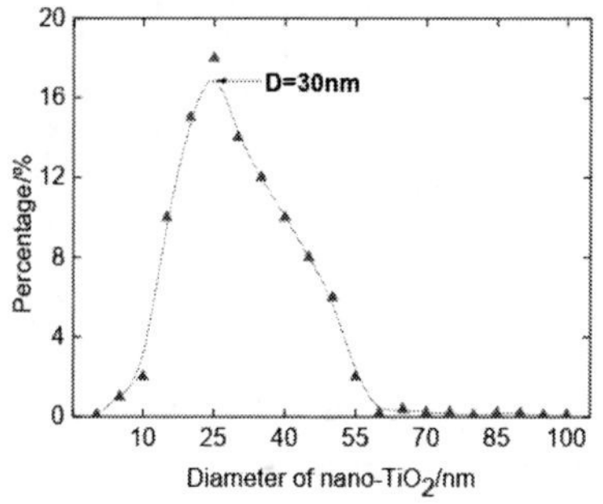


**Fig. 1** Particles-size distribution of nano- $TiO_2$

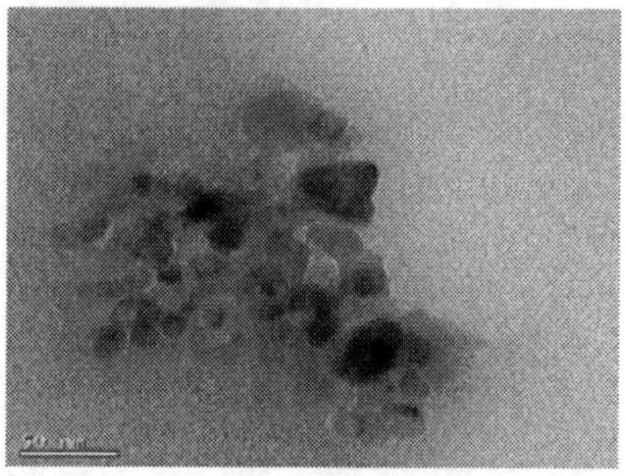

**Fig. 2** TEM images of $TO_2$ nanoparticles

TEM image of nanofluids further demonstrates this phenomenon as presented in Fig. 2. It shows that there are still several agglomerates in the sample, indicates that nanoparticles in the deionized water are difficult to be broken into primary particles completely with the absence of dispersant. However, all nanofluids are synthesized without the participation of dispersant, in addition, there are not obvious agglomeration appear. Thus, it is still effective to investigate the rheological behavior and enhanced viscosity mechanism.

### *3.2. Analysis of rheological behavior*

As one of the most important rheological measurement, dynamic viscosity curve may evaluate the rheological behavior of the fluid intuitively and visually. The dynamic viscosity versus shear rate of $TiO_2$ nanofluids at different volume fraction are shown in Fig. 3. Evidently, the viscosity of nanofluids is little affected by shear rate when volume fraction of $TiO_2$ nanoparticles is lower than 1 vol.%, indicating relatively good Newtonian fluids properties. As the $TiO_2$ volume fraction increases, non-Newtonian behavior (shear thinning) begins to dominate. The phenomenon of shear shinning occurs for higher volume fraction, possibly due to realigning along the flow direction of nanoparticles under the imposed shear stress. An interesting phenomenon is found in Fig. 3, when nanoparticles loading is above 1 vol.%. The dynamic viscosity keeps approximately constant at a higher shear rate from 200 to 1000 $s^{-1}$, acts as Newton behavior. But it decreases obviously when shear rate is lower than 200 $s^{-1}$, behaves as evident shear thinning phenomenon. Previous study of SiC in ethylene glycol nanofluids also confirmed this phenomenon [7].

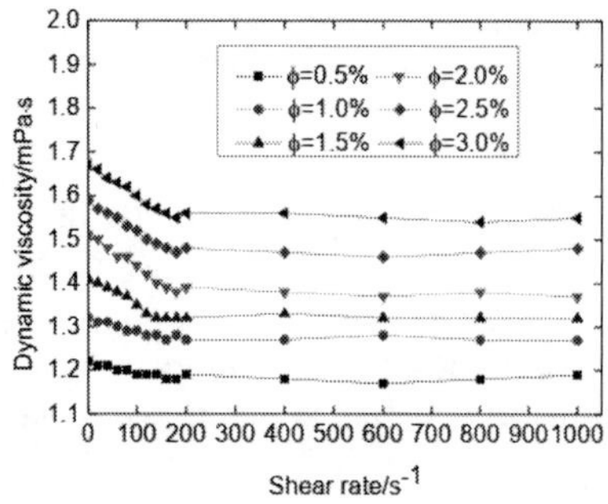


**Fig. 3.** Viscosity of nanofluids as a function of shear rate

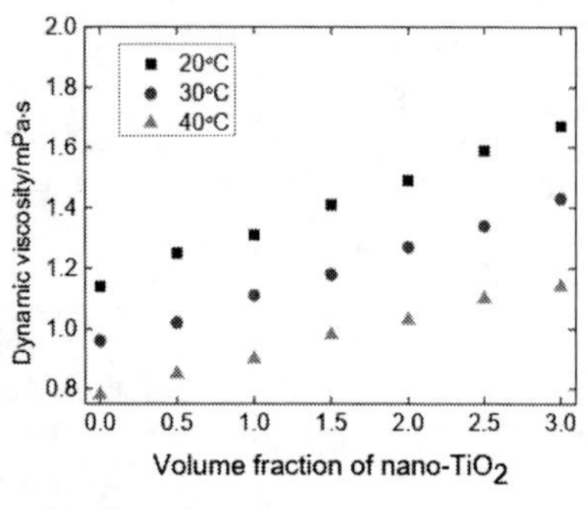


**Fig. 4** Dynamic viscosity of nanofluids as a function of volume fractions

Temperature is one of the important factors that affect the viscosity of nanofluids. Fig. 4 illustrates the viscosity of $TiO_2$ nanofluids as a function of temperature and volume fraction of nanoparticles. It's observed a decrease in viscosity with increased temperature, as the increasing temperature weakens the intermolecular forces of the particles and base fluid itself. For example, dynamic viscosity of nanofluids containning 0.5 vol.% nanoparticles decreases 32% while temperature changed from 20°C to 40°C. Similar trends have been found in several other nanofluids as well. Fig. 4 also represents the viscosity of nanofluids with different solid loading. Due to the addition of nanoparticles, the viscosity of

IOP Conf. Series: Materials Science and Engineering **772** (2020) 012031 doi:10.1088/1757-899X/772/1/012031

nanofluids is higher than that of base fluid, furthermore, as the nanoparticles loading increase, the viscosity of nanofluids is also increased.

If the nanofluids viscosity is expressed by relative viscosity, which is defined as $\eta_{nf}/\eta_{bf}$, relevant data can be concluded in Fig. 5. It represents that the increase in viscosity is a function of particle loading but little dependent of temperature. Compared with the prediction by the well-known Einstein equation and Brinkman equation. It can be seen that both the Einstein and Brinkman equation greatly underestimated the value of viscosity of nanofluids at all volume fraction. Most scholars attributed it to the addition of surfactant or pH value. Anyway, no surfactant was involved in this experiment, pH dependent mechanism also maintains obscure. Thus, some deep connection is urgent to be explored.

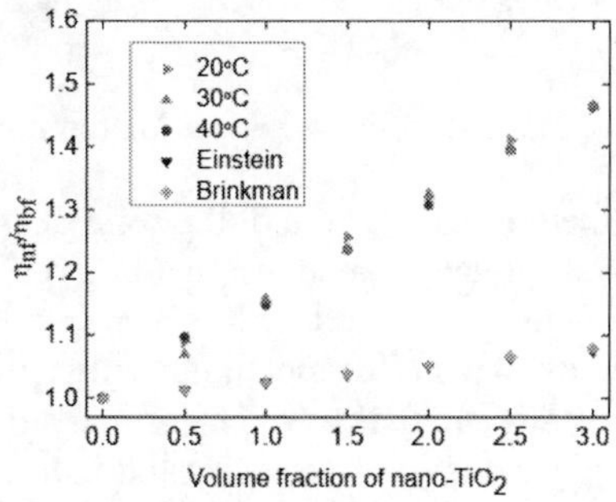


**Fig. 5** Viscosity enhancement of nanofluids as a function of volume fraction of nanoparitcles at given shear rates

### *3.3. Electroviscous effect of nanofluids*

Laminar flow in different conditions can be simplified by Fig. 6. As a pure liquid, laminar flow is only determinate by external factors such as shear rate. When solid particles are added, as shown in Fig. 6b, assume that these particles are neutral in the fluid, they will exert a distortion of laminar flow. In most cases, solid particles in fluid are always charged, thus an extra distortion appears (as Fig.6c), the distortion due to EDL will cause an additional energy dissipation and thus result in further increase in the viscosity. This effect is called electroviscous effect and first considered by Smoluchowski [8]. It can be written as Eqn. (1).

$$\frac{\eta}{\eta_0} = [1 + (2.5 + P)\phi] \tag{1}$$

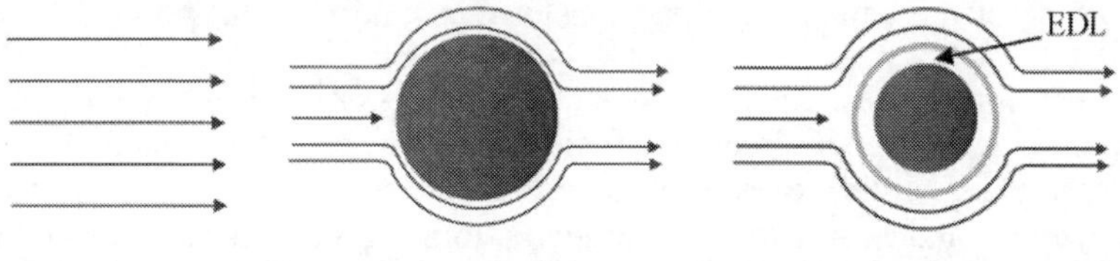


**Fig. 6** Laminar flow sketch of (a) pure fluid; (b) neutral particle in fluid; (c) charged particle in fluid.

Where P is the primary electroviscous coefficient. Obviously, it is not sensible to calculate the sum of all ions. Anyway, measurement of Zeta potential (also called $\zeta$potential) provides a new idea to evaluate the electroviscous effect since the $\kappa^{-1}$ depends on $\zeta$potential in a large extent. It is supposed that at IEP surface charge is completely neutralized or screened and no EDL exist around the particles. There is same distribution at particles surface and into the bulk. Thus, P is declined to 0. Zeta potential of nanofluids with different pH value is analyzed to confirm the IEP, the results are shown in Fig. 7. It illustrates a decrease in Zeta potential with the increased pH, and when pH value is 5.5, Zeta potential declines to 0. Evidently, the IEP of $TiO_2$ nanofluids is pH 5.5.

IOP Conf. Series: Materials Science and Engineering **772** (2020) 012031 doi:10.1088/1757-899X/772/1/012031

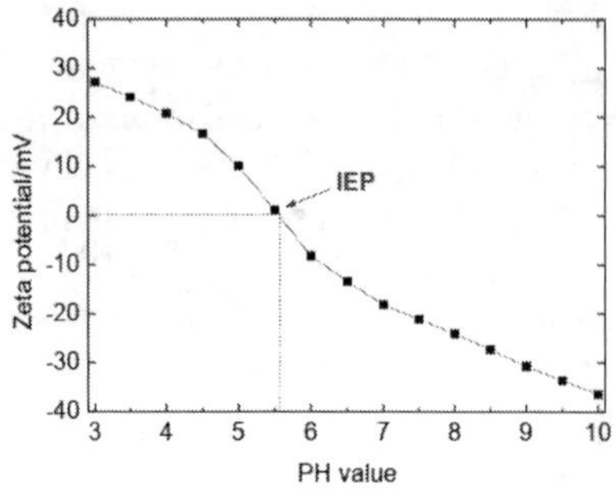


**Fig. 7** Zeta potential of nanofluids at different pH value

Viscosity of nanofluids versus the volume fraction of the solid phase at 20°C is plotted in Fig. 8, for different pH values. Approximatively linear dependence is observed at all pH values, indicating that only primary electroviscous effect is taken into action. It is also found that the slope is large when pH value is far away from the 5.5, possibly due to Zeta potential is higher in that case and consequently the primary electroviscous effect will be more remarkable. The primary electroviscous coefficient at all pH values can be obtained by calculating the ratio of viscosity and volume fraction of solid particles. The experimental results were made a comparison with Booth's model [9]. And they are shown in Fig. 9. As can be seen, there is qualitative agreement between experimental and theoretical model, while the quantitative agreement is poor. Evidently, primary electroviscous effect is severely underestimated by Booth's theory. Attempt is made to explain this disagreement.

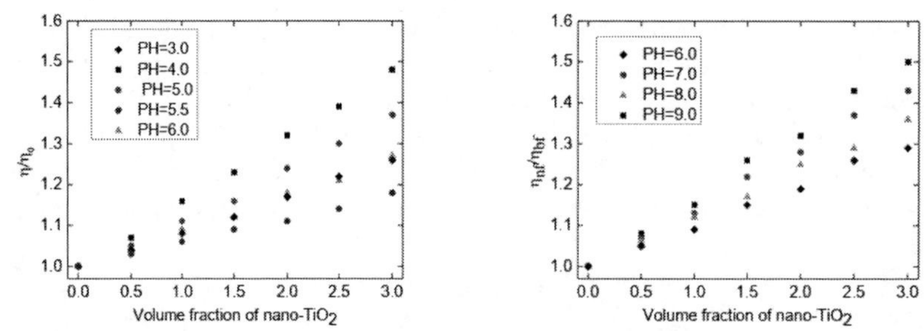


**Fig. 8** Relative viscosity as a function of particles loading at different pH (a) from 3 to 6 (b) from 6 to 9

It is plausible to predict that electroviscous effect is negligible in the vicinity of isoelectric point (pH=5.5), thus the primary electroviscous coefficient should be zero in this point. However, it is not the case in our experiment, as the experiment value P far exceeds to zero. The primary cause may be adsorption of hydrone on the surface of nanoparticles, forming a firm adsorbed film, which consists of an entity with nanoparticles. Volume fraction of nanoparticles is thus increased by this phenomenon indirectly. Besides, some other factors, for examples, instrumental error, also exert an extra influence. However, these factors exist in all case of our experiment. Assume that influence of these factors is constant, actual primary electroviscous coefficient curves should be shifted downwards until the value in pH 5.5 is Zero. As the red line shown in Fig. 9, surprising consensus can be found between modified experimental values with Booth's theory.

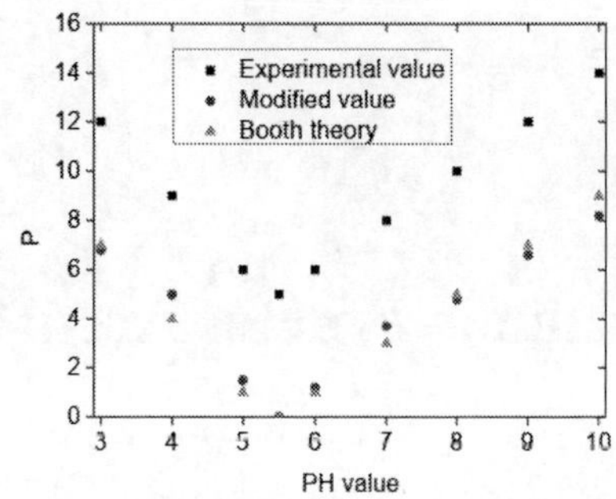


**Fig. 9** Comparison of P between experimental and Booth theory at different pH

## 4. Conclusion

Rheologic behavior of $TiO_2$ nanofluids is determinate by the concentration of $TiO_2$ in base fluid, and there is a transition from Newtonian to Non-Newtonian behavior when the concentration is continuously increased.

Viscosity of $TiO_2$ nanofluids shows strong nanoparitcles content and temperature dependence. It was increased with increasing particles loading and decreased temperature.

There is a huge deviation between experiment value and well-known viscosity model. This deviation is mostly due to electroviscous effect of nanoparticles in base fluids. The primary electroviscous coefficient is evaluated by both experiment and theory calculation. Modified primary electroviscous coefficient is in good agreement with Booth's theory.

**Acknowledgments**

The present study is financially supported by the National Natural Science Foundation of China (51609097)

## References

[1] Jiang H, Li H, Zan C, et al. Temperature dependence of the stability and thermal conductivity of an oil-based nanofluid[J]. Thermochimica Acta, 2014, 579(5):27-30.

[2] Preeti Sati, Rashmi Chandrabhan, Shende, S.Ramaprabhu. An experimental study on thermal conductivity enhancement of DI water-EG based ZnO (CuO)/graphene wrapped carbon nanotubes nanofluids. Thermochimica Acta, 2018, 666(10):75-81.

[3] C T Nguyen, F Desgranges, G Roy, et al. Temperature and particle-size dependent viscosity data for water-based nanofluids – Hysteresis phenomenon. Int. J. Heat Fl., 28(2007) 1492-1506.

[4] SQ Zhou, R Ni, D Funfschilling. Effects of shear rate and temperature on viscosity of alumina polyalphaolefins nanofluids. J. Appl. Phys., 107(2010) 054317-054317-6.

[5] M Baratpour, A Karimipour, M Afrand, et al. Effects of temperature and concentration on the viscosity of nanofluids made of single-wall carbon nanotubes in ethylene glycol. Int. Commun. Heat Mass Transf., 74(2016)108-113.

[6] F J Rubiohernández, A I Gómezmerino, E Ruizreina, et al. An experimental test of Booth's primary electroviscous effect theory. J. Colloid Interf. Sci., 255 (2002) 208.

[7] X Li, C Zou, T Wang, et al. Rheological behavior of ethylene glycol-based SiC nanofluids. Int. J. Heat Mass Transf. 84(2015)925-930.

[8] M V Smoluchowski. Theoretische Bemerkungen über die Viskosität der Kolloide. Colloid Polymer Sci., 18(1916) 190-195.

[9] F Booth. The Cataphoresis of Spherical, Solid Non-Conducting Particles in a Symmetrical Electrolyte. P. Roy. Soc. A , 203(1950) 514-533.

# Determination of Arctiin in Fructus Arctii by Capillary Electrophoresis

**Xiaojing Zhou, Haixing Liu*, Lintong Wang**

Chemistry & Chemical and Environmental Engineering College, Weifang University, Weifang 261061, P.R. China

* Corresponding author e-mail: haixingliu@tom.com

**Abstract.** This paper investigated the determination of arctiin content in Fructus Arctii by high performance capillary electrophoresis (HPCE) method. The borax solution of 37.5 mmol concentration containing 12.5% methanol was chosen as buffer solution. The experiment was performed at a constant voltage of 16 kV and UV detection wavelength of 277 nm. The content of arctiin in Fructus Arctii was 26.45mg/g (RSD=3.34%) (n=7). The recovery was in the range of 87.6%-111.8% (n=5). This method is suitable for the detection of the content of arctiin in Fructus Arctii.

## 1. Introduction

Fructus Arctii is the composite family Arctium Lappa L. dry ripe fruit, pungent, bitter and cold. It is the function of detoxification, detumescence and dispelling wind-heat. It is used treatment of wind-heat type common cold, lung heat cough, sore throat, measles, erysipelas and abscess malignant sore. The Fructus Arctii extractive include lignans, volatile oil, fatty acid, terpenoids, etc. Zhang et al [1] studied the extraction and purification process of total lignans from Arctium lappa L. and to measure the antibacterial activity of total lignans in Arctium lappa L.. Using Arctiin as an indicator, the best extraction process of total Lignan from Arctium lappa L. was investigated by single factor and orthogonal experiments. The purification of the active constituents of Arctium lappa L. was tested by macroporous adsorption resin. The bacteriostasis experiment was performed on the active constituents of the Arctium lappa L. before and after purification by the Oxford Cup method. The best extraction process of Arctium lappa L. was to extract 80 times with 80% ethanol as extraction solvent, and add 12 times of alcohol, 1.5 hours each time. The best purification process was AB-8 type macroporous adsorption resin loading column, 0.05g/mL loading concentration, 6BV distilled water and 5BV 70% ethanol eluting, 70% ethanol eluent eluting part of Arctium lappa L. total wood Lipoprotein. The total lignans of Arctium lappa L. had antibacterial activity against Staphylococcus aureus and Streptococcus pneumoniae. The essential oil from Arctium lappa L. produced in Yunnan was extracted by Hu et al [2] using simultaneous distillation extraction method and its chemical compounds were studied by Gas Chromatograph-mass spectrometry with rate of the carrier gas of 1.0 mL/min. The injection method was split-flow and the ratio of split-flow was 20:1. The temperature of injection port was 280°C and the sample size was 1 μL. The temperature of the programmed route increased. According to computer retrieval by the spectrum analysis, the structures of 89 components (97.55% of the total contents) were identified. The ultrasound-assisted extraction was used by Wang et al [3] to extract arctiin from Frutus arctii, and the effects of particle size, solvent/sample ratio, extraction time, ultrasound power and

Content from this work may be used under the terms of the Creative Commons Attribution 3.0 licence. Any further distribution of this work must maintain attribution to the author(s) and the title of the work, journal citation and DOI.
Published under licence by IOP Publishing Ltd

extraction temperature on the concentration of arctiin were studied. According to the experimental investigated and the Fick's first law, a kinetic model was obtained to describe the ultrasound-assisted extraction process and the calculated values were in good agreement with the experimental data. Xu et al [4] studied extractive of fructus arctii. The fat soluble compounds in the extractive of fructus arctii were removed by ethyl acetate, the aqueous phase drying was isolated by preparative HPLC, and six components were isolated from the extractive of fructus arctii and their structures were identified by physicochemical property analysis, specially by spectral analysis. Their structures were identified as 3,4-dihydroxy cinnamic acid-4-O-β-D-glucoside, 3-ocaffeoylquinic acid, 3,4-dihydroxy cinnamic acid, 3,5-dicaffeoylquinic acid, 1,5-dicaffeoylquinic acid and 3,4-dicaffeoylquinic acid. Fu et al [5] established a HPLC method for the determining Arctiin, Arctigenin and Chlorogenic acid content in Arctium lappa L. and to make a comparison of content of this three components among different parts as well different time that Arctii Fructus fried. The analysis was performed on Welchrom C18 (5μm, 4.6mm × 250mm) column with mobile phase composed of acetonitrile-0.1% phosphoric and flow rate of 1.0mL/min at gradient elution. The detection wavelength was 280nm and the column temperature was 30°C. The UV and IR spectra of petroleum ether, chloroform and ethanol extracts of Fructus Arctii from different places were measured by Shao et al [6]. It was showed that the UV and IR spectra of the same solvent extract of Fructus Arctii from different places had good similarity and characteristics common peak and the UV and IR spectra of different solvent extracts had significant differences. The ultrasonic-assisted extraction was explored by Zhang et al [7] as an alternative technology for the extraction of arctigenin from the hydrolyzed powder of Fructus Arctii. The experimental conditions were investigated by a central composite design, including extraction temperature, ethanol concentration and solvent to solid ratio. The optimum conditions were obtained that the temperature was 52.8°C and ethanol concentration was 80% and solvent to solid ratio was 40 mL/g. Qiao et al [8] established HPLC fingerprints of crude and processed Arctii Fructus, analyzed and compared their chemical constituents, and finally revealed the changes of their chemical constituents after processing of Arctii Fructus. The mobile phase was composed of acetonitrile - 0.2% aqueous formic acid with flow rate of 1.0 ml/min at gradient elution. The column temperature was 30 °C. The detection wavelength was 286 nm. The established chromatographic fingerprints of crude and processed Arctii Fructus were evaluated by similarity analysis and hierarchical cluster analysis. Optimization of the ultrasonic-assisted hot water extraction of polysaccharides from Fructus arctii was investigated by Yu et al [9]. According to single-factor experiments, three independent factors of ratio of water to raw material, extraction time and extraction temperature were tested during extraction. The experiments were arranged by Box-Behnken central composite experiment design. Response surface analysis method was adopted to measure the effect of central composite experiment design. The antioxidant activities of polysaccharides were evaluated by DPPH radical scavenging activity (DPPH·), hydroxyl radical scavenging (·OH), superoxide radical scavenging activity ($O_2^-$) and reducing power. In this paper, the Arctiin content in Fructus Arctii was determined by High Performance Capillary Electrophoresis.

## 2. Experimental section

### *2.1. Instruments and Reagents*

Experimental instruments: CL-1030-type high performance capillary electrophoresis (Beijing Cailu Scientific Instrument Co., Ltd.); HW2000-type chromatography workstation (Nanjing Qianpu Software Ltd.); Capillary (75 μm inner diameter, 52 cm overall length, 44 cm effective length) from Hebei Yongnian Ruifeng Chromatographic Devices Co., Ltd.).

Arctiin (Chinese Drugs and Biological Products); Fructus Arctii (purchase in weifang pharmacy); other reagents used in the experiments were all analytical grade; Double-distilled water was used.

### 2.2. *Experimental Methods*
Before the start of the experiment, capillary was successively washed with 1 mol·L$^{-1}$ hydrochloric acid solution, double-distilled water, 1 mol·L$^{-1}$ sodium hydroxide solution, double-distilled water, buffer solution, each for 5 min. After three times running, capillary was cleaned again using the above method.

Measurements were carried out at 16 kV voltage and experimental temperature at 21°C. UV detection wavelength was 277 nm. Injection time was 10s (7.5 cm height difference).

### 2.3. *Sample Preparation*
Fructus Arctii sample solution: Fructus Arctii was accurately weighed 2.1516 g, added 40 mL water containing 80% methanol, extracted time of 48h at 21°C, filtered, washed and set the volume to 50 mL that was the Fructus Arctii sample solution.

Arctiin standard solution: Arctiin was accurately weighed 0.0046 g, added 2 mL water.

## 3. Results and Discussion

### 3.1. *Selection electrophoresis conditions*
The experiment was carried out at 16 kV voltage. UV detection wavelength was 277 nm.

Based on past experiment experience, 37.5 mmol/L borax solution containing 12.5% methanol was chosen as electrolyte solution.

### 3.2. *Quantitative analysis*

*3.2.1. Standard curve.* First, arctiin standard solution was prepared and its concentrations were 2.3, 1.15, 0.575, 0.2875, 0.1437, 0.0718, 0.0359 mg/mL. Each standard solution was run for three times under the above electrophoresis conditions and the results averaged. The chromatogram of arctiin standard solution was showed in Figure 1. Taking concentration as the abscissa and peak area as the ordinate, the standard curve was drew. Linear regression equation of arctiin (peak area: y μV•s, density: x mg/mL) and the linear range was as follows: y= -2653.6+84674.5x(r=0.998), 0.0359-2.3 mg/mL.

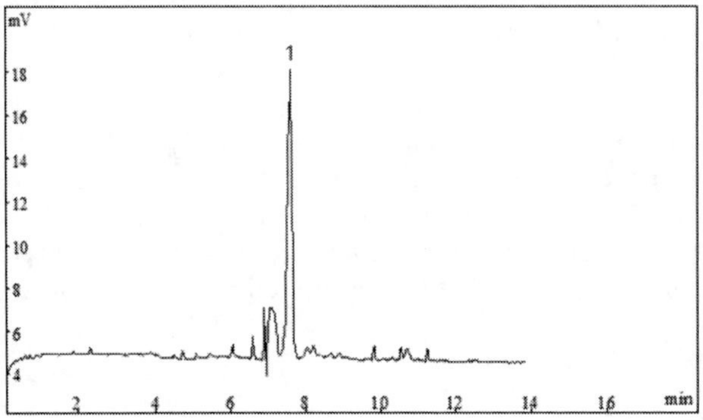


**Fig. 1** Electrophorogram of arctiin standard solution 1-arctiin

*3.2.2. Precision test.* A arctiin standard solution precisely drew and continuously injected for six times under electrophoretic separation conditions, the RSD of arctiin migration time and peak area were 2.25% and 4.91%, indicating good precision.

*3.2.3. Determination of sample content.* Under selected electrophoresis conditions, Fructus Arctii sample solution was run. Separation chromatogram of the Fructus Arctii sample solution was showed in Figure 2. Measured arctiin content in Fructus Arctii was 26.45mg/g(RSD=3.34%)(n=7).

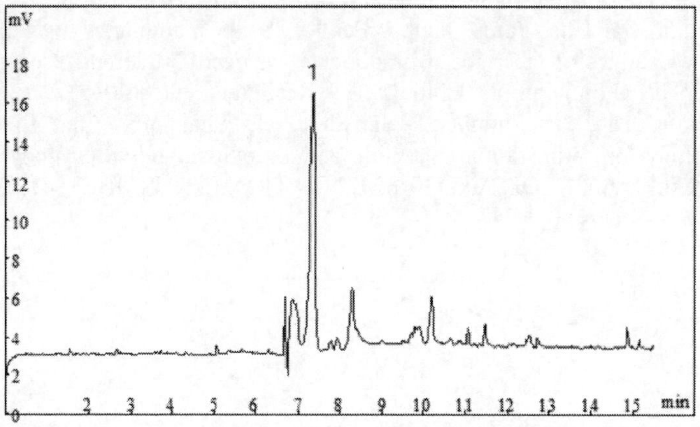


**Fig. 2** Electrophorogram of Fructus Arctii sample solution 1-arctiin

*3.2.4. Recovery.* After determination for five times, the recovery of arctiin in Fructus Arctii sample was in the range of 87.6%-111.8% (n=5). The average recovery was 98.9%.

## 4. Conclusion

This paper investigated the determination of arctiin content in Fructus Arctii by high performance capillary electrophoresis method. Measured arctiin content in Fructus Arctii was 26.45mg/g (RSD=3.34%) (n=7).

**Acknowledgments**

This study were supported by the Natural Science Foundation of Shandong Province (No. ZR2010BL025), Open Project of State Key Laboratory of Supramolecular Structure and Materials (No. sklssm201323) (Jilin University), State Key Laboratory of Inorganic Synthesis and Preparative Chemistry (No. 2011-13) (Jilin University).

## References

[1] Zhang Limin, Zhao Lin, Han Xiao, Zhou Liping, Extraction and purification of total Lignans from Arctium lappa L. and its in vitro antibacterial activity before and after purification, Chemical Engineer, 279(12)(2018), 82-86.

[2] Hu Shanshan, Liang Mengjie, Xu Xingmeng, Bin Jie, Wu Shuangfeng, Lin Yanjun, Guo Yadong, Deng Liang, GC-MS Analysis on the Chemical Components of Arctium Lappa L. Essential Oil by Simultaneous Distillation Extraction, Journal of Kunming University, 40 (3) (2018), 95-99.

[3] Wang Xinyi, Wu Juanxia, Huang Dechun, A Kinetic Model for Ultrasound - Assisted Extraction of Arctiin from Frutus Arctii, Chemical Industry Times, 31(11)(2017), 1-5.

[4] Xu Xiuying, Chen Haiqin, Zheng Yimin, Zhou Zaifu, Isolation and identification of caffeuylquinic acid compounds extractive of fructus arctii, Journal of Southwest University, 39 (11) (2017), 172-176.

[5] Fu Lin, Gu Rui, ZeWeng Yongzhong, TianXi Gele, Simultaneous Determinate the Content of Arctiin, Arctigenin and Chlorogenic acid in Arctium lappa L. by HPLC, Chinese Journal of Ethnomedicine and Ethnopharmacy, 26(11)(2017), 39-42.

[6] Shao Jing, Wen Xiyan, Wang Lanxia, Guo Mei, Study on IR, UV Spectrum of Different Solvent Extract of Fructus Arcti, Journal of Anhui Agri. Sci., 44(7)(2016), 147-150.

[7] Zhang Yong, Wang Xinyi, Gao Zhaohua, Wang Qian, Wang Zhixiang, Optimization of Ultrasonic-Assisted Extraction of Arctigenin from Hydrolyzed Fructus Arctii by Response

Surface Methodology, Chemical Industry Times, 30(3)(2016), 17-20.
[8] Qiao Fengxian, Cai Hao, Fang Qianbo, Pei Ke, Song Xiaoqing, Fan Kailei, Cai Baochang, Study on Changes of Chemical Components Between Crude and Processed Arctii Fructus Based on HPLC Fingerprints, Chin. Med. J. Res. Prac., 29(5)(2015), 9-13.
[9] Yu Jun, Wang Tao, Jia Chunhong, Sun Jiye, Hu Shangqin, Zang Li, Response surface methodology for optimization extraction and antioxidant activities study of polysaccharides from Fructus arctii, FOOD AND FERMENTATION INDUSTRIES, 41(6)(2015), 207-211.

# The promotional effect of sodium chloride on thermophysical properties of nitrate

**Chenge Wang, Yan Li *, Gouyou You, Qunzhi Zhu**
Shanghai University of Electric Power, Shanghai, China

*Corresponding author e-mail: liyandy2000@sina.com

**Abstract.** Nitrate are the excellent phase change materials (PCMs) for thermal energy storage. Nitrate has no phase separation during phase transition and has good chemical stability. The present study concentrates on improving the latent heat of phase change and of heat absorption rate by adding various proportions NaCl into the nitrate/mixed nitrate. Sodium nitrate and sodium chloride were selected to prepare binary mixture by static melting method. The melting point and latent heat of phase change were measured by differential scanning calorimetry (DSC). Compared with pure sodium nitrate, the melting point and solidification point of binary molten salt containing 5% NaCl are reduced by 9℃ and 16℃, respectively. On the other hand, the latent heat of melting and solidification increased by 12J/g and 10J/g, respectively.

## 1. Introduction

The International Energy Agency, based on the latest global Energy consumption projections, predicts that global demand for Energy will continue to increase over the next 20 to 30 years [1]. Solar thermal power generation is an important aspect of solar thermal utilization, which is regarded as the most promising generator of renewable energy equipment [2-4]. Among many inorganic salts, nitrate has excellent heat transfer and fluidity. Its melting point is about 300℃, and the upper limit temperature is about 600℃. When it exceeds 600℃, it will decompose into other substances. At present, the widely used molten salt still has the problems of high melting point, poor thermal stability and high risk of system freezing and plugging, which cannot meet the requirements of advanced solar thermal storage. Therefore, the development of new mixed molten salt with better stability, lower melting point and wider temperature range has become the main research direction and hot spot.

The chloride system has a wide variety of applicable temperature range (150-900℃) [5]. It has high latent heat and specific heat capacity, but high temperature chlorine salt has strong corrosiveness. Sodium chloride (NaCl) could be used as the additive to promote nucleation [6]. This paper aims to study the effect of adding NaCl particles on the thermal properties of binary nitrate and improve the performance of heat storage materials. The thermal properties of mixed molten salt were measured by DSC, and the phase and crystallization of the material were analysed by XRD. Through the experimental results, the performance of nitrate was improved, the application range of nitrate molten salt in phase change heat storage and solar power generation was widened and its reliability was improved.

Content from this work may be used under the terms of the Creative Commons Attribution 3.0 licence. Any further distribution of this work must maintain attribution to the author(s) and the title of the work, journal citation and DOI.
Published under licence by IOP Publishing Ltd

IOP Conf. Series: Materials Science and Engineering **772** (2020) 012033 doi:10.1088/1757-899X/772/1/012033

## 2. Experimental methods and materials

Sodium nitrate(melting point 306.8 ℃ ,latent heat 162.5J/g) and sodium chloride(melting point 801℃,latent heat 482J/g) are provided by Shanghai Feng Shun Fine Chemicals Co China.

### *2.1. Preparation of Materials*

Two kinds of pure salt were dried in a drying oven for 24h at a drying temperature of 120℃, and the crystal water in the sample was removed. Mixed molten salt was prepared by static melting method, sodium chloride solids with different mass fractions of 5%, 10% and 15% were added to pure sodium nitrate to prepare 4 kinds of mixed molten salts with different proportions. Place the 4 groups of mixed salt in a stain steel mortar and grind them until they are completely powdered. Put them into a corundum crucible, then place them in a muffer furnace to heat and melt. The heating time is 2h and the heating temperature is 400℃.After cooling, remove, grind and dry.

### *2.2. Method*

DSC synchronous thermal analyzer (METTLER TGA/DSC2 1600LF) was used to measure the melting point and phase change latent heat of the samples. Samples to be tested after drying were sealed in an aluminum crucible. Place crucible in instrument furnace, at the rate of 10 K/min, the temperature rises from 50℃ to 400℃.The temperature is kept at 400℃for 10 min. Then the temperature is cooled from 400℃to 50℃, and the cooling rate is 10 K/min. Nitrogen was used as experimental protective gas during the measurement, and the flow rate was 25 ml/min.

## 3. Results

### *3.1. Thermal performances of NaCl with various addition proportion*

The DSC curves of samples mixture with various weight rates of NaCl are shown in Fig. 1. It shows that both endothermic and exothermic peaks of the heat flow curve of binary molten salt decreased. The melting point, solidification point and latent heat of sodium nitrate and three mixed molten salts containing different mass fractions of NaCl are listed in Tables 1 and 2. The graph shows that the melting point and solidification point of sodium nitrate can be reduced by adding sodium chloride. However, there is no linear relationship between the amount of sodium chloride added and the value of increase/decrease.

Compared with pure sodium nitrate, the melting point and solidification point of binary molten salt containing 5% NaCl are reduced by 9℃and 16℃ , respectively. On the other hand, the latent heat of melting and solidification increased by 12J/g and 10J/g, respectively. However, the melting point and solidification point of the mixed molten salt decreased, the latent heat also decreased after adding higher mass fraction of NaCl. This is consistent with the known literature results. When adding a small amount of NaCl into PCMs, the latent heat tends to increase . It can be explained that the crystallization of the two salts is always imperfect due to natural reasons. An appropriate amount of NaCl can be uniformly dispersed in the interior of sodium nitrate as a nucleus, which promotes nucleation and contributes to the increase of latent heat. However, latent heat does not increase with the increase of NaCl. There is no linear relationship between them and there is an optimal eutectic ratio.

Fig.2a, b show the starting and ending temperatures of phase change during heating/cooling, and D-value represents the difference between the starting and ending points. The smaller the D-value is, the sharper the endothermic/exothermic peak is, which indicates that the faster the endothermic and exothermic rate of molten salt is, the better the performance of molten salt is. Supercooling is a common phenomenon in the crystallization process of molten salt. Metastable state near the melting point will affect the crystallization process. From Table 1 and 2, it can be seen that in order to increase the latent heat of phase transformation of nitrate, the supercooling degree of molten salt increases slightly. According to nucleation theory [7, 8], NaCl promotes the transformation of molten salts from homogeneous nuclei to heterogeneous nuclei, which improves the latent heat of phase transformation and shortens the time of phase transformation.

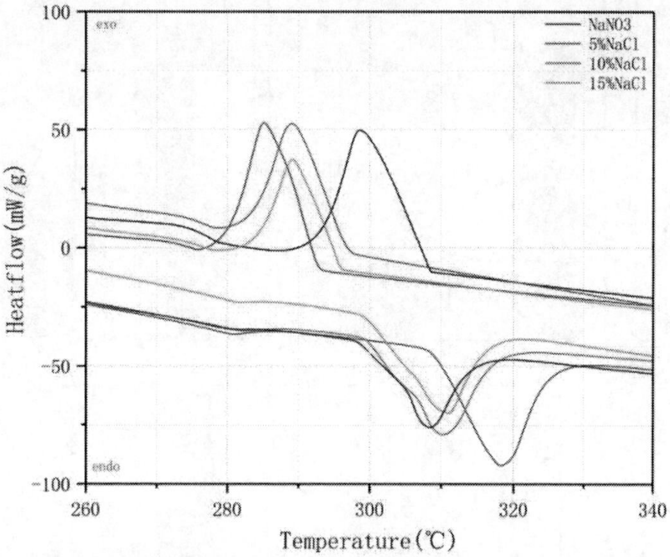


**Figure 1.** DSC curves of samples with various addition rates of NaCl

**Table 1.** Heat of fusion, melting temperature, peak temperature of sodium nitrate and mixed salt

| Heating process | | | | |
|---|---|---|---|---|
| Materials | NaNO3 | 0.05NaCl | 0.1NaCl | 0.15NaCl |
| Enthalpy(J/g) | 224.1 | 236.4 | 198.5 | 174.2 |
| Onset(°C) | 309.0 | 300.7 | 299.4 | 300.8 |
| Peak(°C) | 316.8 | 305.2 | 307.7 | 305.8 |
| End set(°C) | 324.5 | 313.1 | 313.9 | 312.4 |
| D-value | 15.5 | 12.4 | 14.6 | 11.6 |

**Table 2.** Heat of fusion, melting temperature, peak temperature of sodium nitrate and mixed salt

| Cooling process | | | | |
|---|---|---|---|---|
| Materials | NaNO3 | 0.05NaCl | 0.1NaCl | 0.15NaCl |
| Enthalpy(J/g) | 218.3 | 228.7 | 190.6 | 155.6 |
| Onset(°C) | 308.1 | 292.5 | 295.4 | 295.2 |
| Peak(°C) | 300.1 | 286.7 | 289.2 | 289.6 |
| End set(°C) | 293.5 | 280.7 | 282.8 | 284.0 |
| D-value | 14.5 | 11.8 | 12.6 | 11.2 |

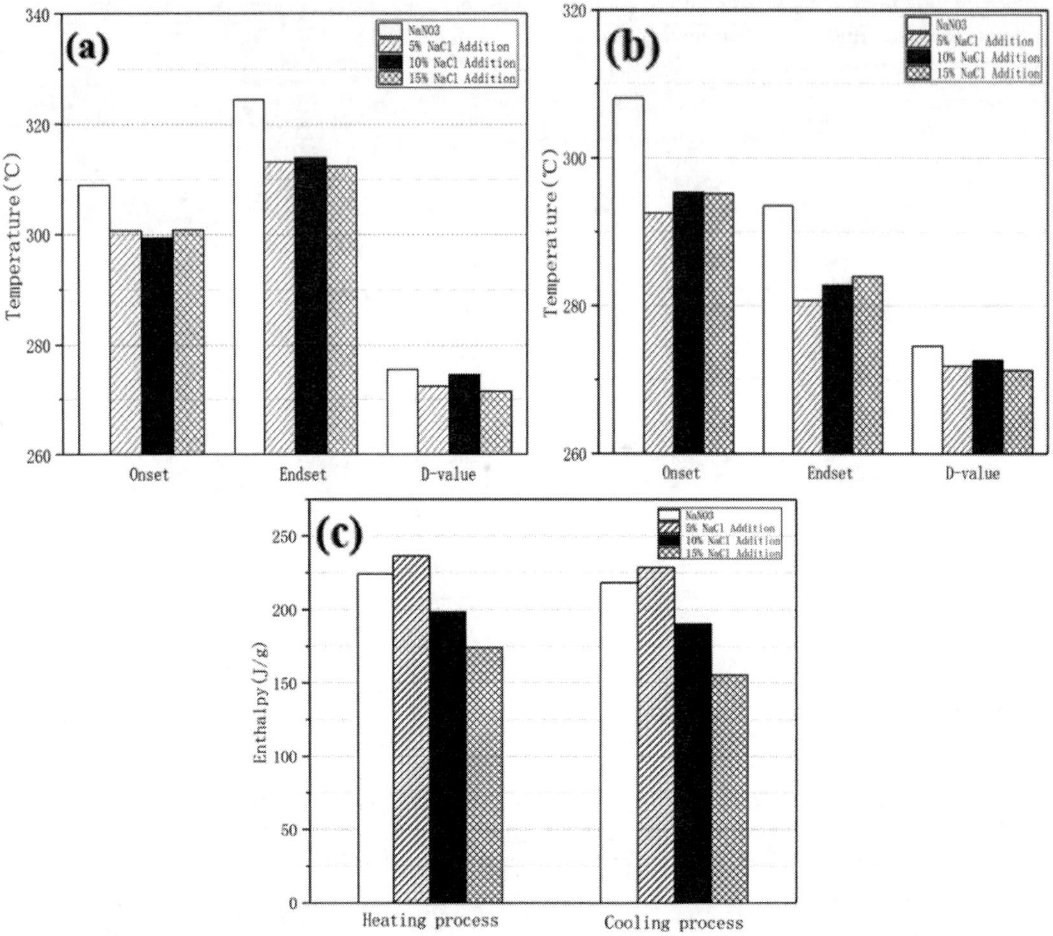


**Figure 2.** (a) phase change temperature during heating process (b) phase change temperature during cooling process (c) latent heat

### *3.2. X-ray diffraction*

Fig. 3 shows the XRD results of sodium nitrate, sodium chloride and mixed salts after five thermal cycles. The spectral peaks of mixed salt correspond to the positions of two pure salt peaks one by one, and there are no irregular peaks. This indicates that the two molten salts are physically mixed and no chemical reaction occurs to produce new substances. The peak strength of each component can indicate the relative crystalline strength of the crystal. Thermal and cold cycles have little effect on the final phase change of phase change materials, showing good thermal stability and chemical compatibility.

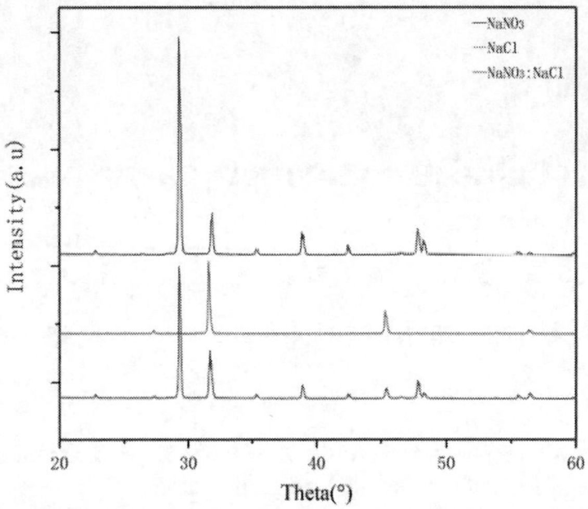


**Figure 3.** XRD pattern of Sodium Nitrate, Sodium Chloride and Mixed Salts

## 4. Conclusion

Appropriate sodium chloride particles can increase the latent heat of phase change of nitrate, accelerate its heat absorption and exothermic rate, and improve the performance of PCMs. However, there is no linear relationship between the amount of NaCl and its properties, and there is an optimum eutectic ratio between sodium chloride and sodium nitrate. The results show that eutectic can be achieved when the ratio of sodium chloride to sodium nitrate is less than 1:9. Compared with pure sodium nitrate, the melting point and solidification point of binary molten salt containing 5% NaCl are reduced by 9℃and 16℃,respectively. On the other hand, the latent heat of melting and solidification increased by 12J/g and 10J/g, respectively.

**Acknowledgments**

This work was financially supported by Shanghai Science and Technology Committee Project (18020501000) fund and National Natural Science Foundation of China (Contract No. 51576119) fund.

**References**

[1] Energy Information Administration.International energy outlook 2016[R].EIA,2016.

[2] Rahman S,Castro A D.Environmental impact ofelectricity generation: A global perspectives [J],IEEE Trans Energy Conv, 1995, 10( 2): 307—313.

[3] HE W Y,YAN Q Y.Molten Salt as a Phase Change Material,Materials reports, 2015, 29: 128-140.

[4] Peng Q,Ding J,Wei X L,et al.The preparation and properties of multi-component molten salts [J]. Apply Energy, 2010, 87(6): 2812.

[5] HE M Z,YANG L W, ZHANG Z T. Supercooling characteristics of inorganic phase change material CaCl2·6H2O.CIESC Journal 2017, 68(11): 4016-4024.

[6] E. Oró, C. Barreneche, M.M. Farid, L.F. Cabeza, Experimental study on the selection of phase change materials for low temperature applications, Renew.Energy 57 (2013) 130–136

[7] M.G. Broadhurst, Extrapolation of the orthorhombic n - paraffin melting properties to very long chain lengths, J. Chem. Phys. 36 (1962) 2578 - 2582.

[8] J. Xu, Y. Ma, W. Hu, M. Rehahn, G. Reiter, Cloning polymer single crystals through self-seeding, Nat. Mater. 8 (2009) 348–353.

# Study on Controllable growth of graphene prepared by CVD

**Xin Zhou**

School of Management, Tianjin University of Technology, Tianjin 300384, China

1228313436@qq.com

**Abstract.** Graphene is a two-dimensional honeycomb crystal of monoatomic thickness composed of hexagonal sp2 hybrid carbon atoms. Since the mechanical separation of Andre Gamm and Konstantin Novoselov in 2004 to obtain a single layer of graphene, this unique two-dimensional structure makes graphene excellent physical, chemical, mechanical, electrical and other properties. It has become the focus of much research in recent years. These properties include exceptionally high charge, high mechanical strength and elasticity, optical clarity and a variety of possible chemical modifications that make graphene promising in micro-nanoelectronics, photodetection and conversion materials, structural and functionally enhanced composites and storage. It can be applied in a wide range of fields, but graphene is different from graphite in nature, and its size is small and the yield is low. So how to prepare high quality graphene is the key to study its application prospects. The emergence of chemical vapor deposition (CVD) technology solves these problems and has evolved into an important method for preparing large-area, large-size, multi-purpose graphene. Large-sized graphite thin single crystals can be obtained on a substrate by controlling the growth mode and different treatments for the substrate. Due to the strong interaction between carbon atoms and the substrate, there are still many difficulties and challenges in directly growing graphite on the insulating substrate. This topic aims to prepare graphene by chemical vapor deposition (CVD) and prepare it. The method is improved to obtain graphene with excellent performance, which provides more feasibility for the practical application of graphite thinning in semiconductor devices.

**Keywords:** graphene, chemical vapor deposition, method control, substrate.

## 1. Graphene overview and characteristics

Graphene, also known as "single-layer graphite sheet", refers to a layer of dense carbon atoms encased in a honeycomb crystal lattice. The carbon atoms are arranged in a two-dimensional structure similar to the graphite monoatomic layer. A new carbon material with a two-dimensional honeycomb lattice structure tightly wrapped by carbon atoms, which can be stacked to form three layers of graphite, which can be crimped to form one-dimensional carbon nanotubes or decomposed into zero-dimensional fullerenes. Early studies have shown that this single-layer atom cannot achieve thermodynamic stability. However, in 2004, the experimental team led by Novoslov, Manchester University, UK, used a simple tape to highly exfoliate graphite, which was used to experiment. The independent two-dimensional graphene crystals studied and won the Nobel Prize in Physics in 2010. This achievement set off a wave

Content from this work may be used under the terms of the Creative Commons Attribution 3.0 licence. Any further distribution of this work must maintain attribution to the author(s) and the title of the work, journal citation and DOI.
Published under licence by IOP Publishing Ltd

IOP Conf. Series: Materials Science and Engineering **772** (2020) 012034 doi:10.1088/1757-899X/772/1/012034

of scholars to study graphene, and various valuable properties of graphene have been discovered successively. Graphene materials have gradually entered the field of industrial production.

## 2. Preparation method of graphene

The first condition for studying graphite thinning is to prepare raw materials. At present, the methods for preparing graphene are mainly divided into physical methods and chemical methods. The physical method refers to a method for preparing graphene using low-cost graphite as a raw material, including a mechanical stripping method, a liquid phase or vapor phase stripping method, an orientation epitaxy method, and the like. Chemical methods include chemical vapor deposition, redox, epitaxial growth, and the like. The chemical method has the advantage of being able to be applied to production on a large scale, and is the main method for preparing graphene so far.

## 3. The research content, methods and means of selection

Although graphene has excellent performance, its performance is often not maximized due to limitations in preparation methods and practical conditions. Even though the specific surface area of graphene is large, chemical conversion graphene (CCG) is liable to be closely stacked due to the $\pi$-$\pi$ bond interaction, and the specific surface area is lost. The main research content of this paper is the controllable preparation of the small layer of graphene material, and some properties of it are simply studied. The micro-mechanical stripping method is used to strip high-oriented pyrolytic graphite into high-quality graphene, even single-layer graphene, and in order to make the produced graphene larger, the silicon wafer is heated in this experiment. Thereafter, the number of layers and the area of the produced graphene were initially judged by an optical microscope, and the number of layers of graphene and some properties were further determined according to Raman spectroscopy. In addition, a series of studies and characterization of graphene strain properties have been carried out.

## 4. Experiment

### *4.1. Experimental instruments and equipment*

Ultrasonic cleaning machine, CVD tube furnace, optical microscope (OM), optical microscope (OM), Raman spectrometer, small ion sputtering instrument.

*4.1.1. Preparation of grapheme.* Cut 25μm thick copper foil into 2cm×2cm, then flatten the copper foil in half, and roll it around. At this time, pay attention to the middle part, do not compact it, put the folded copper foil into the quartz boat, and then use the hook to push the boat. Push into the quartz tube and place it in the center of the furnace. At this time, do not stop the argon gas process and seal the furnace. Open the vacuum pump switch and heating switch, open the angle valve, and run the heating program when the vacuum is reduced to 1-2Pa (1 step, heating time 40 minutes, temperature 1030, hydrogen 10sccm, 2 steps 30 minutes, 1030 degrees, methane 10 scm, hydrogen 10 sccm, 3 steps, 0 degrees, methane 10 sccm, hydrogen 10 sccm). Sampling after cooling, pay attention to vacuum and then argon. The angle valve is closed, the argon gas is 1000sccm, and the vacuum should be heard at 1.13, and the sample should be heard.

*4.1.2. Transfer of grapheme.* After the copper foil sample was taken out, it was carefully spread out, flattened with a glass slide (pad paper towel), and then cut into small pieces of 5 × 5 mm. Spin coating was carried out with a PMMA solution (PMMA 40K or more, 30 g, 500 ml of anisole/toluene/chlorobenzene). Heat on the heating table at 150 degrees for 1 min, remove and mark (or dry in the air for 10 min). Then, the inner layer of the sample was placed in the ammonium persulfate solution upwards, and after 10 minutes, the back surface was rinsed with deionized water, and then placed in an ammonium persulfate solution, and rinsed again after 10 minutes, three times until the copper was completely etched, using a slide. The pmma membrane was transferred to deionized water for 30 min. Then use the cleaned silicon wafer to salvage the pmma membrane and control the moisture.

Here you can use a paper towel to absorb the moisture around the membrane. After the silicon wafer was naturally dried, it was placed in a vacuum drying oven at 150 degrees for 1 hour (30 minutes). Or placed on a heating station 150 degrees for 1-2 minutes. After the sample is taken out, it is placed in acetone, and after 10 minutes, the acetone is exchanged, so that after 3 consecutive times, it is immersed for 1-2 days for a long time. After the sample is taken out, it is immersed in absolute ethanol for 10 minutes, rinsed once with deionized water and then taken out. The transfer flow chart is shown in Figure1:

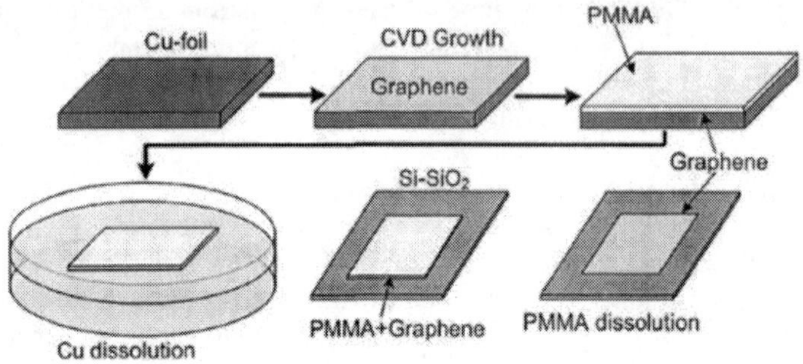


**Figure 1.** Transfer process of graphene prepared by CVD

*4.1.3. Effect of preparation methods on graphem.* In this experiment, a graphene sample was prepared according to a conventional method, and the number of sample layers obtained was large. Through many experiments, the method of heating the silicon wafer was finally selected to realize the controllable preparation of graphene. In the process of preparing graphene, the peeling of graphite was repeated several times at the beginning, and the peeling method was also improved, and the thin graphite was better adhered to the tape by the pressing of the ruler at the time of peeling. The treated silicon wafer was then placed in a box furnace at a temperature controlled at 100 ° C for 2-5 minutes. After the heating was completed, the silicon wafer was taken out and cooled to room temperature, and then the silicon wafer was stuck on a tape with graphene, and the obtained graphene was as shown in Fig. 2. It can be seen from the figure that the graphene produced by this method is not only controlled in the number of layers, but also the area is secured.

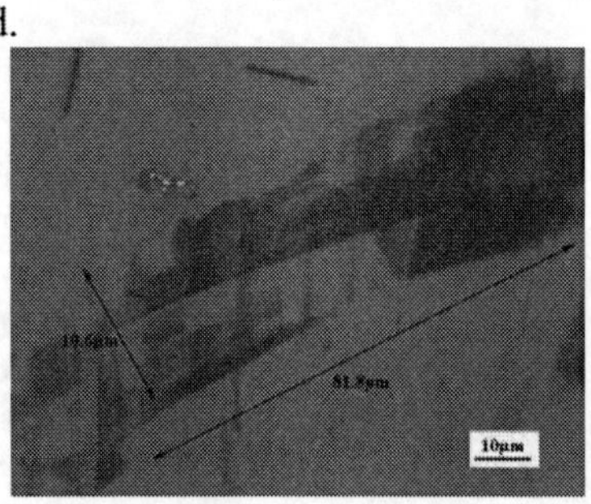


**Figure 2.** Photograph of graphene obtained after heat treatment of silicon wafer

*4.2. graphene characterization results*

*4.2.1. Optical microscope (OM) characterization results.* An OM photograph of graphene prepared under a silicon wafer substrate is shown in Fig. 3.

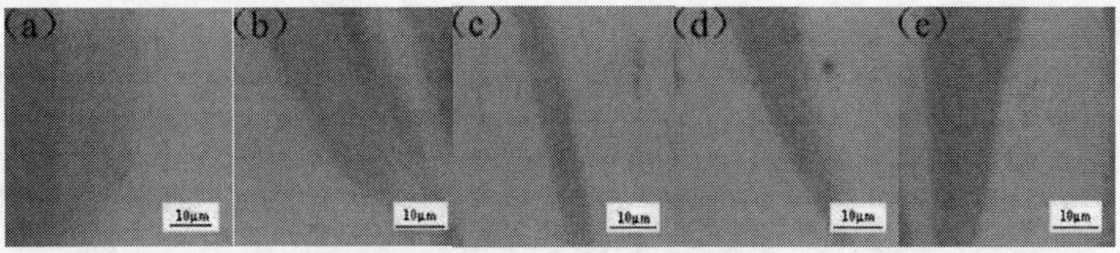


**Figure 3.** A few layers of graphene

By observing the results of optical microscopy characterization, we can know that the preparation method used in this experiment can successfully transfer graphene to the silicon wafer substrate. Among them, 3.5(a) is a single-layer graphene, and its color is very close to that of the silicon substrate, indicating that the transmittance of single-layer graphene is very high. In other graphene samples, as the number of layers increases, the graphene sheets are more and more different from the substrate color. It shows that we can preliminarily discern the number of layers of graphene based on the contrast between the obtained graphene and the substrate color.

*4.2.2. Raman spectrometer (RS) characterization results.*

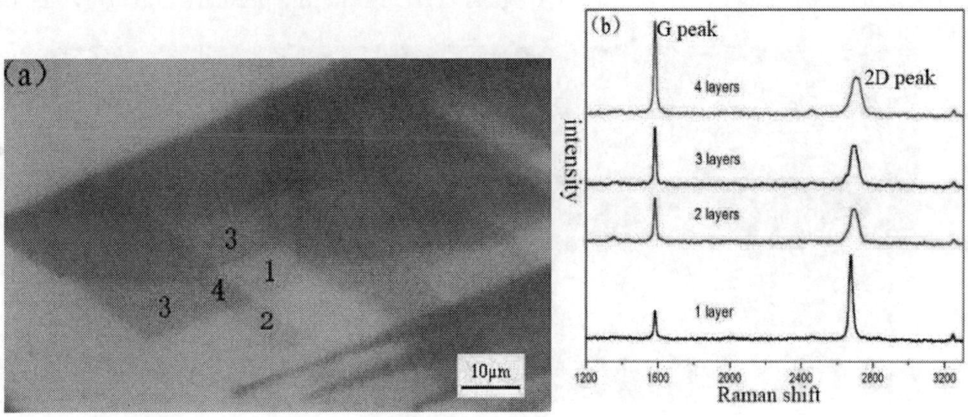


**Figure 4.** (a) OM photo of graphene sample; (b) Raman spectra obtained from single-layer, double-layer, three-layer, and four-layer graphene in a sample

During the test, the laser was focused onto the graphene sample with a scan range of 1200 cm - 1-3500 cm -1. It can be seen from Figure 4 that under the existing experimental conditions, we can prepare a few layers or even a single layer of graphene. Figure3 (b) is a Raman spectrum measured at different layers of graphene based on Figure 4 (a). From Raman spectroscopy, we can know that the number of layers of graphene is different. The Raman spectrum shows different intensity and peak shape, but the common point is that there are two characteristic peaks, namely G peak and 2D peak. As the number of graphene layers increases, the G peak does not shift, and the 2D peak undergoes a Raman shift, which gradually moves toward the long wave direction, that is, a blue shift occurs. Because single-layer graphene does not have interlayer coupling, its 2D peak intensity is very high, but when the number of layers is gradually increased, the interlayer coupling between the layers is manifested. As the number of layers increases, the 2D peak intensity decreases. And the peak shape is widened. The graphene G peaks of different layers have different intensities. As the number of layers increases, the intensity of the G peaks gradually increases, showing a linear law. Therefore, in the Raman spectrum, the single-layer graphene has a lower G-intensity than the other two-layer graphene, and the other layers have a higher G-intensity than the 2D peak, so it can be clearly seen from other The spectra of the samples are resolved, but the 2D peaks of the 2-4 layers are very similar. Based on this, we can distinguish whether the produced graphene is a single layer.

IOP Conf. Series: Materials Science and Engineering **772** (2020) 012034 doi:10.1088/1757-899X/772/1/012034

*4.3. Graphene strain test sample preparation*
First step: In this sample preparation, the pretreatment of the silicon wafer was consistent with the previous method, and was shaken for 15 minutes in deionized water, propanol, absolute ethanol, and deionized water, respectively.

Step 2: The treated silicon wafer is plated with a gold film in a small ion sputtering apparatus with a time setting of 200 s and a current of 20 mA.

The third step: the gold film-coated silicon wafer is placed in a vacuum drying oven for annealing, the temperature is set at 300 ° C, and after heating for 15 minutes, it is cooled with the furnace.

Step 4: Transfer the pre-teared graphene to a gold-plated, annealed silicon wafer.

*4.4. Graphene strain characterization*
In this experiment, the processed samples were mainly subjected to optical microscopy and Raman spectroscopy.

*4.4.1. Optical microscope (OM) characterization results.* As shown in Fig. 5, the area 1 (a) and the area 1 (b) are unsprayed silicon wafers, and the regions (a) and 2 (b) are gold-plated silicon wafers. We can see that after the annealing treatment (a), the original layered gold film becomes granular compared to the previous (b).

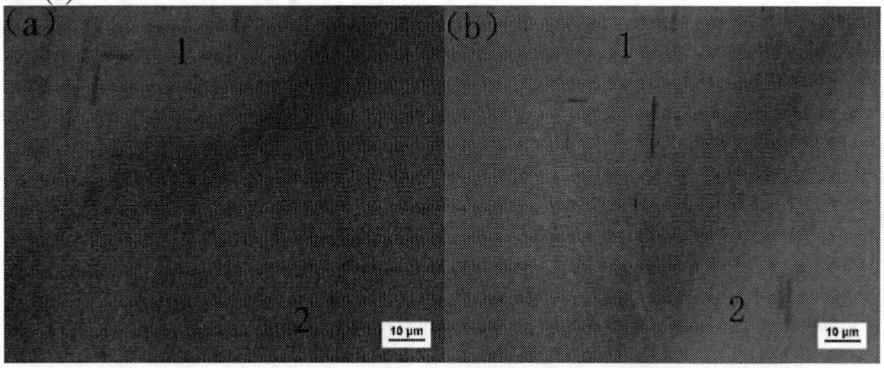


**Figure 5.** (a) Photograph of silicon wafer OM after annealing (b) Photograph of silicon wafer OM before annealing

As shown in Figure 3.8, it is a photomicrograph of graphene after gold annealing of silicon wafer. From the figure, we can see that the small layer of graphene has been transferred to the gold-plated, annealed silicon wafer, and there are obvious grooves at the pick-up point, and the groove width is about 1 µm.

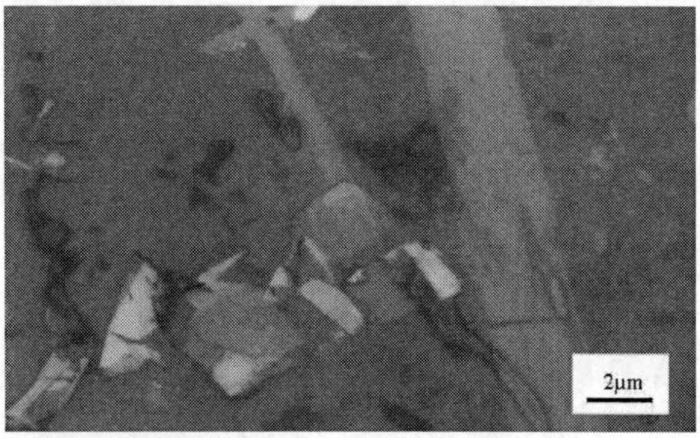


**Figure 6.** Photograph of graphene OM after silicon wafer annealing

*4.4.2. Raman spectroscopy (RS) characterization results.* During the test, the laser is focused to the groove in Figure 3.8, in order to compare the change in the strain of the sprayed gold or not. The scanning range is 1200cm-1-3500cm-1.

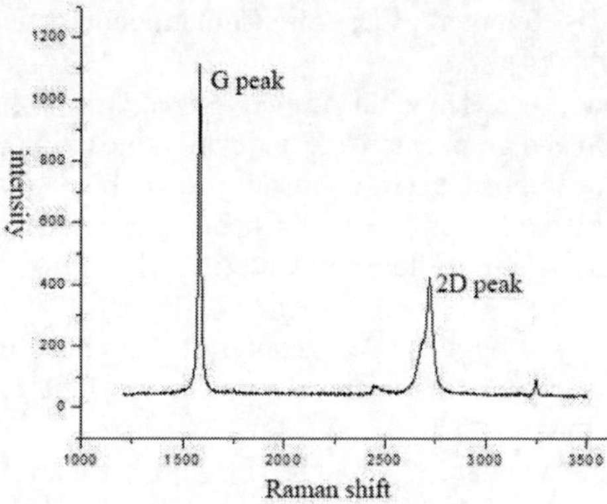


**Figure 7.** Raman spectroscopy of silicon wafers with graphene after gold plating and annealing

In the figure, no D peak was observed, indicating that no defects were found in the graphene-prepared graphene, which proved that graphene may have good performance. The G peak shifted by 2 cm-1 in the long-wave direction, and the long-wave direction of the 2D peak shifted by 39 cm-1, and both of them shifted red, indicating that the strain of graphene changed. However, there are inconsistencies in the references. It is possible that the stress of graphene is different because the peak position of the Raman spectrum is related to the tensile force or tension it receives. Further research on this process can be carried out in future research and study.

## 5. Conclusion

The highly oriented pyrolytic graphite can be peeled off into a single layer or even a small layer of graphene by mechanical stripping method. However, in the course of this experiment, the processing of the silicon wafer is different from the conventional method. In this experiment, the temperature of the silicon wafer is at 100 ° C for a period of time. After 2-5 minutes of heat treatment, the obtained graphene has not only a small number of layers but also an area of several hundred micrometers, indicating that the results of graphene production are affected by temperature and heating time. In this paper, the controllable preparation of graphene is realized. Among them, optical microscope and Raman spectroscopy show that single-layer graphene has different properties from other graphenes. The single-layer graphene has better light transmittance than other graphenes. Many, and as the number of layers increases, the G and 2D peaks in the Raman spectrum exhibit different intensities and peak shapes.

In addition, in the study of the strain characteristics of graphene, it was found that after the gold film was sprayed and annealed, the gold film was changed into a granular shape, and the graphene attached thereto did strain, and the Raman peak occurred. Displacement. In this experiment, the controllable preparation of graphene was studied. Finally, single layer and few layers of graphene were successfully prepared. The single layer, double layer, three layer and four layer graphene were distinguished by characterization method. The true development of graphene has been more than a decade, but its excellent performance has made it a lot of applications, and its related research has also developed rapidly. Although this experiment has gained, there are some shortcomings. I hope that later scholars will improve and improve.

## References

[1] Wang Xinxin,Yin Hongxia,Guan Yuxia,Yang Yuxiang,Huang Yan,Yuan Hongming,Liu Xiangnong, Ni Chaoying. Graphene Oxide Covalently Grafted $Fe_2B@SiO_2$ Nanoparticles for

IOP Conf. Series: Materials Science and Engineering **772** (2020) 012034 doi:10.1088/1757-899X/772/1/012034

Epirubicin Loading and Releasing. [J]. Journal of nanoscience and nanotechnology, 2020, 20(4).

[2] Fan Gaofeng,Huang Xu,Li Dawei,Li Aimin. $TiO_2$-Graphene 3D Hydrogel Supported on Ni Foam for Photoelectrocatalysis Removal of Organic Contaminants. [J]. Journal of nanoscience and nanotechnology, 2020, 20(4).

[3] Zhang Rong,Wang Yihao, Ma Donghui,Ahmed Saeed,Qin Wen,Liu Yaowen. Effects of ultrasonication duration and graphene oxide and nano-zinc oxide contents on the properties of polyvinyl alcohol nanocomposites. [J]. Ultrasonics sonochemistry, 2019, 59.

[4] Kimal Chandula Wasalathilake,Henan Li,Li Xu,Cheng Yan. Recent advances in graphene based materials as anode materials in sodium-ion batteries [J]. Journal of Energy Chemistry, 2020, 42.

[5] Daxiong Wu,Caiyun Wang,Mingguang Wu,Yunfeng Chao,Pengbin He,Jianmin Ma. Porous bowl-shaped VS 2 nanosheets/graphene composite for high-rate lithium-ion storage [J]. Journal of Energy Chemistry, 2020, 43.

[6] Yingqi Xu,Weifeng Zhang,Yaguang Li,Pengfei Lu,Zhong-Shuai Wu. A general bimetal-ion adsorption strategy to prepare nickel single atom catalysts anchored on graphene for efficient oxygen evolution reaction [J]. Journal of Energy Chemistry, 2020, 43.

[7] Parvaneh Eskandari,Zahra Abousalman-Rezvani,Hossein Roghani-Mamaqani,Mehdi Salami-Kalajahi,Hanieh Mardani. Polymer grafting on graphene layers by controlled radical polymerization [J]. Advances in Colloid and Interface Science, 2019, 273.

[8] Chao Wang,Yinxiang Zeng,Xiang Xiao,Shijia Wu,Guobin Zhong,Kaiqi Xu,Zengfu Wei,Wei Su,Xihong Lu. γ-MnO 2 nanorods/graphene composite as efficient cathode for advanced rechargeable aqueous zinc-ion battery [J]. Journal of Energy Chemistry, 2020, 43.

[9] Zang, Xining, Lin, Liwei. Graphene synthesis via droplet CVD AND its photonic applications [P]., 2014.

[10] Lafont, F.,Ribeiro-Palau, R.,Han, Z., Cresti, A.,Cummings, A.W.,Roche, S., Bouchiat, V., Ducourtieux, S., Schopfer, F.,Poirier, W.. Dissipative quantum Hall effect in polycrystalline CVD graphene [P]. Precision Electromagnetic Measurements (CPEM 2014), 2014 Conference on, 2014.

# Research on Evaluation Method of Storage Reliability of a Control Cabin

**Xiaodong Zhao**[1, 2, a], **Xihui Mu**[2, b]

[1] Army Engineering Shijiazhuang Campus, Hebei Shijiazhuang 050003, China;
[2] 32181 Troops of the Chinese People's Liberation Army, Hebei, Shijiazhuang 050003, China

[a] 307794209@qq.com, [b] 1572660524@qq.com

**Abstract.** A problem that is difficult to assess for storage cabin storage reliability. According to the historical data of the control cabin, the life distribution function was assumed, and the model parameters were estimated by the maximum likelihood estimation method and the minimum chi-square estimation method respectively, and different storage life distribution functions were obtained. Then compare and analyze different storage life distribution functions to determine the storage life distribution function of the control cabin. Finally, the Person's chi-square goodness-of-fit test was used to verify the results. The life distribution function of the control cabin is the I-type maximum value distribution, and the storage life is 18.12 years when the confidence is 0.90 and the reliability is 0.95. It provides a basis for subsequent life extension maintenance work and compares different storage reliability assessment methods.

**Keywords:** Control cabin; Storage reliability; Maximum likelihood estimation; Minimal chi-square estimate

## 1. Introduction

As the core component of precision guided munitions, the control cabin is the "brain and central nervous system" of precision guided munitions. Due to its typical characteristics of "long-term storage, one-time use", accurate assessment of its storage life is of great significance for the reliable use of precision guided munitions. In recent years, the evaluation of the storage life of the control cabin had been a hot topic of research [1-2], but most of them were concentrated on the evaluation of the components [3-4], and the evaluation of the control cabin system was less. Based on the historical monitoring data accumulated during the long-term storage of the control cabin, this paper used the maximum likelihood estimation method and the minimum chi-square estimation method to evaluate its storage reliability and determine its life distribution function to provide a basis for subsequent life extension maintenance.

## 2. Historical data

As a special weapon, ammunition has the typical characteristics of "long-term storage, one-time use". After the ammunition was shipped from the factory, it usually needed to be stored in the warehouse for a period of time. Due to the influence of the storage environment, the performance indicators of its components would decrease.

Content from this work may be used under the terms of the Creative Commons Attribution 3.0 licence. Any further distribution of this work must maintain attribution to the author(s) and the title of the work, journal citation and DOI.
Published under licence by IOP Publishing Ltd

In order to identify the quality changes of the control cabin at different stages of the storage period, it was necessary to periodically check the control cabin during storage. The data obtained by the test is the success or failure type data, and since the specific failure time was not known, the detection data of the control cabin is incomplete data. The actual test data is shown in Table 1:

**Tab. 1** Control cabin fault data

| Storage month | Number of samples tested | Number of failures | Failure ratio |
|---|---|---|---|
| 30 | $X_1$ | $Y_1$ | 0.004242 |
| 33 | $X_2$ | $Y_2$ | 0.001038 |
| 36 | $X_3$ | $Y_3$ | 0 |
| 48 | $X_4$ | $Y_4$ | 0 |
| 54 | $X_5$ | $Y_5$ | 0.007568 |
| 59 | $X_6$ | $Y_6$ | 0.003878 |
| 63 | $X_7$ | $Y_7$ | 0.005051 |
| 74 | $X_8$ | $Y_8$ | 0.004337 |
| 78 | $X_9$ | $Y_9$ | 0.005119 |
| 81 | $X_{10}$ | $Y_{10}$ | 0.001001 |
| 85 | $X_{11}$ | $Y_{11}$ | 0.003846 |
| 88 | $X_{12}$ | $Y_{12}$ | 0.011294 |
| 102 | $X_{13}$ | $Y_{13}$ | 0.004114 |
| 113 | $X_{14}$ | $Y_{14}$ | 0.022657 |

It can be seen from Table 1 that the actual fault detection data has zero failure rate at some detection time points. In order to save the calculation cost, the data shown in Table 2 can be obtained.

**Tab. 2** Finished control cabin fault data

| Storage month | Number of samples tested | Number of failures | Failure ratio |
|---|---|---|---|
| 31.4754 | $X_1$ | $Y_1$ | 0.002961 |
| 53.42787 | $X_2$ | $Y_2$ | 0.006846 |
| 60.98718 | $X_3$ | $Y_3$ | 0.004461 |
| 76.1702 | $X_4$ | $Y_4$ | 0.004761 |
| 84.56919 | $X_5$ | $Y_5$ | 0.005616 |
| 102 | $X_6$ | $Y_6$ | 0.004114 |
| 113 | $X_7$ | $Y_7$ | 0.022657 |

The data type can be equivalent to:

Note $t=(t_1,\cdots,t_k)$ is the detection time point. After the storage time $t_i$, the $n_i$ control cabin was taken for testing. Among them, the $X_i$ failed, and the data is obtained:

$$(t_i,n_i,X_i),\quad i=1,2,\cdots,k$$

Among them, $0<t_1<t_2<\cdots<t_k$ , each test is independent of each other.

## 3. Evaluation method

For the incomplete detection of fault data in the control cabin storage process, the reliability of the storage was evaluated by the maximum likelihood estimation method and the minimum chi-square estimation method. The maximum likelihood estimation method gives a conservative estimation of the distribution parameters and the reliable lifetime. The complete detection data is used in the evaluation process, without any deletion, and the distribution life of the control cabin is expected to be obtained under real conditions. The minimum chi-square estimation method combines the goodness-of-fit theory to process the detection data, and gives the estimation of the distribution parameters and the reliable lifetime. At the same time, the distribution function is tested.

### *3.1. Distribution function model*

Considering the physical failure mechanism of the control cabin, it is assumed that the possible forms of life distribution have exponential distribution, Weibull distribution, type I maximum value distribution and type II maximum value distribution. The distribution function expression is as follows:

Index distribution: $F(t)=1-\exp(-\frac{t}{\theta})$

Weibull distribution: $F(t)=1-\exp(-(\frac{t}{\eta})^m)$

Type I maxima distribution: $F(t)=\exp(-\exp(-\frac{t-\mu}{\sigma}))$

Type II maximum value distribution: $F(t)=\exp(-(\frac{t}{\eta})^{-m})$

### *3.2. Maximum likelihood estimation*

The storage time is $t_i$, the number of samples detected at this time is $n_i$, and the number of failures is $x_i$, then the likelihood function is:

$$L=\prod_{i=1}^{k}F(t_i,\vec{\lambda})^{x_i}[1-F(t_i,\vec{\lambda})]^{n_i-x_i}$$

Among them, the exponential distribution $\vec{\lambda}=\theta$, the Weibull distribution $\vec{\lambda}=(\eta,m)$, the I-type maximum value distribution $\vec{\lambda}=(\mu,\sigma)$, and the type II maximum value distribution $\vec{\lambda}=(\eta,m)$.

By solving $\partial\ln L/\partial\lambda_i=0$, the maximum likelihood estimate of the distribution parameters can be obtained.

### *3.3. Pearson goodness of fit test*

Pearson proposed a chi-square statistic to test whether a set of independent samples belong to the same distribution family with specific properties. The variable sample composite Pearson chi-square statistic is of the form [5]:

$$\chi^2(\vec{\lambda})=\sum_{i=1}^{k}(X_i-n_ip_i(\vec{\lambda}))^2/n_ip_i(\vec{\lambda})$$

When $n_i\to\infty$, the limit distribution of $\chi^2(\vec{\lambda})$ is the $\chi^2$ distribution with a degree of freedom $k-1$, that is, $\chi^2(\vec{\lambda})\sim\chi^2_{k-1}$. When the test is performed, since the true value of $\vec{\lambda}$ is unknown, the estimated amount of $\vec{\lambda}$ is used instead of $\vec{\lambda}$ to calculate $\chi^2(\vec{\lambda})$.

Given a significance level $\alpha$, there are:

$$P(\chi^2(\vec{\lambda})\geq\chi^2_{k-s-1}(1-\alpha)\,|\,H_0)\leq\alpha$$

When $\chi^2(\vec{\lambda})\geq\chi^2_{k-s-1}(1-\alpha)$ is established, a small probability event occurs, rejecting the null hypothesis.

### *3.4. Minimal chi-square estimation*

The minimum chi-square estimation refers to the parameter $\tilde{\lambda}$ obtained by minimizing the statistic as the best estimate of the true value $\vec{\lambda}$, ie:

$$\chi^2(\tilde{\lambda}) = \min\{\chi^2(\vec{\lambda}) : \vec{\lambda} \in \Lambda\}$$

A very small chi-square estimate of $\vec{\lambda}$ can be obtained by solving the following equations.

$$\frac{\partial \chi^2(\vec{\lambda})}{\partial \lambda_j} = \sum_{i=1}^{k} \{-\frac{2n_i(X_i - n_i p_i(\vec{\lambda}))}{n_i p_i(\vec{\lambda})} - \frac{(X_i - n_i p_i(\vec{\lambda}))^2}{n_i p_i^2(\vec{\lambda})}\} \cdot \frac{\partial p_i(\vec{\lambda})}{\partial \lambda_j}$$

$$= \sum_{i=1}^{k} \{\frac{-X_i^2 + n_i^2 p_i^2(\vec{\lambda})}{n_i p_i^2(\vec{\lambda})}\} \cdot \frac{\partial p_i(\vec{\lambda})}{\partial \lambda_j}$$

$$= \sum_{i=1}^{k} (1 - (\frac{f_i}{p_i(\vec{\lambda})})^2) \cdot \frac{n_i \cdot \partial p_i(\vec{\lambda})}{\partial \lambda_j}$$

Therefore, the minimum chi-square estimate of $\vec{\lambda}$ is the solution of the following equation:

$$\sum_{i=1}^{k} \{1 - [\frac{f_i}{p_i(\vec{\lambda})}]^2\} \cdot \frac{n_i \cdot \partial p_i(\vec{\lambda})}{\partial \lambda_j} = 0, \quad j = 1, 2, \cdots, s$$

## 4. Storage life assessment

The maximum reliability estimation method and the minimum chi-square estimation method were used to evaluate the storage reliability of the control cabin based on the compiled control cabin fault detection data.

The results of the reliability assessment of the control cabin using the maximum likelihood estimation method are shown in Table 3:

**Tab. 3** Maximum likelihood estimation method control cabin storage reliability evaluation result

| Distribution function | Estimated parameter | Reliable life |
|---|---|---|
| Index distribution | $\hat{\theta}$=999.503 | $t_{0.95}$=51.2678 |
| Weibull distribution | $\hat{\eta}$=3001.0626<br>$\hat{m}$=0.8239 | $t_{0.95}$=81.5877 |
| Type I maxima distribution | $\hat{\mu}$=59.0419<br>$\hat{\sigma}$=32.3060 | $t_{0.95}$=23.5961 |
| Type II maximum value distribution | $\hat{\eta}$=276497.2337<br>$\hat{m}$=0.1520 | $t_{0.95}$=202.6761 |

Comparing the evaluation results in Table 3, it can be concluded that the storage reliability distribution function of the control cabin is a type I maximum value distribution, and the storage life of the control cabin at a confidence level of 0.90 and a reliability of 0.95 is determined to be 23.591 years.

The results of the evaluation of the reliability of the control cabin storage using the minimum chi-square estimation method are shown in Table 4:

**Tab. 4** Minimal chi-square estimation method control cabin storage reliability assessment results

| Distribution function | Estimated parameter | Reliable life |
|---|---|---|
| Index distribution | $\hat{\theta}$=902.525 | t0.95=46.2935 |
| Weibull distribution | $\hat{\eta}$=383.6496<br>$\hat{m}$=1.2062 | t0.95=32.6972 |
| Type I maxima distribution | $\hat{\mu}$=43.2415<br>$\hat{\sigma}$=22.89904 | t0.95=18.1170 |
| Type II maximum value distribution | $\hat{\eta}$=8093.1676<br>$\hat{m}$=0.2233 | t0.95=59.4572 |

Comparing the evaluation results in Table 4, it can be concluded that the storage life distribution function of the control cabin is a type I maximum value distribution, and the storage life of the control cabin at a confidence level of 0.90 and a reliability of 0.95 is 18.1170 years.

By comparing the evaluation results in Tables 3 and 4, it can be determined that the storage reliability of the control cabin obeys the type I maximum value distribution and explains the difference of different evaluation methods.

Finally, the Pearson goodness-of-fit test is carried out to verify the correctness of the minimum chi-square estimation method. A is taken as 0.05, and the goodness-of-fit p values of the four distributions are calculated respectively. The calculation results are shown in Table 5:

**Tab. 5** Goodness of fit p value

| Distribution function | Goodness of fit p value |
|---|---|
| Index distribution | 0.000479 |
| Weibull distribution | 0.000258 |
| Type I maxima distribution | 0.001078 |
| Type II maximum value distribution | 0.0001550 |

The higher the p value, the better the fitting effect. It can be seen from Table 5 that the fitting goodness p value of the type I maximum value distribution is the highest, which verifies that the storage reliability of the control cabin obeys the type I maximum value distribution. The storage life of the control cabin at a confidence level of 0.90 and a reliability of 0.95 was determined to be 18.1170 years.

## 5. Conclusion

According to the fault detection data of the control cabin, the storage reliability of the control cabin was evaluated by the maximum still estimation method and the minimum chi-square estimation method, and the life distribution function was determined as the type I maximum value distribution, and the results were verified by Pearson's goodness of fit theory. The method accurately estimates the storage life under the standard storage environment of the control cabin, determines the storage life distribution function, verifies the rationality of the method, and provides reference for other engineering applications.

## References

[1] WANG Zhao-bin, FU Sai, SHANG Shang, et, al. Storage degradation testing and life prediction for missile electromagnetic relay. Journal of Beijing University of Aeronautics and Astronautics. Vol.42 (2016) No. 12, p. 2610-2619.

[2] SHEN Zheng-guang, YUAN Jing-chun, DONG Jing-yu, et al. Research on acceleration factor estimation method of accelerated life test of missile-borne equipment. Systems Engineering and Electronics. Vol.37 (2015) No. 8,p. 1948-1952.

[3] Takada A, So N, Hajime I. A new acceleration factor decision method for ICCG method based on condition number. IEEE Transactions on Magnetics. Vol.48 (2012) No. 2, p. 519-522.

[4] ZHAO Xiao-dong, MU Xi-hui. Optimization design method of accelerometer,s constant-life test based on sequential statistics. Journal of Chinese Interial Technology. Vol.27 (2019) No. 4, p. 553-560.
[5] Si X S, Wang W, Hu C H, Zhou D H. Remaining useful life estimation-A review on the statistical data driven approaches. European Journal of Operational Research. Vol.213 (2011) No. 1, p. 1-14.

# Reactivity of Metal Boronhydride Derivatives with Amines and Ethanolamines

**Ling Chen*, Liya Yang, Chunxia Wang and Ting Zhu**

School of elementary education, Wanjiang University of Technology, Ma'anshan, Anhui, China

*Corresponding author e-mail: imidazol@163.com

**Abstract.** Reducing agents $LiBH_4$, $NaBH_4$, $NaBH(CO_2CH_3)_3$, $NaBH(OCH_3)_3$ and $NaBH_3CN$ are widely applied in organic synthesis, biochemistry and medical chemistry. In this paper, the structure and the reactivity of metal boronhydride derivatives with amines were studied and the coordination of amines and ethanolamines towards metalboronhydride derivatives was investigated.

## 1. Introduction

$LiBH_4$, $NaBH_4$, $NaBH(CO_2CH_3)_3$, $NaBH(OCH_3)_3$ and $NaBH_3CN$ are all reducing agents. Particularly $LiBH_4$, $NaBH_4$ are commercially produced [1] and have a wide applicability in organic synthesis [2]. $NaBH(CO_2CH_3)_3$, $NaBH(OCH_3)_3$ and $NaBH_3CN$ are also used very popular in biochemistry and medical chemistry [3, 4, 5]. For example, $NaBH(OCH_3)_3$ can be used in reductive amination [6, 7]; $NaBH(OCH_3)_3$ is an effective hydride source in selective reductive displacement of halides [8]; In reductive decyanation with $NaBH_3CN$ excises the cyano group and furnishes unadorned quinolizidies in excellent yield [9].

We are interested in the study of the structure and the reactivity of metal boronhydride derivatives (M = Li, Na, K) with amines. In particular, we investigated the coordination of amines and ethanolamines towards metalboronhydride derivatives.

In this study, we used the following amines: *iso*propyl amine, N, N, N', N'-tetramethylethylendiamine, 2-picolylamine, (N-methyl) ethanolamine, *iso*quinoline and pyridine.

## 2. Results and Discussion

### *2.1. Lone pairs ligands*

There are six reactions in this group. The reactions and their results are described in Table 1. The aim of this research was to prepare metalboronhydrides complexes in order to study their reactivity and structural properties by X-ray diffraction. So N, N, N', N'-tetramethylethylendiamine, *iso*propyl amine and 2-picolylamine were chosen as the ligands because of they have lone pairs.

Content from this work may be used under the terms of the Creative Commons Attribution 3.0 licence. Any further distribution of this work must maintain attribution to the author(s) and the title of the work, journal citation and DOI.
Published under licence by IOP Publishing Ltd

**Table 1.** Preparation of Metalboronhydrides Complexes.

| Number | Reagent 1 | Reagent 2 | NMR-$^{11}$B (ppm) | Results |
|---|---|---|---|---|
| 1 |  | $NaBH_4$ | - 42.5 | no reaction |
| 2 | 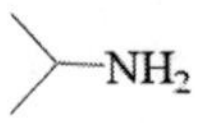 | $KBH_4$ | ----- | compound was insoluble |
| 3 | 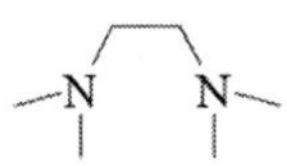 | $NaBH_4$ | - 42.9<br>- 9.5 | reaction |
| 4 | 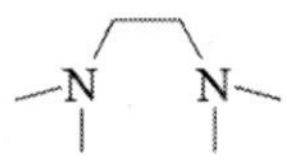 | $LiBH_4$ | - 40.7<br>- 9.5 | reaction |
| 5 | 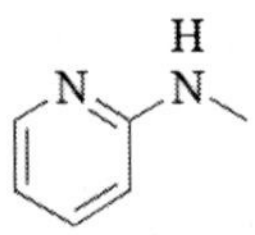 | $NaBH_4$ | - 40.4<br>- 40.6<br>- 40.7 | reaction |
| 6 | 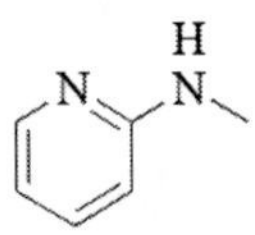 | $LiBH_4$ | - 39.2 | reaction |

The reactions 1 - 6 of $MBH_4$ (M=Li, Na, K) with an excess of amine were performed, its $^{11}$B NMR spectra showed a quintet assigned to the $BH_4^{(-)}$ group. Also, in the reactions 3 and 4, a minor signal (quartet) was observed at - 9.5 ppm, indicating the formation of N→BH3 adduct. The reaction 2 did not proceed because of the scarce solubility of $KBH_4$ in isopropyl amine, the same was observed in the reaction 1.

The used crystallization method was the slow diffusion method. Pentane (5 mL) was diffused into the solution of the corresponding amine and $MBH_4$ (M=Li, Na, K). From the reaction 1, single crystals suitable for X-ray structure analysis were obtained after one week. The X-ray structure showed the lack of the reaction of $NaBH_4$ with isopropyl amine and also showed how the molecule $NaBH_4$ forms an infinite chain in the lattice by the presence of Na-H intermolecular bonds.

### *2.2. Aliphatic and Aromatic Amines*

There are nine reactions in this group. The reactions and their results are described in Table 2. The aim of this research was to prepare the coordination compounds derived from $NaBH_3CN$, $NaBH(OMe)_3$ and $NaBH(OAc)_3$ with aliphatic and aromatic amines, to get single crystals suitable for X-ray diffraction in order to study their structural properties.

IOP Conf. Series: Materials Science and Engineering **772** (2020) 012036 doi:10.1088/1757-899X/772/1/012036

**Table 2.** Preparation of Coordination Compounds.

| Number | Reagent 1 | Reagent 2 | NMR-$^{11}$B (ppm) | Results |
|---|---|---|---|---|
| 7 | 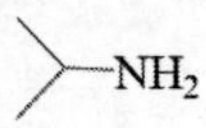 | $NaBH(OAc)_3$ | 0.1 (borate)<br>19.8 ($B(OR)_3$)<br>25.8 | reaction |
| 8 | 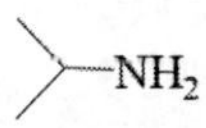 | $NaBH_3CN$ | - 43.5<br>($NaBH_3CN$ complex)<br>-22.0($N\rightarrow BH_3$) | reaction |
| 9 | 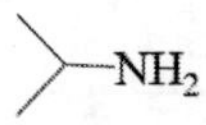 | $NaBH(OMe)_3$ | - 42.0<br>($NaBH(OMe)_3$complex)<br>2.1, 6.6(borate) | reaction |
| 10 | 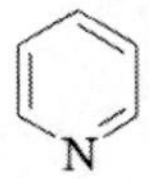 | $NaBH(OAc)_3$ | 1.1 (borate)<br>2.7 (borate) | reaction |
| 11 | 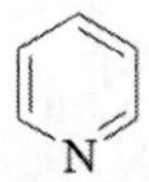 | $NaBH_3CN$ | - 42.4<br>($NaBH_3CN$ complex)<br>-12.1($N\rightarrow BH_3$) | reaction |
| 12 | 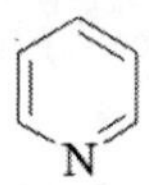 | $NaBH(OMe)_3$ | - 41.0<br>($NaBH(OMe)_3$complex)<br>2.6,7.3 (borate) | reaction |
| 13 | 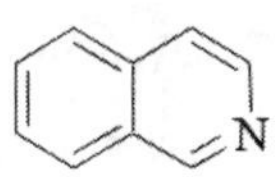 | $NaBH_3CN$ | - 41.1<br>($NaBH_3CN$ complex) | reaction |
| 14 | 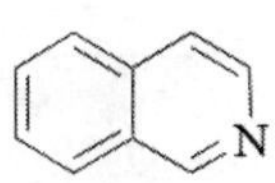 | $NaBH(OMe)_3$ | - 39.6<br>($NaBH(OMe)_3$complex)<br>3.1 (borate) | reaction |
| 15 | 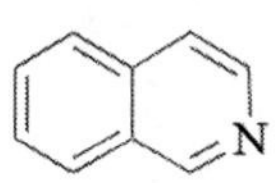 | $NaBH(OAc)_3$ | ----- | compound was insoluble |

The reaction 7 - 14 of sodium boron hydride derivatives with an excess of amine were performed, their $^{11}$B NMR spectra showed a mixture of compounds (as we can see in the Table 2), with the exception of the reaction 13, in which it was observed a signal corresponding to the $BH_3CN$ complex. In this compound, the $\delta$ $^{11}$B is not sensitive to the changes in the electronic density of the sodium atom.

On the other side, $NaBH(OMe)_3$ is more reactive than $NaBH_3CN$ and $NaBH(OAc)_3$, probably because $NaBH(OMe)_3$ is more soluble than the other two.

The used crystallization method was the slow diffusion method. Pentane (5 mL) was diffused into the solution of the corresponding amine and metalboronhydrides derivatives. From the reaction 10 and 12, single crystals were separated from the reaction mixtures after one week. These single crystals were suitable for X-ray structure analysis.

### *2.3. High Temperature Conditions*
The reaction 16 and 17 were performed in the same way as the reactions 7 - 15 under reflux (Table 3). High temperature is good for the reactions, but we can only see one signal in reaction 17, probably because in these compounds, the δ of $^{11}B$ is not sensitive to the electronic environment of the sodium atom.

**Table 3.** Reactions under reflux.

| Number | Reagent 1 | Reagent 2 | NMR-$^{11}B$ (ppm) | Results |
|---|---|---|---|---|
| **16** |  | $NaBH(OMe)_3$ | -44.1<br>2.1 | reaction |
| **17** | 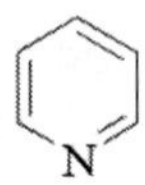 | $NaBH(OMe)_3$ | -40.9 | reaction |

### *2.4. Function of the Ligand's Mass*
There are five reactions in this group. They were performed in different molar ratios (2:1; 1:1; 1:2). Their results are described in Table 4. The aim of this group was to research the function of the ligand's mass in the reaction. At first, ethyl ether was chosen as solvent, but reagent 1 and reagent 2 were not solved on it, so toluene was finally chosen as solvent, although reagent 2 was not solved on it either. After stirring the reaction mixture, it became a suspension. Some minutes later, a kind of white oil-like liquid appeared, it was unsolved in toluene.

**Table 4.** Function of the Ligand's Mass.

| Number | Reagent 1 | Reagent 2 | Solvent | NMR-$^{11}B$ (ppm) | Results |
|---|---|---|---|---|---|
| **18** | 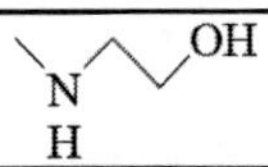 | $NaBH(OMe)_3$ | 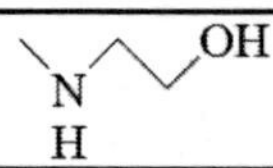 | -42.0<br>1.6 | reaction |
| **19** | 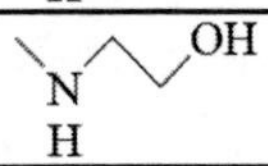 | $NaBH(OAc)_3$ | Toluene | -0.7, 1.6<br>4.2, 7.5 | reaction |
| **20** | 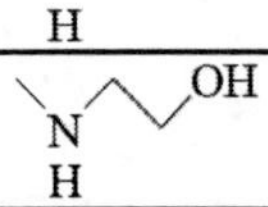 | $NaBH(OAc)_3$ | 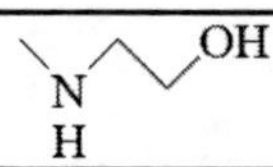 | 4.3 | reaction |
| **21** | 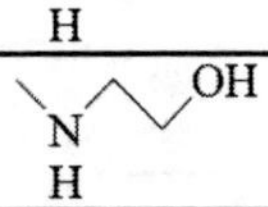 | $NaBH_3CN$ | Toluene | ----- | reaction |
| **22** | 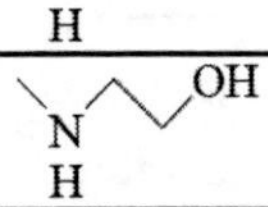 | $NaBH_3CN$ | 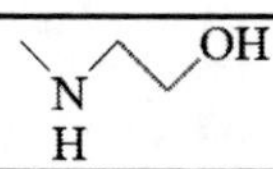 | -43.6<br>2.1 | reaction |

From these reactions, we can see that to choose a good solvent is very important. It can make the reagent diffuse well and has more chance to contact with each other. And the NMR spectra showed also the formation of new borate species in all of the reactions. We observed also that the best reaction conditions were when ethanolamine is used as solvent.

IOP Conf. Series: Materials Science and Engineering **772** (2020) 012036 doi:10.1088/1757-899X/772/1/012036

### 2.5. *Temperature Effect*

There are four reactions in this group. The reactions and their results are described in Table 5. The aim of these reactions was to study how does the reactivity of these boron hydride derivatives in (N-methyl) ethanolamine change with the effect of the temperature.

**Table 5.** Temperature Effect.

| Number | Reagent 1 | Reagent 2 | NMR-$^{11}$B (ppm) | Results |
|---|---|---|---|---|
| **23** | 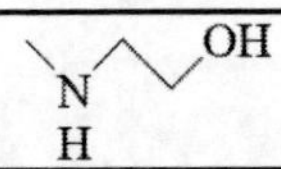 | $NaBH(OMe)_3$ | 1.8 | reaction |
| **24** | 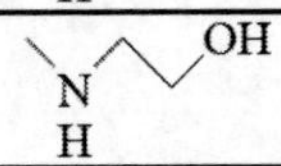 | $NaBH(OAc)_3$ | 4.2 | reaction |
| **25** | 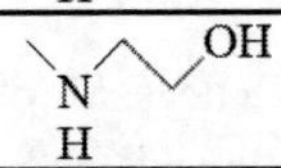 | $NaBH_3CN$ | -43.7<br>2.1 | reaction |
| **26** | 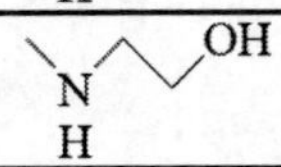 | $LiBH_4$ | -41.4<br>1.7 | reaction |

From the results in the table, the following discussion is made. High temperature is good for the reaction. In reaction 24 and 25, there were no starting materials. That means, reagent 2 are all more reactive in high temperature than in room temperature.

**Acknowledgments**

This work was financially supported by Research Fund of Wanjiang University of Technology (Nr: zl2018017).

**References**

[1] A. A. Hinkley, Encyclopedia of Chemical Technology, 2nd ed., vol. 2, 1966, pp. 210 – 217.
[2] E. Wiberg, E. Amberger, Hydrides of the Elements of Main Group I - IV, 1971.
[3] G. Toth, etc. Helvetica Chimica Acta, 2001, 84, pp. 263 – 270.
[4] C. Rabiller, etc. Eur. J. Org. Chem., 2003, pp. 4810 -4818.
[5] K. Wright, etc. Tetrahedron Letters, 2003, 44, pp. 3381 -3384.
[6] M. Dubber, T. K. Lindhorst, Synthesis, 2001, 2, pp. 327 - 330.
[7] D. C. Beshore, C. J. Dinsmore, Organic Letters, 2002, 4, 7, pp. 1201 - 1204.
[8] R. O. Hutchins, etc. J. Org. Chen., 1978, 43, 11, pp. 2259 - 2267.
[9] R. L. Danheiser, etc. J. Am. Chem. Soc. 2003, 125, pp. 4970 -4971.

IOP Conf. Series: Materials Science and Engineering **772** (2020) 012037 doi:10.1088/1757-899X/772/1/012037

# Study on Molding Process of Ether Anhydride Type Polyimide Materials

**Xiaoqi Chen[1, *], Zhengzheng Wang[2], Haoran Xu[2], Haijun Zhou[1], Mengmeng Zhou[1] and Yantao Li[1]**

[1]Institute of Energy Source, Hebei Academy of Sciences, Shijiazhuang, China
[2]Hebei University of Science and Technology, Shijiazhuang, China

*Corresponding author e-mail: cxq198508@163.com

**Abstract.** The main molding process of ether anhydride type polyimide materials which synthesized by the main meterials of pyromellitic dianhydride (PMDA), 3,3',4,4'-tetracarboxydiphthalic dianhydride (OPDA) and Diamine diphenyl ether (ODA) was researched. The effects of molding temperature, molding time and molding pressure on the tensile properties of molded products were investigated. And the effects of dianhydride ratio on the properties and molding process of ether anhydr ide type polyimide materials was discussed. Results showed that samples molded by polyimide molding powder appeared the best comprehensive performance when the molar ratio of PMDA and OPDA was 2:1, with the tensile strength of 93.5 MPa, the elongation at break of 10.1%, the vitrification transition temperature of 340 ℃ and the thermal decomposition temperature of 5% was 574 ℃. And molding process is under that the molding temperature is 410℃, the molding pressure is 4.0Mpa and the molding time is 15min.

## 1. Introduction

Polyimide (PI) was a kind of special engineering plastics with high application value, which could competed with some kinds of metals. It had excellent properties such as high dielectric property, high strength and high modulus at high temperature [1-2].It was widely used in aviation, aerospace, national defense, military and microelectronics and other high-tech fields [3-6]. Because of the rigid structure of most polyimides chains, the interaction between the chains was so strong that no melt flow or even softening occurs at the decomposition temperature, making it very difficult to form.

In this paper, main materials of pyromellitic dianhydride (PMDA), 3,3', 4,4'-tetracarboxydiphthalic ether dianhydride (OPDA) and 4, 4'-diaminodiphenyl ether(ODA) were used to synthesize ether anhydride polyimide moulding plastics successfully by introducing a certain amount of soft groups into the polyimide molecular chain. On the basis of this research, this paper focused on the study of ether anhydride type polyimide material molding process with different dianhydride ratio, and the tensile properties and thermal stability of molded products were characterized. Then the suitable molding process for ether anhydride polyimide materials were determined.

Content from this work may be used under the terms of the Creative Commons Attribution 3.0 licence. Any further distribution of this work must maintain attribution to the author(s) and the title of the work, journal citation and DOI.
Published under licence by IOP Publishing Ltd

## 2. Experimental

### *2.1. Materials*

The 4, 4'-diaminodiphenyl ether (ODA) were obtained from Nantong Huishun Chemical Co., Ltd. The 3,3',4,4'-tetracarboxydiphthalic ether dianhydride(OPDA) were purchased from Shanghai GuChuang New Chemical Materials Co., Ltd and the pyromelic dianhydride (PMDA) were supplied by Leheng Chemical Co., Ltd. The raw materials were all dried in an vacuum drying oven at 105°C for 6 h. N, N-dimethylacetamide (DMAc) were obtained from Tianjin Yongda Chemical Reagent Co., Ltd.

### *2.2. Preparation of ether anhydride type polyimide molding materials*

ODA was added to the DMAc for completely dissolved, and a mixture of different molar ratios (4:1, 3:1, 2:1 and 1:1) of PMDA and ODPA was put into system at 0°C, respectively. Then a series of ether anhydride type polyamic acid (PAA) solutions were obtained. Through thermal imidization at 160°C~300°C to obtain a series of ether anhydride type polyimide molding powder containing different softening groups, named PI-1, PI-2, PI-3, PI-4. The effects of molding conditions such as molding pressure, molding temperature and molding time on the tensile properties of different ether anhydride type polyimide materials were investigated.

### *2.3. Characterization*

The vitrification transition temperature (Tg) was measured by dynamic thermo-mechanical instrument (DMA) according to ASTM E1640-2013. The thermal stability of PI was measured by TGA. Test conditions: nitrogen atmosphere, temperature rise rate: 20°C/min, test range: 30°C-900°C. Mechanical properties was tested according to the provisions of GB/T 1040.1-2006, and B1 dumbbell test bar was tested at the speed of 2 mm/min on universal tensile testing machine.

## 3. Results and discussion

### *3.1. The vitrification transition temperature analysis*

The vitrification transition temperature (Tg) of the polymer was a transition temperature from the freezing to the movement of the polymer segment, which was an important parameter for polymer processing conditions and performance. The higher the vitrification transition temperature, the higher the temperature at which the material was used, but the more difficult the processing conditions. Due to the high Tg of polyimide, the Tg of ether anhydride type polyimide materials containing different molar ratio of PMDA/OPDA was determined by dynamic thermo-mechanical instrument (DMA). Result were shown on Figure 1.

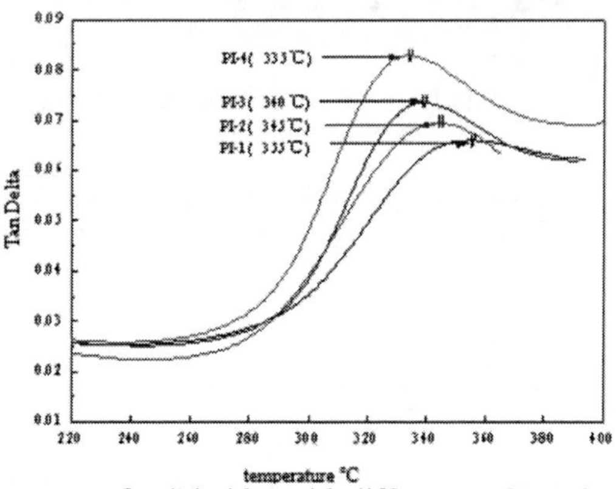


**Figure 1.** DMA spectra of polyimides with different molar ratios of PMDA/OPDA

As can be seen from figure 1, with the increasing of OPDA content in polyimide molding powder, the soft structure in the molecular structure of polyimide increased. And Tg gradually decreased from 355°C to 335°C, which improved the processing performance of PI materials.

### *3.2. Effect of molding process on PI performance*

Taking PI-3 as an example, the effects of different molding pressures, different molding temperatures, different molding times and other factors on the tensile properties of PI materials were investigated. In the case of any two influencing factors remain unchanged, only one influencing factor was changed, a series of PI samples under different molding conditions were obtained, and the tensile properties were measured. Results are shown in Table 1.

**Table 1.** Effect of molding process on tensile properties of PI samples

| | Samples | Elongation at break (%) | Tensile strength (MPa) |
|---|---|---|---|
| Molding Temperatures (°C) | 350 | 5.9 | 55.7 |
| | 380 | 6.2 | 67.4 |
| | 410 | 10.1 | 93.5 |
| | 440 | 7.5 | 88.2 |
| Molding Pressures (MPa) | 2.8 | 6.3 | 60.2 |
| | 3.4 | 9.8 | 88.5 |
| | 4.0 | 10.1 | 93.5 |
| | 4.6 | 10.0 | 93.3 |
| Molding Times (min) | 5 | 3.9 | 61.4 |
| | 10 | 7.3 | 77.3 |
| | 15 | 10.1 | 93.5 |
| | 20 | 6.1 | 75.1 |

It can be seen from Table 1, the PI-3 polyimide molding powder had the best mechanical properties under the condition of the molding temperature of 410°C, the molding time of 15 min, and the molding pressure of 4.0 MPa. The tensile strength and elongation at break can reach 93.5MPa and 10.1%, respectively. Therefore, the optimum molding condition was obtained from above data.

### *3.3. Effect of dianhydride ratio on PI molding process.*

As the content of ether anhydride groups in the molecular chain increased, the Tg of the polyimide decreased, and the fluidity of the melt increased, which was advantageous for the processing of the material. After a series of experiments, the best molding process for polyimide materials with different dianhydride ratios was determined, which was shown in Table 2.

**Table 2.** Best molding process and performance of polyimide materials with different dianhydride ratios

| Sample | Tensile Strength (MPa) | Elongation at Break (%) | Molding Times (min) | Molding Pressures (MPa) | Molding Temperatures (°C) |
|---|---|---|---|---|---|
| PI-1 | 72.0 | 4.1 | 15 | 4.0 | 430 |
| PI-2 | 73.4 | 6.5 | 15 | 4.0 | 420 |
| PI-3 | 93.5 | 10.1 | 15 | 4.0 | 410 |
| PI-4 | 89.5 | 8.1 | 15 | 4.0 | 400 |

As can be seen from table 2, the PIs synthesized with the molar ratio of PMDA/OPDA was 4:1, 3:1, 2:1 and 1:1 under the molding pressure of 4.0Mpa, the molding time of 15 minutes accorded to best molding temperature was 430°C, 420°C, 410°C and 400°C. With the increasing content of ether anhydride groups in the molecular chain, the molding temperature gradually decreased, while the molding time and pressure were almost unchanged. High temperature helped to shorten the molding cycle and improved the comprehensive performance of the samples. However, if the temperature was too high, the surface would become black or even decompose, which degraded the performance of the sample. On the contrary, the molding temperature was too low, the melting was not perfect and the melt fluidity was reduced, which led to the dissatisfaction of the mold filling and affected the tensile properties of the sample. Therefore, the molding temperature was one of the important factors in the molding, and the selection of a suitable molding temperature was very important for the performance of the product.

It can also be seen from Table 2 that the obtained PI-3 when molar ratio of PMDA/OPDA was 2:1, the tensile strength and elongation at break reached a maximum of 93.5 MPa and 10.1%, respectively. Therefore, the elongation at break of PI increased, as the content of ether anhydride groups in the molecular chain , and its flexibility was better.

### *3.4. Effect of dianhydride ratio on the thermal stability of PI*

The thermal weight loss of different ether anhydride type polyimide materials was measured, and the temperature at which the weight loss ratio of the polymer was 5% and 10% was recorded. The results were summarized in Table 3.

**Table 3.** Effect of dianhydride ratio on the thermal stability of PI

| Sample | $n_{PMDA}:n_{OPDA}$ | $Td^5$ (°C) | $Td^{10}$(°C) |
|---|---|---|---|
| PI-1 | 4:1 | 583 | 597 |
| PI-2 | 3:1 | 579 | 591 |
| PI-3 | 2:1 | 574 | 586 |
| PI-4 | 1:1 | 571 | 582 |

As shown in Table 3, comparing the TGA data of different dianhydride ratio polyimide molding powder, it seemed that all four kinds of polyimide molding powder had excellent thermal stability. However, with the increase of ether anhydride groups content, the thermal stability showed a slight decline. The thermal decomposition temperature of PI-1 was the highest and the thermal stability was the best, while the thermal stability of PI-4 was the worst. With the increase of ether anhydride groups content, the content of ether bond in the main chain of polyimide molecules increased, improving the flexibility of macromolecules, so the thermal decomposition temperature gradually decreased, and the thermal stability became worse. This fully showed that ether anhydride groups was the key component to improve the fluidity of molding powder and the flexibility of products. Thus PMDA was an important monomer with high thermal stability of polyimide.

## 4. Conclusion

Ether anhydride polyimide molding powder with different dianhydride ratios was suitable for different molding process. Polyimide molding powder PI-3($n_{PMDA}/n_{OPDA}$= 2:1) had the best comprehensive performance. The suitable molding process was under the molding temperature of 410°C, the molding pressure of 4.0 MPa, and the molding time of 15 min. And the tensile strength of the sample was 93.5Mpa, the elongation at break was up to 10.1%, the vitrification transition temperature was 340°C, and the thermal decomposition temperature of 5% was 574°C, which had good processing performance, resulting in the preparation of mold parts with more complicated structures.

## References

[1] Bershtein V A, David L, Egorova L M, et al. Structure-property relationships in polyimide molecular composites [J]. Materials Research Innovations, 2002, 5(5):230-237.
[2] Wen Dai, Jinhong Yu, Yi Wang, et al. Enhanced thermal and mechanical properties of polyimide/graphene composites [J]. Macromolecular Research. 2014, (9):983-989.
[3] Wilson D. PMR-15 pressing,properties and problems [J]. British Polym, 1998, 20(5): 405~416.
[4] Dong W X , Pei H , Jun S . Press molding conditions of a novel polyimide [J]. Journal of Nanjing

University, 2003.25(5): 9~12.
[5] Zhou H, Liu F, Zhang Y, et al. Novel acetylene-terminated polyisoimides with excellent processability and properties comparison with corresponding polyimides [J]. Journal of Applied Polymer Science, 2011, 122(5): 3493-3503.
[6] Zhang S J, Li Y F, Wang X L, et al. Synthesis and characterization of novel polyimides based on pyridine-containing diamine [J]. Chinese Chemical Letters, 2005, 16(9):1165-1168.

IOP Conf. Series: Materials Science and Engineering **772** (2020) 012038 doi:10.1088/1757-899X/772/1/012038

# Feasibility Study on PRTV of Suppressing insulator Iron Cap from Corrosion

**Ping Tang[1], Jun Zhu[1], You Chen[2], Ming Wei[3,*], Xiaobin Cao[3]**

[1]Sichuan Electric Power Research Institute, Chengdu, China

[2]Skills Training Center of State Grid Chongqing Electric Power Company, Chongqing, China

[3]Electric Engineering School of Southwest Jiaotong University, Chengdu, China

*Corresponding author e-mail: vicweiming@163.com

**Abstract**. UHVDC transmission line porcelain and glass insulator iron caps are seriously corroded, safe and effective anti-corrosion measures are of great significance for the stable operation of the power system. In order to determine the key parameters of PRTV spraying on the surface of insulator iron cap, a three-dimensional finite element model of insulator is established, and the test verified the feasibility of PRTV anti- corrosion of iron cap. The research results show that PRTV can effectively reduce the leakage current density of the insulator iron cap. PRTV can effectively prevent corrosion of the insulator iron cap, and can reduce the surface fouling speed of the insulator, improve the surface resistance of the insulator, and inhibit the insulator flash. The research results can provide reference for the research of anti-corrosion of insulator iron cap.

## 1. Introduction

China has become the country with the largest DC transmission capacity and the longest total transmission line length. [1].With the completion and operation of multiple DC transmission lines, the problem of corrosion of DC insulator iron caps has become increasingly serious. In 2012, more than 2,000 V-shaped string insulators were found to have corrosion of iron caps on the negative side of the DC ±800kV in Hubei province. [2]. The corrosion of the insulator iron cap exhibits a wide distribution range, a large amount of corrosion, and a high corrosion rate. It has a great impact on the mechanical bearing capacity and insulation performance of the insulator, and seriously threatens the safe and stable operation of the power system.

There are few studies on the corrosion of insulator iron caps in foreign countries. The domestic research on the corrosion of insulator iron caps mainly focuses on the accelerated corrosion test method, the calculation of corrosion amount, and the calculation and improvement of the service life of zinc sleeves. Related scholars believe that the main cause of corrosion of insulator iron caps is electrolytic corrosion caused by leakage current [3-12]. However, it can be seen that there is still little research on the corrosion of PRTV coating anti-insulator iron caps.

In this paper, the three-dimensional finite element simulation of the insulator is carried out based on COMSL. The leakage current density and potential of the insulator iron cap when spraying different PRTV coatings are calculated. The influence of PRTV on the leakage current and surface

Content from this work may be used under the terms of the Creative Commons Attribution 3.0 licence. Any further distribution of this work must maintain attribution to the author(s) and the title of the work, journal citation and DOI.
Published under licence by IOP Publishing Ltd

potential of the insulator iron cap is obtained. The anti-corrosion effect of the PRTV coating on the insulator iron cap was verified by an accelerated corrosion test.

## 2. Effects of PRTV on potential and leakage current

PRTV is a common electrical insulation material. Due to its good hydrophobicity, it is often sprayed on the surface of insulator insulation parts to increase the ice-covered flashover voltage and pollution lightning voltage of insulator.

### *2.1. simulation model*

3d modeling of insulators was carried out through Auto CAD, and the basic parameters of edge elements were shown in table 1. Since PRTV coating has basically no effect on the electric field and current density at the ball-and-socket connection, the influence of PRTV coating on the leakage current density and the overall potential distribution on the outer surface of the insulator is mainly analyzed. Therefore, the ball-and-socket connection of the insulator is simplified here. PRTV coating thickness is 1mm. The outer air of the solution domain is a cylinder with the same axis as the insulator. The height of the cylinder is three times the height of the insulator structure, and the bottom radius of the cylinder is three times the radius of the insulator, as shown in figure 1. The Settings of various material parameters are shown in table 2.

**Table 1.** Parameters of insulator

| | Structural height H | Nominal diameter D | Creepage distance L |
|---|---|---|---|
| 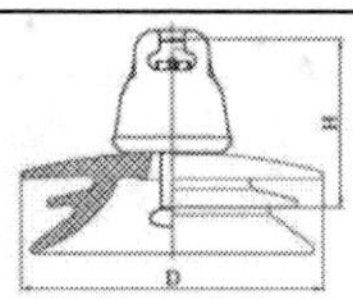 | 195mm | 320mm | 545mm |

**Table 2.** Parameters of various materials

| Material | Iron | Ceramics | Air | Cement | PRTV |
|---|---|---|---|---|---|
| Resistivity/Ω•m | 4E-8 | 1E14 | 1E10 | 1E4 | 1E14 |
| Relative permittivity | 1E10 | 5.7 | 1 | 14 | 6 |

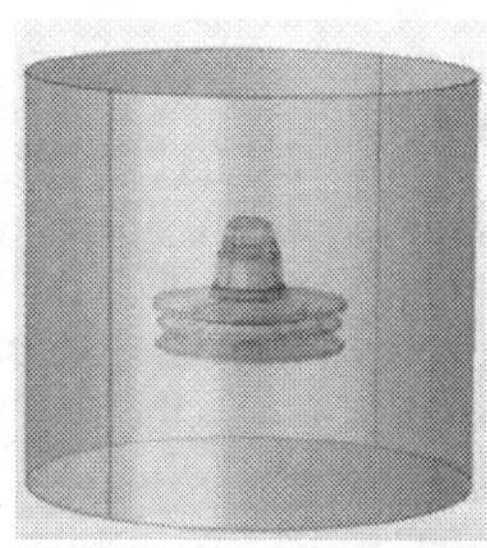

**Fig.1** Three-dimensional computing model

Considering that the potential distribution rule and leakage current density distribution rule of the insulator will not be different with different withstand voltage, only one case under the withstand voltage need to be discussed. In each simulation process, 10000V voltage is added to the iron cap of the insulator while the steel foot is grounded. Additionally, air is set as a natural boundary.

### *2.2. PRTV coating height's influence of leakage current density and potential on iron cap surface*

Four sets of different spraying heights on the iron cap were simulated, that is, the PRTV was not sprayed, the PRTV was sprayed at the gap of the iron cap, the PRTV was sprayed to the middle of the iron cap, and the PRTV was sprayed to the ball joint of the iron cap. The simulation results of spraying PRTV shows the surface potential distribution and leakage current density distribution of the insulator in Fig. 2 and Fig.3. In the picture (a), (b), (c) and (d) respectively represent Simulation results about that the PRTV was not sprayed, the PRTV was sprayed at the gap of the iron cap, the PRTV was sprayed to the middle of the iron cap, and the PRTV was sprayed to the ball joint of the iron cap.

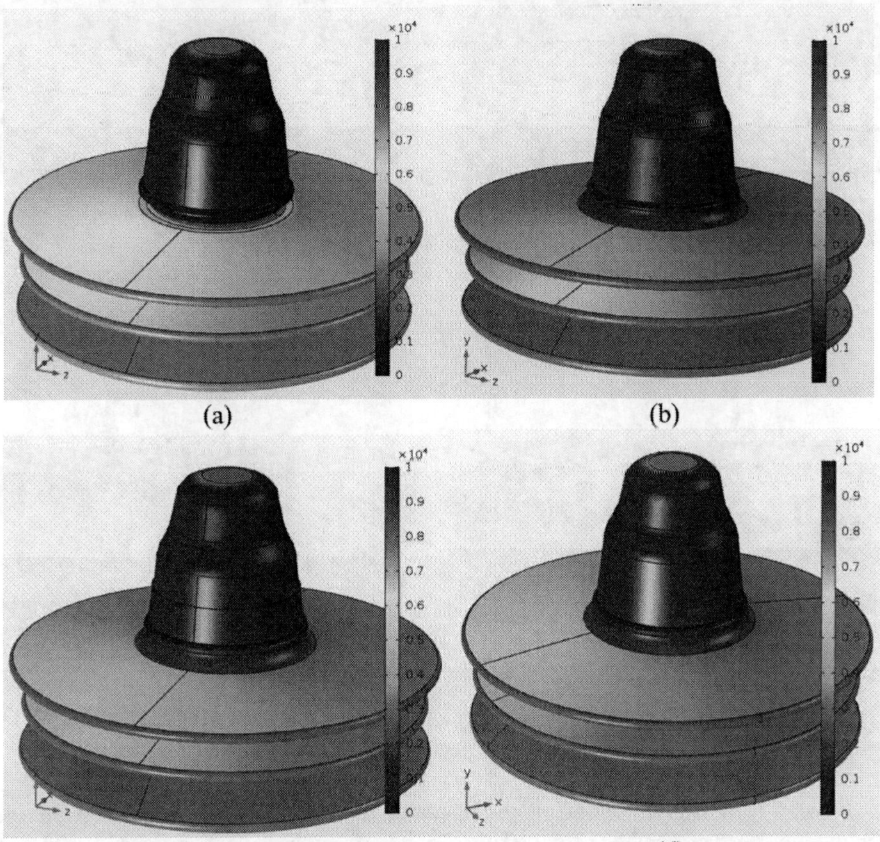

(a) (b)

(c) (d)

**Fig.2** Insulator surface potential distribution

As can be seen from Fig.2, the potential on the insulator ceramics of (b), (c), and (d) is slightly increased compared to (a), but the whole is uniformly distributed, and there is no sudden change in local potential. Comparing (b), (c), (d), it was found that the surface of the insulator iron cap was sprayed with different heights of PRTV coating, and the surface potential of the insulator remained basically unchanged.

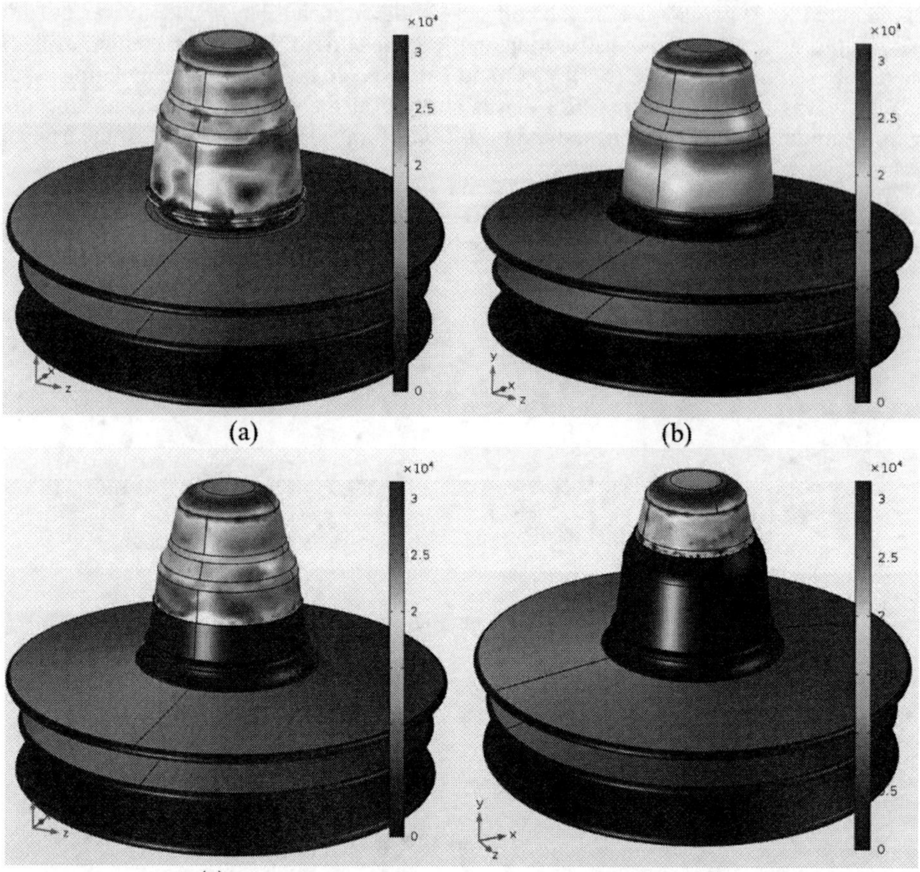

(a) (b) (c) (d)

**Fig.3** Insulator surface leakage current density

As can be seen from Fig.3, the part of the insulator cap is exposed to the air portion , and in this part the maximum current density of (a), (b), (c), and (d) appears near the gap of the iron cap, the middle of the iron cap, the middle of the cap, the end of the PRTV coating. It can be seen that the maximum current density of the insulator iron cap varies depending on the height of the PRTV coating. When the PRTV coating does not cover the gap of the iron cap, the leakage current density at the gap of the iron cap is maximized; when the PRTV coating covers the gap of the iron cap and does not cover the middle portion of the iron cap, at the middle portion of the iron cap and the leakage current density is the highest; when the PRTV coating covers the gap of the iron cap and covers the middle portion of the iron cap, the end leakage current density of the PRTV coating is the largest.

## 3. Accelerated corrosion test

### *3.1. Experiment method*

To simulate the suspension mode of the V-shaped insulator string, test group and control group of insulators are suspended in accordance with the inclination angle of 45°. Reference standard [13] and literature [5], and use the drip method to test for accelerated corrosion test. The 2g of sodium chloride is dissolved in 500mL of water to prepare sodium chloride solution in a certain concentration, and the solution is dropped to the insulator with medical infusion line at a constant speed in order to reduce the resistance of the insulator surface and reduce the duration of the test. For the test group, the sodium chloride solution is added dropwise to the PRTV coating of the insulator's iron cap; for the control group, the sodium chloride solution is added dropwise to the surface of the insulator's iron cap to make sodium chloride solution contact with the iron cap directly. The applied voltage of insulator sample is 25kV in this test, that the iron cap is applied +25kV voltage and the steel feet is grounded. The duration time of voltage application is 60 hours. The test circuit is shown in Fig.4.

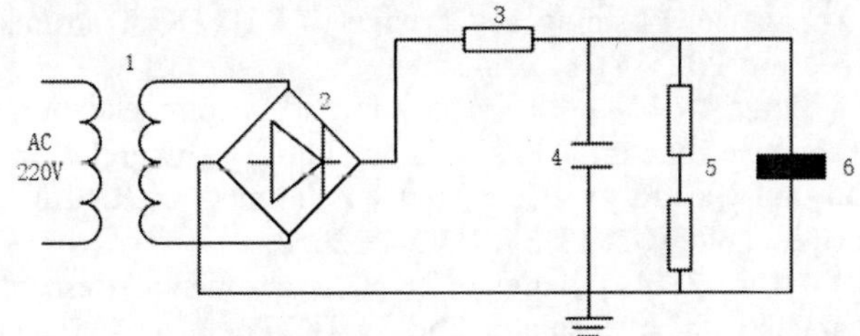


1. Experimental transformer; 2. Fairing; 3. Limiting resistor;
4. Smoothing capacitors; 5. Divider; 6. Test sample

**Fig.4** Test circuit

*3.2. Test results and analysis*
The insulator after accelerated corrosion test is shown in Fig.5.

(a) The control group (b) The test group

**Fig.5** Insulator after accelerated corrosion test

After seen from Fig. 5, the iron cap of the insulator of the control group after accelerated corrosion test appears traces of corrosion, and the porcelain surface of this insulator appears white corrosion products. There was no corrosion mark on the insulator cap of the test group, and there was no corrosion on the porcelain. And the PRTV coating itself does not appear significant local damage and aging phenomenon. From the above, it can be known that the PRTV coating of the iron cap of the insulator can slow down the corrosion of iron cap. Further, the corrosion products of the porcelain surface of insulator obviously decrease when the corrosion of the iron cap of insulator is slowed down, so that this insulator has a surface resistance which is much larger than the surface resistance of the insulator that it's attached by corrosion products, and it makes the flashover voltage of the insulator be raised.

## 4. Conclusion

The following conclusions are drawn from the study of this paper:

1) Spraying PRTV on the surface of the insulator iron cap can reduce the leakage current density of the iron cap surface. The maximum leakage current density on the insulator iron cap increases as the height of the PRTV coating increases;

2) The height of the PRTV coating on the insulator iron cap should take into account two factors such as rainwater dampening and leakage current density reduction. The basic principle is to minimize the height of the PRTV coating while ensuring that the iron cap is not wetted by rain;

3) PRTV can reduce the corrosion of the insulator iron cap, reduce the accumulation of corrosion products on the surface of the insulator, improve the surface resistance of the insulator, and help to improve the flashover voltage of the insulator.

**References**

[1] MA Weimin, FAN Jichao. Planning and Design of UHVDC Transmission System [J]. High Voltage Engineering, 2015, 41(8): 2545-2549.
[2] ZENG Chen, LI Yong, WANG Pin. Analysis and recommendations of insulator gossans corrosion of heavy ice zone in ±800 kV Fufeng line [J]. Hubei Power, 2012, 36(5): 66-67.
[3] SU Zhiyi. Restudy on Electrolytic Corrosion on Fittings of DC Cap and Pin Insulators [J]. Power System Technology, 2013, 37(7): 2047-2052.
[4] GAO Liying, SUN Tao, ZHAO Jiangtao, et al. Experimental Research on Anti-Electrolytic Corrosion Characteristics of Insulator Cap for DC Porcelain Cap and Pin Insulator by Zinc Collar [J]. Power System Technology, 2013, 37(12): 3381-3385.
[5] ZHANG Fuzeng, LIAO Yifan, LUO Ling, et al. Accelerated Electrolytic Erosion Test of Metal Cap of DC Porcelain Insulator for UHVDC Transmission Line [J]. Power System Technology, 2014, 38(2): 311-315.
[6] Luo Ling, Zhang Fuzeng, Li Ruihai, et al. Design for U-Shaped Zinc Ring to Restrain Electrolytic Corrosion of V-String Ceramic Insulators' Caps for DC System [J]. Transactions of China electrotechnical society, 2015, 30(2): 220-228.
[7] LUO Ling, WANG Liming, GUAN Zhicheng, et al. Designing for the Protection Zinc Ring to Suppress Electrolytic Corrosion of Porcelain and Glass Insulators' Iron Caps on HVDC Transmission Lines [J]. Proceedings of the CSEE, 2015, 35(17): 4531-4538.
[8] LUO Ling, ZHANG Fuzeng, REN Guiqing, et al. Feasibility of Using U-shaped Zinc Ring to Restrain Electrolytic Corrosion of Porcelain Insulators' Iron Caps [J]. High Voltage Engineering, 2014, 40(7): 2000-2006.
[9] LI Ruihai, LIAO Yifan, LUO Ling, et al. Research on Hardware Electrolytic Corrosion Problem of Ceramic Insulators on UHVDC Transmission Line [J]. Insulators and Surge Arresters, 2015(1): 1-6.
[10] Ling Luo, Liming Wang, Zhicheng Guan, et al. Influence of Pin Corrosion on Mechanical Characteristic of UHVDC Disc Suspension Insulators and Solutions [J]. IEEE Transactions on Dielectrics and Electrical Insulation, 2015, 22(4): 2242-2251.
[11] Fuzeng Zhang, Bing Luo, Guoli Wang, et al. Study of Test Method for Accelerated Electrolytic Corrosion of Iron Caps of DC Disc Porcelain Insulators [J]. IEEE Transactions on Dielectrics and Electrical Insulation, 2015, 22(5): 2937-2943.
[12] Ling Luo, Liming Wang, Zhicheng Guan, et al. Suppression Method for Electrolytic Corrosion of Porcelain Insulators' Iron Caps on UHVDC Transmission Lines [J]. IEEE Transactions on Dielectrics and Electrical Insulation, 2014, 21(6): 2448-2457.
[13] Accelerated electrolytic corrosion test methods for metal fitting of cap and pin type ceramic or glass insulators for DC systems [S]. Beijing: China electric power press, 2016.

# Synthesis and Electrochemical Performance of Zn-Doped $LiMn_2O_4$ as Cathode Material for Lithium Ion Batteries

**Hao Zheng, Ting Wang, Jinsong Cheng, Lin Li*, Qing Liu and Rongfei Zhao**

Key Laboratory of Functional Materials and Chemistry for Performance and Resource of Guizhou Education Department, Anshun University, Anshun, China

*Corresponding author e-mail: 404003375@qq.com

**Abstract.** The spinel $LiMn_2O_4$ and Zn-doped $LiMn_2O_4$ samples were prepared by a sol-gol method followed by annealing. The structure and morphology of the as-prepared material were analyzed by X-ray diffraction and electron microscopy. The results obtained show that those samples possess a perfect spinel structure. The electrochemical measurements are carried out to study the properties of the $LiMn_2O_4$ and Zn-doped $LiMn_2O_4$. The results indicate that Zn-doped $LiMn_2O_4$ manifests enhanced high capability, rate capability, and superior cycling stability. Especially, the $LiMn_{1.9}Zn_{0.1}O_4$ sample show good electrochemical performance (the reversible discharge capacity of 120 mAh/g at 1C, after 100 cycle).

## 1. Introduction

Spinel-type $LiMn_2O_4$ has been regarded as one of the most prospective cathode materials, due to its low cost, environmental friendliness, high safety, low toxicity and the abundance of manganese resources [1]. However, the application of spinel $LiMn_2O_4$ in high power systems is hindered by problems such as its kinetic problems associated with low electronic conductivity and lithium diffusion coefficient, rapid capacity fading during cycling, especially at elevated temperature [2]. To solve these issues, many attempts have been made to decrease the cation mixing and to strengthen the material structure, such as metal-ion doping and surface modification [3]. Reducing the amount of $Mn^{3+}$ in $LiMn_2O_4$ can depress the Jahn-Teller distortion and therefore improve the cycleability of the materials.

In recent years, several attempts have been made for synthesizing modified $LiMn_2O_4$ doped with various elements to inhibit capacity fading and improve the electrochemical performance [4]. such as B, F, S, Mg, Al, Ti, Zn, Ga, Zr, La, Ce, Nd. The doping method shows better electrochemical performance for it can weaken the Jahn-Teller effect and improve the stability of the structure [5, 6]. The increased stability of the structure and the weakened Jahn-Teller effect can further improve the cycle performance [7].

In this work, we synthesized a series of Zn-doped $LiMn_2O_4$ samples by a sol-gel method followed by annealing. The structural characterization and electrochemical measurements are carried out to study the properties of all the $LiMn_2O_4$ samples. The results indicate that $LiMn_{1.9}Zn_{0.1}O_4$ manifests better electrochemical properties.

Content from this work may be used under the terms of the Creative Commons Attribution 3.0 licence. Any further distribution of this work must maintain attribution to the author(s) and the title of the work, journal citation and DOI.
Published under licence by IOP Publishing Ltd

## 2. Experimental Procedure

The spinel $LiZn_xMn_{2-x}O_4$ (x = 0.00, 0.05, 0.10 and 0.2) were synthesized by sol-gel method. Stoichiometric amount of $LiNO_3$, $Mn(OCOCH_3)_2 \cdot 4H_2O$, $Zn(NO_3)_2 \cdot 6H_2O$, and citric acid were (the ratio of Li:Zn:Mn: citric acid is 1:x:2−x:3.6) dissolved in distilled water. Next, the mixed solution was placed in a water bath as a sol-gel at 80 °C, and when it was stirred to a gel-like substance, it was placed in a drying oven at 120 °C for 12 h to obtain different samples. The obtained samples were ground, placed in a tube furnace, pre-fired at 800 °C for 10 hours in the air to obtain the precursor.

XRD measurements were carried out on a Bruker D8 Advance X-ray diffractometer. The morphologies were characterized by field-emission scanning electron microscopy (SEM, JEOLJSM-7400F, Japan).

Electrochemical performances of these samples were tested in a model test cell system. Positive electrodes were prepared by pressing the mixtures of the active material (80%), acetylene black (10%), and polytetrafluoroethylene (PTFE) (10%) onto a nickel grid. Prior to being used, the positive electrodes were dried at 120 °C in a vacuum furnace for 24 h. The electrolyte was a solution of 1 M $LiPF_6$/ethylene carbonate (EC) and diethyl carbonate (DEC) (1:1 by volume). The separator was Celgard 2400 porous polypropylene. The counter and reference electrodes were lithium foil. The model test cells were assembled in an argon-filled dry box. Charge-discharge tests were carried out at different current densities in the range of 3.0 V to 4.5 V.

## 3. Results and discussion

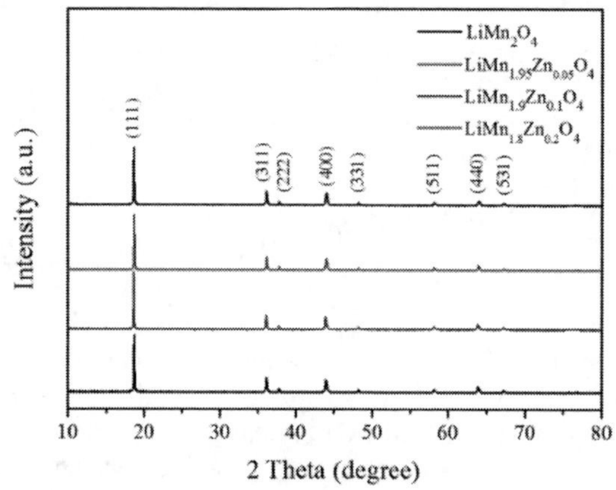


**Figure 1.** Powder XRD patterns of (a) $LiMn_2O_4$, (b) $LiMn_{1.95}Zn_{0.05}O_4$, (c) $LiMn_{1.9}Zn_{0.1}O_4$, and (d) $LiMn_{1.8}Zn_{0.2}O_4$

The X-ray diffraction patterns of the pure $LiMn_2O_4$ and $LiMn_{2-x}Zn_xO_4$ (x = 0.05, 0.1, and 0.2) samples are presented in Fig. 1. The XRD patterns (Fig. 2) of these samples are similar with the structure of spinel $LiMn_2O_4$. It can be clearly seen that all the diffraction peaks are in agreement with those from the JCPDS 35-0782 card, no other phase or impurities were detected.

The morphologies of the pristine $LiMn_2O_4$ and Zn-doped $LiMn_2O_4$ samples are shown in Fig. 2. These samples are composed of many small polycrystalline particles, and agglomerated with the particle size. As shown clearly in Fig. 3a, the sample particles have an octahedral shape. The pristine $LiMn_2O_4$ sample have an un-uniform particle size of 0.1–1.5 μm. It can be seen from Fig. 2b–d that the grain size and crystallinity are smaller with the Zn ion doped. The particles of $LiMn_{1.95}Zn_{0.05}O_4$ sample are agglomerated severely with the particle size from 0.1 to 1.0 μm in Fig. 2b. When the doping amount of 0.1, with a uniform particle size distribution between 0.5 to 1.0 μm of the $LiMn_{1.9}Zn_{0.1}O_4$ sample. However, where the doping amount of 0.2, the particle sizes grow from 0.5 to 1.5 μm in Fig. 2d.

**Figure 2.** SEM images of (a) $LiMn_2O_4$, (b) $LiMn_{1.95}Zn_{0.05}O_4$, (c) $LiMn_{1.9}Zn_{0.1}O_4$, and (d) $LiMn_{1.8}Zn_{0.2}O_4$.

The typical charge and discharge curves of the pristine and doped electrodes in the voltage range between 3.0 and 4.5 V at a low current density of 1 C (148 mA/g) at room temperature, as shown in Fig. 3. From the curves, it can be clearly seen that there are two potential plateaus near 4 V and 4.1 V, which indicates that the insertion and extraction of $Li^+$ that occurs in spinel undergo two stages. One of the voltage plateaus, at about 4.0 V, is associated with the removal of $Li^+$ from half of the tetrahedral sites, in which Li–Li interactions occur; the other, observed at about 4.1 V, is attributed to the removal of lithium ions from the other tetrahedral sites. The pristine $LiMn_2O_4$ delivers an initial discharge capacity of 125 mAh/g, it only had 101 mAh/g after 100 cycles: the loss of the initial value was 19.2 %. But after doping, the initial discharge capacities of the samples $LiMn_{1.95}Zn_{0.05}O_4$, $LiMn_{1.9}Zn_{0.1}O_4$, and $LiMn_{1.8}Zn_{0.2}O_4$ electrodes were 123, 128, and 124 mAh/g, and they can retain 109, 120, and 109 mAh/g, after 100 cycles, respectively. It can be found that $LiMn_{1.9}Zn_{0.1}O_4$ had the best cycling performance among the four samples. It could be concluded that there are three possible reasons that result in improved electrochemical performance for the doped electrodes [4].

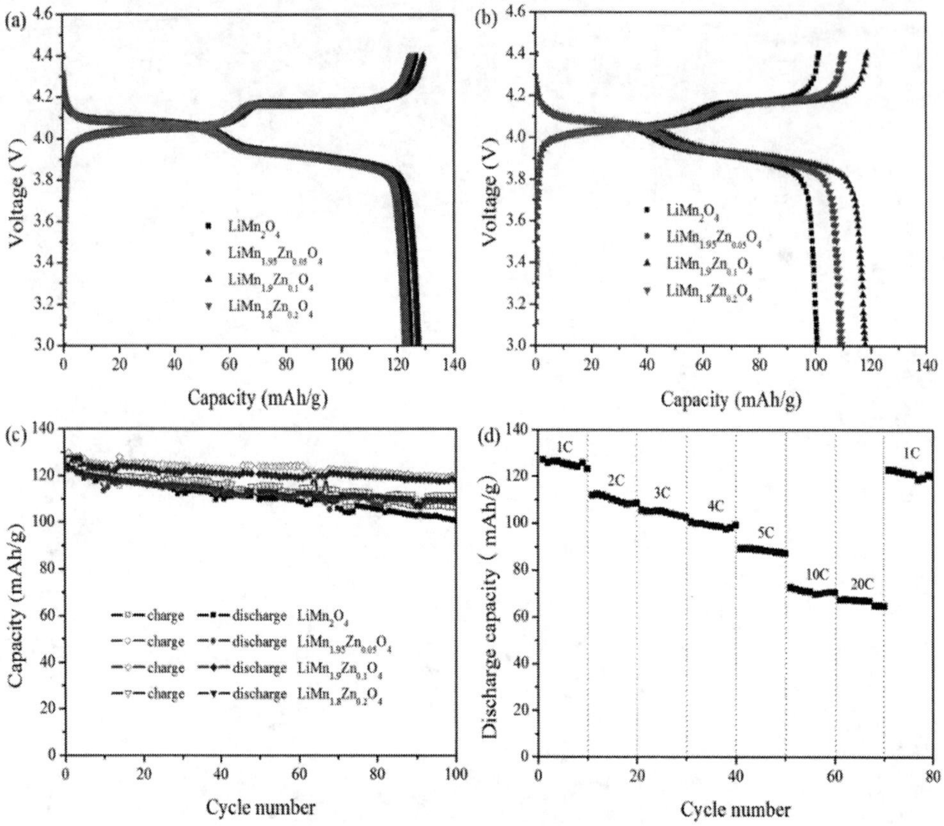


Figure 3. Typical charge/discharge curves for (a) 1st and (b) 100 cycles at a low current density of 1 C, (c) cycling performance, and (d) rate performance of $LiMn_2O_4$, $LiMn_{1.95}Zn_{0.05}O_4$, $LiMn_{1.9}Zn_{0.1}O_4$, and $LiMn_{1.8}Zn_{0.2}O_4$ samples.

Fig.3d shows the cycling performances of the $LiMn_{1.9}Zn_{0.1}O_4$ electrode cycled between 4.5 and 3.0 V at various current densities at room temperature. When the current densities are 1, 2, 3, 4, 5, 10, and 20 C, the discharge capacities of the $LiMn_{1.9}Zn_{0.1}O_4$ electrode are 127, 112, 106, 100, 89, 73, and 67 mAh/g, respectively. When the current density back to 1C, the discharge capacities can retain 123 mAh/g, respectively. Although the discharge capacities decrease with increasing current density, the $LiMn_{1.9}Zn_{0.1}O_4$ electrode displays good capacity retention at different current densities, which indicating its good reversibility and high rate capability.

## 4. Conclusions

The spinel $LiMn_2O_4$ and Zn-doped $LiMn_2O_4$ samples were prepared by a sol-gol method followed by annealing. The structure and morphology of the as-prepared material were analyzed by X-ray diffraction and electron microscopy. The results obtained show that those samples possess a perfect spinel structure. The electrochemical measurements are carried out to study the properties of the $LiMn_2O_4$ and Zn-doped $LiMn_2O_4$. The results indicate that Zn-doped $LiMn_2O_4$ manifests enhanced high capability, rate capability, and superior cycling stability.

**Acknowledgments**

This work was financially supported by the joint science and technology funds of the Guizhou Provincial Education Department (No. KY [2014] 322 and KY [2018] 031).

**References**

[1] F. Cheng, J. Liang, Z. Tao, J. Chen, Functional materials for rechargeable batteries, Adv. Mater. 23 (2011)1695-1715.

[2] M.S. Whittingham, Lithium batteries and cathode naterials, Chem. Rev. 104 (2004) 4271-4302

[3] H. Xia, Z.T. Luo, J.P. Xie, Nanostructured LiMn2O4 and their composites as high-performance cathodes for lithium-ion batteries, Prog. Nat. Sci. Mater. Int. 22 (2012) 572-584.

[4] C.Q. Feng, H. Li, C.F. Zhang, Z.P. Guo, H.M. Wu, J. Tang, Synthesis and electrochemical properties of non-stoichiometric Li–Mn-spinel (Li1.02MxMn1.95O4-yFy) for lithium ion battery application, Electrochim. Acta 61 (2012) 87-93.

[5] C. Peng, H. Bai, M. Xiang, C. Su, G. Liu, J. Guo, Effect of Calcination Temperature on the Electrochemical Properties of Spinel LiMn2O4 Prepared by Solid State Combustion Synthesis. Int. J. Electrochem. Sci. 9 (2014) 1791-1798.

[6] M. Fehse, R. Trócoli, E. Ventosa, E. Hernández, A. Sepúlveda, A. Morata, A. Tarancón, Ultrafast Dischargeable LiMn2O4 Thin-Film Electrodes with Pseudocapacitive Properties for Microbatteries, ACS Appl. Mater. Interfaces 9 (2017) 5295-5301.

[7] L.J. Xi, H.E. Wang, Z.G. Lu, S.L. Yang, R.G. Ma, J.Q. Deng, C.Y. Chung, Facile synthesis of porous LiMn2O4 spheres as positive electrode for high-power lithium ion batterie, J. Power Sources 198 (2012) 251-257.

# Synthesize of $NiMoO_4$/Reduced Graphene Oxide Composite as High Capacity Anode Material for Lithium Ion Batteries

**Lin Li[1,2,*], Jinsong Cheng[1], Qing Liu[1], Ting Wang[1] and Rongfei Zhao[1]**

[1]Key Laboratory of Functional Materials and Chemistry for Performance and Resource of Guizhou Education Department, Anshun University, Anshun, China
[2]School of Chemical Engineering, Guizhou University of Engineering Science, Bijie, China

*Corresponding author e-mail:404003375@qq.com

**Abstract.** $NiMoO_4$/Reduced graphene oxide ($NiMoO_4$/rGO) composites material was prepared by a hydrothermal method. When used as a anode material for lithium-ion batteries, the $NiMoO_4$/rGO composites show superior rate capability and excellent cycling stability. A reversible capacity of approximately 808 mAh $g^{-1}$ is retained even after 100 cycles at 100 mA $g^{-1}$. It is believed that the surface modification by graphene nanosheets contribute to the improved kinetics of lithium-ion diffusion, excellent structural stability and superior electrochemical performance.

## 1. Introduction

Lithium-ion batteries (LIBs) have become the most widely used energy storage systems for portable electronic devices such as laptops, mobile phones, medical microelectronic devices, and electrical vehicles due to their many outstanding features, including high energy density, no memory effect, low maintenance, and little self-discharge [1, 2]. However, the growing requirements for better LIBs requires constant innovation, in terms of improved safety, longer lifetime, smaller size, lighter weight, and lower cost [3]. Transition metal molybdates can be considered as prospective anode materials due to the ability of the metal ions to exist in several oxidation states in these oxides, ranging from 3+ to 6+ for Mo and reversibly reacting with Li delivering high capacity, at potentials lower than 2 V. Molybdenum can form stable oxides with many metals with stoichiometry, $MMoO_4$, where M = Co, Zn, Mn, Cu, Ni, Mg, Pb, Ca, Fe, Cd [4]. $NiMoO_4$ have gradually been considered as promising effective and scalable alternatives to the commercial graphite anode for LIBs in view of their excellent electrochemical performance, low-cost, environment benignity and abundance. Most importantly, $NiMoO_4$ has been reported to demonstrate better performance than single-component metal oxides due to its multiple feasible oxidation states, large cell parameters and comparative higher electrical conductivity ($10^{-6}$ S $cm^{-1}$) [5]. In this work, we have prepared $NiMoO_4$ and $NiMoO_4$/rGO composites using the hydrothermal method. When the prepared materials were evaluated as anode for lithium ion batteries, they shows good electrochemical performance, in terms of rate capability, and cycling stability.

## 2. Experimental

All chemicals were of analytical grade and were used without further purification. GO was made using a modified Hummers method. In a typical experiment, 3 mmol of $NiCl_2 \cdot 6H_2O$ and 3 mmol of

Content from this work may be used under the terms of the Creative Commons Attribution 3.0 licence. Any further distribution of this work must maintain attribution to the author(s) and the title of the work, journal citation and DOI.
Published under licence by IOP Publishing Ltd

$Na_2MoO_4 \cdot 7H_2O$ were dissolved successively in 40 mL GO solution (1 mL $g^{-1}$) with magnetic stirring at room temperature. After treating the above mixture at 180 °C for 12 h in a 50 mL Teflon-lined stainless steel autoclave, it was cooled to room temperature. The product was collected, washed with deionized water and absolute ethanol, and dried in air at 80 °C. For comparison, $NiMoO_4$ sample was also synthesized via the same procedure without the addition of GO.

Powder X-ray diffraction patterns of the samples were obtained with a Rigaku XRD diffractometer. TEM images were taken on a JEOL JEM-2100F microscope. XPS measurements were performed on a KRATOS XSAM800 X-ray photoelectron spectrometer with a Mg Ka excitation source (1253.6 eV).

The electrochemical characterization was performed using CR2032 coin-type test cells. The cell consisted of a cathode with the composition of 70 wt.% active materials, 20 wt.% carbon black, and 10 wt.% PVDF, a lithium metal anode separated by a Celguard 2400 microporous film. The electrolyte was 1 mol $L^{-1}$ $LiPF_6$/ (EC) and (DEC) (1:1). Charge-discharge tests were carried out at different current densities in the range of 0.01 V to 3.0 V. Electrochemical impedance spectroscopy (EIS) experiments were conducted in the frequency range of 0.01-100 kHz on a CHI 600A electrochemical analyzer.

## 3. Results and discussion

Fig. 1a illustrates the XRD patterns of the $NiMoO_4$ and $NiMoO_4$/rGO samples. All diffraction lines can be indexed to the monoclinic $NiMoO_4$, according to the standard XRD data file (JCPDS file no. 12-0348). No peaks of impurities can be detected from the XRD pattern. The chemical composition and the oxidation state of the as-prepared $NiMoO_4$/rGO composites are identified from XPS measurements and the typical survey spectra are given in Fig. 1. As shown in Fig. 1b, the observed peaks correspond to Ni 2p, Mo 3d, C 1s, and O 1s confirm the presence of Ni, Mo, C and O elements within the as-prepared $NiMoO_4$/rGO composites, and the absence of other impurities. The peaks located at 878.6, 871.8, 860.4, and 853.9 eV identified with the binding energy of Ni 2p1/2 and 2p3/2. The peaks appearing at 235.2 and 232.1 eV are attributable to the binding energy of Mo 3d3/2 and Mo 3d1/2. The O 1s orbital spectrum exhibiting a broad peak at 530.7 eV can be fitted by oxygen in the metal–oxygen bonds (Fig. 1e). Moreover, the C 1s orbital spectrum (Fig. 1f) at 284.8 eV further confirmed the existence of the rGO.

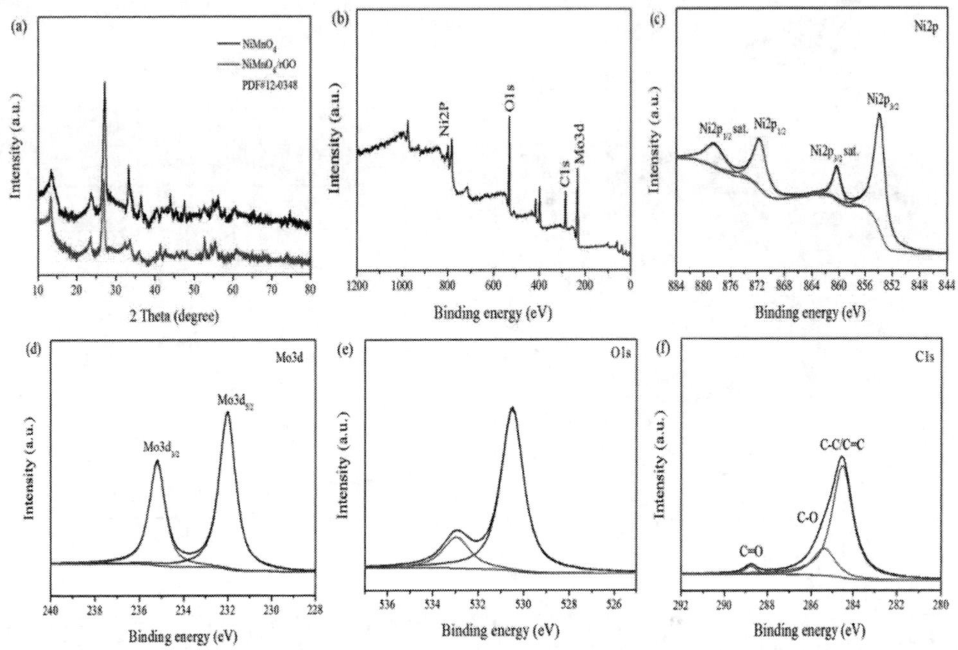


**Figure 1.** (a) XRD patterns of the $NiMoO_4$ and $NiMoO_4$/rGO; XPS spectra of (b) survey spectrum, (c) Ni 2p, (d) Mo 3d, (e) O 1s, and (f) C 1s for the $NiMoO_4$/rGO composites.

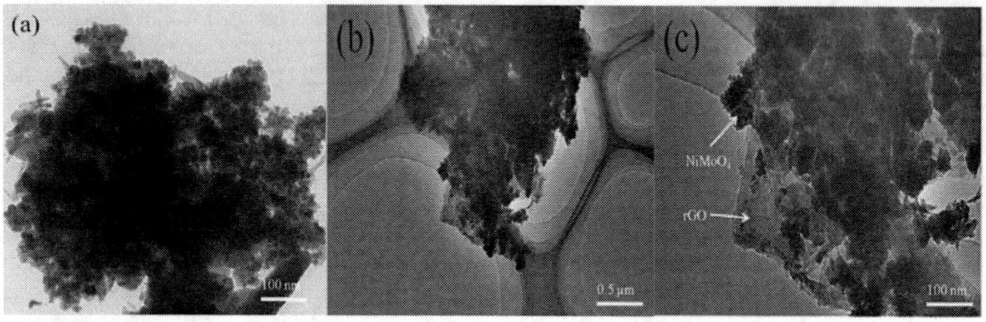


**Figure 2.** Typical TEM images the (a) $NiMoO_4$ and (b, c) $NiMoO_4$/rGO composites.

The surface morphological features of the $NiMoO_4$ and $NiMoO_4$/rGO were investigated by TEM. As shown in Fig.2, they are all composed of non-uniform particles with different sizes. From Fig. 2a, it can be observed that the particles of the bare $NiMoO_4$ sample was agglomerated with the particles sized with ca. 20–100 nm. As shown in Fig. 2b and c, the surfaces of the $NiMoO_4$ were coated with graphene layers. Obviously, the $NiMoO_4$ of the $NiMoO_4$/rGO sample maintained the identical size and morphology with $NiMoO_4$, and the rGO layers are tightly anchored on the surface of the $NiMoO_4$ to form a composite structure. Such hybrid materials can effectively reduce the volume expansion during the lithiation and delithiation processes owing to their better mechanical strength.

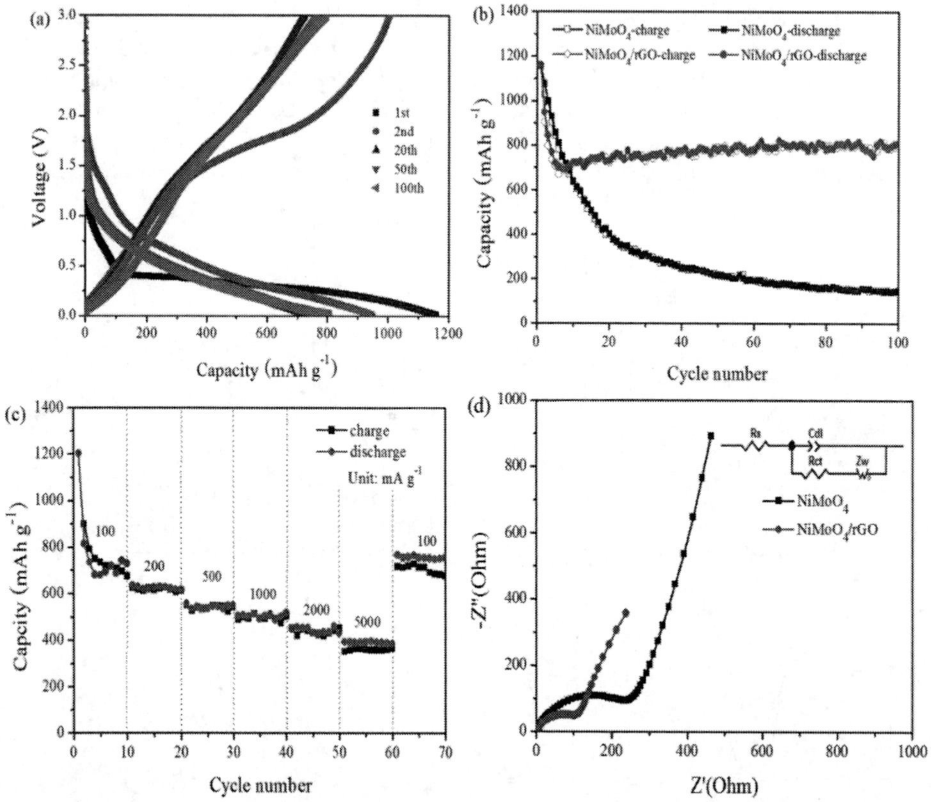


**Figure 3.** (a) Typical discharge/charge curves of $NiMoO_4$/rGO, (b) cycling performance of the $NiMoO_4$ and $NiMoO_4$/rGO electrodes, (c) cycling performance of the $NiMoO_4$ and $NiMoO_4$/rGO electrodes at various current densities, (d) EIS of $NiMoO_4$ and $NiMoO_4$/rGO.

Fig. 3a, b illustrates the electrochemical performance of all the $NiMoO_4$ samples as anode materials for rechargeable lithium batteries at a current density of 100 mA $g^{-1}$ in a potential window between 0.01 and 3.0 V (versus $Li^+/Li$). The specific capacity of the $NiMoO_4$ showed an obvious decrease with cycling, from 1157 mAh $g^{-1}$ for the first cycle to 146 mAh $g^{-1}$ for the 100$^{th}$ cycle, whereas the $NiMoO_4$/rGO delivers a larger initial capacity of 1160 mAh/g and a high reversible capacity of 808 mAh $g^{-1}$ after 100 cycles, which is much higher than what was reported for pure $NiMoO_4$ samples. The excess capacities could be associated with the decomposition of the electrolyte at low voltages generating a solid electrolyte interphase (SEI) layer and the further lithium storage by interfacial charging at the metal/$Li_2O$ interface. It is obvious that the $NiMoO_4$/rGO electrode shows much improved cycling performance with higher specific capacities at the same cycle with the same current density, as compared with the other samples, which may be attributed its bigger porous structure. To better understand the electrochemical behavior of the $NiMoO_4$/rGO electrode, we also investigated its rate performance as shown in Fig.3c. The cell shows good rate capability with average discharge capacity of 1203, 634, 557, 505, 449, and 393 mAh $g^{-1}$ when the current density increased stepwise to 100, 200, 500, 1000, 2000, and 5000 mA $g^{-1}$, respectively. Upon altering the current density back to 100 mA $g^{-1}$, an average discharge capacity as high as 766 mAh $g^{-1}$ could be recovered. These results demonstrate that the $NiMoO_4$/rGO electrode has great potential as high-rate anode material in lithium-ion batteries. To reveal the superior electrochemical performance of $NiMoO_4$/rGO compared with $NiMoO_4$ for lithium energy storage, the charge-transfer resistance was tested by EIS over the frequency domain from0.01 Hz to 100 kHz (Fig.3d). The result showed that the resistance of the $NiMoO_4$/rGO electrode was 105 Ω, which is lower than that of pure $NiMoO_4$ (237 Ω). So, $NiMoO_4$/rGO shows better conductivity than $NiMoO_4$, that is, the $NiMoO_4$/rGO electrode has accommodated the high current density during the cycles. It is why $NiMoO_4$/rGO has a better cycling stability than $NiMoO_4$.

## 4. Conclusion

Herein, the $NiMoO_4$/rGO composites anode material was prepared by a hydrothermal method. When used as a anode material for lithium-ion batteries, the $NiMoO_4$/rGO composites show superior rate capability and excellent cycling stability. A reversible capacity of approximately 808 mAh $g^{-1}$ is retained even after 100 cycles at 100 mA $g^{-1}$, approximately 88.3% of the initial discharge capacity. It is believed that the surface modification by graphene nanosheets contribute to the improved kinetics of lithium-ion diffusion, excellent structural stability and superior electrochemical performance.

## Acknowledgments

This work was financially supported by the joint science and technology funds of the Guizhou Provincial Education Department (No. KY [2014] 322 and KY [2018] 031).

## References

[1] Y. Zhao, X.F. Li, B. Yan, D.B. Xiong, D.J. Li, S. Lawes, X.L. Sun, Recent Developments and Understanding of Novel Mixed Transition-Metal Oxides as Anodes in Lithium Ion Batteries, Adv. Energy Mater. 6 (2016) 1502175.

[2] H. Zheng, X. Chen, Y. Yang, L. Li, G.H. Li, Z.P. Guo, C.Q. Feng, Self-Assembled LiNi1/3Co1/3Mn1/3O2 Nanosheet Cathode with High Electrochemical Performance, ACS Appl. Mater. Interfaces 9 (2017) 39560-39568.

[3] G.D. Park, J.H. Hong, J.K. Lee, Y.C. Kang, Yolk–shell-structured microspheres composed of N-doped-carbon-coated NiMoO4 hollow nanospheres as superior performance anode materials for lithium-ion batteries, Nanoscale 11 (2019) 631-638.

[4] J.H. Ahn, G.D. Park, Y.C. Kang, J.H. Lee, Phase-pure β-NiMoO4 yolk-shell spheres for high-performance anode materials in lithium-ion batteries, Electrochimica. Acta 174 (2015) 102-110.

[5] H. Zheng, S.Q. Wang, J.Z. Wang, J. Wang, L. Li, Y. Yang, C.Q. Feng, Z.Q. Sun, 3D Fe2(MoO4)3 microspheres with nanosheet constituents as high-capacity anode materials for lithium-ion batteries, J Nanopart Res. 17 (2015) 449.

# Design of Control System for 6-DOF Manipulator

**Xianglong Li[1, *], Zikun Quan[2, a] and Dongping Liu[3, b]**

[1]School of Jinan, Qilu normal University, Jinan 250000, China;
[2]School of Weihai, Shandong University, Weihai 264200, China.
[3]CO., LTD of Shenzhen, SHENZHEN FENGPING A.I. CO.,LTD, Shenzhen 580000, China.

*Corresponding author e-mail: 8581998821xl@gmail.com, [a]304916550@qq.com, [b]liud@fengping.tech

**Abstract.** The manipulator is a complex system with high precision, high nonlinearity and strong coupling, the design of the manipulator control system has an important influence on the stable operation of the manipulator. The structural design, drive design and hardware design of the manipulator are the key contents of the manipulator, which needs to be coordinated and controlled as a whole. Through the establishment of a six-degree-of-freedom manipulator dynamics model, and the design of the manipulator control system is conducted in this paper.

**Keywords:** control system; manipulator; six-degree-of-freedom.

## 1. Introduction

The research on the six-degree-of-freedom (6-DOF) manipulator control system is mainly to make the manipulators complete the specified motions in accordance with the established requirements of the people; which require planning the motion trajectory of the manipulators, then the position control is used in each joint axis to drive the manipulator to achieve the desired moving target, and the selection of control algorithm is a key step to ensure the smoothness and accuracy of the manipulator. The selection design of control algorithm generally requires a kinematic model based on the manipulator system, in addition, it should consider both the simplicity of the system control and the motion state of the system, a 6-DOF manipulator dynamics model is established in this paper, based on this, the design and research of the manipulator control system is carried out.

## 2. Technology Development of 6-DOF Manipulator

### *2.1. Research status of 6-DOF manipulator control system*

The 6-DOF manipulator is usually controlled by many motors in coordination, in order to achieve effective communication among different motors, it is necessary to establish the data communication system between the main controller and each motion control unit, at present, the commonly used control system architecture includes the central type and distributed type. The central system architecture is a relatively traditional control method, which can meet the control requirements of most systems, but when the central controller fails, it will have a serious impact on the whole control system. Moreover,

Content from this work may be used under the terms of the Creative Commons Attribution 3.0 licence. Any further distribution of this work must maintain attribution to the author(s) and the title of the work, journal citation and DOI.
Published under licence by IOP Publishing Ltd

due to the complexity of the central system architecture, it is difficult to maintain in the later stage. The distributed control system is a microprocessor that distributes control signals to each control node, under this control system, the main controller only needs to implement the functions of bus connection and manipulator kinematics algorithm, which can simplify the input of control equipment and improve the control stability of the whole system.

### *2.2. Research status of trajectory planning of 6-DOF manipulator*

The trajectory planning of the 6-DOF manipulator is the basis for ensuring the accurate motion and stable control of the manipulator, it is conducive to improve the motion efficiency of the manipulator and improve the tracking error of the system. Under normal conditions, performance optimization indicators are used to describe the trajectory planning performance of manipulator. The research on the manipulator trajectory planning algorithm with time optimization is mature, according to the different trajectory planning; it can be divided into the optimal time motion trajectory algorithm along the pre-set path and the point-to-point processing algorithm under the optimal time. The optimal time motion trajectory algorithm along the pre-set path considers the position, velocity, acceleration and second-order acceleration of the manipulator, the trajectory curve that meets the requirements can be obtained by a high-order polynomial curve connecting the key points in the joint space of the manipulator. The point-to-point processing algorithm under optimal time is mainly to abstract multiple working points in space, and requires the manipulator joints to move at each working point to complete the pre-set task, the motion of each joint is linear and independent and need to have a stay time at the working point, the current common method is to use polynomial interpolation among nodes to conduct trajectory planning.

## 3. Algorithm Design of 6-DOF Manipulator Control System

### *3.1. Establishment of kinematics model of 6-DOF manipulator*

As shown in Fig.1, the mechanical structure of 6-DOF manipulator belongs to 3D open-loop chain structure, which is composed of a base and a rotating arm from top to bottom, these parts are connected by six rotating joints, by controlling these six parts, the full position of the manipulator in the workspace can be achieved.

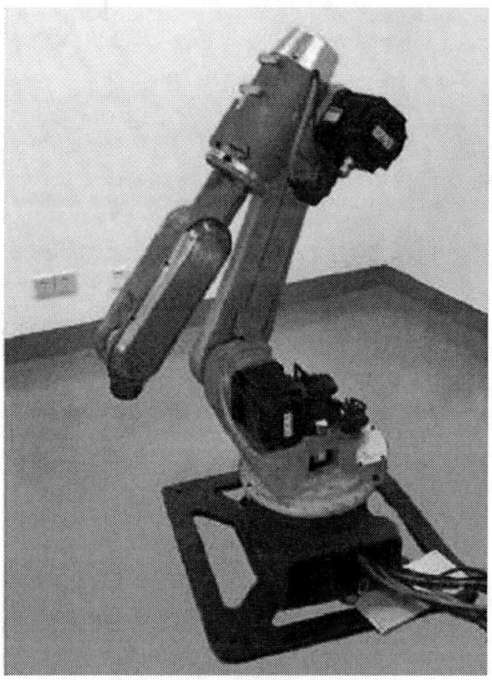

**Fig. 1** manipulator model

IOP Conf. Series: Materials Science and Engineering **772** (2020) 012041 doi:10.1088/1757-899X/772/1/012041

In this paper, the coordinate system is established for the six rotating joints in accordance with the right-hand rule and the D-H principle, according to the spatial geometric relationship among the two adjacent members of manipulator, the positive kinematics equation of the 6-DOF manipulator can be established, the relationship between joint coordinates and mechanical end position can be accurately reflected, but the inverse kinematics mode is more common in practical applications. According to the inverse kinematics model, the manipulator controller can calculate the angle value that each joint needs to reach when completing the ideal motion trajectory, the control system of the manipulator is realized by setting the position of the shutdown. After the selection of the manipulator configuration, the kinematic parameters need to be determined, according to the mechanical arm configuration of Fig.1, the reference coordinate system and joint coordinate system of the manipulator are established according to the manipulator configuration of Fig.1, as shown in Fig.2, because the coordinate system has offsets a1 and a2, therefore, the 1, 2, and 3 coordinate system does not coincide with x-axis, and the spatial relationship of x1, x2, and x3 is shown in Fig.3.

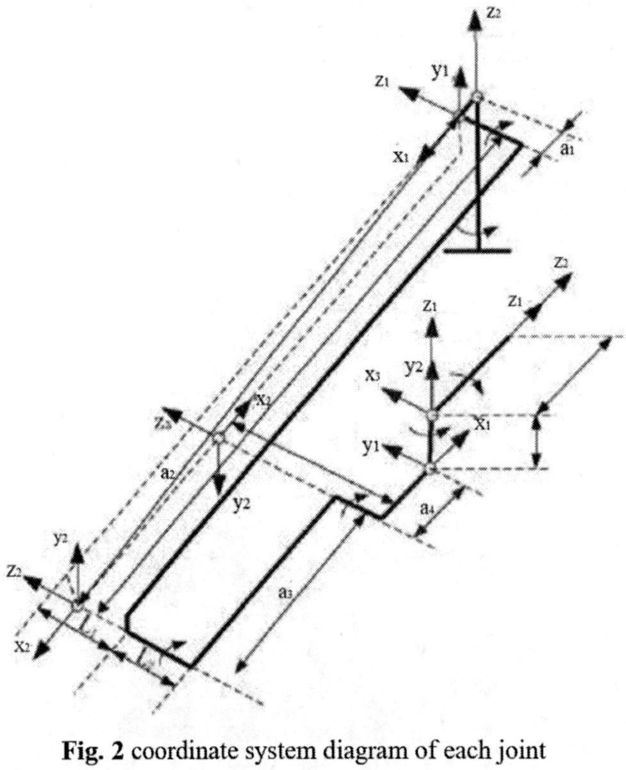

**Fig. 2** coordinate system diagram of each joint

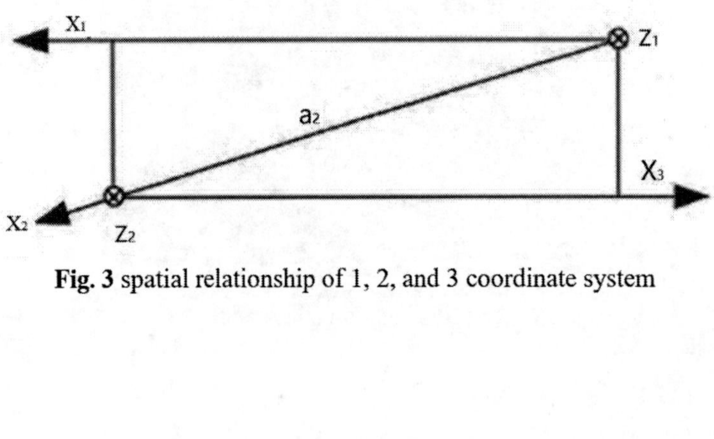


**Fig. 3** spatial relationship of 1, 2, and 3 coordinate system

The kinematic parameters of the 6-DOF manipulator should include the angle of rotation θ around the z-axis, the distance d between two adjacent common perpendiculars on the z-axis, the length of common perpendicular, and the included angle α between two adjacent z-axes. In this paper, the motion parameters are determined by the D-H rule, the homogeneous equation is used to describe the spatial geometric relationship of each connecting rod of manipulator with respect to the reference coordinate system, the 4×4 homogeneous matrix is used to describe the geometric relationship between adjacent connecting rod, then the end position of the manipulator relative to the reference coordinate system can be derived, the D-H parameters of the manipulator in this design are shown in Table 1.

**Table 1.** D-II parameter list

| two adjacent coordinate systems | θ | d | a | α |
|---|---|---|---|---|
| 0 – 1 | θ1 | 0 | a1 | 90° |
| 1 – 2 | θ2 | 0 | a2 | 0 |
| 2 – 3 | θ3 | 0 | a3 | 0 |
| 3 – 4 | θ4 | – d4 | a4 | 90° |
| 4 – 5 | θ5 | d5 | 0 | 90° |
| 5 – 6 | θ6 | 0 | 0 | 0 |

### *3.2. Design of motion planning algorithm for 6-DOF manipulator*

Trajectory planning is an important task in the motion control of manipulators; different motion modes require different trajectory planning algorithms, specifying the appropriate trajectory curve is one of the simplest trajectory specification algorithms, however, this single-type trajectory planning algorithm cannot meet the rigid and flexible impact during the movement of the manipulator, in the subsequent development process, a synthetic trajectory planning curve is designed according to the characteristics of various planning curves, this trajectory planning algorithm usually uses the form of a piecewise function, and the inflection point value of each curve of the function needs to meet the continuity of the motion state parameters. The control algorithm of motion modes of Trial Mode is adopted in this paper, which concentrates the traditional Auto Mode and Teach Mode motion modes; it not only makes the system to move along the established trajectory, but also ensures that the system can be controlled in real time in this process. The motion mode of Trial Mode selects the split trapezoidal velocity profile during the design process and requires segmentation planning during each sampling period. In addition, the interpolation algorithm should also specify the trajectory constraints.

### *3.3. Debugging design of motion planning algorithm of 6-DOF manipulator*

After the design of 6-DOF manipulator control algorithm is completed, before the experimental platform of manipulator is built, the manipulator algorithm debugging software needs to be written, and the algorithm function is solved and verified to ensure the feasibility of the algorithm function and the accuracy of the control logic. In this paper, the manipulator control system is built on the WinCE system platform, C++ language is used to write algorithm code to run within the controller, the MFC framework is used in the selection of software framework, a reasonable program framework is designed, its functions need to include mechanical parameters, tool parameters, operating parameters and coordinate system parameters, and can guarantee the simulation verification of different motion modes.

## 4. Hardware System Design of 6-DOF Manipulator

In the design process of the manipulator, considering the electromechanical coupling relationship of the system, the design process is divided into two parts: mechanical structure design and control system design, in the hardware design process, it is necessary to firstly fit the basic technical parameters of the manipulator to screen components, and then based on the calculation of the shutdown torque, select the motor that meets the power requirements.

IOP Conf. Series: Materials Science and Engineering **772** (2020) 012041 doi:10.1088/1757-899X/772/1/012041

(1) Selection of manipulator configuration

In order to ensure that the manipulator can complete the desired motion requirements in the working space, the object can be grasped to the exact position and the posture movement can be conducted, it is necessary to make the manipulator meet the condition of freedom. The configuration selection of manipulator is the basis of the design of the manipulator, the higher the degree of freedom, the higher the flexibility of the manipulator, but the control structure is also more complicated. Therefore, the freedom of the manipulator should be matched with the task, in this paper, a manipulator with six rotating joints is designed based on the six-degree-of-freedom kinematics model, in order to make the manipulator have a closed solution, it is necessary to make the adjacent three joint axes intersect at one point, therefore, the positioning structure selects a connecting rod with 0° or 90° rotation angle, and the six joints of the manipulator are driven by six torque motors, respectively.

(2) Selection of drive motor of manipulator

There are many kinds of drive motors for the joints of the manipulator; this design selects a DC brush motor with simple configuration and convenient control. In addition, in order to meet the different types of motion of manipulator, it is also necessary to take into account the maximum load torque of the motor and the maximum angular velocity of each joint, under normal circumstances, peak control power at locked-rotor is an important reference indicator in the motor selection process. The dynamic parameter requirements of each joint are the basis for selecting the joint drive components and transmission components; static and dynamic methods can be used to calculate the dynamic parameters of the manipulator, the static method is to ignore the dynamic influence of the load and only calculate the static load; the dynamic method is to design the static load and the dynamic load simultaneously in the calculation process. Assume that the centre of the joint theory node carries the joint focus, and the centre of the connecting rod carries the weight of the connecting rod, and then the required torque of each joint can be calculated.

(3) Selection of manipulator controller

DSP TMS320LF2407 is used as the main control chip in this design, which has 40MHz clock frequency and rich on-chip resources, its built-in event manager module can meet the control requirements of the drive motor, in addition, the chip with small size can save the space of board. In order to improve the dynamic control performance of the system, a double closed-loop control method including a position loop and a speed loop is adopted, the controller can input the expected control scheme of the manipulator in the industrial personal computer, and the controller calculates the ideal trajectory planning curve by algorithm calculation, and converted to the number of pulses of the incremental code disk. The deviation between the actual pulse number and the ideal pulse number is used as the input of the joint controller; the integral separation PID controller solves the position control amount of the system, which can be reflected as the different duty cycle of the PWM waveform, thereby ensuring that the motor driver controls the motor operation.

## 5. Software System Design of 6-DOF manipulator

(1) PID control algorithm

DSP TMS320LF2407 is the main control chip, the speed loop and the position loop use integral separation PID control; in addition, in order to ensure the control stability of the system, it is necessary to debug the control law of the system according to the actual operating environment. When determining the absolute position of the motor, manipulator effectively combines the signal output of the zero-position hall proximity switch and the incremental encoder Z-channel. The PID control method has the advantages of simple principle, convenient use, strong applicability and strong robustness, the relationship between the input e(t) output u(t) of the PID controller is:

$$u(\mathrm{t}) = K_p e(t) + K_I \int_0^1 e(t)dt + K_D \frac{de(t)}{dt}$$

Among them, the function of the proportional adjustment parameter Kp is to adjust the system to reduce the overshoot, moreover, it can control the adjustment speed, but when the parameter is too large, the stability of the system is lowered. The action of the integral adjustment parameter KI is on the elimination of the steady-state error, as long as the error has integral adjustment, it continues until the system reaches the error state. The differential adjustment parameter KD can reflect the change of the system deviation and realize the advanced control of the system, which is beneficial to improve the dynamic performance of the system.

(2) Communication software design

In the control process, the manipulator not only needs to communicate with the six lower computer, but also needs to interact with the I/O board to realize the control of the propulsion device and the hydraulic device, CAN communication is required in these parts, the data frame type and standard frame format are adopted in the CAN communication network, the data transmitted by the upper computer includes the expected angle of each joint, the desired speed, the heavy torque received by each joint, the PID parameter, the I/O control amount, etc., and the accepted data mainly includes the combined angle of each axis of the current manipulator, the current position of the propulsion device and the value of the displacement sensor. The lower computer does not communicate with each other, and the lower computer interacts with the upper computer, therefore, the upper computer can coordinate and manage the seven lower computers by using point-to-point communication mode, the upper computer is the active mode, and the lower computer is the slave mode, only when the lower computer receives the frame data or instruction sent by the upper computer, the data will be returned accordingly.

## 6. Conclusion

The manipulator is a tool that is widely used in modern industry and it is important for ensuring the stability of industrial production. In this design scheme, according to the motion characteristics of the manipulator in the working space, a 6-DOF manipulator with chain joint structure is designed, and by estimating the torque of each joint, the selection of the motor required by the manipulator is achieved. In system modelling, the integral separation PID control method is used to adjust the state performance of the system, achieve improvement in system response speed, overshoot and steady state performance.

## References

[1] Zhang Jinjing, Cui Tao, Pei Dong, Wang Quanzhou, Lu Yan. Design and Implementation of 4-DOF Manipulator Control System Based on FPGA [J], Journal of Machine Design, 2017, 34(7), pp.62-66.

[2] Zhang Xiaoying, Li Dewei, Xi Yugeng, Chen Weidong. Design of Constrained Predictive Control System for 6 Degrees-of-freedom Manipulators [J], Control Theory & Applications, 2014, 31(11), p.1464-1472.

[3] Zhong Jun, Cao Jianshu. Implementation of 6-DOF Robot Arm Control System Based on Kinect V2.0 [J], Machine Tool & Hydraulics, 2018, 46(9), pp.81-85.

[4] Bu Ziling, Yang Ruhan, Peng Yue. System Design of Wireless Controlled six-degree of Freedom Manipulator based on Arduino, [J], China Computer & Communication, 2017, (7), pp.90-93, 97.

[5] Wang Hao, Fan Pingqing, Wang Yansong, Li Cong. Research on Precise Control of Robot Arm Based on ROS Platform, Light Industry Machinery, 2018, 36(6), pp.42-47.

# Calculation of Electrochemical Corrosion Rate at Elbow of Supercritical $CO_2$ Pipeline

**Bin Chen, Qinghua Kang and Rongbin Ju ***

School of Mechanical Engineering, Xi'An Shiyou University, Xi'An, China

*Corresponding author e-mail: 806465856@qq.com

**Abstract.** Based on the CCUS project planned to run in a domestic oil field, the electrochemical corrosion at the bend of supercritical $CO_2$ pipeline was studied by means of high-temperature and high-pressure reaction kettle hanging plate experiment, and the corrosion rate of X80 pipeline steel in unsaturated, saturated and supersaturated supercritical $CO_2$ with water content was obtained. Then through COMSOL simulation of the corrosion rate of 90° bent pipe in the supercritical $CO_2$ corrosion atmosphere, the maximum corrosion rate position of electrochemical corrosion generated under the action of flow accelerated corrosion (FAC) generated by supercritical $CO_2$ fluid in the bent pipe was obtained, and corresponding anticorrosion Suggestions were put forward according to the research results.

## 1. Introduction

Most $CO_2$ mixture captured at the source contains a certain amount of water [1]. When $CO_2$ mixture in carbon steel pipeline is not dehydrated strictly and fully, free water and local high-pressure $CO_2$ combine to form carbonic acid, resulting in a sharp increase in corrosion rate of carbon steel pipeline [2-4]. Different corrosion solutions exist [5].Compared with X65 and X70 pipeline steel [6-7],X80 steel has greater yield strength and corrosion resistance [8-10].The corrosion rate at the pipe bend is much higher than that of the straight pipe, and when the pipe bend is 90°, the corrosion rate is higher than that of the pipe bend at other angles. In other words, X80 pipeline steel is used as the material of the pipe bend at 90°.COMSOL simulation was combined with high temperature and high pressure reaction kettle plate test. Finally, the experimental results and simulation results were compared and analysed.

## 2. Hanging test of X80 pipeline steel in autoclave

X80 pipeline steel was sliced and polished, and the distilled water was subjected to oxygen removal treatment to reduce the influence of oxygen in the distilled water.

The set pressure is 8 MPa, temperature is 40 °C, experimental time is 48 hours. Experimental apparatus: PARR-4578 high temperature and pressure reaction kettle produced by American Parr Company. The corrosion rate was calculated by the weight loss method.

### *2.1. Corrosion morphology analysis*

Figure 1 shows the macroscopic corrosion profile of the pipeline steel coupons used in the supercritical $CO_2$ pipeline. Unsaturated, saturated, supersaturated. In each group, the left image is the coupon before the reaction, and the picture on the right is the coupon after the reaction.

Content from this work may be used under the terms of the Creative Commons Attribution 3.0 licence. Any further distribution of this work must maintain attribution to the author(s) and the title of the work, journal citation and DOI.
Published under licence by IOP Publishing Ltd

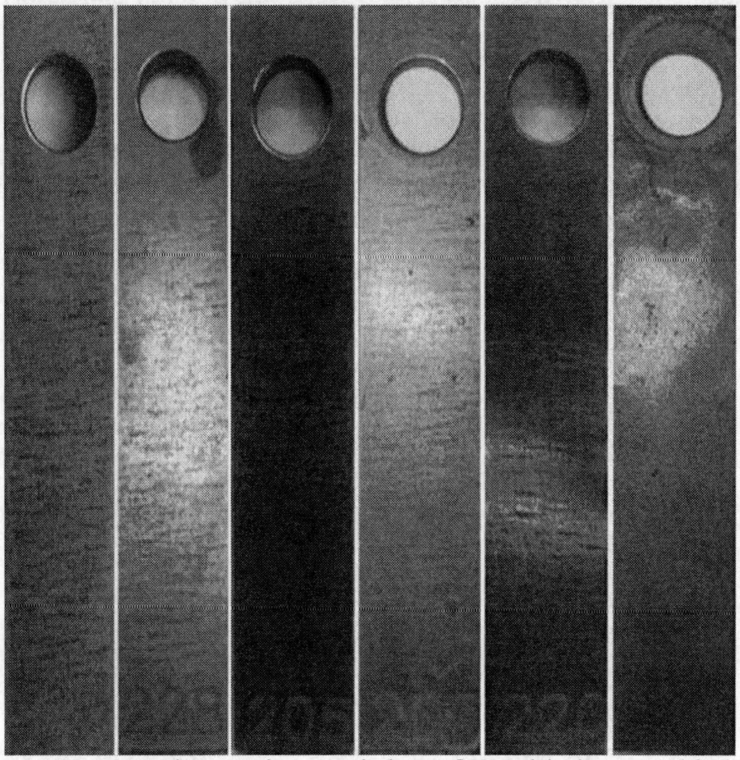

**Figure 1.** Macroscopic corrosion morphology of X80 slabs in supercritical $CO_2$.

It can be found from Fig 1 the hanging pieces after the reaction appear obvious pits, the surface has become uneven, from the color point of view, the film changed from a metallic luster to a blackish gray, and the rust-free trace from the start of the reaction changed to a slight rust, which may be caused by the local corrosion of the salt solution in the reactor [11].

In the case of a hanging piece under the condition that the humidity in the corrosive environment is supersaturated. From the surface leveling degree, the surface of the hanging piece after the reaction becomes rough and uneven, from the color point of view, the rear hangs changed from the original metallic lustrous to the dark brown lustrous [12-13]. By observation, it can be found that the surface has a loose and corrosive product, which can cause the corrosion products to fall off due to slight touch.

*2.2. Experimental results*

The corrosion rate results are shown in Table 1.

**Table 1.** Corrosion rate of slabs at different $CO_2$ phases (mm/a).

| Environment | Material | Unsaturated | Sayurated | Supersaturated |
|---|---|---|---|---|
| Supercritical | X80 | 0.1676 | 0.3653 | 17.2493 |

*2.3. Analysis and discussion*

The corrosion rate of X80 pipeline steel increases with the increase of water content. In supersaturated environment, the corrosion rate increases exponentially and reaches the maximum value. According to the microscopic morphology of the corrosion products, it was found that some of the corroded products were rolled up.The corrosion products are easily detached by the flushing of the fluid, causing further corrosion damage to the pipe wall.

IOP Conf. Series: Materials Science and Engineering **772** (2020) 012042 doi:10.1088/1757-899X/772/1/012042

## 3. Numerical simulation of electrochemical corrosion at the elbow of supercritical $CO_2$ pipeline

### *3.1. Simulation methods and boundary conditions*

COMSOL software was used to model the 90° elbow of the supercritical $CO_2$ pipeline. The turbulence model, the lean mass transfer model and the surface reaction model were coupled, meshed, and mass transfer of $CO_2$ molecules. The solution reacts with the chemical reaction of the inner wall of the $CO_2$ pipeline and the flow of the carbonate solution to construct an electrochemical corrosion rate model with free water in the supercritical $CO_2$ pipeline [14]. The rate of corrosion of the carbonic acid solution on the inner wall of the pipeline steel pipe under the condition that the supercritical $CO_2$ fluid in the pipe is stationary and flowing.

The pipe material is set to X80. The length of the inlet and outlet section of the elbow is both 400mm. The diameter is set to 200mm, and the radius of curvature of the elbow is set to 1.5D.The proposed inlet flow rate of the supercritical $CO_2$ pipeline in China oil field is 2m/s; so, the inlet boundary condition is set to 2m/s for velocity inlet; Set to a pressure outlet of 8 MPa, the temperature is set to 40°C; in the rest state the inlet flow velocity of the 90° elbow of the supercritical $CO_2$ conveying pipeline is set to 0 m/s, and the remaining boundary conditions and dynamic simulation experiments conditions are consistent.

### *3.2. Numerical simulation under static conditions*

The simulation results are shown in Table 2 under the condition that the static condition carbonic acid concentration is 0.05 mol/L.

**Table 2.** Static corrosion rate under the condition of 90 ° elbow.

| Time(s) | 0 | 50 | 100 | 150 |
|---|---|---|---|---|
| Corrosion rate(mm/a) | 0.35 | 0.16 | 0.06 | 0.02 |

According to Table 2, the average corrosion rate in the 90° elbow of the supercritical $CO_2$ conveying pipe of X80 material is about 0.15mm/a. According to scholar Visser[15], under the supercritical $CO_2$ environment at temperature of 40°C,When the carbonic acid concentration is about 0.05mol/L, the water content is about 750000ppmw. Combined with the high temperature and high pressure reactor hanging piece test, the corrosion rate of X80 pipeline steel in unsaturated NaCl solution can reach 0.16mm/a.

Under static conditions, there will be a significant concentration gradient that will provide power for the transport of carbonic acid molecules to the surface of the sample. Carbonate is bent toward 90° due to the consumption of corrosive carbonic acid in the fluid. The rate of mass transfer on the inside surface of the tube is significantly reduced, and the corrosion of the inner surface of the $CO_2$ transport pipe by carbonic acid also shows a downward trend due to the consumption of carbonic acid.

Using COMSOL with the concentration of carbonic acid as the variable, the flow rate of the medium was chosen to be 2m/s, and the concentration gradient of carbonic acid was set to 0.01mol/L, 0.02mol/L, 0.03mol/L, 0.04mol/L. The remaining simulation parameters were in accordance with dynamic $CO_2$. The relationship between the carbonic acid concentration and the corrosion rate shown in Table 3.

**Table 3.** Relationship between carbonic acid concentration and corrosion rate in static conditions.

| Carbonic acid (mol/L) | 0.01 | 0.02 | 0.03 | 0.04 | 0.05 |
|---|---|---|---|---|---|
| Corrosion rate (mm/a) | 0.121 | 0.133 | 0.125 | 0.140 | 0.168 |

It can be seen from table 3 that the corrosion rate of the inner wall surface of the 90° elbow of the supercritical $CO_2$ conveying pipe increases with the increase of the carbonic acid concentration.The corrosion rate of the 90° elbow near wall reached a maximum when the carbonic acid concentration is 0.05 mol/L and value of 0.168 mm/a.

### *3.3. Numerical simulation under dynamic conditions*

When the static condition is the most serious, use COMSOL to model the 90° elbow of the supercritical $CO_2$ pipeline. The inlet speed of the 90° elbow is set to 2m/s, and the other strips remain unchanged. Corrosion rate distribution cloud map, as shown in Figure 2.

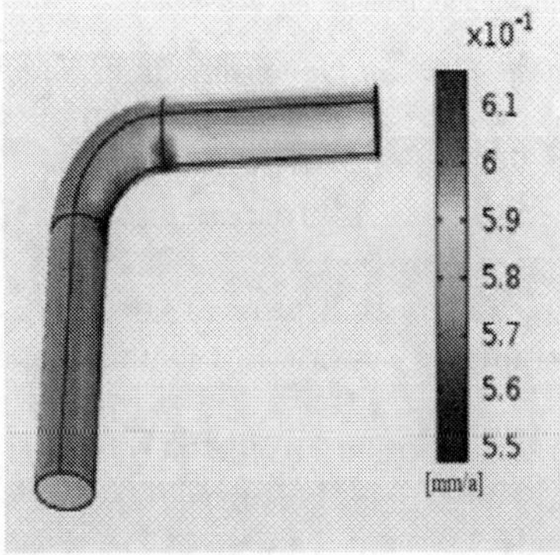


**Figure 2.** Flow under the condition of 90 ° elbow corrosion rate at 0.05 mol/L.

Comparing the static and dynamic results, the corrosion rate obtained by the dynamic simulation results is much larger than the corrosion rate under the static conditions. The reason is caused by different mass transfer modes. Rani [16] performed numerical values on the mass transfer results. The mode shows that the mass transfer coefficient (MTC) of the bend has undergone a circumferential change, and the MTC is the most important parameter for predicting the position of the highly sensitive FAC.

It can be seen from Fig 2 that when the near wall carbonization concentration is 0.05 mol/L and the inlet flow velocity is 2 m/s, the carbonation corrosion is the most serious at the transition point of the elbow and the outlet section of the 90° elbow, which is about 0.6mm/a, The scholar ZHANG [17] combined the array electrode technology with computational fluid dynamics (CFD) simulation to obtain the corrosion rate of the inner wall of the elbow is significantly higher than the outer wall corrosion rate; the maximum corrosion rate appears at the innermost side of the elbow, the minimum corrosion rate appears on the outermost side of the elbow; as time increases, the wall will appear thinner.Kim [18] for the partial wall of the elbow an intuitive visual experiment and numerical analysis were carried out. It was concluded that flow accelerated corrosion (FAC) is the main reason for the thinning of the wall thickness.

### *3.4. Analysis and discussion*

(1) The average corrosion rate obtained by fitting the simulation data under static conditions is 0.15 mm/a. The corrosion rate of the X80 pipeline steel obtained from the high temperature and high pressure reaction reactor coupon test is 0.16 mm/a, which is slightly larger than the simulation result. This may be caused by impurities in the water used in the high temperature and high pressure reactor coupon test.

(2) When the supercritical $CO_2$ fluid flows through the elbow at a certain speed, the turbulent kinetic energy and the wall shearing force are relatively large at the inner wall surface, which accelerates the mass transfer and the corrosion of the corrosive material under dynamic conditions. Therefore, under dynamic conditions, the corrosion rate of the inner wall surface of the elbow is greater than that under static conditions, and the addition of corrosion inhibitor can effectively reduce corrosion.

## 4. Conclusion

(1) The corrosion rate of X80 pipeline steel increases with the increase of water content. In supersaturated environment, the corrosion rate increases exponentially and reaches the maximum value. The water content of $CO_2$ gas source should be strictly controlled;

(2) The corrosion rate inside the 90° elbow is greater than the corrosion rate on the outside. Therefore, when the pipe is corroded and perforated by increasing the wall thickness at the 90° elbow, the thickening of the inner side of the 90° elbow should be slightly higher than the outer side;

(3) In actual conditions, it is far more than a carbonation corrosion. If the control is not good, other water-soluble impurities such as H2S will also cause corrosion of the carbon steel pipeline. Therefore, $CO_2$ should be strictly controlled during pipeline transportation.

**References**

[1] Ju H, Elmoudir W , Aboudheir A , et al. Development of a facile reclaiming process for degraded alkanolamine and glycol solvents used for $CO_2$ capture systems [J]. International Journal of Greenhouse Gas Control, 2018, 74: 174-181.

[2] Hu Y Q, He F, Bao W, et al. Research Progress in Corrosion of $CO_2$ Conveying Pipeline [J]. Surface Technology, 2016, 45(8): 14-21.

[3] Sun C, Wang Y,Sun J B, et al.Research progress on corrosion of supercritical $CO_2$ transport pipeline containing impurities [J]. Journal of Chinese Society for Corrosion and Protection, 2015, 35(5): 379-385.

[4] Zhang Y, Pang X, Qu S, et al. Discussion of the $CO_2$ corrosion mechanism between low partial pressure and supercritical condition [J]. Corrosion Science, 2012, 59: 186-197.

[5] Brown J, Graver B, Gulbrandsen E,et al. Update of DNV Recommended Practice RP-J202 with Focus on $CO_2$ Corrosion with Impurities [J]. Energy Procedia, 2014, 63: 2432-2441.

[6] Cheng Y P, Bai Y, Li Z, et al. Corrosion Characteristics of X65 Steel in $CO_2$/Oil/Water Environment of Gathering Pipeline [J]. Journal of Engineering Science, 2018(5).

[7] Xu M, Zhang Q, Yang X X, et al. Impact of surface roughness and humidity on X70 steel corrosion in supercritical $CO_2$, mixture with $SO_2$, $H_2O$,and $O_2$ [J]. Journal of Supercritical Fluids,2016, 107: 286-297.

[8] Xu M, Li W, Zhou Y, et al.Effect of pressure on corrosion behavior of X60, X65, X70, and X80 carbon steels in water-unsaturated supercritical $CO_2$ environments [J]. International Journal of Greenhouse Gas Control, 2016, 51: 357-368.

[9] Wei A J,Zhao W, Huo F Y, et al. Electrochemical Characteristics of $CO_2$ Corrosion of X80 Steel at Different Temperatures [J]. Corrosion & Protection,2012, 33(4): 281-283.

[10] Mu H P, Yan M C, Yang X, et al. Electrochemical behavior of X80 pipeline steel red soil corrosion in the early stage [J]. Journal of Chinese Society for Corrosion and Protection, 2016, 34(05): 472-476.

[11] Xiang Y, Li C, Hesitao W, et al.Understanding the pitting corrosion mechanism of pipeline steel in an impure supercritical $CO_2$, environment [J]. Journal of Supercritical Fluids, 2018, 138: 132-142.

[12] Farelas F, Galicia M, Brown B, et al. Evolution of dissolution processes at the interface of carbon steel corroding in a $CO_2$ environment studied by EIS [J]. Corrosion Science, 2010, 52(2): 0-517.

[13] Sk M H, Abdullah A M, Ko M ,et al. Local supersaturation and the growth of protective scales during $CO_2$, corrosion of steel:Effect of pH and solution flow [J]. Corrosion Science, 2017, 126: 26-36.

[14] Hu H,Cheng Y F.Modeling by computational fluid dynamics simulation of pipeline corrosion in $CO_2$-containing oil-water two phase flow [J]. Journal of Petroleum Science & Engineering, 2016, 146: 134-141.

[15] Visser E D, Hendriks C, Barrio M, et al. Dynamis CO2 quality recommendations [J]. International Journal of Greenhouse Gas Control, 2008, 2(4): 478-484.

[16] Rani H P, Divya T , Sahaya R R , et al.CFD study of flow accelerated corrosion in 3D elbows [J]. Annals of Nuclear Energy, 2014, 69:344-351.

[17] Zhang G A,Zeng L,Huang H L,et al. A study of flow accelerated corrosion at elbow of carbon steel pipeline by array electrode and computational fluid dynamics simulation[J].Corrosion Science, 2013, 77: 334-341.

[18] Kim H J, Kim K H . Intuitional experiment and numerical analysis of flow characteristics affected by flow accelerated corrosion in elbow pipe system [J]. Nuclear Engineering & Design, 2016, 301: 183-188.

IOP Conf. Series: Materials Science and Engineering **772** (2020) 012043 doi:10.1088/1757-899X/772/1/012043

# Method of Tolerance Design and Verification of Parts Based on Follow-up Crankshaft Grinder with Following Double Grinding Wheel Frame

**Jinwei Fan*, Leilei Huo**

College of Mechanical Engineering and Applied Electronics Technology, Beijing University of Technology, Beijing, China

*Corresponding author e-mail: 710484127@qq.com

**Abstract.** Calculating the parameters of accuracy along each axis by the comprehensive geometric accuracy and then optimizing as well as assigning the accuracy, which is the cutting-edge way of accuracy design of machine tools. However, the guarantee of the comprehensive geometric accuracy of machine tools is ultimately to implement the tolerance of specific moving parts. In view of the present shortage about accuracy design of machine tools, this paper puts forward to a way to design and verifies the tolerance of movement parts based on the comprehensive geometry precision of K2-1080 Crankshaft Grinder with Following Double Grinding Wheel Frame.

**Key Word:** crankshaft grinder; geometric accuracy; precision design; tolerance design.

## 1. Introduction

Crankshaft Grinder with Following Double Grinding Wheel Frame have many motion pairs as well as long kinematic chains, and their comprehensive stiffness and accuracy are easily affected [1]. When compensating the comprehensive errors of Crankshaft Grinder with Following Double Grinding Wheel Frame, it is necessary to measure the errors of machine tools along the direction of each axis, which is more difficult in the actual error compensation of Crankshaft Grinder with Following Double Grinding Wheel Frame. Because tolerances are used to limit errors, the comprehensive geometric errors of machine tools are solved by superimposing random geometric errors of each pair of motion.

Therefore, the geometric tolerance analysis model of five-axis machine tools can be established by giving the geometric tolerance terms of the parts and components of each moving pair to cover the geometric errors in each axis direction. Combined with the randomness of the comprehensive geometric errors of Crankshaft Grinder with Following Double Grinding Wheel Frame, the Monte Carlo simulation method is used to simulate and analyze the three-dimensional errors, so that the final design of the comprehensive geometric accuracy of five-axis machine tools can be achieved. Combining with the randomness of the comprehensive geometric error of five-axis machine tools, the Monte Carlo simulation method is used to simulate and analyze the three-dimensional error, which makes the design of the comprehensive geometric accuracy of Crankshaft Grinder with Following Double Grinding Wheel Frame fall on the geometric tolerance of the components that make up the specific motion pairs

Content from this work may be used under the terms of the Creative Commons Attribution 3.0 licence. Any further distribution of this work must maintain attribution to the author(s) and the title of the work, journal citation and DOI.
Published under licence by IOP Publishing Ltd

[2]. The direction of comprehensive error compensation for Crankshaft Grinder with Following Double Grinding Wheel Frame provides a favorable basis.

## 2. Design of Comprehensive Geometric Accuracy for Crankshaft Grinder

### *2.1. Geometric Error Term of Crankshaft Grinder*

For crankshaft grinder, there are three axes moving parallelly along X, Y and Z directions, each axle will produce three-line displacement errors along X, Y and Z directions, three-line displacement errors around X, Y and Z directions, and three-line displacement errors along X, Y and Z directions of two rotating axes $C_1C_2$ and three-line displacement errors around X, Y and Z directions. In addition, there are three verticality errors between the three linear motion axes. These three verticality errors do not change with the movement of the moving body of the machine. There are 33 geometric errors in the crankshaft grinder. As shown on figure1.

Taking the shifting axis as an example, there are six errors in the movement of the machine tool table along the guide rail, namely three displacement errors and three angle errors. Take X-axis translation motion as an example. There are three translation errors, namely, delta x (X), delta y (X), delta Z (X), and three angular errors, namely, delta x (X), delta y (X), delta Z (X), and three angular errors, namely, delta x (X), delta y (X), delta y (X), and delta Z (X). Delta represents displacement error, epsilon represents angle error, letters in parentheses represent direction of motion, letters in subscripts represent direction of error [3].

Similarly, the errors of translation along the Y axis are: delta x (Y), delta y (Y), delta Z (Y), delta x (Y), delta y (Y), delta y (Y), delta Z (Y). The errors of translation along Z axis are: delta x (Z), delta y (Z), delta Z (Z), delta x (Z), delta y (Z), delta y (Z), delta x (Z), delta y (Z), delta Z (Z).

### *2.2. Geometric Tolerance Term Control of Motion Parts*

Assuming that M is a point on the worktable, because of the existence of both straightness and angle errors. It can be seen that M still fluctuates within a certain limit and does not exceed that limit. This is similar to the concept of contour error in dimensional tolerance.

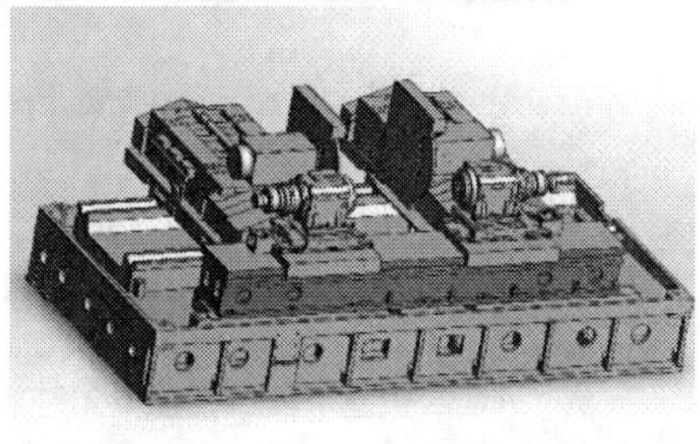

**Figure 1.** K2-1080 Crankshaft Grinder with Following Double Grinding Wheel Frame

According to the above analysis, when the slider moves along the X-axis direction, due to the existence of straightness error and angle error of the guideway, the worktable moves and twists in a certain range of the XY plane and YZ plane [4]. Therefore, for the planar guideway, it can be converted into the surface profile and verticality of the contact surface. Given the flatness of the contact bottom surface, it can cover delta y (X), delta x (X) and delta Z (X), and the verticality of the two contact sides can cover delta Z (X) and delta y (X); For cylindrical guideways, it can be transformed into the straightness and side verticality of V-type guideways [5]. Given the position degree of the axis of V-type guideway, it can cover delta Z (X), delta x (X) and delta y (X), side verticality can cover delta y (X), delta Z (X), and delta x (X) is limited by the accuracy of servo motor along the X axis of machine tool. The initial values of surface contour and verticality are as follows: The factors influencing the flatness of the bottom surface are as follows: delta y (X), delta x (X) and delta y (X).

The resulting flatness magnitude $\Delta_a$ is shown in the formula (1):

$$\Delta_a = \delta_y(X) + \sin(\varepsilon_z(X)) * \frac{L_{aX}}{2} + \sin(\varepsilon_x(X)) * \frac{L_{aY}}{2} \tag{1}$$

The influence factors of lateral verticality are as follows: delta Z (X) and delta x (X). The magnitude $\Delta_b$ of verticality caused by delta Z (X) and delta x (X). See formula (2):

$$\Delta_b = \delta_z(X) + \sin(\varepsilon_y(X)) * \frac{L_{aX}}{2} \tag{2}$$

$L_{aX}$ and $L_{aY}$ respectively indicate the length of the worktable and the machine tool guideway in the X and Y directions, and the angle is radian. The analysis process of the position degree of the cylindrical guideway axis is similar to that above.

### *2.3. Establishing Geometric Tolerance Model of Machine Tool*

In the computer aided tolerance simulation analysis software, the dimension chain is a geometric tolerance model, which is obtained by superimposing geometric tolerances on the components of each moving pair. Therefore, when analyzing the comprehensive geometric errors, it is necessary to convert the geometric errors into the geometric characteristic tolerances of the machine tool assembly parts [5]. This paper takes K2-1080 Crankshaft Grinder with Following Double Grinding Wheel Frame as an example. When building the geometric tolerance analysis model of machine tools, we only consider important parts (such as moving parts) and ignore most of the strengthening parts (such as standard parts). For the parts that constitute the moving pair, the motion errors along the moving direction are limited by the given flatness and verticality of the contact characteristics of the two parts assembly; for the moving pair composed of cylindrical guides, the position of the cylindrical axis needs to be given. [6]. The vertical and parallel errors between the moving directions of the machine tools along the axes require the tolerances of the assembly features and the parallelism of each other to be constrained jointly. Through the above conversion of geometric errors, assembly is carried out according to the actual assembly sequence of machine tools, so the geometric tolerance analysis model of K2-1080 Crankshaft Grinder with Following Double Grinding Wheel Frame can be established. As shown on Figure2

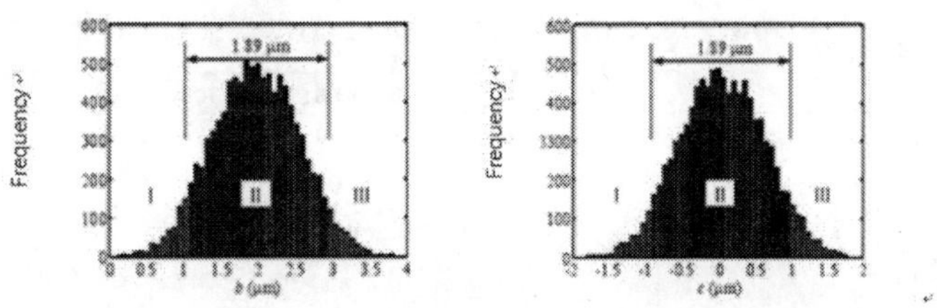

**Figure 2.** Frequency distribution histograms of b and c.

## 3. Comprehensive Geometric Error Analysis

The common methods of tolerance analysis are extreme value method, statistical tolerance method and Monte Carlo method.

The principle of extremum method is simple and the calculation is small. It does not need to consider the size distribution of parts in the tolerance zone. It can guarantee the success rate of assembly and the interchangeability of parts to be 100%. But the design is too conservative, which increases the processing cost. Statistical tolerance analysis method is to describe the change of part size in the form of statistical distribution, and calculate the statistical distribution of part manufacturing capacity and assembly function [7]. This method assumes that the tolerance of parts obeys normal distribution, and the relationship between assembly tolerance and part tolerance is linear. The results are not completely consistent with the actual situation. Monte Carlo method treats the problem of finding the size and tolerance of a closed ring as a statistical problem of finding a random variable. According to the actual distribution of each dimension, a certain algorithm is used to sample in the computer, generate the corresponding random number, and calculate the size and tolerance of the closed ring according to the design function. Under certain conditions, the results obtained by this method are in good agreement with the actual situation.

According to the nature of the synthetic geometric errors of Crankshaft Grinder s, it is based on the structure of Crankshaft Grinder, and the random geometric errors of each pair of motion are transformed and solved. It uses the statistical method of random variables to solve the size and tolerance of the closed loop. The tolerance is a variable with random distribution and limited in a certain range. Its property is consistent with the geometric error of the Crankshaft Grinder l. Therefore, Monte Carlo simulation method can be used to simulate and analyze the three-dimensional geometric errors of Crankshaft Grinder. At present, there are many mature tolerance simulation software using Monte Carlo simulation method, among which Vis VSA developed by Siemens Company is mainly used for spatial dimension chain operation. The assembly errors of Crankshaft Grinder are analyzed by the computer-aided tolerance analysis software Vis VSA. The most important contributing factor to the target size can be found out, and the corresponding parts tolerances can be rationally modified according to the analysis results [8]. On the one hand, the three-dimensional tolerance simulation design verification based on the requirements of the comprehensive geometric accuracy of Crankshaft Grinder can be realized, so that the guarantee of the comprehensive geometric accuracy of Crankshaft Grinder can be realized on the tolerance of each component of the specific motion pair; at the same time, the contribution factor can be used to provide a favorable basis for the follow-up comprehensive error compensation direction of Crankshaft Grinder. On the other hand, the ability to meet the processing quality and the quality stability of products can be evaluated according to the CP and Cpk in the report. In the past research of precision design, the design goal is usually to optimize the accuracy of the whole machine or to minimize the cost [9]. For the precision design of precision horizontal machining centers, we often do not seek some optimal solutions, but to are obtained a set of optimal solutions satisfying certain conditions [10]. Using MATLAB to solve the above optimization problems after 48 iterations, the optimal and design values of design variables and the upper bound of the corresponding attitude error range of each axis moving component are obtained as shown in Table1.

**Table 1.** Results of the optimal error assignment.

| Item | Accuracy Optimization and Allocation Results of Design Variables | | | | | |
|---|---|---|---|---|---|---|
| Installation Error of Guide Rail | $X_1$ | $X_2$ | $X_3$ | $X_4$ | $X_5$ | $X_6$ |
| Optimizesd Value | 4.0254 | 4.9425 | 4.2987 | 9.8535 | 4.6237 | 4.2375 |
| Design Value | 4 | 5 | 4 | 10 | 5 | 4 |
| Trailer Attitude Error | $\varepsilon_1(x)$ | $\varepsilon_1(y)$ | $\varepsilon_1(z)$ | $\varepsilon_2(x)$ | $\varepsilon_2(y)$ | $\varepsilon_2(z)$ |
| Optimizesd Value | 2.755 | 6.405 | 5.124 | 22.39 | 12..77 | 8.956 |
| Design Value | 3 | 6 | 5 | 22 | 13 | 9 |

According to the results of tolerance allocation, 10 000 simulation points are randomly selected in the workspace of machine tools. At each simulation point, a group of manufacturing and installation errors of guideways satisfying tolerance requirements are randomly generated..The position errors of tool relative to workpiece are calculated. Finally, the position errors of x, y and Z directions and corresponding volume errors are obtained. it can be seen that the probability of the position errors of the x, y and Z directions of the end of the machine to meet the design targets is 99.12%, 99.02% and 98.96%, respectively. The rationality of the tolerance design values of the machine tool and the validity of the precision design method are verified. As shown on figure 3.

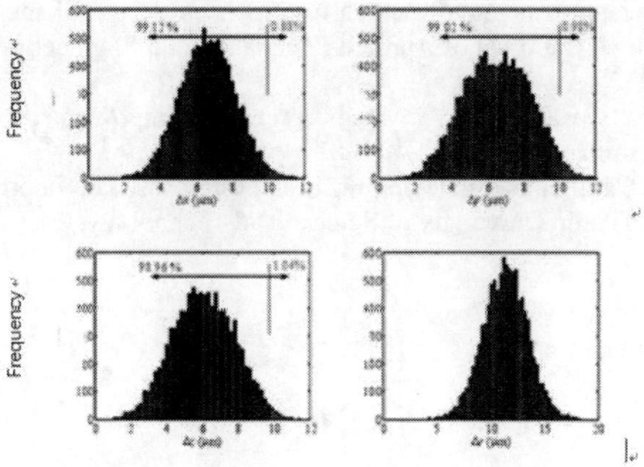


**Figure 3.** Frequency distribution histograms of predicted errors of machine tool.

## 4. Conclusion

The three-dimensional tolerance design verification based on the requirement of the comprehensive geometric accuracy of Crankshaft Grinder is realized. By modifying the corresponding parts tolerance reasonably through the contribution factor, the guarantee of the comprehensive geometric accuracy of Crankshaft Grinder can be realized on the tolerance of each part of the specific motion pair. Available contribution factor can provide a useful basis for the direction of comprehensive error compensation of follow-up Crankshaft Grinder. The error sensitivity analysis model of NC machine tools based on interval analysis theory is established. On this basis, the global maximum sensitivity index is defined to quantitatively describe the above-mentioned influence degree. At the same time, it provides a theoretical basis for the follow-up precision design. The error mapping model between manufacturing and installation errors of guideway and six-dimensional pose errors of moving parts is established. The statistical mapping rules between the two are analyzed by Monte Carlo simulation algorithm, and empirical formulas are formed. It is used to guide the distribution of installation tolerance in subsequent rail manufacturing.

**Acknowledgments**

This work was financially supported by National Major Scientific And Technological Project fund coded: 2016ZX04003-001.

**References**

[1] Li Ming, Felina. Geometric coordinate measurement technology and its application [M]. Beijing: China Quality Inspection Press, China Standard Publication, 2012.
[2] Jiang Qianqian. Theory and Application of New Generation GPS Standard [M]. Beijing: Higher Education Press, 2007.
[3] Lu Qiang, Zhang Youliang. Precision synthesis of 6-leg parallel machine tools by Monte Carlo method [J]. China Mechanical Engineering, 2002, 13 (6): 464-467.
[4] Ngoi BKA,Teck O C.A tolerancing optimization method for product design[J].The International Journal of Advanced Manufacturing Technology,1997,13(4):290-299.
[5] Chaolin, Huangtian, Wang Yang, et al. Precision design of manufacturing-oriented parallel machine tools [J]. China Mechanical Engineering, 1999, 10 (10): 1114-1118.
[6] S utherland G H,Roth B.Mechanism design:Accounting for Manufacturing Tolerances and Costs in Function Generating Problems[J].Journal of Engineering for Industry,1975,97(1):60.
[7] Liu Daming, Zheng Jiaoxing. Monte Carlo simulation application in tolerance analysis of poor assembly rate guidance [J].Science and Engineering Technology, 2006,2(4): 67-82.

[8] Ding Wenzheng, Zhou Minghu. Research on precision design of multi-axis machine tools for remanufacturing [J]. Journal of Applied Foundation and Engineering Sciences, 2007 (4): 559-567.
[9] Dufour P,Groppetti R.Computer Aided Accuracy Improvement in Large NCMachine-Tools[M].Macmillan Education UK,1981.
[10] Jinqiu, Mo Shuai. Parallel tolerance optimization design based on improved cost tolerance model [J]. Journal of Tianjin University of Science and Technology, 2010, 25 (5): 53-56.

# Analysis of Electrostatic Driven Micromachined Gyroscope

**Runyu Jiao, Zhaomin Luo, Jianming Wang and Xingming Xu***
Shandong University of Science and Technology, Taian, China

*Corresponding author e-mail:xxm1601@163.com

**Abstract.** Micromechanical gyroscopes are an important inertial instrument used to measure the angular or angular acceleration of an object. China's research in this field lags behind that of developed countries in the West. The research on the core components of high-performance micromachined gyroscopes is still in its infancy. In this paper, a fully symmetrical multi-ring vibration structure electrostatically driven micromachined gyroscope is proposed, which is studied from the theoretical basis, working principle, structural design, performance simulation, processing technology and device testing, including the following electrostatic drive micro Testing of mechanical resonant devices. The test methods of several main MEMS resonant devices are analyzed. Based on vector network analyzer, vacuum cavity, transimpedance amplifier and other equipment, the frequency sweep test is performed on the device, and the resonant frequency of the drive shaft is 23686.2z. The resonant frequency of the sensitive axis is obtained. For the 23980 Hz based on the principle of electrostatic regulation, a 2.5V electrostatic adjustment voltage is applied to the device, and the frequency cracking of the two modes is reduced from 293.75 Hz to 6.25 Hz, which improves the performance of the device.

## 1. Introduction

A gyroscope is an instrument that measures the angular velocity or angle of an object's rotation in space.The gyroscope's English name "gyroscope" was created by the famous French scientist Foucault in the experiment of measuring the Earth's rotation.It is a combination of the Greek "skopeein" (meaning "see") and "gyros" (meaning "rotation").The work of the gyroscope does not depend on the external reference signal, and can independently detect the change of the attitude of the moving object. In the aerospace, navigation, military guidance and other fields, the traditional gyroscope is widely used, including the mechanical rotor gyroscope, the laser gyroscope, Fiber optic gyroscopes, etc. Mechanical rotor gyroscopes rely on high-speed rotating mechanical rotors. Due to the uneven quality of the rotor itself, it is extremely susceptible to acceleration and the application range is limited. Moreover, the wear caused by the friction of mechanical parts also affects the performance of the gyroscope. And life effects affect the laser gyro, fiber optic gyroscope based on the Sagnac effect, through the interference of light to detect angular velocity; compared to mechanical rotor gyroscope, with no moving parts, simple structure, no need for preheating, long service life, etc. However, such gyroscopes are extremely expensive and are mainly used in some highly sophisticated fields.

Therefore, it is of great significance to study a micro-mechanical gyroscope based on fully symmetric vibration multi-ring structure. This paper studies the physical basis of micro-mechanical gyroscope: analyzes the method of improving the sensitivity of the device; in structural design, micro-machining

Content from this work may be used under the terms of the Creative Commons Attribution 3.0 licence. Any further distribution of this work must maintain attribution to the author(s) and the title of the work, journal citation and DOI.
Published under licence by IOP Publishing Ltd

process, weak signal detection And device testing methods are explored to lay the foundation for the further development of micromachined gyroscopes. There is a lack of systematic research in these areas, and there is a great demand for high performance micromachined gyroscopes. This topic develops micromechanical inertia for China. The device and research mems micro-machining process have great practical significance and application value.

## 2. Design of micro machined gyroscope

### *2.1. Micro-mechanical gyroscope design*

*2.1.1. Structural Design.* The early vibrating ring gyroscope was dominated by a single-ring structure. Later, it evolved from a single-ring structure to a multi-ring structure, also known as a disc structure. The design of multiple rings increases electrode coverage. The area increases the capacitance plate area of the sensitive electrode and the driving electrode, improves the sensitivity of the sensitive signal of the gyro, and increases the effective quality of the device. For the mems resonant device, increasing the effective quality is one of the important means to reduce the noise level. .

The mems gyroscope designed in this paper is based on the main components of multi-ring structure design: vibration ring, drive electrode, sensitive electrode, repair electrode support block.

The vibrating ring is the core of the device and is also used as the main part of the resonator for the angular velocity changes of sensitive systems.The drive electrode is used to apply an excitation signal to drive the device to move.Sensitive electrodes are used to detect the output signal.The trim electrode is used for frequency tuning of the device.The support block functions as a supporting substrate during the thinning process of the device process.

*2.1.2. Working mode as mentioned above, the working mode of the mems vibrating ring type micromachined gyroscope is generally a high-order mode.* The driving electrode and the detecting electrode are both around the vibrating ring structure. Therefore, the driving force acts in parallel with the device. The substrate is perpendicular to the direction of rotation. The vibration of the device is in the horizontal plane, and in theory there is no out-of-plane vibration of the device. Thus, we can obtain the vibration mode of the vibrating ring gyroscope. The vibration amplitude of the two working modes and The angular increase coefficient is different. In order to obtain greater mechanical sensitivity, for the vibrating ring type micromachined gyroscope, the orthogonal mode is 45 degrees.

The vibrating ring gyroscope designed in this paper mainly detects the angle or angular velocity by the vibration of two natural modes.Different gyroscopes can be designed with different modes. In this paper, the angular velocity is detected by two ellipse orthogonal to 45 degrees.In the reciprocating vibration of the vibrating ring from an ellipse to a circle and then to an ellipse, the structure is continuously bent, that is, the capacitor plate constantly moves and the capacitance constantly changes.

The specific working mode is as follows: firstly, the driving electrode is energized, and the excitation signal is applied to make the gyroscope work in a resonance state, the vibration ring changes from a circular shape to an ellipse and continuously vibrates; when an external angular velocity is applied to the object, the energy of the gyroscope is driven. The axis is transferred to the sensitive axis.The vibration modes of the two vibration modes are the same, but the angle of the vibration axis is 45 degrees.When driving the mode, the diameter of the sensitive axis does not change, the capacitance gap at the sensitive electrode does not change, and there is no sensitive signal output.When an angular velocity input occurs, the vibration in the direction of the sensitive axis is excited to generate a displacement.

## 3. Micromachined gyroscope testing

### *3.1. mems resonant device test method*

The micro-nano dynamic mechanical test of the mems resonant device is of great significance to the performance of the device design, process, reliability, etc. The micro-nano dynamic mechanical test generally includes the following parts: vibration excitation, dynamic vibration measurement Modal analysis. The excitation signal is applied to the mems resonant device. After the device generates the vibration response, the response is measured, and the relationship between the excitation and the response is analyzed to determine the resonant frequency and mode shape of the device.

Through the actual micro-nano dynamic response, the theoretical model is verified and the structure and process design optimization of the mems resonant device are guided.

Due to the extremely small size and high resonance frequency of the MEMS device, the conventional conventional contact vibration test method (piezoelectric, photoelastic, strain, etc.) cannot be used. Generally speaking, there are two methods for measuring the mem device: Optical-based non-destructive testing techniques, such as laser Doppler vibrometry, stroboscopic optics, laser interferometry, etc.; electrical-based excitation feedback testing techniques. For mems micromachined gyroscopes, the vibration of the vibrating ring is in-plane Vibration (in-plane vibration), so this section will mainly analyze the in-plane vibration stroboscopic dynamic vision imaging technology and electrical frequency sweep testing technology.

### *3.2. Electrical test based on vector network analyzer*

*3.2.1. Experimental Equipment and Procedures Testing* of MEMS micromachined gyroscope chips requires the design of a simple printed circuit board (PCB), which is used to extract the electrical signals of the chip; on the other hand, as the fixture and carrier of the chip. The platform provides a good environment for testing chips.

The device is mainly subjected to frequency sweep test by means of a vector network analyzer and the like.The test equipment mainly includes a vector network analyzer, a transimpedance amplifier, a vacuum chamber, a DC power supply, etc., and the device resonance frequency is obtained by performing a frequency sweep test on the device.

The specific test steps are as follows:

(1) Place the device in a vacuum chamber, make electrical connections, and test the electrical path with a multimeter.

(2) Using a mechanical pump, pump the vacuum chamber to about 5Pa.

(3) Set up the Agilent E55061B network analyzer: DC Bias port output 10 bias voltage, port1 port output 0dBmV excitation voltage, Port2 port connected transimpedance amplifier, input amplified signal, signal detection equivalent circuit as shown in Figure 1.

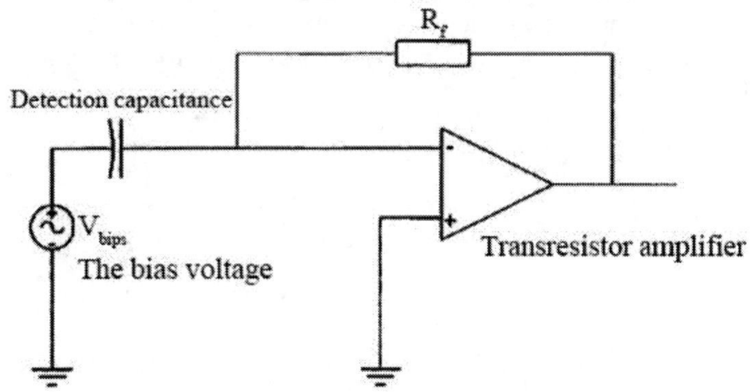


**Figure 1.** Signal Detection Equivalent Circuit

IOP Conf. Series: Materials Science and Engineering **772** (2020) 012044 doi:10.1088/1757-899X/772/1/012044

*3.2.2. Experimental results.* Based on the vector network analyzer and vacuum test environment, the frequency sweeping test of the mems micromachined gyroscope manufactured in this paper is carried out.

(1) The amplitude-frequency response and phase-frequency response curve of the drive shaft, it can be seen that the resonant frequency of the device is 23686.25Hz, as shown in Figure 2.

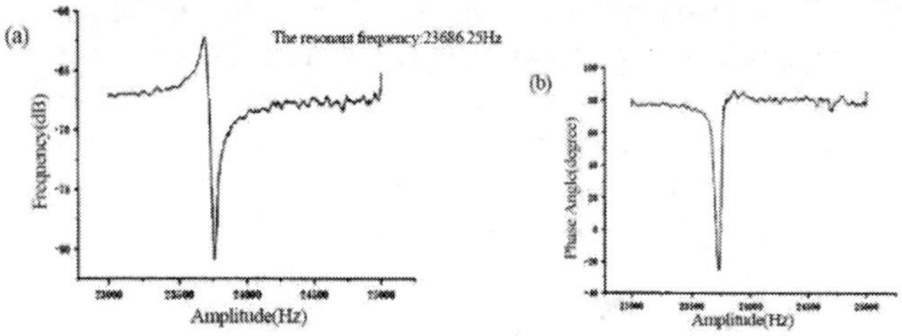


**Figure 2.** Drive Axis Sweep Test (a) Amplitude-Frequency Response Curve (b) Phase-Frequency Response Curve

(2) The amplitude-frequency response and phase-frequency response curve of the sensitive axis, it can be seen that the resonant frequency of the device is 23980 Hz, as shown in Figure 3.

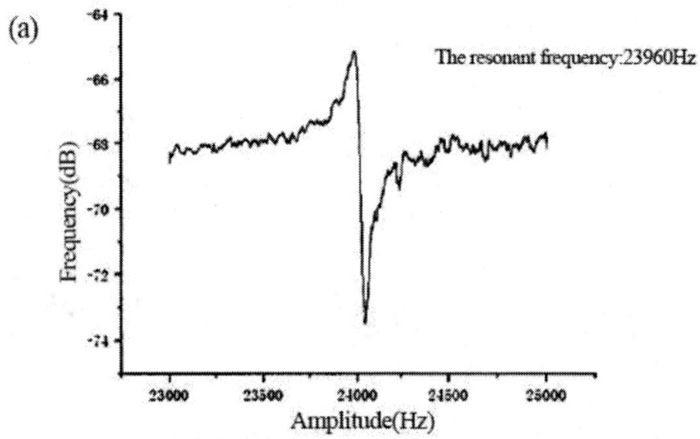


**Figure 3.** Sweep frequency test of sensitive axis (a) amplitude-frequency response curve (b) phasfrequency response curve

Electrostatic adjustment experiment when the 2.5V adjustment voltage is applied to the drive shaft, the amplitude-frequency response curve is shown in Figure 4, and the resonant frequency of the device is 23986.25Hz.The frequency cracking of the two axes has dropped from 293.75z to 6.25hz. From the previous analysis, the resonance frequency cracking of the two axes is greatly reduced, which can greatly improve the sensitivity of the device and lay a foundation for achieving better comprehensive performance of the device.

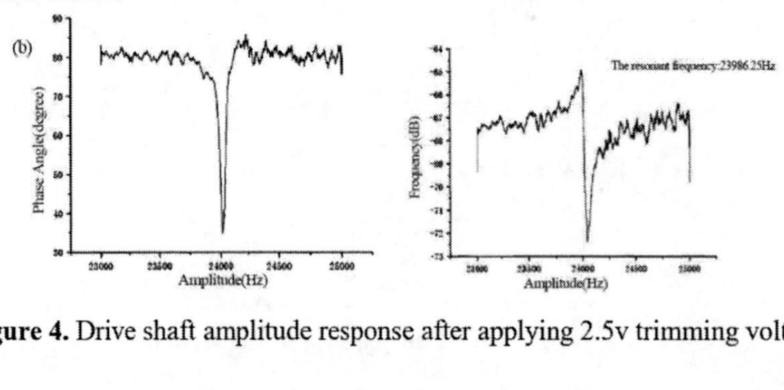


**Figure 4.** Drive shaft amplitude response after applying 2.5v trimming voltage

By comparing the experimental values with the simulated values, it can be found that the simulation frequency of the resonant frequency of the drive shaft is 21052 Hz, which is smaller than the experimental value of 2643.25 Hz; the simulation frequency of the sensitive axis is 21545 Hz, which is 2435 less than the experimental value.

Hz. This is because the deep silicon etching in the process flow sheet has etching (side etching) on the sidewall of the device, and the deep silicon etching has unevenness. In some places, the side etching is large, and in some places, the side etching is small, affecting the device. The resonant frequency.

In the experiment of the relationship between the regulation voltage and the frequency shown in Figure 5.9, the frequency increases with the increase of the adjustment voltage, which is in line with the experimental expectation, and the electrostatic adjustment is realized.Therefore, on the one hand, the improvement of deep silicon etching technology is a key direction of future research; on the other hand, it also proves the importance of frequency adjustment.

## 4. Summary

Micromachined gyroscope is an important inertial measurement device widely used in consumer electronics, automotive, defense guided weapons, aerospace and other fields.This paper presents a micromechanical gyroscope based on multi-ring structure.In the theoretical analysis, structural design, simulation calculation, process design, key micro-nano processing technology test, test analysis and other aspects have been studied.The main research results are as follows:

(1) Combining the application background of this subject, the advantages and disadvantages of common types of micromachined gyroscopes are compared, and the multi-ring structure gyroscope is established as the main design prototype and research object.The basic design parameters were determined. The high aspect ratio design was used. The modal analysis was performed on the device with a capacitance gap of 5 μ m and a depth of 50 m. The resonant frequency of the driving mode was determined to be 21052 Hz. The resonant frequency of the sensitive mode was 21545 hz. The damping was analyzed, especially The effect of air damping on the operation of the device and the establishment of the device operating environment as a vacuum.Under different fixed damping ratios, the harmonic response analysis of the device is carried out to analyze the impact resistance of the device. The X-axis can withstand an acceleration of 8.842×103g, and the E-axis can withstand an acceleration of 1.9× 10g, which meets the design requirements.

(2) Established a test platform for micromachined gyroscopes, using a vector network analyzer, vacuum chamber, transimpedance amplifier and other equipment to perform frequency sweep testing.The resonant frequency of the drive shaft is 23686.25hz and the resonant frequency of the sensitive shaft is 23980Hz.The effect of uneven etching of deep silicon on the device was analyzed.The device was subjected to an electrostatic adjustment experiment. After applying a 2.5V trimming voltage, the frequency cleavage of the two modes was reduced from 293.75 Hz to 6.25 Hz.Verify the effect of different trimming voltages on the resonant frequency of the device

## Acknowledgments

This work was financially supported by the overseas program of Shandong University of science and technology.

## References

[1] ACAR C, SHKEL A M. An approach for increasing drive-mode bandwidth of MEMS vibra-tory gyroscopes [J]. Joumal of microelectromechanica systems, 2005, 14 (3): 520-528.

[2] Lu Hao, Wei Xiaofeng, Pang Xiuzhi Application and Development of Inertial Technology in Precision Guided Weapons [J]. Electro-optical and Control, 2007, 14 (3): 45-47.

[3] Liang Geting, Hui Junjun, Li Yuping Gyro Development and Application. Flying Missile, 2006 (4): 108-110.

[4] SANDERS G, SZAFRANIEC B, LIU, et al. Fiber optic gyros for space, marine, and aviat-ion applications [C]/ Fiber Optic Gyros-20th Anniversary Conference. DENVER, CO:SPIE,1996:

IOP Conf. Series: Materials Science and Engineering **772** (2020) 012044 doi:10.1088/1757-899X/772/1/012044

61-71.
[5] PAVLATH G A. Fiber optic gyros:the vision realized [C]//Optical Fiber Sensors. San Diego, CA:SPIE, 2006: G3140.
[6] BEAN K E, BEAN K E. Anisotropic etching of silicon [J]. IEEE Transactions on electron dvices,1978,25 (10): 1185-1193.
[7] SMITH C S. Piezoresistance effect in germanium and silicon [J]. Physical review 1954, 94 (25): 1185-1193
[8] FRENCH P J, Evans R. Polycrystalline silicon strain sensors [J]. Sensors and actuator s, 1985, 8 (3): 219-225.

IOP Conf. Series: Materials Science and Engineering **772** (2020) 012045 doi:10.1088/1757-899X/772/1/012045

# Microstructures and Mechanical Properties of a new high entropy alloy

**Mingju Chen[1] and Jing Yang[1, 2, 3, a]**

[1]School of Resources and Materials, Northeastern University at Qinhuangdao, Qinhuangdao 066004, PR China
[2]Key Laboratory of advanced metal materials and forming technology in Qinhuangdao, Northeastern University at Qinhuangdao, Qinhuangdao 066004, PR China
[3]State Key Laboratory of Metastable Materials Science and Technology, Yanshan University, Qinhuangdao 066004, PR China

[a] yjwzb@163.com

**Abstract.** All manuscripts must be in English, also the table and figure texts, otherwise we cannot publish your paper. Please keep a second copy of your manuscript in your office. When receiving the paper, we assume that the corresponding authors grant us the copyright to use the paper for the book or journal in question. Should authors use tables or figures from other Publications, they must ask the corresponding publishers to grant them the right to publish this material in their paper.

## 1. Introduction

As is known to all, conventional alloys have been designed to have one or two principal elements with only minor additions of other elements to modify their properties. Generally, adding small amounts of alloying elements is necessary to improve performance or to obtain some special properties [1, 2]. According to traditional experience, too many alloying elements will inevitably lead to the formation of brittle intermetallic compounds, which have greatly limited the development of traditional alloys. However, the appearance of a new class of alloys since 1995, this traditional concept was broken. These alloys have been named as high entropy alloys because the atomic scale mixing entropies are higher than traditional alloys. High entropy alloys generally contain five to thirteen principle elements with each concentration between 5 and 35 at.% [3-5]. Up until now, it has been demonstrated that this explored alloys are feasible to be synthesized, processed and analyzed contrary to the misunderstanding based on traditional experiences. High entropy alloys tend to predominantly consist of simple solid solutions (face-centered cube-FCC, body-centered cube-BCC or mixed) due to high configurational entropy of mixing. Moreover, there are many opportunities in this field for academic studies and industrial applications [6, 7]. Recent developments in the field of high entropy alloys have revealed that they have versatile properties like: ductility, toughness, hardness and corrosion resistance [8-13].

The aim of this paper is to study the microstructures and mechanical properties of a new high entropy alloy, FeCrNiMnAl system. In this high entropy alloy, the cost of alloy elements is lower than that of high entropy alloys studied today. Microstructural analysis and mechanical properties of the FeCrNiMnAl alloy were performed using the as-cast alloy samples.

Content from this work may be used under the terms of the Creative Commons Attribution 3.0 licence. Any further distribution of this work must maintain attribution to the author(s) and the title of the work, journal citation and DOI.
Published under licence by IOP Publishing Ltd

## 2. Experimental Procedure

High-purity (at least 99.9%) powders of the constitutive elements were used as the starting materials, which were melted and processed in an electromagnetic suspension furnace. The melting process was developed in the water-cooled copper shell, sample cooling rate being almost instantaneous. The produced ingot's shape was cylinder with a diameter of 150 mm and height 50 mm. The chemical composition of the FeCrNiMnAl high entropy alloy is presented in Table 1 and reveals the equiatomic percentage for this composition.

**Table 1.** The chemical composition of FeCrNiMnAl high entropy alloy.

| Element | Fe | Cr | Ni | Mn | Al |
|---|---|---|---|---|---|
| At% | 20 | 20 | 20 | 20 | 20 |
| Wt% | 26 | 21 | 24 | 18 | 11 |

The specimens for microstructural characterization and mechanical testing were cut from the produced ingot using an electric discharge machine. Microstructures of samples were examined by optical microscopy (OM, Axiover 200MAT, Zeiss) and X-ray diffractometer (XRD, D/max-2500/PC, Rigaku). The samples for OM and XRD were mechanically ground, polished and etched using 10% nital. The Vicker's hardness was measured using a 5 kg load.

## 3. Results and Discussion

The microstructures of the as-cast alloy sample in low and high magnifications under the optical microscope are shown in Fig. 1(a-c). It can be seen from the figures that the microstructure of the as-cast FeCrNiMnAl high entropy alloy sample is composed of the main phase which is a solid solution with a dendritic structure evenly distributed in the high entropy alloy as can be observed in Fig. 1a.

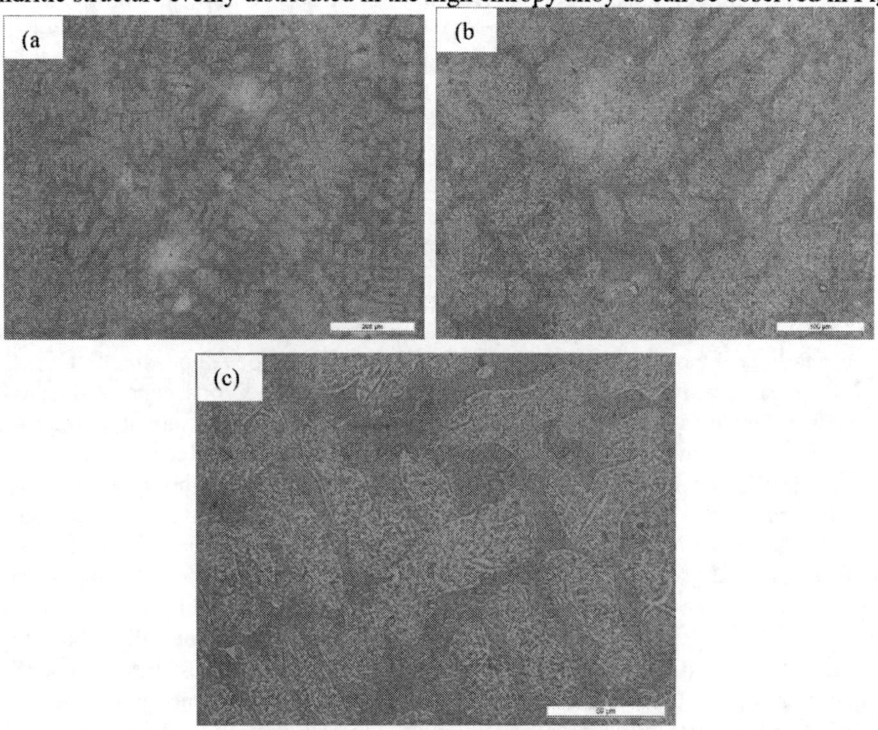


**Fig. 1** OM images of the as-cast FeCrNiMnAl high entropy alloy sample at the magnifications of (a) 100×, (b) 200× and (c) 500×.

Fig. 2 shows the X-ray diffraction pattern of the as-cast alloy sample. The diffraction peaks can be assigned to (110), (200), (211), (220) diffractions, respectively, which represents the typical character of a crystalline body-centered cubic (BCC) phase. Obviously, there is no other distinct diffraction peaks in the as-cast high entropy alloy except BCC peaks.

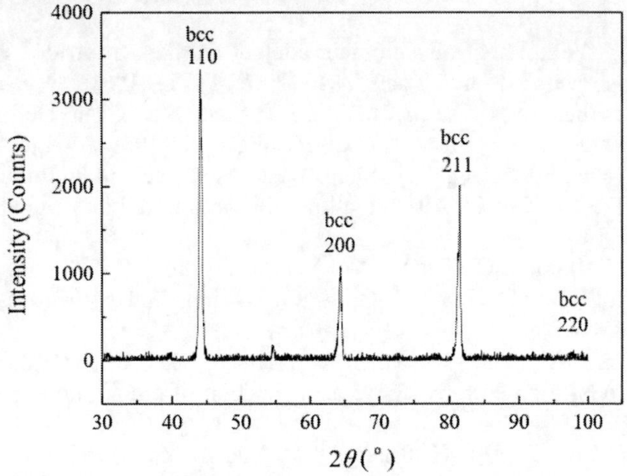


**Fig. 2** XRD pattern of the as-cast FeCrNiMnAl high entropy alloy sample.

Hardness is the property of a material that shows the ability to oppose to an action of permanent shape change when a compressive force is applied. The hardness values for a high entropy alloy are a good indicator of an improvement in mechanical properties. The hardness value for the as-cast FeCrNiMnAl high entropy alloy was determined after a series of 20 measurements along the diameter of the sample and the mean value was calculated. There was no large variation in the hardness values measured along the lines. The hardness mean value for the sample was 422 HV5.

## 4. Conclusion

The high entropy alloy FeCrNiMnAl is a new alloy which was successfully obtained and characterized. Analyzing the as-cast FeCrNiMnAl high-entropy alloy with optical microscopy it is observed that this alloy forms a dentritic structure. The as-cast FeCrNiMnAl high-entropy alloy is single BCC structures. Results for hardness test show that medium value for high entropy alloy FeCrNiMnAl is 422 HV5.

**Acknowledgements**
This work was supported by the National Natural Science Foundation of China (Grant No. 51301033), the Fundamental Research Funds for the Central Universities of Ministry of Education of China (Grant No. N152304001) and the Natural Science Foundation of Hebei Province, China (Grant No. E2017501055).

## References

[1] A.L. Greer, Confusion by design, Nature 366 (1993) 303–304.
[2] H. Baker, Metals Handbook, 9th ed., ASM International, Metals Park, Ohio (1992).
[3] J.W. Yeh, S.K. Chen, S.J. Lin, J.Y. Gan, T.S. Gan, T.S. Chin, T.T. Shun, C.H. Tsau. S.Y. Chang, Formation of simple crystal structures in Cu-Co-Ni-Cr-Al-Fe-Ti-V alloys with multiprincipal metallic elements, Metall. Mater. Trans. A 35 (2004) 2533–2536.
[4] J.W. Yeh, S.K. Chen, J.Y. Gan, S.J. Lin, T.S. Chin, T.T. Shun, C.H. Tsau. S.Y. Chang, Nanostructured high-entropy alloys with multiple principal elements: novel alloy design concepts and outcomes, Adv. Eng. Mater. 6 (2004) 299–303.
[5] B. Cantor, I.T.H. Chang, P. Knight, A.J.B. Vincent, Microstructure development in equiatomic

multicomponent alloys, Mater. Sci. Eng. A 213 (2004) 375–377.
[6] B.S. Murty, J.W. Yeh, S. Ranganathan, High Entropy Alloys, Elsevier (2014).
[7] A. Munitz, M.J. Kaufman, J.P. Chandler, H. Kalaantari, R. Abbaschian, Melt separation phenomena in CoNiCuAlCr high entropy alloy containing silver, Mater. Sci. Eng. A 560 (2013) 633–642.
[8] W.Y. Tang, J.W. Yeh, Effect of aluminum content on plasma-nitrided AlxCoCrCuFeNi high-entropy alloys, Metall. Mater. Trans. A 40 (2009) 1479–1486.
[9] C.W. Tsai, Y.L. Chen, M.H. Tsai, J.W. Yeh, T.T. Shun, S.K. Chen, Deformation and annealing behaviors of high-entropy alloy Al0.5CoCrCuFeNi, J. Alloy. Compd. 486 (2009) 427–435.
[10] C.Y. Hsu, J.W. Yeh, S.K. Chen, T.T. Shun, Wear resistance and high temperature compression strength of FCC CuCoNiCrAl0.5Fe alloy with boron addition, Metall. Mater. Trans. A 35 (2004) 1465–1469.
[11] H.Y. Chen, C.W. Tsai, C.C. Tung, J.W. Yeh, T.T. Shun, C.C. Yang, S.K. Chen, Effect of Substitution of Co by Mn in Al-Cr-Cu-Fe-Co-Ni High-Entropy Alloys, Ann. Chim.-Sci. Mat. 31 (2006) 685–698.
[12] J.M. Wu, S.J. Lin, J.W. Yeh, S.K. Chen, Y.S. Huang, H.C. Chen, Adhesive wear behavior of AlxCoCrCuFeNi high-entropy alloys as a function of aluminum content, Wear 261 (2006) 513–519.
[13] L.H. Wen, H.C. Kou, J.S. Li, H. Chang, X.Y. Xue, L. Zhou, Effect of aging temperature on microstructure and properties of AlCoCrCuFeNi high-entropy alloy, Intermetallics 17 (2009) 266–269.

# Topology Optimization Design of Boat Boom Structure Based on Level Set Method

**Dongyue Qu, Qiang Chen, Jianan Xu *and Tianxin Kou**

College of Mechanical and Electrical Engineering, Harbin Engineering University, Heilongjiang, China

*Corresponding author e-mail: xujianan@hrbeu.edu.cn

**Abstract.** The marine boom is optimized by a level set topology optimization method based on the reaction diffusion equation. In order to overcome the characteristics that the boom is difficult to be topology optimized, the periodic topology optimization theory is introduced to divide the boom into several cycles, and two typical conditions of maximum amplitude and minimum amplitude are selected respectively. Finally, the optimized boom structure is processed and remodelled and finite element analysis is performed, which still meets the original performance index.

## 1. Introduction

In recent years, with the rapid development of marine transportation, the loading and unloading capacity of marine deck cranes is becoming more and more demanding, and cranes with light weight and good performance are more and more popular. Based on this background, the topology optimization of the structure of the marine deck crane is carried out, so as to achieve the purpose of weight reduction on the basis of satisfying the original performance.

Peng Jingjing[1] optimized the structure of the boom by genetic algorithm. Wang Chenglin[2] used SIMP topology optimization method to optimize topology of truss arm top joint structure under single and multi-case conditions. Based on the Lagrange multiplier method and the evolutionary algorithm, Milomir[3] optimized the crane's boom section. Xiong Zhan [4] used the finite element analysis software to optimize the topology of the boom truss of the tower crane under the action of moving load.

This paper mainly optimizes the boom based on the level set topology optimization method of diffusion equation proposed by Takayuki Yamada [5-7]. In the optimization process, the periodic topology optimization theory is applied to the optimization process of the boom. The two conditions of the maximum and minimum amplitude of the crane are selected for research, which makes the optimization more convenient and effective, and the obtained results are also easy to process. Finally, the finite element analysis of the optimized structure is carried out to prove the application value of the level set topology optimization method in engineering practice.

## 2. Structural topology optimization design model of level set method

Based on the level set method of the reaction diffusion equation, the mathematical model of maximizing the stiffness of the boom structure is established, and the structural volume is the mathematical model of the constraint optimization problem. The structure boundary diagram is shown in the following figure:

Content from this work may be used under the terms of the Creative Commons Attribution 3.0 licence. Any further distribution of this work must maintain attribution to the author(s) and the title of the work, journal citation and DOI.
Published under licence by IOP Publishing Ltd

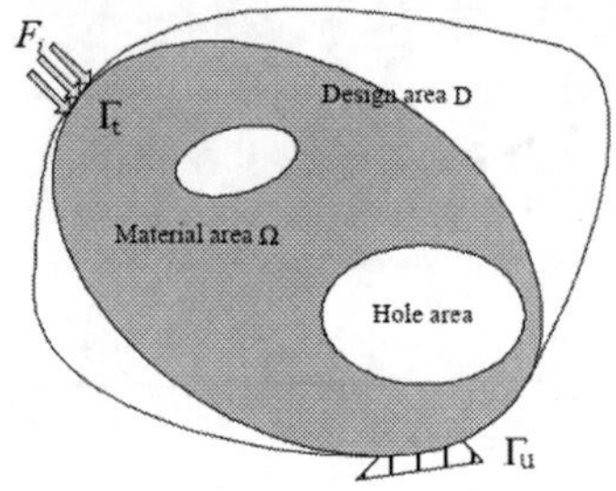


**Figure 1.** Schematic diagram of the level set topology optimization structure boundary.

Where $\Gamma_u$ is a place of displacement constraint. $\Gamma_t$ is the external force constraint. D is the design area and Ω is the material area. The detailed optimization model is shown in equation (1).

$$\begin{aligned} &\inf_{\phi} J = \int_{\Gamma} F_I u_i d\Gamma_t \\ &s.t.G = \int_{\Omega} d\Omega - V_{\max} \leq 0 \\ &div(E_{ijkl} u_{k,l}) = 0 \ on \ \Omega \\ &u_i = \overline{u_i} \quad in \quad \Gamma_u \\ &F_i = \overline{F_i} \quad in \quad \Gamma_t \end{aligned} \tag{1}$$

In the above formula, J is the objective function of the optimization model constructed by the level set topology optimization method, $V_{\max}$ is the largest volume constraint value, $E_{ijkl}$ is the elastic modulus of the material, $u_i$ and $F_i$ are the corresponding displacement and external force constraints, where the displacement of the boundary $\Gamma_u$ is zero, $\overline{F_i}$ represents the given external load, $div(E_{ijkl}u_{k,l}) = 0$ represents no additional external load on material Ω area, and $F_i = \sigma_{ij} n_j = E_{ijkl} u_{k,l} n_j$ represents tension.

The Lagrange multiplier method is used to transform the above constrained optimization problem into an unconstrained problem. The specific expression is shown in equation (2).

$$\overline{J} = \int_{\Gamma} F_i u_i d\Gamma + \int_{\Omega} \tilde{u}_i div(E_{ijkl} u_{k,l}) d\Omega + \lambda(\int_{\Omega} d\Omega - V_{\max}) \tag{2}$$

The level set function is updated by introducing a reaction diffusion equation; the specific formula is as follows:

$$\begin{cases} \dfrac{\partial \phi}{\partial t} = -(-Cd_t \overline{J}_R - \tau \nabla^2 \phi) \ in \ D \\ \phi = 0 \qquad\qquad\qquad\qquad on \ D \end{cases} \tag{3}$$

$\phi$ is a level set function, K is a proportional coefficient greater than zero, and $\tau$ is a regularization parameter, which affects the geometric complexity of the final optimized structure in a certain extent. By reasonably determining the $\tau$ value, an ideal objective function can be obtained, and $\nabla^2 \phi$ is a diffusion term, it ensure the smoothness of the level set function.

## 3. Construction of periodic structure optimization model

The marine deck crane boom belongs to the narrow and long structure type, good optimization results can not obtain through using optimization method directly. The periodic optimization method can make the optimization more convenient and effective, and the optimization result is also convenient for practical engineering applications. The paper use the method of level set topology optimization of reaction diffusion equation to optimize the periodic structure, the optimization process has high efficiency, and the precision of the optimization result is high.

In order to obtain an optimal topology with a periodic structure, the whole structure is divided into an equal number of sub-domains, $n_1$ is the number of sub-domains in the X direction, $n_2$ is the number of sub-domains in the *Y* direction, and $x_{n1,n2,j}$ represents the density value of the *j* unit of the ($n_1$, $n_2$) subfield. For the periodic structure, each sub-domain has the same sub-structure topological configuration as any sub-domain before and after a certain direction, that is the identical unit individual.

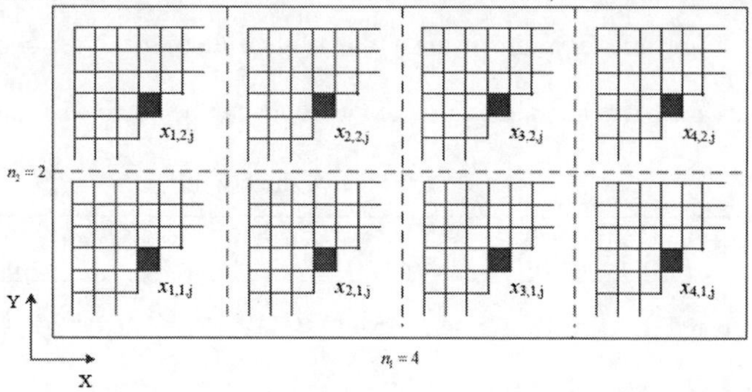


**Figure 2.** Schematic diagramm of the structure dividing sub-domain unit.

At this time, the mathematical model of the topology optimization of the entire periodic structure is:

$$
\begin{aligned}
&\inf_{\phi} J = \sum_{i=1}^{N} x_{i,j} \int_{\Gamma} F_i u_i d\Gamma \\
&s.t. G = \int_{\Omega} d\Omega - \sum_{i=1}^{N}\sum_{j=1}^{M} x_{i,j} V_{i,j} \le 0 \\
&div(E_{ijkl} u_{k,l}) = 0 \ on \ \Omega \\
&u_i = \overline{u_i} \qquad in \ \ \Gamma_u \\
&F_i = \overline{F_i} \qquad in \ \ \Gamma_t \\
&x_{1,j} = x_{2,j} = x_{3,j} = \cdots x_{m,j}
\end{aligned}
\tag{4}
$$

Where $x_{i,j}$ represents the density value of the *i* unit of the a subdomain, $V_{i,j}$ represents the volume of the *j* unit of the *i* subfield, and f denotes that the unit exists in the design domain.

## 4. Optimized design of the boom

### *4.1. Number Determination of boom load*

The boom of the marine deck crane is a box-type boom under double-bar operation, and the movable pulley block is used to realize the lifting and lowering of heavy objects. When working, the load on the boom has variable amplitude load, rotational inertia force, lifting load and wind. Loaded. The side force and restraint state of the boom are shown in Figure 3.

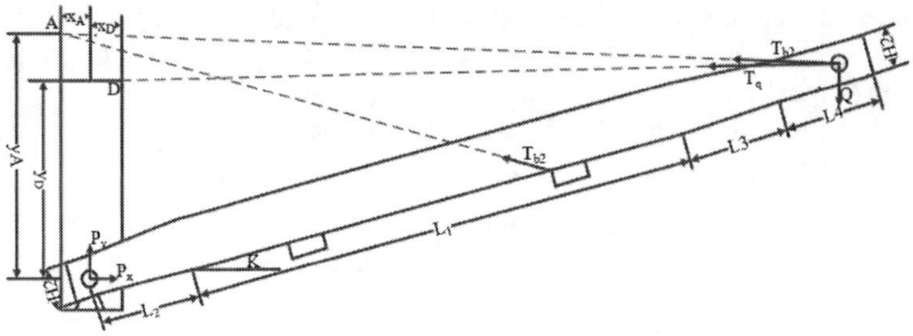


**Figure 3.** Two-dimensional plan of the side frame of the boom

Through analysis and calculation, the force of the boom can be obtained as shown in Table 1.

**Table 1.** Force of the boom.

| Working condition | Lifting wire rope force(t) | Luff wire rope force(t) | Hinge shaft force (t) | Rotating inertial force (kN) | Wind load (t) |
|---|---|---|---|---|---|
| Working condition one ($K=82°$) | 122.8 | $T_{b1}=41$<br>$T_{b2}=21.8$ | $P_x=35.37$<br>$P_y=201$ | Weight of the hanging object 2.36<br>Boom 1.86 | 0.83 |
| Working condition two( $K=20°$ ) | 122.8 | $T_{b1}=5.87$<br>$T_{b2}=3.1$ | $P_x=108$<br>$P_y=55$ | Weight of the hanging object 15.94<br>Boom 10.11 | 0.29 |

### *4.2. Topology optimization of the boom*

The three-dimensional solid model of the boom is shown in Figure 4.

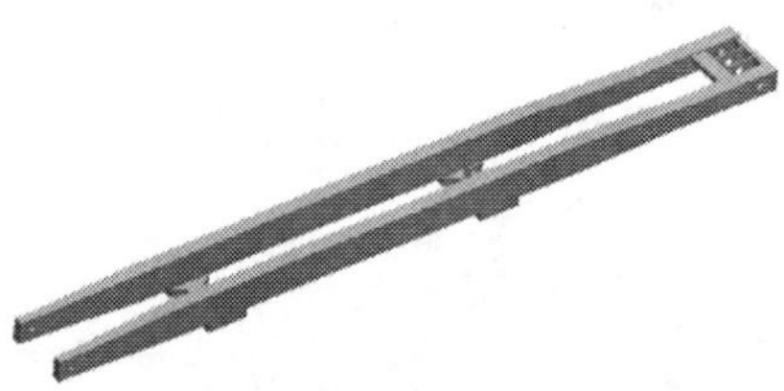

**Figure 4.** 3D solid model of the boom.

In the topology optimization of the boom, combined with the periodic topology optimization theory, the topology optimization of the boom is carried out by using the level set topology optimization method, which not only makes the optimization more convenient and effective, but also makes the obtained results easy to process, the load on the boom and the constraint situation is shown in Figure 3.

When the marine deck crane is working, the elevation angle may change at any time, so it is impossible to study all the working conditions. For the deck crane of this project, the maximum and minimum range of the crane are selected for research, the elevation angle of the boom is 82° and 20°, working condition 3 is the working condition when working condition 1 and working condition 2 correspond to different weight coefficients. In the process of optimization, the above three working conditions are studied in turn, and getting the optimal boom structure by iteration on the basis of satisfying the original performance.

The topology is optimized by the level set method. The optimized result is shown in the figure below.

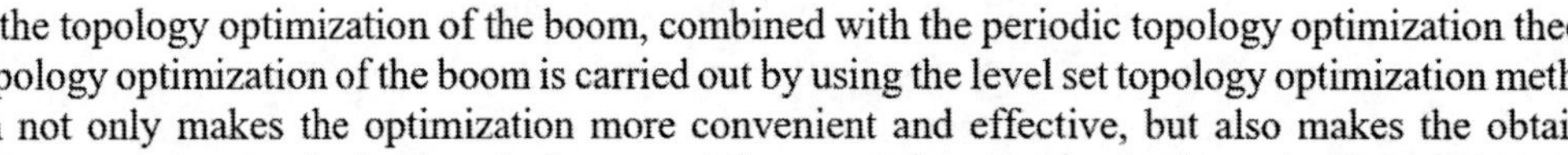

**Figure 5.** Topology optimization results of Boom web about working condition one.

IOP Conf. Series: Materials Science and Engineering **772** (2020) 012046 doi:10.1088/1757-899X/772/1/012046

**Figure 6.** Topology optimization results of Boom web about working condition two.

**Figure 7.** Topology optimization results of Boom web about working condition three.

Re-established the model according to the processed result, the new actual model of the boom is shown in Figure 8.

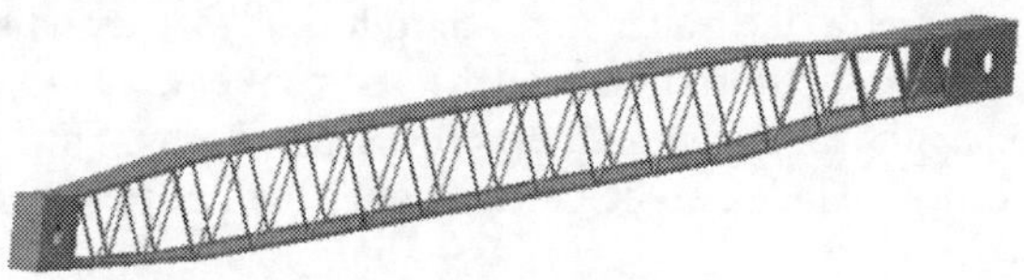

**Figure 8.** Three-dimensional model diagram of the topology optimization of the boom side panel.

After optimization, the weight of the boom was reduced from the original 7t to 5.4t, and the weight of the boom was reduced by 1.6t, achieving the goal of lightweighting the boom.

### *4.3. Simulation Analysis after the Topology Optimization of the Boom*

Remodelling the key components of the optimized boom processed in Section 4.2. The three-dimensional model of the boom topology optimized after assembly, and it is imported into ABAQUS for finite element analysis.

By applying the calculated load and constraints, performing the finite element analysis on the boom position at 82° and 20° respectively, and the obtained stress and deformation results show in the following figure.

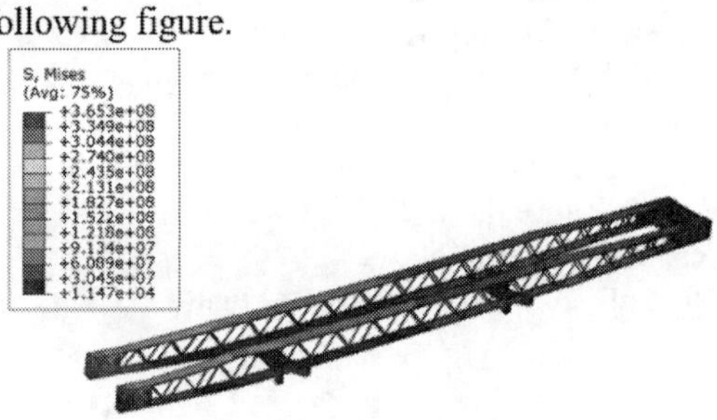


**Figure 9.** Stress distribution of the 82-angle boom

**Figure 10.** Deformation distribution of the 82-angle boom

Through finite element analysis, it can be obtained that the maximum deformation of the boom is about 100mm, and the stress concentration occurs in the hinge shaft. The maximum stress is 365MPa. According to the material properties selected, the position is within the safe range.

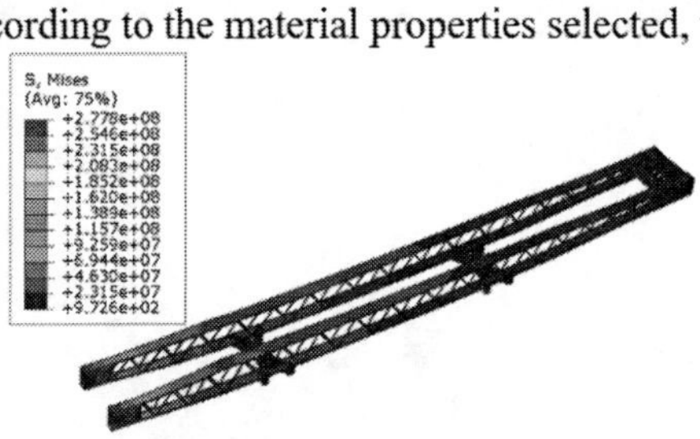


**Figure 11.** Stress distribution of the 20-angle boom.

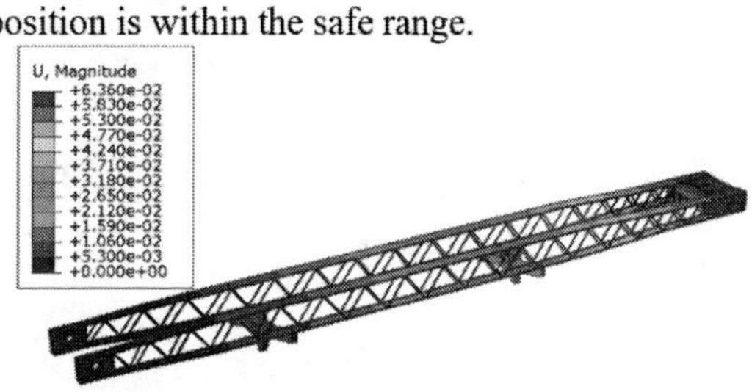


**Figure 12.** Deformation distribution of the 20-angle boom

It can be obtained from the above figure that the maximum deformation of the boom is about 63mm, and the maximum stress of 277MPa appears in the hinged shaft. The boom can meet the actual working conditions. Through the above analysis of the topology optimized boom, it can be seen that, the boom still meets the requirements of the actual working conditions on the basis of the weight reduction of the boom.

## 5. Conclusion

In this paper, the horizontal set topology optimization method of reaction diffusion equation is used to optimize the topology of the ship deck crane boom. Establishing the mathematical model with the maximize structural stiffness as target function and the structural volume as the constraint optimization problem. The equation updates the level set function, which realizes the topology optimization of the shape and size of the boom. At the same time, introducing the periodic theory into the topology optimization, which makes the optimization model more convenient, the obtained boom structure is clear and the engineering application is strong, the processing is easy. Finally, the optimized results are verified to meet the performance index of the boom through the finite element analysis of the optimized boom.

## References

[1] Peng Q Q. Optimization design of the structure of marine deck crane boom based on genetic algorithm [D]. Wuhan University of Technology Master thesis,2018: 38-54.

[2] Wang Chenglin. Research on local topology optimization of crane truss boom section [D]. Master thesis of Dalian University of Technology, 2012:33-65.

[3] Milomir M G, Mile M S, Radovan R B, et al. Optimization of a pentagonal cross section of the truck crane boom using Lagrange's multipliers and differential evolution algorithm [J]. Meccanica, 2011, 46(4): 845-853P.

[4] Xiong Zhan, Mo Jiyun. Topology Optimization Design of Tower Crane Crane Arm Truss [J]. Journal of Wuhan University of Science and Technology, 2014,37(03): 195-198.

[5] Yamada T, Yamasaki S, Nishiwaki S, et al. Design of Compliant Thermal Actuators Using Structural Optimization Based on the Level Set Method [J]. Journal of Computing and Information Science in Engineering, 2008, 11(1): 1217-1225P.

[6] Yamada T, Yamasaki S, Nishiwaki S, et al. A Study of Boundary Settings in the Design Domain for Structural Optimization Based on the Level Set Method [J]. Transactions of the Japan Society for Industrial & Applied Mathematics, 2008, 18:487-505P.

[7] Yamada T, Yamasaki S, Nishiwaki S, et al. Design of Compliant Thermal Actuators Using.

# Dynamic Monitoring of RS and GIS Resources and Ecological Environment Based on High Temperature Materials

**Shuhui Jiang [a], Li Tang [b] and Hui Ju [c]**
Chongqing Jianzhu College, Chongqing 400072, China.

[a] jsh0529@126.com, [b] 398681892@qq.com, [c] 17720444@qq.com

**Abstract.** The dynamic monitoring and analysis of regional ecological environment is an important part of land use research, which can provide support for regional ecological environment and macro regulation of government departments. With the support of remote sensing technology (RS) and geographic information system (GIS), a land use dynamic monitoring system is established to obtain land use change information quickly, accurately and economically. In order to reduce the influence of thermal radiation, noise and vibration of high temperature material machinery on the measurement results in high temperature measurement environment, the paper proposed the measurement method of ultraviolet light high humidity thermal deformation and gray average method. In this paper, the algorithm of digital image correlation method and the establishment of high temperature measurement system are studied. A set of high temperature material mechanical digital image correlation system suitable for high humidity environment measurement is independently established and applied to the high humidity test of materials. The results show that the above comprehensive method can effectively improve the influence of thermal flow disturbance on system imaging in high humidity environment. Comprehensive analysis of regional ecological environment data shows that environmental quality is closely related to regional ecological environment. Meteorological conditions are one of the important factors affecting the environment, and the project of returning farmland to forest plays a vital role in improving the environment.

**Keywords:** High temperature materials; RS and GIS; Dynamic monitoring of ecological environment.

## 1. Introduction

Ecological environment is the organic combination of ecological system and environmental system. It includes biological ecological factors and abiotic ecological factors, such as vegetation, rivers and lakes, land, climate and other physical and geographical conditions and man-made conditions. It is the environmental basis for human survival and development [1]. Environmental quality assessment is a scientific and quantitative description of regional environmental quality. It is to select representative, comparable and operable evaluation indicators and methods according to specific purposes. Correct evaluation of the current situation of ecological environment quality is the basis of regional ecological environment prediction or early warning research, and is an important basis for formulating and planning

Content from this work may be used under the terms of the Creative Commons Attribution 3.0 licence. Any further distribution of this work must maintain attribution to the author(s) and the title of the work, journal citation and DOI.
Published under licence by IOP Publishing Ltd

regional national economic development plans [2]. Remote sensing monitoring of land use in China began in the 1970s. The focus of land use remote sensing monitoring is on the change of cultivated land and the expansion of construction land. However, in recent years, research on global change and regional land use, land use change monitoring and mechanism has been significantly enhanced [3]. It can provide rich resource and environment information quickly and accurately. The geographic information system creates an ideal environment for remote sensing information processing, processing and application. The organic combination of the two provides real-time and dynamic spatio-temporal change information for environmental regulation [4]. Among them, remote sensing (RS), geographic information system (GIS) and their combination provide time-space change information for real-time and dynamic monitoring of ecological environment regulation, as well as information extraction, analysis and processing [5-6]. In order to obtain and manage the dynamic information of land use in a fast, accurate and economical way, scientifically manage the spatio-temporal data of land use, provide advanced management tools for the land management department, and provide decision-making support for regional sustainable development.

## 2. Methodology

The high-temperature digital image correlation method, as its name implies, applies its own digital image correlation method to the high-temperature environment measurement of actual industrial engineering. In recent years, after continuous innovation and optimization by relevant researchers, digital image correlation methods have made great progress in testing the mechanical properties of high temperature materials. The material mechanical measurement under high temperature conditions is very important for the safety, high service life and high temperature of the high temperature resistant materials in the aerospace, military industry, etc., and the deformation and strain field information of the material surface at high temperature is obtained. It is the key and foundation for the study of its mechanical properties [7]. Rigid body displacement experiments at room temperature can be used to verify the accuracy of the measurement system. Because in the experiment, it can be considered that the displacement produced by the displacement platform used in the experiment is standard and accurate. As a space exploration technology, remote sensing has played a significant advantage in land resources investigation. GIS is another advanced science and technology rising after remote sensing technology. It stores, manages, processes, analyzes and displays geographic information under the support of computer software and hardware to provide support for decision-making. Comprehensive application of RS and GIS technology to dynamic monitoring and analysis of land use status can quickly and accurately grasp the information and trends of land use status and dynamic changes, and provide efficient auxiliary means for macro-control of dynamic balance of total land use and ecological environment protection [8].

The biological abundance index is an indicator for assessing the degree of abundance of biodiversity in a region, and is calculated from the proportion of the equivalent area of forests, waters, grasslands and other ecosystems. Table 1 below shows the regional land use structure obtained by using RS and GIS technology.

**Table 1.** Regional land use structure

| Cultivated land | Woodland | Grassland | Building land |
|---|---|---|---|
| 780.36 | 690.12 | 39.92 | 350.11 |
| 661.72 | 249.33 | 8.96 | 159.68 |
| 650.04 | 337.41 | 24.18 | 256.94 |

At present, the organic combination of remote sensing and geographic information system has become a hot spot in the development of high-tech. Remote sensing can provide abundant information of resources and environment quickly and accurately. The high temperature heating and holding system consists of a high temperature furnace with an optical glass observation port on the front and a temperature control platform connected with the high temperature furnace [9]. The system has the function of setting temperature control steps before experiment and automatically adjusting and maintaining temperature. Any object whose surface temperature is higher than absolute zero will emit electromagnetic waves of different frequency bands, and the spectrum distribution of the electromagnetic waves is related to the temperature of the object itself. Remote sensing image processing refers to the geometric correction and image enhancement of the original data acquired by remote sensing. This time we used image processing software to process remote sensing images on the workstation. Because some small maps will inevitably occur during the classification process, these small maps need to be culled and re-classified, and the surrounding small maps are merged into the large categories by clustering, and the non-conforming small patches are removed by filtering [10]. Using database technology and multi-resolution coupling technology to establish a seamless mosaic network information system integrating environment, resources and social economy; this method effectively reduces the interference of vibration and noise on experimental results. The toothed fluctuations are effectively removed. And by comparing the average pre- and post-average data, the obtained data is consistent with the deformation characteristics of the material at high temperatures.

The PSR model is a dynamic model structure that includes three main indicators: stress, state, and response. The three indicators interact and constrain each other. Among them, the urban water ecosystem health PSR model covers the three dimensions of the pressure of the water ecosystem, the maintenance state of the water ecosystem and the dynamic response of human society. Pressure subsystem mainly refers to the negative impact of economic development and water resource consumption on aquatic ecosystem. This model analyses the interaction between human and environment from the perspective of system theory. Figure 1 shows the PSR model of urban aquatic ecosystem health assessment.

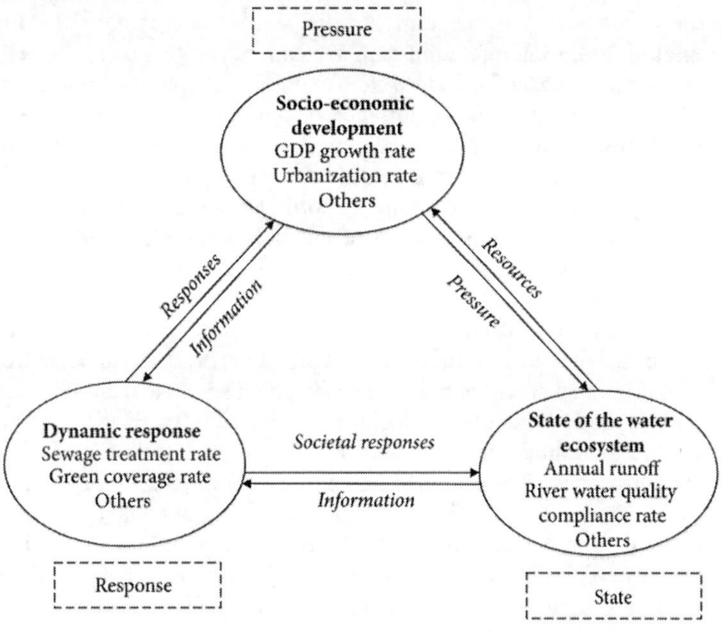


**Figure 1.** PSR model

In order to ensure the measurement of the system in high temperature environment, it is necessary to adopt effective measures to reduce or even eliminate the impact of hot box radiation on imaging. In high temperature measurement, the speckle will fall off or oxidize more or less in long time high temperature environment. Therefore, filling inert gas in the environment of high temperature measurement or pumping the measurement environment into vacuum can improve the measurement accuracy. Taking the patches on the background vector map as the unit, each patch is checked one by one. When the land use type indicated by the image is inconsistent with the background land use type, it is regarded as the land use change area. Using the secondary development technology of GIS, integrating the ecological environment information system and control system, and establishing a decision support system with spatial analysis function and sustainable development discrimination function, it can realize visualization and operation. The ecological comprehensive evaluation index is used to evaluate the quality of the ecological quality. According to the evaluation index values of the evaluation unit and the weight values of each individual indicator, the weighted summation method is used to calculate the comprehensive evaluation index value. Finally, manual visual interpretation is performed on the missing and misclassified pixels, and the undivided pixels are coordinately marked and the global positioning system is used for field sampling. The high-temperature thermal deformation measurement method achieves the original design goal and can be used for high-precision measurement of surface deformation of metal materials and structures in high temperature environments.

## 3. Result Analysis and Discussion

At present, the temperature deformation in the material deformation measurement is mostly in the form of a high temperature furnace. Experiments have shown that in the high temperature environment, a large temperature difference is formed inside and outside the observation window of the high temperature furnace. For the problem of high-temperature imaging affected by blackbody radiation in high temperature environment, the paper compares the illumination of different light sources and uses the corresponding filter to filter out stray light, observes the imaging of different color speckles, and finds an optimal imaging scheme at high temperature. . Due to the existence of the same-spectrum foreign matter and the phenomenon of homosexuality in remote sensing, we need to carry out field verification of the land use change area interpreted indoors. At the same time, the field survey is carried out to compare the differences between remote sensing data and actual data. Error correction and compilation of remote sensing interpretation data are carried out to ensure the reliability of data and various maps. Eco-environmental status index refers to the overall quality of the ecological environment in the evaluated area. According to the importance of biological abundance index, vegetation cover index, water network density index, land degradation index and environmental quality index, different weights are assigned to calculate. Determine the corresponding relationship between the spatial coordinates of the calibration plate and the image coordinates, and calculate the internal and external parameters of the camera. The relative position of the calibration plate and the camera is constantly changing. At this time, the internal parameters of the camera remain unchanged, while the external parameters change with the change of the calibration plate.

The use of remote sensing and GIS means to carry out dynamic monitoring of environmental quality has achieved good results, but there are still a series of problems that need to be improved in future work. Create a database for all data. This is the basis for research on ecological environment regulation and is also a key technical work. The basic steps include: graphical interpretation of digitization, building topological relationships, and building a database. Establishing a spatio-temporal data model of land-use spatio-temporal database is the abstraction and expression of temporal geography and the basis of the land-use dynamic monitoring system. Different spatio-temporal data models will determine the method of spatio-temporal data processing and the organization and management of spatio-temporal databases. The interactive digitalization and database acquisition subsystem mainly carries out thematic information extraction and digitalization. It adopts human-computer interactive input and various data formats (raster data, vector data, attribute data) acquisition methods. Consider combining two kinds of samples into one kind of sample. After multiple sample selection and sample separability calculation, the values of the two parameters are above 1.8, which shows that the separability is good. Snapshot model records land use status at different time of investigation. The model is simple, but the data redundancy is large and the spatial-temporal analysis ability is low. Spatio-temporal composite model records the history and current situation of land use in a map layer. The data redundancy is small, and the spatial-temporal analysis ability is relatively strong.

In the field of industrial engineering, the strength of materials or structures has always been an essential step in verifying whether the materials meet the requirements of working conditions. Therefore, the elastic modulus, as an important index to characterize the stiffness of materials, is one of the

IOP Conf. Series: Materials Science and Engineering **772** (2020) 012047 doi:10.1088/1757-899X/772/1/012047

parameters that must be obtained. There will be errors in the measurement results of any measurement system, and digital image correlation measurement system is no exception. The interference of the external environment and the system errors of the test system itself will affect the accuracy of the test results to a certain extent. The space-time query of land use includes traditional queries on the one hand, such as querying topic attributes from space or querying spatial distribution by topic attributes. Time and space query is a special or comprehensive query on the time, space and thematic attributes of land use, such as land use time and space change query. After data acquisition and preprocessing (including data compression, data conversion, set correction, and windowing of the study area, etc.), data files are generated; Secondly, image data loading, image transformation, image analysis, query, filtering and boundary enhancement are carried out. Ecosystem is a multi-level and multi-factor complex system. Different factors have different impacts on the environment, so it is necessary to determine the weights of the evaluation factors. Because of the harsh high temperature environment and the oxidation of speckles and specimens, it is particularly important to find a new method of speckle making.

## 4. Summary

When measuring high temperature or even ultra-high temperature, some thermal barrier is needed for the camera, light source and other experimental equipment, so that the experimental equipment can be protected. Secondly, the noise effect caused by the temperature rise can be reduced. In the remote sensing land use dynamic monitoring system, the transition between agricultural land and non-agricultural land is relatively easy to identify, but the transition between agricultural land or non-agricultural land types, such as between dry land and irrigated land. There are certain difficulties in transforming recognition. When measuring in a commercial temperature environment, the temperature variation of the system components such as the lens holder also introduces measurement errors. Therefore, a method of blocking the influence is also a well-informed direction for subsequent accuracy. The influence of meteorological conditions on vegetation growth lags behind to a certain extent. Therefore, in future research work, we should properly advance the time of meteorological factors, and when is the most appropriate time to advance is the problem to be solved in the future. It is suggested that environmental protection departments should establish a long-term mechanism for evaluating the ecological environment, and take the trend of dynamic changes in the ecological environment of each district and county as the basis for formulating the policy of ecological function regionalization and land use management, so as to promote the coordinated development of society, economy and ecological livability.

**Acknowledgements**
Chongqing higher education teaching reform research project, project NO. 163270;
Education and teaching research project of Chongqing Jianzhu College, project NO. 16J1

**References**

[1] Zhang R, Morton L D, Smith J D, Gallazzi F, White T A, Ulery B D [J]. Instructive Design of Triblock Peptide Amphiphiles for Structurally Complex Micelle Fabrication. ACS Biomaterials Science & Engineering, 2018, 4(7): 2330 - 2339.
[2] Zhang R, Ulery B D [J]. Synthetic Vaccine Characterization and Design [J]. Journal of Bionanoscience, 2018, 12(1): 1-11.
[3] Zhang R, Smith J D, Allen B N, Kramer J S, Schauflinger M, Ulery B D [J]. Peptide Amphiphile Micelle Vaccine Size and Charge Influence the Host Antibody Response [J]. ACS Biomaterials Science & Engineering, 2018, 4(7): 2463 - 2472.
[4] Zhang R, Kramer J S, Smith J D, Allen B N, Leeper C N, Li X, Morton L D, Gallazzi F, Ulery B D. Vaccine Adjuvant Incorporation Strategy Dictates Peptide Amphiphile Micelle Immunostimulatory Capacity [J]. The AAPS journal, 2018, 20(4): 73
[5] Zhang R, Leeper C N, Wang X, White T A, Ulery B D. Immunomodulatory vasoactive intestinal peptide amphiphile micelles [J]. Biomaterials science, 2018. 6(7): 1717-1722.

[6] Smith J D, Cardwell L N, Porciani D, Nguyen J A, Gallazzi F, Tata R R, Burke D H, Daniels M A, Ulery B D. Aptamer-displaying peptide amphiphile micelles as a cell-targeted delivery vehicle of peptide cargoes [J]. Physical biology, 2018, 15(6): 065006.
[7] Zhang R, Billingsley M M, Mitchell J J. Biomaterials for Vaccine-Based Cancer Immunotherapy [J]. Journal of Controlled Release, 2018, 292(28): 256-276
[8] Qing F, Peng Y. Development and optimization of ecological network of Shunyi District of Beijing based on RS and GIS [J]. Chinese Journal of Applied and Environmental Biology, 2016, 22(6): 1074-1081.
[9] Lu D, Yu H L, Yu G M. Assessing the Land Use Change and Ecological Security Based on RS and GIS: A Case Study of Pingdingshan City, China [J]. Advanced Materials Research, 2014, 905:329-333.
[10] Li X Y, Chen S L, Bian K. Mine Dynamic Monitoring and Integrated Management Based on RS and GIS [J]. Advanced Materials Research, 2014, 955-959: 3840-3844.

# Study on the Synthesis and Photothermal Conversion Property of Polyaniline

**Lihong Bao[1, 2, *], Xin Liu[1]**

[1]School of Materials Science and Engineering, Beijing Institute of Fashion Technology, Beijing 100029, China
[2]Beijing Key Laboratory of Clothing Materials R&D and Assessment, Beijing 100029, China

*Author for correspondence; E-mail: clyblh@bift.edu.cn

**Abstract.** A series of polyaniline (PANI) with different HCl/ aniline and DBSA/ aniline molar ratio was synthesized in this paper. The structure of PANI was characterized using IR spectroscopy. The effect of pH, thermal stability and photothermal conversion properties of PANI were investigated. The results showed that the IR spectrum presented typical absorption peaks of PANI-ES. The pH had a big influence on the optical property of PANI. The PANI had high thermal stability in all HCl/aniline molar ratio. The molar ration of DBSA/aniline can influence the photothermal conversion of PANI, whereas the doped PANI showed better photo thermal conversion efficiency than that of dedoped PANI.

**Keywords:** Polyaniline, photothermal conversion, optical property.

## 1. Introduction

Polyaniline (PANI), the most popular conducting polymer, has gained great attention due to its lower cost, interesting chemical and physical properties, ease of preparation. It has been studied to applied in a wide range of industrial fields, such as rechargeable batteries, gas sensors, polluted water remediation [1], electronic devices, solar cells, and polymer corrosion protection agents.

PANI is a mixed oxidation state polymer consisting of benzoid and oxidized quinoid units [2], the ratio of these two different units change with the synthesis composites and formula, thus endue PANI many unique properties and electronic conduction mechanisms that distinguish it from the other conducting polymers. Emeraldine base (EB), the typical dedoped state of PANI, has a structure consists of equal proportions of amines (–NH–) and imine (=N–) sites, after being protonated by acids HA, resulting in emeraldine salt with the conductivity of a semiconductor level. The electrical properties of PANI have recently been investigated a lot [3-4]. Up to now, only few studies on photo-response of PANI or PANI composite have dealt with. Silvestre B.A. studied the absorption capacity of infrared radiation by PANI nanoparticles dispersion and subsequent conversion to heat, they found under illumination of NIR laser at 780 nm and 100mW power, the temperature of aqueous medium (without PANI nanoparticles) increases only 3 °C after 10 min (600 sec) of laser application, while in presence of PANI nanoparticles the temperature increase 9.5 °C. The temperature change scales up with the concentration of nanoparticles indicating that the nanoparticles absorb strongly the NIR light [5]. PANI-

Content from this work may be used under the terms of the Creative Commons Attribution 3.0 licence. Any further distribution of this work must maintain attribution to the author(s) and the title of the work, journal citation and DOI.
Published under licence by IOP Publishing Ltd

IOP Conf. Series: Materials Science and Engineering **772** (2020) 012048 doi:10.1088/1757-899X/772/1/012048

functionalized graphene quantum dot shows improved photocurrent by enhancing interconnectivity and carrier transport [6]. polyaniline/ reduced graphene oxide (PANI/RGO) composites under visible light irradiation exhibit amazingly improved activity toward the degradation of cationic and anionic dyes in comparison with pristine PANI or RGO [7]. Thus, it would be very interesting to study the intrinsic photo-thermal conversion of PANI materials in the visible light and near infrared region, which can amplify the application of solar energy. Herein, we synthesized two series of PANI with different HCl /aniline and DBSA /aniline molar ratio and investigated the chemical structure, optical properties and thermal stability of these PANI. We also studied the effect of DBSA/ aniline molar ratio on the photo-thermal conversion efficiency of PANI.

## 2. Experiment

### *2.1. Chemicals*

Aniline (AR, ≥ 99.5%) and ammonium peroxydisulfate (APS, initiator) were purchase from Shanghai Aladdin Bio-Chem Technology Co., LTD. dodecylbenzenesulfonic acid (DBSA surfactant, dopant) was from J&K Scientific Ltd. Aniline was purified by distillation in vacuum before use. Other reagents were used as received without purification.

### *2.2. Synthesis of polyaniline (PANI)*

A typical synthesis of polyaniline was carried out as follows: 1.86g of aniline (0.02mol) was dissolved in 80mL of 1M HCl aqueous solution and stirred for 30min at 0°C. An aqueous solution of 3.48g (0.01mol) DBSA dissolute in 30mL water was added in and stirred for 10min. A prepared solution of 4.56g of ammonium persulfate($(NH_4)_2S_2O_8$; 0.02mol) in 20mL of 1M HCl solution, acting as oxidizing agent, was added slowly to the aniline and DBSA solution. The oxidative polymerization proceeded for 3 h at 50°C. The color of polymerization solution changed with polymerization continuing. In the initial stage, aniline/DBSA/ASP solution was transparent. As the polymerization proceeded, the color of the solution changed to yellow, brown and finally to green, which indicates the formation of PANI emeraldine salt (ES). After reaction, the resulting polymer was washed several times with methanol and 1M HCl solution to remove aniline monomers, oligomers, and residual oxidizing agent. The product was then dried in vacuum oven for 24h. Tab. 1 describes the recipes of the respective samples.

### *2.3. Dedoping of polyaniline*

Dedoped PANI was prepared by treatment with ammonia solution. Polyaniline was first dispersed in ammonia solution at room temperature under ultrasonication for 10min. The resulting product was washed with methanol and then dried in vacuum oven for 24h at room temperature.

### *2.4. Characterization*

The UV-Vis spectra of the PANI samples dispersed in N-methyl-2-pyrrolidone were measured on an iS10 UV/Vis spectrophotometer (Scinco, USA). The FT-IR spectra were obtained using a Nicolet iS10 Fourier transform infrared spectrophotometer (Scinco, USA) at a resolution of $4cm^{-1}$ using the attenuated total reflectance technique.

Thermogravimetric experiments were performed on a TGA-105(Shanghai Jiubin Instrument Co., Ltd) analyzer. PANI powder samples ranging from 5 to 7 mg and heated from 25 to 600 °C under $N_2$ atmosphere at a heating rate of 10 °C/min.

Photothermal conversion experiments were performed under simulated sunlight PLS-SXE300C (filter) (Beijing Bofeilai Technology CO., LTD.). The temperature changes over time were recorded using an intelligent auto thermometer with a platinum electrode.

**Tab. 1** Composite of PANI

| Sample | Aniline (mol) | DBSA(mol) | HCl (mol) | APS(mol) |
|---|---|---|---|---|
| PANI1 | 0.02 | 0.01 | 0.02 | 0.02 |
| PANI2 | 0.02 | 0.01 | 0.06 | 0.02 |
| PANI3 | 0.02 | 0.01 | 0.1 | 0.02 |
| PANI4 | 0.02 | 0.01 | 0.14 | 0.02 |
| PANI5 | 0.02 | 0.02 | 0.1 | 0.02 |
| PANI6 | 0.02 | 0.03 | 0.1 | 0.02 |

## 3. Results and discussion

### *3.1. FT-IR*

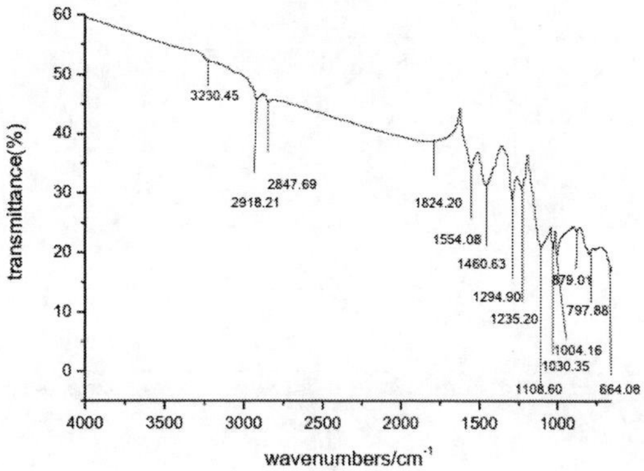


**Fig. 1** FT-IR spectrum of PANI

FT-IR spectra of PANI sample was shown in Fig. 1. The bands of 1460 and 1554cm-1 are assigned to C=C stretching of benzenoid and quinoid rings, respectively [8]. These bands appeared at lower wave number compared to that of dedoped PANI observed near 1500 and 1600cm$^{-1}$. The peak at 1294cm$^{-1}$ is assigned to C-N stretching of secondary amine of PANI backbone. The peak at 1235cm$^{-1}$ is the typical characteristic peak of conducting PANI-ES form [9]. The peak at 1108cm$^{-1}$ is assigned to a plane bending vibration of C-H that formed during protonating process. The peaks ranging from 2800cm$^{-1}$ to 3300cm$^{-1}$, are very small and slightly can be seem, are assigned to aliphatic C-H stretching mode correspondence to the long alkyl tail of DBSA. Therefore, this FT-IR spectra clearly illustrated the well-doped PANI-ES.

### *3.2. Yield and solubility*

**Tab. 2** Yield and solubility of PANI with different composition

| Sample | PANI1 | PANI2 | PANI3 | PANI4 | PANI5 | PANI6 |
|---|---|---|---|---|---|---|
| Yield | 160.6 | 153.1 | 151.04 | 147.8 | 103.16 | 169.15 |
| Solubility | 24.60 | 27.35 | 29.47 | 40.76 | 22.23 | 37.22 |

Tab. 2 showed the yield and solubility of PANI synthesized with different components ratio. The results demonstrated that with the increase of HCl/aniline molar ratio, the yield decreased, and the solubility increased. Which may be due to protonation degree increased with the increase of HCl usage. The yields were higher than 100% for all PANI products was attributed to the existence of sulfonate anion [10]. Tab.2 also showed that the effect of DBSA on the solubility of PANI was more obvious than HCl. This verified the surfactant action of DBSA, which resulted in the better solubility of PANI [11]

### 3.3. *Effect of pH*

Commercially NMP product is alkaline, when PANI dissolve in it, the organic sulfonate acid will interact with NMP and result in the dedoping of PANI in a certain degree and form some emeraldine base. Adjusting the PANI3 solution with buffer solution, we can get PANI3 solution with different pH. The UV-vis spectrum of these solution was recorded and shown in Fig.2. As shown in Fig. 2, when the pH was higher than 5.36, a strong absorption peak appeared in 620nm, which corresponded to a local charge transfer between quinoid ring and the adjacent imine-phenyl-amine unites [12]. At the condition of lower pH, two obvious absorption peaks occur in 430nm and 800nm respectively. Which are due to π-polaron band and polaron-π* band transition of emeraldine salt. What's more, with the pH decrease, the absorption peak increase remarkably, this demonstrated that the amount of PANI emeraldine salt increase gradually.

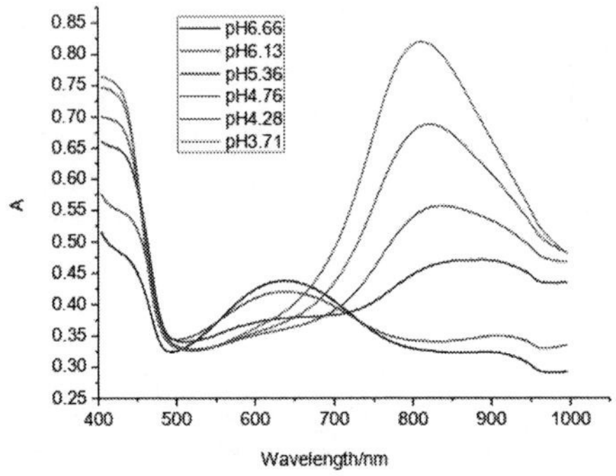


**Fig. 2** UV-Vis spectrum of PANI3 solution with different pH

### 3.4. *TGA*

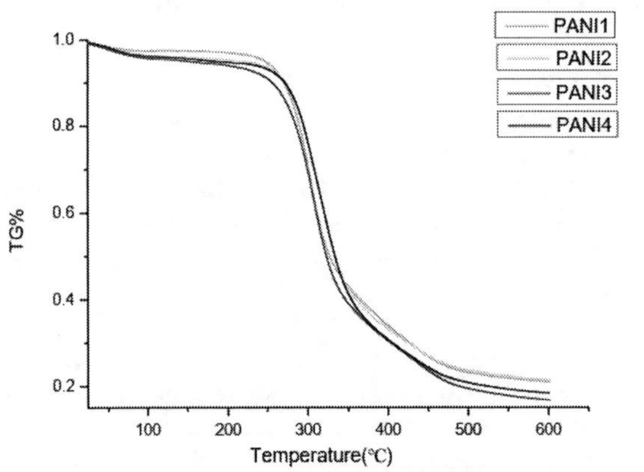


**Fig. 3** TGA curves of PANI

Fig 3 shows the results of TGA measurement for PANI with different aniline/HCl molar ratio. It is clear that the thermal stability of all samples is quite high. The weight loss below 100°C was due to small amount of water contained in the products. The degradation between150 and 200°Cwas probably attributed to the loss of doped DBSA[13]. Whereas, the main weight loss located at about 300 °C is caused by the degradation of PANI main chain. There is no significant difference in the spectra among these samples, which indicated that the amount of HCl has little effect on the thermal stability of PANI polymer.

### *3.5. Photothermal conversion property*

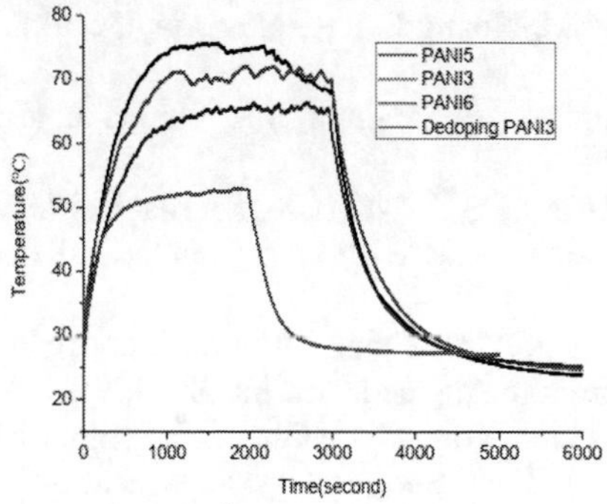


**Fig. 4** Photothermal conversion of PANI with different Aniline/DBSA molar ratio

Photothermal conversion experiments were carried out on the PANI and dedoping PANI solution. Their conversion curves are shown in Fig.4. Tab.3 lists the temperature changes along the illumination time and the highest temperature the solutions can reach. The results indicated that the temperature of the PANI increased faster than that of dedoping PANI as the illumination of the Xenon lamp. This phenomenon can be ascribed to the PANI, which showed green color and had a significant absorption in the infrared radiation region (see Fig. 2b) and can generate heat directly. In contrast, the dedoping PANI, presented blue color, only has a weak absorption in the visible light region and needed to convert the light to thermal energy through a nonradiative decay process. The results suggested that doped PANI could be considered as an effective photothermal agent and pave the way for future solar thermal conversion materials.

**Tab. 3** Dada of photothermal conversion of PANI and dedoping PANI under xenon illumination

| Sample | [a]500s/°C | [b]1000s/°C | [c]1500s/°C | $T_{max}$/°C |
|---|---|---|---|---|
| PANI5 | 65.125 | 74.00 | 75.37 | 75.87 |
| PANI3 | 60.75 | 70.25 | 70.25 | 71.37 |
| PANI6 | 55.12 | 63.50 | 65.37 | 65.87 |
| Dedoping PANI3 | 49.25 | 51.75 | 52.5 | 53.87 |

Note: a, b, c are temperatures of PANI solution at 500s, 1000s, 1500s respectively.

## 4. Conclusion

A series of PANI was synthesized under the present of DBSA. The IR spectrum showed that a typical stretching peak at 1235cm-1 for conducting PANI-ES appeared. The yield and solubility study demonstrated that with the increase of HCl/aniline molar ratio, the yield decreased, and the solubility increased. pH has a big influence in the optical property of PANI and when the pH ranges from 3.71~6.66, the PANI showed absorption peaks in a different wavelength, that means it had different color and attributed to the various exciting process. The thermal stability for all PANI samples of different HCl/aniline molar ration were all quite high and the biggest degradation occurred at 300°C. The photothermal conversion results for PANI and dedoping PANI showed a significant difference. The PANI-ES had a higher efficiency and can reach a higher temperature under illumination at the same time. So, this kind of substance present potential application to future solar thermal conversion materials.

**Acknowledgments**
Authors wants to acknowledge the support of Special Foundation of High-Level Teachers Team Building in Beijing Institute of Fashion Technology for financial support of this research under the grant number of BIFTXZ201808. The authors also declare that they have no conflict of interest.

## References

[1] Ferooze Ahmad Rafiqi, Kowsar Majid, J Mater Sci., 52 (2017):6506–6524.
[2] Kerileng M. Molapo, Peter M. Ndangili, Rachel F. Ajayi, et al, Int. J. Electrochem. Sci., 7 (2012): 11859 – 11875.
[3] Chandra Shekhar Kushwaha, Pratibha Singh, S.K. Shukla, et al, Materials Today: Proceedings, 5 (2018):15253–15260.
[4] ZHAO R., ZHU L.M., QIAN J.F., YANG H.X., Journal of Electrochemistry, 18(2012): 310-313.
[5] Silvestre Bongiovanni Abel, María A Molina, Claudia R Rivarola, et al, Nanotechnology, 25 (2014):495602.
[6] S. K. Lai, C. M. Luk, L. Tang, K. S. Teng, S. P. Lau, Nanoscale, 7(2015):5338.
[7] Mousumi Mitra, Sk. Taheruddin Ahamed, Amrita Ghosh, et al, ACS Omega, 4(2019): 1623−1635.
[8] Moon Gyu Han, Seok KI Cho, Seong Geun Oh, et al, Synthetic Metals, 126(2002)53-60
[9] Jongwan Choi, Sung K.Jang, Felix Sunjoo Kim, Phys. Status Solidi A, 2018, 1701019.
[10] Ma Li, Huang Kelong, Chen Chao, Lu Wei, Chen Fengqiang, J. Cent. South Unic. (Science and Technology), 39(2008):464-468.
[11] Zhou Z., He D., Li X., Li X.T. et al, Acta Polymeric Sinica, 8(2007):757-760.
[12] Kerileng M. Molapo, Peter M. Ndangili, Rachel F. Ajayi, et al, Int. J. Electrochem. Sci., 7(2012):11859-11875.
[13] Moon Gyu Han, Seok KI Cho, Seong Geun Oh, Seung Soon Im, Synthetic Metals, 126(2002)53-60.

# Improved activity for the photo-degradation by organic-inorganic composite photocatalysts

**Lijuan Wan[1, 2, *], Ming Yang[3]**

[1]Nanjing Vocational Institute of Transport Technology, Nanjing, China
[2]Jiangsu Engineering Technology Research Center for Energy Conservation and Emission Reduction of Transportation, Nanjing, China
[3]School of Transportation, Southeast University, Nanjing, China

*Corresponding author e-mail:bartty_ym@163.com

**Abstract.** g-$C_3N_4$ and flowerlike ZnO organic–inorganic composite with varying the content of ZnO has been synthesized through facile mixing and heating method. The as-prepared ZnO samples were characterized by SEM images and BET surface area measurements and etc.The composite catalysts were characterized by BET surface area measurements, X-ray diffraction and etc. Under visible light irradiation, the activity of composite photocatalystg-$C_3N_4$ and flowerlike ZnO organic–inorganic composite with 60% or 80% g-$C_3N_4$ for photo-degradation of MO is higher than that of either single-phase flowerlike ZnO org-$C_3N_4$. The as-prepared g-$C_3N_4$ and flowerlike ZnO organic–inorganic composite exhibits an improved photo-degradation performance for MO due to the enhancement of electron–hole separation.

## 1. Introduction

In recent years for water splitting hydrogen generation or removal of contamination, the photocatalysis technique by using solar energy has a promising application. The photogenerated electron-hole pairs under light irradiation gives rise to the photooxidation-reduction reaction transfer. So far, to improve the photocatalytic performance, various techniques were studied for the photocatalyst, such as increasing surface area[1] and loading cocatalyst[2]. Recently, zinc oxide (ZnO) as a versatile material has attracted considerable attention in a wide range of applications, including catalysis, photodetectors, photoelectrochemical cells, biosensor and so forth.[3] Recently, it has been reported that the microstructures have great influence on these applications, including morphological characters and crystal size.[4] ZnO materials have been prepared with different morphologies, such as rods, cables, wires, and hierarchical nanostructures with high symmetries[5, 6]. In the many existing synthetic strategies[7,8], solvothermal or hydrothermal techniques have been widely used to be the effective approaches due to mild synthesis conditions and the facile operation[9].

An organic polymer photocatalyst, graphite-like carbon nitride (g-$C_3N_4$), has recently been reported [10, 11], which possesses water splitting and the removal of pollutants under visible light irradiation, and in the field of civil engineering, such photocatalysts have potential application for the low-cost advantage. Some composite catalysts such as g-$C_3N_4$/$Bi_2WO_6$ [12], have shown better catalytic performance than that of single component catalyst. Therefore, it is expected to further improve the photocatalytic activity of g-$C_3N_4$ catalyst through forming composite photocatalysts.

Content from this work may be used under the terms of the Creative Commons Attribution 3.0 licence. Any further distribution of this work must maintain attribution to the author(s) and the title of the work, journal citation and DOI.
Published under licence by IOP Publishing Ltd

In this paper, synthesis of the g-$C_3N_4$ and flowerlike ZnO organic–inorganic composite by simple mixing-heating method and the photo-degradation performance of MO by the composite catalysts have been studied. Based on the experimental results for the photo-degradation of MO, it was proposed for the possible mechanism on the improved photocatalytic performance.

## 2. Experimental

The starting materials utilized are zinc chloride, analysis purity grade), 26~28% $NH_3 \cdot H_2O$, absolute ethanol (Nanjing Chemical Reagent Co. Ltd.). Poly(styrene sulfonic acid) sodium salt was purchasedfrom Alfa Aesar and used as received without furtherpurification. Distilled water was used throughout the experiment.

In a typical process, zinc chloride and PSS was dissolved in the solution containing water and ethanol with a proportion, then the $NH_3 \cdot H_2O$ was dropped in it at room temperature. The mixture was stirred for 30 min, then transferred into Teflon-lined stainless steel autoclave, and heated at 130 °C for 6 h. These samples were treated by centrifugation and thoroughly rinsed with water and ethanol several times, and then dried at 70 °C in an oven for subsequent characterization. The photocatalyst of g-$C_3N_4$ was prepared by directly heating melamine at 500 °C (heating rate: 20 °C/min) for 2 h, and the further deammonation treatment was set at 520 °C for 2 h[11], respectively in the semi-closed system to prevent sublimation of melamine. The as-prepared ZnO and g-$C_3N_4$ powder samples were mixed by ball-milling and then calcined at 350 °C for 3 h in a muffle furnace.

Characterization of samples: The specific surface area of the as-prepared powders was obtained on a Micromeritics TriStar 3000 instrument (USA) at 77 K and Brunauer–Emmett–Teller (BET) equation were used to calculate the specific surface area. The products were characterized by X-ray diffraction (XRD) for phase identification on a Rigaku Ultima III diffractometer with Cu $K\alpha$ radiation ($\lambda = 0.154$ nm, 40 kV, 40 mA) and a scan rate of 10 $°\cdot min^{-1}$. Ultraviolet visible (UV-vis) diffuse reflection spectra were measured using a UV-vis spectrophotometer (Shimadzu UV-2550, Japan) and converted from reflection to absorbance by the Kubelka-Munk method.

The photocatalytic reaction was performed in a Pyrex reactor, and the methyl orange (MO) dye was used to test the photocatalytic activities of as-prepared samples. The catalyst (0.1 g) was dispersed in 100 mL MO aqueous solution. The light irradiation system contains a 300 W Xe lamp with a water filter to remove heating effects and cut-off filter L42 was used for visible light. The reaction solutions of the MO photodegradation on catalysts were first stirred in the dark for all experiments for 1 h to reach the adsorption–desorption equilibrium of MO. Using the UV-vis absorption spectra, the MO degradation efficiency was evaluated to measure the peak value of a maximum absorption at wavelength of 463 nm of MO solution.

## 3. Results and discussion

The SEM image provides information on the morphology and crystalline size. It is shown in Figure 1, that ZnO with flower-like morphology is composed of nanocrystalline ZnO rods, and the flower-like ZnO formed by the agglomeration of nanorods. The composite catalyst is composed of flower-like ZnO and g-$C_3N_4$. And from the XRD result, no impurity phases were detected, and the diffraction peaks are in good agreement with a typical wurtzite-type ZnO crystal (hexagonal, $P6_3mc$), JCPDS No.36-1451. The high resolution-SEM image provides further information on the morphology and crystalline size. It is shown in Figure 2, the ZnO nanorods are in the size of about 300-400 nm.

IOP Conf. Series: Materials Science and Engineering **772** (2020) 012049 doi:10.1088/1757-899X/772/1/012049

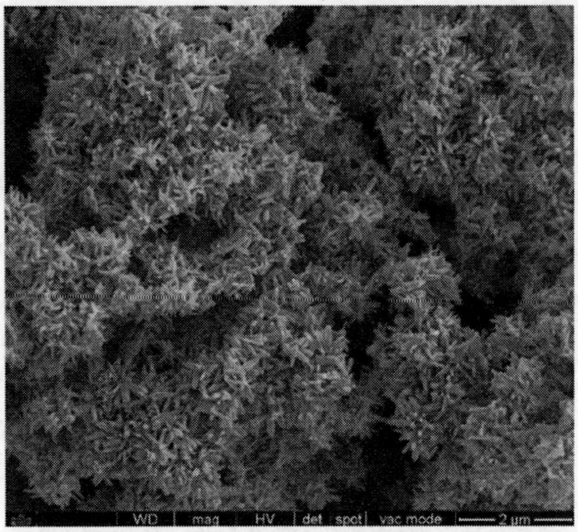


**Figure 1.** SEM image of the as-prepared ZnO sample prepared using zinc chloride as the precursor.

**Figure 2.** High resolution-SEM image of the as-prepared ZnO sample prepared using zinc chloride as the precursor.

For the investigation of the photo-degradation activity of g-$C_3N_4$ and flowerlike ZnO organic–inorganic composite photocatalyst, the pure ZnO, ZnO organic–inorganic composite and pure g-$C_3N_4$ samples were used as photocatalysts for the degradation of MO. It is shown from Figure 3, under visible light irradiation, the results of the degradation activities of MO over different photocatalysts are got. The 60 or 80 wt.% g-$C_3N_4$ and flowerlike ZnO organic–inorganic composite show much higher activity than that of pure g-$C_3N_4$, as shown in Figure 3. This clearly indicates that for the degradation of MO, the g-$C_3N_4$ and flowerlike ZnO composite was determined as an efficient visible-light-driven photocatalyst for the advantage of the formation of chemically bonded interfaces between the two materials by heating treatment.

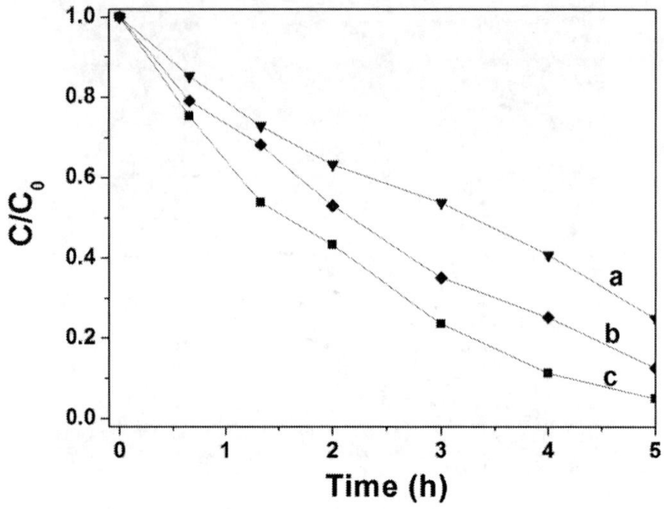


**Figure 3.** Comparison of MO degradation over different photocatalysts: a, g-$C_3N_4$; b, g-$C_3N_4$ and ZnO composite with 80% g-$C_3N_4$; c, g-$C_3N_4$ and ZnO composite with 60% g-$C_3N_4$, under visible light irradiation.

## 4. Conclusion

In summary, g-$C_3N_4$ and flowerlike ZnO organic–inorganic composite photocatalysts with varying the content of g-$C_3N_4$ were synthesized. By coupling ZnO, the photocatalytic degradation result shows that the visible light-induced oxidation performance of MO with g-$C_3N_4$ was remarkably enhanced due to the effective separation of photogenerated electron-hole pairs.

**Acknowledgments**

This work was financially supported by Natural Science Research Projects of Universities in Jiangsu Province (19KJB610014), and High-level Scientific Research Foundation for the Introduction of Talent of Nanjing Vocational Institute of Transport Technology.

## References

[1] F Bosc, A Ayral, P A Albouy, et al. Chemistry of Materials, 2003, 15(12): 2463~2468.
[2] S Sato, J M White. Chemical Physics Letters, 1980, 72(1): 83~86.
[3] Z. W. Zhao, X. J. Chen, B. K. Tay, J. S. Chen, Z. J. Han and K. A. Khor,Biosens. Bioelectron, 2007, 23, 135.
[4] J. G. Yu and X. X. Yu, Environ. Sci. Technol., 2008, 42, 4902.
[5] M. H. Huang, S. Mao, H. Feick, H. Q. Yan, Y. Y. Wu, H. Kind, E. Weber,R. Russo and P. D. Yang, Science, 2001,292, 1897.
[6] Z. W. Pan, Z. Dai and Z. L. Wang, Science, 2001, 291, 1947.
[7] Z. L. Wang, J. Mater. Chem., 2005, 15, 1021.
[8] W. Ogasawara, W. Shenton, S. A. Davis and S. Mann, Chem. Mater., 2000, 12, 2835.
[9] B. Cheng and E. T. Samulski,Chem. Comm.,2004,8,986.
[10] X C Wang, K Maeda, A Thomas, et al. A metal-free polymeric photocatalyst for hydrogen production from water under visible light. Nature Materials, 2009, 8(1): 76~80.
[11] S C Yan, Z S Li,Z G Zou. Langmuir, 2009, 25(17): 10397~10401.
[12] L Ge, C CHan, J Liu, Appl. Catal. B: Environ., 2011, 108–109: 100~107.

# Synthesis of zinc oxide nanorods with different aspect ratios by solvothermal method

**Lijuan Wan[1, 2, *], Ming Yang[3]**

[1]Nanjing Vocational Institute of Transport Technology, Nanjing, China
[2]Jiangsu Engineering Technology Research Center for Energy Conservation and Emission Reduction of Transportation, Nanjing, China
[3]School of Transportation, Southeast University, Nanjing, China

*Corresponding author e-mail:bartty_ym@163.com

**Abstract.** The ZnOnanorods with different aspect ratios have been synthesized by solvothermal method. The as-prepared ZnOsampleswere characterized byBET surface area measurements, SEM and etc. By modulating the different reaction temperatures and the different precursors, the as-prepared ZnO samples with nanorods morphology and different radius were prepared and the influence of the reaction conditions are investigated. Also the different morphologies may influence the potential applications of the as-preparedZnO.

## 1. Introduction

Zinc oxide (ZnO)with different shape and orientation has attractedconsiderable attention as a versatile material due to its potential wide ranging applications, including light-emittingdiodes, dye-sensitized solarcells, catalysis, biosensor, piezoelectric nanogenerators and so forth [1–4]. Recently, ZnO, which is expected as a goodphotocatalyst,has beendocumented that the microstructures have effects on the applications such as orientation andmorphological characters. To get various ZnO nanostructures such as rods, wires, combs,belts, cables, many strategies such as thewet-chemistry route [5] and vapor transport process [6, 7]) have been proposed. Solvothermal or hydrothermal techniques [8, 9] are more economic than vaporphase routeowing to the flexible adjustment of experimentalparameters and mildsynthesis conditions, whichhave beenwidely used as the effective approaches to prepare ZnOwith various morphologies. In previous reports, the synthetic conditions have great effects on the microstructureof ZnO such as reaction conditons or mediumand morphology-directing agents, and the morphologies of ZnO have beenmodulated by some surfactants, such as PSS and etc [10-11]. Particularly, as an anion surfactantfor the morphology modulation, the PSS has beenwidely used in the synthesis of metal oxides with different morphologies [12].

Also,with hydroxylgroups, organic solvents for the synthesis of ZnOhave an important effects on the nucleation and growthof ZnO. With various morphologies,well-defined ZnO nanostructures such as nanowires, rods, tetrapods, stars, flowersand high-symmetry hierarchical nanostructures, for example,under a solvothermal treatment with zinc acetate, ZnOhas beenobtainedwith rod-like morphology and different aspect ratios bychanging organic solvents with different hydroxyl groups, and compared with various low-dimensional ZnO nanostructures, the ZnO with different morphologies such as the core–shell ZnO, ZnOassemblies, those hollow three-dimension, ZnO spheres, may show large

Content from this work may be used under the terms of the Creative Commons Attribution 3.0 licence. Any further distribution of this work must maintain attribution to the author(s) and the title of the work, journal citation and DOI.
Published under licence by IOP Publishing Ltd

light-harvesting efficiencies, high photocatalytic activity,and superior collective properties. [13,14].In this work, the ZnO nanorods with different aspect ratios have been synthesized by solvothermal method.

## 2. Experimental

The starting materials utilized are zinc nitrate hexahydrate($Zn(NO_3)_2 \cdot 6H_2O$, analysis purity grade) or zinc chloride, 26~28% $NH_3 \cdot H_2O$, absolute ethanol (Nanjing Chemical Reagent Co. Ltd.). F127 triblock polymer was purchasedfrom Alfa Aesar and used as received without furtherpurification.Distilled water was usedthroughout the experiment.

In a typical process,zinc nitrate hexahydrateor zinc chloride and F127 was dissolved in the solution containing water and ethanol with a proportion at room temperature, then the $NH_3 \cdot H_2O$ was dropped in it. The mixture was stirred for 30 min, then transferred into Teflon-lined stainless steel autoclave, and heated at different temperatures for 6 h. These samples weretreated by centrifugation and thoroughly rinsed with ethanolandwater for several times, and then dried at 60 °C in an oven, and calcinedat 500 °Cfor 4 h for subsequent characterization.

Characterization of samples: The products were characterized by X-ray diffraction (XRD) for phase identification on a RigakuUltima III diffractometer with Cu $K\alpha$ radiation ($\lambda$ = 0.154 nm, 40 kV, 40 mA) and a scan rate of 10 $^{\circ} \cdot min^{-1}$. The specific surface area of the as-prepared powders was obtained on a Micromeritics TriStar 3000 instrument (USA) at 77 K and Brunauer–Emmett–Teller (BET) equation were used to calculate the specific surface area.The morphology and microstructure were observedusing a field emission scanning electron microscope (FE-SEM;NOVA NanoSEM230, FEI Ltd.) with accelerating voltage of 5 kV.

## 3. Results and discussion

The SEM image provides information on the morphology of as-prepared ZnO samples. From Figure 1, it can be observedthat ZnOnanorodsaresynthesized and the radius of the ZnOnanorods is about 50~100 nm. And, thediffraction peaksfrom the XRD resultare in good agreement with a typical wurtzite-type ZnO crystal,JCPDS No.36-1451, (hexagonal, $P6_3mc$).There were noimpurity phases which were detected.

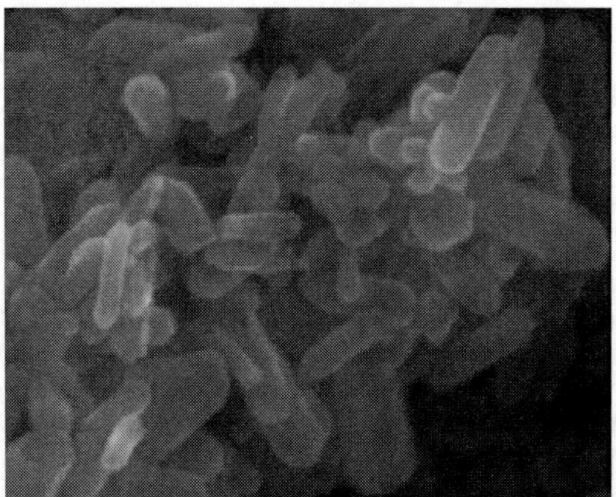


**Figure 1.** SEM image of the as-preparedZnOsample prepared using zinc chloride as the precursor.

To study the influence of the reaction temperature on the morphologies of the as-prepared ZnO samples, the ZnO samples prepared at 160 or 180 °Care synthesized. From Figure 2 and Figure 3, the ZnO samples prepared at 160°C show the nanorod morphology and the radius is about 500 nm, while the ZnO samples prepared at 180 °Cshow the nanorod morphologyand the radius is about 900 nm and the hexagonal morphology is more obvious.

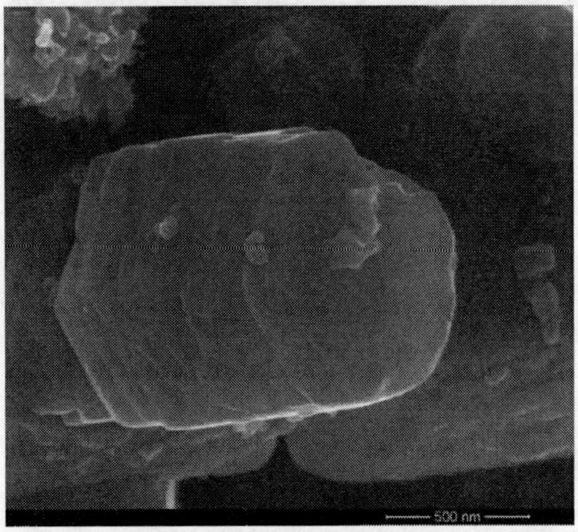


**Figure 2.** SEM image of the as-preparedZnOsample prepared using zinc nitrate hexahydrate as the precursor and heated at 160 °C.

**Figure 3.** SEM image of the as-preparedZnOsample prepared using zinc nitrate hexahydrate as the precursorand heated at 180 °C.

## 4. Conclusion

In summary, the as-prepared ZnOsamples with nanorods morphology and different radius were synthesized due to the different reaction temperatures and the different precursors. The influence of the reaction conditions are studied. And the different morphologies may influence its potential applications.

## Acknowledgments

This work was financially supported by Natural Science Research Projectsof Universities in Jiangsu Province (19KJB610014), and High-level Scientific Research Foundation for the Introduction of Talent of Nanjing Vocational Institute of Transport Technology.

**References**

[1] Z.W. Zhao, X.J. Chen, B.K. Tay, J.S. Chen, Z.J. Han, K.A. Khor, A novel amperometricbiosensor based on ZnO: Co nanoclusters for biosensing glucose, Biosens.Bioelectron. 23 (2007) 135–139.
[2] M.H. Huang, S. Mao, H. Feick, H.Q. Yan, Y.Y. Wu, H. Kind, E. Weber, R. Russo, P.D.Yang, Room-temperature ultraviolet nanowire nanolasers, Science 292 (2001)1897–1899.
[3] S.W. Liu, C. Li, J.G. Yu, Q.J. Xiang, Improved visible-light photocatalytic activity ofporous carbon self-doped ZnO nanosheet-assembled flowers, CrystEngComm13 (2011) 2533–2541.
[4] Y. Qin, X. Wang, Z.L. Wang, Microfibre–nanowire hybrid structure for energyscavenging, Nature 451 (2008) 809–813.
[5] J. Zhang, S.R. Wang, M.J. Xu, Y. Wang, B.L. Zhu, S.M. Zhang, W.P. Huang, S.H.Wu, Hierarchically porous ZnO architectures for gas sensor application, Cryst.Growth Des. 9 (2009) 3532–3537.
[6] Z.L. Wang, Self-assembled nanoarchitectures of polar nanobelts/nanowires, J.Mater. Chem. 15 (2005) 1021–1024.
[7] W. Ogasawara, W. Shenton, S.A. Davis, S. Mann, Template mineralizationof ordered macroporous chitin-silica composites using a cuttlebone-derivedorganic matrix, Chem. Mater. 12 (2000) 2835–2837.
[8] J.G. Yu, X.X. Yu, Hydrothermal synthesis and photocatalytic activity of zincoxide hollow spheres, Environ. Sci. Technol. 42 (2008) 4902–4907.
[9] B. Cheng, E.T. Samulski, Hydrothermal synthesis of one-dimensional ZnOnanostructures with different aspect ratios, Chem. Commun. 8 (2004) 986–987.
[10] X.L. Zhang, R. Qiao, R. Qiu, J.C. Kim, Y.S. Kang, Fabrication of hierarchical ZnO nanostructures via a surfactant-directed process, Cryst. Growth Des. 9 (2009) 2906–2910.
[11] Z. Liu, X.D. Wen, X.L. Wu, Y.J. Gao, H.T. Chen, J. Zhu, P.K. Chu, Intrinsicdipole-field-driven mesoscale crystallization of core–shell ZnOmesocrystalmicrospheres, J. Am. Chem. Soc. 131 (2009) 9405–9412.
[12] J.G. Yu, C. Li, S.W. Liu, Effect of PSS on morphology and optical properties ofZnO, J. Colloid Interface Sci. 326 (2008) 433–438.
[13] Z.W. Pan, Z. Dai, Z.L. Wang, Nanobelts of semiconducting oxides, Science 291(2001) 1947–1949.
[14] S.K.N. Ayudhya, P. Tonto, O. Mekasuwandumrong, V. Pavarajarn, P. Praserthdam, Solvothermal synthesis of ZnO with various aspect ratios using organic solvents, Cryst. Growth Des. 6 (2006) 2446–2450.

# Facile synthesis of ternary oxide and the performance of photo-catalysis

**Lijuan Wan[1, 2, *], Ming Yang[3]**

[1]Nanjing Vocational Institute of Transport Technology, Nanjing, China
[2]Jiangsu Engineering Technology Research Center for Energy Conservation and Emission Reduction of Transportation, Nanjing, China
[3]School of Transportation, Southeast University, Nanjing, China

*Corresponding author e-mail: bartty_ym@163.com

**Abstract.** Nano-structured $Zn_2GeO_4$ photocatalyst was prepared through solution phase route. The as-prepared products were characterized by Brunauer–Emmett–Teller (BET) surface area measurements and scanning electron microscopy (SEM). The experimental results revealed that for photo-degradation of MO, the nano-structured $Zn_2GeO_4$ samples with higher surface area have higher photocatalytic performance than bulk $Zn_2GeO_4$ synthesized through solid-state reaction.

## 1. Introduction

Recently, photo-degradation of organic pollutants as well as water splitting into hydrogen by using photocatalysis method are promising green techniques to resolve the problems of environmental pollution and etc by some metal oxides photocatalysts.[1-3] With chemical and thermal stability, zinc orthogermanate ($Zn_2GeO_4$), has been synthesized and used for deep UV detection,and also is excepted to be a good photocatalyst, and etc.[4-8]

Due to the readily tunable properties of the ternary nanostructures, studies of ternary nanostructures are meaningful. Nano-structured $Zn_2GeO_4$ compared with bulk $Zn_2GeO_4$ exhibit higher photocatalytic activity.[9] The crystallized nano-structured $Zn_2GeO_4$ samples has rarely been reported. However, most of the reported routs had complex process for the preparation of nano-structured $Zn_2GeO_4$. Recently, to synthesize nanostructured $Zn_2GeO_4$, the facile solution route under moderate conditions has been reported. In this paper, at a relatively low temperature under ambient pressure, nano-structured $Zn_2GeO_4$ were successfully prepared through a solution phase route. The as-prepared nanostructured $Zn_2GeO_4$ show the improved photocatalytic performance for the degradation of organic compound due to the larger surface area.

## 2. Experimental

### *2.1. Synthesis of nano-structured $Zn_2GeO_4$*

Isotropic solutions were prepared by dissolving different proportion of $Na_2GeO_3$, $Zn(CH_3COO)_2$, EtOH, $H_2O$ and surfactant F127. The solution was stirred at 60~70 °C for 4~6 h. Then, the as-prepared products were separated by centrifugation and were washed with ethanol and water. Then the samples are dried

Content from this work may be used under the terms of the Creative Commons Attribution 3.0 licence. Any further distribution of this work must maintain attribution to the author(s) and the title of the work, journal citation and DOI.
Published under licence by IOP Publishing Ltd

and calcined at 450 °C for 4 h. A reference sample (SSR-$Zn_2GeO_4$) was prepared at 1300 °C by heating stoichiometric mixture of ZnO and $GeO_2$ for 15 h.

### *2.2. Characterization*

The microstructure and morphology were observed using a field emission scanning electron microscope with accelerating voltage of 15 kV (FE-SEM; NOVA230, FEI Ltd.). The specific surface area of the as-prepared samples was obtained on a Micromeritics TriStar 3000 instrument and to calculate the specific surface area, Brunauer–Emmett–Teller (BET) equation were used.

### *2.3. Photocatalysis test*

The photocatalytic reaction was performed in a Pyrex reactor and the methyl orange (MO) dye was used to test the photocatalytic activities of as-prepared samples. The catalyst (0.1 g) was dispersed in 100 mL MO aqueous solution. The light irradiation system contains a water filter to remove heating effects and a 300 W Xe lamp. To reach the adsorption–desorption equilibrium of MO on catalysts, the reaction solutions of the MO photodegradation for all experiments were first stirred in the dark for 1 h. By using the UV-vis absorption spectra, the MO degradation efficiency was evaluated to measure the peak value at wavelength of 463 nm of a maximum absorption of MO solution.

## 3. Results and discussion

The morphology and microstructure of the as-prepared nano-structured $Zn_2GeO_4$ sample was characterized by SEM. The SEM images of $Zn_2GeO_4$ product synthesized assisted by the surfactant are shown in Figure 1. As shown in Figure 1, nano-structured $Zn_2GeO_4$ can be observed. Figure 2 shows the high-resolution SEM image of the as-prepared $Zn_2GeO_4$ sample. From Figure 2, the morphology of the as-prepared $Zn_2GeO_4$ sample through solution route with ethanol and the surfactant is uniform. To form nano-structured $Zn_2GeO_4$ sample, it is necessary to use the morphology-directing reagent. By the influence of the surfactant, the $Zn_2GeO_4$ nanoparticles are formed.

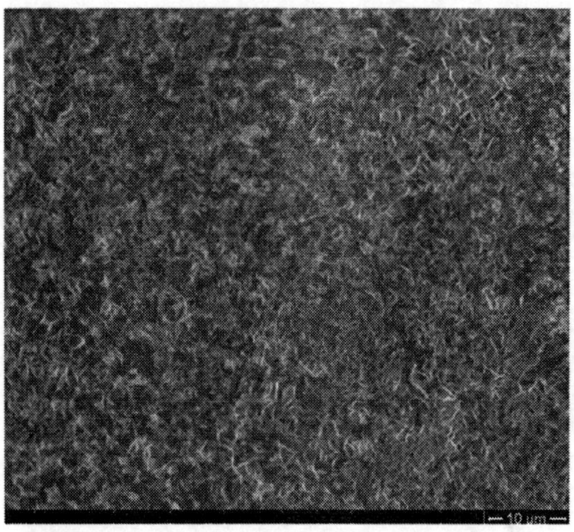


**Figure 1.** SEM image of $Zn_2GeO_4$ sample.

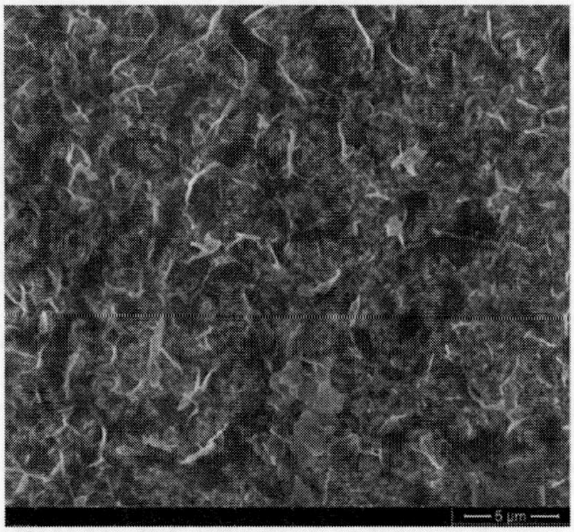

**Figure 2.** High-resolution SEM image of $Zn_2GeO_4$ sample.

The morphology information of the $Zn_2GeO_4$ sample prepared for longer reaction time were further characterized by high-resolution SEM. Figure 3 shows the SEM image of the as-prepared $Zn_2GeO_4$ product synthesized assisted by the surfactant. From Figure 3, the higher magnified SEM image exhibits the cubic $Zn_2GeO_4$ sample with larger size, which may be due to the longer reaction time. The difference in crystallite size between $Zn_2GeO_4$ and SSR-$Zn_2GeO_4$ sample leads to their significant difference in BET surface area: 19.75 $m^2 \cdot g^{-1}$ for nano-structured $Zn_2GeO_4$ sample and 0.86 $m^2 \cdot g^{-1}$ for SSR-$Zn_2GeO_4$ sample.

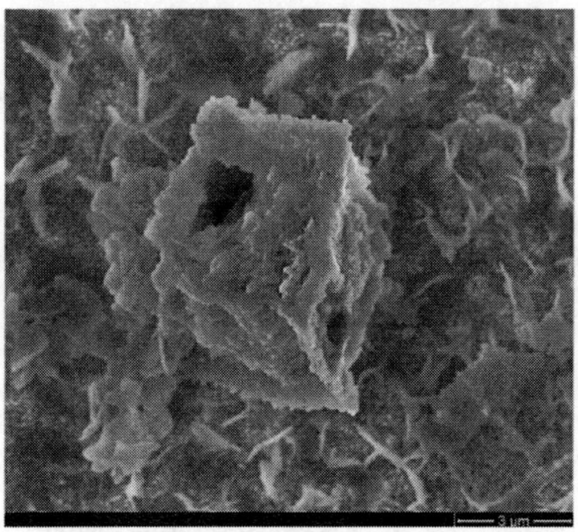

**Figure 3.** High-resolution SEM image of $Zn_2GeO_4$ sample.

In addition, on the as-prepared $Zn_2GeO_4$ sample and SSR-$Zn_2GeO_4$ sample, photocatalytic oxidation of MO is performed. By using the above-mentioned $Zn_2GeO_4$ as photocatalysts, it is shown in Figure 4 that the photocatalytic oxidation of MO is investigated. From Figure 4, the nano-structured $Zn_2GeO_4$ sample obtained by adding the surfactant exhibit much higher activity than SSR-$Zn_2GeO_4$ sample, which is due to the high specific surface area by producing more reaction sites.

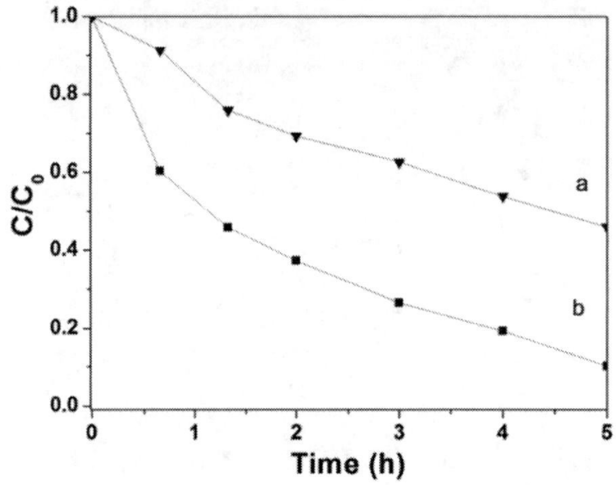


**Figure 4.** Comparison of MO degradation over SSR-Zn2GeO4 (a), as-prepared $Zn_2GeO_4$ sample (b).

## 4. Conclusion

In summary, the nano-structured $Zn_2GeO_4$ samples have been prepared by the solution phase route assisted by surfactant. Compared with the SSR-$Zn_2GeO_4$ sample, for the photodegradation of MO, the nano-structured $Zn_2GeO_4$ samples show higher photocatalytic activity. The high photocatalytic activity of the nano-structured $Zn_2GeO_4$ in the photodegradation of MO may be due to the more reaction sites produced from high specific surface area.

## Acknowledgments

This work was financially supported by Natural Science Research Projects of Universities in Jiangsu Province (19KJB610014), and High-level Scientific Research Foundation for the Introduction of Talent of Nanjing Vocational Institute of Transport Technology.

## References

[1] X. S. Fang, Y. Bando, M. Y. Liao, et al. Adv. Mater. Vol. 21 (2009), 2034.
[2] Z. Q. Liu, D. H. Zhang, S. Han, et al. Adv. Mater. Vol. 15(2003), 1754.
[3] Z. S. Liu, X. P. Jing, L. X. Wang. Luminescence of native defects in Zn2GeO4. J. Electrochem. Soc. Vol. 154 (2007) No. 6, H500-H506.
[4] B. J. Ma, F. Y. Wen, H. F. Jiang, et al. The synergistic effects of two co-catalysts on Zn2GeO4 on photocatalytic water splitting. Catal Lett. Vol. 134 (2010) No. 1, 78-86.
[5] J. H. Huang, K. N. Ding, Y. D. Hou, et al. Synthesis and Photocatalytic Activity of Zn2GeO4 Nanorods for Degradation of organic populltants in water. ChemSusChem. Vol. 1 (2008), 1011-1019.
[6] R. Steven, B. F. Woodfield, J. Boerio-Goates, et al. Chem.Thermodyn. Vol. 36 (2004), 349-357.
[7] J. Huang, X. Wang, Y. Hou, et al. Degradation of benzene over a zinc germanate photocatalyst under ambient conditions. Environ. Sci. Technol. Vol. 42 (2008), 7387–7391.
[8] N. Zhang, S. X. Ouyang, P. Li, et al. Ion-exchange synthesis of a micro/mesoporous Zn2GeO4 photocatalyst at room temperature for photoreduction of CO2. Chem.Commun. Vol. 47 (2011), 2041-2043.
[9] S. C. Yan, L. J. Wan, Z. S. Li, et al. Facile temperature-controlled synthesis of hexagonal Zn2GeO4 nanorods with different aspect ratios toward improved photocatalytic activity for overall water splitting and photoreduction of CO2. Chem Commun. Vol. 47 (2011) No. 19, 5632-5634.

IOP Conf. Series: Materials Science and Engineering **772** (2020) 012052 doi:10.1088/1757-899X/772/1/012052

# A Study on Testing of Rubber Friction Coefficient and Its Wear Surface by UMT-2

**Wei Liu*, Yuqing Zheng, Ting Wang, Lutao Lv, Zonglin Yang**

Key Laboratory of Rubber-Plastics, Ministry of Education/Shandong Provincial Key Laboratory of Rubber and Plastics, Qingdao University of Science and Technology, Qingdao, China

*Corresponding author e-mail: qdlorin@126.com

**Abstract.** For some products as tire, shoes and other wear parts products, friction performance is a very important performance index. A series of theories had been put forward to study the friction properties. With the development of science, the method of rubber friction was more and more abundant, from the friction wear properties to the microstructure, as well as the simulation, all the friction performance of the study provided a reliable basis. UMT - 2 multifunctional friction and wear testing machine, friction properties of the rubber material to make relatively objective evaluation, can provide reliable basis for research on the properties of friction. Due to many factors influencing the rubber materials, rubber friction behaviour of the friction and wear mechanism of exploration would still need a lot of research work.

## 1. Introduction

Friction performance is one of the most important indicators of rubber [1]. According to statistics, friction and wear are wide spread in nature and it has already consumed more than 50% of all primary energy sources in the world [2]. In the process of friction, the deformation by the shear stress happens not only on the rubber surface, but also inside the material under the surface at the same time [3]. It could cause some energy loss when resisting the friction on surface and viscoelastic characterization internal. The study is becoming more and more important on friction behavior between the surface of rubber products and its dual plane, between rubber macromolecule and packing on microcosmic morphology. Then, it could reduce the abrasion, cut down heat of rubber material and improve its fatigue life. And then it could bring considerable economic benefit and social benefit on saving energy, materials, lubricants and other aspects. [4, 5]

By studing the change of the friction coefficient of rubber under different conditions, we could calculate the friction behavior of rubber products under actual-use conditions. We researched the friction behavior of SSBR, and fond the corresponding laws of friction by different sliding velocity, pressure and different dry and wet conditions. On the whole, it provides important reference basis for the performance of rubber research.

Content from this work may be used under the terms of the Creative Commons Attribution 3.0 licence. Any further distribution of this work must maintain attribution to the author(s) and the title of the work, journal citation and DOI.
Published under licence by IOP Publishing Ltd

## 2. Experiments

### *2.1. Materials*
SSBR RC2564S, vinyl content of 65.5%, PetroChina Natural Gas Co. Ltd. , Dushanzi Branch; NR SCRWF, China Hainan Rubber Industry Group Co., Ltd. ;Precipitated silica: Luodiya Silica White (Qingdao) Co., Ltd.; CB N330: Degussa (QD)Co., Ltd.; ZnO: C.P., Tianjin Bodi Chemical Co., Ltd.

### *2.2. Blend recipe*
SSBR 80, NR 20, N330 50 (Silica 50, Si69 2), ZnO 4, SA 1, NS 2.2, S 1.5, NOBS 0.8, 4010NA 2.

### *2.3. Sample preparation*
CB or Silica were blended to the SSBR matrix on the mixer manufactured and other assistants were added in turn at about 130℃.The blends were taken out of the mixer and processed at room temperature with S and NS on an SK-160B two-roll mill, finally being molded into sheets in a VC-150T-FTMO-3RT vacuum press manufactured at 151 ℃ for 20 min.

### *2.4. Main measurements*
The friction coefficients were tested at room temperature (23℃) at different conditions of different pressures and different sliding velocities on a Multi-functional friction and wear tester UMT-2 (as shown in fig.1), Center for Tribology Inc. USA.

Microscope of surface after frictional wear was studied by stereoscopic microscope SMZ1500, Nikon Company, Japan.

SEM of surface after frictional wear was observed using a JEOL-JSM-7500F, JEOL Company, Japan, coated with Pt-Pd.

DMA was carried out using a Netzsch DMA 242 Dynamic Mechanical Analyzer, Netzsch, Germany, from -80 ℃ to 80 ℃, 3 K/min, 1 Hz.

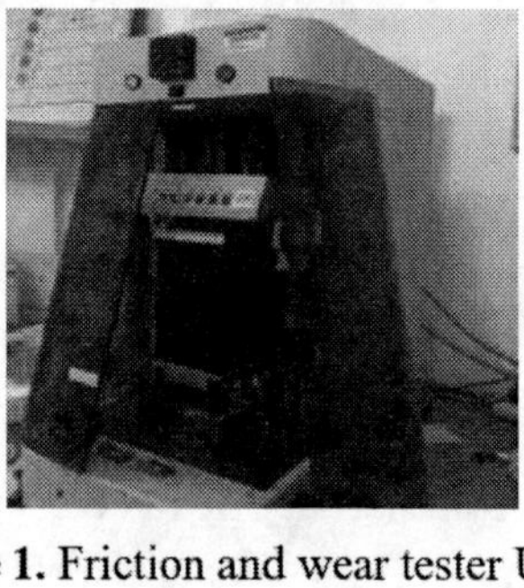

**Figure 1.** Friction and wear tester UMT-2

## 3. Results and discussion

### *3.1. Friction factors*
The experiments were implemented with the steel ball, diameter of 4 mm, at 23 ℃, the relative speed of 15.9mm/s, loading load of 5 N, and different reinforcing agent respectively of CB and Silica.

As shown in figure 2, the different friction factors were tested when the load are respectively 1N, 2N, 5N, 10N, and 20N, at 23 ℃ and the friction velocity is 15.9mm/s.It shown the relationship between the friction coefficient, when the load was small, the friction coefficient was large, which fully explained that most of the rubber friction gave priority to hysteresis friction. When the load was increased gradually, the friction coefficient became small since most of the rubber friction gave priority to sticky friction, and then the pressure recover limited deformation of rubber macromolecules.

IOP Conf. Series: Materials Science and Engineering **772** (2020) 012052 doi:10.1088/1757-899X/772/1/012052

There were two kinds of data listed under the condition of wet and dry friction.It was shown in figure 3 that under water lubricant condition, rubber friction coefficient greatly reduced, and the friction coefficients of CB reinforcing SSBR vulcanizates were lower than that of silica.

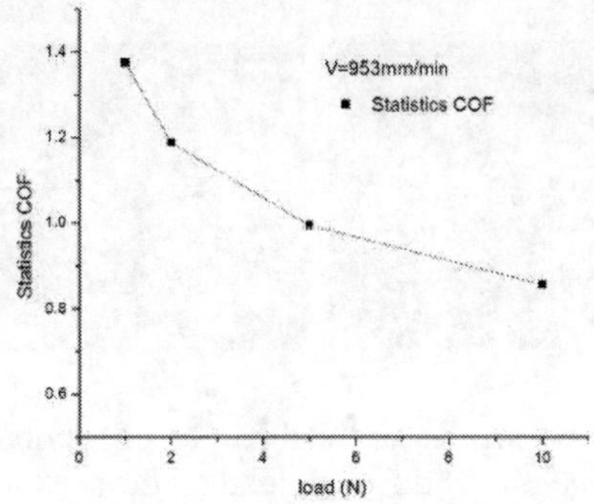


**Figure 2.** The effect of friction coefficient by different friction pressure with steel ball

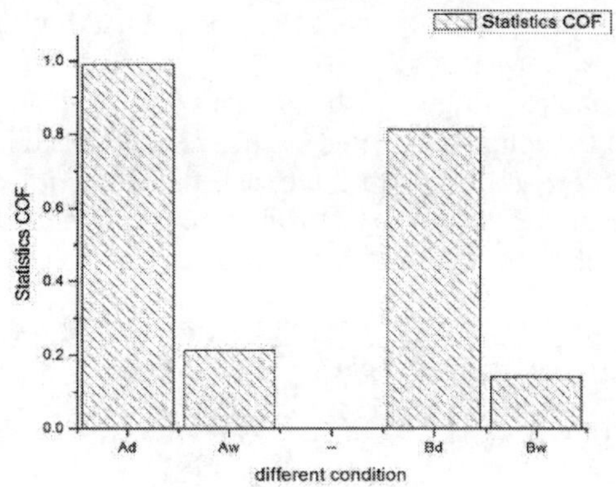


A: CB B: Silicon d: Dry w: Water wet

**Figure 3.** The effect of friction coefficient by water wet condition with steel ball

With other conditions unchanged, the friction velocity was changed to 3.2mm/s, 6.4mm/s, 15.9 mm/s, 31.8 mm/s, 79.4mm/s, in order to get a different friction velocity, as shown in figure 4 .

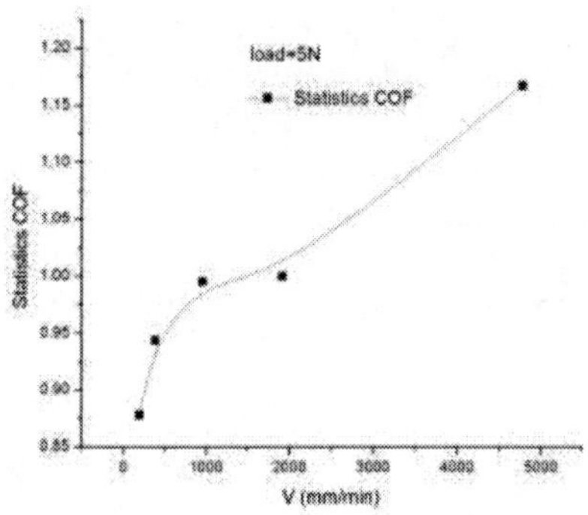


**Figure 4.** The effect of friction coefficient by different fiction speeds with steel ball

IOP Conf. Series: Materials Science and Engineering **772** (2020) 012052 doi:10.1088/1757-899X/772/1/012052

Obtained from the experimental data, the friction coefficient increased with the sliding velocity when the steel ball bearing on the rubber.And under the conditon of hysteresis friction was the main factor, when the relative speed increased, the steel ball forced the greater deformation of rubber and produced the greater friction resistance.As a result, the friction coefficient increased.

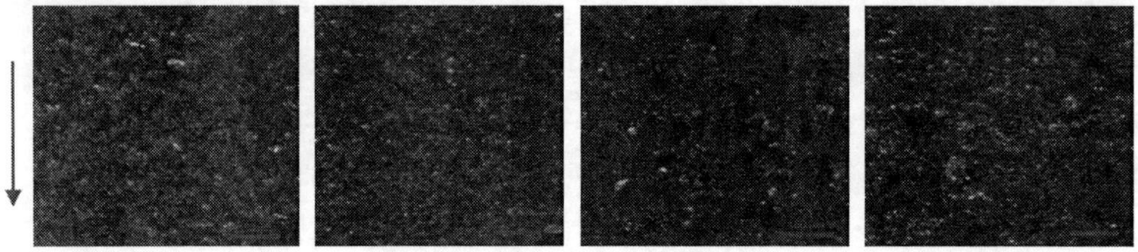

1N 2N 5N 10N

Note: The arrows stands for the direction of friction, 15.8 mm/s

**Figure 5.** The photographs of different load on the steel to the SSBR (100X)

*3.2. Friction surface Microanalysis*

A certain degree of wear and tear occurred on the surface of SSBR after the steel ball slidding on the rubber surface with a constant pressure and speed. Different damage surfaces of SSBR were observed by stereoscopic microscope, as shown in fig.5.

Stereoscopic microscope photos in figure 5 shown that when the load was small, the steel ball slid on the rubber surface and the friction mainly acted as sticky friction, and it would be gradually torn the rubber to scratch.As the friction and pressure increased, the friction acted as adhesive friction and hysteresis friction at the same time, when the steel ball more partial pressure gradually into the rubber, reciprocatingly, then the rubber stack occured.

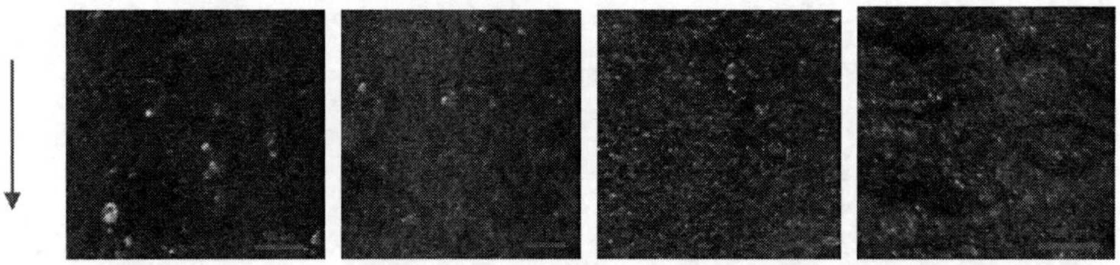

3.2mm/s 15.9 mm/s 31.8 mm/s 79.4mm/s

Note: The arrow for the direction of friction, 5N

**Fig. 6** The Photoshop of different velocity of friction behavior on the SSBR (100X)

Also, fig. 6 shows that through the process of friction between steel ball and rubber materials, there were wear stripes, groove, and the larger of abrasive, volume, etc. On the rubber surface, even the fa

tigue spalling and adhesion occurred . At the same time, fast friction speed was easier to make stack occur on the rubber surface, thus accelerated rubber scratching damage.

When the the friction happened with water lubrication,we found a different result on rubber with different fillers under the same pressure and the relative friction velocity.

With lubricants, the friction coefficient is greatly reduced,then the friction and wear is also changed, as shown in fig.7 for the carbon black reinforcing SSBR with white carbon black reinforcing of SSBR in dry friction and water lubrication friction surface of wet friction conditions.

Ad Aw Bd Bw

Note:A: CB, B:Silica,d:dry, w:wet , the arrow stands for the direction of friction, 15.9 mm/s,5N

**Figure 7.** Influence of Dry and Wet conditions to the friction (100X)

Taking the friction experiment lubricated with water, we found that the friction behavior was significantly different from that without water. And it was also different from the one reinforcing by carbon black and white carbon black, on dry or wet conditions. When the friction took with carbon black reinforcing of SSBR under dry condition, the rubber surface became coarse, with curly phenomenon, while under the condition of water lubricated surface scratches, it was relatively smooth.While the surface became rough after white carbon black reinforcing SSBR friction under dry condition, part surfaces of rubber cracked when water lubricated. Comparison with these two kinds of reinforcing SSBR under the same condition of friction, the wear resistance of white carbon black reinforcing was better than that of carbon black reinforcing SSBR.

Scanning electron microscopy (SEM) analysis was carried out on the friction and wear surface above to show the microstructures (figure 8). Under the condition of dry friction or wet rubbing, different microstructures of friction and wear of surface of carbon black reinforcing SSBR and white carbon black reinforcing SSBR were presented.

After magnified 5000 times and 20000 times respectively, we could see that friction surface under different friction conditions are obviously different. The microstructure of friction surface was stacked and curled on carbon black reinforcing of SSBR under dry condition, while tiny crack occurred under the condition of wet rubbing.While the microstructure of friction surface was curly, but it was appear cracked under the condition of wet friction. From the microstructure, the friction failure mechanisms of these two kinds of reinforcing system of SSBR vulcanizates were similar. Curled up, and the generation of cracks or cracked would accelerate qu torsion for rubber material to damage.

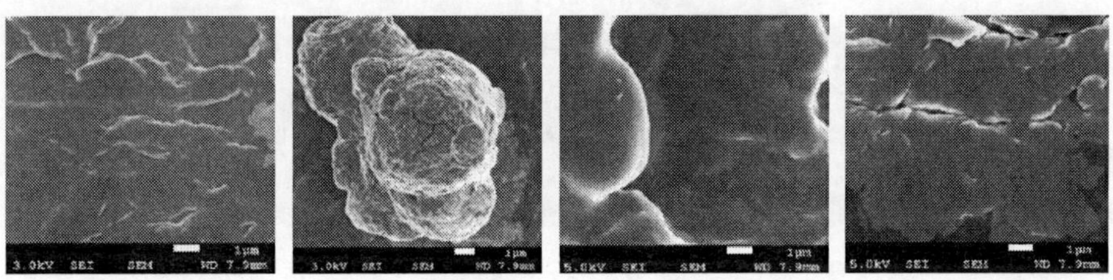

Ad Aw Bd Bw

Note:A: CB, B: Silica, d:dry, w:wet , the arrow stands for the direction of friction, 15.9 mm/s,5N

**Figure 8.** SEM images of different surfaces of friction abrasion of SSBR (5000)

Under the different conditions of the friction above, the situations of friction and damage on the rubber by steel ball were not the same. Broadly speaking, four kinds of mechanisms were obtained from SEM analysis: curl, josephalbert, crack and piling, which could accelerate the rubber to flexible damage, and then the abrasion performance was degraded.

## 4. Conclusion

Obviously, the rubber friction behavior was influenced by the rubber itself inherent nature, in addition to the rubber components. Material hardness and Young's modulus had a great relationship with the relative motion.And the pressure and speed of the friction pair of the relative velocity, the greater the deformation of the rubber to restore, hardness of rubber itself, or flexibility. As a result, the friction wear was generated increasingly.

**Acknowledgments**

This work was financially supported by the fund of Collaborative Innovation of Green Tyres and Rubber.

## References

[1] Zhu Yonggang.Theory of rubber and tire friction on development of valuation, China Elastomerics, 16( 6): 47-50.
[2] Le Gal A, Guy L,Orange G, et a1. Modelling of sliding friction for carbon black and silica filled elastomers on foad traeks1-J3 [J]. Wear, 2008, 264(78): 606-615.
[3] K. G. Gatos, K. Kameo, J. Karger-Kocsis, On the friction and sliding wear of rubber/ layered silicate.
[4] Nanocomposites, eXPRESS Polymer Letters Vol.1, No.1 (2007) 27–31.
[5] Xianhua Chen, B. Steinauer, Dawei Wang, Evolution of aggregate surface texture due to tyre-polishing, J. Cent. South Univ. Technoi. (2011) 18: 259-265.
[6] Jr.N.S. Eiss, H. Czichos. Tribological studies on rubber-modified epoxies:influence of material properties and operating conditions. Wear, 1986, 111(4):347-361.

# Structural Design of Tool Head of Pipe Dredging Device

**Yufei Jia, Mengyu Shang, Haojun Sun, Chuang Pang, Junjie Liu and Xu Yang***

Shenyang Aerospace University, Shenyang 110136, China

*Corresponding author e-mail: 312428849@qq.com

**Abstract.** In the cold season, pipelines are prone to icing-induced clogging, causing great inconvenience to people's production and life. To solve the icing-induced clogging of the pipeline, analysis was made on the reason why the original device fails to break the ice, to provide some guidance for conducting an innovative design of the cutter head of the device. Based on the actual needs for the cutter heads in the course of ice-breaking, a parallel inclined spherical rotary cutter head was proposed, and its 3D model created with the help of a modelling software. Through the operational and arc-traversing performance analyses of the cutter head, the relationship between the radial direction and the axial direction of the cutter head was constructed. The pressure relationship of the device was analysed. The conclusion was drawn as follows: the designed cutter head can improve the effectiveness of the device on ice-breaking and reduce the requirement for the structural intensity, providing a significant basis for the development of its physical prototype.

## 1. Introduction

In the cold season, when the pipe is frozen or the foreign matter is clogged, the pipe is mostly buried in the ground or in a place that is not easily in direct contact, it is relatively difficult to clean up the blockage [1]. Due to the different reasons of pipeline blockage, various types of cleaning devices still have certain limitations in practical applications, and there is still no good solution for pipe icing or hard blockage. In view of the harsh working environment of the pipeline cleaning device and the difficulty in direct intervention, the design of the cleaning device cutter head should meet the requirements of small size, light weight, stable operation and high reliability and so on. Therefore, we propose a parallel tilt ball rotary cutter head.

## 2. Original dredge device cutter work analysis

The existing rotary cutting type cleaning machine has a good application in the clearing of the drainage pipe, and can break the pipe blockage and achieve the purpose of dredging the pipe [2-4]. However, the drawback is that when the principle is applied to the clogging problem caused by icing or hard foreign matter, the linear velocity near the central axis of rotation is small, and it is impossible to effectively cut or break the obstacle in front, and the rear drive device is required to provide a strong thrust. Or a short-term strong impact will hinder the crushing, in order to ensure the advancement of the cutter head. Therefore, the required, power is relatively large, and the structural strength of the device is required to be high. At the same time, the device faces a long structural length and is difficult to cope with the problem of small radius turning, which imposes restrictions on the application of the device.

Content from this work may be used under the terms of the Creative Commons Attribution 3.0 licence. Any further distribution of this work must maintain attribution to the author(s) and the title of the work, journal citation and DOI.
Published under licence by IOP Publishing Ltd

According to the analysis, the designed ice-breaking device cutter can adopt the method of rotary cutting to effectively break the ice, but it needs to effectively solve the shortcomings of low speed and unsatisfactory working effect near the center of rotation, so as to reduce the requirements on the driving device and structural strength, and realize the cutter head that is safe and effective for cornering work to improve its ability to break ice for various types of pipes.

## 3. Dredging device cutter head overall design

The current device head adopts a parallel tilt ball rotary cutter head design, and the design scheme (Fig. 1). The two identical ball cutter heads 5 are rigidly connected with the output shaft of the sub motor 4 to form a rotary cutting body, and the spindles of the cutting body are inclined clockwise by 15° to 20° with respect to the main shaft of the main motor 1, and then are biased toward the main motor main shaft. Move, keeping the outer surface of the head 5 is approximately tangent. The two rotating cutting bodies are rigidly connected to the T-shaped connecting frame 3 and rigidly connected with the output shaft of the main motor. The size of the T-shaped connecting frame is determined by other structures to meet the geometric requirements. The power of the sub motor 4 is input through the overcurrent slip ring 2. The above structure is integrated with the output shaft of the main motor, and realizes the revolving around the main axis while the excavating cutter head realizes the self-rotation cutting

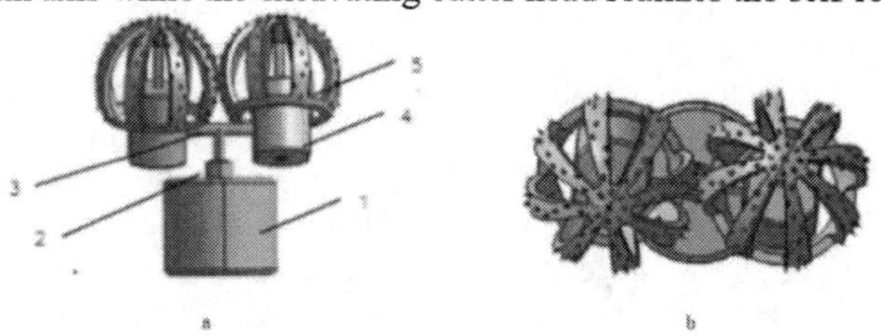


1.main motor; 2.overcurrent slip ring; 3.T-type connector; 4.sub motor; 5.ball cutter head;
a:main view; b: top view magnification

**Figure 1.** cutter head design plan

Through the dynamic simulation of the design scheme through the 3D design software, the design effect of the rotary cutting can be achieved.

### *3.1. Work analysis*

*3.1.1. Work area determination.* In order to illustrate the advantages of the double cutter head design, the double cutter head and the drill bit single cutter head are analyzed and compared (Fig. 2). The area where the two cutter heads block the phase in the pipeline is defined as the work area, and the area where the line speed of the same rotating body is less than 20% of the maximum line speed is defined as the invalid work area. Due to the different curvature radii between the inner and outer sides of the pipeline, the tension of the device on both sides of the neutral line is different, and the relative linear velocity of the invalid working area is too low to effectively cut the front occlusion [5-7]. The structure is required to provide extreme pressure along the speed direction to hinder the front. Crushing can ensure the continuous operation of the device.

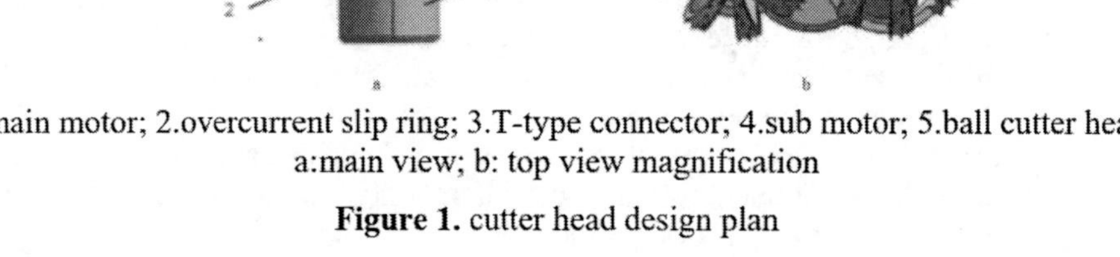


a and b for job analysis c and d for driving force analysis

**Figure 2.** Work area analysis and driving force analysis

In the bit-type single-blade, the ineffective working area accounts for 4% of the working area. The double cutter head makes a tilt of 15° to 20° outside the spindle, effectively avoiding the invalid working area. Therefore, in the working state, there is no invalid working area in the working area of the designed cutter head, and the working efficiency is higher.

*3.1.2. Determination of driving force.* When the two cutter heads are in the same working condition, the driving device gives the same driving force to the two cutter heads, and the driving force F acts on the rear end of the main motor. Since the pipeline working device adopts the cross universal joint shaft connection, the driving force direction along the tangential direction of the pipe center line

The angle between the driving force and the speed direction of the cutter head in the pipe corresponds to the single cutter head and the double cutter head respectively. β1 versus β2. The angle is affected by the length of the two cutter heads and the radius of curvature of the pipe, but is limited by the size requirement of the bend. The size requirement is considered to be cos β1 = cos β2 in the working state. The component force of the driving force along the speed direction of the cutter head reflects the effective force provided by the device for breaking ice, which affects the working efficiency of the cutter head. The component force corresponds to the two cutter heads respectively Fcosβ1 with Fcosβ2. The component force of the driving force along the radial direction of the pipe causes the cutter head to form a pressing friction or even a cutting with the inner wall of the pipe. Therefore, the component force should meet the mechanical performance requirements of the pipe wall design the working area of the two cutter heads is projected in the direction of the speed, and the projected area of the single cutter head is $A_1 = \pi\left(\frac{d}{2}\right)^2 \cos\beta 1$, the double head projection area is $A_2 = \pi\left(\frac{d}{4}\right)^2$, Therefore, the average pressure of the two cutter heads giving the front blockage under the same driving force is:

$$P_1 = \frac{F\cos\beta 1}{A_1} = \frac{4F\cos\beta 1}{\pi d^2}$$

$$P_2 = \frac{F\cos\beta 2}{A_2} = \frac{16F\cos\beta 2}{\pi d^2}$$

Relative pressure:

$$\frac{P_1}{P_2} = \frac{1}{4}$$

It is calculated that under a given driving force, the double cutter head can provide a larger pressure than the drill type single cutter head, so that the double cutter head can reduce the structural strength requirement of the device when faced with the same working demand. Due to the existence of an ineffective working area for a single cutter head, the double cutter head has a more significant advantage.

*3.2. Bending analysis*

For the curve of different diameters and different angles faced by the device during the operation, whether the cutter head can effectively pass the curve will determine its adaptability [8]. Bend analysis of its cornering ability: In order to analyze whether the cutter head can pass the curve, a parametric analysis model is established, and the size needs to meet the size requirement. First, the tool head is simplified (Fig. 3), and the tool head forms a rotating body when working. The radius of curvature of the rotating body is smaller than the radius of the pipe, so that if the rotating body can pass the curve at its maximum contour, the cutting head can pass smoothly.

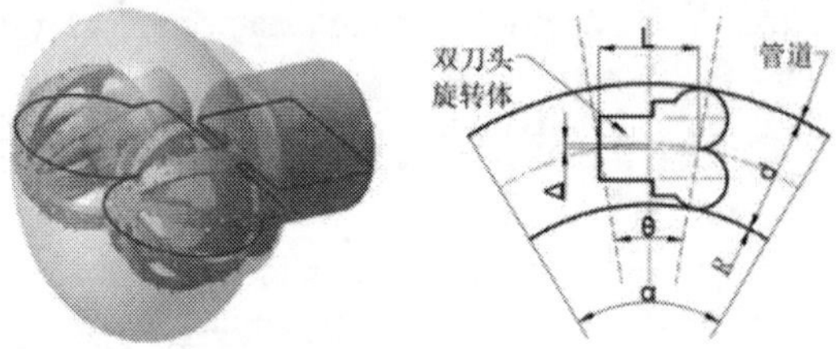


**Figure 3.** Double cutter head rotating body and parameterized analysis model

Since the cutter head keeps spiraling during work, it has an automatic centering capability. In order to ensure the stable advancement and effective cleaning of the cutter head in the pipeline, the tangential point of the two ball cutter heads at the front end of the rotating body and the center point of the rear end of the main motor are on the central axis of the pipeline. At any time, the rotating body is the same as the inner wall of the pipeline. Without interference, the rotating body can pass through the curve, that is, the entire design head can pass. Pipes use relevant design standards in the design process. Figure 3 shows the parametric analysis model. The dimensions of each part include: pipe inner diameter —d, pipe radius of curvature —R, axis offset distance —Δ, axis length—L, main motor diameter—$d_{\text{the Lord}}$, sub motor diameter--$d_{\text{child}}$, the center of the circle —θ, ball cutter radius.—r.

Among them, in order to ensure the effectiveness of the work, the radius of the ball cutter head (r) should satisfy:

$$r = \frac{d}{4}$$

Axis offset distance (Δ) Determined by other structures, the size is:

$$\Delta = \left(R + \frac{d}{2}\right)\left(1 - \cos\frac{\theta}{2}\right)$$

When the cutter head is in the middle of the curve, the dimensions such as length and diameter should satisfy the following relationship:

$$\begin{cases} L = 2\left(R + \frac{d}{2}\right)\sin\frac{\theta}{2} \\ \frac{d_{\text{the Load}}}{2} + \Delta \leq \frac{d}{2} \\ \frac{d_{\text{child}}}{2} + r + \Delta \leq \frac{d}{2} \end{cases}$$

## 4. Conclusion

(1) By establishing an analytical model, the geometric relationship of the dimensions of each part of the double cutter head is established, which satisfies the relationship between the radial direction and the axial direction, and the cutter head can smoothly pass the corner of the curve.

(2) Do work analysis on the double cutter head, reduce the reaction force on the machine body, reduce the structural strength of the device, and improve the work efficiency.

(3) The design of the cutter head device improves the ability to break ice, and has good application value for the development of physical prototype.

**Acknowledgments**
This work was financially supported byNational College Students' innovation and entrepreneurship training program (No. 110418127) fund.

## References

[1] Dianjun Wang, Runping Li, Guangming Huang. Research progress of pipeline robots [J].

Machine Tool Hydraulics, 2008 (04): 185-187.

[2] Weiyuan Zhou, Fei Fang, Lijian Chen, Shou Mao, Ming Xia. Four-way walking intelligent robot system for underground pipelines [J]. Sensors and Microsystems, 2017,36(09): 73-76.

[3] Tao Yu, Huanpeng Guo, Xin Fan, Lingzan Kong. Design and optimization of a new air conditioning ventilation pipe cleaning robot [J]. Mechanical design and manufacturing, 2019 (04): 248-250+255.

[4] Jianqiang Li, Wei Lu, Guangyi Zhao. Application of rotary cutting type cleaning machine in drainage pipe dredging [J]. Water supply and drainage, 2014,40 (03): 86-89.

[5] Lu Yu, Lei Liang, Wei Zhu. Development of pipeline robot for quality inspection of steel pipe [J]. Mechanical Engineer, 2018 (07): 24-26.

[6] Chenzhong Wang, Ting Wang, Jun Liu, Chaolong Zhang, Shaogang Hu. Research on driving characteristics of differential adaptive pipeline robot [J]. Mechanical Transmission, 2017, 41(01): 160-165.

[7] Yaoyong Xue, Jizhong Zhang, Lei Zhang. Design and obstacle analysis of automatic adaptation mechanism for pipeline robots [j]. Manufacturing Automat ion, 2018, 40 (06): 23- 26.

[8] Fengping Xu, Zongquan Deng. Study on the pass ability of pipeline robots in corners [j]. Robotics, 2004 (02): 155- 160.

# Structural optimization design of crane frame for nuclear power waste

**Yu Zhang, Yong Jiang, Yajun Gong *, Baoping Zeng**

Wuhan Second Ship Design and Research Institute, Wuhan, China

*Corresponding author mailbox: 2863535115@qq.com

**Abstract.** The structural stiffness and strength are the basis to ensure the performance of nuclear power waste crane. The structural topology optimization method is an effective method to ensure the structural performance of nuclear power waste crane in the conceptual design stage. In this paper, the topology optimization method is used to establish the finite element topology optimization model of the nuclear waste crane frame, and the structure optimization design is carried out. The optimization results are verified by the finite element method, which proves the effectiveness of the method.

## 1. Introduction

Due to the particularity of the workplace, nuclear waste crane must ensure strict safety, reliability and positioning accuracy. The structural stiffness and strength are the basis to ensure the performance of nuclear power waste crane. The structural topology optimization method is an effective method to ensure the structural performance of nuclear power waste crane in the conceptual design stage.

Structural topology optimization originated from Michell's [1] research on topology optimization design of truss structure in 1904. However, it has attracted a lot of attention since Cheng Gengdong and Olhoff et al. [2] studied the optimization design of the solid plate structure with minimum flexibility. In their research, they innovatively introduced the microstructure into the optimization problem. Later, based on the above ideas, Bendsoe and Kikuchi [3] proposed a topology optimization design method combining microstructure and homogenization. This method transforms the problem of structural topology optimization design into the problem of material distribution. This has played a milestone role in the development of structural topology optimization. In the past 30 years, topology optimization has been developed rapidly.

At present, there are many literatures about topology optimization. According to different optimization objects, topology optimization can be divided into discrete structure topology optimization and continuum structure topology optimization. The topology optimization of discrete structure mainly optimizes the cross-sectional area of members in truss, steel frame and other structures to remove the inefficient members in the structure, so as to obtain the structure with better performance [4,5]. The topology optimization of continuum structure is mainly to optimize the material layout in the solid structure, so as to obtain the structural design scheme meeting the design requirements [6]. The classical topological optimization mathematical model for minimizing the flexibility of continuum structure is shown in formula (1). In the formula, $x_i$is a function representing the material distribution state in the design domain, $x_i = 1$indicating that the point $i$ is filled by solid material, while $x_i = 0$ indicating the change point has no material. $\boldsymbol{u}$ represents the field variable, which satisfies the linear or nonlinear

Content from this work may be used under the terms of the Creative Commons Attribution 3.0 licence. Any further distribution of this work must maintain attribution to the author(s) and the title of the work, journal citation and DOI.
Published under licence by IOP Publishing Ltd

IOP Conf. Series: Materials Science and Engineering **772** (2020) 012054 doi:10.1088/1757-899X/772/1/012054

state equation, $\overline{\boldsymbol{T}}$and$\overline{\boldsymbol{f}}$ represents the surface load and the volume load respectively, $V_0$ represents the initial volume of the design area. Because this paper is mainly about topology optimization of continuum structure, if there is no special explanation, the subsequent content of topology optimization refers to topology optimization of continuum structure.

$$\begin{aligned} &\text{Find:} \quad \mathbf{x}=\{x_1,x_2,\cdots\} \\ &\text{Minimize:} \quad C(\mathbf{x})=\int_{S_\sigma}\mathbf{u}^T\overline{\mathbf{T}}dS+\int_V\mathbf{u}^T\overline{\mathbf{f}}dV \\ &\text{Subject to:} \quad V(\mathbf{x})=\int_\Omega\rho(x)d\Omega\le V_0 \\ &\qquad x_i=0 \text{ or } 1,\ \forall i\in\Omega \end{aligned} \tag{1}$$

In solving the above topological optimization problem (1), a key part is the structural analysis process. Structural analysis mainly refers to the PDE solution corresponding to the physical location of the structure. At present, there are many numerical methods to solve PDE in topology optimization, such as finite element method, meshless method and polynomial mesh method [7]. The theoretical basis of finite element method is very mature, and it is easy to program. Therefore, in the general topology optimization of linear elastic structure, the finite element method is used to solve the PDE. In this paper, the finite element method is also used to solve the PDE in the topology optimization of multiphase materials.

According to the different methods of structural topology description and material interpolation, topology optimization of continuum structure can be divided into homogenization method [3], variable density method [6], level set method [8], progressive structure optimization method, etc. In this paper, the homogenization method is used to optimize the frame structure of nuclear power crane and the finite element method is used to verify.

## 2. Establishment of topology optimization model for crane frame of nuclear power plant

Using the weighted method, the mathematical model of multi-objective topology optimization for nuclear power crane can be generally expressed as follows:

$$\begin{aligned} &\min \Phi=\omega_1\overline{\Phi}_1+\omega_2\overline{\Phi}_2+\ldots+\omega_m\overline{\Phi}_m \\ &\text{s.t.}: \ \mathrm{g}_k(\mathbf{a},\ \rho)\le 0 \quad (k=1,2,\ldots p) \\ &\qquad \mathrm{h}_l(\mathbf{a},\ \rho)=0 \quad (l=1,2,\ldots q) \\ &\rho^{\min}\le\rho\le 1 \end{aligned} \tag{2}$$

Among them, $\Phi$ is the newly constructed weighted objective function, $\overline{\Phi}_m$ represents the normalized m-th objective function, and $\omega_m$ is its corresponding weight coefficient, which satisfying the equation $\sum_{i=1}^{m}\omega_i=1$.

For the optimization problem (2), the gradient based method such as mathematical programming method can be used to solve it. Therefore, a step degree (sensitivity) of all objective functions and constraint functions with respect to design variables $\rho$ must be provided. In this paper, we use the adjoint method to derive the first order analytical sensitivity of the objective function. Consider the m-th objective function:

$$\Phi_m=\int_\Omega\varphi(\mathbf{a},\rho)dV \tag{3}$$

Among them, $\varphi$ is the defined function in the $\Omega$ domain. In order to derive the sensitivity of the objective function $\Phi_m$, we introduce the Lagrange term to construct a new function $\Phi_m^*$.

$$\Phi_m^* = \Phi_m + \boldsymbol{\lambda}^T (\mathbf{Ka} - \mathbf{F}) \tag{4}$$

So:

$$\begin{aligned}\frac{d\Phi_m}{d\rho} &= \frac{\partial \Phi_m}{\partial \rho} + \frac{\partial \Phi_m}{\partial \mathbf{a}}\frac{d\mathbf{a}}{d\rho} + \boldsymbol{\lambda}^T \mathbf{K}\frac{d\mathbf{a}}{d\rho} + \boldsymbol{\lambda}^T \frac{\partial \mathbf{K}}{\partial \rho}\mathbf{a} \\ &= \frac{\partial \Phi_m}{\partial \rho} + \frac{\partial \Phi_m}{\partial \mathbf{a}}\frac{d\mathbf{a}}{d\rho} + \boldsymbol{\lambda}^T \mathbf{K}\frac{d\mathbf{a}}{d\rho} + \boldsymbol{\lambda}^T \frac{\partial \mathbf{K}}{\partial \rho}\mathbf{a} \\ &= \frac{\partial \Phi_m}{\partial \rho} + \left(\frac{\partial \Phi_m}{\partial \mathbf{a}} + \boldsymbol{\lambda}^T \mathbf{K}\right)\frac{d\mathbf{a}}{d\rho} + \boldsymbol{\lambda}^T \frac{\partial \mathbf{K}}{\partial \rho}\mathbf{a}\end{aligned} \tag{5}$$

In order to eliminate the implicit term $\frac{da_i}{d\rho}$, we introduce the adjoint problem as follows:

$$\mathbf{K}\boldsymbol{\lambda} = -\frac{\partial \Phi_m}{\partial \mathbf{a}} \tag{6}$$

Then, bringing it into the formula (6), the sensitivity of the objective function $\Phi_m$ can be obtained:

$$\frac{d\Phi_m}{d\rho} = \frac{\partial \Phi_m}{\partial \rho} + \boldsymbol{\lambda}^T \frac{\partial \mathbf{K}}{\partial \rho}\mathbf{a} \tag{7}$$

Among them, $\frac{\partial \mathbf{K}}{\partial \rho}$ can be directly obtained by the variational method.

In this paper, MMA is used to solve structural topology optimization problems. By introducing the moving asymptote, the implicit optimization problem is transformed into a series of explicit and simpler strictly convex approximation subproblems. The approximation function is determined by the left and right asymptotic points determined in advance and the derivative of the original objective function and constraint function at each point. MMA is the most effective mathematical programming method to solve multi-objective topology optimization.

## 3. Topology optimization results and verification

Figure 1 is the design domain model and boundary conditions of the frame beam structure of nuclear power crane. It is the most important bearing structure of the frame. The topology optimization method is used to optimize the structure and the finite element method is used to verify the optimization results.

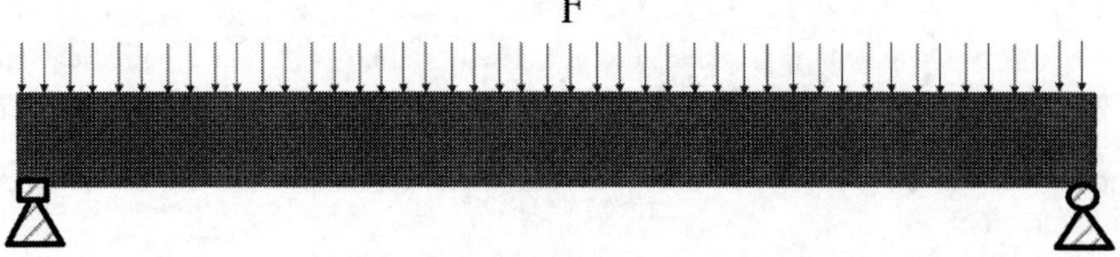


**Figure 1.** Topology optimization design domain and boundary conditions of nuclear waste crane

**Figure 2.** Topology optimization results of frame crossbeam

**Figure 3.** Assembly finite element model of frame beam structure and Z-direction lifting device

**Figure 4.** Deformation cloud chart of the device under the limit working state

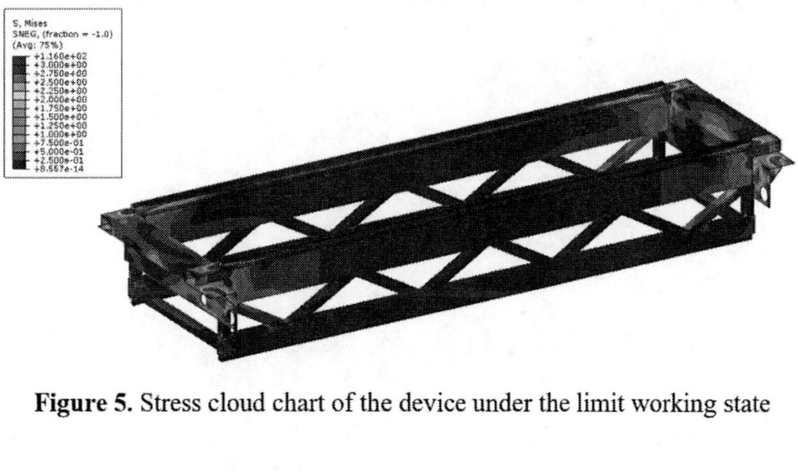


**Figure 5.** Stress cloud chart of the device under the limit working state

Figure 2 is the topological optimization design result of the frame crossbeam structure. According to the optimization result, the structure model of the frame crossbeam is constructed as shown in Figures 3-5. Under the limit condition, the stress and deformation of the whole machine are analyzed by finite element method. The maximum stress and deformation of the structure are 116MPa and 0.095mm respectively. The strength and rigidity meet the design requirements.

## 4. Conclusion

In this paper, the topology optimization method is used to optimize the frame beam structure of nuclear waste crane, and on this basis, the beam structure is designed. Through the finite element simulation analysis under the limit condition, the stiffness and strength indexes of the structure are obtained, and it is found that they can meet the design requirements. The validity of the proposed method is verified.

**Acknowledgments**
National Key R&D Program of China (No. 2017YFC0307800).

## References

[1] Michell A G M. The limits of economy of material in frame-structures [J]. Philosophical Magazine, 1904, Series 6, 8(47): 589-597.
[2] Cheng G D, Olhoff N. An investigation concerning optimal design of solid elastic plates [J]. International Journal of Solids and Structures, 1981, 17(3): 305-323.
[3] Bendsoe M P, Kikuchi N. Generating Optimal Topologies in Structural Design Using a Homogenization Method [J]. Computer Method in Applied Mechanics and Engineering, 1988, 78(1): 197-224.
[4] Stolpe M. Truss optimization with discrete design variables: a critical review [J]. Structural and Multidisciplinary Optimization, 2016, 53(2): 349-374.
[5] Bendsoe M P, Sigmund O. Topology Optimization: Theory, Methods and Applications [M]. Berlin: Springer, 2003.
[6] Talischi C, Paulino G H, Pereira A., et al. PolyTop: a Matlab implementation of a general topology optimization framework using unstructured polygonal finite element meshes [J]. Structural and Multidisciplinary Optimization, 2012, 45(3): 329-357.
[7] Wang M, Wang X, Guo D. A level set method for structural topology optimization [J]. Computer Method in Applied Mechanics and Engineering, 2003, 192(1): 227–246.
[8] Leugering G, Engell S, Griewank A. Constrained Optimization and Optimal Control for Partial Differential Equations [M]. Basel: Springer, 2012.
[9] Xie Y M, Steven G P. Evolutionary structural optimization [M]. London: Springer, 1997.

IOP Conf. Series: Materials Science and Engineering **772** (2020) 012055 doi:10.1088/1757-899X/772/1/012055

# Application of finite element method in strength analysis of ship connecting parts

**Yiou Liu, Zhengjiang Wu *, Yajun Gong, Chao Xiang**

Wuhan Second Ship Design and Research Institute, Wuhan, China

* Corresponding author mailbox: yin@auit.org

**Abstract.** In the design stage, the strength and rigidity of general marine connectors can be calculated according to the relevant ship standard. However, some special nonstandard connections cannot be calculated by ship standard. The finite element method provides us another feasible option. In this paper, the finite element method is used to analyze the strength of the ship connecting flange, and the results are compared with the calculation results of the ship standard.

## 1. Introduction

The finite element method is a numerical method for solving partial differential equations. It approximates the true solution of the element by discretizing the continuous solution area into a finite solution element, and representing the unknown field function to be solved in each element as an approximate function, so that a continuous infinite degree of freedom problem becomes a discrete finite degree of freedom problem [1].

The finite element method has sprouted since the 1940s. In 1943, R. Courant first proposed the application of the principle of minimum potential energy and combined with the piecewise continuous function defined in the triangle region to solve the torsion problem [2]. Its core idea is to introduce shape function to represent unknown field function in triangle region. In 1956, Turner et al. Published the first paper of the finite element method, proposed to use the approximate displacement function in the element to obtain the stiffness matrix of the element and then get the overall stiffness matrix, dealt with the problem of two-dimensional plane and gave the correct answer to solve the stress problem with triangular element, which is considered as the first successful attempt of modern finite element method [3]. The term "finite element method" was first used by clough in 1960 [4]. Then, in 1961, Melosh established the matrix element grid for the two-dimensional plane problem, and established the bending element stiffness matrix of the plane rectangular plate [5]. In 1963, Grafton and Strome studied the axisymmetric and surface bending problems, and established the bending element stiffness matrix of axisymmetric and curved shells [6]. Martin et al. Extended the finite element method to three-dimensional problems, used tetrahedral element mesh and established stiffness matrix to solve the problem [7].

As a scientific calculation method which has been developed for decades, finite element method has many advantages. It can discretize very complex geometric models, and solve problems in a systematic and standardized way. It can develop a flexible and general computer program, which can be widely used in various occasions. So far, a lot of computer programs have been developed that can be applied

Content from this work may be used under the terms of the Creative Commons Attribution 3.0 licence. Any further distribution of this work must maintain attribution to the author(s) and the title of the work, journal citation and DOI.
Published under licence by IOP Publishing Ltd

to various scenes, and it plays an important role in the practical engineering application of all walks of life.

## 2. Strength calculation of marine connecting flange based on shipbuilding standard

Due to the need of installation, it is necessary to enlarge the bolt hole of some flanges by 2mm or 1mm during the design. After the expansion, the stress area of the bolt hole of the flange is correspondingly reduced, and its strength must be reduced. In the case of constant fastener torque, it is necessary to check the local strength through calculation, and check whether the strength of the reamed part meets the requirements under normal tightening torque.

At present, there are standards for checking and calculating the strength of flange, but this standard can only calculate the checking of flange neck strength (circumferential, axial and axial strength as well as torsional strength) and fastener strength under the action of internal fluid pressure of flange. However, according to the calculation standard, it is impossible to check the strength of non-standard connectors after flange reaming.

Figure 1 is a structural diagram of a marine connecting flange. Firstly, according to the strength checking calculation standard of the flange, the axial stress, radial stress and circumferential stress of the flange are calculated, the three-dimensional model of the flange is drawn and the finite element model is established for modal analysis. The stress obtained is compared with the calculated stress to verify the correctness of the method.

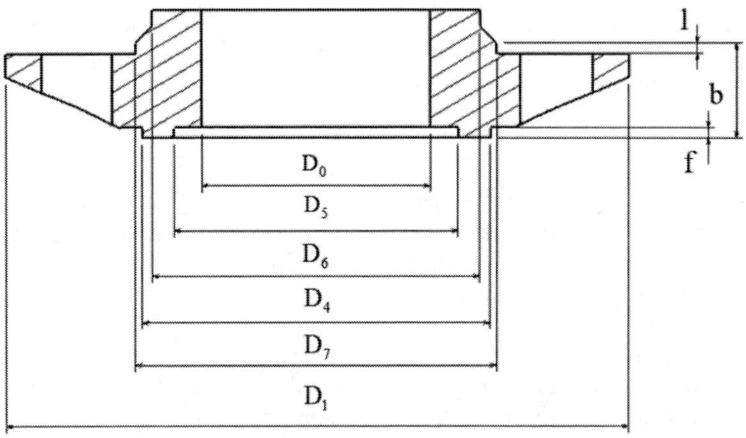


**Figure 1.** Structure of a connecting flange

Flange shape constants are shown in the following table:

$$\delta_0=\delta_1=15,\ \frac{\delta_1}{\delta_0}=1,\ D_1=70,\ D_0=190,\ K=\frac{D_0}{D_1}=\frac{190}{70}=2.71$$

$$h_0=\sqrt{D_1\delta_0}=\sqrt{70\times15}=32.4,\ \text{取}h=1.5\delta_1=22.5,\ \frac{h}{h_0}=\frac{22.5}{32.4}=0.694$$

$$T=1.28,\ U=2.27,\ Y=2.066,\ Z=1.315,\ F_1=0.9,\ V_1=0.55,\ f=1$$

Therefore, according to the calculation formula of flange check, the following results are obtained:

$$e=\frac{F_1}{h_0}=\frac{0.9}{32.4}=0.028,\ d_1=\frac{U}{V_1}h_0\delta_0^2=\frac{2.27}{0.55}\times32.4\times15^2=30087.81,$$

$$\psi = \delta_f e + 1 = (27-3-2)\times 0.028 + 1 = 1.616,\ \beta = \frac{4}{3}\delta_f e + 1 = 1.821,$$

$$\gamma = \frac{\psi}{T} = 1.263,\ \eta = \frac{\delta_f^3}{d_1} = \frac{22^3}{30087.81} = 0.354,\ \lambda = \gamma + \eta = 1.617.$$

Before calculating the allowable stress of flange neck, it is necessary to calculate the design moment $M_0$ of flange in two states of pre-tightening and operation.

$$M_0 = F_D L_D + F_T L_T + F_G L_G$$

Among them:

$$F_D = 0.785 D_1^2 p_c = 0.785 \times 70^2 \times 3.5 = 13462.75N$$

$$L_D = L_A + 0.5\delta_1 = (73-35-15) + 0.5\times 15 = 30.5mm$$

$$F_T = F - F_D = 0.785 D_G^2 p_c - F_D = 0.785 \times \left(110 - 2\times 2.53\sqrt{7.25}\right)^2 \times 3.5 - 13462.75 = 12056.69N$$

$$L_G = \frac{D_b - D_G}{2} = \frac{146-96.37}{2} = 24.81mm,\ L_T = \frac{L_A + \delta_1 + L_G}{2} = \frac{23+15+24.81}{2} = 31.40mm$$

The gasket material is brass, and the coefficient m=3.5 is obtained according to table 9-2 of *GB 150-1998*, therefore:

$$F_G = F_p = 6.28 D_G b m p_c = 6.28 \times 96.37 \times 2.53\sqrt{7.25} \times 3.5 \times 3.5 = 50504.19N$$

Therefore, the design torque of the flange under working condition is:

$$M_0 = 13462.75\times 30.5 + 12056.69 \times 31.40 + 50504.19 \times 24.81 = 2042202.89N\square mm$$

Therefore, the three-dimensional stress of flange neck is obtained:

Axial stress: $\sigma_H = \dfrac{fM_0}{\lambda \delta_1^2 D_1} = \dfrac{1\times 2042202.89}{1.617\times 15^2 \times 70} = 80.18\text{MPa}$

Radial stress: $\sigma_R = \dfrac{\beta M_0}{\lambda \delta_f^2 D_1} = \dfrac{1.821\times 2042202.89}{1.617 \times 22^2 \times 70} = 67.88\text{MPa}$

Tangential stress: $\sigma_T = \dfrac{YM_0}{\delta_f^2 D_1} - Z\sigma_R = \dfrac{2.066 \times 2042202.89}{22^2 \times 70} - 1.315 \times 67.88 = 35.27\text{MPa}$

## 3. Strength check of connector based on finite element method

According to the analysis flow of finite element method. Firstly, the above flange geometric model is meshed in Hyper Mesh software, and the boundary conditions and loads are established in ABAQUS, so as to establish the finite element analysis models of the two flanges, as shown in Figure 2.

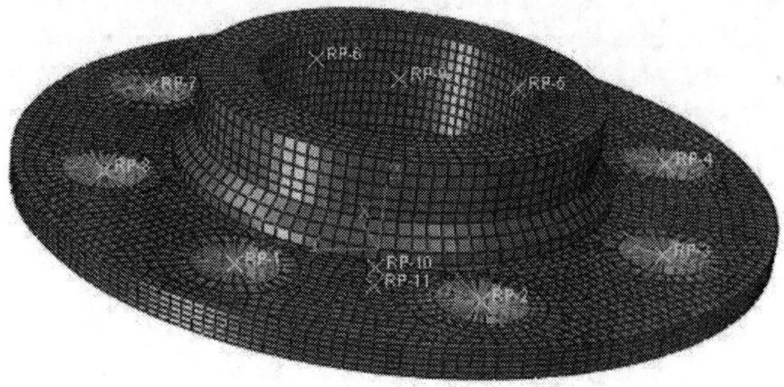

**Figure 2.** Finite element model of a flange

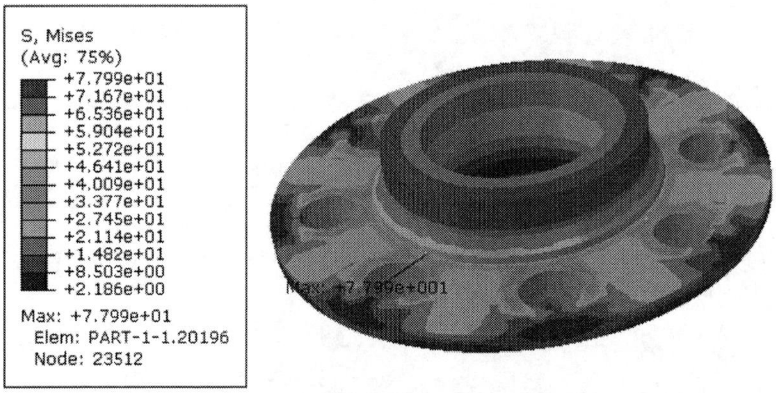


**Figure 3.** Cloud chart of flange stress

It can be seen from figure 3 that the maximum stress of flange obtained by finite element analysis is 77.9MPa respectively. Its position is located in the contact area between the flange neck and the flange plate. The maximum axial stress of the two kinds of flange calculated by theory is 80MPa respectively. The calculated results are basically consistent with each other, which proves the correctness of the above theoretical calculation process.

## 4. Conclusion

In this paper, the finite element method is used to analyze the strength of marine connections, and the calculation results in the ship design standard are compared. Through comparative analysis, it is found that the results of the finite element analysis are in good agreement with the results of the calculation method recommended by the code. The feasibility of the finite element method in the strength analysis of ship connections is verified.

## Acknowledgments

National Key R&D Program of China (No. 2017YFC0307800).

## References

[1] Turner M J. Stiffness and Deflection an alysis of Complex Structures [J]. Journal of Aerosol Sdence, 2015, (23): 805-823.
[2] Clough RW. The Finke Element Method in Plane Stress Analusisi. Proceedings of American Society ofCivil Engineers 1960; 23: 345-37.
[3] Melome J. A Stiffiiess Matrix for the Analysis of Thin Plates in Benng [J]. Journal of the

IOP Conf. Series: Materials Science and Engineering **772** (2020) 012055 doi:10.1088/1757-899X/772/1/012055

Aerospace Sciences, 1961, 28(1): 34-42.
[4] Grafton P E, Stome D R. Analysis of Axisymmetrical Shells by the Direct Stiffness method. Aiaa Journal, 1963, 1(10): 2342-2347.
[5] Martin, H. C. Plane Elasticity Problems and Direct Stiffhe Me od. The Trend in Engineering, 1961, 13, 5-19.
[6] Gallagher Stress Analysis of Heated Complex apes [J]. Ars Journal, 1962, 32 (5): 700-707.
[7] Melosh. J. Structural Analysis of Solids [C]// 1963. S.

IOP Conf. Series: Materials Science and Engineering **772** (2020) 012056 doi:10.1088/1757-899X/772/1/012056

# Surface wettability aging behavior of aramid fiber III after oxygen plasma treatment

**Lu Guo, Jing Wang*, Song Jia, Mingfei Yang, Fuding Sun and Junyang Wang**

School of Materials Science and Engineering & Liaoning Key Laboratory of Advanced Polymer Matrix Composites, Shenyang Aerospace University, Shenyang, China

*Corresponding author e-mail: wally9932@126.com

**Abstract.** Aramid Fiber III was modified by oxygen plasma treatment. The fiber surface wettability before and after different oxygen plasma treatment pressure, and surface wettability aging behavior were observed by Dynamic Contact Angles Analysis (DCAA), respectively. The results showed that aramid fiber surface wettability were increased after oxygen plasma treatment. The total surface free energy increased from 37.78 mJ/m$^2$ for untreated sample to 69.63 mJ/m$^2$ with 40 Pa plasma treated fiber. The fiber surface wettability decreased slowly with storage time in air. The fiber surface wettability decreased from 67.79 mJ/m$^2$ for 20 Pa plasma-treated sample to 63.85 mJ/m$^2$ after 30 days.

## 1. Introduction

The aramid fibers have been widely used in the fields of aerosplace aviation due to its low density and high strength [1, 2]. Aramid fiber III is a recently developed type of para-aramid fibers manufactured in China, which is spun from poly-(polyamide benzimidazole-co-p-phenylene terephthalamide). Due to its surface smooth and chemical inert fiber surfaces, which lead to poor fiber-matrix compatibility and weak adhesion as reinforcement in composite [3-6]. Plasma treatment was proved an effective method to modify the fiber surface for improving fiber surface wettability properties [7].

The purpose of this work is to investigate the effects of oxygen plasma treatment pressure on aramid fiber III surface wettability and its aging behavior. The fiber surface wettability before and after different oxygen plasma treatment pressure, and surface wettability aging behavior were observed by Dynamic Contact Angles Analysis (DCAA), respectively.

## 2. Experimental

### *2.1. Materials and plasma treatment*

Aramid fibers III (polyheteroarylene-co-p-phenyleneterephthalamide) were supplied by China Bluestar Chengrand Chemical Co. Ltd. The fibers were cleaned successively with acetone and distilled water and then dried in a vacuum oven for 3 h at 110 ° C before further analysis. Plasma was excited by an inductive coupling radio frequency generator (13.56 MHz). Oxygen was kept at a flow rate of about 20-30 SCCM. The operation pressure was set at 10Pa, 20Pa, 30 Pa, 40 Pa, 50 Pa, 60 Pa, 70 Pa. The fiber samples were treated for 10 min with plasma treatment power of 200 W. The 20 Pa plasma-treated

Content from this work may be used under the terms of the Creative Commons Attribution 3.0 licence. Any further distribution of this work must maintain attribution to the author(s) and the title of the work, journal citation and DOI.
Published under licence by IOP Publishing Ltd

samples were kept in air for 5, 10, 15, 30 days. After that, the fiber surface wettability was measured by DCAA.

*2.2. Characterization*

Surface wettability of DAF III fiber before and after oxygen plasma treatment was characterized by DCAA (Dataphysics, KRüSS). The fiber samples were cut into about 1 cm in length and fixed directly to a cylindrical rods. Four fibers stick to the rods dispersedly at the same time and then immersed into the testing liquid mediums (water and diiodomethane) by raising the elevating stage at a constant speed of 2 mm/min. The contact angles ($\theta$) were calculated using a computer system from eq. (1). The surface free energy was calculated from Owens-Wendt equation (2) and (3). Each sample are tested three times. The final contact angles with water and diiodomethane are the average value.

$$F = \gamma \cdot p \cdot \cos\theta \quad (1)$$

$$\gamma_l(1+\cos\theta) = 2\cdot\sqrt{\gamma_s^p \gamma_l^p} + 2\cdot\sqrt{\gamma_s^d \gamma_l^d} \quad (2)$$

$$\gamma_s = \gamma_s^p + \gamma_s^d \quad (3)$$

Where F is the wetting force measured by the microbalance; p is the wetted perimeter; θ is the contact angle between fiber and liquid; $\gamma_l$ is the surface tension of the testing liquid; $\gamma_{Total}$ is the surface free energy of the fiber; $\gamma_s^p$ is the polar component and $\gamma_s^d$ is the dispersive component.

## 3. Results and Discussion

*3.1. Effects of plasma treatment pressure on fiber surface wettability*

The results of contact angles with water and diiodomethane before and after oxygen plasma treatment were given in Table 1. For the untreated fiber, the contact angles with water and diiodomethane were 77.47° and 55.02°, respectively. After 10 Pa plasma treatment, the contact angle with water decreased to 32.59° and increased to 65.85°. When the plasma treatment pressure were set between 20 Pa to 60 Pa, the contact angle with water was stable at about 20° and the contact angle with diiodomethane was kept at 67°.The results showed that oxygen plasma treatment can obviously change the fiber surface properties.

**Table 1.** Contact angles with water and diiodomethane before and after oxygen plasma treatment

| Plasma treatment pressure(Pa) | Water (θ) ± s.d. (°) | Diiodomethane (θ) ± s.d. (°) |
|---|---|---|
| Untreated | 77.47 (1.31) | 55.02 (0.78) |
| 10 | 32.59 (1.56) | 65.85 (2.00) |
| 20 | 22.46 (1.13) | 65.31 (1.13) |
| 30 | 19.20 (0.46) | 67.70 (0.83) |
| 40 | 17.15 (1.10) | 70.73 (0.86) |
| 50 | 19.66 (0.33) | 65.48 (2.07) |
| 60 | 21.07 (1.08) | 67.54 (1.46) |
| 70 | 34.25 (0.99) | 67.53 (0.51) |

s.d.: standard deviation

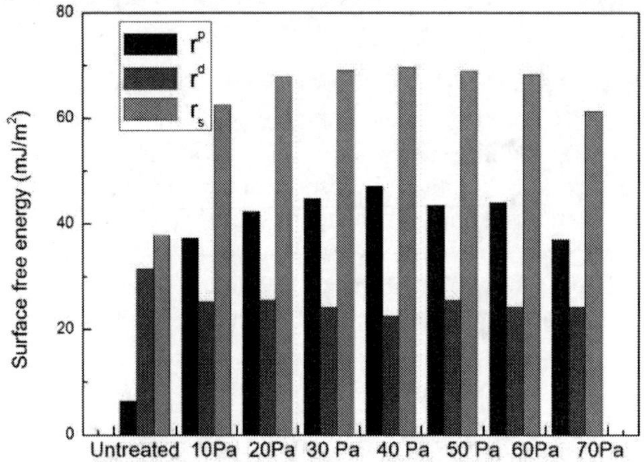


**Figure 1.** Surface free energy of Aramid fiber III via oxygen plasma treatment pressure

According to the equation (1), (2) and (3), we can caculate the fiber surface total free energy, the polar component and dispersive component. Figure 1 showed the results of surface free energy of Aramid fiber III via oxygen plasma treatment pressure. The total surface free energy increased from 37.78 mJ/m$^2$ for untreated sample to 69.63 mJ/m$^2$ with 40 Pa plasma treated fiber. The fiber surface wettability was improved greatly with oxygen plasma treatment.

### *3.2. Fiber surface aging behaviors*

**Table 2.** Contact angles with water and diiodomethane after oxygen plasma treatment with aging time

| Aging days | Water (θ) ± s.d. (°) | Diiodomethane (θ) ± s.d. (°) |
|---|---|---|
| 0 | 22.46 (1.13) | 65.31 (1.13) |
| 5 | 21.33 (1.52) | 65.19 (3.53) |
| 10 | 25.33 (0.53) | 63.97 (1.00) |
| 15 | 26.96 (2.47) | 64.07 (2.01) |
| 30 | 31.72 (1.63) | 59.38 (1.38) |

s.d.: standard deviation

The results of contact angles with water and diiodomethane after oxygen plasma treatment with aging time were given in Table 2. The results of surface free energy of Aramid fiber III after oxygen plasma treatment via aging time was shown in Figure2. For the fresh plasma-treated fiber, the contact angles with water and diiodomethane were 22.46° and 65.31°, respectively. After storage in air for 5 days, the contact angle with water decreased to 21.33°. There is a little slight decrease for short time storage in air. It may be due to the surface chemical structure adjustment after plasma treatment. With the storage time extended to 10 days or longer, the contact angle with water increased slowly. Meantime, the contact angle with diiodomethan decreased. The results showed that oxygen plasma treatment has aging behavior for organic fiber surface. As seen from Figure 2, the fiber surface wettability decreased slowly with storage time in air. The fiber surface wettability decreased from 67.79 mJ/m2 for 20 Pa plasma-treated sample to 63.85 mJ/m2 after 30 days.

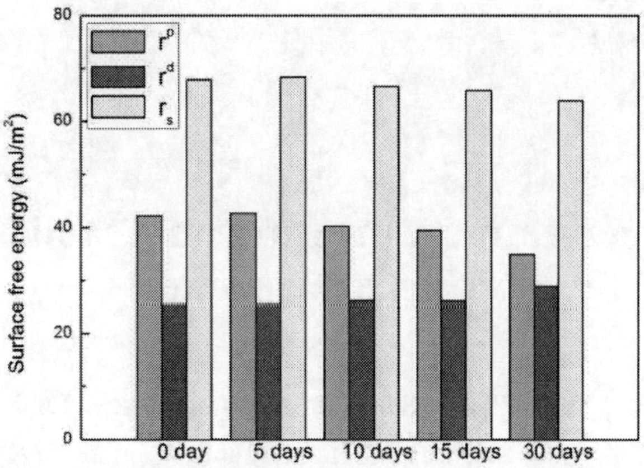


**Figure 2.** Surface free energy of Aramid fiber III after oxygen plasma treatment via aging time

## 4. Conclusion

The dynamic contact angles analysis results showed that oxygen plasma treatment can improve aramid fiber III surface wettablity. The total surface free energy increased from 37.78 mJ/m$^2$ for untreated sample to 69.63 mJ/m$^2$ with 40 Pa plasma treated fiber. The fiber surface wettability decreased slowly with storage time in air. The fiber surface wettability decreased from 67.79 mJ/m$^2$ for 20 Pa plasma-treated sample to 63.85 mJ/m$^2$ after 30 days.

## Acknowledgments

This work was financially supported by National Natural Science Foundation of China (No. 51403129) and Liaoning Natural Science Foundation (20180550705) and University Student Innovation Project (No. 201810143143, 201910143557, 201910143556)

## References

[1] Liu D, Chen P, Mu J J. Improvement and mechanism of interfacial adhesion in PBO fiber/bismaleimide composite by oxygen plasma treatment. Appl Surf Sci.( 2011) 6935–6940.

[2] Gonzalez-Canche N G, Flores-Johnson E A, Carrillo J G. Mechanical characterization of fiber metal laminate based on aramid fiber reinforced polypropylene. Compos Struct. 172(2017)172: 259-266.

[3] Cai R Q, Peng T, Wang F D. Improvement of surface wettability and interfacial adhesion of poly-(p-phenylene terephthalamide) by incorporation of the polyamide benzimidazole segment. Appl Surf Sci, 257(2011) 9562-9567.

[4] Zhang S H, He G Q, Liang G Z. Comparison of F-12 aramid fiber with domestic armid fiber III on surface feature. Appl Surf Sci. 256(2010) 2104-2109.

[5] Xing L X, Liu, L, Huang Y D. Enhanced interfacial properties of domestic aramid fiber-12 via high energy gamma ray irradiation. Composites Part B. 69(2015)50-57.

[6] Liu D D, Hu, YY H, Zhang F F. Interfacial reinforcement of domestic aramid fiber III/PTA composites using azide-containing benzimidazole coupling agent. Acta Mater Composite Sinica, 34( 2017) 336-344.

[7] Jing W, Ping C, Chun L. Improvement of aramid fiber III reinforced bismaleimide composite interfacial adhesion by oxygen plasma treatment. Composite Interfaces, 25(2018):771-783.

# Effects of plasma treatment pressure on Aramid fiber III/BMI composite humidity resistance properties

**Jinyuan Cui, Jing Wang*, Song Jia, Yu Wang, Cheng Long and Yida Kong**

School of Materials Science and Engineering & Liaoning Key Laboratory of Advanced Polymer Matrix Composites, Shenyang Aerospace University, Shenyang, China

*Corresponding author e-mail: wally9932@126.com

**Abstract.** The effects of oxygen plasma treatment pressure on Aramid Fiber III chemical structure and its reinforced bismaleimides (BMI) composite humidity resistance properties were investigated. The aramid fiber III chemical structure under different plasma treatment pressure were measured by FTIR. The composite interlaninar shear strength with different plasma treatment pressure before and after absorption water were tested by short-beam bending test method, respectively. The FTIR results showed that oxygen plasma treatment didn't change the fiber bulk chemical structure. The composite humidity resistance of interlanimar shear strength was enhanced after plasma treatment. The retention rate of composite interlanimar shear strength was above 95%.

## 1. Introduction

The aramid fiber III are widely used in the fields of aviation, automobile due to its excellent properties such as low density, high strength and modulus[1, 2]. However, because of its smooth and chemical inert fiber surfaces, which limit its application in composite system [3, 4]. Thus, the fiber surface need to modify to improve its surface properties. In our previous work, plasma is proved an effective method to enhance the aramid fiber surface wettability and its reinforced composite interfacial properties [5].

The purpose of this work is to investigate the effects of oxygen plasma treatment pressure on aramid fiber III chemical structure and its reinforcing thermosetting bismaleimides (BMI) composite humidity resistance properties. The aramid fiber chemical structure was characterized by FTIR. The aramid fiber III reinforced BMI composite interfacial properties before and after water absorption was evaluated by short beam shear measurement.

## 2. Experimental

### *2.1. Materials*

Aramid fibers III (polyheteroarylene-co-p-phenyleneterephthalamide) were supplied by China Bluestar Chengrand Chemical Co. Ltd. The fibers were cleaned successively with acetone and then dried in a vacuum oven for 3 h at 110 ° C.

The fiber surface was modified by inductive coupling radio frequency (13.56 MHz) plasma reactor. Oxygen was kept at a flow rate of about 20-30 SCCM. The operation pressure was set at 10 Pa, 20 Pa,

Content from this work may be used under the terms of the Creative Commons Attribution 3.0 licence. Any further distribution of this work must maintain attribution to the author(s) and the title of the work, journal citation and DOI.
Published under licence by IOP Publishing Ltd

IOP Conf. Series: Materials Science and Engineering **772** (2020) 012057 doi:10.1088/1757-899X/772/1/012057

30 Pa, 40 Pa and 50 Pa. The fiber samples were treated for 15 min with plasma treatment power of 200 W. BMI resin (QY8911-II) was received from Beijing Aeronautical Manufacturing Technology Research Institute.

### 2.2. *Characterization*

The chemical structure of aramid fiber III were analyzed by infrared spectrometer (Nicolet-20, DXB) at 4 cm-1 resolution and signal-averaged over 32 scans.

Composite water absorption tests were carried as follows: dry, pre-weighed composite specimens (40 mm × 50 mm × 2 mm) were immersed into distilled water (98 ± 2 °C) for 24 h. Then, the composite interlaminar shear strength was tested before and after water absorption.

The composite interlaminar shear strength (ILSS) was measured on short-beam bending test method on a Shimadzu universal testing machine according to ASTM D-2344.

## 3. Results and Discussion

### 3.1. *FTIR*

Figure 1 shows the FTIR results of aramid fiber III with different plasma treatment pressure. It is found that the absorption peak at 3275 $cm^{-1}$ belongs to N - H stretching vibration on the amide bond. the absorption peak at 1634 $cm^{-1}$ is contributed to C = O stretching vibration peaks (amide I band). The absorption peak at 1512 $cm^{-1}$ is N - H group in-plane bending vibration peak (amide I band), 1307 $cm^{-1}$ is C - N groups stretching vibration absorption peak (the amide II band), which are the characteristic absorption peaks of the domestic aramid fiber III[6]. Compared with the untreated fiber sample, the chemical structure of aramid fiber III under different plasma treatment pressure do not change obviously.

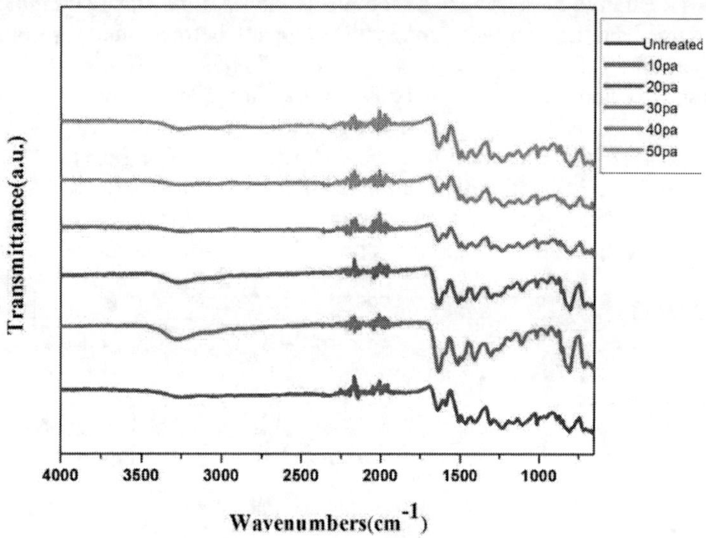


**Figure 1.** FTIR of aramid fiber with different oxygen plasma treatment pressure.

### 3.2. *Water absorption*

Figure 2 shows the water absorption of composites via plasma treatment pressure. It is found that the water absorption of untreated fibers reinforced BMI composite is the largest, while all the other five plasma-treated samples are relatively lower within the beginning two hours. For the untreated sample, the water absorption is 0.22%. For the plasma-treated fiber samples, the water absorption are 0.14%, 0.11%, 0.09%, 0.21% and 0.17%, respectively. The phenomenon maybe due to the composite interface ahesion. If the composite interface adhesion is good, the water molecule can not diffuse into composite inner easily. After being immersed into water for 24 h, the water absorption achieves 0.51% for untreated sample and 0.42%, 0.39% and 0.24% after plasma treated for 10 Pa, 20 Pa, 30 Pa, respectively. However, with the plasma treated pressure increasing to 40 Pa and 50 Pa, the water absorption is larger then the untreated one, which means that the interfacial adhesion of the fiber and the matrix is poor and then the water molecule diffusing into composites along the fiber–matrix interface.

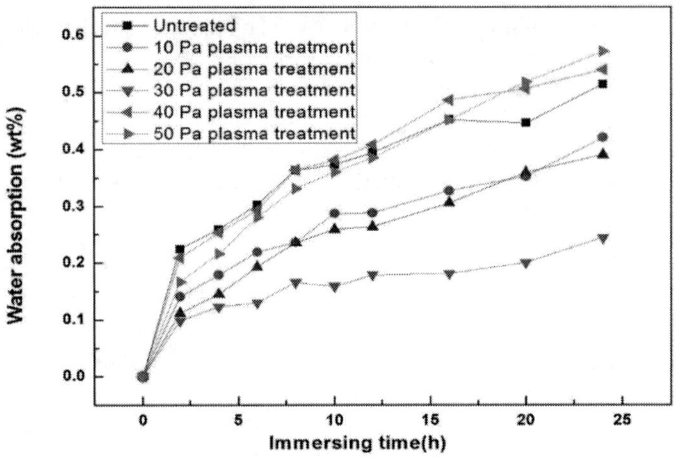


**Figure 2.** Effects of plasma treatment power on humidity resistance of composite bending strength

### *3.3. Humidity resistance*

Figure 3 shows the humidity resistance of composite interlaminar shear strength under different plasma treatment pressure. The ILSS value of untreated composite before water absorption is 49.35 MPa. After plasma treatment, the ILSS values increase to 58.68 MPa, 59.94 MPa, 62.99 MPa, 61.2 MPa and 62.1 MPa after plasma treated pressure set at 10 Pa, 20 Pa, 30 Pa, 40 Pa and 50 Pa, respectively. The result shows plasma treatment can improve composite interfacial adhesion. After water absorption, the ILSS value decrease to 46.2 MPa for the untreated composite. And it decrease to 58.32 MPa, 58.7 MPa, 62.7 MPa, 59.7MPa and 61.1MPa after plasma treated pressure set at 10 Pa, 20 Pa, 30 Pa, 40 Pa and 50 Pa, respectively. The retention rate of composite interlaninar shear strength is about 94% for untreated sample. After plasma treatment, the retention rate of composite interlaninar shear strength exceed 95%.

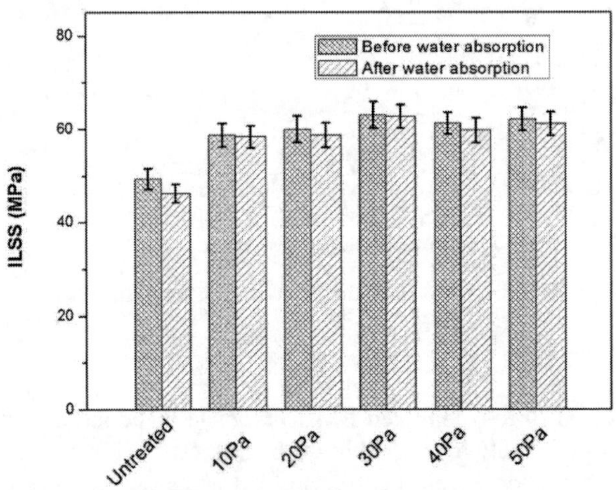


**Figure 3.** Effects of plasma treatment pressure on humidity resistance of composite interlaminar shear strength

## 4. Conclusion

The FTIR results showed that oxygen plasma treatment didn't change the fiber bulk chemical structure. The composite humidity resistance of interlanimar shear strength was enhanced after plasma treatment. The retention rate of composite interlanimar shear strength was above 95%.

## Acknowledgments

This work was financially supported by National Natural Science Foundation of China (No. 51403129) and Liaoning Natural Science Foundation (20180550705) and University Student Innovation Project (No. 201810143144, 201910143517)

## References

[1] Zhang S H, He G Q, Liang G Z. Comparison of F-12 aramid fiber with domestic armid fiber III on surface feature. Appl Surf Sci. 256(2010) 2104-2109.

[2] Liu D D, Hu, YY H, Zhang F F. Interfacial reinforcement of domestic aramid fiber III/PTA composites using azide-containing benzimidazole coupling agent. Acta Mater Composite Sinica, 34( 2017) 336-344.

[3] Wang C X, Du M, Lv J C. Surface modification of aramid fiber by plasma induced vapor phase graft polymerization of acrylic acid. I. Influence of plasma conditions. Appl Surf Sci , 349(2015)333-342.

[4] Wang J, Chen P, Lu C.Improvement of aramid fiber III reinforced bismaleimide composite interfacial adhesion by oxygen plasma treatment. Composite Interfaces, 25(2018):771-783.

[5] Wang J, Chen P, Li H. Surface characteristic of poly (p-phenylene terephthalamide) fibers with oxygen plasma treatment. Surf. Interface Anal., 40(2008) 1299-1303.

[6] Wang H H. Study on Structure and Properties of Aramid Fiber III [D]. Xi'an: Xi' an University of Engineering, 2015

# $SnO_2$ microrods based triethylamine gas sensor

**Lu Wang** [1], **Ruiqin Peng** [2], **Lijie Ci** [2] **and Yuguo Li** [1, *]

[1] School of Physics and Electronics, Shandong Normal University, Jinan 250014, China
[2] SDU & Rice Joint Center for Carbon Nanomaterials, Key Laboratory for Liquid-Solid Structural Evolution & Processing of Materials (Ministry of Education), School of Materials Science and Engineering, Shandong University, Jinan 250061, China

*Corresponding author e-mail: liyuguosd@hotmail.com

**Abstract.** Triethylamine (TEA) gas sensor having high response and stability are in great demand in the actual working environment. In this paper, $SnO_2$ microrods are prepared on $SiO_2$ (300 nm)/Si substrate by thermal evaporation. The $SnO_2$ microrods-based gas sensor is successfully prepared. The gas response and stability of the sensor are tested at an optimal temperature of 200 °C. When the sensor is exposed to 100 ppm TEA gas, the response time is only 4 s. The detection limit is below 1 ppm. The gas-sensing mechanism is further discussed.

## 1. Introduction

In the past few decades, the rapid development of science and technology has promoted the continuous advancement of society, and the living standards of people have been greatly improved. At the same time, it also brings many negative impacts to the whole society, such as increased emissions of harmful gases and pollutants, which poses a serious threat to public health. Therefore, it is imperative to monitor and detect toxic, harmful, flammable and explosive gases in working and living places [1-3]. Among the harmful gases, TEA gas is the most widely concerned. It is often used in the industrial field and is a volatile organic compound with irritating odor, toxic, flammable and explosive, which can cause lung problems, skin burns, and even death. TEA gas can be secreted from dead fish, marine organisms and seashells. Therefore, it poses a huge threat to marine life, water resources, and multiple biological chains. According to the Occupational Safety and Health Administration, the concentration of TEA in the air should not exceed 10 ppm [4-6].

Metal oxide semiconductor (MOS) are widely used because of their reliable sensitivity, low cost, high stability, and long life. Among many kinds of MOSs, $SnO_2$ is the most widely used, and it has a wide band gap ($E_g = 3.6$ eV, T = 300 K), high electron mobility, good chemical, and physical stability, etc [7, 8]. One-dimensional (1D) $SnO_2$ nanomaterials have a large relative specific surface area, the gas sensors based on 1D $SnO_2$ nanomaterials have excellent gas sensitivity performance [9]. Holguin-Momaca J. T et al. [10] successfully prepared $SnO_2$ nanowires by vapor-liquid-solid method, but a mixture of Ar, $O_2$, and $H_2$ was required during the preparation. It is well known that the mixture of $O_2$ and $H_2$ will explode under certain conditions, so the experimental process is dangerous. A. Kolmakov et al. [11] reported that the gas sensing reaction of $SnO_2$ nanowires and nanobelts under the action of Pd catalyst is good, but the preparation process is complicated and not effective for detecting TEA gas.

Content from this work may be used under the terms of the Creative Commons Attribution 3.0 licence. Any further distribution of this work must maintain attribution to the author(s) and the title of the work, journal citation and DOI.
Published under licence by IOP Publishing Ltd

Therefore, it is extremely urgent to develop a 1D $SnO_2$ gas sensor with the simple preparation process, high sensitivity, good stability, and reliability.

In this paper, Sn and SnO mixed powder were used as the source material, and the uniform $SnO_2$ microrods were successfully prepared by a simple thermal evaporation method. This method is not only easy to operate, but also repeatable and reliable. The $SnO_2$ microrods sensor has a fast response to TEA, good stability, and low detection limit.

## 2. Experimental

### *2.1. Preparation of $SnO_2$ microrods*

We successfully prepared a large number of uniform $SnO_2$ microrods on $SiO_2$ (300 nm)/Si substrates by thermal evaporation. Firstly, Sn and SnO mixed powder with a mass ratio of 5:1 were placed at one end of the quartz plate, and several substrates positioned at the other end (about 1 cm apart), which had sputtered with an Au film of about 3-4 nm. Second, the quartz plate is placed in the center of the quartz (the powder is upstream of the gas flow), and the thermal tube is closed, and then vacuumed. After the gas pressure in the furnace is stabilized to 100 Pa, Ar gas (60 sccm) is introduced. Then, it is automatically carried out according to the set program (up to 1000 °C at 10 °C/min, and kept for 60 min). Finally, the samples were taken out after cooling naturally to room temperature. A large number of white wires were found on the substrate surface and were characterized by XRD, Raman, and SEM. Au acts as a catalyst during growth, so the growth mechanism can be ascribed to vapor-liquid-solid growth mechanism [12, 13]. The growing process is shown in Fig. 1.

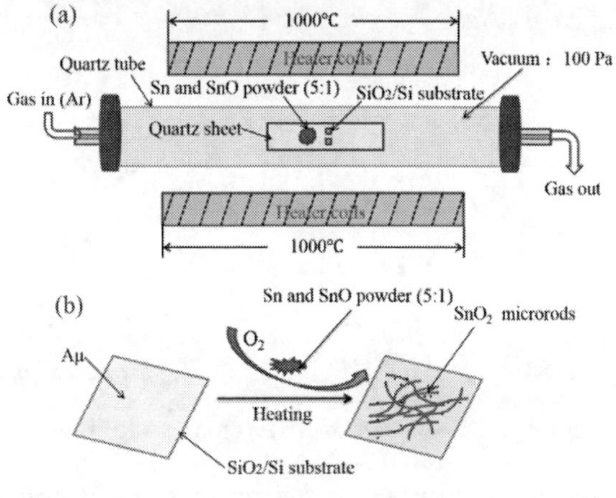


**Figure 1.** (a) Schematic diagram of the growth process of $SnO_2$ microrods; (b) Schematic diagram of the growth of $SnO_2$ microrods on $SiO_2$ /Si substrate.

### *2.2. Gas sensing test*

The sample was sputtered Au film electrodes (7 mm*3 mm*0.1 mm) on the surface. Then, the gas sensor was dried in a vacuum at 100 °C and aged for 24 h to give it better stability and repeatability. The gas-sensing performance was tested using a commercial gas sensing analysis system. The test sensor was placed in a ventilated 18 L sealed chamber. Gas sensitivity experiments (using micro-injectors) were carried out at various temperatures ranging from room temperature to 300 °C for obtaining an optimum temperature. Using the following formula to define the gas sensing response:

$$S = R_a/R_g \quad (1)$$

$R_a$ is the resistance of the test sensor in air, and $R_g$ is the resistance of the test sensor in the target gas. The response and recovery time was calculated as reached a 90% variation in resistance of the difference from the maximum after injecting target gas and backing to air [14].

## 3. Results and discussion

### *3.1. Characterization*

In order to study the sample structure, we performed XRD structural characterization. As shown in Fig. 2(a), the plurality of diffraction peaks were observed. The Au (111) crystal plane was observed at 2θ = 38.50° (Au derived from the catalyst). There are three strong diffraction peaks at 2θ = 26.84°, 34.16°, and 52.04°, corresponding to the $SnO_2$ (110), (101), and (211) crystal planes respectively, which is consistent with the rutile structure $SnO_2$ standard card (JCPDS 41-1445). No other peaks were detected in the XRD pattern, indicating that the Sn and SnO mixed powder were completely converted to $SnO_2$, and the purity of the sample is high. The XRD diffraction peaks of the sample are sharp and the half-height width is narrow, indicating that the prepared $SnO_2$ samples had high crystallinity [15]. To further investigate the sample structure, we also performed Raman test. Fig. 2(b) shows the room temperature Raman spectra of $SnO_2$ microrods. The Raman scattering peaks correspond to the $E_g$, $A_{1g}$, and $B_{2g}$ vibration modes of the rutile structure $SnO_2$ at 474.85 $cm^{-1}$, 633.29 $cm^{-1}$, and 773.5 $cm^{-1}$. It is further confirmed that the prepared $SnO_2$ microrods have the rutile structure, which agrees with the XRD results presented.

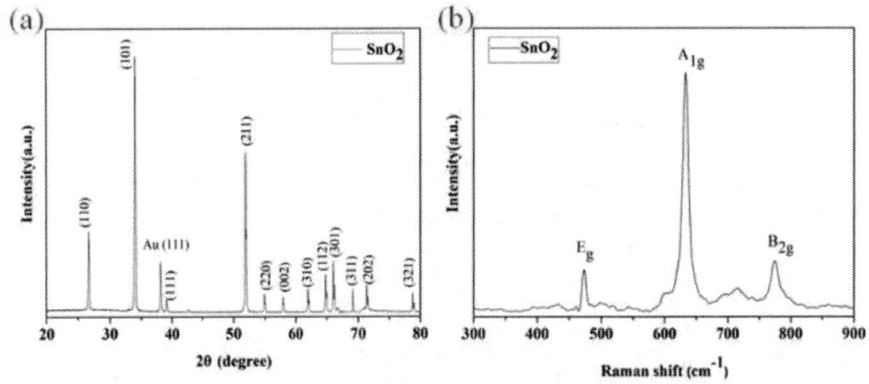


**Figure 2.** (a) XRD pattern of $SnO_2$ microrods; (b) Raman scattering spectra of $SnO_2$ microrods

The morphology of $SnO_2$ microrods was characterized by scanning electron microscope (SEM). As shown in Fig. 3(a), it shows that a large number of rods-like are coated on the surface, and the prepared microrods are uniform and the length is up to 10 μm. In Fig. 3(b), the diameter of the microrods is about 600 nm, the ratio of the length to diameter is up to 16. The microrods surface is rough and some white points are coated on the whole surface.

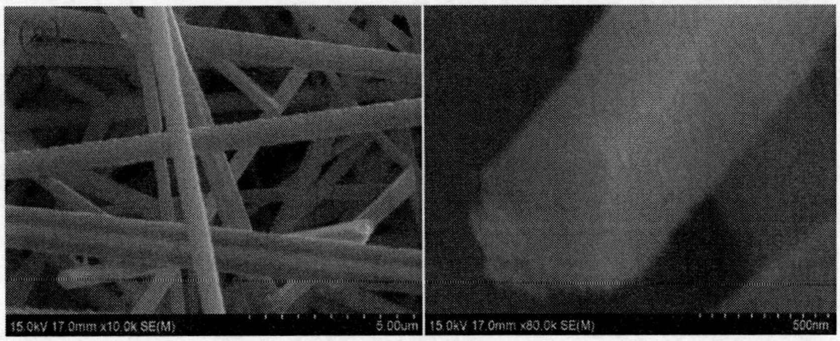


**Figure 3.** (a) Low-resolution SEM image of $SnO_2$ microrods; (b) The high-resolution SEM image of $SnO_2$ microrods.

### *3.2. Gas sensing performance*

The operating temperature of the gas sensor is one of the important factors to evaluate the quality of the device. In order to explore the optimal temperature of gas sensor, we conducted tests to 100 ppm TEA gas at different temperatures (100, 150, 200, 250, and 300 °C), and obtained responses of 1.28, 1.7, 2.6, 1.8, and 1.5 at 100, 150, 200, 250, and 300 °C, respectively. At 200 °C, the sensor got the highest gas response. Thus, the optimum operating temperature can be recognized as 200 °C, which is shown in Fig.4 (a). The following systematically tested gas sensing is all at 200 °C. In Fig. 4(b), it was found that the gas response increased with the increasing concentration ranging from 1 ppm to 500 ppm. When the concentration is 1 ppm, the gas response is 1.6, and when the concentration is 500 ppm, the gas response is up to 4. This gas sensor not only has a good gas-sensitive response to TEA gas but also has a low response limit (less than 1 ppm). We also analyzed the response/recovery property as exposure to the TEA gas concentration of 100 ppm. In Fig. 4(c), the gas sensor has a fast response time (4 s) and a fast recovery time (20 s). Stability is also one of the important factors in describing sensors' performance. We tested the stability of the gas sensor, which is shown in Fig. 4(d). The sensor was tested with 13 cycling times within 2000 s at a TEA gas concentration of 100 ppm, and the cycling response shows no obvious change, indicating that the sensor has good stability.

Based on the above results, the gas mechanism is discussed. Generally, when the TEA gas sensor was exposed to the air, the oxygen molecules in the air would be adsorbed on the surface of the $SnO_2$ microrods, and are ionized to formed $O^{\delta-}$($O_2^-$, $O^-$, and $O^{2-}$) by capturing free electrons from the conduction band of $SnO_2$. Thus, an electron depletion layer is formed by reducing the free electron concentration, and the energy band of the $SnO_2$ microrods is bent upward, the resistance would increase. The detail is shown in Fig. 5(a). When the sensor was exposed to TEA gas, as shown in Fig. 5(b), the TEA gas would react with $O^{\delta-}$ adsorbed on the surface of the $SnO_2$ microrods. The reactions can be calculated by the following equation:

$$(C_2H_5)_3N + O^{\delta-} \rightarrow NO_2 + H_2O + CO_2 + e^- \quad (2)$$

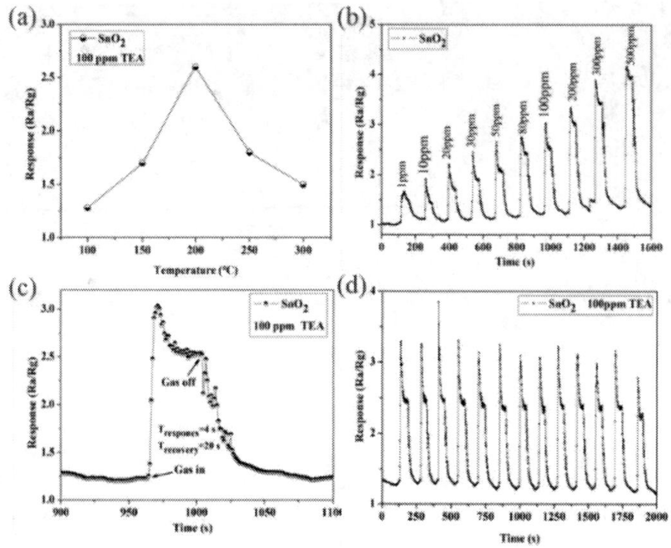

**Figure 4.** (a) The gas response to 100 ppm TEA gas at different temperatures (100-300 °C); (b) The gas response under different TEA gas concentrations; (c) The response time of the gas sensor; (d) The stability of the gas sensor.

The released electrons would return to the $SnO_2$ conduction band, and narrow the energy band bending, the electron depletion layer would be thin, and the resistance reduced accordingly [16, 17], the gas response obtained.

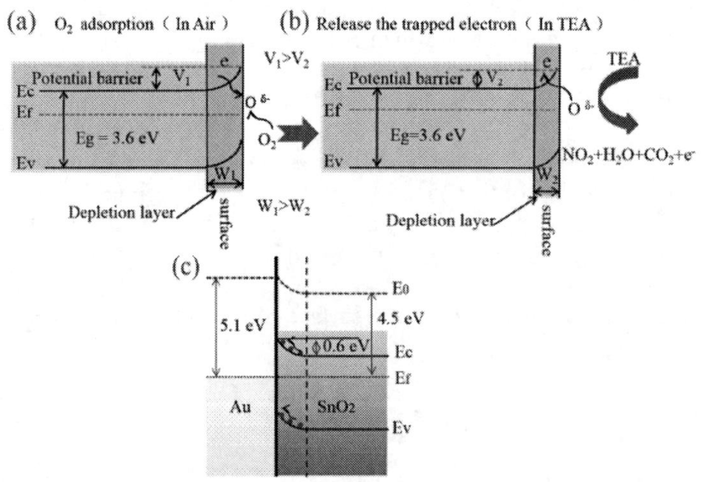


**Figure 5.** Schematic of the $SnO_2$ microrods gas sensing mechanism. (a) The oxygen is adsorbed on the surface of the $SnO_2$ microrods and the depletion layer formed; (b) the depletion layer shown under TEA gas; (b) Energy band diagram of the Au/$SnO_2$

There is another effective factor to contribute to resistance reduction. The XRD pattern revealed that there were Au diffraction peaks, indicating that Au was doped into the $SnO_2$ microrods, they would form Au/$SnO_2$ heterojunctions. As shown in Fig. 5c, For the Au/$SnO_2$ heterojunctions, because of the different work function between Au (5.1 eV) and $SnO_2$ (4.5 eV), after contacting, the electron would transfer from $SnO_2$ to Au and decrease the resistance. In addition, the existence of Au augments the n-type property of $SnO_2$ due to the metal-semiconductor junction. Au induces a great charge depletion, TEA gas molecules are attracted. The adsorbed TEA gas molecules react with the absorbed oxygen ions, and then the captured electrons are released, leading to the thickness of the depletion layer decrease and the resistance are decreased, which is contributing to enhance the gas-sensing performance [18-20].

## 4. Conclusion

In summary, we prepared uniform $SnO_2$ microrods by simple thermal evaporation. The structure property and morphology of $SnO_2$ microrods are confirmed by XRD, Raman, and SEM characterization. The $SnO_2$ microrods-based gas sensors showed the great advantages of fast response, high stability and low detection limit to TEA gas. The gas-sensitive mechanism is also discussed. This work provides a simple and reliable method for preparing $SnO_2$ microrods-based TEA gas.

## References

[1] X. Tian, X. Yang, F. Yang, A visible-light activated gas sensor based on perylenediimide-sensitized $SnO_2$ for $NO_2$ detection at room temperature, [J]. Colloid Surf. A-Physicochem. Eng. Asp. 578 (2019) 123621.

[2] N. Jayababu, M. Poloju, M. V. R. Reddy, Facile synthesis of $SnO_2$-$Fe_2O_3$ core-shell nanostructures and their 2-methoxyethanol gas sensing characteristics, [J]. J. Alloy. Compd. 780 (2019) 523-533.

[3] W. Qi, W. Li, Y. Sun, Influence of low-dimension carbon-based electrodes on the performance of $SnO_2$ nanofiber gas sensors at room temperature, [J]. Nanotech. 30 (2019) 345503.

[4] H. Xu, W. Li, R. Han, Enhanced triethylamine sensing properties by fabricating Au@ $SnO_2$/α-$Fe_2O_3$ core-shell nanoneedles directly on alumina tubes, [J]. Sens. Actuator B: Chem. 262 (2018) 70-78.

[5] S. Bai, C. Liu, R. Luo, Metal organic frameworks-derived sensing material of $SnO_2$/NiO composites for detection of triethylamine, [J]. Appl. Surf. Sci. 437 (2018) 304-313.

[6] R. Peng, Y. Li, J. Chen, P. Si, J. Feng, L. Zhang, L. Ci, Reduced graphene oxide wrapped Au@ZnO core-shell structure for highly selective triethylamine gas sensing application at low temperature, [J]. Sens. Actuator A. 283 (2018) 128-133.

[7] X. Liu, K. Zhao, X. Sun, Electrochemical sensor to environmental pollutant of acetone based on Pd-loaded on mesoporous $In_2O_3$ architecture, [J]. Sens. Actuator B: Chem. 290 (2019) 217-225.

[8] P. G. Choi, N. Izu, N. Shirahata, Improvement of sensing properties for $SnO_2$ gas sensor by tuning of exposed crystal face, [J]. Sens. Actuator B: Chem. (2019) 126655.

[9] J. Zhang, C. Li, B. Wang, Ag-decorated $SnO_2$ nanorods: microwave-assisted green synthesis and enhanced ethanol gas sensing properties, [J]. Micro Nano Lett. 12 (2017) 245-247.

[10] J. T. Holguín-Momaca, F. Espinosa-Magana F, W. Antúnez-Flores, Core to shell switching of the conduction channel on $SnO_2$ nanowire sensors, [J]. Nanotech. 28 (2017) 465501.

[11] A. Kolmakov, D. O. Klenov, Y. Lilach, Enhanced gas sensing by individual $SnO_2$ nanowires and nanobelts functionalized with Pd catalyst particles, [J]. Nano Lett. 5 (2005) 667-673.

[12] T. M. Ngoc, N. Van Duy, C. M. Hung, Self-heated Ag-decorated $SnO_2$ nanowires with low power consumption used as a predictive virtual multi-sensor for $H_2S$-selective sensing, [J]. Anal. Chim. Acta. 1069 (2019) 108-116.

[13] Z. X. Yang, P. Zhang, W. Zhong, Design, growth, and characterization of morphology-tunable $CdxZn_{1-x}S$ nanostructures generated by a one-step thermal evaporation process, [J]. Crystengcomm. 14(2012) 4298-4305.

[14] S. Zhu, Y. Liu, G. Wu, Mechanism study on extraordinary room-temperature CO sensing capabilities of Pd-$SnO_2$ composite nanoceramics, [J]. Sens. Actuator B: Chem. 285 (2019) 49-55.

[15] D. Xue, P. Wang, Z. Zhang, Enhanced methane sensing property of flower-like $SnO_2$ doped by Pt nanoparticles: A combined experimental and first-principle study, [J]. Sens. Actuator B:

Chem. (2019) 126710.
[16] B. Nam, W. Lee, T. K. Ko, Influence of the distribution of nanoparticles on the $NO_2$ sensing properties of $SnO_2$ nanorods decorated with CaO and Pt, [J]. J. Alloy. Compd. 2019.
[17] T. Y. Lai, T. H. Fang, Y. J. Hsiao, Characteristics of Au-doped $SnO_2$-ZnO hetero-nanostructures for gas sensing applications, [J]. Vacuum, 166 (2019) 155-161.
[18] S. M. Yousefi, S. Rahbarpour, H. Ghafoorifard, Describing the effect of Ag/Au modification on operating temperature and gas sensing properties of thick film $SnO_2$ gas sensors by gas diffusion theory, [J]. Mater. Chem. Phys. 227 (2019)148-156.
[19] G. Tofighi, D. Degler, B. Junker, Microfluidically synthesized Au, Pd and Au-Pd nanoparticles supported on $SnO_2$ for gas sensing applications, [J]. Sens. Actuator B: Chem. 292 (2019) 48-56.
[20] R. Abdelghani, H. S. Hassan, I. Morsi, Nano-architecture of highly sensitive $SnO_2$-based gas sensors for acetone and ammonia using molecular imprinting technique [J]. Sens. Actuator B: Chem. 29 (2019) 126668.

IOP Conf. Series: Materials Science and Engineering **772** (2020) 012059 doi:10.1088/1757-899X/772/1/012059

# Experimental investigation on the dynamics of cathode spots in arc plasma descaling

**Zhongliang Tang, Haibin Li, Shulong Liu, Yi Liu, Jing Hu***

College of Physics and Electronic Information, Huaibei Normal University, Anhui Huaibei, China

*Corresponding author e-mail: hujing200407696@126.com

**Abstract.** In the ferrous metallurgical industry, removal of the oxide layer on steel surface (i.e. descaling) with low pressure arc plasma is an environmentally friendly method. In this paper, high speed camera and oscilloscope were employed to diagnose the process of arc decaling. The evolution of cathode spots morphologies and the property of arc voltage fluctuations were investigated for different discharge conditions. The motion and lifetime of the cathode spots are affected obviously by discharge parameters such as pressure, electrode gap and arc current. The lifetime and velocity of the cathode spots are essentially depended on the properties of the substructure of the cathode spot. The fluctuations of arc voltage are directly associated with the dynamics of the cathode spots, and there are large voltage fluctuations for short spot lifetime, large velocity of the spots and poor conditions for spot ignition.

## 1. Introduction

The vacuum arc is widely used in many fields, such as vacuum circuit breaker and vacuum arc coatings, etc [1-4]. Cathode spots are the most predominant and most studied feature of the vacuum arc. The most common form of the cathode spots consists of an ensemble of luminous spots that move fast and randomly over the cathode surface. The materials under the spots are almost simultaneously evaporated and subsequently ionized. Each cathode spot consists of smaller substructures, called ectons or cells [5, 6]. The motion of the cathode spots on the surface is a direct consequence of the extinguishing and ignition of the ectons, which further results in the fluctuation of arc voltage and almost all other significant parameter fluctuations.

In recent years, low pressure arc have been employed to remove of the oxide layer on the steel surface [7-9]. In early work, researchers concentrated their studies on the treatment processes, e.g. surface roughness and arc voltage varied with the thickness of the oxide layer and the electrode gap, respectively [7-10]. Only limited attention was paid in the dynamics of the cathode spots. Shi et al. found retrograde motion of the cathode spots [11]. In our previous works, the role of the thickness of the oxide layers in affecting the spot dynamics was studied as well [12].

In this contribution, the dynamics of the cathode spots in low pressure arc descaling were further investigated, employing high speed camera and oscilloscope for diagnostic of the arc discharge. Experimental studies for the dynamics of the cathode spots were carried out via the evolution of the morphologies of the cathode spots and fluctuations of the arc voltage.

Content from this work may be used under the terms of the Creative Commons Attribution 3.0 licence. Any further distribution of this work must maintain attribution to the author(s) and the title of the work, journal citation and DOI.
Published under licence by IOP Publishing Ltd

IOP Conf. Series: Materials Science and Engineering **772** (2020) 012059 doi:10.1088/1757-899X/772/1/012059

## 2. Experimental arrangement

The schematic arrangements of the experimental apparatus are shown in Fig. 1. The experiments were conducted in a glass vacuum chamber with a rotary pump. A copper cylinder with a diameter of 40 mm severed as the anode. A low carbon steel (99.672% of Fe, 0.04% of C, 0.25% of Mn, 0.038% of Al) of 30×30×3 mm3 was connected with the negative polarity of a dc power supply with a no-load voltage of 450 V and served as the cathode. The oxide layer with a thickness of 6.4 μm was formed in the process of hot roll in air atmosphere at a temperature of 1250 ℃. A high-voltage pulse igniter with a voltage of 8 kV was used to ignite the arc discharge. Then, the arc discharge was maintained by the dc power supply. The plasma is produced through the discharge between anode and cathode. The arc current, electrode gap and work gas pressure were adjusted in the range of 100 to 1000 Pa, 10 to 50 mm and 10 to 60 A, respectively. After the chamber was evacuated to 5 Pa, the argon was introduced. The pictures of the cathode spots were captured by a high-speed camera (Phantom v 12.1) with an exposure time of 1 μs. The arc voltage was monitored with an oscilloscope (Tektronix TDS2014).

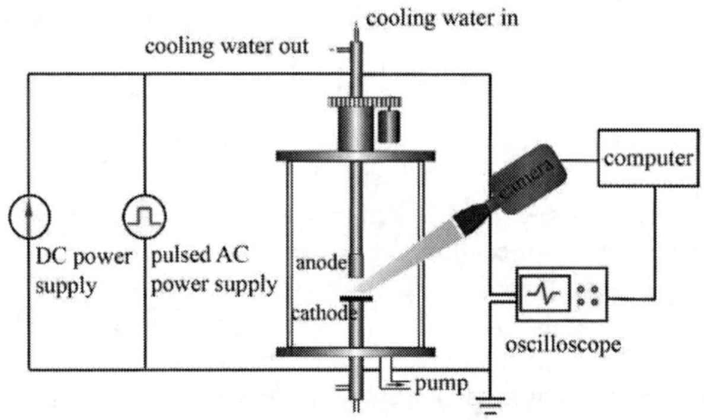


**Fig. 1** Schematic illustration of the experimental setup for arc descaling

## 3. Results and discussion

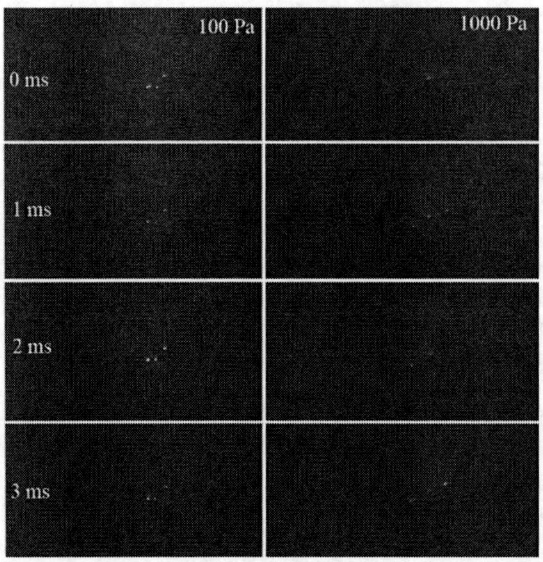


**Fig. 2** Morphologies of cathode spots for different pressure at an arc current of 30 A and an electrode distance of 15 mm.

IOP Conf. Series: Materials Science and Engineering **772** (2020) 012059 doi:10.1088/1757-899X/772/1/012059

Figure 2 shows the evolution of the morphologies of the cathode spots at 30 A, 15 mm while the pressure are 100 Pa and 1000 Pa. At the lower pressure, the cathode spots with large diameter and high brightness tend to agglomerate into several large spot groups. At 1000 Pa, the amount of the cathode spots becomes larger because of the splitting of the cathode spots. The current filaments formed by the ectons attract each other in the mutual self-magnetic fields due to the Lorentz force, while external work gas and the plasma pressure provide repelling forces. Therefore, the increase of the repelling forces between ectons with increasing the gas pressure makes the splitting of the spots more frequent [13].

The threshold current $i_{th}$ of an ecton is the same for a certain cathode material [6]. The ecton will extinguish when the current lower than $2i_{th}$, and split when current larger than $2i_{th}$. The ecton number is almost constant when the total arc current is unchanged. Consequently the amount of the ectons contained in one spot decrease at high pressure, and meanwhile, the spot number increase. The average lifetime of a spot

$$\theta = t_c / q^L$$

Where $t_c$ is the cycle duration of the ecton, $q<1$, $L$ is the total ecton number simultaneously operating during the spot operation [14, 15]. Hence, the average lifetime $\theta$ of the spot will shorten with the decrease of $L$.

Compared with other physical quantities of cathodic arcs, the arc voltage is intimately related to cathode processes, with very little distortion or superposition by secondary parameters of the system. The fluctuations of the arc voltage are essentially associated with the non-stationary process and are resulted from the extinguishing and ignition of the spots on the cathode surface. The fluctuations of the arc voltage also directly reflect the spot dynamics and the interaction between spots and cathode material. Tang et al. investigated the power spectrum of the arc voltage fluctuations and the evolution of the cathode spots, revealed the relationship between the spot dynamics and the nature of the cathode surface [12].

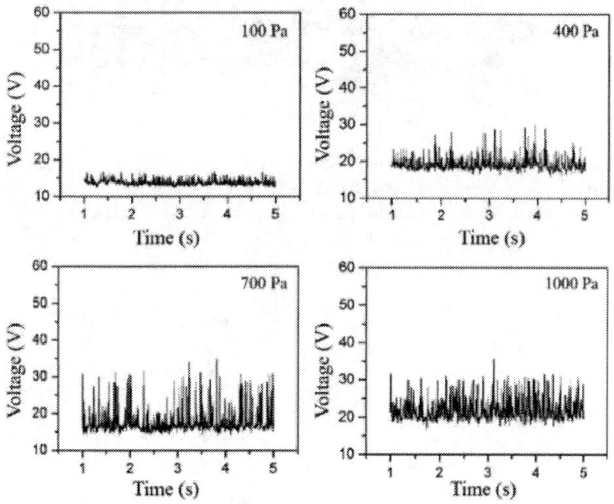


**Fig. 3** Fluctuations of arc voltage for different pressure at an arc current of 30 A and an electrode distance of 15 mm.

The waveforms of the arc voltage were recorded at different pressure, as shown in Fig. 3. The arc voltage and its fluctuations increase significantly with gas pressure. From the view of the spot dynamics, the decrease in the spot lifetime and increase in the ignition frequency cause the rise of the voltage fluctuations, which corresponds with spot dynamics.

IOP Conf. Series: Materials Science and Engineering **772** (2020) 012059 doi:10.1088/1757-899X/772/1/012059

Electrode gap width plays an important role in controlling descaling rate in low pressure arc descaling. The descaling rate decreases with increasing the gap width at a constant current. Figure 4 represents the evolution of the cathode spots at 100 Pa and 30 A, while the gap width is 15 mm and 50 mm. When the arc operates with a large electrode gap width, the number of the cathode spots increases, but the brightness falls off. The length of the current filaments increase with electrode gap width, consequently the magnetic interaction between ectons is enhanced. The increase of the distance between the spots results in the decline of the plasma density. Meanwhile, the reduction of the electric field between the electrodes increases the velocity of the positive space charge. The local strong electric fields formed between the positive space charge and the cathode is beneficial to the ignition of the new ectons, and hence the spot velocity increases [16]. Spot motion is not motion of matter but a process associated with extinguishing and ignition of ecton sites, which is composed of a series of discrete processes.

Figure 5 shows the fluctuations of the arc voltage with a variable electrode gap width from 10 mm to 50 mm. When the electrode gap width increases, the voltage fluctuation grows. The reduction of the plasma density makes the ignition of the new ectons more difficult, which causes the increase of the voltage fluctuation. The higher spot velocity will lead to the increase of the voltage fluctuation as well.

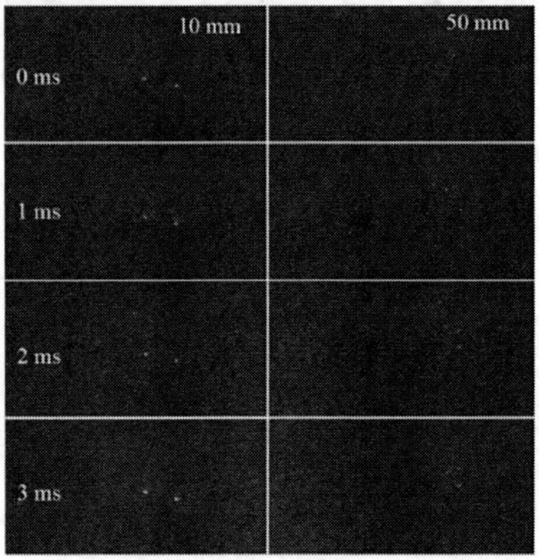


**Fig. 4** Morphologies of cathode spots for different electrode distance at a pressure of 100 Pa and an arc current of 30 A.

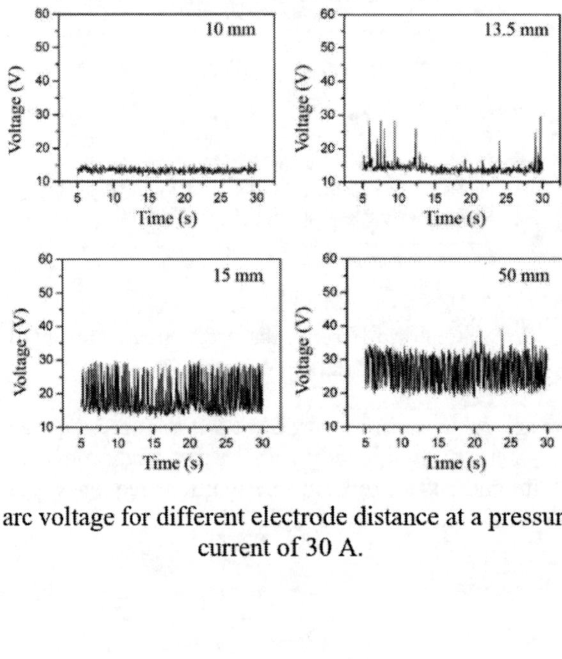


**Fig. 5** Fluctuations of arc voltage for different electrode distance at a pressure of 100 Pa and an arc current of 30 A.

IOP Conf. Series: Materials Science and Engineering **772** (2020) 012059 doi:10.1088/1757-899X/772/1/012059

In the arc descaling, the descaling rate is approximately proportional to the arc current [9]. Figure 6 shows the spot morphologies with different arc current at 100 Pa and 15 mm. The spot number is found to increase linearly with arc current due to the splitting of the spots.

Figure 7 shows the change of the voltage fluctuation with arc current. It can be seen that the voltage fluctuation decrease with increasing the arc current. The ecton number and the plasma density both increase with arc current. The ignition of new ectons is facilitated by the existence of plasma generated from already active ectons [17, 18]. Additionally, there is long spot lifetime for high plasma density and thus the voltage fluctuation decreases.

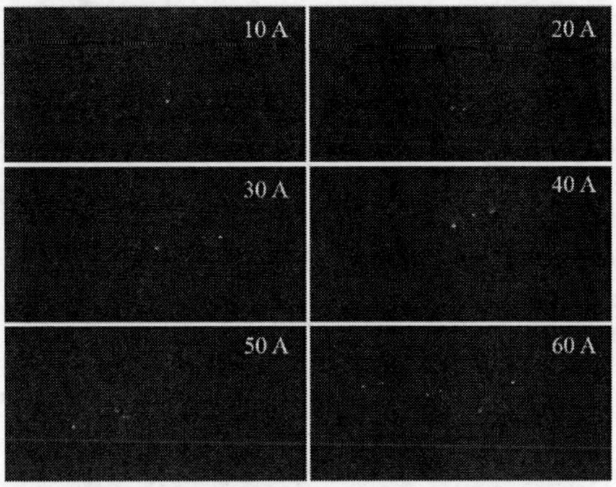


**Fig. 6** Morphologies of cathode spots for different current at a pressure of 100 Pa and an electrode distance of 15 mm.

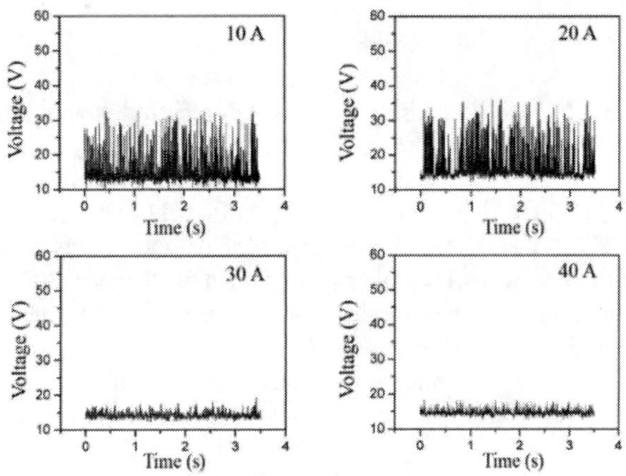


**Fig. 7** Fluctuations of arc voltage for different arc current at a pressure of 100 Pa and an electrode distance of 15 mm.

## 4. Conclusion

The experimental results show that the dynamics of the cathode spots is affected by the discharge conditions in the process of low pressure arc descaling. The motion and the lifetime of the cathode spots depend on the nature of the quasi-periodical mutual repulsion and attraction between the ectons within a spot. The increase of the pressure can enhance the repelling force between the ectons, which reduces the number of the ectons within a spot and consequently shortens the lifetime of the spots. The magnetic repelling force between ectons is enhanced due to the increase of the electrode gap width; therefore, the number of the cathode spots increases, but the brightness falls off. The spot number is found to increase linearly with arc current as well. The fluctuation of the arc voltage is associated with the dynamics of the cathode spots, and there are large voltage fluctuations for short spot lifetime, large velocity of the spots and poor conditions for spot ignition.

**Acknowledgments**
This work was financially supported by the Natural Science Foundation of Anhui Province [No. 1708085ME96]

**References**

[1] Malkin P. The vacuum arc and vacuum interruption [J]. Journal of Physics D Applied Physics, 1989, 22.
[2] Boxman R L , Goldsmith S . Principles and applications of vacuum arc coatings [J]. IEEE Transactions on Plasma Science, 1989, 17(5):0-712..
[3] Niwa Y, Funahashi T, Yokokura K, et al. Basic investigation of a high-speed vacuum circuit breaker and its vacuum arc characteristics [J]. IEE Proceedings-Generation, Transmission and Distribution, 2006, 153(1):11-0..
[4] Pat S, Şadan Korkmaz, Özen S, et al. GaN thin film deposition on glass and PET substrates by thermionic vacuum arc (TVA)[J]. Materials Chemistry & Physics, 2015, 159:1-5.
[5] Beilis I I. State of the theory of vacuum arcs [J]. Plasma Science IEEE Transactions on, 2001, 29(5):657-670.
[6] Mesyats G A. Ectons and their role in plasma processes [J]. Plasma Physics & Controlled Fusion, 2005, 47(47):A109.
[7] Iwao T, Inagaki Y, Sato A, et al. Oxide-Layer Thickness Effect for Surface Roughness Using Low-Pressure Arc [J]. IEEE Transactions on Plasma Science, 2006, 34(4):1223-1228.
[8] Takeda K. Dry cleaning of metal surfaces by a vacuum arc [J]. Surface & Coatings Technology, 2000, 131(1–3):234-238.
[9] Takeda K, Takeuchi S. Removal of Oxide Layer on Metal Surface by Vacuum Arc [J]. JIM, 2007, 63(7):636-642.
[10] Takahashi H, Sugimoto M, Takeda K, et al. Effect of electrode gap width in vacuum arc cleaning [J]. Thin Solid Films, 2006, s 506–507(-):346-349.
[11] Shi Z, Xiao J, Jia S, et al. The influence of gap distance on the random walk of cathode spot in vacuum arc [J]. JOURNAL OF PHYSICS D-APPLIED PHYSICS, 2007, 40(19):5924.
[12] Tang Z L, Yang K, Liu H X, et al. Dynamics of cathode spots in low-pressure arc plasma removing oxide layer on steel surfaces [J]. Physics of Plasmas, 2016, 23(3):1168-256.
[13] Jüttner B. The dynamics of arc cathode spots in vacuum: new measurements [J]. Journal of Physics D Applied Physics, 1999, 30(2):221.
[14] Barengolts S A, Mesyats G A, Shmelev D L. Structure and time behavior of vacuum arc cathode spots [J]. IEEE Transactions on Plasma Science, 2003, 31(5):809-816.
[15] Barengolts S A, Mesyats G A, Tsventoukh M M. The ecton mechanism of unipolar arcing in magnetic confinement fusion devices [J]. Nuclear Fusion, 2010, 50(12): 125004.
[16] Fang D Y. Cathode spot velocity of vacuum arcs [J]. Journal of Physics D Applied Physics, 1982, 15(15):833.
[17] Rosen J, Anders A. Time and material dependence of the voltage noise generated by cathodic vacuum arcs [J]. Journal of Physics D Applied Physics, 2005, 38(23): 4184.
[18] Anders A, Oks E M, Yushkov G Y. Cathodic arcs: Fractal voltage and cohesive energy rule [J]. Applied Physics Letters, 2005, 86(21): R103.

IOP Conf. Series: Materials Science and Engineering **772** (2020) 012060 doi:10.1088/1757-899X/772/1/012060

# The Effect of Photoconductive Semiconductor Materials in Improving the Resolution of Femtosecond Streak Camera

**Xiaorong Liu[1,] *, Song Li[2]**

[1] School of Management, Tianjin University of Technology, Tianjin, China
[2] Tianjin Zhonghuan Electronic Information Group, Tianjin, China

*Corresponding author e-mail: 2365861939@qq.com

**Abstract.** In this study, through the research on the effect of photoconductive semiconductor materials in improving the resolution of femtosecond streak camera, the effective value of applying photoconductive semiconductor materials to femtosecond streak camera is discussed and analyzed, and the effectiveness of photoconductive semiconductor materials in improving the resolution of femtosecond streak camera is verified. Method: In this study, first of all, the generation and recombination of carriers in semiconductors are studied. Solving the continuity equation is applied to further explore the law of the motion of carriers. Later, photoconductive semiconductor materials are used in the preparation of switching devices and scanning circuits. After that, the prepared circuit module is tested by dark current. The main purpose is to realize the resistance of the circuit switch in the intrinsic state and the dark state withstand voltage ability of the circuit switch. Results: It is found that the dark state withstand voltage of each switch can exceed 6000V. In addition, it is found that the current value of the switch under different light is different. When in the natural light, the current value is large. When it is in the state of shading, its current value is relatively small. The efficiency of all three switches is over 50%. The efficiency of nonlinear model is obviously higher than that of linear model, which can be more than 90%. By applying the semiconductor material to the femtosecond streak camera, the scanning circuit can be optimized effectively to meet the high standard requirements of the femtosecond streak camera for the scanning circuit, which has certain advantages, and has certain value and significance for improving the resolution of the second streak camera.

## 1. Introduction

Ultrafast diagnosis technology is a fruitful and enduring research topic. In the research process of more than 100 years, there are many kinds of ultrafast diagnostic equipment developed by people, and the scope of diagnosis ranges from the microsecond level to the femtosecond level [1-3]. According to the working mechanism, three kinds of ultrafast diagnostic equipment have been developed. The first one is the optical mechanical high-speed camera, through the high-speed movement of the mechanical components in the camera to achieve the instantaneous capture and imaging of the super fast target. The second is electro-optical or magneto-optical shutter high-speed camera, which is based on the electro-optical or magneto-optical modulation effect of materials, a shutter type high-speed camera, such as a Kerr box camera, is made. The third is a high-speed camera with image converter. A high-speed camera

Content from this work may be used under the terms of the Creative Commons Attribution 3.0 licence. Any further distribution of this work must maintain attribution to the author(s) and the title of the work, journal citation and DOI.
Published under licence by IOP Publishing Ltd

based on the vacuum image converter technology senses the ultrafast light signal through the photocathode and converts it into an electrical signal, which is either deflected by scanning or directly imaged on the fluorescent screen through pulse gating [4-6]. Among the three types of high-speed camera, the high-speed camera with image converter has the widest diagnosis range, the best performance and the widest application, which is the research hotspot of the ultra fast diagnosis technology [7].

As the best indicator to measure the technical level of streak camera, time resolution has always been a hot competition in the field of ultrafast diagnosis in various countries [8]. As early as the 1980s, the temporal resolution of the streak image tube has entered the femtosecond level. However, there are not many countries that can develop the femtosecond streak camera according to the design results. The Institute of general physics of Russian Academy of Sciences has been in the leading position in the development of streak camera. In terms of the development of the femtosecond time-resolved streak camera, the institute has been successful since the 1990s (the former Soviet Union), with a time resolution of 200fs [9] for the FV-FS-M type streak tube. Hamamatsu company of Japan is the research institute with the highest level of standardization and commercialization of streak camera. Similarly, in the 1990s, Hamamatsu company has reported a streak image converter with a time resolution of 180fs. In recent years, the company has launched a commercial femtosecond streak camera with a time resolution of 200fs [10]. In China, the professional development units of streak camera include Xi'an Institute of Optics and mechanics of Chinese Academy of Sciences and Shenzhen University. Among them, the research and development of the streak camera of Xi'an Institute of Optics and mechanics began in 1963. In the past 50 years, Xi'an Institute of Optics and mechanics has developed many kinds of streak cameras, whose diagnosis range is from nanosecond to picosecond.

In this study, photoconductive semiconductor materials are used in the preparation of scanning circuit because of the high requirement of scanning circuit for femtosecond streak camera. First of all, the principles and approaches of the generation and recombination of carriers in semiconductors are further analyzed. After that, the numerical simulation and simplification of linear model are carried out. Then, on this basis, circuit modules and devices are prepared. Finally, the dark current test method is applied to verify the effectiveness of photoconductive semiconductor materials. It is found that the application of photoconductive semiconductor materials can meet the high requirements of scanning circuit to some extent, and can improve the performance of optimized scanning circuit. Whether it is in the nonlinear mode or in the linear mode, the efficiency of voltage conversion and the efficiency of circuit switch have been improved. It is proved that the photoconductive semiconductor material can meet the requirements of femtosecond streak camera.

## 2. Method

### *2.1. Generation and recombination of carriers in semiconductors*

The core of semiconductor devices is the generation and recombination of carriers. The most important basis of its operation is the energy band theory of solid, which includes two kinds. The first is the guide band and the second is the price band. Only electrons in the conduction band and holes in the valence band can promote the generation of carriers. When it is at absolute zero, the valence band of semiconductor will become full band. On the contrary, the conduction band will become empty. However, there is a disadvantage that the electrons in the full band do not have conductivity and no carrier. As shown in figure 1, the states of semiconductors are respectively at room temperature of absolute zero, in the intrinsic state and in the excited state. When it is at absolute zero, the semiconductor can be in the intrinsic state without any doping, and some electrons will migrate to the conduction band and leave holes. When it is in this state, the carrier concentration of the conductor will decrease, but the resistivity will not change, and it is still in a very high state. When excited, the electrons move all the way to the conduction band, leaving a large number of holes behind, and the concentration will rise quickly, and finally change to the state of conductor.

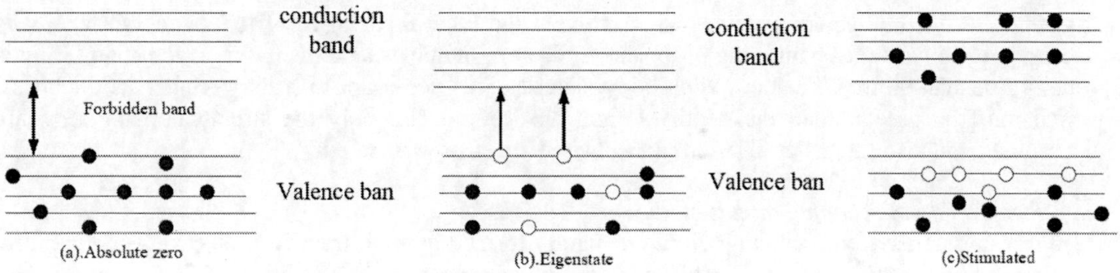


**Figure 1.** Generation of carriers in semiconductors

There are two ways semiconductor can generate carriers. The first is electrical injection. The second is light injection. The main way of application in this study is light injection. Its principle is to modulate the conductivity of semiconductor by using light.

There are four mechanisms of carrier recombination: direct recombination, indirect recombination, surface recombination and Auger recombination. Among them, direct recombination means that the electrons in the conduction band will fall directly into the valence band and can be combined with holes, so the direct coincidence rate of carriers can be calculated. The equation is as follows:

$$U_d = k_d(pn - n_i^2) \tag{1}$$

Among them, $k_d$ represents the direct compound probability of electron and hole, the concentration of hole is represented by p, and the concentration of electron is represented by n.

Indirect recombination means that the electrons in the conduction band and the holes in the valence band can be compounded by other auxiliary means. In this process, the most important auxiliary body is the third level in the band gap. Therefore, the equation for calculating the indirect recombination rate is as follows:

$$U_1 = \frac{(pn - n_i^2)}{\tau_n(p + n_i) + \tau_p(n + n_i)} \tag{2}$$

In the equation, $\tau_n$ represents the lifetime of electron, $\tau_p$ represents the lifetime of hole, the concentration of hole is represented by p, and the concentration of electron is represented by n.

Auger recombination is the recombination of carriers in the process of transfer from high energy level to low energy level. At this time, the remaining energy will be transferred to other carriers. Carriers move to higher energy levels. When transferred to the lower energy level, it will release energy again. The calculation equation of the composite rate is as follows:

$$U_E = (\gamma_e n + \gamma_n p)(pn - n_i^2) \tag{3}$$

The consciousness of surface recombination is that the recombination central energy level of forbidden band will be produced in the process of recombination. In general, the speed of composite velocity is expressed by surface composite velocity. The equation is as follows:

$$S_p = \sigma + v_T N_{st} \tag{4}$$

In the equation, the total number of composite centers is represented by $N_{st}$.

### 2.2. *Preparation of photoconductive switching devices*

For the photoconductive switch device, its main structural parameters are electrode gap G, electrode length L, switch width and switch thickness, electrode fillet radius, etc. Among these parameters, the withstand voltage of the switch is determined by the electrode gap, but it should also meet the requirements of the circuit. In addition, the electrode gap should be combined with the spot size of the trigger pulse. In this experiment, in rder to prevent the local breakdown of the switch surface, it is necessary to convert the chamfering of the electrode at right angles, and the radius of the chamfering is usually less than 0.5mm. For the switch body material, it is necessary to strictly require its thickness, which must be greater than the depth of light absorption. Not only the structural parameters are determined, but also the materials used in the switch are strictly selected.

### *2.3. Preparation of scanning circuit module*

As a transmission system, scanning circuit is mainly formed by high frequency electromagnetic pulse. It is very fast for the scanning circuit, and the signal spectrum is relatively wide. Therefore, two principles should be followed in the preparation of circuit modules. The first principle is to spread all the high frequency signals in the microstrip line or coaxial line. PCB should be made into double-layer board. One side is in contact with the ground. The characteristic impedance of the microstrip line is designed to be 50 ohms. The second principle is to select the components with relatively small parasitic parameters. In this way, the bandwidth of the circuit can be improved. Moreover, the voltage withstand value should be high enough to meet the conditions. The voltage withstand value of the chip element used in this experiment is 3000V. During circuit scanning, two synchronous pulses are connected to the 50 ohm coaxial line through the SMA interface. GaAs has a certain sensitivity to light, so it is necessary to pay attention to protect it from light when preparing it. In addition, 705 transparent insulating adhesive can be coated on the surface in order to prevent the same surface photoconductive switch from discharging along the surface.

### *2.4. Definition and simplification of continuity equation*

There are two ways to study the motion of carriers. The first is to solve the continuity equation, and the second is Monte Carlo simulation. The semi quantitative research object of this study is mainly the generation, diffusion and compound law of downcomer in linear mode of photoconductive switch, the method of solving continuity equation is chosen.

In order to analyze the internal photoelectric effect of semiconductor, the continuity equation is simplified. Before that, it is necessary to simplify the physical model in four cases. The first case is that the external light can irradiate the surface of the semiconductor uniformly, and the carriers will only diffuse in the same direction. The second case is that this experiment only studies the generation, recombination and diffusion of carriers. In the third case, other possible ways of carrier generation are not considered, only focus on the carrier generated by the cause. The fourth situation is that this experiment will take into account the four composite approaches. If the light beam is injected along the z-axis, the carrier generation rate is expressed in g(z, t). The simplified continuity equation is as follows:

$$\frac{\partial p(z,t)}{\partial t} = D_p \frac{\partial^2 p}{\partial z^2} - k_d(pn - n_i^2) - \frac{(pn - n_i^2)}{\tau_n(p+n_i) + \tau_p(n+n_i)} + g(z,t) \tag{5}$$

$$\frac{\partial n(z,t)}{\partial t} = D_n \frac{\partial^2 p}{\partial z^2} - k_d(pn - n_i^2) - \frac{(pn - n_i^2)}{\tau_n(p+n_i) + \tau_p(n+n_i)} + g(z,t) \tag{6}$$

In the equation, p and n represent the concentration of electrons and holes. Dp and Dn represent the diffusion coefficients of holes and electrons. The equations of the boundary conditions are as follows;

$$P(0,t) - n_i = \frac{D_p}{s_p}\frac{\partial p}{\partial z}|z = 0, p(\infty,t) = n_i \tag{7}$$

$$n(0,t) - n_i = \frac{D_p}{s_n}\frac{\partial_n}{\partial z}|z = 0, p(\infty,t) = n_i \tag{8}$$

IOP Conf. Series: Materials Science and Engineering **772** (2020) 012060 doi:10.1088/1757-899X/772/1/012060

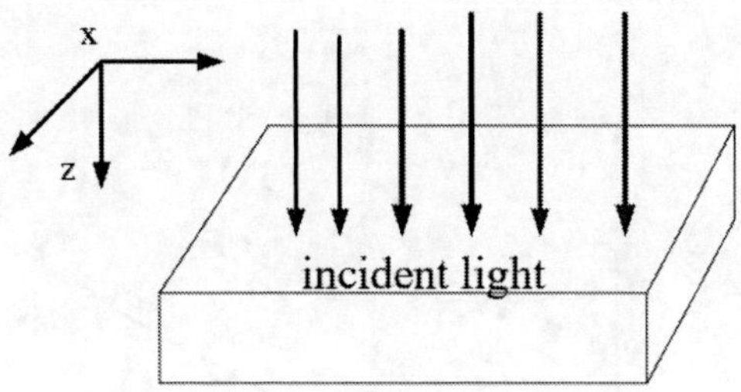


**Figure 2.** Simplified physical model

## 3. Results and discussion

In this study, the prepared circuit modules are tested for dark current. Using this method, not only the resistance of the circuit switch when it is in the eigenstate can be known, but also the dark state withstand voltage ability of the circuit switch can be clearly realized. The schematic diagram of dark current test is shown in figure 3. The main test object is the dark current of several switches. During the test, the + HV and - HV are adjusted synchronously. Among them, the main function of R1 and R2 is to limit current, and its resistance value is 470K Ω. When the switch is in the intrinsic state, its resistance value will far exceed R1 and R2, so the voltage drop on R1 and R2 will not be included in the test. The voltage difference between + HV and - HV is taken as the voltage at both ends of the switch, and then the acute current of microammeter is recorded. After testing, it can be concluded that the dark state withstand voltage of each switch can exceed 6000V. In addition, it is found that the current value of the switch under different light is different. When in the natural light, the current value is large. When it is in the state of shading, its current value is relatively small. The final test results are shown in figure 4 and figure 5. It can be seen from the figure that the current degrees of the three switches are very small when they are in the blackout state, especially No. 1 and No. 3. The current of switch 3 is kept at about 1 μ A. For switch 2, its dark state resistance is very large, indicating that the current is also very small. When the switch is in the non shading state, it can be seen that the current of the three switches will increase. Among them, the current of switch 3 is the largest. In addition, when the bias voltage exceeds 6000V, the other two switches will discharge along the surface.

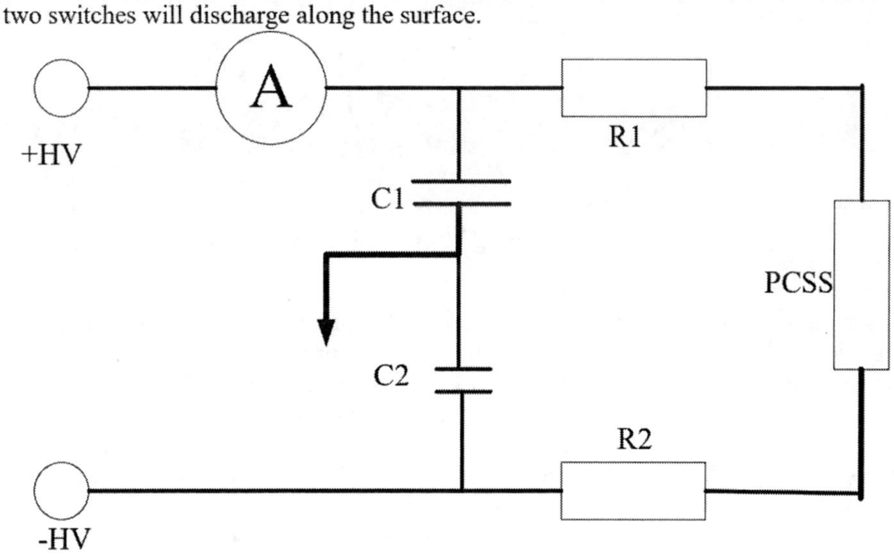


**Figure 3.** Dark current test schematic diagram

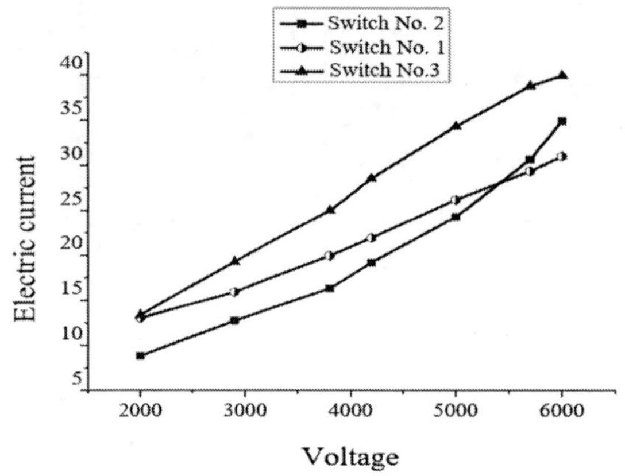


**Figure 4.** Dark current test results of photoconductive switch in non shading state

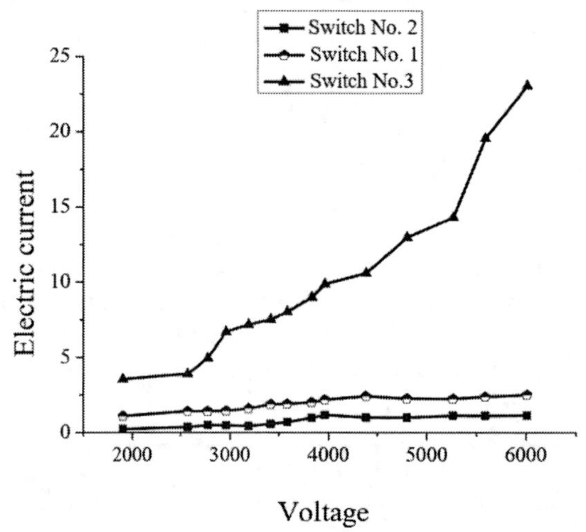


**Figure 5.** Dark current test results of photoconductive switch under shading state

With the development of science and technology, the appearance of femtosecond streak camera has brought a lot of benefits to people. As the fastest time resolution diagnostic equipment, its resolution can even reach femtosecond level. Among them, the femtosecond streak tube has a decisive influence on the time resolution of the femtosecond streak camera, but the femtosecond streak tube has a high demand and standard for the slope of the scanning voltage and the triggering shake. Conventional scanning circuit has many disadvantages. In recent years, many experts and scholars have carried out in-depth research based on this, but there is still no significant breakthrough. Therefore, it is very important and necessary to study the scanning voltage in femtosecond streak tube.

In this study, photoconductive semiconductor materials are used in the preparation of scanning circuit to optimize the scanning voltage. First of all, the principles and approaches of the generation and recombination of carriers in semiconductors are analyzed. After that, the circuit is optimized by applying the numerical simulation and simplification of linear mode. Finally, the effectiveness of photoconductive semiconductor materials is verified by dark current test. After testing, it is found that the dark state withstand voltage of each switch can exceed 6000V. The current value of the switch under different light is different. When the current value is larger in the natural light than in the dark state, the current value is smaller, and the efficiency of the three switches is more than 50%. The efficiency of nonlinear model is obviously higher than that of linear model, which can be more than 90%. It can be seen that the application of photoconductive semiconductor materials can further optimize the performance of scanning circuit. No matter in any mode, the efficiency of voltage conversion and the efficiency of circuit switch have been improved. It is proved that photoconductive semiconductor material is of great value to optimize the resolution of full femtosecond streak camera.

## 4. Conclusion

In this experiment, through the use of photoconductive semiconductor materials for the preparation of scanning circuit, it is found that photoconductive semiconductor materials can avoid the shortcomings of traditional scanning circuit, optimize and improve its performance. This is mainly reflected in the application of photoconductive semiconductor materials. The voltage of the scanning circuit and the efficiency of the switch have been effectively improved, which can further improve the time resolution of the femtosecond streak camera by optimizing the femtosecond streak tube. However, all the tests in this study belong to single trigger mode, but femtosecond streak camera requires the repetition frequency of scanning circuit to be 1kHz ~ 5KHz. Therefore, in the future research, it is necessary to pay attention to this point. This study is still in the experimental exploration stage, without engineering design, which will also be the work to be carried out in the follow-up study.

## References

[1] Shaoqiang Wang, Wei Shi. The Role of the Photo-generated Carrier in Surface Flashover of the GaAs Photoconductive Semiconductor Switch [J]. IEEE Journal of the Electron Devices Society, 2017, (99). pp. 1-1.
[2] Zhang Yongping, Chen Zhizhan, Shi Wangzhou, et al. Effect of laser excitation energy on resistance of lateral geometry 4H-SiC photoconductive semiconductor switches [J]. High Power Laser & Particle Beams, 2015, 27(5), pp. 233-237.
[3] Wei Shi, Yu Ji, Ming Xu, et al. Comparison on the synchronization of 2 parallel GaAs photoconductive semiconductor switches excited by laser diodes [J]. 2017, (99), pp. 1-1.
[4] María Vicente Prieto, Ana Luisa Caseiro Gomes, Javier Montero Martín, et al. The Effect of Femtosecond Laser Treatment on the Effectiveness of Resin-Zirconia Adhesive: An In Vitro Study [J]. Journal of Lasers in Medical Sciences, 2016, 7(4), pp. 214-219.
[5] K. Usha, P. Kumbhakar, B. Mondal. Effect of Ag-doped TiO2 thin film passive layers on the performance of photo-anodes for dye-sensitized solar cells [J]. Materials Science in Semiconductor Processing, 2016, 43, pp. 17-24.
[6] Ehsan Rahmati, Mehdi Ahmadi-Boroujeni. Improving the efficiency and directivity of THz photoconductive antennas by using a defective photonic crystal substrate [J]. Optics Communications, 2018, 412, pp. 74-79.
[7] Tianshi Wang, Zhigang Gui, Anderson Janotti, et al. Strong effect of electron-phonon interaction on the lattice thermal conductivity in 3C-SiC [J]. Physical Review Materials, 2017, 1(3), pp. 034601.
[8] J Wei Wang, Liansheng Xia, Yi Chen, et al. Research on synchronization of 15 parallel high gain photoconductive semiconductor switches triggered by high power pulse laser diodes [J]. Applied Physics Letters, 2015, 106(2), pp. 022108.
[9] AbdulJabbar, N, Forrest, T, Gronsky, R, et al. Effect of stoichiometric vacancies on the structure and properties of, the Ga2SeTe2 compound semiconductor [J]. 2018, 108(52), pp. 21276-21281.
[10] Ferdinand Grüneis. An Intermittent Generation–Recombination Process as a Possible Origin of 1/f Fluctuations in Semiconductor Materials [J]. Fluctuation and Noise Letters, 2017, 16(43), pp. 1750034.

# Robustness interval of apparent air bubbles of fair-faced concrete

**Jian Zhang**[1, 2, 3, a], **Chunxiang Qian**[1, 2, 3, *], **Yanqiang Chen**[1, 2, 3, b]

[1]School of Materials Science and Engineering, Southeast University, Nanjing, China
[2]Key Laboratory of Materials Microbiology, Southeast University, Nanjing, China
[3]Jiangsu Key Laboratory for Construction Materials, Southeast University, Nanjing, China

*Corresponding author e-mail: cxqian@seu.edu.cn, [a]2364963216@qq.com, [b]seucyq@163.com

**Abstract**. In this paper, a method of controlling the apparent quality of fair-faced concrete is adopted. By adjusting the rheological parameters of fair-faced concrete slurry, the robustness interval of repeatedly stable preparation of fair-faced concrete with excellent apparent quality is obtained. It has a good economy and has a good guiding significance for practical engineering. The results show that the plastic viscosity coefficient of fair-faced concrete slurry is 5.67~9.69 Pa·s, and the yield stress is between 112.25~195.68 Pa. The fair-faced concrete with good apparent quality can be obtained.

## 1. Introduction

Fair-faced concrete is a special type of concrete which is formed in one step and is decorated directly with the original concrete surface or the concrete surface treated with transparent protective agent [1, 2]. Its smooth surface, no obvious color difference, Sharp Angle, no damage and other quality issues, through its own texture to achieve the clear water concrete "inside and outside the actual beauty" of the technical requirements.

Zerbino [3] found that a reasonable viscosity range can not only improve the rheological properties of self-compacting concrete, but also improve the apparent quality of fair-faced concrete. Technical Specification for application of self-compacting concrete [4] mentioned that although self-compacting fair-faced concrete has the advantages of both self-compacting and fair-faced water in theory, self-compacting concrete has great fluidity and the amount of cementitious material used in slurry is large In order to prevent the segregation bleeding, the slurry must have a large viscosity, which makes it difficult to remove the air bubbles on the surface of concrete and affects the apparent quality.

Fair-faced concrete has gradually become a trend of development. It has many advantages [5, 6] , such as being able to avoid environmental pollution caused by building decoration materials, and having simple, plain and true architectural performance effects, etc. However, the surface quality of fair-faced concrete is still a practical problem which is difficult to solve in engineering [7, 8] . For the problem of air hole, there are many methods to deal with it at present, but most of them are based on experience and have great randomicity. Based on the plastic viscosity Coefficient and yield stress of the paste, the control method of the surface blowhole of fair-faced concrete is put forward in this paper.

Content from this work may be used under the terms of the Creative Commons Attribution 3.0 licence. Any further distribution of this work must maintain attribution to the author(s) and the title of the work, journal citation and DOI.
Published under licence by IOP Publishing Ltd

IOP Conf. Series: Materials Science and Engineering **772** (2020) 012061 doi:10.1088/1757-899X/772/1/012061

## 2. Raw materials and test methods

### *2.1. Raw materials*
PO 42.5 ordinary portland cement produced by Nanjing Conch Cement (CA), with a density of 3.16 g/cm$^3$; PO 42.5 ordinary portland cement produced by Nanjing Jiangnan Xiaoyetian cement (CB) , with a density of 3.13 g/cm$^3$; grade I fly ash produced by Changshu Power Plant; grade S95 mineral powder produced by Zhangjiagang Lianfeng iron and steel; medium coarse sand from Dongting Lake in Hubei Province with a sieve of 4.75mm; 5-25mm stone produced by Kangcheng is used; polycarboxylate lipid superplasticizer ($PS_A$) provided by Sika additive company is used, with water reduction rate of 30% and polycarboxylate ether superplasticizer ($PS_B$), with water reduction rate of 30%; drinking tap water is used for mixing water.

### *2.2. Specimen preparation*
The strength grades of fair-faced concrete commonly used in engineering are C40, C45, C50, C55 and C60 respectively, and the cross preparation tests of fair-faced concrete with the above two kinds of cement and two kinds of admixtures are carried out Specific test combinations are shown in Table 1.

**Table 1.** Mix proportion of fair-faced concrete of five strength grades

| Strength grade | W/B | Water/ kg·m$^{-3}$ | Cement/ kg·m$^{-3}$ | FlyAsh/ kg·m$^{-3}$ | Mineral powder/ kg·m$^{-3}$ | Sand/ kg·m$^{-3}$ | Stone/ kg·m$^{-3}$ | Water-reducing admixture/% |
|---|---|---|---|---|---|---|---|---|
| C40 | 0.33 | 154 | 341 | 47 | 80 | 712 | 1068 | 1.2 |
| C45 | 0.31 | 152 | 358 | 49 | 83 | 668 | 1090 | 1.2 |
| C50 | 0.29 | 152 | 398 | 42 | 84 | 639 | 1135 | 1.2 |
| C55 | 0.27 | 150 | 406 | 55.6 | 94.5 | 662.7 | 1081.3 | 1.2 |
| C60 | 0.25 | 148 | 432.2 | 59.2 | 100.6 | 649.8 | 1060.2 | 1.2 |

For the convenience and requirement of the experiment, the 50cm×10cm×50cm wooden mould designed by myself is used, as shown in figure 1.

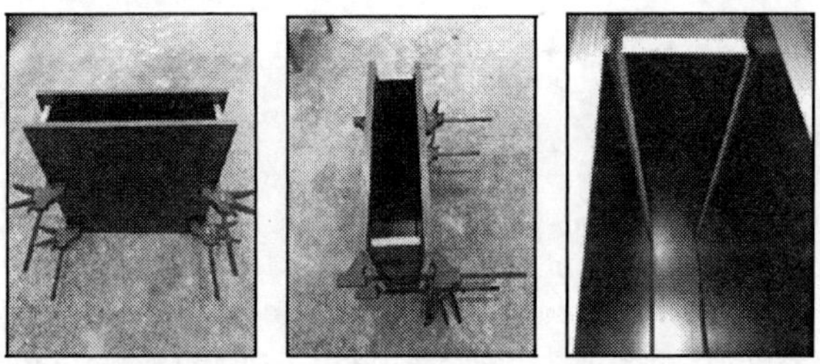

**Figure 1.** Fair-faced concrete mould

### *2.3. Determination and adjustment of yield stress and plastic viscosity coefficient*

*2.3.1. Determination of yield stress and plastic viscosity coefficient.* The Brookfield RST Rotational viscometer is used to determine the yield stress and the plastic viscosity coefficient of fair-faced concrete pastes. Due to the rotation of the rotors in the pastes, the coarse aggregates in the concrete pastes will touch the rotors Rotor damage and affect the accuracy of the data measured, so the paste in the coarse aggregate after leaching to determine.

The program of the Rheologic Property Tester of concrete can be designed into four stages: 1 slow running stage: rotational speed 0.1 RPM, duration 60s; 2 acceleration stage: Duration 100s, rotational speed is divided into 10 gradient, each gradient increase speed 10RPM, duration 10s; 3 constant speed

IOP Conf. Series: Materials Science and Engineering **772** (2020) 012061 doi:10.1088/1757-899X/772/1/012061

stage: Duration 60s, speed 100RPM; 4 Deceleration stage: Duration 100s, speed divided into 10 gradient, each gradient deceleration 10RPM, duration 10s;

Bingham model was used to fit the measured data. Bingham rheological equation was as follows:

$$\tau = \tau_0 + \eta_p \cdot \gamma \quad (1)$$

In this equation, $\tau$ is the the shear stress of Bingham Fluid, Pa; $\tau_0$ is the yield stress of Bingham Fluid, Pa; $\eta_p$ is the plastic viscosity of Bingham Fluid, Pa·s; and $\gamma$ is the shear rate of Bingham fluid, $s^{-1}$.

The intercept and slope of the equation are the yield stress and the plastic viscosity coefficient of the concrete paste.

*2.3.2. Adjustment of yield stress and plastic viscosity coefficient.* The yield stress and plastic viscosity coefficient of fair-faced concrete paste can be adjusted by the dosage of water reducing agent, which is 0.3%, 0.6%, 0.9%, 1.2% and 1.5% respectively.

### *2.4. Evaluation of apparent performance*

Three parameters, such as surface porosity, pore size distribution and maximum pore size, can be used to describe the surface porosity of fair-faced concrete in a more comprehensive way.

**Table 2.** Evaluation criteria for surface porosity of fair-faced concrete

| Order of evaluation | Excellence (1-2) | Good (2-4) | Disqualification (4-6) |
|---|---|---|---|
| Evaluation criterion | The maximum diameter is less than 4mm, and the surface stomatal area ratio is less than 0.2% | The maximum diameter is greater than 4 mm and less than 8 mm and the number is less than $5/m^2$. The surface stomatal area ratio is less than 0.2% | The maximum diameter is more than 8 mm or the surface stomatal area ratio is more than 0.2% |

## 3. Results and Discussion

### *3.1. Yield stress and plastic viscosity coefficient*

Based on the same initial air content, the influence of yield stress and plastic viscosity coefficient on the apparent quality robustness of fair-faced concrete is tested. Figure 3 shows the relationship between the plastic viscosity coefficient, yield stress and apparent mass of fair-faced concrete of different strength grades, with the increase of plastic viscosity coefficient The fluctuation of yield stress increases correspondingly, and the apparent quality of fair-faced concrete becomes worse and worse correspondingly. The points of plastic viscosity coefficient and yield stress with good apparent quality are taken out of the four charts, and the overlapping intervals of these points. The plastic viscosity Coefficient and yield stress robust intervals with excellent apparent quality, are taken out The plastic viscosity coefficient of these points overlaps between 5.67 Pa·s and 9.69 Pa·s, and the yield stress overlaps between 112.25 Pa and 195.68 Pa.

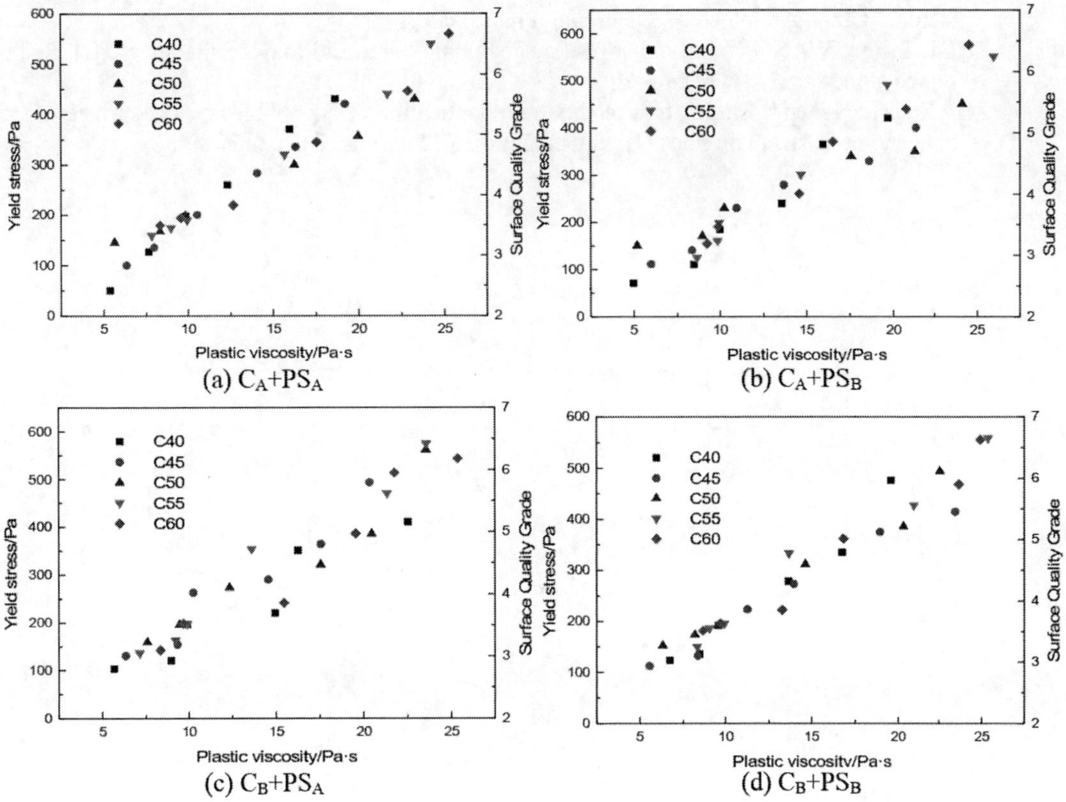


**Figure 2.** Effect of plastic viscosity and yield stress on apparent quality

## 4. Conclusion

Based on the same initial air content range, when the plastic viscosity coefficient is between 5.67 Pa·s and 9.69 Pa·s, and the yield stress is between 112.25 Pa and 195.68 Pa, the apparent quality of as-cast-finish concrete is excellent.

## Acknowledgments

The authors appreciate the financial support from the National Natural Science Foundation of China (Grant No.51572047), the National Natural Science Foundation of China (Grant No. 51738003).

## References

[1] JGJ169. Technical specification for application of fair-faced concrete [S]. Beijing: Ready-mixed Concrete, 2009, (5): 6-7.

[2] LI Qiang, LI Xin-min, MENG Wen-yuan, et al. Development status of finish concrete technology in our country and its existing problems and countermeasures [J]. Architecture Technology, 2007, 38 (1): 2-3.

[3] Zerbino, R., Barragán, B., Garcia, T., et al. (2009) Workability Tests and Rheological Parameters in Self-Compacting Concrete. Materials and Structures, 42, 947-960.

[4] Ministry of housing and urban rural development of the people's Republic of China. JGJ / T 283-2012 self-compacting concrete application technical specification [S]. Beijing: China Construction Industry Press, 2012: 75.

[5] GONG Shi-jie. Defect Prevention for Fair- Faced Concrete and Prospective Research on Construction Techniques and Materials [J]. Building Construction, 2004, 26 (4): 333-335.

[6] GU Yong-xin. Construction technology and technology of fair-faced concrete project [M].

Beijing: China Architecture & Building Press, 2006:1-12.
[7] ZHOU Li-jun, WANG Xiao-bing. Appearance bubble control of fair-faced concrete [J]. Ready-mixed Concrete, 2016, (4): 68-70.
[8] ZHANG Zhao-sheng. Elimination method and application of air bubbles on the surface of C50 concrete precast square piles [J]. Fujian Building Materials, 2015, (12):80-81.

# Analysis and Optimization Design of Cantilever Structure of Shear Wall Corner

**Di Hu[1,*], Xuezhong Peng[2] and Tinfeng Shi[2]**

[1]School School of Civil Engineering, Central South University, Changsha, China
[2]Guangzhou Port Group Co., Ltd., Guangzhou, China

*Corresponding author e-mail: hudibridge@csu.edu.com

**Abstract.** In the frame-shear wall high-rise building, it is necessary to design a special cantilever structure to connect shear wall with two mutually perpendicular frame beams when they do not intersect. In the paper an innovative simplified approach to design special cantilever structure with steel reinforced concrete is presented, which includes two steps as design of feasible steel structure and design of reinforced concrete based on structural analysis and optimization. The demonstrated design idea and design method provide valuable reference for innovative design of complex cantilever structure of shear wall corner with frame beams.

## 1. Introduction

In the frame-shear wall high-rise building, it is common that the frame beams are connected with the shear wall directly so that the load of frame beams is safely transmitted to the shear wall. However, for some special structures, or due to optimizing the layout of the floor space, it is encountered that the two mutually perpendicular frame beams do not intersect with shear wall at corner area. In this case, it is necessary to design a special cantilever structure to connect shear wall corner to frame beams, which bears the moments and vertical shear forces from the frame beams at two perpendicular directions, resulting in complicated forces with vertical shear force, bending moment and torque. There are no calculation methods and related design regulations for this kind of cantilever structure in the current specifications.

In view of the collapse of reinforced concrete cantilevers occurred in recent years, Yang [1] analyzed the cause of the failure and proposed design countermeasures. Duan [2] probed into the mechanism of the collapses of some cantilever structures from the perspective of design and construction, and proposed preventive measures. Gu [3] completed the structural analysis of reinforced concrete (RC) cantilever beam with span of 7.5m and 8.7m in high intensity area and discussed the structural design. Li [4] made a conclusion that steel reinforced concrete (SRC) structure is the best choice in design of large cantilever after comparative analysis.

According to the position relationship between the shear wall and two mutually perpendicular frame beams of a high-rise building structure, the preliminary design and structural analysis of a special cantilever structure are carried out, then the optimized design is obtained and a feasible design method is proposed.

Content from this work may be used under the terms of the Creative Commons Attribution 3.0 licence. Any further distribution of this work must maintain attribution to the author(s) and the title of the work, journal citation and DOI.
Published under licence by IOP Publishing Ltd

IOP Conf. Series: Materials Science and Engineering **772** (2020) 012062 doi:10.1088/1757-899X/772/1/012062

## 2. Preliminary Design of Cantilever structure

The South China International Port Service Center is a frame-shear wall structure with 52 floors above ground (H=250m). The floor is made of large RC slab supported by beams with depth 600mm. At the area of elevator shear wall, two mutually perpendicular frame beams do not intersect with shear wall, as shown in Figure 1.The height of cantilever structure should be the same as that of the frame beam, that is h=600mm. Estimation indicates that the reinforcement cannot meet the bearing capacity requirement if the cantilever structure is a RC structure, while it is difficult to connect with RC frame beam if steel structure is adopted. Therefore, the steel reinforced concrete (SRC) is used.

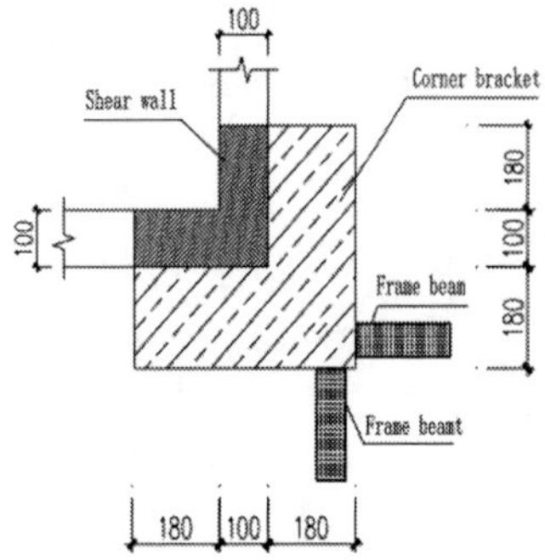


**Figure 1.** Shear wall and beams/mm

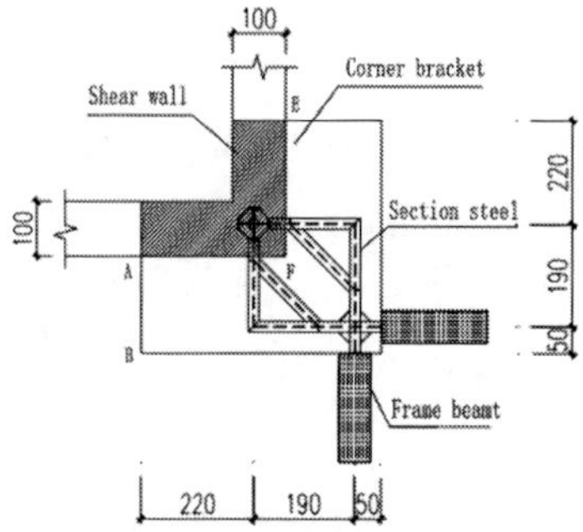


**Figure 2.** Scheme 1/ mm

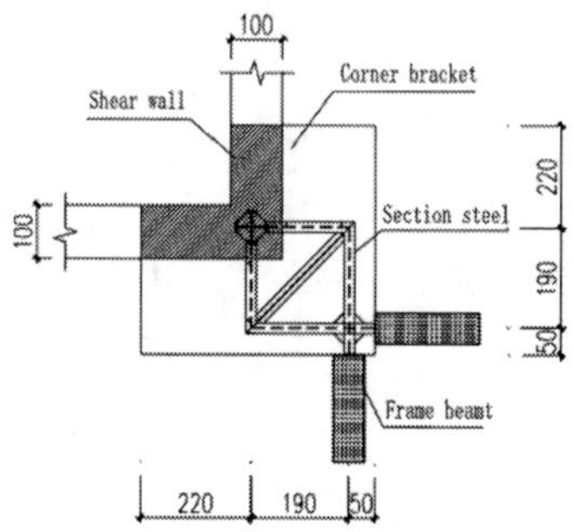


**Figure 3.** Scheme 1/ mm

In order to transfer the loads of the frame beam safely, reliably and quickly to the shear wall, the steel structure should be connected the frame beam end and the shear wall, as shown in Fig.2 and Fig.3, which two alternative designs are feasible. The section of the section steel is H400×200×18×24 (mm) and the steel material adopts Q345.

## 3. Calculation of Internal Force and Design of Reinforcement

Under the action of the loads from frame beams, the steel structure in Fig.4 is symmetric, so it can be detatced into half to structural analysis. Set the boundary point of member AB to the shear wall to be origin to establish the coordinate system along the AB direction and the vertical direction. Boundary section O is fixed by shear wall, so the semi-steel structure is a statically indeterminate system. The internal force of each member, $M_k$ (free torsional torque), $M_\omega$ (restrained torsional torque), $B$ (bending-tortional moment), $M_y$ (bending moment) and $Q_z$ (shear force), can be calculated by force method or displacement method underthe action of vertical force at ends of frame beams.

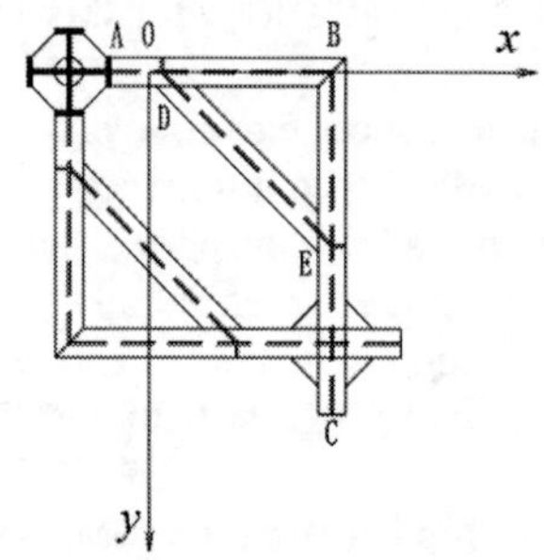


**Figure 4.** Steel structure

For a restrained torsion member, the H section has sectorial shear stress $\tau_\omega$, pure torsional shear stress $\tau_k$ and sectorial principal stress $\sigma_\omega$. When a vertical force $N$ is applied at point C, the stresses of section are given by

$$\sigma=\sigma_{M_y}+\sigma_{\omega}=\frac{M_y Z}{I_y}+\frac{B\omega}{I_\omega} \tag{1}$$

$$\tau=\tau_{Qz}+\tau_k+\tau_\omega=-\frac{Q_z S^0}{I_y\delta}+\frac{M_k\delta}{I_k}-\frac{M_\omega S_\omega^0}{I_\omega\delta} \tag{2}$$

Where $I_k$ is free torsional moment of inertia, $I_\omega$ is restrained torsional moment of inertia, $I_y$ is bending moment, $S_\omega^0$ is sectorial area moment, $S^0$ is area moment, $\delta$ is thickness of web.

According to the fourth strength theory [5], the strength yield criterion (strength condition) of section steel can be obtained by

$$\sigma_{eq}=\sqrt{\sigma^2+3\tau^2}\le[\sigma] \tag{3}$$

Where $\sigma_{eq}$ is an equivalent stress.

A vertical force $N_g$ applied at point P and Q (end of frame beam) corresponds an equivalent stress $\sigma_{eq}$, the better scheme means that the value of $\sigma_{eq}$ under the same $N_g$ is smaller. Comparison of structural analysis shows that scheme 1 in Fig. 2 is better than scheme 2 in Fig.3, so scheme 1 of steel strucrure is adopted for further design of reinforcement.

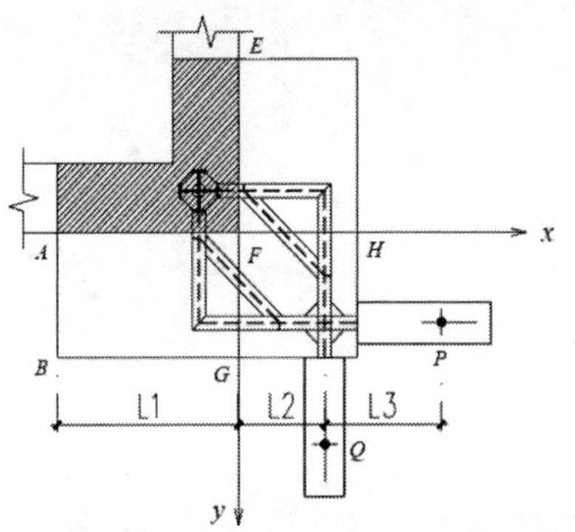


**Figure 5.** Computing internal force in AF

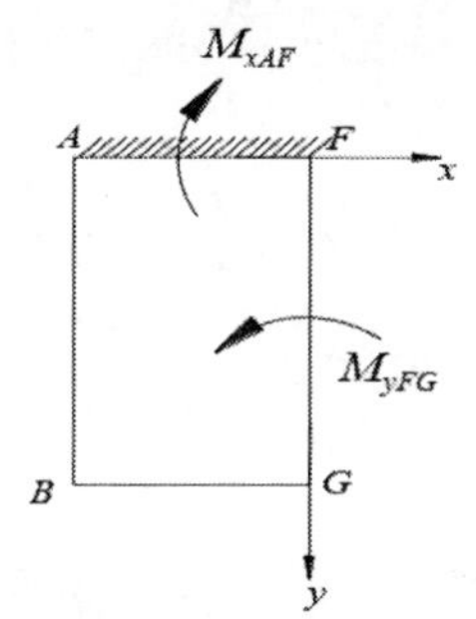


**Figure 6.** Detached ABGF concrete slab

Take out ABGF from cantilever structure in Fig. 5, as shown in Fig.6. At ultimate state of bearing capacity, it is assumed that the steel strucrure achieves it's bearing capacity, so the loads that reinforced concrete bears in cantilever structure are those deducting $N_g$ from loads of frame beames.

The internal force in section AF can be obtained from Fig.5, as

$$N_{AF}=N_0-N_g \tag{4}$$

$$M_{xAF}=M_{yFG}=\frac{L_1}{L_1+L_2}\left[(N_0-N_g)(L_2+L_3)+(N_0-N_g)L_2+M_0\right] \tag{5}$$

Due to complex internal force in section AF, steel bars in upper and lower edges are arrranged (Fig.7). When the cantilever structure fails, the compressive steel bars may yield and not, in this case, the stress of compressive steel bars can be obtained by plane hypothesis. According to conditions for compatibility of cross-section deformation and internal force balance, set $A_s=A_s'$, we have

$$\sigma_s^{'} = E_s\left(1-\frac{\beta_1 a_s^{'}}{x}\right)\varepsilon_{cu} \tag{6}$$

$$\alpha_1 f_c bx + \sigma_s^{'} A_s^{'} = f_y A_s \tag{7}$$

$$M_{xAF} = \alpha_1 f_c bx\left(h_0 - \frac{x}{2}\right) + \sigma_s^{'} A_s^{'}\left(h_0 - a_s^{'}\right) \tag{8}$$

Where $x$ is the height of the concrete slab compression zone, $\varepsilon_{cu}$ is the ultimate compressive strain of concrete, $a_s^{'}$ is cover depth of compressive steel bars, $f_y$ is strength of steel, $\beta_1$ is a coefficient, when the concrete strength grade does not exceed C50, it is taken as 0.8.

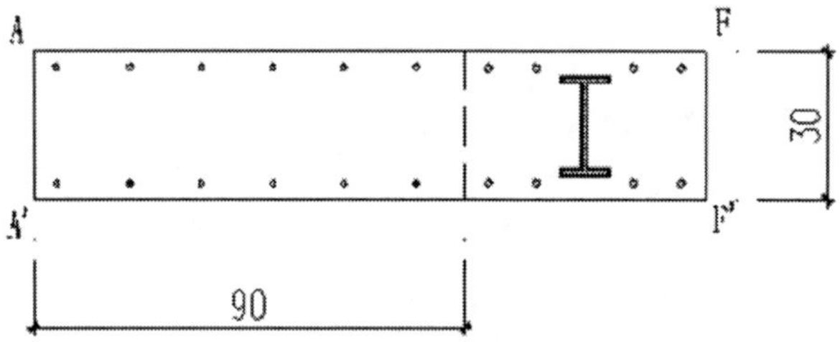


**Figure 7.** Schematic diagram of section AF/cm

Equations 6-8 result in following formula:

$$Ax^3 + Bx^2 + Cx + D = 0 \tag{9}$$

Where $A = -\frac{\alpha_1 f_c b\left(f_y - \sigma_{s0}\right)}{2}$, $B = \left(f_y - \sigma_{s0}\right)\alpha_1 f_c bh_0 + \alpha_1 f_c b\sigma_{s0}\left(h_0 - a_s^{'}\right) - \frac{\alpha_1 f_c b\sigma_{s0}\beta_1 a_s^{'}}{2}$,

$C = \alpha_1 f_c bh_0\sigma_{s0}\beta_1 a_s^{'} - \left(f_y - \sigma_{s0}\right)M_{xAF} - \alpha_1 f_c b\sigma_{s0}\left(h_0 - a_s^{'}\right)\beta_1 a_s^{'}$, $D = -\sigma_{s0}\beta_1 a_s^{'} M_{xAF}$, $\sigma_{s0} = E_s\varepsilon_{cu}$.

After the height of the concrete compression zone is obtained from Equation 9, the required area $A_s$ of steel bars in section AF can be determined by

$$A_s = \frac{\alpha_1 f_c bx}{f_y - E_s\varepsilon_{cu}\left(1-\frac{\beta_1 a_s^{'}}{x}\right)} \tag{10}$$

## 4. Examples and Analysis

### *4.1. Comparison of Steel Structure Scheme*

In order to verify the correctness of the design method of the cantilever structure proposed in this paper, the finite element analysis, check of bearing capacity and model test are carried out for the designed scheme. In consideration of the fact that the full-scale test model (plane dimension 460cm × 460cm) is not easy to hoist and transport, a 1:2 scale model is adopted (model test is discussed in another paper). Thus, the 230cm × 230cm coner corbel with section steel H200×100×9×12 (Q345) under the action of vertical force $N_0 = 320kN$ and bending moment $M_0 = 162.5kN \cdot m$ at each end of frame beam (point P and Q in Fig.5) is used to optimal design and design of steel bars.

ANSYS is used to structural analysis for steel structure, elememt solid95 is adopted for steel. When the equivalent stress of steel at a certain position reaches $f_y = 345MPa$ approximately, $N_g$ corresponds the bearing capacity of the steel structure. Calculations show that $N_g = 85kN$ in scheme 1 and $N_g = 25kN$ in scheme 2, respectively, and the equivalent stress contours are shown in Fig. 9. The comparison of results by analytical method and by ANSYS is listed Table 1.

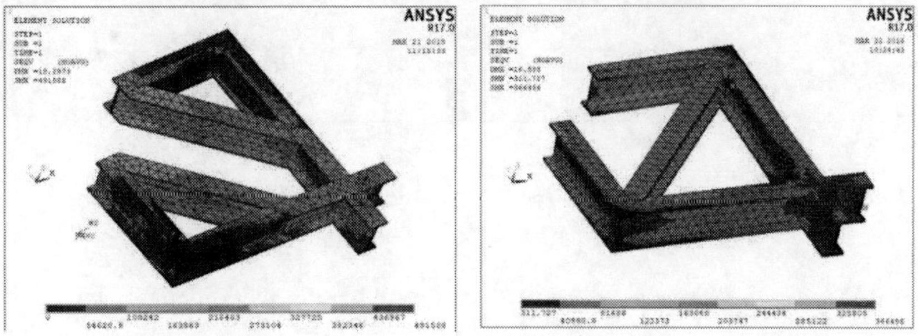

**Figure 8.** Equivalent stress contours (ANSYS) in streel structures

**Table 1.** Comparison of theoretical values of steel structure

| Scheme | Computing method | Ng (kN) | Equivalent stress (MPa) |
|---|---|---|---|
| I | Analytical method | 85.0 | 348.2 |
| | ANSYS | 85.0 | 346.6 |
| II | Analytical method | 25.0 | 348.5 |
| | ANSYS | 25.0 | 347.9 |

Above analysis shows that the steel structure of scheme 1 has a larger bearing capacity, which the arrangement of section steel demonstrates the clear path and high efficiency of transmitting frame beam loads to shear wall.

*4.2. Design of Reinforcement and Check of Bearing Capacity*

The height of concrete corbel is 30cm, the concrete adopts C35 and the steel bar adopts Grade II. $f_c$=16.7$MPa$ , $f_y = f_y'$=360$MPa$ , $\varepsilon_{cu}$=0.0033 , $E_s$=2×10$^5$$MPa$ , $\alpha_1$=1.0 $\beta_1$=0.8 .Deducting the vertical force the steel structure bears, the vertical load exerted on cantilever structure by each frame beam is $N_b = N_0 - N_g = N_0 - 85kN$ .

According to Equation 5, $M_{xAF} = 432.8kN \cdot m$ is obtained, and the area of steel bars in section AF is $A_s = 6110.8\ mm^2$ (tension zone) is obtained based on Equation 10. The steel bars are arranged as 10Φ28@150mm ( $A_s = 6158\ mm^2$ ) in tension zone and compression zone (Fig.9), respectively. The stirrup is Φ10@100mm. The arrangement of steel bars in section FH is the same as in AF.

Calculation shows that the bending and shearing control the design of cantilever structure. According to Technical Specification of Steel-Reinforced Concrete Structures [6], the bearing capacity of the fixed end section AF provided by reinforced concrete structures is

$$M_b^{rc} = M_b^{st} + A_s f_y \left( h_0 - \frac{f_y A_s - f_y' A_s'}{2\alpha_1 f_c b} \right) \quad (11)$$

$$V_{cs} = \alpha_{cv} f_t b h_0 + f_{yv} \frac{A_{sv}}{s} h_0 \quad (12)$$

Substituting parameters into above Equations, we have

$M_b^{rc} = 490.2kN \cdot m > M_{xAF} = 432.8kN \cdot m$, $V_u = 631.1kN > V_0 - N_g = 320kN - 85kN = 245kN$

Thus, the designed scheme of steel reinforced concrete cantilever structure meets the specification.

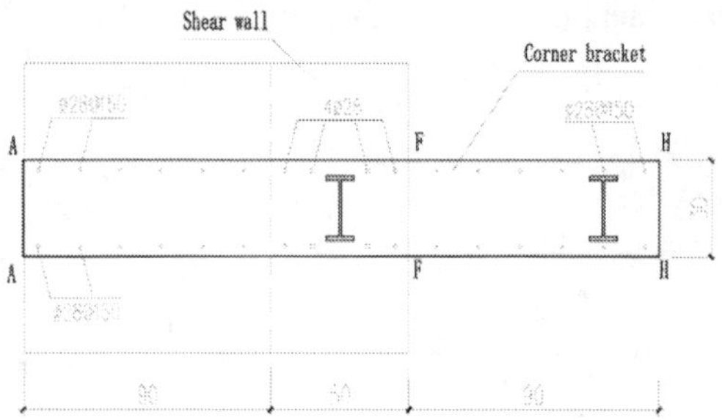


**Figure 9.** Arrangement of steel bars in section/cm

## 5. Conclusion

It is common occurred that two mutually perpendicular frame beams do not intersect with shear wall in frame-shear wall high-rise buildings. In this paper, an innovative two-step approach to design cantilever structures with steel reinforced concrete is discussed, a feasible and reasonable cantilever structure for a real high-rise building is designed along with structural analysis and optimization. The presented approach of designing safe and economic unconventional cantilever structures provides a meaningful reference.

## References

[1] W. Yang, The destruction causes of cantilever structure ofreinforced concrete and its counter - measure, Coal Eng., 1 (2005):20-21.
[2] J. Duan, A.Q. Hou, S.F. Qi,J.S. Qiao, The causes of the overhanging structure collaping and its countermeasure,J. of Jiaozuo Inst. of Tech., 04 (1999): 278-280.
[3] W.J. Gu, L.Feng, J.Wang, A.Zhang, S.L.Zhou, Structural design of large-span reinforced concrete cantilever beam in high intensity areas. Building Structure, S1 (2012):154-156.
[4] W.G. Li. Structure design and analysis of a large cantilever construction, 09(2011):7-10.
[5] S.J. Foster, R.E. Powell, H.S. Selim. Performance of high-strength concrete brackets,ACI Struct. J., 93(1996):555-563.
[6] Technical Specification of Steel-Reinforced Concrete Structures (YB 9082-2006), Metallurgical industry press, 2006
[7] S.H.Zhang, B.L.Yuan. Steel ration and strength calculation of reinforced concrete brackets of small shear span ratio under vertical static load,J.of Qingdao Inst. of Arch. and Eng., 03 (1990):1-11.

# Deflection Calculation of Steel-Concrete Composite Beams Considering Effects of Shear Lag and Slip

**Di Hu[1], Yafei Wang[2,] * and Faxing Ding[1]**

[1]School School of Civil Engineering, Central South University, Changsha, China
[2]College of Civil Engineering, Hunan University, Changsha, China

*Corresponding author e-mail: 854634352@qq.com

**Abstract.** Based on the proposed longitudinal deformation equation considering the slip between steel and concrete and the shear lag, the energy variation function is deduced, then the formula to calculate the deflection of steel-concrete composite beams considering the effects of shear lag and slip under the action of uniformly distributed load is derived. A group of experimental steel-concrete composite beams with different space of stud is made and the experimental data are compared with theoretical values showing that the formula holds high calculation accuracy.

## 1. Introduction

Steel-concrete composite material is widely used in large-span structures and high-rise buildings for their good working performance and high economy. A steel-concrete composite beam is composed of concrete slab, steel beam and shear connectors. The shear lag of flange slab, shear deformation of shear connectors and relative slip between steel and concrete affect the deflection of the composite beams [2-6]. Up-to-present, two group approaches to calculate the deflection as analytical methods [2-9] and numerical methods [10-12] are broadly used, including conversion section method [2], modified conversion section method [3], interpolation method based on conversion section method [4], reduction stiffness method [5-7], energy principle method and so on.

In order to simplify the analytical calculation of composite beams, some studies consider inter laminar slip but omit the effect of shear lag [8, 9], while some studies omit inter laminar slip but consider the effect of shear lag [10]. The experimental and theoretical studies show that the interlayer slip obviously reduces the overall stiffness of composite beams resulting in increasing the deflection of the beams [6, 8-9]. Therefore, the comprehensive effects of shear lag and slip should be considered in the calculation of deflection of steel-concrete composite beams.

Studies show that the spacing of studs, the shear strength of single studs and the cross-section characteristics affect the slip [5]. However, the existing analytical formulae fail to consider the influence of the above parameters comprehensively. In this paper, the energy variation function of steel-concrete composite box girder is established based on the longitudinal displacement equation considering the effect of interlayer slip and shear lag, then the formula for calculating the deflection of simply supported steel-concrete composite box girder under the action of uniformly distributed load is derived by using energy variation method.

Content from this work may be used under the terms of the Creative Commons Attribution 3.0 licence. Any further distribution of this work must maintain attribution to the author(s) and the title of the work, journal citation and DOI.
Published under licence by IOP Publishing Ltd

## 2. Derivation of Basic Differential Equations

### *2.1. Basic Consumptions*

Fig.1 is the elevation coordinate schematic diagram of the composite beam, and Fig. 2 is the cross-section schematic diagram of the composite beam. Set the maximum width among the width of webs of the concrete slabs, the width of the cantilever, and the width of the upper and lower flanges of the steel gird as "b", then the other widths are expressed as $\xi_i b$ ($i=1,2,3,4$).Unless otherwise stated, the subscript "c" stands for concrete and "s" stands for steel.

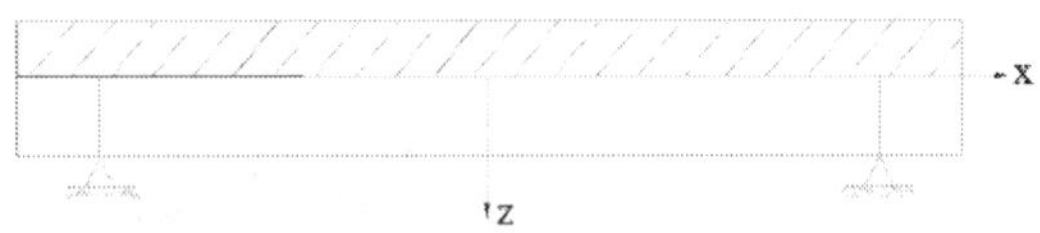


**Figure 1.** Schematic diagram of elevation coordinate

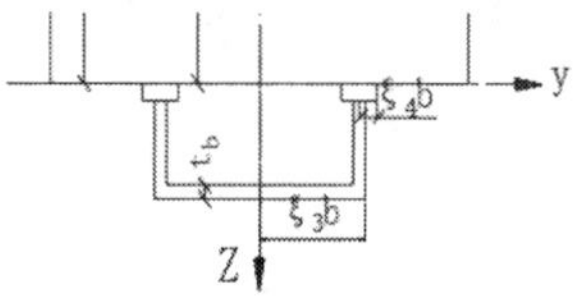


**Figure 2.** Schematic diagram of section

By introducing three generalized displacements, the vertical deflection, the longitudinal displacement and the generalized slip between layers of the beam, the longitudinal displacement mode of any point at any section of the composite beam can be written as follows:

$$U_{ci}(x,y) = -Z_c\left\{\frac{dw}{dx} + \left[1-\left(\frac{y}{\xi_i b}\right)^3\right]u(x)\right\} - g_c \cdot S(x) \tag{1}$$

$$U_{si}(x,y) = +Z_{st(b)}\left\{\frac{dw}{dx} + \left[1-\left(\frac{y}{\xi_i b}\right)^3\right]u(x)\right\} + g_s \cdot S(x) \tag{2}$$

Where $u(x)$ is the maximum difference of shear angle, $Z_c$ and $Z_{st(b)}$ respectively are the distances from the middle surface of the concrete and the upper and lower flanges of the profiled steel to the cross-sectional centroid axis of composite beam, $S(x)$ is the relative slip.

For a simply supported steel-concrete composite beams under the action of uniformly distributed load, the relative slip between steel and concrete can be deduced as:

$$S(x) = \beta q\left[\frac{e^{-\frac{\alpha l}{2}-\alpha x} - e^{-\frac{\alpha l}{2}+\alpha x}}{\alpha\left(1+e^{-\alpha l}\right)} + x\right] \tag{3}$$

Where $\alpha = \sqrt{\dfrac{K_0\left[\left(I_s + I_c E_c/E_s\right)/\left(1/A_s + E_s/A_c E_c\right) + (0.5h)^2\right]}{E_s d\left(I_s + I_c E_c/E_s\right)}}$ ; $\beta = \dfrac{hd}{0.58 n_s A_s \sqrt{E_c f_c A_1}}$ ; $I_s$ and $I_c$ are inertia moments of steel and concrete slabs, respectively; $A_s$ and $A_c$ are cross-sectional areas of steel and concrete slabs, respectively; $n_s$ is the number of studs in the same cross-section, $q$ is the uniform load; $l$ is the calculation span; $d$ is longitudinal spacing of studs.

When establishing the energy equation, it is assumed that the steel section and the concrete section satisfy the plane hypothesis at the section of the webs, the beam body works elastically.

### *2.2. Establishment of Function of Potential Energy*

According to the principle of virtual work, the expression of total potential energy of composite beams under the action of vertical uniformly distributed load is

$$\Pi = \overline{V_b} - \overline{W} \tag{4}$$

Where $\overline{V_b}$ is the deformation potential energy of steel webs, concrete slab, upper and lower flanges of section steel, and $\overline{W}$ is the load potential energy of composite beams.

In consideration of $\varepsilon_{ci} = \frac{\partial U_{ci}(x,y)}{\partial x}$, $\gamma_{ci} = \frac{\partial U_{ci}(x,y)}{\partial y}$, $\varepsilon_{si} = \frac{\partial U_{si}(x,y)}{\partial x}$, $\gamma_{si} = \frac{\partial U_{si}(x,y)}{\partial y}$, the total potential energy is obtained as:

$$\overline{V_b} = 2\times 2\times \frac{1}{2}\sum_{i=1}^{2}\int_0^{\frac{l}{2}}\int_0^{\xi_i b}\left(E_c\varepsilon_{ci}^2 + G_c\gamma_{ci}^2\right)t_{ui}dxdy + 2\times 2\times \frac{1}{2}\int_0^{\frac{l}{2}}\int_0^{\xi_3 b}\left(E_s\varepsilon_{s3}^2 + G_s\gamma_{s3}^2\right)t_b dxdy$$

$$+4\times 2\times \frac{1}{2}\int_0^{\frac{l}{2}}\int_0^{\xi_4 b}\left(E_s\varepsilon_{s4}^2 + G_s\gamma_{s4}^2\right)t_t dxdy + 2\times \frac{1}{2}\int_0^{\frac{l}{2}} E_s I_w \left(w''\right)^2 dx \tag{5}$$

$$\overline{W} = -2\int_0^{\frac{l}{2}} M(x)w''(x)dx \tag{6}$$

Where $I_w$ is inertia moment of steel web and $G$ is shear modulus.

Substitute Equations 5 and 6 into Equation 4, the total potential energy is obtained.

## 3. Deflection and Stress

When the beam is under the action of uniformly distributed load $q$, according to the principle of minimum potential energy, $\delta\Pi = \delta\left(\overline{V_b} - \overline{W}\right) = 0$, the following formulae are derived:

$$u'' - k^2 u = \frac{7n}{6}\left[\frac{Q(x)}{E_c I_4} + \frac{\beta q h_c}{Z_c h} I_3 \left(\frac{1}{I_4} - \frac{1}{I_1}\right) g''(x)\right] \tag{7}$$

$$w'' = -\frac{1}{E_c I_4}\left[M(x) + \frac{3}{4}E_c I_1 u' + \frac{\beta q h_c}{Z_c h} E_c I_3 g(x)\right] - \frac{3}{4}w''' - \frac{9}{14}u'' + \frac{9G_c I_2}{5E_c b^2 I_1}u - \frac{3}{4}\frac{\beta q h_c I_1}{Z_c h I_3}g'' = 0 \tag{8}$$

Where $n = \left(1 - 0.875 I_1/I_4\right)^{-1}$, $k = 14nI_3G_c/\left(5E_c b^2 I_1\right)$, $I_1 = I_{cu1} + I_{cu2} + n_1 I_{sb} + n_1 I_{st}$,

$I_2 = \frac{1}{\xi_1^2}I_{cu1} + \frac{1}{\xi_2^2}I_{cu2} + \frac{n_2}{\xi_3^2}I_{sb} + \frac{n_2}{\xi_4^2}I_{st}$, $I_3 = I_{cu1} + I_{cu2} + n_1 n_3 I_{sb} + n_1 n_4 I_{st}$; $I_4 = n_1 I_{sw} + I_1$,

$I_{cu1} = 2t_{u1}\xi_1 bZ_c^2$; $I_{cu2} = 2t_{u2}\xi_2 bZ_c^2$, $I_{sb} = 2t_b\xi_3 bZ_{sb}^2$; $I_{st} = 4t_t\xi_4 bZ_{st}^2$,

$g(x) = 1 - e^{-\frac{\alpha l}{2} - \alpha x}/\left(1 + e^{-\alpha l}\right) - e^{-\frac{\alpha l}{2} + \alpha x}/\left(1 + e^{-\alpha l}\right)$.

Substituting $M(x) = \frac{1}{8}ql^2 - \frac{1}{2}qx^2$ and $Q(x) = qx$ into Equation 7, and considering boundary conditions $u'(x = l/2) = 0$ and $u'(x = l/2) = 0$, $u(x)$ is obtained, hence the deflection of beam can be derived from Equation 8 as:

$$w(x) = \frac{5}{384E_c I_4}ql^4 + \frac{\beta q h_c}{Z_c h}\frac{I_3}{I_4}\left(\frac{l^2}{8} - \frac{1}{\alpha^2}\right) + \frac{3I_1}{4I_4}\left[\frac{C_0}{k^2} - \frac{7nq}{6k^2}\left(\frac{l^2}{8E_c I_4} + \frac{\beta h_c}{Z_c h}\cdot\frac{I_1 I_3 - I_3 I_4}{I_1 I_4}\right)\right]$$

$$+\frac{2qx^4 - 3ql^2x^2}{48E_c I_4} + \frac{\beta q h_c}{Z_c h}\frac{I_3}{I_4}\left\{\frac{e^{-\frac{\alpha l}{2} - \alpha x} + e^{-\frac{\alpha l}{2} + \alpha x}}{\alpha^2\left(1 + e^{-\alpha l}\right)} - \frac{1}{2}x^2\right\}$$

$$+\frac{3I_1}{4I_4}\left\{\frac{7nq}{6k^2}\left[\frac{x^2}{2E_cI_4}+\frac{\beta h_c g}{Z_c h}\cdot\frac{I_1I_3-I_3I_4}{I_1I_4}\right]-\left(\frac{1}{E_cI_4}-\frac{\alpha^2\beta h_c}{Z_c h}\cdot\frac{I_1I_3-I_3I_4}{I_1I_4}\right)\frac{14nqsh\frac{kl}{2}}{6k^4shkl}\cdot chkx\right\} \quad (9)$$

Concrete stress can be calculated by $\sigma_s = E_s\frac{\partial U(x,y)}{\partial x}$ and steel stress can be calculated by $\sigma_c = E_c\frac{\partial U(x,y)}{\partial x}$.

## 4. Examples and Analysis

### *4.1. Test Beams*

In order to verify the accuracy of the Equation 9, three simply supported steel-concrete composite beams with bolts spacing of 120 mm, 180 mm and 300 mm were fabricated, in which the shear connectivity degrees are1.33, 0.88 and 0.5 (calculated according to reference [1]). The dimension of the test beams are shown in Fig. 3 and Fig.4. The uniformly distributed load is $8.9kN/m$; the concrete slab is made of C40 concrete, $E_c = 34.5Gpa$;the steel beam is made of Q235, $E_s = 206Gpa$. The shear connector is made of cylindrical head welding nails with the diameter of 12.8 mm and the length of 45 mm.

In this test, the deflections at mid-span are measured.

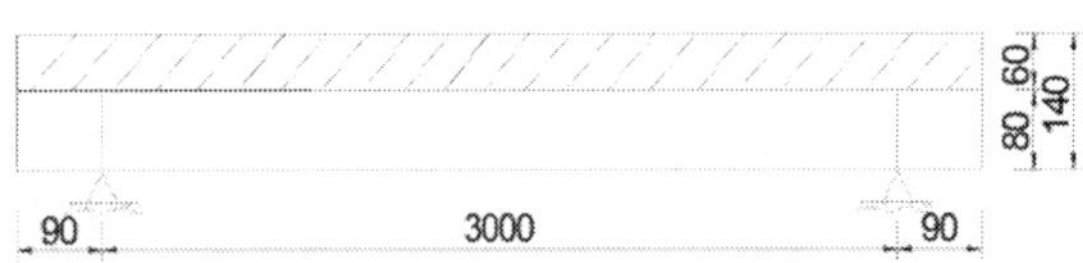

**Figure 3.** Longitudinal arrangement of the test beam/mm

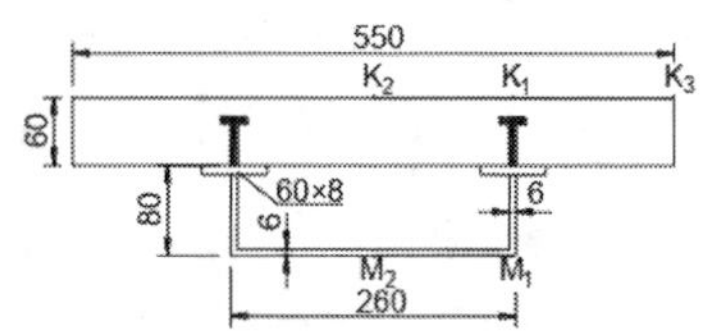

**Figure 4.** Transverse section /mm

### *4.2. Comparison of Results*

Comparisons between the experiment data of mid-span deflection and the theoretical results of Equation 9 are listed in Table 1, where the relative differences are compared with the experimental data.

**Table 1.** Comparison of deflections under different longitudinal spacing of stud

| Bolt spacing/mm | Deflections/mm | | |
|---|---|---|---|
| | Experimental data | Values by Formula 9 | Relative difference |
| d=120 | 4.12 | 4.31 | 4.61% |
| d=180 | 4.62 | 4.83 | 4.55% |
| d=300 | 5.23 | 5.59 | 6.88% |

The following conclusions can be drawn from Table 1:

(1) When the bolt spacing is 120 mm, 180 mm and 300 mm respectively, the relative differences between the calculated values by Formula 9 and the experimental data are 4.61%, 4.55% and 6.88%, respectively. The calculated values are close to the experimental value, showing that the calculation accuracy of Formula 9 is satisfactory.

(2) As the spacing of bolts increases, the deflection at mid-span increases.

In order to understand further the effect of degree of shear connection between concrete slab and steel structure to deflection of steel-concrete composite beams, the theoretical deflections at mid-span of beam varying with the spacing of bolts are calculated by Equation 9, which are drawn in Fig. 5. For the purpose of verify the computation accuracy of Equation 9, ANSYS is used to calculate the deflection in case of two extreme conditions, as no slip (steel and concrete is completely connected) and full free slipping between concrete slab and steel structure. The finite element theoretical values are also drawn in Fig. 5

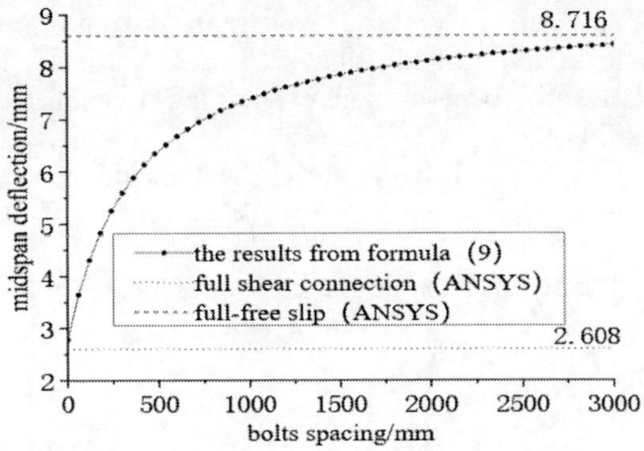


**Figure 5.** The relationship between mid-span deflection and longitudinal space of stud

From Fig.5 we can make following conclusions:

(1) The mid-span deflection increases with the increase of the bolt spacing, while the rate of the increase reduces gradually.

(2) In two extreme conditions, there is no slip and fullly free slipping between concrete slab and steel structure, the calculated values by Formula 9 are close to the finite element theoretical values (ANSYS). When the steel and the concrete is completely connected (no slipping), the calculated value by Formula 9 is 2.78 mm and the result by ANSYS is 2.608 mm, its relative difference is 6.59%; when the shear connection between steel and concrete is complete free, the calculated value by Formula 9 is 8.62 mm and the result by ANSYS is 8.716 mm, its relative difference between them is -1.10%, showing that Formula 9 gives reliable deflection of steel-concrete composite beams.

## 5. Conclusion

Based on the proposed longitudinal displacement model considering the effect of interlayer slip, the analytical formula for calculating the deflection of simply supported steel-concrete composite beams under the action of uniformly distributed load considering the effect of shear lag and slip is established by using the energy variation method. The calculated values by the established formulae are agreeable with the experimental data showing that the formulae hold high calculation accuracy. The presented formulae can easily consider the influence of the number, spacing and the shear capacity of the bolts, and the characteristics of the cross-sections.

## References

[1] Code for design of steel and concrete composite bridge (GB 50917), China Planning Press, Beijing, 2013.

[2] Code for design of steel structures (GB 50017-2003), Beijing: China Planning Press, 2003.

[3] Y.J. Shao, P.R. Zhu, Z.H. Chen, Modified transformed-section method of deflection calculation of steel-concrete composite beams, J. of building struct., 29(2008):99-103.

[4] R.P. Johnson, I.M. May, Partial-interaction design of composite beams, Struct. Eng.,53(1975): 305-311.

[5] J.G. Nie, J.M. Shen, Y.S. Yuan, A general formula for predicting the deflection of simply supported composite steel-concrete beams with the consideration of slip effect, Eng. Mech., 11(1994):21-27.

[6] J.G. Nie, Z.W. Yu, A reduced deformation rigidity method for calculating of composite steel-concrete beams, China Civil Eng. J., 6 (1995):11-17.

[7] A. Gjelsvik, Analog-beam method for determining shear-lag effects, J. Eng. Mech., 117(1991):

1575-1594.

[8] L. Miao, D.W. Chen, Closed-form solution of composite beam considering interfacial slip effects, J. Tongji Univ., 39(2011 ):1113-1119.

[9] L.Z. Jiang, Z.W. Yu, J. Li, Theoretical analysis of slip and deformation of steel concrete composite beam under uniformly distributed loads, Eng. Mech., 20(2003):133-137.

[10] F.X. Li, J.G. Nie, Elastic analytical solutions of shear lag effect of steel-concrete composite beam, Eng. Mech., 28(2011):1-8.

[11] F. Gara, G. Leoni, L. Dezi, A beam finite element including shear lag effect for the time-dependent analysis of steel–concrete composite decks, Steel Construction, , 31(2009):1888-1902.

[12] F.F. Sun, G.Q. Li, Beam formulation for composite beams with slip, shear lag and shear deformation, J. Tongji Univ., 32(2004):436-440.

# Effects of Alloying Elements on Mechanical and Electronic Properties of Cu by First-Principles Calculations

**Cheng Peng [a], Lijuan Zeng [b], Fuxiang Huang *, Mingjun Zhong [c], Shuang Liang [d], Xiaojie Ran [e] and Lu Zhou [f]**

College of Materials Science & Engineering, Chongqing University of Technology, Chongqing 400054, China

*Corresponding author email: hfx@cqut.edu.cn, [a] pc123@2017.cqut.edu.cn, [b] 1422173163@qq.com, [c] 463064227@qq.com, [d] ls@2017.cqut.edu.cn, [e] 1071602004@qq.com, [f] 1137662948@qq.com

**Abstract.** The effects of alloying elements (Mn, Cr, Sn, P, Be and Ti) on the mechanical and electronic properties of Cu have been investigated performing first-principles calculations. A supercell consisting of 107 Cu atoms and one solute atom is used. The calculated results show that the binding energy values of $Cu_{107}X$ alloys are negative, indicating they have thermodynamic stable structures. The ductility properties and hardness of Cu and $Cu_{107}X$ alloys are also determined based on the calculated elastic constants ($c_{ij}$'s). It is found that the ductility properties of $Cu_{107}X$ alloys decrease due to the addition of alloying elements, but they still exhibit plasticity. However, the Vickers hardness of $Cu_{107}X$ alloy is improved, and the $Cu_{107}Be$ has the largest calculated value of 5.62 GPa. In addition, the electrical conductivity of $Cu_{107}X$ alloy is reduced, among them, $Cu_{107}Be$ and $Cu_{107}Mn$ have relatively high electrical conductivity of 2.58 and $2.53*10^7$ S/m, respectively.

**Keywords:** Copper Alloys; First-Principles Calculations; Mechanical Properties; Electrical Conductivity.

## 1. Introduction

Copper and its alloys are widely used in machinery manufacturing, transportation, electric power and other fields [1-2]. Especially in the electronic information industry, copper wire is regarded as an important basic material for electronic packaging due to their excellent electrical conductivity [3-4]. Nowadays, with the rapid development of the microelectronics industry, it is required that the bonding copper wire has more excellent mechanical properties and electrical conductivity. However, high strength and high electrical conductivity are mutually exclusive in nature; therefore, the contradictory relationship must be considered between the properties in the design of the alloy [5]. Research suggests that the solid solution of alloying elements into the Cu matrix usually cause lattice distortion, which improves the mechanical properties of the alloy, but increases the scattering of conducting electrons and impairing the electrical conductivity of Cu [6]. So, how to improve the strength of copper alloys while ensuring good conductivity is one of the important research directions of copper alloys [7]. In order to solve the above problems, various strengthening processes such as alloying, deformation strengthening,

Content from this work may be used under the terms of the Creative Commons Attribution 3.0 licence. Any further distribution of this work must maintain attribution to the author(s) and the title of the work, journal citation and DOI.
Published under licence by IOP Publishing Ltd

and composite materials method, etc, have been proposed, thus, alloys having high strength and high conductivity such as Cu-Zr, Cu-Cr, and Cu-Re were prepared [8-10].

Although very reliable experimental results can be obtained according to conventional research methods, but it takes a lot of time and cost. Therefore, theoretical prediction for the effects of alloying additions on Cu from the first principles can provide essential guidance in identifying materials with desired mechanical and electronic properties. In the present work, the first-principles calculations are conducted to systematically study the mechanical properties and electrical conductivity of Cu dilute alloys with the addition of elements X (Mn, Cr, Sn, P, Be and Ti).

## 2. Details of the calculations

In this paper, the first-principles calculations of trace element doping of copper were performed with CASTEP (Cambridge Sequential Total Energy Package) package based on the density functional theory (DFT) ultra-soft potential plane wave method [11]. In the calculation process, the Perdew-Burke-Emzerhof (PBE) potential function in generalized gradient approximation (GGA) is applied to describe the electronic exchange correlation energy part [12], and the convergence calculation is performed by self-consistent field (SCF) method. The accuracy is set to $5.0\times10^{-7}$ eV/atom. In order to suffice the accuracy of the calculation and increase the calculation speed, all calculations are performed in the reciprocal K space. After a series of convergence tests, the Monkhorst Pack [13] 4×4×4 mesh was used in Brillouin zone (BZ) integration, and the plane wave truncation can be set to 450 eV. The crystal structure was geometrically optimized using the Brodyden-Fletcher-Goldfarb-Shanno (BFGS) algorithm [14], before all calculations were performed to obtain the most stable structure in the local area. The convergence parameters are setting as follows: The convergence accuracy of the total energy of the system is $5.0\times10^{-6}$ eV/atom, the atomic average force is less than 0.1 eV/nm, the tolerance offset is less than $5.0\times10^{-7}$ nm, and the maximum stress deviation is 0.02 GPa.

Cu belongs to a face-centered cubic (FCC) structure with space group of Fm-3m, a lattice constant of a=b=c=3.6146 Å, and α=β=γ=90° [15]. Subsequently, supercell model consisting of 3×3×3 arrangement of Cu crystal and having 108 atoms was used to simulate the doping of trace X (Mn, Cr, Sn, P, Be, Ti) atoms in Cu for the current calculations, in which the alloying element X replaces the top-corner sites. After the substitution of the X atom, the doped alloy is represented by $Cu_{107}X$, wherein the X content is 0.926 at%, and the crystal structure is as shown in Fig. 1.

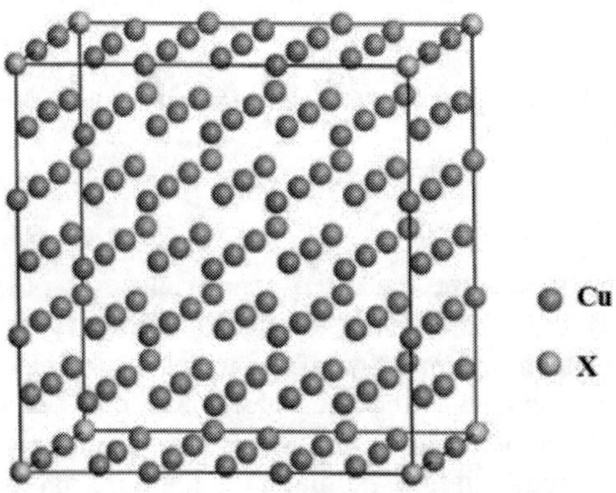


**Figure 1.** Supercell model of atom doping Cu with X (Mn, Cr, Sn, P, Be, Ti)

## 3. Results and discussions

### *3.1. Thermodynamic Stability*

In order to evaluate the stability of doped $Cu_{107}X$ alloys with respect to the $Cu_{108}$ solid, we study their thermodynamic properties by calculating the formation enthalpy and binding energy. Here, the formation enthalpy is used to characterize the energy released/absorbed by different types of atoms from

IOP Conf. Series: Materials Science and Engineering **772** (2020) 012064 doi:10.1088/1757-899X/772/1/012064

the elemental state-generating compound, and the easier the formation of the compound if it has the smaller value. The resulting is expressed as ΔHf and is calculated as follows [16]:

$$\Delta H_f = \frac{E_{tot}^{Cu_aX_b} - aE_{solid}^{Cu} - bE_{solid}^{X}}{a+b} \tag{1}$$

where $E_{tot}$ is the total energy of the supercell, a (or b) refers to the number of Cu or X atoms in the supercell, respectively. $E_{solid}^{Cu}$ and $E_{solid}^{X}$ denote the energies per atom of Al and X in the solid states, respectively.

The binding energy is interpreted as the energy released when free atoms are combined into a crystal. The larger the absolute value, the stronger the binding force between atoms in the crystal, indicating that the structure of the compound is more stable. The binding energy is expressed as $\Delta E_{coh}$ and is defined as [16]:

$$\Delta E_{coh} = \frac{E_{tot}^{Cu_aX_b} - aE_{atom}^{Cu} - bE_{atom}^{X}}{a+b} \tag{2}$$

Where $E_{atom}^{Cu}$ and $E_{atom}^{X}$ are the energies of Cu and X atom in the isolate state, respectively. According to the Eq. 1-2, the calculation results of the formation enthalpy and binding energy of $Cu_{108}$ and $Cu_{107}X$ are shown in Table 1. It can be readily seen that $Cu_{108}$ is a simple substance, and the $\Delta H_f$ is 0 eV·atom$^{-1}$. After the X atom is doped, the $Cu_{107}X$ alloy has a small enthalpy of formation, and they are easy to form. Among them, the formation enthalpy of $Cu_{107}Mn$ and $Cu_{107}Be$ alloys is negative, expressing that they are easier to form than $Cu_{108}$ and other $Cu_{107}X$ (Cr, Sn, P, Ti) alloys. The calculated binding energy of $Cu_{108}$ is -3.6644 eV·atom$^{-1}$, which is only 5% deviation compared with experimental value of -3.49 eV·atom$^{-1}$ [15], suggesting that the calculations in the present work are reasonable. Meanwhile, the calculated binding energy of $Cu_{107}X$ alloy is negative, revealing that they all have thermodynamic stability. It is notice that the calculated binding energies of $Cu_{107}Mn$, $Cu_{107}P$ and $Cu_{107}Ti$ alloys are -3.6729, -3.6765 and -3.6786 eV·atom$^{-1}$, respectively, which are smaller than $Cu_{108}$, so they have a more stable crystal structure than $Cu_{108}$, and $Cu_{107}Ti$ is the most stable. On the contrary, the calculated binding energies of $Cu_{107}Cr$, $Cu_{107}Sn$ and $Cu_{107}Be$ alloys are all larger than $Cu_{108}$, so their crystal structure stability is slightly worse than $Cu_{108}$. The absolute value of the experimental values of the binding energies of pure metals Mn, Cr, Sn, P, Be and Ti is ranked as Ti>Cr>P>Be>Sn>Mn [15], and it can be found that the structural stability of $Cu_{107}X$ is largely affected by the stability of the metal X itself. However, the doping of Cr and Mn elements does not conform to the above rule, which probably caused by the different crystal structure of Cr and Cu. Doping causes Cu to generate large lattice distortion, which requires more energy to be absorbed during the formation. Mn and Cu are both face-centered cubic structures, so the thermodynamic stability of $Cu_{107}Mn$ is better.

**Table 1.** The calculated results of $Cu_{107}X$ alloys unit cell total energy, $\Delta H_f$ and $\Delta E_{coh}$, solid atom and isolated atom average energy

| Phase | Unit cell energy (eV) | Solid atom energy (eV) | Atom energy (eV) | $\Delta H_f$ *10$^{-2}$ (eV·atom$^{-1}$) | $\Delta E_{coh}$ (eV·atom$^{-1}$) |
|---|---|---|---|---|---|
| $Cu_{108}$ | -181540.7658 | -1680.9330 | -1677.2686 | 0.00 | -3.6644 |
| $Cu_{107}Mn$ | -180483.798 | -623.0098 | -619.3843 | -0.88 | -3.6729 |
| $Cu_{107}Cr$ | -182258.2712 | -2399.4492 | -2395.1292 | 0.94 | -3.6611 |
| $Cu_{107}Sn$ | -181908.0577 | -2048.7549 | -2044.999 | 0.49 | -3.6604 |
| $Cu_{107}P$ | -180079.1137 | -218.3445 | -214.3107 | 0.87 | -3.6765 |
| $Cu_{107}Be$ | -180252.7426 | -392.2348 | -389.6937 | -0.63 | -3.6603 |
| $Cu_{107}Ti$ | -181453.3308 | -1593.8376 | -1588.2996 | 0.31 | -3.6786 |

### *3.2. Mechanical Properties*

The mechanical properties of the material are closely related to the elastic constant of the solid [17]. For instance, according to calculating the elastic constant $C_{ij}$ of $Cu_{108}$ and $Cu_{107}X$, combined with the Voig-Reuss-Hill approximation [18], the bulk modulus (B), shear modulus (G), Young's modulus (E), Poisson's ratio (v) of $Cu_{107}X$ alloy can be derived. Moreover, according to the Paugh empirical scheme B/G to evaluate the brittleness of the material, and to predict the Vickers hardness (Hv) of the $Cu_{107}X$ alloy by semi-empirical formula [19]. The computational formulas are expressed as follows:

$$B = \frac{C_{11} + 2C_{12}}{3} \tag{3}$$

$$G = \frac{1}{2}\left[\frac{C_{11} - C_{12} + 3C_{44}}{5} + \frac{5(C_{11} - C_{12})C_{44}}{3(C_{11} - C_{12}) + 4C_{44}}\right] \tag{4}$$

$$E = \frac{9BG}{3B + G} \tag{5}$$

$$v = \frac{3B - 2G}{2(3B + G)} \tag{6}$$

$$Hv = \frac{1 - 2v}{6(1 + v)}E \tag{7}$$

According to the above formulas, the calculation results are lists in Table 2. It is noteworthy that the calculated values of the elastic constant of $Cu_{108}$ are in good agreement with experimental results [20]. According to Born-Huang's elastic stability theory, there are three independent elastic constants for the cubic crystal. The restriction of the mechanical stability are given as follows [21]: $C_{11}>0$, $C_{12}>0$, $C_{44}>0$, $C_{11} - C_{12}>0$, $C_{11} + 2C_{12}>0$. As shown in Table 2, all of the elastic constants for $Cu_{107}X$ alloys satisfy these criteria mentioned above, indicating that they are mechanically stable.

**Table 2.** Elastic properties of $Cu_{108}$ and $Cu_{107}X$ alloys

| Phase | Cij | | | B | G | E | B/G | v | $H_v$ (GPa) | |
|---|---|---|---|---|---|---|---|---|---|---|
| | $C_{11}$ | $C_{12}$ | $C_{44}$ | (GPa) | (GPa) | (GPa) | | | | |
| $Cu_{108}$ | 172.0 | 123.2 | 78.5 | 135.5 | 41.4 | 113.0 | 3.27 | 0.365 | 3.73 | |
| | 168.4 | 121.4 | 75.4 | 137.0 | 48.3 | 129.8 | - | - | | [20] |
| $Cu_{107}Mn$ | 151.0 | 113.4 | 77.2 | 125.9 | 44.1 | 118.5 | 2.85 | 0.343 | 4.62 | |
| $Cu_{107}Cr$ | 155.7 | 120.3 | 75.2 | 132.1 | 42.4 | 115.0 | 3.11 | 0.355 | 4.11 | |
| $Cu_{107}Sn$ | 149.1 | 107.9 | 72.5 | 121.6 | 43.9 | 117.6 | 2.77 | 0.339 | 4.72 | |
| $Cu_{107}P$ | 147.7 | 109.3 | 71.9 | 122.1 | 42.6 | 114.4 | 2.87 | 0.344 | 4.43 | |
| $Cu_{107}Be$ | 154.5 | 107.8 | 78.8 | 123.3 | 48.5 | 128.7 | 2.54 | 0.326 | 5.62 | |
| $Cu_{107}Ti$ | 150.4 | 110.3 | 73.9 | 123.7 | 44.0 | 118.0 | 2.81 | 0.341 | 4.66 | |

The ductility and brittleness of the material can be evaluated by the B/G ratio and Poisson's ratio v [22]. The material exhibits a plastic property when the B/G value is greater than 1.75, and the bigger the value of ratio is, the better the ductility of the material is. If the ratio is less than 1.75, it is a brittle material. For the Poisson's ratio v, it is reflect the toughness and brittleness of the material. When the value of v is greater than 0.3, the material exhibits plasticity, and the larger the ratio, the better the plasticity of the material. Comparing the B/G value and Poisson's ratio v of $Cu_{107}X$ alloy in Table 2, on the one hand, it can be seen that the B/G value of the $Cu_{107}X$ alloy doped with trace elements are between 2.54–3.1, both greater than 1.75. On the another hand, the Poisson's ratio v is between 0.326 and 0.355 and are greater than 0.3. At the same time, it is easy to compare the B/G value and v value of $Cu_{107}X$, and their influence on the ductile and brittleness of the doped alloy is consistent. In general, the doping of Be element greatly reduces the B/G value and Poisson's ratio v of Cu, so the plasticity of $Cu_{107}Be$

IOP Conf. Series: Materials Science and Engineering **772** (2020) 012064 doi:10.1088/1757-899X/772/1/012064

alloy is worse than that of other doped alloys. The doping of the other elements has not obviously effect on the B/G value and Poisson's ratio v of Cu. Therefore, The $Cu_{107}Mn$, $Cu_{107}P$ and $Cu_{107}Ti$ have good plasticity.

The hardness of a material always plays an important role in its applications, thus it is necessary to investigate the hardness of material. Table 2 lists the calculated Vickers hardness values for $Cu_{108}$ and $Cu_{107}X$ alloys. Since the $Cu_{108}$ used in the calculation is an ideal model, and the defects of the crystal are not considered, the calculated pure copper Vickers hardness value of 3.73 GPa is larger than the experimental value of the annealing state at room temperature in the Ref. [23] of 1.53 GPa. In addition, Zeng [24] reported that the incorporation of Sn, Mn, P, and Be elements will increase the hardness of pure Cu, which is consistent with the calculation results. So, the doping of the Be element greatly increases the hardness of Cu, and the doping of other elements also appropriately increases the hardness of Cu.

### *3.3. Electrical Properties*

In order to study the electrical conductivity of $Cu_{108}$ and $Cu_{107}X$ alloys, the kinetic relaxation including NVE and NVT of $Cu_{108}$ and $Cu_{107}X$ was first carried out, and the optical conductivity of several frame structures in the kinetic trajectory was calculated. Finally, the Kubo-Greenwood formula was used to calculate the conductivities of $Cu_{108}$ and $Cu_{107}X$. The specific form of the Kubo-Greenwood formula is as follows [25]:

$$\sigma_1(k, \omega)=\frac{2\pi}{3\omega\Omega}\sum_{j=1}^{n_b}\sum_{i=1}^{n_b}\sum_{\alpha=1}^{3}\left(F(\varepsilon_{i,k})-F(\varepsilon_{j,k})\right)\times\left|\langle\Psi_{j,k}|\nabla\alpha|\psi_{i,k}\rangle\right|^2\delta(\varepsilon_{j,k}-\varepsilon_{i,k}-\omega) \quad (8)$$

Here, the parameters in Formula (8) have been described in previous reports [25], so this paper will not introduce. The optical conductivity of $Cu_{108}$ and $Cu_{107}X$ alloys is shown in Figure 2. The optical conductivity is used to describe the behavior of the material to excite free electrons after absorbing photons [26]. From Fig. 2, the optical conductivity of $Cu_{108}$ and $Cu_{107}X$ alloys is greater than zero when the incident light energy is 0 eV. This is because the metal has no band gap, and even if there is no incident light, a transition from the valence band to the conduction band can occur. The optical conductivity of $Cu_{108}$ is the largest, and the optical conductivity of $Cu_{107}X$ alloy is consistent with the change trend of $Cu_{108}$. The optical conductivity of $Cu_{107}Mn$ and $Cu_{107}Be$ is relatively large, while the optical conductivity of $Cu_{107}P$ and $Cu_{107}Sn$ is relatively small.

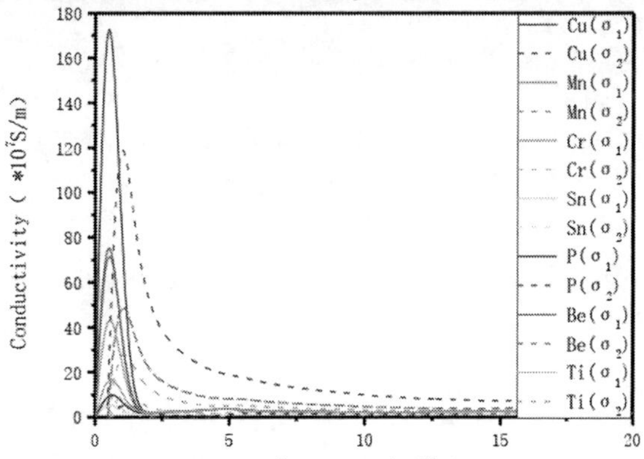


**Figure 2.** Photoelectric conductivity of $Cu_{108}$ and $Cu_{107}X$ alloys

According to the Ref. [25,27], the optical conductivity of the $Cu_{108}$ and $Cu_{107}X$ alloys in Fig. 2 at the frequency of 0eV is extracted, which is the electronic conductivity of $Cu_{108}$ and $Cu_{107}X$ alloy, as shown in Fig. 3. The calculated value of $Cu_{108}$ conductivity is $5.39*10^7$ S/m, and the error of pure copper conductivity 5.8*$10^7$ S/m [28] is 7%. The calculated value is lower than the experimental value due to the selection of correlation function and pseudopotential. From Fig. 2, we noted that the conductivity of Cu is reduced after the X element is doped, which is mainly due to lattice distortion caused by element doping. Among them, the conductivity of $Cu_{107}Sn$ is relatively small compared with other alloys, and its value is 1.91*$10^7$ S/m, which is similar to the conductivity of 2.1*$10^7$ S/m when the Sn doping content is about 1wt% [29]. Furthermore, $Cu_{107}Be$ and $Cu_{107}Mn$ have relatively high electrical conductivity of 2.58 and 2.53*$10^7$ S/m, respectively.

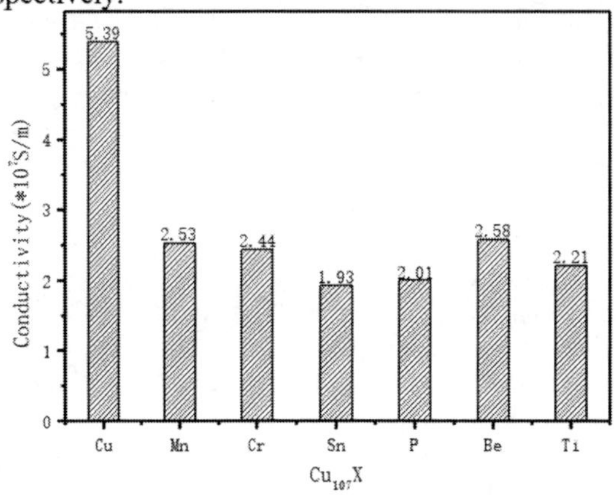


**Figure 3.** Conductivity of $Cu_{108}$ and $Cu_{107}X$

## 4. Conclusion

In the present work, we have systematically investigated the thermodynamic stability, mechanical properties as well as electronic properties of different alloyed $Cu_{107}X$ (X=Mn, Cr, Sn, P, Be and Ti) systems by employed first principles calculation. From the present study, the following conclusions are obtained:

1). The binding energies of $Cu_{107}X$ alloys are negative, indicating that they all have a thermodynamic stable structure. Among them, the binding energy of $Cu_{107}Ti$, $Cu_{107}P$ and $Cu_{107}Mn$ are -3.6786, -3.6765 and -3.6729 eV·atom$^{-1}$, respectively, which have a more stable crystal structure than $Cu_{108}$.

2). The ductility of $Cu_{107}X$ alloy decreases due to the addition of alloying elements, but they still exhibit plasticity, among which $Cu_{107}Mn$, $Cu_{107}P$ and $Cu_{107}Ti$ have good plasticity. On the contrary, the Vickers hardness of the $Cu_{107}X$ alloy after doping was improved, and the $Cu_{107}Be$ has the largest calculated value of 5.62 GPa.

3). The conductivity of $Cu_{107}X$ alloy decreased after doping. Among them, $Cu_{107}Be$ and $Cu_{107}Mn$ have relatively high electrical conductivity, and their values are 2.58 and 2.53*$10^7$ S/m, respectively.

4). Considering comprehensively, the doping of Be element can improve the mechanical properties of Cu alloy and have less influence on its conductivity.

**Acknowledgements**

This work was financially supported by the National Key Research and Development Program of China (Grant No. 2016YFB0402602).

**References**

[1] M.P. Wang, Advanced Copper Alloy With High Strength And Conductivity (Central South University Press, China, 2015), pp. 1–3.
[2] G.R. Yang, Y. Hao and W.M. Song, Foundry Technology, 13–15 (2003).
[3] Y.T. Ding, J. Cao and G.J. Xu, Foundry Technology, 971–974 (2006).
[4] M. Mayer and Y.S. Lai, Microelectronics Reliability, 1–2 (2011).
[5] R.G. Li, H.J. Kang and Z.N. Chen, Scientific Reports, 20799–20808 (2016).
[6] L. Lu, Y.F. Shen and X.H. Chen, Science, 422–426 (2004).

[7] J. Yuan, Y.S. Long and S.H. Zhao, Special Casting & Nonferrous Alloys, 828–832 (2015).
[8] F. Jang, X.B. Chen and L. Jang, Hot Working Technology,1–5 (2009).
[9] R.R. Liu, H.T. Zhou and X. Zhou, Materials Review, 100–105 (2012).
[10] J. Zhang, B. Xie and L.Z. Zhao, Hot Working Technology, 21–27 (2014).
[11] T.H. Fischer and J. Almlof, Journal of Physical Chemistry, 9768–9774 (1992).
[12] G. Kresse and J. Hafner, Journal of Physics Condensed Matter, 8245–8257 (1994).
[13] D.J. Chadi, Physical Review B Condensed Matter, 5188–5192 (1977).
[14] G.F. Li, S.Q. Lu and X.J. Dong, Journal of Alloys and Compounds, 170–176 (2012).
[15] D. Wei, Solid State Physical (Tsinghua University Press, China, 2003), pp. 8–10.
[16] W.C. Hu, Y. Liu and D.J. Li, Computational Materials Science, 381–389 (2015).
[17] D.E. Kim, S.L. Shang and Z. K. Liu, Computational Materials Science, 254–260 (2009).
[18] Y. Lu, D.F. Li and B.T. Wang, Journal of Nuclear Materials, 136–141 (2011).
[19] Z. Hong, H.W. Huang and G. Li, Journal of Materials Science & Engineering, 482–485 (2016).
[20] J.H. Li, X.D. Dai and S.H. Liang, Physics Reports, 1–134 (2008).
[21] M. Born and K. Huang, American Journal of Physics, 113–127 (1954).
[22] S.F. Pugh, Philosophical Magazine, 823–843 (1954).
[23] Z. Han, B. Yao, and K.Lu, Acta Metallurgica Sinica Chinese Edition, 238–244 (2014).
[24] Z.M. Zeng, Practical Nonferrous Metal Materials Handbook (Mechanical Industry Press, China, 2015), pp. 63–79.
[25] S. Mazevet, M. Torrent and V. Recoules, High Energy Density Physics, 84–88 (2010).
[26] H.Y. Guo, L. Hou and J.W. Zhang, Applied Physics Letters, 191905–191907 (2007).
[27] D.W. Zhou, J.F. Song and G.Q. Li, Journal of Nanyang Normal University, 27–30 (2011).
[28] L.B. Li, Metal Materials Physical Properties Handbook (Mechanical Industry Press, China, 2011), pp. 136–142.
[29] D.M. Zhao, Q.M. Dong and P. Liu: Gongneng Cailiao/Journal of Functional Materials, 609-611 (2001).

# Preparation and research of brush-coated solvent-free polyurethane anti-corrosion coatings

**Zhongzhi Han[a], Jiafeng Shi, Chao Ding, Shi Li, Cancan Cui, Shaoming Duan, Lei Wang and Xiaojun Guo***

CNPC Engineering Technology Research Co., LTD, Tianjin, China

*Corresponding author e-mail: guoxj@cnpc.com.cn.com,
[a]hanzz01@cnpc.com.cn.com

**Abstract**. In order to solve the problem of short gel time and difficulty in brushing of solvent-free polyurethane coatings, the coatings choose polyether polyester polyol and polyisocyanate prepolymer as the main film-forming materials, adding spherical ceramic powder to improve the wear resistance of the coating, combination of molecular sieve physical dehydration and oxazolidine chemical dehydration, effectively solves the problem of water sensitivity of paint, a solvent-free polyurethane anticorrosive coating for brush application was prepared. The coating also has good adhesion, excellent wear resistance and corrosion resistance.

## 1. Introduction

Polyurethane coatings have strong corrosion resistance, good weather resistance, excellent performance, especially the emergence of cross-linked two-component polyurethane coatings, combined with environmentally friendly functions, making polyurethane coatings leapfrog. Polyurethane coatings are widely used in wood, construction, plastics, anti-corrosion, machinery and other fields, becoming one of the fastest growing coating varieties [1, 2].

With the gradual strengthening of environmental protection requirements in China, reducing the VOC content and emissions in coatings, the use of solvent-free and water-based environmentally friendly coatings instead of solvent coatings has become the development direction of anti-corrosion coatings. Solvent-free polyurethane coating is a solvent-free, non-flammable, 100% solids environmentally friendly coating with excellent oil resistance, acid and alkali resistance, low temperature curability, and ductility [3-5].

At present, the high viscosity of the solvent-free polyurethane coating makes it need to be heated and sprayed. The construction requires special high-pressure airless impact spraying equipment. The surface of the coating is easy to form pitting, orange peel, the appearance is poor, the wettability to the substrate is not good, the reaction is exothermic, the internal stress is large, and will have negative effect on adhesion. The fast curing speed of the coating also results in a short pot life of the coating, which is not conducive to small-area construction and repair work on site [6-8].

To this end, this study develop a solvent-free polyurethane anticorrosive coating that can be brushed by studying the polyol and curing system in the polyurethane coating film. The coating has excellent physical properties, mechanical properties and corrosion resistance, making it suitable for small-area construction and repair work.

Content from this work may be used under the terms of the Creative Commons Attribution 3.0 licence. Any further distribution of this work must maintain attribution to the author(s) and the title of the work, journal citation and DOI.
Published under licence by IOP Publishing Ltd

## 2. Experimental part

Preparation of paint: step one, The molecular sieve activated powder, the antifoaming agent and the dispersing agent are added to a certain amount of the polyether polyester polyol, and dispersed at a high speed for 1 hour to obtain a dehydrating agent;

Step two, The ceramic powder, the talc powder, the titanium dioxide powder, the barium sulfate and the carbon black are heated in a 150 ° C high-temperature cabinet for 12 hours to obtain a coating pigment filler;

Step three, the polyether polyester polyol, the water removing agent, the defoaming agent and the dispersing agent are mixed. The mixture is stirred at a high speed and room temperature. The coating pigment and filler are added under stirring, and the high-speed stirring is continued for 30 to 40 minutes to uniformly disperse the components. And then grind with a grinder for 0.5 to 1 hour to obtain a mixture.

Step four, a dehydrating agent and an antifoaming agent obtained in the step one are added to the mixture obtained in the step three, and dispersed at a high speed for 30 minutes to obtain a component A of the solventless polyurethane coating.

Step five, dispersing polymethylene polyphenyl isocyanate PM200 to form component B.

Step six, mixing the A component and the B component uniformly to obtain the Solvent-free polyurethane coating.

## 3. Results and discussion

### *3.1. Study on polyether polyols*

Among the polyurethane raw materials, a hydroxyl group (resin) mainly composed of a polyol is an important component. As a soft segment structure of polyurethane coatings, polyols are also an important factor in determining the properties of polyurethanes. In polyurethane raw materials, there are two main categories: polyether polyols and polyester polyols [9].

Polyether polyols are cheaper than polyester polyols and have a wide range of raw materials. Polyester polyurethane has more polar groups such as ester groups and urethane groups, has strong cohesive strength and adhesion, and has high strength and wear resistance. Therefore, after the urethane bond is embedded in the molecular chain of the low molecular weight polyether polyol, the strength of the paint film can be greatly improved, and the paint film has a high hardness.

The polyether polyols based on ether bonds were used in this test, and four different types of polyether polyols were used for basic performance testing. The test results are shown in Table 1.

**Table 1.** Basic properties of different types of polyether polyols

| Polyether polyol type | surface drying time (min) | hard drying time (h) | Impact Strength (J) | Adhesion (MPa) | Anti-bending (1.5°) | Hardness (ShoreD) |
|---|---|---|---|---|---|---|
| Branched polyol | 120 | 10 | 16 | 9.2 | No cracks | 58 |
| Branched polyether/polyester primary hydroxyl polyol | 90 | 6 | 14 | 13.2 | No cracks | 71 |
| Castor oil modified polyol | 60 | 5 | 16 | 9.5 | No cracks | 55 |
| Cashew nut shell modified polyol | 40 | 4 | 10 | 12.8 | No cracks | 85 |

As can be seen from Table 1, the branched polyol has good impact resistance and chemical resistance, but the hardness of the coating is low and the Shore D is only 58. Through comparative analysis, the branched polyether/polyester primary hydroxyl polyol has good physical and mechanical properties and chemical resistance, and the basic properties meet the requirements.

### *3.2. Study on isocyanate curing agent*

Two-component polyurethane coating polyisocyanate prepolymer as a polyurethane curing agent has many advantages such as high thermal stability, corrosion resistance, radiation resistance, fast curing and high hardness. There are very few free isocyanate monomers in the polyisocyanate prepolymer, which is beneficial to the chain extension and crosslinking reaction to be carried out quickly and thoroughly. It can reduce the occurrence of side reactions, improve the compatibility of the raw material system, and improve the physical properties of the coating [10].

The suitable curing agent is selected to make the coating have a long pot life, so that the solvent-free polyurethane coating can be brushed. This experiment selected four prepolymers from three manufacturers on the market to test the properties of coatings. Branched polyether/polyester primary hydroxyl polyol for polyol component. The test results are shown in Tables 2 and 3.

**Table 2.** Basic properties of different types of isocyanates

| type | surface drying (min) | hard drying (h) | pot life (h) | Impact Strength (J) | Anti-bending (1.5°) | Adhesion (MPa) | Hardness (ShoreD) |
|---|---|---|---|---|---|---|---|
| PM1 | 90 | 6 | 3 | 14 | No cracks | 13.2 | 71 |
| PM2 | 30 | 5 | 1 | 12 | No cracks | 14.3 | 82 |
| PM3 | 100 | 8 | 3.5 | 16 | No cracks | 12.8 | 59 |
| PM4 | 90 | 6 | 3 | 15 | No cracks | 13.6 | 75 |

**Table 3.** Chemical resistance of different types of isocyanates

| type | $H_2SO_4$ (10%, 30d) | NaOH (10%, 30d) | NaCl (30%, 30d) | Diesel resistance (2#, 30d) | Cathodic disbanding (65℃, -1.5V, 48h) |
|---|---|---|---|---|---|
| PM1 | Coating intact | Coating intact | Coating intact | Coating intact | 12mm |
| PM2 | Coating intact | Coating intact | Coating intact | Coating intact | 14mm |
| PM3 | Coating intact | Coating intact | Coating intact | Coating intact | 15mm |
| PM4 | Coating intact | Coating intact | Coating intact | Coating intact | 10mm |

As can be seen from Tables 2 and 3, PM2 isocyanate has good impact resistance and chemical resistance, but the coating has a short pot life. PM3 isocyanate has a longer surface dry time, lower hardness and a Shore D hardness of only 59. PM4 and PM1 isocyanate have good physical and mechanical properties and chemical resistance, and PM4 is better than PM1 in cathodic resistance. Therefore, PM4 polyisocyanate prepolymer was used as polyurethane coating isocyanate component confirmed by the above study.

### *3.3. Study on wear-resisting filler*

Solvent-free polyurethane coatings require coatings with good abrasion resistance. The addition of wear-resistant fillers can improve the wear resistance of epoxy coatings, and ceramic powders are the most common in wear-resistant fillers. Mainly because ceramic powder itself has high hardness and good wear resistance.

Ceramic powder is used as a wear-resistant filler. Under the conditions of the film-forming resin, the coating is prepared according to the different proportions of the ceramic powder in the pigment filler. The performance comparison data of the coating is shown in Table 4.

**Table 4.** Effect of different ratios of ceramic powder and pigment filler on coating properties

| Ceramic powder proportion (%) | Wear resistance (CS17, 1kg, 1000r) mg | Anti-bending (1.5°) | Flexibility (mm) | Adhesion (Grade) | Impact Strength (J) | Hardness (ShoreD) |
|---|---|---|---|---|---|---|
| 5 | 55.3 | No cracks | 1 | 1 | 15 | 79 |
| 10 | 52.4 | No cracks | 1 | 1 | 16 | 81 |
| 15 | 59.8 | No cracks | 1 | 1 | 16 | 82 |
| 20 | 48.2 | No cracks | 1 | 1 | 17 | 84 |
| 25 | 46.7 | cracks appear | 1 | 1 | 15 | 85 |
| 30 | 46.4 | cracks appear | 2 | 1 | 14 | 85 |

With the increase of the amount of ceramic powder, the wear resistance of the coating is enhanced, the flexibility is decreased, the impact resistance is first increased and then decreased, the overall change of adhesion is not large, and the hardness gradually Increase. Ceramic powder was used to add 20% of pigment and filler.

### *3.4. Study on coating additive system*

In the polyurethane paint, the reaction between NCO and water produces $CO_2$, which is the main source of bubbles in the paint film. By adding a dehydrating agent, the water in the solvent can be effectively removed, bubbles and pinholes are eliminated [11].

The system uses physical dehydration combined with chemical water removal. The physical dewatering of polyurethane coatings mainly uses molecular sieves. The molecular sieves are aluminosilicates, which are mainly composed of silica-alumina and oxygen bridges to form an open skeleton structure. The chemical dehydrating agent has an effect on inhibiting the bubbles of the polyurethane paint, wherein the effect of the oxazolidine is most obvious, and the effect of moisture can be effectively suppressed, bubbles and pinholes are avoided. This is because when the oxazolidine encounters a trace amount of moisture in the formulation raw material or is exposed to moisture, the oxazolidine preferentially reacts with water, thereby preventing the isocyanate from reacting with moisture to generate carbon dioxide.

During the construction process, the isocyanate in the prepolymer will produce a certain amount of bubbles in the coating when it comes into contact with a trace amount of moisture in the air. Therefore, defoaming agents should be added to eliminate or inhibit. Therefore, when a defoaming agent is used, a suds suppressor is often used in combination to suppress the generation of bubbles. The polyurethane coating system uses an antifoaming agent produced by BYK Chemical Company of Germany, and its dosage is usually 0.3~0.5% of the coating content.

The dispersing agent has the functions of wetting, reducing viscosity, improving leveling, preventing floating color, sagging, sedimentation and so on. The coating selected a polyacrylate dispersant, which has good compatibility with the film-forming resin, and has a high molecular weight to effectively separate the pigment particles, and the dispersant is used in an amount of 0.4% of the total formulation.

### *3.5. Physical and chemical properties of coatings*

The physical and chemical properties of the coatings were tested in accordance with relevant standards. The test results are shown in Table 5.

IOP Conf. Series: Materials Science and Engineering **772** (2020) 012065 doi:10.1088/1757-899X/772/1/012065

**Table 5.** Performance of coatings

| type | actual measurement |
|---|---|
| Solid content (%) | 98.6 |
| surface drying time (25°C, min) | 18 |
| hard drying time (25°C, min) | 80 |
| pot life (h) | 3 |
| shore hardness (Shore D) | 84 |
| Anti-bending (1.5°) | Coating intact |
| Impact strength (J) | 15 |
| Adhesion (MPa) | 12.8 |
| Wear resistance (Cs17, 1kg, 1000r)mg | 48.6 |
| Cathodic disbanding (65°C, 48h)mm | 9.3 |
| $H_2SO_4$, NaOH and NaCl (10%, 30d) | Coating intact |
| Salt spray resistance (1500h) | Coating intact |

As can be seen from Table 4, the developed coating has a suitable pot life, making it suitable for small-area construction and repair work. Coatings have excellent physical and mechanical properties and corrosion resistance.

## 4. Conclusion

Through the study of polyether polyols and isocyanates, solvent-free polyurethane coatings have a suitable pot life and can be applied by brushing. The addition of spherical ceramic powder as a wear-resistant filler in the coating improves the wear resistance of the coating. The combination of physical dehydration and chemical dehydration effectively solves the problem of water sensitivity of the coating, eliminates bubbles and pinholes of the coating, and reduces loss of light and haze.

## References


[1] Zhang W Y, Gao W, Lin F. Preparation and Performance Test of Solvent-Free Heavy-Duty Polyurethane Coating [J]. Advanced Materials Research, 2014, 915-916:816-820.

[2] Mobin M , Malik A U, Al-Muaili F, et al. Performance Evaluation of a Commercial Polyurethane Coating in Marine Environment [J]. Journal of Materials Engineering and Performance, 2012, 21(7): 1292-1299.

[3] Heidarian M, Shishesaz M R. Study on effect of duration of the ultrasonication process on solvent-free polyurethane/organoclay nanocomposite coatings: Structural characteristics and barrier performance analysis [J]. Journal of Applied Polymer Science, 2012, 126(6): 2035---2048.

[4] Doley S, Dolui S K . Solvent and catalyst free synthesis of sunflower oil based polyurethane through non-isocyanate route and its coatings properties [J]. European Polymer Journal, 2018: S0014305717316749.

[5] Zhang W Y, Gao W, Lin F. Preparation of Solvent-Free Moisture Curable Heavy Anti Corrosion PU Coating [J]. Advanced Materials Research, 2014, 898:80-83.

[6] Chen S L, Chen G X , Chen Q F, et al. The Preparation of Solvent-Free Polyurethane Laminating Adhesive with Mixed-Polyols [J]. Advanced Materials Research, 2012, 549: 177-182.

[7] Wang Q, Chen G X, Chen S L. Preparation of contamination free polyurethane adhesive based on biopolymer castor oil [J]. Materials Research Innovations, 2013, 17(S1): 162-165.

[8] Chattopadhyay D K, Raju K V S N. Structural engineering of polyurethane coatings for high performance applications [J]. Progress in Polymer Science, 2007, 32(3): 352-418.

[9] Liu F C, Hao Y S, Wang Z Y, et al. Flaking and degradation of polyurethane coatings after 2 years of outdoor exposure in Lhasa [J]. Chinese Science Bulletin, 2010, 55(7): 650-655.

[10] Yang X F, Vang C, Tallman D E, et al. Weathering degradation of a polyurethane coating [J]. Polymer Degradation & Stability, 2001, 74(2): 341-351.

[11] Y. González-García, S. González, Souto R M . Electrochemical and structural properties of a polyurethane coating on steel substrates for corrosion protection [J]. Corrosion Science, 2007, 49(9): 0-3526.

IOP Conf. Series: Materials Science and Engineering **772** (2020) 012066 doi:10.1088/1757-899X/772/1/012066

# The study of spacing influence on the performance of Contra-Rotation-Propeller in stratospheric airship

**Cong Xie, Guoquan Tao * and Zhe Wu**

School of Aeronautics Science and Engineering, Beihang University (BUAA), Beijing, China

*Corresponding author e-mail: taogq_buaa@163.com

**Abstract.** According to the requirements of a foreign high-altitude airship project, the Single-Rotation-Propeller (SRP) was designed under the maximum endurance factor criterion. The spacing influence on the Contra-Rotation-Propeller (CRP) was studied by numerical simulation analysis, the performance of front and rear propeller was compared with SRP. The result indicated that CRP was more efficient than SPR, the CRP has enormous potential in enhancing the aerodynamic performance.

## 1. Introduction

The long-endurance, low-energy consumption, and high security make the stratospheric airship extensive application in both military and civilian fields. Efficiency propelling propeller is beneficial to improve the maneuverability and sustainability of airship, which cannot be matched by a turbojet or turbofan.

However, the environment at the stratosphere, which above 20 km, has a much smaller air density. Also, the advance velocity of stratospheric airship is 10 m/s which are extremely slow. The low Reynolds number and small advance ratio make it a big challenge to design an efficient propelling propeller. The Counter-Rotation-Propeller (CRP) is used to improve the efficiency.

The airfoil of high altitude vehicle was researched by prof. Michael S Selig [1], [2], the aerodynamic characteristics of six typical low Reynolds number, high lift, and low resistance airfoils were analyzed and wind tunnel experiments are carried out. Angleo[3] proposed a method by design the chord length and twist Angle to maximize propeller efficiency under design requirements, Xiang [4] improved the Angleo's method to make it more content with the requirement of engineering application.

Many experimental investigations of CRP have been conducted during the last century, Biermann and Hartman [5] found that the performance of counter-rotating propellers was significantly improved at lower advance ratio. Naiman [6] developed a method for calculating the performance of a CRP from the section characteristics of the blade elements. Tang's [7] research indicated the performance of CRP is different with variation of Reynolds number, and the efficiency of CRP achieved appears to be a few percent greater than that for a standard conventional propulsion system. Chen [9], Liu [10] and Wang [11] studied other aspects of CRP.

This paper designed a SRP which all airfoils along the radial direction of the propeller worked under the maximum endurance factor $(C_l^{1.5}/C_d)$, the CRP was composed of two coaxial SRP, aerodynamics of was analyzed and compared with SRP.

Content from this work may be used under the terms of the Creative Commons Attribution 3.0 licence. Any further distribution of this work must maintain attribution to the author(s) and the title of the work, journal citation and DOI.
Published under licence by IOP Publishing Ltd

## 2. The Performance of Single-Rotation-Propeller

According to the requirements of a foreign high-altitude airship project, the propeller design indexes were proposed as follows:

**Table 1.** The design indexes of CRP

| Altitude | Velocity | Rated speed | Rated load | Diameter | Power |
|---|---|---|---|---|---|
| 20 km | 10 m/s | 500 RPM | ≥ 180 N | ≤ 4 m | ≤ 2500 |

### *2.1. the design of SRP*

The two-blade propeller was selected due to the low speed and large diameter of the SRP, the propeller-tips Mach number should be restricted as follows[1]:

$$\frac{\sqrt{(\pi D n_s)^2 + V_0^2}}{\text{a}} \leq 0.7 \tag{1}$$

$$D \leq \frac{\sqrt{(0.7\text{a})^2 - V_0^2}}{\pi n_s} = 7.88 \text{ m} \tag{2}$$

Where the D was the diameter of the SRP, $n_s$ was the rotation speed per second, $V_0$ was the forward velocity, a was the local sound velocity which can be calculated by the local temperature. The diameter of SRP was chosen 4 m, due to the restriction of the design indexes. The hub of the propeller was chosen 25% of the diameter.

Prandtl simplified method was adopted to calculate the thrust distribution of the minimum energy dissipation and accordingly calculated the appropriate propeller chord length to minimize the energy loss of the propeller. The induction Angle of each section of the propeller was calculated, so that all airfoils along the radial direction of the propeller worked under the maximum endurance factor ($C_l^{1.5}/C_d$), made the propeller had the highest efficiency [12].

The shape parameters of the propeller were shown in Figure 1. Where the radial location was the location related to the radius, and the relative chord length was related to the diameter.

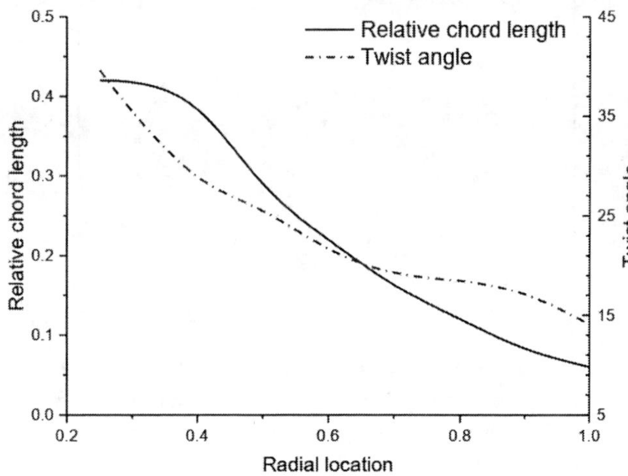


**Figure 1.** The shape parameters of the propeller

The 3D model of the SRP was constructed by the CAD software according to the design, which was shown in Figure 2.

**Figure 2.** The 3D model of the SRP

*2.2. Numberical simulation anlysis of Single-Rotation-Propeller*

STAR-CCM+ was used for numerical simulation analysis of the SRP, the control equation was discretized by the finite volume method, the external flow field was separated by three-dimensional steady constant density, Reynolds average N-S equation and k-ω turbulence model was adopted.

**Table 2.** The rated atmospheric conditions for numerical simulation

| T/K | ρ/(kg/m³) | P/ Pa | a/(m/s) | μ/(kg/m·s) |
|---|---|---|---|---|
| 216.7 | 0.08805 | 5474.9 | 295.07 | $1.4216\times 10^{-5}$ |

The performance of the SRP at different rotation speed was analyzed under the rated atmospheric condition which was shown in Table 2. As shown in Figure 3, the thrust, torque, and power were increased as the rotation speed increases, the efficiency increased a little firstly, reaching the peak 86.21% at 300 RPM, then decline gradually. The thrust and power at 500RPM were 98.98 N and 1298.71 W, which were satisfied the design indexes of the SRP.

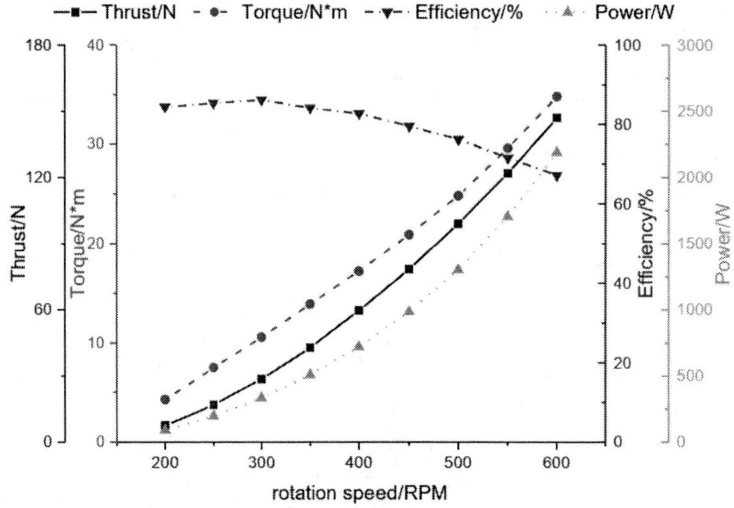


**Figure 3.** The performance of the SRP

## 3. The Performance of Counter-Rotation-Propeller

The front propeller (FP) and rear propeller (RP) were components of the CRP, which were exactly the same except the rotational directions. The spacing of 0.8m, 1.2m, 1.6m, 2m, 2.4m were calculated at the rotation speed of 500RPM.

STAR-CCM+ was used for numerical simulation analysis of the CRP, which the control equation and turbulence model were consistent with the SRP.

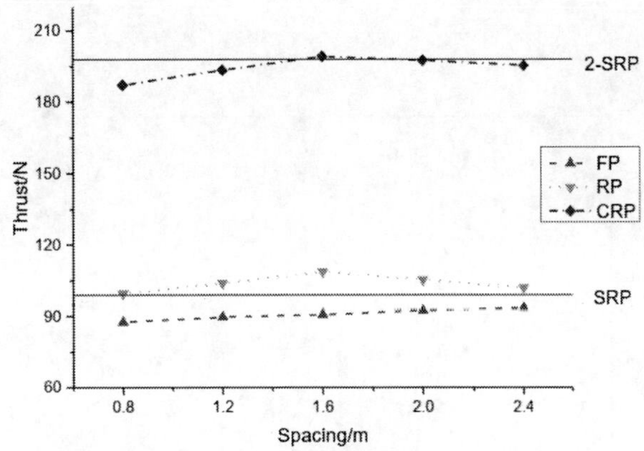


**Figure 4.** The thrust of the CRP compare to SRP

As shown in Figure 4, the thrust of the front propeller was increased as the spacing increased, the rear propeller was climbed up first and then decline, the maximum thrust was occurred at the spacing of 1.6m. The total thrust tendency of CRP was consistent to the rear propeller, the total thrust beyond the two independent SRPs only at the spacing of 1.6m.

The differences between front propeller, rear propeller and SRP prove the existence of interference between front and rear propellers. The total thrust of CRP was mostly less than two independent SRPs, only in certain spacing had a higher total thrust, which the interference was disadvantage to the total thrust. The thrust of the front propeller was less than SRP, and the rear propeller was more, it indicated that the rear propeller was beneficial from the CRP while the front propeller was not conducive to.

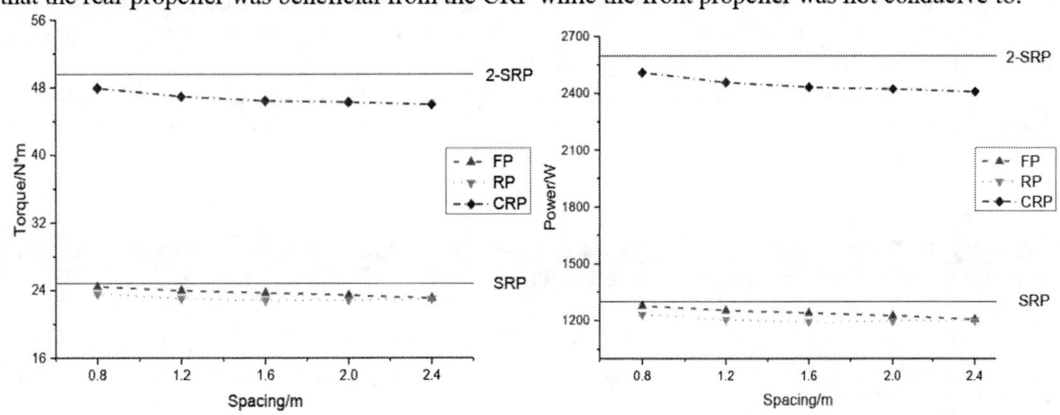


**Figure 5.** The torque and power of the CRP compare to SRP

The torque of front propeller and CRP were decreased as the spacing increased, which was shown in Figure 5, the rear propeller was decline firstly and then climb up. The torque of the front propeller and rear propeller were less than SRP, which made the total torque of CRP was less than two independent SRPs. The rear propeller impacted more from CRP which torque was less than front propeller.

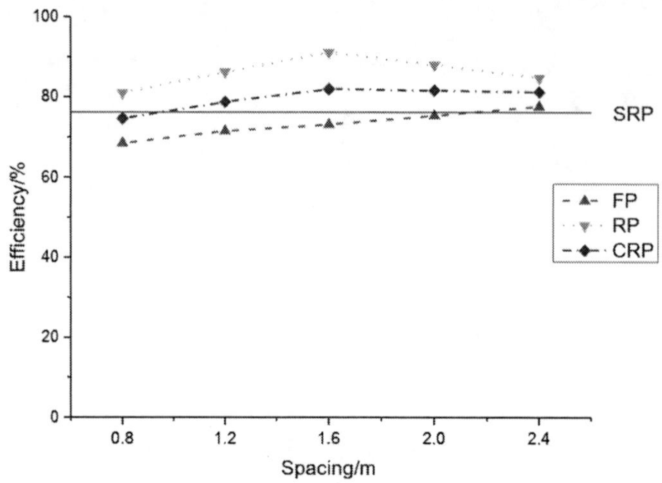


**Figure 6.** The efficiency of the CRP compare to SRP

As shown in Figure 6, the efficiency of the front propeller was increased as the spacing increased, and the rear propeller and CRP were climbed up first and then decline. The efficiency of the front propeller increased more than the loss of the rear propeller, making the efficiency of CRP exceeded the SRP.

## 4. Conclusion

The thrust, torque, and power of SRP were increased as the rotation speed increases, the efficiency increased a little firstly then decline gradually. The thrust and power at 500RPM were satisfied the design indexes, which could be used to constitute the CRP.

The differences between front propeller, rear propeller and SRP prove the existence of interference between front and rear propellers, which was beneficial to rear propeller and not conducive to front propeller. The interference increased the thrust of the rear propeller and decreased the thrust of the front propeller, which reduced the torque and power of both, making the efficiency of the rear propeller exceed the SRP.

The thrust, torque, and power of CRP were less than two independent SPRs. The thrust reduction of CRP was not significant, but there was a certain drop in torque and power, which made the efficiency of CRP beyond the SPR.

The CRP has enormous potential in enhancing the aerodynamic performance, by optimizing the geometric parameters of the front and rear propeller, the thrust force can be effectively improved while improving the efficiency.

## References

[1] Selig M S, Guglielmo J J, High-Lift Low Reynolds Number Airfoil Design, J. Journal of Aircraft, 1997, 34(1):72-79.

[2] Selig M S, Mcgranahan B D, Wind Tunnel Aerodynamic Tests of Six Airfoils for Use on Small Wind Turbine, J. Journal of Solar Energy Engineering, 2004, 126(4):986.

[3] Salvo D'Angelo, Berardi F , Minisci E, Aerodynamic Performance of Propellers with Parametric Considerations on the Optimal Design, J. Aeronautical Journal -New Series-, 2002, 106(1060):313-320.

[4] Xiang Song, Wang Ji, Zhang Liguo, A design method for high efficiency propeller, J. Journal of Aerospace Power, 2015, 30(1):136-141.

[5] Biermann D , Hartman E P, Wind-tunnel tests of four- and six-blade single- and dual-rotating tractor propellers, J. Technical Report Archive & Image Library, 1942.

[6] Naiman I, Method of Calculating Performance of Dual-Rotating Propellers From Airfoil Characteristics, J. Technical Report Archive & Image Library, 1943.
[7] Tang Z, Liu P, Sun J, Performance of Contra-Rotating Propellers for Stratospheric Airships, J. International Journal of Aeronautical & Space Sciences, 2015, 16(4):485-492.
[8] Liu Peiqing, Theory and Application of The Air Porpeller, M. Beihang University Press,2006.
[9] Chen Y, Liu P, Tang Z, et al. Wind tunnel tests of stratospheric airship counter rotating propellers[J]. Theoretical & Applied Mechanics Letters, 2015, 5(1):58-61.
[10] Liu, Z, Liu, P, Effect of Advance Ratio and Blade Planform on the Propeller Performance of a High Altitude Airship, J. Journal of Applied Fluid Mechanics, 2016, 9(6):2993-3000.
[11] Wang Yufu, Liu Zhenguo, Tao Guoquan, Numerical simulation of high altitude propeller's aerodynamic characteristics and wind tunnel test, J. Journal of Beijing University of Aeronautics and Astronautics, 2013, 39(8):1102-1105.
[12] Su Yang, Tao Guoquan, Design and numerical simulation of high altitude propeller, C. National conference on industrial fluid mechanics, 2018, 8-12

# Study on an angle measuring device based on mechanical gyroscope

**Zilin Zhang[1], Huimin Zhao[1], Xiyuan Li[1] and Xu Zhao[1], Jingchao Sun[2], ZhouYan[3,*]**

[1] Shenyang Aerospace University, Shenyang, China
[2] Science College of Shenyang Aerospace University, Shenyang, China
[3] Engineering Training Center of Shenyang Aerospace University, Shenyang, China

*Corresponding author e-mail: sdyanzhou16888@163.com

**Abstract.** The circular staggered magnetic field is studied and built. Then combining with the mechanical structure of the mechanical gyroscope, a tilt angle measurement instrument which cannot be interfered by the external conditions is designed. A stable working magnetic field is generated in the system itself. Based on the principle of electromagnetic induction, the tilt angle change is transformed into the electrical signal change output, and then the tilt angle is measured.

## 1. Introduction

The measurement of tilt angle plays an important role in many fields, such as landslide monitoring [1], building tilt, navigation system [2], etc. In war or in life, we can easily get the geographical coordinate position and find the direction accurately by compass, modern radar or electronic gyroscope. But in some strong magnetic field or emtirely blocked environment, radio communication and electronic gyroscope will be weaken or even lose function, which will make measurement get in trouble. [3]

In order to deal with the situation, this paper presents a real-time measuring device [6-7] of direction and angle, which is based on the principle of electromagnetic induction and mechanical gyroscope [4] [5], and can be free from external interference.

## 2. Principle and Design Scheme

### *2.1. Principle of Electromagnetic Induction*

Electromagnetic induction method is based on Faraday's law of electromagnetic induction. Place a circular coil with n turns and s cross-section in an alternating magnetic field with

$$B = B_m \sin(\omega t)$$

Magnetic field strength. If the angle between the normal of the coil and the magnetic induction line is $\theta$, the induced electromotive force generated in the coil is [8-9]:

Content from this work may be used under the terms of the Creative Commons Attribution 3.0 licence. Any further distribution of this work must maintain attribution to the author(s) and the title of the work, journal citation and DOI.
Published under licence by IOP Publishing Ltd

$$\varepsilon = -\frac{d\Phi}{dt}$$
$$= \text{-NS}\omega \text{B}_{\text{m}} \cos\theta \bullet \cos(\omega t)$$
$$= -\varepsilon_m \cos(\omega t)$$

Where: $\varepsilon_m = NS\omega B_m \cos\theta$ is the amplitude of induced electromotive force.

It can be seen from the formula that the alternating magnetic field is established and the induced potential is obtained through the electromagnetic induction coil, and then the electrical signal change is obtained. The device mainly uses the electromagnetic induction law to convert the angle change of any surface into the current pulse output, and obtains the angle change through Arduino microprocessor processing. At the same time, the gyroscope is combined with the above principles to determine the reference plane in the early stage of measurement and play an auxiliary role in measuring the angle.

The measurement of the change of the angle is transformed into the measurement of the change of the magnetic flux, which produces the induced potential, and then causes the change of the current. The current signal is amplified, and the final angle value is obtained after the program processing.

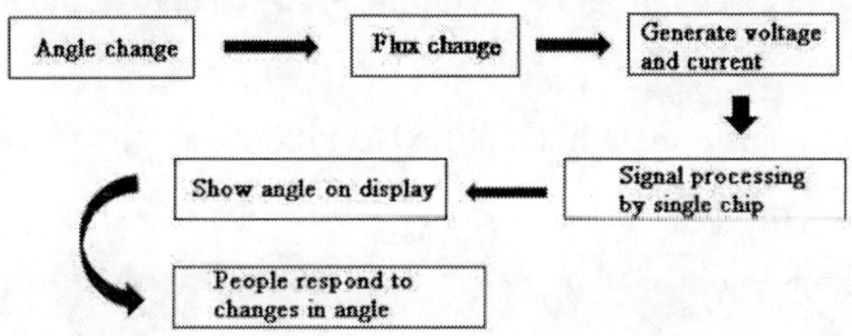


**Figure 1.** Principle of angle measurement.

*2.2. Establishment of Circular Alternating Magnetic Field*
An annular alternating magnetic field is established by fixing 358 micro strong magnetic fields on the annular device, as shown in Figure 2.

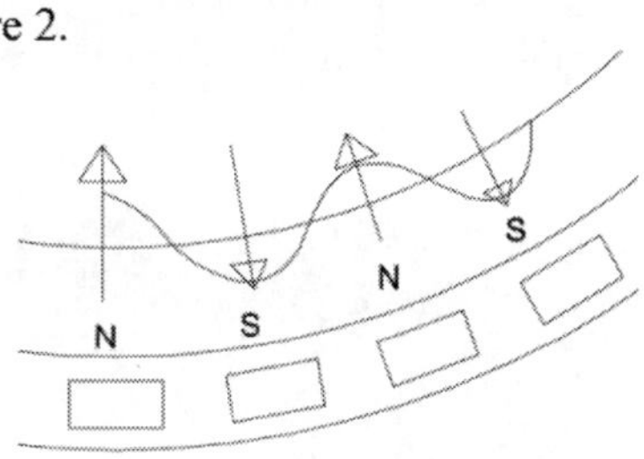


**Figure 2.** Cyclic alternating magnetic field.

The internal magnetic field is similar to the sine curve, and the specific situation should be obtained through MATLAB analysis and fitting after experimental measurement, as shown in Figure 3, the maximum value of the magnetic field is about 180mT.

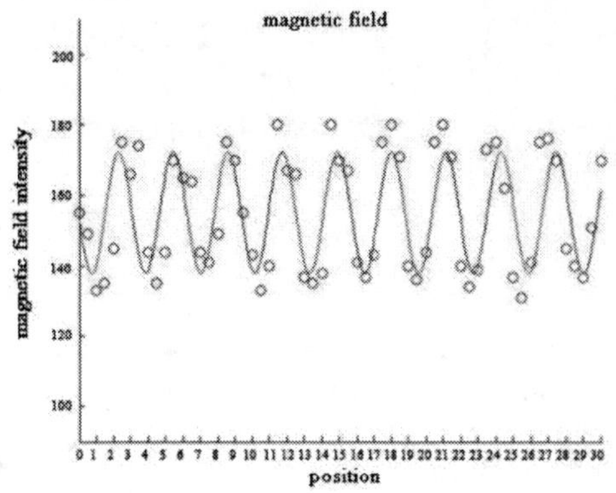


**Figure 3.** Fitting diagram of magnetic field strength.

Combining the function of magnetic field with the movement of measuring arm, the accuracy of measuring angle can be adjusted by program according to the change of induced potential.

### *2.3. Principle of Measuring Machinery*

The measurement part of the instrument is simplified as follows, and its function diagram is as follows:

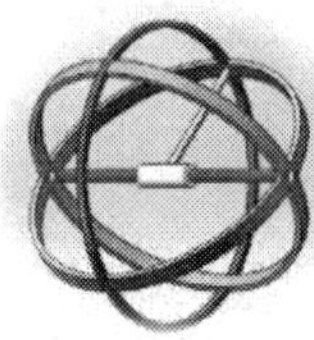

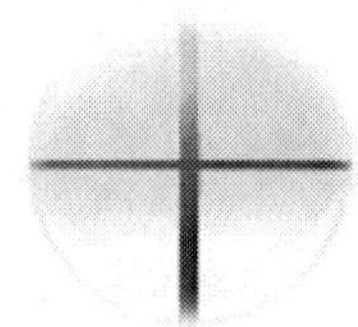

**Figure 4.** Function diagram I **Figure 5.** Function diagram II

As shown in Figure 4, the blue part divides a space into four areas, as shown in Figure 5; therefore, any angle change in the space can be represented by using the two mutually perpendicular planes where the two blue circles in Figure 4 are located.

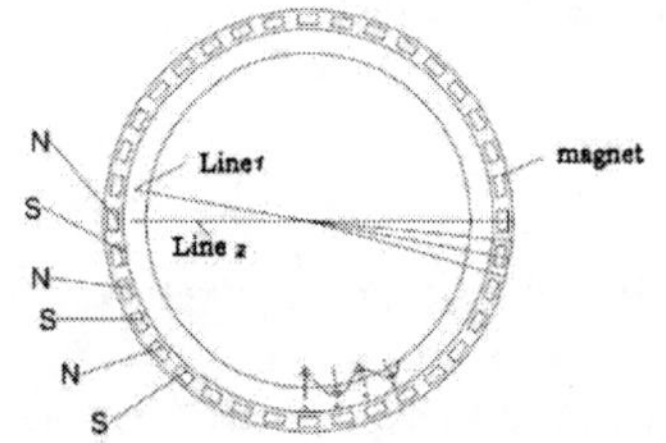


**Figure 6.** Measurement diagram.

In Figure 6, using the principle of gyroscope, the measuring arm is in the position of line 2 on one side, but the external magnet rotates with the different plane. After analysis, assuming that the external magnet does not move, the measuring arm has two degrees of freedom in three-dimensional coordinates, that is, the shell where the magnet is located can rotate freely in 360 plane. Only angle measurements in one of the directions are analyzed below.

When the outer shell rotates, the position of the measuring arm moves from 2 to 1. In this process, the two ends of the measuring arm will experience the same number of magnets in turn, and the two sides will generate reverse current of equal times. This is to eliminate the error and make the measurement more accurate.

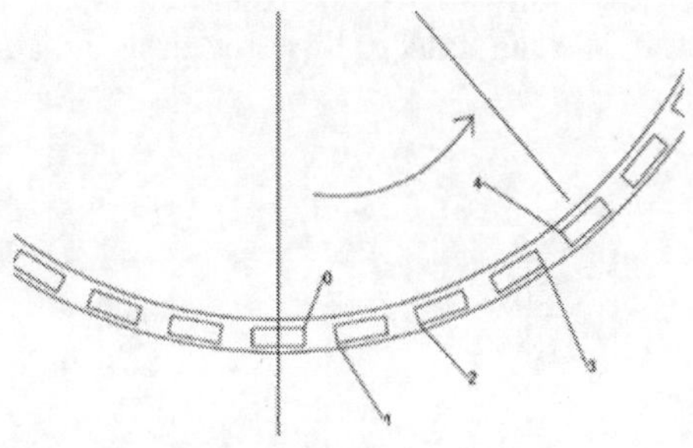

**Figure 7.** Simplified diagram of measurement process.

When the two ends of the measuring arm are relatively static with the last magnet, the last electrical signal is generated. After being processed by the processor, the corresponding angle of the coil is displayed on the display screen.

The overall instrument measurement process is shown in figure 8.

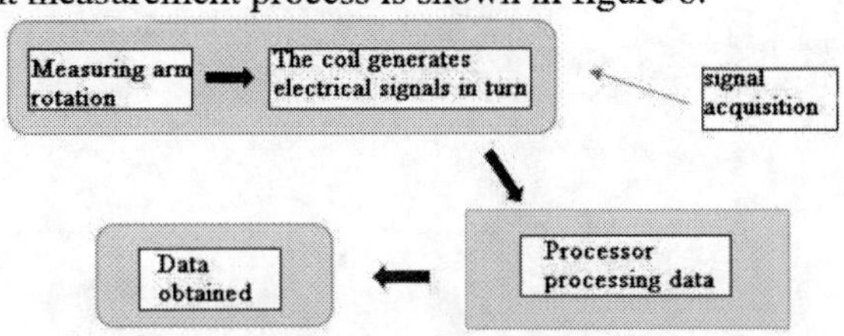


**Figure 8.** Flow chart of measurement process.

*2.4. Signal Processing and Algorithm Processing*

In the process of signal processing, the current generated by the coil under the action of the induced potential is processed, and the current is transmitted to the processor through the amplification circuit in the middle after the current is equal to that of the example amplifier. The process is as follows.

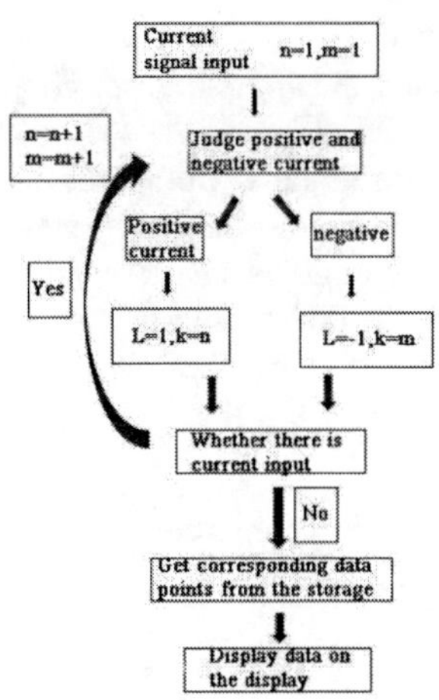


**Figure 9.** Schematic diagram of data processing process.

The data operation principle inside the single chip microcomputer is shown in Figure 10. Simply speaking, it is to assign the angle of the coil on the outer shell, according to the corresponding current value of each point, use the Eviews software to establish the database, so as to get the corresponding different angle value of each coil. We can think of this process as a mapping.

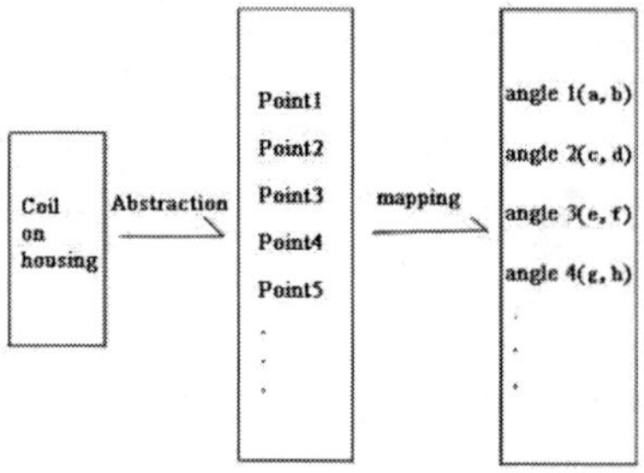


**Figure 10.** Diagram of data operation process.

The module structure of electronic part is as follows

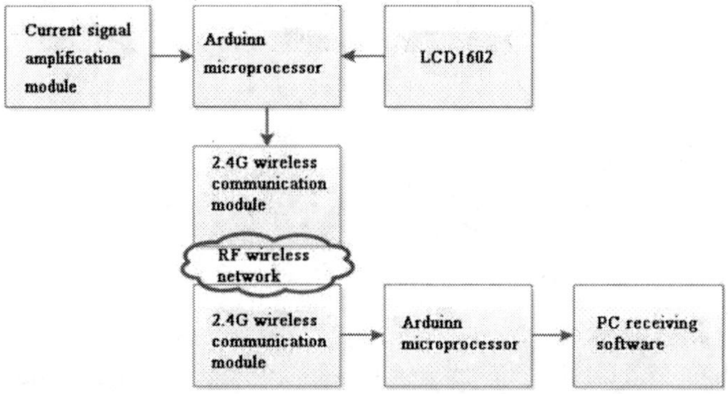


**Figure 11.** Module composition diagram.

## 3. Data processing design

The whole idea of angle measurement system is to decompose any angle in the space into the virtual space coordinate system. After measuring the angle in three coordinate planes, the space angle is given through the calculation of the system.

The measurement results are shown in Table 1. Compared with the measurement results of dxl360s level meter, it can be seen that when measuring the horizontal inclination angle of 3-10 degrees, the absolute error of measurement with large angle is smaller, and the measurement error with small angle is larger, which is related to the fact that the device is larger and the bottom surface cannot be completely flat. And the overall average error is 0.02875 degrees, which can meet the standards in construction, civil engineering and daily life.

**Table 1.** Comparison results of instrument measurement (unit: degree)

| Inclination angle measured by instrument | Dxl360s level measurement results | absolute error |
|---|---|---|
| 9.98 | 10.00 | 0.02 |
| 8.99 | 9.01 | 0.02 |
| 7.98 | 8.00 | 0.02 |
| 6.99 | 7.00 | 0.01 |
| 5.98 | 6.01 | 0.03 |
| 4.97 | 5.01 | 0.04 |
| 3.98 | 4.00 | 0.02 |
| 2.95 | 3.02 | 0.07 |

## 4. Conclusion

In this paper, a new horizontal angle measuring device based on electromagnetic induction principle and mechanical gyroscope is designed and implemented. The system structure and working principle are described in detail, and the calibration, contrast experiment, repeatability experiment and error analysis of the instrument are carried out. The experimental results show that the system has good repeatability and accuracy. Compared with dxl360s level, the average absolute error of measurement is within 0.05 degree, which can be used in construction, civil engineering and daily life. The main advantage of the device is that it is not interfered by the external environment, and the algorithm analysis of the angle measurement data is simple, which provides a widely used scheme for measuring the angle of inclination.

## Acknowledgments

This work was financially supported by Shenyang Aerospace University Student's Platform for Innovation and Entrepreneurship Training Program fund.

First author: Zilin Zhang (2000.01--), female, sophomore, the main research direction is mechanical design, manufacturing and automation.

Corresponding author: Zhou Yan (1982.12--), male, engineer, main research direction is the application of numerical control technology and mechanical design research.

## References

[1] Joel S. Marciano,Calvin G. Hilario,Mary Ann B. Zabanal etal,Monitoring system for deep-seated landslides usinglocally-developed tilt and moisture sensors: System improvements and experiences from real world deployment [C], Global Humanitarian Technology Conference, 2014, pp.263-270.

[2] Villeneuve, Emma, Harwin, William, Holderbaum, William etal,Signal quality and compactness of a dual-accelerometer system for gyro-free human motion analysis [J], IEEE Sensors Journal, 2016, pp.6261-6269.

[3] Pengfei Gui,Liqiong Tang,Subhas Mukhopadhyay,MEMS Based IMU for tilting measurement: Comparison of complementary and kalman filter based data fusion [C], Industrial Electronics and Applications, 2015, pp. 2004-2009.

[4] Lu Kai, Principle and Application of Gyroscope National Defense Industry Press, July 1, 1981

[5] Almenise,New Gyroscope Technology, National Defense Industry Press, July 1, 2013.

[6] Jiang Zuoming, Angle Measurement, March 1, 1995.

[7] Mao Zhenfu, Magnetic Field Measurement, Beijing Atomic Energy Press, May 1, 1985.

[8] Tang Hongwei, Zhao Ruiming, Theoretical Basis of Electromagnetic Field, China Electric Power Press, April 2010.

[9] Zhang Yu,Development of space magnetic field detection device [D], Xi'an University of technology, 2013.

IOP Conf. Series: Materials Science and Engineering **772** (2020) 012068 doi:10.1088/1757-899X/772/1/012068

# Analysis of Fatigue Fracture and Strength Improvement of Backup Rolls

**Jiamin Zhang[a], Xiaochen Wang*, Huaqiang Liu[b], Quan Yang[c] and Dong Xu[d]**

National Engineering Research Center of Flat Rolling Equipment, University of Science and Technology Beijing, Beijing 100083, China

*Corresponding author e-mail: wangxiaochen@ustb.edu.cn,
[a]zhangjm_1993@163.com, [b]Lhuaqiang@163.com, [c]yangquan@nercar.ustb.edu.cn,
[d]xudong@ustb.edu.cn

**Abstract**. As an important part of the cold rolling mill, the backup rolls are always in service under complicated stress conditions. Once the roll fails, it will bring huge economic losses. In this paper, the problem of fracture failure of the backup roll neck position in an 1150mm six-roll reversible cold rolling mill of a company is analyzed. The stress distribution of the barrels under working condition is calculated by the finite element software ABAQUS, and the stress value at the dangerous section of the backup roll is obtained. The fatigue strength safety factor of the backup roll is obtained by analytical calculation. Combined with the actual production situation, from the perspective of structure and size design, improved schemes can be proposed to improve the fatigue strength of the backup roll. This analysis can provide reference for related enterprises.

## 1. Introduction

During rolling process, the backup rolls are subject to the combined effects of bending stress, shear stress, contact stress between the rolls, and residual stresses formed during the roller manufacturing process[1].Common forms of failure include roll surface wear, spalling, cracks, fracture, etc. Among them, the fracture of the backup roll not only increases the cost of production, but also greatly affects the normal production of the unit. Figure 1 shows the crack and fracture failure of a company's backup rolls.

(a) (b)

**Figure 1.** Crack and fracture of Backup roll (a) Crack (b) Fracture

Content from this work may be used under the terms of the Creative Commons Attribution 3.0 licence. Any further distribution of this work must maintain attribution to the author(s) and the title of the work, journal citation and DOI.
Published under licence by IOP Publishing Ltd

IOP Conf. Series: Materials Science and Engineering **772** (2020) 012068 doi:10.1088/1757-899X/772/1/012068

In recent years, Yan [2] calculated the bending stress of the backup roll neck and barrel by bending strength analytical calculation and finite element simulation to determine the weak point of the backup roll at the roll neck. Wang [3] analyzed the failure modes of the CSP production line backup roll, such as the crack of the barrel, the small drop; the shoulder; the cause of the crack at the arc of the roller neck, and proposed countermeasures. Sun [4] by comparing the common structures of several roll neck transitions, the final choice is to increase the radius of the arc to further reduce the stress level, thereby improving the bending strength of the backup roll of the cold rolling mill.Ouyang [5] through the analytical method to check the fatigue strength of high-speed finishing roller shaft, found that rolling force, rolling torque and steel temperature are the main factors affecting the fatigue strength of the roller. The fatigue strength of the roller shaft is improved by increasing the surface finish of the fracture portion of the roll and increasing the diameter of the fracture portion of the roll.

In this paper, the combination of finite element numerical simulation and analytical calculation is used to analyze the causes of fracture of the backup roll. Quantitative study of the relationship between different structures and dimensions and the safety factor of fatigue strength provides a reference for improving the fatigue strength of the backup rolls.

## 2. Analysis of the reasons for the fracture of the backup roll

### *2.1. Establishing a Finite Element Model of Roller System*

There are many factors that cause roll fracture failure. It is confirmed that these rolls have no defects in the manufacturing process, and the working load is also smaller than the maximum rolling force. Therefore, the causes of roll fracture are analyzed from the viewpoint of structure and size design. In order to be closer to the actual force of the backup roll, half geometric model of the roller system is established during the simulation.

The structure of the backup roll is guaranteed to be the same as that given by the company. Part of the roll structure and key dimensions are shown in Figure 2. The center distance of the bearing is 1860mm. The connection between the barrel and the roll neck is a transition circle with a radius of 75mm. The structure of the work roll and the intermediate roll is simplified, and the barrel and roll neck size are retained. The key dimensions of each roll are shown in Table 1.

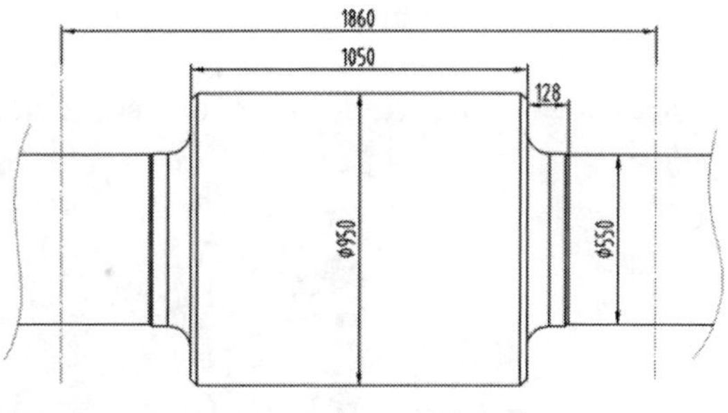


**Figure 2.** Part of the structure of the backup roll

**Table 1.** Key dimensions of rolls

| Roll Name | Barrel Length(mm) | Barrel Diameter(mm) | Neck Diameter(mm) |
|---|---|---|---|
| Work Roll | 1150 | 310 | 160 |
| Intermediate Roll | 1150 | 370 | 190 |
| Backup Roll | 1050 | 950 | 550 |

The backup roll material is the hypereutectoid alloy steel with Cr content of about 5%. The Cr5 steel has good hardenability and hardenability. The Young's modulus is 236,000 and the Poisson's ratio is 0.3.

IOP Conf. Series: Materials Science and Engineering **772** (2020) 012068 doi:10.1088/1757-899X/772/1/012068

The mesh is refined at the contact portion between the rollers and the portion where the dangerous section appears, that is, the transition portion between the roll body and the roll neck of the backup roll. The contact between the rollers is wedge-shaped mesh, and the rest is a hexahedral mesh. The mesh type is C3D8R. The overall roller meshing and local refinement are shown in Figure 3.

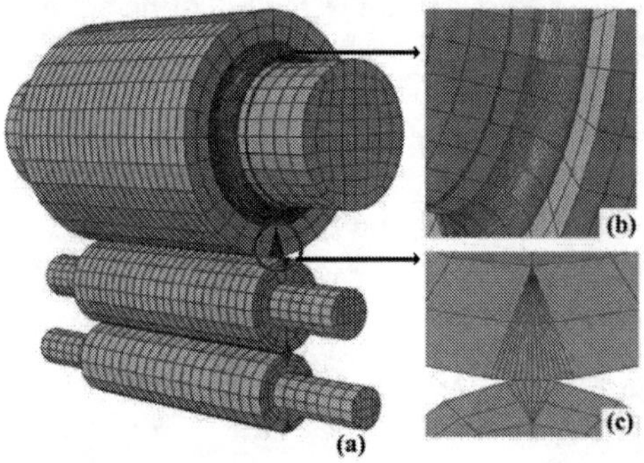


**Figure 3.** Mesh elements for the model (a) Whole structure of the model (b) Partcial enlarged view of the roll neck (c) Partcial enlarged view of the roller contact

Since the bending moment during the rolling of the six-high mill is mainly carried by the backup roll, when the strength of the backup roll is checked, it is considered to be subjected to the full rolling force. Set 75% of the maximum rolling force for the working load, that is, 9000KN.The backup roll sets a displacement constraint at the cross section of the bearing center. Both the work roll and the intermediate roll side are free ends.

## *2.2. Finite Element Simulation Calculation Results*

Figure 4 shows the stress distribution obtained by the finite element analysis of the static load of the roller system. Figure 4 (a) is the stress distribution of the roll system, and Figure 4 (b) is the stress distribution at the top of the backup roll. In order to facilitate the observation of the deformation of the roller system, the deformation scale factor is 150.It is known from the calculation results that the stress concentration occurs at the root of the backup roll neck, which is the most dangerous point of the full roller system. When the rolling force is 9000KN, the maximum tensile stress at the root of the backup roll is 148.34 MPa.

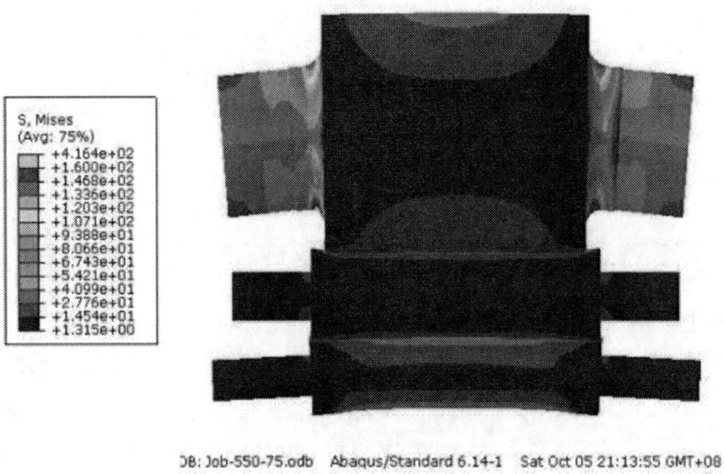


(a)

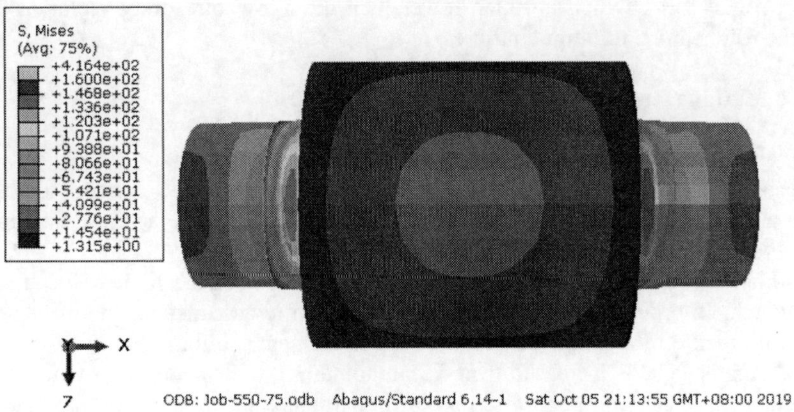


(b)

**Figure 4.** Stress distribution

### *2.3. Calculation of Safety Factor for Fatigue Strength of Backup Roll*

When structure or size of the roll is improper, the stress concentration at the root of the roll neck may cause fatigue cracks. During the high-speed rotation of the rolling, the alternating load subjected to compressive stress and tensile stress acts for a long time, crack propagation will cause the roll to fracture. In this paper, the fatigue strength of the backup roll is checked by calculating the fatigue strength safety factor.

According to the formula (1) and combined with the finite element simulation calculation of the stress value at the neck of the backup roll, the fatigue strength safety factor can be estimated [6].

$$n = \frac{\varepsilon_\sigma \beta \sigma_{-1}}{K_\sigma \sigma_{\max}} \tag{1}$$

$$\sigma_{-1} = 0.4\sigma_b \tag{2}$$

n–The safety factor ;$\sigma_{-1}$–The permanent limit of the standard sample material; $\sigma_{max}$–Maximum working stress;$\varepsilon_\sigma$–Dimensional coefficient; β–The surface quality coefficient, the surface of the backup roller is processed by grinding; $K_\sigma$– The stress concentration factor, the corresponding values of different roll sizes are different [7]. The values of these influence factors are shown in Table 2 [8].

**Table 2.** Influence coefficient value

| $K_\sigma$ | $\varepsilon_\sigma$ | $\beta$ | $\sigma_b$ **(MPa)** | $\sigma_{-1}$ **(MPa)** | $\sigma_{\max 1}$ **(MPa)** |
|---|---|---|---|---|---|
| 1.764 | 0.6 | 1 | 1500 | 600 | 148 |

Formula (3) is the safety factor of the roll according to the fatigue strength:

$$n_1 = \frac{\varepsilon_\sigma \beta \sigma_{-1}}{K_\sigma \sigma_{\max 1}} = \frac{0.6 \times 1 \times 600}{1.764 \times 148} \approx 1.38 \tag{3}$$

When designing shaft parts, if the shaft diameter is greater than 200mm, the allowable safety factor according to the fatigue strength calculation should be between 1.8 and 2.5. However, the safety factor of the support roll is 1.38, which is much lower than the allowable safety factor of 1.8. From this perspective, the roll is prone to fatigue failure.

## 3. Research on fatigue strength improvement of backup roll

### *3.1. Increasing the Fillet Radius*

With reference to the design of shaft parts, this section studies three options for reducing stress concentration and improving fatigue strength by changing the structure at the transition between the roll body and the roll neck or increasing the size.

Backup roll has a compact structure, and the excessive radius of the fillet will change the existing structure too much. In order to keep the original roll structure as constant, the influence of the fillet radius at the roll neck on the stress distribution under the same rolling conditions is studied. In this paper, the fillet radius value is increased from 75mm to 85mm of the original structure.

The finite element calculation and analysis of the structure shows that the maximum stress at the root of the roll neck is 139.7MPa, and the stress concentration factor is 1.714.According to formula (1), the fatigue strength safety factor of the backup roll with a fillet radius of 85 mm is calculated:

$$n_1 = \frac{\varepsilon_\sigma \beta \sigma_{-1}}{K_\sigma \sigma_{\max 2}} = \frac{0.6\times 1\times 600}{1.714\times 139.7} \approx 1.50 \tag{4}$$

When the transition fillet radius is increased by 10 mm, the safety factor is increased by 8.7%, but the safety factor is still not achieve the allowable safety factor. Therefore, under the premise of keeping the existing structure and function unchanged, the method has limited improvement on the bearing capacity of the roll.

### *3.2. Increasing the Neck Diameter*

The series of rolls with fatigue fracture, the roll neck diameter d is 550mm and the roll diameter D is 950mm, the ratio d/D is 0.58. In this part, the finite element calculation is carried out for the stress distribution of the barrel when the roll neck diameter d is 570mm, 590mm, that is, the d/D is 0.6, 0.62 and the transition fillet radius is 75mm and 85mm respectively. The corresponding maximum stress value and the calculated safety factor are shown in Table 3.

**Table 3.** Fatigue strength safety factor for different structural dimensions

| d/D | Fillet Radius(mm) | Stress Concentration Factor | $\sigma_{max}$ (Mpa) | Safety Factor |
|---|---|---|---|---|
| 570/950=0.6 | 75 | 1.772 | 134.7 | 1.51 |
| 590/950=0.62 | 75 | 1.797 | 118.8 | 1.68 |
| 570/950=0.6 | 85 | 1.702 | 126.9 | 1.67 |
| 590/950=0.62 | 85 | 1.712 | 113.2 | 1.85 |

It can be obtained from the calculation results that when the roll neck diameter is 590 mm and the transition fillet radius is 85 mm, the safety factor is greater than the allowable safety factor of 1.8. The fatigue strength condition is satisfied, and the roll neck is less prone to fatigue fracture.

The backup roll is mainly subjected to radial force during rolling. The bearing is usually selected from four-row cylindrical roller bearings. The outer dimensions of the bearing are large. Therefore, when designing the journal size, the bearing selection should also be considered. Through the above calculations, it is recommended to adopt a measure of appropriately increasing the diameter and the fillet radius of the backup roll neck to reduce the stress concentration at the root of the roll neck to improve the fatigue strength.

IOP Conf. Series: Materials Science and Engineering **772** (2020) 012068 doi:10.1088/1757-899X/772/1/012068

### *3.3. Changing the Structure of the Transition*
In addition to the rounded structure, the barrel and the roll neck may be combined with a bevel and a rounded corner. Proper structural design can effectively increase the radial dimension of the transition, thereby reducing stress concentration. Figure 5 is a structural comparison diagram of two structures, wherein the structure 1 represents a rounded structure and the structure 2 is a structure with a bevelled and rounded corner.

In order to make as few changes as possible to the original structure, the fillet radius is 50mm, and the stress distribution at the top of the roll body is calculated by finite element calculation as shown in Fig. 6. The maximum stress value at the transition is 123.1 MPa and the safety factor is about 1.88, also greater than the allowable safety factor of 1.8.

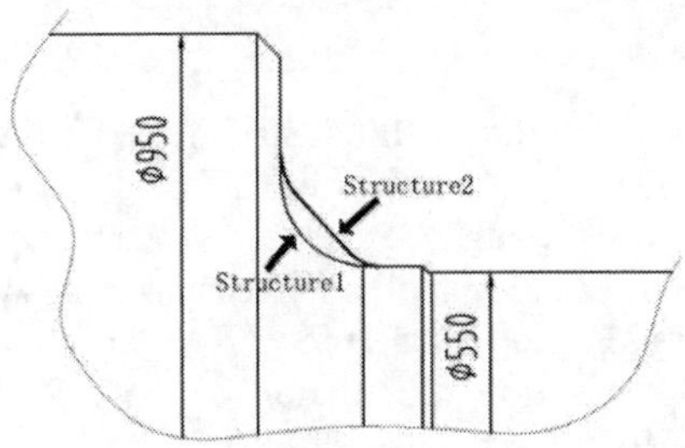


**Figure 5.** Structure comparison diagram

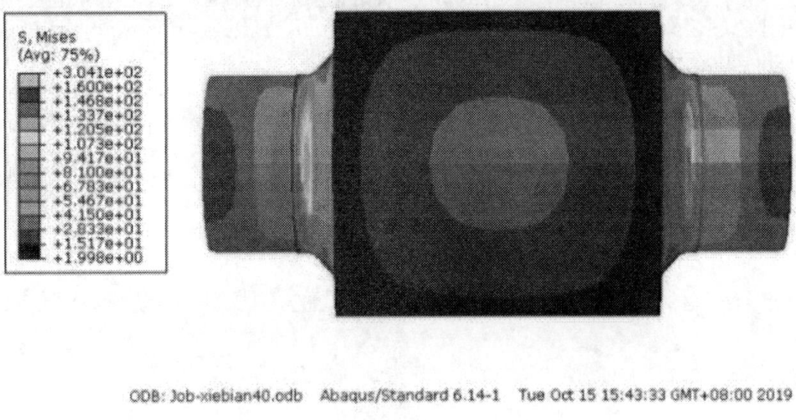


**Figure 6.** Stress distribution of structure 2

It can be seen from the above analysis that a reasonable structure can effectively reduce the stress concentration and improve the fatigue strength of the roll.

## 4. Conclusion
In this paper, the finite element software is used to calculate the stress distribution of the support roller in which the company has fracture failure, and the safety factor calculated from the fatigue strength at the stress concentration. Then, three structural size designs that can improve the fatigue strength of the support rolls are discussed. The conclusions are as follows:

(1) If the rounded transition structure is still used, the fillet radius is increased by 5 mm and the safety factor is increased by 9%-12%. The diameter of the roll neck is increased by 20 mm, the maximum stress at the dangerous section can be reduced by 12 MPa to 15 MPa, and the safety factor

is increased by about 10%. The ratio d/D of the diameter of the journal journal d to the diameter D of the roll body should be greater than or equal to 0.62 to ensure sufficient fatigue strength.

(2) If the rounded corner structure of the transition is changed to a bevelled and rounded structure, the fillet radius is 50mm, and the angle of the bevel is 45°, the safety factor can be satisfied if the roll neck diameter is constant.

## Acknowledgments

This work was supported by Beijing Natural Science Foundation (Grant No.3182026), National Natural Science Foundation of China (Grant No.51975043 and No.51604024), Guangxi Special Funding Program for Innovation-Driven Development (Grant No.GKAA17202008) and Fundamental Research Funds for the Central Universities (Grant No. FRF-TP-19-002A3).

## References

[1] J. Zou, *Rolling Machinery*. (Metallurgical Industry Press, Beijing, 2014).
[2] H. Yan and A.M. Wu," Bending strength check of cold rolling mill support roll " in The 7th (2009) China Iron and Steel Annual Conference Proceedings (below). 2009.
[3] Q.Y. Wang, Failure Form of Back-up Roll and Reasons Analysis in Liangang CSP Plant. Metal Materials and Metallurgy Engineering, 2009, 37(4): 32-34.
[4] X.Z.Sun and Z.L.Liu. Optimization Calculation of Cold Rolled Flat Roller Structure Large Castings, 2019(5): 54-55.
[5] G.Ouyang,. High Speed Wire Finishing Roller Shaft Fatigue Fracture Analysis and Processing. Jiangxi Metallurgy, 2016, 36(3): 35-40.
[6] R.H. Liu and X.Z. Guo, *Material Mechanics* (Chemical Industry Press, Beijing ,2017 ).
[7] W.Sa and Z.G.Zhang, *Stress Concentration Handbook.* (Heilongjiang Science and Technology Press, Heilongjiang,1983).
[8] Y.Yuan. "Analysis of Ultimate Heat Treatment on Microstructure and Property of Cry Steel Back-up Rolls". M.S. thesis, Henan University of Science and Tecnology.2015.

IOP Conf. Series: Materials Science and Engineering **772** (2020) 012069 doi:10.1088/1757-899X/772/1/012069

# Analysis on calculating process of mechanical precision of devices on satellite

**Bijie Xue*, Lei Zhang, Hongxian Zhong and Fa Jing**

Institute of Telecommunications Satellite, China Academy of Space Technology, Beijing, China

*Corresponding author e-mail: xuebijie@126.com

**Abstract.** With the gradual complication of the devices on satellite, the mechanical precision of some devices with accuracy requirements on satellite should be predicted after the assembly, in order to conduct the previous design of devices and accuracy assignment of satellite. This paper will discuss main factors which influence the mechanical precision of devices, process design and calculating method. And the position and orientation matrix is used to analyse the influence of error at each stage on mechanical precision of devices after installation. One thruster on satellite is chosen as an example to specify the calculating process and feasibility of this calculating method. This paper proves that the satellite structure assembly has a small impact on mechanical precision. The control method of mechanical precision is put forward at design stage and assembly stage of satellite, and the mechanical precision and error distribution method of devices on satellite is predicted through position and orientation matrix at previous work stage.

## 1. Introduction

The attitude control of satellite on orbit is collectively completed by gyroscopes, sensors and thrusters. The premise of these devices which operates stable and efficient is that those devices still keeps high quality of mechanical precision after installation. In order to ensure the normal work of satellite, some key devices in satellite have the strict requirement of mechanical precision [1]. Before the installation of those devices asked for mechanical precision, the satellite will experience some stages such as structure assembly, cabins disassembling. These repeated disassembly, coordinate transfer and large-scale test will produce effects on the mechanical precision of device after installation. With the development of satellite assembly technology and the gradual complication of devices on satellite, more and more typical multibody systems appear in satellite [2]. The installation process of those devices is complex and non-testable links are numerous, and how to calculate the mechanical precision of these devices becomes a problem that should be solved urgently. [3]

This paper discusses main factors which influence the mechanical precision of devices, process design and calculating method. From the very beginning of satellite manufacturer, the position and orientation matrix is used to analyze the influence of error at each stage, such as satellite structure assembly and satellite precision basement transfer, on mechanical precision of devices after installation. And one thruster on satellite is chosen as an example to specify the calculating process and feasibility of this calculating method.

Content from this work may be used under the terms of the Creative Commons Attribution 3.0 licence. Any further distribution of this work must maintain attribution to the author(s) and the title of the work, journal citation and DOI.
Published under licence by IOP Publishing Ltd

## 2. Calculating process of mechanical precision

*2.1. Satellite structure assembly error*

Satellite of central cylinder will firstly experience structure assembly, then the device assembly. The errors from structure assembly accumulate to device assembly stage, which will influence the mechanical precision of devices. The structure assembly includes propulsion module assembly and service module assembly, the installation process of main panels is shown as follows.

(1) Central cylinder assembly

Take the positioning pin on satellite-rocket separation interface as a standard, the central cylinder is connected to foundation support of structure assembly through bolts. Z-axis of the central cylinder is vertically upward to establish the precision basement of structure assembly.

(2) Horizontal panels assembly

Two horizontal panels are installed to central cylinder which is shown in Fig.1. A holder is fixed to the foundation support of structure assembly, which can assign the parallelism of two horizontal panels with satellite-rocket separation interface. It is assumed that M and N are two installation holes of some device on horizontal panel, because of the leaning within the limit of error, that is to say, two coordinate values of Z-axis under satellite coordinate are different, which will result in the leaning of device after installation. So in the device assembly stage, the mechanical precision of device should consider the inherent errors in the stage of structure assembly stage.

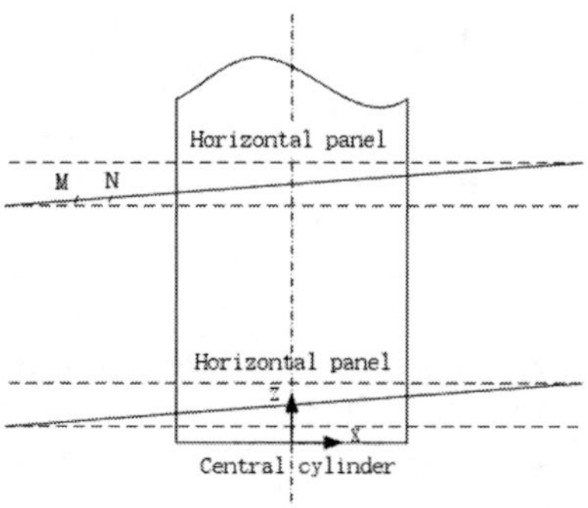


**Figure 1.** Installation diagram of two horizontal panels.

Then other panels are installed into central cylinder or two horizontal panels. The precision requirements of every panel are presented according to the satellite coordinate in the structure assembly, so the precision relationships between satellite panels are decoupled.

*2.2. Satellite precision basement transfer error*

In the stage of satellite structure assembly, the satellite mechanical coordinate is determined by three positioning pin in precision turntable. Before the satellite turned into the stage of devices assembly, the basement of satellite should be transferred to reference mirror in satellite. This process is called satellite precision basement transfer, which will use four mounting theodolites to realize, and this process will introduce a certain mechanical error.

The auto collimation measuring error of one mounting theodolite is ∂_1=0.5", the leveling error of one mounting theodolite is ∂_2=0.7", the calibration error and milling error of reference mirror are ∂_3= 〖3.6〗^" and ∂_4=1". Mounting theodolite system will measure for three times, and RSS method is used to calculate error. Angular measurement error at every time is as follows.

$$\sigma_1 = \sqrt{\partial_1^{\ 2} + \partial_2^{\ 2}} \tag{1}$$

$$\sigma_2 = \sqrt{\partial_1^{\ 2} + \partial_2^{\ 2} + \partial_3^{\ 2} + \partial_4^{\ 2}} \tag{2}$$

IOP Conf. Series: Materials Science and Engineering **772** (2020) 012069 doi:10.1088/1757-899X/772/1/012069

$$\sigma_1 = \sqrt{\partial_1^{\ 2} + \partial_2^{\ 2}} \tag{3}$$

3σ rule is chosen and measuring accuracy is rounded [4], the angular measurement errors are 0.0008°, 0.004°and 0.0008°. In the practical basement transfer process, mounting theodolite system will measure for three times, and RSS method is used again to calculate error. Angular measurement error is as follow.

$$\sigma = \sqrt{\sigma_1^{\ 2} + \sigma_2^{\ 2} + \sigma_3^{\ 2}} = 0.004157° = 14.965'' \tag{4}$$

The angle value above is the worst case in the stage of satellite precision basement transfer. When structure assembly and precision basement transfer completed, satellite turns into the stage of device assembly. The error from structure assembly will also reflect in the mechanical precision of devices.

## 3. Calculating process design of mechanical precision of devices

Before devices installation, satellite will experience stages of structure assembly and precision basement transfer, as described in the previous section. The errors of these two stages will reflect in the mechanical precisions of device. So they will be seen as two point errors and added into the mechanical precision links of devices, shown as Fig.2.

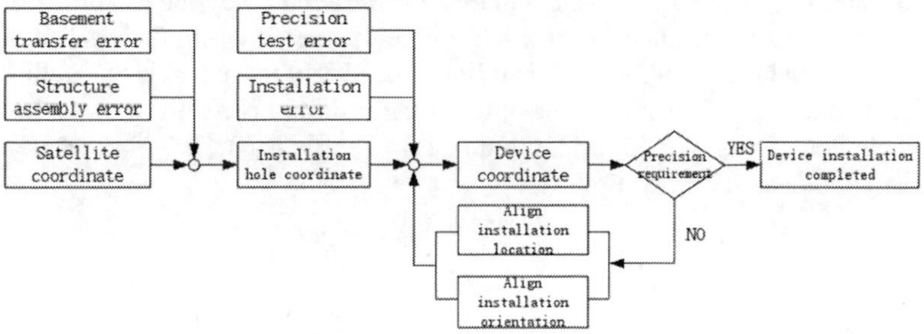


**Figure 2.** Calculating process of mechanical precision of devices on satellite.

Shown as Fig.2, when mechanical precisions of devices are calculated, there are three coordinate should be considered, satellite coordinate, installation hole coordinate and device coordinate. Build the relationship between these three coordinate through location and orientation matrix and reflect in the mechanical precisions of device.

Location and orientation matrix is a block matrix of 3×4. The first three columns present the orientation relationship between two coordinates. The last one column presents the location relationship between two coordinates. When a device is installed to a satellite, the relationship between device coordinate and satellite coordinate will be obtained through mounting theodolite or laser tracker, and the location and orientation matrix is obtained then.

Location and orientation matrix describes the location and orientation relationships between two coordinate [5]. Lead the location and orientation matrix method into the calculating of mechanical precisions of device, and treat the structure assembly error and basement transfer error as two errors and add them into the location and orientation matrix between satellite coordinate and installation coordinate in the form of continuous multiplication. So the errors of lifecycle of satellite will be considered into mechanical precisions of device. One thruster is taken as an example to explain the mechanical precision method of device based on location and orientation matrix.

## 4. Example

It is known that one thruster is fixed upon the edge +X+Y of one panel. From structure assembly stage, the influence of installation error, satellite basement transfer error and bracket installation error on mechanical precision of thruster is considered. This process involves three coordinates, there are satellite coordinate, panel installation hole coordinate and thruster coordinate, shown as Fig.3.

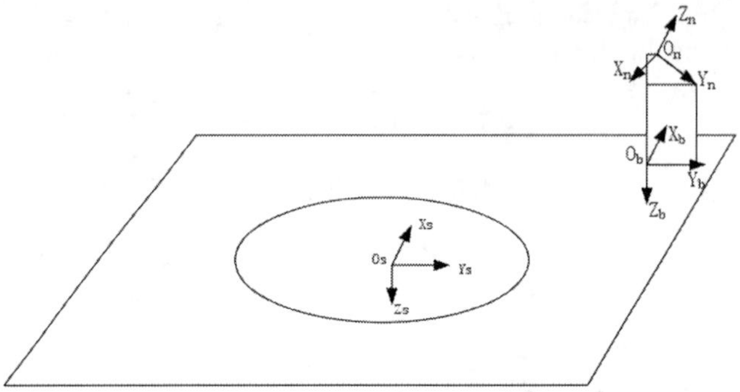


**Figure 3.** Installation diagram of one thruster.

(1) Relationship between satellite coordinate and panel installation hole coordinate

The origin of satellite coordinate is located in the center of central cylinder, while the origin of panel installation hole coordinate is located in the +X+Y edge of one panel. In the case of theory, the three axes of panel installation hole coordinate are parallel respectively to each axis of satellite coordinate. So it is assumed that in the case of theory, the position of the origin of panel installation hole coordinate in satellite coordinate is (1180mm, 1050mm, 100mm), Euler angle is (0°, 0°, 0°), the theory position and orientation matrix between these two is:

$$C^{\mathrm{T}}_{b\to s}=\begin{pmatrix}1 & 0 & 0 & 1180\\ 0 & 1 & 0 & 1050\\ 0 & 0 & 1 & 100\end{pmatrix} \tag{5}$$

In practice, it is assumed that the maximum error of structure assembly of one panel is $\pm 1$mm, and the satellite basement transfer error is 15″. Then the position changes to (1179.999941mm, 1049.999947mm, 101mm). Let the norm of the axes of satellite coordinate be 1, there are:

$$\begin{array}{ll}\vec{x}_s=(1,0,0) & \vec{x}_b=(0.99999995,0,0.00031654)\\ \vec{y}_s=(0,1,0) & \vec{y}_b=(0,0.99999995,0.00031654)\\ \vec{z}_s=(0,0,1) & \vec{z}_b=(0.00031654,0.00031654,0.99999995)\end{array} \tag{6}$$

Then the actual angle matrix between satellite coordinate and panel installation hole coordinate is obtained. The position and orientation matrix and Euler angle are obtained through cosine of that actual angle matrix. Then add the satellite basement transfer error of 15″ to this Euler angle, the actual position and orientation matrix is obtained.

$$C^{\mathrm{R}}_{b\to s}=\begin{pmatrix}0.99999997 & -0.00007272 & -0.00024378 & 1179.999941\\ 0.00007278 & 0.999999968 & 0.000243775 & 1049.999947\\ 0.00024376 & 0.000243793 & 0.999999941 & 101\end{pmatrix} \tag{7}$$

(2) Relationship between panel installation hole coordinate and thruster coordinate

Thruster coordinate is located in the top of its bracket. Local enlarge Fig.3, the relationship between panel installation hole coordinate and thruster coordinate is obtained, as Fig.4 shown. It is assumed that in the case of theory, the position of the origin of thruster coordinate in panel installation hole coordinate is (0mm, 64.65mm, -224.64mm). Let the norm of the axes of satellite coordinate be 1, there are:

$$\begin{array}{ll} \vec{x}_b = (1,0,0) & \vec{x}_n = (0,0.25881905,0.96592583) \\ \vec{y}_b = (0,1,0) & \vec{y}_n = (1,0,0) \\ \vec{z}_b = (0,0,1) & \vec{z}_n = (0,0.96592583,-0.25881905) \end{array} \tag{8}$$

The theory angle matrix is obtained through the included angles of these six vectors:

$$A^{\mathrm{T}}_{n\to b} = \begin{pmatrix} 90 & 0 & 90 \\ 74.99999971 & 90 & 14.99999918 \\ 14.99999918 & 90 & 105 \end{pmatrix} \tag{9}$$

Then the theory Euler angle is (-105°, 0°, -90°), the theory position and orientation matrix is:

$$C^{\mathrm{T}}_{n\to b} = \begin{pmatrix} 0 & 1 & 0 & 0 \\ 0.258819 & 0 & 0.96592583 & 64.65 \\ 0.96592583 & 0 & -0.258819 & -244.64 \end{pmatrix} \tag{10}$$

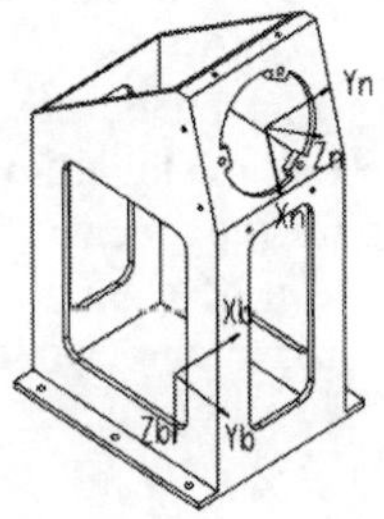


**Figure 4.** Two coordinates of panel installation hole and thruster.

(3) Relationship between satellite coordinate and thruster coordinate

The position and orientation matrix $C^{\mathrm{T}}_{b\to s}$, $C^{\mathrm{T}}_{b\to s}$ between satellite coordinate and panel installation coordinate are obtained from step (2), and the position and orientation matrix $C^{\mathrm{T}}_{n\to b}$ between panel installation coordinate and thruster coordinate is also obtained. Then the theory position and orientation matrix between satellite coordinate and thruster coordinate is gained as $C^{\mathrm{T}}_{n\to s} = C^{\mathrm{T}}_{b\to s} \cdot C^{\mathrm{T}}_{n\to b}$. And the actual position and orientation matrix $C^{\mathrm{R}}_{n\to s} = C^{\mathrm{R}}_{b\to s} \cdot C^{\mathrm{R}}_{n\to b}$ considering errors from structure assembly and satellite basement transfer is

$$C^{\mathrm{T}}_{n\to s} = C^{\mathrm{T}}_{b\to s} \cdot C^{\mathrm{T}}_{n\to b} = \begin{pmatrix} 0 & 1 & 0 & 1180 \\ 0.258819 & 0 & 0.96592583 & 1114.65 \\ 0.96592583 & 0 & -0.258819 & -144.64 \end{pmatrix} \tag{11}$$

$$C^{\mathrm{R}}_{n\to s} = C^{\mathrm{R}}_{b\to s} \cdot C^{\mathrm{R}}_{n\to b} = \begin{pmatrix} -0.00025429 & 0.99999997 & 0.0000071 & 1179.999941 \\ 0.25905446 & 0.00007278 & 0.96586271 & 1114.649947 \\ 0.96598887 & 0.00024376 & -0.25858350 & -143.64 \end{pmatrix} \tag{12}$$

$$\Delta = \left|C^{\mathrm{R}}_{n\to s} - C^{\mathrm{T}}_{n\to s}\right| = \left|C^{\mathrm{R}}_{b\to s} \cdot C^{\mathrm{R}}_{n\to b} - C^{\mathrm{T}}_{b\to s} \cdot C^{\mathrm{T}}_{n\to b}\right| = \begin{pmatrix} 0.00025429 & 0.00000003 & 0.0000071 & 0.000059 \\ 0.00023546 & 0.00007278 & 0.00006312 & 0.000053 \\ 0.00006304 & 0.00024376 & 0.00023550 & 1 \end{pmatrix} \tag{13}$$

From the value Δ of theoretical subtracting actual position and orientation matrix, it can be seen that the influence of error from structure assembly on mechanical precisions of device is very small. If the errors from structure assembly and satellite basement transfer are only considered, while the error from device assembly is not considered, the maximum angle error of thruster is only 0.00025429°, and the maximum position error is 1mm. Taking one thruster as an example and using the position and orientation method, not only specify the feasibility of this calculating method, but also demonstrate the rationality of precision allocation in the stages of structure assembly and satellite basement transfer.

## 5. Conclusion

This paper discusses main factors which influence the mechanical precision of devices, process design and calculating method. The position and orientation matrix is used to analyze the influence of errors at each stage, such as satellite structure assembly and satellite precision basement transfer, on mechanical precision of devices after installation. And one thruster on satellite is taken as an example to specify the calculating process and feasibility of this calculating method. This paper proves that the satellite structure assembly has a small impact on mechanical precision of device. The control methods of mechanical precision are put forward at design stage and assembly stage of satellite, and the mechanical precision and error distribution methods of devices on satellite are predicted through position and orientation matrix at previous work stage.

(1) Mechanical precision of device after installation is thought ahead, which may reduce the cost of devices and conserve resources of machine work.

(2) If it is known that the design precision of some device on satellite is lower, the installation location of device on satellite can be aligned at the design stage. Some location on satellite which has a higher mechanical precision is chosen to make up for the mechanical precision of device.

(3) The mechanical precision of device can be calculated ahead through the method of this paper, and compare with test value after installation, which verifies the correctness of design process.

## References

[1] Han Dong, Zhong Shisheng, Xia Pingjun. Modeling & development of assembly precision adjusting system for devices on satellite [J], Computer integrated manufacturing systems, 2007, 13(12).

[2] You Fudi, Analysis and control of dynamic pointing accuracy for satellite antenna [C], Harbin Institute of Technology, 2011

[3] Dirk L, Martin G, Micheal B J. In flight simulation in support of an aircraft certification process[C], AIAA Modeling and simulation technologies conference and exhibit, Austion, Texas, 2003.

[4] Zaihua Yang. Measurement of orientations of optic axis of space camera with theodolite [J], Spacecraft environment engineering, 2010, 27(1).

[5] Bhatta S, Agrawal S K, Spherical rolling robot: A design and motion planning studies[J], IEEE Transaction on robotics and automation, 2000,16(3).

# The Measurement to the Pillar Fillet Weld of the Differential Settlement Spherical Tank by the ACFM

**Zemin Wu [a], Guang Dai, Guoqiang Zhou and Jiancheng Leng**

The Northeast Petroleum University, Daqing 163318, China.

[a] jaywu1980@163.com

**Abstract.** Present the method that using the ACFM to test the spherical tank which had appeared the differential settlement. First introduced the principle and the system composition of the ACFM, by using which, the pillar fillet weld of the LPG spherical tank had been tested, and then the quantitative evaluation of the weld cracks had been get out. After analyzed deeply, the reliability of the ACFM when testing the pillar fillet weld of the spherical tank had been proved, in additionally, the causes of how the cracks come out had been given, too.

**Keywords:** Spherical tank; Differential settlement; Pillar fillet weld; Alternating current field measurement.

## 1. Introduction

The Spherical tank is one of the most common pressure vessel in petroleum and chemical industries, which was mainly used in storing of the material, like LPG and LNG, etc. , which included the shells, the supports and the accessories, and the pillars could be fell into pillar type and apron type. At present, most of the spherical tanks use the support type of pillar with equatorial tangent. To the spherical tank with pillar support, the welding connection between the pillar and the shell would support the weight of the whole sphere and the material.

Differential settlement is a common problem to the spherical tank, which should lead the stress unbalance in each pillar of the spherical tank, and could cause additional stress, which would easily cause the crack of the pillar fillet weld to appeared out, and evenly impact the safety in using of the spherical tank [1-3]. As to the special structure style of the pillar fillet weld, normal testing methods would cost much more time, and there might be some chances of mis-detection. Thus, this text used ACFM to do the test, which had many disadvantages like non-contacted measurement, high speed acquisition, high accuracy, qualified and quantified quickly, and the testing result could be shown with computer graphics, which could make a good effect to detect the pillar fillet weld in test.

## 2. The ACFM Technique

### *2.1. The Detecting Principle*

The ACFM technique was founded and developed on the basis of ACPD, of which, the theoretical basis was the electromagnetic induction principle[4-5], that is to say, when an inductive probe with alternating current approached the work piece, the alternating current would produce a alternating magnetic field

Content from this work may be used under the terms of the Creative Commons Attribution 3.0 licence. Any further distribution of this work must maintain attribution to the author(s) and the title of the work, journal citation and DOI.
Published under licence by IOP Publishing Ltd

around itself, and according to the test of the intensity variation, people could judge whether there was a crack or not, as shown in the Figure 1(in which, the X axial was in the direction of the crack , and the Y axial perpendicular to the plane of the crack, and meanwhile, paralleled to the current direction, the Z axial perpendicular to the surface of the conductor, and pointed to outside). When there wasn't flaw on the surface of the work piece, when the induction density Bx, which paralleled to the surface of the work piece, paralleled to the direction of the X axial, then the induction density Bz, which perpendicular to the surface of the work piece will be zero, and if the flaw existed, with the increasing of the crack's resistivity index, the current field and the induced magnetic field would be distorted, in which, if the scattered electric streamlines corresponded to the crossing of the Bx, as the magnetic induction intensity, then the depth of the crack could be tested, if the clockwise and anti clockwise electric streamlines respectively corresponded to the positive and negative peak of the Bz, as the magnetic induction intensity, then, the length of the crack could be tested. Thus, according to the variation of two mutually orthogonal magnetics, people could quantitatively check out the size of the flaw.

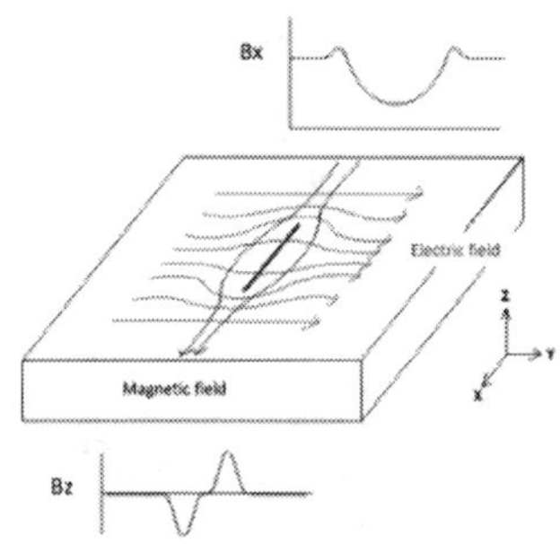


**Figure 1.** The schematic diagram of the ACFM.

### *2.2. Detecting System*

The ACFM testing system included two parts, such as the hardware and the software [6]. The hardware part, is made up of the scanner, the ACFM mainframe, and the computer, etc. as shown in Figure 2, and the scanner drived out the uniform alternating current on the surface of the workpiece, then get the magnetic signal's changes from the surface of the workpiece in the same time, and analyzed and processed the signal with crack detection analysis software, could intelligently analyzed and identified the crack of the workpiece in testing.

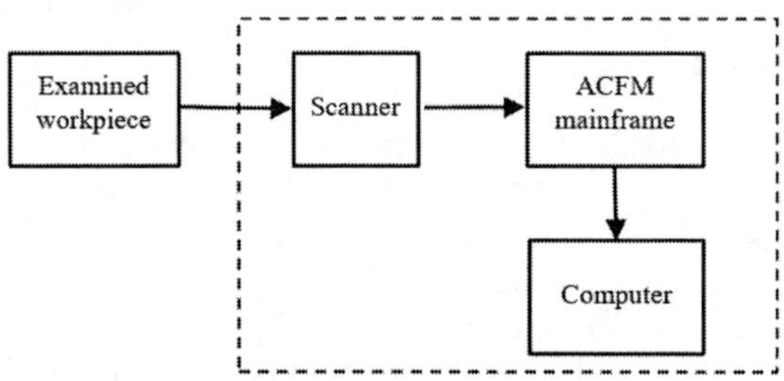


**Figure 2.** ACFM testing system

## 3. The ACFM Testing To the Pillar Fillet Weld

A LPG spherical tank belonged to a petrochemical company was built and went into operation in 2006, whose volume was 2500m³, who had the structure with the pillar of equatorial tangent with 10 pillars. During the regular checking in 2016, the spherical tank was found out that there was a differential settlement problem in the tank, which is, the settling differential value between the adjacent pillars was too big, because of which the standard request of GB12337 was exceeded [7]. Thus, it was very necessary to measure the pillar fillet weld by the ACFM to ensure the spherical tank operate safety.

### *3.1. The Testing Method*

In order to record and issued the report the 10 pillars of the tank was in ordered with number, as the figure 3. In thinking of the working place of the pillar fillet weld area was limited and the length of the weld was long, the testing area was divided into 4 parts (a testing part, b testing part, c testing part and d testing part), the testing could began, as shown in Figure 4.

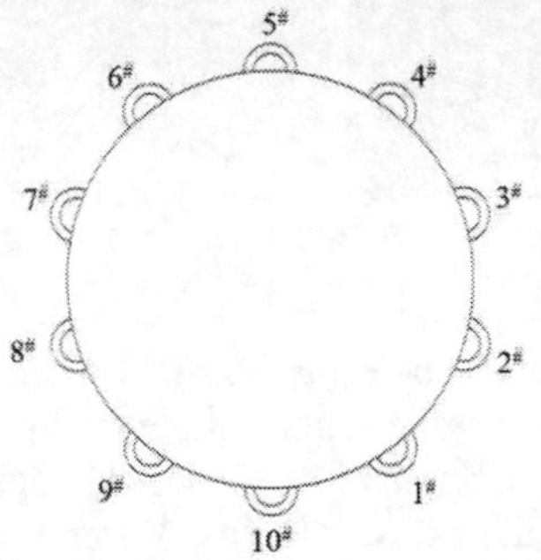


**Figure 3.** The number of the spherical tank's pillars

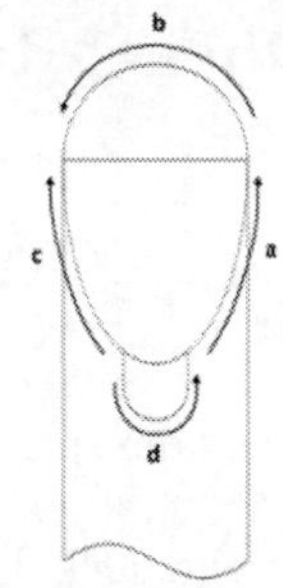


**Figure 4.** The testing to the pillar fillet weld

According to the structure of the pillar fillet weld area, the pencil-type probe was used, after which was connected with the main engine, the first step was to checkout the equipment and the probe by using the standard testing panel, in order to ensure the ACFM system operate normally to meet the requirements of the detection precision, the curve of the testing signal was shown in figure 5.

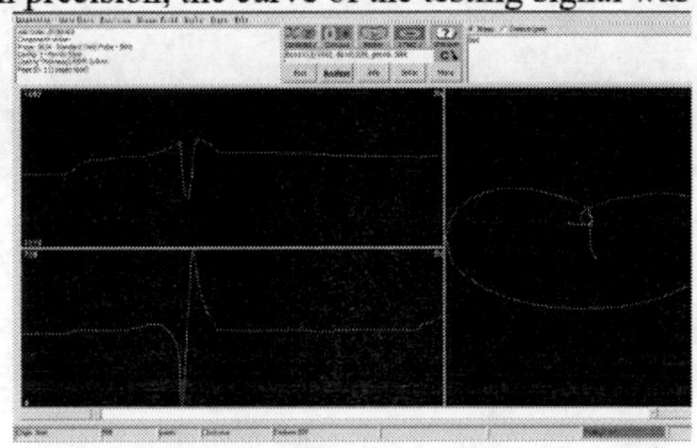

**Figure 5.** The testing signals of the ACFM's standard panel

During testing, the probe always moved along with the direction which paralleled with the welded joints, because the scanning speed would affect the strength of the testing signals directly, so the scanning speed should be steady, so the probe should maintain the state to fit into the weld and the weld-toe. The width of the pillar fillet weld was about 40mm, the probe would scan 3 times in the parallel directions to the weld to ensure covering to whole width of the weld. When scanning the sectorial weld, for fear of the missing test caused by fringe effect and the lift-off effect, every part must be scanned overlapped by the length of 50mm. And scanned the transversal crack of the weld in a Z-shape [8], then recorded the marked position in the weld corresponded to the curve peak of Bx, after which, the strike of the crack could be confirmed by connecting the marked positions.

During the testing process, the software could store the signal curve scanned by the probe automatically, and also had the function of replaying, which could be checked and analyzed lately.

### *3.2. The Testing Result*

When testing the 10 pillar fillet welds of the spherical tank with differential settlement by the ACFM, it was found that there was a suspicious crack in the 6th site of the pillar 5# from which, as shown in figure 6.

IOP Conf. Series: Materials Science and Engineering **772** (2020) 012070 doi:10.1088/1757-899X/772/1/012070

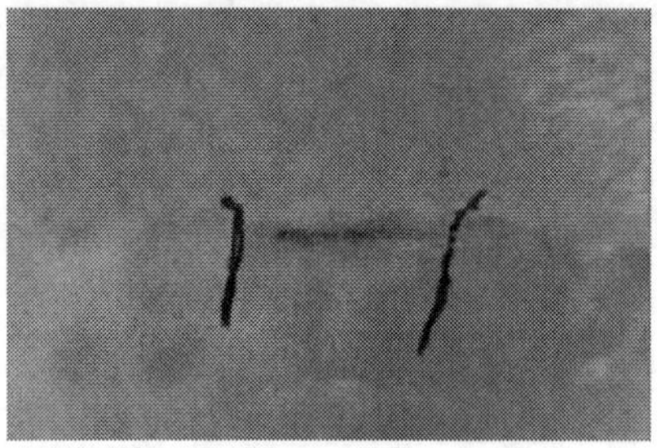

**Figure 6.** The position of the suspicious crack

When quantitatively evaluate to the suspicious crack, after measuring the mark point corresponded to the trough and the crest of the Bz signal according to the original curve, the estimated value of the crack length was get, from which, the system software could compensate to the crack length on the basis of the theoretical model, then size of the crack could be calculated, which was, 13mm long and 2.2mm deep, as shown in figure 7.

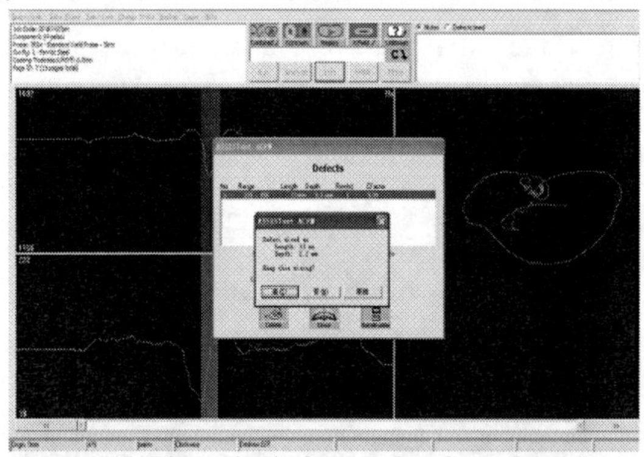

**Figure 7.** The calculation to the size of the flaw

After the treatment of depainting had been done to the surface of the suspicious cracking area, by using the MPT method, the existence of the crack could be proved, and according to the result of MPT, it was found that, the crack's length was about 12~13 mm, and after polished and smooth transited to the crack, it was also found that the depth of the crack was about 2~2.5 mm, and the flaw didn't appear anymore when rechecked.

## 4. The Cause Analysis of the Crack in Pillar Fillet Weld

### *4.1. Regarding to the welding material*

After looking up the construct data of the spherical tank, it was informed that, the tank was installed not only with the rods Ø3.2 as the backing weld, but also the rods Ø4.0 in other levels whose content of diffusible hydrogen were 6.46ml/100g, which had exceeded the claim in GB12337 as the content should less than or equal to 6ml/100g. If the diffusible hydrogen in the weld was too much would cause hydrogen brittleness and delay cracks, which would cause the low stress crack in the structure.

### *4.2. Regarding to the quality of the welding process*

Because of the structure of the fillet weld of the connection between the pillars and the shell of the spherical tank, affected by which, there must be a V-type groove on the pillar in sideways, which had used on side welding Bothe sides formation technology. This kind of welding process could lead to the non-fusion flaws on the backside, such as, the pore inside the welds and the slag inclusions, etc., which would become to the pillar fillet weld cracks under alternate loading over a long period of time.

### *4.3. Regarding to the differential settlement*

This spherical tank was built on the coastal backfill foundation, which would lead to the differential settlement after a long-time running. During the regular checking in 2016, the value of the differential settlement of the pillars as shown in table 1(take pillar 1# as the measurement standard), could be 12.6mm in Maximo between two adjacent pillars.

**Table 1.** The differential settling volume of the pillar

| Number of the pillar | Settlement value (mm) |
|---|---|
| 1# | 0 |
| 2# | -7.2 |
| 3# | -11.8 |
| 4# | -19.7 |
| 5# | -14.0 |
| 6# | -24.6 |
| 7# | -12.0 |
| 8# | -7.6 |
| 9# | -4.9 |
| 10# | -3.2 |

After setting up the finite element model by using of the ABAQUS software, as shown in Figure 8. Recording to the working condition of the spherical tank, when applied load to the finite element model, including that, the gravitational load— the gravity of the shell, pillars, pull-rods and the accessories, the gas load— the oil gas, pressure over LPG, the hydraulic load— the gravity action of the LGP to the internal surface of the spherical tank. The boundary condition of the finite element model was the value of the differential settlement. By setting the downward displacement along the axial direction with the bottom of the spherical tank's pillars to carry out the value of the settlement between the pillars. From the cloud atlas of the stress coming from the pillars and the shell of the tank, as shown in Figure 9 and 10, it could be seen that, when differential settlement appeared, the stress distributed very unevenly on the pillars, that is, the pillars 1#, 3#, 5# and 8# had raised bigger stress and the value of the stress in the fillet weld area connected to the shell was big too, to which, the out load (like wind, snow, etc.) overlaid, the stress would exceed the allowable strength of the material, which would lead the structural damage in parts of the tank.

**Figure 8.** The finite model of the spherical tank.

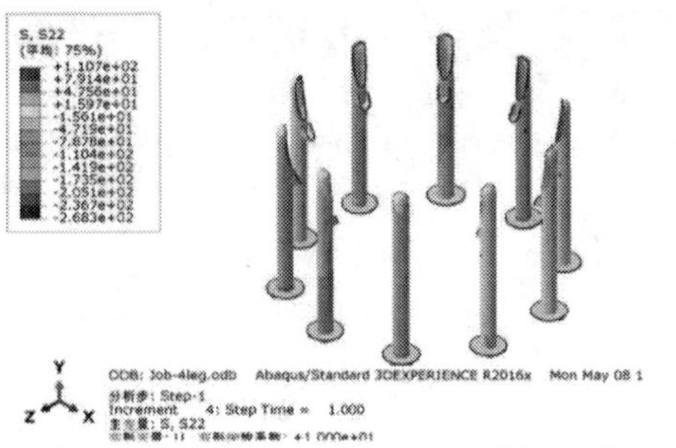

**Figure 9.** Stress nephogram of the pillars

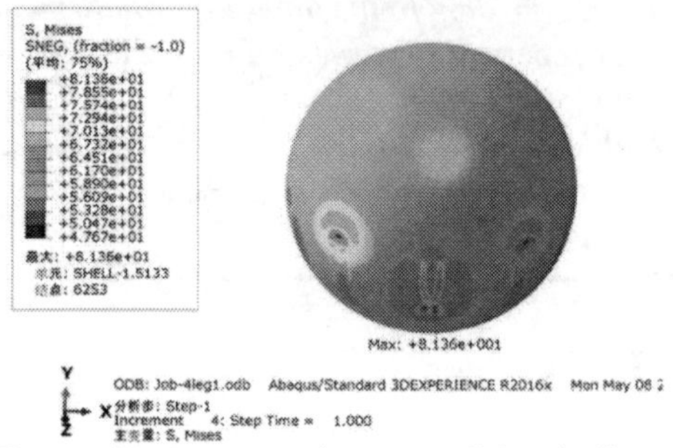

**Figure 10.** Stress nephogram of the shell

## 5. The conclusion

(1) When measuring the pillar fillet weld of a spherical tank which had differential settlement with the ACFM, the crack had been found out from the testing position in stage d of the pillar 5#, to which, processed the evaluation, the existence of the crack could be varied with MT method.

(2) When the differential settlement appears in the spherical tank, there would be a additional stress produced from the fillet welds between the pillars or between the pillars and the spherical shell, which is, added the effected of external factors and load conditions, and would produce flaws such as crack easily, which should be the important testing part in NDT.

(3) According to the local testing and application results, it was shown that, to test the pillar fillet weld of the spherical tank by using the ACFM method, can meet the site operation demand such as no need to polish, high testing speed, the quantitative measurement to the flaw, low effect from external force, etc. From those, the ACFM method is proved to be an effective method of testing to pillar fillet welds of the spherical tank.

## References

[1] Hongli GAO, Weijun Li, Zhihai Li. Stress Analysis of A LPG Spherical Tank on Multiple Differential Settled Groud. Pressure vessel Technology, Vol.30 (2013), No.1, p.40-44.

[2] Yongwei WANG, Wanchun FAN, Xia YIN. FEM analysis of stress of the connection of support to shell of spherical tank. Petro-chemical Equipment, Vol.36 (2007), No.6, p.69-72.

[3] Jun CHEN, Jiangying SHI, Daoqin ZHAO. Analysis of Cracks in Fillet Welds of Spherical Shell Columns. Special Equipment Safety Technology, Vol.2 (2016), p.26-27.

[4] Jianwei ZHOU, M.C.LUGG, R.A.COLLINS. Non-uni-form Model for Alternating Current Field Measurement of Fatigue Cracks in Metals. International Journal of Applied Electromagnetics and Mechanics, Vol.10 (1999), No.3, p.221-235.

[5] B BLAKELEY, M LUGG. Application of ACFM for Inspection through Metal Coating. Insight, Vol.52 (2010), No.6, p.310-314.

[6] Yatian GAO, Lihua WANG, Xiujie MIAO, et al. Application of Empirical Mode Decomposition to ACFM Signal. Advanced Materials Research, Vol.889-890 (2014), p.761-765.

[7] GB12337-2014.Steel spherical tanks. Edited by National quality technology supervision bureau (2014), p.59.

[8] Yong CHEN, Dongmin PAN, Ping DENG. Influencing Factors of the Signal for ACFM and Recognition Determination of Cracks. Nondestructive Testing, Vol.35 (2013), No.9, p.61-65.

IOP Conf. Series: Materials Science and Engineering **772** (2020) 012071 doi:10.1088/1757-899X/772/1/012071

# Research on Vibration Reduction Compatibility for Multiple Transportation Conditions

**Botao Xu [1, 2], Zhikai Fang [1, 2], Yugang Liu [1, 2, *] and Xingyue Liu [1]**

[1] Beijing Institute of Spacecraft Environment Engineering, Beijing, 100094, China
[2] Beijing Engineering Research Centre of the Intelligent Assembly Technology and Equipment for Aerospace Product, Beijing, 100094, China

*Corresponding author's e-mail: liuyugang46497@sina.com

**Abstract.** Vibrations in the process of transportation are often neglected in product design. In order to avoid damage to acceleration sensitive equipment or components of spacecraft due to destructive transport load, vibration absorbers are usually installed in the container during spacecraft transportation. How to determine the vibration reduction scheme suitable for different payload weight has become the key point and technical difficulty in the design of vibration reduction in transportation. Aiming at this problem, the size and weight of the target product are analysed, and the vibration features of different sections of the satellite are analysed and tested. On this basis, a general vibration reduction subsystem of the container is designed, which is suitable for three kinds of satellite sections and four types of transportation conditions, and the loading distribution parameters are optimized. Transportation test data show that the vibration reduction effect of the new system for aerospace products meets the design requirements.

## 1. Introduction

Transportation is an important part of the life cycle of spacecraft. When the transport excitation frequency is equal to the natural frequency of the satellite-container, resonance will occur, which will damage the vibration sensitive components on the satellite. Spacecraft transportation experiences continuous mechanical loads [1-3]. This kind of load is characterized by random vibration (excitation frequency ranges from 5Hz to 300Hz) and periodic load [4]. In addition, impact load is inevitable. Referring to the mechanical environment conditions of satellite design, the vibration at the interface of satellite is generally no more than 0.6g and the impact is no more than 1G. In actual highway transportation, the vibration requirements are generally satisfied, but there are usually several shocks exceeding 1G.

The dynamic environment of launch vehicle is the main design basis of satellite structure design. Vibration in the process of satellite ground transportation are easily neglected in satellite design. Satellite container system usually adopts fixed fundamental frequency vibration reduction system because of its reliability, size and other constraints. [5]

How to determine the vibration reduction scheme suitable for different payload weight has become the key point and technical difficulty in the design of vibration reduction in transportation. Aiming at this problem, the size and weight of the target product are analysed, and the vibration features of different sections of the satellite are analysed and tested. On this basis, a general vibration reduction subsystem

Content from this work may be used under the terms of the Creative Commons Attribution 3.0 licence. Any further distribution of this work must maintain attribution to the author(s) and the title of the work, journal citation and DOI.
Published under licence by IOP Publishing Ltd

of the container is designed, which is suitable for three kinds of satellite sections and four types of transportation conditions, and the loading distribution parameters are optimized in this paper.

## 2. Platform size and layout design

In order to ensure and expand the universality of the container, it is required that the payload adapter in the packing cabin should be equipped with a universal support interface to meet the mechanical constraints of all kinds of satellite and meet the matching requirements of the above three types of transport conditions. Moreover, for the smaller MEO, the task requires both single transport and two at the same time.

Based on the above conditions, the adapter should also satisfy the following constraints:

- The size should not exceed the transport limit.
- Easy to assemble and disassemble.
- The overall stiffness of the adapter is greater than that of the payload.

In addition, the transporting vibration sensor should be located in the location with better local stiffness when the size of the shock absorber loading table is large. The structural design forms satisfying the above conditions are shown in Figure 1.

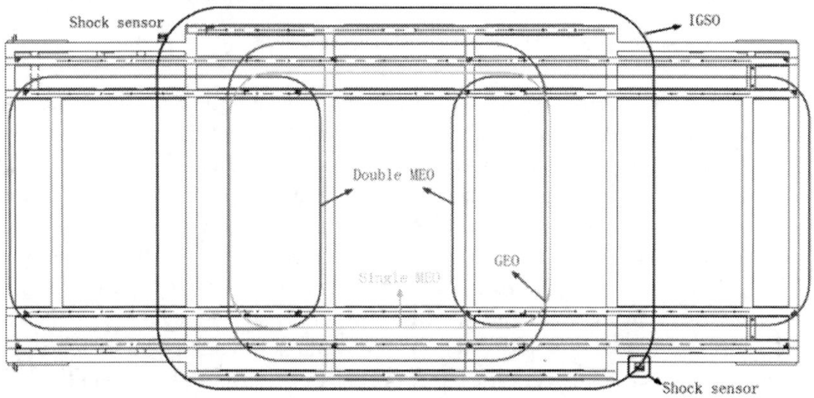


**Figure 1.** Compatible Transportation Support Interface

The basic parameters of various transport loads are shown in Table 1.

**Table 1.** Comparison table of basic parameters of various types of transport loads

| Load type | | MEO | IGSO | GEO | Double MEO | Max. Envelope |
|---|---|---|---|---|---|---|
| Transportation weight (kg) | | 590 | 1450 | 650 | 1250 | 1500 |
| Size (mm) | X | 2506 | 3056 | 2630 | 5412 | 3056 |
| | Y | 1900 | 2428 | 1950 | 2428 | 2428 |
| | Z | 1942 | 2882 | 1870 | 1942 | 2882 |
| Transverse screw spacing (mm) | | 1470 | 2324 | 1910 | 1470 | / |
| Longitudinal screw spacing (mm) | | 200 | 200 | 200 | 200/400 | 200 |

## 3. Design of Vibration Absorption Scheme

### *3.1. Statics analysis*

Constraints and Load Applying: Fixed constraints in six areas below the loading platform, and loaded 1.6t in the black area. As shown in Figure 2

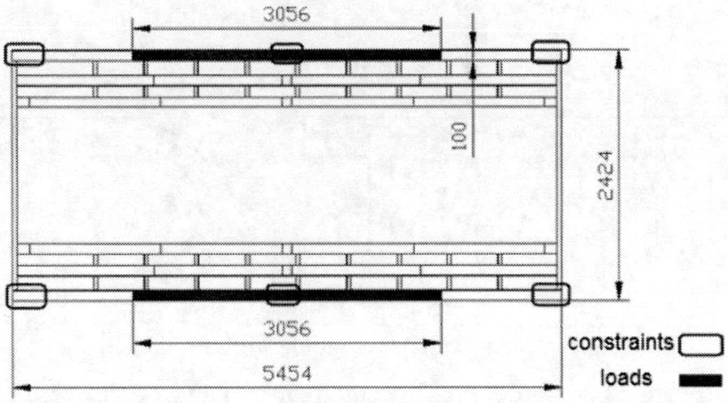


**Figure 2.** Mechanical analysis constraints and load sketch of loading platform

The deformation analysis shows a maximum deformation of 0.12 mm and less than the required 5 mm. The maximum stress is 92.2 MPa at the upper middle area and less than steel's breaking limit (520 MPa).

### *3.2. Design of vibration absorption system*

In order to reduce the impact of vibration and shock on the product during transportation, the selection of shock absorber should avoid approaching the first fundamental frequency of the product, so as to ensure that the vibration and shock acceleration transmitted to the product after vibration reduction meet the established requirements.

On the premise of inheriting mature vibration reduction schemes, we should also consider saving the space occupied in the container, so we choose the way of vibration reduction inside the container. According to the different configurations of the loaded products, the number of vibration absorption modules can be adjusted flexibly.

The vibration magnitude of aerospace products in highway transportation is generally between (0,0.2) and less than the damage threshold required by the products. To assess the impact on products, the focus is on the occasional impact events in transportation, such as the impact caused by crossing railway crossings, deceleration belts, deceleration prompt marking, etc. It can be seen from the data of field transportation test that all the impact exceeding limits occur in the vertical direction (Z direction). The data are shown in Table 2. Therefore, the vertical vibration reduction is preferred in the transport vibration reduction scheme.

**Table 2.** Shocklog RD298 recorded overload data

| Serial no. | Sampling rate/Hz | X/g | Y/g | Z/g |
|---|---|---|---|---|
| 1 | 256 | 0.246 | 0.366 | 1.05 |
| 2 | 32 | 0.84 | 0.48 | 1.36 |
| 3 | 64 | 0.201 | 0.366 | 1.13 |
| 4 | 256 | 0.402 | 0.456 | 1.72 |
| 5 | 256 | 0.231 | 0.291 | 1.22 |
| 6 | 128 | 0.21 | 0.357 | 1.19 |

Figure 3 is a sketch of vertical preferential vibration isolator.

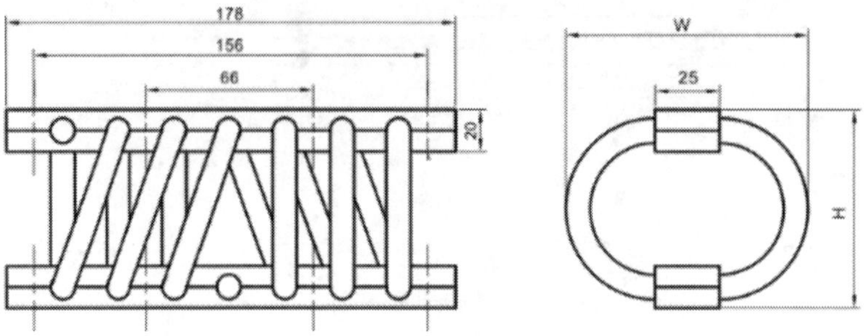


**Figure 3.** Vertical Preferential Vibration Isolator

The calculation of vertical first-order fundamental frequency Fnz of wire rope isolator is as follows

$$2\pi f_{nz} = \sqrt{k_z \bullet \alpha / M} \tag{1}$$

In the formula:

*kz*: Static stiffness (N/mm) under load.

$\alpha$: Dynamic stiffness coefficient (average dynamic stiffness coefficient a = 1.5)

*M*: Loading mass per assembly (kg)

Formula (1) shows that increasing M can reduce the first-order fundamental frequency of the vibration reduction system without reaching the maximum deformation. Therefore, the actual load of the isolator is generally larger than the nominal load. Ring wire rope isolator stiffness and damping are both non-linear. The static stiffness of working state should be calculated by first-order fundamental frequency. The use of average static stiffness will lead to inaccurate results.

The shock absorber should be arranged symmetrically under the loading platform plane according to the configuration of Figure 4. Considering that the vibration reduction effect for each satellite should be close to each other, two alternatives are proposed as options, eighteen stronger isolators (GGT220-108) or twenty weaker isolators (GGT250-137). According to the maximum envelope calculation, the total weight G = 2220 kg. The first order frequencies of the two schemes can be obtained respectively, as shown in Table 3.

**Table 3.** Optional isolator model parameters

| Vibration Isolator Model | GGT220-108 | GGT250-137 |
|---|---|---|
| Number of isolators | 18 | 20 |
| Nominal load (kg) | 60 | 66 |
| Working Static Stiffness (N/mm) | 190 | 100 |
| Maximum dynamic deformation(mm) | 58 | 90 |
| Maximum impact force(kN) | 4.5 | 4.8 |
| Capacitance (J) | 220 | 250 |
| Height (mm) | 108 | 137 |
| Calculated *Fnz* (Hz) | ~ 10.6 | ~ 5.8 |

And the system vibration transmission rate:

$$T = \frac{Output}{Input} = \sqrt{\frac{1+(2\zeta\lambda)^2}{(1-\lambda^2)^2+(2\zeta\lambda)^2}} \tag{2}$$

Where:

$\lambda$: vibration frequency ratio, λ = fi/fn.

$f_i$: Frequency of external road disturbance, fi:= ω /(2 π )

$\zeta$: The average damping ratio of the system is 0.15.

It can be seen that the applicability of scheme 2 is better than that of scheme 1. In scheme 2, when the input excitation is greater than 8.2Hz, the system enters the vibration absorption region; when the input excitation frequency is not less than 17.4Hz, the system isolation effect is more than 80%, and the system vibration isolation design meets the design requirements. The vibration reduction arrangement of the loading platform under the maximum envelope is determined as follows: the loading platform is connected to the container through 20 wire rope dampers, and is equipped with two real-time monitoring acceleration sensors, a ShockLog298 vibration and shock recorder, which monitors the vibration state of the whole transportation process of the product, as shown in Figure 4.

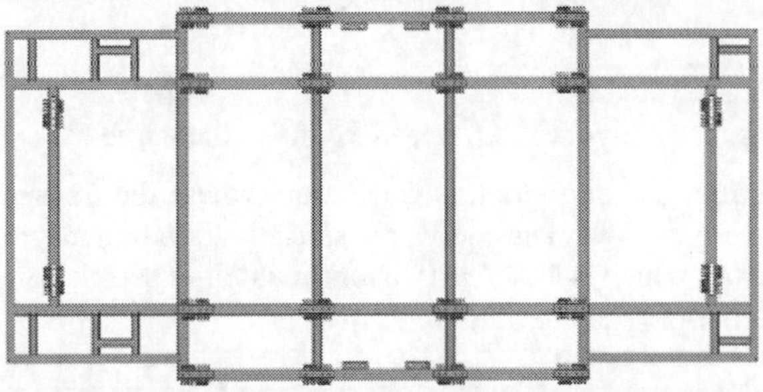

**Figure 4.** Envelope Vibration Absorption Arrangement Scheme

Referring to the loads and constraints imposed on the system in the section of static analysis, the vibration transmission rate of the loading platform is analysed. The simulation results show that the vibration isolation performance of the system has reached more than 80% when the vibration transmission rate is about 15 Hz. The response data is the response at the vibration measuring point in Figure 2. From the curve of Figure 5, it can be seen that the first-order frequency of the vertical direction of the vibration absorber system is about 5Hz, which is slightly smaller than the calculation result (5.8 Hz) in Table 3. This is mainly due to the larger size of the loading platform, the weakened stiffness, and the reduction of the number of equivalent isolators under the specific excitation, resulting in the increase of the equivalent load of a single isolator.

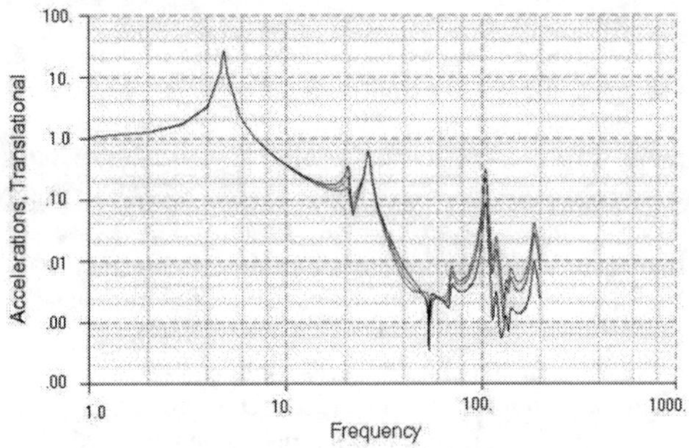


**Figure 5.** Vibration Transmittance Simulation Curve

According to the above calculation, on the basis of the maximum equivalent shock absorber load meeting the above requirements, the isolator layout can be quickly and conveniently adjusted according to four typical loading modes.

## 4. Test Verification

Transportation tests were carried out using 1450 kg counterweight based on IGSO satellite. After the end of test, the data in the shock recorder are exported and the time domain waveform of the data larger than 0.9g is analysed. By comparing the shock events before and after vibration reduction, it can be found that the vibration amplitude decreases significantly. See Figures 6.

IOP Conf. Series: Materials Science and Engineering **772** (2020) 012071 doi:10.1088/1757-899X/772/1/012071

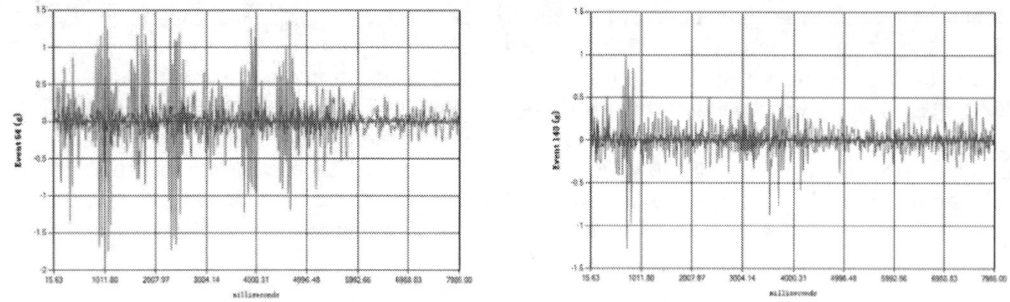

a) Before Vibration Reduction b) After Vibration Reduction

**Figure 6.** Overload Points in the curve

Shock response spectrum transformation is carried out where the transportation exceeds the limit point. The conclusions are as follows: The equivalent sinusoid vibration magnitude is no more than 0.3g (0Hz~100Hz accumulative), which is less than 0.6g of the satellite permission level, so the shock signal does no harm to the satellite. As shown in figure 7.

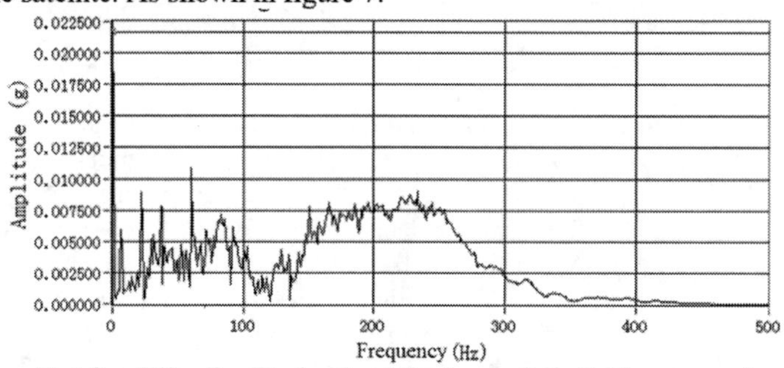


**Figure 7.** After Vibration Reduction Maximum Point's Spectrum Analysis

## 5. Conclusion

In this paper, a technical scheme of loading multiple products in a container is proposed. Through statics analysis and dynamics analysis, the type of isolator is determined, the optimal scheme is defined, and the problem of effective vibration reduction of multiple transport loads is solved. The results of transportation test prove the validity of the scheme.

## References

[1] Liu Chen, Zhu Jiantao, et al. Dynamics parameter measurement and analysis of certain satellite platform during launch and in orbit [J]. Spacecraft Environment Engineering, 2017, 34(3): 270-276.

[2] CGWIC. LM-3A Series Launch Vehicles User's Manual [R]. Issue2011.

[3] GJB 1027A-2005 the requirement for launch, upper-stage, and space vehicle[S]. Beijing: The committee of technology and industry of national defence, 2006.

[4] Shi Lixia, et al. Dynamics environment excitations during spacecraft transportations [J]. Spacecraft Environment Engineering, 2013(3): 250-255.

[5] PENG Chengrong. System design for spacecraft [M]. Beijing: China Science and Technology Press, 2011.

IOP Conf. Series: Materials Science and Engineering **772** (2020) 012072 doi:10.1088/1757-899X/772/1/012072

# Exploration on Extracting Geometric Model by Topology Optimization Design

**Jiaqi Miao**

School of Chemical Machinery and Safety Engineering, Dalian University of Technology, Dalian, China

**Abstract.** Topology optimization design is a way to achieve more efficient design with a minimal priori decision. However, due to the complexity of the solution obtained, topology optimization is usually limited to theoretical researches and can not be applied into actual manufacturing. As a new manufacturing method that is currently developing rapidly, additive manufacturing has made great progress in many fields. Additive manufacturing has fewer limitations on the shape and complexity of the design. And it is currently moving toward higher precision and larger dimensions. The combination of these two ways can solve the problem of topology optimization design well. However, the topology optimization design's principle determines that it is difficult to transform into a 3D model that can be processed and manufactured pefectly. Based on the topology-optimized structure, this paper uses MATLAB to extract the geometric model and smooth it, which provides an effective method for extracting 3D models from topology optimization. Also, this paper puts forward an efficient method to optimize the structure and it gets a pretty good result in the simulation.

## 1. Introduction

Topology optimization is a mathematical method that optimizes material distribution over a given area based on given load conditions, constraints, and performance metrics. Seeing material as the optimization object, topology optimization can find the best distribution scheme in the design space. Compared with other traditional structural optimization methods, it has more degree of freedom and can obtain much more design space. At present, common topology optimization methods include homogenization method [1], variable density method [2], progressive structure optimization method [3] and so on. As for additive manufacturing, great progress has been made in recent years. Jacob Koffler from the University of California, San Diego, used additive manufacturing techniques to mimic the spinal cord scaffold of the central nervous system structure, which successfully helped the rat to recover its motor function [4]. Researchers at Tel Aviv University in Israel use the patient's own tissue as a raw material to complete heart building. Building complete heart structure is the world's first. Both of these technologies are at the forefront of their respective fields, and additive manufacturing can be well solved for the problem of topology optimization. However, the process of transforming the topology optimization structure into a manufacturable 3D model is still not well implemented. This is because the topology optimization result is probabilistic information, which is theoretically optimal, but can not give the actual processing and manufacturing model [5]. To this problem, this paper obtains its geometric model by a method of boundary optimization and extracting the processed contours. At the end of the

Content from this work may be used under the terms of the Creative Commons Attribution 3.0 licence. Any further distribution of this work must maintain attribution to the author(s) and the title of the work, journal citation and DOI.
Published under licence by IOP Publishing Ltd

model, the smooth processing and analysis of the model also have been carried out. What is more, the paper talks something about the future work.

## 2. Extraction of geometric models

This paper extracts the geometric model based on the results obtained by Erik Andreassen et al.'s topology optimization program [6]. Fig.1 shows the results of topology optimization under the conditions of Table 1. The Fig.1 has a fuzzy probabilistic part, and the fuzzy value at the model boundary is between 0 and 1 (1 means there must be, 0 means no). Firstly, the digital information of the model boundary is preprocessed to ensure that the boundary value is optimized without affecting the optimization performance. At the same time, the influence of the surrounding structure points is considered. Secondly, the geometric model is extracted by selecting the contour line at 0.5, and the selection of the interpolation method is explained.

**Table 1.** Parameter setting.

| %% MATERIAL PROPERTIES<br>%top88(nelx, nely, volfrac, penal, rmin, ft) | %% MATERIAL PROPERTIES<br>%top88(nelx, nely, volfrac, penal, rmin, ft) |
|---|---|
| nelx = 90; | nelx = 15; |
| nely = 90; | nely = 30; |
| volfrac = 0.5; | volfrac = 0.5; |
| penal = 3; | penal = 3; |
| rmin = 2; | rmin = 2; |
| ft = 2; | ft = 2; |
| tic | tic |
| E0 = 1; | E0 = 1; |
| Emin = 1e-9; | Emin = 1e-9; |
| nu = 0.3; | nu = 0.3; |

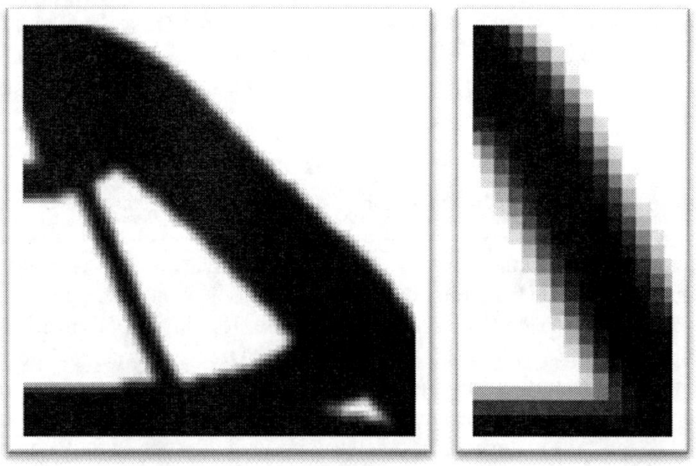

**Figure 1.** Topologically optimized graph.

### *2.1. Boundary value optimization*

In the optimization of the boundary value, the influence of the surrounding point values should be considered. The specific method is as follows: each element value is changed into a matrix of n*n, and each number in the matrix is the value of this element. Taking the influence of the surrounding point values into account, each element takes the value of its own sum with the surrounding values, and the result is shown in Fig.2.

IOP Conf. Series: Materials Science and Engineering **772** (2020) 012072 doi:10.1088/1757-899X/772/1/012072

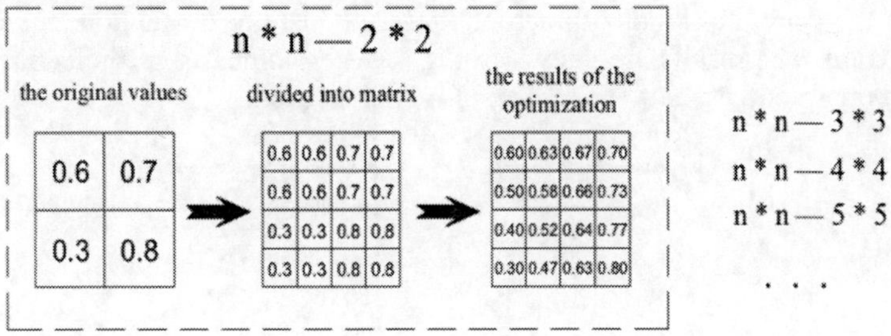


**Figure 2.** Schematic diagram of boundary value optimization

### *2.2. Contour extraction*

First adjust the results of the topology optimization (matrix), and use the linspace function and the griddata function to divide and interpolate. The specific code is shown in Table 2.

**Table 2.** Extract geometric model.

| %% geometric model |
|---|
| xPhys_1 = xPhys'; |
| [a, b] = size(xPhys_1); |
| x = ones(a,1); |
| y = [1: b]; |
| y = y'; |
| X = kron (x, y); Y = kron (y, x); Z = xPhys_1 (:); |
| xx = linspace (1, a, 100); yy = linspace (1, b, 100); |
| yy = yy'; |
| zz = griddata (X, Y, Z, xx, yy, 'cubic'); |
| Figure |
| contour (xx, yy, zz, [0.5,0.50001]); |
| set (gca, 'ydir', 'reverse', 'FontSize', 16) |
| title ('the geometric model'); |
| axis equal; |

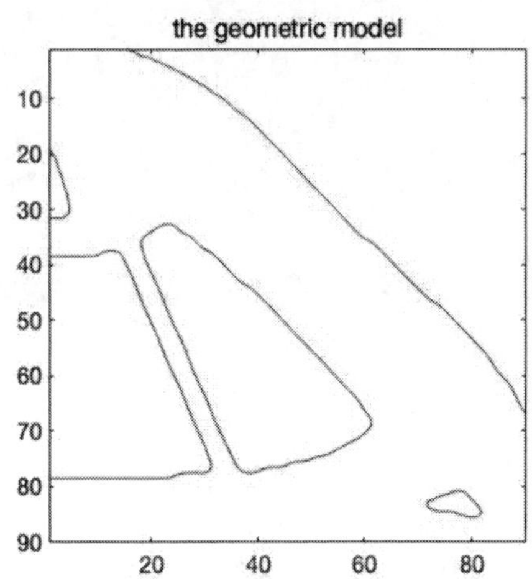


**Figure 3.** Geometric model diagram.

Among them, there are four kinds of interpolation methods: linear function, cubic function, nearest function, v4 function. The results are shown in Fig.4. After comparison, the cubic function with the shortest time and the smoothest curve is selected for interpolation.

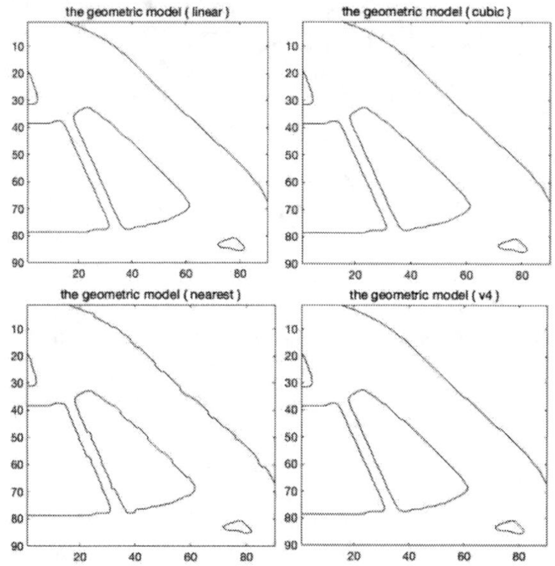


**Figure 4.** Comparisons of different interpolation methods.

## 3. Optimization results and analysis

### *3.1. Optimized results*

According to the optimization method in the section "Boundary Value Optimization", we use different n values for optimization analysis, and the results are shown in Fig.5.

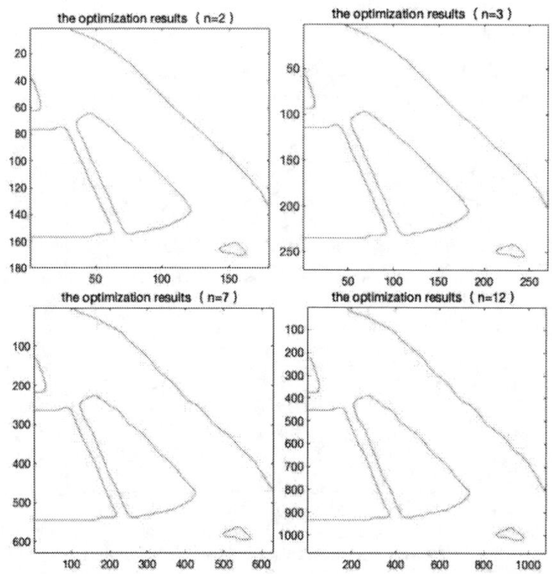


**Figure 5.** Optimization results with different values of n.

### 3.2. *Analysis of optimization methods*

It can be seen from Fig.6 that the value of n is not as large as possible, but instead presents a trend of superiority and inferiority. To further verify the universality of the conclusions, different topological optimization results are used for analysis.

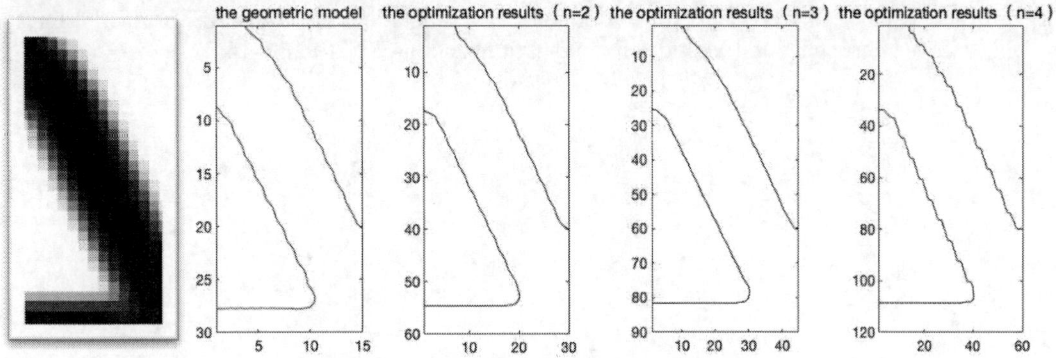


**Figure 6.** Results at different values of n.

The processed results show that our optimization method does solve the problem of non-smoothness very well, but it seems that each different size model has a corresponding optimal n value. In the structure of Fig.6 above, the optimum state is clearly reached when n=3. If the value of n is greater than or less than this value, the output model will become unsmooth, which will cause processing difficulties for the output model. Approximate in this model is as follows: the correspondence between the unit length and the optimal n value (seeing in Table 3).

**Table 3.** Width and its optimal n value.

| | n = 2 | n = 3 |
|---|---|---|
| **Width = 0~3.7** | Suitable | Not suitable |
| **Width = 3.7~30** | Not suitable | Suitable |
| **Width > 30** | Not suitable | Not suitable |

## 4. Conclusion

This paper extracts the geometric model of the topology-optimized structure by its contour through the MATLAB software. Aiming at the boundary value considering the influence of the surrounding points, a method of mean value is proposed, which can be seen in the simulation results to make the boundary smoother. On this basis, the optimization effect of the expanded matrix dimension n under different conditional models is studied. Finally, an optimal n-value approximation map of different width structures is given. Through the exploration of geometric model extraction, it is better to extend the model building of complex structures in the future. Therefore, the topology optimization method and the additive manufacturing technology are contacted, and the theoretical optimal topology optimization result is truly applied to the actual processing and production. In the future work, we can widely move this method to the 3D model, and perhaps artificial intelligence recognition can be applied to make the surface smoother and easier to manufacture. There is no doubt that these jobs will provide the basis for the perfect combination of topology optimization and additive manufacturing.

## References

[1] Bends on, M. P, "Generating optimal topologies in structural design using a homogenization method." Computer Methods in Applied Mechanics and Engineering 71(1988).
[2] Bendson, M. P, "Optimal shape design as a material distribution problem." Struct. Optim. 1(1989).
[3] Xie, Y. M and G. P. Steven, "A simple evolutionary procedure for structural

optimization."Computers & Structure 49.5(1993): 885-896.
[4] Koffler J, Zhu W, Qu X, Platoshyn O. Dulin. J. N. Brock, J., et al. "Biomimetic 3D-printed scaffolds for spinal cord injury repair." Nature Medicine. 2019, 25(2): 263–269.
[5] Zegard, Tomás, and G. H. Paulino, "Bridging topology optimization and additive manufacturing." Structural and Multidisciplinary Optimization 53.1(2016): 175-192.
[6] Andreassen, Erik, et al. "Efficient topology optimization in MATLAB using 88 lines of code." Structural and Multidisciplinary Optimization43.1(2011): 1-16.

IOP Conf. Series: Materials Science and Engineering **772** (2020) 012073 doi:10.1088/1757-899X/772/1/012073

# Study on the Lubrication Performances of Rolling Oil for Cold Rolled Laminated Cu/Al Composite Sheet

**Sang Xiong[1, 2, 3, *], Zhengping Shang[1, a], Huan Gao[1, b] and Qijie Zhai[2, c]**

[1] Jiangsuzhongse Composite Materials Co., LTD., Wuxi 214000, China
[2] School of Materials Science and Engineering, Shanghai University, Shanghai 200000, China
[3] College of Materials Science and Engineering, Nanjing Institute of Technology, Nanjing Jiang Su 211167, China

* Corresponding author: jxbjxs2008@163.com, [a] 641483235@qq.com, [b] 411622074@qq.com, [c] 1271800877@ qq.com

**Abstract.** Three kinds of ester base oils (D85, D100, D115), dodecyl alcohol, dodecyl acid, butyl stearate as oiliness agents, and thiophosphate, fatty alcohol phosphates and dibutyl phosphate as extreme pressure agents were used to prepare and investigate the tribological properties of rolling oils for cold rolled laminated Cu/Al composite sheet by an orthogonal experiment, and lubrication performances of the rolling oils were studied through rolling experiments. Surface roughness and the interface of Cu/Al composite sheet were analyzed. The results show that the optimum tribological properties of the rolling oil is compose of D100 as basic oil, butyl stearate as oiliness agent and the compound ratio of fatty alcohol phosphates and thiophosphate of 3:1 as extreme pressure antiwar agent. The PB value reaches 1020 N, the coefficient of friction is 0.0795. Basic oil is the main factor of rolling capacity. The rolling oil contented D115, showed the best rolling capacity while the minimum thickness was only 0.045mm. The addition of butyl stearate improved the surface quality of both copper and aluminum side of the rolled stripes and make the sheet surface roughness decreased. After cold rolling, new composite interface, which is clear and tightly bound replaced the brittle mesophase.

**Key words:** Cu-Al composite sheet; lubrication; cold rolling; surface topography.

## 1. Introduction

This document was prepared using the AIP Proceedings template for Microsoft Word. It provides a simple example of a paper and offers guidelines for preparing your article. Here we introduce the paragraph styles for Level 1, Level 2, and Level 3 headings. Cu-Al composite sheet have attracted much attention in recent years, such as high conductivity, high thermal conductivity, low contact resistance and light weight and corrosion resistance of aluminum [1, 2]. It is widely used in aviation, aerospace, machinery, medical treatment, food, civil and other fields [3]. In the rolling process of Cu-Al composite sheet, effective lubrication could reduce the minimum rolling gauge, improve the surface quality of the product, prolong the service life of the casting roll and reduce energy consumption [4]. In the process

Content from this work may be used under the terms of the Creative Commons Attribution 3.0 licence. Any further distribution of this work must maintain attribution to the author(s) and the title of the work, journal citation and DOI.
Published under licence by IOP Publishing Ltd

of cold rolling, due to the close contact between the roller and the workpiece, friction and wear will inevitably occur, increasing the energy loss. The lubricant containing phosphorus and sulfur can form a lubricating film on the workpiece and reduce the friction and wear in the rolling process [5]. With the emergence of high-precision Cu-Al composite sheet technology, an opportunity is provided to develop a new kind of cold rolling oil and to solve the above-mentioned problems [6]. The reasonable selection of base oil and additives is crucial for rolling of Cu-Al composite sheet under the same condition.

The present work focuses on the assessment of the effect of different base oils (D85, D100, D115), oiliness agents (dodecyl alcohol, dodecyl acid, butyl stearate) and three ratios of the various extreme pressure agent on the rolling oil lubrication property of Cu-Al composite sheet. Therefore, the tribological properties of rolling oils, the lubrication performance and the surface quality of the rolled Cu-Al composite sheet were studied. Furthermore, this work can offer a reference for the development of high-precision Cu-Al composite sheet technology in the future

## 2. Experiment and Method

### *2.1. Materials*

Here is how to display a pop-up window from which to select and apply the AIP Proceedings template paragraph styles. Pure copper (T2) and aluminum (1060) were used in the experiment and the chemical compositions are listed in Table 1. Three kinds of ester base oils (D85, D100, D115), dodecyl alcohol, dodecyl acid, butyl stearate as oiliness agents, and thiophosphate, fatty alcohol phosphates and dibutyl phosphate as extreme pressure agents were used. The flash points of the D85, D100 and D115 are 85, 100 and 115°C, respectively. At 40°C, kinematic viscosities of D85, D100 and D115 are 2.0 $mm^2 \cdot s^{-1}$, 2.5 $mm^2 \cdot s^{-1}$ and 3.0 $mm^2 \cdot s^{-1}$. Orthogonal design of four factors and three levels was taken to obtain the proper agents and given in Table 2 and Table 3. The rolling oils were consisting of base oil 90%, oiliness agents 9.6% and extreme pressure agents 0.4%. Heat the base oil to about 40°C, add oily agent and extreme pressure agent and stir, make rolling oil mix evenly.

**Table 1.** Chemical compositions of Cu-Al composite sheet (Atom%)

| **Elements** | **Cu** | **Pb** | **S** | **Cd** | **P** | **Fe** | | |
|---|---|---|---|---|---|---|---|---|
| T2 | 99.9900 | 0.0003 | 0.0016 | 0.0002 | 0.0013 | 0.0009 | - | - |
| Elements | Fe | Mn | Mg | Si | Zn | Ti | Cu | Al |
| 1060 | 0.35 | 0.03 | 0.03 | 0.25 | 0.05 | 0.03 | 0.05 | 99.21 |

**Table 2.** Orthogonal design of four factors and three levels of types of base oils, oiliness agents, extreme pressure agents

| **Levels** | **Factors** | | | |
|---|---|---|---|---|
| | **A Types of base oils** | **B Types of oiliness agents** | **C Types of extreme pressure agents** | **D Extreme pressure agents ratios** |
| 1 | D85 | Dodecyl alcohol | Thiophosphate + fatty alcohol phosphates | 1:1 |
| 2 | D100 | Dodecyl acid | Fatty alcohol phosphates + dibutyl phosphate | 1:3 |
| 3 | D115 | Butyl stearate | Thiophosphate + dibutyl phosphate | 3:1 |

**Table 3.** Orthogonal test table L9 ($3^4$)

| Samples | Factors | | | |
|---|---|---|---|---|
| | **A Types of base oils** | **B Types of oiliness agents** | **C Types of extreme pressure agents** | **D Extreme pressure agents ratios** |
| 1# | 1 | 1 | 1 | 1 |
| 2# | 1 | 2 | 2 | 2 |
| 3# | 1 | 3 | 3 | 3 |
| 4# | 2 | 1 | 2 | 3 |
| 5# | 2 | 2 | 3 | 1 |
| 6# | 2 | 3 | 1 | 2 |
| 7# | 3 | 1 | 3 | 2 |
| 8# | 3 | 2 | 1 | 3 |
| 9# | 3 | 3 | 2 | 1 |

### *2.2. Test on Tribological Properties*
An MRS-10A four ball testing machine was used according to the national standard GB/T 12583—1998 to measure the maximum non-seizure load ($P_B$ value) adopted for estimating the load carrying capacity of the rolling oils. The average coefficient of friction of different rolling oil samples was obtained from experiments conducted in 30 minutes of testing at room temperature under a loading force of 147 N, a rotational speed of 1 200 revolutions per minute. Each experiment was measured three times.

### *2.3. Cold Rolling Test*
A four-high rolling mill of Φ95/200 mm×200 mm with a velocity of 60 r/min and a rolling power of 35 kW was used to evaluate the lubricity of the rolling oils. The workpiece thickness was 1.0 mm and the pass-reduction was 20%. The surface morphology analyses of rolled copper foil were conducted by an FTS-S3c roughness instrument and LEICA metalloscopy. The interface of Cu/Al composite sheet was analyzed by a scanning electron microscope (SEM).

## 3. Results and Discussions

### *3.1. Tribological Properties*
$P_B$ is an important parameter that reflects the load carrying capacity of rolling oil, which is determined by the rolling oil viscosity, the adsorption film toughness and the sliding speed [5]. $P_B$ and the coefficient of friction are two important indexes of extreme pressure and antiwear performance of the rolling oil. The bigger the $P_B$ value is, the stronger oil film and better anti-wear and friction-reducing performance of the rolling oil will be. Similarly, the smaller the coefficient of friction is, the better the lubricity will be. However, the two factors are not always coincident. For the convenience of the analysis, an extreme pressure and anti-wear coefficient (ω) that can combine the $P_B$ value and the coefficient of friction with the same standards to reflect the tribological properties of rolling oil directly, as shown in Eq. (1).

$$\omega = \log_{10}(\frac{P_B}{\mu}) \tag{1}$$

Where $P_B$ is the maximum non-seizure load, and $\mu$ is the coefficient of friction.

The values of $P_B$, μ and ω from the tribological results of orthogonal experiment are given in Table 4. Employing the method of orthogonal trial with mixed types of agents and processing the data with range analyze and listed in Table 5.

**Table 4.** Tribological test of orthogonal experiment

| Samples | Effects | | | | $P_B$/N | μ | ω |
|---|---|---|---|---|---|---|---|
| | A | B | C | D | | | |
| 1# | 1 | 1 | 1 | 1 | 932 | 0.0602 | 4.190 |
| 2# | 1 | 2 | 2 | 2 | 412 | 0.0753 | 3.738 |
| 3# | 1 | 3 | 3 | 3 | 932 | 0.0697 | 4.126 |
| 4# | 2 | 1 | 2 | 3 | 883 | 0.0803 | 4.041 |
| 5# | 2 | 2 | 3 | 1 | 696 | 0.0779 | 3.951 |
| 6# | 2 | 3 | 1 | 2 | 1020 | 0.0795 | 4.108 |
| 7# | 3 | 1 | 3 | 2 | 412 | 0.0768 | 3.730 |
| 8# | 3 | 2 | 1 | 3 | 784 | 0.0756 | 4.016 |
| 9# | 3 | 3 | 2 | 1 | 559 | 0.0788 | 3.851 |

**Table 5.** Range analyzes of orthogonal experiment

| Samples | A Types of base oils | B Types of oiliness agents | C Types of extreme pressure agents | D Extreme pressure agents' ratios |
|---|---|---|---|---|
| k1j | 4.018 | 3.987 | 4.105 | 3.977 |
| k2j | 4.033 | 3.902 | 3.877 | 4.061 |
| k3j | 3.866 | 4.028 | 3.936 | 3.859 |
| Rj | 0.167 | 0.126 | 0.228 | 0.202 |
| Order | C>D>A>B | | | |
| Optimal level | $A_2$ | $B_3$ | $C_1$ | $D_2$ |
| Optimal portfolio | $A_2B_3C_1D_2$ | | | |

In Table 5, Kij refers to the cumulative results of experimental data processing of different j factors at the same level in each group. j factors include A, B, C and D factors, and i levels include 1, 2 and 3. Rj represents range, that is, the difference between the maximum value and minimum value of each Kj value in factor j. The larger the Rj, the more obvious the influence of j factor in the orthogonal table on the experimental results. It can be seen from the table that the types of extreme pressure agents play a major role in increasing oil film strength and reducing wear of sheet and strip. From the influence of extreme pressure agent ratio D, it can be seen that the ratio of different extreme pressure agent has a certain effect on the rolling oil lubrication properties, presumably related to the type of extreme pressure agent. After the compound of thiophosphate and fatty alcohol phosphates, the oil film strength was increased to the highest level, and the compound synergistic effect was played in the base oil. From the above, the optimum tribological properties of the rolling oil is compose of D100 as basic oil, butyl stearate as oiliness agent and the compound ratio of fatty alcohol phosphates and thiophosphate of 3:1 as extreme pressure antiwear agent. The PB value reaches 1020 N, the coefficient of friction is 0.0795. Basic oil is the main factor of rolling capacity.

### *3.2. Lubrication Performance of Rolling Oil*

The rolling thickness of Cu-Al composite sheet with 20% reduction per pass using different lubricating oils are shown in Fig. 1. It can be clearly seen from the Fig.1 that the rolling oil of sample 2# and sample 3# has the maximum rolled thickness, the minimum draft, while the rolling oil of sample 6# and sample7# has the minimum rolling gauge and the maximum draft is 0.955 mm. The rolled thickness is calculated to the formula,

$$h = \frac{4.148K\mu D\sqrt{\varepsilon}}{E} \tag{2}$$

where E is elasticity modulus, N/mm2, K is resistance to deformation, N/mm2, μ is coefficient of friction, D is the roll diameter, mm and is relative pressure, %. It can be concluded that sample 6# and sample 7# have the best lubricity during rolling.

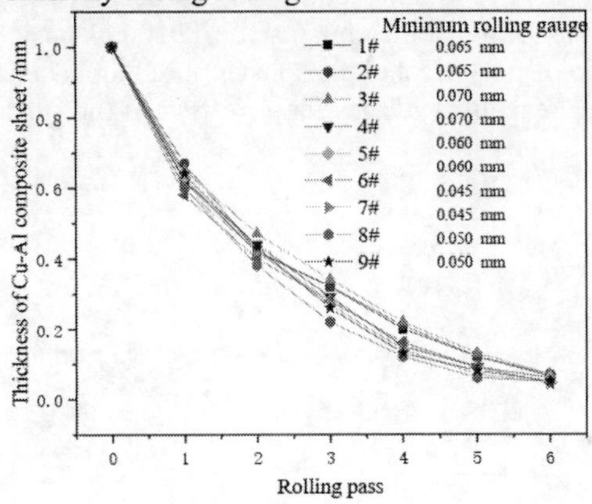


**Figure 1.** Thickness of Cu-Al composite sheet under different lubrication conditions

The parameter *Ra* (mean height of profile irregularities) is used to characterized the surface roughness of rolled Cu-Al composite sheet and Fig. 2. It is confirmed that the surface roughness of Cu-Al composite sheet is affected by the types of oiliness agents, and ester oil agents have a greater influence on the surface of Al side and Cu side. The rolling oil mixed with the oily agent butyl stearate has good lubrication effect on the copper and aluminum layers, that is to say, it plays an excellent role in wetting and friction reduction during the lubrication process. The surface roughness $R_a$ of Cu side lubricated with the sample 6#, 7# and 9# containing butyl stearate as oiliness agents were only 0.199 μm, 0.427 μm and 0.143 μm, respectively.

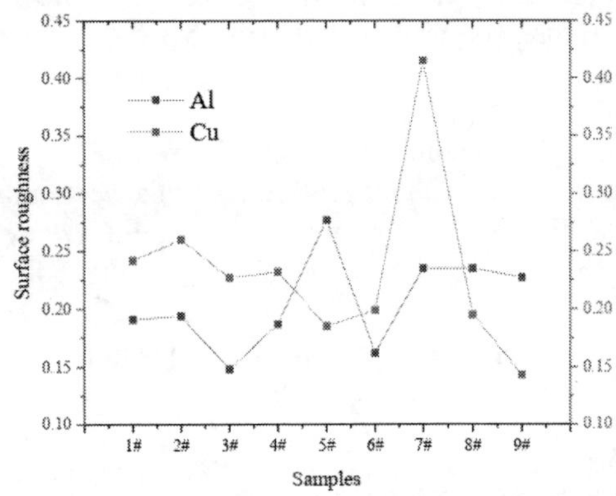


**Figure 2.** Rolled surface roughness curve lubricated with the rolling oils

Surface morphology of Cu side lubricated with the rolling oils and given in Fig. 3. It is obvious that the rolled surface has defects and deep scratches with the sample 7# lubrication, which are consistent with the results of surface roughness. A small amount of metal falling off and copper layer tearing perpendicular to the rolling direction can be observed. Sample 6# can effectively reduce the friction and wear of roller and Cu-Al composite sheet, which play a good role in lubrication, and ensure the smooth surface and uniform rolling texture after rolling. Sample 7# of the rolling oil has a higher viscosity and the oil film strength is higher than that of sample 6#, if the oil film thickness formed in the deformation zone exceeds the sum of the roughness of the roll surface and composite sheet, then the lubrication form in the deformation zone will not be the typical boundary lubrication, but the fluid lubrication form. In this way, the rolling oil film will isolate the roller and rolling piece surface, there is no direct bite of the surface micro-convex body, but the rolled product surface becomes rough. Therefore, the optimum lubrication performance of the rolling oil is composed of D100 as basic oil, butyl stearate as oiliness agent and the compound ratio of fatty alcohol phosphates and thiophosphate of 3:1 as extreme pressure antiwear agent, which has the excellent lubrication effects, and the surface quality of rolled sheets is good.

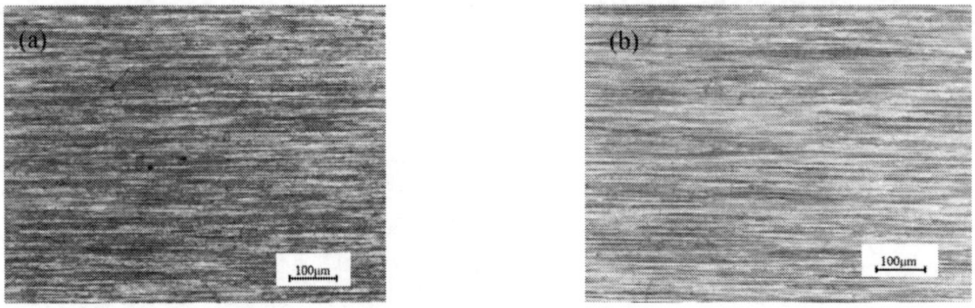


**Figure 3.** Surface morphology of Cu side lubricated with the rolling oil sample 6# (a) and 7# (b)

Cu-Al intermetallic compounds of a certain thickness are easily formed at the composite interface after rolling. According to the Cu-Al phase diagram, the composition of this layer of intermetallic compounds are $Al_2Cu$, AlCu, $Al_4Cu_9$. The transition layer has an important effect on the composite quality of Cu-Al composite sheet. Generally, the transition layer at the interface should not be too thick, otherwise, the brittle intermetallic compound with low hardness at the interface will inevitably reduce the bonding strength and overall electrical conductivity of the Cu-Al composite sheet. Interfacial SEM images and interfacial composition curve of Cu-Al composite sheet are shown in Fig.4. There are obvious intermetallic compound layers at the Cu-Al interface of Cu-Al composite plate before rolling. The thickness of the intermetallic compound layer is about 1.9 μm. The hardness and brittleness of the intermetallic compound layer are relatively high. In the cold rolling process of Cu-Al composite plate, due to the low internal temperature, the diffusion velocity between copper and aluminum elements is very low, and it is difficult to form intermediate phase near the interface. Therefore, brittle Cu-Al intermetallic compound is effectively prevented, and the interface composite quality is improved to some extent. The relatively thick Cu-Al composite interface layer of the original plate was damaged and torn in the process of large deformation, and the Cu-Al materials were embedded into each other again. In Fig. 4b, there is an irregular light-colored narrow strip at the interface of the composite plate after rolling, which is caused by deformation heat generated by the cold rolling process with large deformation rate, resulting in a slight local diffusion process of copper and aluminum elements at the interface of the copper and aluminum composite plate after rolling, forming metallurgical bonding. Especially, The addition of butyl stearate improved the surface quality of both copper and aluminum side of the rolled stripes and make the sheet surface roughness decreased. After cold rolling, the brittle intermediate phase was replaced by a new composite interface with clear interface and tight binding.

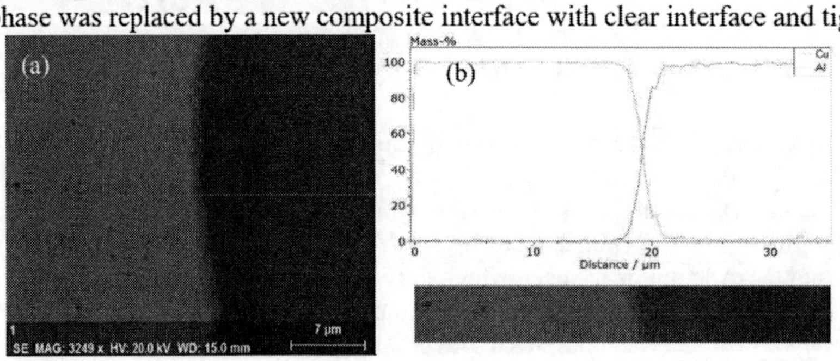


Raw material

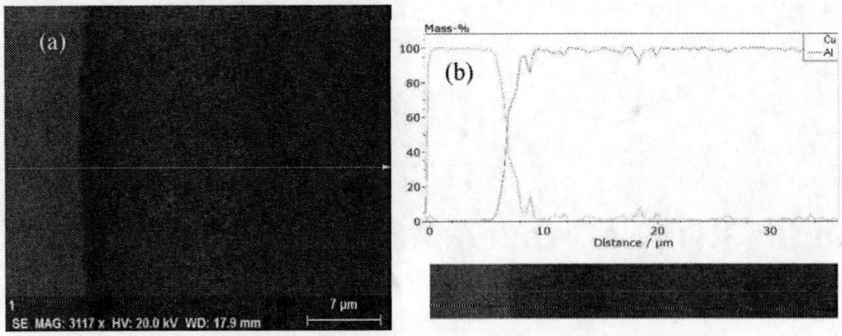


Rolled interface lubricated with sample 6#

**Figure 4.** Interfacial SEM images and interfacial composition curve of Cu-Al composite sheet

## 4. Conclusion

Three kinds of ester base oils (D85, D100, D115), dodecyl alcohol, dodecyl acid, butyl stearate as oiliness agents, and thiophosphate, fatty alcohol phosphates and dibutyl phosphate as extreme pressure agents were used to prepare and investigate the tribological properties of rolling oils for cold rolled laminated Cu/Al composite sheet by an orthogonal experiment. The optimum lubrication performance of the rolling oil is composed of D100 as basic oil, butyl stearate as oiliness agent and the compound ratio of fatty alcohol phosphates and thiophosphate of 3:1 as extreme pressure antiwear agent, which has the excellent tribological properties. Lubrication performances of the rolling oils were studied through rolling experiments. Types of base oil is the main affecting factor to determine the rolling capacity of composite sheet among the parameters, D115 base oil with the highest viscosity exhibit the strongest ability of rolling, the optimal sample minimum rolling gauge is 0.045 mm. The rolling oil containing butyl stearate as oiliness agent has good lubrication effect on surface roughness of Al-Cu surface. It plays an excellent role in reducing friction during lubrication, and effectively improves the surface quality and reduces the surface roughness of the rolled Cu-Al sheet. Before cold rolling, Cu-Al composite sheet has brittle intermediate phase with thickness of 1.9μm. However, mesophase disappears at low temperature after cold rolling, forming a clear boundary and tight interface.

## Acknowledgments

The authors gratefully acknowledge the financial assistance provided by Natural Science Foundation of Jiangsu Province (No. BK20181026), Scientific Foundation of Nanjing Institute of Technology (No. YKJ201706).

## References

[1] F.Moisy, A.Gueydan, X. Sauvage, A. Guillet and C. Keller, Materials & Design 155, 366-374(2018).
[2] Y. J.Su, X. H. Liu, Y. F.Wu, H. Y. Huang, and J. X. Xie, International Journal of Minerals, Metallurgy, and Materials 20, 684-692(2013).
[3] Rhee K Y, Han W Y and Park H J, Materials Science and Engineering: A 384,70-76(2004).
[4] H. Xu, C. Liu, V. V. Silberschmidt, S. S.Pramana, T. J. White, and Z. Chen, Acta Materialia 59, 5661-5673(2011).
[5] Q. Huang, J. Zhang, L. Zhu, L. Jiang, and X.Gao, Rare Metal Materials & Engineering 46,1749-175 (2017)5.
[6] J. W. Kaczmar, K. Pietrzak, and W. Wlosiński, Journal of Materials Processing Technology 106, 58-67(2000).

# Study on the Reuse of Abandoned Shared Bicycle Materials

**Xingfei Shui, Xinhui Lu *, Xinyu Lin**

College of Forestry, Sichuan Agricultural University, Sichuan, China

*Corresponding author's e-mail: 278102491@qq.com

**Abstract.** The abandoned shared bicycle materials was redesigned and reused to manufacture fitness bikes, in basis of the analysis of its function and structure, to boost its recyclability. This work was referred from the mature re-manufacturing technology. The structure and dimensions including lever length, seat height and crank length of fitness bike were analyzed using ergonomics method.

**Key Words:** Shared Bicycle; Recycling; Fitness bike.

## 1. Introduction

Mobile internet technology has built a platform for the development of Shared Bicycle because of the users of Shared Bicycle have experienced explosive growth. Unfortunately,in operation appeared the phenomenon of "bicycle cemetery" in cities, which caused waste of resources and environmental pollution. Recycling system is immature, such as low recycling price resulting in recycling barriers. At present, most of the existing researches are about sharing economic and piled public bicycle system [1, 2]. It ignores the importance of recycling whereas how to effectively recycle Shared Bicycle has become an urgent problem. The way of re-manufacturing in Germany is useful for the redesign of Shared Bicycles. whereas how to effectively recycle Shared Bicycle has become an urgent problem.The components of the Shared Bicycle were split and the relevant specific dimensions were determined under the premise of ergonomics. An exercise bike was redesigned through the 3Dmax expression.

## 2. Shared Bicycle recycling method

German Automobile Recycling Experimental Center classifies used cars. Determine the best dismantling step. After inspection, parts with separate scrap levels will be refurbished and refurbished, improve the recycling rate of components to 95% [4]. There are certain similarities between the recycling of Shared Bicycle and cars. Based on the operation flow chart of the dismantling of ALBA vehicles in Berlin, this paper discusses recycling and utilization of Shared Bicycle. During the disassembly process, the fixing device is first removed. The reusable components are removed, and the basic components are finally removed. Considering the principle of minimum energy consumption, the types of materials should be cut in the development and recycling design, so that parts can be easily dismantled and maintained, and energy consumption can be reduced in the life cycle. Therefore, the raw materials of the fitness bike design mostly follow the Shared Bicycle parts.

Content from this work may be used under the terms of the Creative Commons Attribution 3.0 licence. Any further distribution of this work must maintain attribution to the author(s) and the title of the work, journal citation and DOI.
Published under licence by IOP Publishing Ltd

## 3. Shared Bicycle function structure

The Shared Bicycle has 25 parts, including a frame, tires and chain. Depending to the working characteristics of each component, its structure can be roughly divided into guiding system, driving system and braking system [3]. (1) Guide system: Composed of handlebars, front fork, front wheel and other components. The handlebars have a different texture and are made of rubber; the upper end of the front fork is connected with the handlebar component and the lower end is connected with the front axle component; the front tires are 26 inches, and the tires are butyl rubber inner tubes. (2) Drive systems: Composed of pedal, chain, flywheel and other components. The pedal assembly is mounted on the left and right crank on the axle assembly to convert flat power into power; adopt a chain drive, turn the crank, the middle shaft and the chain flywheel of the pedal force to make the rear wheel rotation. (3) Brake systems: Composed of brake parts. OFO Shared Bicycle mainly adopts wire-pull disc brake, through the tension of the wire, drive the brake pad in the calipers to rub against the disc. This sets up a braking effect.

**Figure 1.** OFO Shared Bicycle explosion map.

## 4. Results and analysis

Fitness bike movement way: The pedal drive sprocket is driven by the chain to the left wheel and then to the right wheel, and then finally transmitted to the seat through the crank rocker mechanism. The fitness bike is elevated, and the front wheel axles are on the same horizontal line as the rocker bracket fulcrums, which together form a crank rocker mechanism. The fitness bike is elevated, and the front wheel axles are on the same horizontal line as the rocker bracket fulcrums, which together form a crank rocker mechanism. The child seat is fixed by riveting and rocker bar. The rocker bar is fixed on the bracket by two bearings, which can freely go up and down. The connecting rod is inserted into the rocker and driven up and down by the connecting rod under the action of the crank of the opposite wheel. The child seat follows the seat of Shared Bicycle. From the perspective of safety, the handle is adjusted to fit the grasp size of children, and the front pedal of children's seat following the stents of Shared Bicycle. The front fork of Shared Bicycle is employed to supporting parts on both sides of connecting rod. The body adopts a single front fork to facilitate the circular motion of the wheel. Draw the fitness bike in 3Dmax software, like this is shown in Fig.2. The feasibility of seesaw function is illustrated by selecting four state comparison diagrams. When the wheel drives the connecting rod to turn, showing the rising state, the seat under the lever principle and the gravity action, the seat will show the downward trend. Similarly, when the state is falling, the seat will rise.

IOP Conf. Series: Materials Science and Engineering **772** (2020) 012074 doi:10.1088/1757-899X/772/1/012074

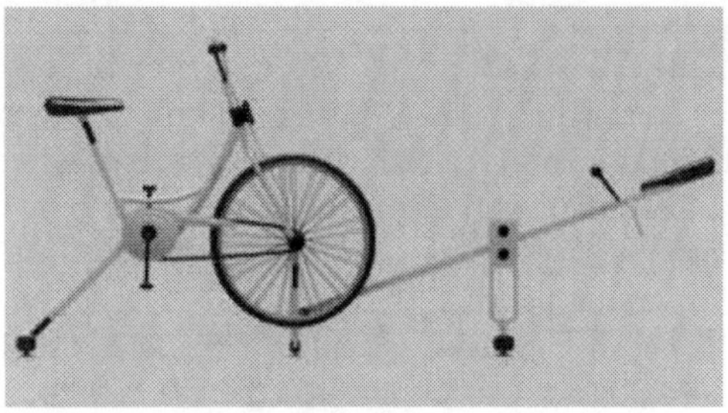

A. Connecting rod with wheel nadir state.

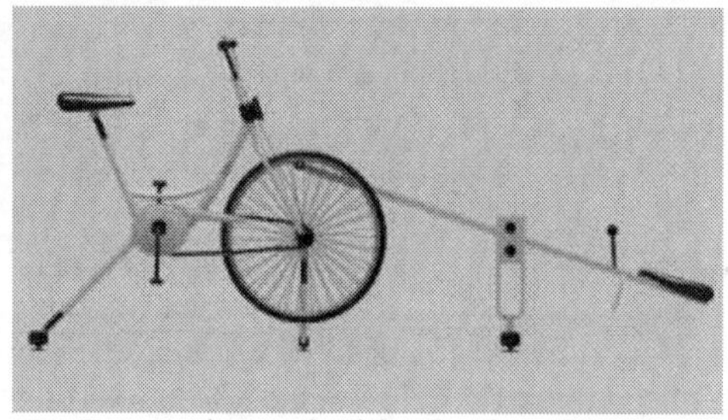

B. Connecting rod with wheel peak state.

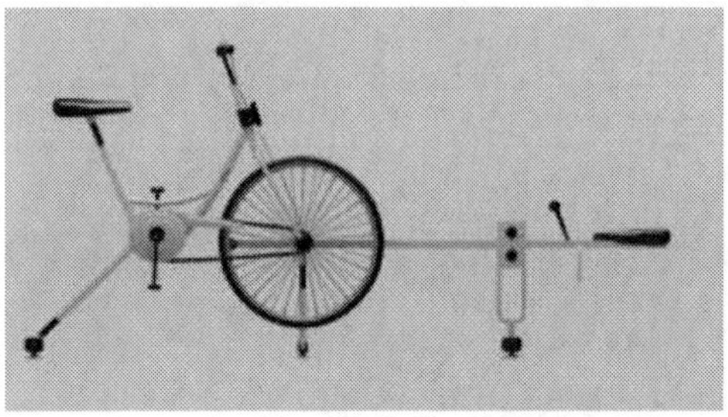

C. Connecting rod with wheel closest point state.

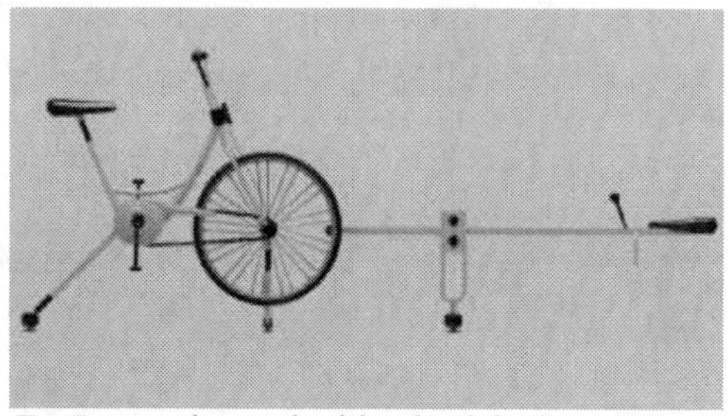

D. Connecting rod with wheel farthest point state.

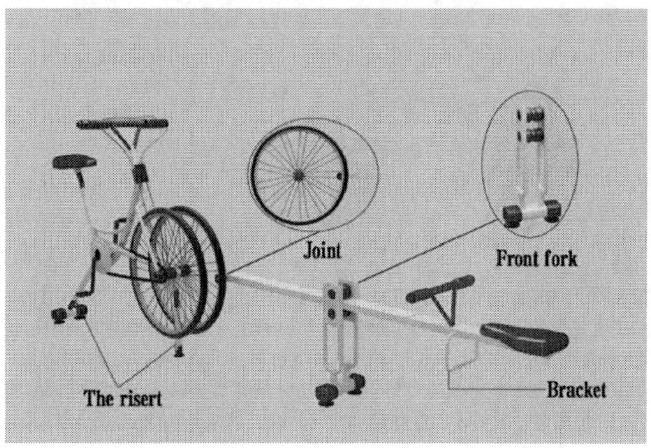


**Figure 2.** General structure of exercise bike.

### *4.1. Size Specification*

According to the size analysis of OFO Shared Bicycle, the optimal length range of connecting rod of fitness bike is studied. Among them, the value range of the seat height and the position relationship between the three nodes are very important for comfort, and the reason crank length is also an important site of the fitness bike.

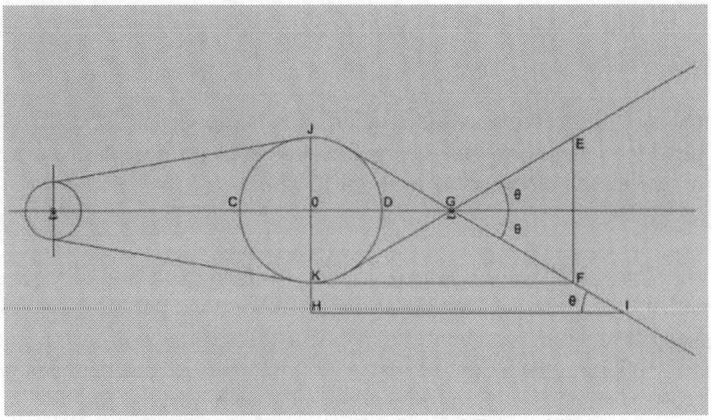


**Figure 3.** Schematic diagram of fitness bike

*4.1.1. Lever length.* In order to ensure the safety of mechanism movement, the preset Angle of theta is 30°, and the motion Angle of the seat is 60°. Tire size of the 2.0 version OFO Shared Bicycle is 26 inches, while the diameter of JK is 640mm and the radius of DO is 320mm. HK is 200mm above ground height of fitness bike. According to figure 3, JK=EF, so JG=GE. The minimum length of the lever isGC=CO+OG=320+ $\frac{320}{tan\theta}$≈874mm,the longest length of the lever is JI=GJ+GI=$\frac{320}{sin\theta}$ +$\frac{320+200}{sin\theta}$ ≈ 1680mm.T Therefore, the value range of the lever is between 874mm~1680mm, take the intermediate range is 1200mm.

*4.1.2. Location of three nodes.* Three points is handlebars, seat, pedal. The relative position between the handlebar and the seat determines the rider's upper body position, and the height of the handlebar is too high or too low to cause discomfort. The position of the seat relative to the axis of the central axis is a key to determine whether the muscle strength of the lower limb muscles can be effectively utilized [5] Under the guidance of ergonomics theory, men's P95 and women's P5 are used as valves to adjust the key size of the fitness bike according to the actual situation. The size data [6] involved are shown in Table 1.

**Table 1.** Body size of fitness bike. (Unit: mm)

| Measuring project | Male P95 | WomenP5 | Measuring project | Male P95 | WomenP5 |
|---|---|---|---|---|---|
| Height | 1775 | 184 | Sit two elbows and shoulders wide | 489 | 378 |
| Leg add foot | 448 | 342 | Body foot wide | 105 | 80 |
| Thick chest | 245 | 170 | Shoulder breadth | 403 | 320 |
| Thigh length | 505 | 402 | Hip width | 334 | 290 |
| Standing shoulder height | 1455 | 1195 | Sitting elbow height | 1096 | 899 |
| Standing position cubits high | 1096 | 899 | Sitting thigh thick | 151 | 113 |
| Upper limb length | 569 | 455 | lower limb length | 1063 | 851 |

The optimal operating height of the upper limbs is 76mm lower than the elbow and is self-binding [7]. Handlebar height should be less than P5 women's elbow height, assuming that the average sole thickness is 25mm. The height range of handlebars is 899-76+25mm≤h≤1195-25mm, 848 mm≤h≤1170mm, take an intermediate range 1000mm; seat height shall meet the minimum percentile woman's template, h seat = calf height + thigh length + sole correction =342+402+25=769mm, take 770 mm; the distance between handlebars and seats should take into account the arm length and design margin of the maximum percentile male and the minimum percentile female, the maximum value is upper arm + forearm length =455mm, the minimum value is 245+18=263mm, 263mm≤h≤455mm, take an intermediate range 360mm.

*4.1.3. Crank length.* Now the common way to determine the length of the crank is to take 1/10 of the height [8]. According to the human body template, the average height of Chinese adults is about 1700mm. Therefore, the length of the crank is about 170mm.

## 5. Conclusion

The abandoned OFO Shared Bicycle the form of motion of the bicycle and the seesaw combined by the crank rocker mechanism. The Shared Bicycle after the full service period is recycled, dismantled and reused. Through the man-machine size analysis of fitness bike, the length of the lever is 1200mm, the height of handlebar is 1000mm, the height of seat is 770mm, the distance between handlebar and seat is 360mm, and the length of crank is 170mm. In addition, the author of the fitness bike size design is still in a preliminary exploration, to deeper research, but also needs to carry out ergonomic experiments.

**Acknowledgments**

This work was financially supported by Sichuan Provincial Social Science Key Research Base 'Cultural Industry Development Research Center' funded project (WHCY2019B08) and Sichuan Agricultural University 2019 Art Special Research Project fund.

## References

[1] Fishman E and Washington S, "Factors influencing bike share membership:an analysis of Melbourne and Brisbane," Transportation Research Part A: Policy and Practice,2015,pp. 17-22.
[2] Fishman E, "Bike share a review of recent literature,"in Transport Reviews,2016, pp. 92-110.
[3] Xiandun Song, "Bicycle design,"in Chinese Bicycle,1989, pp. 45.
[4] Shu Pei and Xiuming Tian, "Research and analysis of vehicle recovery and remanufacturing,"in China Makes Comprehensive Use Of Resources, 2001,pp. 23-26.
[5] Chengde Feng and Zuoda GUO, "Principles and applications of ergonomic,"in Sichuan University,2000, pp. 29-30.
[6] Yulan Ding, Ergonomics(Yulan Ding, Beijing,2000), pp. 82.
[7] Naying Zhang,"Application of ergonomics in bicycle product design,"in The Development of Science and Technology,2001, pp. 26.
[8] Hongmei Kuang and Suixian Yang,"Ergonomic bicycle riding posture improvement design,"in Design And Research,2007, pp. 9-10.

# Effect of aging treatment on corrosion resistance of Inconel 718 fabricated by selective laser melting

**Xuenan Liu [1], Runyu Zhang [1], Sengnan Feng [1], Huixiang Li [1], Yizhe Wang [1], Jianhong Gong [1,*] and Hui Sun [2,3]**

[1] School of Electrical, Mechanical and Information Engineering, Shan Dong University, Shandong, China.
[2] School of Space Science and Physics, Shandong University, Weihai 264209, China
[3] Shandong Provincial Key Laboratory of Optical Astronomy and Solar-Terrestrial Environment, Shandong University, Weihai, 264209, China

*Corresponding author e-mail: gongjh@sdu.edu.cn

**Abstract.** The microstructure and corrosion resistance of Inconel 718 alloy made by selective laser melting (SLM) by aging heat treatment (AT) were studied. After the solution treatment at 1050°C, the aging treatment temperature ranges from 600°C to 800°C. Scanning electron microscopy showed that γ" was precipitated out of the γ austenite matrix, with σ decrease in delta phase. The morphology, distribution and formation mechanism of precipitates in these microstructures were analyzed and discussed. Open circuit voltage curve (OCPT), alternating current impedance spectrum (IMP) and TAFEL curve (TAFEL) were tested by electrochemical workstation. Electrochemical corrosion results and corrosion resistance of samples after four different heat treatments were analyzed, and the best aging heat treatment mode and multistage aging treatment were discussed.

## 1. Introduction

SLM manufacturing technology can make up for the defects of traditional manufacturing process in high-tech fields such as aerospace, which are difficult to produce or have substandard performance, and has gradually become one of the mainstream development directions of additive manufacturing[1].High quality metal powder materials has been avtechnical problem that selective laser melting technology development needs to overcome, and nickel based high temperature alloy has the high temperature strength, oxidation resistance, corrosion resistance and other high temperature under the condition of the use of the advantages. Inconel 718 nickel-based alloy since its launch has experienced more than 50 years of continuous improvement, already mature application in the aerospace, petrochemical and other fields [2], and is good raw materials of selective laser melting.

Although SLM has many advantages, the additive manufacturing technology is in the development stage and the technology is not mature. There are still many defects in the parts produced by SLM, such as incomplete powder melting, poor metallurgical bonding and hole formation [3].The heat treatment process of Inconel 718 alloy need high requirements, and its microstructure is greatly affected by heat treatment. Different heat treatments will cause huge differences in morphology and quantity of strengthened phase and precipitated phase [4], resulting in differences in mechanical properties. The

Content from this work may be used under the terms of the Creative Commons Attribution 3.0 licence. Any further distribution of this work must maintain attribution to the author(s) and the title of the work, journal citation and DOI.
Published under licence by IOP Publishing Ltd

immature selective laser melting manufacturing technology may aggravate these defects. The solution treatment + aging treatment is the most common heat treatment method for strengthening Inconel 718 alloy. Among them, the solution treatment is the heat treatment of Inconel 718 at 900°C~2000°C for 1h~2h. During the experiment, the time and temperature can be adjusted and matched. Aging treatment is to conduct heat treatment for 4h~8h at 600°C~800°C. Similarly, three data curves are selected for comparison, and the experimental results are tested by an electrochemical workstation. The parts are easy to be oxidized and corroded under high temperature and high strength working conditions, so it is necessary to study the corrosion resistance of Inconel 718.

The research on Inconel 718 alloy started late in China [5]. There are few studies on the corrosion properties of Inconel 718 alloy materials manufactured by additive at home and abroad. Therefore, this paper chooses to study the influence of aging treatment on the corrosion resistance of Inconel 718 alloy prepared by selective laser melting, compares the heat treatment results, and explores the corrosion resistance rules of Inconel 718 alloy.

## 2. Experimental procedures

### *2.1. Material*

The wire cutting machine was used to divide Inconel 718 alloy (ingredients as shown in table 1) into four rectangular blocks of 10 mm× 45 mm× 30 mm, and then the long strip material was cut into four pentahedral blocks with side length of 50 mm, 50 mm, 70 mm and height of 30 mm.

**Table 1.** Three Scheme comparing.

| Fe | Cr | Nb | Mo | Ti | Al | C | B | Ni |
|---|---|---|---|---|---|---|---|---|
| 18.50 | 18.00 | 5.30 | 3.00 | 0.95 | 0.50 | 0.003 | 0.003 | Balance |

### *2.2. Heat Treatment*

The box type resistance furnace of gwl-1400°C manufactured by guoju experimental electric furnace factory in luoyang was used for heat treatment. The following four groups of samples are heat-treated with 300°C/h of heating rate according to table 2 respectively.

**Table 2.** Heat treatment

| Number | Heat Treatment |
|---|---|
| A | 1050°C×1h, AC |
| B | 1050°C×1h+620°C×8h, AC |
| C | 1050°C×1h+620°C×8h+720°C×8h, AC |
| D | 1050°C×1h+820°C×8h, AC |

### *2.3. Characterization Analysis*

The microstructure of metallographic surface was observed by optical electron microscope. Scanning electron microscope (SEM) was used to analyze the original and aging microstructure of the alloy.

### *2.4. Electrochemical testing*

Firstly, NaCl powder and water were used to prepare 3.5wt% sodium chloride solution as electrolyte, and electromagnetic mixer was used to mix it evenly. Then platinum electrode, mercury electrode and sample electrode were placed on the same plane and inserted into the foam plate to put the electrolyte. Finally, Shanghai chenhua CHI660E electrochemical workstation was used for testing. During the test, the initial potential OCPT of four samples obtained from 300s was measured after the point position was stabilized for a while, and the initial potential median value was set to obtain IMP and TAFEL curves.

## 3. Results and Discussion

Electron microscope images of the microstructure and delta phase morphology of Inconel718 alloy after solution heat treatment and aging heat treatment at three different temperatures are shown in Fig.1 (a), (b), (c) and (d).

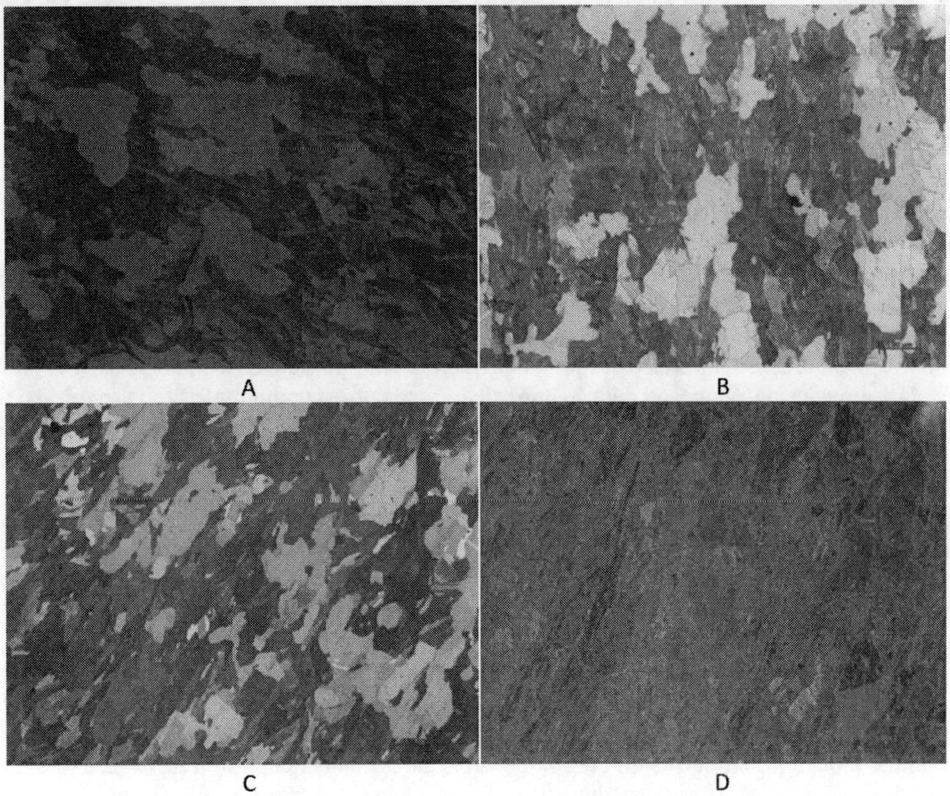


**Figure 1.** Metallographic microscope ×50 times (100μm)

After three heat treatment processes, the grain size of Inconel718 alloy is not significantly different from that of the solid solution treatment. It can be found from the observation of metallography that the only solid solution treatment Inconel718 alloy is mainly composed of γ-phase matrix and a small amount of carbides, while the size of other phases is too small to be observed [6]. After solid solution and aging treatment, many spherical γ'phase, disc γ"phase, rod σ phase and a large amount of carbides were separated from the dendritic γ matrix, and the strengthening γ' phase and γ" phase were the main factors to improve the hardness of the materials.γ"phase appeared in large quantities after heat treatment, and γ" was the main strengthening phase of the materials after heat treatment, which was a kind of metastable phase, appearing as a face-centered cubic phase and dispersing out in the shape of a ball, which played a part in strengthening the alloy[7]. γ" phase is metastable phase, during long-term use, γ" phase tends to be σ phase.

**Table 3.** Dynamic potential scanning parameters

| Electrochemical parameter | Ecorr/ (V) | Icorr ($\mu A\cdot cm^{-2}$) | Rp ($K\Omega\cdot cm^{-2}$) |
|---|---|---|---|
| A | -0.542 | 2.996 | 180.908 |
| B | -0.310 | 0.387 | 1400.517 |
| C | -0.310 | 0.3156 | 1717.364 |
| D | -0.160 | 0.17248 | 3142.393 |

The curve was fitted by Tafel epitaxial method, and the polarization parameters were obtained. Icorr was the self-corrosion current density of the alloy, and Ecorr was the self-corrosion potential of the alloy. The corrosion rate was the thickness of the alloy corrosion every year, and the corrosion rate was proportional to the self-corrosion current density. From the perspective of corrosion dynamics, the higher the density of self-corrosion current, the faster the corrosion rate and the worse the corrosion resistance of the alloy. Alloy treated at 820°C had the smallest Icorr, followed by dual aging treatment at 720 °C, and only solid solution treatment had the largest Icorr. The self-corrosion current density of alloys of different states varies from large to small: solid solution > 620°C aging > 720°C dual aging > 820°C aging.

The main factors affecting the corrosion performance of materials include chemical composition and organizational structure. The corrosion resistance of Inconel 718 alloy fabricated by SLM may be affected by the porosity defects in SLM parent material samples [8]. The strengthening methods of nickel-based superalloy mainly include solution strengthening, precipitation strengthening and grain boundary strengthening. The main strengthening phase in Inconel 718 alloy is γ phase (Ni3Nb), and the stability of γ phase directly affects the properties of the alloy. γ phase → γ' phase → γ"phase → σ phase transition occurs during solid solution treatment. σ phase precipitation of Inconel 718 alloy after solid solution treatment reduced the concentration of Nb element in γ phase which could be used to precipitate Inconel 718 alloy during aging, and the γ in grain decreased correspondingly[9], and the volume fraction of alloy strengthened phase decreased. Characteristic curves such as self - corrosion potential - corrosion time, ac impedance spectrum and tafell polarization curve are usually used to characterize the corrosion resistance of materials. The self-corrosion potential represents the corrosion tendency of the material in the corrosion solution, and the more positive the self-corrosion potential is, the weaker the corrosion tendency is[10]. When the electrode system is a sine wave voltage (current) communication signal disturbance, it produces a corresponding response signal current (voltage), and these signals can be obtained by the impedance of the electrode or admittance. The impedance spectrum generated by the sinusoidal signal of a series of frequencies is called electrochemical impedance spectrum. The greater the arc radius of the electrochemical impedance spectrum, the stronger the corrosion resistance of the material. The tuffell curve is a section of a polarization curve in which the electrode potential is changed in order to achieve a certain current, TAFEL is obtained by logarithm of current density and overpotential plot. The self-corrosion current density represents the corrosion resistance of the sample in the corrosion solution. The lower the self-corrosion current density is, the stronger the corrosion resistance of the material will be [11].

## 4. Conclusion

The conclusions are as follows:

(1)After 720°C dual aging and 820°C high-temperature aging heat treatment, σ phase will be precipitated in the grain boundary in the form of short bars or grains, and 820°C high-temperature aging treatment is conducive to the transformation of the metastable γ" phase to stable σ phase, and more σ phase will be precipitated in the grain and grain boundary than 720°C dual aging treatment.

(2)Since the corrosion current density changes from large to small, it is solid solution > 620°C aging >720°C dual aging >820°C aging and the order of corrosion resistance tendency is 820°C aging >720°C dual aging >620°C aging > solid solution. Therefore, 820°C high-temperature aging alloy has the best corrosion resistance, while only the alloy treated with solid solution has the worst corrosion resistance.

(3)The high-temperature aging heat treatment temperature of 820°C is higher than the dual aging temperature of 720°C, which increases the number and size of σ phase. The corrosion resistance of samples treated with solid solution + aging is better than that treated with solid solution only. The fractional aging has little impact on the corrosion resistance of the alloy, and the corrosion resistance of samples treated with high aging is better.

**Acknowledgments**
We gratefully acknowledge the Shandong Province Natural Science Foundation (ZR2018QEM002) and the Young Scholars Program of Shandong University, Weihai for their financial support. We also thank the laboratory teacher for the article experiment guidance, at the same time, thanks to palace of the teacher for the article content guidance.

**References**
[1] R. Wang, W. Q. Yang. Selective laser melting technology and topology optimization for aerospace components [J]. Modern manufacturing engineering,2018(12):24-30.
[2] R. Y. Liu. Study on microstructure and properties of GH4169 superalloy [D]. Donghua university,2013.
[3] X. K. Sun. Study on laser melting pool morphology and porosity defects of GH4169 alloy selection [D]. Harbin Institute of Technology,2018.
[4] Y. J. Chen, C. G. Zhang, X. T. Zhang, C. Wang, A. Wang. Effect of solid solution + aging on alternating corrosion fatigue properties of GH4169 alloy [J]. Thermal processing technology, 2008,47(08):158-162+165.
[5] Y. Wang. Effect of delta alloy on high temperature deformation and recrystallization [D]. Harbin Institute of Technology,2008.
[6] R. B. Li, M. Yao, W. C. Liu et al. Isolation and determination for δ, γ' and γ' phases in Inconel 718 alloy [J]. Scripta Mater, 2002, 46(9): 635-638.
[7] J. T. Chen, H. Q. Jiangg, H. Y. Pan, H. D. Xia, Z. R Yang, Q. Jiang, L. Q. Fang. Effect of heat treatment process on microstructure and properties of GH4169 alloy [J]. Steel vanadium and titanium, 2008,39(06):168-172.
[8] Sun Y, Moroz A, Alrbaey K. Sliding Wear Characteristics and Corrosion Behaviour of Selective Laser Melted 316L Stainless Steel. Journal of Materials Engineering and Performance, 2014, 23(2): 518-526.
[9] Yun. Research on welding characteristics of typical materials after laser selection melting forming [D]. Huazhong university of science and technology,2017.
[10] C. Yang. Delta phase dissolution and deformation behavior of GH4169 alloy [D]. Harbin Institute of Technology,2007.
[11] Gao Z, Ojo O A. Modeling analysis of hybrid laser-arc welding of single-crystal nickel-base superalloys. ACTA MATERIALIA, 2012, 60(6-7): 3153-3167.

IOP Conf. Series: Materials Science and Engineering **772** (2020) 012076 doi:10.1088/1757-899X/772/1/012076

# Effect of Three Heat Treatments on High Temperature Oxidation Properties of GH4169 Alloy Produced by SLM Additive Technique

**Xuenan Liu [1], Siyuan Liu [1], Sengnan Feng [1], Huixiang Li [1], Yizhe Wang [1], Jianhong Gong [1,]* and Hui Sun [2, 3]**

[1] School of Electrical, Mechanical and Information Engineering, Shan Dong University, Shandong, China.
[2] School of Space Science and Physics, Shandong University, Weihai 264209, China
[3] Shandong Provincial Key Laboratory of Optical Astronomy and Solar-Terrestrial Environment, Shandong University, Weihai, 264209, China

*Corresponding author e-mail: gongjh@sdu.edu.cn

**Abstract.** In this paper, GH4169 nickel-based superalloy molded samples produced by Selective Laser Melting (SLM) were subjected to aging treatment and solution and aging treatment to study the high temperature oxidation behavior under different heat treatments. It is found that the high temperature oxidation products of GH4169 samples produced by SLM is mainly Cr2O3. The alloy sample under heat treatment has a higher high temperature oxidation rate. The oxide film of GH4169 alloy sample treated by solution and aging treatment has a high forming speed, and the surface of the oxide film is continuous and dense, and the quality is the best, and the protection effect on the alloy matrix is the best.

## 1. Introduction

GH4169 is a nickel-based superalloy with excellent comprehensive mechanical properties. It is widely used in the manufacture of important parts in aerospace, petrochemical and other fields [1]. It is an important basic alloy material, and its mechanical properties have been extensively studied [2, 3]. Since high-temperature nickel-based alloys have sufficient oxidation resistance at the service temperature (<700°C), there are few studies on their high temperature oxidation problems. However, the GH4169 alloy has many oxidizable elements such as Nb, Mo and Fe and a spot of Al, the alloy parts therefore may be oxidized on the surface during hot working or high temperature service, and excessively thick oxidation may be formed in severe cases, which in turn affects the normal and safe use of parts, and even leads to scrap [4].

In addition, therefore, the application of SLM rapid prototyping technology to produce complex solid superalloy parts with excellent performance can solve the problem of low forming efficiency of parts due to various factors in the traditional manufacturing process. Therefore, this study has a great theoretical and practical significance to study on the high temperature oxidation performance of GH4169 alloy manufactured by SLM and the influence of different heat treatments on its high temperature oxidation performance. The main research contents of this topic are as follows:

Content from this work may be used under the terms of the Creative Commons Attribution 3.0 licence. Any further distribution of this work must maintain attribution to the author(s) and the title of the work, journal citation and DOI.
Published under licence by IOP Publishing Ltd

(1) Study the results of alloys under different heat treatments after heating and oxidizing experiment, and analyze the law of oxidation mass gain;

(2) Study the surface morphology and phase composition of the oxide film after isothermal temperature oxidation, analyze the oxidation process, and summarize the effect of heat treatment on the high temperature oxidation performance of GH4169 alloy and its mechanism.

## 2. Experimental procedures

### *2.1. Experimental Materials*

The alloy powder was prepared into GH4169 alloy samples by a EOSINT-M280 selective laser melt forming system. In the SLM process, the laser power is 300W, the laser diameter is 0.1mm, the scanning speed is 1200mm/s, the scanning distance is 0.12mm, and the thickness of each layer is 0.03mm. Its chemical composition is shown in Table. 1.

**Table 1.** Chemical composition of GH4169 alloy (in weight percent, wt%).

| C | Si | B | Mn | Cr | Ni | Mo | Al | Nb | Ti | Cu | Co | Fe |
|---|---|---|---|---|---|---|---|---|---|---|---|---|
| 0.03 | 0.05 | 0.003 | 0.03 | 19.01 | 52.30 | 3.06 | 0.57 | 5.07 | 1.00 | 0.02 | 0.02 | Bal. |

### *2.2. Heat Treatment*

According to the TTT curve of the alloy and the conventional heat treatment system, three different heat treatment processes were developed in this experiment.

**Table 2.** Heat treatment scheme for GH4169 alloy samples

| Number | Heating Treatment | Name |
|---|---|---|
| A | 720°C× 3h， 50°C/h FC to 620°C× 3h， AC | DA |
| B | 950°C× 1h， AC+ 720°C× 3h, 50°C/h FC to 620°C× 3h, AC | ST |
| C | 1050°C× 1h， AC+ 720°C× 3h, 50°C/h FC to 620°C× 3h, AC | GST |
| D | Untreated | AS |

The heat treatment experiment of the GH4169 alloy sample was carried out in GWL-1400 electric furnace at a heating rate of about 300°C / h. The samples' number, heat treatment and its name are shown in Table. 2.

### *2.3. High Temperature Oxidation Experiment*

The polished and cleaned samples were oxidized for 48h in a GWL-1400 box-type electric furnace at 850°C. Another set of samples is placed in a differential scanning calorimetry (DSC) to perform a heating and oxidizing experiment. The experimental atmosphere is pure oxygen, the oxidation temperature is gradually increased from 25°C to 1200°C, the heating rate is 0.25°C / s, and the total heating time is 4700 s.

The oxidation resistance of the GH4169 alloy is analyzed by the percentage of oxidation mass gain of samples. To calculate the percentage increase in weight of high temperature oxidation samples, refer to the following formula:

$$P^{+} = \frac{W_t - W_0}{W_0} \times 100\% \tag{1}$$

Where: $P^{+}$ is the percentage of oxidation mass gain of the sample; $W_t$ is the mass of the sample after the oxidation experiment (mg); $W_0$ is the mass (mg) of the sample before the oxidation experiment.

### 2.4. *Characterization Analysis*

The metallographic structure is observed using an optical microscope (OM). The phase composition of the alloy sample and its oxidized oxide film is analyzed by Rigaku Ultima IV X-ray diffractometer (XRD). The scanning range is 2θ=20-80°. The FEI Nova Nano SEM 450 scanning electron microscope is used to observe and analyze the morphology of the surface oxide film of the alloy samples under different heat treatments.

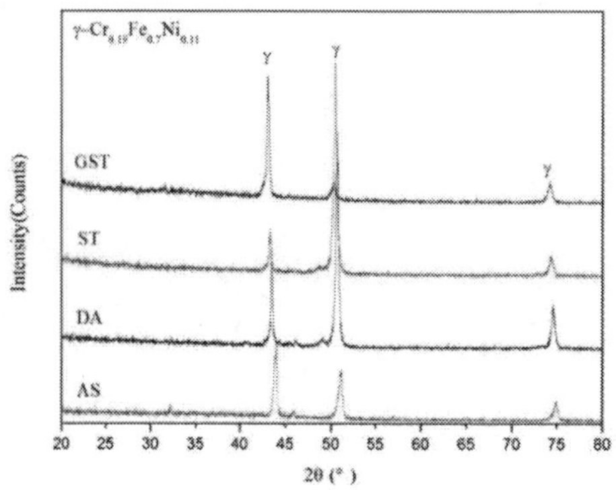


**Figure 1.** XRD patterns of GH4169 alloy samples under different heat treatments

## 3. Results and discussion

### 3.1. *Unoxidized Phase*

Figure. 1 shows that the diffraction peak of the matrix phase can be monitored under all conditions.The γ' phase and the γ" phase are the two main strengthening phases of the GH4169 alloy, but the two strengthening phases are co-precipitated with the γ matrix phase, so the diffraction peaks of the γ' phase and the γ" phase are covered up by γ phase, which results in the inability to observe individual diffraction peaks of the γ' phase and the γ" phase from the diffraction pattern. In addition, since the volume fraction of the σ phase and the carbide phase in the alloy are small, it is difficult to detect the diffraction peaks of them. These results are basically consistent with the findings of Song et al [5].

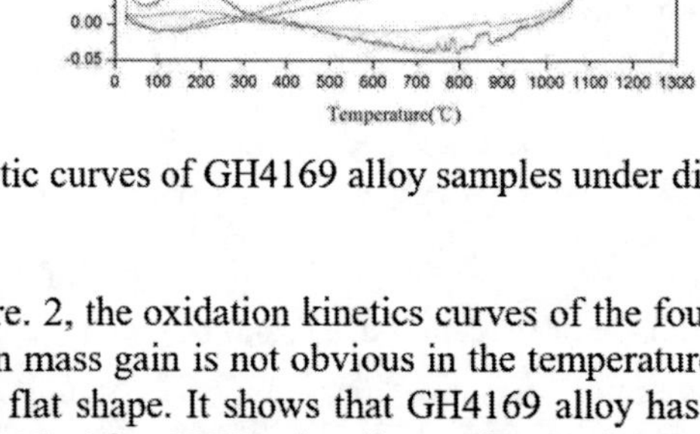


**Figure 2.** Oxidation kinetic curves of GH4169 alloy samples under different heat treatments

### 3.2. *Oxidation Kinetics*

As can be seen from the Figure. 2, the oxidation kinetics curves of the four samples showed the same trend as a whole: the oxidation mass gain is not obvious in the temperature range of 25°C-800°C, and the curve showed a relatively flat shape. It shows that GH4169 alloy has excellent high temperature oxidation resistance below 800°C. After 800°C, the alloy undergoes intense oxidation reaction and the weight increases rapidly. This rule accords with the conclusions of C. Wagner and Tomashov [6-8].The

AS sample does not show a significant tendency to oxidize before 900°C and shows a rapid increase after 900°C. The DA and ST samples both have small oxidation mass gains and low oxidation rates before 1000°C. But the rates increase significantly after 1000°C, and the percentage of mass gain of the ST sample is higher than that of the DA sample in the same case. Before 1100°C, the general trend of mass gain of GST sample is about the same as the curve of AS sample.

### *3.3. Oxide Film Analysis*

*3.3.1. Phases.* It can be seen from the Figure. 3 that after isothermal temperature oxidation, the diffraction peaks of γ matrix phase and Cr2O3 are observed in each sample, and the difference in heat treatments also makes the intensities of characteristic peaks of each phase different. A small amount of Fe2O3 is also observed in the oxide film of each oxide film. Besides, a NiCr2O4 spinel phase is observed both in the GST sample and the AS sample.

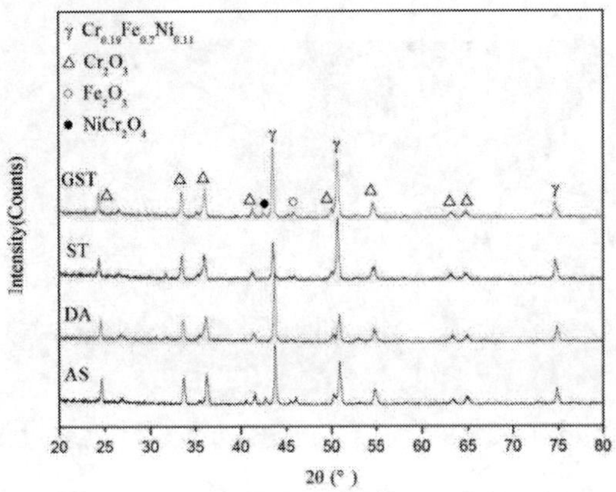


**Figure 3.** XRD patterns of oxide film of GH4169 alloy samples under different heat treatments

In the initial stage of oxidation, the surface of the nickel-based superalloy will react to form a large amount of Cr2O3 oxide layer [9], and the weight of the oxidation product of the alloy increases rapidly. Since Cr rapidly diffuses to the surface of the alloy for oxidation, a depleted area of Cr appears under the oxide layer, which is called depletion of Cr [10-11]. Under the action of the chemical element concentration gradient, Ni element reacts with oxygen to form NiO. At the same time, at a higher temperature, NiO can react with Cr2O3 to form NiCr2O4 spinel phase. The formed Cr2O3 oxide film is uniform and dense, which can prevent further oxidation of the inner alloy matrix, and NiCr2O4 relatively does not have such protection.

Due to the depletion of Cr, Fe element is also diffused to the surface of the alloy by the concentration gradient of chemical elements [12], and reacts with oxygen in the surrounding environment to form Fe2O3. NiO and Cr2O3 can also interact to further form a NiFe2O4 spinel phase.

In this experiment, the oxidation products Cr2O3 are formed in the four samples, and no obvious NiO characteristic peaks and NiCr2O4, NiFe2O4 spinel phase characteristic peaks are observed. It is proposed that the formed oxide film Cr2O3 is continuously dense and the depletion of Cr of matrix surface is not significant, which hinders the formation of NiCr2O4 and NiFe2O4 spinel phases.

*3.3.2. Surface Morphology.* It can be seen from Fig. 4 that the surface structure of the oxide film of the AS sample is relatively uneven, and the grains of oxidation structure are sparse (Fig. 4a), and a large number of pores and cracks can be observed. The pore size of the oxide film on the surface of the DA sample is reduced, but severe folds occurred (Fig. 4b). Besides, plenty of cracks are found at the edges of the folds (Fig. 4b). The surface oxide films of the ST and GST samples are both dense and uniform, and there are fewer holes and cracks (Fig. 4c, 4d), indicating that the solution treatment can improve quality of oxide film of GH4169 alloy. Furthermore, as the solution temperature increases, the pores and cracks of oxide film of alloy reduce, making the compactness of oxide film better.

In addition, it can be seen that the surface of the oxide film of the ST and GST samples under solution treatment even shows a distinct continuous clustered rod-like oxidation structure (Fig. 4c, 4d), which are more clear on the surface of the GST oxidized sample (Fig. 4d ). It is speculated that the formation position is at grain boundary because the grain boundary can provide the nucleation position and nucleation energy required for the oxide nucleation [13]. The nucleation oxide grows initially perpendicular to the surface of the alloy. Then it follows by re-nucleation at the interface to form new oxides and interconnects to grow between the oxides, ultimately forming a continuous oxide film structure. The oxide firstly develop and grow at the grain boundary, and it is known from the foregoing results that the grain size and boundary size become large after solution treatment, so that the rod-shaped oxide structure is formed at the large grain boundary during the oxidation process and can be observed easily.

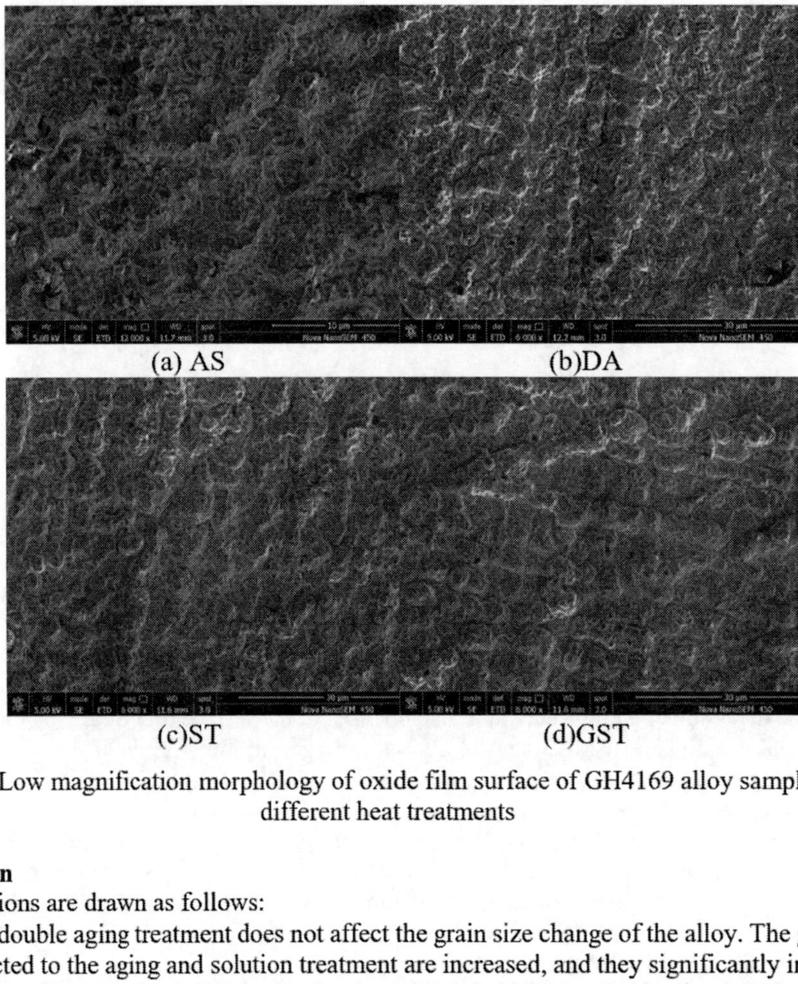

(a) AS (b)DA

(c)ST (d)GST

**Figure 4.** Low magnification morphology of oxide film surface of GH4169 alloy samples under different heat treatments

## 4. Conclusion

Main conclusions are drawn as follows:

(1) Direct double aging treatment does not affect the grain size change of the alloy. The grains of the sample subjected to the aging and solution treatment are increased, and they significantly increase after solid solution at 1050°C and aging treatment because the σ phase is completely dissolved into the matrix at 1050°C, its limiting effect on grain growth disappears, resulting in grain coarsening and grain boundary flattening.

(2) The three samples treated by heat treatment formed a continuous dense oxide film during oxidation, and the surface defects were not obvious. The surface of the oxide film of GST sample is

smooth, continuous and dense, few cracks existed, and the number of holes is the least. In addition, obvious rod-like oxidation structure is observed in the oxide film morphology both of ST and GST samples. This is because the rod-shaped oxidized structure should be an oxide cluster structure formed at the grain boundary, and can grow rapidly and be clearly observed on the surface of the oxide film of solution-treated alloy with large grain and grain boundary.

**Acknowledgments**

We gratefully acknowledge the Shandong Province Natural Science Foundation (ZR2018QEM002) and the Young Scholars Program of Shandong University, Weihai for their financial support. We also thank the laboratory teacher for the article experiment guidance, at thc same time, thanks to palace of the teacher for the article content guidance.

**References**

[1] J. T. Guo. High Temperature Alloy Materials Science (Volume 1) [M]. Beijing: Science Press, 2008.

[2] J. T. Chen, H. Q. Jiang, H. Y. Pan, H. D. Xia, Z. R. Yang, Q. Jiang, L. Q. Fang. Effect of Heat Treatments on the Microstructure and Mechanical Properties of GH4169 Superalloy[J]. Iron Steel Vanadium Titanium, 2018, 39(06):168-172.

[3] Y. J Chen, C. T. Zhang, X. T. Zhang, C. Wang, A. WANG. Effect of Solid Solution and Aging on Alternating Corrosion Fatigue Properties of GH4169 Alloy [J]. Hot Working Technology, 2018, 47 (08): 158-162+165.

[4] S. H. Li, J. Z. Ba, Y. X. Jiang, F. X. Yan, X. Gao. A Research Progress of High Temperature Oxidation Behaviors in IN718 [J]. Chemical Defence on Ships, 2008(02): 8-14.

[5] SONG Kan YU Kai LIN Xin CHEN Jing YANG Haiou HUANG Weidong. Microstructure And Mechanical Properties Of Heat Treatment Laser Solid Forming Superalloy Inconel 718 [J]. Acta Metallurgica Sinica, 2015, 51(08): 935-942.

[6] Wagner C. Diffusion and high temperature oxidation of metals [J]. Atom movements, 1951:153-173.

[7] Tomashov N D. Theory of corrosion and protection of metals [M]. 60 Fifth Avenue,New York 10011:The Macmillan Company, 1965.

[8] Tomashov N D. Theory of Corrosion of Metals [M]. Moscow: Mashgiz, 1952.

[9] W. X. Zhao, L. H. Tan, Y. You, X. G. Shi, S. YOU, Qifan.High Temperature Oxidation Behavior of Electron Beam Refined Inconel718 Alloy [J]. Materials for Mechanical Engineering, 2018, 42(05): 1-8+13.

[10] X. D. Lu, T. Chen. Isothermal oxidation behaviour of a Ni-base superalloy at 900 and 1000℃ [J]. Journal of Functional Materials, 2015, 46(04): 4025-4030.

[11] X. D. Lu, S. G. Tian, Z. D. Sun. Isothermal oxidation behaviour of a Ni-base superalloy at 850 and 950℃ [J]. Corrosion Science and Protection Technology, 2011, 23(04): 298-302.

[12] Huntz A M, Reckmann A, Haut C, et al. Oxidation of AISI 304 and AISI 439 stainless steels [J]. Materials Science and Engineering: A, 2007, 447(1-2): 266-276.

[13] F. J. Liu, M. C. Zhang, J. X. Dong, et al. High-temperature oxidation of FGH96 P/M superalloy [J]. Acta Metallurgica Sinica, 2007, 20(2): 102-110.

# One-Step Synthesis of Au Nanoplates with the Assistance of Formaldehyde

**Jing Xia** *

National Institute of Clean and Low Carbon Energy, Beijing, 102211, China

*Corresponding author e-mail: Jing.xia.e@chnenergy.com.cn

**Abstract.** In this work, a facile method has been developed to fabricate Au nanoplates in one step through heating formaldehyde and $HAuCl_4$ aqueous solution. The obtained Au nanoplates are single crystal with an average edge length of approximately 160 nm. And the formation of plate-like Au nanostructures is because of the adsorption of formaldehyde on the surfaces of Au nanostructures, leading to a similar surface energy of {111} and {100} facets.

## 1. Introduction

In the past decades, gold nanostructures have been widely studied due to their unique size and morphology dependent electronic, chemical, physical, and optical properties [1]. To date, a myriad of gold nanostructures, such as nanoparticles, nanorods, nanorings, nanowires, nanodisks, nanocubes and branched nanocrystals, have been synthesized [2, 3], and they are widely used in the fields of catalysis, biomedicine, biosensing, imaging and photoelectronic devices [4, 5].

Two-dimensional (2D) Au nanoplates have recently attracted much attention among those gold nanostructures, mainly due to a strong plasmon band of these 2D nanostructures in the Near Infrared Ray (NIR) region. Nanoplates with plasmon bands in the NIR region are very sensitive to the changes of small morphology and close environment, providing possible applications in cancer hyperthermia, sensing, imaging, and biological assays [6]. Meanwhile, 2D nanoplates with long edges and sharp vertices present enhanced electric field at the vertices and anisotropic electrical conductivity [7]. Therefore, they are excellent candidates of information storage, photonics and optoelectronics, metal enhanced fluorescence and surface enhanced Raman scattering [8]. Additionally, Au nanoplates are also good substrates for tip-enhanced Raman scattering measurements, because of the atomically flat surfaces of Au nanoplates, easy deposition of analytes on surfaces, and higher enhanced factor between gold surfaces and metal tips. It is also reported that Au nanoplates provides new possibilities for fundamental studies and potential applications in catalyzing, and new nanodevices [9].

Nowadays, a variety of methods have been exploited to get these anisotropic Au nanoplates. The most common method for the formation of Au nanoplates is the seed-mediated approach [10]. In this method, nanoparticles with the diameter of 3-6 nm are prepared firstly, and then they are mixed with a solution of $HAuCl_4$, reducing agent and surfactant. Another conventional method is the polyol process, in which the Au nanoplates are generated by heating a mixed solution of $HAuCl_4$, surfactant and ethylene polyol above 120 °C for several hours. Aldehyde group is formed in this method when the reaction temperature is higher than 100 °C.[11] It is reported that this aldehyde group can be used to reduce $AuCl_4^-$ and the intermediate induced by oxidized aldehyde group plays an important role in the formation

Content from this work may be used under the terms of the Creative Commons Attribution 3.0 licence. Any further distribution of this work must maintain attribution to the author(s) and the title of the work, journal citation and DOI.
Published under licence by IOP Publishing Ltd

of Au nanoplates[12]. However, the above seed-mediated method is complex, in which two separate steps are needed, including formation of Au nanoparticles and growth of Au nanoplates. As for the polyol process, the reaction temperature is high, which is energy-intensive and unsafe.

Here, we developed a simple one-step method to fabricate Au nanoplates with the assistance of the formaldehyde. The morphology and crystal structure of obtained Au nanoplates were analyzed by transmission electron microscopy (TEM) and scanning electron microscopy (SEM). Also, the formation mechanism was discussed.

## 2. Experimental Section

### *2.1. Materials and Chemicals*
Formaldehyde (36.5-38%) was purchased from Sigma-Aldrich. Hydrogen tetrachloroaurate trihydrate ($HAuCl_4 \cdot 3H_2O$) was purchased from Alfa Aesar. All the chemicals were used without further purification.

### *2.2. Preparation of Au Nanoplates*
The reaction solution was prepared by adding 1 mL formaldehyde and 0.5 mL of 10 mM $HAuCl_4 \cdot 3H_2O$ into 9 mL deionized water. Then the reaction solution was heated at 40 °C for 5 h. After the heat treatment, the final product was obtained by centrifugation at 6000 rpm for 30 min.

### *2.3. Characterization*
Au nanoplates obtained from previous steps were first ultrasonicated for 3 min as the pre-process for characterization, making sure that the Au nanoplates were well dispersed in solution. The ultrasonic cleaner used in our research was KS-120D. Before the characterization of TEM, 10 μL of the well-dispersed products was dripped on carbon-coated copper grids. Then TEM and High-resolution TEM (HRTEM) images were taken by using a JEOL JEM ARM200F. For the characterization of SEM, a drop of these nanoplates was placed on a 1 x 1 $cm^2$ silicon slide. After the samples dried, SEM images were collected by a FEI Nova NanoSEM 450.

## 3. Results and Discussion

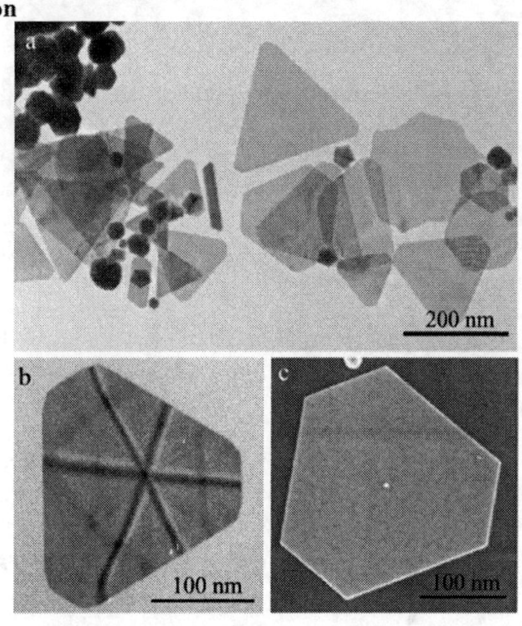


**Figure 1.** a) The typical TEM image of Au nanoplates. b) TEM and c) SEM images of an individual Au nanoplate.

In this work, Au nanoplates were fabricated in one step with the help of formaldehyde. Fig. 1 shows the morphology of Au nanoplates. As shown in Fig. 1a, the average edge length of Au nanoplates is about 160 nm and the shapes of them include triangle, truncated triangle and hexagon. In Fig.1b, the TEM image of an individual Au nanoplate exhibits contrast in its surface. This is because of the existence of the internal stress, which is caused by the buckling of the ultrathin nanoplate [13].

Fig.2a shows the HRTEM image of the Au nanoplate. It demonstrates that the Au nanoplate is single-crystalline. And the fringe spacing is 0.14 nm, which is indexed to {220} reflections of face center cubic (fcc) Au [14]. Its corresponding Fourier transform pattern is shown in Fig. 2b. The hexagonal Fourier transform pattern demonstrates that the flat surface of the Au nanoplate is {111} face, as reported previously [15]. And each nanoplate is bounded by two {111} planes as the top and bottom faces and a mixture of {111} and {100} planes as the side faces [15].

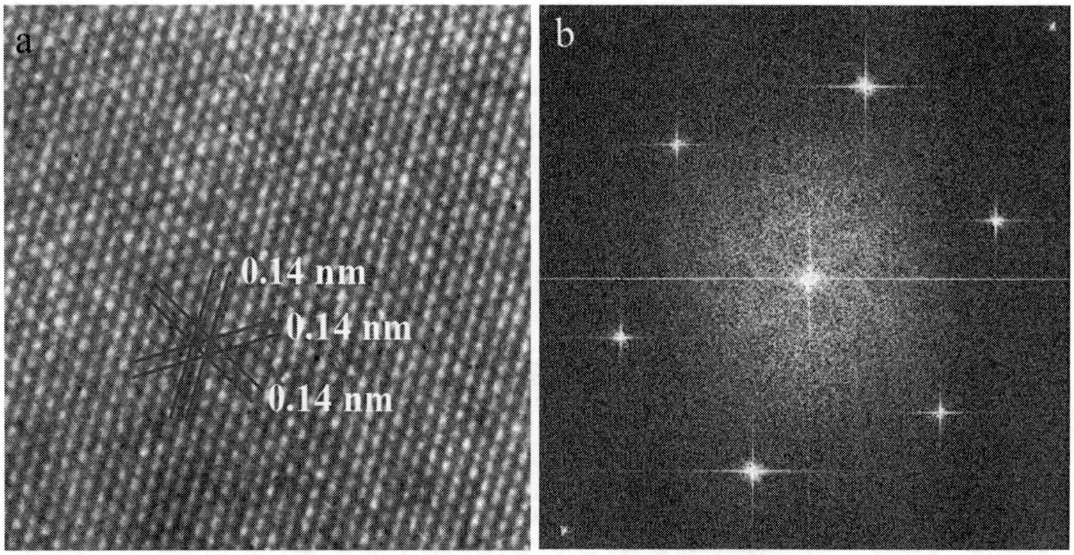


**Figure 2.** a) HRTEM image and b) corresponding Fourier transform pattern of the Au nanoplate.

In the formation of Au nanoplates, Au precursors are firstly reduced to Au atoms by formaldehyde. Then Au atoms grow to Au nanostructures with plate-like shape[16]. In the shape control of Au nanostructures, it is reported that the adsorption of an appropriate ligand can control the morphology of nanostructures through interaction between ligands and metal facets[17]. In this research, the surface energy of {111} planes is lower than that of {100} and {110} planes after the adsorption of formaldehyde on surfaces of Au nanoplates. Meanwhile, the surface energy of {100} planes is close to that of {111} planes. Thus, Au atoms grow to Au nanostructures with plate-like shape finally, as reported in previous research[16]. And further study on the formation of Au nanoplates is still in progress.

## 4. Conclusion

A facile one-step method has been developed to fabricate Au nanoplates through heating formaldehyde and $HAuCl_4$ aqueous solution at 40 °C. The obtained Au nanoplates are single crystal and their average edge length is approximately 160 nm. And the formation of plate-like Au nanostructures is because of the adsorption of formaldehyde on the surfaces of Au nanostructures, leading to a similar surface energy of {111} and {100} planes.

**Acknowledgments**

The authors are very grateful for the financial support from the China Energy.

**References**

[1] Grzelczak M, Perez-Juste J, Mulvaney P, Liz-Marzan LM. Chemical Society reviews 2008;37:1783-91.
[2] Chithrani BD, Ghazani AA, Chan WCW. Nano Letters 2006;6:662-8.
[3] Murphy CJ, Gole AM, Stone JW, Sisco PN, Alkilany AM, Goldsmith EC, et al. Accounts of Chemical Research 2008;41:1721-30.
[4] Guan G, Liu S, Cai Y, Low M, Bharathi MS, Zhang S, et al. Advanced Materials 2014; 26:3427-32.
[5] Ye E, Win KY, Tan HR, Lin M, Teng CP, Mlayah A, et al. Journal of the American Chemical Society 2011;133:8506-9.
[6] Shankar SS, Rai A, Ahmad A, Sastry M. Chemistry of Materials 2005;17:566-72.
[7] Amit S, Minakshi C, Murali S. Nanotechnology 2006;17:2399.
[8] Aslan K, Lakowicz JR, Geddes CD. The Journal of Physical Chemistry B 2005;109:6247-51.
[9] Hsu S-J, Lin IJB. Journal of the Chinese Chemical Society 2009;56:98-106.
[10] Millstone JE, Métraux GS, Mirkin CA. Advanced Functional Materials 2006;16:1209-14.
[11] Tsuji M, Hashimoto M, Nishizawa Y, Tsuji T. Chemistry Letters 2003;32:1114-5.
[12] Li CC, Cai WP, Cao BQ, Sun FQ, Li Y, Kan CX, et al. Advanced Functional Materials 2006;16:83-90.
[13] Shankar SS, Rai A, Ankamwar B, Singh A, Ahmad A, Sastry M, Nat. Mater. 2004; 3, 482.
[14] Lin WH, Lu YH, Hsu YJ, Journal of Colloid and Interface Science 2014; 418: 87-94.
[15] Xia Y, Gilroy KD, Peng HC, Xia X, Angewandte Chemie International Edition 2017; 56:60-95.
[16] Xia J, Dong ZL, Cai YQ, et al. Chemistry-A European Journal 2018; 58:15589-15595.
[17] Sun Y, Mayers B, Herricks T, Xia Y, Nano Letters 2003;3:955-960.

IOP Conf. Series: Materials Science and Engineering **772** (2020) 012078 doi:10.1088/1757-899X/772/1/012078

# Calculation on Refloating a damaged ship with salvage pontoon

**Dewei Pan, Yuqiang Sun, Zhaoxin Zhou*, Kangdi Zhao and Yancai Chen**

School of Navigation, Shandong Jiaotong University, Weihai, China

*Corresponding author e-mail: 15065135036@126.com

**Abstract.** Refloating a damaged ship with salvage pontoon can be very complicated. In the calculation process of traditional pontoon salvage for shipwrecks, the calculation of inflation time, analysis of pontoon motion and intelligent control are mainly studied. When the pontoon is inflated, the volume of gas in the pontoon will change with the change of time and temperature, which will affect the accuracy of salvage measurement. In this paper, According to the theory of heat conduction, a mathematical model of gas volume in pontoon varying with time and temperature is established, which is beneficial to getting the buoyancy and stability of ships.Through the simulation, the refloating process of a damaged ship with salvage pontoon was researched. Computation result shows that shipwreck with many pontoons sunk in the water totally of the preliminary stage, since the stability of the shipwreck is poor, the aeration rate in the pontoon should be calculated accurately to prevent the capsizing of the ship. As the draft of the wreck decreases, the stability of the hull increases gradually, which is beneficial to the process of lifting and floating. The change of pitch angle is small in the course of buoyancy, but it has obvious influence on the buoyancy distribution of ship hull. The position of pontoon has a great influence on the distribution of shear force and bending moment of ship hull. Relevant measures should be taken to reduce the damage to ship hull in the course of buoyancy.

## 1. Introduction

Generally, there are few major ship accidents. Because the design and construction phases of ships will be completed strictly in accordance with the relevant classification societies and international regulations.

Man-made and natural factors are the main reasons for the wreck. Man-made factors include carelessness of drivers, rule-breaking operation of management staff, low professional skills of technicians, etc., which damages the wrecked ship and causes it to sink in the water. Natural factors include bad weather (storm, typhoon, fog), underwater collision (reef, iceberg), collision of marine organisms (whale, shark, flag and other large organisms) and corrosion damage caused by marine organisms attaching to the surface of ships[1,2].

There are many ways to salvage shipwrecks. Each method can be used independently or in combination, depending on the specific needs. These methods are respectively sealed pumping salvage method, pontoon salvage method, ship lifting salvage method, foam plastic salvage method, cofferdam salvage method, inflatable drainage salvage method and so on. Among them, the pontoon salvage

Content from this work may be used under the terms of the Creative Commons Attribution 3.0 licence. Any further distribution of this work must maintain attribution to the author(s) and the title of the work, journal citation and DOI.
Published under licence by IOP Publishing Ltd

method is the one most commonly used [3-5]. The selection of a wreck removal method for a particular case involves taking into account a multitude of parameters. The most important are environmental parameters, such as prevailing weather conditions and the type of the seabed, and the weight of the shipwreck, which will determine the required lifting capacity. pontoon salvage method was more often used in practical engineering[6].

At present, there is short of research on the pontoon salvage of shipwrecks. The research focus on the influence of pontoon on the hull, the mechanical analysis of pontoon, the analysis of pontoon motion and the motion control of pontoon.

In the salvage project of SEWOL, engineers found that pontoons of different heights could not be inflated synchronously, which resulted in low efficiency[7]. In order to reduce the lifting force of the bow, the bending moment and shear force of the hull girder and the overturn of the hull to the deck side after inflating the sunken ship cabin, it is necessary to allocate sufficient external buoyancy for the sunken ship. Pontoons were used in the bow lifting project of SEWOL, which provided 1630 tons of lifting buoyancy. According to the internal and external space and structure of the ship, additional buoyancy can be provided by installing airbags, rubber pontoons and steel salvage pontoons in the ship[8].

Ilias Zilakos and Michael Toulios studied the force of the air bag in the inflatable state and the lifting state, which provided a theoretical basis for the design of the air bag for rescue of damaged ships [9]. The mechanics changes of hull during the rescue of damaged ship by airbag are studied, but the influence of floating motion on structure is not involved [10]. In order to improve the launching speed and safety, reduce the difficulty of operation and reduce the time of launching, Lu Heng et al. put forward a novel floating structure [11].

LU et al. designed a new automatic control salvage pontoon based on Siemens, which can realize remote operation and reduce labor costs. The bottom of the pontoon is equipped with a submersible pump and a pressure sensor B, and the sensor A is installed inside the pontoon. Sensors A and B are connected to computer on the water surface. Workers on surface can use Siemens to obtain the internal pressure of the pontoon and decide whether to turn on or off the submersible pump to control the pontoon's descent and rise [12]. In order to improve the anti-sinking performance of damaged ships, Hee et al. designed a buoyancy support system. The system uses fire fighting system as the inflatable structure of air cavity. In this paper, the ship hull with complex structure is tested, and the results are good, but it is difficult to control the water inflow [13].

A.K.D. Velayudhan designed an automatic control method for underwater lifting of airbag sediment, but the effect of air bag intake and exhaust on salvage process has not been studied in detail [14]. Chen Xiayin et al. systematically calculate buoyancy of pontoons by using multi-physical field simulation and piecewise curved-to- straight algorithm, which reduces the calculation error [15]. Yang Xin and Yebo introduced a set of automatic monitoring system with obvious advantages for pontoon salvage controlled by computer and PLC bus. The control mode of this system is mainly controlled by communication between upper computer and lower computer, which has very good stability [16].

Liu et al. introduced the application of safety analysis method of fuzzy logic to safety analysis of pontoon salvage of shipwrecks. This method is suitable for dealing with uncertainty in safety assessment [17].

At present, there are few references about the salvage pontoon. Researches mainly about the pontoon inflation calculation, hydrodynamic calculation and intelligent control, ignoring the change of temperature after air inflation of the pontoon. When the pontoon is used to lift a sunken ship, the buoyancy of the pontoon will change because of the heat dissipation after the pontoon is inflated, which will affect the accuracy of the calculation of the buoyancy. In this paper, the influence of temperature change on pressure after pontoon inflation is theoretically deduced. The process of pontoon lifting and sinking is further simulated, and the influence of different aeration rate on the state of sinking ship is analyzed.

## 2. Calculation of gas in pontoon

When the heat dissipation area of pontoon inflation space is $S_p$, the heat conductivity $h_P$, the constant-volume specific heat of gas $C_P$, and the volume of gas $V$, the heat conduction constant of pontoon inflation space is obtained as follows[18]:

$$H = \frac{S_p \cdot h_P}{C_p \cdot \rho_0 \cdot V} \tag{1}$$

In the process of blowing up, the speed of inflation is much higher than that of temperature change, assuming that the inflation stage can be regarded as an adiabatic process. The density and temperature of cabin gases can be approximated as linear variations, which are presented as follows [19]:

$$\rho(t) = \rho_0 + \frac{\rho(t_0) - \rho_0}{t_0} \cdot t \tag{2}$$

Here, $\rho_0$ is density of air; $\rho(t_0)$ is density of cabin gas after the process of blowing up; $t_0$ is time of the process of blowing up; $t$ is any time of the process of blowing up.

$$T(t) = T_0 + \frac{T(t_0) - T_0}{t_0} \cdot t \tag{3}$$

Where, $T_0$ is temperature of air; $T(t_0)$ is temperature of cabin gas after the process of blowing up;

In practical engineering, when the temperature of the cabin decreases $T^{\Delta}(t)$ due to the heat dissipation in the pontoon, the energy conservation equation is as follows [20]:

$$H[T(t) - T^{\Delta}(t) - T_0] = \frac{\rho(t)}{\rho_0} \cdot \frac{dT^{\Delta}(t)}{dt} \tag{4}$$

By the equation of (2), (3), (4), equation (4) is obtained:

$$T^{\Delta}(t_0) = \left\{ H \cdot t_0 - 1 + [1 + \frac{P_s}{P_0} - \left(\frac{P_s}{P_0}\right)^{\frac{k-1}{k}}]^{\frac{-H \cdot t_0}{\frac{P_s}{P_0} - \left(\frac{P_s}{P_0}\right)^{\frac{k-1}{k}}}} \right\} \cdot \frac{\left(\frac{P_s}{P_0}\right)^{\frac{1}{k}} - \frac{P_s}{P_0}}{H \cdot t_0 + \frac{P_s}{P_0} - \left(\frac{P_s}{P_0}\right)^{\frac{k-1}{k}}} \cdot T(t_0) \tag{5}$$

At the end of inflation, the actual temperature and density of the gas in the cabin can be presented as follows:

$$T(t_{0a}) \approx T(t_0) - T^{\Delta}(t_0) \tag{6}$$

$$\rho(t_{0a}) \approx \frac{P}{R \cdot T(t_{0a})} \tag{7}$$

Here, $P$ is air pressure before the process of blowing up. $R$ is gas constant.

After inflation, the cabin temperature gradually decreases due to heat dissipation. The cabin temperature is obtained from equation (4).

$$T(t_i) = T_0 - [T_0 - T(t_0) + T^{\Delta}(t_0)] \cdot e^{\frac{P_0}{P} \cdot H \cdot t} \tag{8}$$

The pressure of the gas in the cabin can be obtained:

$$P(t_{0a}) = \rho(t_{0a}) \cdot T(t_i) \cdot R \tag{9}$$

The volume of gas in the cabin is:

$$V(t_{0a}) = \frac{P \cdot V}{P(t_{0a})} \tag{10}$$

According to equation (10), the inflow volume, weight and center of gravity of the pontoon and the buoyancy of the pontoon and the state of the sunken ship can be obtained.

## 3. Analysis of examples

Above is the theoretical analysis of the floating process of the damaged sunken ship. This example simulates and analyses the floating process of a barge by pontoon. The layout of pontoons is shown in Fig. 1, and the main dimensions of the ship are shown in Table 1. The above is the theoretical analysis on refloating a shipwreck.

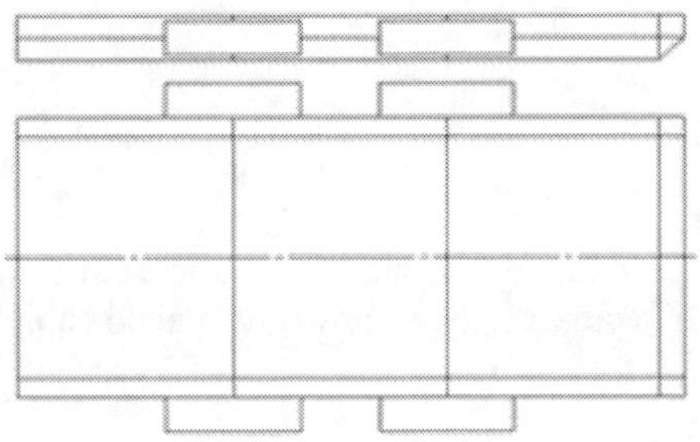

**Figure 1.** Arrangement plan of pontoons.

**Table 1.** The main dimensions of the ship.

| Overall length /m | Moulded breadth /m | Moulded depth /m | Weight W/kN |
|---|---|---|---|
| 79.54 | 32.4 | 5.2 | 26998.48 |

The ship's hold is broken causing ship sinking during the voyage due to an accident. Through underwater inspection of the damaged ship, it is found that the bow cabin and side cabins of the damaged ship are broken. Here, the lifting process of the shipwrecks is calculated.

The ship was sank in an accident during the voyage. Through underwater detection of the damaged ship, it is found that the bow part of the cabin and cabins on either side are broken.

Case A, liquid load rate is 0.5, draft is 5.241, local effluent of stern, tail inclination is 0.47;

Case B, liquid load rate is 0.25, draft is 4.725, local effluent of stern, tail inclination is 0.39;

Case C, liquid load rate is 0.05, draft is 4.314, local effluent of stern, tail inclination is 0.33;

Case D, liquid load rate is 0, draft is 4.198, local effluent of stern, tail inclination is 0.3.

## 4. Results and discussions

### *4.1. Ship stability*

In the process of lowering and floating, we should not only pay attention to the buoyancy of the hull, but also calculate the stability of the hull to prevent accidents. Fig. 2 shows the stability of the wreck during its lowering and lifting. During the simulation process, with the decrease of hull draft, the stability of the ship increases gradually. The maximum stability arm heels of the four working conditions are 0.209 m, 1.191 m, 2.93 m and 3.391 m, respectively. The stability distances are 5.29 degrees, 102.15 degrees, 97.74 degrees and 97.26 degrees, which fully meet the requirements of lifting and floating engineering. The transverse metacentric radius of the hull is 10.500 m, 12.100 m, 13.203 m, 22.712 m, and the transverse metacentric height is 10.700 m, 12.167 m, 13.115 m and 22.576 M. If the angle of the hull is not properly controlled at the beginning of the floating, the hull will appear negative stability, and it is easy to tilt to one side. At this time, due to the small overall stability, if a larger overturning moment is received, the hull will overturn, which will increase the difficulty of salvage.

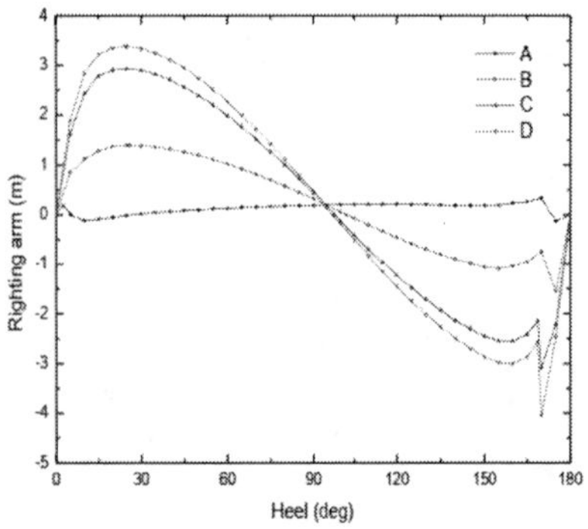


**Figure 2.** The static stability curves of ship.

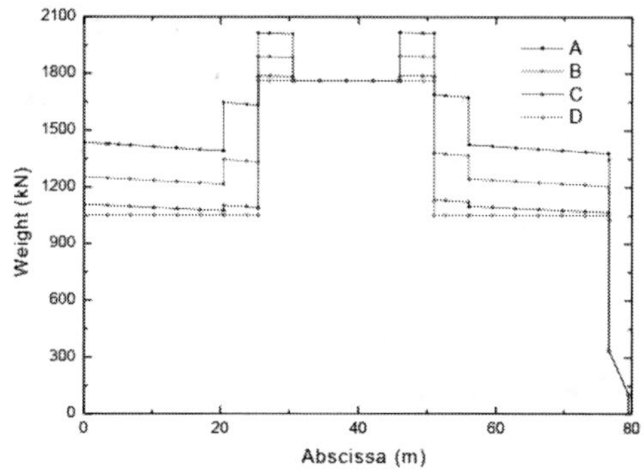


**Figure 3.** Weight distribution of ship.

*4.2. Weight distribution*

Fig. 3 shows the weight distribution along the length direction of the hull, including the inflow into the pontoon. As the draft of the hull decreases, the inflow of damaged cabin decreases gradually.

*4.3. Buoyancy distribution*

As the draft of the hull decreases, the buoyancy of the hull decreases gradually. Fig. 4 shows the buoyancy distribution along the hull length, including buoyancy of pontoons. Because the hull inclines toward the stern, the buoyancy distribution gradually decreases from stern to bow.

*4.4. Shear distribution*

Fig. 5 shows the shear distribution along the length direction of the hull, which is similar to that of the normal hull. However, due to the effect of pontoons, the maximum/minimum shear force is located in the middle of the pontoon rather than in the first quarter of the hull. With the decrease of hull draft, the distribution of shear force under four working conditions gradually increases.

*4.5. Bending moment distribution*

Fig. 6 shows the bending moment distribution along the length direction of the hull, which is the same as that of the normal hull, but with with a critic condition. Measures should be taken to reduce the load in the middle cabin of the hull during lifting. The bending moment of the hull increases gradually during the lifting process, which indicates that the influence of the inflow of broken cabin on the bending moment of the hull is weaker than that of the load. It is necessary to take weight reduction measures for the successful lifting of the hull. The maximum shear force of working condition A is 42.075 m, and that of the other three working conditions is 38.25 m, which corresponds to the change of hull pitch angle.

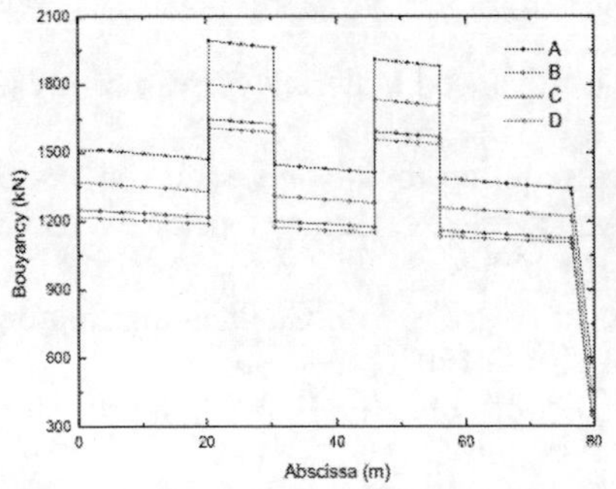


**Figure 4.** Buoyancy distribution of ship.

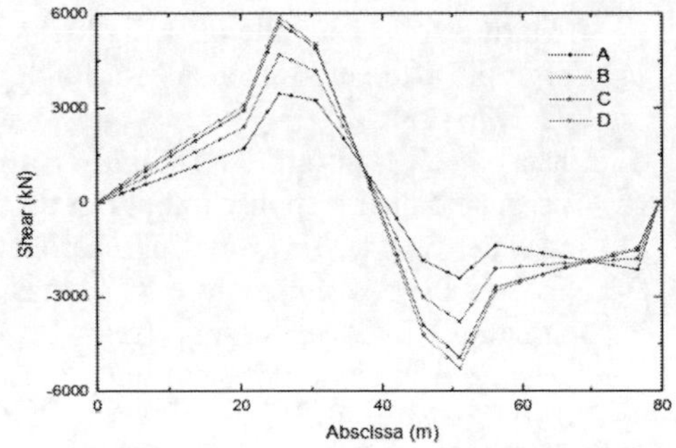


**Figure 5.** Shear force distribution of ship.

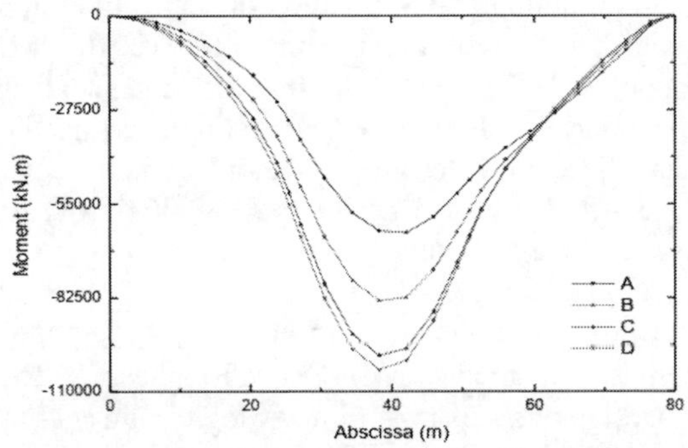


**Figure 6.** Bending moment distribution of ship.

## 5. Conclusion

Refloating a damaged ship with salvage pontoon is a complex engineering. When the pontoon is used to lift a sunken ship, after the inflation of pontoon, the pressure in the pontoon will change causing the change of the buoyancy of the pontoon, which will affect the accuracy of the calculation of the buoyancy. In this paper, the influence of temperature change on pressure after pontoon inflation is deduced theoretically, and the variation of gas volume along time in buoy is further obtained. According to the principle of ship statics, buoyancy of pontoon can be obtained. In the preliminary stage of shipwreck with many pontoons sunk in the water totally, since the stability of the shipwreck is poor, the aeration ratein the pontoon should be calculated accurately to prevent the capsizing of the ship. As the draft of the wreck decreases, the stability of the hull increases gradually, which is beneficial to the process of lifting and floating. The change of pitch angle is small in the course of buoyancy, but it has obvious influence on the buoyancy distribution of ship hull. The position of pontoon has a great influence on the distribution of shear force and bending moment of ship hull. Relevant measures should be taken.

**Acknowledgments**

The authors thank the anonymous reviewers for their valuable remarks and comments.
This paper is funded by the Shandong Jiaotong University Scientific Research Foundation (Z201715).

**References**

[1] Y. Wang, E. Zio, X.Y. Wei, A resilience perspective on water transport systems: The case of Eastern Star, J. International Journal of Disaster Risk Reduction. 33 (2019) 343-354.
[2] R. Szlapczynski, P. Krata, Determining and visualizing safe motion parameters of a ship navigating in severe weather conditions, J. Ocean Engineering. 158 (2018) 63-274.
[3] F. Chihhao, H. Chengjui, J.Y. Lin, Taiwan's legal framework for marine pollution control and responses to marine oil spills and its implementation on T.S. Taipei cargo shipwreck salvage,

J. Marine pollution bulletin. 136 (2018) 84-91.
[4] A. Varsami, Different Methods of Refloating a Ship, J. Advanced Materials Research. 837 (2014) 792-797.
[5] L. Chansoo, Y. Hansam, K. Dongha, Lifting forces required to salvage a sunken vessel and caisson and their response to bottom friction, buoyancy release, surface tension, water capture and water release, J. Ocean Engineering. 125 (2016) 82-89.
[6] V. Nikolaos P, A. Koimtzoglou, K. Louzis, Shipwreck: A Crisis with Challenging Solutions, J. Journal of Risk Analysis and Crisis Response. 4.3 (2014) 160-174.
[7] C.C. Zhang, Application of steel buoy in salvage project of "Shi Yue Hao", J. World Shipping. 40.10 (2017) 31-36.
[8] S.H. Chen, W.P. Wang, Y. Jiang, Design and implementation of "Shi Yue Hao" salvage project, J. World Shipping. 40.10 (2017) 6-23.
[9] Z.Ilias, T. Michael, On Modeling and Simulation of Innovative Ship Rescue System, J. Journal of Offshore Mechanics and Arctic Engineering. 140.6 (2018) 061303-.
[10] I. Zilakos, E. Chatzidouros, N. Tsouvalis, An innovative ship salvage concept and its effect on the hull structural response, J. Journal of Marine Engineering & Technology. (2018) 1-12.
[11] H. Lu, T.Y. Li, Y. Zhao, Structural Design of Cylindrical Shell Buoys in the Buoyancy Supported Risers of FPSO, J. Journal of Ocean Technology. 33.06 (2014) 103-107.
[12] Z. Zha, New automatic salvage pontoon used in deep water, J. American Society of Civil Engineers. (2011) 2744-2750.
[13] H.J. Kang, I. Kim, J. Choi, A concept study for the buoyancy support system based on the fixed fire-fighting system for damaged ships, J. Ocean Engineering. 155 (2018) 361 - 370.
[14] A.K.D. Velayudhan, Design of a supervisory fuzzy logic controller for monitoring the inflow and purging of gas through lift bags for a safe and viable salvaging operation, J. Ocean Engineering. 171 (2019) 193-201.
[15] X.Y. Chen, J.H. Peng, Z. Cheng, et al., "Design of monitoring system of salvaged pontoons based on FameView, J. Application of Electronic Technique. 45.4 (2019) 69-73.
[16] X. Yang, B. Ye, Design of Salvage Buoy Monitoring, J. Information & Communications. 5 (2019) 118-119.
[17] Y.H. Liu, L.Y. Chen, X.K. Huang, "Safety Analysis Methodology in Marine Salvage System Design," in Proceedings of the 3rd International Conference on Mechatronics, Robotics and Automation, 2015, Proceedings of the 3rd International Conference on Mechatronics, Robotics and Automation, edited by xxxx, Shenzhen, China, 2015, 155-159.
[18] H.P. Hu, Rechuandao Lilun, first ed., China University of Science and Technology Press, Hefei, 2010.
[19] B. Wu, Ye ya yu qi ya chuan dong yuan li ji ying yong, first ed. China Electric Power Press, 2010.
[20] Department of Teaching, Fudan University, Liuti Lixue(Shi yong ben), first ed. Science and Technology Press, 1960.

# Exploration on the Structure Parameters of Nozzle-to-Throat Clearance in Vacuum Generator

**Xilong Guo *, Yong Li**

School of Mechanical and Electrical Engineering, Qingdao, China

*Corresponding author email: 961616153@qq.com

**Abstract.** Nozzle-to-throat clearance is one of the main factors affecting the injection coefficient of the vacuum generator. The increase of nozzle-to-throat clearance will lead to the aggrandizement of the annular duct area which the driving fluid passes, thereby amplifying the ejector flow rate of ejector tube and the injection coefficient. On the contrary, Keenan [1] obtained the result through the experiment that the injection coefficient decreased as nozzle-to-throat clearance increases, because the accretion in nozzle-to-throat clearance would increase the pressure of inlet plane of the mixing chamber, resulting in the decrease of the driving pressure difference of the driving fluid, hence the injection coefficient diminished. In this paper, the numerical simulation method is used to obtain the optimal nozzle-to-throat clearance parameter size.

## 1. Introduction

With the recovery of the global economy, the manufacturing industry has also gained unprecedented opportunities. In the vacuum industry and the pneumatic industry, the vacuum generator has gradually received extensive attention at home and abroad because of its small size, simple structure, maintenance-free, low noise and convenient control. It can effectively put an end to the delamination of conveying materials and ensure the homogeneity of the composition of the mixture. At present, it is widely used in tablet presses, capsule filling machines, dry granulators, packaging machines, vibrating screens and other machinery for automatic feeding.

However, various types of vacuum generators on the market are currently mixed. Moreover, the theoretical research on the structure of vacuum generators is relatively lacking at home and abroad. There is no corresponding reference standard for the influence of the structural dimensions of different parts of the vacuum generator on its working efficiency. Therefore, in order to meet the transportation requirements in production, the vacuum generator with different specifications is generally adopted for final selection of experiments, but this method is cumbersome and inefficient. This paper mainly discusses the influence of the structural parameters of nozzle-to-throat clearance on the working efficiency of the vacuum generator. By using the method of numerical simulation, the variation curve of different parameters is obtained, and an optimized structure size is found.

## 2. Effect of Nozzle-to-Throat Clearance on Injection Coefficient

In this paper, under the condition of 0.45MPa/1.5 (working pressure 0.45 MPA, gas compression ratio 1.5), the geometric structural parameters of vacuum generator are kept consistent, and the vacuum

Content from this work may be used under the terms of the Creative Commons Attribution 3.0 licence. Any further distribution of this work must maintain attribution to the author(s) and the title of the work, journal citation and DOI.
Published under licence by IOP Publishing Ltd

generators with nozzle-to-throat clearance of 5.4, 9.4, 17.4, 21.4, 29.4, 33.4, 37.4 and 37.4 are simulated to explore the effect of nozzle-to-throat clearance on injection coefficient.

Figure 1 shows the variation process of the injection coefficient of the vacuum generator with the change of nozzle-to-throat clearance. It can be seen that the injection coefficient increases at first and then decreases with the increase of nozzle-to-throat clearance, and reaches the maximum at the peak of the curve, so the corresponding nozzle-to-throat clearance is the optimal clearance.

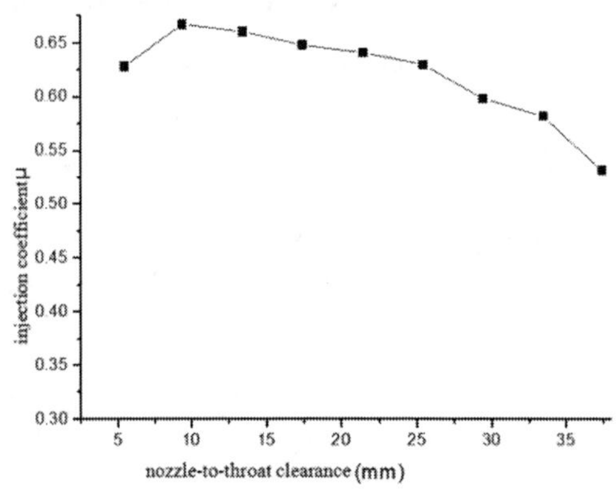


**Figure 1.** Injection coefficient of vacuum generator with different nozzle-to-throat clearance

According to the simulation results, the maximum injection coefficient of the vacuum generator designed under the condition of 0.45MPa/1.5 is 0.667, which corresponds to the nozzle-to-throat clearance of 9.4mm. The range of throat nozzle-to-throat clearance calculated by Sokolov[2] based on steam ejector is 25mm ⩽ 33mm, which is very different from the simulation results. Therefore, the working fluid medium is different (the working fluid calculated by Sokolov is steam, and the simulation process uses air as working fluid), which has a great influence on the result. Wang Xiuhui[3] studied the application of the ejector in gas recovery, and obtained the maximum negative pressure when nozzle-to-throat clearance was equal to nozzle throat diameter. The larger the negative pressure, the larger the driving pressure difference, and the higher the ejection fluid flow rate, so the injection coefficient is larger. The nozzle throat diameter of this vacuum generator model is 12mm, so when the nozzle-to-throat clearance is close to the nozzle throat diameter, a larger injection coefficient can be obtained, which is consistent with the conclusion of Wang Xiuhui's experimental research.

Figure 2 is a Mach number cloud diagram of a vacuum generator with nozzle-to-throat clearances (mm) of 5.4, 9.4, 13.4, 17.4, 21.4, 25.4, 29.4, 33.4, 37.4. It can be seen that as the nozzle-to-throat clearance increases, the distance between the Laval nozzle outlet and the inlet of the mixing chamber becomes larger and larger, and the ejector fluid is prematurely mixed with the working fluid in the receiving chamber of the vacuum generator, which lead to the loss of energy. By observing the Mach number cloud diagram of a vacuum generator with nozzle-to-throat clearances of 5.4mm and 37.4mm, it can be seen that when nozzle-to-throat clearance is too large, the supersonic region generated by the working airflow is significantly reduced, and the jet wake is shortened. Although increasing nozzle-to-throat clearance can increase the area of the annular passage through which the ejector fluid passes [4] (ie, the effective cross-sectional area through which the ejector fluid can pass), the injection coefficient becomes lower due to premature loss of energy in the receiving chamber.

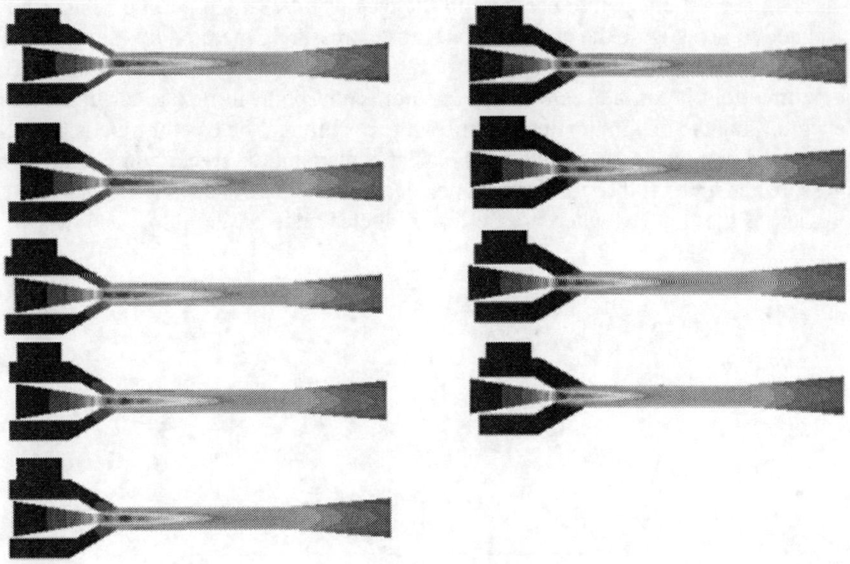

**Figure 2.** Mach number cloud diagram with different nozzle-to-throat clearance

## 3. The Influence of Nozzle-to-Throat Clearance on Vacuum Degree

Combined with the injection coefficient variation diagram (Fig.1), although the increase of nozzle-to-throat clearance will increase the annular passage which the ejector fluid passes, the ejector flow rate does not increase continuously, which indicates that the annular channel area is not the only factor that determines the flow of the ejected fluid under the conditions of this paper. According to Wang Xiuhui's research on the ejector, the vacuum degree of the vacuum generator and the vacuum degree of the ejector tube are statistically analyzed.

Figure 3 shows the double-Y coordinate diagram of the injection coefficient and vacuum degree at different nozzle-to-throat clearance. From the vacuum degree, the vacuum degree has no obvious law as nozzle-to-throat clearance increases when the maximum vacuum is generated at 17.4mm, and nozzle-to-throat clearance is in the range of 5-21.4mm. However, the degree of vacuum drops significantly after the throat distance is greater than 25.4. It can be seen that the injection coefficient is inevitably reduced when the vacuum degree of the vacuum generator is small, but the maximum vacuum degree and the injection coefficient are not significantly consistent.

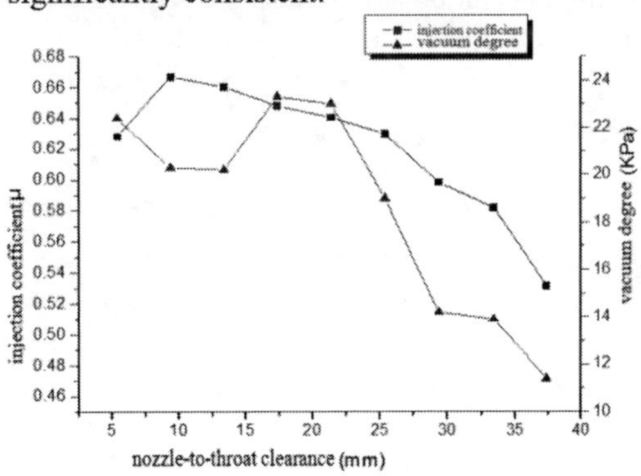


**Figure 3.** Injection coefficient and maximum vacuum at different nozzle-to-throat clearance

The maximum vacuum of the vacuum generator should be generated at the shock of the nozzle outlet, and due to the instability of the shock, the negative pressure of the injected fluid cannot be visually reflected. In this paper, the negative pressure in the ejector tube (the pipe into which the ejector fluid enters) is counted to investigate the amount of negative pressure generated by the vacuum generator of different nozzle-to-throat clearances at the ejector tube. Figure 4 shows the pressure curve of the ejector tube in the Y direction. It can be seen that there is different point in nozzle-to-throat clearance of the vacuum generator, but the maximum negative pressure position in the ejector tube is basically the same. This position is the entrance position where the ejector tube and the receiving chamber intersect. The negative pressure generated is the largest when nozzle-to-throat clearance is 9.4 mm, while the generated negative pressure is the smallest when nozzle-to-throat clearance is 37.4 mm.

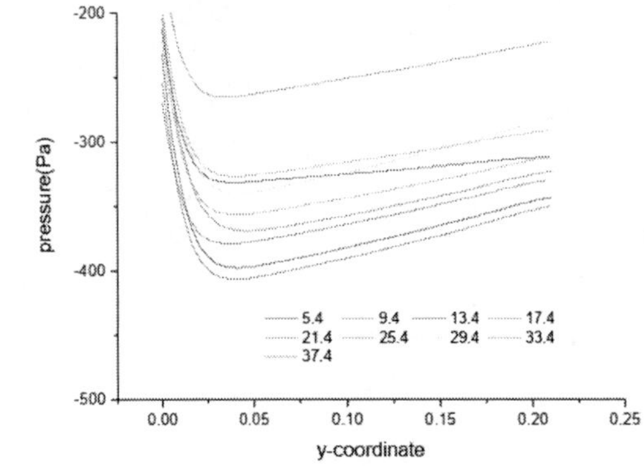


**Figure 4.** Y direction pressure curve of ejector tube

Figure 5 shows the double Y coordinate diagram of the injection coefficient and the vacuum degree of the ejector tube at different nozzle-to-throat clearance. It can be seen from the figure that the variation trend of the injection coefficient is basically the same as the vacuum degree in the ejector tube. As the degree of vacuum becomes larger, the injection coefficient also becomes larger, and the injection coefficient of the vacuum is decreased. Therefore, it can be concluded that there is a closer relationship between the injection coefficient and the vacuum of the ejector tube. On the other hand, the ejector fluid is subsonic and its flow velocity is smaller than that of the working fluid, so the greater the vacuum, the larger the negative pressure, the larger the driving pressure difference of the ejector fluid, and the larger the flow velocity of the ejector fluid. Therefore, the flow rate of the ejector fluid becomes larger as the maximum degree of vacuum in the ejector tube. The vacuum degree of the ejector tube is also inseparable from the area of the annular passage. Especially when nozzle-to-throat clearance is too small, the area of the annular passage (the passage of the ejector fluid into the mixing chamber) is reduced. The jet fluid cannot enter the mixing chamber, and the injection coefficient suddenly drops, resulting in a sharp drop in the maximum vacuum within the ejector tube.

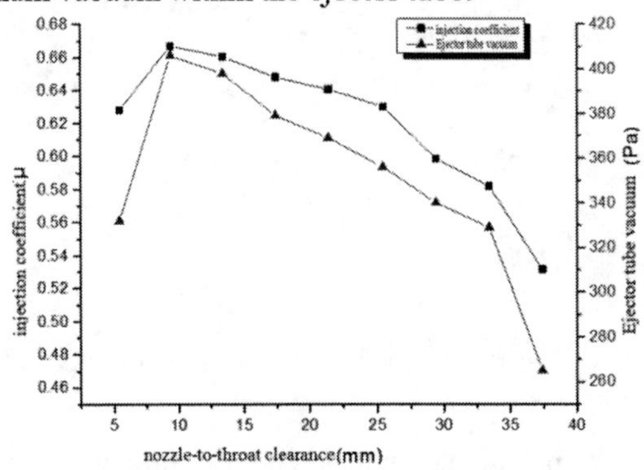


**Figure 5.** Injection coefficient and maximum vacuum in ejector tube at different nozzle-to-throat clearance

## 4. Conclusion

This paper mainly studies the influence of the geometric parameters of nozzle-to-throat clearance of the vacuum generator on the injection coefficient and vacuum degree. By changing the geometric parameters, the data of each group were simulated separately, and finally the optimal nozzle-to-throat clearance of the vacuum generator designed under the working condition of 0.45MPa/1.5 was obtained. And the relationship between nozzle-to-throat clearance and injection coefficient is preliminarily obtained, which provides reference for the structural size design of vacuum generators of other specifications, and plays a certain role in promoting the development of the pneumatic conveying industry.

## References

[1] Keenan JH. An Investigation of Ejector Design by Analysis and Experiment [J]. Journal of Applied Mechanics, 1950, 17(3): 299-309.
[2] Sokolov et al. Ejector [M]. Science Press, 1977: 1-71.
[3] Wang Xiuhui, Zhao Junyou, Zuo Peng et al. Experimental Study on the Influence of Nozzle-to-Throat Clearance on the Performance of Jet-Guide [J]. Petroleum Mine Machinery, 2014 (10): 35-38.
[4] Chen Weixiong, Chen Huiqiang, Shi Chaoxi, et al. Experimental Study on the Influence of Nozzle-to-Throat Clearance on Injector Performance [J]. Journal of University of Chinese Academy of Sciences, 2016, 33(2): 253-257.

IOP Conf. Series: Materials Science and Engineering **772** (2020) 012080 doi:10.1088/1757-899X/772/1/012080

# Structural Optimization of Machine Tool Column Based on Dynamic Characteristic Analysis and Response Surface

**Zhongfu Bao *, Dejiang Zeng and Darao Wu**

Guangdong Mechanical & Electrical Polytechnic, Guangzhou 510515, China

*Corresponding author e-mail: bzfjixie@163.com

**Abstract.** Based on digital simulation technology, the dynamic characteristics of a NC machine tool column are studied, including modal analysis and harmonic response analysis. Then the response surface of design parameters and indicators is obtained based on the experimental design. Finally, the optimal design is obtained by using genetic algorithm based on response surface.

## 1. Introduction

An important development trend of modern mechanical equipment is precision and high speed, especially all kinds of machine tools and processing centers. Because of the elasticity of machine tool components, the real-time change of cutting force and the influence of environment in the process of machine tool processing, the machine tool will inevitably produce vibration. Vibration not only reduces the machining accuracy and surface quality of the workpiece, but also brings many hazards to the machine tool itself, such as accelerating tool wear, accelerating fatigue of transmission parts and so on. Dynamic research can effectively alleviate vibration problems and improve the accuracy and life of machinery manufacturing equipment.

In this study, the dynamic characteristics of a NC machine tool column are analyzed based on dynamic numerical simulation technology, and the response surface is obtained based on experimental design. Finally, the structure is optimized.

## 2. Technological process

The research is based on ANSYS Workbench platform. Because of the use of parametric analysis technology, design variables need to be specified in three-dimensional modeling. The three-dimensional model of the main body of the NC vertical machining tool is shown in Figure 1.

The machine tool bed is mainly composed of base, pillar, spindle box and so on. The pillar is the support of the spindle components, which directly affects the feed rate and cutting depth. Therefore, the stiffness of the pillar has a great impact on the accuracy of the machine. The pillars should be matched with the bed and spindle box, and the relevant dimensions should not be modified as far as possible. Therefore, the following three design variables are determined: the span under the column is marked as a; the wall thickness of the column is marked as b; and the wall thickness of the reinforcement is marked as c, as shown in Figure 2.

Static structure analysis, modal analysis and harmonic response analysis tasks are established in Workbench, which share the same three-dimensional model. Considering that harmonic response analysis is based on modal analysis, it is necessary to import the results of modal analysis into harmonic

Content from this work may be used under the terms of the Creative Commons Attribution 3.0 licence. Any further distribution of this work must maintain attribution to the author(s) and the title of the work, journal citation and DOI.
Published under licence by IOP Publishing Ltd

response analysis. In each analysis task, the result data focused on are designated as variables, including the maximum static stress value, the first natural frequency in modal analysis, and the response value of the specified position in the harmonic response. These variables are the most important performance indicators for evaluating the dynamic and static characteristics of machine tool columns. On the basis of three analysis tasks, experimental design is carried out. The experimental samples are determined according to the central composite method, and response surface is created according to the specified performance parameters as response values. Optimum design based on response surface can avoid multiple simulation analysis and improve design efficiency.

The model was preprocessed. In order to reduce the amount of calculation, only the column is taken out as the object of analysis. In order to fit the actual load-bearing situation as much as possible, the remote load is adopted, that is, the position of the spindle is taken as the stress point, and the remote load is added to the area where the column and the spindle box cooperate, as shown in Figure 3. The bolt holes are fixed at the bottom of the column. Automated partition of entity grid is adopted.

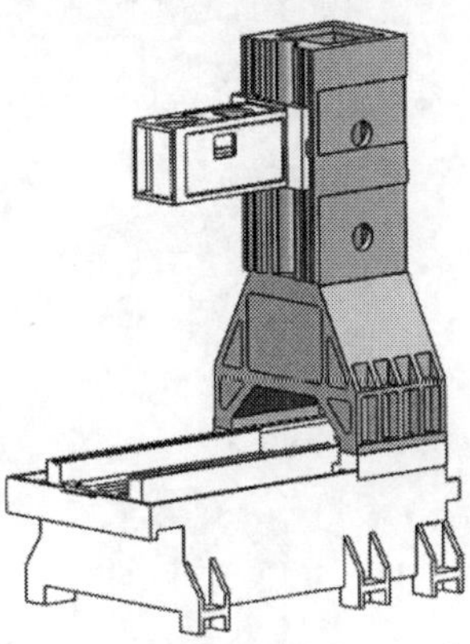
**Figure 1.** Machining tool

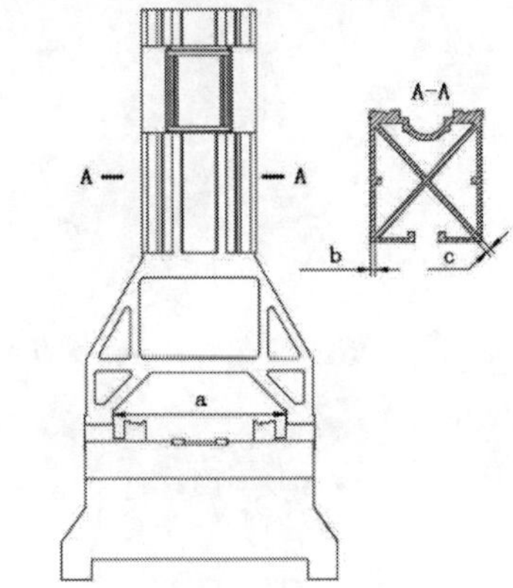

**Figure 2.** Design Variables

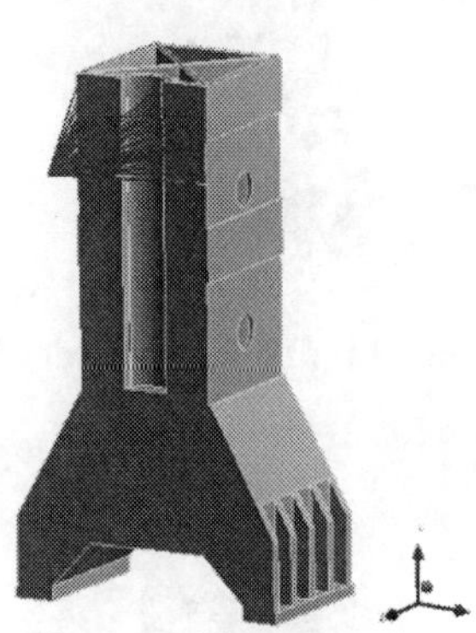
**Figure 3.** Preprocessing

## 3. Dynamic characteristic analysis

### *3.1. Modal analysis*

Modal and Harmonious Response Analysis is the Main Method of Dynamics Research. Modal analysis is a numerical technique for calculating the vibration characteristics of structures. Its main purpose is to obtain the natural frequencies and modes of structures. Modal analysis can effectively avoid resonance or vibration at a specific frequency.

The dynamic equation for undamped modal analysis is

$$[M]\{X''\}+[K]\{X\}=\{0\} \tag{1}$$

The free vibration of the structure is simple harmonic vibration, so the displacement is

$$x=x\sin(\omega t) \tag{2}$$

The equation (3) is obtained by introducing (2) into (1).

$$([K]-\omega^2[M])\{X\}=\{0\} \tag{3}$$

From equation (3), eigenvalues and corresponding eigenvectors can be calculated, in which eigenvalues $\omega_i$ are natural circular frequencies and eigenvectors $\{x\}_i$ are vibration modes.

In ANSYS Workbench, the modal analysis of the column can be performed by calling the Modal module. Although objects have infinite order natural frequencies, from the point of view of mechanical design, we need to focus on the natural frequencies of low frequency band, because it is very resonance-inducing. Therefore, the results of the first six orders are queried, as shown in Table 1 and Figure 4-9.

**Table1.** The first six natural frequencies

| Order | Natural Frequencies /Hz | Mode Shapes |
|---|---|---|
| 1 | 73.93 | The column swings around the x-axis |
| 2 | 124.38 | The column swings around the z-axis |
| 3 | 277.78 | The column swings around the y-axis |
| 4 | 328.16 | The upper part of the column moves along the x-axis |
| 5 | 342.31 | Column twisting around Y axis |
| 6 | 392.67 | Column twisting around z-axis |

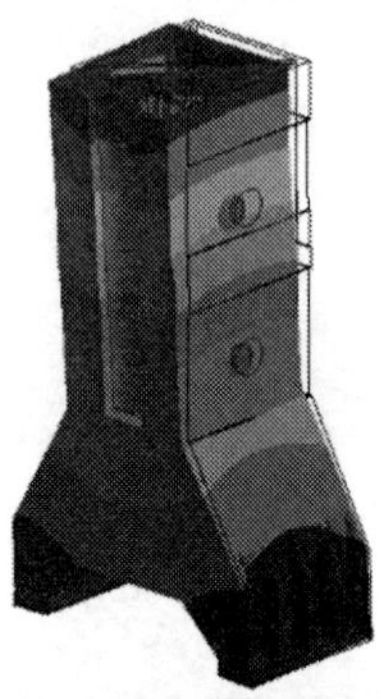

**Figure 4.** First-order

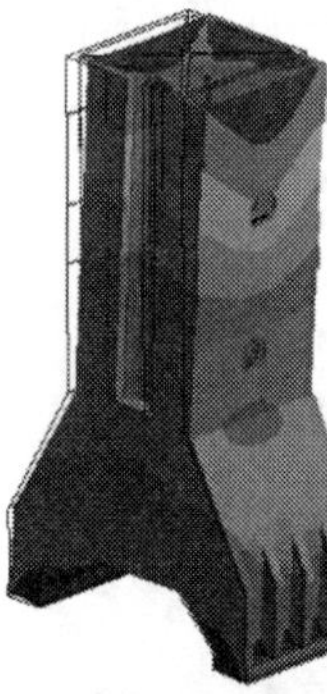

**Figure 5.** Second-order

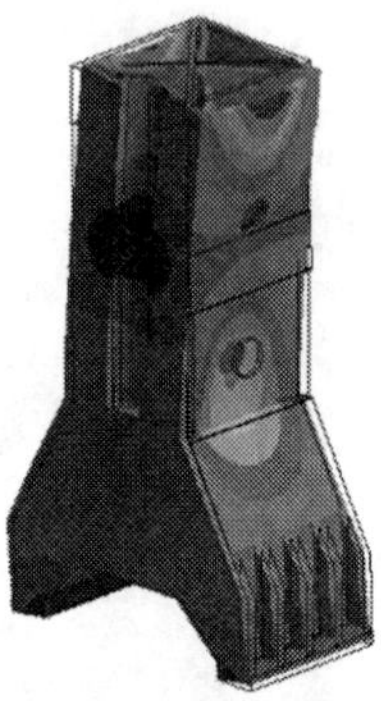

**Figure 6.** Third-order

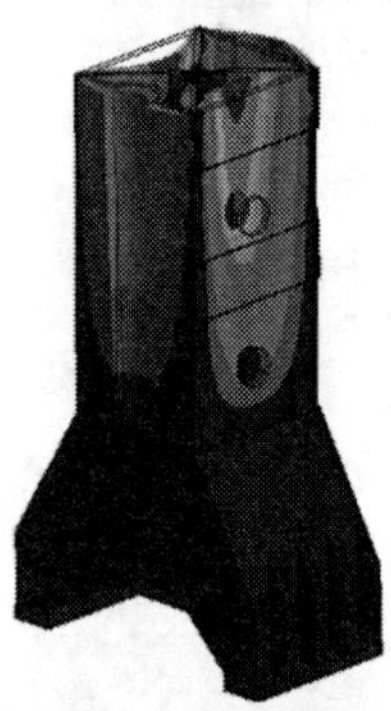

**Figure 7.** Fourth-order

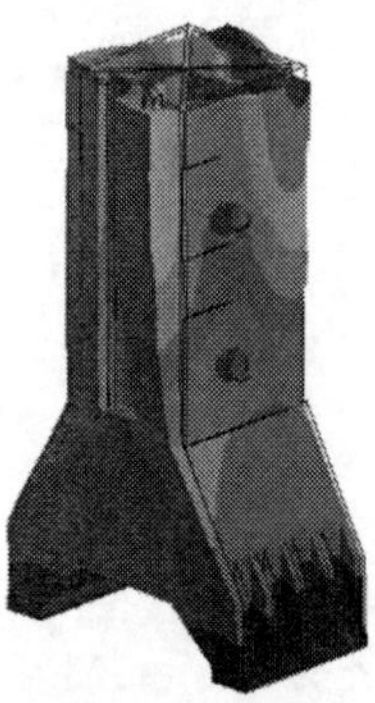

**Figure 8.** Fifth-order

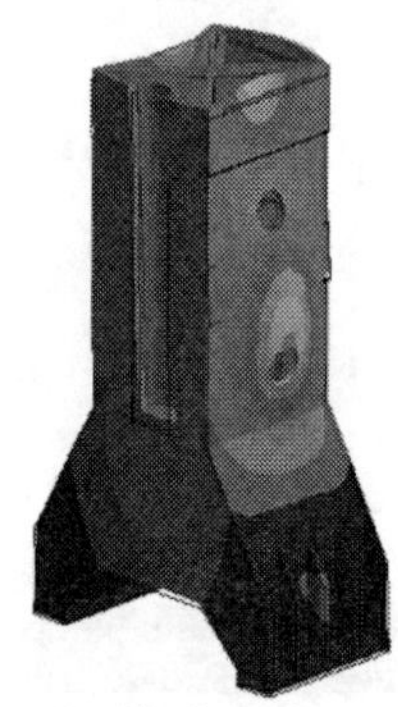

**Figure 9.** Sixth-order

### *3.2. Harmonic response analysis*

Harmonic response analysis is to analyze the response of the structure under harmonic loads of different frequencies and amplitudes. By detecting the resonance, the weak stiffness of the structure can be found, so as to achieve targeted structural optimization and improvement. The natural frequencies and modes of the structure are obtained by modal analysis, and the influence of external forces at different frequencies on the structure can be obtained by harmonic response analysis.

When the load of the structure changes with time according to the harmonic law, its equation is as follows:

$$[M]\{x''\}+[C]\{x'\}+[K]\{x\}=\{F\} \tag{4}$$

Since $\{F\}$ and $\{x\}$ represent harmonic load and displacement respectively, the following relations can be obtained:

$$\{F\}=\{F_{\max}e^{j\psi}\}e^{j\omega}=(\{F_1\}+i\{F_2\})e^{j\omega} \tag{5}$$

$$\{x\}=\{x_{\max}e^{j\psi}\}e^{j\omega}=(\{x_1\}+i\{x_2\})e^{j\omega} \tag{6}$$

Where, $\omega$ is Frequency of Excitation Load; $\psi$ Phase angle of exciting load; $\{F_1\}$ is The Real Part of the Load; $\{F_2\}$ is Imaginary part of load; $\{x_1\}$ is Real parts of displacement; $\{x_2\}$ is Imaginary parts of displacement.

Equations (5) and (6) are introduced into equation (4) to obtain the characteristic equation of harmonic response analysis.

$$([K]-\omega^2[M]+i\omega[C])(\{x_1\}+i\{x_2\})=(\{F_1\}+i\{F_2\}) \tag{7}$$

Because of the structure and constraints of the machine tool column, the response of each point on it is different. Considering that the area matched with the spindle box directly affects the processing accuracy, the displacement response value of this area is focused on, as shown in Figure 10. Figure 11 shows that the x-direction displacement has greater response when the excitation frequency is about 124 Hz and 328 Hz. Figure 12 shows that the y-direction displacement has greater response when the excitation frequency is about 74 Hz, 124 Hz, 277 and 392. Figure 13 shows that the z-direction displacement has greater response when the excitation frequency is about 74 Hz, 124 Hz and 328 Hz. Comprehensive comparison shows that the structural stiffness of columns in Y and Z directions is relatively weak.

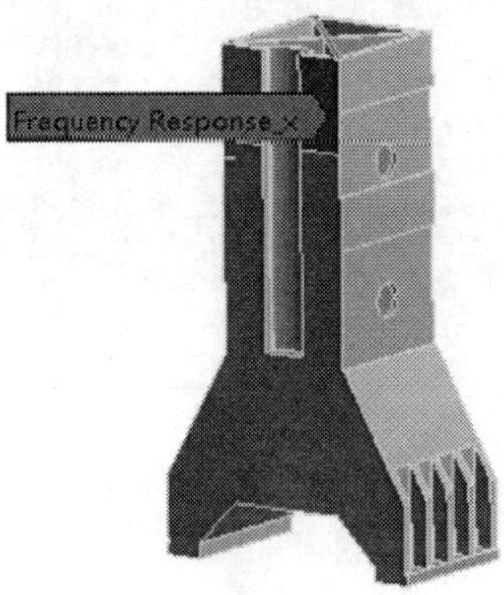


**Figure 10.** Areas of focus

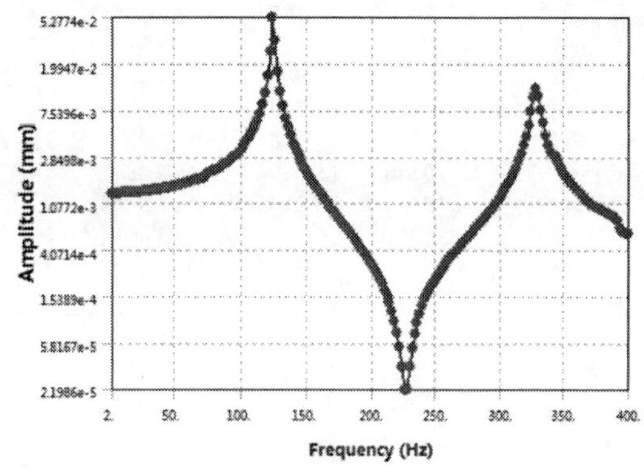


**Figure 11.** X-direction displacement response

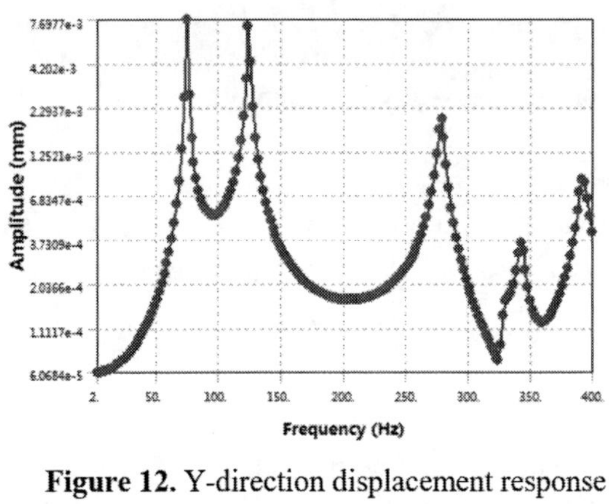


**Figure 12.** Y-direction displacement response

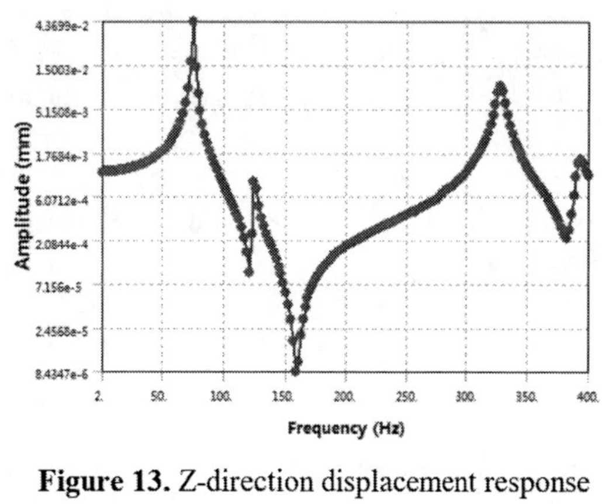


**Figure 13.** Z-direction displacement response

## 4. Experimental design and response surface construction

Response Surface Design (RSD) is an effective method to establish the relationship between test indices and continuous variables through a small amount of experimental data. Its basic idea is to approximate a polynomial model with explicit expression in a small area to express a complex unknown function relationship. This method is based on experimental design and mathematical statistics. It not only considers the random error of experiment, but also calculates simply. It is an effective means to solve the problem.

In this experiment, the design variables are span a, column wall thickness B and reinforcement wall thickness c, which are recorded as P1, P2 and P3 respectively. The results of analysis which need to be focused on are taken as the response values, including the column mass, the first natural frequency, the maximum static stress and the displacement response values of the specified position in the harmonic response, which are recorded as P4-P9 in turn.

This method combines the traditional distribution of interpolation nodes with the design of full or partial factors. The experimental data are shown in Table 2.

**Table 2.** Three-factor Central Compound Experiments Table

| | P1/mm | P2/mm | P3/mm | P4/kg | P5/Hz | P6/ MPa | P7/mm | P8/mm | P9/mm |
|---|---|---|---|---|---|---|---|---|---|
| 1 | 560 | 18 | 15 | 474.7429 | 73.93009 | 66.74668 | 0.052774 | 0.007698 | 0.043699 |
| 2 | 504 | 18 | 15 | 531.0856 | 77.01912 | 66.66999 | 0.0497 | 0.006585 | 0.025106 |
| 3 | 616 | 18 | 15 | 425.9552 | 75.6173 | 62.09837 | 0.048061 | 0.006493 | 0.036552 |
| 4 | 560 | 16.2 | 15 | 546.6722 | 74.03675 | 65.00346 | 0.048319 | 0.007133 | 0.040311 |
| 5 | 560 | 19.8 | 15 | 499.7859 | 72.06995 | 61.02433 | 0.049724 | 0.007537 | 0.042858 |
| 6 | 560 | 18 | 13.5 | 465.5954 | 75.1682 | 72.49665 | 0.054976 | 0.007245 | 0.029157 |
| 7 | 560 | 18 | 16.5 | 483.8331 | 72.7864 | 61.0033 | 0.046217 | 0.006089 | 0.030193 |
| 8 | 514.4701 | 16.53654 | 13.78045 | 493.0992 | 79.2702 | 79.18216 | 0.052986 | 0.007084 | 0.030268 |
| 9 | 605.5299 | 16.53654 | 13.78045 | 405.6569 | 80.68867 | 78.919 | 0.054862 | 0.007475 | 0.03091 |
| 10 | 514.4701 | 19.46346 | 13.78045 | 531.9726 | 75.88097 | 62.89568 | 0.043244 | 0.007081 | 0.039864 |
| 11 | 605.5299 | 19.46346 | 13.78045 | 448.4342 | 77.28036 | 62.84295 | 0.045714 | 0.006138 | 0.02801 |
| 12 | 514.4701 | 16.53654 | 16.21955 | 508.0554 | 77.13119 | 66.21922 | 0.042199 | 0.005662 | 0.027081 |
| 13 | 605.5299 | 16.53654 | 16.21955 | 420.6132 | 78.45866 | 65.58773 | 0.043865 | 0.005999 | 0.034738 |
| 14 | 514.4701 | 19.46346 | 16.21955 | 546.6722 | 74.03675 | 65.00346 | 0.048319 | 0.007133 | 0.040311 |
| 15 | 605.5299 | 19.46346 | 16.21955 | 463.1337 | 75.38895 | 64.74619 | 0.043115 | 0.005767 | 0.029631 |

The expression of response surface is as follows.

$$y(x)=\alpha_0+\sum_{i=1}^{k}\alpha_i x_i+\sum_{i=1}^{k}\alpha_{ii}x_i^2+\sum_{i<j}^{k}\alpha_{ij}x_i x_j+\varepsilon \tag{8}$$

Where, $k$ is number of design variables; $x_i$ is design variables; $\alpha_i,\alpha_{ii},\alpha_{ij}$ are undetermined coefficients of polynomials; $\varepsilon$ is error.

After n experiments, the response surface model can be expressed as equation (9).

$$Y=X\mathrm{A}+\varepsilon=Y+\varepsilon \tag{9}$$

Where, $Y$ is AMESIM simulation value; $y$ is response surface approximation function value; $\varepsilon$ is fitting error. The parameters of quadratic polynomial can be estimated by least square method.

$$L=\sum\varepsilon_j^2=(Y-XA)^T(Y-XA) \tag{10}$$

$$a=(X'X)^{-1}X'Y \tag{11}$$

Based on the above equation, the response surface model can be obtained. Two variables with significant influence on each response value can be displayed as three-dimensional surface, as shown in Figure 14-19.

Through the response surface model, we can more intuitively understand the impact of design variables on indicators. Figure 14 shows that span a has a significant effect on mass. The larger the parameter is, the smaller the mass is. The function shows a parabolic approximation. Other variables have little effect on mass. Figure 15 shows that the stiffener thickness has a significant effect on the static stress value in a certain range. When the variable is greater than 15, the stress value does not change significantly; when the variable is less than 15, the stress rises sharply.

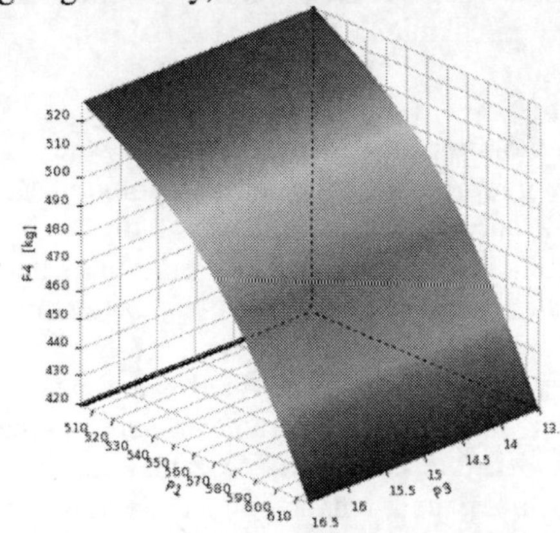


**Figure 14.** Response of P4 to P1 and P3

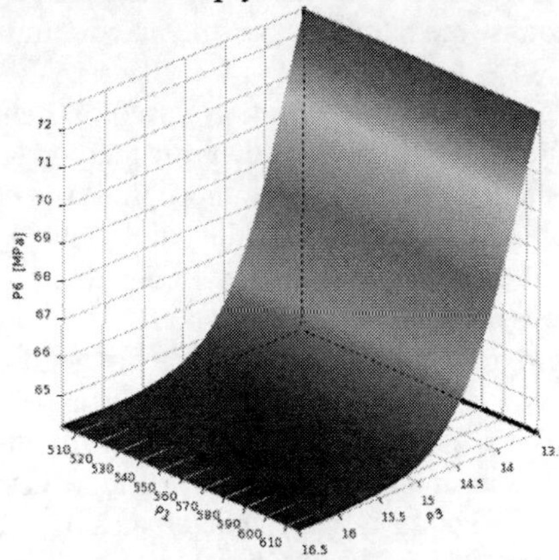


**Figure 15.** Response of P6 to P1 and P3

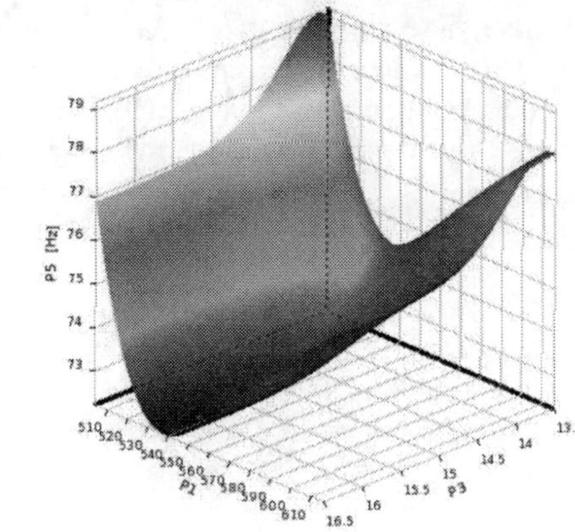


**Figure 16.** Response of P5 to P1 and P3

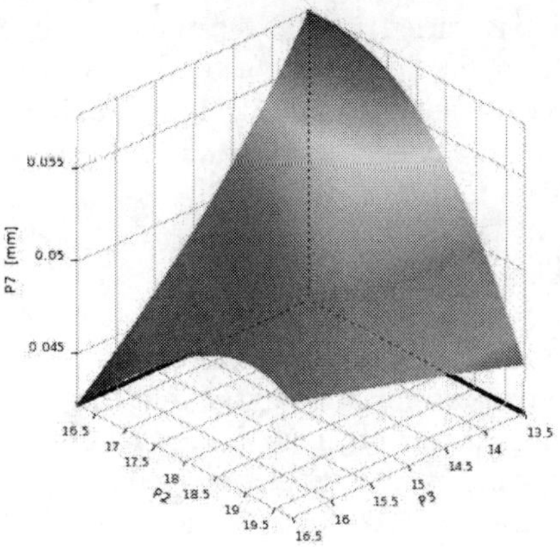


**Figure 17.** Response of P7 to P1 and P3

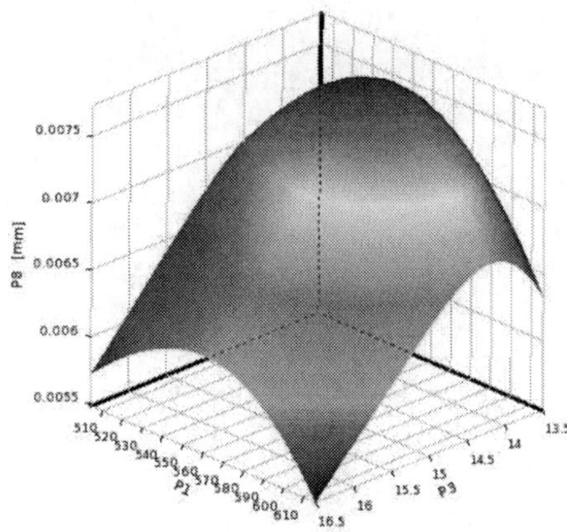


**Figure 18.** Response of P8 to P1 and P3

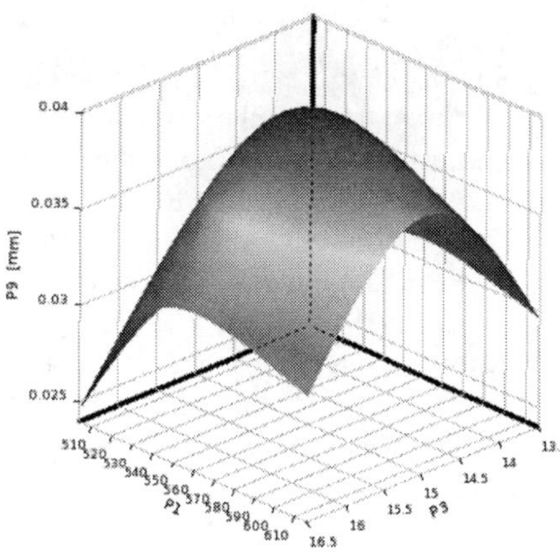


**Figure 19.** Response of P9 to P1 and P3

Figure 16 shows that the influence of span a on the first natural frequency presents a parabolic law. When A is about 550, the natural frequency reaches the lowest. The smaller the reinforcement thickness C is, the higher the first natural frequency value is. From Figure 17, it can be seen that the regularity of the responses of column wall thickness B and reinforcement thickness C to the x-direction harmonic values is similar to the saddle surface. From Figure 18 and 19, it can be seen that the response values of span a and reinforcement wall thickness C to Y-direction harmonic response and Z-direction harmonic response are parabolic, i.e. the maximum value is reached in the middle region of variables.

## 5. Structural optimization based on genetic algorithm

Genetic algorithm is a stochastic method to simulate the evolution process of organisms, which is based on natural selection. The algorithm starts with a population representing the potential solution set of the problem. In each generation, individuals are selected according to the fitness of individuals in the problem domain, and genetic operators are used to combine, cross and mutate to produce a population representing the new solution set. According to the principle of survival of the fittest, better and better approximate solutions are gradually generated.

The outstanding advantage of genetic algorithm is that it can find the optimal value in the whole search space, avoiding the problem of getting the optimal value when the parameters fall into the local optimum in the optimization process. Based on the established response surface model, it can be used to optimize the target characteristics of the genetic algorithm population, and no simulation solution is needed, which can greatly improve the computational efficiency.

In order to achieve lightweight design and ensure good comprehensive performance, the optimal design is carried out with the minimum quality as the objective, the maximum static stress not exceeding 80 MPa, the first section natural frequency not less than 80 Hz, and the displacement response in three directions not exceeding 0.05 mm as constraints. Finally, the optimum design parameters are obtained. The span a under the column is 506 mm, the wall thickness B is 16.616 mm, and the wall thickness C is 14.42 mm.

## 6. Conclusion

Based on the digital simulation technology and experimental design, the dynamic characteristics of the column of a NC machine tool are studied. First, the natural frequencies and modes of the structure are obtained by modal analysis, and then the responses of the structure to external forces at different frequencies are obtained by harmonic response analysis. The response surface was obtained by central composite experiment design, and the genetic algorithm was used to optimize the design based on the response surface. Finally, the lightweight design of the column is realized, and good dynamic characteristics are guaranteed.

## References

[1] Sun, Li Qing , et al, Simulation and Test Study on Dynamic Characteristic of Air Spring with Auxiliary Chamber, Applied Mechanics and Materials 341-342.1 (2013):391-394..

[2] Sun, Li Qin , et al, Simulation and Test of ASAC Dynamic Characteristic Based on ABAQUS. Advanced Materials Research.765 (2013) ,331-335.

[3] Faria Neto, Antonio , A. F. B. Costa , and M. F. De Lima, Use of Factorial Designs and the Response Surface Methodology to Optimize a Heat Staking Process, Experimental Techniques. 42 (2018) 319-331.

[4] Patel, Siddhant B. , and J. V. Gohel , Enhanced solar cell performance by optimization of spray coated CZTS thin film using Taguchi and response surface method, Journal of Materials Science Materials in Electronics 29 (2018) 5613-5623.

# Finite element analysis of building block ark

# Strength Analysis and Design of Building Block Ark Based on Finite Element

**Hongyan Li, Yu Wang***

School of Qingdao University of science and technology, Qingdao 266061, China.

*Corresponding author e-mail: 4570536@qq.com

**Abstract.** The research on the building block ark can be divided into two key issues: first, the hydrodynamic performance analysis of the building block ark, the main content is to study the hull shape and main scale of the building block ark. The second is the square of blocks.The strength analysis of the boat is mainly to study the overall structure and local strength of the building block ark. This article will explore the Lord.The content is the structural strength analysis of building block ark.

**Keywords:** block ark, strength analysis, finite element analysis.

## 1. Introduction

The building block ark is the earliest splicing catamaran built at home and abroad. A catamaran is a ship connected laterally, such as the middle deck. A long time ago, it was found in shipbuilding that the two hulls, connected horizontally, could sail more smoothly on the sea or Inland River [1]. The discovery led to the earliest use of the hull for the construction of sailboats, which can withstand higher winds and waves due to its unique hull. Compared with ordinary ships, catamarans of the same tonnage have larger deck area and larger cabin capacity, and have more stable characteristics. In addition, if the design of the catamaran is reasonable, it can also consume less energy and greatly improve the navigation speed [2]. It can be said that the emergence of catamaran is an important milestone in the history of ship development [3]. There are two sides of the hull of the catamaran. When the hull shakes due to wave impact and fluctuation, the stability of the catamaran is several times that of the ordinary monohull. Its excellent stability makes the catamaran quickly become a new favorite in the field of ship design and research [4].

## 2. Building block Ark design scheme

The structural strength of the building block ark is more complex than that of the single ship or even the catamaran. Therefore, the research and design of building block ark is less than that of ordinary single ship, and there are a lot of ship models and real ship data. In view of this situation, the structural strength of the building block ark is calculated. The design scheme of building block deck the ship is composed of a sharp floating body, a floating body, a connecting body, a L-shaped deck, a Z-shaped deck, a U-

Content from this work may be used under the terms of the Creative Commons Attribution 3.0 licence. Any further distribution of this work must maintain attribution to the author(s) and the title of the work, journal citation and DOI.
Published under licence by IOP Publishing Ltd

shaped deck, a baffle, a guard board and a cover plate. Building block ark is a new type of ship, similar to catamaran. Length can be freely changed by the number of joints. It is very flexible and can be used as a pontoon. The structure of the ark and the external load under the design condition are very complex. By using the direct calculation method based on the finite element theory, the stress state of the whole ship can be calculated effectively and quickly on the basis of the reasonable establishment of the finite element model. Therefore, the ship strength calculation method based on the finite element theory is an improvement of the traditional beam theory. In this paper, the finite element calculation of the building block ark is carried out under the above premise. The main contents are as follows:

(1) Based on the development of catamaran, the research status of catamaran is discussed, and the structural characteristics and overall performance of catamaran are analyzed in detail. The finite element simulation software is explained, and the application of the finite element method in the strength calculation of ship structure is described.

(2) Measurement of the size and structure of each component, three-dimensional model of each component of the hull, assembly and overall design and drawing of some components and SolidWorks software.

(3) Analyze the strength finite element analysis of the first part of the component structure loaded by the building block ark under the forced state, meshing, imposed load and constraint using the finite element software. The calculation results are compared with the strength standards, which do not conform to the strength of the standard components and the continuous improvement of the structure, and finally meet the strength requirements.

(4) Finally, the overall strength of the hull under five different working conditions, namely, the middle arch, the middle concave and the wave surface, is calculated by ANSYS software. It includes five kinds of working condition analysis, external load analysis, load formula derivation and calculation, solution model simplification, whole ship model grid division and optimization, constraint and load application.

The deformation and stress of the hull structure are solved. The strength of the hull is judged by comparing with the corresponding strength standards, and the components and structures that do not meet the strength standards are improved. After calculation and improvement, the structure of the block cabinet finally meets the strength conditions under the design conditions, which provides a reference for further structural design of the block cabinet in the future. Through the verification and summary of launching test of physical model of building block ark, the feasibility of assembling hull design is verified. Refer to the initial simplified model building block cabinet and the continuously improved 3D model building ark, preliminarily determine the structure and size of each component, and SolidWorks software is used to design and draw the building block ark of the 3D model, as shown in Figure 1:

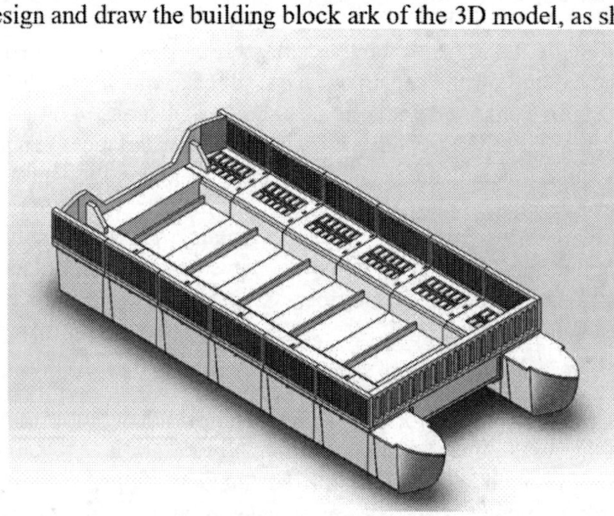

**Fig. 1** Block Ark Outline Drawing

### *2.1. Stress analysis of building block Ark*

The finite element analysis of the total strength of the building block ark under five working conditions is carried out by using ANSYS software. Through the solution of the finite element software, the equivalent stress nephogram and the total deformation diagram are obtained. Due to the different load action under different working conditions, the calculated stress and deformation of the building block ark are different. The total deformation of the hull is relatively small, and the deformation will be magnified and displayed, mainly referring to the data of the deformation size. According to the calculated stress size and distribution, compared with the mechanical properties of PP material, we can judge whether the structural strength meets the requirements.

In the following, the total longitudinal strength of the building block ark is evaluated by solving the results of each working condition.

Working condition 1

The condition that the bow of the building block ark is impacted by waves.

A total of 2940N load weight is applied to the U-shaped deck between the floating bodies of deck 3, 4 and 5; the water resistance and wave force are applied to the upstream of the bow, with the size of 2520N, and the uniform load is applied to the upstream; the buoyancy is 4410N, and the uniform load is applied to the bottom of the hull. In order to constrain the displacement of the stern in z-direction, the bow is subjected to wave impact and water resistance to simulate the dynamic propulsion. Take the x-direction sideline of the upper surface of the stern joint floating body, constrain its X-and Z-direction displacement, take the upper surface of the stern joint floating body, and constrain the y-direction displacement.

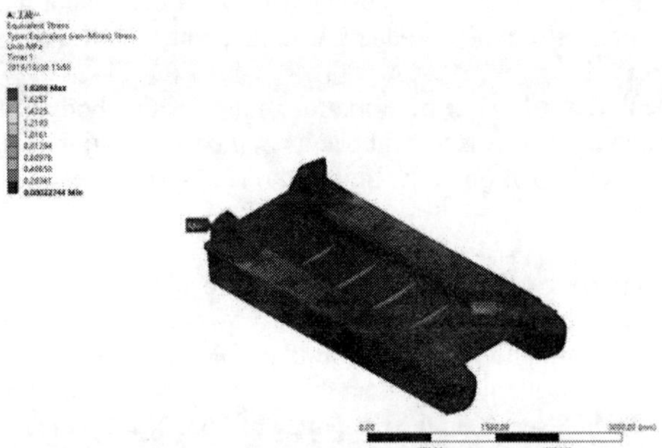

**Fig. 2** Equivalent stress diagram of condition 1

In this case, the inclination angle of the hull is small and the stress distribution of the deck is symmetrical on the starboard and starboard sides. Due to the wave impact load on the bow of the ark and the design driving conditions of the building block ark, i.e. under the condition of small waves, driving at low speed, the wave impact load is limited. The maximum stress of the deck is about 2MPa, and the stress of the impacted part of the bow is slightly higher than that of the middle section. The total deformation is small. Due to the small load imposed according to the design form conditions, the maximum deformation of the hull deck is 6.2mm, and the deformation of other positions is small, which can be ignored.

Working condition 2

The condition that the building ark is impacted by the wave side.

In this case, the ship's direction of travel is perpendicular to the direction of wave propulsion. Therefore, the water resistance is applied to the upstream of the bow, which is 1210N; the wave force is applied to the side wave impact surface, which is 2760N; the U-shaped deck between 3, 4 and 5 floating bodies is applied with a total load of 2940N; the buoyancy is 4410N, which is applied to the bottom of the hull with a uniform load. The X and Y-direction displacement of the outer edge line of the upper surface of the floating body on the uninfluenced side is taken to constrain the z-axis displacement of the stern on the uninfluenced side, and the bow is subjected to wave impact and water resistance to simulate the dynamic propulsion.

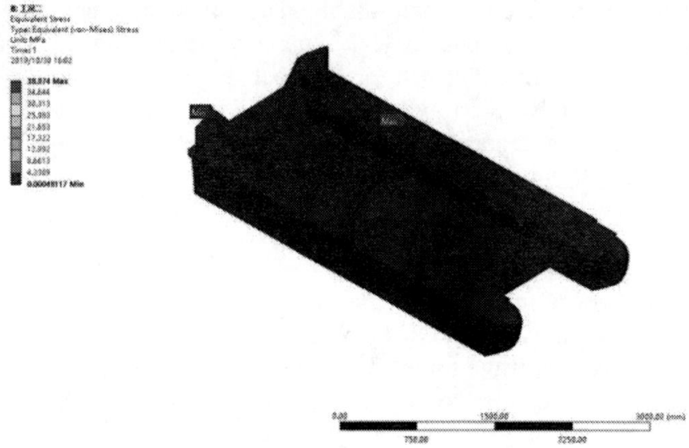

**Fig. 3** Equivalent stress diagram of condition 2

In this case, the inclination angle of the hull is larger than that of case I. The design driving condition of the building block ark is that the impact load of the wave is limited under the condition of small wave and low speed. The maximum stress of U-shaped deck is about 39 MPa, and the stress of ship side in wave impact area is increased.

Deformation: first of all, the transverse deformation of the floating body is larger than that of the first working condition. The maximum displacement occurs at the first floating body connecting deck on the impacted side, and the total deformation of the hull is about 244mm. The deformation in other locations is small and can be ignored.

Working condition 3

The moment of arching wave in the design of building block ark.

Under no-load condition, the middle part is arched upward, and both ends of bow and stern are sagging. The buoyancy of the middle part of the building block ark is greater than that of the gravity, and the buoyancy of the bow and stern is less than that of the gravity. The buoyancy of 550N is applied to the bottom of the 1 and 6 hull sections; the buoyancy of 1655N is applied to the bottom of the 2 and 5 hull sections; the buoyancy of 2205N is applied to the bottom of the 3 and 4 hull sections; the U-shaped deck between the 3, 4 and 5 floating bodies is applied with a total load of 2940N. The load is distributed symmetrically in the middle section of the hull, so the z-axis displacement of the middle section of the hull is constrained, and the x-axis sidelines at both ends of the bow and stern restrict the Y-axis and z-axis displacement.

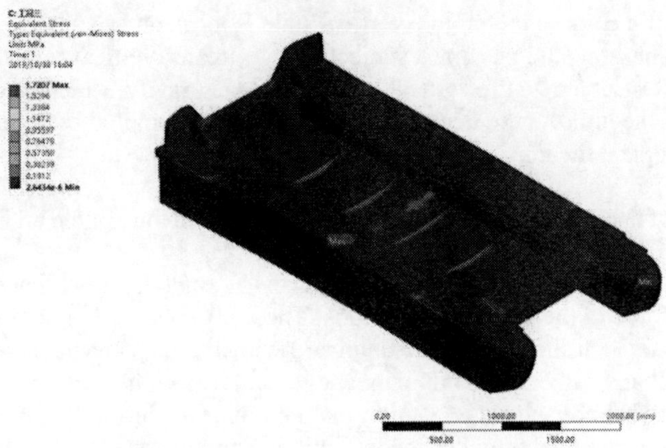

**Fig. 4** Equivalent stress diagram of condition 3

Theoretically, when the wave length is 1.05-1.1 times of the length of the ark, the wave bending moment of the building block ark is the largest. In the actual calculation, the wave length is always equal to the length of the ark. The driving conditions of the design of the building block ark, i.e. driving at a low speed under a small wave condition, are often taken. Under this condition, the middle part of the hull is subjected to a large vertical upward buoyancy and a vertical downward load weight, and the deformation of the ark is small. The maximum stress is about 9Mpa.

Working condition 4

Under the condition of the bending moment of the vertical wave in the design of the building block ark.

In this working condition, under no-load condition, both ends of the bow and stern are arched upward and the middle part is sagging. The buoyancy of the fore and aft parts of the building block ark is greater than that of the gravity, and the buoyancy of the middle part is less than that of the gravity. The buoyancy of 2205N is applied to the bottom of 1,6 two hull sections; 1655N is applied to the bottom of 2,5 two hull sections; 550n is applied to the bottom of 3,4 two hull sections; a total of 2940N is applied to the U-shaped deck between 3,4,5 floating bodies. The x-axis displacement of the middle section of the hull is constrained, and the x-axis sidelines at both ends of the bow and stern restrict the Y-axis and z-axis displacement.

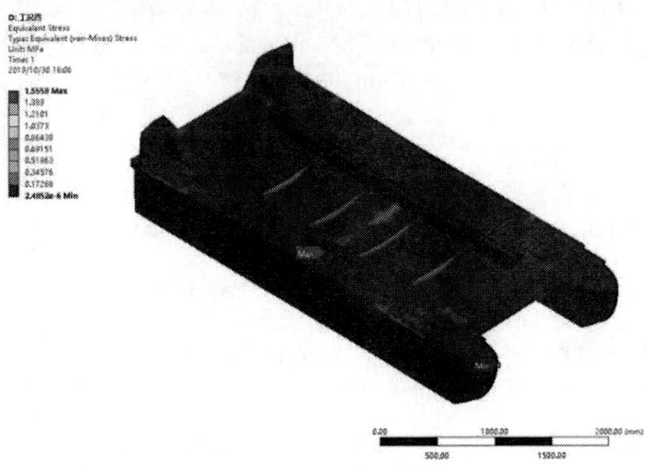

**Fig. 5** Equivalent stress diagram of condition 4

The wave trough of the ark in the middle vertical state is located in the midship cross section of the building block ark. The bow and stern are subject to the greater vertical upward buoyancy, and the middle part of the hull is subject to the vertical downward load gravity, which causes the hull to deform along the direction of the hull. As the deck is not a whole, but assembled by several pieces, there is no large bending moment, and the maximum stress is about 1.5MPa.

Working condition 5

The combination of bending moment and torque of the wave in the middle arch of the building block ark.

Under this condition, the ship is sailing at an angle of 45 °, and the wave front is subjected to water resistance and wave force in the Z and X directions. The buoyancy load is not uniform, and the wave peak position in the middle of the hull is arched upward with a large buoyancy, and the buoyancy of the bow and stern is small and sagging. 2205N buoyancy is applied to the bottom of the left hull 2, 3 and the right hull 4, 5; 1655N buoyancy is applied to the bottom of the left hull 1, 4 and the right hull 3, 6; 550N buoyancy is applied to the bottom of the left hull 5, 6 and the right hull 1, 2; a total of 2940n load weight is applied to the U-shaped deck between the 3, 4 and 5 floating bodies. The z-axis displacement of the middle section of the hull is constrained, and the X-and Y-direction displacement is constrained by the side line of the right bow section of the left stern section.

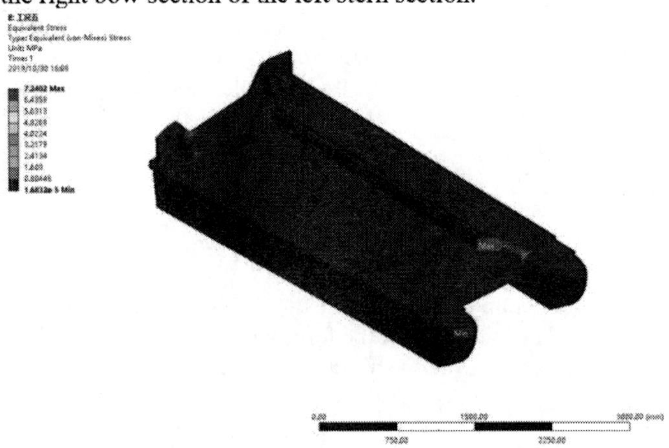


**Fig. 6** Equivalent stress diagram of condition 5

Under this condition, the hull is subjected to both bending moment and torque. The stress distribution changes greatly along the boat width direction. The stress in the middle of the U-shaped deck is larger, and gradually decreases to both sides. The maximum stress value is about 7.2MPa.

## 3. Summary

The finite element analysis of the strength of the building block ark under different working conditions is as follows:

(1) Analyze the external load of the hull, determine the load conditions of the hull under different working conditions, and calculate the load;

(2) Analysis of working conditions: the working conditions are divided into five working conditions: the bow and stern are impacted by waves, the side is impacted by waves, the middle arch, the middle sag, the combination of wave bending moment and torque, and the transverse bending moment.

(3) The optimized building block ark is meshed as a whole, and the grid is adjusted and optimized to get a better quality grid.

(4) Analyze the strength of the ark under five working conditions. The maximum stress is 39 MPa, which is lower than the strength of PP material (45MPa), so the strength of building block ark meets the strength requirements.

## References

[1] HuI Xu, Zaojian Zou. Numerical simulation of viscous flow of small waterplane catamaran based on FLUENT software [J]. Journal of Wuhan University of Technology (traffic science and Engineering Edition), 2004, 28 (1): 8-10.

[2] Jie Zheng, Wei Xie , Wei Luo, et al. Finite element calculation of transverse strength and torsional strength of Wave Piercing Catamaran [J]. Chinese naval research, 20105 (1): 14-18.

[3] Xiaoping Lu, Yan Huang, Zhongkun Shi. Study on wave making resistance of Wave Piercing Catamaran [J]. Ship engineering, 1999 (4): 5-9.

[4] Wenqiang Zuo, Wencai Dong ,Xiang Xia, et al. Experimental study on the influence of stern spoiler on the resistance of Wave Piercing Catamaran [J]. Chinese naval research, 20061 (4): 52-55.

# Preparation of polyacrylic acid and its activation on macrophages

**Mengmeng Zhou[1, 2], Xin Li[1], Zhengzheng Wang[3], Haijun Zhou[1], Xiaoqi Chen[1], Shulan Yang[1], Yantao Li[1, *]**

[1]Institute of Energy Source, Hebei Academy of Sciences, Shijiazhuang, 050000, China
[2]School of materials science and engineering, Hebei University of Technology, Tianjin, 300401, China
[3]Hebei University of Science and Technology, Shijiazhuang 050018, China

*Corresponding author e-mail:13903116163@163.com

**Abstract.** With the rapid and efficient development of animal husbandry industry, intensive a nd large-scale has become the main body, and the corresponding use of vaccines and vaccine adjuvants has gradually increased. Polyacrylic acid has been widely used in vaccine adjuvants because of its safety and low toxicity. In this paper, polyacrylic acid was been prepared by aqueous solution polymerization, taking redox system as catalyst. The molecular structure of polymer was characterized by FT-IR. Viscosity average molecular weight was tested on a ubbelohde viscometer. Moreover, activation of polyacrylic acid on macrophages was analyzed by cell culture method. Results showed that the vaccine prepared by polyacrylic acid as adjuvant and chicken ovalbumin could stimulate macrophages to produce higher IL-1 β.

## 1. Introduction

Animal vaccines are biological preparations made by bacteria, viruses and so on, which can make the body produce specific immunity. With the decrease of immunogenicity, adjuvants are needed to enhance the immune response. Adjuvant is a kind of non-specific immune enhancer [1-2]. When injected into the body together with antigen, it can enhance the immune response of the body to the antigen and improve the immunogenicity of the antigen. With the continuous progress of drug research and development, many kinds of new adjuvants have emerged at home and abroad in recent years [3]. As the most classical adjuvant, aluminum salt adjuvant has been used in clinical practice for 80 years, which can effectively induce humoral immune response, but has no effect on cellular immunity. The emulsion adjuvants include oil/water and oil/water adjuvants, such as FIA, Montanide 65 adjuvant, etc. Most of the commonly vaccines are made up of such adjuvants and antigens. But this kind of emulsion contains mineral oil, which can't be metabolized and remain in the injection site to cause some side effects, which has certain toxicity. However, polyanionic polymers have better performance in vaccine adjuvants, which has been confirmed by most researchers [4]. In 1988, l. A. TH. Hilgers esterified polyacrylic acid with weight average molecular weight of 450000 with butanol or octanol to obtain polyacrylate. It was found that the alkyl polyacrylate showed a good secondary immune response, and the study showed that the relationship between the immune performance and esterification rate of the polymer [5-6]. In this paper, polyacrylic acid polymer with small molecular weight was synthesized using redox catalyst. The

Content from this work may be used under the terms of the Creative Commons Attribution 3.0 licence. Any further distribution of this work must maintain attribution to the author(s) and the title of the work, journal citation and DOI.
Published under licence by IOP Publishing Ltd

IOP Conf. Series: Materials Science and Engineering **772** (2020) 012082 doi:10.1088/1757-899X/772/1/012082

adjuvant prepared from this kind of polymer has low viscosity, easy injection and good safety performance. The activation of macrophages was also studied in macrophage culture in mice.

## 2. Experimental

### *2.1. Materials*

The raw materials are as follows: Acrylic acid and sodium bisulfite, Tianjin Damao Chemical Reagent Factory; Ammonium persulfate, Tianjin Yongda Chemical Reagent Co., Ltd; Sodium hydroxide, Tianjin Feng Chuan chemical reagent Technology Co., Ltd; PBS and OVA were supplied by Institution of biology, Hebei Academy of Sciences. The water was distilled.

### *2.2. Characterizations*

The structure of polyacrylic acid was tested on Fourier Transform Infrared Spectrometer (FT-IR); Viscosity average molecular weight was measured by Ubbelohde viscometer with inner diameter of 0.5-0.6mm. The viscosity of the product was measured at 30 ± 0.05 °C with NaOH of 2mol / L as solvent; the relationship between intrinsic viscosity and viscosity average molecular weight of polymer can be expressed by mark houdwink equation:

$$\eta = 3.38 \times 10\text{-}3 \times M0.43 \quad (1)$$

Where η represents viscosity (Pa.s) and M represents viscosity average molecular weight.

The activation of polyacrylic on macrophages was tested by cell culture. The macrophages of 1.0×106 cells/mL in a 24-well plate were cultured for 24 h and for experiment.

### *2.3. Preparation of Polyacrylic acid*

Half mass of acrylic acid with concentration of 40% and ammonium persulfate/ sodium bisulfite was added into a flask equipped with a stirrer, a thermometer and a dropping funnel. The system temperature rises to 40 °C, and the remaining monomers and initiators with a drop funnel was slowly add. Polyacrylic acid polymer was obtained after 3 hours of reaction. The solid polyacrylic acid can be obtained by vacuum drying and dehydration. The experimental process is as follows.

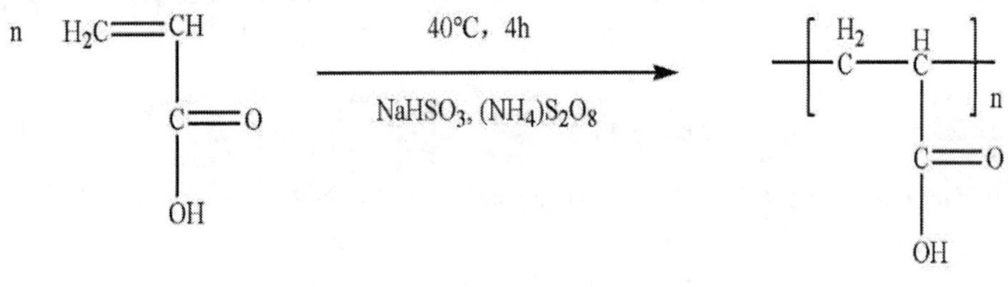


**Figure 1.** Synthesis route of polyacrylic acid

### *2.4. Preparation of vaccine*

Dilute polyacrylic acid with PBS to a concentration of 125ug / ml, and neutralize it with sodium hydroxide solution to a pH value of 7.4. The concentration of OVA was 5ug/ml. Vaccine was prepared in the ratio of 1:1 with above polyacrylic acid solution and OVA. This sample was used for the experiment of activation of macrophages.

## 3. Results

### *3.1. FT-IR analysis of molecular structure of polyacrylic acid*

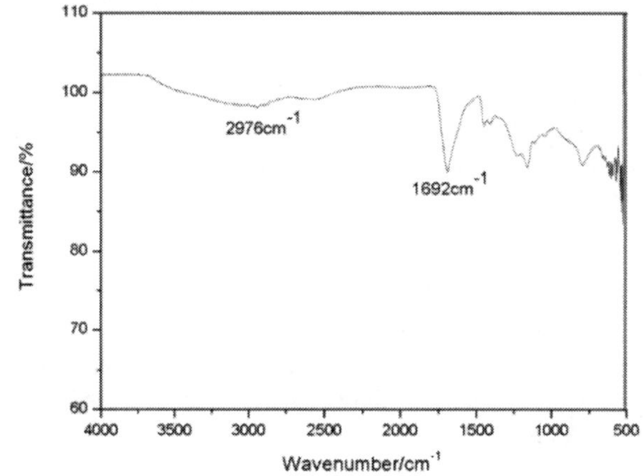


**Figure 2.** FT-IR spectra of polyacrylic acid

The molecular structure of polyacrylic acid is shown in Fig.1. It can be seen that absorption peak of methylene and methylene in the molecular structure of polyacrylic acid was 2976cm-1. 1692cm-1 is the characteristic absorption peak of C = O bond. While No Characteristic absorption peak was appeared near 1640cm-1, which proves the successful preparation of polyacrylic acid polymer.

### *3.2. Effect of polymerization temperature on molecular weight of polyacrylic acid*

The polymerization temperature in redox system has a great influence on the molecular weight of the polymer. Too high temperature will lead to violent reaction of the system, which will reduce the molecular weight. The initiator decomposes rapidly and generates more free radicals at high temperature, leading to the increasing of the rate of continuous initiation and chain termination and decreasing of molecular. When the reaction temperature is too low, the initiator's initiation temperature cannot be reached and the polymerization reaction is slow [7]. Therefore, the oxidation-reduction reaction temperature selected in the bulk system is 40 °C.

### *3.3. Effect of initiator content on viscosity average molecular weight of polyacrylic acid*

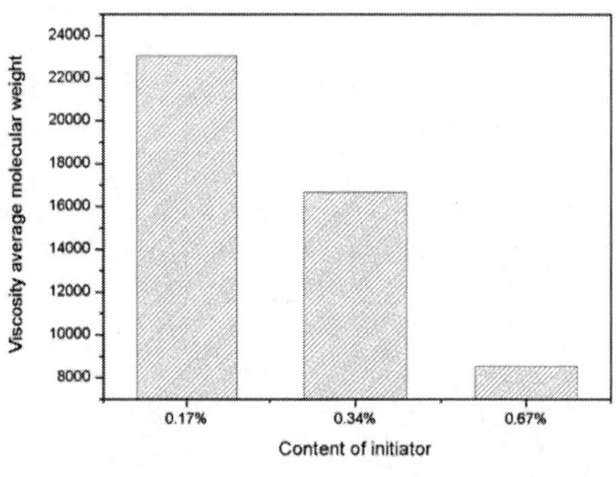


**Figure 3.** Effect of initiator content on viscosity average molecular weight of polyacrylic acid

With the increase of initiator dosage, the molecular weight of polymer decreased, which was in accordance with the rule that the length of kinetic chain was inversely proportional to the square root of initiator concentration. The higher the concentration of initiator, the higher the concentration of free radicals. Therefore, the more active centers of polymerization can be formed, resulted more polymer chains. Then the chain length of each molecular chain is shortened, which leads to the decrease of molecular weight.

### *3.4. Activation of macrophages by polyacrylic acid*

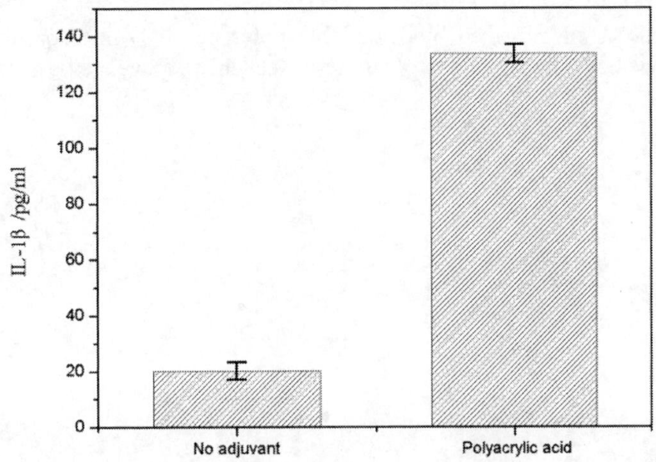


**Figure 4.** The activation effect of polyacrylic acid on macrophages

Proinflammatory cytokines can mediate the activity of antigen-presenting cells, and the level of proinflammatory cytokines is one of the markers of the activation of antigen-presenting cells. IL-1β is an important pro-inflammatory cytokine. Macrophages can secrete IL-1β. The IL-1β secretion of macrophages stimulated by adjuvants can reflect the activation level of macrophages [8]. We added OVA-PAA to macrophage culture medium and taking PBS/OVA as blank control group. The results showed that the concentration of IL-1 β in macrophage culture medium added ova / PAA was as high as 132pg / ml, while that in blank control group was 21pg / ml. It is concluded that PAA can induce macrophages to secrete IL-1 β and has a good activation effect on macrophages.

## 4. Conclusion

In this paper, polyacrylic acid polymer was successfully prepared by redox catalysis system, and its structure was characterized. The effect of initiator amount on viscosity average molecular weight was studied. In this paper, OVA was selected as antigen and polyacrylic acid as adjuvant to prepare vaccine. PAA has excellent ability to activate macrophages through the culture of macrophages in vitro. On the basis of this work, polyacrylic acid will have a broader application prospect in cellular immunology. The application of polyacrylic acid in vaccine needs to be explored deeply in the future.

## References

[1] A.S. Mckee, P. Marrack, Old and new adjuvants, Current Opinion in Immunology 47 (2017) 44-51.

[2] Audibert A.C, Lise L.D. Adjuvants: current status, clinical perspectives and future prospects. Immunol Today [J], 1993, 14(6): 281-284.

[3] R. Vialle, L. Dupuis, S. Deville. Microgel particulate adjuvant: characterization and mechanisms of action. Procedia in Vaccinology [J], 2010.2: 12-16.

[4] T. Diamantstein, B. Wagner, Inge Beyse. Stimulation of humoral antibody formation by

polyanions. I.The effect of polyacrylic acid on the primary immune response in mice immunized with sheep red blood cells. European Journal of Immunology [J], 1971.I: 335-340.
[5] L.A.Th. Hilgers. Alkyl-esters of polyacrylic acid as vaccine adjuvants. Vaccine [J], 1998, 16(16): 1575-1581.
[6] L.A.Th. Hilgers, L. Ghenne, I. Nicolas. Alkyl-polyacrylate esters are strong mucosal adjuvants, Vaccine, 2000.18: 3319-3325.
[7] Wan Fuzhong, Chen Changjing, Zhu Yanmei, Synthesis of high molecular weight sodium polyacrylate at room temperature and determination of its molecular weight, Hubei Chemical Industry [J] 1994,3:28-30.
[8] Zhang Weifeng, Uniform-sized biodegradable polymer-based microparticles as adjuvants for subunit vaccines. The university of Chinese Academy of Science [D], Beijing, 2015.

# Effect of MWCNTs' Content on the Properties of Polyimide Composite Molding Materials

**Xiaoqi Chen[1, *], Zhengzheng Wang[2], Haoran Xu[2], Haijun Zhou[1], Mengmeng Zhou[1] and Yantao Li[1]**

[1]Institute of Energy Source, Hebei Academy of Sciences, Shijiazhuang, China
[2]Hebei University of Science and Technology, Shijiazhuang, China

*Corresponding author e-mail: cxq198508@163.com

**Abstract.** A series of MWCNTs/PI composite molding powders with different MWCNTs contents were prepared by ultrasonic dispersion in situ polymerization and blending method, and the samples were prepared by molding. In this experiment, 3,3', 4,4' - biphenyltetracarboxylic dianhydride (BPDA), pyromellitic dianhydride (PMDA) and 4,4 ' - diaminodiphenyl ether (ODA) are taken as the main raw materials and MWCNTs as reinforcement materials. The molecular structure, thermal stability, mechanical properties, wear resistance and hardness of the composites were characterized. Results showed that mechanical properties of the composite molding materials prepared by ultrasonic dispersion in situ polymerization are better than those prepared by dry mixing method. The composite molding materials showed the best properties at the MWCNTs content of 0.5wt%. What's more, the tensile strength and elongation at break are 108.9MPa and 4.5%, which are 26% and 23.6% higher than those of pure PI.

## 1. Introduction

Polyimide (PI), as a kind of high performance engineering plastics, has excellent mechanical properties, electrical properties, chemical stability, high radiation resistance and high temperature and low temperature resistance.Various application fields have placed higher demands on polyimide materials, and conventional polyimide materials have been unable to meet the demand. In order to meet higher requirements, more and more researchers focus on the performance modification of PI. Multi-walled carbon nanotubes (MWCNTs) as the tubular structure, with excellent mechanical, electrical and chemical properties and specific self-lubricating function, etc., have been developed into an ideal modified material for many materials [1-5].In recent years, the combination of MWCNTs with high-polymers such as polytetrafluoroethylene [6], ultrahigh molecular weight polyethylene, polyurethane [7] and bismaleimide [8] has become the research focus.

Therefore, in this paper, functionalized MWCNTs was used to modify PI composite molding materials. Effects of ultrasonic in-situ polymerization method and blending method on the properties of MWCNTs/PI composite molding materials were studied, and the effects of MWCNTs' content on the mechanical property, thermal stability, wear resistance and hardness of PI composite molding materials were discussed.

Content from this work may be used under the terms of the Creative Commons Attribution 3.0 licence. Any further distribution of this work must maintain attribution to the author(s) and the title of the work, journal citation and DOI.
Published under licence by IOP Publishing Ltd

## 2. Experimental

### *2.1. Materials*

Carboxylated multi-walled carbon nanotubes (MWCNTs) was supplied by Beijing Boyu High-tech New Materials Technology Company with carboxyl group content of 0.2 wt% to 1 .0 wt%, the diameter of 20 nm to 30 nm, the length of 10 to 30 um, and purity of 95%. The 4, 4'-diaminodiphenyl ether (ODA) were obtained from Nantong Huishun Chemical Co., Ltd, the 3,3'4, 4'-biphenyltetracarboxylic dianhydride (BPDA) were obtained from Shijiazhuang haili Chemical Co., Ltd and the pyromelic dianhydride (PMDA) were obtained from Rugao Leheng Chemical Co., Ltd. The raw materials were all dried in an vacuum drying oven at 105℃ for 6 h. N, N-dimethylacetamide (DMAc) were obtained from Tianjin Yongda Chemical Reagent Co., Ltd.

### *2.2. Preparation of PI molding materials*

10 wt% ODA/DMAC solution was added to a four-neck flask equipped with mechanical stirring. Then an equimolar mixture of BTDA/PMDA (nBPDA/nPMDA=2/1) at 20℃ was added into system. 4 hours' later, a kind of viscous yellow polyamic acid (PAA) solution was obtained. Then xylene and catalyst were added. The mixture was stirred uniformly and heated up. After refluxing for 3 hours, the mixture was cooled and filtered to obtain powders. The powders were washed three times with acetone, suction filtered and placed in an oven for thermal imidization to obtain a pure PI molding powder. Lastly, a certain amount of PI molding powder was weighed, put into the mold, and PI material sample was obtained by hot pressing machine.

### *2.3. Synthesis of MWCNTs/PI composite molding materials by ultrasonic in-situ polymerization*

Firstly, the MWCNTs were added to a quantitative DMAC and ultrasonically dispersed for 1h at a power of 400 W, then ODA was added to the dispersion. After the ODA was completely dissolved, an equimolar mixture of BTDA/PMDA(nBPDA/nPMDA=2/1) at 20℃ was added into system. After reacting for 4 hours, a viscous black MWCNTs/PAA solution was obtained. Then xylene and catalyst were added, the mixture was stirred uniformly and heated up. After refluxing for 3 hours, the mixture was cooled and filtered to obtain powder. The powder was washed three times with acetone, suction filtered and placed in an oven for thermal imidization. Finally, a series of MWCNTs/PI composite molding powders with different MWCNTs content were obtained. Lastly, a certain amount of PI molding powders were weighed and placed in a mold, and samples of the MWCNTs/PI composite molding material was obtained by hot press molding machine.

### *2.4. Synthesis Process of Carbon MWCNTs/PI Composite Molding Materials by Blending Method*

The MWCNTs and pure PI molding powders were added to a high-speed mixer in different mass ratios, and mixed at a high speed of 25000 r/min for 1 min to obtain a series of PI composite molding powders. A certain amount of powder was placed into a mold, and a sample of MWCNTs/PI composite molding material was obtained by a hot press molding machine.

### *2.5. Characterization*

The molecular structure of PI and MWCNTs/PI composites was tested on Fourier transform infrared spectroscopy (FT IR,Spectrum 100 infrared spectrometer test). The thermal stability of composites was tested by TGA under nitrogen atmosphere with heating rate of 20℃/min and Test range of 30℃-900℃. The mechanical properties of composites were tested by universal tensile testing machine according to GB/T 1447-2005. Rockwell hardness of composite materials was tested with a plastic Rockwell hardness tester according to GB/T 3398.2-2008. And the wear resistance of composite materials are tested by the plastic sliding friction tester according to the provisions of GB/T 3960-2016.

## 3. Results and Discussions

### *3.1. FT IR analysis of PI and MWCNTs/PI composite materials*

The molecular structures of PI molding powder and MWCNTs/PI composite molding powders were determined by infrared spectrometer in Figure 1.

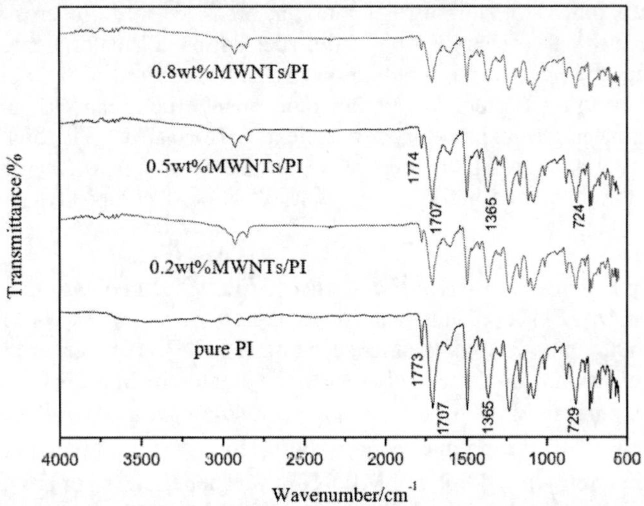


**Figure 1.** FI IR spectrum of pure PI and MWCNTs/PI composites materials.

As was known in Figure 1, in infrared spectrum of pure PI, 1773cm-1and 1707cm-1 is the asymmetric and symmetric stretching vibration peaks of C=O on the five-membered imine ring, and 1365cm-1 is the stretching vibration peaks of C-N, and 729cm-1 is the bending vibration peak of C=O, and 1234cm-1 is the stretching vibration peaks of C-O-C. Compared with Infrared spectrum of MWCNTs/PI composites materials, the characteristic absorption peak of MWCNTs/PI composite molding materials is basically the same as that of pure PI, and the addition of MWCNTs has little effect on the imamination of polyamide acid.

### *3.2. Effects of different preparation processes on mechanical property of MWCNTs/PI composite molding materials*

In this paper, ultrasonic in-situ polymerization method and blending method were used, and two different MWCNTs/PI composite molding materials were prepared separately. Effects of different MWCNTs content on their mechanical property were discussed in Table 1.

**Table 1.** The effect of MWCNTs content on mechanical properties

| MWCNTs content/ (wt%) | Ultrasonic in Situ Polymerization Method | | Blending Method | |
|---|---|---|---|---|
| | Tensile strength/MPa | Elongation at break % | Tensile strength/MPa | Elongation at break/% |
| 0 | 86.4 | 3.6 | 86.4 | 3.6 |
| 0.2 | 91.6 | 2.6 | 89.6 | 1.73 |
| 0.5 | 108.9 | 4.5 | 92.9 | 1.97 |
| 0.8 | 106.6 | 4.4 | 90.5 | 1.69 |
| 1.0 | 95.9 | 3.7 | 86.9 | 1.50 |

IOP Conf. Series: Materials Science and Engineering **772** (2020) 012083 doi:10.1088/1757-899X/772/1/012083

As shown in Table 1, the tensile strength and elongation at break of MWCNTs/PI composite molding materials prepared by ultrasonic in-situ polymerization increased first and then decreased with the increasing of MWCNTs content. The tensile strength (108.9MPa) and elongation at break (4.5%)of the composites reached the highest, which was 26% and 23.6% higher than pure PI when the content of MWCNTs is 0.5 wt%. Comparing with pure PI, the tensile strength (93MPa) of MWCNTs/PI composite molding materials prepared by blending method had a small improvement, reaching the maximum when MWCNTs content was 0.5wt%. While its elongation at break showed a decreasing trend. The above results showed that the MWCNTs can not be well dispersed into PI material simply by blending, and even became a defect in the composite molding materials. Through ultrasonic in-situ polymerization, MWCNTs can be uniformly dispersed into the molecular structure of PI, and the MWCNTs/PI composite molding materials prepared had excellent performance. Therefore, the MWCNTs/PI composite molding materials were finally prepared by ultrasonic in-situ polymerization, and the thermal stability, friction performance and hardness of the MWCNTs/PI composite molding materials were discussed.

### *3.3. Effects of MWCNTs content on thermal stability of MWCNTs/PI composite molding materials*
The effects of different MWCNTs content on the thermal properties of MWCNTs/PI composite molding materials were shown in Figure 2. With the increasing of MWCNTs content the thermal decomposition temperature of composites increased gradually. When the amount of MWCNTs was 1wt%, the thermal decomposition temperatures of MWCNTs/PI composite molding materials on Td5and Td10 were 580℃ and 600℃ respectively, while the thermal decomposition temperatures of pure PI on Td5and Td10 were 574℃ and 596℃. Therefore, the addition of MWCNTs was beneficial to improve the thermal stability of MWCNTs/PI composite molding materials, but the low MWCNTs content had little effect on the thermal stability of MWCNTs/PI composite molding materials.

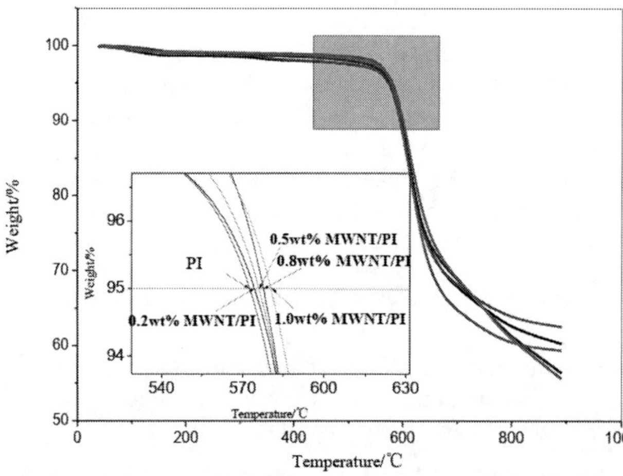


**Figure 2.** Thermogravimetric curve of pure PI materials and MWCNTs/PI composite molding materials

### *3.4. Effects of MWCNTs content on wear resistance of MWCNTs/PI composite molding materials*
The influence of different MWCNTs content on the wear resistance of MWCNTs/PI composite molding materials was studied. As can be seen from table 2, it was beneficial to improving the wear resistance of MWCNTs/PI composite molding materials by adding MWCNTs. When the MWCNTs content was 0.5wt%, the effect was most obvious, and the friction coefficient of the composite was only 0.22. That was because MWCNTs was a tubular structure with a good load transfer capacity. During the sliding process, a transfer film can be better formed on the friction surface and the wear resistance of MWCNTs/PI composite material can be improved.

**Table 2.** The effect of MWCNTs content on wear resistance of composite molding materials

| Sample | MWCNTs Content/ (wt%) | Friction Coefficient |
|---|---|---|
| 1 | 0 | 0.32 |
| 2 | 0.2 | 0.30 |
| 3 | 0.5 | 0.24 |
| 4 | 0.8 | 0.27 |
| 5 | 1.0 | 0.27 |

### *3.5. Effects of MWCNTs Content on the hardness of MWCNTs/PI Composite Molding Materials*

**Table 3.** The effect of MWCNTs content on the hardness of MWCNTs/PI composite molding materials

| Sample | MWCNTs Content/ (wt%) | Rockwell Hardness |
|---|---|---|
| 1 | 0 | 105.4 |
| 2 | 0.2 | 104.5 |
| 3 | 0.5 | 106.5 |
| 4 | 0.8 | 107.7 |
| 5 | 1.0 | 107.4 |

Table 3 showed the influence of different MWCNTs content on the Rockwell Hardness of MWCNTs/PI composite molding materials. The addition of MWCNTs was beneficial to increase the hardness of MWCNTs/PI composite molding materials. When the addition amount of MWCNTs was 1.0 wt%, the hardness of MWCNTs/PI composite molding materials reached 108.4, which increased slightly comparing with pure PI.

## 4. Conclusion

Comparing with ultrasonic in-situ polymerization and blending method, the MWCNTs/PI composite molding materials which were prepared by ultrasonic in-situ polymerization had a better mechanical property. The tensile strength and elongation at break of MWCNTs/PI composite molding materials were 108.9 MPa and 4.5% respectively when the MWCNTs content was 0.5 wt%. And comparing with pure PI materials, those were increased by 26% and 23.6% respectively. Due to the small amount of MWNTs added, the thermal stability, wear resistance and hardness of PI materials are improved, but the improvement range is small.

## References

[1] Bin Y Z, Oishi K, Koganemaru A, et al. Catalytic effect of nickel under carbonization of polyimide films [J]. Carbon, 2005, 43( 8): 1617-1627.

[2] Gooding J J. Nanostructuring electrodes with carbon nanotubes: A review on electrochemistry and applications for sensing [J]. Electrochim Acta, 2005, 50 ( 15): 3049-3060.

[3] Li Zhenlian, Zhang Xinrui, et al. Tribological properties of complex filled polyimide lubricating composites [J]. Lubrucation Engineering, 2016, 41(10): 82-85.

[4] Zhang Mingyan, Cui Yu, Wu Qiong, et al. Mechanical properties of polyimide/multi-walled carbon nanotubes hybrid film. Journal of Functional Materials, 2014, 45(24): 24035-24040.

[5] So H H, Cho J W, Sahoo N G. Effect of carbon nanotubes on mechanical and electrical properties of polyimide/carbon nanotubes nanocomposites [J]. European Polymer Journal, 2007, 43(9): 3750-3756.

[6] Shi Y, Feng X, Wang H, et al. Tribological and mechanical properties of carbon-nanofiber-filled polytetrafluoroethylene composites [J]. Journal of Applied Polymer Science, 2010, 104(4): 2430-2437.

[7] Song H J, Zhang Z Z, Men X H. Surface-modified carbon nanotubes and the effect of their addition on the tribological behavior of a polyurethane coating [J]. European Polymer Journal, 2007, 43(10): 4092-4102.

# Research on Preparation Technology for Continuous Carbon Fiber Reinforced Printing Filaments

**Xiangjun Bi [1], *, Hongsheng Tan [2], Zonghui Li [3], Yuqin Li [1] and Tengfei Liu [4]**

[1] Aerospace Research Institute of Special Material and Processing Technology, Beijing, China
[2] Shandong University of Technology, Zibo, China
[3] Hua Dian Electric Power Research Institute Co., LTD., Beijing, China
[4] Xi`an Jiaotong University, Xi`an,China

*Corresponding author e-mail: bxiangj@163.com

**Abstract.** Continuous carbon fiber reinforced thermoplastic composites with high mechanical strength can make up for the low-bearing faults of pure resin material and particle reinforced material. At the same time, the physical properties of fiber conduction can meet the structure damage detection and conductive functional requirements. It is seen as one of the advanced technology in thermoplastic 3D printing field. At present, the material has been gradually used in robot manufacturing, aerospace components manufacturing and so on. This paper introduces the preparation process of continuous fiber reinforced printing filaments and the characterization of mechanical properties, morphology and printing results. It points out that the printing filaments can meet the basic requirements of continuous fiber reinforced printing, which with high fiber orientation is affected by prepreg tension function. In addition, the small diameter filaments have obvious advantages in transportation, printing fabrications and printing effects. It is suggested that continuous fiber reinforced printing is suitable for choosing filaments with the diameter of 0.4mm or less.

**Keywords:** Additive manufacturing, 3D printing, FFF, filament fabrication

## 1. Introduction

With the development trend of low cost and intelligentization of composite materials and some requirements (short cycle development, rapid prototyping, simple preparation of complex components and environment-friendly manufacturing), a new composite material manufacturing concept, additive manufacturing (i.e. 3D printing), is proposed. It is a new manufacturing method by adding materials and manufacturing layer by layer, and a quick manufacturing way of physical objects consistent with the three-dimensional model. At the same time, it is a best way to solve the contradiction between low cost, quick molding and the requirements of products high performance, which has an important impact on product design, manufacturing and maintenance mode. And it is known as the core technology of the "third industrial revolution". [1] 3D printing and specifically fused filament fabrication (FFF) have brought about a new dimension to the fabrication of PCs. [2, 3] Now the United States has issued and supported several programs which integrate 3D printing into the country's manufacturing strategy, such

Content from this work may be used under the terms of the Creative Commons Attribution 3.0 licence. Any further distribution of this work must maintain attribution to the author(s) and the title of the work, journal citation and DOI.
Published under licence by IOP Publishing Ltd

as NGMTI (in 2006) [4], NNMI (in 2012), A 21st Century Science, Technology and Innovation Strategy for America's National Security, Emerging Science and Technology Trends: 2016-2045--A Synthesis of Leading Forecasts (APRIL 2016) [5]. The research institutes and enterprises involved include Binstate University, University of Alabama, Massachusetts institute of technology, Oak Ridge National Laboratory, Lockheed Martin, Boeing, and so on [6]. In China, 3D printing technology has also been promoted to a very important position, and issue the Plans: the Development and Promotion of the National Additive Manufacturing Industry 2015-2016, the Plan for the Development and Action of the National Additive Manufacturing Industry 2017-2020 [7, 8, 9].

Filaments for 3D printing, especially for thermoplastic resin matrix composites, have become a bottleneck in the development of printing technology, which limits the stability and safety of 3D printing products. At present, filaments reinforced with small particles or chopped fiber is given priority, such as chopped glass fiber reinforced PP, ABS or PLA; they have low mechanical strength, tensile strength under 100 MPa), low weather resistance, low heat resistance and poor ultraviolet resistance performance, far less than the traditional fiber reinforced composite material. So these did not meet the requirements of aerospace structural use.[10] In this paper, the preparation of continuous fiber-reinforced nylon filaments was taken as the research object, and introduced the morphology, usage and printing status of two materials with different diameters were compared.

## 2. Experimental

### *2.1. Material preparation*
Nylon, PA6, TP-4208, ZIG SHENG INDUSTRIAL CO., LTD. Carbon fiber, T700-12K, Toray. Carbon fiber, T700-1K, Jiangsu Hengshen Co. LTD. Filament Forming Equipment, self-made equipment; Continuous Fiber Printing Equipment, COMBOT-200, Shanxi Fibertech Co. LTD.

### *2.2. Technological Process*
In order to ensure the continuity of fiber, the equipment of continuous fiber reinforced thermoplastic filaments is crossing form in prepreg process, which is different from linear flow design of discontinuous fiber wire [11], and has different prepreg form, mixed state, extrusion state and sizing technology.The following is a brief description of the molding process:

- ◆ Material pretreatment:

Dry fiber: PA6 was placed in a 90°C air-blast drying box and dried for 8h.

Dry plastic: The fiber was placed in a 90°C air-blast drying box and dried for 1h.

- ◆ Preimpregnated fibre, Screw extrusion: In this process, three sets of prepreg rolls were used to infiltrate the fiber into the nylon melt under the action of pressure.
- ◆ Wire sizing: Use special sizing equipment to control the final filament diameter.
- ◆ Air cooling: After passing through the high-temperature sizing equipment, the wire was cooled at room temperature and 2m away from the wire.
- ◆ Windig for use: Use a specified diameter plastic reel.

In this paper, two kinds of filaments with diameters of 1.75mm and 0.4mm were prepared. The fiber materials selected were one bundle of 12k carbon fiber and one bundle of 1k carbon fiber respectively.

### *2.3. Testing and Characterization*

*2.3.1. The tensile test.* It is difficult to determine the bending performance of a filament with a diameter of 0.4mm. Here, a wire with a diameter of 1.75mm was used as the test object. The testing was conducted on the basis of ASTM D3039/D3039M Standard Test Method for Tensile Properties of Polymer Matrix Composite Materials and ISO 3597-2:2003 Textile-glass-reinforced plastics-Determination of mechanical properties on rods made of roving-reinforced resin-Part 2: Determination of flexural strength. These tests were carried out using the universal testing machine (INSTRON Co., Instron 5969), as shown in figure 1.

**Fig. 1** filament tensile and bending test diagram

*2.3.2. Surface topography observation.* The section morphologies of monofilament were measured and analyzed with the micro optical microscope (ANMO electronics corporation, AN413T).

## 3. discussion

### *3.1. Mechanical test*

The measured data were respectively 820MPa (tensile strength) and 42GPa (tensile modulus). The strength data is close to the body of carbon fiber, which indicates that the fiber is consolidated and formed under the action of traction, and maintains a good fiber collimation.

### *3.2. Surface morphology analysis*

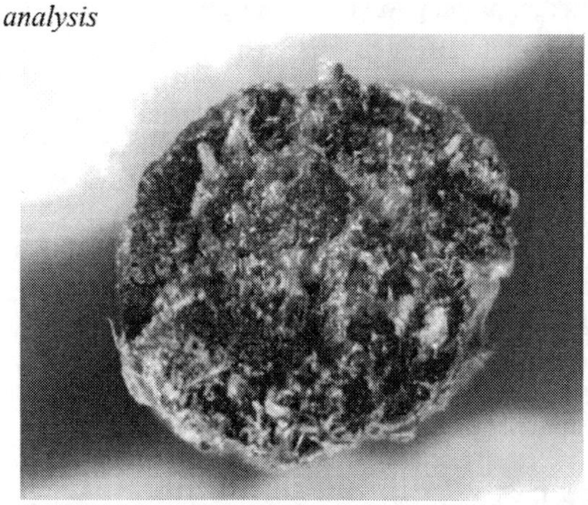

**Fig. 2** the fiber section morphology (200×)

Through the analysis of the fiber section morphology, it is found that there is no core-shell structure [12] in the fiber prepared by traction, but the resin and fiber are coated with each other and prepreg, forming a relatively uniform term. This may be because the fiber content is too high, which causes the fiber to occupy the volume of the resin. However, the mutual coating state between the fiber and the resin also indicates that the roll prepreg preparation process achieves the effect of complete prepreg of the fiber.

IOP Conf. Series: Materials Science and Engineering **772** (2020) 012084 doi:10.1088/1757-899X/772/1/012084

### *3.3. Effect of fiber diameter on winding state*

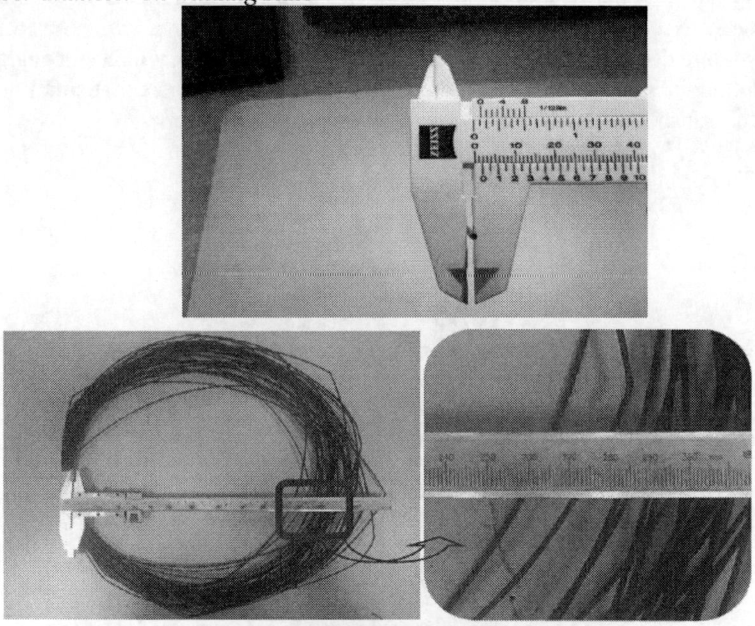

a) A diameter of 1.75mm

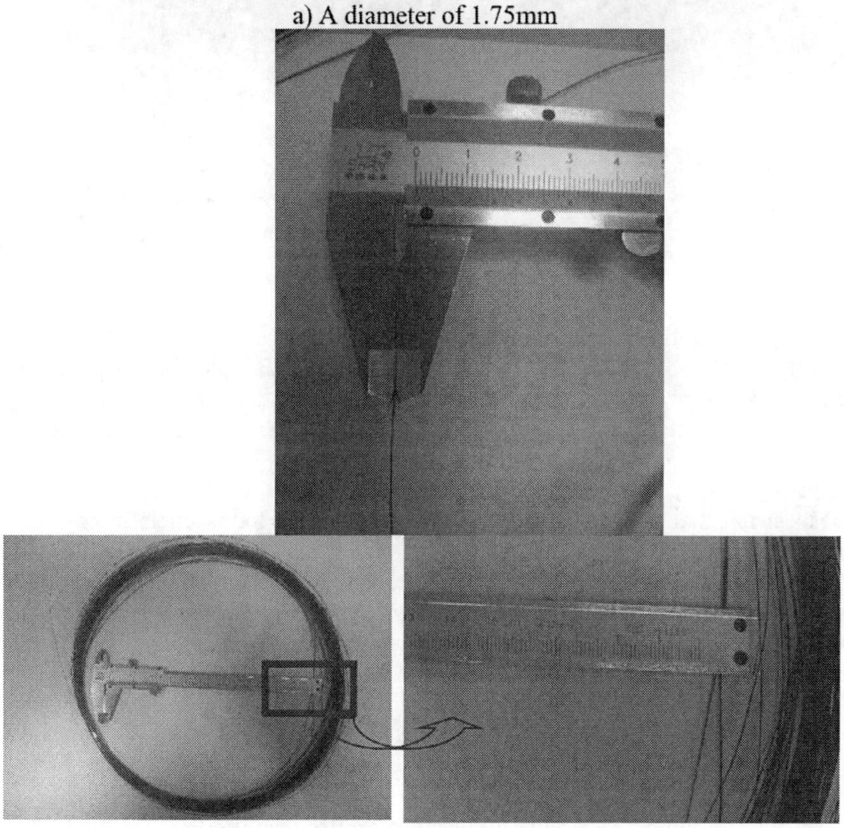

b) A diameter of 0.4mm

**Fig. 3** Wire with different diameters

According to figure 3, the winding diameter of the wire with a diameter of 1.75mm is very large. When the diameter exceeds 300mm, part of the filament is still broken, which has affected the use of printing. The winding diameter of 0.4mm wire is no more than 200mm, which can guarantee the original smooth and continuous state without any sign of breaking. This indicates that small diameter wires are more suitable for transportation and ensure the effective length of fibers.

*3.4. Effect of fiber diameter on printing*

a) Square sample with side length of 20mm (diameter of 1.75mm)

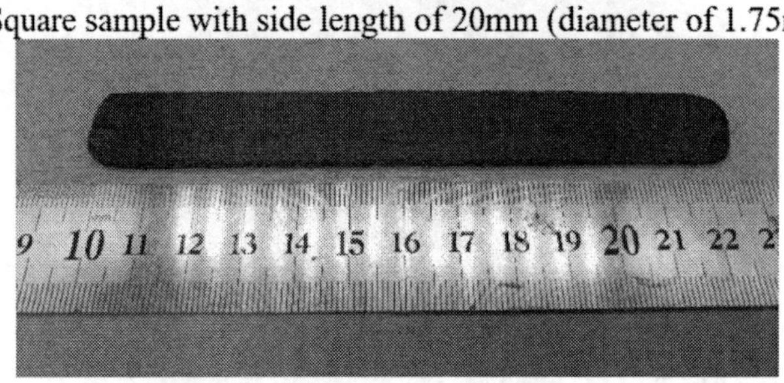


b) Print splines with length of 120mm (diameter of 0.4mm)

**Figure 4.** Comparison of printing effects of wires with different diameters

By comparing the printing effects of the two wires as shown in figure 4, it is found that compared with the filaments with 1.75mm, the sample with 0.4mm filaments has a small rotation return phenomenon and the space between adjacent fibers, and it can be guaranteed to turn back and print at a small Angle, which is more conducive to precision control and the printing of high-precision parts.

## 4. conclusion

Through the mechanical measurement, it can be predicted that the force action of the traction equipment can guarantee the consistency of the fiber direction, and the obtained wire with high orientation can meet the requirement of high strength of wire. At the same time, the fiber and resin in the fiber bundle were coated with each other, which indicated that the fiber was fully preimpregnated. The above shows that the preparation process is suitable for the preparation of continuous fiber reinforced nylon wire. By comparing the indexes of two kinds of filament with diameters (1.75mm and 0.4mm), it is shown that the small diameter wire is easy to be stored and used, and the printing defects (such as filaments and surface smoothness) are well guaranteed.

## References

[1] Fabrizio Regina, Fulvio Lavecchia, Luigi Maria Galantucci. Preliminary study for a full colour low cost open source 3D printer, based on the combination of fused depositionmodelling (FDM) or fused filament fabrication (FFF) and inkjet printing. International Journal on

Interactive Design and Manufacturing (IJIDeM), 2018, 12:979–993.
[2] Yehia Ibrahim, Garrett W. Melenka, Roger Kempers. Additive manufacturing of Continuous Wire Polymer Composites. Manufacturing Letters, 2018, 16: 49–51.
[3] LUAN Cong-cong, YAO Xin-hua, LIU Cheng-zhe, FU Jian-zhong. Carbon fiber-thermoplastic composite 3D printing technology and its self-monitoring. Journal of Zhejiang University (Engineering Science), 2017, 09: 1808-1814.
[4] WANG Jianhua, LI Rupeng. Latest Development of Aircraft Digital Design and Manufacturing Technology. Aeronautical Manufacturing Technology, 2016, 05: 78-82.
[5] ZHU Hongkang, JIA Yudong. Research on Policies of American Manufacturing Innovation. Materials China, 2017, 05: 395-400.
[6] ZHANG Xuejun, TANG Siyi, ZHAO Hengyue, GUO Shaoqing, LI Neng, SUN Bingbing, CHEN Bingqing. Research Status and key technologies of 3D Printing. Journal of Materials Engineering, 2016, 44(2): 122-128.
[7] http://www.most.gov.cn/tztg/201304/t20130416_100885.htm.
[8] Xue Fengtong, Liu Haibin. The overview of friction stir additive manufacturing. Modern Manufacturing Engineering, 2019, 04: 33-40.
[9] AN Guojin. Application and prospect of metal additive manufacturing technology in aerospace. Modern Machinery, 2019, 03: 39-43.
[10] [10] MING Yueke, DUAN Yugang, WANG Ben, XIAO Hong, ZHANG Xiaohui. 3D Printing for High Performance Fiber Reinforced Polymer Composites, Aeronautical Manufacturing Technology, 2019: 34-38
[11] Wu Hai-hua, Wang Jian, Cai Yu, Peng Jian-hui, Zhong Ji-hong, Wei Zheng-ying. Preparations and properties of graphene/PLA composite wire for 3D printing. CARBON TECHNIQUES, 2018, 6(37): 61-65.
[12] XIAO Jian-hua. Extrusion of Carbon Fiber Reinforced Thermoplastic Plastics for 3D-Printing. CHINA PLASTICS INDUSTRY, 2016, 44(6): 46-48.

# Hot-pack Accumulative Roll-bonding (HP-ARB) Process of Ti-44Al-5Nb-1Mo-(V, B) Alloys

**Tianrui Li, Guohuai Liu*, Ruiqi Guo, Xiuqi Wang and Zhaodong Wang**

State Key Laboratory of Rolling and Automation, Northeastern University, Shenyang 110819, China

* Corresponding author: liugh@ral.neu.edu.cn

**Abstract.** Ti-44Al-5Nb-1Mo-(V, B) alloy preforms were processed under hot-pack accumulative roll-bonding (HP-ARB) process directly from HIPed ingot. A thin crack-free HP-ARB sheet was obtained. The corresponding microstructure evolution, alloy element distribution, and tensile properties were systematically studied. The alloy matrix were measured as ~703-844 μm, while the defect-free bonding interfaces were measured as ~96 and 221 μm. Near-lamellae and fine near-duplex microstructure were characterized in the bonding interface and the sheet matrix, respectively. The differences of stress state between the alloy matrix and the plate edge, and large deformation strain during HP-ARB are considered to be the main factors to form the microstructure characteristics. After HP-ARB, the sheet exhibited high performance at room-temperature (RT) as well as high-temperature. Ultimate tensile stress (UTS) of 863 Mpa and ductility of 1.71% were exhibited at RT. The 800 oC and 850 oC UTS/ductility were 754 Mpa/7.24% and 599 Mpa/67.5%, respectively. The pure bonding interface and fine homogeneous microstructure contributed positively to the RT and high-temperature tensile properties.

## 1. Introduction

TiAl alloys are being considered to be the potential new structural materials due to their attractive properties, such as low density, excellent high-temperature strength, good oxidation and creep resistance [1-3]. However, their industrial applications are limited because of their low-temperature brittleness, poor workability and narrow processing window. Significant effort has been made, including alloying, hot working process, heat treatment, etc., to improve their inherent ductility. Therefore, β-solidifying γ-TiAl alloys which primarily consist of α2, γ and B2 phases were developed to obtain superior hot workability. Some grain refiners such as B, C, Y, etc. are also added which could efficiently increase the applicability of β-solidified γ-TiAl alloys.

TiAl cast ingot generally has inhomogeneous deformation induced by the multiphase structure and complex microstructure during thermal-mechanical processing. The deformation incompatibility could aggravate flow localization and easily induce crack initiation and propagation. Considerable effort has been devoted to microstructure control and grain refinement for TiAl alloys by hot working, such as forging, extruding and rolling [4]. Direct hot pack rolling process by as-cast or as-HIPed TiAl alloys, which is considered as an effective low-cost method, has been recently proposed. Furthermore, accumulative roll-bonding (ARB) is widely considered to be an intense straining process for bulk

Content from this work may be used under the terms of the Creative Commons Attribution 3.0 licence. Any further distribution of this work must maintain attribution to the author(s) and the title of the work, journal citation and DOI.
Published under licence by IOP Publishing Ltd

materials using rolling deformation, which has been applied in various materials, such as aluminum, titanium alloy and steels [5-7]. To date, the strength and ductility at room or elevated temperatures of these alloy sheets are significant improved [8, 9]. However, the ARB process, corresponding microstructure evolution, and mechanical properties remain elusive for TiAl alloys.

In the present study, ARB process was introduced in Ti-44Al-5Nb-1Mo-(V, B) alloys via direct hot-pack rolling. Next, attention was focused on microstructure characteristics and element distributions of the sheet matrix and bonding interface. Finally, the room- and high-temperature tensile properties of the as-rolled alloy were investigated.

## 2. Materials and experimental procedure

The alloy ingot (about Φ110 mm×420 mm) with nominal composition of Ti-44Al-5Nb-1Mo-2V-0.2B (at. %) was prepared using induction skull melting (ISM) under argon atmosphere. Subsequently, hot isostatic pressing (HIP) was carried out at 1260 °C in Ar atmosphere with a pressure of 175 MPa for 4 h followed by furnace cooling. The preforms for hot-pack accumulative roll-bonding (HP-ARB) were cut by electric-discharge machining from the center zone of the HIPed ingot and cleaned by grinding and acetone. Stainless steel cans were used to encapsulate the several pieces of preforms. Oxidation resistant coating as well as insulation materials between cans and TiAl alloys were necessary to avoid Fe-Ti reaction and oxidation at rolling temperature. The preform preparation and rolling process were manifested by a diagrammatic drawing as shown in Fig. 1a. Fig. 1b shows the crack-free HP-ARB sheet.

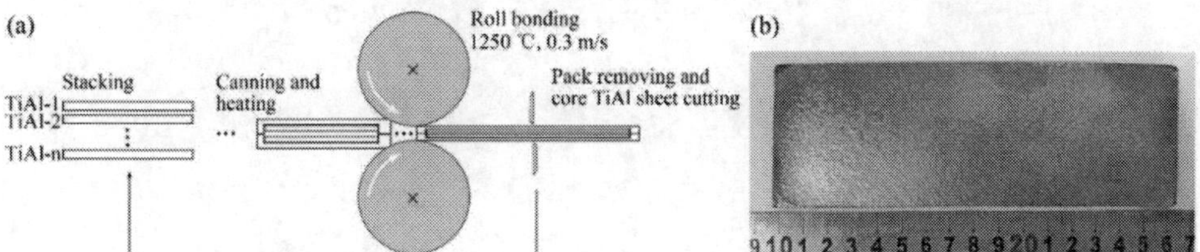


**Figure 1.** (a) Schematics showing the HP-ARB process; (b) macro appearance of the accumulative-rolled Ti-44Al-5Nb-1Mo-(V, B) alloy sheet.

Microstructure, fracture surface, and element distribution analysis were carried out using a Zeiss Ultra 55 scanning electron microscopy (SEM) and a JEOL 8530F Electron Probe Micro-analyzer (EPMA). Both SEM and EPMA specimens were prepared by electro-polishing procedure. The room-temperature (RT) and high-temperature (800 and 850 °C) tensile tests for rolling sheets (along the rolling direction, RD) were conducted using AG-X100kN material testing machine at a speed of 0.5 mm/min.

## 3. Results and discussions

After hot-pack accumulative roll-bonding (HP-ARB) process, the macro morphology of the sheet on normal direction-rolling direction (ND-RD) plane was characterized in Fig. 2a. The alloy matrix are measured as ~703-844 μm, while the bonding interfaces can be clearly observed with size of ~96 μm and 221 μm which are indicated by red dotted lines. Ti-44Al-5Nb-1Mo-(V, B) alloy preforms are combined together without any impurity or defect.

The microstructure and their higher magnification morphologies of the interface and matrix for HP-ARB sheet are shown in Fig. 2b and c, respectively [10]. There is no clear boundaries between the matrix and the interface, except the different contents of γ phase, B2 phase and lamellar colonies. The microstructure of the bonding interface is near-lamellae, while near-duplex microstructure is obtained in the sheet matrix. Coarse colonies, which are ~10-30 μm in size, with fine γ grains (~5-10 μm) at colony boundaries are indicated in Fig. 2b. Moreover, fewer B2 phase can be seen compared with the sheet matrix. Homogeneous microstructure consisting of fine equiaxed colonies, γ and B2 grains with mean size of ~10 μm is shown in Fig. 2c.

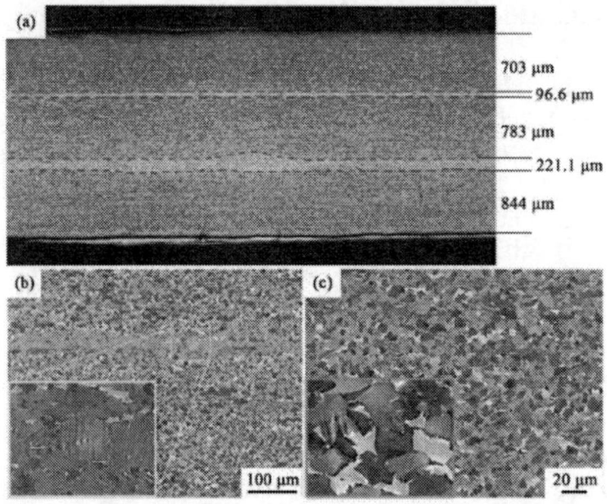


**Figure 2.** SEM/BSE images of the accumulative-rolled Ti-44Al-5Nb-1Mo-(V, B) alloy sheet observed on ND-RD plane: (a) macro morphology of the sheet, microstructure of (b) the bonding interface, and (c) the alloy matrix; The instes in (b) and (c) are the higher magnification morphologies.

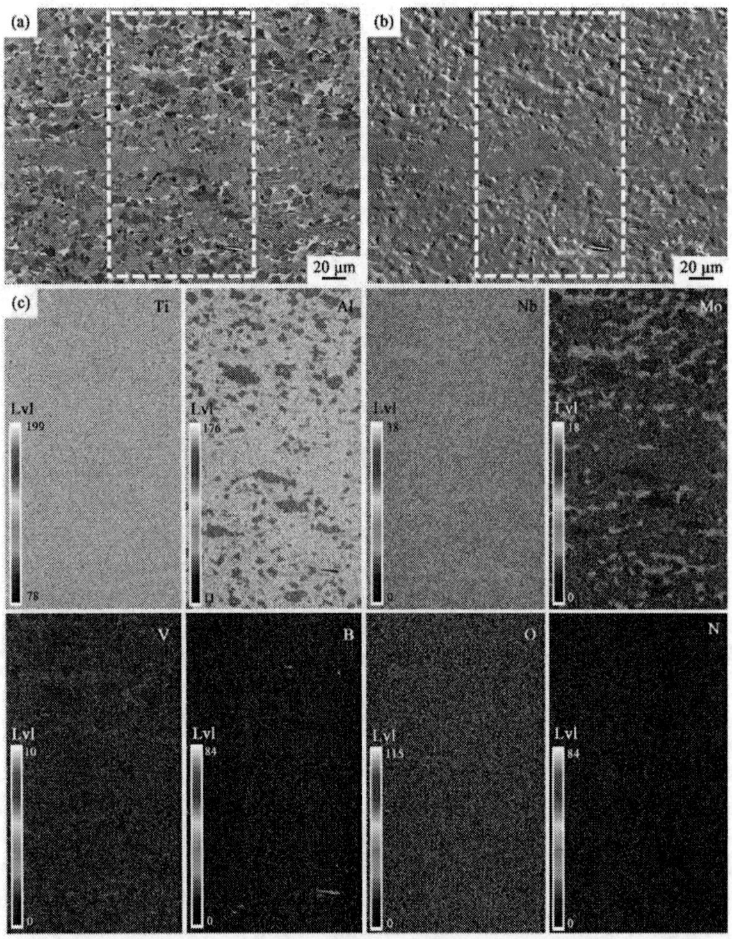


**Figure 3.** EPMA analysis of the accumulative rolling Ti-44Al-5Nb-1Mo-(V, B) alloy sheet interface: (a) EPMA/BSE and (b) EPMA/SE images for the bonding interface; (c) distribution maps for different elements.

To explain the causes of the interface formation, the accumulative-rolled Ti-44Al-5Nb-1Mo-(V, B) alloy sheet was studied by EPMA [11] as shown in Fig. 3. Here, all the relevant alloy elements as well as O, and N which may have effects on the interface bonding results were scanned and discussed.

Fig. 3a and b show the EPMA scanned BSE and SE images, in which the white dotted box was selected for element analysis. It can be clearly seen in Fig. 3b that a band exist at the image center which is identified as bonding interface under EPMA observation. It is distinctly smoother than the surrounding matrix ascribed to its better corrosion resistance during electro-polishing procedure. Obvious element segregation and distribution differences between interface and matrix can be observed in Fig. 3c, especially for Ti, Al and Mo elements. The Al, Nb, Mo, and V elements tended to migrate into the TiAl matrix during the intense straining process, while Ti migrated to the opposite direction. There is nearly no visible element enrich at the interface comparing with the matrix, that is to say higher homogeneities have been obtained. As strong β-stabilizers, blocky enrichments of Mo and V elements are shown in the sheet matrix indicating the high contents of B2. Furthermore, several borides, enriched by Nb and B, are observed at the upper part and bottom in the vision field. The HP-ARB process would reduce the boride particle quantity within bonding interface.

In order to rule out the effects of O and N elements on the formation of these microstructure characteristics, their distributions were also acquired. Completely homogeneous element distribution mapping were shown in Fig. 3c-O, and N. Therefore, the differences of stress state between the alloy matrix and the plate edge, and large deformation strain during hot-working are deemed to be the main influence factors [12].

The room- and high-temperature tensile properties of the HP-ARB Ti-44Al-5Nb-1Mo-(V, B) sheet were tested using samples with a gauge section of 15 × 5 × 2 mm3. The specimens for high-temperature tensile tests were homogenized for 20 min at the target temperatures (800 or 850 °C) to ensure the temperature homogeneity. The following Table 1 shows the tensile property results.

**Table 1.** Tensile properties of the HP-ARB Ti-44Al-5Nb-1Mo-(V, B) sheet.

| | YS / Mpa | UTS / Mpa | δ / % |
|---|---|---|---|
| RT | -- | 862.63 | 1.71 |
| 800 °C | 422.85 | 745.20 | 7.24 |
| 850 °C | 368.82 | 598.99 | 67.50 |

NOTE: RT represents room temperature, YS represents yield strength, UTS represents ultimate tensile strength, and δ represents breaking elongation.

The UTS at RT of the as-ARBed sheet was 862.6 MPa with an elongation of 1.71% which performed well. At high temperatures, the sheet still exhibited excellent properties. When the tensile temperature was 800 oC, the YS, UTS and elongation for the ARB sheet were 422.8 MPa, 745.2 MPa and 7.24%, respectively. As the test temperature increased to 850 oC, the YS and UTS decreased to 368.8 MPa and 599 MPa, while superior performance of ductility of 67.5% exhibited which is more excellent than the majority of β-γ TiAl alloys.

## 4. Conclusion

In this study, hot-pack accumulative roll-bonding (HP-ARB) directly from HIPed Ti-44Al-5Nb-1Mo-(V, B) alloy ingot was progressed. The corresponding microstructure characteristics were discussed. RT, 800 and 850 °C tensile properties were tested. The following are the main conclusions:

Thin crack-free sheet was obtained by HP-ARB process. The alloy matrix were measured as ~703-844 μm, while the defect-free bonding interfaces were measured as ~96 and 221 μm. The preforms were combined well without any impurity or defect.

Near-lamellae and fine near-duplex microstructure were characterized in the bonding interface and the sheet matrix, respectively. Fine and homogeneous microstructure was observed. The differences between the sheet matrix and interface were ascribed to the different stress state between the alloy matrix and the plate edge.

After HP-ARB, the sheet exhibited good performance at room and high temperatures. At RT, the UTS and ductility were 863 Mpa and 1.71%. The 800 and 850 oC UTS/ductility were 754 Mpa/7.24% and 599 Mpa/67.5%, respectively. The pure bonding interface and fine homogeneous microstructure contributed positively to the tensile properties.

**Acknowledgements**
This project was supported by the National Key Research and Development Program of China (2016YFB0301201), National Natural Science Foundation of China (51504060, 51301140) and Fundamental Research Funds for the Central Universities (N160713001).

**References**
[1] T. R. Li, G. H. Liu, M. Xu, et al. Flow Stress Prediction and Hot Deformation Mechanisms in Ti-44Al-5Nb-(Mo, V, B) Alloy, Materials. 11 (2018) 2044-2060.
[2] G.H Liu, X. Z. Li, Y. Q. et al. Microstructure, microsegregation pattern and the formation of B2 phase in directionally solidified Ti–46Al–8Nb alloy, J. Alloy Compd. 541 (2012), 275-282.
[3] Q. Wang, R. R. Chen, X. Gong, et al. Microstructure, Mechanical Properties, and Crack Propagation Behavior in High-Nb TiAl Alloys by Directional Solidification, Metall. Mater. Trans. A. 46 (2018), 4555-4564.
[4] M. Xu, G. H. Liu, T. R. Li, et al. Microstructure characteristics of Ti–43Al alloy during twin-roll strip casting and heat treatment, Trans. Nonferrous Met. Soc. China. 29 (2019), 1017-1025.
[5] N. Tsuji, R. Ueji, and Y. Minamino. Nanoscale crystallographic analysis of ultrafine grained IF steel fabricated by ARB process, Scr. Mater. 47 (2002), 69-76.
[6] N. Tsuji, Y. Ito, Y. Saito, et al. Strength and ductility of ultrafine grained aluminum and iron produced by ARB and annealing, Scr. Mater. 47 (2002),893-899.
[7] G.H Liu, T. R. Li, M. Xu, et al. Microstructural Evolution and Mechanical Properties of TC4 Titanium Alloy During Acculative Roll Bonding Process, Acta Metall. 53 (2017), 1038-1046.
[8] S. Pasebani, and M. R. Toroghinejad. Nano-grained 70/30 brass strip produced by accumulative roll-bonding (ARB) process, Mater. Sci. Eng., A. 527 (2010), 491-497.
[9] Y. Saito, H. Utsunomiya, N. Tsuji, et al. Novel ultra-high straining process for bulk materials—development of the accumulative roll-bonding (ARB) process, Acta Mater. 47 (1998), 579-583.
[10] T. R. Li, G. H. Liu, M. Xu, et al. High temperature deformation and control of homogeneous microstructure during hot pack rolling of Ti-44Al-5Nb-(Mo, V, B) alloys: The impact on mechanical properties, Mater. Sci. Eng., A. 751 (2019), 1-9.
[11] Y. J. Li, J. Kang, W. N. Zhang, et al. A novel phase transition behavior during dynamic partitioning and analysis of retained austenite in quenched and partitioned steels, Mater. Sci. Eng., A. 710 (2018), 181-191.
[12] S. H. Lee, Y. Saito, N. Tsuji, et al. Role of shear strain in ultragrain refinement by accumulative roll-bonding (ARB) process, Scr. Mater. 46 (2002), 281-285.

# The Static Characteristic Analysis on Key Components of Boring-milling Machining Center

**Lin Tao[1, 2], Yuan Zhao[1, 2], Yue Zhao[1, 2], Lili Zhang[1, 2]**

[1] Liaoning Key Laboratory of Information Physics Fusion and Intelligent Manufacturing for CNC Machine, Fushun Liaoning 113122 China
[2] Shenyang Institute of Technology College of Mechanical and Vehicle Engineering, Fushun Liaoning 113122 China

**Abstract.** Taking TH5650X50 Boring-milling Machining Center as the research object, analyze the static characteristics of the two key components of the milling system. Use finite element software to create 3D model of the two key components of the milling system and design the whole virtual assembly of milling system. Based on the theory of cutting deformation, consider the application of its forces and non-forces in the way of a remote payload and remote quality. The finite element model of spindle box-principle axis and crossbeam are obtained by circular grid division method, and all the parameters of the finite element model are solved by FFEPlus solver in Solidworks Simulation module. The study verifies the accuracy of the finite element mode, and obtains the deformation of the two key components under the most adverse conditions.

**Key words:** Boring-milling machining center, Key components, Static structural analysis, Finite element analysis, Solidworks.

## 1. Introduction

As the continuous improvement of machining accuracy of machine tools, the technology development of boring-milling machining center has gradually obtained the characteristics of intelligence, high speed, precision, modularization and complexity [1-2]. The static and dynamic characteristics of machine tools are important factors affecting the machining accuracy and stability of the equipment. Therefore, it is problem demanding prompt solution to analyze the static and dynamic characteristics and explore measures to improve them[3-5].

In this paper, static characteristic analysis was carried out for spindle box and crossbeam, two key components of boring-milling center milling system, using Solidworks software. A series of useful conclusions were obtained through the establishment of finite element model and the analysis of the static comprehensive displacement, which laid a foundation for the of the whole structure's dynamic characteristic analysis and the whole machine's optimization design and electromechanical simulation.

## 2. Whole structure

As shown in Fig. 1, TH5650X50 boring-milling machining center includes two spindles of boring spindle and milling spindle, with the structural characteristics of both boring machines and milling machines. For the processing of large complex box parts, the boring, milling, drilling and other

Content from this work may be used under the terms of the Creative Commons Attribution 3.0 licence. Any further distribution of this work must maintain attribution to the author(s) and the title of the work, journal citation and DOI.
Published under licence by IOP Publishing Ltd

processing procedures on multiple surfaces of the parts can be completed after one clamping, which effectively avoids the problem of accuracy decline caused by multiple clamping.

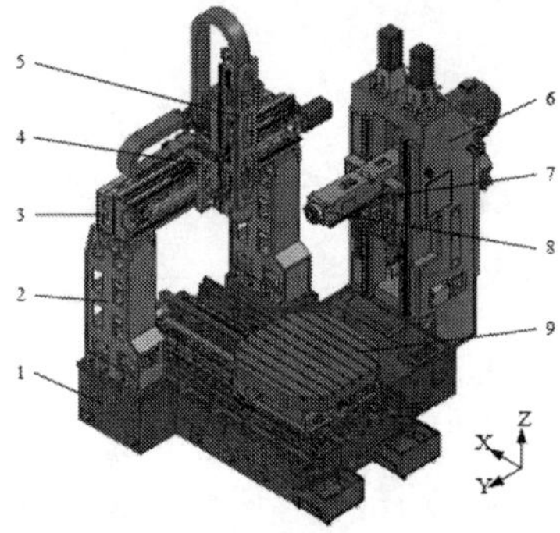


1. Bed 2. Milling pillar 3. Crossbeam 4. Milling slide table 5. Spindle box 6. Boring pillar 7. Boring slide table 8. Ram 9. Workbench

**Fig. 1** TH5650X50 boring-milling center model

## 3. FEA parametric modeling

### *3.1. Definition of material properties*

For TH5650X50 boring-milling machining center, spindle box-principle axis is made of gray cast iron HT250 and crossbeam is made of gray cast iron HT300. The characteristic parameters of each material are shown in table 1.

**Table 1.** Characteristic parameters of each material

| Materials | Elasticity modulus(GPa) | Density(kg/m$^3$) | Poisson ratio |
|---|---|---|---|
| HT250 | 135 | 7350 | 0.23 |
| HT300 | 157 | 7800 | 0.28 |

### *3.2. Load and constraint*

(1) Determination of cutting force

From the perspective of mechanical manufacturing process, the boring process of cemented carbide tools can be divided into two main processes of rough machining and finish machining[6]. By analyzing the values of various parameters under rough milling and finish milling conditions, it can be seen that the machining accuracy of boring is negatively correlated with its power, and positively correlated with its boring speed[7-8]. Therefore, the total milling force should decrease with the increase of machining accuracy. In order to ensure the accuracy of the analysis results, the cutting force under rough milling condition is selected as the nonpotential force affected by the components in the static simulation analysis of key components.

(2) Application of loads and constraints

After analysis, when spindle box was in the lower limit position, its displacement reached the maximum. At this time, the six surfaces in contact with the guide and the slide table should be fully fixed; the nut of the driving screw on the spindle box was fully fixed on the slide table. In addition, the gravity of the spindle box was balanced by the hydraulic cylinder fixed on the slide table, so the influence of its gravity could be ignored in the analysis. The effect of the cutting force on the spindle box was applied to the lower end surface of the spindle box and to the two ring surfaces that are in contact with the motorized spindle through a remote load.

In the analysis of the crossbeam, because the four faces in the negative direction of the Z axis of the crossbeam were all fixed on the two stand columns with bolts during installation, the four faces in the negative direction of the Z axis of the crossbeam were subject to the fully fixed constraint. On behalf of the force of gravity on it, the gravitational field applied to the entire crossbeam was applied to the six contact surfaces between the slide table and the guide rail on the crossbeam through a remote load; finally, the milling force received by the milling cutter was also applied to the six contact surfaces between the slide table and the rail through a remote load.

### *3.3. Establishment of finite element model*

Finite element grid division was carried out for spindle box and crossbeam in Solidworks Simulation module [9-10]. A more reasonable grid accuracy was obtained using the cyclic grid division method. In other words, the element size was set in advance according to the size of finite element model. After solving a set of results, the element size was reduced to half of the previous size for calculation; if the difference between the two calculated results was less than 5%, the previous element size was selected; if the difference between two results was greater than 5%, the element size was reduced continuously until a satisfactory result was obtained. Finite element model of spindle box and crossbeam is shown in Fig. 2-3.

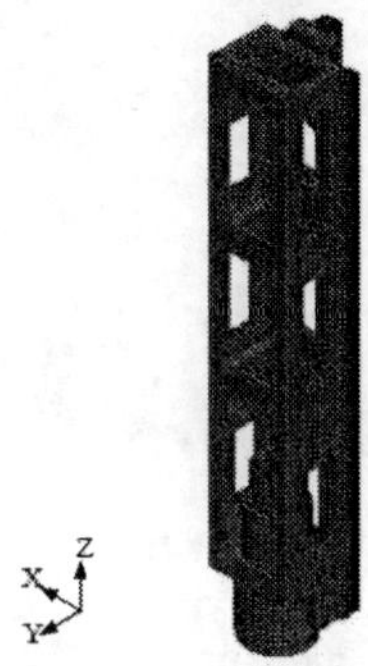


**Fig. 2** Finite element model of spindle box

**Fig. 3** Finite element model of crossbeam

## 4. Static characteristics analysis of key components

The finite element models of spindle box-principle axis and crossbeam were solved using FFEPlus solver in Solidworks Simulation module to obtain their comprehensive deformation and their displacement and deformation along the X, Y and Z directions, as shown in Fig. 4-5.

According to the analysis of Fig. 4, under the influence of cutting force, the maximum displacement of the spindle appears at the lower end of the spindle box, with a value of $1.189\times10^{-2}$mm; the maximum displacement of the spindle box in the X-axis direction appears at the lower end of the spindle box, with a value of $9.462\times10^{-3}$ mm; the maximum displacement of the spindle box in the Y-axis direction appears at the lower end of the spindle box, with a value of $7.869\times10^{-3}$mm; the maximum displacement of the spindle box in the z-axis direction appears at the lower end of the spindle box, with a value of $1.986\times10^{-3}$mm.

According to the analysis of Fig. 5, under the influence of the total force, the maximum displacement of the crossbeam appears in the middle of the crossbeam, with a value of $1.853\times10^{-2}$mm; the maximum displacement of the crossbeam in the X-axis direction appears in the middle of the crossbeam, with a value of $1.180\times10^{-2}$mm; the maximum displacement of the crossbeam in the Y-axis direction appears in the middle section between the middle and end of the crossbeam, with a value of $2.846\times10^{-3}$mm; the maximum displacement of the crossbeam in the z-axis direction appears in the middle of the crossbeam, with a value of $1.612\times10^{-2}$mm.

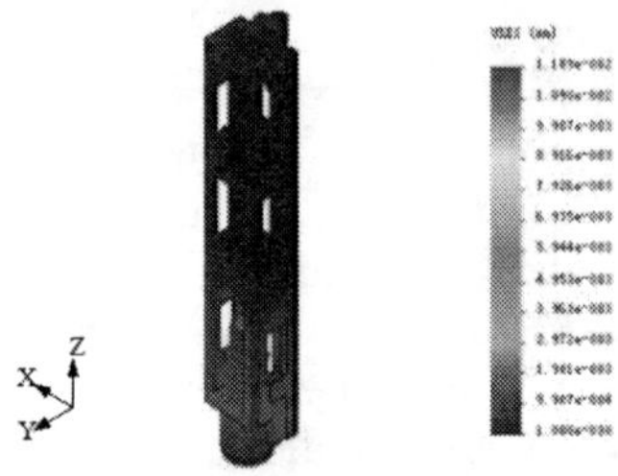

(a) Total displacement

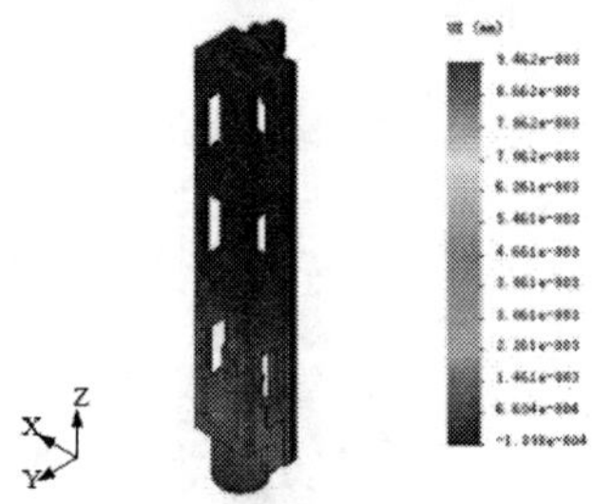

(b) X direction

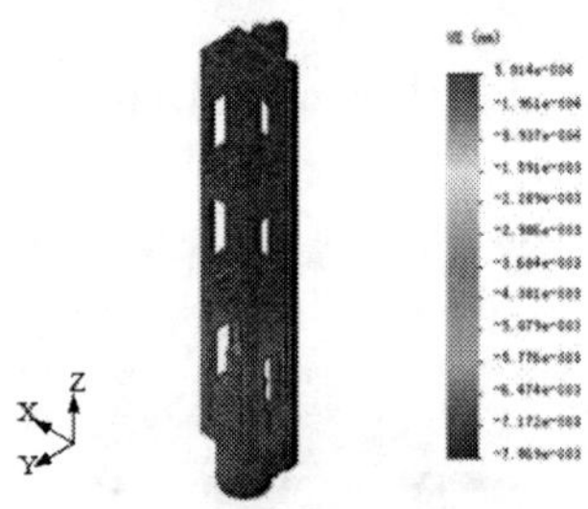

(c) Y direction

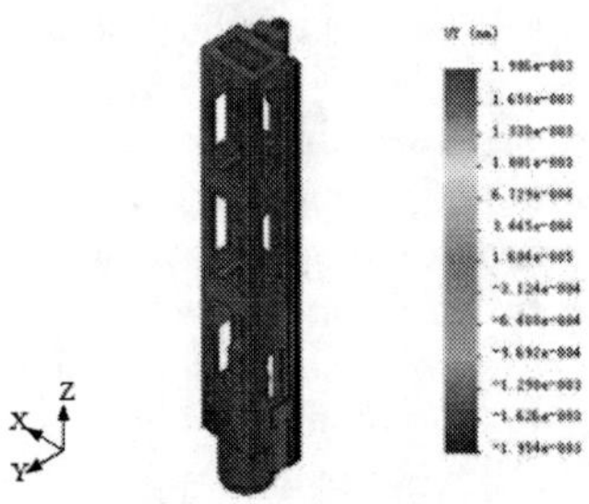

(d) Z direction

**Fig. 4** Deformation nephogram of spindle box in each direction

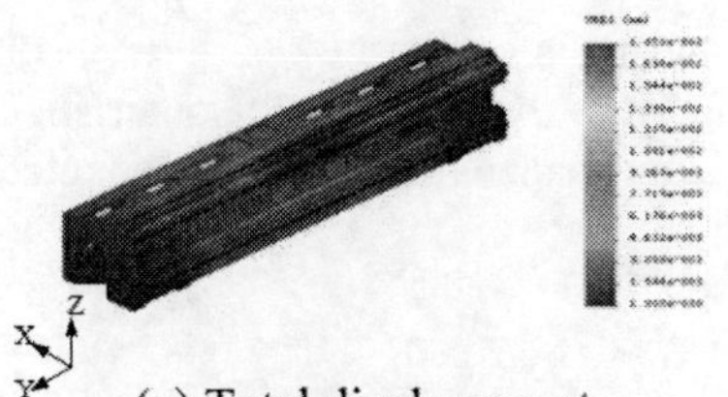

(a) Total displacement

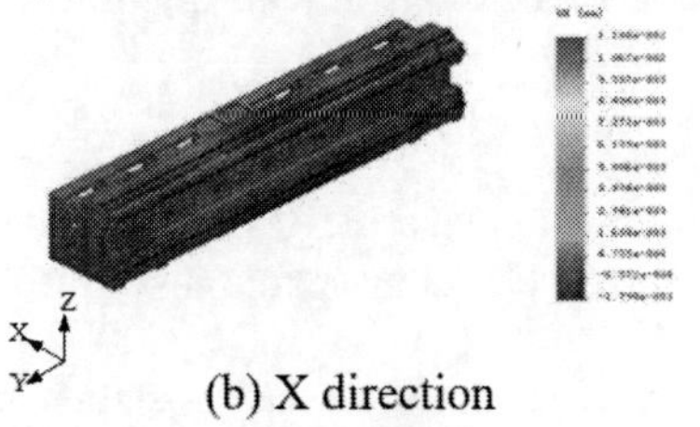

(b) X direction

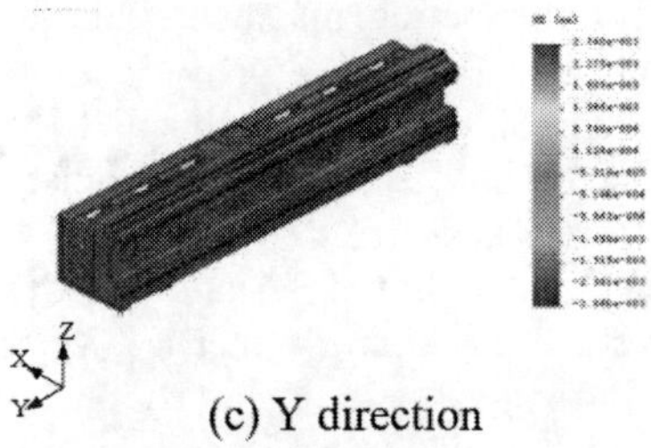

(c) Y direction

(d) Z direction

**Fig. 5** Deformation nephogram of crossbeam in each direction

## 5. Conclusion

The spindle box-principle axis and crossbeam of TH5650X50 boring-milling center milling system were taken as the research object, and the application of the force and non-force load of the two key components was considered in the way of remote load and remote mass, and the static and static finite element analysis were conducted on them in turn. Finally, finite element analysis data of spindle box-principle axis and crossbeam, and the following main conclusions were obtained:

(1) The finite element model and various parameters of spindle box-principle axis and crossbeam were obtained by the cyclic grid division method, and the above finite element model was calculated using FFEPlus solver in Solidworks Simulation module, so as to obtain its static comprehensive displacement. The remote load method, remote mass method and cyclic grid method in the simulation process can provide feasible approaches for the analysis of similar problems.

(2) Whether the design of various component meets its accuracy requirements was verified using the static comprehensive displacement results. The results show that the comprehensive displacement deformation of milling crossbeam is the largest. Under the influence of gravity and cutting force, the maximum displacement of the crossbeam appears in the middle of the crossbeam, with a value of $1.853\times10^{-2}$mm. The comprehensive displacement does not exceed $2.0\times10^{-2}$mm, meeting the requirement of TH5650X50 boring-milling center that the static displacement error of various key part should not exceed 20 μm.

(3) By performing the statics analysis on the key parts of TH5650X50 boring-milling machining center, the research lays a foundation for the dynamic characteristic analysis and rigidity analysis of the milling system, which is of great significance for the subsequent electromechanical joint simulation and modeling of the whole machine.

**Acknowledgments**
This work was financially supported by NFSC (61603262, 61403071), Liaoning Natural Science Foundation (20180550418), Foundation of i5 Institute of Intelligent Manufacturing, Shenyang Institute of Technology (i5201702).

**References**

[1] Gao Feng, Guo Weizhong, Song Qingyu et al. Research and development of heavy manufacturing equipment at home and abroad [J]. Chinese journal of mechanical engineering, 2010 (19): 92-107.
[2] Li Tao, Kong Lulu, Zhang Hongchao. Research status and development trend of energy consumption model of typical cutting machine tools [J]. Chinese journal of mechanical engineering, 2014,07: 102-111.
[3] Ji Min, Zhang Jinsheng, Wang Shuda et al. Static and dynamic characteristic analysis and optimization of open machining equipment combination machine tool and automatic processing technology, 2017(09): 62-65.
[4] Gao Dongqiang, Chen Chaoqun, Ma Jinfeng. Static and dynamic characteristics analysis and optimization of bed in vertical machining center [J]. Mechanical design and manufacturing, 2013(12): 221-227.
[5] Liu Aoxiang, Wang Jun, Kang Junxian et al. Dynamic characteristics analysis of plano milling system of TX1600G boring-milling center[J]. Integrated machine tool and automatic machining technology, 2014, (05): 50-55.
[6] Wang Xiankui. Milling and sawing of machining process manual (single edition) [M]. Beijing: China Machine Press, 2008.
[7] Zhang Yifang. Metal cutting manual (fourth edition) [M]. Shanghai: Shanghai science and technology press, 2011.
[8] Zhu Lida, Yu Tianbiao, Wang Wanshan. Simulation analysis of cutting force of orthogonal milling machining [J]. Journal of China Ordnance, 2012, 33(4): 419-424.
[9] Zhang Lei, Xie Zhikun, Li Yan et al. Dynamic and static optimization design of high-speed plano five-axis machining center based on finite element [J]. Machinery design and manufacturing, 2011(12): 36-38.
[10] He Ning, Zhang Dongsheng,Tong Guorong et al. Static characteristics analysis of the crossbeam of plano machine tool based on finite element method [J]. Journal of shaanxi institute of technology (natural science edition), 2014, 30(3): 1-4.

# Size Effect of Ultra Pure Ti Thin Strip under Asymmetrical Rolling

**Meng Song[1, 2, *], Libo Yin[1, 2], Fengmei Bai[1, 2], Zhenyi Huang[1, 2], and Xianghua Liu[3]**

[1]Key Laboratory of Metallurgical Emission Reduction & Resources Recycling (Anhui University of Technology), Ministry of Education, Maanshan 243032, China.
[2]School of Metallurgical Engineering, Anhui University of Technology, Maanshan 243032, China.
[3]State Key Laboratory of Rolling and Automation, Northeastern University, Shenyang 110819, China.

*Corresponding author e-mail: songmeng@ahut.edu.cn

**Abstract.** In order to study the size effect on properties of ultra pure Ti tensile tests were performed with ultra pure Ti foil (thickness range from 30μm to 280μm) by asymmetrical rolling. A pronounced size effect was observed at thickness of 34μm. The results were shown that the ultimate strength were increased as sample thickness down from 280μm to 34μm, however, were decreased as the thickness from 34μm to 30μm with ultrahigh strain. These features seemed to be link to the increasing of fraction of surface grain to volume, which led to the less districting on dislocation slip.

## 1. Introduction

In recent years, many researchers interested in the mechanical properties of materials in micro scale, as significantly increasing demand of miniaturization production and multi-functional integration. To fabricate micro scaled parts efficiently and accurately, the mechanical behavior in micro scaled deformation is one of the most significant issues to be explored and in-depth understanding of those behaviors is of great necessity to be established [1].

When the components are in the micrometer or below, the plastics deformation behaviors and the properties are no longer the same as those of bulk materials, which known as the "size effect". Many studies have focused on the effect of different size parameters, such as specimen size, grain size and surface topography, on the mechanical properties including ultimate strength and yield strength [2-4]. It has been found that, when the ratio of the thickness to grain size is below 5, the yield strength of Ti increases with the decrease in thickness due to the large volume fraction of oxides layer on the surface [5].

Although the size dependent mechanical behavior has been reported in many materials, the size effect on the deformation of ultra pure Ti foils, which in asymmetrical rolling process, have been paid little attention. Therefore, the present work studies the size effects on the deformation bahaviors of ultra pure Ti with thickness ranging from 30 μm to 280 μm via tensile testing at room temperature.

Content from this work may be used under the terms of the Creative Commons Attribution 3.0 licence. Any further distribution of this work must maintain attribution to the author(s) and the title of the work, journal citation and DOI.
Published under licence by IOP Publishing Ltd

IOP Conf. Series: Materials Science and Engineering **772** (2020) 012087 doi:10.1088/1757-899X/772/1/012087

## 2. Materials and methods

The study was carried out on ultra pure Titanium in the form of plate with dimensions of 11.2×40×60 mm (thickness×width×length), which cut from round cakes with diameter of 385 mm, that was shown in Fig.1. The chemical composition (wt%) of ultra pure Titanium utilized is 99.999% Ti.

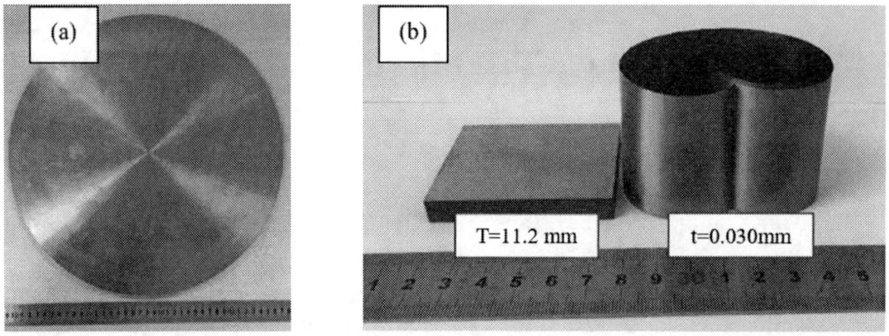


**Figure 1.** The industrial pure copper specimens (a) before and (b) after rolling

The material in the initial state was subjected to a main cold rolling to a final thickness 30μm obtained in the two stages without intermediate annealing at room temperature. At first, it was rolled to 1500μm on a laboratory two high mill with the work roll diameter of 180 mm, then, rolled to final thickness 30μm by Micro Metal forming Mill (3M mill).

The 3M mill was a new kind of rolling mill with ultra-thin rolling capacity developed by our research group. It was shown in Fig.2. The mainly parameters and characters were as follows: backup roll diameter 120 mm, work roll diameter 50 mm, roll barrel length 130 mm, maximum roll force 200kN. The speed ratio between up and down rolls can be changed from 1 to 1.3 continuously.

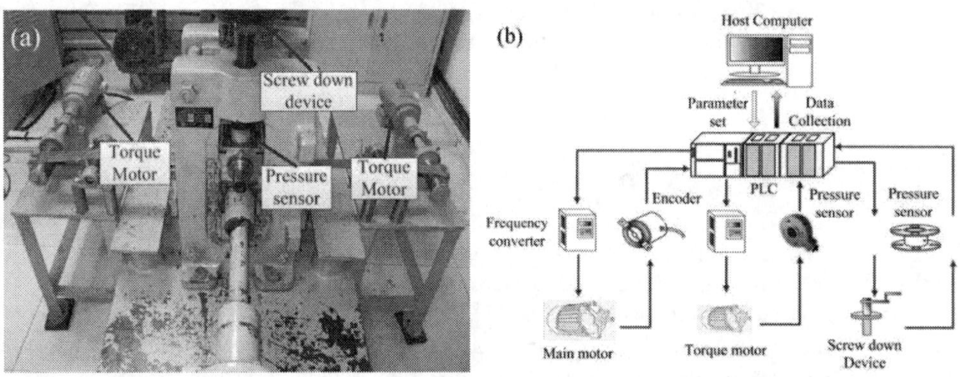


**Figure 2.** Illustration of (a) 3M-Iasymmetric mill and (b) the control system.

## 3. Results and discussions

True stress-strain curves of samples with thickness from 30μm to 280μm were depicted in Fig.3. When the ultra pure Ti sample was deformed by micro rolling, dislocations were generated, move and stored; the storage cause the materials to work harden.

The stress strain curves almost covered each other. However, a pronounced size effect was found when thinner than 34μm: the ultimate strength decreased as thickness decrease from 34μm to 30μm with further deformation, that is, the thinner foils had lower ultimate tensile strength. As the same time, the elongation decreased in the asymmetrical rolling process.

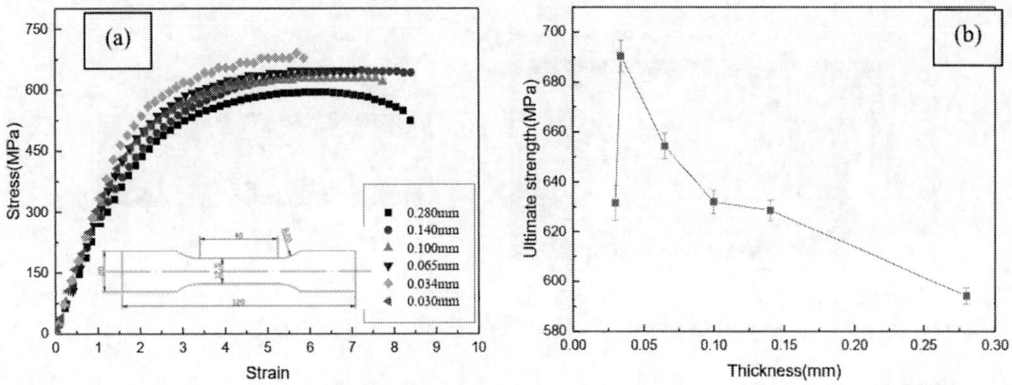


**Figure 3.** (a)Tensile true stress-strain curves; and (b) ultimate strength vs. thickness for ultra pure Ti of thickness from 30μm to 280μm

The elastic deformation was affected by the size effect, that is, a surface grain ration change in deformed metal was found when thickness of ultra pure Ti samples down from 280μm to 30μm by asymmetrical rolling. Due to the micro rolling, the geometry scale of ultra pure Ti was spanned from macro to micro, in which it was rolled from 280μm to 30μm without intermediate annealing at room temperature. The notable character of micro rolling was the negative roll gap, that is, the elastic contact deformation between work roll, which outside the loaded roll gap, was remained. Fig.4 was shown the difference of normal gap rolling and negative gap rolling. The different rolling process, the friction was increased as the thickness varying from macro to micro, which in agreement with literature [6], that could lead to the effective grain refinement.

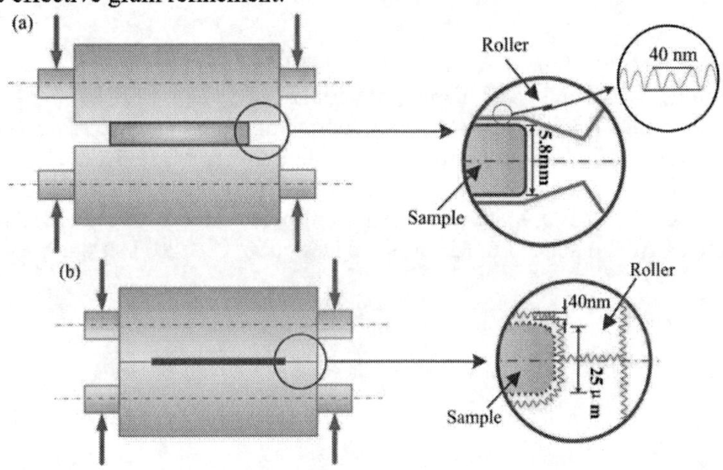


**Figure 4.** Difference of (a) normal gap rolling and (b) negative gap rolling

As asymmetrical rolling process, the surface grains significantly increased with thickness decreasing. The constraints along the grain boundaries of ultra pure Ti on the surface grains are relaxed. As result, the size effect was found. A remarkable work softening from 34μm to 30μm was displayed clearly in Fig.5, which are rarely observed in other heavily cold worked materials [1].

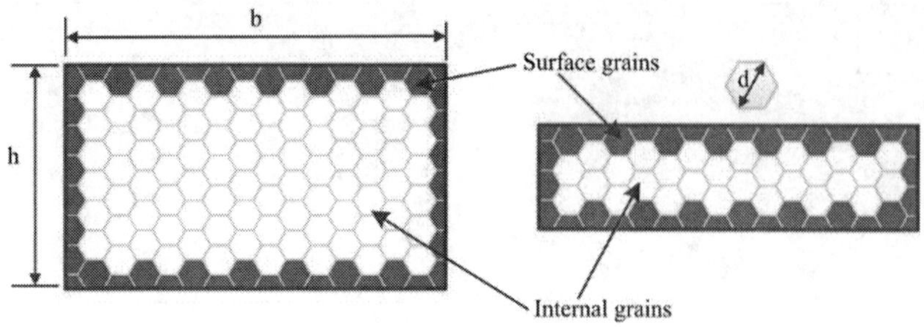


**Figure 5.** Surface grains and internal grains in sample

## 4. Conclusion

The size effect on mechanical properties in ultra pure Ti foil during compound deformation, was systematically investigated in the present work. The following conclusions can be drawn. The uniaxial tensile tests were carried out with different thickness of 280μm, 140μm, 100μm, 65μm, 34μm and 30μm. The ultimate strength of ultra pure Ti foil at first increased from 280μm to 34μm, however, decreased from 34μm to 30μm. Decreasing thickness increased the surface ratio, which an important factor in constitutive model considered size dependent. High surface ration means more surface grain in sample, which less strict in dislocation slip, led to ultimate strength decrease.

## Acknowledgments

The authors would like to thank the Key project of the Natural Science Foundation of the Anhui Higher Education Institutions of China (Grant No. KJ2019A0059) and National Natural Science Foundation of China (Grant No. 51374069) for their financial supports for the current study.

## References

[1] M. Song, X. Liu and L. Liu, Materials 10 (5), (2017)538 .
[2] S. V. Sajadifar, C. Atli and G. G. Yapici, Materials Letters 244, (2019)100-103 ..
[3] B. Meng, W. Wang, Y. Zhang and M. Wan, Journal of Materials Processing Technology (2019).
[4] M. Song, X. Liu and L. Liu, Joural of central south university 24 (11), (2017)2783-2792 .
[5] S. Wang, L. Niu, C. Chen, Y. Pang, B. Liao, Z. H. Zhong, P.Lu, P.Li, X. D. Wu, J. W. Coenen, L. F. Cao and Y. C. Wu, Materials Science & Engineering A 730, (2018).244-261
[6] W. L. Chan, M. W. Fu and J. Lu, Materials & Design 32 (1), (2011)198-206 .

# Effect of Deformation on Precipitation Behaviors of Carbide in Ti-Nb-Mo Steel

**Yong Yang[1, 2, *], Kang An[2], Xiaofeng Zhang[2], Zhenyi Huang[2] and Yanmei Li[3]**

[1]Anhui Province Key Laboratory of Metallurgical Engineering & Resources Recycling, Anhui University of Technology, Maanshan, China
[2]School of Metallurgical Engineering, Anhui University of Technology, Maanshan, China
[3]State Key Laboratory of Rolling and Automation, Northeastern University, Shenyang, China

*Corresponding author e-mail: yangyongboxian@163.com

**Abstract.** The precipitation behaviors of (Ti, Nb, Mo)C in specimens of Ti-Nb-Mo steel hold at 700 °C under no deformation condition and deformation condition with engineering strain 0.4 imposed in austenite region have been studied. The results indicate: under no deformation condition, the rows of interphase precipitation show straight and/or curve morphology, and the precipitation behaviors include interphase precipitation, random precipitation and fibrous precipitation; under deformation condition, the rows of interphase precipitation are straight, and the precipitation behaviors include interphase precipitation and strain-induced precipitation. Furthermore, the dispersion of precipitate particles formed under deformation condition is more uniform than those formed under no deformation condition.

## 1. Introduction

Based on the characteristics of moderate strength, good weldability and economic applicability, high strength low alloy (HSLA) steels are widely used in automobiles, bridges and other fields [1]. Alloying elements can play important roles in the controlling of microstructures and properties of steels. Since the inception of 'Nanohiten' steel [2], HSLA steels, such as Ti-alloyed steels, Nb-alloyed steels, Ti-Nb-alloyed steels, Ti-V-alloyed steels, in which nano-scale precipitates, in forms of interphase precipitation and/or random precipitation, can provide 200 ~ 400 MPa strength increment to yield strength, have been widely investigated for more than ten years [3]. For example, the effects of chemical composition of steels[4], cooling rate[5], final cooling temperature, holding temperature and holding time[6] on precipitation behaviors of nano-scale particles have been widely studied. Based on the reported results, it can be concluded that increasing the addition amounts of alloyed elements could attribute to obtain interphase precipitation; decreasing the final cooling temperature can increase the possibility of getting supersaturated precipitation; addition of Mo could effectively reduce the growth rate and coarsening rate of precipitation particles[7]; using continuous cooling method is favourable to get coexisting of interphase precipitation and supersaturated precipitation. Though the effects of different factors on precipitation behaviors have been studied, there is little attention focused on the effect of deformation on precipitation behaviors of (Ti,Nb,Mo)C for Ti-Nb-Mo steel. Therefore, the precipitation behaviors

Content from this work may be used under the terms of the Creative Commons Attribution 3.0 licence. Any further distribution of this work must maintain attribution to the author(s) and the title of the work, journal citation and DOI.
Published under licence by IOP Publishing Ltd

of (Ti,Nb,Mo)C under conditions of deformation and no deformation were investigated in the present work.

## 2. Material and experiments

The experimental steel was melted using a vacuum induction furnace and cast as an ingot with 50 kg and subsequently forged into 80 mm × 80 mm billet. The chemical composition in weight percent of the steel is Fe-0.085C-1.8Mn-0.2Si-0.1Ti-0.1Nb-0.2Mo. A 80 mm × 80 mm × 100 mm block was cut from the billet and then homogenized at 1200 °C for 1.5 h and subsequently hot-rolled into steel plate of 15 mm thickness via six passes on the Φ450 mm two-roller reversible rolling mill.

The cylindrical specimens of diameter 5 mm and length 10 mm for experiments were machined from the center of the billet. The hot compression tests were performed on DIL805A/D and the detail experimental technologies can be seen in Fig.1. Hereafter, the specimens subjected to strains of 0 and 4mm were referred as specimen A and specimen B, respectively. The microstructures of specimens were examined by transmission electron microscopy (TEM – TECNAI-G2). Thin foils for TEM were prepared using twin-jet electrolytic thinning. The electrolyte was composed of 10% (volume fraction) perchloric acid and 90% alcohol, maintained at -25 °C, 30 V, and current of 45 mA.

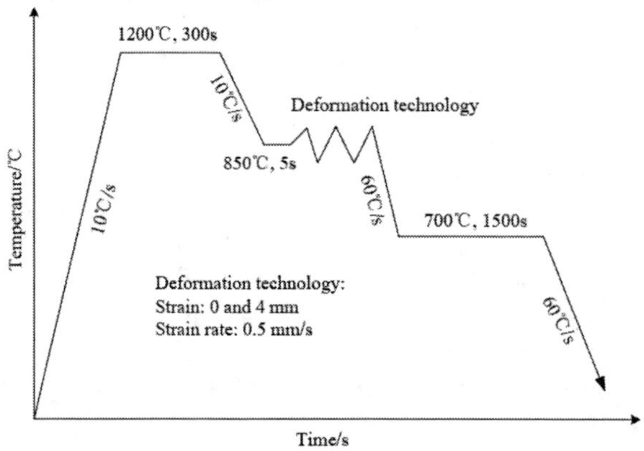


**Figure 1.** Schematic representation of experimental technology.

## 3. Results and discussions

### *3.1. Precipitation Behaviors of (Ti,Nb,Mo)C under the Condition of no Deformation*

The characteristics of interphase precipitation formed in specimen A with no deformation were presented in Fig. 2. From Fig. 2(a), it can be seen that precipitates are aligned in row regularly and this phenomenon of precipitation is referred to as interphase precipitation [8], which can be formed by ledge mechanism[9] or bowing mechanism[10]. Fig. 2(b) is the dark-field image of interphase precipitation shown in Fig. 2(a). It can be seen that the rows of interphase precipitates are not straight entirely, i.e. the rows in local areas are curved, which indicates that the interphase precipitates presented in this work are formed through both ledge mechanism and bowing mechanism of ferrite phase transformation. Bowing mechanism indicates that the driving force for ferrite phase transformation at 700 °C under no deformation condition is not sufficient enough, because the phase transformation of ferrite is keen to adopt ledge mechanism if phase transformation driving force is large enough. The sheet spacing of interphase precipitation was measured and exhibited in Fig. 2(c), from which it can be seen that sheet spacing of interphase precipitation is mainly distributed in the range of 30 nm to 38 nm, and the mean value of sheet spacing is 33.6 nm.

In addition to interphase precipitation, there are also two kinds of precipitation behaviors in specimen A, i.e. random precipitation and fibrous precipitation [11], which are shown in Fig. 3(a) and Fig. 3(b), respectively. Random precipitates, which produced in supersaturated matrix, showed aggregation behavior in red circles in figure 3(a) and no uniform dispersion behavior in the red circle in Fig. 3(b). Fibrous precipitates, which usually appears in V-alloyed steels [11], were observed in Ti-Nb-Mo steel and showed in the red square in Fig. 3(b).

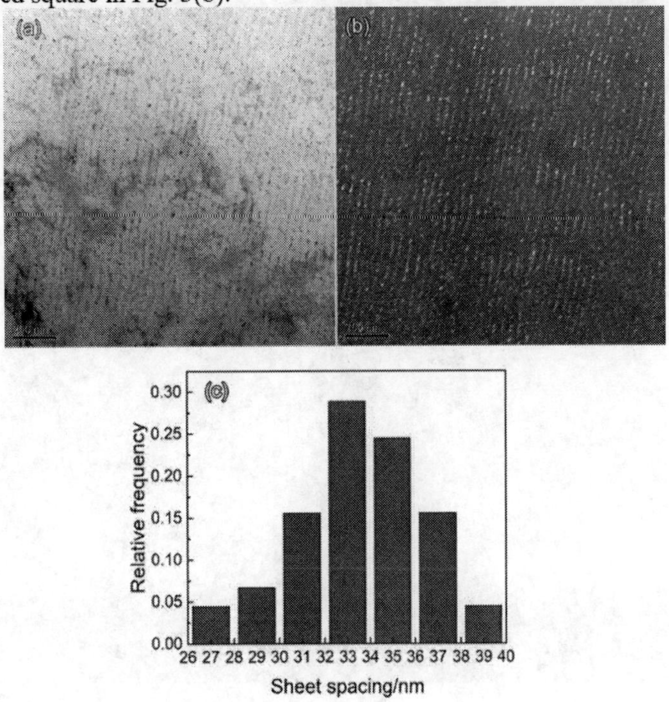


**Figure 2.** Interphase precipitation behaviors of (Ti, Nb, Mo) C at 700 °C under no deformation condition: (a) bright-field image of interphase precipitation, (b) corresponding dark-field image of (a) and (c) frequency distribution of sheet spacing of interphase precipitation.

### *3.2. Precipitation Behaviors of (Ti,Nb,Mo)C under Deformation Condition*

The precipitation behaviors of (Ti,Nb,Mo)C in specimen B subjected to compression deformation of 4 mm were shown in Fig. 4, and Fig. 4(a) and 4(b) presented the bright-field image and dark-field image of interphase precipitation, respectively. It can be seen that the rows of interphase precipitation are perfectly straight, which indicates that the interphase precipitates presented in this work are formed through ledge mechanism, which indicates that the driving force for ferrite phase transformation is large enough, i.e. deformation increases the driving force. The frequency distribution of sheet spacing of interphase precipitation is shown in Fig. 4(c), from which it can be seen that the sheet spacing is mainly located in the range of 22 nm to 26 nm, which is smaller than that showed in Fig. 2(c) for interphase precipitation formed under no deformation condition, and the mean value of sheet spacing is 25.3 nm. This phenomenon also indicates that the driving force for ferrite phase transformation is higher under deformation condition than that under no deformation. This deduction is supported by Bhadeshia's theory [12]. Apart from interphase precipitation, strain-induced precipitation [13] also observed in specimen B, which is shown in the ellipse in Fig. 4(d). Compared with the precipitation behaviors of carbide particles in specimen A, precipitates in specimen B disperse more uniformly.

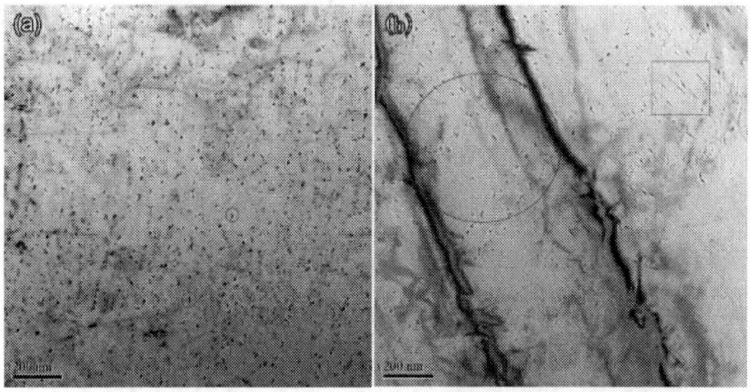


**Figure 3.** Precipitation behaviors of random precipitation (a) and fibrous precipitation (b) of (Ti, Nb, Mo) C at 700 °C under no deformation condition.

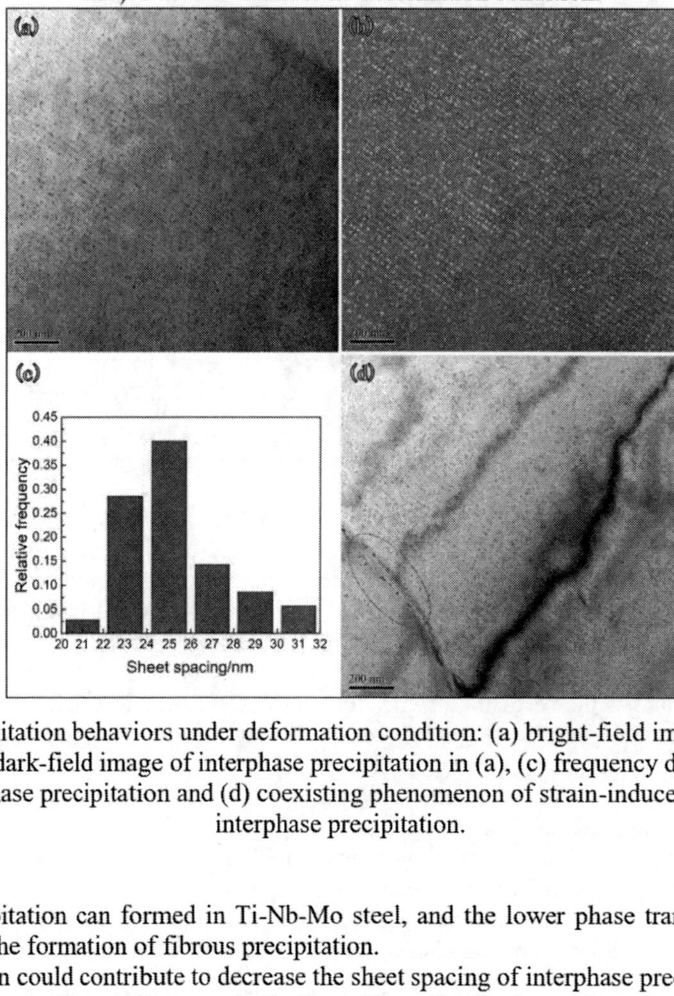


**Figure 4.** Precipitation behaviors under deformation condition: (a) bright-field image of interphase precipitation, (b) dark-field image of interphase precipitation in (a), (c) frequency distribution of sheet spacing of interphase precipitation and (d) coexisting phenomenon of strain-induced precipitation and interphase precipitation.

## 4. Conclusion

(1) Fibrous precipitation can formed in Ti-Nb-Mo steel, and the lower phase transformation rate of ferrite, the easier the formation of fibrous precipitation.

(2) Deformation could contribute to decrease the sheet spacing of interphase precipitation.

(3) Deformation could contribute to obtain the uniform distribution of interphase precipitation and obtain strain-induced precipitation.

**Acknowledgments**

This work is financially supported by National Key R&D Program of China from the Ministry of Science and Technology of China under grant No. 2017YFB0305300 and the Opening Foundation of Anhui

Province Key Laboratory of Metallurgical Engineering & Resources Recycling under grant No. SKF19-02.

**References**

[1] J. Zhang, F.M. Wang, Z.B. Yang, C.R. Li. Metall. Mater. Trans. A 47, 6621-6631 (2016).
[2] Y. Funakawa, T. Shiozaki, K. Tomita, T. Yamamoto, E. Maeda. ISIJ Int. 44, 1945-1951 (2004).
[3] I.B. Timokhina, P.D. Hodgson, S.P. Ringer, R.K. Zheng, E.V. Pereloma. Scripta Mater. 56, 601-604 (2007).
[4] S.P. Tsai, T.C. Su, J.R. Yang, C.Y. Chen, Y.T. Wang, C.Y. Huang. Mater. Design 119, 319-325 (2017).
[5] C.Y. Chen, H.W. Yen, F.H. Kao, W.C. Li, C.Y. Huang, J.R. Yang, S.H. Wang. Mater. Sci. Eng. A 499, 162-166 (2009).
[6] C.C. Yuan, C.S. Fan, C.C. Chon, Y.J. Ren. Mater. Sci. Eng. A 634, 123-133 (2015).
[7] Z. Wang, H. Zhang, C. Guo, Z. Leng, Z. Yang, X. Sun, C. Yao, Z. Zhang, F. Jiang. Mater. Design 109, 361-366 (2016).
[8] R. Okamoto, A. Borgenstam, J. Ågren. Acta Mater. 58, 4783-4790 (2010).
[9] H.W. Yen, P.Y. Chen, C.Y. Huang, J.R. Yang. Acta Mater. 59, 6264-6274 (2011).
[10] R.A. Ricks, P.R. Howell. Met. Sci. 16, 317-322 (2013).
[11] M.Y. Chen, H.W. Yen, J.R. Yang. Scripta Mater. 68, 829-832 (2013).
[12] H.K.D.H. Bhadeshia. Phys. Status Solidi 69, 745-750 (1982)
[13] Y.B. Cao, F.R. Xiao, G.Y. Qiao, C.J. Huang, X.B. Zhang, Z.X. Wu, B. Liao. Mater. Sci. Eng. A 552, 502-513 (2012).

# An Algorithm for Leakage Detection of the Heating Pipeline

**Huijie Guo *, Yulong Li and Fan Yang**

Beijing Institute of Radio Metrology and Measurement, Beijing, China

*Corresponding author e-mail: ghj203@126.com

**Abstract.** The existing heating pipeline in our country has the serious problem of bubbling, dropping and leaking, which causes the huge waste of energy and the loss of pipeline investment. An algorithm for leakage detection of the heating pipeline is proposed to solve the difficult problem of heating pipeline leakage monitoring by establishing the background noise database, capturing effective noise signals, screening suspected leakage noise signals and recognizing the leakage noise signal with an advanced coarse-to-fine process. Once the pipeline leakage is detected, professionals will be informed to locate and repair the leakage point to minimize the impact of the accident. Meanwhile, the background noise database and the parameters of leak detection algorithm are updated by using false alarm signals, which can continuously improve the accuracy of heating pipeline leakage detection.

## 1. Introduction

With the acceleration of urbanization in China, central heating has developed vigorously. However, the existing heating pipeline construction level is low, with serious leakage, which results in a huge waste of energy and pipeline investment loss [1]. Due to poor construction quality of heating network, it is difficult to find pipeline leakage. The detection of medium leakage fault in the pipeline basically relies on booster pressure, segmenting valve shutoff and other very inefficient and backward methods to find the leakage pipeline segment, and then find the leakage point through correlator, sound bar and ultrasonic detector [2].

Therefore, it is necessary to install simple leakage detection equipment, and use a method that can detect and identify the leakage of heating pipeline in real time, in order to conduct real-time monitoring of the leakage state of the pipeline, detect the leakage as soon as possible and repair it in time, so as to reduce the operation and maintenance cost of the pipeline, reduce energy loss and improve the service life of the pipeline.

## 2. Algorithm implementation

The purpose of the algorithm proposed is to provide a leakage detection method for heating pipeline. The noise recorder is used to collect the noise signal of the pipeline, and then the leakage noise of the pipeline is detected and identified in real time, in order to detect the leakage of the pipeline in time and alarm it, so as to provide a basis for scientific maintenance of the pipeline operation.

To achieve the above purposes, the proposed algorithm adopts the following technical scheme, with the specific implementation as follows.

Content from this work may be used under the terms of the Creative Commons Attribution 3.0 licence. Any further distribution of this work must maintain attribution to the author(s) and the title of the work, journal citation and DOI.
Published under licence by IOP Publishing Ltd

### *2.1. Build the background noise database*

The noise recorder will be installed on the outer wall of the pipeline in the key monitoring section before the heating season, and it is ensured that there is no leakage point in the pipeline to be monitored before installation. Background noise signals will be collected at regular intervals during 15 days before and 15 days after the beginning of heating season. The collection interval is 10 minutes and the collection duration is 10 seconds every time, with the sampling frequency of 6000 points per second [3]. The collection will last for 30 days.

Background noise database is built according to Manhattan distance. The Manhattan distance between signal X and signal Y is calculated as follows.

$$\begin{cases} md(X,Y) = \sum_{k=1}^{n} |x_k - y_k| \\ \forall X = (x_1, \cdots, x_n), Y = (y_1, \cdots, y_n) \end{cases} \tag{1}$$

Where, $md(X, Y)$ is the Manhattan distance between signal $X$ and signal $Y$. Given that the Manhattan distance threshold of background noise signals is $T_{md}$, which is typically equal to 0.4. Given the newly acquired background noise signal $N$, Manhattan distances between $N$ and all signals in the background noise database are calculated respectively to obtain the maximum value $M_{md}$. If $M_{md} > T_{md}$, the signal $N$ is added to the background noise database to update the database, otherwise, discard $N$ and wait for the next time to collect the background noise signal, until the end of the background noise signal collection period, and then the background noise database is constructed [4].

### *2.2. Capture effective noise signals*

The effective noise signal is captured in accordance with the maximum time-frequency vibromotive ratio of the background noise database.

First, the time-domain oscillation peak $op(X)$ of signal $X$ is defined as follows.

$$\begin{cases} op(X) = \max(X) - \min(X) \\ \forall X = (x_1, \cdots, x_n) \end{cases} \tag{2}$$

Where, max ( ) is the function to take the maximum value, and min ( ) is the function to take the minimum value. The frequency domain energy density $ed(X)$ of signal $X$ is defined as follows.

$$ed(X) = \mathrm{sum}(\mathrm{fft}(X).\wedge 2)/F_m \tag{3}$$

Where, fft ( ) is the fast Fourier transform function, sum ( ) is the summation function, and $F_m$ is the maximum frequency value of the spectrum of $X$.

Then, the ratio of time-domain oscillation peak and frequency domain energy density of all signals in the background noise database are calculated respectively, by which the maximum time-frequency vibromotive ratio $OM_{tf}$ of the background noise database is obtained as follows.

$$OM_{tf} = max\left(\frac{op(X_1)}{ed(X_1)}, \cdots, \frac{op(X_k)}{ed(X_k)}, \cdots, \frac{op(X_n)}{ed(X_n)}\right) \tag{4}$$

Where, $X_k$ represents the $k$th signal in the background noise database, and $n$ represents the number of signals in the database.

In practice, the noise signal $Y$ of the pipeline is collected in real time, and $op(Y)/ed(Y)$ is calculated. If $op(Y)/ed(Y) > a \times OM_{tf}$, $Y$ is determined to be an effective noise signal and subsequent discrimination is required. Otherwise, $Y$ is discarded and noise signals are collected again for detection [5]. Where, $a$ is the weighted decision coefficient. Typically, the value of $a$ is from 0.6 to 0.75.

### *2.3. Screen suspected leakage noise signals*

The suspected leakage noise signal is screened by noise clustering and category enclosing.

First, the mean deviation algorithm is used to aggregate all signals in the background noise database into $m$ classes, and the Euclidean distance between the centroid of every two classes of signals is denoted as $D_k$. Thus, there are $K = m(m-1)/2$ class spacing in total, that is $k = 1, 2, \ldots, K$. Therefore, the minimum category spacing $D_{mi}$ of the background noise database is calculated as follows.

$$D_{mi} = \min(D_k), k = 1, 2, \ldots, K \tag{5}$$

Then, the Euclidean distances between the effective noise signal $\boldsymbol{Y}$ obtained in the previous step and the centroid of signals of $m$ classes in the background noise database are calculated respectively, and the maximum value among them is denoted as $D_{te}$. If $D_{te} > D_{mi}$, $\boldsymbol{Y}$ is determined to be a suspected leakage noise signal, and subsequent discrimination is required. Otherwise, $\boldsymbol{Y}$ is determined as background interference noise to discard, and noise signals are collected again for detection [6].

### *2.4. Recognize leakage noise signals*

According to the Mahalanobis distance of joint feature similarity in time-frequency domain, the noise signal is classified as a leakage noise signal or a background noise signal.

First, empirical mode decomposition is performed for the suspected leakage noise signal $\boldsymbol{Y}$ obtained in the previous step.

$$\mathrm{emd}(\boldsymbol{Y}) = \sum_{k=1}^{n} imf_k + ref \tag{6}$$

Where, emd( ) represents the empirical mode decomposition function, $imf_k$ represents the natural mode components with number of $n$, and $ref$ represents the residual component [7].

Then, the time-frequency joint feature vector $\boldsymbol{tf}(\boldsymbol{Y})$ of signal $\boldsymbol{Y}$ is defined as follows.

$$\boldsymbol{tf}(\boldsymbol{Y}) = \mathrm{cat}([\mathrm{mean}(imf_k), \mathrm{median}(imf_k), \mathrm{fem}(imf_k), \mathrm{ffm}(imf_k)],\ l) \tag{7}$$

Where, $k = 1, 2, \ldots, l$, mean( ) represents taking the mean function, median( ) represents taking the median function, fem( ) represents the function of taking the frequency value of maximum spectrum energy, ffm( ) represents taking the maximum frequency spectrum value function, and cat(•, $l$) represents concatenating $l$ vectors [8].

And then, the time-frequency domain joint feature vectors of all signals in the background noise database are calculated respectively to form the time-frequency domain joint feature sample set $\boldsymbol{TF} = (\boldsymbol{tf}_1, \ldots, \boldsymbol{tf}_n)$ of the background noise database. Given the mean vector $\boldsymbol{\mu}$ of $\boldsymbol{TF}$ and the covariance matrix $\boldsymbol{S}$, the Mahalanobis distance $M_{ij}$ between the two vectors in $\boldsymbol{TF}$ is calculated as follows.

$$M_{ij} = \sqrt{(\boldsymbol{tf}_i - \boldsymbol{tf}_j)^{\mathrm{T}} \boldsymbol{S}^{-1} (\boldsymbol{tf}_i - \boldsymbol{tf}_j)} \tag{8}$$

Where, $(\ )^{\mathrm{T}}$ represents the vector transpose operation, $i, j = 1, 2, \ldots$, n, and the maximum of $M_{ij}$ is denoted as $MS$. The Mahalanobis distance $MD$ of signal $\boldsymbol{Y}$ to $\boldsymbol{TF}$ is defined as follows.

$$MD = \sqrt{(\boldsymbol{Y} - \boldsymbol{\mu})^{\mathrm{T}} \boldsymbol{S}^{-1} (\boldsymbol{Y} - \boldsymbol{\mu})} \tag{9}$$

If $MD > \beta \times MS$, $\boldsymbol{Y}$ is identified as the leakage noise signal, and immediately sent an alarm to inform professionals to locate and repair the leakage point in the current monitoring section. Otherwise, $\boldsymbol{Y}$ is identified as background interference noise to discard, and noise signals are collected again for detection [9]. Where, $\beta$ is the weighted classification coefficient. Typically, the value of $\beta$ is from 0.9 to 1.15.

### *2.5. Update the database and model*

In practical application, in order to improve the accuracy and robustness of the aforementioned leak detection algorithm and reduce the disturbance caused by a large number of background noise, the dynamic updating mechanism of the model should be added to gradually reduce the detection error rate of the algorithm, so as to adapt to different detection environments.

In a real system, if background interference noise is identified as leakage noise signal, the background interference noise is added to the background noise database to update the database, and then $OM_{tf}$, $D_{mi}$, $MS$ and other discriminant parameters are updated accordingly, so as to optimize the above leakage detection algorithm of the heating pipeline [10].

The above steps can effectively solve the difficult problem of heating pipeline leakage monitoring through the establishment of background noise database, collection of effective noise signals, screening of suspected leakage noise and determination of leakage noise signals by the coarse-to-fine process. At the same time, the background noise database and parameters of leakage detection algorithm are updated by using false alarm signals, which can continuously improve the accuracy of heating pipeline leakage detection.

## 3. Summary

The above algorithm starts from the background noise of heating pipeline which is easy to collect, and gradually discriminates whether the noise signal collected in practice is leakage noise signal from coarse to fine by establishing background noise database. Moreover, with the continuous updating of background noise database, the accuracy of the leakage detection model of heating pipeline will get higher.

The algorithm proposed in this paper has a strong applicability for the leakage detection of the heating pipeline, and can monitor the leakage status of the pipeline in real time and give a timely alarm when the leakage occurs, which is conducive to improving the operation and maintenance level of the heating pipeline.

## References

[1] Y. Liu, Study of design highlights about heating pipeline in comprehensive pipeline trench, Shandong Transportation Technology, 6 (2018) 83-87.
[2] C. Xu, X. Ma, J. He, et al, Status' problem and solution of urban underground pipeline safety and development, Urban Studies, 20(3) (2013) 108-112.
[3] K. Liu, Y. Jiang, N. Deng, et al, Analysis on main risk factors of urban underground pipelines, Geotechnical Investigation & Surveying, 9 (2013) 51-55.
[4] L. Zhang, Y. Wu, L. Guo, et al, Design and implementation of leak acoustic signal correlator for water pipelines, Information Technology Journal, 12(11) (2013) 2195-2200.
[5] K. Sun, F. Luo, Pipeline leakage detection and positioning system research based on DSP and MCU, Computing Technology and Automation, 33(1) (2014) 29-32.
[6] T. Wu, B. Wang, Y. Zhao, et al, Renovated fault analysis method for ammonia leakage based on the fault tree analysis, Journal of Safety and Environment, 14(4) (2014) 15-20.
[7] G. Dougherty, Pattern recognition and classification: an introduction, China Machine Press, Beijing, 2019.
[8] H. Zhang, Y. Ni, A connected domain marker algorithm for object feature extraction, Computer & Internet, 7 (2015) 58-61.
[9] J. Ramirez, F. Carbone, E. Pastor, et al, Assessing the consequence of pipeline accidents to support land-used planning, Safety Science, 97 (2017) 34-42.
[10] S. Liu, H. Wang, Location of water leakage in water supply network based on intelligent optimization algorithm, Journal of Tongji University (Natural Science), 42(5) (2014) 740-744.

# Evaluation Research on Green Degree of Green Packaging Materials

**Yan Li [a], Zhenqing Sun**

School of economics and management, Tianjin University of Science & Technology, Tianjin 300222, China;

[a] 1198221091@qq.com

**Abstract.** Green packaging materials are packaging materials which are harmless to human health, have a good protective effect on ecological environment and can be recycled in packaging materials that people produce, manufacture, use and recycle. Green packaging materials can save resources and energy; they can be naturally degraded or reused without breaking ecological balance after being discarded, moreover, green packaging materials have a wide range of sources, low energy consumption, easy recovery and high recycling efficiency. Green packaging materials are the premise and foundation of green packaging design. Based on the analysis of the shortcomings of traditional material selection methods, this paper establishes the selection principles of materials in green packaging design, and elaborates on how to use theoretical method of pan-environmental function to evaluate the green degree of green packaging materials, and finally select the best green packaging materials.

**Keywords:** green packaging; material selection; green degree evaluation.

## 1. Introduction

In recent years, with the improvement of people's living standards, the packaging requirement for products have become higher and higher, it not only requires the products to be beautifully packaged, economical, practical, but also conform to environmental protection requirements. Packaging not only makes the product easy to transport and carry, but also protects and beautifies the goods, and plays a role in extending the product shelf life and promoting sales, however, packaging brings convenience to people and also causes serious harm to the environment in which human beings live, the environmental pollution caused by packaging waste is increasing, in order to protect the environment, people have proposed green packaging concept.

## 2. Material Elements of Green Packaging

The material elements of green packaging include basic materials (paper materials, plastic materials, glass materials, metal materials, ceramic materials, bamboo materials and other composite materials etc.) and auxiliary materials (adhesives, coatings, inks, etc.), they are the material basis for realizing three functions of packaging (protection, convenience and sales), which is directly related to the overall functions and economic costs of packaging, the processing ways and the recycling of packaging waste. Material selection in green packaging design should follow the following principles:

Content from this work may be used under the terms of the Creative Commons Attribution 3.0 licence. Any further distribution of this work must maintain attribution to the author(s) and the title of the work, journal citation and DOI.
Published under licence by IOP Publishing Ltd

(1) Lightweight, thin, easy to separate, high-performance packaging materials.
(2) Recyclable and renewable packaging materials.
(3) Edible packaging materials.
(4) Degradable packaging materials.
(5) Natural packaging materials developed by using natural resources.
(6) Try to use paper packaging.

## 3. Green Packaging Design

The harmless materials of green packaging means that the materials used should strictly limit the levels of harmful heavy metals such as lead, mercury, chromium and other harmful elements to the human body, try to select materials that are not toxic to humans and the environment, and non-exhaustive, non-rare resource in the environment. When designing green packaging, people should try to use a single type of packaging material and use less heterogeneous or composite materials, this is because considering packaging entering the waste recycling stage, different packaging materials are mixed together, and the separation requires a large amount of work.

The energy saving of green packaging design means that people can select simple printing and manufacturing methods to save costs and save energy. The improvement of design and production methods can reduce waste in production, such as making similar products with less material, or reducing the waste of cutting. The pre-contamination and air pollution are eliminated in production by selecting environmentally friendly materials and improving production methods.

The cyclic design of the packaging structure means that the packaging structure needs to be simplified. The succinct structure and bright graphic design are a simple modern style packaging. People should try to create packaging methods that save materials and process steps. The single material is used for smart internal separation and buffer design, replace the buffer harmful to the environment, and form a unified design style. We can promote consumers' sustainable consumption habits and use design to help them build sustainable lifestyles. Glue-free structural style can be used when packaging lighter products to avoid contamination of the adhesive and facilitate recycling, and make recycling to be a habit rather than a responsibility.

## 4. Pan-environmental Function Evaluation Method of Green Degree of Green Packaging Materials

### *4.1. Pan-environmental function*

The production process of materials is a complex process involving many specific material parameters. In order to simplify the problem, the material production process is treated as a black box, without regard to the specific reactions in the production process, only the relationship between input and output is studied. After simplification, the input parameters include resources and energy; the output parameters include various forms of emissions in addition to the product, as shown in Fig.1.

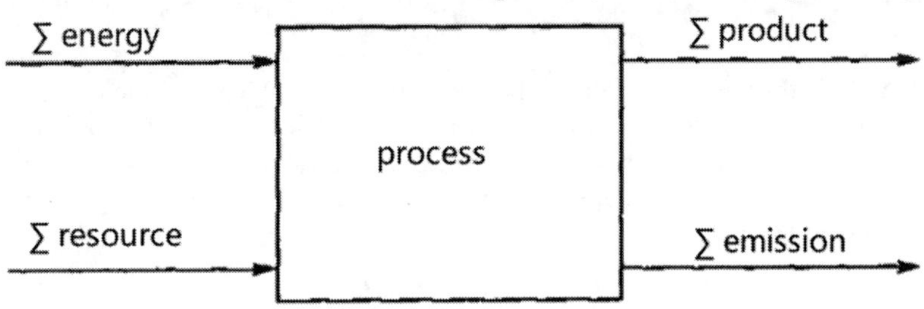


**Fig.1** Input output parameters of a process

In the material production process, there are four parameters for interaction between the system and the outside world: resources, energy, emissions, and products (as shown in Fig.1). These four parameters have an effect on the outside world, in which the product has a positive effect on the outside world, and the other three have a direct or indirect negative effect on the outside world. Therefore, in the material evaluation, it is necessary not only to consider the direct hazard of pollutants, but also consider the indirect hazards of resource consumption and energy consumption. The scope of these hazards and impacts is called the pan environment. It is described with function expression to get a pan-environmental function loop. The value of pan-environmental function is defined as a measure of the comprehensive degree of resource consumption, energy consumption, and environmental impacts and pollutant emissions in the process.

Obviously, the general mathematical expression of pan-environment function is:

$$ELV=\psi(n)=f(R(n), E(n), P(n)) \tag{1}$$

In the formula, *ELV* is a pan-environmental function, whose value is also called a pan-environmental load, which is a function of the material process number *n*. *R* is the resource environmental factor of the material, *E* is the energy environmental factor of the material, and *P* is the pollutant factor of the material. *R, E, and P* are functions of the process number n. For the whole process of materials, the pan-environment function is:

$$ELV = \int_0^n f(R,E,P)dn \tag{2}$$

Here, the pan-environmental function does not study the single environmental factors *R*, *E* or P in isolation, it combines the resources, energy and waste involved in the material and conducts comprehensive research.

$$R(n) = y_1(a_i) \tag{3}$$

$$E(n) = y_2(b_j) \tag{4}$$

$$P(n) = y_3(c_k) \tag{5}$$

In the formula, $a_i$ represents the superposition of various resource consumptions throughout the life of the material, $b_j$ represents the superposition of various energy consumptions, and $C_k$ represents the superposition of various wastes. Since the environmental impact factors of resource environmental factors, energy environmental factors and pollutant environmental factors are relatively independent, their contribution to the pan-environment function can be superimposed. The pan-environmental function *ELV* is added and processed

$$ELV=a'R+b'E+c'P \tag{6}$$

In the formula, a', b', and c' are weighting coefficients. Since the environmental factors of *R*, *E*, and *P* have different dimensions, in order to can superimpose, it is necessary to conduct non-quantization processing separately. Here the concept of environmental equivalent index is introduced, and the environmental equivalent index is defined as:

$$I_{ij} = \frac{Cij}{Sij} \quad i=1, 2, 3,..., n \ ; \ j=1, 2, 3 \tag{7}$$

In the formula, $C_{ij}$ is the measured data of the i-th item in the j-th environmental factor, and its unit is the occurrence amount per unit of product, $S_{ij}$ is the national or industry standard of the i-th item in the j-th environmental factor. Obviously the environmental equivalent index is a non-quantitative value. *J*=1 is the resource environment factor item, *j*=2 is the energy environment factor item, and *j*=3 is the pollutant environmental factor item.

$$R=\sum I_{i1} \qquad i=1,2,3\ldots,n \tag{8}$$

$$E=\sum I_{i2} \qquad i=1,2,3\ldots,n \tag{9}$$

$$P=\sum I_{i3} \qquad i=1,2,3\ldots,n \tag{10}$$

### *4.2. Evaluation of material green degree*

The green degree of the material is an important basis for green products. When selecting green packaging materials, first, the materials that meet the basic requirements of the packaging from the candidate materials are selected; then according to the material selection principle of the green packaging design, the green degree of the selected materials is evaluated, and finally the materials with the highest green degree are selected, that is the best material.

According to the life cycle theory, the life cycle of the materials includes the following processes: raw material-finished material - finished product – use, the evaluation of the green degree of waste materials is the evaluation for the entire life cycle of materials, namely in the whole process of the life cycle, the materials are comprehensively evaluated in energy, resources, and environmental protection.

The green degree of materials can be expressed with pan-environment function: $ELV=\psi(n)=f(R(n), E(n), P(n))$

The steps to green degree evaluation of the material:

(1) Calculate $I_{i1}$, $I_{i2}$, *Ii3* according to formula (7). Specifically, formula (7) is to calculate a dimensionless comparison value, if the $S_{IJ}$ data is insufficient, any one of several materials to be evaluated and compared may be selected, and its $C_{IJ}$ is used as the standard data $S_{IJ}$ for use, $C_{IJ}$ of other materials is compared. (2) The calculation results in step (1) are put into the formulas (8), (9), and (10), and the values of *R*, *E*, and *P* are obtained. (3) The result in step (2) is put into formula (6), calculate the value of ELV. Here, the weight is determined according to the product and the environment in which it is located. In areas where energy and resources are relatively scarce, designers tend to choose larger weight values; while areas with relatively abundant energy and resources are the opposite. (4) Compare the calculated *ELV* values of various materials, the lower the *ELV* value, the higher the green degree value of the material, and the better the green performance of the material.

## 5. Application Examples

The above method is further discussed below by specific examples. For example, in the choice of green packaging materials, resin high-density polyethylene and polypropylene all conform to the basic requirements of the design, then the level of green level is the decisive factor in the choice of the best material. The green degree of the two materials is evaluated by using the above method.

**Table 1.** Quantitative data of Ikg plastic LCA

| inputs and outputs of energy consumption and emissions | | unit | HDPE | PP |
|---|---|---|---|---|
| energy consumption | | MJ | 16.94 | 18.47 |
| resource | crude oil | kg | 1.16 | 1.27 |
| | water | kg | 9.50 | 3.10 |
| | others | g | 4.70 | 5.90 |
| waste gas | coal dust | g | 2.00 | 2.00 |
| | $CO_2$ | kg | 0.94 | 1.10 |
| | $SO_x$ | g | 6.00 | 11.00 |
| | $NO_x$ | g | 10.00 | 10.00 |
| | hydrocarbon | g | 21.00 | 13.00 |
| | others | g | 0.60 | 0.70 |
| waste water | BOD | mg | 100.00 | 60.00 |
| | COD | mg | 200.00 | 400.00 |
| | suspended particles | g | 2.10 | 1.90 |
| waste residue | sludge, dust, etc. | g | 5.00 | 5.00 |
| | waste | g | 27.00 | 26.00 |
| data origin | | PWMI | PWMI | |

According to the above table, the data of PP is used as the reference data, and the data of HDPE is compared and calculated, the calculation result of HDPE is as follows:

**Table 2.** HDPE calculation results

| inputs and outputs of energy consumption, emissions, etc. | *I* | *E* | *R* | *P* | weight | ELV |
|---|---|---|---|---|---|---|
| energy consumption | | 0.917 | 0.917 | | (0.3,0.3, 0.4) | 6.1806 |
| resource | crude oil | 0.916 | | 4.777 | | |
| | water | 3.065 | | | | |
| | others | 0.796 | | | | |
| waste gas | coal dust | 1 | | | | |
| | coal dust | 1 | | | | |
| | $SO_2$ | 0.855 | | | | |
| | hydrocarbon | 1.615 | | | | |
| | others | 0.857 | | | | |
| waste water | BOD | 1.667 | | | | |
| | COD | 0.5 | | | | |
| | suspended particles | 1.105 | | | | |
| waste residue | sludge, dust, etc. | 1 | | | | |
| | waste | 1.038 | | | | |

Resources: Crude oil 0.9164.777, water 3.065, others 0.796, waste gas dust 111.181, $CO_2$ 0.854, $SO_2$ 0.545, $NO_x$ l, hydrocarbon 1.615, others 0.85, waste water BOD 1.667, COD 0.5, suspended particulates 1.105, waste sludge, dust, etc. 1, waste 1.03.

According to the selected weight above, the values of PP are taken as 1, and the *ELV* value of PP is calculated to be 5, 6. It can be seen that under this weight, the green degree of PP is better than that of HDPE. If the weights are selected differently, the evaluation results of the material green degree will also be different.

## 6. Conclusion

The theoretical method of pan-environmental function is applied to the evaluation of material green degree in the selection of green packaging materials in this paper, and the problem of material selection under the basic design requirements is successfully solved, which is scientific, objective and workable. The method is simple and easy to implement, and has strong practical value.

## References

[1] Shen Yanfang. Research on the Application and Development of Green Packaging materials [D]. Nanchang University, 2012, 11(5): 23-25.
[2] Lin Chaoping. Study of Selection for Green Packing Material on Design of Material [J]. Light Industry Machinery, 2011, 03 (02): 8-10.
[3] Li Gang, Huang Wenxiu. The Research And Discussion Of The Green Packaging Design [J]. Sheji, 2015, (1), 146-148.
[4] Yang Jie, Li Huilin. Evaluation of Green Degree for Packaging Industry in China Based on Fuzzy Analytic Hierarchy Process [J]. Ecological Economy, 2018, (1), 78-82.
[5] Li Hongming Zhou Yi. An Evaluation Method for Green Package Design [J]. Ecological Economy, 2017, 38(13), 208-212.

# Study on Positive and Negative Materials of High Performance Ion Batteries

**Jialong Song, Zijia Sun, Mingfen Sun**

Zhuhai College of Jilin University, Guangdong, Zhuhai, 519041, China

**Abstract**. Study on positive and negative materials of high performance Ion batteries is presented in this manuscript. In recent years, sodium ion batteries have developed rapidly, and a number of key materials with electrochemical activity have emerged. Among them, the transition metal oxide as a core sodium ion battery positive electrode shows considerable sodium-embedded capacity, but the capacity decays severely during continuous cycling. This paper studies the novel findings on the mentioned issues to provide the new perspectives for assisting the future research.

## 1. Introduction

The structured battery is a unitary body of a battery and a structural material. In order to obtain an efficient structural battery, the battery generally requires a high operating voltage, a long life cycle, and a large specific energy; the material is required to have a high compressive and shear resistance. Its development has experienced hydrogen nickel type, lithium ion battery type and polymer lithium ion battery type, which may be applied in the form of fiber battery or the fully homomorphic thin film battery in the future. Recently, research on lithium ion secondary batteries has continued. Several new systems for improved lithium ion secondary batteries have emerged, such as the polymer lithium ion secondary batteries using in-situ polymerization processes, which simplify the process and improve safety and specific energy; while all-solid-state thin film lithium-ion batteries or fibers Lithium-ion battery is a hot spot at home and abroad. It is a new generation of lithium-ion battery developed on the basis of liquid lithium ion secondary battery. Based on the structure of a common lithium ion battery, it is deposited on the substrate by physical coating.

According to the current research progress, modern sodium ion batteries have three outstanding advantages compared with lithium ion batteries.

- Sodium batteries have relatively stable electrochemical properties and are safer to use. At the same time, sodium-ion batteries also have defects, such as the relative atomic mass of sodium is much higher than that of lithium while also then resulting in a small theoretical specific capacity.
- The raw material resource is rich, the cost is low and the distribution is widespread.
- The half-cell potential of the sodium ion battery is 0.3-0.4v higher than that of lithium ion battery, that is to say, the electrolyte solvent and electrolyte salt with lower decomposition potential can be used, and the selection range of electrolyte is wider.

In recent years, sodium ion batteries have developed rapidly, and a number of key materials with electrochemical activity have emerged. Among them, the transition metal oxide as a core sodium ion battery positive electrode shows considerable sodium-embedded capacity, but the capacity decays severely during continuous cycling. Doping non-transition metals improves the structural stability of

Content from this work may be used under the terms of the Creative Commons Attribution 3.0 licence. Any further distribution of this work must maintain attribution to the author(s) and the title of the work, journal citation and DOI.
Published under licence by IOP Publishing Ltd

the material, but it will sacrifice the specific capacity of the core material. Therefore, the equilibrium capacity and cycling performance are the main research bottlenecks of this kind of the materials.

Phosphate compounds have the characteristics of the high voltage and high cycling stability. The research on these materials is mainly to solve the problem of the poor conductivity. The design and optimization of carbon content and its structure is an important method to improve the comprehensive performance of these electrodes. As the sample, figure 1 gives the demonstration.

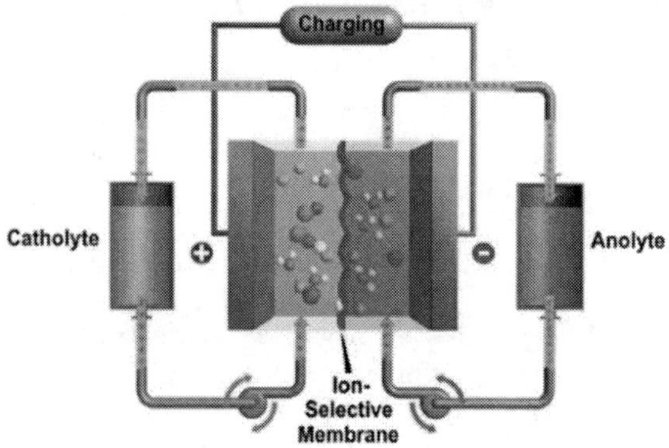


**Figure 1.** The Inner Organization of the High Performance Batteries

## 2. The Proposed Methodology

### *2.1. High Performance Batteries Overview*

The structure battery is called the multi-purpose structure battery is in order to overcome the present commercial massive battery quality big, the volume big shortcoming, but unifies an organic entity the battery and the load-carrying structure which forms.

The structural battery can provide the energy and can be used as a load-bearing structural material, which can effectively reduce the quality of the system; the fork can save system space and increase its payload by dispersing the battery in the structure of the system. With the emergence of lithium-ion batteries, lithium-ion batteries with long service life, high working voltage, large specific energy and strong load resistance are used as the core of the lightweight honeycomb plate of the structural battery. In this way, the total mass and volume of the system are further reduced, and the specific energy and power of the structured battery are greatly improved. Some experts predict that the new technologies, new methodologies, new materials and new structures will replace the current lead-acid batteries. At present, the electrode active material used by most lead-acid battery manufacturers is a mixture of fine particles, that is, lead powder. Lead powder is the starting material for the formation of active materials in the lead-acid battery. Its properties and the quality control play an important role in the performance of the entire lead-acid battery. In order to develop high-performance lead-acid batteries with high energy, long cycle life and high rate discharge performance, domestic and foreign scholars have carried out a large number of ultra-fine lead powder synthesis and battery performance research.

For each application, each battery has the different performance requirements. Adding additives or creating a porous environment, or increasing the efficiency of use, or strengthening the strength of the paste, or improving the conductivity, but the addition of an additive often cannot meet improvement of multi-faceted performance at the same time. In addition, an all-solid-state thin film lithium ion secondary battery using a flexible substrate (stainless steel or polyimide) is easily curled into various shapes and can be combined with a thin film solar cell to form an integral thin film battery.

### *2.2. Ion Battery*

Sodium is abundant in the earth and widely distributed, and it belongs to the same family element as lithium. Therefore, the development of sodium ion battery technology is of great significance.

However, amount of sodium is larger than that of lithium, and standard electrochemical potential is lower than that of lithium.

Therefore, the energy density of sodium ion batteries is difficult to match that of the lithium ion batteries. However, when large-scale energy storage applications require large amounts of the alkali metals, the cost advantage of the sodium-ion batteries is revealed. Sodium ion battery and lithium ion battery are essentially the same that are a rocking chair battery. Based on the research experience of lithium ion battery, sodium ion battery has made significant progress in electrochemical activity of its key material after some rapid development in recent years. A lithium ion battery using an organic electrolyte has a wide electrochemical window of 3V or more, and thus has a higher energy density than a water ion battery. However, organic electrolytes are flammable and toxic. If they are not used properly, they will bring serious safety and environmental problems, which restricts the application of the lithium-ion batteries in large-scale energy storage.

In the battery structure, aluminum foil or the foamed nickel is usually used as a gasket between the negative electrode and the battery case to reduce the gap to reduce the battery resistance; the negative electrode usually uses the core battery-grade lithium metal plate; PC (Li2O); GC and PC (BN) are Li2O-loaded polymer film, glass-ceramic separator and the boron nitride-loaded polymer film. These three films form a battery separator for the easy ion exchange. The solvent is dimethyl Sulfone or the Tetraethylene glycol dimethyl ether, solute is electrolyte of lithium Bistrifluoromethanesulfonimide battery positive electrode is a catalyst-supported glassy carbon electrode, usually using foamed nickel or the like between the battery case and the positive electrode to enhance conductivity feature. In the Li-CO2 battery that has been reported so far, a glass carbon fiber membrane is often used as a battery separator to ensure rapid ion exchange. Figure 2 demonstrates the mentioned ideas.

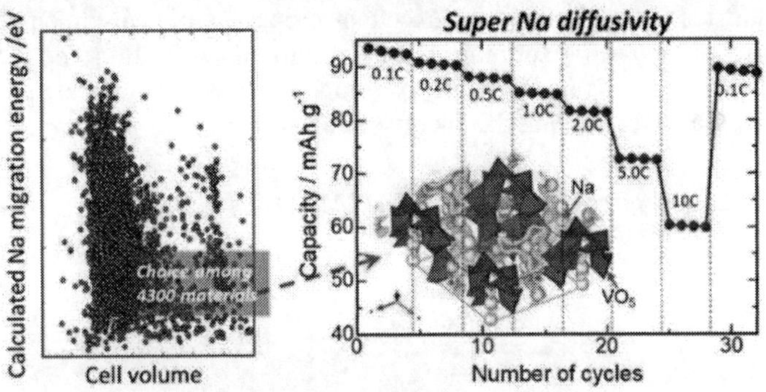


**Figure 2.** The Performance Simulation of the Ion Battery

### 2.3. *Ion Battery Positive and Negative Materials*

In order to obtain a sodium ion battery electrode material with high specific capacity and good battery stability, it is particularly important to analyze the mechanism of insertion and removal of sodium ions, the structure of the electrode material and electrochemical performance during charge and discharge. At this stage, the electrode materials of the sodium ion batteries are mainly inorganic compounds based on transition metal elements such as Fe, Co, Ni, Mn, etc., which are limited in resources, expensive, and difficult to recycle. In addition, the lattice size of inorganic metal compounds is fixed while making large-scale sodium ion insertion and extraction are limited, so capacity is limited, and there are not many types of inorganic compounds that can be used as electrode materials for sodium ion batteries.

Therefore, we present the samples as the follows.

(1) Positive Materials list

- The earliest research on the polyanionic cathode materials was applied to the positive electrode of lithium ion batteries, which was later applied to the sodium ion batteries. The research of such materials is relatively early, mainly representing a class of olivine crystals such as

NaMnPO4 and NaFePO4. This electrode material has good rate performance and cycle stability as a positive electrode for sodium ion batteries.

- The rise of two-dimensional materials came after the discovery of graphene, which showed for the first time that two-dimensional structures could be stable as crystals. Since then, the two-dimensional materials have been developed. Among them, a series of transition metal two-dimensional materials such as general molybdenum disulfide, transition metal bimetal hydroxide, transition metal sulfide have been then gradually synthesized and applied to the positive electrode of sodium ion battery with good results.
- Transition metal phosphate is also a positive electrode material used in general lithium ion batteries. It has crystal structures and mature synthesis process. Therefore, hydrothermal synthesis has become a new way to then synthesize sodium iron phosphate. Currently, the research of layered cathode materials mainly focuses on the P2 phase and 03 phase and the specific capacity of P2 phase is generally higher and the stability is better. This is mainly because the coordination space of the triangular prism occupied by Na+ in the P2 phase is larger than the octahedral coordination space in the 03 phase, which is favorable for the diffusion of Na+.

(2) Negative Materials list

- Graphite anode materials have been used in applications in lithium-ion batteries, and also experimental and theoretical studies have shown sodium ions are difficult to intercalate between graphite layers, limiting the application of the graphite in sodium-ion batteries. Interpretation of Na can be explained by first-principles calculations. The mechanism of the extraction, and then find ways to improve the performance of graphite storage.
- As the cathode of sodium ion battery, metal compounds are mainly two-dimensional metal carbide and nitride. In addition to the excellent properties of two-dimensional materials, these materials can not only store sodium ions by means of adsorption and intercalation, but also combine with sodium ions to store energy through pseudo-capacitance generated by chemical reactions, thus greatly enhancing the energy storage effect.

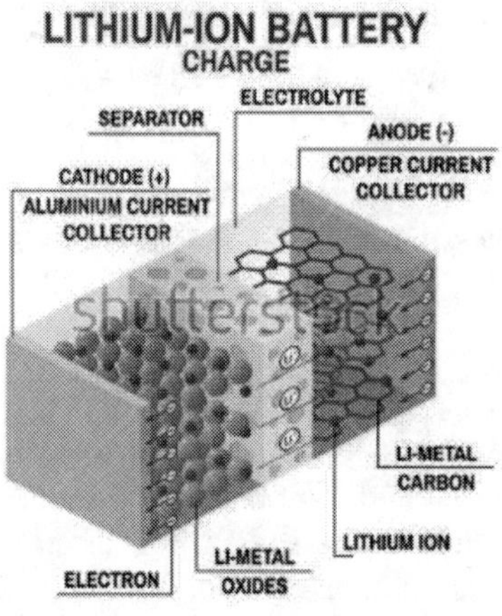


**Figure 3.** The Ion Battery Positive and Negative Materials

## 3. Conclusion

Study on the positive and negative materials of high performance Ion batteries is presented in this manuscript. The structured battery is a unitary body of a battery and a structural material. In order to obtain an efficient structural battery, the battery generally requires a high operating voltage, a long life cycle, and a large specific energy; the material is required to have a high compressive and shear resistance. This paper proposes the novel perspectives on the future advancement of the related work. The ideas are novel and in the next research, we will conduct the experimental analysis.

**Acknowledgements**
This research is supported by the Guangdong Province Key Platforms and Major Research Projects (No. 2018KQNCX347): Preparation and Modification Positive Materials (NCM523) of Lithium-ion Battery and Electrochemical Properties Study.

**References**

[1] Fang, Y., Yu, X.Y. and Lou, X.W., 2018. Formation of Hierarchical Cu - Doped CoSe2 Microboxes via Sequential Ion Exchange for High - Performance Sodium - Ion Batteries. Advanced Materials, 30(21), p.1706668.
[2] Fan, L., Li, X., Song, X., Hu, N., Xiong, D., Koo, A. and Sun, X., 2018. Promising dual-doped graphene aerogel/SnS2 nanocrystal building high performance sodium ion batteries. ACS applied materials & interfaces, 10(3), pp.2637-2648.
[3] Li, L., Zheng, Y., Zhang, S., Yang, J., Shao, Z. and Guo, Z., 2018. Recent progress on sodium ion batteries: Potential high-performance anodes. Energy & Environmental Science, 11(9), pp.2310-2340.
[4] Plageras, A.P., Psannis, K.E., Stergiou, C., Wang, H. and Gupta, B.B., 2018. Efficient IoT-based sensor BIG Data collection–processing and analysis in smart buildings. Future Generation Computer Systems, 82, pp.349-357.
[5] Wang, Q., Zhu, X., Liu, Y., Fang, Y., Zhou, X. and Bao, J., 2018. Rice husk-derived hard carbons as high-performance anode materials for sodium-ion batteries. Carbon, 127, pp.658-666.

IOP Conf. Series: Materials Science and Engineering **772** (2020) 012092 doi:10.1088/1757-899X/772/1/012092

# Experiments Research on Electrical Discharge Grinding Polycrystalline Diamond Tool based on Surface Quality Analysis

**Yunhai Jia[1, a], Jianmei Guo[1], Qinjian Zhang[2] and Yue Sun[1]**

[1]Research and development center, Beijing Institute of Electro-Machining, Beijing 100191, China
[2]School of mechanical and electrical engineering, Beijing Information Science and Technology University, Beijing 100192, China

[a]jyh308401@sina.com.cn

**Abstract**. As the hardest material in the field of synthetic materials, polycrystalline diamond has attracted more and more attention in the field of tools, which reflects the basic characteristics of 'high efficiency, precision and flexibility' of modern advanced cutting technology and clean production. Electrical discharge grinding (EDG) is often used in polycrystalline diamond tools, especially in special shape, thin and fine edge cutting tools. Taking two kinds of grain size of polycrystalline diamond as the research object, taking the machining test and material test as the analysis means, this paper simply compares the surface quality of disc electrode and wire electrode machining, and studies the influence of diamond particle size, electrode polarity and electrode rotation speed on the surface quality and material removal rate of precision electrical discharge grinding polycrystalline diamond. The results show that there is no porous structure, good surface quality and no selectivity in discharge removal when using the negative polarity machining. With the increase of electrode rotation speed, the removal amount of polycrystalline diamond material increases gradually, and the surface roughness of polycrystalline diamond material decreases first, then increases. When the electrode rotation speed reaches 80 m/min, the surface roughness of the polycrystalline diamond sample reaches the minimum value, and the removal amount of polycrystalline diamond material also tends to be stable.

## 1. Introduction

Polycrystalline diamond (PCD) is a synthetic diamond powder grown by catalyst under high temperature and high pressure. Its structure is very similar to natural diamond, which is formed by C-C bond and has good toughness. Polycrystalline Diamond tool has a higher hardness than HSS cutting tools, carbide cutting tools and better wear resistance, better thermal stability, chemical stability and thermal conductivity, excellent embodies the modern advanced cutting technology [1, 2].Developed countries in aerospace, automobile manufacturing, machinery processing, such as precision machining industry, especially the processing of non-ferrous metals and their alloys, the widespread adoption of PCD cutting tool instead of carbide cutting tools, processing efficiency more than 2 times [1], the

Content from this work may be used under the terms of the Creative Commons Attribution 3.0 licence. Any further distribution of this work must maintain attribution to the author(s) and the title of the work, journal citation and DOI.
Published under licence by IOP Publishing Ltd

cutting tool service life can be improved to ten times the traditional carbide cutting tools, reduce the tool wear and failure rate [3-4].

Because the hardness and abrasion resistance of polycrystalline diamond are very high, it is a common method to use electrical discharge machining (EDM) to achieve material removal by using soft and rigid materials. EDM is a self-excited discharge process based on the principle of pulsed discharge erosion. The physical process of discharge erosion is a comprehensive process of electromagnetism, thermodynamics, fluid dynamics, etc. [5, 6].Different from mechanical machining, EDM belongs to non-contact machining, without mechanical cutting force, with the advantages of simple tool electrode forming, low relative loss, etc., it can be effectively applied to the PCD cutting tool machining with sharp edge and complex shape edge. However, there are also differences between disc electrode and wire electrode processing, suitable for processing tools are also different, the cutting edge of the processing tools also shows different morphology.

Experts have made many explorations in PCD materials such as EDG and EDM wire cutting, and obtained much valuable research result. These studies can be divided into three categories, basic type of research is around the processing mechanism of PCD material removal of [2, 6-9], the processing and analysis of a combination of materials testing, SEM observation before and after by PCD material processing, XRD analysis, energy spectrum analysis and Raman spectrum analysis and carry on the processing element before and after the contrast and the analysis of the components, the analysis of the grain boundary, to judge the PCD material removal mechanism in EDM. Another kind is the study of PCD material processing technology [10-12], also basic use test is given priority to, to the processing surface quality of workpiece, such as surface roughness, straightness, etc.), processing efficiency, material removal rate as the evaluation index, such as establishing process parameters (such as open circuit voltage, peak current, pulse width, pulse interval, etc.) and the relationship between evaluation index model, search process. The third category is the selection of electrode materials for EDM [2, 13, 14], and the influence of white copper, red copper and graphite electrodes on EDM and the effect of rotating electrode on machining efficiency. These studies all focus on important factors affecting discharge machining and ignore seemingly unimportant processing parameters, such as electrode polarity, electrode rotation speed and diamond particle size. For EDM precision machining, these "minor factors" also have important influences on surface quality and machining efficiency. In addition, the particle size of polycrystalline diamond also affects the quality and efficiency of processing to a certain extent.

In this study two kinds of grain size of polycrystalline diamond material as the research object, through processing according to the method of analysis of the material science, simple analysis compares the surface quality of plate electrode and wire electrode processing differences, explore the diamond granularity, electrode and electrode polarity speed of polycrystalline diamond spark discharge grinding surface quality and the influence of material removal rate. This provides an experimental basis for making reasonable electrical discharge grinding process of polycrystalline diamond.

## 2. Experiments method and equipment

**Table 1.** Physical property of PCD tool blanks.

| Crystal granularity size (μm) | Diamond content (%) | Cobalt content (%) | Density (g/(cm)$^3$) | Elastic Modulus (GPa) | Thermal Conductivity (W/m.k°) |
|---|---|---|---|---|---|
| 2 | 84.5 | 15 | 4.2 | 930 | 520 |
| 10 | 90 | 10 | 4.0 | 1050 | 580 |

IOP Conf. Series: Materials Science and Engineering **772** (2020) 012092 doi:10.1088/1757-899X/772/1/012092

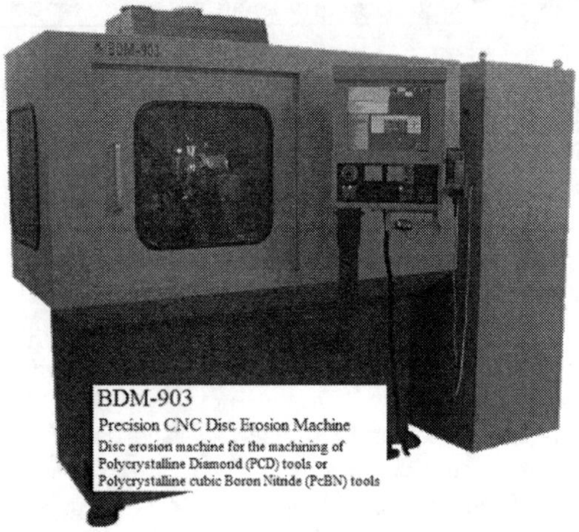


**Fig. 1** Experimental equipment BDM-903

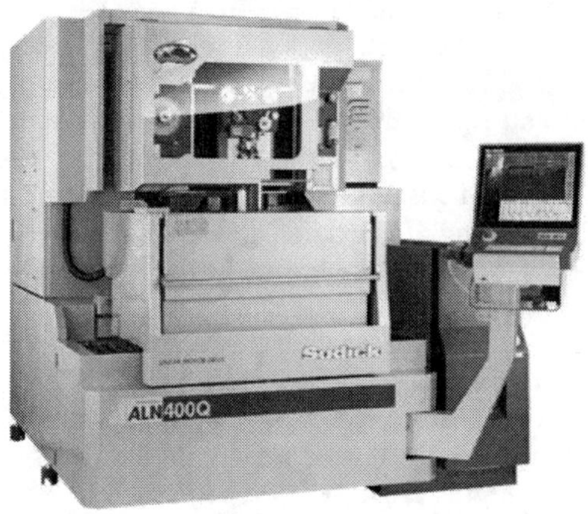


**Fig. 2** Experimental equipment ALN400Q

Polycrystalline diamond samples were selected with diamond PCD composite tablets with particle sizes of 2 and 10 microns respectively (their material properties are shown in table 1).The electrode adopts disc-shaped copper electrode and copper wire, the end surface of which is used for grinding PCD layer, and the line electrode is used for cutting PCD cutter edge. The electrical discharge machine named BDM-903 (as showed in Fig. 1) produced by Beijing Institute of Electro-machining is selected as experimental machining equipment. The machine tool with a wire electrode selected the ALN400Q of Sodick co. (as shown in Fig. 2). HITACHI's s-4800 scanning electron microscope, D/max-ra X-ray diffraction analyzer and domestic TR240 surface roughness meter were selected as the experimental analysis instruments.

## 3. Surface quality analysis of EDG disk electrode and wire electrode

The cutting edge of PCD sample with particle size of 10 microns processed by 0.2mm brass wire is shown in Fig.3. Three modes of rough processing, semi-finishing processing and finishing processing are respectively adopted. It can be easily seen from Fig.3 that after PCD processing, the cutting edge was in an obvious melting state, and there were different degrees of corrosion pits on the cutting edge, and the cutting edge had a good straightness. The size and number of each crater are related to the processing technology.

**Fig. 3** SEM of wire-EDM (a) rough machining; (b) semi-precision machining; (c) precision machining;

Figure 4 shows the surface morphology of PCD processed with 200mm diameter copper disc electrode, as well as rough machining, semi-finishing and finishing. From the figure, it can be easily found that the processed surface also has molten materials and craters of different sizes, but the straightness of the cutting edge is general, and a curved surface appears at the cutting edge.

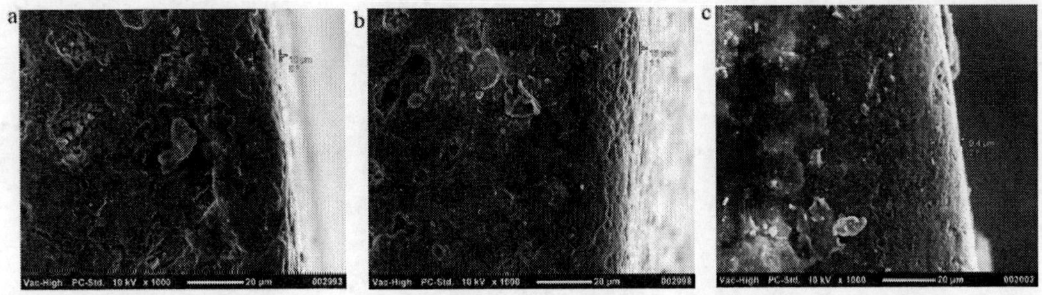

**Fig. 4** SEM of EDG (a) rough machining; (b) semi-precision machining; (c) precision machining;

Compared with Fig.3 and Fig.4, it is obvious that the straightness of the cutting edge processed by disc electrode is worse than that processed by wire electrode, and the size of the etching craters is basically the same. From the surface roughness, disc electrode processing is better than line electrode processing.

## 4. Influence of electrode rotation speed and diamond particle size

Fig.5 and Fig.6 are the curves of the relationship between the electrode speed, diamond and the quality and efficiency of the discharge grinding process under the unified discharge machining process parameters test. As can be seen from Fig.5, with the increase of electrode rotation speed, the surface roughness of PCD samples first decreases, and then increases. In particular, when electrode speed is at 60 m/min, the surface roughness of samples reaches the minimum value. Similarly, about 10 microns granularity PCD material samples of the surface roughness value is reduced, first after the increase, but the sample surface roughness value reaches the minimum value is in electrode speed 80 m/min, it also suggests that, in the same discharge processing, particle size is different, the surface roughness of the workpiece are different, the optimization of processing technology changed correspondingly adjusted according to the material particle size.

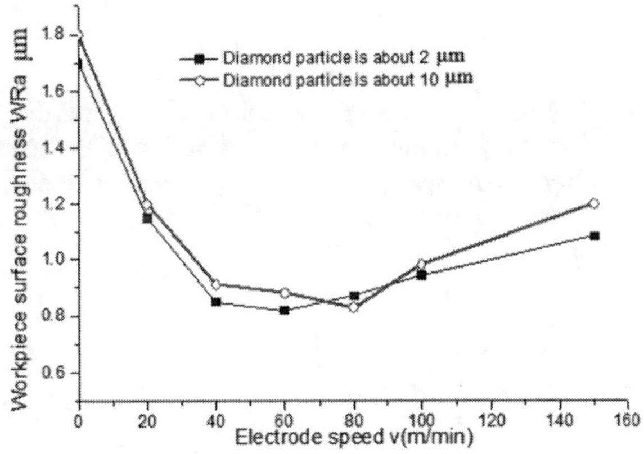


**Fig. 5** The relationship between the speed of the electrode, diamond particle size and the surface quality of the workpiece

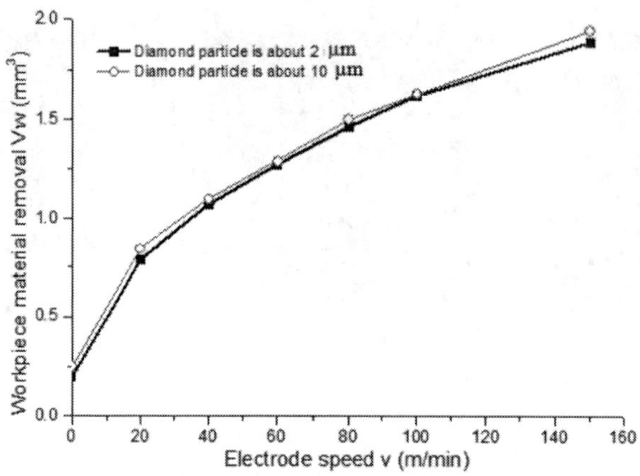


**Fig. 6** The relation curve of the electrode speed, diamond particle size and the workpiece material removal

As can be seen from Fig.6, when the electrode is not rotated, the removal amount of PCD material is the lowest. With the increase of the rotation speed of the electrode, the removal amount of PCD material gradually increased. When the electrode speed reached 80m/min, the removal amount of PCD material gradually slowed down. This is because the rotation of the electrode improves the discharge machining conditions and accelerates the material removal speed. However, when the linear velocity of the electrode rotation matches the discharge parameters, the effect of electrode rotation speed on material removal speed is not obvious. It can also be seen from Fig.6 that, with the increase of the rotation speed of the electrode, the removal speed of large-size diamond is faster, but the change trend is the same.

## 5. Conclusions

From the above analysis, we can get the following conclusion:

(1) The straightness of the cutting edge processed by disc electrode is worse than that processed by wire electrode, and the size of the etching craters is basically the same.

(2) With the increase of the speed of electrode rotation, the surface roughness of PCD material first decreases and then increases. When electrode speed is at 60~80 m/min, the surface roughness of the sample reaches the minimum value. With the increase of particle size, the speed of the electrode which obtained the best surface quality also increased under the same discharge machining process.

(3) With the increase of the electrode rotation speed, the removal amount of PCD material gradually increases. When electrode speed reaches 80m/min, the removal amount of PCD material gradually slows down.

## Acknowledgments

This research is supported by Beijing Natural Science Foundation the Grant No. 3162013 and Young Scholar Program of Beijing Academy of Science and Technology the Grant No. YS201905. The authors would also like to thank the anonymous reviewers whose comments greatly helped in making this paper better organized and more readability.

## References

[1] M.W.Cook, P.K.Bossom. Trends and recent developments in the material manufacture and cutting tool application of polycrystalline diamond and polycrystalline cubic boron nitride. International Journal of Refractory Metals & Hard Materials. 18 (2000) 147-156.

[2] Yunhai JIA, Jiangang LI. Impact Analysis of Electrode Material on Electrical Discharge Grinding Polycrystalline Diamond Cutting Tools. Procedia CIRP. 68 (2018), p. 643.

[3] Michiko Ota, Juny Okida, et al. High Speed Cutting of Titanium Alloy with PCD Tools. Key Engineering Materials. 389 (2009), p. 157.

[4] Ezugwu E, Bonney J and Yamane Y. An overview of the machinability of aero engine alloys. Journal of Materials Processing Technology. 134 (2003) 233-253.

[5] Yu Jiashan. Theory of electrical discharge machining. Beijing: National Defence Industry Press; 2011. p. 55-71.

[6] M. Kunieda. Fundamentals of electrical discharge machining and future prospects [J]. Journal of the Japan Society for Precision Engineering. 71(2005):189-194.

[7] Liu, J, Deng, F, Lu, X, Zhang, P. A study on structural evolution of metamorphic layer on the surface of PCD in electrical discharge machining. Diamond and Related Materials. 91(2019) 46-53.

[8] Yan, M, Cheng, Y, Luo, S. Improvement of Wire Electrical Discharge Machining Characteristics in Machining Boron-doped Polycrystalline Diamond Using a Novel Iso-pulse Generator. International Journal of Precision Engineering and Manufacturing. 20(2019) 159-166

[9] Jia, Y.H, Guo, J.M, Guo, Y, Yu, F. Research on micro-size electrical discharge machining polycrystalline diamond. Materials Science Forum. 943(2018) 14-19.

[10] Tang, J, Zhang, H, Ye, P. Research on electronic discharge grinding of polycrystalline diamond based on response surface method. Key Engineering Materials, 764(2018) 123-132.

[11] Yunhai Jia, Jiangang Li and Lin Chen. Research on Electrical Discharge Grinding Technics of Polycrystalline Diamond Compact Cutting Tool. Advanced Materials Research. 1004-1005 (2014) 1182-1185.

[12] M.Boujelbene, E.Bayraktar, W.Tebni. Influence of machining parameters on the surface integrity in electrical discharge machining. Archives of Materials Science and Engineering. 39 (2009) 110-116.

[13] Jiwang Yan, Kazunori Watanabe, Tojiro Aoyama. Micro-electrical discharge machining of polycrystalline diamond using rotary cupronickel electrode. CIRP Annals – Manufacturing Technology 63 (2014) 209-212.

[14] M. Zulafif Rahim, Songlin Ding and John Mo, Electrical Discharge Grinding of Polycrystalline Diamond—Effect of Machining Parameters and Finishing In Feed. Journal of Manufacturing Science and Engineering. 137(2015) 1-11.

# Research and design of CSA-510 color analyzer

**Yunlong Hu*, Yingjie Jiang, Wenzhong Li and Jun Yao**

Zhejiang University Kunshan Innovation Institute, Suzhou, China

*Corresponding author e-mail: 3257063992@qq.com

**Abstract.** This paper introduces the measurement principle of color analyzer firstly, and designs a kind detection equipment of color analyzer which can be used for accurate measurement and analysis of color, An experimental comparative study was carried out with foreign mature product CA-310 at last. The results show that When the brightness is greater ⩾ 0.1cd/m2, the results of brightness test keep a linear relationship, and the range of error with competitor brightness is less than 2%. The error range of color coordinate accuracy is basically controlled within ±0.0050. Compared to the high prices of abroad equipment, it has certain competitive advantages.

## 1. Introduction

With the continuous development of society, the brightness and chroma accuracy of the display screen is higher demanded increasingly by the display equipment industry which is also the key technology of the core competitiveness of manufacturers. Color analyzer is one of the most commonly used detection devices in the field of color detection. The domestic equipment is more poor than mature foreign products with poor accuracy, low range and inability to accurately detect low brightness. The main causes include: 1. inaccurate design of optical path; 2. imperfect filter matching; 3. rough design of circuit; 4. unreasonable designs [1].Therefore, this paper carries out a detailed experimental to solve the existing problems of the color analyzer, while it puts forward its design scheme, and designs a new kind of detection equipment of color analyzer which has certain reference value in the field of color and brightness detection.

## 2. The working theory of color analyzer

The color analyzer is a photoelectric integral colorimetric measuring instrument [2]. Which covering the photoelectric detector with a filter to modify the relative spectral sensitivity S (λ) of the three X Y Z detectors to be the spectral stimulus functions x (λ), y (λ), Z (λ) recommended by the International Lighting Committee for 1931 standard chroma observers, as shown in Figure 1 [3]. Using the three detectors, the tristimulus values X, Y and Z of the target can be measured at one time.

$$X=\begin{cases} k\int_{380}^{500}\varphi(\lambda)\bar{x}(\lambda)d\lambda = c_x\int_{380}^{500}\varphi(\lambda)S(\lambda)\tau_{xb}(\lambda)d\lambda \\ k\int_{500}^{780}\varphi(\lambda)\bar{x}(\lambda)d\lambda = c_x\int_{500}^{780}\varphi(\lambda)S(\lambda)\tau_{xr}(\lambda)d\lambda \end{cases} \quad (1)$$

Content from this work may be used under the terms of the Creative Commons Attribution 3.0 licence. Any further distribution of this work must maintain attribution to the author(s) and the title of the work, journal citation and DOI.
Published under licence by IOP Publishing Ltd

$$Y = k\int_{380}^{780}\varphi(\lambda)\bar{y}(\lambda)d\lambda = c_y\int_{380}^{780}\varphi(\lambda)S(\lambda)\tau_y(\lambda)d\lambda \quad (2)$$

$$Z = k\int_{380}^{780}\varphi(\lambda)\bar{z}(\lambda)d\lambda = c_z\int_{380}^{780}\varphi(\lambda)S(\lambda)\tau_z(\lambda)d\lambda \quad (3)$$

Among the k, $C_x$, $C_y$ and $C_z$ are constant $\tau_{xy}(\lambda)$ , $\tau_{xb}(\lambda)$ and $\tau_z(\lambda)$ are the spectral transmittance of each filter, $S(\lambda)$ is the relative spectral sensitivity of the detector, $\lambda$ is the wavelength.

$$\tau_{xb}(\lambda) = \frac{\bar{x}(\lambda)}{S(\lambda)}, (380nm \le \lambda \le 500nm) \quad (4)$$

$$\tau_{xr}(\lambda) = \frac{\bar{x}(\lambda)}{S(\lambda)}, (500nm \le \lambda \le 780nm) \quad (5)$$

$$\tau_y(\lambda) = \frac{\bar{y}(\lambda)}{S(\lambda)} \quad (6)$$

$$\tau_z(\lambda) = \frac{\bar{z}(\lambda)}{S(\lambda)} \quad (7)$$

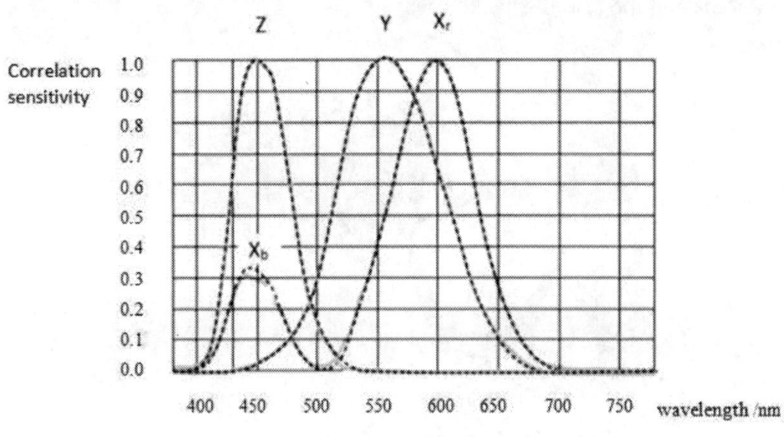


**Figure 1.** Tristimulus curve of spectrum of CIE-XYZ standard chroma observer.

Through the tristimulus values X, Y and Z of the measured object to measure the color coordinates of the measured object. Because the $X_b$ are similar to Z curve, So $X_b$=0.151Z are taken as the short wave part of X curve.

## 3. Design of color analyzer

The color analyzer system is mainly divided into two parts: probe and control panel. The probe part includes the design of optical path, mechanical structure and signal acquisition and processing circuit board. The core of the control panel is circuit board. The composition flow chart of system is shown in Figure. 2.

IOP Conf. Series: Materials Science and Engineering **772** (2020) 012093 doi:10.1088/1757-899X/772/1/012093

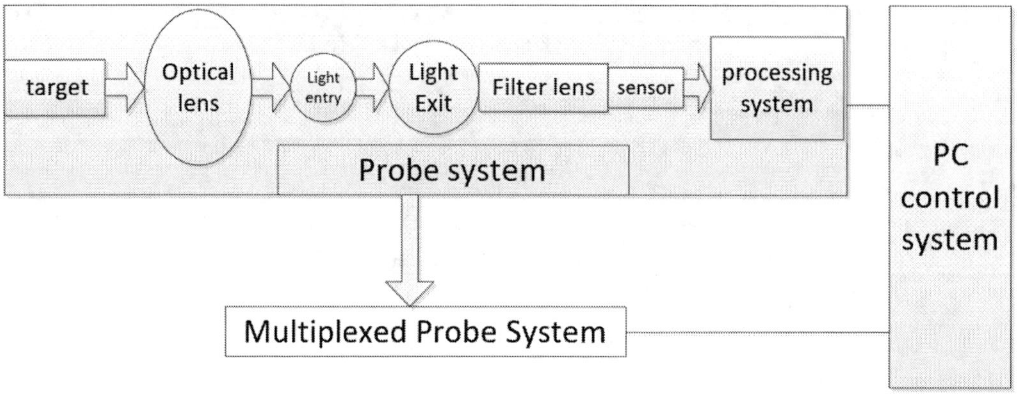


**Figure 2.** The flow chart of system.

The optical system of the probe mainly includes objective lens, optical fiber module, filter lens and sensor. It's working principle: the light of source to be measured gathers on the receiving window surface of the optical fiber module through the objective lens, and guides to the sensor through the optical fiber module and the filter lens. Especially To obtain accurate actual values in time of testing low brightness, the most important thing is that reduces the light loss to transmitted to the sensor, and prevents the light from focusing on the outside area of the receiving window of the optical fiber module, In such a case to reduce the light loss and could obtain accurate measurement. The following Figs are respectively the appearance of prototype and the test interface.

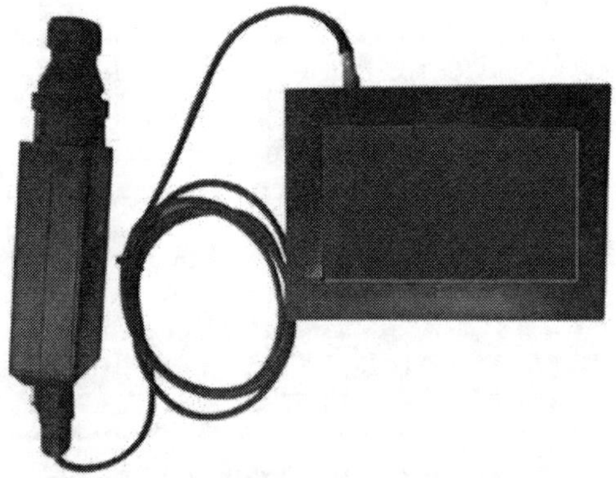

**Figure 3.** The appearance of prototype.

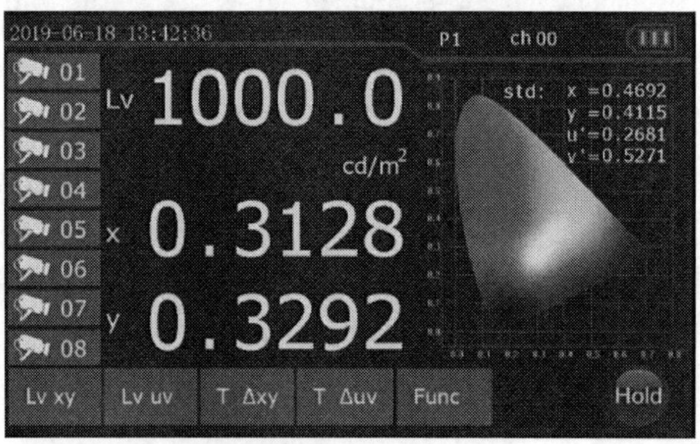


**Figure 4.** The test interface.

Experimental results and analysis: The CA-310 color analyzer is a mature product. As a competitive product, it will be tested and compared under the same conditions to verify the accuracy and reliability of the product. Experimental conditions: standard lamp, CA-310, self-made CSA-510 and other auxiliary equipment, etc. Brightness and color coordinates were measured. The test results are as follows.

**Table 1.** Comparison of test results.

| CA-310 | | | CSA-510 | | |
|---|---|---|---|---|---|
| lv | x | y | lv | x | y |
| 458.00 | 0.4715 | 0.4156 | 452.10 | 0.4741 | 0.4109 |
| 299.10 | 0.4717 | 0.4154 | 294.30 | 0.4740 | 0.4109 |
| 199.50 | 0.4716 | 0.4154 | 197.30 | 0.4731 | 0.4147 |
| 100.60 | 0.4710 | 0.4154 | 99.70 | 0.4724 | 0.4130 |
| 50.40 | 0.4708 | 0.4156 | 51.40 | 0.4730 | 0.4126 |
| 29.50 | 0.4713 | 0.4154 | 30.10 | 0.4729 | 0.4141 |
| 15.00 | 0.4709 | 0.4154 | 15.20 | 0.4730 | 0.4140 |
| 10.07 | 0.4712 | 0.4156 | 10.10 | 0.4722 | 0.4177 |
| 5.00 | 0.4714 | 0.4162 | 5.10 | 0.4748 | 0.4167 |
| 4.00 | 0.4713 | 0.4164 | 4.00 | 0.4743 | 0.4177 |
| 3.20 | 0.4713 | 0.4164 | 3.20 | 0.4766 | 0.4131 |
| 2.50 | 0.4819 | 0.4239 | 2.45 | 0.4857 | 0.4228 |
| 2.07 | 0.4829 | 0.4243 | 2.00 | 0.4859 | 0.4255 |
| 1.50 | 0.4822 | 0.4258 | 1.48 | 0.4823 | 0.4277 |
| 1.00 | 0.4837 | 0.4263 | 1.02 | 0.4847 | 0.4254 |
| 0.50 | 0.4829 | 0.4272 | 0.50 | 0.4836 | 0.4265 |
| 0.10 | 0.4850 | 0.4284 | 0.10 | 0.4871 | 0.4281 |

According to the data in the table: Under the same test conditions, the brightness of test results basically maintain the same linear relationship with the calibration results. The error is less than 2%. the color coordinates keeps good consistency. The error is within ±0.0050. At the same time, the same result can be maintained with CA-310 for low brightness measurement.

## 4. Conclusion

In this paper, according to the principle of color measurement, The CSA-510 color analyzer is designed. Proved by experiment: the brightness of test results basically maintain the same linear relationship with the calibration results. The error is less than 2%. The color coordinates keeps good consistency. The error is within ±0.0050. At the same time, the same result can be maintained with CA-310 for low brightness measurement. It can meet the requirement of color measurement. It has certain reference significance for the design of color testing products.

**Acknowledgments**
This work was financially supported by Zhejiang University Kunshan Innovation Institute fund.

**References**

[1] Jun Yao. Study on the Way of Quickly Setting up the Standard White Field of TV [J]. Chinese Journal of Scientific Instrument,2002,323(3):79-81.
[2] Jun Yao,Guan rong Ye. Calibration of color analyze[J]. Measurement Technique ,1997(8):30-32.
[3] Jun Zhang,Guo kui li,Hua qing Sheng. Development of calibration system for color analyzer[J]. Advanced Measurement and Laboratory Managemen,2013(1):12-14.

# Noise Prediction of Traction Gear of High-Speed EMU Base on Acoustic-Structural Coupled Method

**Zhaoping Tang[1, *], Min Wang[1, a], Yi Wu[2, b], Min Zhao[3, c] and Xianliang Zheng[3, d]**

[1]School of Information Engineering, East China Jiaotong University, Nanchang Jiangxi, 330013, China.
[2]College of Science and Technology, Nanchang University, Nanchang, 330029, China
[3]CSR Qishuyan Locomotive & Rolling Stock Technology Research Institute Co., Ltd, Changzhou Jiangsu, 213011, China.

* Corresponding author email: tzp@ecjtu.edu.cn, [a] 1291966861@qq.com, [b] 1443059097@qq.com, [c] zhaomin.qs@crrcgc.cc, [d] zhengxianl@163.com

**Abstract.** The dynamic characteristics of traction gear transmission system have great influence on the safety, comfort and reliability of EMU. Base on acoustic-structural coupled theory, the noise radiation characteristics of gear transmission system in high-speed train CRH380A are studied by finite element-boundary element method. The acoustic radiation analysis model is built by using acoustic BEM. The radiated noise is predicted by Helmholtz boundary integral equation. The research results can provide a theoretical basis for the optimization design of the low noise gear in EMU.

**Keywords:** traction gear of EMU, dynamic characteristics, acoustic BEM, noise prediction.

## 1. Introduction

Traction transmission system is the core of power source for high-speed electric multiple unit (EMU) running. To study the dynamic mechanical characteristics and acoustic response characteristics in gear meshing process, furthermore to analyze and predict the magnitude and the origin of noise is the key for improving the transmission performance of traction gear and reducing vibration and noise.

The internal excitation is the characteristic that distinguishes gear transmission from other mechanical systems, also the origin of gear transmission system noise. Xue et al. [1] established the Spur Planetary Gear model by using the parametric method, simulated and analyzed the gear meshing process, and obtained the contact pressure and stress in a meshing period. Tang et al. [2] established finite element model of traction gear transmission system for the high-speed train CRH380A, and based on transient dynamics, analyzed the distribution and variation of equivalent stress and contact pressure on the gear tooth surface. Taking the dynamic meshing force along the meshing line, away from the meshing line, the dynamic tooth force and the bearing force as measure, Ghosh and Chakraborty [3] evaluated the vibration and noise of gears under different conditions.

The noise of gear transmission is not only about transmission error fluctuation, tooth surface stress distribution, dynamic meshing force change and other mechanical characteristics, but also about the structural characteristics, vibration characteristics of gear and transmission characteristics of meshing

Content from this work may be used under the terms of the Creative Commons Attribution 3.0 licence. Any further distribution of this work must maintain attribution to the author(s) and the title of the work, journal citation and DOI.
Published under licence by IOP Publishing Ltd

force. Hou[4] studied the test vibration data of CRH2 EMU gearbox, made some comparison on the vibration frequency characteristics of gearbox under different working conditions, and explored the standard for vibration test and analysis of EMU gearbox. Through modeling and simulation test of the carriage of inter-city EMU CJ-1, Jia et al.[5] analyzed the generation, propagation, as well as acoustic-vibration relationship between propagation rule and bodywork, and found the main noise origin parts of EMU. Li[6] constructed the finite element simulation model for EMU gearbox, avoided resonance by modal analysis, meanwhile, measured the transmission noise by load test method.

The noise analysis and prediction of gear transmission system was achieved indirectly by studying the impact vibration and dynamic load on gear teeth [7-8], however, further acoustic study on gear transmission system is rare. The main reason is that it is difficult to build a noise analysis model for system, which is also much more complicated [9]. In this paper, based on acoustic-structural coupled system, by the finite-element boundary method, the dynamic characteristics and vibration response characteristics of the gear meshing has be analyzed for EMU (high-speed train CRH380A) under continuous traction.

## 2. Parametric Modeling of Gear Transmission System in Emu

In this paper, using software PRO/E, a parametric solid model is constructed for helical gear transmission system.

### *2.1. The Basic Parameters of Gear Pair*

The basic geometric parameters of traction gear pair of high-speed train CRH380A are shown in Table 1.

**Table 1.** Basic geometric parameters of gear pair

| Name | Driving gear | Driven gear |
|---|---|---|
| Teeth $z$ | 29 | 69 |
| Modulus $m_n$ / (mm) | 7 | 7 |
| Helix Angle $\beta$ / (°) | 20 | 20 |
| Tooth face width $B$ | 70 | 70 |

### *2.2. Establishment of Parametric Model*

**Figure 1.** Simplified model of gear transmission system

Considering the workload of computation in simulation, to ignore parts such as the transmission shaft, bearing and spline, the simplified gear pair is taken as the research object model (Fig.1).

## 3. Dynamic Characteristics Analysis of Gear Transmission System

By analyzing and solving the dynamic characteristics of gear pair, the accurate origin of noise and acoustic boundary conditions are provided for the acoustic radiation study of the system.

### *3.1. Dynamic Simulation in Gear Meshing Based on RecurDyn*

The built model of gear pair in PRO/E is imported into the multi-body system dynamics simulation software RecurDyn and the dynamic meshing force of gear pair is solved. The main steps are as follows:

1. To define the transmission mode.
2. To define material properties. These properties are shown in Table 2.

**Table 2.** Material properties of gear pair

| Material properties | Driving gear | Driven gear |
|---|---|---|
| Elastic modulus E / (MPa) | $2.06 \times 10^{11}$ | $2 \times 10^{11}$ |
| Poisson's ratio μ | 0.31 | 0.29 |
| Density ρ / $(kg/m^3)$ | 7850 | 7800 |

3. To define boundary conditions. The speed and torque are respectively 841.5N.m and 434rad/s for traction gear pair of high-speed train CRH380A under the continuous working condition.
4. To simulate and solve. The dynamic meshing force is solved by finite element method.

It can be seen from Fig. 2 and Fig. 3, the overall trend of the meshing force in each meshing period is stable and presents a periodic variation law although the meshing force of the gear pair fluctuates with time.

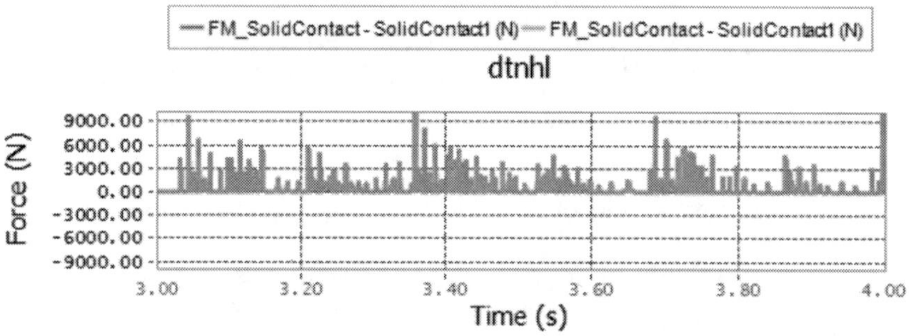


**Figure 2.** Time domain diagram of gear pair meshing force

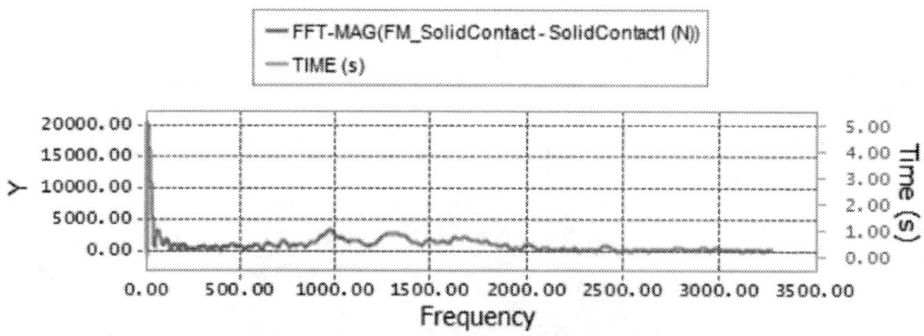


**Figure 3.** Frequency domain diagram of gear pair meshing force

### *3.2. Modal Analysis*

By the modal analysis, the any ordered natural frequencies and modal characteristics of the transmission gear pair in a certain frequency range can be solved. On this basis, the vibration response of the gear pair under various excitations in a certain frequency range can be predicted.

The modal analysis can be realized through the workbench module in ANSYS. The finite element structures were solved, and the results are shown in Table 3. The mode of the first order is shown in Fig. 4, and the others are omitted.

**Table 3.** Natural frequency of gear pair in the first six order

| Order | Frequency (Hz) |
|---|---|
| 1 | 507.34 |
| 2 | 769.1 |
| 3 | 882.65 |
| 4 | 1457.6 |
| 5 | 2186.7 |
| 6 | 2603 |

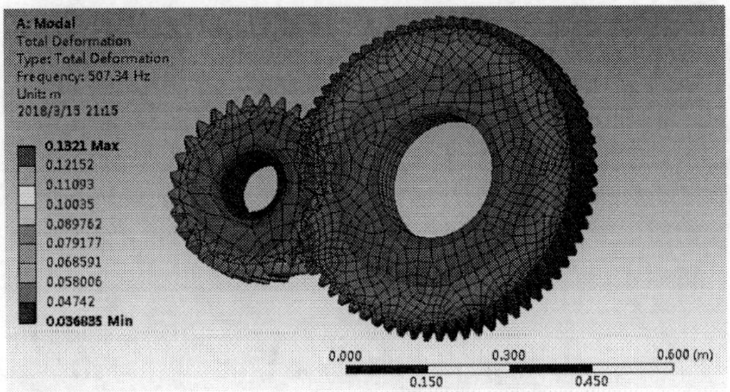


**Figure 4.** The first order natural mode of vibration of gear pair

### 3.3. *Dynamic Harmonic Response Analysis of Transmission Gear*

The harmonic response analysis include vibration displacement, vibration velocity and vibration acceleration. In the analysis, the peak value and the corresponding frequency should be emphatically observed as to avoid fatigue and resonance. The forced vibration equation of undamped system is:

$$[\mathbf{M}]\{\ddot{u}\}+[\mathbf{K}]\{u\}=\{\mathbf{F}\} \tag{1}$$

In the equation, $u$ is vibration displacement. $\ddot{u}$ is vibration acceleration, $M$ represents systematic mass matrix. $K$ represents stiffness matrix.

According to formula 1, the vibration response characteristics are calculated and the frequency response output curve is gotten (Fig. 5).

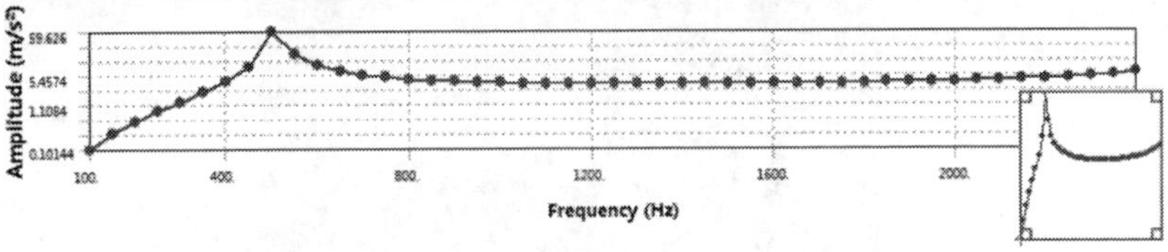


**Figure 5.** Gear pair frequency response curve

According to Fig.5, the frequency response curve has a peak at frequency 500Hz,and there is no significant peak near the following order natural frequencies, it shows that the vibration generated in the process of gear transmission which is mostly concentrated in the low frequency band. It is completely consistent with the actual situation, and the reliability of the model and method is verified.

## 4. Noise Analysis of Gear Transmission

The above dynamic characteristics analysis can provide boundary conditions for acoustic noise analysis of gear transmission system.

### 4.1. *To Select the Numerical Computation Methods for Acoustic*

In dealing with the acoustic radiation issue in unbounded sound field, if the whole model entity is meshed by the finite element method, the computational complexity is often too high to calculate. However, by the boundary element method based on Helmholtz boundary integral equation, the acoustic radiation can be calculated with one pattern by extracting the surface mesh of the structure[10]. The expression is as follows:

$$p=[\mathbf{A}_v(w)]^{\mathrm{T}}v_n(w) \tag{2}$$

$$p(r_a)=\sum_{i=1}^{n_e}N_i^e(r_a)\cdot a_{pi},\quad r_a\in\Omega_{ae} \tag{3}$$

$$v_n(r_a)=\sum_{i=1}^{n_e} N_i^e(r_a)\cdot a_{vi},\quad r_a\in\Omega_{ae} \tag{4}$$

In these equations, $p$ is sound pressure. $\mathbf{A}_v(w)$ is acoustic transfer vector. $v_n(w)$ is the normal velocity of vibration on structural surface unit. $n_e$ is the number of nodes on the $\Omega_{ae}$ unit. $a_{pi}$ is sound pressure on the boundary element nodes. $a_{vi}$ is normal velocity on the unit nodes. $N_i^e$ is shape function of unit.

Then, the normal velocity $v_n$ and sound pressure $p$ at a distance of $r$ from the origin of noise in the gear transmission system are satisfied as equation (3) and equation (4).

*4.2. Noise Analysis Based on Boundary Element Method*

The obtained results of the vibration response from the analysis in Section 3 are imported into the noise vibration analysis software LMS, the noise of the gear transmission system can be simulated and predicted.

The sound field medium is defined as the air fluid property, and the action of the field point is similar to the acoustic sensor, where it needs to calculate the noise level in the sound field [11].The distance is about 1m from each point to the center of the gear pair on the ISO spherical acoustic field. Taking the positions of passengers are right above the gear pair into consideration, in the sound field, as marked in Fig.6, the four appropriate points is selected. To check the sound pressure level at those points, and the corresponding sound pressure level curves are plotted (Fig. 7).

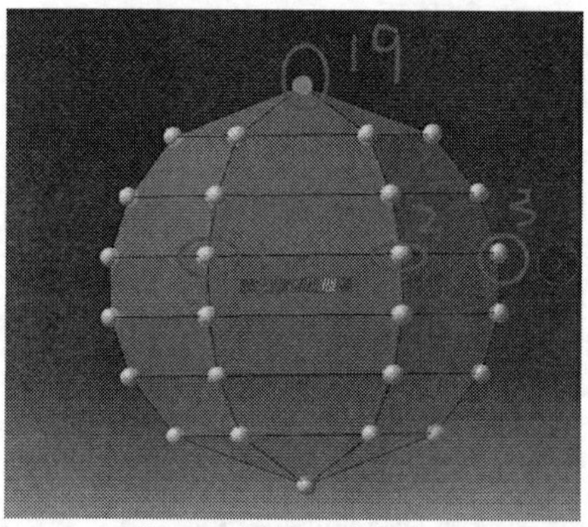


**Figure 6.** Pictorial view

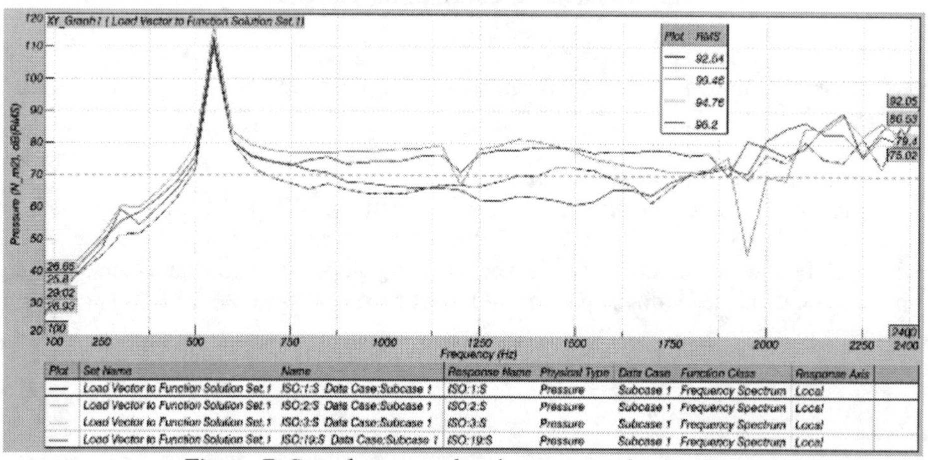


**Figure 7.** Sound pressure level curveat various points

### *4.3. To Predict the Noise*

In the paper, the sound power level of the radiated noise will be solved, and it will reflect the radiated energy of the origin of noise in a time unit. The average sound power is the average sound energy passing through the plane and being perpendicular to the plane in the unit time. The expression is shown as follows:

$$W = \xi c_0 S \tag{5}$$

$$\xi = \frac{E}{V_0} = \frac{1}{2}\rho_0\left(v^2 + \frac{1}{2\rho_0 c_0^2}p^2\right) \tag{6}$$

In the equation, $c_0$ is the speed of sound propagation. $S$ is the area of acoustic diffusion direction. $\xi$ is energy density.

Through the sound pressure level contour obtained in the upper section (Fig. 8), according to the equation (5)-(6), the sound power level curve of the transmission gear pair and the root mean square (RMS) of the sound power level of the transmission noise can be obtained as Fig. 8. It can be seen from the figure that the corresponding RMS in the whole scanning frequency band is 114.44 dB.

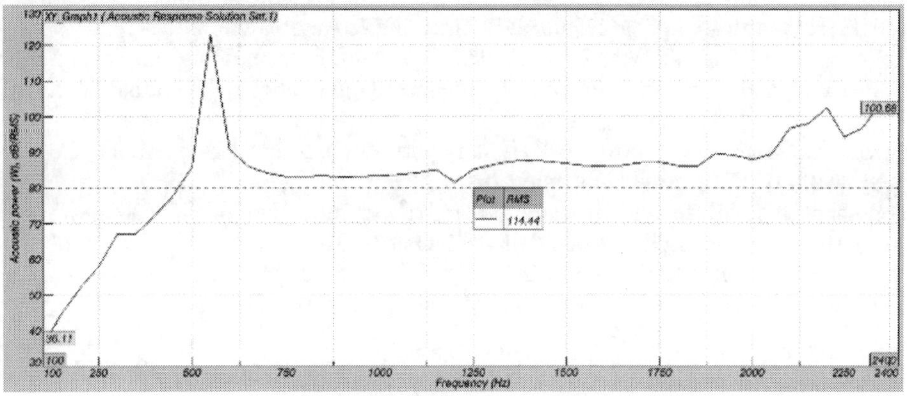


**Figure 8.** Sound power curve

## 5. Conclusion

Through the simulation analysis of the dynamic mechanical properties and acoustic response characteristics of the traction gear transmission system in high-speed train CRH380A, the noise of the gear transmission system is predicted by using of the finite element-boundary element method, which can provide reference for the optimization design of low noise gears for high-speed EMUs and the modification design of vibration and noise reduction. The main conclusions are shown as follows:

(1) Based on the parametric model of traction gear transmission system of high-speed EMU, the dynamic characteristics are analyzed, including dynamic meshing force, natural vibration characteristics and dynamic vibration response. The research results show that the generated vibration is mainly concentrated in the middle and low frequency band in the running process of the gears, and the peak value of frequency response curve appears at 500 Hz, which is just at the first order natural frequency.

(2) The vibration characteristic curve obtained from harmonic response analysis of gear transmission system, can establish boundary conditions for solving the noise response by boundary element method. The results show that the corresponding RMS of power level of the traction gear in the whole scanning frequency band is 114.44 dB.

**Acknowledgements**

This project is supported by National Natural Science Foundation of China (Grant No.51765015) and supported by the Natural Science Foundation Project of Jiangxi Province (Grant No.20171BAB206027).

**References**

[1] Xue S, Entwistle R, Mazhar I, et al. "The spur planetary gear torsional stiffness and its crack sensitivity under quasi-static conditions [J],"Engineering Failure Analysis, 2016, pp. 63: 106-120.
[2] Tang Z P, Sun J P, Yan L, et al. "Dynamic Contact Analysis and Tooth Modification Design for EMU Traction Gear [J],"International Journal of Simulation Modelling, 2017, pp. 16: 742-753.
[3] Ghosh S S, Chakraborty G, "On optimal tooth profile modification for reduction of vibration and noise in spur gear pairs [J],"Mechanism and Machine Theory, 2016, pp.105:145-163.
[4] Youzhong HOU, "Analysis and research of the vibration data during the running-in test of the gearbox for CRH2 EMU [J], "Rail Transportation Equipment and Technology, 2013, pp. 3: 42-44.
[5] Huanying JIA, Yanqiang CAI, Fanshuai MENG, et al. "Vibration and noise reduction scheme of CJ-1 EMU [J],"Railway Locomotive & Car, 2017,pp.39:49-53.
[6] Feng LI, "Simulation analysis and experimental study for gear transmission of EMU bogie [D]," Shanghai Jiaotong University, 2015.
[7] Xiaojing ZHU, "Optimization design of tooth profile of gear in motor car gearbox based on KISSsoft software [J]," Mechanical Engineering & Automation, 2016, pp. 92-93.
[8] Kunming CHUN, Jiaxu WANG, Wei PU, et al. "Study on meshing impact and modification method of filtering reducer considering mixed lubrication [J],"Journal of Xi'an Jiaotong University, 2015, pp.49:140-148
[9] Zeyin HE, "Transmission error analysis and optimization research on vibration and noise of gear system [D],"Chongqing University, 2015.
[10] Yongfang MA, "Research on vibration and noise reduction of SEM30 gearbox based on constrained damping [D]," Shandong University, 2016.
[11] Xin MENG, Liangjun LI, Xuyong WANG, "A new method for indoor sound power measurement of compressor's noise [J]," Noise and Vibration Control, 2015, pp.35:202-206.

IOP Conf. Series: Materials Science and Engineering **772** (2020) 012095 doi:10.1088/1757-899X/772/1/012095

# Study on Wear Performance of GS-18NiMoCr36/42CrMo Friction Pair

**Guo Xu [1, 2, *], Bo Chen [1, 2], Kun Feng [1, 2], Dongya Zhang[1, 2], Pengxiao Zhu[1, 2]**

[1] Jiangsu XCMG Construction Machinery Research Institute Ltd, Xuzhou Construction Machinery Group, Xuzhou, China

[2] State Key Laboratory of Intelligent Manufacturing of Advanced Construction Machinery, Xuzhou Construction Machinery Group, Xuzhou, China

*Corresponding author e-mail:1454014925@qq.com

**Abstract.** GS-18NiMoCr36 steel was used to prepare fixed pin with different hardness. The rotating ring with hardness of 53HRC was prepared by 42CrMo steel. Under the different load conditions, MR-H3B high-speed ring on block wear tester was used to study the influence of different hardness matching friction pair wear performance. The results show that with the increase of load, the weight loss of wear has the law of increasing-decreasing-increasing. The form of wear is oxidation wear - adhesive wear - oxidation wear - adhesive wear. As the hardness of the fixed pin increases, the weight loss of wear tends to decrease, and the hardness of the material can be improved within a certain hardness range, which can significantly improve the wear resistance.

## 1. Introduction

The large tonnage four-wheeled belt travel mechanism mainly includes a roller, a drive wheel, a guide wheel, a carrier roller and a track shoe. The main failure mode of the four-wheel belt is wear, concentrated in the following contact parts: the guide wheel and the track link section; the roller and track section; the sprocket and track section [1, 2]. These contact points are typical rolling friction wear. With the increase of surface friction, the rolling friction resistance increases rapidly [3]. It is especially important to study the effects of different loads on the wear characteristics of friction pairs. In recent years, GS-18NiMoCr36 (hereinafter referred to as G18) has been widely used in large-tonnage track shoes due to its good heat treatment properties and toughness, and 42CrMo is used in the manufacture of wheel bodies [4, 5]. The author simulates the influence of different loads on the G18/42CrMo friction pair on the MR-H3B high-speed ring block wear tester, and provides data support for the performance study of the whole four-wheel system.

## 2. Materials and methods

G18 round steel were cut into fixed pins of size 12mm×12mm×19mm by wire-cutting. It is quenched and tempered by different temperatures before grinding. The hardness is 26HRC, 36HRC and 46HRC respectively.

For the test, 42CrMo steel is a round bar with a size of Φ60mm. After machining, a circular ring grinding sample (hereinafter referred to as rotating ring) is obtained, the thickness is 13mm, the outer

Content from this work may be used under the terms of the Creative Commons Attribution 3.0 licence. Any further distribution of this work must maintain attribution to the author(s) and the title of the work, journal citation and DOI.
Published under licence by IOP Publishing Ltd

diameter is 49mm, and the inner diameter is 30mm.The composition of G18 steel and 42CrMo steel used in the test is shown in Table 1.

The ring block wear test was performed using a MR-H3B high speed ring block wear tester (the device schematic is shown in Figure 1.). The load is selected from 100N, 300N, 600N and 1000N, the rotation speed is 100r/min, and the time is 60min. The fixing pin and the rotating ring were ultrasonically cleaned with absolute ethanol before and after the test, and weighed using an electronic balance with an accuracy of 0.1 mg to calculate the weight loss of the worn sample. Three groups were matched for each set of matching tests, and the weight loss of wear was averaged using the results of the three groups. The worn surface of the worn sample was cleaned and dried, the worn surface was analyzed by a scanning electron microscope, and the micro-area composition of the worn surface was analyzed by using EDS.

**Table 1.**The composition of G18 steel and 42CrMo steel

| Grade | C | Si | Mn | Cr | Ni | Mo |
|---|---|---|---|---|---|---|
| G18 | 0.21 | 0.31 | 0.83 | 0.52 | 0.74 | 0.58 |
| 42CrMo | 0.415 | 0.248 | 0.572 | 0.97 | / | 0.16 |

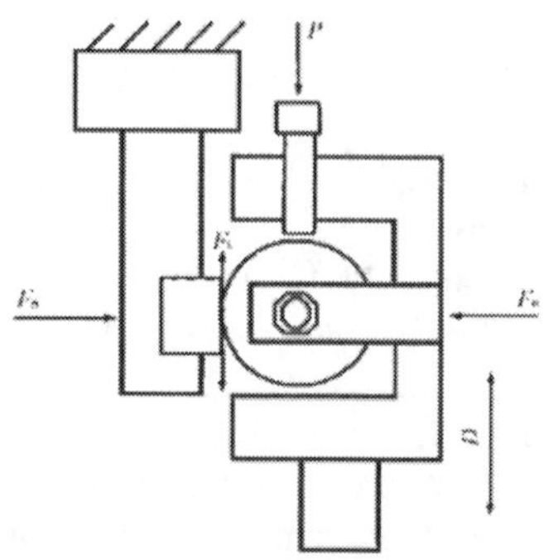

**Figure 1.** Schematic of MR-H3B high speed ring block wear tester

## 3. Results and discussion

### *3.1. Effect of load on wear performance of friction pair*

The test results of the ring block wear of different hardness pin and ring friction pair samples under different load conditions are shown in Table 2.

According to the results in Table 2, the weight change curves of the fixed pin and the rotating ring are wearing shown in Figure 2 and Figure 3 under different load conditions. At 100N~300N, the wear weight loss of pin and ring increases significantly when the load increases, taking the hardness of fixed pin (26HRC) as an example, the wear weight loss increased from76.1mg to 189.6mg, which increased by 2.5 times. When the load is 600N, the weight loss of the fixed pin (36HRC) is reduced from 185.6mg to 89.7mg, wear resistance increased by 51%. When the load is increased to 1000N, the wear loss of the friction pair is significantly reduced. It can be seen from Table 2 that the wear weight loss of the fixing pin is always higher than the rotating ring's. In general, the wear resistance of high hardness materials is higher than that of low hardness materials. This is consistent with the soft abrasive wear law. For the low and medium hardness fixed pins of 26HRC and 36HRC, the fluctuation curve of the grinding loss weight changes greatly. The magnitude of change was 115.5mg (26HRC) and 130.9mg (36HRC), indicating that the load has a significant effect in the wear process. And for the higher hardness sample of 46HRC, the weight loss of the pin changes slowly, the change range is only 44.2mg.

Ring-block wear simulates a relative sliding frictional wear process of a line contact, usually explained by the "adhesion-mechanical meshing" theory. Since the microscopic state of the friction surface between the friction pairs is uneven, the friction only occurs where the convex peak and the convex peak between the friction surfaces are exactly opposite each other [6, 7]. In general, when the fixing pin (26HRC, 36HRC) with low hardness and good plasticity is ground against a high-hardness rotating ring (46HRC), the hard micro-protrusion on the ring presses and cuts the pin to cause plastic deformation of the pin contact surface, and eventually the deformation of the deformed area causes wear. Therefore, the wear resistance of the material can be improved by increasing the hardness of the fixing pin.

**Table 2.** Grinding loss weight of friction pairs under different loads

| Load/N | The hardness of pin/HRC | The hardness of ring /HRC | Weight loss of pin/mg | Weight loss of ring /mg |
|---|---|---|---|---|
| | 26 | | 76.1 | 25.7 |
| 100 | 36 | 53 | 54.7 | 21.7 |
| | 46 | | 29.7 | 21.3 |
| | 26 | | 189.6 | 107.4 |
| 300 | 36 | 53 | 185.6 | 107.7 |
| | 46 | | 73.9 | 49.6 |
| | 26 | | 139.6 | 37.0 |
| 600 | 36 | 53 | 89.7 | 41.7 |
| | 46 | | 51.2 | 24.8 |
| | 26 | | 173.5 | 82.9 |
| 1000 | 36 | 53 | 165.2 | 84.4 |
| | 46 | | 68.8 | 60.8 |

In this test, when the load is increased from 300N to 600N, the wear of the friction pair decreases greatly due to the formation of oxidation layer with high hardness. When the load continues to rise to 1000N, the temperature rise of the friction pair intensifies. In addition to the formation of oxidation layer between the contact surfaces of the friction pair, a high-temperature softening layer will also be formed under the oxidation layer. At this time, the stress between the friction pairs is extremely large, and the bearing capacity of the softened layer is insufficient, which leads to the peeling of the oxidation layer with high hardness and brittleness on the surface due to brittle cracking, and the wear amount is increased.

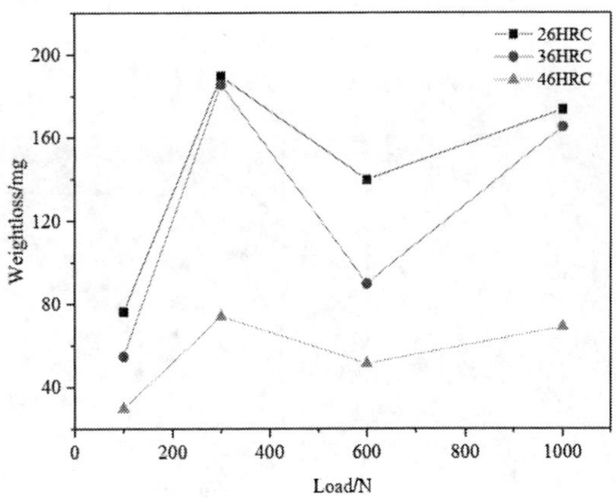


**Figure 2.** Effect of different loads on the wear performance of fixed pins

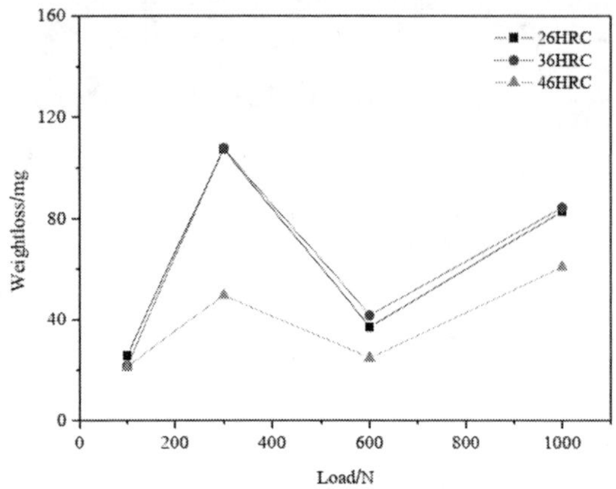


**Figure 3.** Effect of different loads on the wear performance of the rotating ring

*3.2. Analysis of wear trace morphology of ring-block wear*

Figure 4 shows the wear scar morphology of the fixed pin under different loads. The EDS energy spectrum analysis results of different regions in Figure 4 are shown in Table 3. It can be seen from Figure 4(a) that only a small amount of the surface of the wear scar is peeled off, and the wear debris is formed by the external force. From the EDS analysis results of the two regions A and B, it can be seen that the particles are oxides, and the surface of the wear scar is also an oxide. At this time, the oxidation wear is mainly dominant. With the EDS results of regions C and D, it can be seen from Figure 4(b) that the load is increased from 100N to 300N, the oxide film is peeled off, and it is broken into oxide particles under the action of the load. Oxide particles of different sizes are distributed on the surface of the wear scar, resulting in an increase in the roughness of the wear surface and aggravation of wear. At this time, the adhesive wear is dominant. As shown in Figure 4(c), when the load is 600N, combined with the EDS analysis result (the region of E and F ), the surface re-generates the oxide film due to the increase in temperature, reduces the friction coefficient, and reduces the wear amount. There are shallow furrows on the surface of the wear scar and typical adhesive wear morphology. It can be seen from Figure 4(d) that when the load is 1000N, cracks appear on the surface of the wear scar. The oxide particles formed by the rupture of the oxide film are pressed into the crack.

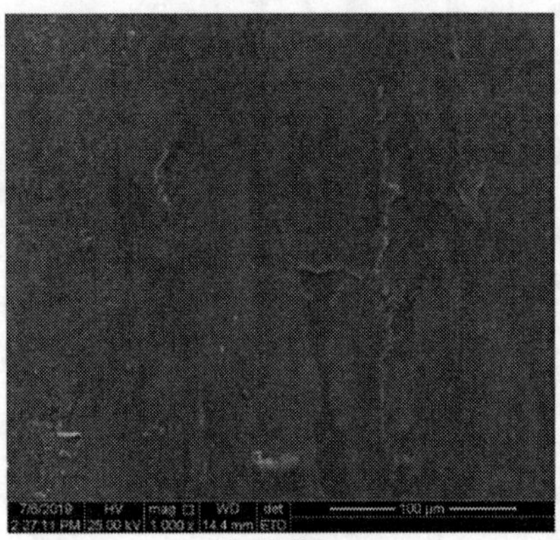

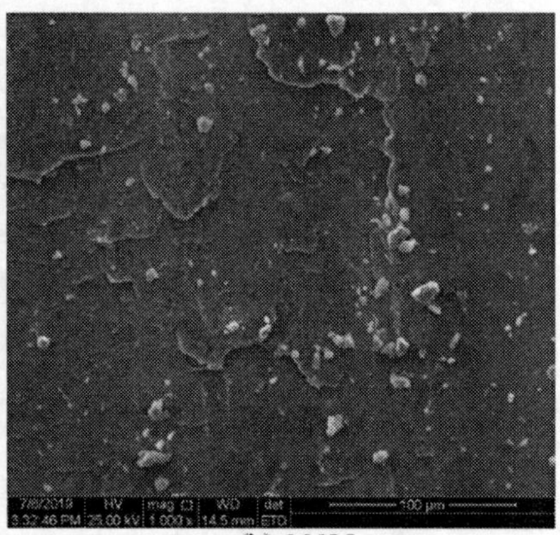

(a) 100N (b) 300N

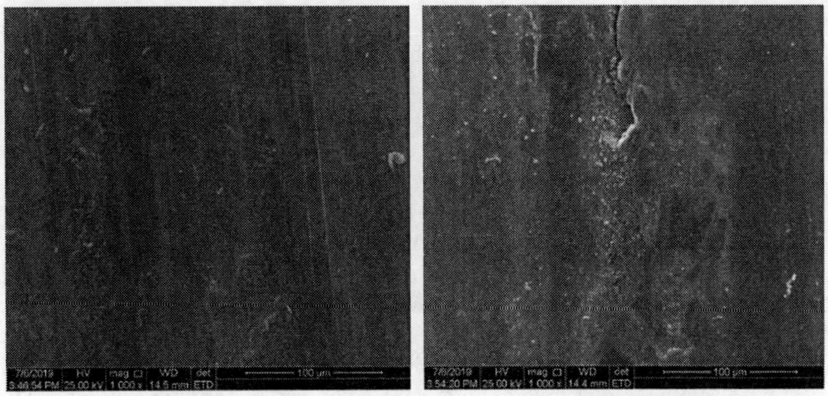

(c) 600N (d) 1000N

**Figure 4.** Wear scar morphology of the fixed pin under different loads

**Table 3.** EDS results for different regions in Figure 4

| Element | O | | Cr | | Fe | |
|---|---|---|---|---|---|---|
| Region | wt./% | at/% | wt./% | at/% | wt./% | at/% |
| A | 41.63 | 71.35 | / | / | 58.37 | 28.65 |
| B | 34.51 | 64.76 | 0.84 | 0.48 | 64.65 | 34.76 |
| C | 45.02 | 74.06 | 0.77 | 0.39 | 54.21 | 25.55 |
| D | / | / | 1.02 | 1.09 | 98.98 | 98.91 |
| E | 14.26 | 36.72 | 1.03 | 0.81 | 84.71 | 62.74 |
| F | 41.91 | 71.55 | 0.83 | 0.43 | 57.27 | 28.01 |
| G | / | / | 1.11 | 1.19 | 98.89 | 98.81 |
| H | 44.28 | 73.47 | 0.95 | 0.48 | 54.78 | 26.04 |

During the increase of the load, due to the conversion of mechanical energy, a temperature rise occurs between the contact surfaces of the friction pair. When the load increases to a certain extent, the temperature between the contact surfaces of the friction pair raises rapidly, and the oxide layer is formed due to the high temperature between the contact surfaces. The thermal conductivity of the oxide layer is lower than that of the metal matrix, which exacerbates the temperature rise of the contact surface. A denser, thicker dense oxide layer is formed between the faces. Studies have shown that the instantaneous temperature (flash temperature) between the metal contact surfaces under high stress can be as high as 1000℃ [8]. The high temperature on the one hand exacerbates the oxidation of the contact surface, increases the thickness of the oxide layer, and on the other hand, it softens the metal in contact areas due to high temperature. Due to the high hardness of the oxide layer, the substrate can be well protected and has a certain anti-wear effect. The ability of the softened metal to resist stress deformation is significantly reduced, resulting in rapid wear. Therefore, the influence of temperature rise on wear resistance of friction pair is complex, which is generally a combination of the two different wear mechanisms mentioned above.

## 4. Conclusion

(1) As the load increases, the weight loss of wear has the law of increasing-decreasing-increasing. When the hardness of the fixing pin is 36HRC, the wear weight loss is the lowest at 100kN, only 54.7mg, and the wear weight loss is 89.7mg when the applied load is 600kN, the wear resistance is increased by 52.6% compared with 300kN, and the wear resistance is increased by 45.7% compared to 1000kN.

(2) As the hardness of the fixed pin increases, the weight loss of wear tends to decrease. In all the results, the wear weight loss of the rotating ring is smaller than that of the fixed pin.

(3) At 600kN, the oxide film is re-formed due to temperature. The oxidation film reduces the friction coefficient, which results in the loss of wear weight of the friction pair.

**References**

[1] XIAO Qing. Wear and Preventive Measures of Track Wheel Dozer "Four Wheels and One Belt"[J]. China Equipment Engineering, 2007 (2): 35-36.

[2] ZHAO Ming, YAN Wei. Wear Limit and Maintenance Points of C42000 Crawler Crane "Four Wheels and One Belt" [J]. Construction Machinery and Maintenance, 2014 (10):128-129.

[3] LIU Qi-yue, ZHOU Zhong-rong, SHI Xin-yu, et al. Rolling Wear Characteristics of 45# Steel/GCr15 Steel Friction Pairs [J]. Journal of Tribology, 2001, 21 (1): 33-36.

[4] FAN Hui-ji, Li Yi, Jin Xin, et al. Effect of Tempering Temperature on Microstructure and Properties of G18NiMoCr3-6 Steel [J]. Heat Treatment of Metals, 2017, 42 (11): 163-165.

[5] ZHAI Jin-hui, Liu Shiyu. Optimization of Process Parameters for Quenching and Heat Treatment of 42CrMo High Hardness Roller [J]. Large Castings, 2007 (4):6-8.

[6] WANG Wen-jian, ZHANG Xiang-long, ZHANG Ji-wang, et al. Effect of carbon content on rollng friction and wear properties of wheel steel [J]. Journal of the China Railway Society, 2012, 34 (2): 32-35.

[7] BAI Wan-zhen, WEI Shi-zhong, LONG Rui, et al. Effect of heat treatment process on rolling wear performance of high vanadium high speed steel [J]. Foundry Technology, 2008, 30 (7): 864-868.

[8] LIU Yu-wei, ZHANG Wei-zheng, YE Fu-hao. Study on Flash Point Temperature in Single Microprotrusion Contact Process [J]. Journal of Xi'an Jiaotong University, 2016, 50 (3): 90-93.

IOP Conf. Series: Materials Science and Engineering **772** (2020) 012096 doi:10.1088/1757-899X/772/1/012096

# Study of a fatigue damage model for flange considering the friction coefficient of the faying face

**Yuhao Zhao[1,*], Jing Wei[1], Wencan Yin[2], Shijie Zhang[1]**

[1] School of Chongqing University, Chongqing, China
[2] Jiangsu Kwoyuen New Energy Technology Group Co., Ltd., Huaian 211600, China

*Corresponding author e-mail: zhaoyuhao1992@126.com

**Abstract.** In the flange connections, part of the load is transmitted by the friction moment between the faying faces, thus the distribution of the input load in the connection is changing with the friction coefficient. Current models usually calculate the fatigue damage directly according to the load spectrum without considering the variation of the load due to friction coefficient. In the paper, a fatigue damage model for flange plate considering friction coefficient is proposed, and the influence of friction coefficient on the load distribution of the connection is deduced. Based on Palmgren-Miner linear damage theory, the algorithmic expressions of fatigue damage at varying friction coefficient are derived, which not only assesses the fatigue strength of flange under variable friction coefficient, but obtains the friction coefficient range which meets the fatigue strength requirement of flange connections.

## 1. Introduction

Flange connections are often applied to connect pipes or to transmit loads. In a transmission system, the working condition is usually dynamic [1], which can easily lead to fatigue failure. The flange connection and transmission mechanism of a wind power system is shown in Fig. 1. As a key component of transmission system, the fatigue life of flange is directly related to the performance of the whole system. After Fatigue failure, maintenance also causes a lot of manual labor and financial losses [2]. Thus, it is of great significance to assess the fatigue damage and check the fatigue strength of flanges during mechanical design [3, 4].

At present work, numerous fatigue damage calculation approaches were proposed based on different elements, among which time domain analysis approach is extensively used [5]. The fatigue damage is modeled mainly based on nominal stress method, local stress-strain method, energy method. Although there are various fatigue damage models for various systems, the fatigue calculation based on friction coefficient and considering the load distribution is still unprecedented. Thus, a fatigue damage model of flange considering friction coefficient is presented in this work. Based on a 3.0MW wind turbine, the fatigue damage and fatigue strength of the flange based on variable friction coefficient is calculated, and the influence of friction coefficient on the stress distribution and fatigue strength of the flange is obtained.

Content from this work may be used under the terms of the Creative Commons Attribution 3.0 licence. Any further distribution of this work must maintain attribution to the author(s) and the title of the work, journal citation and DOI.
Published under licence by IOP Publishing Ltd

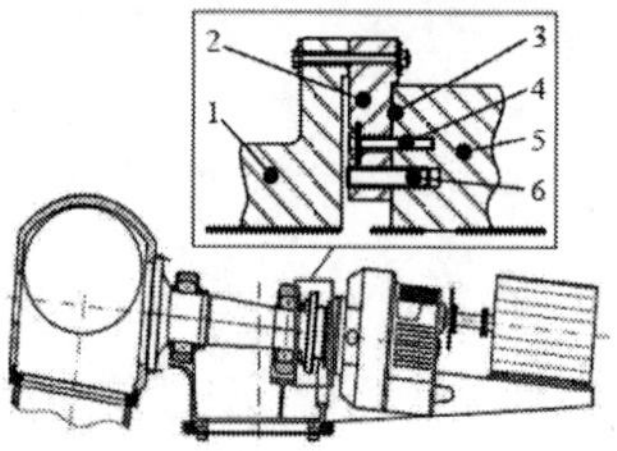


Figure 1. Inputting flange connection

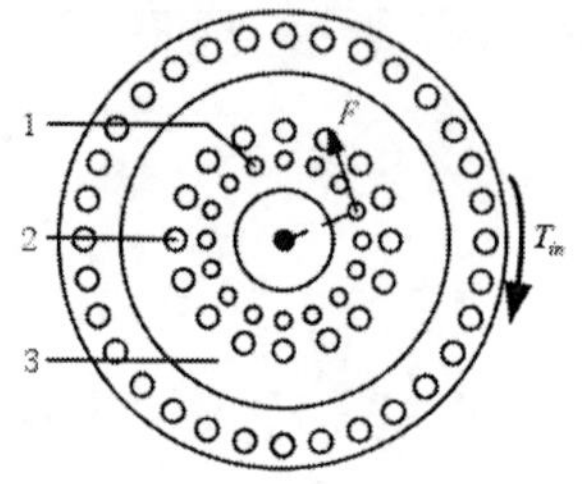


**Figure 2.** Force acting on flange

## 2. Theory model

### *2.1. Fatigue load considering friction coefficient*

During operating, the faying face of the flange is subjected to frictional force and the pin holes of the flange are subjected to circumferential positive pressure, as shown in Fig. 2. And the wind turbine is subjected to random torsional loads during operation, which is shown in Fig. 3, so the torque transmitted by the flange is not equal to the total load. The magnitude is determined by the area of the faying surface and the friction factor. According to the Palmgren-Miner linear damage theory, the fatigue damage accumulates with the increase of stress cycle, and fatigue failure is led when the total fatigue damage D is greater than the allowable value $D_{\max}$, which is given according to particular criteria [6, 7]. Assume the faying face is a complete ring, then the maximum static friction moment acted on the flange is:

$$T_{f\max} = 2\pi pf\int_r^R x^2 dx \tag{1}$$

In Eq. (1), $p$ is the unit positive pressure of the faying face, $f$ is the friction coefficient, $r$ and $R$ are the inner circle radius and outer circle radius of the faying face, respectively.

According to the nature of static friction, when the input torque $T$ is less than $T_{f\max}$, the friction moment is equal to the input torque, means that the load is transmitted by pins alone; otherwise the load is transmitted paratactic by frictional force and pin hole force. Based on Eq. (1), the pinhole force is calculated as:

$$\begin{cases} T_{pin} = 0 & T_{in} \le T_{f\max} \\ T_{pin} = T_{in} - T_{f\max} & T_{in} > T_{f\max} \end{cases} \tag{2}$$

In Eq. (2), $T_{pin}$ is the torque transmitted by pins, $T_{in}$ is the input torque. Eq. (2) implies that when the input torque increases from 0, there is a non-linear relationship between pin torque and the input torque. It is further deduced that the circumferential force acting on each pin hole of the flange is:

$$\begin{cases} F_p = 0 & T \le T_{f\max} \\ F_p = \dfrac{T_{pin}}{r_p \times n} & T > T_{f\max} \end{cases} \tag{3}$$

In Eq. (3), $r_p$ is the radius of the pinholes' position on the flange, $n$ is the quantity of pin hole.

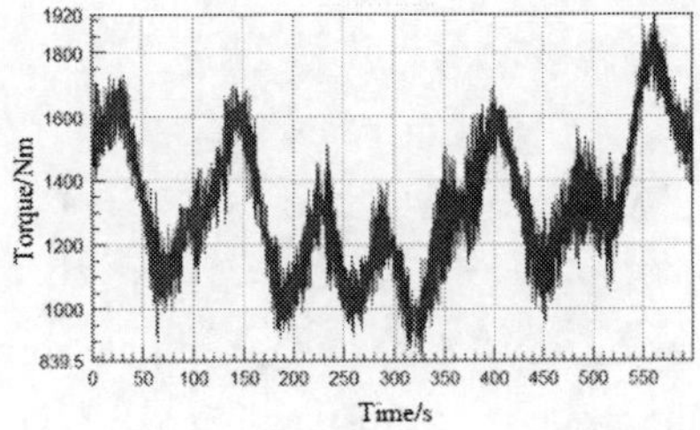


**Figure 3.** Random torsional load of wind turbine

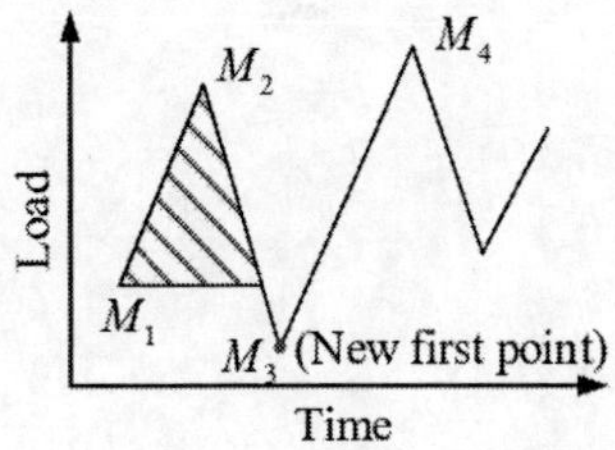


**Figure 4.** The RFC method

### *2.2. Rainflow cycle counting*

As for the counting of load time histories, there are three main approaches to count the number of load cycles: level crossing approach [8], peak cycle approach [9] and rainflow cycle method (RFC) approach [10].Over the years, ASTM [11] has proposed many cycle counting methods to solve the fatigue damage of wind turbines and supporting structures, which usually based on the time history of stress or strain. Among these counting methods, the most widely applied is RFC. In 1968, Matsuishi et al. [12] first proposed the RFC based on the principle of raindrops dropping from eaves. Then this method was given the mathematical definition by Dowling [13]. After improvement by Rychlik [14], a simplified RFC for the periodic amplitude of rainflow in random loading process was obtained. In 1993, Frendahl [15] discussed RFC under fixed load and the linear fatigue damage accumulation theory, and then proved the accuracy of it. RFC has been improved and developed in different applications [16, 17] among which the three-point counting method, benefits from its high accuracy, has been widely used.

The three-point RFC method use three successive points to form a cycle. As shown in Fig. 4, The method is based on three successive stress points ($T_1$, $T_2$, $T_3$) that defined the successive ranges $\Delta T_1 = T_1 - T_2$ and $\Delta T_2 = T_2 - T_3$. If $\Delta T_1 \le \Delta T_2$, $dT_1 \le dT_2$, one cycle from $T_1$ to $T_2$ is extracted and the point $T_3$ will be regarded as the new first point, otherwise no cycle is counted. After counted by RFC method, the load will be translated to a symmetric cyclic one, which is shown in Fig. 5.

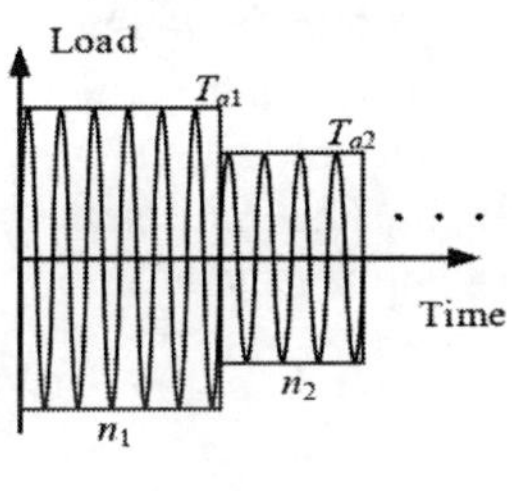


**Figure 5.** Symmetric cyclic loads

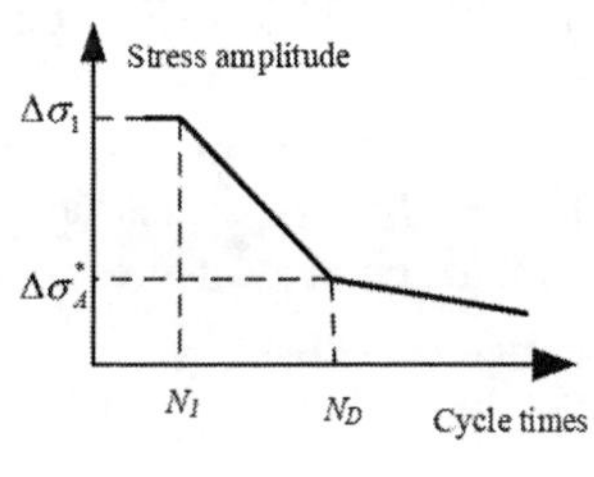


**Figure 6.** The S-N curve

### *2.3. Linear fatigue damage theory*

At present, it is a common practice to apply the Palmgren-Miner linear damage theory to the calculation of the total fatigue damage. Hashin et al. [11] proposed the method of calculating fatigue damage of specimens with S-N curve and Palmgren-Miner linear damage theory and proved its reliability through a series of experiments. Subsequently, Leipholz [12] proposed the concept of modified S-N curve which takes more load factors into account, which is widely used in the present work. The modified S-N curve is shown as Fig. 6 and is expressed as follow:

$$\begin{cases} S_a = \Delta\sigma_A^*\left(\frac{N}{N_D}\right)^{b_2} & 0 < S_a \le \Delta\sigma_A^* \\ S_a = \Delta\sigma_A^*\left(\frac{N}{N_D}\right)^{b_1} & \Delta\sigma_A^* < S_a \le \Delta\sigma_1 \end{cases} \tag{4}$$

In Eq. (4), $S_a$ is the stress amplitude in the loading history, $\Delta\sigma_A^*$ is the ordinate corresponding to S-N curve low cycle fatigue inflection point, $N_D$ is the abscissae corresponding to S-N curve low cycle fatigue inflection point, $b_1$ and $b_2$ are the slopes of high-cycle fatigue curve and low-cycle fatigue curve on S-N curve, respectively.

$$D = \sum_{i=1}^{n} \frac{n_i}{N_i} \tag{5}$$

In Eq. (5), $n_i$ is the number of load cycles in the loading history, $N_i$ is the fatigue life. As for asymmetrical stress, the damage caused by mean stress should also be taken into account [13]. According to Goodman's model [14], it is assumed that the mean stress is linearly related to the stress amplitude as:

$$S_a = (1 - \frac{S_m}{S_u})\mathrm{S}_e \tag{6}$$

In Eq. (6), $S_a$ is the stress amplitude, $S_m$ is the mean stress, $S_u$ is the ultimate tensile strength of the material, $\mathrm{S}_e$ is the fatigue limit of the material. With a combination of Eq. (4), Eq. (5) and Eq. (6), the total fatigue damage is:

$$D = \sum_{i=1}^{n} n_i / [\mathrm{N}_D(\frac{S_{ai}S_u}{(\mathrm{S}_u - \mathrm{S}_m)\Delta\sigma_A^*})^{1/b_j}] \tag{7}$$

In Eq. (7) $j = 1$ or 2, depending on the S-N curve. In order to obtain the fatigue damage of flange based on loading history, the input loading history must be transformed into the stress history of the pin hole. In the present practical application, the procedure is: calculate the equivalent stress under unit torque first, then apply a linear relationship to obtain the stress history. The finite element method (FEM) is used to obtain the equivalent stress of flange under unit torque, which is deduced as:

$$\sigma_0 = \frac{1}{\sqrt{2}}[(\varepsilon_1 - \varepsilon_2)^2 + (\varepsilon_2 - \varepsilon_3)^2 + (\varepsilon_3 - \varepsilon_1)^2]^{\frac{1}{2}} \tag{8}$$

The stress amplitude and mean stress in the pin hole is deduced as follow:

$$\begin{cases} S_{ai} = T_{ai}\sigma_0 \\ S_{mi} = T_{mi}\sigma_0 \end{cases} T_i > T_{f\max} \tag{9}$$

With a combination of Eq. (5) to Eq. (9), the total fatigue damage of the flange is derived:

$$D' = \sum_{i=1}^{n} n_i / [\mathrm{N}_D(\frac{T_{ai}\sigma_0 S_u}{(\mathrm{S}_u - \mathrm{S}_m)\Delta\sigma_A^*})^{1/b_j}] \tag{10}$$

Thus the factor of stress (FOS) of the flange, through which the fatigue strength of flange can be evaluated, can be deduced as:

$$FOS = \frac{1}{D'} \tag{11}$$

IOP Conf. Series: Materials Science and Engineering **772** (2020) 012096 doi:10.1088/1757-899X/772/1/012096

## 3. Simulation

### *3.1. Static Analysis of the flange*

The detailed parameters of the input flange are shown in Table 1. As shown in Fig. 7, the model has a total of 774960 elements and 1122603 nodes. According to the actual force of the flange, the full constraint is applied to the outer ring, and the unit torque is applied in the pinholes.

**Table 1.** Analysis parameters

| Parameters | E(MPa) | μ | $\Delta\sigma_A^*$ (MPa) | $N_D$ (MPa) | P(N/mm) | r(mm) | R(mm) |
|---|---|---|---|---|---|---|---|
| value | 20600 | 0.3 | 257.89 | 1002000 | 40 | 175 | 425 |

According to Eq. (1), the maximum static friction moment is 2674.6 kNm. The equivalent stress under the unit load (1kNm), calculated using by ANSYS, is shown in Fig. 8.

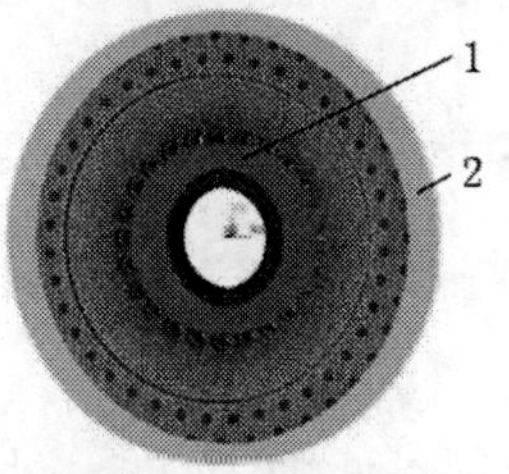


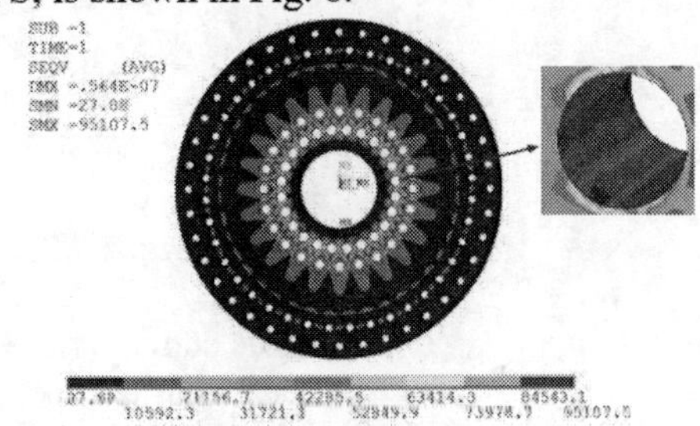

**Figure 7.** Finite element model **Figure 8.** Equivalent stress of flange

It is shown in Fig. 8 that the maximum equivalent stress of the flange is 95107.5 MPa.

### *3.2. Transformation of the fatigue load*

In the 3MW wind turbine, the unit positive pressure of the faying face is 40N/mm, radius of outer circle of contact surface is 425mm, radius of the inner circle of contact surface is 175mm. When the friction coefficient is 0, 0.1, 0.2, 0.3, 0.4 and 0.5 respectively, relationship between the pin torque and the input torque is shown in Fig. 9.

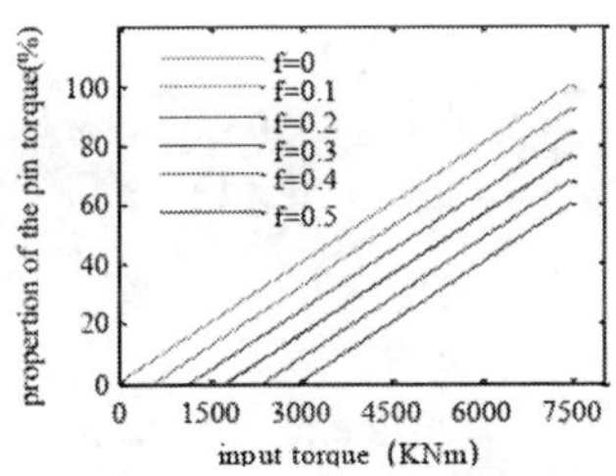


**Figure 9.** Loads based on different friction coefficient

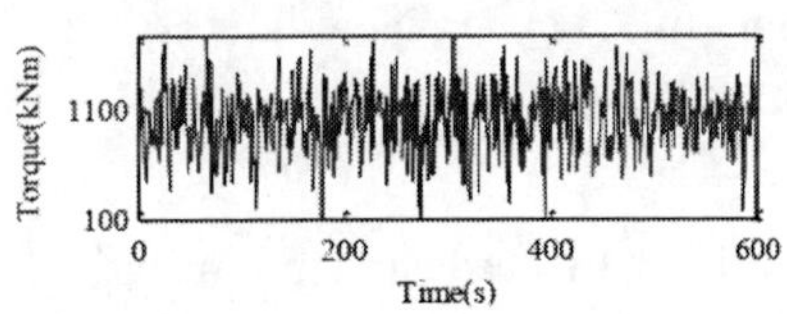


**Figure 10.** Pinhole load of DLC1.2-18-0( $f = 0.4$ )

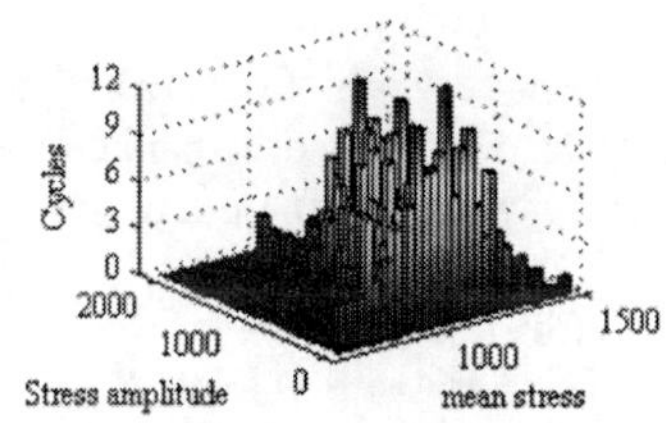


**Figure 11.** Rainflow matrix of load DLC1.2-18-0 ( $f = 0.4$ )

It is shown in Fig. 9 that when the friction coefficient is constant, the pin torque increases with the increase of the input torque; and when the input torque is constant, the pin load decreases with the increase of the friction coefficient. Particularly, the pin torque equals to the input torque when the friction coefficient is 0, which means that the input torque is entirely transmitted by the pin joint.

It can be seen is Fig. 10 that the fatigue load of the working condition DLC1.2-18-0 is translated to the pinhole load using a three-point, which is then counted by RFC method as the Fig. 11 shows..

### *3.3. Fatigue life Assessment of the flange*

Combined Eq. (12), Eq. (13) and Eq. (14), and FOS of flange under different friction coefficient are calculated. The results are shown in Table 2 and the trend of and FOS is shown in Figure 12.

**Table 2.** Result of D and FOS

| *f* | *D* | *FOS* |
|---|---|---|
| 0 | 2.963 | 0.338 |
| 0.1 | 1.827 | 0.547 |
| 0.2 | 1.218 | 0.821 |
| 0.3 | 0.304 | 3.292 |
| 0.4 | 0.110 | 9.113 |
| 0.5 | 0.074 | 13.48 |

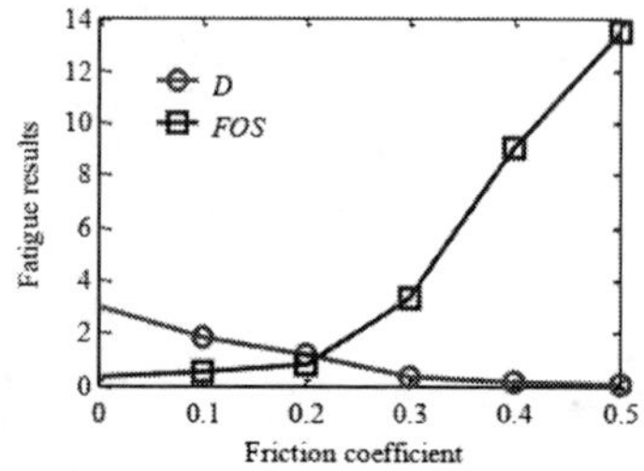


**Fig. 12** The trend of D and FOS

As shown in Table (2), fatigue damage is greater than 1 without considering friction moment (f = 0), while it decreases with the increase of friction coefficient. It can be obtained by interpolation approach that the friction coefficient is 0.224 when fatigue damage reaches the value of 1. Therefore, to avoid fatigue failure, the friction coefficient must be greater than 0.224.

## 4. Conclusion

The total fatigue damage of flange is negatively correlated with the friction coefficient. That is, the larger the friction coefficient, the smaller the fatigue damage and the higher the fatigue strength of flange. Therefore, the fatigue strength of flange can be improved by increasing the friction coefficient.

Through the fatigue analysis, the very friction coefficient satisfying fatigue strength can be obtained, thus providing a theoretical basis for the mechanical design of flange connections.

## Acknowledgements

This study was supported by the National Natural Science Foundation of China (51775058). We also wish to express our thanks to Jiangsu Kwoyuen New Energy Technology Group Co., Ltd., China for sharing the parameters of the wind turbine.

## References

[1] Marino E , Giusti A , Manuel L . Offshore wind turbine fatigue loads: The influence of alternative wave modeling for different turbulent and mean winds[J]. Renewable Energy, 2017, 102:157-169.

[2] Fuglsang P, Thomsen K. Cost optimization of wind turbine for large scale offshore wind farms, technical report. Risø National Laboratory; 1998.

[3] Yeter B, Garbatov Y, Soares C G.. Fatigue damage assessment of fixed offshore wind turbine tripod support structures[J]. Engineering Structures, 2015, 101:518-528.

[4] Christoph W. Kensche. Fatigue of composites for wind turbines[J]. International Journal of Fatigue,2006,28(10).

[5] Yeter B, Garbatov Y, Soares C G. Evaluation of fatigue damage model predictions for fixed offshore wind turbine support structures [J]. International Journal of Fatigue, 2016, 87: S0142112316000086.

[6] Socie D. Multiaxial Fatigue Damage Models [J]. Key Engineering Materials, 2006, 324-325(4):747-750.

[7] Dong W , Moan T , Gao Z . Long-term fatigue analysis of multi-planar tubular joints for jacket-type offshore wind turbine in time domain [J]. Engineering Structures, 2011, 33(6): 2002-2014.
[8] Rice R , editor, Leis B , et al. Fatigue design handbook [M].
[9] Veers, P.S, Winterstein, et al. Variable amplitude load models for fatigue damage and crack growth [J]. Astm Special Technical Publication, 1987(1006):26.
[10] Dowling N E . Fatigue Failure Predictions for Complicated Stress-Strain Histories [J]. Journal of Materials, 1971, 7(1): 71.
[11] Standard practices for cycle counting in fatigue analysis [J]. Astm, 1985.
[12] M. Matsuishi, T. EndoFatigue of metals subjected to varying stress. Jpn Soc Mech Eng (1968), pp. 37-40
[13] Downing S D, Socie D F. Simple rainflow counting algorithms [J]. International Journal of Fatigue, 1982, 4(1): 31-40.
[14] Rychlik I. A new definition of the rainflow cycle counting method [J]. International Journal of Fatigue, 1987, 9(2): 119-121.
[15] Frendahl M , Rychlik I . Rainflow analysis: Markov method [J]. International Journal of Fatigue, 1993, 15(4): 265-272.
[16] Amzallag C , Gerey J P , Robert J L , et al. Standardization of the rainflow counting method for fatigue analysis [J]. International Journal of Fatigue, 1994, 16(4): 287-293.
[17] Glinka G , Kam J C P . Rainflow counting algorithm for very long stress histories [J]. International Journal of Fatigue, 1987, 9(4): 223-228.

IOP Conf. Series: Materials Science and Engineering **772** (2020) 012097 doi:10.1088/1757-899X/772/1/012097

# Optimization of a cantilever steel frame structure of bottom platform by finite element method

**Weihai Sun[1], Guiling Ju[2,*], Limei Ju[1] and Jingwen Pan[1]**

[1] Vehicle Engineering Department, Army Academy of Armored Forces, Beijing 100072, China
[2] Basic Courses Department, Army Academy of Armored Forces, Beijing 100072, China

*Corresponding author e-mail: guilingju@163.com

**Abstract.** The finite element analysis of a cantilever steel frame structure of the bottom platform in civil engineering is carried out. Based on ANSYS and the basic characteristics of steel frame structure, parametric modelling is carried out, and the stress and deformation of the structure under extreme load conditions are analysed. The first-order optimization algorithm is used to optimize the beam cross-section size, which significantly reduces the total weight of the structure, thus providing a basis for the optimal design of the engineering structure.

## 1. Introduction

In the design of industrial buildings, warehouses, stadiums and ports, steel structures have become a major structural form, which has the advantages of lower cost, simple processing and manufacturing, and convenient transportation and installation [1, 2]. In order to further save steel and reduce costs, the total amount of steel used in steel structures needs to be well controlled [3-6]. Based on the engineering practice, the finite element method is used to analyze the cantilever steel frame structure of the bottom platform. The stress and deformation distribution of the structure are obtained, and then the structural optimization design is carried out.

## 2. Computational models

The bottom platform is a left-right symmetrical structure. According to the symmetry, the left part is taken for modeling and calculation. For the inverted cantilever steel frame structure of the bottom platform, the mechanical model of the steel truss structure is established and analyzed, and the load generated by the auxiliary mass such as the lifted member, the spreader, the keel and the patterned steel plate is applied to the steel truss structure to simulate the application of the load. The stress and deformation of the steel truss structure are obtained after the loads are applied. Then, the optimal design of the structure is performed based on the computational model.

The material of the frame structure is Q345-C steel. Bilinear hardening model is adopted for the steel. Material constants for Q345-C steel are as follows: yield stress: 350 MPa, elastic modulus: 210 GPa, Poisson's ratio: 0.3, density: 7850 kg/m3, linear hardening modulus: 2.1 Gpa.

According to technical performance requirements: single side maximum 10T vertical load and 7 or less staff, the 11 T load is actually loaded when the calculation is performed. Since the exact position of

Content from this work may be used under the terms of the Creative Commons Attribution 3.0 licence. Any further distribution of this work must maintain attribution to the author(s) and the title of the work, journal citation and DOI.
Published under licence by IOP Publishing Ltd

the load is undetermined and the load may move during the working process, the load is applied to the end of the cantilever platform in the worst case, and the calculation result is conservative compared with the actual situation.

The parametric model is established with ANSYS's APDL command stream. In the modeling process, the coordinates of the key points are first established, then the line is established according to the drawing, and finally the cross section of the beam is defined and the ID number is associated. For square tube 160 mm × 160 mm × 6 mm, section ID number is defined as 1; for 14a channel steel, section ID number is defined as 2; for 100 mm × 100 mm × 5 mm square tube, section ID number is defined as 4; for rectangular tube 100 mm × 50 mm × 4 mm, section ID number is defined as 5; for the rectangular tube 40 mm × 60 mm × 3 mm, the section ID number is defined as 7. The parameter definition for the cross section is shown in Fig. 1. According to the drawing the geometric model of cantilever steel frame structure of bottom platform is shown in Fig. 2.

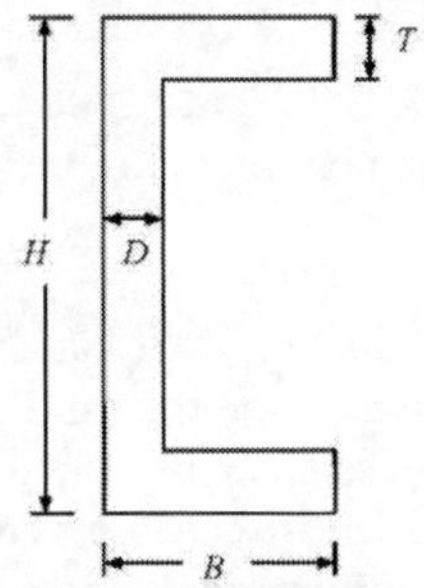


(a) Parameter definition of channel cross section

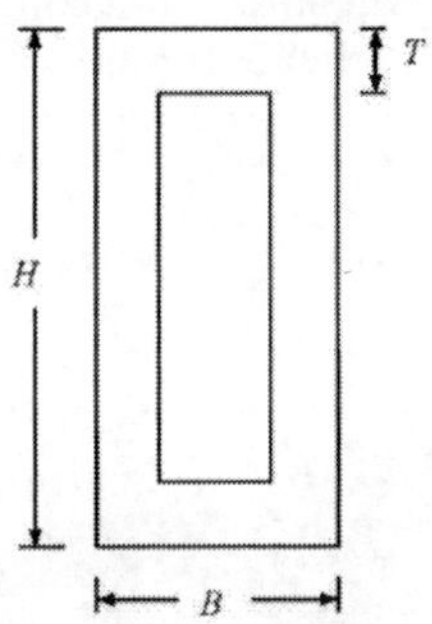


(b) Parameter definition of rectangular tube cross section

**Figure 1.** Parameter definition of the cross section of the beam

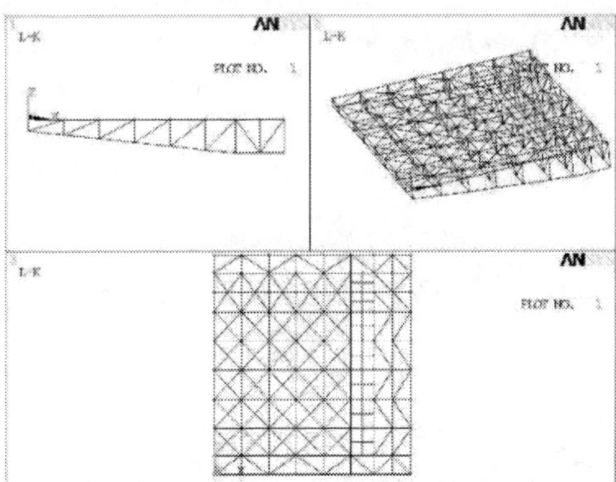

**Figure 2.** Geometric model of the steel frame structure

The steel frame structure is modeled using BEAM188 elements, and a total of 2941 elements are obtained. The finite element model showing the beam size is shown in Fig. 3. Parametric model with ANSYS's APDL command stream makes it easy to modify the model and optimize the structure.

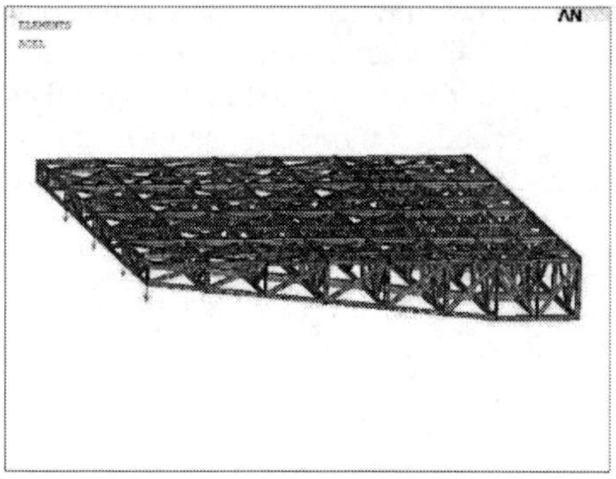

**Figure 3.** Finite element model of the steel frame structure

## 3. Results and discussion

The Von Mises stress contour is shown in Fig. 4. It can be seen that the main beam of the bottom platform and the joint portion of the fixed part and the cantilever part have a higher stress level, and the maximum stress is 61 MPa, which is less than the yield stress of the material of 350 MPa. The deformation contour is shown in Fig. 5. It can be observed that the deformation near the end of the cantilever is larger, and the maximum deformation is 1.36 cm at the end.

The distribution law of the stress and deformation of the steel frame structure is similar with the cantilever beam. That is, the closer to the end of the cantilever, the larger the displacement and the smaller the stress; while the root of the cantilever (at the fixed end) has the smallest displacement and the maximum stress.

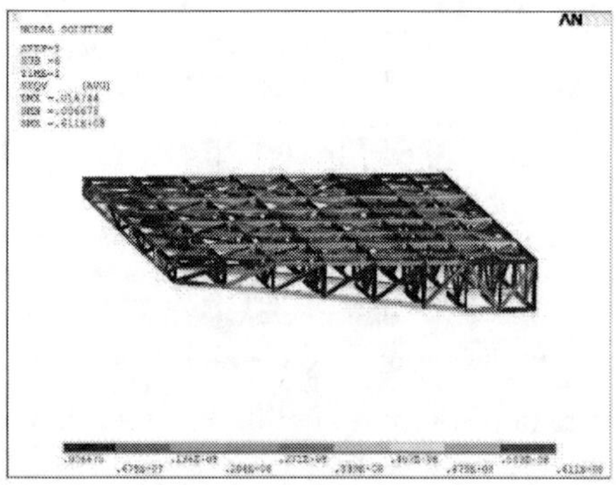

**Figure 4.** Von Mises stress contour before optimization

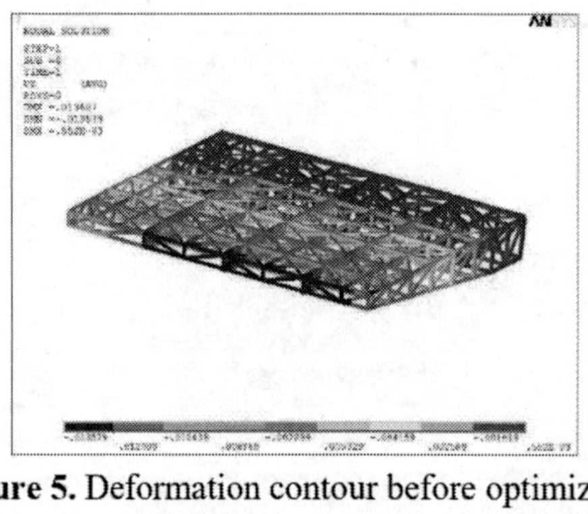

**Figure 5.** Deformation contour before optimization

## 4. Structural optimization design

### *4.1. Process of optimization design*

*4.1.1. Basic element statement.* Design Variables (DVS): Design variable parameters that need to be constantly adjusted during the design process. Each design variable may have upper and lower limits to specify the range of values for the design variables.

Here, dimensions $B_i$, $H_i$, $T_i$ ($i$ = 1, 4, 5, 7) having a cross-sectional ID of $i$ and dimensions $B_2$, $H_2$, $T_2$, and $D_2$ having a cross-sectional ID of 2 are design variables. The upper and lower limits are taken as: the original value is added and subtracted by 50% of the original value.

State Variable (SVS): A variable parameter specifies the constraints that the design requires. It is a dependent variable of the design and a function of the design variable.

The equivalent Mises stress is taken as the state variable, and the maximum equivalent Mises stress is kept constant during the optimization process.

Objective Function: The variable parameter that is minimized in the design. It must also be a function of the design variable. Changing the value of the design variable will change the value of the objective function. In the ANSYS optimizer, only one objective function can be set.

Take the total volume $V_{TOT}$ as the objective function, and the minimum total volume is equivalent to the minimum mass.First-order optimization algorithm is used in optimization process.

*4.1.2. Basic analysis process.* Specify analysis files; declare optimization variables; select optimization tools or optimization methods; specify optimization cycle control methods; conduct optimization analysis; view design sequence results.

### *4.2. Discussion of optimization results*

Keeping the maximum equivalent Mises stress value unchanged, the size of the cross section of the beam is optimized. The optimized design sequence obtained by the calculation is as follows, there are eight groups, of which the eighth group is the optimal design sequence. The relationship between VTOT and the optimized design serial number is shown in Fig. 6. The convergence of the optimized design can be observed. The stress and deformation contours after the optimized design are shown in Figs. 7 and 8, respectively.

The total volume of the cantilever steel frame before optimization is 1.4031 m3 and that of the cantilever steel frame after optimization is 1.2890 m3. With the maximum equivalent Mises stress unchanged, the total weight of the cantilever steel frame is reduced by 10.1% by optimizing the cross-section size of the beam.

However, it can be seen from Fig. 8 that the maximum deformation of the cantilever steel frame after optimization is 1.67 cm, and the maximum deformation before optimization is 1.36 cm. Therefore, the maximum deformation of the cantilever steel frame after the optimized design is increased by 0.31 cm. If there is a requirement for stiffness, the maximum deformation of the platform can be taken as a state variable to satisfy the stiffness condition.

**Table 1.** Section dimensions before and after optimization

| Variables | $B_1$/mm | $H_1$/mm | $T_1$/mm | $H_2$/mm | $B_2$/mm | $D_2$/mm | $T_2$/mm | $B_4$/mm | $H_4$/mm |
|---|---|---|---|---|---|---|---|---|---|
| Before optimization | 160 | 160 | 6 | 140 | 58 | 6 | 9.5 | 100 | 100 |
| After optimization | 158.99 | 155.2 | 5.95 | 122.45 | 54.61 | 5.79 | 8.98 | 101.16 | 99.77 |

| Variables | $T_4$/mm | $B_5$/mm | $H_5$/mm | $T_5$/mm | $B_7$/mm | $H_7$/mm | $T_7$/mm | $V_{TOT}$/m$^3$ |
|---|---|---|---|---|---|---|---|---|
| Before optimization | 5 | 50 | 100 | 4 | 40 | 60 | 3 | 1.40 |
| After optimization | 4.92 | 46.0 | 92.69 | 3.60 | 39.56 | 59.0 | 2.93 | 1.29 |

IOP Conf. Series: Materials Science and Engineering **772** (2020) 012097 doi:10.1088/1757-899X/772/1/012097

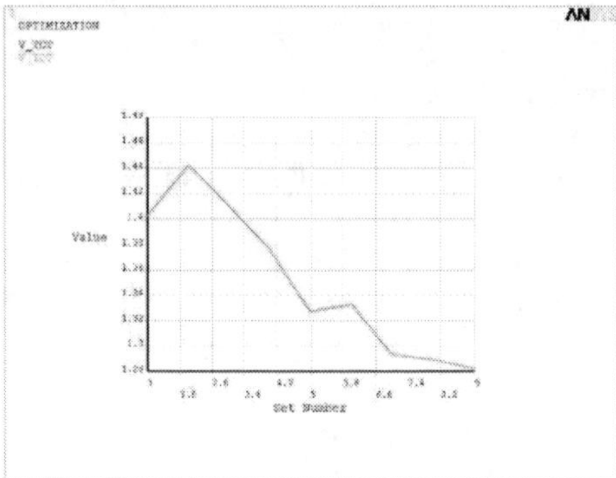


**Figure 6.** The Variation of VTOT with optimized design serial number

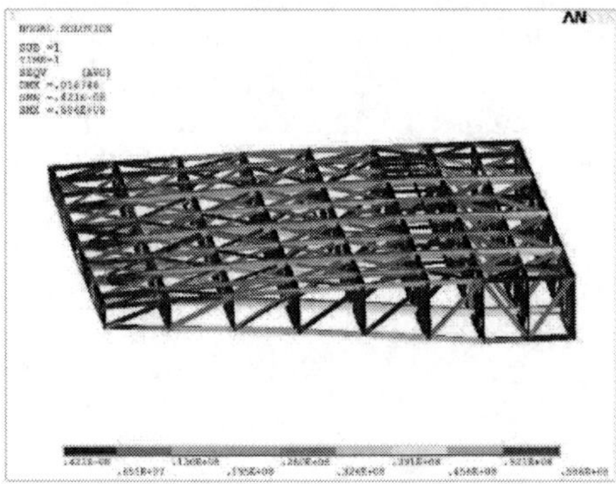

**Figure 7.** Von Mises stress contour after optimization

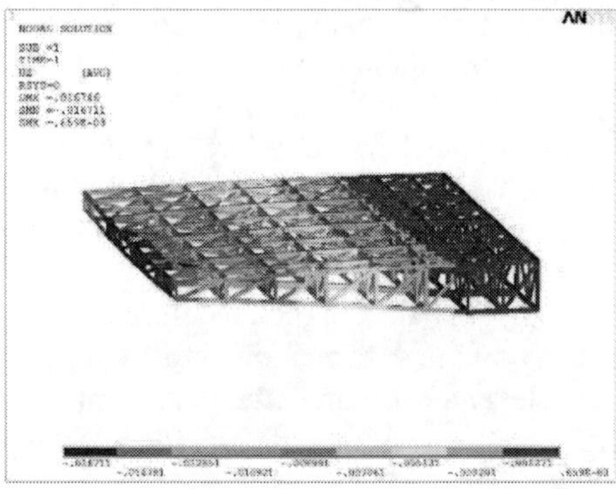

**Figure 8.** Deformation contour after optimization

## 5. Conclusion

The finite element model of the cantilever steel frame structure of the bottom platform is established from the engineering practice, and the stress and deformation of the steel frame structure under the limit load condition are analyzed. The analysis results show that the cantilever steel frame structure is safe. The first-order optimization algorithm is used to optimize the cross-section size of the beam, keeping the maximum equivalent Mises stress unchanged. Within the scope of the study, the total weight of the structure is reduced by 10.1%. The research method in this paper provides a reference for the design and optimization analysis of the cantilever steel structure.

## References

[1] F. Mohammad, M. Mohsen, C. V. Camp, et al., School based optimization algorithm for design of steel frames, Engineering Structures, 171(2018) 326-335.

[2] M. Stolpe. Truss optimization with discrete design variables: a critical review, Structural & Multidisciplinary Optimization, 53(2016) 1-26.

[3] M. Liu, S. A. Burns, Multiple fully stressed designs of steel frame structures with semi-rigid connections. International Journal for Numerical Methods in Engineering, 58(2003) 821-838.

[4] L. Gil, A. Andreu. Shape and cross-section optimisation of a truss structure, Computers & Structures, 79(2001) 681-689.

[5] B. Xu, J. Jiang, W. Tong, et al., Topology group concept for truss topology optimization with frequency constraints, Journal of Sound and Vibration, 261(2003) 911-925.

[6] M. S. Abdulsalam, B. M. Z. M. Z. Affandi, B. A. H. M. Hadli, Cross Section Optimization of Plane Truss among Different Spans, Applied Mechanics and Materials, 679(2014) 1-5.

IOP Conf. Series: Materials Science and Engineering **772** (2020) 012098 doi:10.1088/1757-899X/772/1/012098

# Study on the Preparation Technology of Hollow Yarn with Hollow Stability

**yifei Li***

School of Wuhan Textile University, Wuhan, China

*Corresponding author e-mail: 18986072242@163.com

**Abstract.** At present, there are few researches on hollow yarn. This paper mainly focuses on the strength of hollow yarn. The purpose is to prepare a hollow stable hollow yarn. Several different spinning processes were used in the experiment. Control group and experimental group were used in the experiment. Among them, the control group was prepared by spinning ordinary hollow yarn; the experimental group was prepared by soluble vinylon and hot-melt polyester filament. According to the difference of melting point and solubility in water of these materials, two groups of hollow yarns are formed by high temperature heating and water bath heating respectively. After spinning, the stability of hollow yarn was tested by strength test.

## 1. Introduction

Hollow fiber consists of water-soluble fiber prepared by mixed spinning and solvent wet gel method. The outer layer forms cotton fiber, and the inner layer of double layer structure forms water soluble fiber. Cotton, wool, silk, cannabis and other natural fibers are mixed and blended, encapsulated in human and synthetic fibers to change function, which is the main direction in the 1980s and 1990s [1-3]. In the 21st century, however, there is a yarn product that has not been twisted into knee products. Cotton towels are crepe-free and soft. It is used to clean the face, soft and soft. This is called a beauty towel [4].

## 2. Cotton Fiber Cohesion

### *2.1. PVA fibers*

The object of the present invention is to provide a yarn forming method of hollow yarn, which can improve the water absorption and softness of the fabric. The hollow yarn of the invention is a core-spun yarn with skin core structure formed by water-soluble fibers and cotton fibers, and then the water-soluble fibers located at the core of the core of the core-spun yarn are dissolved to form hollow yarn with hollow core structure. A kind of hollow yarn forming method is as follows: water-soluble polyvinyl alcohol fiber as core yarn, combed cotton sliver with outer fiber 1214g/10m (Table 1 and Table 2).

Content from this work may be used under the terms of the Creative Commons Attribution 3.0 licence. Any further distribution of this work must maintain attribution to the author(s) and the title of the work, journal citation and DOI.
Published under licence by IOP Publishing Ltd

**Table 1.** Experimental group

| number | Strength/cN | Elongation rate/% |
|---|---|---|
| 1 | 574 | 6.2 |
| 2 | 488 | 6.8 |
| 3 | 500 | 6.8 |
| 4 | 554 | 6.6 |
| 5 | 498 | 6.0 |
| 6 | 625 | 5.4 |
| 7 | 468 | 5.8 |
| 8 | 504 | 7.0 |
| 9 | 535 | 7.8 |
| 10 | 571 | 8.2 |
| average | 541.7 | 6.66 |

**Table 2.** Control group

| number | Strength/cN | Elongation rate/% |
|---|---|---|
| 1 | 574 | 6.2 |
| 2 | 488 | 6.8 |
| 3 | 500 | 6.8 |
| 4 | 554 | 6.6 |
| 5 | 498 | 6.0 |
| 6 | 625 | 5.4 |
| 7 | 468 | 5.8 |
| 8 | 504 | 7.0 |
| 9 | 535 | 7.8 |
| 10 | 571 | 8.2 |
| average | 528 | 5.62 |

### 2.2. *Water-soluble PVA*

Core-spun yarn is made by spinning equipment, then core-spun yarn is dissolved in water at 5-100 degree C for 10-30 minutes, after polyvinyl alcohol is completely dissolved from the core-spun yarn, 45s-10s hollow yarn is obtained. In the core-spun yarn, the weight ratio of core yarn is less than 40%, and that of combed cotton is more than 60%. The coring process of cotton/PVA core-spun yarn is actually the process of dissolving core-spun yarn completely under certain conditions, thus forming pure cotton hollow and fluffy yarn. Temperature is the most important factor in the process of core-unwinding of cotton/PVA core-spun yarn. Choosing the appropriate temperature can improve the core-dissolving efficiency, and can dissolve the core layer PVA to form a good cotton hollow yarn. Secondly, time, under the optimum temperature conditions, water-soluble PVA dissolves linearly and rapidly into fine particles. After reaching the dissolution equilibrium, the PVA particles in the yarn should be placed for a period of time to dissolve completely from the yarn into water and reach the state of complete dissolution of PVA. Otherwise, after drying the fabrics, some powders will be produced from time to time, which will affect the quality of the fabrics. Water-soluble PVA can only dissolve at a certain water temperature. Although PVA will swell before dissolving in the process of dissolution, the water needed is limited, and excessive use of water will only waste water resources and increase costs. Water-soluble PVA will completely dissolve after a certain time in a certain water temperature, so long as it is selected reasonably. Dissolve it at a temperature for a period of time. The degree of polymerization and saponification of PVA fibers have significant effects on their water solubility. The longer the molecular weight of PVA is, the more perfect the aggregate structure is. Water-soluble PVA generally has a degree of polymerization less than 1700, and the degree of saponification DS mainly determines the characteristics of PVA. If the saponification DS of PVA is adjusted to 88-99.5%, the water-soluble temperature of PVA fibers can be adjusted in the range of 5-100 degrees C. The PVA fibers dissolved in ambient temperature water can be obtained by gel spinning. The degree of polymerization and saponification of PVA fibers produced by wet spinning are less than 1700 [5] and 99.5% respectively. The water-soluble temperature of PVA fibers can be controlled at 80 C, 50 C and 25 C, respectively.

## 3. Core-spun yarn

Core-spun yarn, also known as synthetic yarn or coated yarn, is a new type of yarn composed of two or more kinds of fibers. The original core yarn was staple yarn and filament yarn, which were used as cotton and polyester yarns [6, 7]. The new bead line uses polyester coated polyester core. Core-spun yarns are usually made by twisting staple fibers (such as cotton, wool or bonded fibers) with synthetic filaments with strength and elasticity as core yarns. Core-spun yarn combines the excellent properties of filament yarn and external staple fibers. More common core yarns are polyester-cotton core-spun yarns made of polyester filament yarns and wrapped in cotton fibers. There are also spandex core-spun yarns made of Spandex filaments and other fibers. Knitted or knitted fabrics made of these core-spun yarns are flexible when stretched and can be worn comfortably and comfortably. The main purpose is to strengthen the cotton canvas, and keep the waterproof of cotton fibers when inflated with water, and polyester has tensile, tear and shrinkage resistance when raining. At present, many types of core-spun yarns have been developed, which can be summarized as staple and staple fiber core-spun yarn, chemical filament and staple fiber core-spun yarn, chemical filament and chemical filament core-spun yarn. At present, core-spun yarns are usually made of core yarns with chemical filaments, and various staple fibers are formed to form the unique structure of core-spun yarns. Chemical filaments commonly used in core yarns are polyester filaments, nylon filaments and spandex filaments. The external staple fibers are cotton, polyester cotton, polyester, nylon, acrylic acid and wool. In addition to the advantages of core-spun yarns, many defects can be supplemented by utilizing the excellent physical properties of special structures, the properties of external fabrics and the surface properties of core yarns and maximizing the properties of the two fibers. For example, polyester-based core-spun yarn can fully demonstrate the advantages of polyester filament, easy to cool, fold, wash and dry quickly, while absorbing water, generating less static electricity, can show special faces. Core-spun yarn is mainly composed of core yarn for garments, core yarn for elastic fabrics, core yarn for decorative fabrics and core yarn for sewing yarns. Core spinning has a variety of spinning methods, such as ring spinning, electrostatic spinning, eddy current spinning and magnetic twisting [8]. At present, China's textile industry mainly uses cotton spinning system to process core-spun yarns. Spinning process. There are two processes for spinning core-spun yarn at present, taking polyester-cotton core-spun yarn as an example. Carding system: Cotton: Cleaning, carding, roving and spinning (core-spun). Combing system: Cotton: Cleaning, carding, cotton, coil, combing, roving and spinning. In the actual manufacturing process, the choice of core-spun yarn process depends on yarn demand and spinning quantity. The general principle is that combing can be used without combing. When the combing system is used, the process can be shortened and shortened to reduce production costs. If some manufacturers wrap their fragments with thousands of irregularities, the fibers align well with the straightness of the fibers, and once they pass, the strips can meet the coating requirements. The process can be adopted as follows: Cotton: Cleaning, carding, winding, combing, roving and spinning (core-spun). Yarn Treatment: A hollow and stable hollow yarn can be obtained by melting the spun yarn at 110 °C and dissolving the soluble polyvinyl alcohol at 70 °C.

## 4. Conclusion

From the theoretical point of view, there are many research results on hollow yarn at present, but the research results on the stability of hollow yarn are relatively few, and the research results on a specific kind of hollow yarn are insufficient. Therefore, the research results of this paper can complement the relevant theories to a certain extent, and provide a theoretical basis for further research on this issue for follow-up researchers.

**References**

[1] Sudhir Kumar Saw, Chandan Datta. Thermomechanical Properties Of Jute/Bagasse Hybrid Fibre Reinforced Epoxy Thermoset Composites [J]. BioResources, 2009, 4(4).

[2] Henrikki Liimatainen, Sanna Haavisto, Antti Haapala et al.. Influence Of Adsorbed And Dissolved Carboxymethyl Cellulose On Fibre Suspension Dispersing, Dewaterability, And Fines Retention [J]. BioResources, 2009, 4 (1).

[3] Kalina Maćkowiak, Natalia Torlińska-Walkowiak, Barbara Torlińska. Dietary fibre as an important constituent of the diet [J]. Postępy Higieny i Medycyny Doświadczalnej, 2016, 70.

[4] Leonard Y. Mwaikambo. TENSILE PROPERTIES OF ALKALISED JUTE FIBRES[J]. BioResources, 2009, 4(2).

[5] P Kaliappan, R Kesavan, B Vijaya Ramnath. Investigation on effect of fibre hybridization and orientation on mechanical behaviour of natural fibre epoxy composite [J]. Bulletin of Materials Science, 2017, 40(4).

[6] Hollow Yarn in Friction Spinning Part 2: Yarn Structure and Deformation Under Axial Tension and Lateral Forces. Ali Akbar Merati and Masaaki Okamura.

[7] Maria Morissa Lu, Aart Willem Van Vuure. Improving moisture durability of flax fibre composites by using non-dry fibres [J]. Composites Part A, 2019.

[8] O. Olivieri, M. Antonini, G. Acuti et al.. Fibre characteristics and follicle activity in Angora kids as affected by diet [J]. Italian Journal of Animal Science, 2011, 2(1S).

IOP Conf. Series: Materials Science and Engineering **772** (2020) 012099 doi:10.1088/1757-899X/772/1/012099

# Comparative Study on Molecular Dynamics Simulation Methods of Mechanical Properties of Tricalcium Silicate and Dicalcium Silicate

**Yuhuan Bu, Wenxiang Du ***

School of petroleum engineering, China University of Petroleum, Qindao, China

*Corresponding author e-mail: wenxiang_du973@163.com

**Abstract.** A comparative study on the mechanical properties of tricalcium silicate ($C_3S$) and dicalcium silicate ($C_2S$) under different force fields and different sizes by molecular dynamics method. The $C_3S$ and $C_2S$ single crystal and supercell molecular model were established respectively. The molecular dynamics simulation was carried out under the three force fields of COMPASS, COMPASS II and Universal to obtain the mechanical properties of the nanostructure. It is found that the COMPASS II force field is suitable for $C_3S$ and $C_2S$ mechanical simulation performance under the same size, which is closer to the literature value. Under the same force field, the large-size unit cell mechanical performance simulation is more stable and reliable than the small-size simulation result. Mechanical properties are one of the most important properties of cement stone. The hydration products of $C_3S$ and $C_2S$ are responsible for the mechanical strength of the cement stone. Therefore, studying the mechanical properties of $C_3S$ and $C_2S$ is of great significance for the development of cement materials.

## 1. Introduction

Cement is the most widely used building material in the world, and Portland cement is the most widely used [1]. With the development of computer technology, molecular dynamics simulation based on classical mechanics is widely used in the fields of biological materials, composite materials, polymers and polymer structures, and is one of the best simulation methods at the nanoscale. This will certainly provide a new perspective for the study of the mechanical properties of cement clinker. There are many components of Portland cement clinker, $C_3S$ and $C_2S$ are the main bodies, accounting for 50-70% and 15-30% of the total content respectively [2].

$C_3S$ consists of three calcium oxides and one silica. The molecular formula is $3CaO{\cdot}SiO_2$, also known as Alite. It reacts rapidly with water and is responsible for the formation and development of early strength within 28 days of cement stone. The $C_3S$ crystal is composed of independent tetrahedrons, the four corners are $Ca^{2+}$, the center is the oxygen atom, and the P-1 space group. Figure 1(a) shows its single crystal model [3].

$C_2S$ is a compound formed by the combination of two calcium oxides and one silica. It is also called belite, and its molecular formula is $2CaO{\cdot}SiO_2$. There are many crystal forms such as $\alpha$-$C_2S$, $\beta$-$C_2S$, $\gamma$-$C_2S$. $\beta$-$C_2S$ is the most common. It is a P21/N space group, and the silicon tetrahedron and calcium atoms are ordered. $\beta$-$C_2S$ molecular model is created in this study. The single crystal is shown in Figure

Content from this work may be used under the terms of the Creative Commons Attribution 3.0 licence. Any further distribution of this work must maintain attribution to the author(s) and the title of the work, journal citation and DOI.
Published under licence by IOP Publishing Ltd

1(d). $C_2S$ hydration rate is slow, which plays a major role in the long-term strength of cement stone after 28d [4].

In molecular dynamics simulations (MD), the choice of force field is critical to the accuracy of the simulated results. Among the many force fields, COMPASS is the first force field based on the ab initio calculation method, which can accurately calculate meteorological properties (structure, conformation, vibration, etc.) and condensed phase properties (condensation energy, etc.), and obtain covalent and non-covalent bonds. Force [5]; COMPASS II force field is developed on the COMPASS force field, can calculate more atom types and types, has a richer functional expression; Universal force field [6] is from the atomic point of view, its calculation range covers the entire periodic table of elements. The mechanical properties of $C_3S$ and $C_2S$ can be simulated in all three force fields, but the adaptability of the model needs to be explored in order to provide basis for subsequent research. The choice of simulated dimensions is equally important for molecular dynamics simulations, given the computational efficiency and accuracy requirements. If the simulation size is too large, more computational resources are required, and the calculation efficiency is low; if the simulation size is too small, the calculation accuracy is difficult to guarantee.

In summary, this study uses a computer for molecular dynamics simulation, which aims to consider the effects of different force field settings, simulation size selection on the mechanical properties of $C_3S$ and $C_2S$.

## 2. Simulation part

### *2.1. Portland cement clinker composition and crystal structure*

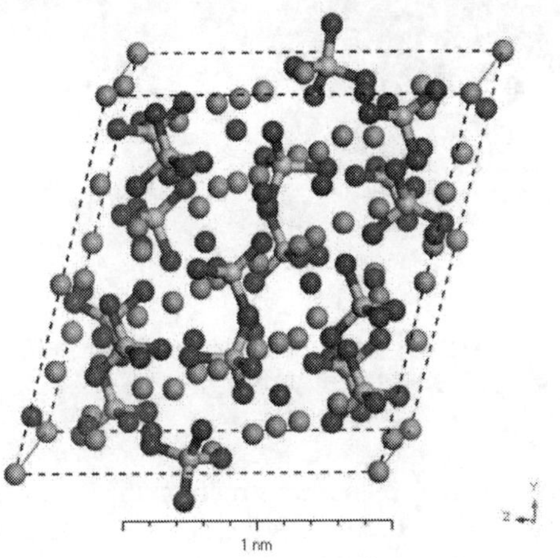


(a) a×b×c $C_3S$ single crystal cell

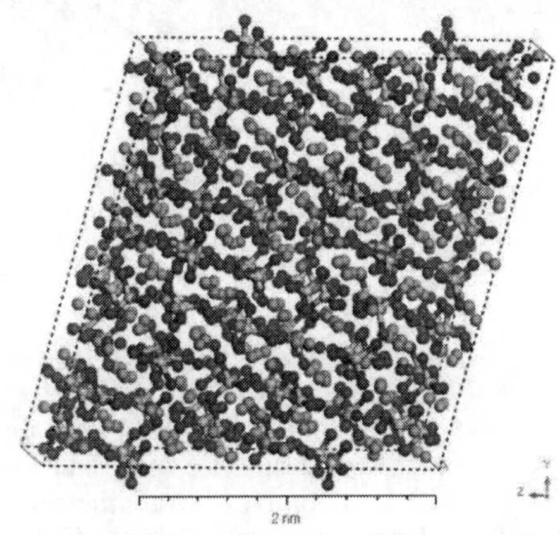


(b) 2a×2b×2c $C_3S$ supercell

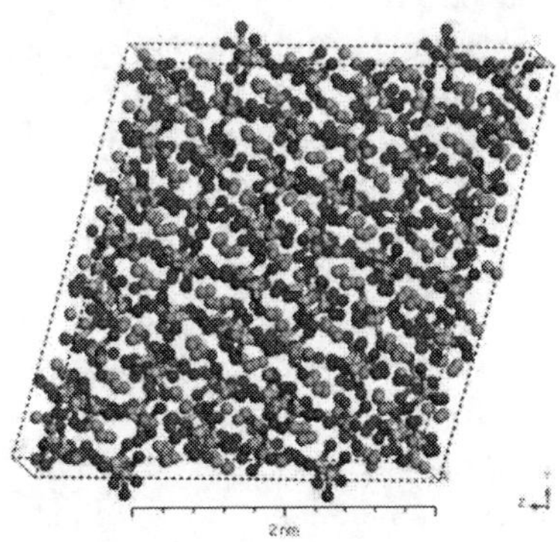


(c) 3a×3b×3c $C_3S$ supercell

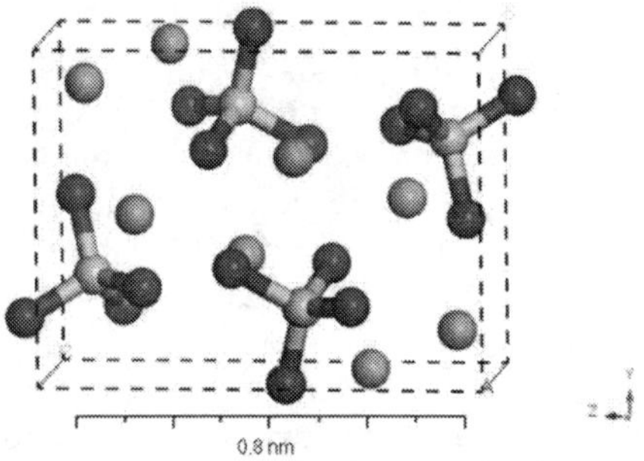


(c) a×b×c $C_2S$ single crystal cell

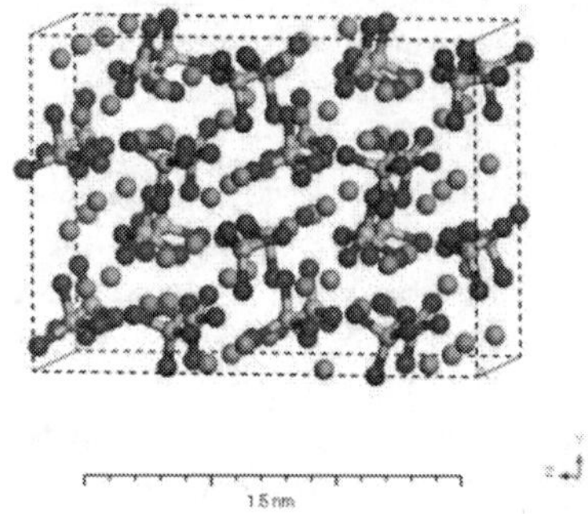


(e) 2a×2b×2c $C_2S$ supercell

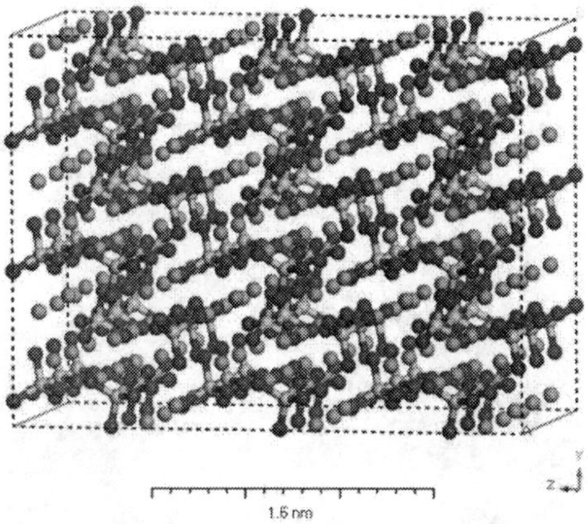


(f) 3a×3b×3c $C_2S$ supercell

**Figure 1.** $C_3S$ and $C_2S$ crystal single crystal and super crystal model (red ball: oxygen atom; yellow ball: silicon atom; green ball: calcium atom)

### *2.2. Simulation details*

Taking the established single crystal cell and super unit cell as the initial structure, the Smart algorithm is first selected to optimize the geometric structure of the atomic structure. This method is the combination of the steepest descent method, the conjugate gradient method and the Newton Lapson method. Then the COMPASS, COMPASS II and Universal force fields were used to carry out molecular dynamics simulation under the NPT ensemble with constant atomic number, pressure and temperature. The whole system adopts Nose pressure control and Berendsen temperature control method, and the temperature and pressure are controlled respectively 298K and 0.0001GPa, simulating the room temperature environment; in order to fully relax and ensure the system reaches equilibrium state, the whole simulation duration is set to 100ps, the time step is set to 0.1fs, the integral of the motion equation is calculated by Verlet algorithm, Ewald summation method was used to calculate the electrostatic interaction and van der Waals interaction. Finally, the mechanical properties of the equilibrium crystal are calculated. To ensure that the deformation is within the elastic range, the maximum strain amplitude in each direction is set to 0.003. All simulations were performed in the Materials Studio software.

The bulk modulus ( $K$ ) and shear modulus ( $G$ ) of the structure are obtained according to Voigt-Reuss-Hill (VRH) [7] method. The elastic modulus ( $E$ ) and Poisson's ratio ( $\nu$ ) are solved according to the following equation:

$$E = \frac{9KG}{G + 3K} \quad (1)$$

$$v = \frac{3K - 2G}{6K + 2G} \quad (2)$$

## 3. Results and discussion

The simulation results are shown in Table 1, Table 2:

**Table 1.** MD simulation results of $C_3S$

| Force field type | Size | K(GPa) | G(GPa) | E(GPa) | v |
|---|---|---|---|---|---|
| COMPASS | single cell | 46.95 | 25.76 | 65.32 | 0.27 |
| | 2a×2b×2c supercell | 46.57 | 26.45 | 66.71 | 0.26 |
| | 3a×3b×3c supercell | 52.45 | 30.28 | 76.18 | 0.25 |
| COMPASS II | single cell | 49.68 | 28.00 | 70.71 | 0.26 |
| | 2a×2b×2c supercell | 49.17 | 27.81 | 70.19 | 0.26 |
| | 3a×3b×3c supercell | 63.35 | 27.87 | 70.98 | 0.30 |
| Universal | single cell | 20.64 | 21.32 | 47.59 | 0.12 |
| | 2a×2b×2c supercell | 37.81 | 23.49 | 58.38 | 0.24 |
| | 3a×3b×3c supercell | 42.14 | 23.85 | 60.20 | 0.26 |
| Ref.value [7-8] | | 59.8-84 | 25.2-58 | 63.3-141 | 0.22-0.32 |

**Table 2.** MD simulation results of $C_2S$

| Force field type | Size | K(GPa) | G(GPa) | E(GPa) | v |
|---|---|---|---|---|---|
| COMPASS | single cell | 39.78 | 22.32 | 56.42 | 0.26 |
| | 2a×2b×2c supercell | 40.01 | 22.73 | 57.34 | 0.26 |
| | 3a×3b×3c supercell | 42.42 | 25.31 | 63.33 | 0.26 |
| COMPASS II | single cell | 36.82 | 24.42 | 60.00 | 0.23 |
| | 2a×2b×2c supercell | 48.87 | 27.92 | 67.91 | 0.22 |
| | 3a×3b×3c supercell | 65.94 | 30.85 | 80.06 | 0.29 |
| Universal | single cell | 37.09 | 21.09 | 53.19 | 0.26 |
| | 2a×2b×2c supercell | 35.40 | 20.15 | 50.80 | 0.26 |
| | 3a×3b×3c supercell | 34.19 | 22.45 | 71.15 | 0.15 |
| Ref.value[9-10] | | 63.2-177 | 21.4-116.0 | 58-285 | 0.19-0.23 |

It can be seen from the table that when the simulated size are the same, the bulk modulus and shear modulus calculated by the Universal force field are smaller than those of the COMPASS and COMPASS II force fields, resulting in a significant difference between the elastic modulus and the Poisson's ratio. The Poisson's ratio of $C_3S$ single crystal is only 46% of the COMPASS II force field calculation, which deviates from the literature value. The Universal force field is not suitable for simulating the mechanical properties of $C_3S$ and $C_2S$. Compared with the literature value, the bulk modulus calculated by the COMPASS force field. The $C_3S$ and $C_2S$ bulk modulus values calculated by the COMPASS force field are smaller than the literature values in different size, but more consistent with the literature value when calculated by COMPASS II force field. So the COMPASS II force field is suitable for simulating the mechanical properties of $C_3S$ and $C_2S$.

Under the COMPASS II force field, the bulk modulus of $C_3S$ and $C_2S$ single crystal and 2a×2b×2c supercell is smaller than the experimental value, but the bulk modulus, shear modulus and elastic modulus of 3a×3b×3c supercell are in good agreement with the literature values, indicating that when the simulation size is large, the calculation error caused by accidental error and size effect can be effectively reduced. This result is consistent with the general perception.

## 4. Conclusion

This study compares the molecular dynamics simulation methods of $C_3S$ and $C_2S$ and obtains the following conclusions:

(1) Molecular dynamics simulation study the mechanical properties of $C_3S$ and $C_2S$ are feasible, and the simulation results are in good agreement with literature values.

(2) COMPASS II force field is more suitable for the study of mechanical properties of $C_3S$ and $C_2S$. Comparing the simulation results of COMPASS, COMPASS II and Universal force field, the simulation results of the Universal force field are generally smaller than those of the COMPASS and COMPASSII force fields, and the deviation from the literature value is large; The COMPASS compared with COMPASS II force field, COMPASSII force field on the mechanical properties of $C_3S$ and $C_2S$ simulation is more applicable. It is suitable for studying the mechanical properties of $C_3S$ and $C_2S$ ion-covalent bonds.

(3) The mechanical properties of large-size simulation calculations are stable and reliable. The calculation results of bulk modulus, shear modulus and elastic modulus of the $3a\times3b\times3c$ supercell model are more stable and reliable. Hence, the large size model can avoid the calculation deviation caused by size effect and accidental error when the calculation resources are allowed.

## Acknowledgments

This work was financially supported by National Natural Science Foundation of China (51974355).

## References

[1] Bacarji, E., et al. "Sustainability perspective of marble and granite residues as concrete fillers. Constr Build Mater." Construction & Building Materials 45.13(2013): 1-10.
[2] Goñi, Sara, and A. Guerrero. "Study of alkaline hydrothermal activation of belite cements by thermal analysis." Journal of Thermal Analysis & Calorimetry 99.2(2010): 471-477.
[3] Juilland, Patrick , et al. "Dissolution theory applied to the induction period in alite hydration." Cement and Concrete Research 40.6(2010): 831-844.
[4] Cuberos, Antonio J. M. , et al. "Active Iron-Rich Belite Sulfoaluminate Cements: Clinkering and Hydration." Environmental Science & Technology 44.17(2010):6855-6862.
[5] Sun, H. "COMPASS: An ab Initio Force-Field Optimized for Condensed-Phase Applications Overview with Details on Alkane and Benzene Compounds." Journal of Physical Chemistry B 102.38(1998): 7338-7364.
[6] Rappe, A. K., K. S. Colwell, and C. J. Casewit. "Application of a universal force field to metal complexes." Inorganic Chemistry 32.16(1993): 3438-3450.
[7] Hill, R. "The Elastic Behaviour of a Crystalline Aggregate." Proceedings of the Physical Society 65.5(1952): 349-354.
[8] Velez, Karine , et al. "Determination by nanoindentation of elastic modulus and hardness of pure constituents of Portland cement clinker." Cement & Concrete Research 31.4(2001):555-561.
[9] Manzano, H., J. S. Dolado, and A. Ayuela . "Elastic properties of the main species present in Portland cement pastes." Acta Materialia 57.5(2009): 1666-1674.
[10] Wang, Qianqian, et al. "Hydration Mechanism of Reactive and Passive Dicalcium Silicate Polymorphs from Molecular Simulations." The Journal of Physical Chemistry C (2015): 150817090050007.

# Research on the Integration of Industrial Design and Mechanical Product Design

**Lijuan Zhang, Qin Yang *, Kewei Zhang, Jiayue Liu, Caicai Feng, Wen Hou and Xin He**

School of Science and Technology Liaoning, Liaoning, China

*Corresponding author e-mail: 1535569532@qq.com

**Abstract.** The purpose of this paper is to integrate industrial design into mechanical design, improve the core competitiveness of mechanical products, and become a benign development path for the transformation and upgrading of traditional machinery enterprises. To provide perfect and feasible ideas, theories and concrete methods and means for traditional enterprises; Through the research and analysis of the present situation of the development of traditional machinery industry, this paper finds out a series of problems existing in the upgrading and transformation of traditional manufacturing industry in the emerging stage, combined with the core ideas of informatization and industrialization under the background of industrial design 4.0, taking Anshan old industrial base as an example. An innovative development model suitable for the integration of industrial design and regional machinery manufacturing industry is developed. Relying on the platform of Anshan Industrial Design Center, the development model of "industrial design industrial alliance + association + university" led by the government is established. And take the industrial design public service as the support, adjust measures to the local conditions. Focusing on "public service, platform construction, design competition, talent training, publicity and promotion, brand activities, standard formulation, basic research" and other aspects, to achieve the integration of new industrial design industry ecological model of machinery manufacturing industry.

## 1. The present situation of the development of machinery industry

The traditional mechanical design method pays attention to the realization of functional structure, in order to meet the mechanical conditions as the main design criterion. With the development of computer technology, more and more auxiliary design software helps designers to complete product design. At the same time of improving the design efficiency, the homogeneity of the results of the machinery industry is serious, this traditional portable design method has been gradually unable to meet the needs of users and market demand. Most of the traditional small and medium-sized enterprises and old industrial design bases are faced with a series of problems to be solved, such as difficult transformation and upgrading, low industrial value chain and limited output of industrial achievements. In the face of such a situation, Many domestic experts and scholars put forward the concept of "integrating industrial design theory and method into mechanical innovation and development", which has greatly improved the brand recognition, product modeling, color painting, ergonomics and other problems of mechanical

Content from this work may be used under the terms of the Creative Commons Attribution 3.0 licence. Any further distribution of this work must maintain attribution to the author(s) and the title of the work, journal citation and DOI.
Published under licence by IOP Publishing Ltd

products. Based on this concept, the value of traditional mechanical products is gradually returning to spring. However, on the one hand, designers limited by mechanical theory can not completely solve the defects in the current mechanical design, and some of the designs still remain in the category of structure and function. Based on the background of industrial design 4.0, mechanical products and industrial design will move towards the development trend of "innovation-driven, shared integration". On the other hand, although the concept of "combining the two and keeping pace with each other" has been put forward, most of the studies have only stayed at the superficial theoretical level. Failing to make an in-depth study and practical combination of the characteristics of the development of the regional machinery industry and the state's guidance on industrial design, the advantages between the two have not been maximized, and there is a lack of substantive innovation models and practical and effective guidance. As a result, there are some problems, such as "no cure, slow effect" and so on, in the transformation and upgrading of the traditional machinery industry.

## 2. The leading role of Industrial Design

### *2.1. Overview of Industrial Design*

Design is a discipline that studies the changes and laws of form or style. Industrial design is defined as "a special service work." A creative activity that optimizes the shape, function, and use value of products and product families for the benefit of both users (consumers) and producers, and properly designs all human environments, It is a comprehensive applied discipline.

In the design activities, industrial design is more inclined to the study of product concept, shape, color collocation, visual interface and man-machine relationship. The core of industrial design is to "create a better life for mankind and create a better world." Through creative activities in industrial design, respond to economic, social, environmental and ethical issues, propose solutions, visualize and operationalize them, and build better product, system, service experiences or business networks.

### *2.2. The significance of the Integration of Industrial Design and Machinery Industry*

The real economy of the development of machinery manufacturing industry has always been the mainstay of the healthy development of the national economy. With the concept of German industrial design 4.0 put forward, The United States, Britain, France and other developed countries have also put forward the strategy of "reindustrialization" one after another. In order to develop faster and better, We should speed up the planning and layout of manufacturing upgrading, put forward "made in China 2025" to strengthen the deep integration of industrialization and informatization, and vigorously promote China's transformation from a manufacturing power to a manufacturing power. The development of China's manufacturing industry is inseparable from industrial design. On the one hand, the upgrading of manufacturing brand has an urgent need for industrial design. On the other hand, the cooperation between professional industrial design companies and manufacturing enterprises is getting closer and closer, the mode of oriented service manufacturing enterprises has become more common, and some industrial design companies have the ability to produce customized products independently. Therefore, combined with the national conditions and the development of the design industry, the traditional machinery industry in the design level is in urgent need of industrial design theory to improve the grade of design products and increase the competitiveness of products in the market!

## 3. Research on Integration Mode of Industrial Design

### *3.1. Analysis on the background of Innovation and Development of Regional Machinery Industry in Industrial Design Integration*

Since the 19th CPC National Congress, the development of machinery industry is facing new opportunities. In order to promote China's industry to move towards the middle and high end of the global value chain, cultivate a number of world-class advanced manufacturing clusters; In response to the call of the state, we should help to complete the advanced manufacturing cluster, which provides a new development direction for the traditional manufacturing industry. All localities should adjust measures to local conditions and seek a new development model suitable for the region. Taking Anshan City, Liaoning Province as an example, this paper studies the innovative development model of industrial design and integration of regional machinery manufacturing industry. Anshan City relying on the iron and steel industry, the development of iron and steel as the center, other industries for iron and steel services industrial development cluster! How to better promote the economic development of Anshan, complete the transformation and upgrading of traditional enterprises, how to efficiently integrate industrial design into the traditional machinery industry, coordinate the use of resources, and help traditional enterprises to complete the upgrading and transformation, At present, it is an urgent problem for traditional manufacturing enterprises to solve.

### *3.2. concrete schemes for the innovative development of regional machinery manufacturing industry in the integration of industrial design*

In order to give full play to the innovative and leading role of industrial design in the development of local industrial economy, and to promote the transformation and upgrading of manufacturing industry in Anshan, Anshan has set up an "Anshan Industrial Design Center" to rely on this platform. To establish a government-led "industrial design industry alliance + association + colleges and universities" form of development model. With the public service of industrial design as the support, adjusting measures to local conditions, focusing on "public service, platform construction, design competition, personnel training, publicity and promotion, brand activities, basic research on standard formulation", To realize the ecological model of machinery manufacturing industry with the integration of new industrial design industry.

In terms of public services, we should provide industrial design services for equipment manufacturing enterprises in Anshan area, improve the industrial design deployment of the manufacturing industry, achieve practical economic benefits, and gradually complete the new industrialization transformation of enterprises. Reshape the brand image of Anshan equipment manufacturing industry. It also creates research industrial design projects for heavy industrial enterprises, creates industrial design projects for product planning and modeling improvement design for manufacturing enterprises, and provides industrial design consultation and comprehensive services for small and medium-sized enterprises. Promote the formation of industrial design service market system in Anshan area.

*3.2.1. Public service.* 1. Mainly provide industrial design services for equipment manufacturing enterprises in Anshan area, use 5-10 years to improve the industrial design and deployment of the manufacturing industry, achieve actual economic benefits, and gradually complete the new industrialization transformation of enterprises. Reshape the brand image of Anshan equipment manufacturing industry. 2. To create research industrial design projects for heavy industrial enterprises, to create industrial design projects for product planning and modeling improvement design for manufacturing enterprises, and to provide industrial design consultation and comprehensive services for small and medium-sized enterprises. Promote the formation of industrial design service market system in Anshan area.

*3.2.2. Platform construction.* 1. Vigorously developing the public service platform for industrial design under the new situation is the practical demand for the adjustment of China's industrial structure and industrial transformation and upgrading. Facing the "Anshan characteristics" to change the scattered, weak and imprecise situation of the Anshan industrial design industry, Concentrate the resources of all parties, optimize integration, improve design quality, strengthen market competitiveness, reduce vicious competition, and create a municipal industrial design service platform. To enhance the guidance and driving operation of industrial design to the development of local industry. Impact on the direction of provincial industrial design center. In order to change the current situation of "provincial industrial design center" in Anshan City, so that the center can really play a practical role. In line with the principle of "giving priority to characteristics and industrial agglomeration", condensing Anshan industrial design industry resources and characteristic advantage industries to form an effective interaction, so that advanced manufacturing industry and modern design service industry can integrate, promote and develop together. Work together to reach the provincial industrial design center. 3. According to the requirements of the "guidelines for the Establishment of the National Industrial Design and Research Institute" of the Ministry of Industry and Information Technology, the Anshan Industrial Design and Research Institute is established. Outstanding responsibilities and tasks include basic research, technical support, achievement transformation, consulting services, personnel training, exchange and cooperation of the key functions.

*3.2.3. Design Competition.* 1. Enterprise orientation competition. In order to promote the deep integration of industrial design and manufacturing industry, promote the development of industrial design to scale, specialization, high-end, internationalization and characteristics, and solicit design themes for "characteristic culture, characteristic industries and key products" in Anshan, Take the form of "enterprise registration + visit and investigation" to hold a special competition in the form of naming. 2. Set up the Anshan Industrial Design Award led by the government. To achieve a breakthrough in the zero industrial design award in Anshan City, and become the second city in Liaoning Province to have a government-led industrial design award, promoting the commercialization of the design of the award-winning works. And recommended to participate in the China Outstanding Industrial Design Award, China Red Star Design Award, China good Design, Germany IF Award, Germany red dot Award, the United States IDEA Award and other famous design competitions.

*3.2.4. Personnel training.* 1. Skills training and vocational certification. Including 3D printing advanced college training and certification, Solid Works global certification examination (CSWA) training and certification, two-dimensional graphic designer training, hand-drawn product effect map training and other skills training; We will start the pilot work of professional qualification certification for industrial design practitioners led by the municipal government, further broaden the scope of evaluation of industrial design practitioners, improve the evaluation system, innovate evaluation methods, and speed up the professionalization of industrial design personnel. 2. Special training. Support well-known design experts, excellent industrial designers, university teachers for manufacturing enterprises, design institutions, scientific research institutes, industry associations and other units related personnel, by category, field, level of industrial design innovation personnel training. 3. Establish a training base. Strengthen cooperation with industrial parks, leading enterprises, well-known design institutions and industrial design colleges and universities inside and outside the province, jointly build industrial design experiments and training bases inside and outside the university, and employ outstanding industrial designers to teach part-time in colleges and universities. Train compound applied industrial design talents. 4. Training Institute. Strengthen cooperation with famous design colleges and universities at home and abroad, introduce the concept of advanced design education, establish industrial design research institute, aim at the training of high-end and compound talents in the field of design, and establish a regional and industrial design talent training base. Strengthen cooperation with relevant colleges and universities and scientific research institutions to carry out professional training such as knowledge popularization and skills upgrading according to the characteristics of the industry. We should explore the open mode of talent training and promote the international and domestic two-way exchange and joint training of design talents.

*3.2.5. Publicity and promotion.* Organize and carry out all kinds of publicity and promotion activities, improve the understanding of industrial design in the whole society, shape the overall image of Anshan design, create "Anshan characteristics", enhance the competitiveness of Anshan industrial design, and set up a new business card of "Anshan design". 1. Publish design magazines and Anshan Industrial Design Yearbook in Anshan area. Matched with the "Design" magazine of the China Industrial Design Association, prepared to set up the "Anshan Design" magazine. Regular publication, regular release of industry design information, explore the establishment of Anshan local economy design service system. At the end of the year, the Anshan Industrial Design Yearbook was published to establish the benchmark and example of industrial design in Anshan area. 2. To establish the official micro of Anshan Industrial Design Center. Set up a special person to be responsible for the establishment of the "Anshan Industrial Design Center" official WeChat and official Weibo, regularly push "design news, design cutting edge, design information 'related information, to create a" Anshan design "atmosphere.

*3.2.6. Brand activities.* 1. Forum exhibition. Strengthen cooperation with famous design colleges and universities, industry design associations, design industry parks and well-known design institutions at home and abroad to hold national or regional design forums and design exhibitions to create the brand influence of "Anshan Design". To the high-end, international, open direction of development. 2. Design docking. Relying on Liaoning Province Industrial Design Association, various Industry Design Associations, Anshan Industrial Design Association, Liao Anshan Industrial Design Center, related Industrial Design institutions and Manufacturing Enterprises, Co-organized "Design Forum, Design Exhibition, into the Industrial Cluster" and other forms of "Anshan innovative Design week" series of activities to enhance the brand influence of "Anshan Design".

*3.2.7. Standard formulation.* In order to strengthen the protection and supervision of intellectual property rights, design standards and design norms, and establish a good ecological environment for the design industry, a research group was set up. To formulate policies and systems related to design standards, design evaluation, market transactions, intellectual property protection, supervision and management, so as to provide fundamental protection for the development of creative and industrial design industries.

*3.2.8. Basic research.* Set up a research group to face the fields of "brand planning and image promotion, product design and strategy consultation, marketing planning and market operation", Carry out research and services in the direction of "industry research, operation analysis, concept promotion, planning research, policy publicity, knowledge popularization and education", and regularly publish white papers on industry design and development to provide support for government decision-making.

## 4. Conclusion

Based on the analysis of the development of traditional machinery industry at the present stage, combined with local characteristics, this paper expounds in detail the methods of the integration of industrial design and machinery industry in Anshan City, combining local characteristics and adjusting measures to local conditions to promote the development of regional economy. Anshan to find a suitable for their own development of the old industrial transformation of the road, the future of the traditional machinery industry will coruscate a new competitiveness.

**Acknowledgments**

The work of this paper is completed under the careful guidance of my mentor, Professor Yang Qin. Professor Yang Qin's rigorous attitude and scientific working methods have greatly helped and influenced me. In the study and life have given me a lot of concern and help, I would like to express my heartfelt thanks.

## References

[1] Zhang Ye, Jiang Liang, Xiao Jianghao. The idea of Industrial Design in Mechanical products [J]. Design, 2018 (08): 102-104.
[2] G. N. Eremin. Improved standards regarding electrical steel and precision alloys [J]. Steel in Translation, 2017, 47 (2).
[3] Chen Yan. Research on innovative Industrial Design promoting the Development of Regional economy in Henan Province [J]. Journal of Hubei Academy of Fine Arts, 2015 (01): 95-97.

IOP Conf. Series: Materials Science and Engineering **772** (2020) 012100 doi:10.1088/1757-899X/772/1/012100

[4] Hua Zhengwei. Research on Creative Industry Cluster and Regional Economic Development in China [D]. Northeast normal University, 2012.
[5] Zhang Fanglan, Zhang Zhiyu. Research on Construction and Regional Integration of Industrial Design Innovation platform [J]. Teaching Research, 2011, 34 (05): 50-51. 57.

# Polystyrene-supported nickel complex catalysed deprotection of allylic tertiary amine with sodium borohydride

**Lingxiang Bao, Honggang Liang, Haonan Zhang, Zhiyuan Yan, Chenghui Sun * and Siping Pang**

School of Materials Science & Engineering, Beijing Institute of Technology, Beijing, China

*Corresponding author e-mail: sunch@bit.edu.cn

**Abstract.** This work represents a green and facile approach for the polystyrene supported nickel catalysed deprotection of allylic tertiary amine. The careful instrumental investigations with XPS, TEM, TGA, XRD and ICP-AES precisely characterized the developed heterogeneous catalyst. Cheap metal, mild reaction conditions, good yields as well as the reusability of the catalyst made this method more environmentally friendly.

## 1. Introduction

Finding efficient protecting groups and selective deprotection methods is still major motif in organic chemistry [1]. The characteristic of allyl protecting group, such as high selectivity and stability under acidic and basic conditions, which made its wide use of the protection of heteroatoms (nitrogen and oxygen atoms) [2, 3].The transition metal catalysed methods are the most widely and commonly used for allyl group deprotection. In recent decades, some homogeneous catalysts based on Pd [4, 5], Ru [6], Rh [7], Ni [8] and Co [9] have been used for the deprotection reaction and are capable of obtaining the desired product in high yield. However, the homogeneous catalytic system have drawbacks such as difficulty in separation and recovery of the catalyst which limits its application in the industrial field. Furthermore, the requirement of stoichiometric amounts of nucleophilic compound which act as the allyl group scavenger in the deprotection reaction bring environmental problem. Therefore, it is meaningful that develop a green, economical and simple method to remove allyl groups.

The use of heterogeneous catalysts for organic synthesis has proven to be a good way to minimize waste production and optimize catalyst efficiency. Among those, the heterogenization of homogeneous catalysts onto polymer supports has been well investigated and proven to be more attractive in industrial applications [10]. Polystyrene supported transition metal catalysts are an integral part of this large family and are evoking a renewed interest [11]. They can be easily removed from the reaction mixture, have excellent thermal stability, and can be often reused and regenerated. Polystyrene supported transition metal catalysts has been found to show high catalytic activity towards a wide range of organic reactions, including hydrogenation [12], epoxidation [13], coupling [14, 15] and so on. Even though the efficiency and selectivity of polystyrene supported transition metal, there remains an absence of any reports using them for deallylation of allylic amine.

Herein, we report a new methodologies for the synthesis of a polystyrene supported Ni catalyst that exhibits excellent activity for the deallylation of tertiary amine derivatives with sodium borohydride and

Content from this work may be used under the terms of the Creative Commons Attribution 3.0 licence. Any further distribution of this work must maintain attribution to the author(s) and the title of the work, journal citation and DOI.
Published under licence by IOP Publishing Ltd

methanol. Reactions were conducted under mild conditions and didn't produce any hazardous waste, which can meet the requirements of environmentally friendly synthesis.

## 2. Experimental

### *2.1. Materials*
Chloromethylated polystyrene and Nickel precursor $NiCl_2$ were obtained from TCI Development Co., Ltd. Lithium diphenyl phosphate was purchased from Sigma-Aldrich (USA). All the solvents used here were purchased from Sinopharm Chemical Reagent Co., Ltd. Spectrochem.

### *2.2. Preparation of catalyst*
The desired catalyst was prepared cautiously in two step process. Initially, 3.46 g of chloromethylated polystyrene in 30mL tetrahydrofuran was swelled under ultrasound for 1h. Then, swelled chloromethylated polystyrene was mixed with lithium diphenyl phosphate in 45 mL tetrahydrofuran solution in argon atmosphere. Then the solution was continuously stirred under ultrasound until the color of the solution changed from red to colourless. Then resultant mixture was filtered and washed with methanol, ether, toluene and tetrahydrofuran, respectively. Freshly prepared polymer was taken in 50 mL ethanol/toluene mixed solvent (volume ratio 2:1) and continuously stirred under ultrasound for 2 h. The resultant polystyrene was filtered, repeat impregnation and filter for 4 times to remove impure component absolutely and dried in vacuum at 60 °C.

In ethanol/toluene mixed solvent (volume ratio 2:1), the diphenylphosphine methyl polystyrene (PS-DP) (0.5g) and nickel chloride (0.15g) was stirred for 1h in argon atmosphere. The mixture was continuously stirred under ultrasound for another 1h and filtered. The product was washed by ethanol, acetone and ether respectively. Then solid was dried at 60 °C under vacuum. Preparation scheme of the Polystyrene-supported nickel complex is outlined in Scheme 1.

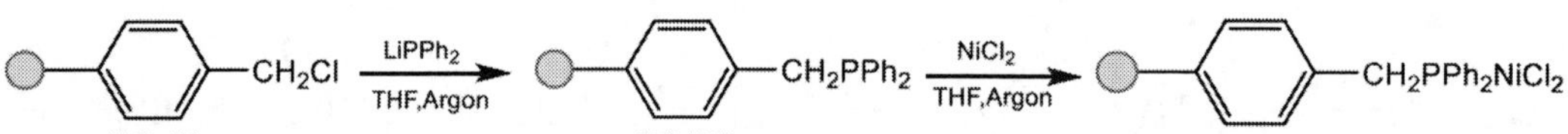


**Scheme 1.** Preparation of polymer incorporated nickel catalyst.

### *2.3. General procedure for deprotection of allylic tertiary amine*
In a 25 mL glass vial, allylic tertiary amine (1 mmol), PS-DP-Ni catalyst (5% mmol), sodium borohydride (1 mmol) were mixed in 10 mL of methanol at −20°C. After that the vial was sealed and stirred at room temperature (23−28 °C). The progress of the reaction was monitored by TLC. Upon completion of the reaction, the catalyst was filtered off and the reaction mixture was purified by silica gel chromatography using an appropriate eluent. The solid product was analyzed by NMR, IR and MS.

### *2.4. Characterization*
Transmission electron microscope (TEM) images were performed on FEI Tecnai G2 F30. Ni loading was determined by Thermo ICAP6300 inductively coupled plasma atomic emission spectrometry (ICP-AES). Powder X-ray diffractometer (PXRD) patterns was analyzed by a Bruker D8 Advance X-ray diffractometer equipped with a Cu Kα radiation. The scanning 2θ range was 10 – 80°. X-ray photoelectron spectroscopy (XPS) with an AXIS Ultra (Kratos, England) was used to examine the electronic properties of Ni active sites. The binding energies were calibrated using a C1s binding energy of 284.6 eV. Thermal stability of PS-DP-Ni catalyst was measured by a TA-DSC Q2000 differential scanning calorimeter under $N_2$ atmosphere at a heating rate of 5 °C /min from 30°C to 600 °C.

## 3. Results and discussion

### *3.1. Characterization of the catalyst*

*3.1.1. Transmission electron microscope (TEM).* Image 1(A) and 1(B) are the TEM images of PS-DP-Ni catalyst in two different magnification, where black spherical balls are well dispersed over the whole specimen. These black balls confirm coordination of Ni with polymer supported ligand.

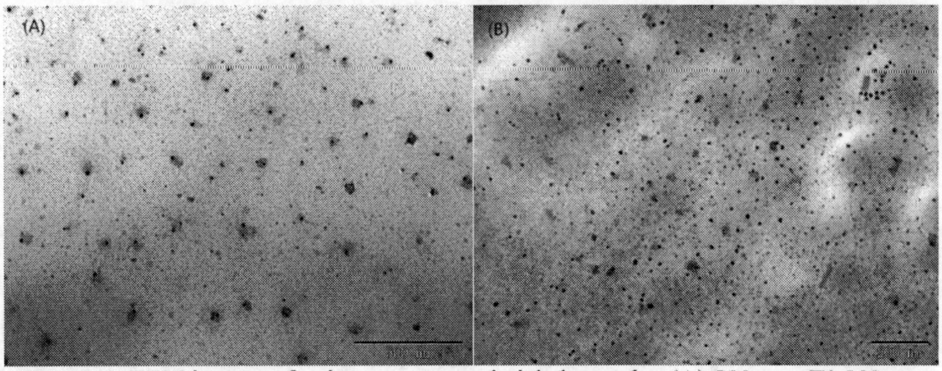


**Figure 1.** TEM images of polymer supported nickel complex (A) 500 nm, (B) 200 nm

*3.1.2. X-ray photoelectron spectroscopy (XPS).* The XPS survey spectrum for PS-DP-Ni catalyst is depicted in Fig.2 (A). The spectrum showed a strong existence of C 1s (284.6 eV) core level binding energy peaks along with Ni $2p_{3/2}$ (860 eV). The Ni $2p_{3/2}$ spectra of the PS-DP-Ni catalyst were characterized by a main peak with a satellite at higher binding energy (BE).The BE of Ni $2p_{3/2}$ in PS-DP-Ni of 854.3 eV was close to that of the Ni species in $[CH_3PPh_3]_2[NiCl_4]$. Its satellite peak due to 860.1 eV (curve-fitted value) also was observed in the XPS spectra shown in Fig. 2 (B).

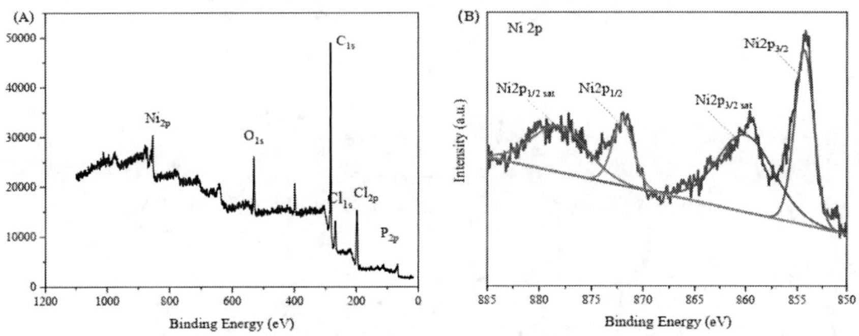


**Figure 2.** XPS spectrum of the as-synthesized PS-DP-Ni (A) full range (B) Ni 2p

*3.1.3. Powder X-ray diffractometer (PXRD).* X-ray diffraction measurement was carried out to determine the structure of the polystyrene supported nickel complex (Fig. 3). The broad peaks at 2θ values between 10° and 30° are attributed to the amorphous structure of the polystyrene. It has been observed that the PS-DP show new characteristic peaks at 32°, 45.7°, 56.7° in the XRD measurement than chloromethylated polystyrene. Additionally, after react with $NiCl_2$, the characteristic peaks at 16.8° and 36° can be found in the patterns of PS-DP-Ni. The loading of Ni in PS-DP-Ni was 0.84wt% by ICP-AES, but in XRD spectrum, no characteristic diffraction peak belonging to Ni were found, indicating that Ni may exist in the form of complexes. This is consistent with the phenomenon that most Ni species is +2 price from XPS, which might be helpful to improve the catalytic performance of the supported catalyst.

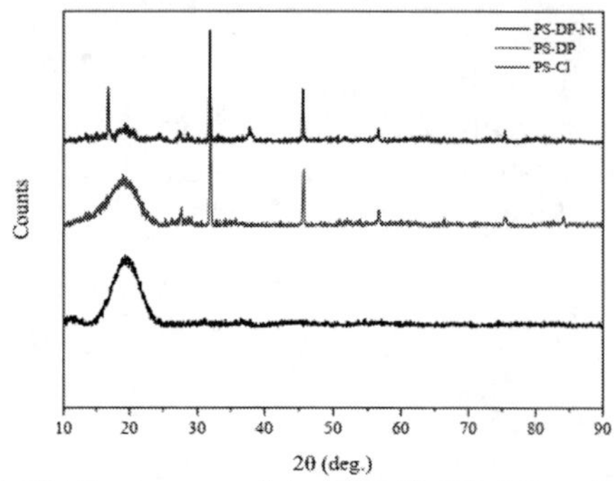


**Figure 3.** XRD patterns of the PS-Cl, PS-DP and PS-DP-Ni

*3.1.4. Thermogravimetric analysis (TGA).* Thermal properties of the supported complex was confirmed by TGA over a temperature between 30°C and 600°C at a rate of heating 5°C/min in $N_2$ atmosphere. A TGA-plot of the chloromethylated polystyrene, PS-DP and PS-DP-Ni catalyst is shown in Fig. 4. The thermal stability of the PS-DP-Ni catalyst is slightly lower compared to PS-DP and chloromethylated polystyrene, which shown that the Ni loading to the surface of the polymer. Additionally, from the TGA curve it can be outlined that thermal degradation of the heterogeneous catalyst occurred after 315°C and the decomposition of the organic structure mainly occurred in one step from 350–480°C. All the above results, suggest that the polystyrene supported nickel complex is thermally stable and can be utilized in a wide range of temperatures.

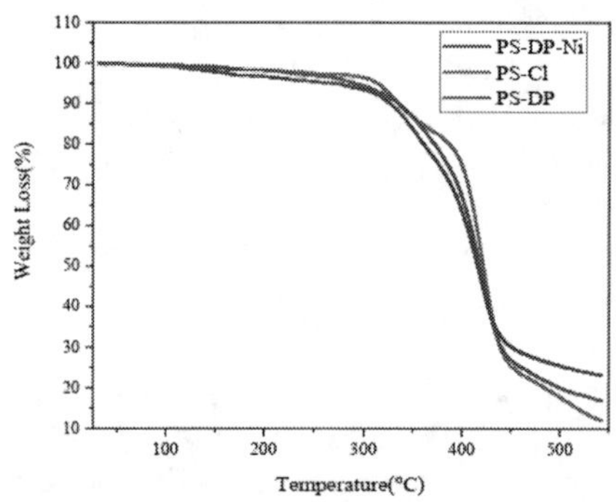


**Figure 4.** TGA plots of the PS-Cl, PS-DP and PS-DP-Ni

## *3.2. Catalytic activity of PS-DP-Ni*

To evaluate the activity of the polystyrene resin supported nickel complex, it was used in the deallylation of allylic tertiary amine reaction. Initial studies were performed upon the deprotection of N-allyl-N-methylaniline to optimize the reaction parameters including the solvent, catalyst loading and the reaction temperature. The results are summarized in Table 1. The effects of different temperature on the model reaction were first surveyed. It was found that the reaction didn't occur at low temperature while increase the reaction temperature can't improve the yield (Table 1, entry 4). The influence of the solvent on the allyl removal reaction was also investigated. Among the solvents tested, methanol was the most suitable reaction medium for the cleavage of allyl group (Table 1, entry 3). THF was so poor that give the desired product in a yield of 7% (Table 1, entry 6). When the reaction was carried out in toluene or DMF, no conversion of the starting material was observed, probably because the solvents were aprotic solvent (Table 1, entries 7 and 8).

The scope of the N-deallylation reaction under the above-derived optimized reaction conditions was investigated using 0.3mol% of the PS-DP-Ni catalyst in methanol as the solvent at 25 °C. An array of tertiary amine derivatives with different structure were surveyed and the results are outlined in Table 2.

The effectiveness of catalyst was examined with N-allyl-N-methylaniline, 1-allyl-1, 2, 3, 4-tetrahydroquinoline, 2-allyl-1-methyl-1, 2, 3, 4-tetrahydroisoquinoline, 4-allylthiomorpholine, 1-allyl-4-ethylpiperazine. Among them, 4-allylthiomorpholine correspond to the highest yield. However, 1-allylpiperdin-4-ol and 1-allyl-1H-imidazole provides a poor yield of tertiary amine.

**Table 1.** Optimization of reaction parameter for the .deprotection of allylic tertiary amine

| Entry | Solvent | Catalyst (mol % of Ni) | Temp (°C) | Time (h) | Yield[b] (%) |
|---|---|---|---|---|---|
| 1 | MeOH | 0.5 | -25 | 24 | 5 |
| 2 | MeOH | 0.5 | 0 | 24 | 12 |
| 3 | MeOH | 0.5 | 25 | 24 | 52 |
| 4 | MeOH | 0.5 | 50 | 24 | 52 |
| 5 | EtOH | 0.5 | 25 | 24 | 10 |
| 6 | THF | 0.5 | 25 | 24 | 7 |
| 7 | DMF | 0.5 | 25 | 24 | trace |
| 8 | Toluene | 0.5 | 25 | 24 | trace |
| 9 | MeOH | 0.5 | 25 | 16 | 51 |
| 10 | MeOH | 0.3 | 25 | 16 | 51 |

[a] Reaction conditions: allylic tertiary amine (1 mmol), sodium borohydride(1 mmol), PS-DP-Ni catalyst (0.5% mmol), solvent 10 mL. [b] Determined by GC using n-hexadecane as an internal standard;.

**Table 2.** Deprotection of allylic tertiary amine.

| Entry | Substrate | Time (h) | Yield[b] (%) |
|---|---|---|---|
| 1 | 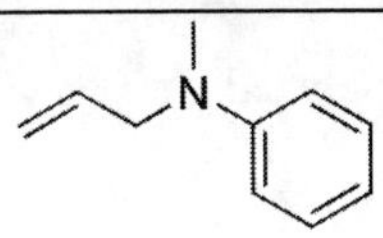 | 16 | 52 |
| 2 | 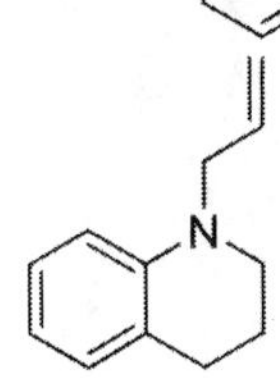 | 18 | 35 |
| 3 | 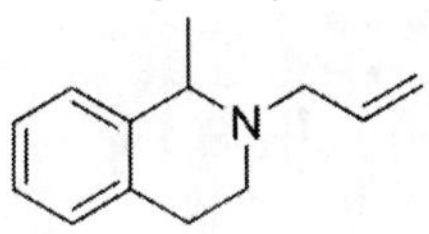 | 18 | 51 |
| 4 | 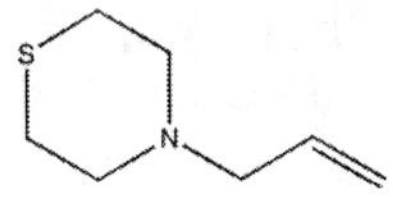 | 15 | 73 |
| 5 | 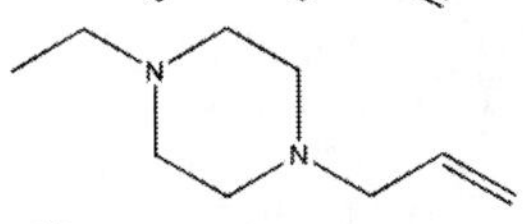 | 24 | 55 |
| 6 | 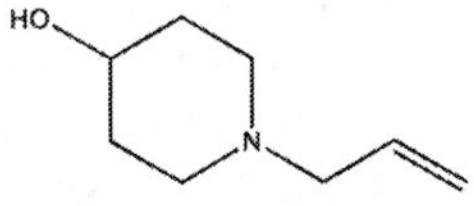 | 13 | trace |
| 7 | 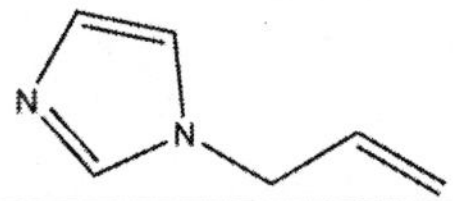 | 9 | Trace |

[a] Reaction conditions: allylic tertiary amine (2 mmol), sodium borohydride (2 mmol), PS-DP-Ni catalyst (0.3% mmol), solvent 10 mL. [b] Isolated yields.

## 4. Conclusion

In Conclusion, we have developed a novel polystyrene supported Ni (II) complex as an efficient catalyst in the deprotection reaction of various allylic tertiary amine. The allylic tertiary amines are cleaved in the presence of catalyst, sodium borohydride and methanol with moderate to high yields. This heterogeneous catalyst was easily recovered by simple filtration and reused without significant loss of its catalytic activity.

**Acknowledgments**
This work was financially supported by the National Natural Science Foundation of China (21576026).

## References

[1] P. Kumar, S.K. Cherian, R. Jain, K. Show, Chemoselective deprotection of N-allylic amines using DDQ. Tetrahedron letters 2014, 55 (52), 7172-7176.
[2] P. Nandi, J.L. Dye, J.E. Jackson, Reductive amine deallyl- and debenzylation with alkali metal in Silica Gel (M-SG). Tetrahedron letters 2009, 50 (27), 3864-3866.
[3] T. Taniguchi, K. Ogasawara, Facile and Specific Nickel-Catalyzed De-N-allylation. Tetrahedron letters 1998, 39, 4679-4682.
[4] S.Lemaire-Audoire, M. Savignac., J. Pierre Genet, Selective Deprotection of Allyl Amines using Palladium. Tetrahedron letters 1995, 36 (8), 1267-1270;
[5] Z. Dong, S. Pyne, Synthesis of Chiral Allylic Amines via Palladium (0) Catalysed Allylations Of Allylic Carbonates With Chiral Sulfinamide Anions. Sulfur Letters 2002, 25 (1), 37-43.
[6] K. Kajihara, M Arisawa, S. Shuto, A Method for Cleaving an Allyl Protecting Group at the amide nitrogen of peptides by One-pot Olefin isomerization-Oxidation. The Journal of organic chemistry 2008, 73, 9494-9496.
[7] M.J. Zacuto, X. Feng, One-Step RhCl3-Catalyzed Deprotection of Acylic N-allyl Amides. The Journal of organic chemistry 2007, 72, 6298-6300.
[8] S. Kamijo, Z. Huo, T. Jin, C. Kanazawa, Y. Yamamoto, Facile Deallylation Protocols for the Preparation of N-Unsubstituted Triazoles and Tetrazoles. The Journal of organic chemistry 2005, 70 (16), 6389-6397.
[9] H. Kumobayashi, S. Akutagawa, S. Otsuka, Metal-assisted terpenoid syntheses. 6. Enantioselective hydrogen migration in prochiral allylamine systems by chiral cobalt catalysts. Journal of the American Chemical Society 1978, 100 (12), 3949-3950.
[10] Y. Zhang, S. N. Riduan, Functional porous organic polymers for heterogeneous catalysis. Chemical Society reviews 2012, 41 (6), 2083-94.
[11] N.E. Leadbeater; M. Marco, Preparation of Polymer-Supported Ligands and Metal Complexes for Use in Catalysis. Chemical Reviews 2002, 102 (10), 3217-3274.
[12] A.S. Sharma; H. Kaur, Microwave assisted hydrogenation of olefins by Pd NPs@polystyrene resin using a gas addition kit: a robust and sustainable protocol. New Journal of Chemistry 2018, 42 (23), 18935-18941.
[13] J. Bartáček, P. Drabina, J. Váňa, M. Sedlák, Recoverable polystyrene-supported catalysts for Sharpless allylic alcohols epoxidations. Reactive and Functional Polymers 2019, 137, 123-132.
[14] B. Movassagh, F. Hajizadeh, E. Mohammadi, Polystyrene-supported Pd(II)-N-heterocyclic carbene complex as a heterogeneous and recyclable precatalyst for cross-coupling of acyl chlorides with arylboronic acids. Applied Organometallic Chemistry 2018, 32 (1), e3982;
[15] A. Ohtaka, A. Sakon, A. Yasui, T. Kawaguchi, G. Hamasaka, Y. Uozumi; T. Shinagawa; O.Shimomura; Catalytic specificity of linear polystyrene-stabilized Pd nanoparticles during Ullmann coupling reaction in water and the associated mechanism.2018,854,87-93

# Failure analysis of the low-pressure blade lacing wire in steam turbine

**Lijie Qiao *, Keyun Feng**

Huadian Electric Power Research Institute Co., Ltd, Hangzhou, China

*Corresponding author e-mail: qiaolj@126.com

**Abstract.** The failure analysis of the low-pressure last stage blade lacing wire in steam turbine made of 0Cr17Ni4Cu4Nb(S17400) is presented. The failure of the low-pressure last stage blade occurred at 7 operating years. Several examinations were carried out to identify the blade lacing wire failure's root cause: macroscopic inspection, chemical composition analysis, hardness test, metallographic analysis, microscopic examination and inspection of maintenance records. It is found that there is local displacement of the roots of the low-pressure rotor, and the collision between the lacing and the lacing holes is the direct cause and external mechanical incentive of the lacing fracture.the high content of P element and the high hardness reduce the plasticity and toughness of the material which is the intrinsic factor of this tensile fracture, without obvious plastic deformation, the low-pressure last stage blade lacing wire breaks rapidly. It is suggested to find out the local displacement of blade root and fix it in a reliable way, check the strength after eliminating the harmful defects such as cracks, and replace the blade if necessary, strictly according to DL/T438-2016 and the manufacturer's standard requirements for the inspection of the lacing wire before use.

**Keywords:** Turbine blade; lacing wire; fracture; failure analysis.

## 1. Introduction

Steam turbine blades are critical components in power plants, which convert the linear motion of high temperature and high pressure steam flowing down a pressure gradient into a rotary motion of the turbine shaft. The low pressure turbine blades, designed to extract the final remnant of energy from the passing steam flow, are relatively large scale rotating airfoils due to the significant centrifugal forces experienced during normal operation. four blades are grouped together by a lacing wire. Low pressure (LP) blades of a steam turbine are generally found to be more susceptible to failure than intermediate pressure (IP) and high pressure (HP) blades. [1-4]

0Cr17Ni4Cu4Nb (S17400) steel is a martensitic precipitation hardening stainless steel. The steel has high corrosion resistance, good attenuation performance, easy adjustment of strength level, simple welding process, especially corrosion resistance and water drop resistance than 12% Cr steel, and has good comprehensive mechanical properties. It can be used for turbine low pressure final stage moving blades and structural parts with working temperature below 300℃ in corrosive environment.[5-8]

Content from this work may be used under the terms of the Creative Commons Attribution 3.0 licence. Any further distribution of this work must maintain attribution to the author(s) and the title of the work, journal citation and DOI.
Published under licence by IOP Publishing Ltd

The gas turbine, steam turbine and generator are rigidly connected in series on a long axis, the steam turbine model is HTC158# (D10 optimized type), three-pressure, intermediate reheat, single-shaft, impulsive non-extraction pure solidification unit. The high-pressure cylinder of the steam turbine is arranged in reverse flow. At the final stage of the medium-pressure cylinder, there is intake steam from the waste heat boiler low-pressure superheated steam. The medium pressure cylinders are mixed with the steam and flow into the low pressure cylinder. The low pressure cylinder is a double-flow down-flow type with 6 pressure stages for a total of 12 stages. The parameters of low pressure steam are shown in Table 1.

**Table 1.** The parameters of low pressure steam.

| | Extraction condition | Pure coagulation condition |
|---|---|---|
| Steam inlet pressure(MPa) | 0.37 | 0.40 |
| Steam inlet temperature(°C) | 305.05 | 304.54 |
| Steam flow(t/h) | 38.28 | 44.24 |
| Exhaust steam pressure(KPa) | 3.67 | 4.90 |
| Exhaust steam temperature(°C) | 27.51 | 32.54 |
| Exhaust flow(t/h) | 211.16 | 362.42 |

During the No.7 steam turbine set overhaul in April 2019, it was found that there were lacing wire fracture and leaf blade hole damage of the last stage low-pressure blade, and the material was 0Cr17Ni4Cu4Nb(S17400) with 13 mm diameter. A series of tests were carried out to analyze the cause of failure.

## 2. Experimental procedure

The objects to investigate into were lacing wire fracture and leaf blade hole damage (8.090 h service exposed) obtained from the plant. After an extensive visual examination, chemical analysis of the lacing wire material was carried out by spectral analysis method. A portion of the failed lacing wire fractured surface was cut for metallography. The metallographic samples were prepared by using standard metallographic techniques and etched with reagent Ferric chloride and hydrochloric acid during 15s at room temperature. The brinell hardness test and vickers hardness test were carried out on the cross section of the lacing wire. Brinell Hardness Testing Machine with bearing of 2.5mm in diameter under a load of 1840 N.Micro-hardness Vickers tests were performed with a 200 g load and an indentation time of 15 s. Micro-indentation tests were also carried out on defined areas of the sample. The applied load was 1960 mN. The microstructure of the lacing wire material was analyzed using an optical microscope and a TESCAN VEGA3 LM scanning electron microscope (SEM).

## 3. Results

### *3.1. Visual inspection*

Low pressure blade lacing wire and brace hole site inspection situation as shown in fig.1.There is a gap between lacing wire with brace hole, accidental abnormal factors may lead to the collision of lacing wire and brace hole in operation, local defects were first produced on the outer surface or the inner surface of the pull rib hole, local defects subsequently develop into microcracks in high-speed rotation or re-collision, macroscopic defects caused by microcrack growth are related to collision energy, frequency and contact area, etc. local defects can form Local crack of blade hole (b),gap of blade hole(c),severe crack of blade hole(c).In addition, it can be seen from (b) that the collision at different locations can form multiple cracks at different locations (9 o 'clock and 12 o 'clock), extending along the inner wall to the outer wall.(e) visible fracture in overall level, no obvious plastic deformation, the characteristics of brittle fracture, fracture surface local oxidation corrosion, show the maintenance when the brace has broken the oxidation, fracture surface smooth metallic parts, may be associated with cracking section between local touch after grinding, according to the fracture extended morphology, visible inspection

IOP Conf. Series: Materials Science and Engineering **772** (2020) 012102 doi:10.1088/1757-899X/772/1/012102

blade fracture source of brace in the A1 and A2 position, extension area located at the position B, instantaneous fault zone located at the position C.

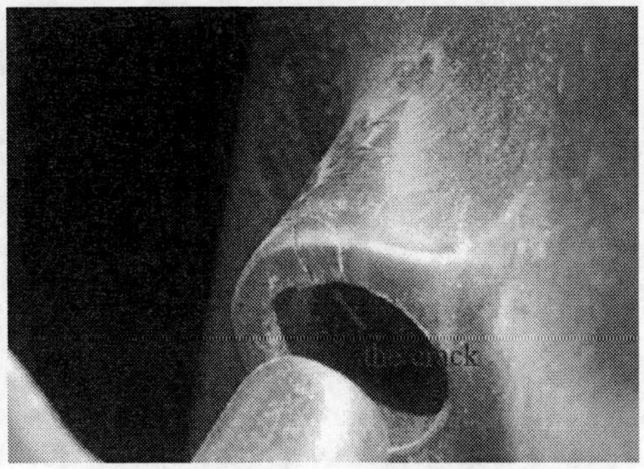


(a) Fit between lacing wire and blade hole (b) Local crack of blade hole

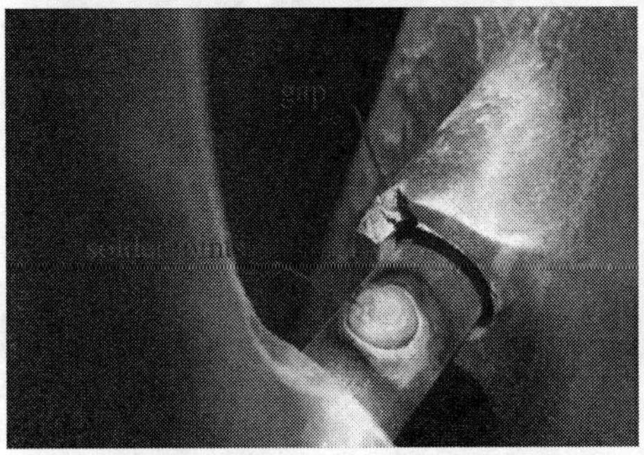


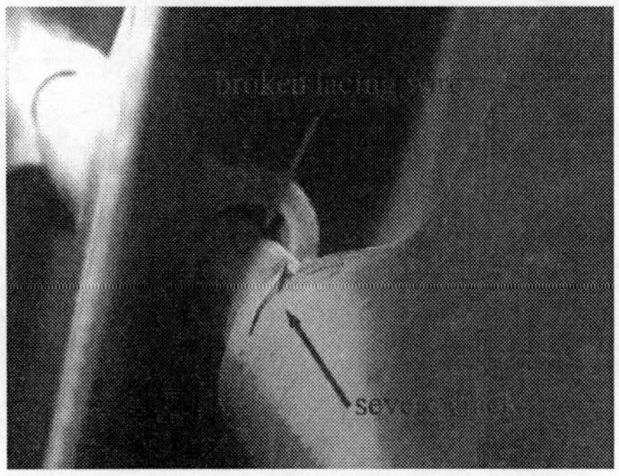


(c) Gap of blade hole (d) severe crack of blade hole

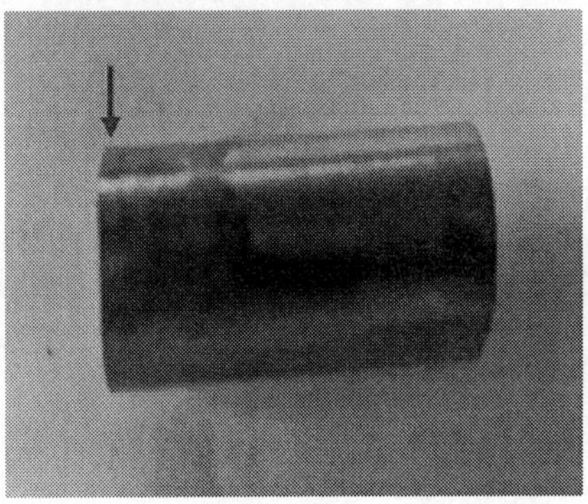

(e) Fracture of lacing wire

**Figure 1.** Morphology of failed low-pressure blade

### *3.2. Chemical analysis*

Thermo ARL8860 spectrometer was used for spectral analysis, and the analysis results were the average of the three test results, as shown in table.2.It can be seen that the content of P exceeds the requirements of manufacturer's standard, and the content of other major elements meets the requirements. Generally, P element is considered as the harmful element in steel, which can reduce the plasticity and toughness of steel, and increase the brittle transition temperature of steel, leading to the cold brittleness of steel.

IOP Conf. Series: Materials Science and Engineering **772** (2020) 012102 doi:10.1088/1757-899X/772/1/012102

**Table 2.** Spectral analysis results (wt%).

| Elements | C | Si | Mn | P | S | Cr | Ni | Cu | Nb+Ta |
|---|---|---|---|---|---|---|---|---|---|
| Result | 0.040 | 0.38 | 0.23 | 0.035 | 0.004 | 15.84 | 4.41 | 3.37 | 0.28 |
| Manufacturer's standard | ≤0.055 | ≤1.00 | ≤0.50 | ≤0.030 | ≤0.020 | 15.00~16.00 | 3.80~4.50 | 3.00~3.70 | 0.15~0.35 |

### *3.3. Hardness test*

Brinell hardness test and vickers hardness test were carried out on the cross section of the lacing wire . The hardness test location was shown in fig.2, and the results of brinell hardness test were shown in table.3, and vickers hardness test results were shown in fig.3.It can be seen that the detected brinell hardness value exceeds the manufacturer's standard requirements, and the distribution trend of vickers hardness of the tensile rib cross-section of the blade shows an upward trend from the central part to the edge part. The measured minimum value is 308HV, and the measured maximum value is 326HV, both of which exceed the manufacturer's standard requirements.

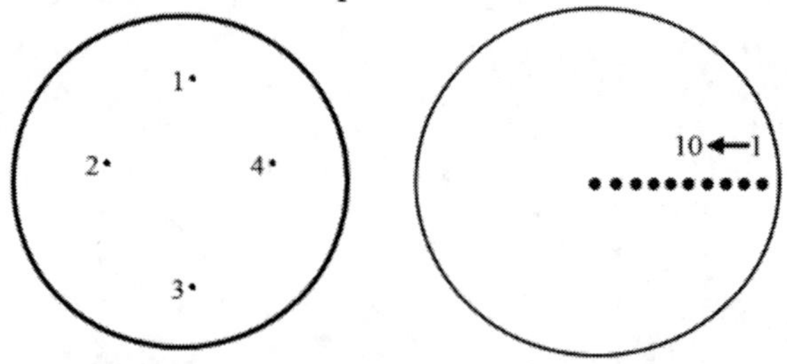


(a) Brinell hardness test (b) vickers hardness test

**Figure 2.** Location of hardness measuring point

**Table 3.** Brinell hardness test results (HBW)

| Measuring point | 1 | 2 | 3 | 4 | manufacturer's standard |
|---|---|---|---|---|---|
| Result | 320 | 314 | 315 | 318 | 262-302 |

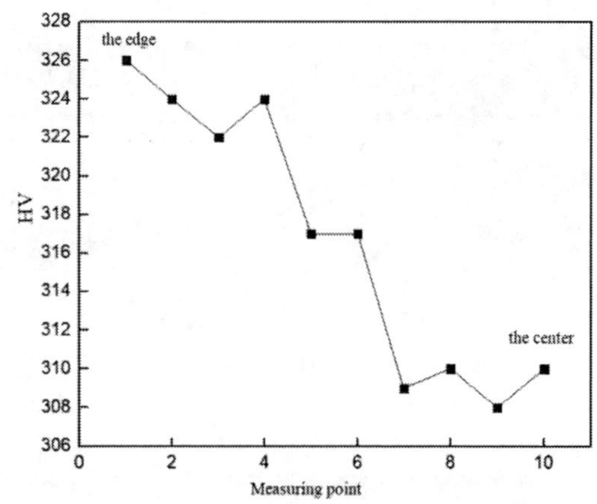


**Figure 3.** Vickers hardness test results

### *3.4. Microscopical investigation*

The metallographic structure of the outer edge and the central part is martensite without obvious abnormality, which was shown in fig.4.

IOP Conf. Series: Materials Science and Engineering **772** (2020) 012102 doi:10.1088/1757-899X/772/1/012102

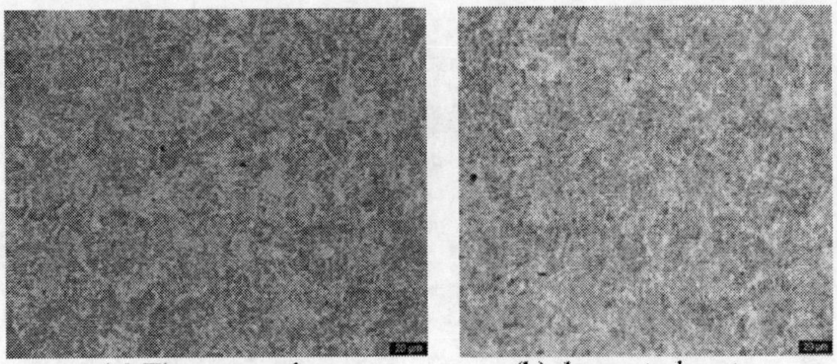

(a) The outer edge (b) the central part

**Figure 4.** microstructure of lacing wire

The source area, extension area and transient fracture area of lacing wire were analyzed by sem respectively, and the results are shown in fig.5.Visible fracture source (A1) local pit, for external surface of the lacing wire after the collision, the follow-up development in high speed rotating or collision again as the crack, and gradually expand to the central part, micro has the obvious river pattern, spreading area and transient breaking cleavage fracture morphology, found no toughening nest morphology, as a typical brittle fracture. There are smooth grinding surface and directional grinding mark in the fracture, which may be related to the friction between the sections after lacing wire fracture. The local pore morphology of the fracture may be the original defect of the material.

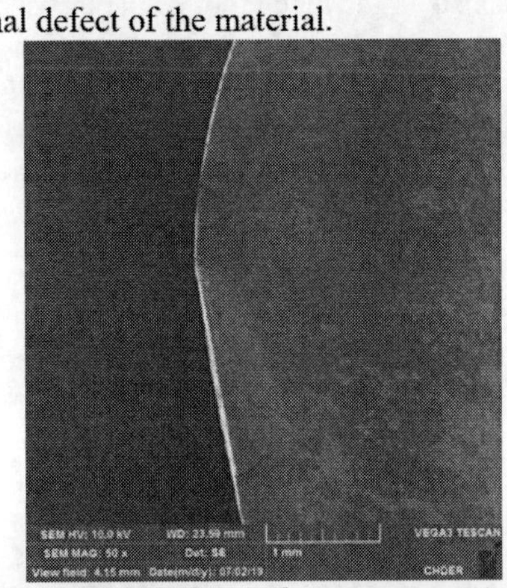

(a) Morphology of source area A1

(b) Local rubbing marks in extension area

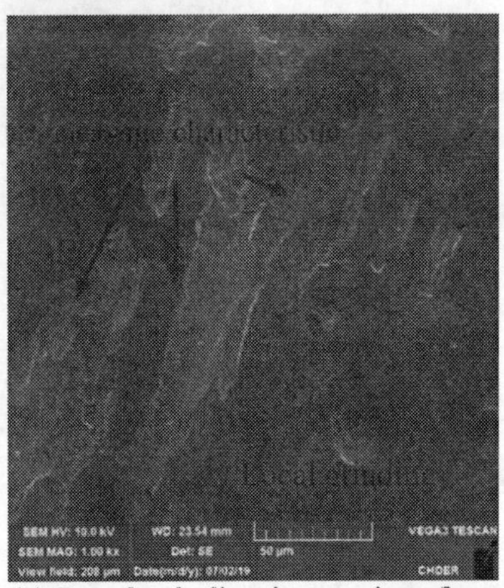

(c) Local grinding in transient fracture area

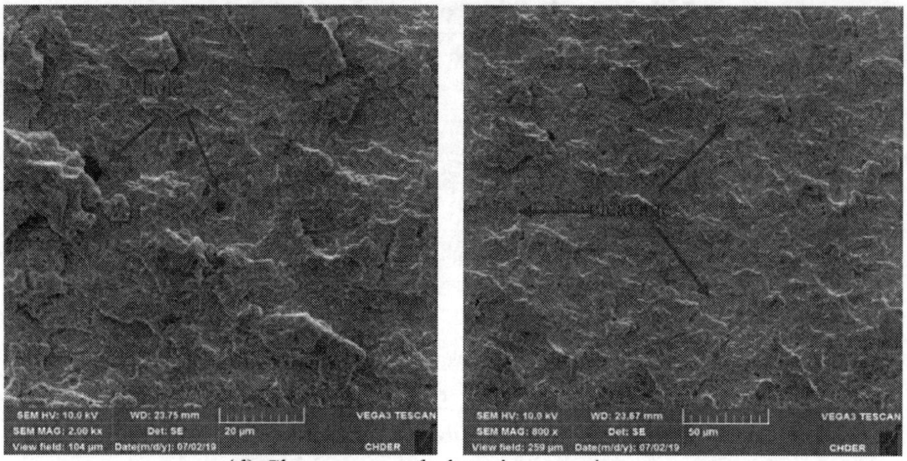

(d) Cleavage morphology in extension area

**Figure 5.** SEM morphology of fracture

## 4. Conclusion

In summary, local displacement exists in the blade root of the last stage of low-pressure rotor, and the collision between the lacing wire and brace hole during operation is the direct cause and external mechanical inducement of the failure. The high content P element and hardness of lacing wire reduce the plasticity and toughness of the material, which are the internal factors of this lacing wire fracture. Brittle fracture occurs rapidly without obvious plastic deformation of the steel.

## References

[1] Mukhopadhay NK, Ghosh Chowdhury S, Das G, Chattoraj I, Das SK, Bhattacharya DK. Eng Failure Analysis 1998;5(3):181.

[2] Atrens A, Meyer H, Faber G, Schneider K. In: Speidel MO, Atrens A, editors. Corrosion in power generating equipment. Metals Park, Ohio: Plenum Press;

[3] 1983. p. 299.

[4] Viswanathan R. Damage mechanisms and life assessment of high temperature components. Metals Park, Ohio: ASM International; 1989. 313.

[5] ASM handbook, Vol. 1., ASM Metals Park (OH), 1993. p. 284.

[6] Thermal Power Plant Metal Materials Handbook, China Electric Power Press,2001. p. 284.

[7] A. Mokaberi, R. Derakhshandeh-Haghighi, Y. Abbaszadeh,Engineering Failure Analysis, Volume 58, Part 1, December 2015, Pages 1-7.

[8] D. Ziegler, M. Puccinelli, B. Bergallo, A. Picasso,Case Studies in Engineering Failure Analysis, Volume 1, Issue 3, July 2013, Pages 192-199.

[9] Goutam Das, Sandip Ghosh Chowdhury, Ashok Kumar Ray, Swapan Kumar Das, Deepak Kumar Bhattacharya,Engineering Failure Analysis, Volume 10, Issue 1, February 2003, Pages 85-91.

# Microstructure Evolution Simulation of GH4169 Fasteners in Forging Process

**Yanju Wang*, Jiaying Jiang, Tao Wang, Yukai Yang, Aixue Sha and Xingwu Li**

Aviation Engine Corporation of China, Beijing Institute of Aeronautical Materials, Beijing, 100095, China

*Corresponding author e-mail: wyjbiam@163.com

**Abstract.** This paper mainly calculated the microstructure evolution characteristics of GH4169 alloy fasteners during hot forging deformation. The constitutive equation and the microstructure evolution models of the material were embedded in the finite element software. By analysis, the grain size and dynamic recrystallization volume fraction were calculated, and the key points on the fastener in forging process were analyzed. As a result, the microstructure evolution of the GH4169 alloy fastener during the hot forging deformation process was obtained. The results show that, through the plastic deformation by the hot forging process, the grain size on the fastener's head is gradually refined, the minimum average grain size is reduced from the initial 100 μ m to less than 10 μ m. During the deformation process, the splinted part of the fastener begins to undergo dynamic recrystallization first. After that, dynamic recrystallization volume fraction gradually increases from the central region to the edge, with the highest value of 60%. These results provide a basic data reference for GH4169 fastener forging process development and parameters optimization.

**Key words:** GH4169; fastener; forging; microstructure; numerical simulation.

## 1. Introduction

Superalloy GH4169 is a main material as wrought superalloys due to its high strength, oxidation resistance and corrosion resistance. it is widely used in advanced aeroengines and aircraft at home and abroad. In terms of aero-engine fasteners, GH4169 alloy bolt fasteners account for more than 70% of the total bolt demand, and is also an indispensable high-strength load-bearing fastener material for aircraft. The main purpose of the forging process is to form and modify. Forming is a basic requirement, and the performance of the forgings should be maximized as far as the shape requirements are reached. GH4169 alloy fasteners work under high temperature and high pressure, which have extremely high requirements on microstructure and mechanical properties of the material, and need to withstand bending and shear stress caused by alternating load during service. Therefore, so great performance is also required such as fatigue resistance and wear resistance. However, GH4169 alloy fasteners may undergo a variety of microstructure evolution during the hot forging process, including dynamic recrystallization, grain growth and so on. The recrystallization process directly affects the change of the grain size in the forging process, which could affects the mechanical and structural properties of the

Content from this work may be used under the terms of the Creative Commons Attribution 3.0 licence. Any further distribution of this work must maintain attribution to the author(s) and the title of the work, journal citation and DOI.
Published under licence by IOP Publishing Ltd

material. By numerical simulation, the microstructure evolution of the GH4169 fasteners in forging process could be studied, and the parameters of the forging process could be optimized. Therefore, this work has great significance for optimizing process parameters and improving product quality and performance.

In this paper, the microstructure evolution characteristic of GH4169 alloy fasteners during hot forging deformation was calculated. The constitutive equation and the tissue evolution model of the material were embedded in the finite element software, and the grain size and dynamic recrystallization volume fraction of forging process were obtained. The key points on the fasteners were analyzed, and the microstructure evolution law of GH4169 alloy fasteners during hot forging deformation was obtained, which provided the basic data reference for the formulation and parameters optimization of GH4169 fastener in forging process.

## 2. Numerical modelling of the fasteners in forging process

For the hot forging process simulation of GH4169 alloy fasteners, firstly, the shape and size of the initial blank and the forging dies need to be determined, and the three-dimensional assembly model should be established. As shown in Fig. 1(a), according to the spatial geometric relationship between the blank and the upper and lower dies, the model is initially positioned and assembled. The upper and lower dies are set as rigid bodies, and the blank is elastic-plastic. And the initial mesh is discredited by absolute meshing method. Finally, there are 61,969 nodes, 45,600 polygons and 225,364 tetrahedral on the initial blank, as shown in Fig.1 (b). During the simulation analysis, the meshes were attached to the material, and the material is easily deformed due to excessive deformation of the corresponding mesh shape during the flow process. The calculation process may be interrupted by the distortion. Therefore, it is very important to ensure that the analysis can continue after the material has undergone a large deformation in the simulation process, but the results still have enough accuracy. After the mesh distortion reaches to a certain level, the distorted mesh is automatically redevised, and the new high-quality meshes are generated.

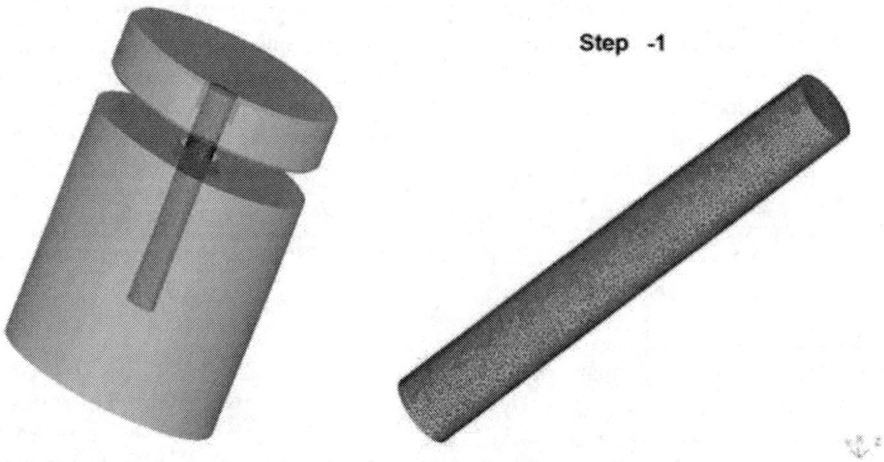


(a) (b)

**Fig. 1** 3D modeling of the fastener in forging process (a-assembly, b-blank mesh)

## 3. Material definitions

Since the forging process is a plastic deformation process, involving the flow and plastic deformation of the material, the GH4169material model should be defined in the numerical simulation model .And al l the material properties including the high temperature flow stress parameters, elastic modulus, Poisson's ratio, thermal expansion coefficient, thermal conductivity, specific heat are all need to be input.

In order to study the microstructure evolution of the fastener in forging process, the constitutive and microstructure evolution models of the material should also be defined as following:

- Critical strain and peak strain models:

$$\varepsilon_c = 0.83\varepsilon_p \tag{1}$$

$$\varepsilon_p = 0.004659\,\dot{\varepsilon}^{0.1238}\exp(49520 / RT) \tag{2}$$

- Static recrystallization models:

$$X_{srex} = 1 - \exp[-0.693(\frac{t}{t_{0.5}})^{0.3}] \tag{3}$$

$$t_{0.5} = 3.16\varepsilon^{-0.75}\exp(74790 / RT) \tag{4}$$

$$d_{srex} = 678\exp(-31694 / RT) \tag{5}$$

- Meta-dynamic recrystallization models:

$$X_{mrex} = 1 - \exp[-0.693(\frac{t}{t_{0.5}})] \tag{6}$$

$$t_{0.5} = 5.04E - 9\varepsilon^{-1.42}\dot{\varepsilon}^{-0.408}\exp(196000 / RT) \tag{7}$$

$$d_{mrex} = 4.85E10\varepsilon^{-0.41}\dot{\varepsilon}^{-0.028}\exp(-240000 / RT) \tag{8}$$

- Dynamic recrystallization models:

$$X_{drex} = 1 - \exp[-0.693(\frac{\varepsilon - 0.8\varepsilon_p}{\varepsilon_{0.5}})] \tag{9}$$

$$\varepsilon_{0.5} = 5.043E - 9\varepsilon^{-1.42}\dot{\varepsilon}^{-0.408}\exp(196000 / RT) \tag{10}$$

$$d_{mrex} = 4.85E10\varepsilon^{-0.41}\dot{\varepsilon}^{-0.028}\exp(-240000 / RT) \tag{11}$$

- Grain growth models:

$$d_g = d_0^2 1.58E16t\exp(-390753 / RT) \tag{12}$$

The equations of (1)~(12) are all embedded into the numerical analysis model for the stress, strain and microstructure evolution calculation.

## 4. Grain size calculation results

The calculation results of the average grain size change during the forging process of GH4169 alloy fasteners are shown in Fig. 2. The initial billet is homogenized, and the average grain size is 100 μm. By the hot forging plastic deformation process, the grain on the fastener head is gradually refined, and the minimum average grain size is reduced from the initial 100 μm to less than 10 μm. Due to the small plastic deformation of the central area, the grain size is thicker than other areas, and more than 85% of the deformation area is refined by hot forging plastic deformation, which helps to improve the application performance of fasteners.

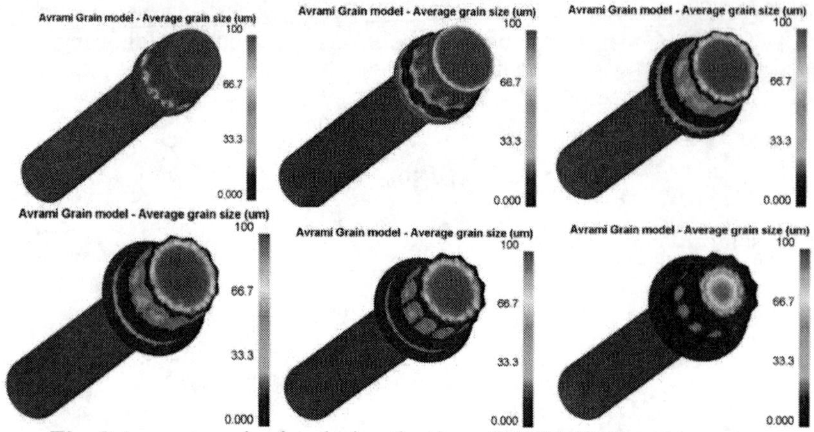


**Fig. 2** Average grain size during forging of GH4169 alloy fasteners

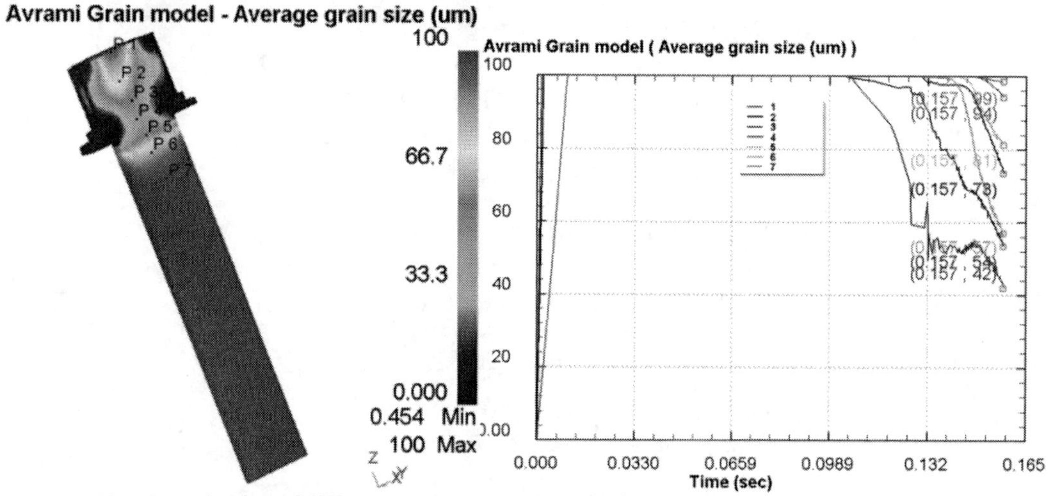


**Fig. 3** Grain size of different regions on GH4169 alloy fastener in forging process

Select the key points ( P1~P6) in different areas of the central axis section on the fastener to analyze the average grain size. As shown in Fig.3.With the deformation of the material during the forging process, the grain size of each point are gradually refined. Among them, the P3, P4, and P5 first enter the grain refining process. When the forging is completed, the average grain size value of the P4 is the smallest, which is refined from 100 μm to 42 μm.

## 5. Dynamic recrystallization calculation results

The calculation results of the dynamic recrystallization volume fraction of the GH4169 alloy fastener in forging process are shown in Fig. 4. During the deformation process, the spine portion of the fastener head begins to undergo dynamic recrystallization first, and then the dynamic recrystallization volume fraction increases with the deformation process from the central region to the edge, with a maximum of 60%.

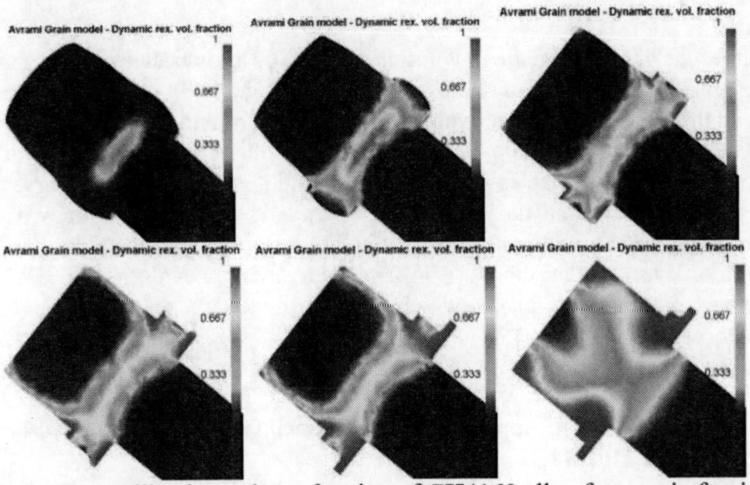


**Fig. 4** Dynamic recrystallization volume fraction of GH4169 alloy fastener in forging process

The dynamic recrystallization volume fraction of different regions in the forging process of GH4169 alloy fasteners was selected as shown in Fig. 5. The calculation results show that the fasteners complete the dynamic recrystallization softening after the forging process is completed, and the softening degree of the peripheral region is higher than that of the core. The highest value of the dynamic recrystallization volume fraction is approximately 60%.

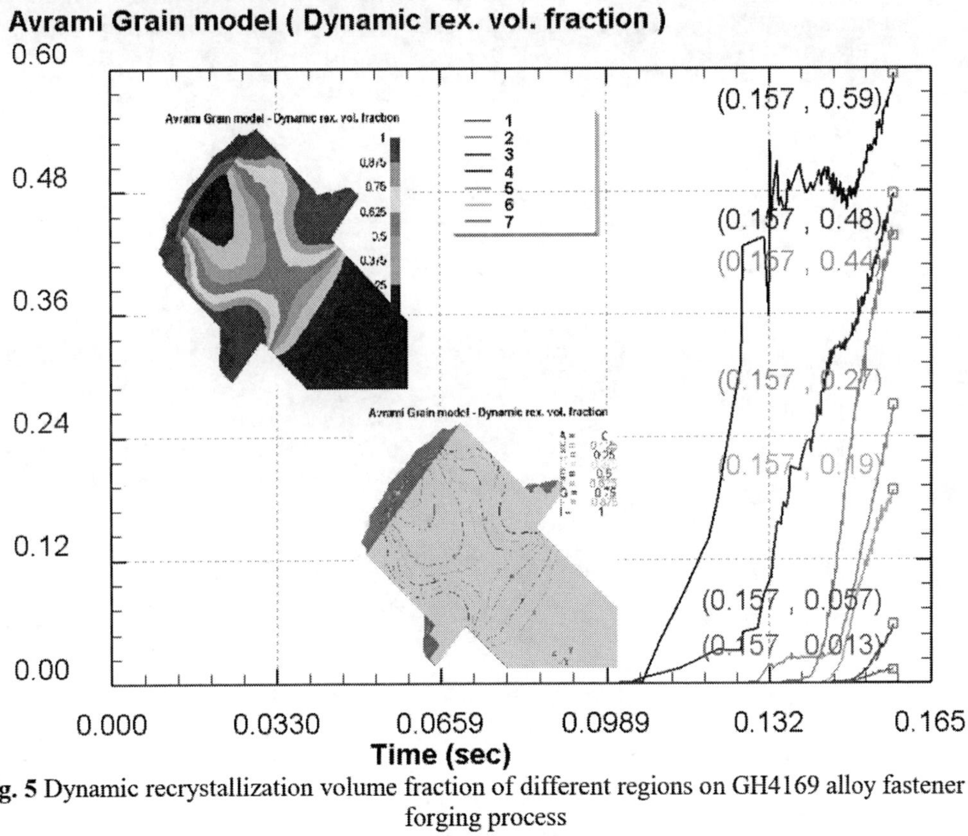


**Fig. 5** Dynamic recrystallization volume fraction of different regions on GH4169 alloy fastener in forging process

## 6. Conclusion

The calculation results of microstructure evolution of GH4169 alloy fasteners during the forging process show that the grain deformation on the fastener head is gradually refined by hot forging plastic deformation, and the minimum average grain size is reduced from the initial 100μm to less than 10μm. The calculation results of dynamic recrystallization volume fraction show that the dynamic recrystallization on the head of the faster occurs first, and then the dynamic recrystallization volume fraction gradually increases from the center area to the edge, with a maximum value of 60%. The microstructure evolution characteristics of different parts on the fastener head during forging deformation were analyzed by data comparison. The microstructure evolution of GH4169 alloy fasteners during the forming process were given, which could provide data reference for the forging process parameters optimization.

## References

[1] L.Y.Zhu, T.Xie,H.Zhang. Application of Material GH4169&TB8 in Fastener. Mechanical Engineer, 2013 (10): 41-42.
[2] Gao L, Pang Y H, Sun L, et al. Numerical Simulation of GH4169 Alloy Section of Hot Roll Forming. Advanced Materials Research, 2014: 479-482.
[3] Yongchun Xu, Derong Feng. Analysis for heat heading cracking of GH4169 nuts. Fasteners Technology, 2013(1):33-38.
[4] Ping Luo. Dynamic recrystallization simulation of GH4169 alloy during thermal deformation. Northeastern University,2014.
[5] Liang Yang, Jian Yu, Jianxin Dong. Modeling of microstructure control and forging process simulation of GH4169 alloy. Journal of Iron and Steel Research. 2011(S2):166-169.

# The explosion resistance analysis of the safety evacuation device

**Yong Wang, Xuezhi Feng*, Lizhuang Cui, Gengyu Sun, Yongyan Wang and Xiumin Jing**

School of Qingdao University of science and technology, Qingdao, China

*Corresponding author e-mail: 2500875646@qq.com

**Abstract.** The explosion resistance is one of the main functions of the safe evacuation device, and the explosion resistance of the safety evacuation device is mainly determined by the safe evacuation cabin, and therefore it requires the building structure to have sufficient strength and stiffness to resist the blast wave application to the cabin. In this paper, a new model of the similar triangular cabin structure is established. The explosion resistance impact analysis of the similar triangular cabin structure is carried out by AUTODYNA and ABAQUS software. The similar triangular cabin structure is placed in the simulated residential building, and the pressure cloud in the room during the explosion and the change of the pressure at each time in the single room can be obtained by TNT explosion test simulation. The pressure curve is used to analyze explosion resistance of the safe evacuation device, and the stress and deformation of the new cabin structure can be obtained. At the same time, the stress concentration are located at the corners of the interior of the cabin and the maximum deformation is located on the front and back walls of the pod which can meets the safety evacuation shock resistance.

## 1. Introduction

The so-called safety refuge device is a kind of airtight cabin structure which can provide emergency refuge, protect life safety and have a certain volume for the evacuees who are affected or suffered in sudden disasters or disasters. When explosion occurs, it is impossible to escape from the fire passage smoothly, which requires that the safety refuge device must have explosion-resistant ability to provide security for the lives of trapped people.

Abroad, Brzedzislav et al. used Auto Rea Gas to simulate the whole process of gas explosion in a mine with 9.5% TNT concentration. By comparing the results of simulation and test, it was found that the results were basically similar [2]. Dmitry Khokhlov et al. simulated and analyzed the whole process of mixed gas explosion, and considered the explosion comprehensively. There is a coupling relationship between combustion factors and physical quantities [3]. In China, Yu Yicheng of Harbin University of Technology and others used numerical simulation method to obtain the diffusion law of indoor gas under different influencing factors, and analyzed and calculated the consequences of explosion [4]. Gao Erxin and others used Fluent fluid software to simulate the movement law of TNT explosion shock wave flow. The distribution law of pressure field and temperature field in Coal Mine Roadway after TNT explosion

Content from this work may be used under the terms of the Creative Commons Attribution 3.0 licence. Any further distribution of this work must maintain attribution to the author(s) and the title of the work, journal citation and DOI.
Published under licence by IOP Publishing Ltd

is obtained. Dou Hongying and Baibo of Taiyuan University of Technology re-studied the anti-explosion analysis of movable lifesaving cabin.

In this paper, the advantages and disadvantages of the existing shelter structures are analyzed, and a new triangle-like shelter structure is proposed. In addition, AUTODYN is used to simulate the propagation of shock wave in a single room, and the explosion pressure curve is obtained. On this basis, the ABAQUS software is used to load the explosion pressure curve on the surface of the cabin to obtain the stress and deformation of the cabin, and to verify the anti-explosion performance of the cabin structure.

## 2. Cabin design of safety refuge device

According to the design requirements of pressure resistance [7], the pressure resistance of the round cabin is the best, but higher requirements are put forward for the placement and space utilization. The pressure resistance of the square cabin is the worst, but its placement is flexible and suitable for placement. On the basis of this, combined with the characteristics of circle and square, a new design scheme, i.e. triangle-like cabin structure, is put forward, which can improve the compressive performance, reduce the deformation and stress concentration as much as possible under the premise of guaranteeing the high utilization of space. Appearance. Fig.1 is a triangle-like cabin structure model.

**Fig. 1** The structural model of similar triangle cabin

## 3. Analysis of Cabin Explosion Based on AUTODYN

### *3.1. Material Model Setting*

The materials used in Explosion Analysis in this paper include TNT explosive, air and material of cabin structure.

(1) According to the relevant literature, HIGH_EXPLOSIVE_BURN model is selected as the material model of explosives, while JWL equation of state is used for the equation of explosives [8]:

$$P = A\left(1-\frac{\omega}{R_1 V}\right)e^{-R_1 V} + B\left(1-\frac{\omega}{R_2 V}\right)e^{-R_2 V} + \frac{\omega E}{V} \quad (1)$$

(2) The material model of air is NULL material model, and LINEAR_POLYNOMIAL equation of state [8]:

$$P = C_0 + C_1\mu + C_2\mu^2 + C_3\mu^3 + \left(C_4 + C_5\mu + C_6\mu^2\right)E \quad (2)$$

$$\mu = \frac{1}{V} - 1 \quad (3)$$

And: C0 = C1 = C2 = C3 = C6 = 0; C4 = C5 = 0.4; air density p = 1.225kg/m3; initial relative volume V0 = 1.0

(3) The material used in the cabin structure is 35CrMo steel plate, the corresponding material model is MAT_JOHN_COOK model, and the equation of state is John Cook [8]:

$$\sigma_y = \left(A' + B'\bar{\varepsilon}^{p^{n'}}\right)\left(1 + C'\ln\dot{\varepsilon}^*\right)\left(1 - T^{*m'}\right) \quad (4)$$

### *3.2. Establishment and Setting of Cabin Model*

The space size of each residential building is uncertain and the decoration methods are different, which makes it difficult to calculate the model. This paper chooses a single room in general to study. It is assumed that the structure of the safe shelter cabin is placed on one corner of the room, the equivalent TNT explosive is placed on the diagonal line of the cabin position, and the whole room is equipped with ventilated windows and exit doors, etc. The plane layout of the model is obtained as follows. Figure 2 shows.

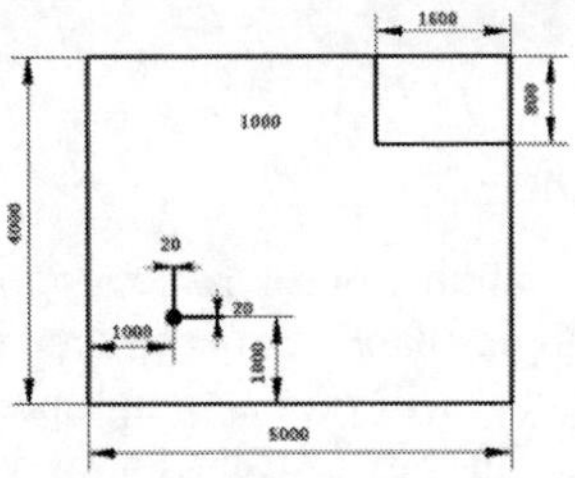


**Fig. 2** Plane layout in a single room

The equivalent TNT is put into a single chamber and filled with air to mix the air with the equivalent TNT. The cabin structure of the safety refuge device is placed in a corner of the single chamber. The explosion point is located at the center of the TNT explosive, and the TNT explosive is placed on the diagonal line of the safety refuge device, and the boundary condition is added. In AUTODYN software, the boundary of air area automatically defaults to rigid wall [9], while for the cabin structure, the boundary condition is fixed, and the system will automatically identify. The density of meshing should not be too dense, otherwise it will be difficult for computer to realize because of the huge amount of calculation. Therefore, the cabin structure model is calculated by unstructured grid, and the Eulerian multi-material field grid is used in the air domain [10]. The model diagram after adding material is obtained, as shown in Figure 3.After the pretreatment of the whole Explosion Analysis target, 40 Gaussian monitoring points are applied to the walls of the single chamber and the surface of the safety refuge device, in order to further analyze and observe the pressure changes of the explosion in each part of the single chamber. The distribution of Gauss points in the whole single room is shown in Figure. 4.

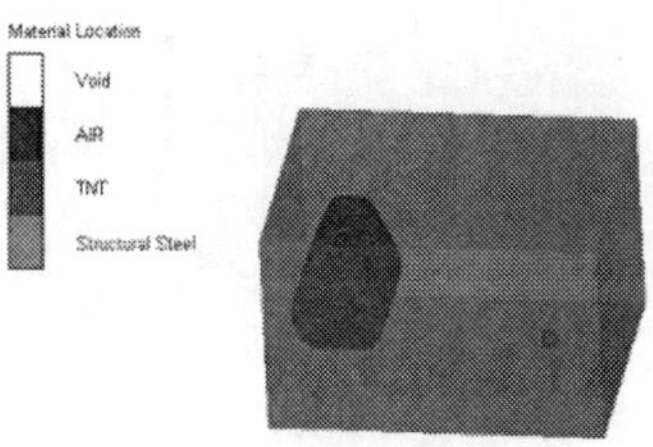


**Fig. 3** Single chamber calculation model

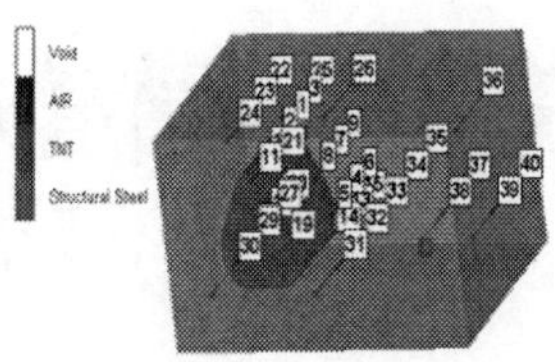


**Fig. 4** Single chamber Gaussian point distribution

### *3.3. Result Analysis*

The explosion time of a single chamber is set at 100 ms. After analysis, the pressure cloud of a single chamber is obtained from the beginning of explosion to 100 ms, as shown in Fig. 5.

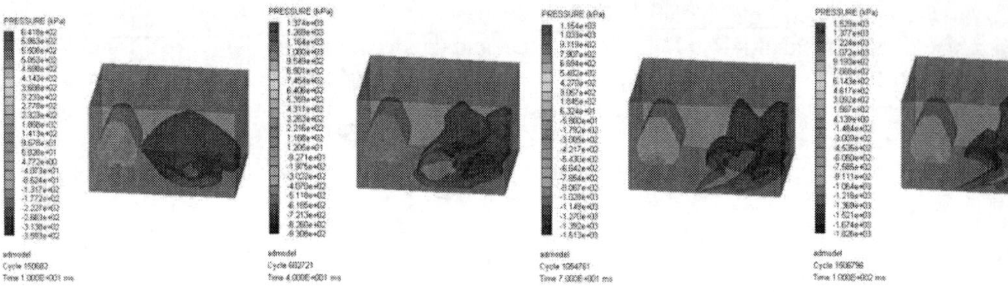

(a) 10ms (b) 40ms. (c) 70ms. (d) 100ms.

**Fig. 5** Single chamber pressure nephogram at all times

From the pressure nephogram of single chamber at each time in Fig. 5, it can be clearly observed that once the explosion source is detonated, it will produce high pressure. After detonation for 10 ms, the maximum pressure in single chamber reaches 641.8 KPa, and the shock wave after detonation first begins to spread around the detonation, and then it propagates to a little farther from the opposite side. Until it reaches the walls of a single room.

Take several observation points from a single chamber and observe the pressure change curve of the observation points during the whole explosion process, as shown in Fig. 6 and Fig. 7.

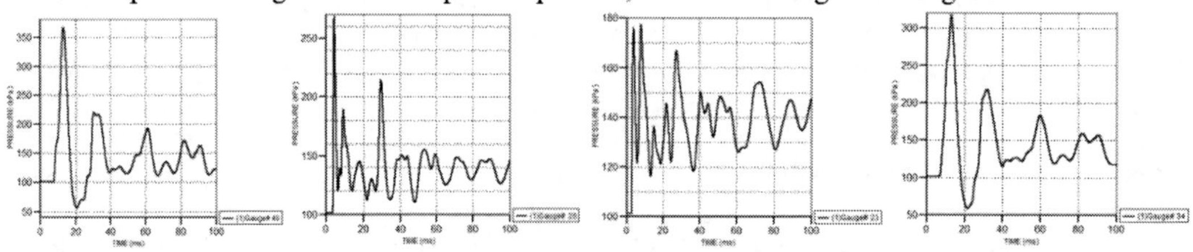

(a) 23# (b) 40# (c) 26# (d) 34#

**Fig. 6** The pressure curve of the points on each side wall of a single chamber

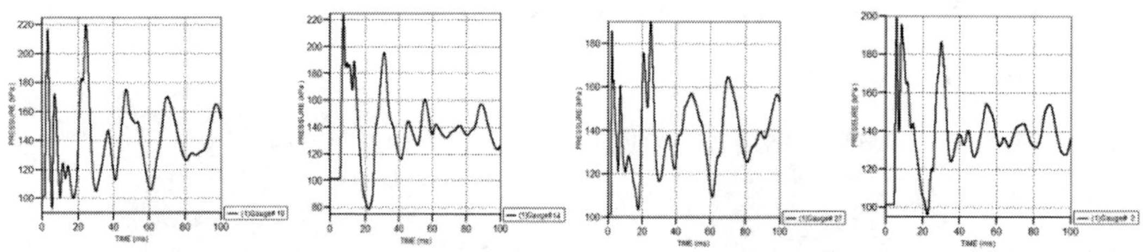

(a) 10# (b) 14# (c) 27# (d) 2#

**Fig. 7** Pressure curve of points on the surface of the cabin

Fig. 6 the pressure curves of the walls in the single room are taken from the left wall 23, the right wall 40, the back wall 26, and 34, respectively. From the analysis, it can be seen that the shock waves in a single chamber reach the walls with similar degree of vibration, and the closer to the explosion source, the greater the peak pressure, and its value reaches 350 KPa.

Fig. 7 Pressure curves on the surface of the cabin. The left side of the cabin is 14, the right side is 10, the top surface is 27, and the door is 2. From the analysis chart, it can be seen that the waveform of the monitoring points on the surface of the cabin is less oscillating than that near the explosion source in a single chamber, and the peak pressure on the left cabin is higher than that on the right cabin. The peak pressure on the left cabin has exceeded 220KPa, which is due to the reflection overpressure of the side walls in a single chamber. The peak pressure is applied to the deck near the side of the wall, while the peak pressure is in the range of 180-200 KPa at other locations on the surface of the deck.

## 4. Explosive Shock Resistance Analysis Based on ABAQUS

### *4.1. Analytical Pretreatment Device*

Model import: The cabin model in Creo is imported into ABAQUS.

Additional materials: The materials used in this paper are 35CrMo steel plate, 213GPa for Young's Modulus and 0.286 for Poisson's Ratio.

Mesh generation: Free mesh generation is adopted.

Applying boundary conditions: Because the safety refuge device is affected by the installation location, all other aspects of the safety refuge device are impacted by explosion except the bottom. According to the propagation law of shock wave produced by gas explosion studied by scholars in document [11], this paper simplifies the load conditions applied in the finite element analysis of anti-gas explosion shock performance of safety refuge device. Based on the analysis results of shock wave in the above section, ABAQUS software is used in the cabin. By applying the required waveform, the shock wave is simplified to a triangular shock wave. Since the pressure on the cabin does not exceed 0.4 MPa, the peak pressure can be set to 0.4 MPa. The shock wave can be retained 60 ms before the shock wave, and the shock wave can be reduced from 60 ms to 400 ms. The dynamic response of the anti-gas explosion shock performance of the safety refuge device is calculated. The applied shock wave waveform is shown in Fig. 8.

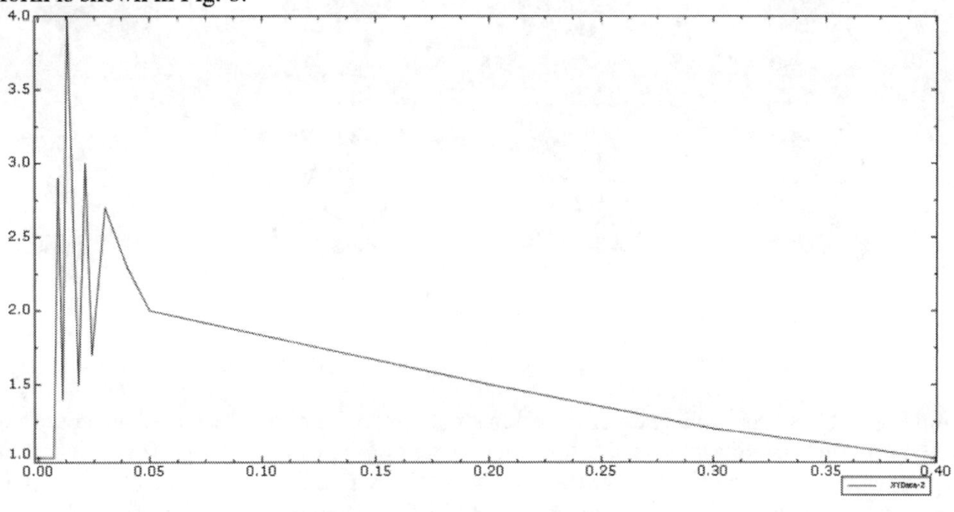


**Fig. 8** The explosive shock waveform

### *4.2. Result Analysis*

The calculation time of the whole cabin analysis is 0.4s. Considering the impact of explosion inside the cabin, after the whole cabin analysis, the cabin structure is sliced along the XOZ plane by the create command under view cut in the Tools menu, and the cabin interior situation is observed. The post-processing of the cabin is checked by using ABAQUS/Explicit solver. As a result, the stress and deformation nephograms of the cabin structure at each time are obtained, as shown in figs. 9 and 10.

IOP Conf. Series: Materials Science and Engineering **772** (2020) 012104 doi:10.1088/1757-899X/772/1/012104

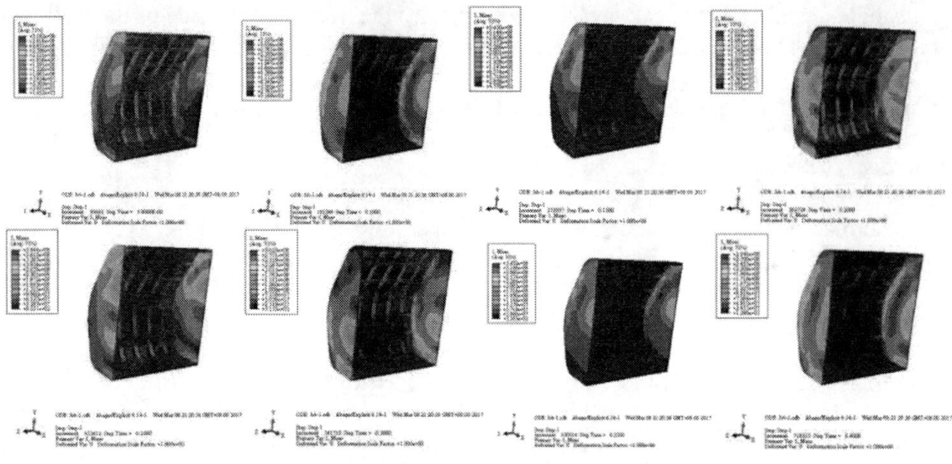

**Fig. 9** Stress nephogram of cabin at 50ms-400ms time

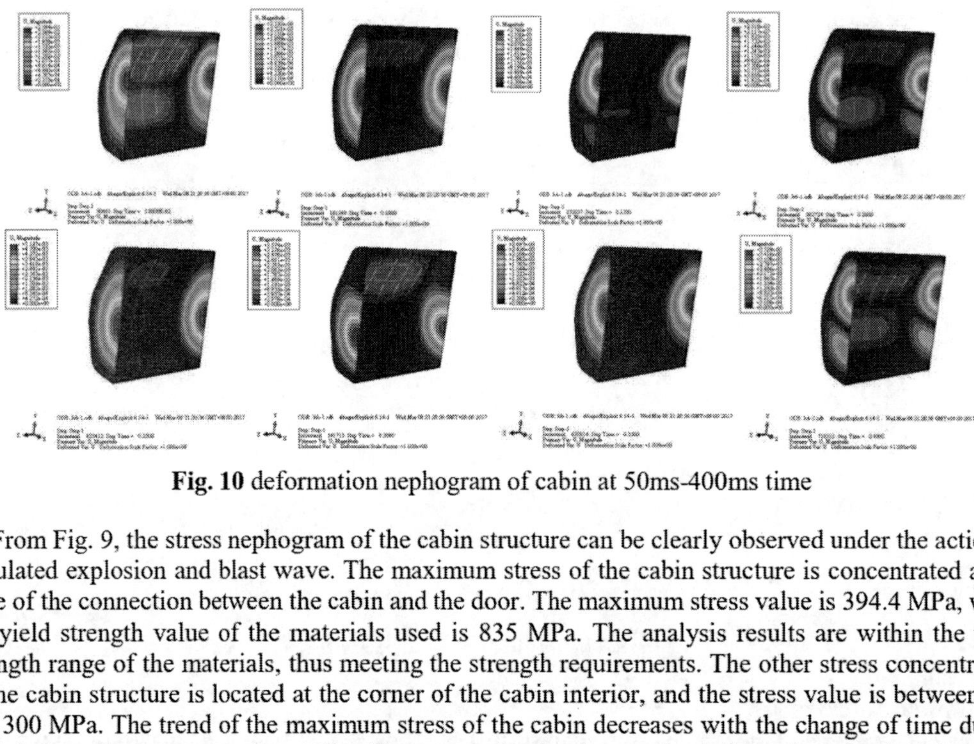

**Fig. 10** deformation nephogram of cabin at 50ms-400ms time

From Fig. 9, the stress nephogram of the cabin structure can be clearly observed under the action of simulated explosion and blast wave. The maximum stress of the cabin structure is concentrated at the edge of the connection between the cabin and the door. The maximum stress value is 394.4 MPa, while the yield strength value of the materials used is 835 MPa. The analysis results are within the yield strength range of the materials, thus meeting the strength requirements. The other stress concentration of the cabin structure is located at the corner of the cabin interior, and the stress value is between 200 and 300 MPa. The trend of the maximum stress of the cabin decreases with the change of time during the whole analysis process. From Fig. 10, we can clearly observe the deformation cloud of the cabin structure under the action of simulated explosion and blast wave. The maximum deformation of the cabin structure occurs on the front and rear walls of the cabin, and its maximum deformation is 5.127mm. This is because there is no reinforcement on the front and back walls of the cabin, so the deformation is relatively large here, limited by the area of residential buildings, and the installation location should be relatively reasonable, the cabin structure should be based on the wall. According to the actual situation, the impact load on the rear wall of the cabin should be smaller than that of the simulation analysis, and the deformation of the other positions of the cabin should be about 2-3 mm.

## 5. Conclusion

(1) Based on the structure of the original safety refuge device, a new triangle-like safety refuge device is proposed in this paper, and on this basis, a model is built by Creo.

(2) The triangle-like cabin structure is placed in the simulated residential building, and the TNT is used to simulate the explosion test. The pressure cloud in the room during the explosion and the pressure variation curve in each time in the single room are obtained.

(3) The pressure curve is used to analyze the explosion resistance of the safety refuge device, and the stress and deformation nephogram of the new cabin structure is obtained. The stress concentration is located at the corner of the cabin interior, and the maximum deformation is located on the front and back walls of the cabin, which meets the explosion shock resistance performance of the new safety refuge device.

## Acknowledgments

This work was supported by the Natural Science Foundation of Shandong Province (ZR2019MEE082).

## References

[1] GAO Guang-wei, ZHANG Lu-hua. Design of coal mine underground life-saving cabin [J]. China Safety Science and Technology, 2009, (8): 162-164.

[2] Janovsky B. Selesovsky P, Horkel J. Vented confined explosins in Stramberk experimental mine and Auto Rea Gas simulation [J]. Journal of Loss Prevention in the Process Industies, 2006, 19(23): 280-287.

[3] Samanta A, Mukhopadhyay M. Free vibration of stiffened shells by the finite element technique. Eur J Mech A Solids, 2004, 23: 159-79.

[4] YU Yicheng. Study on Simulation of Indoor Gas Leakage and Explosion Result [J]; Harbin Institute of Technology;

[5] GAO Er-xin, DING Chen-xi, HAN Shuang-liang, MENG Sha-sha, ZHANG Xiao-rui. Parameter distribution of gas explosion in a coal mine [J]. Blasting, 2013, (01): 5-7.

[6] DONG Hong-ying. Study on the whole structure design and anti-blast performance of mine life-saving cabinets [D]. Taiyuan University of Technology, 2016.

[7] WANG Bi-hao. Structural strength calculation of safety chamber in coal mine [D]. Harbin Institute of Technology, 2013.

[8] FAN Ya-fu, DUAN Zhu-ping. Experimental Study on Johnson-Cook Material Model Parameters [J]. Mechanics and Practice, 2003, (05): 40-43.

[9] FANG Wen-feng. Study on dynamic behavior of life-saving and chamber protection in coal mine [M]. China University of Mining and Technology, 2012.

[10] Ma - dong, PAN Hai - yan, WANG Yun, MENG Zhi - juan. Numerical simulation of the impact of explosion protection in underground life - saving cabin [J]. Vibration and Shock, 2012, (20): 172-176.

[11] Li - qiang, BAI Bo, XIE Qinghai. Numerical simulation of dynamic response of mobile life - saving cabin under impact loading [J]. Vibration and shock, 2013,32 (16): 146-150.

# Parameter Simulation and Analysis of Rotary Feeder

**Fang Dong and Yong Li**

School of Mechanical and Electrical Engineering, Qingdao University of Science and Technology, Qingdao 266061, Shandong Province, China

**Abstract.** As a kind of continuous conveying equipment for powder and pellets, rotary feeder is widely used in the processing of non-staple food products, chemical industry production, metal smelting, wine brewing and terminal transportation because of its convenient control, high reliability and stable transportation. Such industries have an important position in the bulk material conveying industry. Different process parameters and structural parameters of the rotary feeder will have an impact on the conveying effect. The process parameters affecting the conveying effect mainly include rotor speed and material filling coefficient. In this paper, the dismantling rotary feeder with the rotor diameter of Φ62 is studied by simulation and test, and the influence of rotor speed and material filling coefficient on the conveying effect is analyzed.

## 1. Introduction

At this stage, the rotary feeder has become a commonly used feeder in material conveying systems, and its use is widespread and popular. With the development of science and technology and the needs of industrial production, rotating machinery is constantly developing in the direction of large-scale, complicated, integrated, and high-speed. At present, for the continuous transportation machinery, the optimization research and design results for the rotary feeder in the material conveying system are relatively small, so it is necessary to carry out research work in this area to solve practical problems for the project.

## 2. Establishment of a rotary feeder model

The structure of the rotary feeder is relatively complicated, and the rotary feeder needs to be simplified, and unnecessary structures such as motors and bearings are omitted, so that the calculation speed is accelerated, and the simulation results are more clear and targeted.

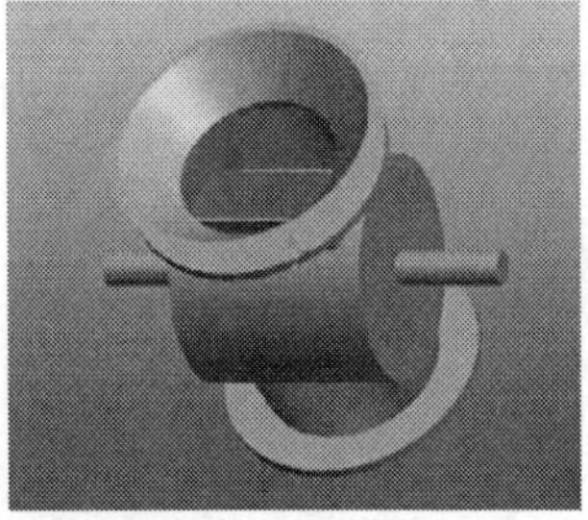

**Fig. 1** Simplified model of rotary feeder

Content from this work may be used under the terms of the Creative Commons Attribution 3.0 licence. Any further distribution of this work must maintain attribution to the author(s) and the title of the work, journal citation and DOI.
Published under licence by IOP Publishing Ltd

As shown in Figure 1, a simplified rotary feeder model is built in Proe to scale the model of the rotary feeder to a certain scale.

The size of the model, the accuracy of the mesh being divided, and the computational power of the computer all have an important impact on the simulation experiment. Therefore, in the modeling, we take the rotor shaft with a diameter of 10 mm, a blade diameter of 62 mm, and a blade width of 60 mm. The number of blades is 9 and distributed on the rotor shaft, and the model is introduced into the EDEM.

## 3. Simulation and results analysis

### *3.1. The effect of speed on the amount of discharge*

The main problem to be studied in this subject is the effect of the rotor speed of the feeder and the filling rate of the conveying material on the conveying effect of the rotary feeder. In this experiment, four particles with a rotation speed of 20r/min are selected for analysis. Four simulated particles were randomly selected from all simulated particles, and four particles were marked and their motion trajectories were tracked. For rotary vane feeders, the user typically selects the appropriate feeder specifications by calculating the feed capacity of the feeder. Formula (2-1) is the conveying capacity of the rotary vane feeder:

$$Q = 60\rho\beta V_0 KN \tag{1}$$

In the formula:

Q——The conveying capacity of the rotary vane feeder, T/h;

P——The actual pouring density entering the feeder, kg/m3;

η——Filler coefficient of the feeder;

K——The number of upper wheel chambers of the rotor;

n——Rotor's speed, r/min;

$V_n$——The volume of a single wheel chamber, m3.

In the actual production application, the rotor speed of the feeder plays a vital role in the control of the material conveying volume during the conveying process. Under normal circumstances, the larger the rotor speed, the larger the material conveying amount, but when the speed is too high, The blade particles of the discharge port have not yet fallen, and the particles follow the blade to return to the feed port, and the discharge is not realized, so there is a maximum rotation speed. When the rotation speed exceeds the rotation speed, the material conveyance amount decreases. In this experiment, quantitative analysis of the discharge amount was carried out by observing the discharge amount under different rotor speeds.

(a) (b)

**Fig. 2** Particle motion diagram

IOP Conf. Series: Materials Science and Engineering **772** (2020) 012105 doi:10.1088/1757-899X/772/1/012105

According to past experience and existing data, the feed rate of the rotary valve feeder is generally controlled at 15 ~ 30r / min, so in this simulation scheme, the rotor speed is set to 15, 20, 25, 30, 35r / Min. Record the mass of the material passing through the unit box every 0.2s, and export the data to draw the blanking diagram under the rotor speed, as shown in Figure 3. The figure is the data derived from the initial time 0 seconds when the rotation speed is 15r/min. Since 0~2 seconds is the time when the material starts to fall, the material has not reached the position of the discharge port at this time, and the drop of the material is unstable. . Therefore, in the next analysis, only the stable blanking section (the blanking amount after 2 s) will be analyzed to observe the law of the blanking amount.

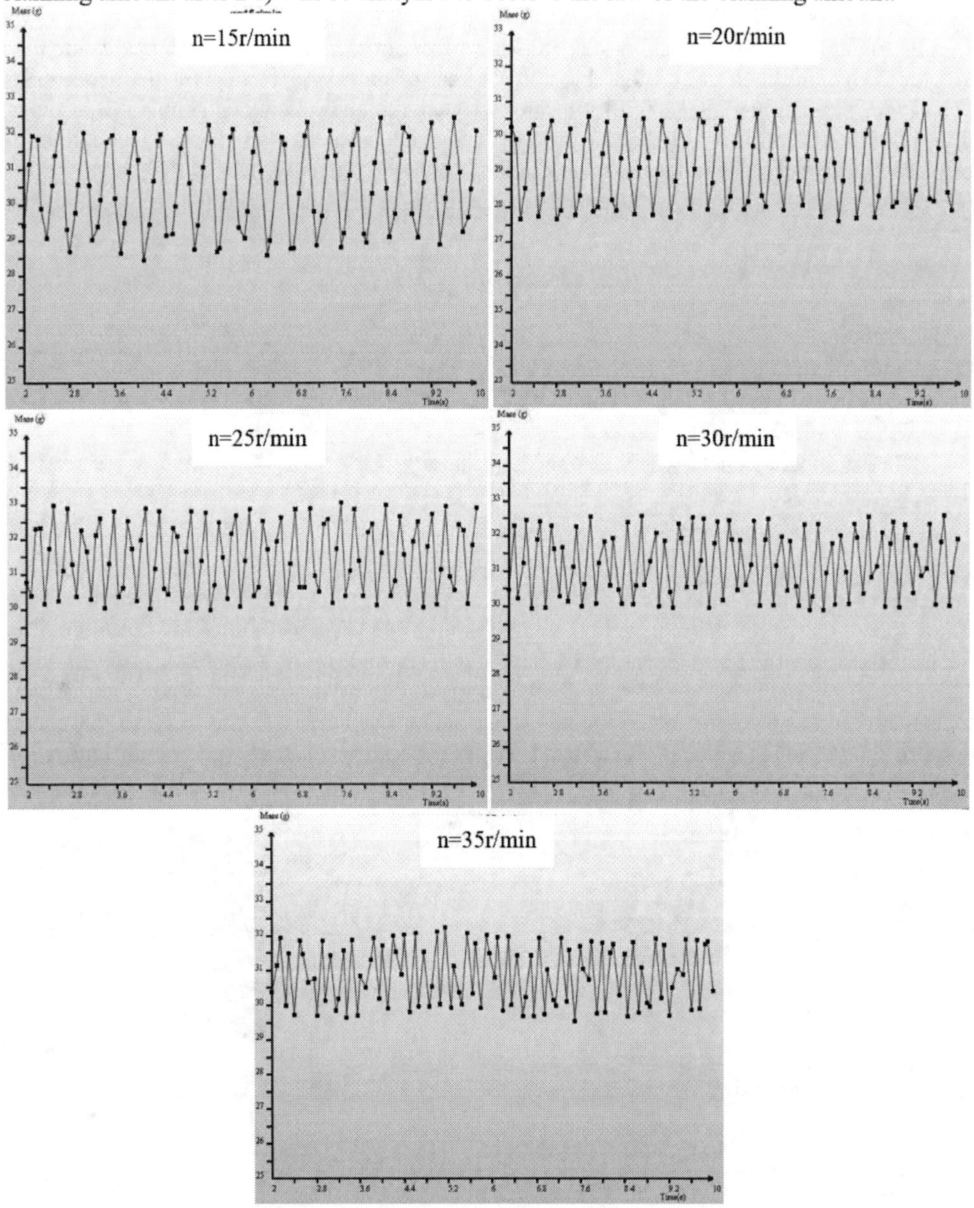


**Fig. 3** Rotary feeder blanking diagram

Figure 3 shows the blank feeder blanking diagram. The vertical coordinate is the blanking mass and the abscissa is the time. From the figure, we can see that the feeding amount of the feeder changes with the speed. The mass of the output material of the feeder is periodically changed with time. This phenomenon is the pulsation phenomenon. The pulsation phenomenon occurring during the conveying process is directly related to the rotation speed of the rotary feeder rotor. The faster the rotation speed of the rotor, the shorter the pulsation cycle. Now we will get the blanking amount obtained under different rotor speeds. The maximum, minimum, amplitude, and rotation periods are averaged and organized as shown in Table 3-1 below.

**Tab. 1** The variation of the quality with the rotational speed

| n (r/min) | Max (g) | Min (g) | Amplitude (g) | Cycle (s) |
|---|---|---|---|---|
| 15 | 32.457 | 29.716 | 2.74 | 0.42 |
| 20 | 32.573 | 29.832 | 2.94 | 0.31 |
| 25 | 32.893 | 29.909 | 2.98 | 0.28 |
| 30 | 32.593 | 29.906 | 2.69 | 0.21 |
| 35 | 32.252 | 29.550 | 2.70 | 0.18 |

From Table 1, we can see that as the rotor speed increases, when the rotor speed is 15~25r/min, the maximum and minimum values of the blanking amount increase with the increase of the rotor speed. Maximize at 25r/min. Now the maximum and minimum values of the blanking amount are plotted, as shown in Figure 4. It can be found through observation that the change of the blanking amount in a certain speed range is basically linear, when the rotor speed is 15~25r/ At min, the maximum and minimum values of the blanking amount increase with the increase of the rotational speed. However, when the rotor speed reaches 30r/min and 35r/min, the maximum and minimum values of the blanking amount are reduced. This is because when the rotor speed is too large, the material will be brought back completely if it fails to fall. As shown in 3-11(b), it can be clearly seen that some of the particles have not fallen, indicating that there is an optimum speed, and the optimum speed of 25 r/min is very difficult to approach the actual application speed, which is significantly lower than the theoretically calculated optimum speed. It shows that the simulation method is indeed more reliable than the theoretical calculation. At the same time, the numerical simulation is very intuitive to reflect the pulsation of the rotary feeder feeding capacity.

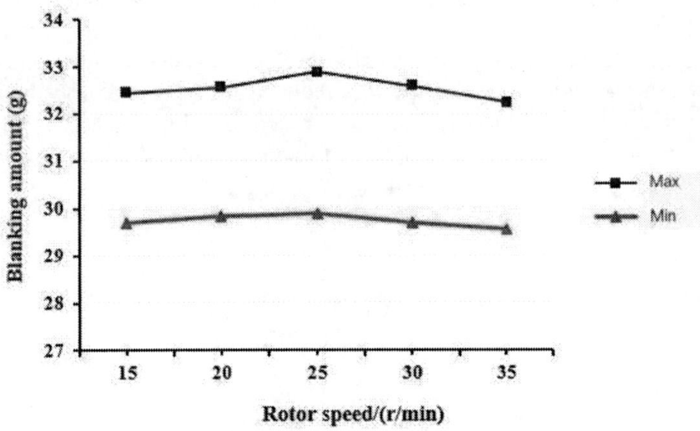


**Fig. 4** Blanking amount atdiffrent rotation rate

It is known that the calculation formula of the rotation period of the rotary feeder rotor is:

$$T = \frac{2\pi}{w} \tag{2}$$

According to the formula, when the rotor speed is 15, 20, 25, 30, 35r/min, the rotation time of the blade is 3.93s, 2.99s, 2.42s, 2.01s. In the experiment, the rotor has a total of 9 blades. It can be seen that the rotation period of each blade is 0.44s, 0.33s, 0.27s, 0.22s, which is consistent with the rotation period in the simulation experiment, so that the experimental rotor rotation period can be consistent with the theoretical value.

It can be seen from the analysis of Fig. 4 that in the material transportation process of one cycle, the material flow rate conveyed in each wheel chamber is not completely consistent, and there is unevenness. The relatively large influence on the whole feeding process is the rotation speed of the rotor. It can be known that in order to control the amount of material to be conveyed during the material conveying process, the first thing to consider is the rotor speed of the feeder. Through this simulation experiment, when the rotor speed is about 25r/min, the blanking amount reaches the maximum value, that is, the conveying efficiency is the highest when the rotor speed is about 25r/min.

### *3.2. The effect of speed on mass flow rate*

The mass flow rate of the medium is a parameter that quantifies the flow pattern of the particles in the feeder by speed. The measurement of this parameter is performed by selecting a plane parallel to the rotor axis and recording the number of all particles flowing through the plane. In this simulation experiment, a plane parallel to the exit is selected as the detection plane of the mass flow rate.

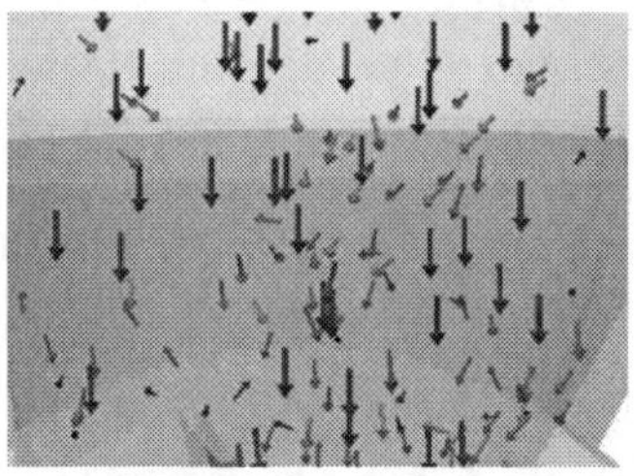
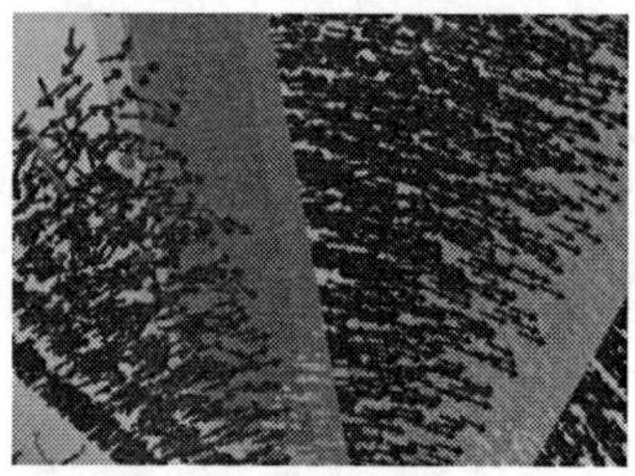
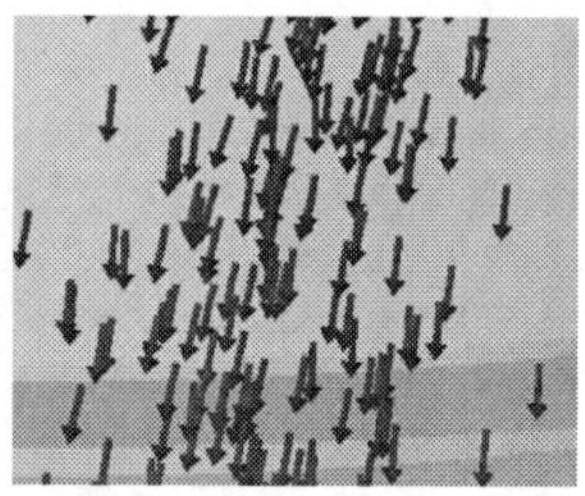

**Fig. 5** Feeding mouth speed vector **Fig.6** Wheel indoor speed vector **Fig. 7** Outlet speed vector

Figure 5 shows the speed vector at the feed port. The difference in particle velocity at the feed port is due to the fact that the particles are not simply free-falling movements before entering the feeder inlet. The actual situation is that the particles have different initial velocities when feeding the material. It can be seen from Figure 6 that after the particles enter the feeder wheel chamber, the initial kinetic energy is partially offset by the rotor force, the speed is reduced, and the particle velocity entering the rotor wheel chamber varies with the rotor speed. . It can be seen from Fig. 7 that the material of the discharge port gradually becomes larger due to the rotation of the rotor and the action of gravity.

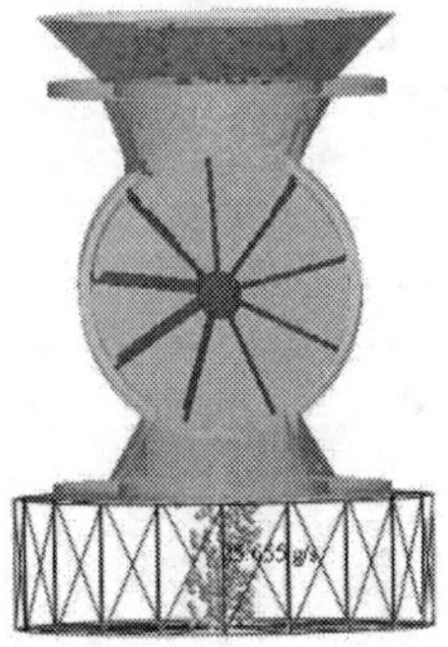

**Fig. 8** Mass flow rate chart

Figure 8 shows the mass flow rate detection diagram. Select a plane parallel to the rotor axis at the material exit, establish a mass flow velocity sensing cell grid, perform real-time detection of the mass flow rate, and extract the mass flow rate to average and analyze the difference. Fill factor and mass flow rate at different speeds.

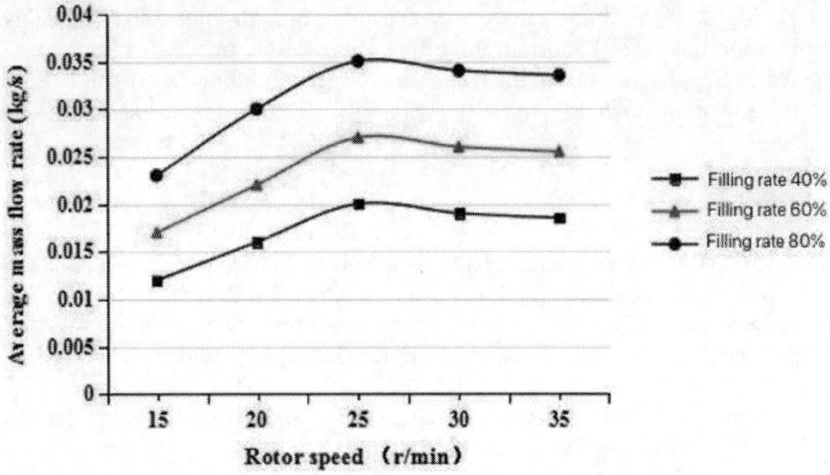


**Fig. 9** Relationship between rotor speed and mass flow rate

Figure 9 shows the mass flow rate at different rotor speeds. When the material fill factor is set to 40%, the rotor speed is 15r/min, 20r/min, 25r/min, 30r/min, 35r/min. The average mass flow rates were 0.012 kg/s, 0.016 kg/s, 0.020 kg/s, 0.023 kg/s, and 0.025 kg/s, respectively. When the material filling factor is set to 45% and the rotor speed is 15r/min, 20r/min, 25r/min, 30r/min, 35r/min, the average mass flow rate of the material is 0.017kg/s, 0.022kg/s. , 0.027 kg / s, 0.031 kg / s, 0.034 kg / s. When the material filling factor is set to 60% and the rotation speed is 15r/min, 20r/min, 25r/min, 30r/min, 35r/min, the average mass flow rate of the materials is 0.023kg/s and 0.030kg/s, respectively. 0.035 kg/s, 0.040 kg/s, 0.043 kg/s. When the filling factor of the material is increased from 40% to 80%, the mass flow rate of the material at different rotor speeds increases by 1.92 times, 1.88 times, 1.75 times, 1.73 times, and 1.72 times, respectively. It can be seen from the data that when the filling factor of the material is doubled, the increase in the mass flow rate of the material is not doubled. At the same time, as the rotational speed of the rotor increases, the increase in the mass flow rate of the material is decreasing. Analysis of Figure 4-16 shows that under the same material fill factor, the mass flow rate increases with the increase of the rotor speed, and the material mass flow rate value is basically linearly related. In addition, it can be found that at the same rotor speed, the filling factor of the material increases, and the average mass flow rate of the material also increases correspondingly, and when the rotational speed of the rotor increases, in the case of different material filling factors of the same rotor speed, The magnitude of the increase in the average mass flow rate of the material has increased. In addition, it can be found that under the same material filling factor, the increase in the mass flow rate of the material is reduced after the rotor speed is increased to 30 r/min. Through simulation analysis, it is found that if the mass flow rate of the material is to be changed simply by changing the filling factor and cannot be quantitatively changed by the material filling coefficient, the material mass flow rate can be adjusted by the combination of the rotor speed and the filling coefficient.

## 4. Conclusion

Five rotor speeds and three material fill factors were set in the experiment, and 15 sets of simulation experiments were carried out. Through 15 sets of simulation experiments, the data such as material drop amount and average mass flow rate during the transportation process were obtained, and the optimal value of material transportation was obtained by analyzing the data, and the empirical parameter values selected in the previous transportation were optimized. Through experiments, it is found that to achieve the same mass flow rate, the filling factor required for different speeds is also different, the filling factor is reduced, and the rotor speed to be set should be increased accordingly. Through simulation experiments, the transmission efficiency is highest when the rotor speed is about 25r/min. When the rotor speed is higher than 30r/min, the power consumption will increase significantly. Therefore, the rotor speed should not exceed 30r/min; the filling factor of the material is improved. The smaller the moving speed of the material inside the feeder, the larger the filling factor can be selected as much as possible under the premise of the conveying conditions.

**References**

[1] Hirota M, Sogo Y, etc. Effect of mechanical properties of power on pnenumatic conveying in inclined pipe. Power Technology[J]. 2002, 122 (1): 150-155
[2] Roberts, A. W. The influence of granular vortex motion on the volumetric performance of enclosed screwconveyors[J]. Powder Technology, 1999, 1(5): 56-57
[3] Bauverlag GmbH. The perfect solution for screw conveying equipment WAM ES-type cement screw conveyor[J]. Betonwerkund Fertigteil-Technik /Concrete Plant and Precast Technology, 2008, (4): 48-51
[4] Wada, Kenzou, Ebisui, Masahide. Mechanism of transportation inside the screw conveyer[J]. Powder Technology, 2000, (5): 57-63
[5] Use of Cas-Solid-Ejectors as Inward Transfer Units for Pneumatic conveying of BulkSolids,Bulk Solids Handling [M]. 2006
[6] B Gupta, AK Nayak, TK Kandar, S Nair. Investigation of air–water two phase flow through a venturi [J]. Experimental Thermal & Fluid Science, 2016, 70: 148-154
[7] John Favier. Industrial application of DEM: Opportunities and Challenges [J]. DEM Solutions Ltd, 2008

# Design of an ultra-thin magnetic radar absorber embedded with double period frequency selective surface for entire L-band microwave absorption

**Houdi Xiao *, Lingdong Zhang, Mingyun Lv, Zhipeng Qu, Chuanzhi Wang, Ruiru Qin**

School of Aeronautic Science and Engineering, Beihang University, Beijing, 100191, P. R. China

*Corresponding author e-mail: xhdbuaa@126.com

**Abstract.** In this paper, a thin, broadband and low frequency with high absorption (more than 90% absorption) radar absorber (RA) consisting of magnetic-type material substrate and double period frequency selective surface (FSS) is demonstrated. The geometric parameters were investigated for its contribution towards absorption by the finite-difference time domain (FDTD) method. The effects of incidence angles of electromagnetic radiation were also discussed. The simulation indicates that the use of the double period FSS makes a remarkable increase of operational bandwidth and moves the resonant frequency to the lower frequency comparison with single magnetic radar absorption materials (SMRA). Due to the excellent symmetric nature of the proposed structure, the reflectivity is insensitive to incident angle from 0° to 30°. The reflectance of RA lower than -10 dB (over 90% absorption) is obtained in the whole frequency range of L-band which is 1.5 times wider than the bandwidth of the same thickness SMRA when the structure thickness is only 2.62 mm. The simulation method is verified by comparison with the experiment in the literature. Thus, the presented RA breaks the limitation of the thickness (λ/4) for the traditional low-frequency band absorbing material and has wide application prospect in electromagnetic shielding and low-frequency stealth.

## 1. Introduction

The stealth technology of weapon equipment has been the key technology to determine its survivability. Therefore, many scholars carry out research on stealth technology [1, 2]. Among them, the research of radar absorbing material (RAM) and radar absorbing structure (RAS) are the key points [3]. For traditional RAM, the absorbing performance is strongly due to permeability and permittivity of composites consequently, many researchers focus on the preparation of absorbent [4, 5]. The mixed type of microwave absorbing materials that consolidated magnetic and dielectric losses are researching [6]. Many researchers use frequency selective surface(FSS) or period structure of metallic patches in the modern design, because of the absorption performance can be easily improved and controlled by adjusting the geometric dimensions and patterns [7, 8]. However, most of the absorption band in the study is mainly concentrated in the C-band or above C-band, like X band, terahertz band, visible light and infrared band, etc. There are few ultra-thin absorbing materials or absorbing structures involved in

Content from this work may be used under the terms of the Creative Commons Attribution 3.0 licence. Any further distribution of this work must maintain attribution to the author(s) and the title of the work, journal citation and DOI.
Published under licence by IOP Publishing Ltd

L-band [9]. Although an RAS that had a absorption performance in L-band had designed by Haiyan Chen, it did not cover the entire L band [9].

In this letter, a novel thin, broadband and low frequency with high absorption (more than 90% absorption) radar absorber (RA) based on magnetic-type material substrate and double period frequency selective surface (FSS) is proposed and numerically characterized. The effects of design parameters on the reflectivity of RA were investigated. The microwave absorbing performance of SMRA was improved by FSS in the frequency range of 1-2 GHz (L-band)

## 2. Design

As shown in Figure 1, the structure of the considered RA in this letter consists of a double periods FSS and magnetic-type absorbing material based on perfect conductor ground. Here, the double periods FSS described in Figure 1(b). The FSS layer contains two orders of two-dimension periodic array of square ring cell. The cycles of the FSS layer are $d_0$ and $d_1 = d_0 / 2$ ,respectively. The magnetic-type absorbing material is the material in reference [9], and its relative permittivity $\varepsilon$ and relative permeability $\mu$ are shown in Figure 2. $\varepsilon'$ and $\varepsilon''$ represent the real part and imaginary part of relative permittivity of the materials. $\mu'$ and $\mu''$ represent the real part and imaginary part of relative permeability of the materials. We performed computer simulations of our design using the commercial finite-difference time domain (FDTD) solver microwave studio by CST that has been proved to be successful in similar types of analysis [10]. The program simulated a single unit cell with appropriate boundary conditions, as unit cell boundary on the side of the modal. The Floquet port was used to simulate the plane wave. The material properties and structure parameters are also defined in the simulation system. We simulated the electromagnetic parameter of RA at normal and oblique incidence condition.

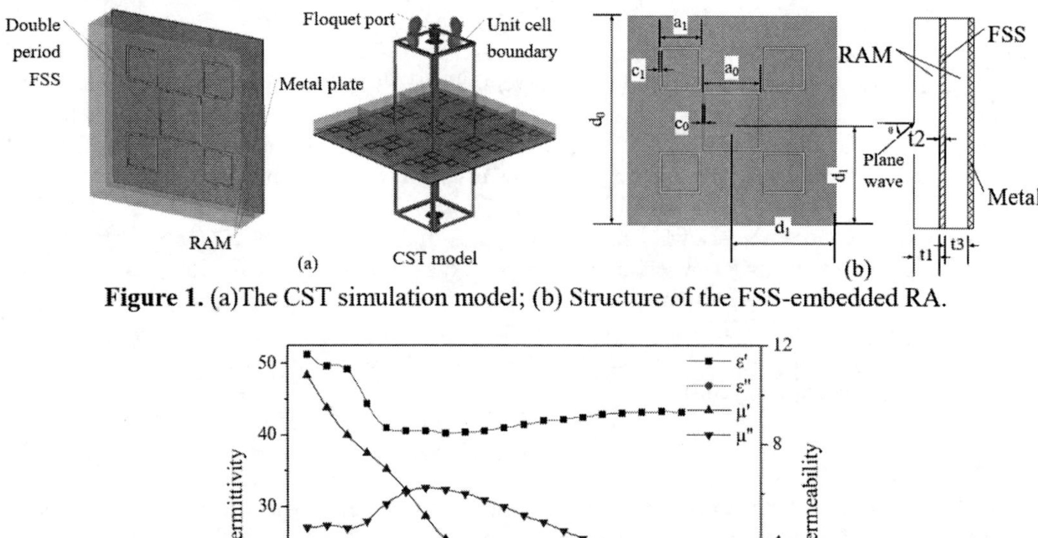


**Figure 1.** (a)The CST simulation model; (b) Structure of the FSS-embedded RA.

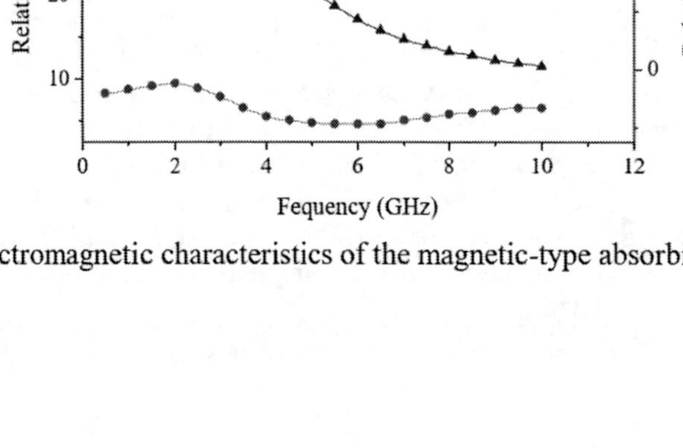


**Figure 2.** electromagnetic characteristics of the magnetic-type absorbing materials [9]

IOP Conf. Series: Materials Science and Engineering **772** (2020) 012106 doi:10.1088/1757-899X/772/1/012106

## 3. Results and discussion

The geometry and size of the unit cell is optimized by numerical simulation to obtain a good microwave absorption performance. The optimized detailed parameter dimensions of the absorber are listed in Table 1.

**Table 1.** Dimensions of the absorber

| Parameter | Value | Parameter | Value |
|---|---|---|---|
| d0 | 10 mm | t2 | 0.02 mm |
| d1 | 5 mm | t3 | 0.9 mm |
| a0 | 2.8 mm | c0 | 0.1 mm |
| a1 | 2 mm | c1 | 0.1 mm |
| t1 | 1.7 mm | | |

The reflectivity properties of the double period FSS-embedded RA are proposed in Figure 3. At the same time, the magnetic-type absorbing materials are also simulated and compared with the experimental data in reference [9], that is proved that our simulation results are basically consistent with the experimental results. The reflectivity of the magnetic-type absorbing materials is agreed with the experimental data. The operating bandwidths of 0.9~2.1 GHz of the double period FSS-embedded RA is obtained, but the operating bandwidth of the magnetic-type absorbing materials is only 0.97~1.75 GHz. Just like the results in Figure 3, we can observe that the operating frequency bandwidths for the double period FSS-embedded RA can be greatly increased and shifted to low frequency as compared to the magnetic-type absorbing materials only.

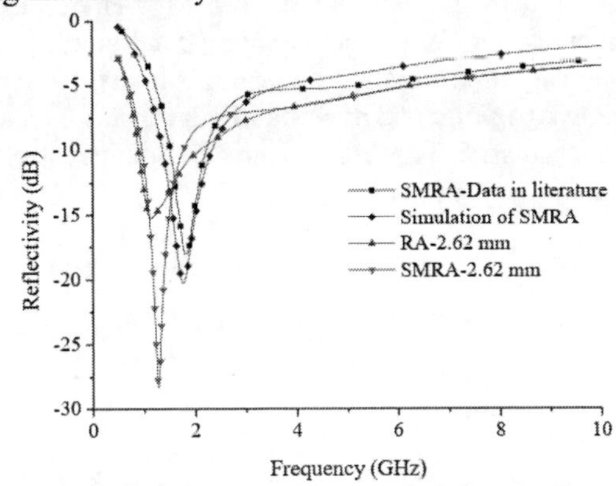


**Figure 3.** Reflectivity of RA and single magnetic-type absorbing materials

To understand the source of resonant absorption, seven main dimensional parameters, d0, a0, a1, c0, c1, t1, t3, are studied to examine their effects on the reflectivity of the absorber. The influences of period d0 and d1 of the double period FSS on the reflectivity of RA are given in Figure 4, while other parameters are kept unchanged. From the simulated results, two resonance peaks were observed at 0.7 GHz and 1.7 GHz when d0 is equal to 6 mm, and the resonance absorbing peak appear at different frequencies with variable d0. With the increasing of the period of double period FSS unit cell, the value of left-resonance peak at about 0.7 GHz gradually decrease, whereas the right-resonance peak gradually decreases to the minimum and then increase slightly. The intensity of low-frequency resonance peak gradually decreases, whereas the high-frequency resonance peak gradually increases firstly and decrease secondly and then increase slightly. This result mean that we can change the period of the FSS to adjust the absorptive bandwidth range. As observed, there is a single resonance peak at about 1.1 GHz with increasing d0 above 10 mm. At the same time, a broad bandwidth is obtained with a reflectance lower than -10 dB in the frequency range from 0.9~2.1 GHz which cover the entire L band.

IOP Conf. Series: Materials Science and Engineering **772** (2020) 012106 doi:10.1088/1757-899X/772/1/012106

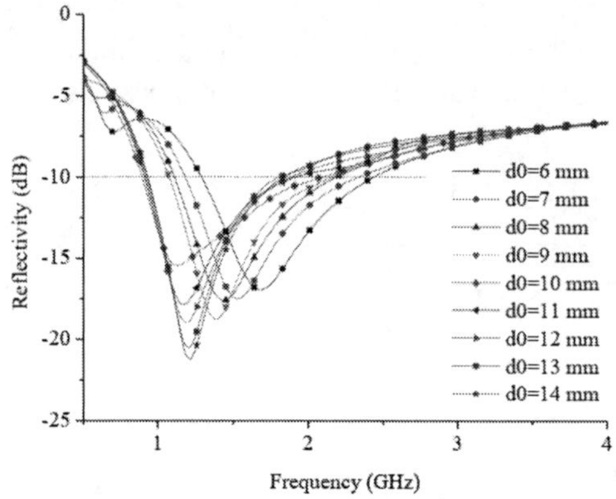


**Figure 4.** Simulated reflectivity of the RA versus frequency for different d0.

The calculated reflectivity of RA with different values of thickness for the magnetic-type material t1 and t2 are presented in Figure 5, respectively, while other parameters are kept unchanged. As observed, there is one resonant peak at about 1.14 GHz when t1=1.7 mm, t3=0.9 mm, which indicates that the impedance of RA matches with the free space at this resonance point. Hence, the absorber can absorb the electromagnetic wave energy at this point. As observed in Figure 5(a), with the increase of t1, the resonance frequency absorption peak gradually shifts to the lower frequency, and the intensity of the absorption peak gradually increases, whereas the operating bandwidths gradually decrease. Unlike the results of Figure 5(b), the intensity of the absorption peak gradually decreases firstly, and then increase with increasing t3. The reflectivity in Figure 5 have significance variations trend, which resonance point and intensity were changed regularly. This result can be explained by the principle of impedance matching in transmission line.

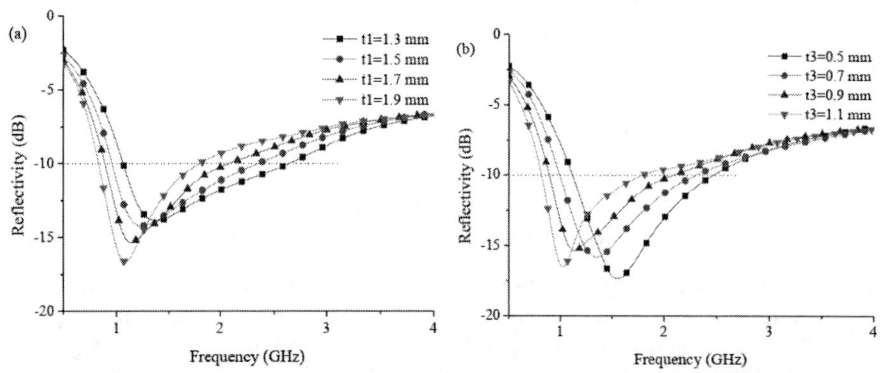


**Figure 5.** Simulated reflectivity of the RA versus frequency for different t1, t3.

The simulated reflectivity of the RA with different values of a0 and a1 are shown in Figure 6. As observed in Figure 6(a), the resonant absorption peak shifts toward low-frequency region firstly and then to high-frequency region with the increase of a0. The intensity of the resonance peak gradually decreases to the minimum and then increase, whereas the absorption bandwidth with a reflection coefficient lower than -10 dB in the frequency range from 0.9 GHz~2.1 GHz covers the entire L band when a0 is equal to 2.8 mm. The calculated reflection spectra of the absorber with different a1 are presented in Figure 6(b). From the simulated results, one resonance peak was observed at about 1.22 GHz when a1 is quale to 1 mm, and the resonance absorbing peak appear at different frequencies with variable a1. With increasing a1, the value of resonance peak at about 1.22 GHz gradually decrease to the minimum then increase to the maximum and finally increase to about 1.5 GHz. With the increase of a1, the intensity of resonance peak gradually decreases to the minimum then increased as the a1 increase from 1 mm to 5 mm. The absorber has the widest absorption bandwidth which cover the entire L band when a1 is quale to 2 mm.

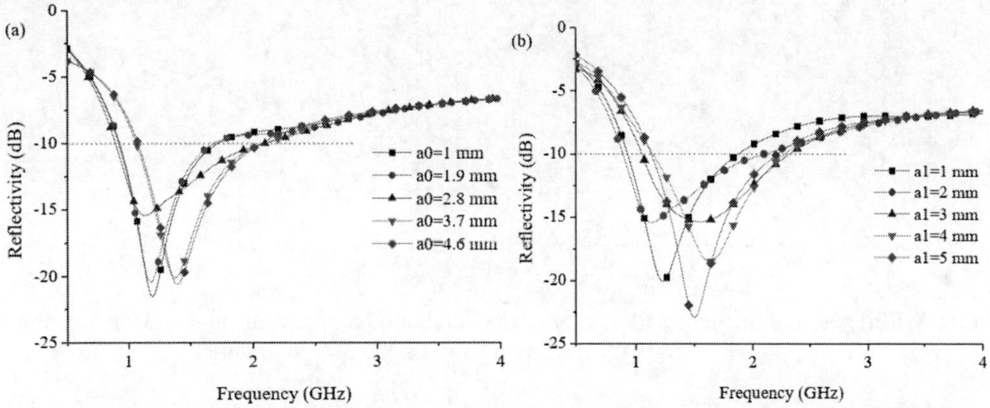


**Figure 6.** Simulated reflectivity of the RA versus frequency for different a0 and a1.

Figure 7 shows the reflection coefficient of the RA when the line width of the zero FSS c0 and the line width of the first FSS c1 are variable and other parameters are kept unchanged. As shown in Figure 7, the resonant absorption peak and the low-frequency position don't move with the increase of c0 and c1. However, the high-frequency position small shifts toward high-frequency region with the increase of c0, and small shifts toward low-frequency region with the increase of c1.

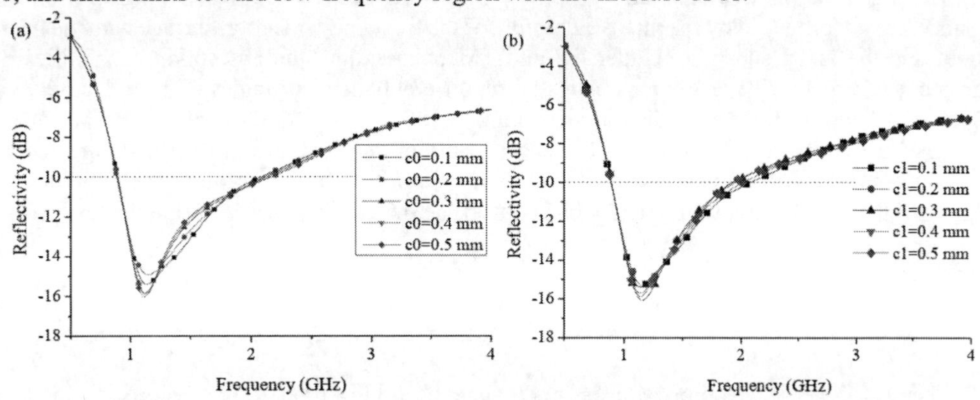


**Figure 7.** Simulated reflectivity of the RA versus frequency for different c0 and c1.

In order to make better use of the design, the reflection for TE-polarized and TM-polarized waves at different incident angles from 0° to 30° were given in Figure 8. As shown in Figure 8, in case of TE-polarized waves, the bandwidth with reflection coefficient below -10 dB gradually reduces with the increase of incident angles, whereas the bandwidth slowly increases with the incident angles from 0° to 30° under TM-polarized waves. As the incident angle increases, the minimum reflection coefficient for the TE polarization slightly increases, whereas that for the TM polarization smoothly decreases. The resonant frequencies at 1.13 GHz is stable with respect to the angle variation. This confirms that the resonant peak of designed RA is insensitive to incident angle from 0° to 30°.

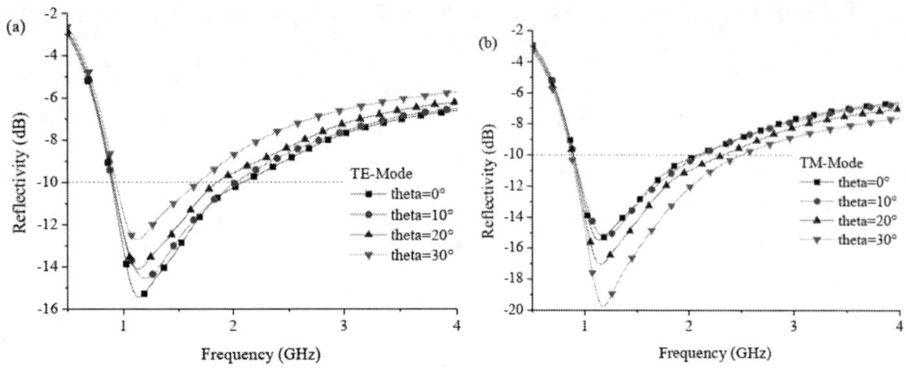


**Figure 8.** Reflection of the designed RA for normal and oblique incidence of TE-polarized and TM-polarized waves, (a) TE polarization, (b) TM polarization

## 4. Conclusion

A thin and broadband low frequency RA consisting of magnetic-type material substrate and double period FSS has been presented. The influences of geometric parameters on absorption properties are investigated. The structure research shows that we can adjust the absorption frequency and bandwidth lightly by changing the geometry of the structure. The frequency response of the proposed structure provides a thin (2.62 mm thick), broadband (0.9 GHz~2.1 GHz) and low frequency (cover the entire L band) with high absorption (more than 90% absorption) which resulted from the strong electromagnetic resonance that are produced by the intense coupling of the frequency selection surface with the magnetic medium and the metal substrate. Under TE and TM polarization, the absorption capabilities of the suggested RA structure have been investigated at different incident angles. The reflectivity of the designed RA is insensitive to incident angle from 0° to 30°. Moreover, the validity of the simulation method is verified by comparison with the experiments in the literature. Thus, this work reported a design for a new low frequency and broadband absorber which breaks through the limitation of the thickness of the traditional low-frequency band absorbing material and has wide application prospect.

## References

[1] Wang, X.K., et al., Synthesis of tremella-like graphene@SiC nano-structure for electromagnetic wave absorbing material application. Journal of Alloys and Compounds, 2018. 741: p. 205-210.

[2] Chen, M.D., et al., Optimization on microwave absorbing properties of carbon nanotubes and magnetic oxide composite materials. Applied Surface Science, 2018. 434: p. 1321-1326.

[3] Lee, S.E., et al., Broadband all fiber-reinforced composite radar absorbing structure integrated by inductive frequency selective carbon fiber fabric and carbon-nanotube-loaded glass fabrics. Carbon, 2016. 107: p. 564-572.

[4] Zhang, Y.C., S.T. Gao, and H.L. Xing, Reduced graphene oxide wrapped cube-like ZnSnO3 : As a high-performance microwave absorber. Journal of Alloys and Compounds, 2019. 777: p. 544-553.

[5] Sozeri, H., et al., Cr3+-substituted Ba nanohexaferrites as high-quality microwave absorber in X band. Journal of Alloys and Compounds, 2019. 779: p. 420-426.

[6] Yang, Z.N., et al., Design of a thin and broadband microwave absorber using double layer frequency selective surface. Journal of Alloys and Compounds, 2017. 699: p. 534-539.

[7] Ranjan, P., A. Choubey, and S.K. Mahto, A novel approach for optimal design of multilayer wideband microwave absorber using wind driven optimization technique. Aeu-International Journal of Electronics and Communications, 2018. 83: p. 81-87.

[8] Tuong, P.V., et al., Perfect-absorber metamaterial based on flower-shaped structure. Photonics and Nanostructures-Fundamentals and Applications, 2013. 11(1): p. 89-94.

[9] Chen, H.Y., H.B. Zhang, and L.J. Deng, Design of an Ultra-Thin Magnetic-Type Radar Absorber Embedded With FSS. Ieee Antennas and Wireless Propagation Letters, 2010. 9: p. 899-901.
[10] Yang, Z.N., et al., Design of a broadband electromagnetic absorbers based on TiO2/Al2O3 ceramic coatings with metamaterial surfaces. Journal of Alloys and Compounds, 2016. 687: p. 384-388.

# Potential Transparent Supercapacitor Electrode Materials

**Ziyan Fan**

Irvine High School, 4231 Walnut Ave, Irvine, CA, 62604, USA

20fanziyan@gmail.com

**Abstract.** Supercapacitor utilizes the nano-scaled pores to increase the surface area indefinitely and shortens the distance between the positive and negative plate, on the account that some larger particles cannot fit into the pores. According to the capacitance equation "capacitance = εrε0 (area/distance)," as the distance decreasing and area increasing, the magnitude of the capacitance is tremendously magnified. A transparent supercapacitor describes a supercapacitor that has a high optical transmittance. If the intensity of light that transmits across the substance is higher than the rate that it is blocked or reflected. The substance possesses a transparency. The flexibility of transparent supercapacitors is another important area of studies. It generalizes a combination of stretchability, bendability, twistibility, or compressibility. In real life, it's important to obtain the optimal optical transmittance, flexibility, and electrical conductivity for future devices such as smart glasses. These characteristics enable us to develop something more complex and advanced in the future.

## 1. Introduction

Transparent supercapacitor is a distinct type of supercapacitor that revolutionizes the electrochemical industry. It is developing toward computer glasses, wearing and portable electronics, or other highly advanced technological products. The transparency of an electrode material is directly proportional to its optical transmittance. Light waves with certain frequency strikes an object. If the object has the same frequency as the light, photons are absorbed and transformed into thermal energy of vibrational movement. The light waves pass through the object with minimum reflection, and re-emits on the other side of the object. This particular light wave is called transmitted light wave. Reports have shown that an underlying transparent indium - tin oxide layer [1] could provide efficient current collection and transparency. However, the element indium is expensive and unlikely to apply in the printed electronics in the future. Furthermore, its brittleness prevents applications where the flexibility of a supercapacitor is required. Graphene is a two-dimensional layer with sp2 hybridized structure. It has a strong mechanical strength, excellent flexibility, and good optical transparency. The implementation of chemical vapor deposition graphene film shows great stretchability and transparency. And the inexpensive synthesis of graphene film of inexpensive Ni and Cu foils is a desirable approach. A wrinkled structure of graphene [2] onto the polydimethylsiloxane (PDMS) showed less transparency and flexibility than that of the directly chemical-vapor-deposited layers of graphene film on the PDMS. Thus, the arrangement of graphene films also affect the transparency and flexibility of a supercapacitor. Carbon nanomaterials, generally have great flexibility and electric double-layer capacitance (EDLC) behaviours, showing wide applications in flexible supercapacitors. However, their optical transmittance is often limited by their low conductivity. It turns out its other replacement poly(3,4-

Content from this work may be used under the terms of the Creative Commons Attribution 3.0 licence. Any further distribution of this work must maintain attribution to the author(s) and the title of the work, journal citation and DOI.
Published under licence by IOP Publishing Ltd

ethylenedioxythiophene) polystyrene sulfonate (PEDOT:PSS), a sprayed conductive macromolecular polymer, show more potential in the future mass production of transparent supercapacitors. PEDOT:PSS [3] is a versatile electrode film. It not only fits the biosensor supercapacitor (in which it traps the virus and tests if the sample from a patient is contaminated by cancer cells or viruses), but it is also suit, and commercially available for a transparent supercapacitor. The PEDOT:PSS's volume alters the film transmittance, and the transparency increases with the film transmittance. Experiments show that PEDOT:PSS is essentially better than most graphene electrodes, on the account that its superior performance attributes as a transparent supercapacitor charge storage film to its low optical conductivity. The PEDOT:PSS on the electrode is spin-coated from a doped PEDOT:PSS solution. Various asymmetric supercapacitors electrodes or hybrid electrodes composed of PEDOT:PSS films show superior electrochemical behavior. Nevertheless, the thick PEDOT:PSS film layer conveys lower optical transmittance that is contrary to a transparent supercapacitor. The triangular Galvanic Charge-Discharge (GCD) shape of PEDOT:PSS indicates that it shows great capacitive behaviors as well. Polyvinylpyrrolidone (PVP) and $LiClO_4$ electrolytes can also serve as the coating of a transparent supercapacitor. PVP is an amorphous polymer. It has good stability and displays moderate ionic conductivity. Whereas the properties of PVP can change due to external factors, such as the increase of humidity and loss of residual solvents. These films have been employed in the fabrication of a symmetrical supercapacitor, which shows good transparency, capacitive behaviors, and some flexibility. The new devices illustrate the continued improvement in the development of transparent and flexible supercapacitors. In the future, researchers plan to further improve the energy density, flexibility, and durability, and stretchability of a supercapacitor.

## 2. Results and Discussion

### *2.1. Material*

Material is the key indicator of the transparency, capacitance, and flexibility of a transparent supercapacitor electrode (TSE) material. Even though indium tin oxide (ITO) is still the most widely used TSE material nowadays, the material itself is brittle and expensive, which is not the most desirable material we want. Therefore, more materials emerge from studies of TSE around the globe. In general, the materials are categorized into two big groups. And these two groups extend many branches. The first big group is the distinct materials. This section is divided into five large groups: 1. ITO/FTO; 2. conductive polymers; 3. Metallic nanostructures; 4. Mxenes; 5. carbon materials. The second big group is categorized by the different chemical configurations of a material. The most usual material configuration that is implemented onto the TSC are the one-dimensional material (1D material), and two-dimensional material (2D material). Three dimensional materials (3D materials) are rarely used, but some of them convey really high flexibility and optical transmittance due to their unique mechanical and optical properties. For technological advancement in the future, we seek better and more robust TSC material every day.

**Table 1.** Properties of Different Transparent Supercapacitor Materials

| Electrode | Electrolyte | Optical Transmittance | Voltage Window | Specific Capacitance | Cycling Stability | Energy Density | Power Density |
|---|---|---|---|---|---|---|---|
| ITO | [BMIM][Cl] RTIL | 91.80% | -0.5 - 0.5 V | 22 F/g at 5 mV/s | | | |
| Graphene films | KCl | 70% | 0 - 1 V | 111 F/g at 10 mV/s | | 15.4 Wh/kg | 7200 W/kg |
| CNC | PVA-$H_3PO_4$ gel | 71% | 0 - 1 V | 0.33 F/cm$^2$ at 10-500 mV/s | 10,000 cycles | 47 mWh/cm$^3$ | 19 mW/cm$^3$ |
| PEDOT | PVP-$LiCO_4$ | 50% | -1 - 1 V | 15 F/g at 100 mV/s | 1,400 cycles | | |
| Highly aligned CNT | PVA | 75% | 0 - 0.8 V | 7.3 F/g at 0.1 V/s | | 1.0 Wh/kg | 1.1 kW/kg |
| $RuO_2$–ITO | $H_2SO_4$ | | 0.3 - 0.8 V | 1235 F/g at 50 mV/s | 4,000 cycles | | |
| Freestanding and flexible graphene membrane | PVA | 59% | 0 - 1 V | 55 F/g | 20,000 cycles | 430 μWh cm$^{-3}$ | 190 mW cm$^{-3}$ |
| AACS nanowires | Polymer electrolyte | 85% | 0 - 0.8 V | 136.5 μF/cm$^2$ | | | |
| PSC/MAM | PVA/$H_2SO_4$ | | | 459.6 F/m$^2$ | | 35.9 mWh/m$^2$ | 461.5 mW/m$^2$ |
| $MoO_3$ nanobelts | LiCl | 90% | 0 - 0.8 V | 1198 F/g at 2 mV/s | 20,000 cycles (96.5%) | 22.89 Wh/kg | 686.84 W/kg |
| $Co_3O_4$ nanocrystal | PVA/$H_3PO_4$ | 51% at 550nm | 0.2 - 0.6 V | 177 F/g at 1 mV/s | 20,000 cycles (100%) | 3.01 Wh/kg | 1.152 kW/kg |
| MnO2@Ni | $Na_2SO_4$ | > 84% | 0 - 0.8 V | 131.5 mF cm-2 at 100 mV/s | 10,000 cycles (96.3%) | | |
| Inject printed graphene | $H_3PO_4$ | 87% at 550 nm | 0 - 1.0 V | 8.7 μF cm$^{-2}$ at 50 mV/s | 10,000 cycles (91.3%) | | |

ITO is widely used due to its high optical transmittance and electrical conductivity. It comprises of 74% In, 18% O2, and 8% Sn by weight. Its conductivity comes from the metallic bonds of indium doped tin oxide. On the other hand, its transmittance is high owing to its high light-scattering haze factor, which is the ratio of transmitted light and reflected light. The material looks like a piece of glass. Fluorine tin oxide (FTO) shares similar properties with the ITO. In further lab research, a solid-state flexible supercapacitor with room temperature ionic liquid gel, and ITO electrodes is found by researchers to be safe, environmentally friendly, flexible, and transparent. [BMIM][Cl] is used as the solvent, and ITO is coated onto the polyethylene naphthalate (PEN) [4]. Moreover, ITO is immensely used as the material for the touch screen of a smartphone. Conductive polymers are another important material in the fabrication of TSE. They are characterized as robust, transparent, and highly conductive. Polymers are long organic chains. Picture a polymer as countless joints connected, since these are monomers with only a few nanometers. These joints give it multiple pivot points that enable it to move around quite easily. Therefore, Young's modulus of the polymer materials is relatively high, which means it has high flexibility. Polymers are transparent because the light can transmit through certain plastic materials. For example, polyethylene terephthalate (PET) is a common material to fabricate plastic bottles. Some polymers take thousands of years to decompose, which is environmentally unfriendly, so it's essential to choose the polymer materials we want wisely. The potential use of PEDOT: PSS conductive polymer films on a TSE intrigues many researchers. The PEDOT: PSS coated transparent supercapacitor has electrodes that are 95% transparent, and a capacitance of 12 mF at a scan rate of 50 mV/s. Its capacitance is much higher than any previously reported transparent supercapacitor. PEDOT: PSS polymers are commercially available, unlike the expensive ITO. These polymers have good transmittance due to their low optical conductivity. They are easily adjustable through doping. For example, their sheet resistance decreases dramatically after a small modification. In the meantime, its capacitance is maximized. An aqueous dispersion of PEDOT: PSS is used to create PEDOT: PSS films on PET substrates. Films are prepared after a formic acid post-treatment. A free-standing PEDOT: PSS electrodes not only have superior electrochemical performance and high flexibility but also excellent electrochemical performance in a flexible and transparent all-solid-state supercapacitor. Its 2 vol% surfactant (Triton-X 100) [5] enhances the electrical conductivity, improves the wettability, and reduces the surface tension, which is favorable for the multilayer spin coating of the PEDOT: PSS films to form a smooth and uniform surface morphology. PEDOT: PSS electrodes with one layer were fabricated via spin coating the resulting formulation on PET substrates at 500 rpm for 6s. The gel electrolyte was then coated on the PEDOT: PSS/PET. Most polymers are not electrically conductive by itself. However, if a thin sheet of metal is usually deposited onto the polymer, it enables them to convey both high flexibility and high electrical conductivity. This is where the third material nanostructures come in place. We create metallic nanowires on the surface of a substrate. The substrate is usually polymer to maximize its conductance. The thickness of the substance is inversely proportional to its reluctance. Thus, its thickness must be right to convey the maximum conductivity and transparency at the same time. The polymer can exist as either the current collector or the electrolyte to maximize the conductivity of a TSE. For instance, a supercapacitor based on electrochemically stable Ag–Au-core–shell nanowire percolation network electrode [6] has an excellent electrical conductivity as well as greatly enhanced chemical and electrochemical stabilities. Ag–Au-core–shell is synthesized through a simple Au coating process in the solution environment. Polyvinyl alcohol (PVA) serves as its electrolyte, and this sort of supercapacitor exhibits no degradation upon repeating the charge-discharge cycle. Nevertheless, the substrate can be other materials such as manganese oxide, which also conveys good transparency and flexibility. The researchers developed a novel Au@MnO2 core-shell nanomesh structure [7] on a flexible polymeric substrate through nanosphere lithography combined with electro-deposition processing. The direct growth of MnO2 on gold nanomesh increases the transparency and achieves an efficient contact between the current collector and active materials. As a result, it has a high areal capacitance of 4.72 mF/cm2. Ultimately, the research of metallic nanostructures is another way to approach a perfect TSE. MXenes is different from metallic nanostructures. Even though Mxenes contain metal compounds and are electrically conductive, it is, in fact, a 2D transition metal carbides and nitrides which have displayed promising properties in numerous applications, such as energy storage, electromagnetic interference shielding, and catalysis. A transparent, flexible, and conductive 2D titanium carbide (MXene) films display high volumetric capacitance [8]. Its layered structure creates a good condition for intercalation reaction to take place. Therefore, it has a high energy density due to its high pseudocapacitance, and meanwhile a high power density. Last but not least, carbon materials are the second most abundant TSE materials due to their practicability. They are characterized as highly flexible and conductive. Furthermore, 1D and 2D carbon materials show how different arrangements of materials can affect the property of that material. Carbon nanotube (CNT) [9] is a common 1D carbon material. As its name suggests, it's a tubular structure. When trillions of these tubular structures are interconnected, it forms a giant network. As a result, when the network is disturbed by a force, it can disperse the pressure into multiple units, and therefore, very flexible. Single-walled carbon nanotubes (SWCNTs) [10] are particularly attractive replacement of ITO. A solution-based deposition is the main method for fabricating SWCNTs transparent and conductive films (TCF) because it is low-cost and scaled up to large areas. Chemical vapor deposition technique fabricated transparent and conducting SWCNTs films. It decreases the surface resistance and increases power density. On the other hand, 2D material is a flexible material by itself. Graphene is a good example. It has multiple ring structures and shows high tensile strength. Each ring acts as a joint that disperses the outside force, as the CNTs. Moreover, it has free electrons all over the place due to its resonance structure. Therefore, it conveys a high electrical conductivity on the account that currents can flow across its surface or interior easily. A laminated ultrathin chemical vapor deposition graphene films based [11] stretchable and transparent high-rate supercapacitor is stretchable and transparent. Its electrode is made of four-layer laminated ultrathin graphene films. Specifically, It has an optical transmittance of 72.9%, and a stretchability of 40%. Its specific capacitance (5.33 uF/cm2 at 200 mV/s) showed no degradation and even increased slightly over 10,000 cycles. The weak D band of these layers of graphenes suggests that neither stretching or processing induces mechanical damages to the buckled graphene films. Furthermore, a wavelength of 550 nm (within the range of visible lights) gives the material its transparency. A single layer graphene film was synthesized using chemical vapor deposition methods. The laminating procedure stacked single-layered graphene layers into polymethyl methacrylate (PMMA). The H2SO4 electrolyte is then prepared and coated on the electrode.

### *2.2. All in one device*

One way to further improve the TSE is to make a hybrid electrode material that takes the advantages of different materials. The reason that we add metallic nanowires onto a thin polymer sheet is that we want the optimal electrical conductivity, optical transmittance, and flexibility altogether. Vertically aligned ZnO@CuS@PEDOT core@shellnanorod arrays [12] decorated with MnO2 nanoparticles for a high-performance and semi-transparent supercapacitor electrode is a good example. Hybrid nano-architectures with high electrochemical performance for supercapacitors have been designed by growing hierarchical ZnO NRs@CuS@PEDOT@MnO2 core@shell nanorod arrays on ITO/glass substrates. These 1D semiconducting ZnO structures have a high aspect ratio, high surface area, and very high crystalline quality and can provide a direct conduction path for electrons. These advantages contribute to an areal capacitance of 9.85 mF/cm2 at 100 mV/s, and a power density of 10 kW/ kg.

## 3. Conclusion

The two main characteristics of a future supercapacitors are flexibility and transparency. In particular, its flexibility comes from flexible materials such as conductive polymers and carbon materials. Even though conductive polymers are usually not stable under high temperature or high humidity, which leads to short cycle lives, they can be readily manufactured, and chemical doping strengthens their stability. Moreover, these conductive polymers have long chains and strong bonding. As a result, conductive polymers are typically more elastic than other materials. On the other hand, there are multiple kinds of polymer materials. They can be either amorphous or crystalline. Amorphous polymers are generally more transparent than crystalline polymers due to their near-zero light refraction. Nevertheless, it's possible to make these crystalline polymers transparent. One way to achieve that goal is to create ultrathin film of these polymers. In the meantime, it's important to make sure the crystals inside these crystalline polymers are oriented in a way that the size of the crystals become smaller, on the account that smaller crystals do not cause the refraction of light rays. Furthermore, an additive technology also improves the transparency of crystalline polymers. The injection of a metallocene catalysts can potentially increase the transparency of a polymer film due to a decrease in the degree of crystallinity. Still one thing to notice here, at a temperature of about 120 - 130 °C, the sample will begin to form a haze caused by the crystal formation. It's still under investigation how to preserve the properties of conductive polymers we want under extreme conditions. Carbon materials are also an excellent

IOP Conf. Series: Materials Science and Engineering **772** (2020) 012107 doi:10.1088/1757-899X/772/1/012107

candidate for a transparent flexible supercapacitor. 1D and 2D carbon materials both convey high flexibility. CNTs are the most common flexible supercapacitor electrode material in 1D carbon. One CNT is neither flexible nor a good material for an electrode material, but if you "stitch" trillions of these CNTs together. A CNT network forms. Each of the CNTs act as a mechanical joint, therefore, the Young's modulus of CNT network material is particularly big. Likewise, 2D carbon materials utilize this mechanical property as well. For example, the hexagonal ring structures on graphene molecular configuration also act as joints. In contrast, 2D carbon graphene is already a flexible material by itself. Graphene also has a large surface area in contact with the electrolyte. Therefore, graphene electrode material conveys a large capacitance [13]. However, these two materials are not perfect. Both of them are difficult to be processed in solution because of their low solubility in most solvents with no treatment. Moreover, graphene is susceptible to re-stack over cycles [14]. As a result, its active surface area is limited and it lowers the actual capacitance of graphene electrode in comparison to its theoretical capacitance. MXene is a 2D transition metal carbide/nitride. It's another excellent 2D electrode material for a flexible supercapacitor. Graphene and MXene both have good electrical conductivity, on the account that graphene has free electrons moving around, and MXenes have metallic intermolecular bonding. The high conductivity enables these two materials to have stronger currents, and thus, higher power density. In addition, MXenes have solid's layered structures. Because of that, more intercalation reactions can take place within the supercapacitor, which leads to a high pseudocapacitance. Even though carbon materials and MXene are so desirable flexible materials, their thickness must be thin enough to make these materials transparent. And problems come along. When these materials are much smaller than they were before, their tensile stress limit decreases and resistance increases owing to the decrease in thickness. The same issue falls on another candidate of transparent supercapacitor electrode. Metallic nanostructures make full use of ultrathin metals (nanowires and nanopores) to tremendously increase the electrode surface area, and it has high electrical conductance as MXenes. Nevertheless, thin metals are less reactive, and get oxidized quickly when it's exposed to air. To tackle the conundrum, a possible solution is to a tolerant polymer film such as PET under these materials. It will prevent the materials from losing the quality they had before, and increase the optical transmittance of the material.

It's now time to take real consideration of the all in one devices. Almost no materials are perfect alone, but if you choose the proper materials and combine them together. Their advantages will compensate their disadvantages. Hybrid supercapacitor electrode is the future of supercapacitor businesses. It will enable us to build better interactive devices such as smartphones, smart windows, smart glasses, tablets, and other touchable devices. In other words, the assembly of different materials maximize the functionality of a transparent supercapacitor devices. It merges the electrical conductivity, optical transmittance, and mechanical properties.

- Compositing electrode materials make the best combination of high performance
- Highly trasnparent and flesible electrolyte with safe and ionic conductive are to be studied,
- Metrics of quantifying transparency and flexibility of supercapacitors are to be developed.
- Lowering the cost;miniaturizing the device

IOP Conf. Series: Materials Science and Engineering **772** (2020) 012107 doi:10.1088/1757-899X/772/1/012107

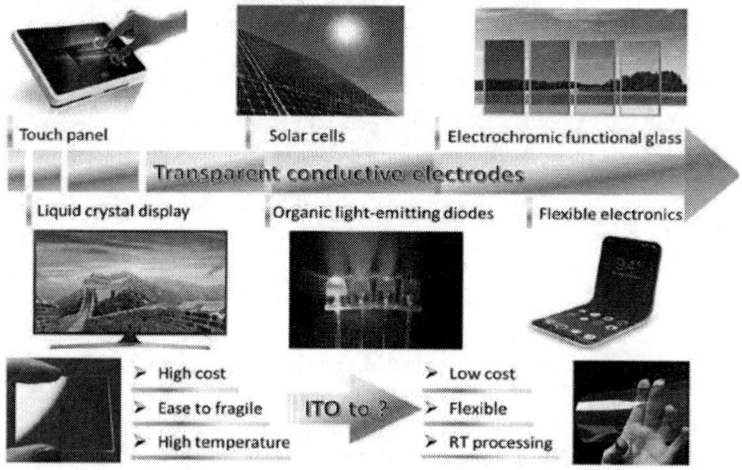


Applications of transparent conductive electrodes. Also included are the current challenges of ITO and expected features in the ITO substitutes. Reprinted with permission from Ref. [1]. Copyright 2019 Elsevier.

(1)Zhang, Chuanfang John, and Valeria Nicolosi. "Graphene and MXene-based transparent conductive electrodes and supercapacitors." Energy Storage Materials 16 (2019): 102-125.

**Fig. 1** Transparent Conductive Electrdes

In that way, future technologies like hologram will show up in the near future because of the fabrication of an almost perfect transparent supercapacitor.

## References

[1] Ilhwan Ryu, MinHo Yang, Hyemin Kwon, Hoo Keun Park, Young Rag Do, Sang Bok Lee, and Sanggyu Yim, Coaxial RuO2−ITO Nanopillars for Transparent Supercapacitor Application. *Langmuir* **2014** *30* (6), 1704-09

[2] Tao Chen, Yuhua Xue, Ajit K. Roy and Liming Dai, Transparent and Stretchable High-Performance Supercapacitors Based on Wrinkled Graphene Electrodes. *ACS Nano* **2014** *8* (1), 1039-46.

[3] Thomas M. Higgins and Jonathan N. Coleman, Avoiding Resistance Limitations in High-Performance Transparent Supercapacitor Electrodes Based on Large-Area, High-Conductivity PEDOT: PSS Films. ACS Applied Materials and Interfaces, **2015** *7* (30), 16495-506

[4] Di Wei, Steve J. Wakeham, Tin Wing Ng, Mike J. Thwaites, Hayley Brown, Paul Beecher, Transparent, flexible and solid-state supercapacitors based on room temperature ionic liquid gel. Electrochemistry Communications, 2009, 11, 2285-87

[5] Tao Cheng,a Yi-Zhou Zhang, Jian-Dong Zhang, Wen-Yong Lai, and Wei Huangab, High-performance free-standing PEDOT:PSS electrodes for flexible and transparent all-solid state supercapacitors. **J. Mater. Chem. A**, 2016, **4** , 10493-99

[6] Habeom Lee, Sukjoon Hong, Jinhwan Lee, Young Duk Suh, Jinhyeong Kwon, Hyunjin Moon, Hyeonseok Kim, Junyeob Yeo, and Seung Hwan Ko, Highly Stretchable and Transparent Supercapacitor by Ag–Au Core–Shell Nanowire Network with High Electrochemical Stability. *ACS Applied Materials & Interfaces,***2016** *8* (24), 15449-58

[7] Tengfei Qiu, Bin Luo, Michael Giersig, Eser Metin Akinoglu, Long Hao, Au@MnO2 Core–Shell Nanomesh Electrodes for Transparent Flexible Supercapacitors, Wiley-VCH Verlag GmbH &

Co. KGaA, Weinheim, 2014, 10.

[8] Chuanfang (John) Zhang, Babak Anasori, Andrés Seral-Ascaso, Sang-Hoon Park, Transparent, Flexible, and Conductive 2D Titanium Carbide (MXene) Films with High Volumetric Capacitance. WILEY-VCH Verlag GmbH & Co. KGaA, Weinheim, 2017.

[9] Tao Chen1, Huisheng Peng2, Michael Durstock3 & Liming Dai, High-performance transparent and stretchable all-solid supercapacitors based on highly aligned carbon nanotube sheets. Scientific Reports, 3, 2013.

[10] Zhiqiang Niu, Weiya Zhou, Jun Chen, Guoxing Feng, Hong Li, Yongsheng Hu, A Repeated Halving Approach to Fabricate Ultrathin, Single-Walled Carbon Nanotube Films for Transparent supercapacitors. Wiley-VCH Verlag GmbH & Co. KGaA, Weinheim, 2013, 9.

[11] Ping Xu, Junmo Kang, Jae-Boong Choi, Laminated Ultrathin Chemical Vapor Deposition Graphene Films Based Stretchable and Transparent High-Rate Supercapacitor. Acs nano, 2014, 8.

[12] Jorge Rodriguez-Moreno, Elena Navarrete-Astorga, Enrique A. Dalchiele, Vertically aligned ZnO@CuS@PEDOT core@shell nanorod arrays decorated with MnO2 nanoparticles for a high-performance and semi-transparent supercapacitor electrode. The Royal Society of Chemistry, 2014.

[13] Aiping Yu, Isaac Roes, Aaron Davies, and Zhongwei Chen, Ultrathin, transparent, and flexible graphene films for supercapacitor application. APL, 5, 2010.

[14] Na Li, Gongzheng Yang, Yong Sun, Huawei Song, Hao Cui, Guowei Yang, and Chengxin Wang, Free-Standing and Transparent Graphene Membrane of Polyhedron Box-Shaped Basic Building Units Directly Grown Using a NaCl Template for Flexible Transparent and Stretchable Solid-State Supercapacitors. *Nano Letters* **2015** *15* (5), 3195-203

# Damage identification of bridge structures based on curvature mod

**Zongjun Sun, Kai Li***

Shandong University of Science and Technology, College of transportation, Qingdao, China

*Corresponding author e-mail: 825840462@qq.com

**Abstract.** Based on the modal analysis theory of bridge dynamic characteristics, a pre-damaged reinforced concrete model beam is established by ANSYS as the object of analysis. The regularity of modal parameters such as displac, ement mode and curvature mode of the model beam at the preset damage location and different damage degree and the identification of modal parameters in structure damage are studied by finite element numerical analysis. Sensitivity. The results show that the displacement mode is difficult to identify the damage of the structure effectively, but the curvature mode can reflect the damage of the structure more accurately. The damage identification method based on the curvature mode can be applied to the damage diagnosis of the structure in practical engineering.

## 1. Introduction

During the operation and use of bridges, the long-term impact of various loads will continue to occur various types of damage, different damage gradually accumulated to a certain extent will have an impact on the safety of bridge structures [1]. Therefore, the detection of bridge structure location and the identification of damage degree have important research significance. The finite element numerical analysis is used to restore the original state information of the structure as much as possible, and the damage of the structure is identified by comparing it with the measured information. In this paper, ANSYS finite element analysis software is used to establish reinforced concrete beam bridge models with different crack damage conditions [2-3]. Curvature modal analysis method of bridge dynamic characteristics is used to locate structural damage location and identify damage degree. The curvature mode based on modal analysis can not only locate the damage location effectively, but also be more sensitive to local damage than displacement mode.

## 2. Strain mode and curvature mode theory

The displacement mode of beam body is an important parameter in the analysis of bridge dynamic characteristics. Strain mode is the first derivative of displacement mode. The distribution of strain mode corresponds to that of displacement mode one by one. Curvature mode is the deformation mode in the middle of the structure, which is the same as strain mode. According to the theory of material mechanics, the bending curvature of the beam is as follows:

$$\rho = \frac{M}{EI} \tag{1}$$

Content from this work may be used under the terms of the Creative Commons Attribution 3.0 licence. Any further distribution of this work must maintain attribution to the author(s) and the title of the work, journal citation and DOI.
Published under licence by IOP Publishing Ltd

Formula: $\rho$ is curvature; $M$ is bending moment; $EI$ is Flexural rigidity.

The differential equation of beam vibration is:

$$\frac{\partial^2}{\partial x^2}\left[EI(x)\left(\frac{\partial v(x,t)}{\partial x^2}+a_1\frac{\partial v^3(x,t)}{\partial x^2 \partial t}\right)\right]+m(x)\frac{\partial v^2(x,t)}{\partial x^2}+c(x)\frac{\partial v(x,t)}{\partial t}=f(x,t) \tag{2}$$

Formula: $v(x,t)$ is transverse displacement of t axis section x at time; $a_1$ is stiffness proportional coefficient; $EI(x)$ is flexural stiffness at $x$; $m(x)$ is quality of $x$; $c(x)$ is damping at $x$.

The stress $\sigma$ after change is:

$$\sigma = E(\varepsilon + a_1\dot{\varepsilon}) \tag{3}$$

Formula: $E$ is modulus of elasticity; $\varepsilon$ is initial strain; $\dot{\varepsilon}$ is variation of strain.

According to the modal theory of structural dynamic characteristics, the solution $v(x,t)$ of formula (3):

$$v(x,t)=\sum_{i=1}^{\infty}\varphi_i(x)q_i(t) \tag{4}$$

Formula: $\varphi_i(x)$ is displacement mode shapes; $q_i(t)$ is Modal coordinates.

In the theory of material mechanics, the curvature function of an elastic beam is obtained by calculating the second derivative of the displacement function, and then the curvature change at any section $x$ is as follows:

$$\rho(x)=\frac{\partial^2 v(x,t)}{\partial x^2}=v''(x)=\sum_{i=1}^{N}\varphi_r''(x)q_i(x) \tag{5}$$

Formula: $\varphi_r''(x)$ is curvature mode shape.

When the bending deformation occurs, the curvature formula of the neutral plane in the beam body is as follows:

$$\rho(x)=\frac{v''(x)}{\left(1+v'(x)\right)^{\frac{3}{2}}}\approx v''(x) \tag{6}$$

It can be seen from the formula that when cracks and other local damages occur in the beam body, the dynamic structural parameters such as elastic modulus of the corresponding damaged beam body decrease, the local bending stiffness of the structure decreases, and the increase of curvature causes a sudden change in the smooth curvature modal curve without damage. According to the observed curvature modal curve, the damage of the beam body can be identified location[4-7].

The steps of solving the curvature mode of the beam are as follows:

(1)The displacement modes $\varphi(r)$ of each measuring point of the beam body are measured to form the displacement mode of the beam body:

$$\varphi(r)=\left[\varphi(1),\varphi(2),\cdots,\varphi(r)\right] \tag{7}$$

(2)Solving the corresponding curvature mode $\rho(r)$ by central difference method for each measuring point:

$$\rho(r)=\frac{\varphi(r-1)-2\varphi(r)+\varphi(r+1)}{d_{r,r+1}\cdot d_{r,r-1}} \tag{8}$$

Formula: $r$ is measure point;; $d_{r,r+1}$ and $d_{r,r-1}$ represents the distances between adjacent measuring points $r$, $r+1$ and $r-1$, respectively.

(3) The curvature modes of the beam are as follows:

$$P(r)=\left[\rho(1),\rho(2),\cdots,\rho(r)\right] \tag{9}$$

## 3. Finite Element Model of Pre-Damaged Reinforced Concrete Beams

The pre-damaged reinforced concrete model beams are 7 meters long, 0.3 meters wide and 0.5 meters high, respectively. The strength grade of concrete is C60, and the density is 3 x 103kg/m3. The thickness of concrete protective layer is 0.5 meters, two compressive longitudinal bars are 25, three tension longitudinal bars are 25, stirrups are 9, the spacing of stirrups is 0.15 meters, the elastic modulus of steel bars is 200GPa, and the density is 8 *103kg/m3. Poisson's ratio is 0.25.

Beam damage is a transverse crack with a width of 2 mm precast along the beam height at the bottom of 1/4 span, 1/8 span, 1/2 span, 3/4 span and 3/8 span. The crack depth is 5%, 10%, 15%, 20%, 25%, 30%, respectively, i.e. 0.025 m, 0.05m, 0.075 m, 0.1 m, 0.125 m and 0.15m (case 1-case 5), finite element model and beam damage. The cross section is shown in Figures 1 and 2.

**Fig. 1** Finite element model of beam body

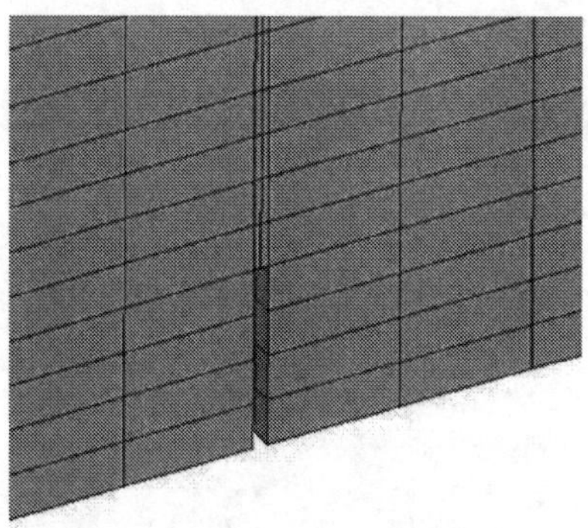

**Fig. 2** Cracks in Damaged Section

### *3.1. Displacement Modal*

The first five displacement modes of reinforced concrete beams under non-destructive condition are shown in figure 3.

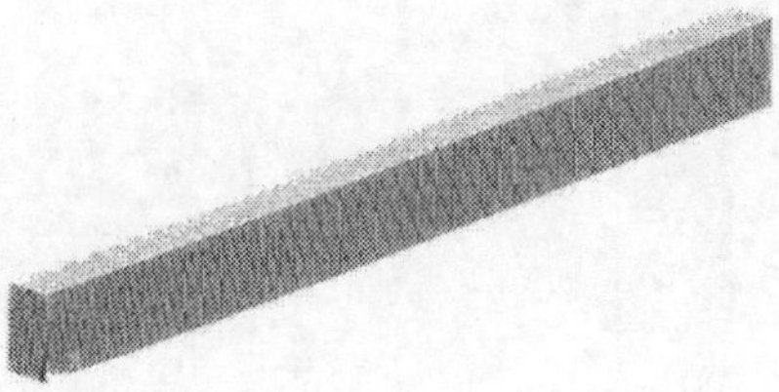

(a) Frist-order displacement mode shapes

(b) Second-order displacement mode shapes

(c) Third-order displacement mode shapes

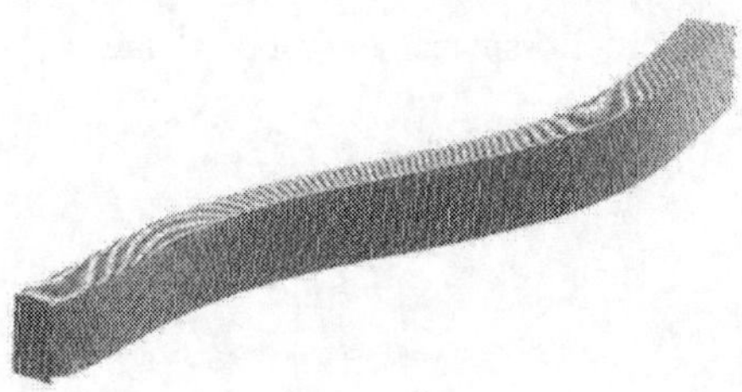

(d) Fourth-order displacement mode shapes

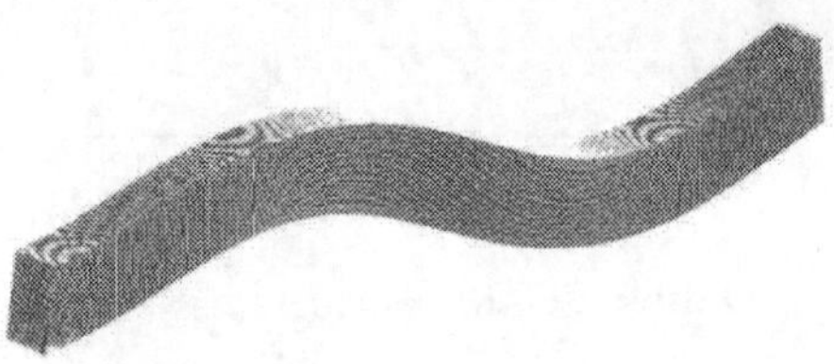

(e) Fifth-order displacement mode shapes

**Fig. 3** The displacement modes

In the displacement mode diagram shown in Figure 3, the displacement mode deformation of each node at the same height and the same horizontal position is the same. The minimum preset damage of the model beam is 5% of the beam height. Therefore, the displacement mode is extracted as the displacement variation of the second layer of the model layer, that is, the displacement of 49 nodes at 0.0125 position from the bottom of the beam. The distance between the two nodes in the model is 0.15 meters. The displacement modes under different damage states are shown in Fig. 4.

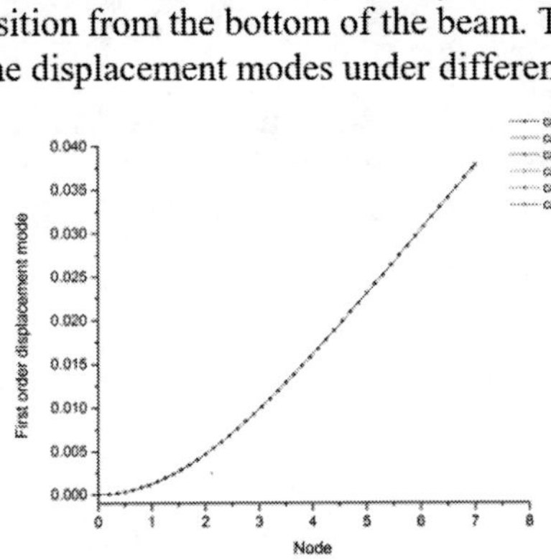


(a) First-order displacement mode shapes

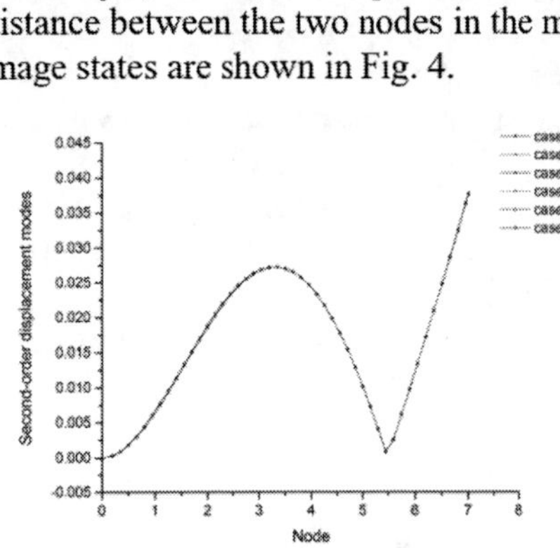


(b) Second-order displacement mode shapes

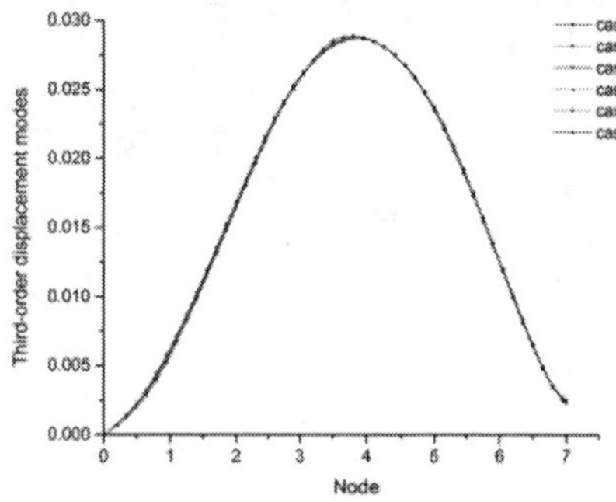


(c) Third-order displacement mode shapes

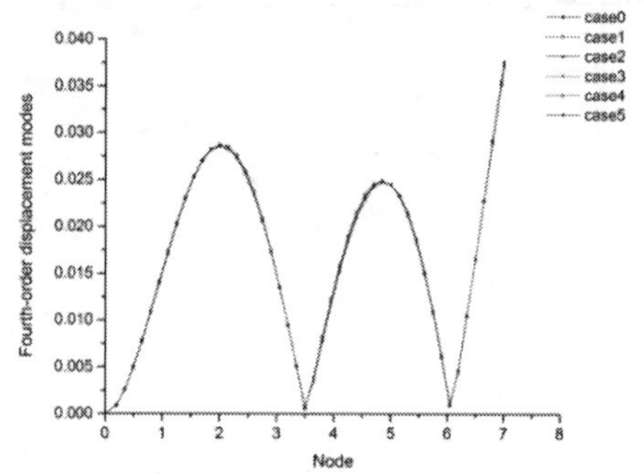


(d) Fourth-order displacement mode shapes

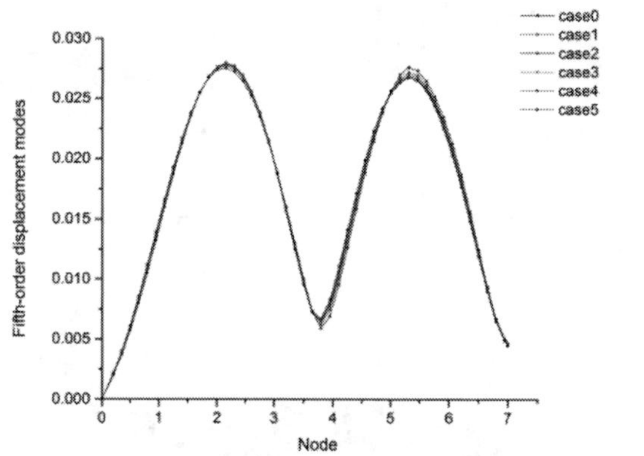


(e) Fifth-order displacement mode shapes

**Fig. 4** Displacement mode shapes with different damage degrees

From the displacement mode shapes of different damage states in Fig. 4, it can be seen that the coincidence degree of displacement mode shapes is better. Only in the fifth-order displacement mode shapes, slight changes of displacement mode and non-destructive state can be seen under different damage conditions, but it can't accurately reflect the existence of structural damage, nor can the location of damage be judged. Displacement modes can't reflect the structural damage sensitively, nor can they accurately identify the structural damage of reinforced concrete girder bridges by relying solely on displacement modes.

### *3.2. Curvature Mode Shape*

The displacement mode shapes obtained from the above finite element analysis are calculated according to the formula (7) ~ (9) and the corresponding curvature mode shapes under different damage degrees are calculated according to the solving steps of the curvature mode of the beam. The fifth order curvature mode shapes under five damage degrees are shown in Fig. 5.

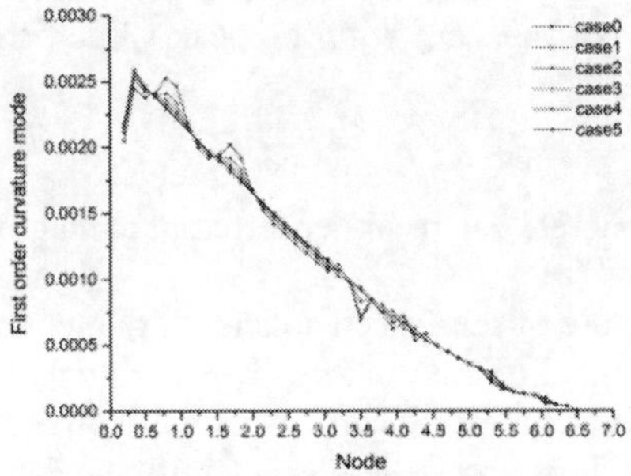


(a) First-order curvature modes

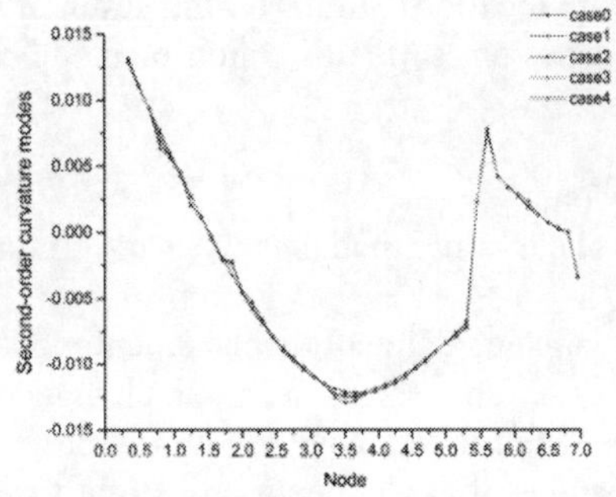


(b) Second-order curvature modes

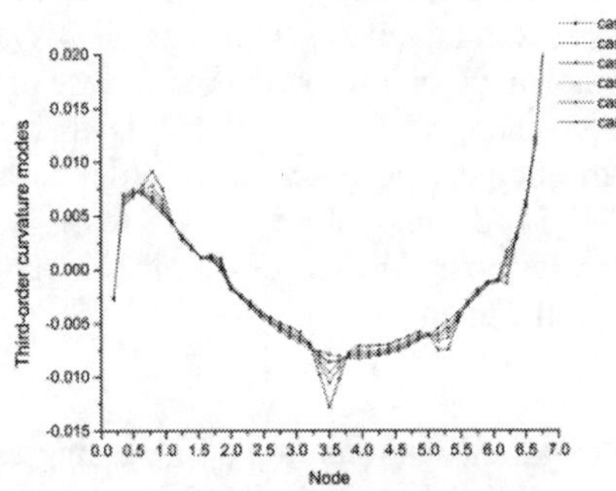


(c) Third-order curvature modes

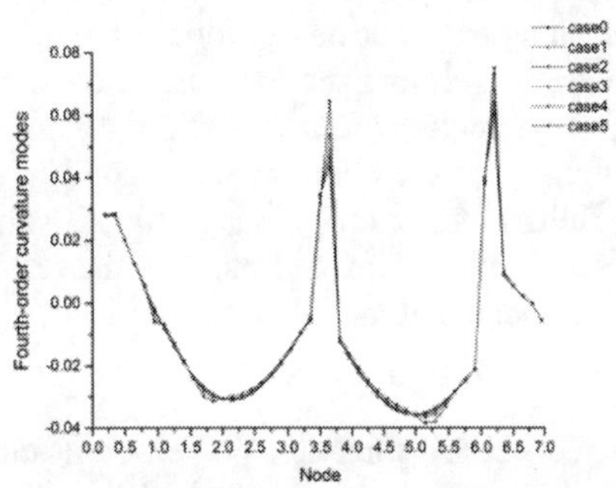


(d) Fourth-order curvature modes

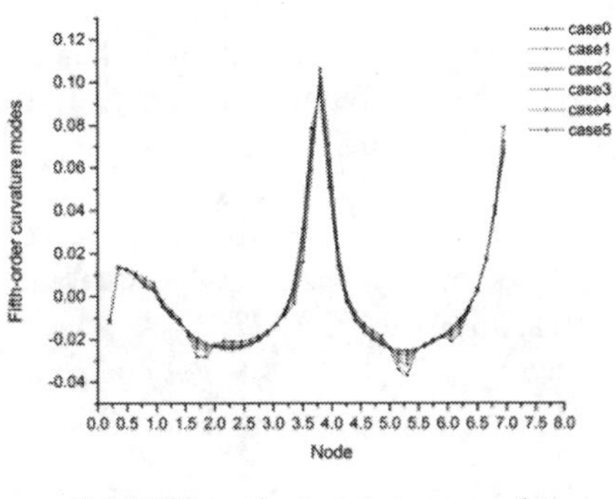


(e) Fifth-order curvature modes

**Fig. 5** Curvature mode shapes with different damage degrees

From the curvature mode shapes obtained by the above finite element numerical simulation, it can be seen that: (1) the curvature mode shapes at different damage locations have obvious abrupt changes on both sides of the damage location. The abrupt change of the mode shapes at the damage location reaches the maximum, and the abrupt change increases with the deepening of the damage degree, and the curvature mode energy. It can better reflect the damage location and degree of structural damage. (2) The first-order curvature modes fluctuate dramatically in the midspan, while the higher-order curvature modes do not. It can be seen that the order selection of modal modes is very important for judging the damage of structures, besides considering the arrangement of measuring points, in the damage diagnosis of vibration systems of bridge structures. Important role. (3) The sudden change direction of the curvature mode is consistent with the peak direction of the order mode. The local damage of the structure results in the decrease of the local stiffness of the structure, which leads to the increase of the curvature of the damage location. Similarly, the strain at the damage location increases significantly due to the stress concentration, so the direction of mode change is consistent with the peak value change at the mutation location.

## 4. Conclusion

The results of curvature modal analysis of reinforced concrete beams under different damage conditions show that:

(1) The damage degree of displacement modal structure of reinforced concrete finite element model beams increases and shows a small change rule. It is difficult to measure in actual engineering monitoring. When the damage amount is large enough, some displacement modal modes at different damage locations of model beams are slightly distorted. It is also difficult to identify the damage of the bridge structure, which further proves that the displacement modes are insensitive to the damage of the structure, that is, the structural damage can't be effectively identified by the displacement modes.

(2) The curvature mode has a sudden change at the structural damage, which can identify the location of the preset damage more accurately, and the mode distortion increases with the increase of the damage degree of the structure. The direction of the sudden change of mode shape is the same as the corresponding peak direction of the second mode. In engineering practice, in order to obtain more accurate results in structural damage identification, high-order modal modes are reasonably selected according to different damaged parts of the tested structure. As far as possible, it is avoided to select modal modes located at modal nodes to identify structural damage.

## References

[1] Wu Weiwei, Li Hongyan, Ye Fei. Finite element numerical simulation of pre-damaged reinforced concrete beams [J]. Journal of Systems Simulation, 2018, 30 (9): 3437-3444.

[2] Wu Duo, Liu Laijun, Zhang Xiaoyu, Liu Hongyan. Bridge damage identification based on curvature modal curve change [J]. Journal of Architectural Science and Engineering, 2018, 35 (2): 119-124.

[3] Wang Chao, Zhu Hongping. Design and application of health monitoring system for Junshan Yangtze River Bridge [J]. Journal of Wuhan University of Technology, 2015, 37 (5): 76-80.

[4] Zhu Hongping, Zhang Junbing. Health Monitoring of Bridge Structures [J]. China Journal of Highway, 2013, 13 (2).

[5] Sun Jie. Research on structural damage identification method based on multi-modal neural network [J]. Journal of Wuhan University of Technology (Transportation Science and Engineering Edition), 2012, 36 (6): 1240-1242.

[6] Wu Xiangnan, Xu Yue, Liang Peng. Present situation and Prospect of bridge structural damage identification [J]. Journal of Chang'an University (Natural Science Edition), 2013, 33 (6): 49-55.

[7] Wu Fusheng, Gu Peiying, Deng Chang. Structural modal analysis and damage identification [M]. Nanjing: Southeast University Press, 2008.

IOP Conf. Series: Materials Science and Engineering **772** (2020) 012109 doi:10.1088/1757-899X/772/1/012109

# Effect of Lignin Removal on Moisture Absorption of Wood

**He Zhang[1, a], Yuchen Chen[2, *], Yongkuo Wang[1, b]**

[1]Ministry of Civil Affairs of the People's Republic of China, Beijing, China
[2]School of Laval University, Quebec

*Corresponding author e-mail: carmenchen24@outlook.com, [a]mishanxixi@126.com, [b]yongkuow@163.com

**Abstract**. In the study, we carried out heartwood and sapwood of fast-growing Chinese fir as research objects, the lignin was removed by sodium chlorite method, then the changes of wood chemical composition were detected by Raman spectra, the moisture absorption changes of wood were studied by saturated salt solution method, and the H-H model fitting data was used to analyze the effect of lignin removal. It aims to provide reference for the research and development of new wood modification technologies or wood-based products [4]. The results showed that there was no significant difference between sapwood and heartwood when the relative humidity was lower than 30%. With the increase of relative humidity in the environment, the water absorption capacity of sapwood cells was higher than that of heartwood. The water absorption capacity of heartwood and heartwood after delignification was significantly stronger than that of control wood, especially when the relative humidity was 95%, the equilibrium water content reached about 60%.

## 1. Introduction

Wood is one of the only four renewable materials in the world, and its processing and utilization have attracted more and more attention from the world. However, in order to meet the people's growing yearning for a better life, the pressure on China's timber supply is increasing. International timber trade bears more ecological risks and policy risks. Reducing dependence on tropical hardwood has become the consensus of many countries including China. China has also implemented a national reserve forest construction project, and forests play an important role in the construction of the ecological system of mountains, rivers, forests, fields, lakes and grasses. Therefore, the wood industry must accelerate two changes, one is the continuous acceleration of the transformation of raw material supply to fast-growing plantations, and the other is the acceleration of the transformation of utilization mode to high-quality utilization. However, these developments and changes have always been unable to avoid the fundamental problem of unsatisfactory wood properties of artificial forests. Due to the common problems of low density, poor dimensional stability, low mechanical strength and poor durability, this greatly limits the application of plantation wood, especially the use of high-quality solid wood.

In recent years, modification technologies based on wood chemical properties, or chemical-based materials derived from wood, have also attracted much attention from academia. The more advanced technologies include transparent wood, super-strong wood, sponge wood and so on. These studies have provided new angles and new ideas for the utilization of high quality wood. These studies are all

Content from this work may be used under the terms of the Creative Commons Attribution 3.0 licence. Any further distribution of this work must maintain attribution to the author(s) and the title of the work, journal citation and DOI.
Published under licence by IOP Publishing Ltd

based on the composition relationship of the basic chemical components of wood. This requires further exploration of the physical or chemical properties of cellulose, hemicellulose, lignin and extracts. This has become the focus and hotspot of wood scientists at home and abroad. Based on this, this paper takes fast-growing Chinese fir as the research object. The effect of lignin removal on moisture absorption of fast-growing wood was discussed. The physical and chemical properties of wood are mainly studied by Raman spectrometer. The effect of saturated salt solution method on hygroscopicity of heartwood and sapwood was studied. It aims to provide reference for the research and development of new wood modification technologies or wood-based products.

## 2. Materials and Methods

### *2.1. Materials*

The tested fir wood is a clone of fir wood collected from Kaihua County Forest Farm in Zhejiang Province. Dam 8 planted in 1998, no CL-8-4. Cut 3.5m to 5.5m timber from the felled sample along the tree height direction, according to the sapwood and heartwood two parts respectively cut specimens, processing to 20 mm×20 mm×5 mm(R× T× L) for standby. The chemical reagent is sodium chlorite (98.0%, 500g), Chengdu aikeda chemical reagent limited company produces, Glacial acetic acid (analytically pure, mass fraction ≥99.5%), Guangdong Guanghua Sci-Tech Co.,Ltd produce. Self-made distilled water.

### *2.2. Method*

*2.2.1. Lignin Removal.* Selecting the prepared sample, Drying at 103 ℃for 24 hours, Obtaining the absolute dry quality before delignification, Based on every 150ml distilled water, 1:2 mixed sodium chlorite and glacial acetic acid solution, To configure, Put the test piece into the mixed solution. Water bath heating at 85℃ for 12 hours, and then put into a drying box after air drying for several days. Dry at 103℃ for 24h, the absolute dry quality after delignification is obtained, the weight loss rate (%) of delignification treatment was calculated. The removal of lignin is specifically carried out according to the method in "Chemical Components Effect on Mechanical Properties of Wood Cell Wall" [1].

*2.2.2. Hygroscopicity experiment.* First, the sample was dried for 24 hours. Weigh, Then 100ml of magnesium chloride, sodium bromide, potassium chloride and potassium nitrate are added respectively at 20℃ to prepare saturated salt solution. The ambient humidity is 30%, 60%, 80% and 95% respectively. The quality is weighed after the time interval of 24 hours and 48 hours. Wait until equilibrium is reached and remove it. Data processing and calculation methods refers to "Wood Drying Science" [5].

On the premise of a certain temperature, the equilibrium water content in the sample cells is:

$$M_S = (W_{moist} - W_{dry})/W_{dry} \tag{1}$$

In the formula:
$M_S$——Equilibrium moisture content of samples;
$W_{moist}$——The sample weight at a certain relative humidity;
$W_{dry}$——Absolute dry weight of sample.

## 3. Results and Discussion

### *3.1. Weight Loss Rate after Treatment*

Table 1 shows the changes of heartwood and sapwood before and after delignification. After being treated by sodium chlorite method, Lignin is effectively removed.

**Table 1.** Sample quality before and after treatment

| Type | Sample number | Pre-treatment quality (g) | Quality after treatment (g) | Weight Loss Rate (%) |
|---|---|---|---|---|
| heartwood | X1 | 0.4845 | 0.3341 | 31.04% |
| | X2 | 0.8021 | 0.5506 | 31.36% |
| | X3 | 0.6332 | 0.4333 | 31.57% |
| | X4 | 0.6189 | 0.4421 | 28.57% |
| | X5 | 0.6056 | 0.4073 | 32.74% |
| | Average (standard deviation) | - | - | 31.06%(0.0137) |
| sapwood | B1 | 0.8278 | 0.5927 | 28.40% |
| | B2 | 0.6812 | 0.4865 | 28.58% |
| | B3 | 0.5346 | 0.3653 | 31.67% |
| | B4 | 0.6547 | 0.4658 | 28.85% |
| | B5 | 0.6453 | 0.4367 | 32.33% |
| | Average (standard deviation) | - | - | 29.97%(0.0168) |

*3.2. Raman Spectrum Analysis*

In this study, Raman spectroscopy was used to further analyze the effect of delignification on wood spectroscopy. Table 2 shows the Raman spectrum characteristic peaks of Chinese fir cell wall in 4000-400 $cm^{-1}$ band. The Raman spectrum after treatment is shown in Figure 1.

**Table 2.** Characteristic peaks of Raman spectra of cell wall of Chinese fir

| | Wavelength($cm^{-1}$) | Spectral peak attribution |
|---|---|---|
| Lignin | 1273 | Telescopic vibration of benzene ring |
| | 1604 | Telescopic vibration of benzene ring |
| | 1655 | C=C bond and γ-C=O bond |
| | 2942 | Methoxy C-H bond |
| Cellulose | 917 | C-H deformation |
| | 1095 | CC and CO vibrations |
| | 1463 | HCH and HOC plane bending vibration |
| | 2889 | CH and $CH_2$ Telescopic Vibration |
| glucomannan | 1340 | Mannose skeleton vibration |

The characteristic peak representing lignin at 1604 $cm^{-1}$ in Figure 1 is very significant, after acid treatment of lignin, it is found that the characteristic peak at this location is obviously lower than that of the control material. Almost disappeared, indicating that lignin is removed, further observe the spectrum near 1604 $cm^{-1}$ (Figure 1), it can be seen that there is a peak at 1655 $cm^{-1}$. Corresponding to C=C bond on coniferyl alcohol and γ-C=O bond on coniferyl aldehyde in lignin unit, this peak disappeared almost completely after delignification treatment. The results showed that sodium chlorite method did cause the cleavage of carbonyl groups on lignin units. The peaks at 1273 $cm^{-1}$and 2942 $cm^{-1}$ respectively correspond to the hydroxyl or C-H bond on methoxy and methoxy groups connected to lignin and phenyl, after delignification, the peak heights of the two positions are obviously reduced, It shows that sodium chlorite method can remove methoxy group from lignin. 1340 $cm^{-1}$ represents the characteristic peak of hemicellulose and 917 $cm^{-1}$, 1095 $cm^{-1}$, 1463 $cm^{-1}$, 2889 $cm^{-1}$ represents the characteristic peak of cellulose without obvious reduction. It shows that after sodium chlorite treatment, Hemicellulose and cellulose have not been largely removed, based on the data analysis in

Table 1, it can be concluded that: After being treated by sodium chlorite method, almost all lignin was removed from the specimen [2].

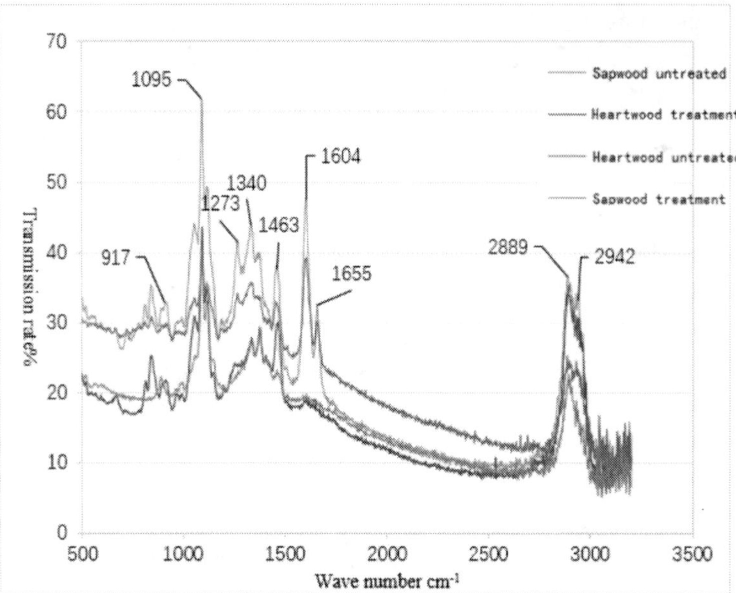


**Figure 1.** Raman spectroscopic local map

As can be seen from the results of Raman spectrum, the chemical composition between heartwood and sapwood is the same but the content is not the same. The heartwood and sapwood after delignification treatment, the characteristic peaks of lignin at 1510 $cm^{-1}$and 1423 $cm^{-1}$ are obviously lower than those before treatment, the characteristic peak representing lignin at 1510 $cm^{-1}$ even disappears, however, there are still differences between heartwood and sapwood after treatment. Presumptive reason: During xylem formation of trees, various types of cells in cambium region and xylem have undergone obvious changes in cell morphology, structure and function. This also leads to different contents of heartwood and sapwood although their chemical compositions are the same. However, the samples treated by sodium chlorite method mainly change the chemical structure of lignin and remove it, with little influence on other chemical components.

### *3.3. Hygroscopicity Analysis*

Figure 2 and Table 3 show the moisture absorption equilibrium moisture content of the treated material and the control material at different relative humidity at room temperature. As can be seen from figure 2, when the humidity of heartwood of fir wood is 30%, there is little difference in moisture absorption performance. However, as the humidity increases, sapwood absorbs more water than heartwood, and after lignin is removed, the moisture absorption capacity of wood cells has little difference at lower humidity (30%), When in an environment of higher humidity (higher than 60%),The acid treated sample has significantly higher hygroscopicity than the untreated sample. But when the humidity reaches 95%, with surprising results, samples treated by sodium chlorite method, the hygroscopic capacity is much higher than that of untreated samples. The equilibrium moisture content is as high as 60%. However, only about 22% of them were untreated. This shows that: Firstly, delignification treatment improves the hygroscopicity of wood, with the increase of relative humidity, the hygroscopic capacity becomes more obvious. Second, when it is not treated, the hygroscopicity of heartwood is lower than that of sapwood. This may be mainly due to the difference in contents between the two [3]. After treatment, the moisture absorption rate of heartwood is higher than that of sapwood, especially at high relative humidity.

**Table 3.** Average hygroscopicity before and after treatment

| Relative humidity | heartwood | | sapwood | |
|---|---|---|---|---|
| | treatment | Contrast | treatment | Contrast |
| 30% | 5.76±0.23 | 5.33±0.23 | 5.80±0.41 | 6.14±0.27 |
| 60% | 10.07±1.48 | 9.46±0.22 | 9.19±0.58 | 10.03±0.22 |
| 85% | 15.34±0.96 | 12.46±0.38 | 14.82±0.54 | 13.79±0.14 |
| 95% | 69.83±6.75 | 21.32±0.64 | 51.79±3.86 | 24.11±0.79 |

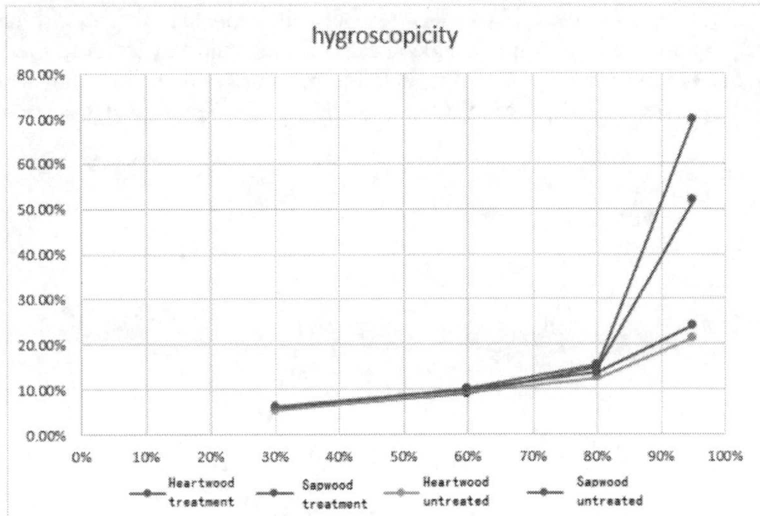


**Figure 2.** Wood hygroscopicity before and after delignification

(1) Evaluation of fitting effect

The fitting effect of the model was evaluated by determining coefficient ($R^2$) and average relative error [6]. Among them:

$$R^2 = 1 - \frac{SSE}{SST} \tag{2}$$

In the formula: SSE—-residual sum of squares; SST—-total sum of squares.

$R^2$ represents the degree of curve fitting between experimental data and model functions. The closer its value is to 1, the better the fitting effect of the curve is [6].

The relationship between wood and water is the basic theory in wood science. It has always been the focus, hotspot and difficulty of research in the industry. However, with the progress of related technologies, Not only does wood as a whole pay attention to water properties and apply mathematical models to analyze water adsorption properties, new methods such as low-field nuclear magnetic resonance and DSC are used to explore the wood-water relationship. Thybring (2013) in the review, it is suggested that the fiber saturation point FSP can even be as high as 40%.

In order to better analyze the moisture absorption of Chinese fir heartwood and sapwood after delignification, Using Hailwood-Horrobin moisture absorption model (H-H model), According to the following formula, the sorption isotherm is constructed to visually represent the changes before and after treatment.

IOP Conf. Series: Materials Science and Engineering **772** (2020) 012109 doi:10.1088/1757-899X/772/1/012109

$$m = \left[\frac{A}{a_w} + B - Ca_w\right]^{-1} \tag{3}$$

$$A = \frac{W}{18} \times \frac{1}{k_2(k_1+1)} \tag{4}$$

$$B = \frac{W}{18} \times \frac{k_1-1}{k_1+1} \tag{5}$$

$$C = \frac{W}{18} \times \frac{k_1 k_2}{k_1+1} \tag{6}$$

In the formula, $a_w$ is relative humidity (%), A, B, C, $k_1$, $k_2$, w are constants, and m is equilibrium moisture content (%).Among them, $k_1$ and $k_2$ respectively represent the equilibrium constants of water-wood system during adsorption of single and multi-molecular layers. And the larger the value, the stronger the adsorption capacity. 18/W reveals the moisture content of wood when the effective sorption group is saturated with hydration water. The larger W, the less water adsorption sites on the polymer surface.

(2) Results Analysis

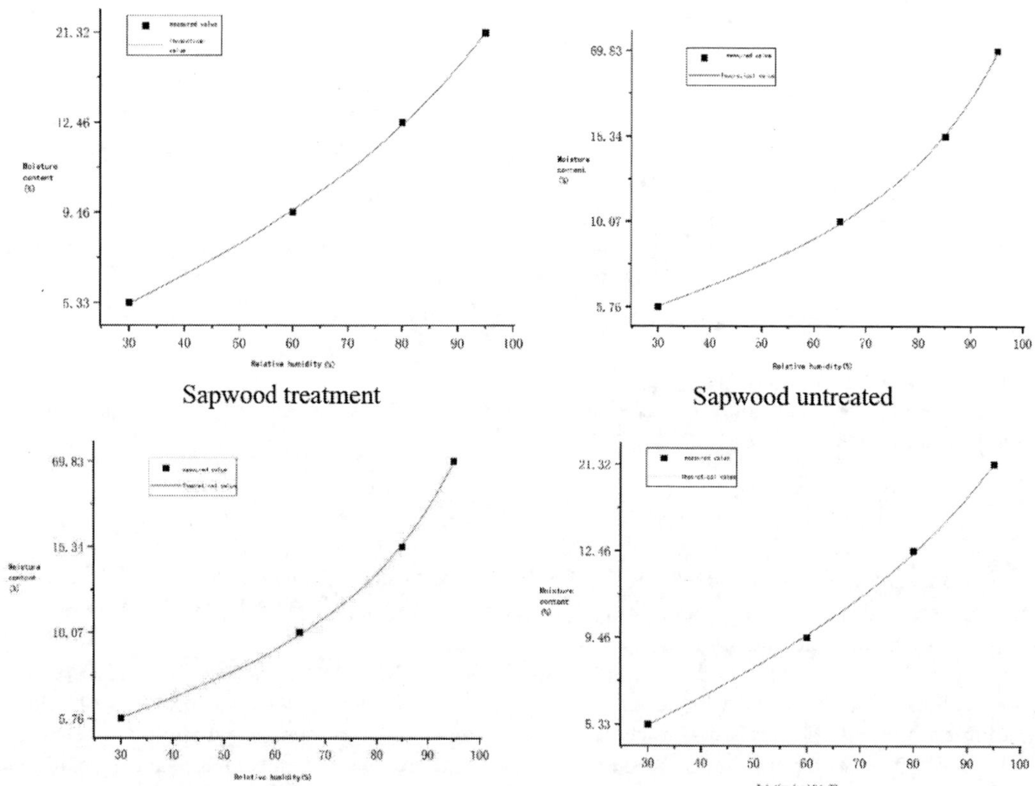


Sapwood treatment Sapwood untreated

Heartwood treatment Heartwood untreated

**Figure 3.** Hygroscopicity isotherm of wood before and after Delignification

**Table 4.** Fitting parameters of curve before and after treatment

| Parameter | Sapwood treatment | Heartwood treatment | Sapwood untreated | Heartwood untreated |
|---|---|---|---|---|
| A | 15.322 | 15.201 | 17.225 | 18.321 |
| B | 0.668 | 0.668 | 0.768 | 0.765 |
| C | 0.006 | 0.006 | 0.004 | 0.003 |
| $k_1$ | 0.048 | 0.048 | 0.045 | 0.037 |
| $k_2$ | 0.763 | 0.764 | 0.471 | 0.378 |
| W | 16.261 | 16.239 | 16.745 | 16.261 |
| 18/W | 1.107 | 1.108 | 1.074 | 1.107 |
| $R^2$ | 0.9743 | 0.9656 | 0.9833 | 0.9835 |

As can be seen from Figure 3 and Table 4, At different relative humidity, Both treated and untreated materials conform to the H-H model of $m = \left[\frac{A}{a_w} + B - Ca_w\right]^{-1}$ . It can be seen from the Figure that the moisture absorption equilibrium moisture content of the treated and untreated materials almost completely coincides with the theoretical value of the H-H model. The fitting degree $R^2$ is above 0.96, The curve type is similar to the typical "S(II)" isothermal adsorption line. This shows that the adsorption process of water on the specimen and control material treated by sodium chlorite method includes both monolayer adsorption and monolayer adsorption, there is also multi-molecular layer adsorption, In addition, in the area where the relative humidity is 80%-95%, The sample treated by sodium chlorite method, The slope of the curve is close to 1. Indicates that within this relative humidity range, after sodium chlorite treatment of part of the water vapor in the specimen gathered into liquid water.

As can be seen from Table 4, the W value after delignification is smaller than that of the control material. The results showed that the number of water adsorption sites in delignified wood was higher than that of control wood. The moisture adsorption content of the specimens treated by sodium chlorite method increased. The values of $k_1$ *and* $k_2$ are positive values, this means that the adsorption capacity of the multi-molecular layer is greater than that of the single-molecular layer [7]. From the analysis in Table 4, The $k_1$ value of the treated specimen is larger than that of the control material. It means that the moisture absorption capacity of the single molecular layer of fast-growing Chinese fir after delignification is stronger than that of the control material. Therefore, after lignin removal, in the lower humidity range (below 30%), the difference between the samples before and after treatment is not obvious. However, in the higher humidity range (above 60%), there are obvious changes. The phenomenon that the treated sample absorbs more water than the sample before treatment. The water adsorption capacity of wood monolayer is mainly due to the existence of free hydroxyl groups, which are relatively hydrophilic. However, the existence of lignin previously hindered the hygroscopic ability of hydrophilic groups. With the removal of lignin, the proportion of effective hydroxyl in the treated specimen increases. Therefore, the moisture absorption capacity of the treated test piece is improved, this also explains the difference in hygroscopicity before and after delignification. In Table 4, it can be found that the $k_2$ value of the specimen treated by sodium chlorite method is greater than that of the control material. This indicates that the adsorption capacity of the multi-molecular layer of the sample after delignification treatment is enhanced compared with that of the control material. To sum up, after delignification treatment, the adsorption capacity of the monolayer of the specimen to water is improved, and at the same time, the secondary adsorption sites in the specimen are also increased.

The hygroscopicity of wood is not only affected by the accessibility of hydroxyl groups, but also by the microscopic pores of wood. Vitas (2019) study shows that delignification treatment can increase the proportion of macro pores, that is, Nakatani (2008) study points out that the pores in lignin are mainly micro pores below 0.6nm. There are three reasons why delignification treatment increases hygroscopicity: First, the removal of hydrophobic substance lignin improves the accessibility of water molecules of wood cell wall substance hemicellulose and cellulose amorphous region. Secondly, the removal of lignin increases the proportion of macro pores in the distribution of wood pore system, making the condensation effect in macro pores significantly increase under the condition of high relative humidity. Thirdly, the quality of wood per unit volume decreases and the content of hygroscopic substances (hydroxyl) in cell wall components increases significantly.

## 4. Conclusion

Fast-growing wood is the main material for wood processing in China. However, due to the limitation of experimental conditions, only the influence of delignification treatment on moisture absorption performance of wood is discussed in this paper. The research results in this paper are of high value for exploring the microstructure and properties of cell walls, providing a better research perspective and some interesting entry points for subsequent research. The main research conclusions are as follows:

1) Through spectral analysis, it can be found that the average weight loss rate after sodium chlorite treatment is 32%. It is presumed that most lignin has been removed, and there may be a small amount of hemicellulose removed, which is confirmed by Raman spectra.

2) hygroscopicity test shows that: When the relative humidity is lower than 30%, the hygroscopicity change is not obvious. With the increase of relative humidity, the exposure of hydrophilic groups, the increase of cell wall microscopic voids and the increase of the mass ratio of hygroscopic substances caused by delignification treatment lead to a significant increase of moisture absorption equilibrium moisture content. Even at 95% high relative humidity, the equilibrium moisture content reaches an amazing 60%. It is believed that it is caused by condensation of microcapillary, but it needs further verification.

3) There are some differences in hygroscopicity between heartwood and sapwood. Combined with H-H model fitting of hygroscopicity data, the reason is inferred: When the humidity is high, the difference of moisture absorption ability between sapwood and heartwood mainly comes from the moisture absorption of multi-molecular layer (adsorbed water and condensed water). However, the moisture absorption capacity of heartwood and sapwood after delignification is obviously enhanced. Mainly because lignin is a relatively hydrophobic chemical component, mainly because lignin is a relatively hydrophobic chemical component. The lignin content directly affects the moisture absorption capacity of wood. After lignin removal, the number of hydrophilic groups in the cell wall increases and the moisture absorption capacity of wood increases.

## References

[1] Zhang Shuangyan. Chemical Components Effect on Mechanical Properties of Wood Cell Wall [D]. Chinese Academy of Forestry, 2011.

[2] Ding Tao, Wang Changjun, Peng Wenwen. A theoretical study of moisture sorpition behavior of heat-treated pin wood using Raman spectroscopic analysis [J]. Journal Forestry Engineering, 2016, 1(5): 15-19.

[3] Song Kunlin.Changes of Structure and Performance of Cell Walls during the Transformation from Sapwood to Heartwood of Cunninghamia lanceolate [D]. Chinese Academy of Forestry, 2012.

[4] Feng Xiangyuan, Cheng Lifeng, Duan Shengwen. Composition of monosaccharides in hydrolyzed ramie detected by high performance liquid chromatography with pre-column derivatization [J]. Journal of Textile Research, 2015-12.

[5] Wei Lu, Sun Zhaobin, Chen Fengyi. Influence of pretreatment on shrinkage-swelling characteristic for black walnut woood [J]. Journal of Central South University of Forestry & Technology, 2017, 37(3): 104-110.

[6] Liu Jijian,Yu Jianfang,Wang Ximing. Study on Simulation of Wood Moisture Absorption Based on Finite Difference Method [J]. Wood Processing Machinery, 2016, 27(03): 34-37+8.

[7] Jiang Jiali, Lǚ Jianxion.Hydrothermal Efffects on Dynamic Viscoelastic Properties of Wood.Scientia Sillvae Sinicae, 2014, 50(12), 104-111.

# Microstructure characterization and micro-hardness of Fe-9Cr ODS alloy produced by laser powder bed fusion

**Yingnan Shi, Zheng Lu * and Rui Xie**

Key Laboratory for Anisotropy and Texture of Materials (Ministry of Education), School of Materials Science and Engineering, Northeastern University, Shenyang 110819, P. R. China

*Corresponding author e-mail: luz@atm.neu.edu.cn

**Abstract.** Oxide dispersion strengthened (ODS) alloys have been considered as the promising candidates for advance Generation-IV fission reactor application due to excellent radiation tolerance and high temperature mechanical properties but traditional manufacturing process is complex. A 9Cr-ODS alloy with nominal composition of Fe-9Cr-1.5W-0.3Ti-0.3$Y_2O_3$ (wt %) was produced by laser powder bed fusion (LPBF) which is a laser additive manufacturing technique. The microstructure and nanoscale oxide particles were characterized by using optical microscope (OM), electron backscattered diffraction (EBSD) and transmission electron microscopy (TEM). The micro-hardness was tested by hardness tester. The results showed that the equiaxed grains with an average size of 3.8 μm were obtained. The dispersed oxide particles with 10-20 nm were observed in LPBFed 9Cr-ODS alloy. The micro-hardness of built sample was about 460 HV.

## 1. Introduction

The harsh work environments of the advanced Generation-IV fusion reactors require structural materials with outstanding performances. Nanostructured oxide dispersion strengthened (ODS) alloys present excellent irradiation resistance and mechanical properties because of the high-density of nano-sized oxides, which is considered as one of the most promising candidate materials for fuel cladding in fast reactors [1-3]. At present, powder metallurgy (PM, such as mechanical alloy + high temperature consolidation) is a conventional process to produce ODS alloys [4-7]. Although the powder metallurgy is a ripe and effective method to produce ODS alloys with high-density of nano-sized oxide particles and outstanding performances, the long manufacturing period, high cost and less flexibility are still insurmountable.

Recently, developments in laser additive manufacturing (LAM) have offered a potential method for structural materials application in the future advanced reactors. The LAM has unique advantages compared to the conventional PM such as, short manufacturing cycle, low cost and great flexibility. The ODS alloys with high chromium content produced by LAM have been explored in recent years. T. Boegelein et al have successfully manufactured the high chromium content and Al-containing ODS alloy named PM2000 by selective laser melting (SLM) process [8, 9]. The fine precipitates with high-density were obtained. A Fe-18Cr Al-free ODS alloy which was manufactured by electron beam selective melting (EBSM) presented high-density precipitates and tensile strength of 1226 MPa [10].

Content from this work may be used under the terms of the Creative Commons Attribution 3.0 licence. Any further distribution of this work must maintain attribution to the author(s) and the title of the work, journal citation and DOI.
Published under licence by IOP Publishing Ltd

However, the aging embrittlement and irradiation embrittlement are inevitable for the ODS alloys with high chromium content [11]. Therefore, the low-Cr (9-12Cr) ODS alloys are considered for application in reactors with higher temperature and irradiation dose.

In this study, a 9Cr-ODS alloy was produced by laser powder bed fusion (LPBF) which is a powder-bed metal printing system. The microstructure and nano-sized particles were characterized and examined carefully.

## 2. Experimental

The high-purity elemental powders (Fe, Cr, W, Ti) and nano-scale $Y_2O_3$ powder were mixed according to the ratio of Fe-9Cr-1.5W-0.3Ti-0.3$Y_2O_3$ (wt%). The mechanical alloying (MA) was carried out in Fritsch P-5 planetary mill under high purity argon (99.99%) atmosphere for 50 hours. Detailed MA parameters for producing the supersaturated alloy powder are detailed in a previous publication [12].

The LPBF machine was a self-assembly apparatus and mainly composed of $CO_2$ laser, numerical control system, coaxial powder feeder and inert gas protection system. Scanning strategy was zigzag scan vectors. The substrate was AISI 1045 steel. The schematic of process is shown in Fig.1. In this process a thin layer of powder is first deposited onto the substrate. The laser beam is scanned onto the powder bed to form solidified layers. Scanning path is generated by the CAD and sliced software. The parts can be built up layer-by-layer into a net-shape. The process parameters used in this study are listed in Table 1. The direction of deposition is along the Z direction, X direction is the scanning direction.

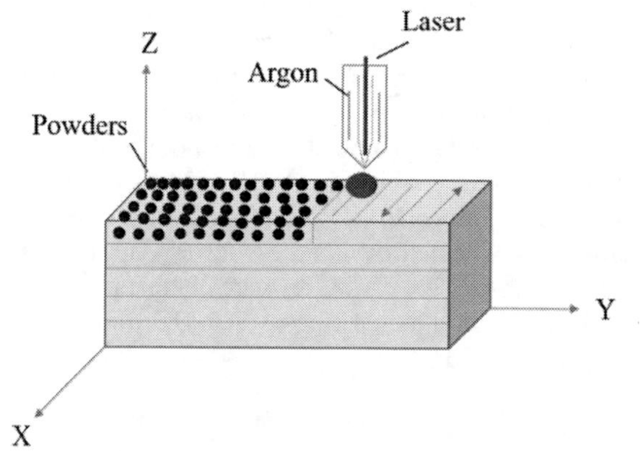


**Figure 1.** The schematic of process

**Table 1.** The process parameters in this study

| Laser power | Scanning speed | Layer thickness | Hatch distance |
|---|---|---|---|
| 1500 W | 6 mm/s | 0.8 mm | 1.5mm |

The microstructure of the YOZ section was characterized by an optical microscope (OM, OLYMPUS GX71) and a JSM-7001F scanning electron microscope (SEM) with an electron backscattered diffraction (EBSD). The sample was etched by using 5% nital solution for observation of OM. The EBSD specimens were prepared by mechanical ground and electrolytic polishing. The electrolytic polishing was performed in 10% $HClO_4$ + 90% $C_2H_5OH$ solution with 20 V voltage at room temperature. The microstructure was further investigated using a Titan $G^2$60-300 transmission electron microscope (TEM). The TEM thin foil specimens were prepared by a JIB-4600F focused ion beam (FIB). The Vickers micro-hardness tests were carried out by a 41MVDtm digital micro Vickers hardness tester. 5 indentation tests were made for the built sample to obtain the average value of hardness. The load was 0.98 N and loading time was 10 s.

## 3. Results and discussion

The Fig.2 (a) shows the OM image of LPBFed 9Cr-ODS alloy in vertical section (YOZ). The equiaxed grains are observed. The EBSD also presents the similar feature (see the Fig.2 (b)) and there is a fine grain size distribution (Fig. 2 (c)). More than 90% of grains are less 5μm and a main peak at ~0.5-1.0μm is obvious. The fine grains are attribute to the higher cooling rate. The cooling rate can reach up about $10^5$-$10^6$ K/s during the laser additive manufacturing [13]. The high cooling rate can result in the rapid solidification as well as inhibits the growth of grains. The average grain size of 3.8 μm is smaller than the SLMed PM2000 (~20μm) and EBSMed ODS alloy (~5-10μm) [8, 10]. This could be associated with the differences of alloy composition and LAM process.

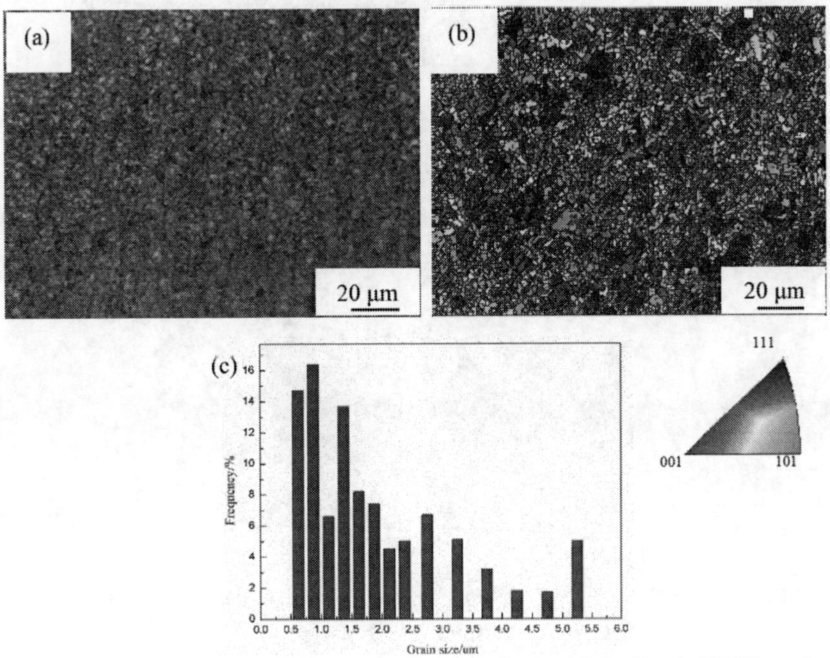


**Figure 2.** The (a) OM and (b) EBSD images of LPBFed 9Cr-ODS alloy in YOZ section, (c) the corresponding size distribution of grains.

The lath martensites are observed in the LPBF built 9Cr-ODS alloy (see the Fig.3). A typical feature of 9Cr-ODS alloy is that the phase transformation from ferrite/martensite (α) to austenite (γ) can be controlled by heat treatment [14]. Therefore, some lath martensites formed during the LPBF because of the higher cooling rate.

In order to clarify the morphology of oxide particles, the high angle annular dark field (HAADF) images of STEM analysis were conducted (Fig. 4). These oxide particles exhibit darker contrast than the matrix in the HAADF image, which are attribute to lower mas compared to the matrix. A large amount of oxide particles with ~10-20 nm are found and these particles are homogeneously dispersed in the matrix.

Cr, Y and Ti et al elements were supersaturated solution in Fe matrix after mechanical alloying. During the laser additive manufacturing, the loss of Y and Ti was found due to some micron-sized oxides formed in the melting pool [9, 15]. And these oxides in melting pool would float and agglomerate driven due to continuous stirring and lower mass density than ODS alloy matrix. However, Y, Ti et al elements diffuse slowly in the α-Fe matrix [16]. Thus, a certain amount of alloy elements appears to still be atomic supersaturated solution in matrix. The finer precipitates (seen in the Fig. 4) may form in the solidified matrix after deposition during the repeated heating cycles caused by laser powder bed fusion.

**Figure 3.** The TEM micrograph of as LPBF built 9Cr-ODS alloy

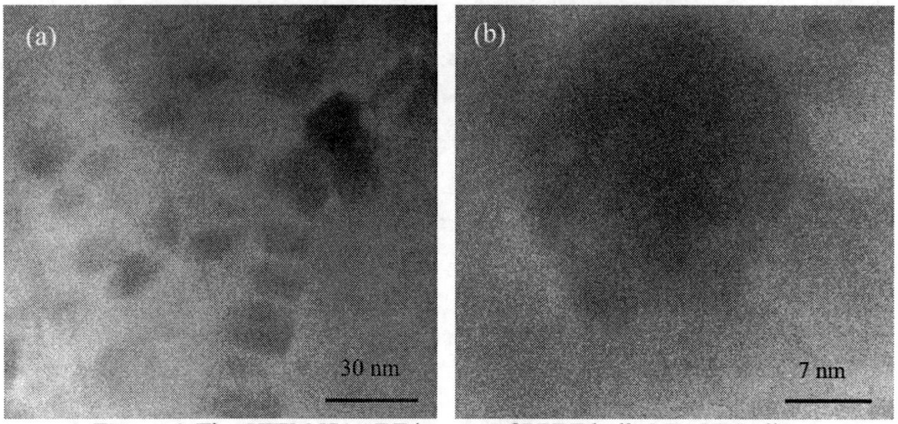


**Figure 4.** The STEM HAADF images of LPBF built 9Cr-ODS alloy

The Vickers micro-hardness of built sample is about 463$\pm$15 HV which is similar with the 9Cr-ODS Fe-matrix alloy produced by conventional process [17].

## 4. Conclusion

The 9Cr-ODS alloy was successfully produced by laser powder bed fusion. The characterization of grain morphology and nano-sized oxide particles of LPBFed 9Cr-ODS alloy was studied. The fine grains (average size about 3.8 μm) are formed in built 9Cr-ODS alloy. Some lath martensites are introduced due to the high cooling rate. Dispersed nanoscale particles (~10-20 μm) are obtained in the LPBFed 9Cr-ODS alloy. The micro-hardness of this sample is up ~460 HV.

**Acknowledgments**
This work was financially supported by the National Natural Science Foundation of China (51871044 and 51471049).

## References

[1] G. R. Odette, M. J. Alinger, B. D. Wirth, Recent Developments in Irradiation-Resistant Steels, Annu. Rev. Mater. Res. 38 (2008) 471-503.
[2] S. Ukai, T. Nishida, H. Okada, et al, Development of Oxide Dispersion Strengthened Ferritic Steels for FBR Core Application, J. Nucl. Sci. Technol. 34 (1997) 256-263.
[3] R. L. Klueh, J. P. Shingledecker, R. W. Swindeman, et al, Oxide dispersion-strengthened steels: A comparison of some commercial and experimental alloys, J. Nucl. Mater. 341 (2005) 103-114.
[4] M. J. Alinger, G. R. Odette, D. T. Hoelzer, On the role of alloy composition and processing

parameters in nanocluster formation and dispersion strengthening in nanostructured ferritic alloys, Acta. Mater. 57 (2009) 392-406.
[5] S. Ukai, S. Mizuta, T. Yoshitake, et al, Tube manufacturing and characterization of oxide dispersion strengthened ferritic steels, J. Nucl. Mater. 283-287 (2000) 702-706.
[6] C. Y. Lu, Z. Lu, C. M. Liu, Microstructure of nano-structrued ODS CLAM steel by mechanical alloying and hot isostatic pressing, J. Nucl. Mater. 442 (2013) S148-S152.
[7] Z. Y. Li, Z. Lu, R. Xie, et al, Effect of spark plasma sintering temperature on microstrucutre and mechanical properties of 14Cr-ODS ferritic steels, Mater. Sci. Eng. A 660 (2016) 52-60.
[8] T. Boegelein, S. N. Dryepondt, A. Pandey, et al, Mechanical response and deformation mechanisms of ferritic oxide dispersion strengthened steel structures produced by selective laser melting, Acta. Mater. 87 (2015) 201-215.
[9] T. Boegelein, E. Louvis, K. Dawson, et al, Characterisation of a complex thin walled strucutre fabricated by selective laser melting using a ferritic oxide dispersion strengthened steel, Mater. Char. 112 (2016) 30-40.
[10] R. Gao, L, Zeng, Q. F. Fang, et al, Characterization of oxide dispersion strengthened ferritic steel fabricated by electron beam selective melting, Mater. Des. 89 (2016) 1171-1180.
[11] A. Kimura, R. Kasada, N. Iwata, et al, Development of Al added high-Cr ODS steels for fuel cladding of next generation nuclear systems, J. Nucl. Mater. 417 (2011) 176-179.
[12] R. Xie, Z. Lu, C. Lu, et al, Microstructure and mechanical properties of 9Cr oxide dispersion strengthened steel produced by spark plasma sintering, Fusion. Eng. Des. 115 (2017) 67-73.
[13] K. Yang, P. Rometsch, C. Davies, et al, Effect of heat treatment on the microstructure and anisotropy in mechanical properties of A357 alloy produced by selective laser melting, Mater. Des. 154 (2018) 275-290.
[14] S. Ukai, S. Mizuta, M. Fujiwara, et al, Development of 9Cr-ODS martensitic steel claddings for fuel pins by means of ferritic to austenite phase transformation, J. Nucl. Sci. Technol. 39 (2002) 778-788.
[15] Y. Shi, Z. Lu, H. Xu, et al, Microstrucutre characterization and mechanical properties of laser additive manufactured oxide dispersion strengthened Fe-9Cr alloy, J. Alloys. Compd. 791 (2019) 121-133.
[16] C. Hin, B. Wirth, J. Neaton, Formation of Y2O3 nanoclusters in nanostrucutred ferritic alloys during isothermal and anisothermal heat treatment: A Kinetic Monte Carlo study, Phys. Rev. B. 80 (2009) 134118.
[17] H. R. Z. Sandim, R. A. Renzetti, A. F. Padilha, et al, Annealing behavior of ferritic-martensitic 9%Cr-ODS-Eurofer steel, Mater. Sci. Eng. A. 527 (2010) 3602-3608.

# Numerical Simulation Study on Bolted Joints of Steel-wood Structures

**Yang Zheng, Chunmei Xu ***

School of Civil engineering, Southwest Forestry University, Kunming, China

*Corresponding author e-mail: 2219139473@qq.com

**Abstract.** As one of the most widely used rigid connections, bolted joints have the advantages of convenient loading and unloading, increased prestressing to prevent loosening, and no change in material composition at the joints. In this paper, six sets of test pieces are designed, and the monotonic loading is performed. The bearing capacity of the joints under the two parameters of bolt margin and wood characteristics is analyzed, and compared with the theoretical value, the reliability of the finite element simulation results is increased.

## 1. Introduction

The steel-wood structure is a new type of structure that effectively combines the steel structure and the wood structure. It has the durability of the traditional steel structure and the good compressive and compressive performance of the wooden structure. It is easy to be applied to the support rod of the large-span structure with Pieces, light weight, environmental protection, low energy consumption, insulation and corrosion resistance [1].

At present, many researchers have conducted in-depth research on the joint test of steel-wood structural bolting: Li Feng [2] found that the screw force in the screw group is uneven, and the screw near the centroid of the screw group can resist seismic energy consumption; Enzhen Zhu [3] proposed a calculation formula for the bearing capacity of bolted joints by carrying out the bearing pressure test and bolting performance test of the larch and Pinus sylvestris var. mongolica; Kezhen Liu et al [4] proposed the force failure mode includes the tensile failure of the steel plate and the compression of the wall of the hole.

In this paper, six sets of test pieces are designed, and the monotonic loading is carried out. The bearing capacity of the joints under the two parameters of the bolt margin (45mm, 52.5mm, 60mm) and the wood characteristics (red pine, glulam) are analyzed. And compared with the theoretical value, increase the reliability of the finite element simulation results.

## 2. Finite element model establishment

### *2.1. Material parameter input*

(1) The steel is made of Q235. The volume of H-shaped steel is 300mm*300mm*18mm*24mm, the yield strength is 235Mpa, the Poisson's ratio is 0.3, the elastic modulus is 210Gpa, and the tangential modulus is 21Gpa.

Content from this work may be used under the terms of the Creative Commons Attribution 3.0 licence. Any further distribution of this work must maintain attribution to the author(s) and the title of the work, journal citation and DOI.
Published under licence by IOP Publishing Ltd

(2) Wood: The test uses Korean pine and glulam, both of which are anisotropic materials. The linear elastic parameters of the materials are based on the experimental data in [5], as shown in Table 1 below:

**Table 1.** Wood line elastic parameter.

| Numble | $E_L$ (Mpa) | $E_R$ (Mpa) | $E_T$ (Mpa) | $\mu_{RT}$ | $\mu_{LR}$ | $\mu_{LT}$ | $G_{RT}$ (Mpa) | $G_{LR}$ (Mpa) | $G_{LT}$ (Mpa) |
|---|---|---|---|---|---|---|---|---|---|
| Korean pine | 13743 | 434 | 434 | 0.341 | 0.341 | 0.528 | 935 | 765 | 935 |
| Glued wood | 10000 | 497.51 | 497.51 | 0.337 | 0.337 | 0.372 | 675 | 675 | 675 |

Note: The yield strength of glulam is 25Mpa, the tangent modulus is 2.5Mpa, and the density of Korean pine is 0.42g/cm3.

(3) Bolt: 30mm bolt is used, and the bolt strength grade is 10.9.

### *2.2. Test design*

This test design 6 sets of test pieces, Table 2 is the sample parameter table: where L1, L2, L3 are the influence of bolt margin on the joint connection when the wood is all red pine; L4, L5, L6 are when When the wood is glulam, the influence of bolt margin on the joint connection; when L1, L4 are 45mm, the influence of wood characteristics on the joint connection; L2, L5 when the bolt margin is 52.5mm, , the influence of wood characteristics on the joint connection; L3, L6 when the bolt margin is 60mm, the influence of wood characteristics on the joint connection.

**Table 2.** Sample parameter table.

| Numble | H-beam length (mm) | Board size (mm) | Wood properties (Mpa) | Bolt margin (mm) | Bolt diameter (mm) |
|---|---|---|---|---|---|
| L1 | 1500 | 1500*300*150 | Korean pine | 45 | 30 |
| L2 | 1500 | 1500*300*150 | Korean pine | 52.5 | 30 |
| L3 | 1500 | 1500*300*150 | Korean pine | 60 | 30 |
| L4 | 1500 | 1500*300*150 | Glued wood | 45 | 30 |
| L5 | 1500 | 1500*300*150 | Glued wood | 52.5 | 30 |
| L6 | 1500 | 1500*300*150 | Glued wood | 60 | 30 |

### *2.3. Modeling method and unit selection*

This experiment uses the method of symmetrical analysis, using the workbench software to make the model. The default unit used is solid186, and only 1/4 of the model is used, which facilitates the subsequent calculations to save computational complexity, making the results of the model calculation highly symmetrical. The constitutive model of the steel adopts a bilinear equivalent strengthening model.

Using 3D solid modeling, the model is shown in Figure 1. The width direction is the X direction, the length direction is the Y direction, and the thickness direction is the Z direction. All the freedom is constrained at the bottom of the test piece.

Degree.

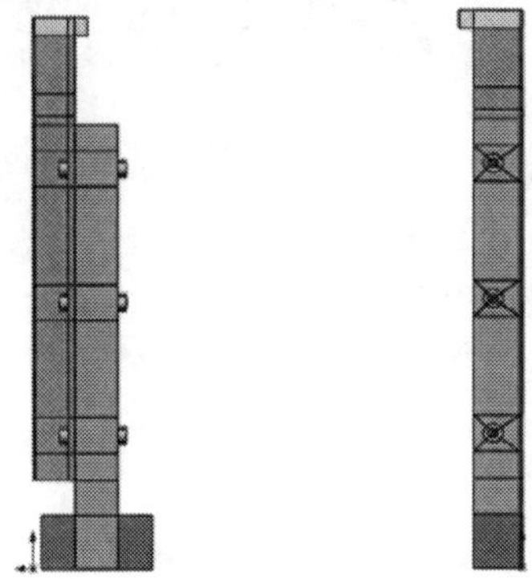

**Figure 1.** (a) 1/4 model front view (b) 1/4 model left view

## 3. Analysis settings

### *3.1. Contact and loading method*

There are five types of contact surfaces:

(1) The contact between the wood and the steel plate is due to the large difference in stiffness between the two, the H-shaped steel with high rigidity is taken as the target surface, the wood with low rigidity is used as the contact surface, and the asymmetric contact is adopted, and the friction coefficient is 0.1.

(2) The contact between the bolt column and the steel plate, both of which are Q235 steel, directly adopts symmetrical contact and is easy to converge.

(3) The contact between the nut and the steel has a small difference in stiffness and uses symmetrical contact.

(4) The contact between the nut and the wood board has a large difference in stiffness. The asymmetric contact is used, the bolt is used as the target surface, and the wood board is used as the contact surface.

(5) The contact between the bolt column and the wood hole has a large difference in stiffness. The asymmetric contact is adopted, the bolt is used as the target surface, and the wood board is used as the contact surface.

The loading method is applying a load of 250KN downward to the top of the 1/4 model, and the current load step end time is 1000s, that is, loading 1000KN within 1000s.

### *3.2. Meshing*

For the bolt, that is, the edge portion of the contact surface adopts a 10 mm mesh, and the non-contact surface edge adopts a 40 mm mesh, which can greatly reduce the number of nodes, can simplify the calculation process, and make the calculation result more accurate.

## 4. Numerical simulation results

### *4.1. Load-displacement curve*

Figure 2 shows the load-displacement curves of the six groups of tests. The horizontal axis is the displacement and the vertical axis represents the load. It can be seen from the figure 2 that the curve is smooth and the node is actually in the initial stage of loading. The force loading is not enough, the displacement is not obvious, because the time is proportional to the load, at the beginning, the time does not move, so the load does not change, only a slight displacement. Because the grid has a discrete effect, there is a gap with the bolt hole, there is no displacement without applying pressure, and then the load is increased, and the displacement is also greatly increased.

The woods of the L1, L2, and L3 groups are all red pine, and the bolt margins are 45 mm, 52.5 mm, and 60 mm, respectively. In the elastic phase, as the bolt margin increases, the corresponding load increment decreases as the same displacement increases.

The wood of L4, L5 and L6 groups are all glued wood, and the bolt margins are 45mm, 52.5mm, 60 mm, respectively. In the elastic phase, as the bolt and bolt margin increases, the corresponding load increment increases as the same displacement increases.

The bolts of the L1 and L4 groups are all 45mm. The wood is made of Korean pine and glulam. In the elastic phase, when the same displacement is increased, the corresponding load increment is reduced.

The bolt spacing of L2 and L5 groups is 52.5mm, and the wood is taken from red pine and glulam. In the elastic phase, when the same displacement is increased, the corresponding load increment is unchanged.

The bolts of the L3 and L6 groups are all 60mm. The wood is made of Korean pine and glulam. In the elastic phase, when the same displacement is increased, the corresponding load increment is unchanged.

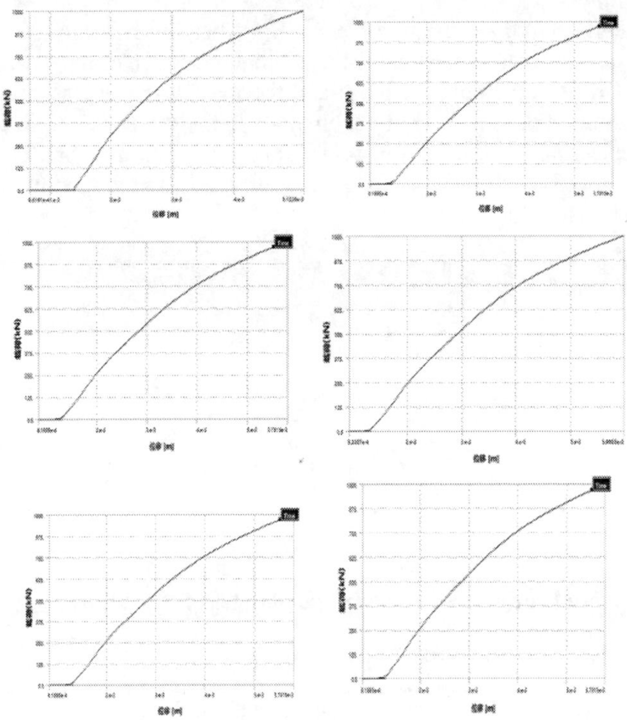

**Figure 2.** Load displacement curve

*4.2. Stress cloud*

From the figure 3, it can be seen that the wood is an anisotropic material with a large elastic modulus in the longitudinal direction of the fiber, mainly by the length of the fiber supporting the bolt, and the transverse fiber is weak. The inner side of the bolt on the wood has large stress and the upper stress is small; the upper side has large stress and the lower stress is small because the H-beam is pressed downward, the bolt is pressed downward, the lower stress is large, and the upper stress is small; the outer side of the bolt The bolt head is pressed upwards, the upper stress is large, and the stress below is small.

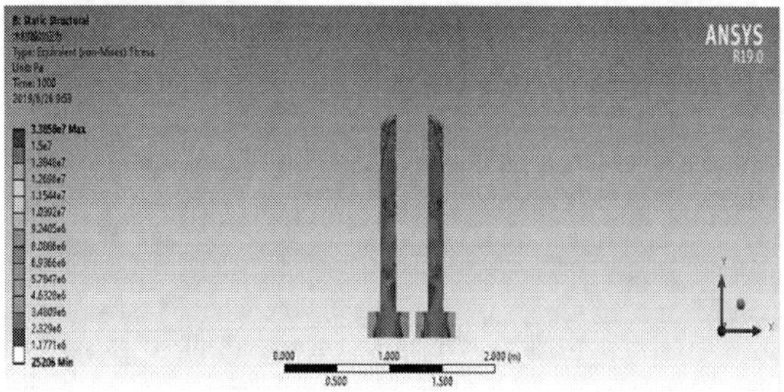


**Figure 3.** Equivalent stress map of wood

### *4.3. Load capacity analysis*

According to the American Wood Structure Design Specification NDS [6], the yield strength of the multi-bolt joint specimen is taken as the load value at the intersection of the horizontal offset line of the initial stiffness of 5% d (d is the bolt diameter) and the load-displacement curve, ASTM D 176 [ 7] stipulated that the maximum deformation of the pin node in the test does not exceed 15mm. This paper only explores the damage of the bolt joint in 1000KN, and the deformation does not exceed 6mm, so the ultimate bearing capacity is 1000KN. The method of yield strength and ultimate load in the paper by Chao Fang [8] is shown in Figure 4 below:

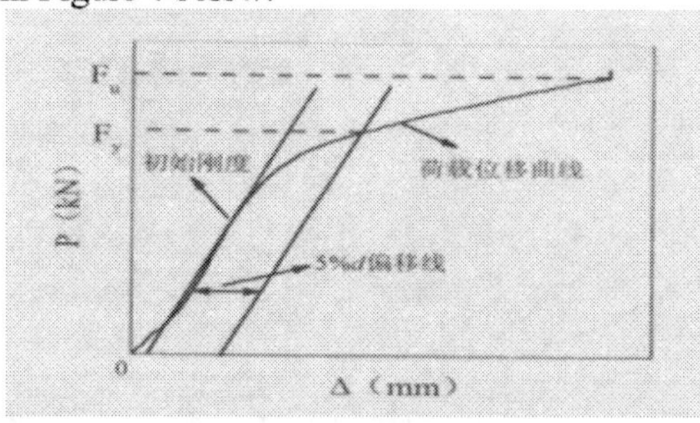


**Figure 4.** Method for determining yield strength and ultimate load

The static analysis results of the test pieces of the joint test are as follows:

**Table 3.** Carrying capacity finite element analysis results

| Numble | Yield strength (KN) | Ultimate strength (KN) | Initial stiffness (KN/mm) |
|---|---|---|---|
| $L_1$ | 956.63 | 1000 | 543.48 |
| $L_2$ | 776.13 | 1000 | 403.23 |
| $L_3$ | 776.13 | 1000 | 403.23 |
| $L_4$ | 865.63 | 1000 | 304.88 |
| $L_6$ | 776.13 | 1000 | 403.23 |
| $L_6$ | 776.13 | 1000 | 403.23 |

## 5. Numerical simulation results compared with theoretical values

### *5.1. Calculation of bolt bearing capacity by "wood structure design code"*

The "Code for Design of Wood Structures" (GB-50005-2003) [9] stipulates that the design bearing capacity of each shear plane of a single bolt connection is determined as a formula (1).

$$N_V = K_V d^2 \sqrt{f_c} \tag{1}$$

Wherein, $N_V$ the bearing capacity design value (KN) of each shear surface of the bolt
$K_V$ Bearing capacity coefficient of bolt connection, check the table, the value is 5
d Bolt diameter, take 30mm
fc Wood design value of compressive bearing capacity (MPA)

From the above formula, it can be seen that when the bolt diameter and the bearing capacity coefficient are constant, the bolt bearing capacity is only related to the design value of the grain compressive bearing capacity of the wood. According to the test results of Jia You [10], the design value of the bearing capacity of Pinus koraiensis is 26.21-30.21 MPA, and the value of this paper is 27 MPA. According to the technical specification of the glulam structure [11], the design value of the bearing capacity of the glulam TCT24 is 14.8MPA.

### *5.2. Numerical simulation results and calculated values*

The finite element simulation results and the theoretical value results are as shown in Table 4. The 5% yield strength is 1.04-1.6 times larger than the value of the wood structure design specification. The Chinese wood structure design specification has a certain conservative estimation range, so the finite element the simulation results are relatively reasonable.

**Table 4.** Comparison table between finite element simulation results and theoretical value results

| Numble | 5%Yield load(KN) | Ultimate load (KN) | Wood structure specification (KN/mm) |
|---|---|---|---|
| $L_1$ | 956.63 | 1000 | 739.43 |
| $L_2$ | 776.13 | 1000 | 739.43 |
| $L_3$ | 776.13 | 1000 | 739.43 |
| $L_4$ | 865.63 | 1000 | 547.49 |
| $L_6$ | 776.13 | 1000 | 547.49 |
| $L_6$ | 776.13 | 1000 | 547.49 |

## 6. Conclusion

In this paper, the bearing capacity of the pine and glulam at 45mm, 52.5mm and 60mm bolt margins is studied and compared with the theoretical value. The following conclusions were drawn:

(1) The wood mainly supports the bolts in the direction of the longitudinal fibers.

(2) The yield strength of Pinus koraiensis decreases with the increase of bolt margin; the yield strength of glulam decreases with the increase of bolt margin. At a bolt margin of 45 mm, the yield strength of Korean pine is greater than that of glulam. At 52.5 mm and 60 mm bolt margins, wood properties have little effect on yield strength.

(3) The ultimate strength has little effect on bolt margins and wood properties.

(4) The initial stiffness of Korean pine decreases with the increase of bolt margin; the initial stiffness of cemented wood increases with the increase of bolt margin. At a bolt margin of 45 mm, the initial stiffness of the pine is greater than that of the glulam. At 52.5 mm and 60 mm bolt margins, wood properties have little effect on initial stiffness.

## References

[1] Feng Qian et al. Steel-wood mixed string shell structure swimming pool practice-Shanghai Chongming Sports Training Base swimming pool design. [J] Architectural Journal. 2017. Special Feature: Towards the performance integration of architecture and structural design: 40-43.

[2] Li Feng et al. Experimental study on bearing capacity of bolt-joint joints of glued bamboo structures. [J] Journal of Building Structures. 2014.35(4): 230-235.

[3] En chun Zhu et al.Experimental study on bolting connection of timber structure and determination of bearing capacity design value. [J] Journal of Building Structures.2016. 37(4): 54-63

[4] Kezhen Liu et al.Research progress in bolting of wood structures[J].Wood Processing Machinery.2011(1): 40-42.
[5] Kun Kang et al.Finite element analysis of the influence of the gap between the two layers on the bearing capacity of the dovetail joints of ancient buildings. [J] Chinese scientific papers.2016.11(1): 38-42
[6] National Forest Products Association.NDS 1997 National design specification for wood construction [s]. Washington.DC.: National Forest Products Association
[7] Amercian Society of Mechanical Engineering.ASTM D1761-88 standard test methods for mechanical fasteners in wood [s]. Washington, USA, 2008.
[8] Chao Fang.Study on static performance of steel-wood composite connection [D]. Beijing: Beijing Jiaotong University, 2015.
[9] GB50005-2003. Wood structure design specification [S]. Beijing: China Building Industry Press, 2003.
[10] Jia You et al.Experimental study on bearing strength of Yujiahonghong pinning groove. [J]Wood processing machinery, 2018, 29(2): 24-28.
[11] GB50708-2012. Technical specification for glued wood structure [S]. Beijing: China Building Industry Press, 2012.

# Effect of Post-weld Heat Treatment on Microstructure and Mechanical Properties of Al-Zn-Mg-Cu Alloy Welded Joint

**Xiang Zhang[1, 2, *], Dejie Xie[1, 2] and Bo Chen[1, 2]**

[1] Jiangsu Xuzhou Construction Machinery Research Institute, Xuzhou Construction Machinery Group, Xuzhou, Jiangsu, China
[2] State Key Laboratory of Intelligent Manufacturing of Advanced Construction Machinery, Xuzhou Construction Machinery Group, Xuzhou, Jiangsu, China

*Corresponding author e-mail: vip525@163.com

**Abstract.** Metal inert-gas welding (MIG) test was carried out using ER5183 welding wire as filler metal and Al-4.8Zn-1.6Mg-0.16Cu alloy profile as base material. Different post-weld heat treatment processes were adopted for Al-4.8Zn-1.6Mg-0.16Cu alloy welded joints. The effects of different post-weld heat treatment processes on the microstructures and mechanical properties of welded joints were studied by means of hardness test, microstructure observation and scanning electron microscopy analysis. The results show that the mechanical properties of welded joints can be improved by artificial aging treatment after welding. At the same time, the comprehensive mechanical properties of re-solution and aging after welding will be significantly improved. The inhomogeneity of welded joints and the distribution of strengthening phases will also be improved, the softening of heat affected zone will almost disappear, and the strength of welded joints will be restored.

## 1. Introduction

With the rapid development of rail transit industry, energy saving, emission reduction and high-speed safety have become the focus and development direction of major rail transit enterprises and the best solution to this problem is applications of lightweight car body materials [1-3]. Medium and high strength Al-Zn-Mg-Cu alloys have excellent mechanical and technological properties after solution and aging (usually in T6 state) [4-7], so they are ideal materials for lightweight body. Aluminum alloy body manufacturing is mostly based on MIG or TIG arc welding with large heat input [8-11]. The joint softening after welding is serious, which directly affects vehicle safety and service life. Due to the influence of welding heat input, the weld is prone to second phase segregation, intergranular corrosion and stress corrosion [12]. Usually, the welding performance can be improved by post-weld heat treatment [13, 14].

Therefore, the effect of different post-weld heat treatment processes on the microstructures and mechanical properties of MIG welded joints is studied in this paper. These studies will provide theoretical basis for expanding the engineering application of Al-Zn-Mg-Cu alloys, and have important significance for promoting the development, design and in-depth study of post-weld heat treatment technology of Al-Zn-Mg-Cu alloys.

Content from this work may be used under the terms of the Creative Commons Attribution 3.0 licence. Any further distribution of this work must maintain attribution to the author(s) and the title of the work, journal citation and DOI.
Published under licence by IOP Publishing Ltd

IOP Conf. Series: Materials Science and Engineering **772** (2020) 012112 doi:10.1088/1757-899X/772/1/012112

## 2. Experimental materials and methods

The experimental material is Al-4.8Zn-1.6Mg-0.16Cu alloy. The raw material is melted in smelting furnace and semi-continuous casting machine is used for semi-continuous casting. The Al-Zn-Mg-Cu alloy profile (75×45×5mm) is extruded by 2000 t extruder. T6-treated alloy is obtained by on-line water-cooled quenching. The 100mm profile parallel to the extrusion direction was intercepted as the welding base material, and the filler wire is ER5183 with a 1.2 mm diameter. The chemical composition of base metal and welding wire was determined by ARL 3460 spark direct reading spectrometer. The result of the base metal is: Zn 4.8%, Mg 1.6%, Cu 0.16%, Si 0.09%, Fe 0.15%, Mn 0.41%, Cr 0.12%, and the residual amount is Al. The result of the filler wire is: Zn 0.11%, Mg 4.88%, Cu 0.03%, Si 0.05%, Fe 0.13%, Mn 0.52%, Cr 0.11%, and the residual amount is Al.

The weld joint type and groove size (t1=t2=5mm, =70°, b=0~0.5mm) are shown in Fig. 1. Before welding, pneumatic steel wire brush was used to polish the welded parts on the base metal surface to expose the metallic luster. MIG welding was carried out on the base metal by using the Fornis TPS 5000 welding machine. The welding parameters are as follows: the shielding gas is high purity argon (purity > 99.99%); welding current is 125A; welding speed is 60cm/min; argon flow rate is 20L/min.

The weld joint type and groove size (t1=t2=5mm, =70°, b=0~0.5mm) are shown in Figure 1. Before welding, pneumatic steel wire brush was used to polish the welded parts on the base metal surface to expose the metallic luster. MIG welding was carried out on the base metal by using the Fornis TPS 5000 welding machine. The welding parameters are as follows: shielding gas is high purity argon with purity > 99.99%; the welding current is 125A; the welding speed is 60cm/min; and the argon flow rate is 20L/min.

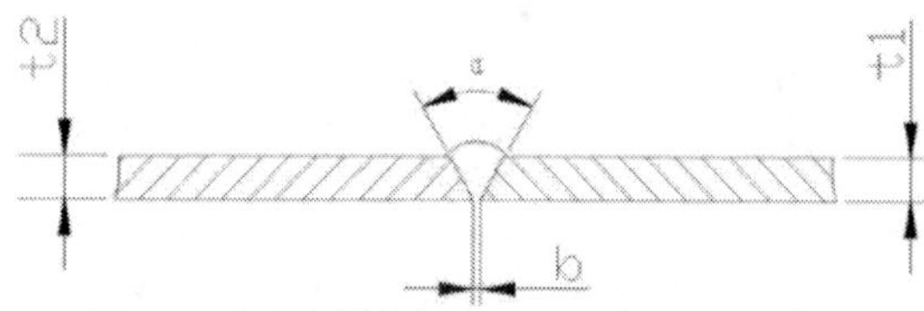


**Figure 1.** Weld joint type and groove size

The penetration test of the welded samples is carried out according to NB/T 47013.5-2015 "Non-destructive Testing of Pressure Equipment Part 5: Permeation Testing". The results meet the requirements. The welding samples and penetration test results are shown in Figure 2. The samples are heat treated after welding. The post-weld heat treatment process is shown in Table 1.

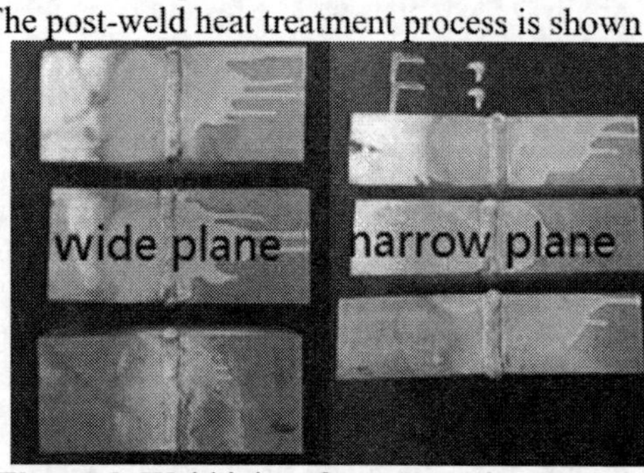


**Figure 2.** Weld joint after penetration detection

Tensile tests were carried out with SHIMADZU 100 kN electronic universal testing machine at a loading rate of 10mm/min. Samples were prepared according to the standard requirements of GB/T 2651-2008 tensile test method for welded joints. Three samples were tested in each group, and the average value was taken as the test result. The sampling position of tensile samples was shown in Figure 3. Tensile properties of wide and narrow surfaces of rectangular tubes were measured separately. FEI Inspect S50 type scanning electron microscope (SEM) was used to observe the tensile fracture morphology. Microstructure was observed by means of Leica DMI 5000M metallographic microscope.

The KB30S type Vickers hardness tester is used to test the hardness. The testing position is one-half of the wall thickness in the horizontal direction. The testing load is 0.5kg, and the distance between testing points is 0.5mm. The hardness testing position is shown in Figure 4.

**Table 1.** Post-weld heat treatment process

| Welding condition | post-weld heat treatment process | parameters |
|---|---|---|
| S1 | without heat treatment | - |
| S2 | Aging treatment | 150°C for 6 h |
| S3 | solid solution + ageing treatment | 520°C for 30 min+150°C for 8 h |

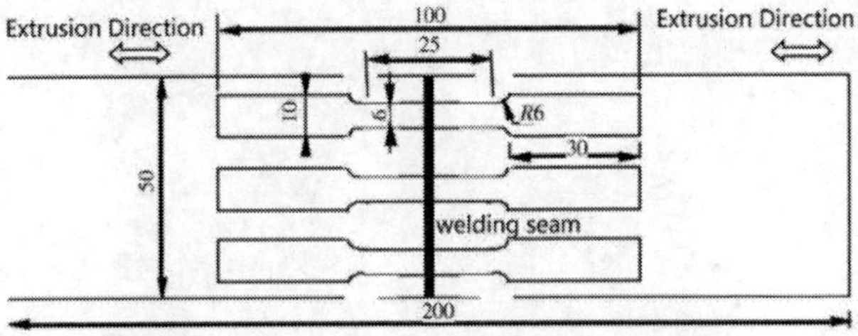


**Figure 3.** Schematic diagram of welding and sampling location (mm)

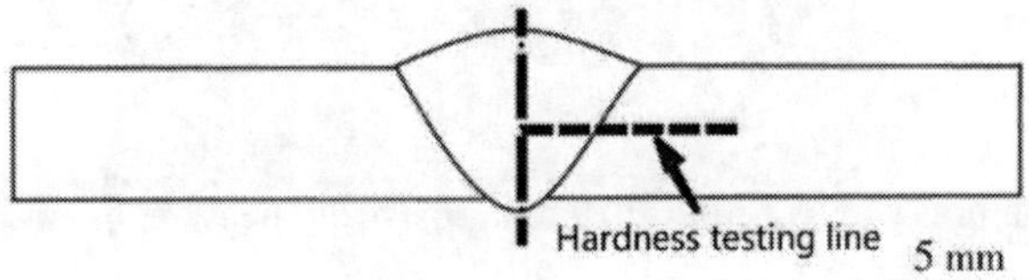


**Figure 4.** Location of hardness test points

## 3. Results and Discussions

### *3.1. Microstructure*

The microstructure of weld zone is typical as-cast microstructure of aluminum alloy, and there is no obvious difference before and after heat treatment. Fig. 4 shows the microstructure of HAZ in different post-weld heat treatment regimes. It can be seen that the heat affected zones of S1 and S2 samples are recrystallized under the influence of welding thermal cycle, and the grains grow up, and there are growing black $Mg_2Si$ phases. After re-solution and aging, the coarse $Mg_2Si$ phase in the heat-affected zone dissolves back, and the dispersed strengthening phase is precipitated again, and the homogeneity of microstructure is improved.

### *3.2. Hardness*

The hardness of different post-weld heat treatment specimens was tested, and the test results are shown in Fig. 6. As can be seen from figure 6, it can be seen that the base metal is less affected by welding heat cycle, and the hardness change trend of S1 and S2 welding specimens is similar, that is, with the increase of distance from the weld, the hardness first increases, then decreases, then increases to the base metal hardness level and remains stable. The lowest hardness of S1 sample is at the weld, the hardness value is 64.0HV, the lowest hardness of S2 sample is 1 mm away from the center of the weld, and the hardness value is 69HV. With the increase of the distance from the weld, the influence of welding heat cycle on the material decreases, and the hardness of HAZ begins to recover to the hardness of base metal. Compared with the untreated sample (S1), the hardness of the weld and HAZ of the S2 sample after post-weld heat treatment has been improved to a certain extent, and the width of HAZ has been narrowed obviously. After re-solution aging treatment, the change trend of hardness of S3 welded joint is obviously different. In addition to the increase of weld hardness, the softening of heat affected zone is significantly improved, and the hardness is obviously restored. The lowest hardness value is 82HV.

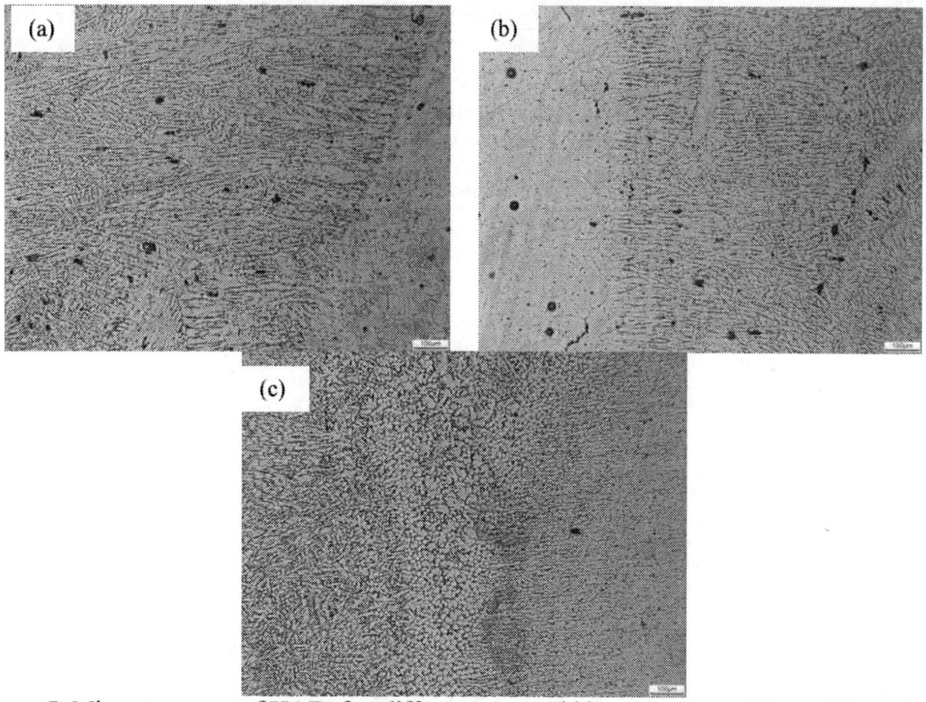

**Figure 5.** Microstructures of HAZ after different post-weld heat treatment: (a) S1, (b) S2, (c) S3

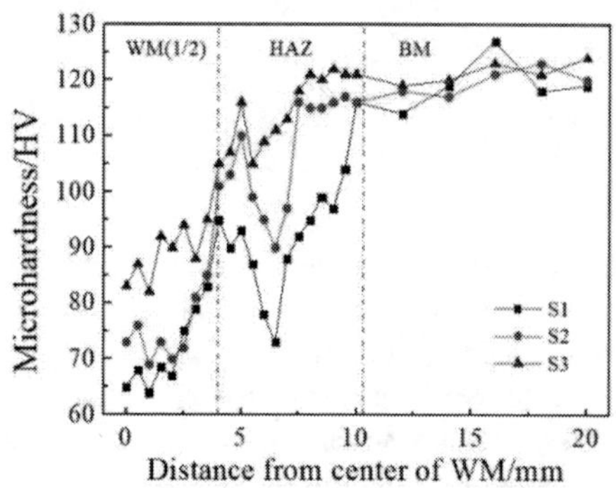

**Figure 6.** Hardness of welding joints after different post-weld heat treatment

*3.3. Tensile test*

Table 2 shows the test results of mechanical properties of different post-weld heat treatment samples. Fig. 6 shows the trend of the mean values of the wide, narrow and overall tensile strength of different post-weld heat treatment samples. As shown in Table 2, the overall average tensile strength of base metal is 405 MPa, and the wide side is close to the narrow side. The overall average tensile strength of welded joints without post-weld heat treatment is 263 MPa, and the welding coefficient (joint tensile strength/base material tensile strength) is 64.9%. Tensile strength increased to 285 MPa and the strength coefficient was 70.4% after ageing at 175 C for 6 h. After re-solution and re-aging after welding, the tensile strength reaches 307 MPa and the welding coefficient reaches 75.8%. Compared with the direct aging treatment after welding, the tensile strength and yield strength of the joint after re-solution and re-aging treatment have been improved. The trend of yield strength in Table 2 was further observed, and it was found that the trend coincided with that of tensile strength. However, it can also be seen from Fig. 6 that the strength difference between wide and narrow faces of S3 specimen increases with the increase of tensile strength. This is mainly due to the shorter welds on narrow surfaces, its welding defects are always higher than the wide surface, resulting in the overall effect of subsequent heat treatment is still not as good as the wide surface.

**Table 2.** Post-weld heat treatment process

| Welding condition | Position | UTS /MPa | E(UTS) /MPa | E(UTS) /MPa | YS /MPa | E(YS) /MPa | E(YS) /MPa |
|---|---|---|---|---|---|---|---|
| WM | Wide plane | 401 | 402 | 405 | 339 | 341 | 339 |
| | | 406 | | | 343 | | |
| | | 400 | | | 340 | | |
| | Narrow plane | 400 | 407 | | 334 | 338 | |
| | | 413 | | | 339 | | |
| | | 409 | | | 340 | | |
| S1 | Wide plane | 288 | 256 | 263 | 191 | 187 | 189 |
| | | 226 | | | 179 | | |
| | | 253 | | | 190 | | |
| | Narrow plane | 317 | 269 | | 209 | 192 | |
| | | 264 | | | 188 | | |
| | | 227 | | | 179 | | |
| S2 | Wide plane | 278 | 279 | 285 | 201 | 201 | 209 |
| | | 292 | | | 203 | | |
| | | 268 | | | 199 | | |
| | Narrow plane | 310 | 291 | | 225 | 216 | |
| | | 256 | | | 211 | | |
| | | 308 | | | 213 | | |
| S3 | Wide plane | 279 | 295 | 307 | 197 | 209 | 220 |
| | | 307 | | | 220 | | |
| | | 298 | | | 209 | | |
| | Narrow plane | 327 | 319 | | 239 | 232 | |
| | | 333 | | | 234 | | |
| | | 296 | | | 222 | | |

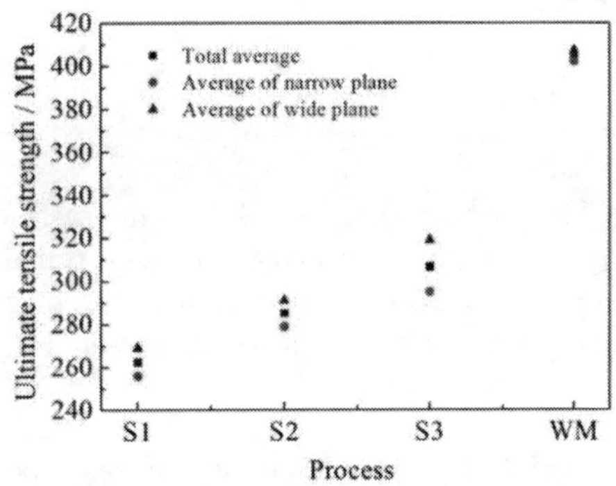


**Figure 7.** Tensile strength of welding joints after different post-weld heat treatment

### *3.4. Fracture Analysis*

Fracture morphology of tensile specimens after different post-weld heat treatment was observed by SEM, and the results were shown in Fig. 8. The fracture morphology of S1 and S2 specimens is a mixed mode of dimple fracture and cleavage fracture, but the number and size of dimples are quite different. The dimple size of S1 is larger and less than that of S2, and the cleavage surface area is larger. This indicates that there are more second phases in the thick fracture position of S1, which is easy to cause stress concentration cracking and promote crack propagation. S3 sample fracture morphology dimple size is uniform, dimple aggregation to form holes, tear edge is obvious, there are obvious holes, is formed by welding seam pore or dimple aggregation. The dimples size of S3 sample was uniform, and the dimples aggregated to form holes with obvious tearing edges. And there are obvious holes in the fracture, which are formed by the coalescence of weld pore or dimple. Samples S1 and S2 are subdivided into solution zone and over-aging zone by welding heat cycle. The temperature of heat affected zone near the weld is higher than that of Mg and Si atoms, and the Mg2Si strengthening phase dissolves back, that is to say, the solid solution zone is formed. However, the temperature in the heat affected zone far from the weld is lower than the solution temperature of Mg and Si atoms. The agglomeration and growth of Mg2Si strengthening phases occur, that is, the over-aging zone is formed, and the mechanical properties and plasticity of the joints decrease. However, after the post-weld heat treatment of S2 sample, the inhomogeneity of microstructure and distribution of strengthening phases are improved, so the tensile strength is improved. Relative to the microstructure of heat affected zone of S1 and S2 samples, Mg and Si elements almost completely re-solution after Solution-Aging, the softening zone of heat affected zone almost disappears, and the strength of joints increases greatly.

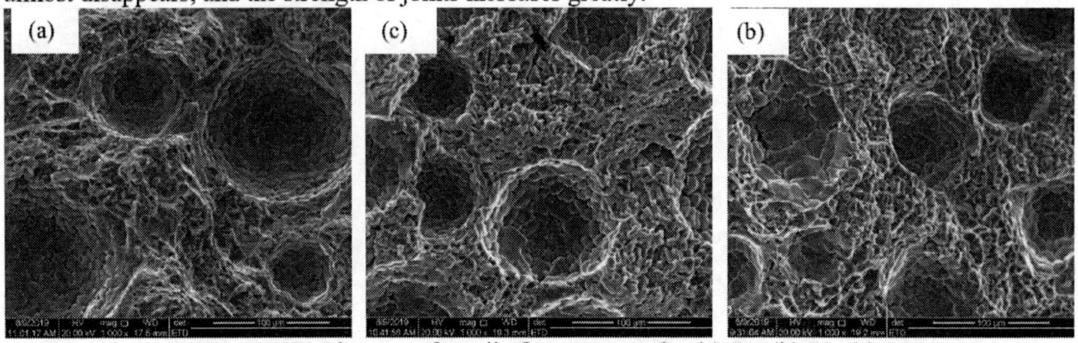


**Figure 8.** SEM images of tensile fracture sample: (a) S1, (b) S2, (c) S3

## 4. Conclusion

(1) The minimum hardness of the weld increased from 64 HV to 69 HV, the tensile strength increased from 263 MPa to 285 MPa, and the yield strength increased from 189 MPa to 209 MPa after post-weld heat heat treatment at 150℃ for 6 h.

(2) After post-weld heat treatment at 520℃ for 30 min & 150℃ for 8 h, the strength and hardness of the joints were greatly improved, the minimum hardness of the weld increased from 64 HV to 84 HV, the tensile strength increased from 263 MPa to 307 MPa, and the yield strength increased from 189 MPa to 220 MPa.

(3) After aging treatment, the inhomogeneity of the microstructure and the distribution of strengthening phases are improved. After re-solution and aging treatment, the Mg2Si phase is dissolved and the dispersed strengthening phases are precipitated again. Therefore, post-weld heat treatment is an effective way to improve the strength of welded joints of heat-treated strengthened aluminium alloy.

## References

[1] ZIJIE F, LIANGJIN G, RUIYI S. Research and development of automotive lightweight technology [J]. J Automotive Safety and Energy, 2014, 5(1): 1-16.

[2] CHUAN-FU L, LI-PING Z. Discussion on the Aluminum Alloy Materials in Automotive Lightweight Application and Development in the Future [J]. Equipment Manufacturing Technology, 2015, 4): 143-5.

[3] YUQIN G, XINFENG Z, YAN Y, et al. Research state of lightweight material and manufacture processes in automotive industry [J]. Forging and Stamping Technology, 2015, 40(3): 1-6.

[4] Yun-Lai Deng, Li Wan, Yun-Ya Zhang,. Microstructural evolution of Al－Zn－Mg－Cu alloy during homogenization [J]. Journal of Materials Science, 2011, 46(4):875-881.

[5] LI H, CHEN P, WANG Z, et al. Tensile properties, microstructures and fracture behaviors of an Al-Zn-Mg-Cu alloy during ageing after solution treating and cold-rolling [J]. Materials

Science and Engineering: A, 2019, 742: 798-812.

[6] ZHANG Y, JIN S, TRIMBY P W, et al. Dynamic precipitation, segregation and strengthening of an Al-Zn-Mg-Cu alloy (AA7075) processed by high-pressure torsion [J]. Acta Materialia, 2019, 162: 19-32.

[7] WANG W, PAN Q, WANG X, et al. Mechanical properties and microstructure evolution of ultra-high strength Al-Zn-Mg-Cu alloy processed by room temperature ECAP with post aging [J]. Materials Science and Engineering: A, 2018, 731: 195-208.

[8] RAVISANKAR V, BALASUBRAMANIAN V. Optimising pulsed current TIG welding parameters to refine the fusion zone [J]. Science & Technology of Welding & Joining, 2006, 11(1): 57-60.

[9] ZHANG L, LI X, NIE Z, et al. Softening Behavior of a New Al-Zn-Mg-Cu Alloy Due to TIG Welding [J]. Journal of Materials Engineering & Performance, 2016, 25(5): 1870-9.

[10] ZHANG L, LI X, NIE Z, et al. Microstructure and mechanical properties of joints of a new Al-Zn-Mg-Cu alloy welded by TIG [J]. Rare Metal Materials & Engineering, 2016: 334-338

[11] MILANI A M, PAIDAR M, KHODABANDEH A, et al. Influence of filler wire and wire feed speed on metallurgical and mechanical properties of MIG welding–brazing of automotive galvanized steel/5754 aluminum alloy in a lap joint configuration [J]. The International Journal of Advanced Manufacturing Technology, 2016, 82(9-12): 1495-1506

[12] ZHENG M, FU-LIANG Y, XIAO-LI D, et al. Heat Treatment Process of Al-Mg-Si-Cu Alloy [J]. Heat Treatment of Metals, 2011, 36(5): 97-102.

[13] Srinivas B , Krishna N M , Cheepu M , et al. Studies on post weld heat treatment of dissimilar aluminum alloys by laser beam welding technique[J]. IOP Conference Series Materials Science and Engineering, 2018, 330.

[14] ZHANG D-K, WANG G-Q, WU A-P, et al. Effects of Post-weld Heat Treatment on Microstructure, Mechanical Properties and the Role of Weld Reinforcement in 2219 Aluminum Alloy TIG-Welded Joints [J]. Acta Metallurgica Sinica (English Letters), 2019, 2: 1-11.

IOP Conf. Series: Materials Science and Engineering **772** (2020) 012113 doi:10.1088/1757-899X/772/1/012113

# Comparative analysis of the fracture splitting deformation of the big end for C70S6, 36MnVS4 and 46MnVS5 connecting rods

**Jingsha Liu[1, a], Shuqing Kou[2, b] and Dongye He[2, *]**

[1]College of Materials Science and Engineering, Jilin University, Changchun, China
[2]Roll-Forging Research Institute, Jilin University, Changchun, China

* Corresponding author e-mail:JilinUHL@163.com, [a]jsliu36@163.com, [b]kousq@jlu.edu.cn

**Abstract.** During the fracture splitting process of the connecting rod, the plastic deformation, especially the plastic deformation before cracking, not only causes the big end hole of the connecting rod to be out of roundness, but also increases the residual stress, affecting the assembly precision and even the service life of the connecting rod. In this paper, the mechanical properties of microalloyed steel C70S6, 36MnVS4 and 46MnVS5 connecting rod were determined by tensile test. The displacement field, plastic strain and the deformation principle of the big end before the crack initiation of the fracture splitting of the connecting rod were numerically simulated. The results show that the yield strength, tensile strength and plasticity index of the new medium carbon microalloyed steels 46MnVS5 and 36MnVS4 are significantly higher than those of C70S6. However, the C70S6 connecting rod has a long loading time before crack initiation, and the plastic deformation area of the groove root front is large and the strain is high. The plastic deformation of 36MnVS4 and 46MnVS5 connecting rods is small and the area is concentrated. Among the three, the 46MnVS5 connecting rod has the smallest deformation and the C70S6 connecting rod has the largest deformation.

## 1. Introduction

Fracture splitting has become the main way for the machining of connecting rod. The three key processes of connecting rod machining include prefabricated incision, directional crack and bolt assembly [1], among which directional splitting has the largest impact on the deformation of the big end of connecting rod. The roundness loss of the big end hole is one of the important detected items for the machining of connecting rods. The large deformation not only causes the roundness loss of the big end hole, but also reduces the assembly accuracy and even affects the performance of the connecting rod.

C70S6 is the first mature case of micro-alloy non-quenched and tempered steel for the fracture splitting of connecting rod [2], which is characterized with good purity and excellent splitting performance [3,4]. However, its low yield strength, poor cutting performance and unsatisfactory fatigue strength make it difficult to achieve the design for structural weight reduction [5, 6]. In order to adapt to the lightweight development of automobiles, new connecting rod materials of 36MnVS4, 38MnV5 and 46MnVS5 have been developed in recent years [7]. Their comprehensive mechanical properties

Content from this work may be used under the terms of the Creative Commons Attribution 3.0 licence. Any further distribution of this work must maintain attribution to the author(s) and the title of the work, journal citation and DOI.
Published under licence by IOP Publishing Ltd

have been significantly improved compared with C70S6 connecting rod, and the fatigue strength can be increased by about 10%, which has a great potential of development and application.

In view of the transient nature and the complexity of the dynamic fracture splitting process, it is difficult to dynamically capture the parameters. In this paper, the numerical fracture splitting simulation analysis of three typical microalloyed non-quenched and tempered connecting rods of C70S6, 36MnVS4 and 46MnVS5 is carried out. The mechanical properties of the three materials were obtained by the tensile strength test of the forged sample, and the fracture splitting deformation of the big end of the connecting rods were numerically analyzed.

## 2. Material properties

The chemical compositions of C70S6, 36MnVS4 and 46MnVS5 microalloyed non-quenched and tempered steels are shown in Table 1. It can be seen that compared with C70S6, the carbon contents of 36MnVS4 and 46MnVS5 are significantly reduced, the content of S, Mn, Cr, V increases; The reduction of C is conducive to improving machinability and toughness. The increase of the content of the above alloy elements can make grain refined and get higher strenghth, enhance the toughness of materials, and make up for the influence of the reduction of C on mechanical properties. The reduction of S and the increase of Mn can improve the thermal processability of materials.

**Table 1.** Chemical composition of three materials

| Material | C | Mn | Si | S | Cr |
|---|---|---|---|---|---|
| C70S6 | 0.71 | 0.50 | 0.20 | 0.07 | 0.12 |
| 36MnVS4 | 0.36 | 0.99 | 0.70 | 0.05 | 0.25 |
| 46MnVS5 | 0.47 | 1.10 | 0.75 | 0.06 | 0.30 |

The forging of the connecting rod is carried out at 1150~1250 °C, and the air-controlled cooling is procssed after forging according to the CCT curve of the three materials (first rapid air cooling and heap of cooling) to obtain good pearlite and ferrite structure, suitable hardness and mechanical properties. The forging rod of the connecting rod was sampled to prepare a sample with a diameter of Φ 5 mm and a length of 25 mm. The mechanical properties obtained by the tensile test are shown in Table 2.

**Table 2.** Results of tensile test of three materials

| Material | Sample Number | Yield Strength(σs)/ MPa | Tensile Strength(σb)/ MPa | rate of reduction /% | Elongation /% |
|---|---|---|---|---|---|
| C70S6 | 1#: | 605 | 990 | 25.5 | 14.0 |
| | 2#: | 590 | 985 | 28.0 | 16.5 |
| | 3#: | 585 | 980 | 25.0 | 17.5 |
| | average | 593 | 985 | 26.2 | 16 |
| 36MnVS4 | 1# | 895 | 1100 | 36.0 | 22.0 |
| | 2# | 920 | 1130 | 42.0 | 23.0 |
| | 3# | 895 | 1130 | 39.0 | 19.5 |
| | average | 903 | 1120 | 39 | 21.5 |
| 46MnVS5 | 1# | 795 | 1130 | 29.5 | 18.5 |
| | 2# | 775 | 1110 | 33.0 | 16.5 |
| | 3# | 830 | 1120 | 29.5 | 18.0 |
| | average | 800 | 1120 | 30.7 | 17.7 |

It can be seen that the high carbon microalloyed steel C70S6 connecting rod has the lowest yield and tensile strength, the lowest rate of reduction and elongation, the 36MnVS4 connecting rod is the highest, and the 46MnVS5 is in the middle. The yield ratio of C70S6 is only 0.602, which is significantly lower than that of medium carbon microalloyed steel.

## 3. Numerical analysis model construction

### *3.1. Principle of fracture splitting mechanics of connecting rod*

The fracture splitting process of the connecting rod is shown in Figure 1. The wedge-shaped pull rod moves downward. The wedge-shaped surface of the pull rod pushes the horizontal movement of the moving sleeve to continuously apply tensile load to the large hole of the connecting rod. Stress concentration occurs at the root of the large hole of the connecting rod, and the slit is cracked. And the crack spreads rapidly until the fracture is separated into the connecting rod body and the connecting rod cover.

The fracture splitting essence of the connecting rod is the I-type quasi-brittle fracture of the elastoplastic material after a small yield. The theoretical basis is that the external stress is perpendicular to the fracture surface, and the low stress brittle fracture is easy to occur, the plastic deformation is small and the fracture is flat.And the maximum tensile stress is adopted as the basic criterion in the finite element numerical simulation [8].

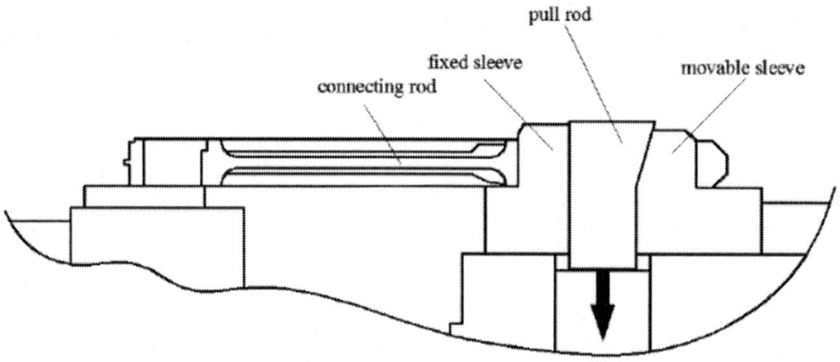


**Figure 1.** Fracture splitting diagram of the connecting rod

### *3.2. Parameters of connecting rod and groove root*

One connecting rod of a car engine is selected, the main dimensions are shown in Figure 2(a).The diameter of the big end hole is 41.80 mm, the center distance between the big end hole and small end hole is 147.50mm, the bolt hole diameter is 6.50 mm, and the distance between the bolt holes and the outer edge is 2.10mm. The thickness of the connecting rod is 17.90mm.

According to the previous research experience on the fracture splitting of C70S6 connecting rod, 0.4-0.6mm for the groove depth, 0.1-0.3mm for the radius of curvature are chosen. To prevent the influence of different geometric parameters on the modeling results, the groove root selects the same parameters.The size of the groove root was as Figure 2(b): the groove length L=17.9 mm, the groove depth h=0.5 mm, the opening angle=60° , and the radius of curvature = 0.2 mm.

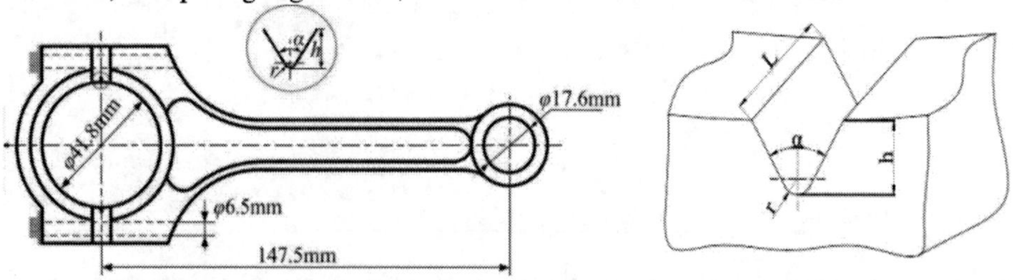


(a) Main dimensions of the connecting rod (b) Parameters of groove root

**Figure 2.** Parameters of the connecting rod and the groove root

IOP Conf. Series: Materials Science and Engineering **772** (2020) 012113 doi:10.1088/1757-899X/772/1/012113

### *3.3. Geometric modeling*

The geometry, constraint and force of the connecting rod are symmetrical about the mid-surface and the central axis , while the connecting rod small end, the connecting rod, the boss on the connecting rod, the outer round corner have little influence on the fracture splitting process, and thus the structure such as the small end and the rod are neglected.

Based on the geometry of the connecting rod and the symmetry of the restraint force, and in order to improve the calculation efficiency and reduce the difficulty of meshing, the quarter end of the connecting rod is modeled, and the moving and fixed block are simplified to one quarter.The model is shown as Figure 3. The simplified model has a plane of symmetry along the thickness direction (middle plane, plane of symmetry 1) and a plane of symmetry along the central axis (symmetric plane 2).

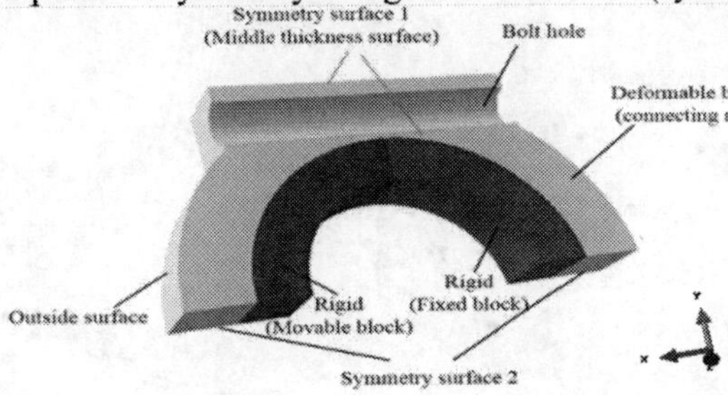


**Figure 3.** Geometric model

### *3.4. Boundary conditions and load handling*

The model is simplified to 1/4, which has two symmetry planes. The displacement Uz=0 and Ux=0 are applied on the nodes in the symmetry plane 1 (thickness midplane) and the symmetry plane 2 as shown in figure 3.

There are two contact pairs in the model, which are the contact between the big end and the fixed sleeve and the big end and the movable sleeve. A fully fixed constraint is applied to the fixed sleeve, which limits all rotations and Uz = Ux = 0. The speed load is applied to the moving sleeve along the X direction of the connecting rod.

## 4. Numerical simulation results and analysis

### *4.1. Displacement field before crack initiation*

The displacement of the big end before the cracking of the connecting rod is shown in Figure 4, where U1 stands for the displacement in the X direction (the direction of the connecting rod axis), U2 stands for the displacement in the Y direction (the direction of the vertical axis and parallel to the predetermined fracture surface), and U3 stands for the displacement in the Z direction (the thickness direction), U stands for the total displacement.

Table 3 shows the maximum displacement of the big end of the connecting rod in three directions. It can be seen that the displacement of the big end of the C70S6 connecting rod in three directions and the total displacement are all the largest, 36MnVS5 is the second, and 46MnVS5 is the smallest. Since the tensile stress is applied to the fixed sleeve and the movable sleeve in the direction of the axis X in the big end of the connecting rod, the big end of the connecting rod exhibits an elongation in the X direction and the maximum displacement in the X direction occurs in the inner wall of big end ; The largest displacement in the Y direction occurs near the groove root and the predetermined fracture surface, and produces a Y-direction compression deformation; the displacement in the Z direction is small during the fracture splitting of the connecting rod. This indicates that the fracture splitting of the big end mainly shows the appearance of out of roundness in the X-Y direction, in which the C70S6 ellipse is the most obvious, and 36MnVS5 and 46MnVS5 are sequentially decreased.

IOP Conf. Series: Materials Science and Engineering **772** (2020) 012113 doi:10.1088/1757-899X/772/1/012113

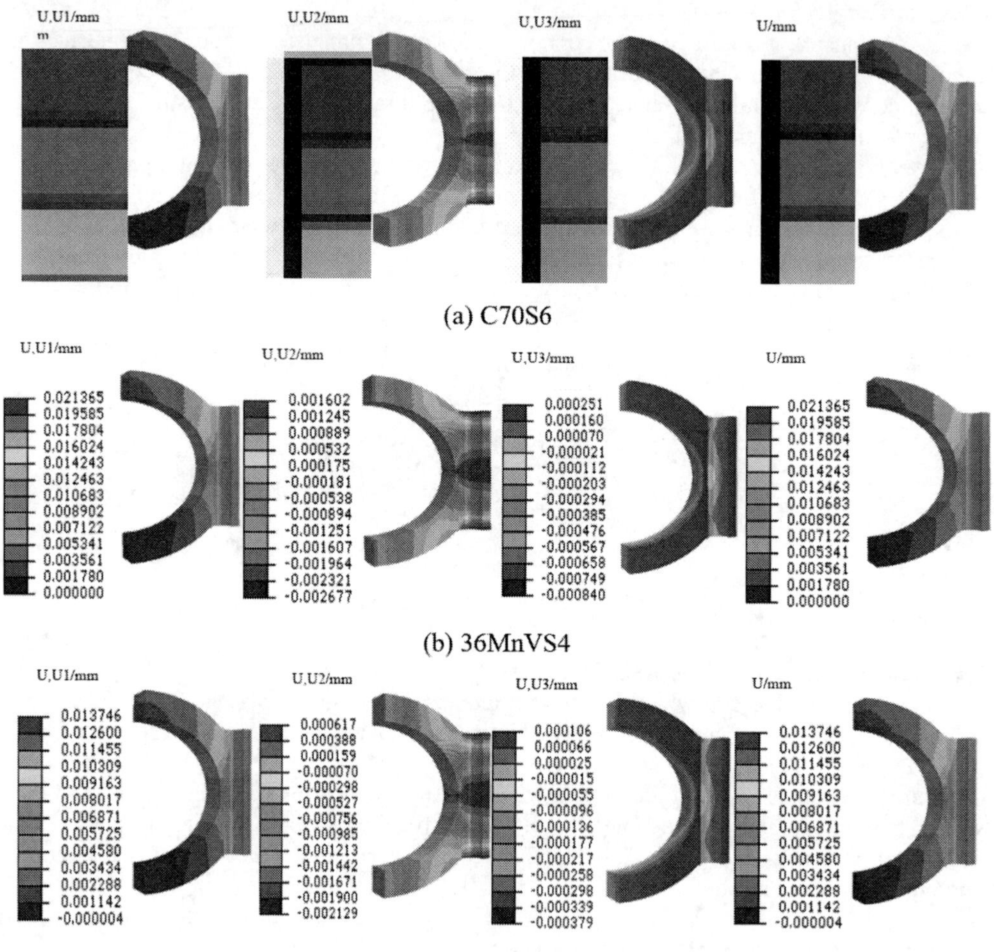

(a) C70S6

(b) 36MnVS4

(c) 46MnVS5

**Figure 4.** Displacement of the big end before crack starts

**Table 3.** Displacement change of the big end of the connecting rod

| Material / Displacement | C70S6 | 36MnVS4 | 46MnVS5 |
|---|---|---|---|
| U1,Magnitude | 0.032672 | 0.021365 | 0.013746 |
| U2,Magnitude | 0.003554 | 0.001602 | 0.000617 |
| U3,Magnitude | 0.000459 | 0.000251 | 0.000106 |
| U,Magnitude | 0.032672 | 0.021365 | 0.013746 |

### *4.2. The deformation of the big end*

Figure 5 and Figure 6 respectively show the maximum, minimum and intermediate displacement of the connecting rod of the three materials in the X and Y directions with the increase of time. The x-direction displacement is the largest at the contact point between the moving sleeve and the inner wall of the big hole along the axis of the connecting rod, and the X-axis displacement is the smallest. According to Figure 4, the maximum displacement in the Y direction is located in the groove root and the predetermined fracture surface region, and the minimum displacement in the Y direction is in the region of 20 to 30 degrees from the X-axis and Y-axis.

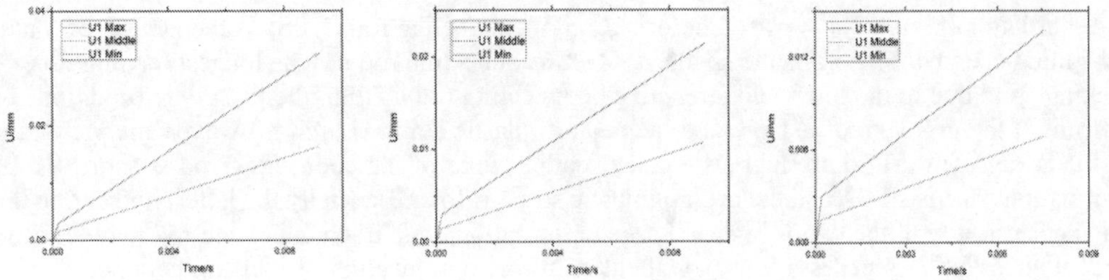


**Figure 5.** Change of X-direction displacement of big end with time

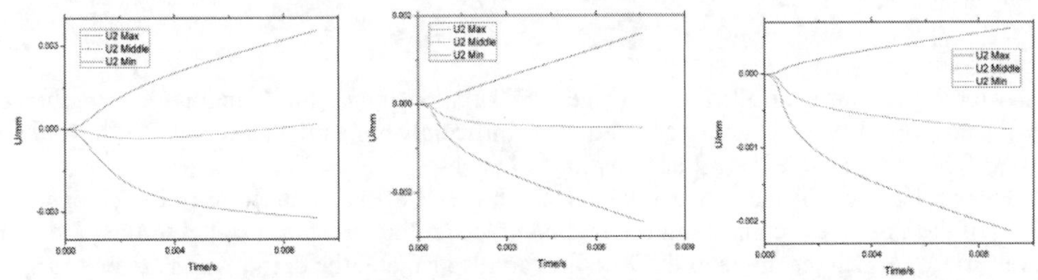


**Figure 6.** Change of Y-direction displacement of big end of connecting rod with time

It can be seen from Figure 5 that the x-direction displacement of the big end of the connecting rod of the three materials changes in the same pattern with time. The maximum displacement increases linearly with the increase of time, and the minimum displacement is always 0. It indicates that with the increase of time, the displacement of each point in the big end along the X direction varies unevenly.

It can be seen from Figure 6 that the Y-direction displacement of the connecting rod of the three materials changes with time basically in the same rule. The maximum displacement and minimum displacement increase with time, and the intermediate displacement changes little. The displacement of the big end of the connecting rod in Y direction is uneven.

### *4.3. Plastic strain near the groove root*

Figure 7 shows the plastic strain of connecting rod before cracking. It can be seen that the plastic deformation of connecting rod is completely concentrated near the groove root, and the plastic strain of other areas is 0, which further indicates that the fracture splitting of connecting rod is a quasi-brittle fracture with yield within a small range near the groove root. As shown in Figure 4, the big end of the connecting rod extends in the X direction and contracts in the Y direction in the predetermined fracture surface area. After unloading, a large amount of elastic deformation can be released, and the big end only retains a small amount of ellipticity, which can be completely eliminated by subsequent fine boring.

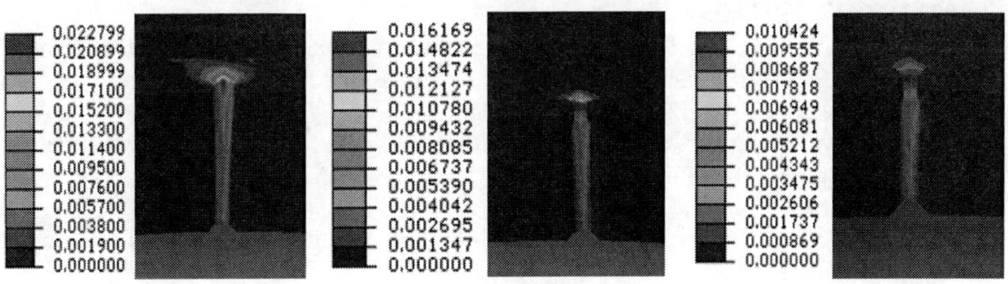


(a) C70S6 (b) 36MVS4 (c) 46MnVS5

**Figure 7.** Plastic strain near groove root

According to Figure 7, the plastic deformation of connecting rod C70S6 is the largest, 36MnVS4 is in the middle, and 46MnVS5 is the smallest. C70S6 connecting rod has the longest loading time before cracking. And due to the low yield strength of connecting rod C70S6, the plastic deformation zone of the front of the groove root is large and the peak of plastic strain is high. However, the yield ratios of 36MnVS4 and 46MnVS5 are high, the crack loading time of the connecting rod is short, the plastic deformation zone is small, and the peak of plastic strain is low (especially the latter), indicating that the incision sensitivity of the two is higher. Large deformation of C70S6 connecting rod may lead to large residual stress, so it is necessary to pay attention to the optimization of incision parameters, splitting fixture design and reasonable selection of force and energy parameters to avoid the impact on the assembly accuracy.

## 5. Conclusion

(1) After forging and air-controlled cooling, the yield strength and tensile strength of the medium carbon microalloyed steels 36MnVS4 and 46MnVS5 are significantly higher than those of C70S6.And with the decrease of the carbon content, the plasticity index increases.

(2) Before the crack of the connecting rod, the big end of the connecting rod is elongated in the direction of the axis of the connecting rod and contracts in the direction parallel to the predetermined fracture surface, so that the circle in the XY direction is lost; and the deformation amount increases as the loading time increases; The ovality of C70S6 connecting rod is the most obvious, 46MnVS5 is the lowest. The deformation of the connecting rod in the thickness direction is extremely small.

(3) The loading time of C70S6 connecting rod is the longest before cracking, the plastic deformation zone of the front of the groove root is the largest, and the peak of the plastic strain is the highest; while the 36MnVS4, 46MnVS5 connnecting rod has a high yield ratio, the loading time of the connecting rod is short before cracking, and the plastic deformation area is small. And the plastic strain of the 46MnVS5 connecting rod significantly decreases.

## References

[1] Fukuda S, Eto H. Development of fracture splitting connecting rod [J]. Jsae Review, 2002, 23(1):101-104.

[2] Wen X L , Jiang B, Zhou L Y, et al. Splitting Property Analysis of Automotive Con-Rod Steel C70S6[J]. 2015.

[3] Gu Z, Yang S, Ku S, et al. Fracture splitting technology of automobile engine connecting rod [J]. The International Journal of Advanced Manufacturing Technology, 2005, 25(9-10): 883-887.

[4] Kou S Q , Gao Y, Gao W Q. The influence of auxiliary gases in the optimized analysis of pulsed laser grooving of a C70S6 connecting rod for fracture splitting [J]. Results in Physics, 2017, 7:628-635.

[5] Zhang D , Harris S J, Mccartney D G, et al. The effect of laser transformation notching on the controlled fracture of a high carbon (C70S6) steel [J]. Materials Science and Engineering: A, 2008, 489(1-2):273-284.

[6] Kim H S , Kim T G, Chung T J , et al. Fatigue characteristics of high strength C70S6 and SMA40 steels[J]. Materials Science & Engineering: A (Structural Materials: Properties, Microstructure and Processing), 2010, 527(12):2813-2818.

[7] Wei Y, Fang W, Zhang C L, et al. Effect of Different Microstructures on the Performance of Air-Cooled Forging Steel 46MnVS5 Fracture Splitting Connecting Rod [C]. Materials Science Forum. 2018.

[8] Mo, S., Yu, Y., and Wang, J., Engineering Material Mechanical Properties (Beijing: Peking University Press, 2013.

# Numerical Calculation of Composite Water Lubricated Bearing Considering Effect of Elastic Deformation

**Changjiang Liu [1], Bin Yao [2], Genyuan Gao [3], Yanzhen Wang [3], Yang Mao [3, *]**

[1] China Shipbuilding Industry Corp, 712 research institute, Hubei Wuhan, China
[2] Zhejiang Zhuji Bearing Factory Co., Ltd., Zhejiang Zhuji, China
[3] Shanghai Jiaotong University, School of mechanical engineering, Shanghai, China

*Corresponding author e-mail: maoyang90@sjtu.edu.cn

**Abstract.** Traditional bearing design neglects the influence of elastic deformation of materials on bearing lubrication characteristics, based on a new type of polymer composite water-lubricated bearing materials, the influence of elastic deformation of bearing material is introduced into the simulation. The lubrication performance of the water-lubricated bearings is studied with respect to material parameters and structural parameters. The effects of the elastic deformation on the lubrication characteristics such as the maximum pressure value and the minimum film thickness were compared and analyzed under different bearing clearance structures of a certain type of stern bearings. The results show that, the maximum water film pressure of bearing increases with the clearance ratio, and the maximum water film pressure of elastic bearing is less than that of rigid bearing. With the increase of clearance ratio, the elastic deformation of bearing increases linearly. The increase of bearing elastic deformation results in the difference between minimum water film thickness and maximum water pressure between the elastic bearing and rigid bearing.

## 1. Introduction

As a critical component of the ship propulsion system, the friction and lubrication characteristics of stern tube bearings have an important impact on the service performance and life reliability of the ship. Due to the low-speed and heavy-load operating conditions of water-lubricated bearings combined with the low viscosity of the water medium, bearings usually run in mixed lubrication or boundary lubrication, which will inevitably lead to direct contact between the shaft and the bearing friction surface and wear of bearing material. Therefore, compared with other types of sliding bearings, water-lubricated bearings have requirements on the wear resistance of materials. In addition, the elastic modulus of water-lubricated bearing material is only a few hundredths of that of metal material, which makes the bearing elastic deformation and the water film thickness in the same order of magnitude. Hence, the influence of elastic deformation on bearing lubrication performance and structural design cannot be ignored.

At present, new water-lubricated bearing materials are mostly made of high molecular polymer [1]. The low hardness and small elastic modulus of high molecular polymer combined with the low speed and heavy load operating characteristics usually make high molecular polymer bearings generate significant elastic deformation [2-3] during operation. Simultaneously, due to the low viscosity of the water medium and poor dynamic pressure bearing capacity, such bearings are more likely to run in

Content from this work may be used under the terms of the Creative Commons Attribution 3.0 licence. Any further distribution of this work must maintain attribution to the author(s) and the title of the work, journal citation and DOI.
Published under licence by IOP Publishing Ltd

mixed lubrication under low speed, heavy load and start-stop conditions, which results in severe wear of bearing material [4-6]. The matching scale has an important influence on the friction and wear performance of the friction pair [7-8]. The traditional bearing simulation method does not consider the influence of the elastic deformation of the bearing on the bearing friction and lubrication characteristics. The simulation analysis results have a large error compared with the engineering practice, which cannot meet the demand of the current water-lubricated bearing simulation analysis. Therefore, based on the operation characteristics of water-lubricated high molecular polymer bearings, it is necessary to introduce the elastic deformation simulation method into the water-lubricated bearing simulation model to establish a suitable numerical analysis method for the calculation of the lubrication characteristics of ship water-lubricated bearings.

In this paper, based on the modified UHMWPE bearing material parameters [9-10], the related lubrication theory of water-lubricated bearings is analyzed firstly. The boundary conditions and the flow state of the lubricating medium in the numerical calculation process of water-lubricated bearings are studied. The bearing lubrication performance analysis method considering elastic deformation is established. And considering the elastic modulus of the polymer composite material, the water-lubricated bearing pressure, minimum film thickness, deformation, friction coefficient and eccentricity under different clearances are calculated. Finally, the calculation results are compared and analyzed with the rigid bearing.

## 2. Water-lubricated bearing model considering elastic deformation

### 2.1. Calculation method

For the water-lubricated bearings, the principle of the dynamic pressure formation is shown in Figure 1. Under the effect of the external load, the certain eccentric distance e will be generated between the center of the rotating shaft ($O_j$) and the center of the bearing ($O_b$), thereby forming a wedge-shaped gap in the water film convergence region, while the rotating shaft drives the water medium with a certain viscosity relative to the bearing. The movement meets three conditions of fluid dynamic pressure lubrication formation, hence establishing dynamic pressure lubrication and supporting the external load.

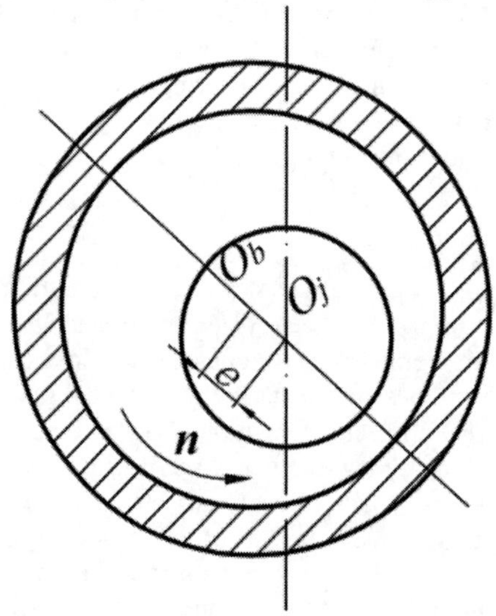


**Figure 1.** Principle of Bearing Dynamic Pressure Formation

The fluid particle motion equation is expressed as:

$$\begin{cases} \rho(\frac{\partial v_x}{\partial t} + \mathbf{v}\cdot\nabla v_x) = \rho X + \frac{\partial \sigma_x}{\partial x} + \frac{\partial \tau_{yx}}{\partial y} + \frac{\partial \tau_{zx}}{\partial z} \\ \rho(\frac{\partial v_y}{\partial t} + \mathbf{v}\cdot\nabla v_y) = \rho Y + \frac{\partial \tau_{xy}}{\partial x} + \frac{\partial \sigma_y}{\partial y} + \frac{\partial \tau_{zy}}{\partial z} \\ \rho(\frac{\partial v_z}{\partial t} + \mathbf{v}\cdot\nabla v_z) = \rho Z + \frac{\partial \tau_{xz}}{\partial x} + \frac{\partial \tau_{yz}}{\partial y} + \frac{\partial \sigma_z}{\partial z} \end{cases} \tag{1}$$

The mass conservation equation:

$$\frac{\partial \rho}{\partial t} + \frac{\partial(\rho v_x)}{\partial x} + \frac{\partial(\rho v_y)}{\partial y} + \frac{\partial(\rho v_z)}{\partial z} = 0 \tag{2}$$

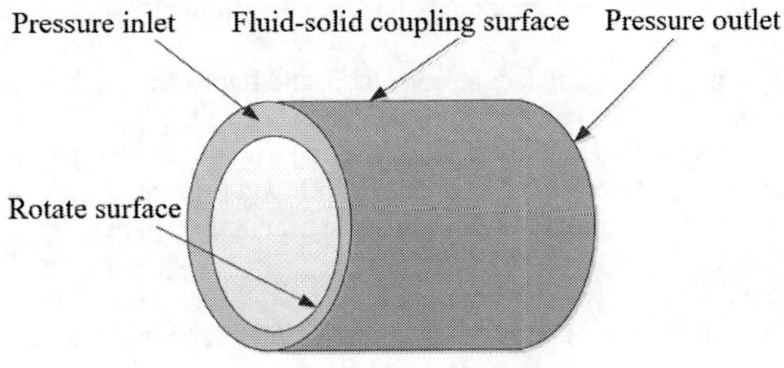


**Figure 2.** Water Film Boundary Conditions

### *2.2. Model design and analysis*

The water film boundary conditions are set as shown in Figure 2. The left and right end faces of the water film are the pressure inlet and the pressure outlet respectively. The outer surface of the water film is in contact with the bearing, which is a fluid-solid coupling surface. The inner surface of the water film is in contact with the shaft, which is a rotating wall surface. Due to the special structure of the water film, the radial dimension is several tens of micrometers or even several micrometers, which is much smaller than the axial and circumferential dimensions. In order to accelerate the convergence of the calculation results and increase the stability of the calculation process, the structured grid is used for the fluid in this calculation. The solid mesh is divided in the ANSYS Workbench, and the material parameters and boundary conditions are set at the same time, finally the *.inp file containing the solid domain information is output. The fluid grid and the *.inp file of the solid are imported into CFX for fluid-solid coupling calculations.

The bearing solid elements are deformed and displaced under the effect of the water film pressure, and the motion follows the Newton's Second law.

$$\rho_s \ddot{\mathrm{d}}_s = \nabla \cdot \sigma_s + f_s \tag{3}$$

Where $\rho_s$ is the solid density, $\ddot{\mathrm{d}}_s$ represents the acceleration of the solid element, $\sigma_s$ is the solid stress tensor and $f_s$ is the volume force vector. At the interface between the fluid and the solid, the fluid pressure deforms the solid, which changes the boundary shape of the fluid in turn. The elements at the interface between the fluid and the solid follows the equality of variables such as displacement, stress, heat flow and temperature. This paper neglects the temperature variation of the water medium in the calculation, so the change in heat flow and temperature is ignored. Therefore, the equilibrium equations are expressed as:

$$\begin{cases} d_f = d_s \\ \boldsymbol{n} \cdot \tau_f = \boldsymbol{n} \cdot \tau_s \end{cases} \tag{4}$$

Where $d_f$ and $d_s$ are fluid displacement and solid displacement respectively, $\tau_f$ and $\tau_s$ are the stress of fluid particles and solid elements respectively. The governing equations of the fluid and solid domains are coupled to the same system matrix for solution:

$$\begin{bmatrix} \mathbf{A}_{ff} & \mathbf{A}_{fs} \\ \mathbf{A}_{sf} & \mathbf{A}_{ss} \end{bmatrix} \begin{bmatrix} \Delta \boldsymbol{X}_f^k \\ \Delta \boldsymbol{X}_s^k \end{bmatrix} = \begin{bmatrix} \boldsymbol{F}_f \\ \boldsymbol{F}_s \end{bmatrix} \tag{5}$$

Where $\mathbf{A}_{ff}$ and $\mathbf{A}_{ss}$ are system matrices of the fluid and solid domains respectively, $\mathbf{A}_{fs}$ and $\mathbf{A}_{sf}$ are fluid-solid coupling matrices, $\Delta X_f^k$ and$\Delta X_s^k$ are the pending solutions of the fluid and solid domains respectively, $\boldsymbol{F}_f$ and $\boldsymbol{F}_s$ are the external loads acting on the fluid and solid domains respectively, $k$ is the number of iteration steps.

All manuscripts must be in English, also the table and figure texts, otherwise we cannot publish your paper. Please keep a second copy of your manuscript in your office. When receiving the paper, we assume that the corresponding authors grant us the copyright to use the paper for the book or journal in question. Should authors use tables or figures from other Publications, they must ask the corresponding publishers to grant them the right to publish this material in their paper.

## 3. Results analyze

The clearance has an important influence on the bearing carrying capacity and friction coefficient. Therefore, it is important to study the effect of different clearances on the bearing load and lubrication performance. According to the compression test, the modified UHMWPE bearing material has the compressive elastic modulus of 1000MPa and the Poisson's ratio of 0.44, which are used as the material parameters when studying the influence of structural parameters on the lubrication performance of the bearing. In Figure 3(a)~(e), the water-lubricated stern bearing with the diameter of 80mm and the length-diameter ratio of 1.0 was taken as the research object to study the influence of different clearance ratios on maximum water film pressure, minimum water film thickness, deformation, friction coefficient and eccentricity with and without elastic deformation.

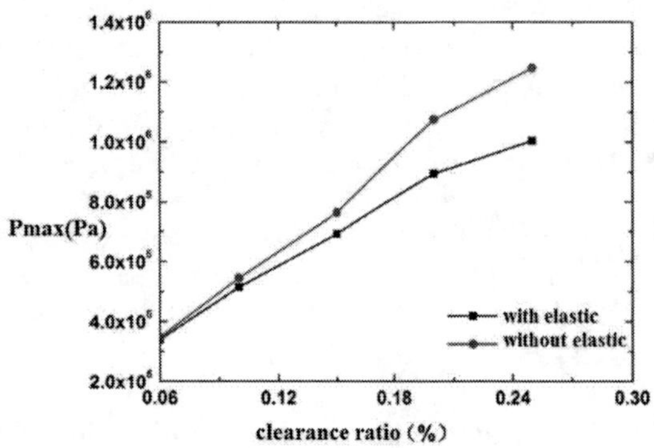


(a) Maximum Water Film Pressure Comparison

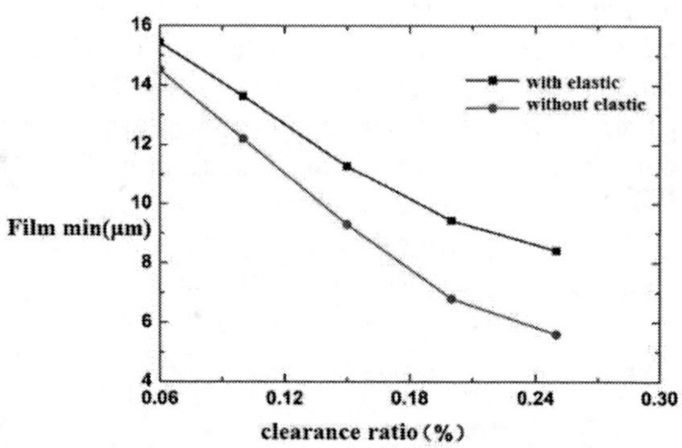


(b) Minimum Water Film Thickness Comparison

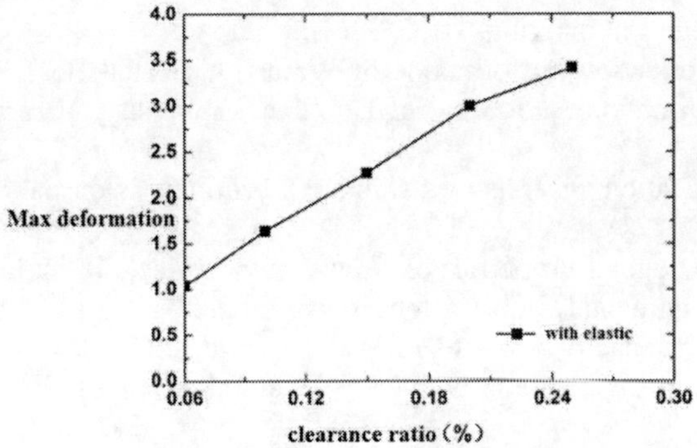


(c) Maximum Deformation Comparison

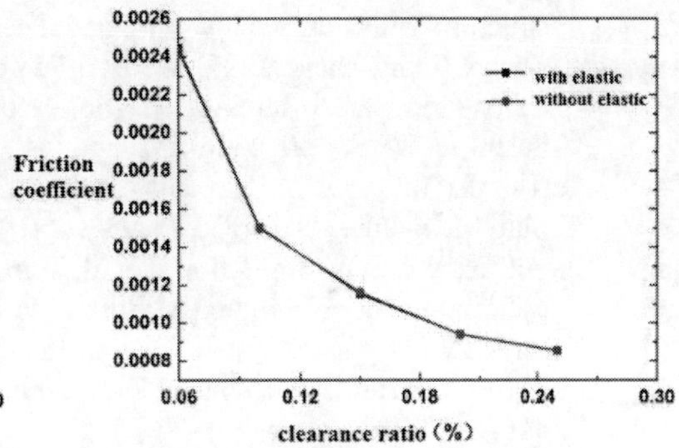


(d) Friction Coefficient Comparison

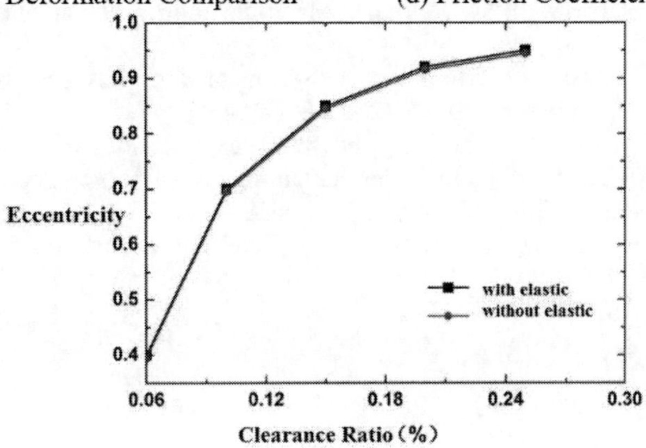


(e) Eccentricity Comparison

**Figure 3.** Effect of Different Clearance Ratio on Lubrication Performance of Bearings

## 4. Conclusion

The results show that when the bearing clearance ratio increases from 0.06% to 0.25%, the bearing eccentricity increases from 0.4 to 0.95 under constant load and speed conditions, which indicates that the bearing carrying capacity decreases and the change of the eccentricity gradually slows down with the increase of the clearance ratio, and the influence of the clearance on the bearing carrying capacity is gradually reduced at the same time. The maximum water film pressure increases with the increase of the clearance ratio. The maximum water film pressure of the elastic bearing is smaller than that of the rigid bearing, and the difference between them increases with the increase of the clearance ratio. In addition, the minimum water film thickness and friction coefficient reduce with the increase of the clearance ratio. The minimum water film thickness of the elastic bearing is larger than that of the rigid bearing, and the difference between them increases with the increase of the clearance ratio. Simultaneously, with the increase of the clearance ratio, the bearing elastic deformation increases linearly. After the bearing clearance ratio increases from 0.06% to 0.25%, the bearing elastic deformation increases by 3.5 times. It is also because of the increase of the bearing elastic deformation that the difference between the minimum water film thickness and the maximum water film pressure of the elastic bearing and the rigid bearing gets larger.

## References

[1] Li J H, Li L C, Liu Yu. Experimental Study on Tribological Properties of Polymer Lining of Water Lubricated Stern Bearing[J]. Ship Ocean Engineering, 2013, 42(2):67-69.
[2] Qu Bo, Liu Tao, Xu D S. Optimal parametric design of bulkhead vibration control for underwater

structure[J]. Naval Architecture and Ocean Engineering, 2016, 32(3):34-38.

[3] Li Zheng, Jiang Dan, Yin Z W, et al. Lubrication Performance of Water Lubricated Bearing Considering the Effects of Elastic Deformation[J]. Journal of Donghua University, Natural Sciences, 2018, 44(4):526-531.

[4] He Tao, Li W Y, Lu X Q, et al.Mixed lubrication analysis of a ship stern bearing[J]. Journal of Ship Mechanics, 2014(8):973-981.

[5] Shen C X, Wu Z W, Jiang Song, et al. Tribological Properties of New Water-lubricated Bearing Materials under Simulated Marine Environment[J].Lubrication Engineering, 2017, 42(4):1-5.

[6] Wang Y Z, Jiang Dan, Yin Z W, et al.Load Capacity Analysis of Water Lubricated Hydrostatic Thrust Bearing Based on CFD[J]. Journal of Donghua University, Natural Sciences, 2015, 41(4):428-432.

[7] El-Domiaty A, El-Fadaly M, Nassef A E. Wear characteristics of ultrahigh molecular weight polyethylene (UHMWPE)[J]. Journal of Materials Engineering and Performance, 2002, 11(5): 577-583.

[8] Wang J Z, Yan F Y, Xue Q J. Tribological behaviors of some polymeric materials in sea water[J]. Chinese Science Bulletin, 2009, 54(24): 4541-4548.

[9] Wang Y, Yin Z, Li H, et al. Friction and wear characteristics of ultrahigh molecular weight polyethylene (UHMWPE) composites containing glass fibers and carbon fibers under dry and water-lubricated conditions[J]. Wear, 2017, s 380–381:42-51.

[10] Wu Z, Sheng C, Guo Z, et al. Surface texture processing for tribological performance improvement of UHMWPE-based water-lubricated bearings[J]. Industrial Lubrication and Tribology, 2018, 70(7):1341-1349.

IOP Conf. Series: Materials Science and Engineering **772** (2020) 012115 doi:10.1088/1757-899X/772/1/012115

# Preparation of Ferrite $Fe_3O_4$ and Its Electromagnetic Wave Absorption Properties

**Fuli Yang[1], Xingzhe Hou[1], Lirui Wang[2], Youjiang Li[2] and Miao Yu[2,*]**

[1]Electric power research institute, Chongqing Electric Power Corporation, State Grid, Chongqing China
[2]College of Optoelectronic Engineering, Chongqing University, Chongqing China

*Corresponding author's e-mail: yumiao@cqu.edu.cn

**Abstract.** Ferrite materials have good frequency characteristics. Their relative permeability is large and relative permittivity is small. These make it suitable for matching layer and have a good application prospect in broadening the absorption bandwidth. In this paper, ferrite was prepared by a simple chemical precipitation-dehydration method, and the ferrite $Fe_3O_4$ particles with 40% mass fraction were mixed with paraffin to form a ring test sample. The experimental results show that the as-prepared sample at higher temperatures has more stable nucleation and fewer defects, so it has more significant electromagnetic wave absorption performance. The minimum reflection loss is about - 44 dB, and the electromagnetic absorption band with the reflection loss RL less than - 10 dB is located at 12.5-14.1 GHz (3 mm) and 6.5-8.1 GHz (5 mm), respectively.

## 1. Introduction

Excessive electromagnetic waves cause serious damage to human organs such as human eyes and central nervous system [1, 2]. United Nations Conference on the human environment has listed electromagnetic radiation as one of the major pollutants. In particular, the proliferation of electronic and communication equipment has increased the density of electromagnetic energy in confined spaces, and electromagnetic interference and pollution have become more and more serious [3]. At the same time, radar detection technology seriously threatens the survivability and combat effectiveness of battlefield weapons [4, 5]. Therefore, both electromagnetic protection and electromagnetic stealth technology are urgent issues that we are currently facing, and the core issue is the development of electromagnetic wave absorbing materials.

Ferrite is a kind of double complex dielectric material with both magnetic and dielectric properties, which has both electrical loss and magnetic loss for electromagnetic wave loss. Due to high resistivity ($10^8$-$10^{12}$), ferrite can avoid skin effect of metal conductor at high frequency, so it can still maintain high permeability at high frequency. At the same time, ferrite also has a wide range of raw materials and low price characteristics. The research on ferrite absorbing materials has achieved outstanding results and has also been applied to some practical engineering fields. However, the absorbing properties (such as absorption strength, bandwidth, etc.) of the ferrite material and the preparation process still need to be further improved.

Content from this work may be used under the terms of the Creative Commons Attribution 3.0 licence. Any further distribution of this work must maintain attribution to the author(s) and the title of the work, journal citation and DOI.
Published under licence by IOP Publishing Ltd

In this paper, ferrite Fe3O4 particles were prepared by simple chemical precipitation-dehydration method. Ferrous chloride (FeCl2) and ferric chloride (FeCl3) were iron sources. Compared with other common methods, such as sol-gel method, this method has the advantages of simple operation, short preparation period and low cost. Meanwhile, ferrite Fe3O4, with thin absorbing coating thickness, can achieve significant electromagnetic wave absorption in a wide frequency band. The loss mechanism of ferrite $Fe_3O_4$ was also studied in-depth.

## 2. Experiment

### *2.1. Raw material*

Ferrous chloride ($FeCl_2$) and ferric chloride ($FeCl_3$), analytical grade, Tianjin kemio Chemical Reagent Co., Ltd.; Sodium hydroxide (NaOH), analytical grade, Chengdu Kelong Chemical Reagent plant; deionized water, were self-made in the laboratory.

### *2.2. Preparation of ferrite $Fe_3O_4$*

First, Excess NaOH powder was dissolved in 100 ml of deionized water. Then, ferrous chloride (FeCl2) and ferric chloride (FeCl3) were weighed according to a molar ratio of n(FeCl3): n(FeCl2)=1:2, and they were poured into 200 ml of deionized water to prepare a homogeneous solution. Thereafter, the NaOH solution and the mixed solution containing iron ions were simultaneously slowly dropped into the reaction tank, and the water bath control temperature was maintained at 25 ° C (sample S1) or 80 ° C (sample S2), respectively. The reaction system was then gently stirred. After the reaction was completed, it was allowed to stand for 10 h. The obtained product was washed several times with deionized water and absolute ethanol, suction filtered, and dried in a vacuum oven at 80 ° C for 6 h. Finally, the final product Fe3O4 particles are obtained.

### *2.3. Testing and characterization*

The elemental composition, crystal structure and morphology of the particles are detected by vega2 X-ray spectrometer (EDS), X-ray diffraction XRD and scanning electron microscope (SEM), respectively.

Ring-shaped samples having an outer diameter of 7 mm, an inner diameter of 3.04 mm, and a thickness of 2 to 5 mm were pressed by paraffin-bonding 40 wt% of ferrite Fe3O4. The relative permeability and permittivity of the samples were tested by coaxial reflection/transmission technique using an Agilent 85071E vector network analyzer. The test range is from 2 to 18 GHz.

### *2.4. Calculation of absorption performance*

According to the transmission line theory, for single-layer absorbing materials with metal as the back plate, the reflection loss (RL) can usually be calculated by the following formula [6-8]:

$$Z_{in} = \sqrt{\frac{\mu_0 \mu_r}{\varepsilon_0 \varepsilon_r}} \tanh\left( j\left(\frac{2\pi}{c}\right) fd\sqrt{\mu_r \varepsilon_r} \right) \quad (1)$$

$$L = 20\log_{10}\left|(Z_{in} - Z_0)/(Z_{in} + Z_0)\right| \quad (2)$$

Where $Z_0$ is the impedance of free space, $Z_{in}$ is the input impedance between free space and the material interface, $c$ is the speed of light, $f$ is the frequency of the microwave, and $d$ is the thickness of the absorbing material.

## 3. Results and discussion

### *3.1. Characterization of the samples*

It can be seen from the figure 1 that the diffraction peaks are located at 2θ=30.1°, 35.6°, 43.1°, 53.5°, 57.1°, 62.7°, respectively, corresponding to (220), (311), (400), (422), (511), (440) crystal planes. This matches the cubic spinel structure magnetite (JCPDS card 19-0629) [2].

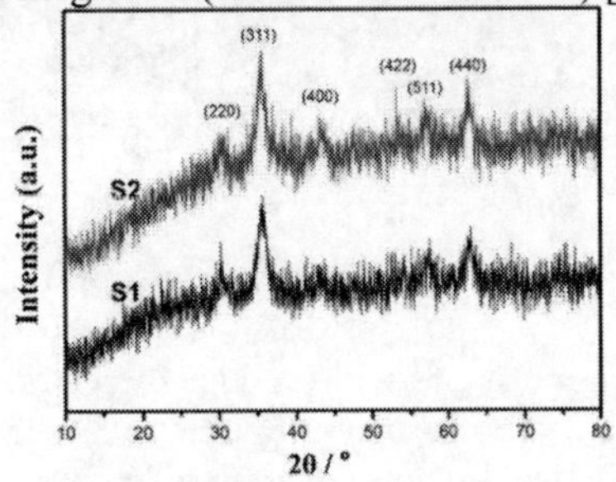


**Figure 1.** XRD image of ferrite $Fe_3O_4$

From SEM, the surface of ferrite $Fe_3O_4$ particles is rough and uneven, and has an irregular circle shape with a diameter of about 1-3 μm. The main elements are Fe and O. Cl and Na are reagents remaining and C might be mixed during scanning electron microscope operation.

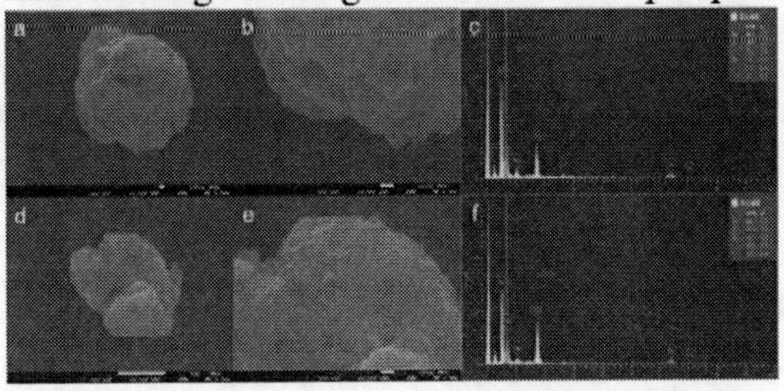


**Figure 2.** SEM and EDS images of ferrite $Fe_3O_4$ (a-c is a different multiple of SEM and EDS images for sample S1; d-f is a different SEM and EDS images for sample S2).

### *3.2. Electromagnetic parameters and microwave absorption properties*

The electromagnetic losses are closely related to complex permittivity and permeability. Figure 3a showed the relationship of complex permittivity of ferrite $Fe_3O_4$ with frequency. In 2-12 GHz, the complex permittivity of sample S1 and sample S2 is almost unchanged. Above 16 GHz, the complex permittivity of the samples fluctuated slightly with the frequency. However, in 12-16 GHz, the complex permittivity fluctuated greatly, and the valley and peak values appeared in 13-14 GHz. According to dielectric loss tangent ($\tan\sigma_E=\varepsilon''/\varepsilon'$, as shown in figure 4a), ferrite had a strong dielectric loss capability at 13-15 GHz. Due to the presence of Fe2+ and Fe3+ valence ions in the ferrite, electric dipole polarization loss was formed in the alternating electromagnetic field.

From figure 3b, the real part of complex permeability $\mu'$=0.72～1.2, the imaginary part of complex permeability $\mu''$=-0.04～0.46. The peak and valley value of the complex permeability appeared near 4 GHz, 7 GHz and 14 GHz. Based on magnetic loss tangent ($\tan\sigma_M=\mu''/\mu'$, as shown in figure 4a), there was strong magnetic loss capacity near the corresponding frequency, which was mainly caused by the natural resonance loss of ferrite.

IOP Conf. Series: Materials Science and Engineering **772** (2020) 012115 doi:10.1088/1757-899X/772/1/012115

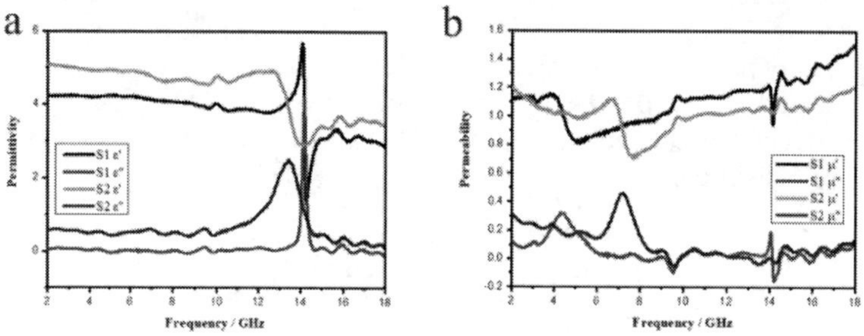


Figure 3. Complex dielectric constant and complex permeability u of $Fe_3O_4$ (samples S1 and S2)

Polarization hysteresis between interfaces will produce dielectric loss with the change of frequency [9]. However, there were $Fe^{2+}$, $Fe^{3+}$ valence changing ions in ferrite, so it was believed that the ferrite $Fe_3O_4$ had better dielectric loss. Debye dipole relaxation is an important mechanism for dielectric loss [10]. According to the transmission line theory, permittiviity can be expressed by the following formula [11,12]:

$$\left(\varepsilon'-\frac{\varepsilon_s+\varepsilon_\infty}{2}\right)^2+\left(\varepsilon''\right)^2=\left(\frac{\varepsilon_s-\varepsilon_\infty}{2}\right)^2 \tag{3}$$

According to formula (3), the curve of the real and imaginary parts of permittivity is a semicircle called Cole-Cole semicircle. Each Cole-Cole semicircle corresponds to a Debye relaxation process [13,14]. Both samples had a Cole-Cole semicircle. However, the Cole-Cole semicircle of sample S1 was relatively smooth, while the Cole-Cole semicircle of sample S2 was twisted obviously. This indicates that other loss mechanisms (such as electron polarization, conductivity loss and oxygen defect) also contribute greatly to sample S2.

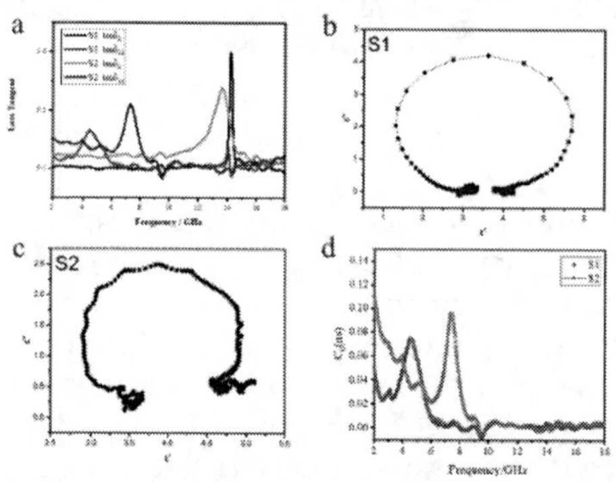


**Figure 4.** Electromagnetic loss tangent of ferrite $Fe_3O_4$ a; Cole-Cole semicircle b, c; with the value of $\mu''(\mu')^2f^{-1}$ varies with frequency in 2~18GHz.

The magnetic loss of ferrite $Fe_3O_4$ mainly comes from hysteresis, eddy current loss and natural resonance in 2-18GHz [15]. The hysteresis is caused by irreversible magnetization and is negligible under weak external magnetic field [16]. As can be seen from formula $\mu''\approx2\pi\mu_0(\mu')^2\sigma d^2f/3$, eddy current loss is related to the diameter $d$ and conductivity $\sigma$ of metal particles, here $\mu_0$ is the vacuum permeability [17,18]. If the magnetic loss of the absorbing material was only due to the eddy current loss, the value of $C_0=\mu''(\mu')^2f^{-1}$ should remain constant when the frequency changed [19]. However, the value of $C_0$ of ferrite fluctuated in 2-18 GHz. In other words, the effect of eddy current on their permeability can be ignored, and the magnetic loss of prepared ferrite mainly comed from natural resonance.

The reflection loss (RL) curve of the samples with different thickness was shown in figure 5. As is known to all, the reflection loss value RL< -10 dB (corresponding to 90% microwave absorption) is the target value of EM absorber in practical application. The reflection loss value and absorption bandwidth of sample S2 were better than that of sample S1. This might be because the nucleation of sample S2 was relatively more stable and the crystal nucleus defects were less. For sample S2, double absorption peaks occurred at 6-8 GHz and 13-15 GHz. When the thickness of the sample S2 was 2mm, the minimum reflection loss reached -44dB with bandwidth of 1.6 GHz at 6.5-8.1 GHz. While the thickness of the sample S2 continued to increase to 5 mm, the minimum reflection loss reached -43 Db with bandwidth of 1.6 GHz at 12.5-14.1 GHz.

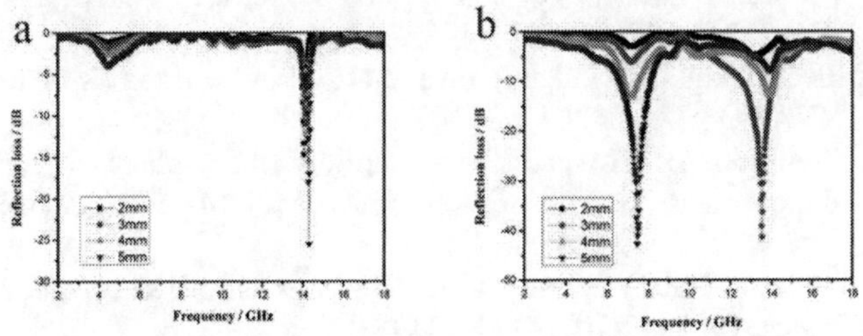


**Figure 5.** Reflection loss of ferrite $Fe_3O_4$ with thickness of 2-5mm (a. Sample S1; b. Sample S2)

## 4. Conclusion

Ferrite Fe3O4 absorber particles were successfully prepared by chemical precipitation-dehydration method in this paper. The prepared ferrite has both magnetic loss and dielectric loss. From the electromagnetic loss angle, Debye relaxation and natural resonance, the principle of absorption loss was systematically studied, which would provide a basis for future research on ferrite absorbing materials. The ferrite sample S2 prepared at 80 ° C had good electromagnetic wave absorption with the minimum reflection loss of -44 dB. Moreover, its absorption bandwidth reached 1.6 GHz. The high absorption ability and wide absorption band indicated that the effectiveness of the ferrite $Fe_3O_4$ absorbing particles in the field of electromagnetic stealth and electromagnetic protection. In addition, the flexible and convenient preparation method provided a new strategy for the preparation of other ferrite particles.

**Acknowledgments**

The financial support from Electric power research institute of Chongqing Electric Power Corporation of State Grid (52202317001D) is highly acknowledged.

## References

[1] Lv H, et al. Porous Three-Dimensional Flower-like Co/CoO and Its Excellent Electromagnetic Absorption Properties. [J]. ACS Applied Materials & Interfaces, 2015, 7(18):9776-9783

[2] Wang Z, et al. Enhanced microwave absorption of Fe3O4 nanocrystals after heterogeneously growing with ZnO nanoshell. [J]. RSC Advances, 2013, 3(10):3309-3315

[3] Zhao B, et al.Preparation and electromagnetic wave absorption properties of novel dendrite-like NiCu alloy composite [J].RSC Adv, 2015, 5(53): 42587-42590

[4] Zhao B, et al.Facile synthesis and novel microwave electromagnetic properties of flower-like Ni structures by a solvothermal method.[J].Journal of Materials Science: Materials in Electronics, 2014, 25(8):3614-3621

[5] Liu T, et al.Microporous Co@CoO nanoparticles with superior microwave absorption properties[J].Nanoscale, 2014, 6(4): 2447-2454

[6] Shi X, et al. Dual nonlinear dielectric resonance and nesting microwave absorption peaks of hollow cobalt nanochains composites with negative permeability[J]. Applied Physics Letters, 2009, 95(16):163108-1 -163108 -3

[7] Beshkar F, J.H.M.M, et al. Dendritic a-Fe2O3 nanostructures: facile hydrothermal synthesis,

characterization and microwave absorption. Journal of Materials Science Materials in Electronics, 2016. 12(27): 1-7

[8] Biswas S, G P Kar and S Bose, et al. Attenuating microwave radiation by absorption through controlled nanoparticle localization in PC/PVDF blends. Phys Chem Chem Phys, 2015. 17(41): 27698-712

[9] Zhuo R.F, et al. Multistep Synthesis, Growth Mechanism, Optical, and Microwave Absorption Properties of ZnO Dendritic Nanostructures. The Journal of Physical Chemistry CJ. Phys. Chem. C, 2008, 112(31): 11767-11775

[10] Zhao B, et al. Synthesis of flower-like CuS hollow microspheres based on nanoflakes self-assembly and their microwave absorption properties. JOURNAL OF MATERIALS CHEMISTRY A, 2015, 3(19): 10345-10352

[11] Zhu W, et al. Electromagnetic and microwave-absorbing properties of magnetic nickel ferrite nanocrystals. NANOSCALE, 2011, 3(7): 2862-2864

[12] Zhao B, et al. Synthesis of flower-like CuS hollow microspheres based on nanoflakes self-assembly and their microwave absorption properties. J. Mater. Chem. A, 2015, 3(19): 10345-10352

[13] Fang P H, et al. Cole—Cole Diagram and the Distribution of Relaxation Times. The Journal of Chemical Physics, 1965, 42(10): 3411-3413

[14] Du Y, et al. Shell Thickness-Dependent Microwave Absorption of Core–Shell Fe3O4@C Composites. ACS Applied Materials & Interfaces, 2014, 6(15): 12997-13006

[15] Yu M, et al. Flower-like carbonyl iron powder modified by nanoflakes: Preparation and microwave absorption properties. Applied Physics Letters, 2015, 106(16): 161904

[16] Wu M, et al. Microwave magnetic properties of Co50/(SiO2)50 nanoparticles. Applied Physics Letters, 2002, 80(23): 4404-4406.

[17] Murali K P, et al. Structure-magnetic property correlations in nickel-polymer nanocomposites. Journal of Materials Science: Materials in Electronics, 2016, 27(1): 154-162

[18] Wu M, et al. Microwave magnetic properties of Co[sub 50]/(SiO[sub 2])[sub 50] nanoparticles. Applied Physics Letters, 2002, 80(23): 4404.

[19] Wang Z, et al. Magnetic and microwave absorption properties of self-assemblies composed of core-shell cobalt-cobalt oxide nanocrystals. Phys Chem Chem Phys, 2015. 17(5): 3796-801.

# Study on Electrochemistry of Composite Electrode of Nano Reticular $MoS_2/RuO_2$

**Yun Lu, Jiancheng Ke, Wei Wang, Ruonan Wang, Li Wang**

School of Electronic Science and Engineering, University of Electronic Science and Technology of China, Chengdu 610054, China

**Abstract.** Nano Reticular $MoS_2$ is prepared in the hydrothermal method. By means of hydrothermal method, the cathode material of composite electrode of $MoS_2/RuO_2$ is prepared. For all kinds of prepared samples, physical characterization analysis, such as SEM and XRD, and electrochemistry tests, such as cyclic voltammetry, alternating-current impedance and galvanostatic charge-discharge, are conducted. The test results show that the addition of a small amount of molybdenum disulfide improves the specific capacity of composite electrode, lowers alternating equivalent impedance and improves conservation ratio of specific capacity. When the doping content of molybdenum disulfide is 25%, specific capacity of binary composite electrode can reach 413 F/g and equivalent alternating-current impedance is equal to 0.665 Ω. By means of 800 charge-discharge cycles, specific capacity still remains around 87%. Key words: Molybdenum Disulfide; Hydrothermal Method; Binary Composite Electrode Materials; Equivalent Alternating-current Impedance; Specific Capacity.

In recent years, most researches focus on improving the energy production and the performance of storage device. The super capacitor is considered to be a very promising power storage device and compared with similar power storage device, has superior performance, which means battery. Different from batteries, charging and discharging process of super capacitors is faster. Compared with batteries, it has higher power density and longer service life [1]. In the next several years, it can play more important role in energy storage field. Super capacitors of transition metal oxides, such as Sn, Mn, Ti, Co, and Ru, are explored to be applied to this application [2-5].Transition metal oxides, such as $RuO_2$, have high theoretical specific capacity, but have poor conductive property, poor cycling stability and other shortcomings, which have limited its application.

Due to high surface volume ratio and surface charge ratio, two-dimension nano materials have attracted the interests of researchers with its unique physical, chemical and biological functions. From the perspective of crystallography, two-dimensional crystals commonly have highly anisotropic physical property [6-9]. Molybdenum disulfide has the merits, such as less gathering, larger than superficial area, network interconnection and vertical channels' effectively exposing active sites. Molybdenum disulfide is an important 2D material and the supplementary material of grapheme and is very attractive. $MoS_2$ has the merits, such as compared with superficial area, larger area, good chemical stability, high ionic conductivity and laminated structure, and has certain advantages in electrochemical devices [10, 11]. Intrinsic ionic conductivity is higher than the one of oxides. Theoretical capacitance is higher than graphite specific conductance [12-14]. Excellent storage capacitance makes it have multitudes of excellent mechanical, optical and electronic performance and has it applied in many electronic circuits,

Content from this work may be used under the terms of the Creative Commons Attribution 3.0 licence. Any further distribution of this work must maintain attribution to the author(s) and the title of the work, journal citation and DOI.
Published under licence by IOP Publishing Ltd

which need low standby power, as well as effective photocatalysis, solar energy funnel, electronics and excitonic integrated circuit [15, 16]. In the semiconductor prepared by $MoS_2$, it can be observed the change from indirect band gap type to direct band gap type [17, 18].

Molybdenum disulfide has these merits, such as less gathering, compared with superficial area, lager area, network interconnection, vertical channels' effectively exposing active sites. Transition metal -- ruthenium oxide has the features, such as high specific capacity, easy agglomeration and poor cycling performance. This article has prepared reticular nano molybdenum disulfide/$RuO_2$ composite electrode material and improved the shortcoming of ruthenium oxide by using the advantage of molybdenum disulfide, which is not easily condensed, and then further improved the performance of composite electrode material.

## 1. Experiment

### *1.1. Preparation of $MoS_2$*
Weigh appropriate amount of high-purity molybdenum powder and disperse it into the deionized water (50ml); add in appropriate hydrogen peroxide and drip dispersion liquid of molybdenum trioxide into thiourea solution; mix it with magnetic force for 60 min, disperse it with the ultrasound of ultrasonic wave for 30 min, transfer the mixed solution into high pressure reactor (100ml) and seal it. And then put it into the constant temperature drying box (200°C) for about 10 hours and cool it till the room temperature. Filter the solution in the reactor and clean it with deionized water and absolute ethyl alcohol. Put the product after being filtered into the vacuum (80°C) to dry for a whole day. In the end, the reticular nano molybdenum disulfide can be achieved.

### *1.2. Synthesis of $MoS_2/RuO_2$ Composite Electrode Material*
Weigh $RuCl_3.H_2O$ (2.7g) and dissolve it in water (100ml). Add in appropriate amount of nano molybdenum disulfide into it. Stir it with magnetic stirrer for 30 min and conduct the ultrasonic dispersion for 30 min. Pour the mixed liquors of $RuCl_3/MoS_2$ into the high-pressure reactor under normal pressure and around 200°C. After 12hours, it can be naturally cooled till room temperature. And then repeatedly wash it with deionized water and ethyl alcohol. In the end, put the product into the vacuum drying box (150°C) to dry for 10 hours.

### *1.3. Preparation of Composite Electrode*
Weigh polyvinylidene fluoride (0.020g) and dissolve it in NMP solution (1.5g). Stir it with magnetic stirrers till it is transparent. Weigh mixed composite material of $MoS_2/RuO_2$ (0.16g) and mix it with conductive additive - acetylene black (0.02g). Add the mixed materials into the NMP solution mixed in polyvinylidene fluoride and sufficiently mix it till the proper viscosity. Use the applicator to evenly apply electrode slurry onto the tantalum sheet, which is handled properly. And then put the tantalum sheet into the vacuum drying box (60°C) to dry for 6 hours, then dry for 8 hours under the temperature of 110°C, as well as dry for 3 hours under the temperature of 170°C. Then composite electrode is achieved.

### *1.4. Testing Instruments*
Use the JSM-6490LV Scanning Electron Microscope from JEOL, D8 ADVANCEX ray diffractometer from Rigaku Industrial Corporation, CS2350 double-cell electro-chemical workstation from Wuhan Corrtest Instruments Corporation and CT2001A super-capacitor test system from Wuhan LAND Electronic Co., Ltd. The test process is completed under room temperature.

## 2. Results and Discussion

### *2.1. Scanning Electron Microscopy (SEM) Analysis*

Figure (a) is the nano ruthenium oxide synthesized by liquid phase precipitation method. It is globular aggregation and arrangement among ruthenium oxide. Figure (b) is the reticular nano molybdenum disulfide, which is produced by using high-purity molybdenum powder and thiourea under hydrothermal reaction. Figure (c) is composite electrode material prepared by hydrothermal method of $MoS_2/RuO_2$. From the fiture, we can see clearly that reticular nano molybdenum disulfide can be better embeded in ruthenium oxide, which is accumulated and arranged. Figure (d) is EDS chart of composite electrode material of $MoS_2/RuO_2$. From thc figure, we can see the main components of composite electrode material are the elements -- O, Mo and Ru.

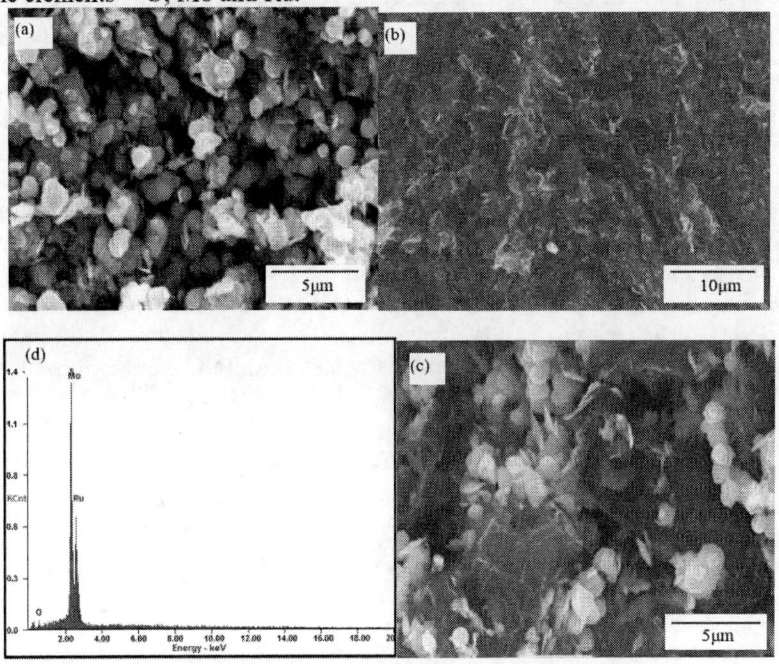


**Figure 1.** XRD and EDS of Electrode Material
(a) XRD of Granular Ruthenium Oxide Synthesized by Liquid Precipitation Method; (b) XRD of Reticular Nano Molybdenum Disulfide Generated by High-purity Molybdenum Powder and Thiourea under Hydrothermal Reaction; (c) XRD of Composite Electrode Materials of $MoS_2/RuO_2$; (d) EDS of Composite Electrode Materials of $MoS_2/RuO_2$

### *2.2. XRD Analysis*

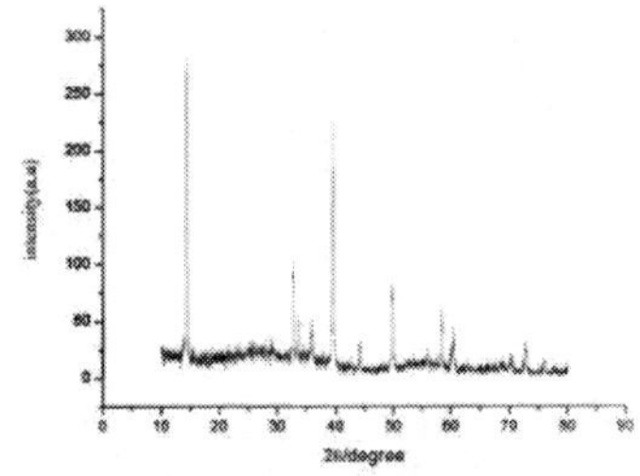

**Figure 2.** XRD Diffraction Pattern of Composite Materials of $MoS_2/RuO_2$ Prepared and Synthesized with Hydrothermal Method

IOP Conf. Series: Materials Science and Engineering **772** (2020) 012116 doi:10.1088/1757-899X/772/1/012116

Figure 2 is XRD of composite electrode materials of $MoS_2/RuO_2$. In the figure, 14.2°, 33.2°, 39.6°and 59°are the diffraction peaks of MoS2, which respectively respond to crystal faces of (002), (100), (103) and (110) of MoS2. 34.51°, 50.76°and 62.5°are the crystal faces of (101), (211) and (310) of RuO2. In the figure, the diffraction peak at 14.2° is very pointed. The diffraction peak corresponds to 14.2°of $MoS_2$, which means along the direction of C axis, reticular nanostructure is good. At the same time, at 39.6°of diffraction peak, it respectively corresponds to the crystal face of (103) of $MoS_2$. $MoS_2$ is better embedded in $RuO_2$, which is accumulated.

## *2.3. Cyclic Voltammetry Features*

The electrode materials are performed the test of electrochemical cycling performance. The electrolytic tank is put in 38% sulfuric acid solution, Pt electrode is applied for the auxiliary electrode, mercury - mercurous sulfate electrode is applied for reference electrode (saturated $K_2SO_4$). From the figure, it can be clearly seen that composite electrode material of $MoS_2/RuO_2$ has very good pseudocapacitance feature. From the figure, we can see with the improvement of the content of molybdenum disulfide, the area of rectangle presents the trend of increasing first and reducing later. And rectangle area is the representation of material specific capacity. When the content of molybdenum disulfide is 25%, the area of rectangle is the largest. When the content of molybdenum disulfide is within the range from 0% to 25%, specific capacity presents the trend of improvement. Because laminated molybdenum disulfide is embedded in the accumulated ruthenium oxide, it makes the superficial area of composite electrode larger and improves specific conductance, and then further improves specific capacity. And when the content of molybdenum disulfide keeps improving, specific capacity of electrode material presents the trend of declining. The reason is that the content of molybdenum disulfide improves, and correspondingly, the content of ruthenium oxide is reduced. And the feature of pseudocapacitance is also reduced.

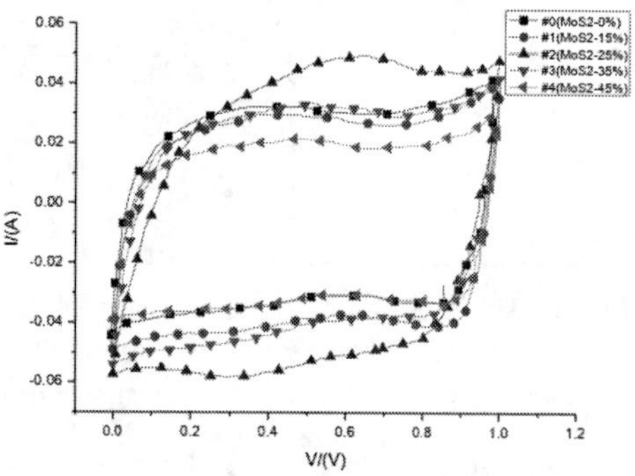


**Figure 3.** Cyclic Voltammetry Features of Composite Electrode of $MoS_2/RuO_2$ under different contents of $MoS_2$ (10mV/s)

## *2.4. Measurement of Specific Capacitance*

For the computational formula of specific capacitance, refer to formula 1. Thereof, *I* is the discharge current (unit: A), $\Delta t$ is the discharge time (unit: s), *m* is the electrode material quality (unit: g), $\Delta V$ is the voltage range and *C* is the capacity of supercapacitor (unit: F/g)

$$C = \frac{I\Delta t}{m\Delta V} \quad (1)$$

Figure 4 is the charge-discharge curve of current under different contents of $MoS_2$. Discharge current is $50mA/cm^2$, electrolysis is the sulfuric acid solution, with the concentration of 38%. The voltage range of charge-discharge is 0 ~ 1 V. Table 1 is the discharge time and specific capacity under different concentrations of $MoS_2$.

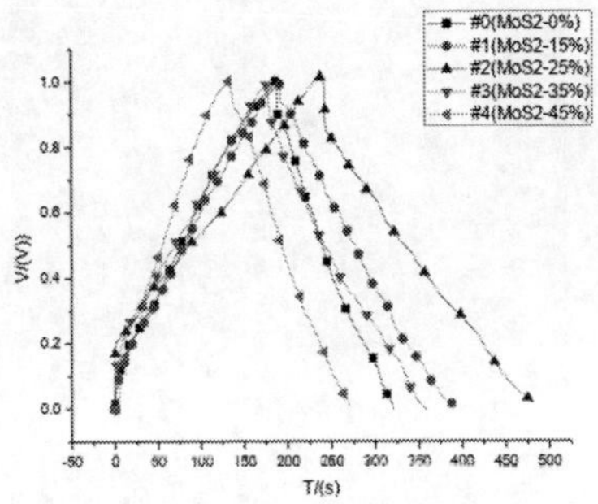


**Figure 4.** Galvanostatic Charge-Discharge Curve of Composite Electrode of $MoS_2/RuO_2$ under Different Contents of $MoS_2$

**Table 1.** Specific Capacitance of Composite Sample under Different Contents of $MoS_2$

| MoS2 content | 0% | 15% | 25% | 35% | 45% |
|---|---|---|---|---|---|
| Discharge time(s) | 326s | 393s | 478s | 356s | 273s |
| Specific capacitance (F/g) | 282 | 340 | 413 | 308 | 236 |

### *2.5. AC Impedance Features*

Figure 5 sees AC equivalent impedance of electrode materials under different contents of $MoS_2$. AC equivalent impedances under 0%, 15%, 25%, 35% and 45% are respectively 0.934Ω, 0.864Ω, 0.665Ω, 0.64Ω and 0.358Ω. From the figure and data, we can see when $MoS_2$ is added in, AC equivalent impedance is reduced constantly. The reason is nanometer $MoS_2$ is embedded in globose ruthenium oxide, which has piled and arranged. This improveds the electric conductivity among ruthenium oxide and then further improves resistance properties.

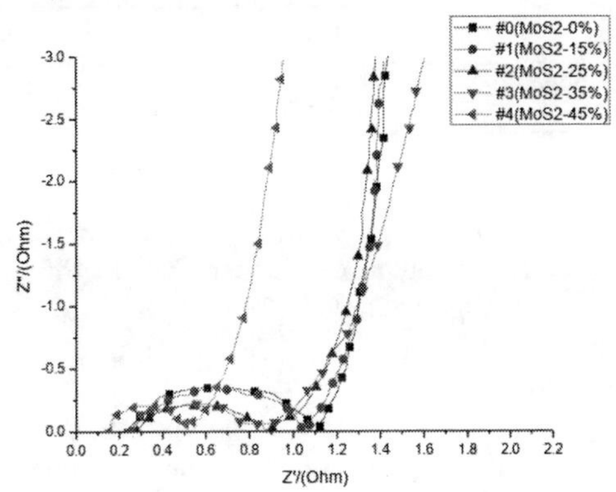


**Figure 5.** Electrochemical Impedance Spectroscopy of Composite Electrode of $MoS_2/RuO_2$ under Different Contents of $MoS_2$

### *2.6. Cycle Life*

Cycle performance of composite electrode materials is the important index of the service life of super-capacitor, which makes the measurement system of LAND battery have the cycle performance of charging for 800 times under the voltage range of 0~1V and charge - discharge current density of $50mA/cm^2$.

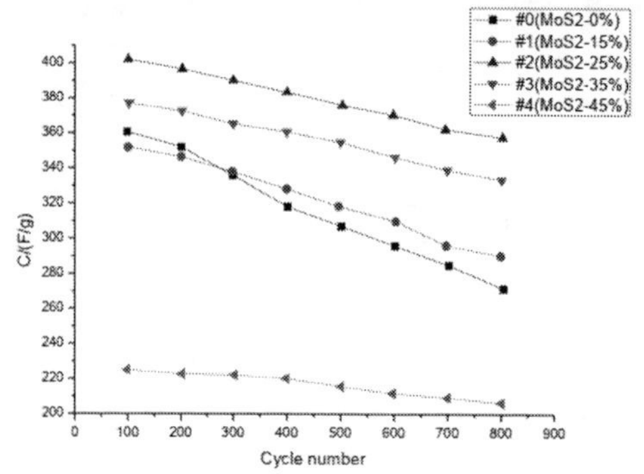


**Figure 6.** The Curve of Specific Capacitance with the Number of Charges and Discharges

From the figure, we can see that under 0% content of $MoS_2$, with more charge-discharge times, the specific capacity of electrode material largely declines. The reason is although the transition metal - ruthenium oxide has higher specific capacity, agglomeration leads to poor dispersity and then further leads to the fast decline of ruthenium oxide capacity. From the fugure, the content of $MoS_2$ is improved, which can improve the cycle performance of ruthenium oxide and the conservation rate of specific capacitance. The reason is reticular nano - molybdenum disulfide has the merits, such as less gathering, lager area (compared with superficial area), network interconnection, vertical channels' effectively exposing active sites, which make up for the shortcoming that agglomeration of transit metal leads to poor cycle performance.

**Table 2.** Conversation Rate of Specific Capacitance of Composite Samples under Different Contents of $MoS_2$

| $MoS_2$ content | 0% | 15% | 25% | 35% | 45% |
|---|---|---|---|---|---|
| Conversation Rate of Specific Capacitance (%) | 75.2% | 82.6% | 87.1% | 88.5% | 91.5% |

## 3. Conclusion

(1) The addition of reticular nanometer molybdenum disulfide enlarges the superficial area of $MoS_2/RuO_2$ and improves specific conductance. Under the electric current density of 50mA/cm$^2$, specific capacity is improved to 413F/g and impedance is reduced to less than 0.665Ω.

(2)By means of 800 times of galvanostatic charge-discharge, specific capacity of composite electrode of $MoS_2/RuO_2$ remains around 87%.

(3) By utilizing the merits of reticular nanometer molybdenum disulfide--difficult agglomeration, larger area (compared with superficial area) and high ionic conductivity, the composite electrode materials of $MoS_2/RuO_2$ synthesized can effectively improve pseudocapacitance feature of ruthenium oxide.

## References

[1] Choi J H , Lee C, Cho S, et al. High capacitance and energy density supercapacitor based on biomass-derived activated carbons with reduced graphene oxide binder [J]. Carbon, 2018: S0008622318301143.

[2] Bonu V, Gupta B , Chandra S , et al. Electrochemical supercapacitor performance of SnO2 quantum dots[J]. Electrochimica Acta, 2016:S0013468616307241.

[3] Aravinda L S, Bhat K U, Bhat B R. Nano CeO2 /activated carbon based composite electrodes for high performance supercapacitor [J]. Materials Letters, 2013, 112(12): 158-161.

[4] Aravinda L S , Nagaraja K K , Nagaraja H S , et al. Fabrication and performance evaluation of hybrid supercapacitor electrodes based on carbon nanotubes and sputtered TiO2. [J].

Nanotechnology, 2016, 27(31): 314001.
[5] Liang D , Tian Z , Liu J , et al. MoS2 nanosheets decorated with ultrafine Co3O4 nanoparticles for high-performance electrochemical capacitors [J]. Electrochimica Acta, 2015.
[6] Feng Z S, Kang Z H, Wu F G, et al. SHG in doped GaSe: In crystals [J]. Optics Express, 2008, 16(13):9978.
[7] V.V. Atuchin, T.A. Gavrilova, T.I. Grigorieva,et al. Sublimation growth and vibrational microspectrometry of α-MoO3 single crystals [J]. Journal of Crystal Growth, 2011, 318 (1): 987-990.
[8] Atuchin V V, Golyashov V A, Kokh K A, et al. Formation of Inert $Bi_2Se_3$ (0001) Cleaved Surface[J]. Crystal Growth & Design, 2011, 11(12):5507-5514.
[9] Kokh K A, Atuchin V V, Gavrilova T A, et al. Microstructural and vibrational properties of PVT grown $Sb_2Te_3$ crystals [J]. Solid State Communications, 2014, 177(1): 16-19.
[10] Jie X, Choi D, Cosimbescu L, et al. Exfoliated MoS2 Nanocomposite as an Anode Material for Lithium Ion Batteries [J]. Chemistry of Materials, 2010, 22(16):4522-4524.
[11] HUANG, KeJing, Wang L, et al. Layered MoS2-graphene composites for supercapacitor applications with enhanced capacitive performance [J]. International Journal of Hydrogen Energy, 2013, 38(32):14027-14034.
[12] Thangappan R, Kalaiselvam S, Elayaperumal A, et al. Graphene decorated with MoS2 nanosheets: a synergetic energy storage composite electrode for supercapacitor applications. [J]. Dalton Transactions, 2016, 45(6): 2637-2646.
[13] Verble J L, Wieting T J. Lattice Mode Degeneracy in MoS2 and Other Layer Compounds [J]. Phys.rev.lett, 1970, 25(6): 362-365.
[14] Verbrugge M W. Adaptive Characterization and Modeling of Electrochemical Energy Storage Devices for Hybrid Electric Vehicle Applications [M]// Modeling and Numerical Simulations. 2008.
[15] Braibanti A, Ostacoli G, Paoletti P, et al. Recommended procedure for testing the potentiometric apparatus and technique for the pH-metric measurement of metal-complex equilibrium constants [J]. Pure & Applied Chemistry, 1987, 59(12): 1721-1728.
[16] Fortin E, Sears W M. Photovoltaic effect and optical absorption in MoS 2[J]. Journal of Physics & Chemistry of Solids, 1982, 43(9): 881-884.
[17] Mak K F, Lee C , Hone J , et al. Atomically thin MoS2: A new direct-gap semiconductor[J]. Physical Review Letters, 2010, 105 (13): 136805.
[18] Fu L, Wan Y, Tang N, et al. K-Λ crossover transition in the conduction band of monolayer MoS2under hydrostatic pressure: [J]. Sci Adv, 2017, 3(11): e1700162.

# Study on Strain Test Law of the Semi-rigid base Asphalt Pavement

**Feng Liu [1, 2], Chuanhai Wu [1, 2], Shanqiang Li [1, 2] and Xinquan Xu [1, 2]**

[1] Research and Development Center on Road Transport Safety and Emergency Support Technology & Equipment, Ministry of Transport, PRC, Guangzhou 510420, Guangdong, China
[2] Guangdong Hualu Transport Technology Co., LTD, Guangzhou 510420, Guangdong, China

**Abstract.** In order to grasp the similarities and differences in the experimental strain law of different types of semi-rigid base asphalt pavement, based on the Yunluo highway durability test pavement, the strain monitoring system , temperature monitoring system and humidity monitoring system of the full-scale pavement were constructed, and corresponding test plans were formulated. Through the test scheme, the strain response laws of different structural layers of two asphalt pavements were obtained. Through the comparative analysis with the relevant theoretical analysis results, the rationality and credibility of the strain layout scheme and test scheme of the Yunluo durability test pavement were confirmed. Through the analysis of the measured strain data, it was verified that the graded crushed rock layer did have a decentralized effect on the load. At the same time, it was also confirmed that when the same pavement material was in different structural layers, its mechanical response under load was different.

**Keywords:** Road Engineering, Graded Crushed Rock, Cement Bound Granular, Layout and Test Plan.

Different from the research on the actual mechanical response of AASHTO, WASHO (Western Association of State Highway Officials) and other experimental roads developed by foreign developed countries for full-thick asphalt pavement structure[1-3], because China is subject to construction funds and high-quality asphalt resources in the early stage of expressway construction, the semi-rigid base asphalt pavement represented by cement stabilized graded gravel has developed rapidly and occupied a dominant position in China's highway construction[4]. Therefore, the practical experience of the fully flexible asphalt pavement structure based on the sensing system cannot be fully applied to the mechanical response analysis and design of the semi-rigid base asphalt pavement structure. In order to reveal the early damage mechanism of the semi-rigid base asphalt pavement and the factors affecting its mechanical behavior, the Highway Research Institute of the Ministry of Transport has built a full-scale test site in Beijing, and arranged corresponding sensors to study different pavements[5]. Hu Peng[6] carried out research on the response and damage of asphalt pavement under dynamic load. Based on the durability test road project of Yunluo Expressway in Guangdong Province, this paper will greatly enrich

Content from this work may be used under the terms of the Creative Commons Attribution 3.0 licence. Any further distribution of this work must maintain attribution to the author(s) and the title of the work, journal citation and DOI.
Published under licence by IOP Publishing Ltd

the theory and practical experience of semi-rigid base asphalt pavement structure by studying the measured strain law of each type of semi-rigid base asphalt pavement.

## 1. Overview of Yunluo Expressway Test Section

Yunluo Expressway started in Luoding, Guangdong Province, and the research team paved three test sections with a total length of about 3km[7]. Among them, two types of semi-rigid base asphalt roads of 2km are paved, as shown in Figure 1. The base layer and the lower base layer of the structure 1 are composed of cement stabilized graded gravel, and only the strength of the material is different, and the structure is called a reinforced semi-rigid base asphalt pavement. Considering the dispersion of the flexible material to the stress, the lower base layer of the structure 3 is paved with graded gravel, which is called a flip-chip semi-rigid base asphalt pavement.

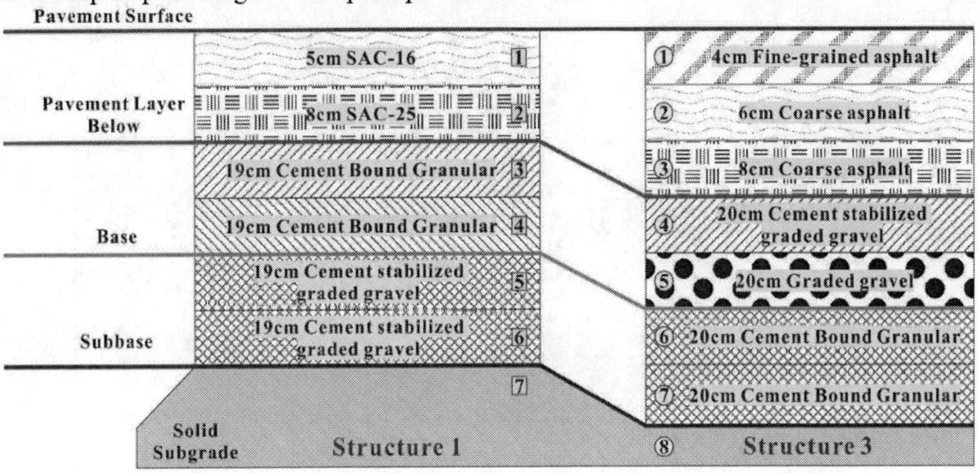


**Fig. 1** Schematic diagram of two types of semi-rigid base asphalt pavement

## 2. Strain Sensor Layout and Test Plan

### *2.1. Strain Sensor Layout*

Structural layer strain test was carried out on the Yunluo expressway test section (Fig. 2 is the CTL strain sensor site layout diagram), temperature field test, humidity field test and various structural layer pressure tests. However, due to space limitations, this article can only analyzed the strain field law of two rigid base asphalt pavement structure.

**Fig. 2** Field installation of the US Building Technology Laboratory (CTL) resistance strain sensor

Figure 3 shows a schematic diagram of two structural sensor arrangements. In the floor plan, the blue "I" and "H" are the locations of the strain sensors. According to the literature[8], two strain sensors adjacent to the same structural layer are arranged at a pitch of 60 cm. Figure 4 is a layout view of the strain sensor corresponding to Figure 3 in each structural layer position. Among them, S1-1 represents the first sensor of structure 1, and S3-1 in parentheses represents the first sensor of structure three, and the others are the same.

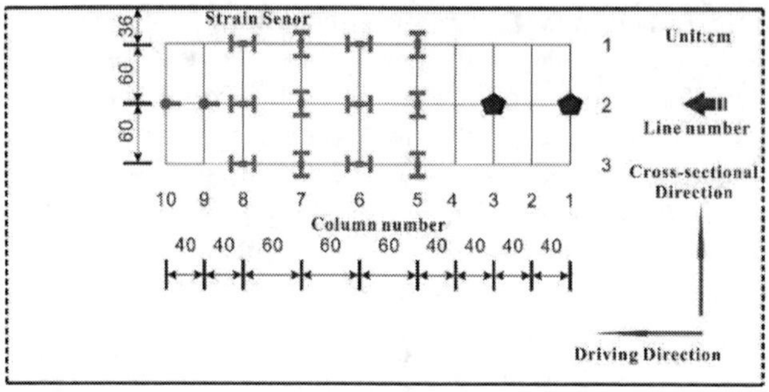


(a) Layout plan

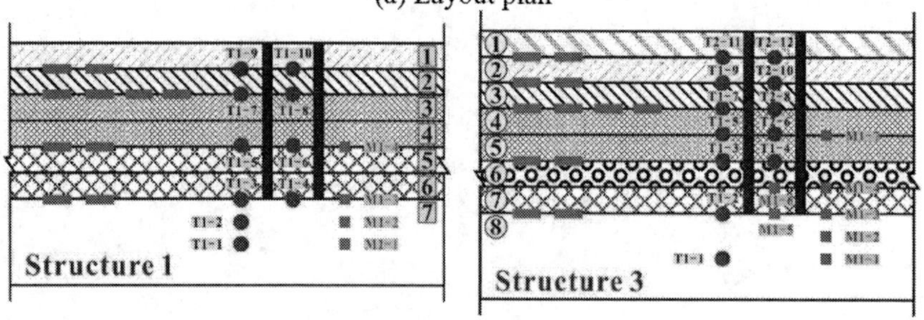


(b) Elevation layout

**Fig. 3** Two kinds of sensor layout of semi-rigid base asphalt pavement structure

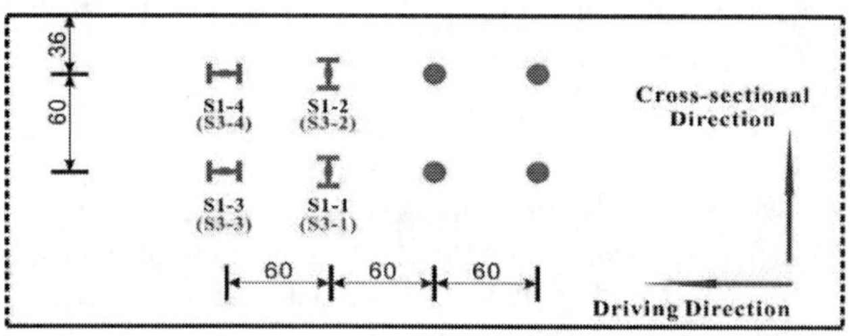


**Note: There is no strain sensor at the blue origin in the figure.。**

(a) Lower base layer bottom sensor layout

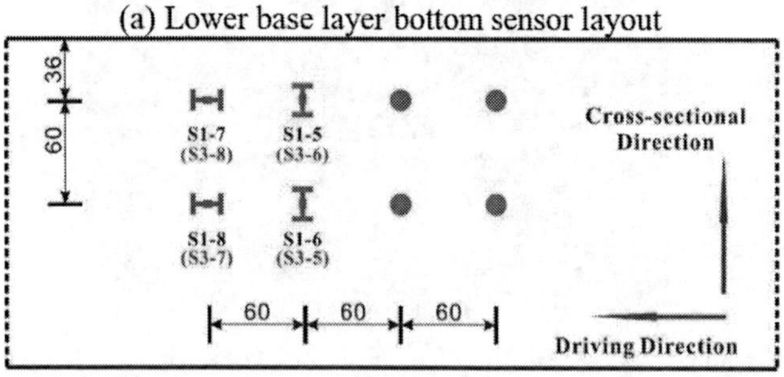


(b) Lower base layer sensor layout

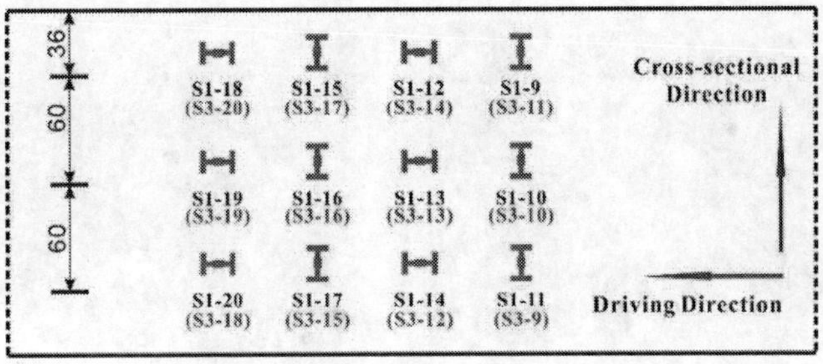


(c) Lower layer bottom sensor layout

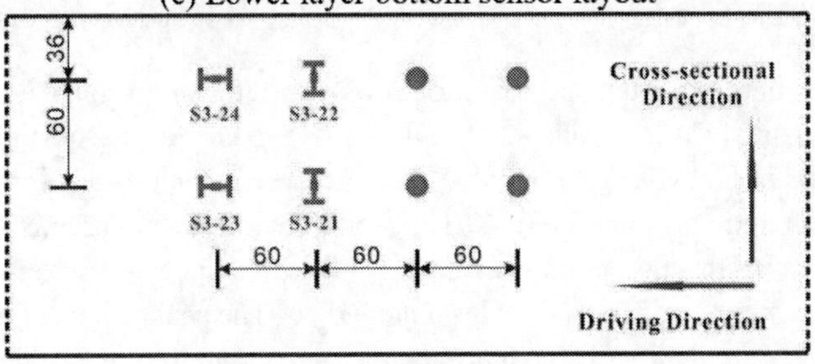


(d) Middle-layer bottom sensor layout

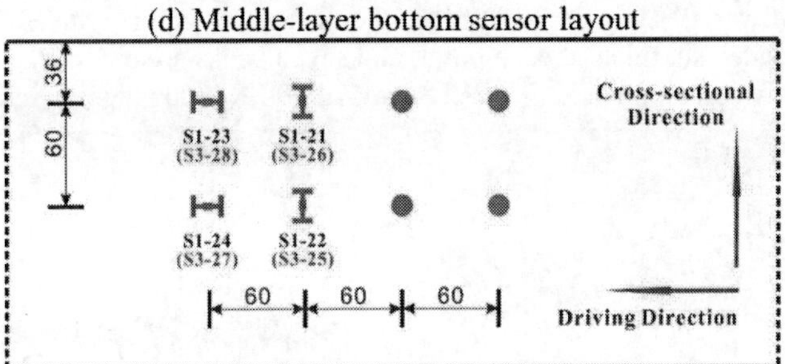


(e) Upper layer bottom sensor layout

**Fig. 4** Test road pavement structure scheme sensor marking diagram

*2.2. Mechanical Response Test Program*

The mechanical response test of the Yunluo test section uses two loading modes: FWD (Fig. 5) and real vehicle load (Fig. 6). FWD is a pulse load. The real vehicle load is loaded along the line number 2 in the direction of the blue arrow in the plane layout diagram of Figure 3. However, due to space limitations, only the strain response law of asphalt pavement under FWD load is analyzed here. Loading test is using FWD at the blue "I" and "H" of the strain sensor layout scheme of Figure 3. After loading, use the Donghua DH3817F dynamic data acquisition instrument to collect the strain sensor signals of each structural layer and save them accordingly. The saved strain data format is as follows: FWD + load action position (structure type + line number + column number) + sensor number. FWD118S1-23 indicates that the FWD load acts at the intersection of line number 1 and column number 8 of structure 1 of Figure 3a, while S1-23 represents the strain sensor of the structure-upper layer bottom.

**Fig. 5** FWD test

**Fig. 6** Driving load test

In engineering practice, it is often necessary to process a useful strain signal containing noise (Figure 7b). The signal processing flow is as follows: ①DH3817F dynamic data acquisition instrument collects raw data, and converts it into MATLAB data. ②The measured strain data is imported into MATLAB, the data is filtered (filtered and smoothed). ③The strain measurement curve is zeroed at the baseline, and the data before the strain strain point is averaged.④The average value is zeroed, the signal curve is translated, and the true strain waveform signal is obtained. ⑤The peak value of the relevant strain signal is extracted. Due to the complexity of signal processing, the research team developed the corresponding signal processing software[9]. Figure 7(b) shows the processed strain signal. It can be seen from the figure that FWD is loaded six times at each point, and seven peaks are obtained. The second peak to the sixth peak respectively corresponds to the FWD loads of 50 kN, 80 kN and 100 kN.

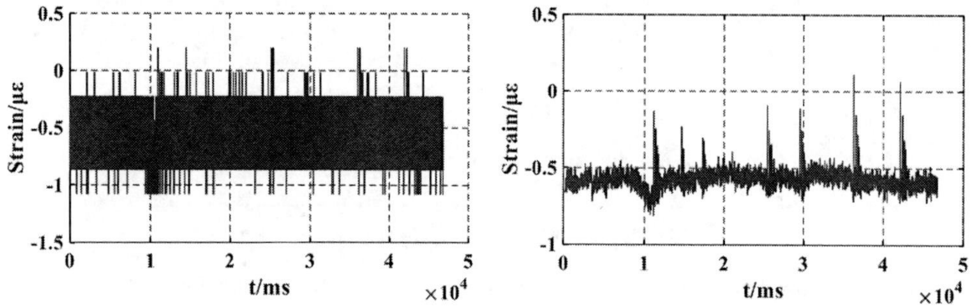


(a) Original strain signal (b) Treated strain signal

**Fig. 7** Asphalt pavement FWD118S1-8 test point strain curve

## 3. Analysis of Actual Measurement Strain Law of Asphalt Pavement

Figure 8 to Figure 12 are the measured strain curves of the driving directions of the structural layers of the two semi-rigid base asphalt pavements. Among them, Fig. 9 is the measured strain curve of the cross-sectional direction and the driving direction of the three middle layers of the structure. The following is divided into two aspects: asphalt surface layer and (bottom) base.

### *3.1. Analysis of Strain Law of Asphalt Surface*

From Figure 8, it can be seen that the upper layer bottom of two-layer asphalt pavement and three-layer asphalt pavement structure are under the compressive strain under load, which is beneficial to the stress of the structure. It can be seen from Fig. 10 that the underlying layers of the two structures are subjected to tensile strain, which is in good agreement with the theoretical analysis results of the elastic layered system of asphalt pavement. This indicates that there is the possibility of fatigue cracking at the bottom of the asphalt layer, and it is unreasonable to use the asphalt layer bottom stress to control the cracking to design the asphalt pavement. Under the same 80kN load, the tensile strain of structure three (about 20με) is about 12με than that of structure one (about 8με), which means that the risk of structural three-

IOP Conf. Series: Materials Science and Engineering **772** (2020) 012117 doi:10.1088/1757-899X/772/1/012117

fatigue cracking is larger than that of structure one. It can be seen from the strain measurement curves in the two directions of the middle layer of the structure of Fig. 9 that the middle layer of the structure is subjected to tensile strain in both directions, and the tensile strain in the driving direction is larger than that in the cross-sectional direction. Therefore, this kind of structure has a higher proportion of fatigue lateral cracks in the middle layer than in the longitudinal direction. At the same time, comparing the tensile strain in the driving direction of the middle layer (Fig. 9a) and the lower layer (Fig. 10 right), the tensile strain of the lower layer is smaller than that of the middle layer, and the middle layer of the structure 3 is less favorable.

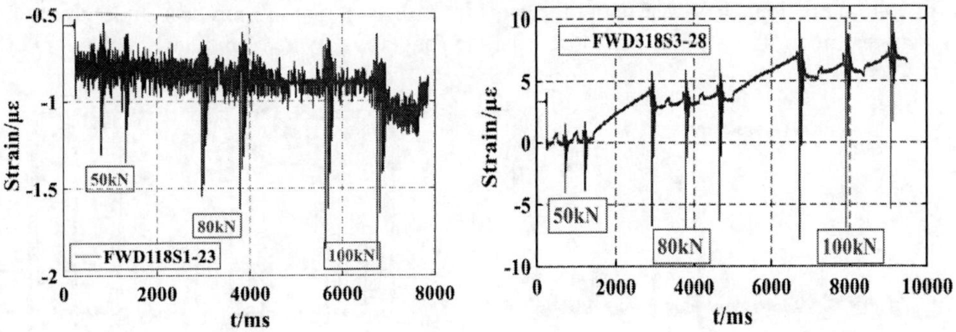


**Fig. 8** Strain curve of the corresponding measuring point at the bottom of the upper layer

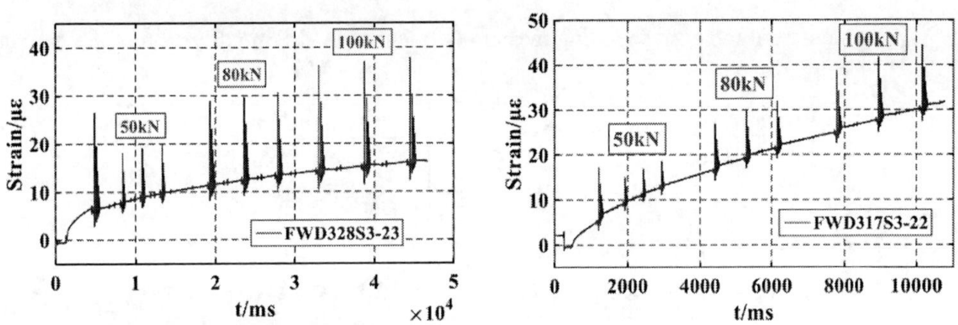


(a) Driving direction strain (b) Cross-sectional direction strain

**Fig. 9** Strain curve of the measuring point corresponding to the driving direction of the middle layer in structure 3

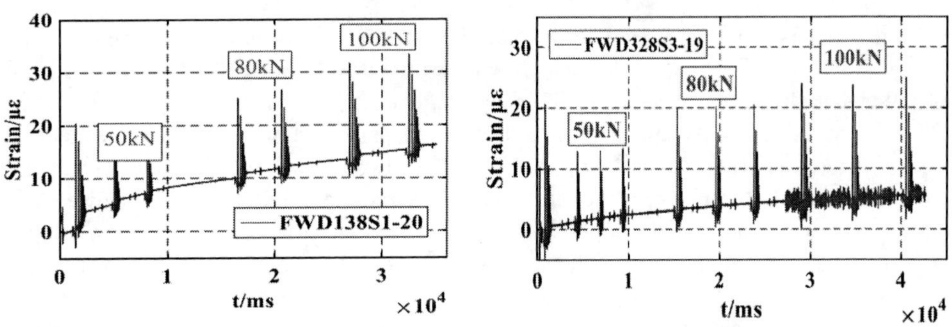


**Fig. 10** The driving direction of the lower layer corresponds to the strain curve of the measuring point

### *3.2. Analysis of Measured Strain Law of Lower (Bottom) Base Layer*

By analyzing the measured strain curves of the base layer and the lower base layer of the two structures, both structural layers are subjected to tensile strain, and the tensile strains tend to increase with the increase of the load. However, in the same layer, the same semi-rigid material under the same level of FWD load (100kN), the tensile strain at the bottom of the lower base layer containing the graded gravel structure 3 is about 5με compared with the structure one (Figure 11 ). According to Figure 12, the tensile strain of the bottom base of the structure 3 is smaller than that of the structure 1. This shows that graded gravel layer has dispersing effect on load, and the existence of graded gravel layer is beneficial to the stress of soil foundation. However, under the same load, the strain of the sub-base under the structure is larger than that under the sub-base. It shows that the dangerous layers of fatigue cracking are different between them.

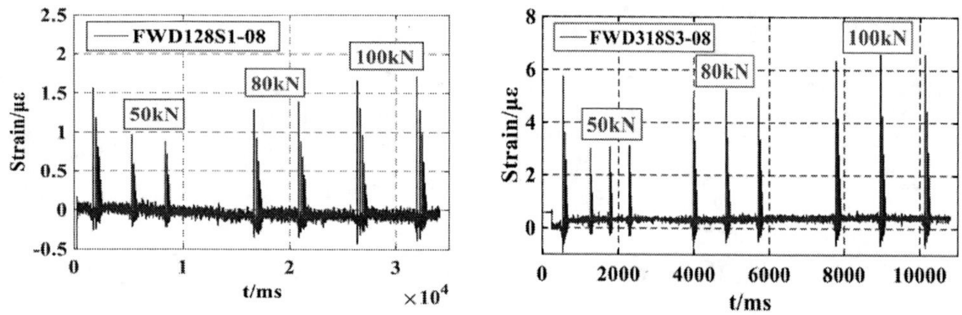


**Fig. 11** Strain curve of the corresponding measuring point in the driving direction of the base layer

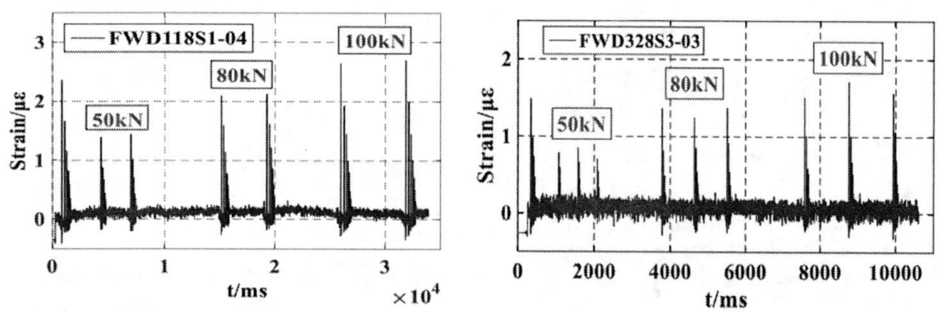


**Fig. 12** Strain curve of the corresponding measuring point at the bottom of the lower base layer

## 4. Conclusion

Through the analysis of the measured strain data of each structural layer of the durability test road of Yunluo Expressway in Guangdong Province, the measured strain law of two types of semi-rigid base asphalt pavement structures is obtained. Research indicates: ①It is feasible to use the strain sensor embedded in the full-scale pavement to measure the mechanical response of the pavement and then guide the structural design of the asphalt pavement. ②To two types of asphalt pavement structure, the upper layer is subjected to compressive strain, and the structure is relatively safe. However, the two layers may have different levels of fatigue cracking. The structure 1 is located in the lower layer, while the structure 3 is located in the middle layer. ③Semi-rigid materials with the same mechanical properties show different mechanical behaviors when applied to different structural layers, and there are some differences in mechanical properties. ④The measured results show that the graded gravel layer has a certain dispersing effect on the load, and it is a more favorable conclusion for the overall force of the structure.

Due to the differences of climate, geological conditions and traffic volume in actual projects, the two types of semi-rigid base asphalt pavement based on this research can not represent the mechanical response law of all semi-rigid base asphalt pavement. At the same time, because of construction and other reasons, the survival rate of strain sensor can not reach 100%, which leads to the failure of some strain sensor output. All these have brought obstacles to the improvement of the design method of semi-rigid base asphalt pavement. However, with the development of science and technology, the field measurement sensor technology will surely enter the field of pavement design more and more, and will gradually reveal the dynamic response characteristics of the real pavement once considered as "black box".

**Acknowledgments**
The authors appreciate the support of the science and technology project of Guangdong Provincial Communications Department (Project number: Science and technology -2012-02-011).

**References**

[1] Wen Z G, Yang Y H, Wei L X, et al. Research on the Effect of Repair Modulus on Asphalt Pavement Potholes Joint Interface Mechanics Behavior [J]. Highway, 2017.
[2] Sachs S G, Vandenbossche J M, Li Z, et al. Accounting for Temperature Susceptibility of Asphalt Stiffness When Designing Bonded Concrete Overlays of Asphalt Pavements [J]. Journal of Transportation Engineering, 2016, 142(10): 04016040.
[3] Li A, Zhao Q, Wang T, et al. The Research Summarization of Road Simulation for Automobile Performance Test [J]. Agricultural Equipment & Vehicle Engineering, 2016.
[4] Wu S, Chen H, Zhang J, et al. Effects of interlayer bonding conditions between semi-rigid base layer and asphalt layer on mechanical responses of asphalt pavement structure [J]. International Journal of Pavement Research & Technology, 2017, 10(3): 274-281.
[5] Lu Pengmin, Dong Zhonghong. Mechanical Analysis of Vehicle-Asphalt Pavement System[M]. Beijing : People's Communications Publishing House, 2010.
[6] Hu Peng. Response and Failure Analysis of Asphalt Pavement under Dynamic Load[M]. Beijing : People's Communications Publishing House Co., Ltd., 2018.
[7] Wang Jiachen, Li Hao, Wu Chuanhai, etc.. Study on Dynamic Response Measurement System of Long Life Asphalt Pavement in Heavy-Load Area [J]. Chinese and foreign highway, 2016, 36(2): 58-62.
[8] Tan Zhenyu, Li Hao, Xu Xinquan, etc.. Study on Layout Scheme of Mechanical Response Monitoring System for Asphalt Pavement [J]. Journal of Wuhan University of Technology (Transportation Science and Engineering Edition), 2017, 41(3):528-532.
[9] Liu F, Wu C, Xu X, et al. Mechanical Response Analysis of Long-life Asphalt Pavement Structure of Yunluo Highway on the Semi-rigid Base[C]// IOP Conference Series: Earth and Environmental Science. IOP Conference Series: Earth and Environmental Science, 2018:022002.

IOP Conf. Series: Materials Science and Engineering **772** (2020) 012118 doi:10.1088/1757-899X/772/1/012118

# Analysis of Pavement Mechanical Properties under Different Parameters of Asphalt Materials

**Liu Feng[1, 2], Li Hao[1, 2], Liu Gui[1, 2] and Zhou Yong[1, 2]**

[1] Guangdong Hualu Transport Technology Co., LTD, Guangzhou 510420, Guangdong, China
[2] Research and Development Center on Road Transport Safety and Emergency Support Technology & Equipment, Ministry of Transport, PRC, Guangzhou 510420, Guangdong, China

**Abstract.** In view of the differences between the 2006 Chinese version of the asphalt pavement design specification and the 2017 version of the asphalt pavement design specification, taking the road structure of Renxin Expressway in Guangdong Province as an example, the finite element analysis method was used to establish the finite element model of the road structure. The mechanical response of asphalt pavement with different surface materials is analyzed. Then the mechanical response of asphalt pavement to asphalt pavement structure with different thickness is analyzed under the condition that the asphalt pavement remains unchanged in both the base layer and the subgrade. Finally, according to the above analysis results, the problems that need attention in asphalt pavement design are obtained.

**Keywords:** Road Engineering, Material Parameters, Mechanical Properties.

Unlike the full-thickness asphalt pavement structure in developed countries, the development of China's highways was more based on the semi-rigid base of the stable grade and gravel of inorganic binder because of the constraints of high-quality asphalt resources and economic conditions in the early China [1]. The previous research shows that the mechanical response law of the full-thickness asphalt pavement structure is very different from that of the semi-rigid asphalt pavement structure [2]. Through the continuous monitoring of the test section of Yunluo Expressway in Guangdong Province, the research team can see that the asphalt pavement on the semi-rigid base layer designed with strong foundation, thin surface and stable soil foundation can realize the long life of asphalt pavement [3,4]. The road surface condition is shown in Figure 1. However, the monitoring results also show that the dynamic response law of asphalt pavement under vehicle load is different from that of asphalt pavement under static load [5]. Therefore, in order to consider this difference, the 2017 Chinese asphalt pavement design specification changes the static modulus of the material to a dynamic modulus. This is a major step in the design of Chinese asphalt pavement. However, with the change of material parameters, whether the mechanical response law of asphalt pavement structure under the action of the original static parameters will change with the different material parameters is worth studying.

Content from this work may be used under the terms of the Creative Commons Attribution 3.0 licence. Any further distribution of this work must maintain attribution to the author(s) and the title of the work, journal citation and DOI.
Published under licence by IOP Publishing Ltd

**Fig. 1** Current status of Yunluo highway long-life test road in Guangdong Province

## 1. Renxin Highway Overview

Xinbo expressway is one of the sections of the Wushen expressway. The starting point of the project is located in Xinfeng County, Shaoguan City, and is connected to Boshen expressway in Boluo County. From the south to the north, the Guanghui expressway, the Guanghe expressway and the Daguang expressway are connected in series, and the northbound direction of Hunan is connected to the Renxin Expressway through the Daguang expressway. The basic structure of Renxin expressway pavement is shown in Figure 2. In order to test the influence of aggregates with different mechanical properties on the asphalt surface layer and the overall pavement structure, the research team paved four test roads. And the corresponding theoretical analysis was carried out before the paving of the test road. The theoretical analysis part mainly consists of two aspects. Firstly, the overall mechanical response of asphalt pavement under different surface layers is analyzed. Secondly, Based on the basic structure of Renxin expressway, the mechanical response of pavement structure under different asphalt layer thickness is studied under the premise of the same thickness of base and subgrade.

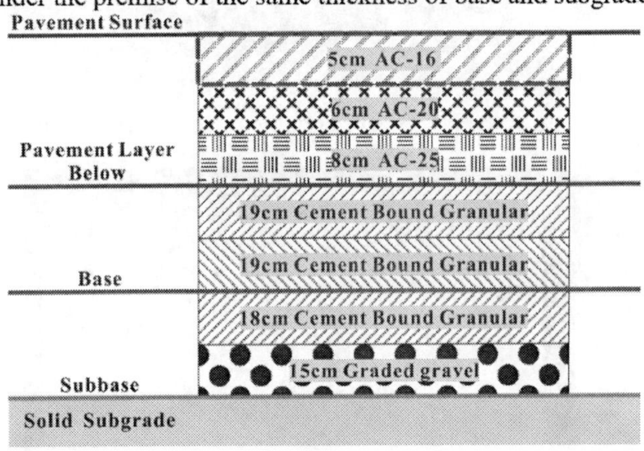


**Fig. 2** Schematic diagram of asphalt pavement structure of Renxin highway

## 2. Finite Element Model of Road Structure

### *2.1. Structure model*

In the mechanical analysis of road structures, the basic assumptions are as follows:

1) The materials of each structural layer are assumed to be linear elastic materials and all conform to the generalized Hooke theorem.

2) The thickness of each structural layer of the road is limited. According to the relevant literature, when the ground structure is subjected to static load and dynamic load analysis, when the thickness of

the foundation is greater than 6m and greater than 12m, the measured results are close to the simulation results and can meet the calculation requirements.

3) Complete contact between the structural layers of the road.

4) The load acts perpendicular to the pavement surface layer, and the stress, strain and displacement of the road structure at the infinite depth of the load are zero.

The asphalt pavement was analyzed using ANSYS finite element analysis software. In the ANSYS analysis, the foundation and each structural layer are blocky. However, the size of the foundation model is larger than the size of each structural layer above the foundation. The specific foundation size is 16m×9m×10.5m (ie. x×y×z, where x is the driving direction; y is the depth direction; z is the cross-sectional direction). The planar dimensions of each structural layer other than the foundation are 10m × 4.5m. Before dividing the grid, the solid subgrade and the structure layers are connected using the GLUE command. Ensure continuous contact between the soil subgrade and each structural layer and each structural layer. SOLID 45 element is used to divide the solid model grid in soil subgrade and each structure layer. And establish the finite element model as shown in Figure 3. The selection of the calculated points is shown in Figure 4.

**Fig. 3** Finite element model of Renxin highway asphalt pavement structure under double circular load

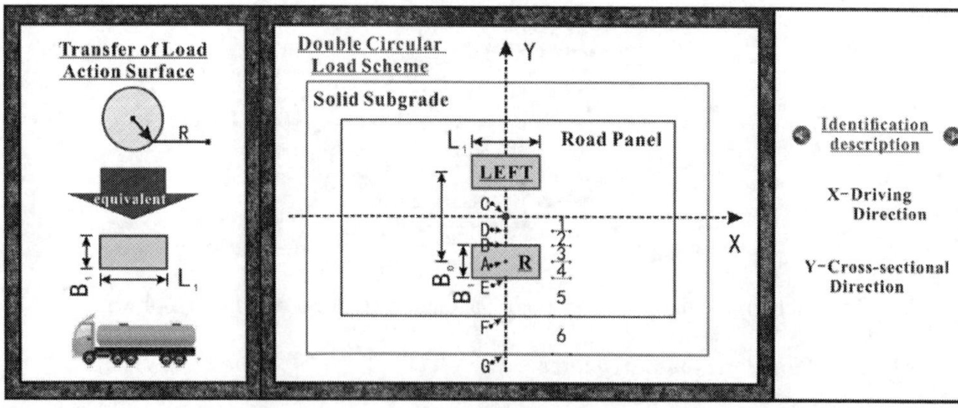


**Fig. 4** Double circular load contact surface parameter setting

## 2.2. *Material Parameters*

Before carrying out the structural structural response analysis of the road, the material parameters and structural parameters of the asphalt pavement structure are determined according to Table 1.

**Table 1.** Material parameters of Renxin expressway for analysis of mechanical properties of asphalt pavement

| NO. | Material Name | Thickness /mm | Elastic Modulus / MPa Version 06[6] | Version 17[7] | Cracking strength /MPa | Poisson's ratio |
|---|---|---|---|---|---|---|
| 1 | GAC-16 | 50 | 1200~1800 | 8000~12000 | 1.2 | 0.25 |
| 2 | GAC-20 | 60 | 1200 | 10000 | 1.0 | |
| 3 | GAC-25 | 80 | 1000 | 9000 | 0.8 | |
| 4 | Cement stabilized macadam | 190+190+180 | 1600 | 9000 | 0.5 | |
| 5 | Graded gravel | 150 | 350 | 350 | -- | 0.35 |
| 6 | Soil base | -- | 70 | 70 | -- | 0.40 |

## 3. Analysis of Law of Road Mechanics Response

### *3.1. Analysis of the Variation of Mechanical Response Components with the Modulus of the Upper Layer*

Because different materials are used in the surface layer in Renxin expressway, this part will analyze the mechanical response law of asphalt pavement structure when the elastic modulus of the upper layer changes under the condition that the base modulus, the middle layer modulus and the lower layer modulus remain unchanged. Analyze according to 06 edition specification and 17 edition specification respectively. Calculation parameters as shown in table 1. The calculation point is taken as point C of Fig. 4. Among them, the results of the deflection calculation are shown in Figure 5. The results of stress and strain response at the bottom of each structural layer are shown in Table 2.

**Table 2.** Analysis results of the mechanical response components of Renxin expressway with the change of the elastic modulus of the upper layer

| Mechanical response parameters | Calculatebasis | Elastic modulus of the upper layer /MPa | Upper Layer | Middle Layer | Lower Layer | Upper base | Upper sub-base |
|---|---|---|---|---|---|---|---|
| Structural bottom strain /με | Version 06 | 1200 | -23.57 | 46.97 | 24.23 | 19.86 | 37.14 |
| | | 1400 | -17.30 | 49.02 | 24.81 | 19.89 | 36.82 |
| | | 1600 | -11.32 | 50.81 | 23.34 | 19.93 | 36.53 |
| | | 1800 | -6.36 | 52.39 | 25.83 | 19.98 | 36.26 |
| | Version 17 | 8000 | -8.68 | 2.02 | 4.13 | 3.82 | 10.49 |
| | | 9000 | -7.89 | 2.32 | 4.23 | 3.83 | 10.40 |
| | | 10000 | -7.18 | 2.60 | 4.33 | 3.85 | 10.32 |
| | | 12000 | -5.92 | 3.10 | 4.51 | 3.88 | 10.17 |
| Structural bottom stress /MPa | Version 06 | 1200 | -0.098 | -0.035 | -0.044 | 0.0050 | 0.0739 |
| | | 1400 | -0.102 | -0.032 | -0.043 | 0.0055 | 0.0733 |
| | | 1600 | -0.105 | -0.029 | -0.042 | 0.0060 | 0.0728 |
| | | 1800 | -0.107 | -0.027 | -0.041 | 0.0064 | 0.0722 |
| | Version 17 | 8000 | -0.1368 | -0.0767 | -0.0324 | 0.0126 | 0.123 |
| | | 9000 | -0.1451 | -0.0729 | -0.0306 | 0.0131 | 0.122 |
| | | 10000 | -0.1525 | -0.0692 | -0.0289 | 0.0136 | 0.121 |
| | | 12000 | -0.1643 | -0.0627 | -0.0257 | 0.0150 | 0.119 |

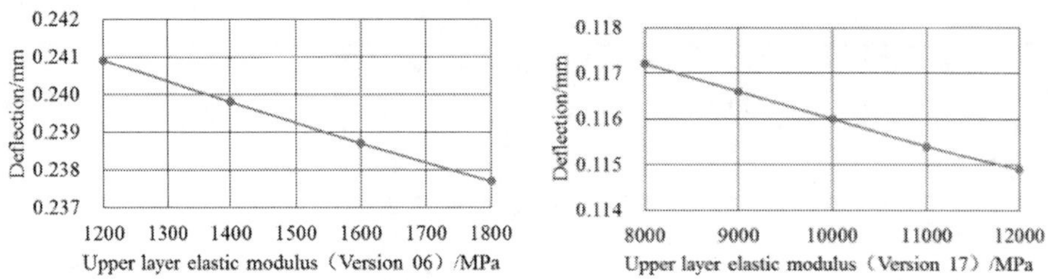


**Fig. 5** Trend diagram of road surface deflection with upper layer elastic modulus

As can be seen from Figure 5 and Table 2. ①As the elastic modulus of the upper layer increases, the deflection value of the road surface shows a downward trend. ②The stress and strain at the bottom of the upper layer are both negative. The bottom layer is subjected to compressive stress and compressive strain. ③The stress of the middle layer and the bottom layer is consistent with the upper layer. But the strain at the bottom of the two layers is in a pull state. The three layers above the graded rock layer are subjected to tensile strain or tensile stress. The stress and strain at the base of the upper layer increase with the increase of the elastic modulus of the upper layer. However, the stress and strain of the lower and upper basement layers decrease slightly with the increase of the elastic modulus of the upper layer. However, the reduction is not large, and the strain value of the upper basement layer is the largest.

### *3.2. Analysis of the Variation of Mechanical Response Components with the Thickness of the Surface Layer*

This section will take the existing pavement structure of Renxin expressway as the benchmark, change the thickness of the surface layer, and investigate the mechanical response laws of the pavement structure after the thickness increases or decreases. The material composition is shown in Table 1. The surface layer structure is divided into the following five types. ①4+5+7(Upper layer + Middle layer + Lower layer). This structure is abbreviated as S1. ②5+6+8(Upper layer + Middle layer + Lower layer, Renxin expressway structure). This structure is abbreviated as S2. ③5+8+8(Upper layer + Middle layer + Lower layer). This structure is abbreviated as S3. The total thickness of the top layer is 21cm. ④ 5+6+8+6(Upper layer + Middle layer + Lower layer + ATB-25). This structure is abbreviated as S4. The total thickness of the top layer is 25cm. ⑤5+6+8+9(Upper layer + Middle layer + Lower layer + ATB-25). This structure is abbreviated as S5. The total thickness of the top layer is 28cm.

**Table 3.** Analysis Results of the Mechanical Response Components of Renxin Expressway with the Variation of the Thickness of the Surface Layer

| Mechanical response parameters | structure type | Upper Layer | Middle Layer | Lower Layer | ATB-25 | Upper base | Upper sub-base |
|---|---|---|---|---|---|---|---|
| Structural bottom strain /με | S1 | -39.62 | 27.78 | 26.84 | -- | 21.82 | 38.59 |
| | S2 | -20.04 | 38.86 | 27.82 | -- | 20.23 | 36.03 |
| | S3 | -20.32 | 43.57 | 28.39 | -- | 19.24 | 34.46 |
| | S4 | -19.00 | 42.57 | 40.91 | 25.03 | 17.39 | 31.74 |
| | S5 | -18.63 | 43.53 | 43.39 | 23.04 | 16.18 | 29.88 |
| Structural bottom stress /MPa | S1 | -0.1310 | -0.0544 | -0.0418 | -- | 0.0059 | 0.0767 |
| | S2 | -0.1105 | -0.0357 | -0.0332 | -- | 0.0073 | 0.0718 |
| | S3 | -0.1103 | -0.0237 | -0.0287 | -- | 0.0081 | 0.0688 |
| | S4 | -0.1077 | -0.0290 | -0.0173 | -0.0225 | 0.0089 | 0.0635 |
| | S5 | -0.1067 | -0.0271 | -0.0136 | -0.0187 | 0.0094 | 0.0598 |

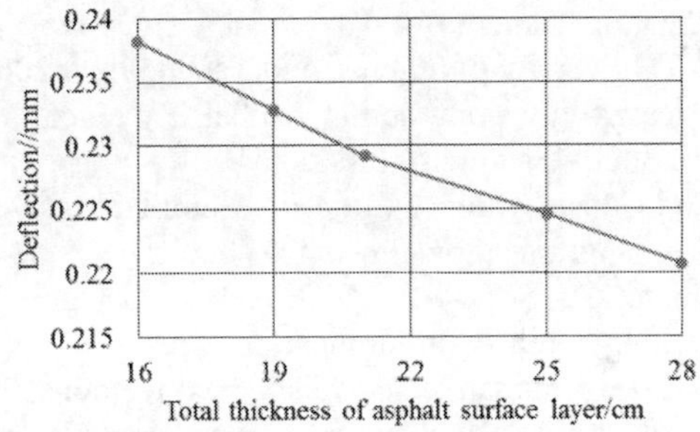


**Fig. 6** Trend diagram of total thickness variation of road surface deflection

It can be seen from Table 3 and Figure 6. ①The pavement structure with different surface thickness shows a decreasing trend with the increase of surface thickness. ②The surface layer stress of the five asphalt pavement structures (including the ATB-25 structural layer, the same below) are compressive stresses, which have little effect on the structure. It can be ignored. ③In addition to the upper layer, the bottom strain of other asphalt-containing structural layers is tensile strain. However, the pull strain of the bottom layer increases with the increase of the total thickness of the road surface layer. ④The strain law of the upper base, lower base and upper base layer is more obvious. Both decrease with the increase of the total thickness of the surface layer. It can be seen that with the increase of the total thickness of the surface layer, the asphalt layer bears more loads. The load of the inorganic composite material layer has been reduced. This is also more consistent with the concept of durable asphalt pavement. ⑤At the same time, by analyzing the bottom layer stress data of the base layer and the sub-base layer in the table, it can be known that the stress at the bottom of the lower base layer and the upper base layer decreases as the total thickness of the asphalt pavement structure increases. However, the case of the upper base layer is opposite to that of the upper base layer, indicating that as the total thickness of the surface layer increases, the upper base layer will be a weak layer of the entire structural layer, which may crack under the load before the other layers. In short, the increase in the thickness of the asphalt surface layer is beneficial to the force of the inorganic composite material layer. But it will also have a negative impact on the asphalt layer. However, from the point of view of easy maintenance of the entire pavement structure, it is advantageous to increase the thickness of the asphalt layer to the overall force of the pavement.

## 4. Conclusion

According to the results of the analysis above, whether the asphalt pavement is designed with static parameters or the asphalt pavement is designed with dynamic parameters, the road surface bending of asphalt pavement is consistent with the stress and strain laws of each structure. However, there are certain differences in the calculation values. In the structure of three layers of asphalt, the middle layer is most unfavorable, and its strain value is about 1 times higher than the strain value of the lower layer. The inorganic binder layer is the most unfavorable for the upper base layer. It is subjected to tensile stress and tensile strain and has a large value. The above descriptions show that in the asphalt pavement design, in the three-layer asphalt pavement structure on the semi-rigid base layer, the middle surface and the upper base are the most unfavorable layers. The corresponding mechanical indexes should be controlled, otherwise it is easy to shorten the overall life of the asphalt pavement structure.

**Acknowledgments**
The authors appreciate the support of the science and technology project of Guangdong Provincial Communications Department (Project number: Science and technology -2012-02-011).

IOP Conf. Series: Materials Science and Engineering **772** (2020) 012118 doi:10.1088/1757-899X/772/1/012118

**References**

[1] Sha Qinglin. The Premature Damage of Pavement Expressway and the Countermeasure [J]. Journal of Changsha University of Science and Technology, 2006, 3(3):1-5.
[2] Xie Jin. Research on Heavy-duty Long-acting Asphalt Pavement Material and Structure Design [M]. Hunan: Central South University Press, 2019.
[3] Sha Qinglin. Design and Construction of Long-life Semi-rigid Pavement for Heavy-duty Traffic [M]. Beijing: China Communications Press, 2011.
[4] Wei Jinyu, Wang Lin, et al. Semi-rigid Base Layer Damage Evaluation and Engineering Application [M]. Beijing: China Communications Press, 2019.
[5] Liu Feng, Wu Chuanhai, Xu Xinquan, et al. Mechanical Response Analysis of Long-life Asphalt Pavement Structure of Yunluo Highway on the Semi-rigid Base[C]// IOP Conference Series: Earth and Environmental Science. IOP Conference Series: Earth and Environmental Science, 2018:022002.
[6] JTG D50-2006 Specification for Design of Highway Asphalt Pavement[S]. Beijing: China Communications Press, 2006.
[7] JTG D50-2017 Specification for Design of Highway Asphalt Pavement[S]. Beijing: China Communications Press, 2017.

IOP Conf. Series: Materials Science and Engineering **772** (2020) 012119 doi:10.1088/1757-899X/772/1/012119

# Modeling of "Concentric Drum" Tilt Angle Model Based on Mechanical Analysis

**Zihao Yang[1, a], Shiwei Zhang[2, *] and Shanshan Cheng[2, b]**

[1]School of Information and Control Engineering, Xi'an University of Architecture and Technology, Xi'an 71005, China;
[2]School of Metallurgical Engineering, Xi'an University of Architecture and Technology, Xi'an 710055, China;

*Corresponding author e-mail: 2234789895@qq.com, [a]1105453436@qq.com, [b]1961265970@qq.com

**Abstract.** " Concentric drums" is a group cooperation project, which requires better tacit understanding between the team members. At the same time, the force exerted on the rope can be accurately controlled, which contains a more complex mechanical relationship. In reality, it is impossible for the players to make force synchronously and the force is difficult to control accurately, which can be attributed to the following two situations: first, there is no situation in which the team has no force in advance but the size is not uniform when the force is applied, so the drum surface is at the level position at the moment when the players start to make force; second, there are individual players who make force in advance, and the strength of each player is different, resulting in the tilted of the drum surface when the players are exerting force. In the first case, the vertical component forces of two opposite members in the same straight line are combined into one force, then the moment of each resultant force is calculated and the vector sum is performed, the angular acceleration is solved according to the rigid body mechanics formula, and finally the tilt angle is calculated by using the time brought in by double integration. In the second case, firstly, consider that some members of the team will start the force ahead of time, and those members who do not start the force will be brought into the model of case I according to the force value of 0 to obtain tilt angle 1, Secondly, the force of the remaining team members is considered. At this time, based on the inclined plane, the force of each member after treatment is brought into the model of case I again to get the inclined angle tilt angle 2 caused by different forces. After two inclinations, three planes with different angles can be obtained. The angle between the final plane and the original plane can be calculated by using the space coordinate normal vector, that is, the bevel angle of horizontal inclination of the drum surface.

**Key words:** Force analysis; Space vector; Elastic collision.

Content from this work may be used under the terms of the Creative Commons Attribution 3.0 licence. Any further distribution of this work must maintain attribution to the author(s) and the title of the work, journal citation and DOI.
Published under licence by IOP Publishing Ltd

## 1. Introduction

In this problem, the diameter of the drum surface of the concentric drum d = 40 cm, the height of the drum body n = 22 cm, the mass of the drum $m_1$ = 3.6kg, and the mass of the volleyball used $m_2$ = 0.27kg. The number of members pulling the rope around the drum body n ≥ 8, and the minimum distance between members P ≥ 60cm. The drum surface is horizontal at the initial time, and its initial position is 11 cm lower than that of the rope at the horizontal time. Under actual circumstances, because the players can not precisely control the timing and strength of the force, the drum surface may be tilted, and a mathematical model is established to describe the relationship between the timing and strength of the player's force and the angle of the drum surface at a particular moment. According to the different timing and strength of the players, the inclination angle of the drum surface is obtained at 0.1 s.

## 2. Construction of tilt angle model based on mechanical analysis

### 2.1. Problem analysis

In the first case, the spatial rectangular coordinate system I is established, and the pulling force $f_n$ n of each rope is decomposed to the vertical direction, and the z-direction component $f_n \sin\theta_n$ of each rope pulling force is obtained. The symmetrical pulling force in the horizontal direction is synthesized in the z-direction component, and four pulling force components along the z-direction are obtained, representing the moment $\sum M_n$ and the moment direction of each group of forces; establish the rectangular coordinate system in the horizontal direction II. Four sets of moments are decomposed into x-axis and Y-axis of coordinate system II, and the moment components on each coordinate axis are synthesized into $\sum M_X$ and $\sum M_Y$ to obtain the total resultant moment $\sum M_{total}$. The total moment causes the drum to rotate. The drum is simplified as a thin-walled drum. The expression of the drum rotation angular acceleration α is listed. The angular acceleration α is integrated to get the drum rotation angle δ. The program 2 is written with MATLAB software to solve the drum tilt angle. In the second case, when n (n ≤ 8) team members exert force $f_n$ in advance time of t s, the drum inclines at an angle of $\beta_1$. The force is decomposed to the direction perpendicular to the drum surface, and the moment is calculated. According to the moment, the angle $\beta_1$ of n team members exert force in advance of t s, and the angle between each tension and the horizontal direction is $\theta_n$. according to the solution method in the first case, after the force is applied, the angle $\beta$ of the drum caused by all the pulling forces after 8-n teams are exerted can be calculated by the improved program 2. Using the above two procedures, we can get the tilt angle of the drum surface at 0.1s under different starting time and force of Table 1.

**Table 1.** symbol description

| symbol | description |
|---|---|
| $f_n$ | Tension of each rope |
| $\theta_0$ | Angle between rope and horizontal plane when drum is suspension still |
| $\theta_n$ | Angle between tension of each rope and horizontal direction, |
| n | Number of players pulling ropes around drum |
| $m_1$ | Drum quality |
| $m_2$ | The quality of Volleyball |

### 2.2. Establishment and solution of the model

It is specified that the number of members who pull the rope around the drum body in an ideal state is n = 8, and the force f of each person to control the rope is the same in size and direction, and the drum surface remains horizontal, as shown in Figure 1. The problem can be considered as one-dimensional if only the ball is moving vertically.

The multiple reciprocation of the ball is regarded as multiple motion periods, and I, II, III and IV are recorded as one motion period, which is divided into four stages: static stage I before the movement of drum and ball, motion stage II before the collision, collision stage III and motion stage IV after the collision, as shown in Figure 1.

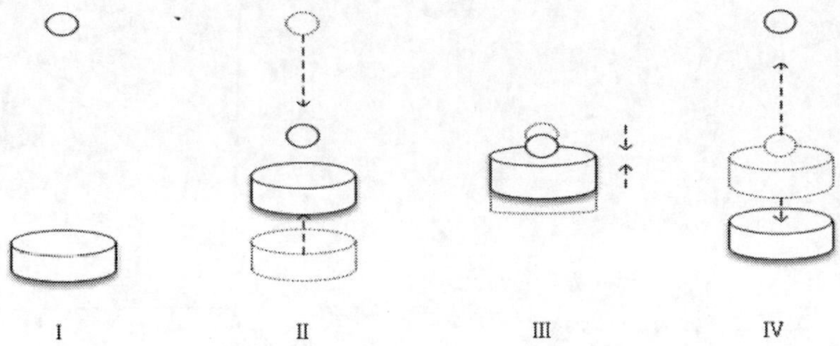


**Fig. 1** Four motion stages I, II, III and IV in a cycle

Static stage I: The conduct force analysis on the state of drum in static stage I before drum and ball movement is as shown in Fig. 2 and Fig. 3. The number of players is n = 8, the distance between players is p = 60cm, and the horizontal projection of each rope length is l (make $\theta_0 = 45°$):

$$l = \frac{60}{2\sin(\pi/4)} = 78.4$$

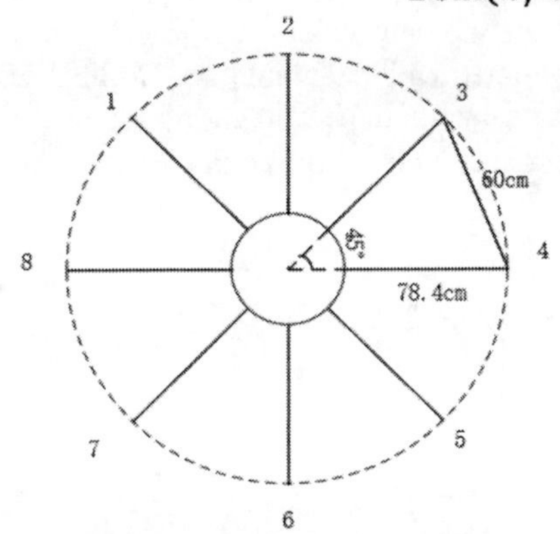


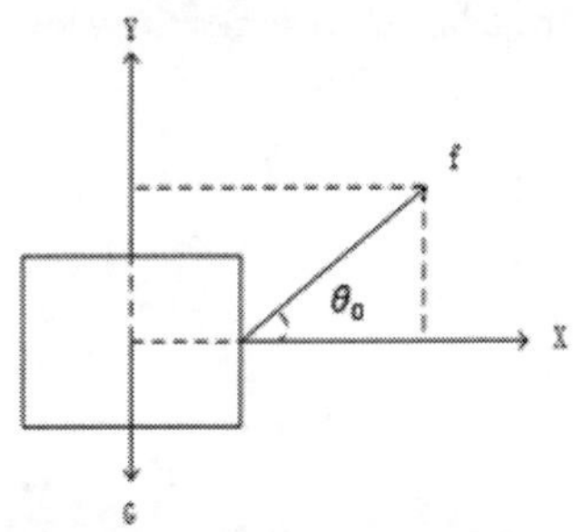


**Fig. 2** Top view of drum surface in ideal state **Fig. 3** Lateral force analysis of drum surface

For the first type of situation, the drum is horizontal at the beginning, but after the team members exert force, due to the different magnitude and direction of the force, the drum surface inclines, and a space rectangular coordinate system I is established. The angle of each team member's pulling rope and the applied force are shown in Figure 4, that is, $\varphi=(\theta_n, f_n)$. The specific conditions of 8 team members are as follows: $\varphi_1=(\theta_1, f_1)$, $\varphi_2=(\theta_2, f_2)$,…,$\varphi_8=(\theta_8, f_8)$

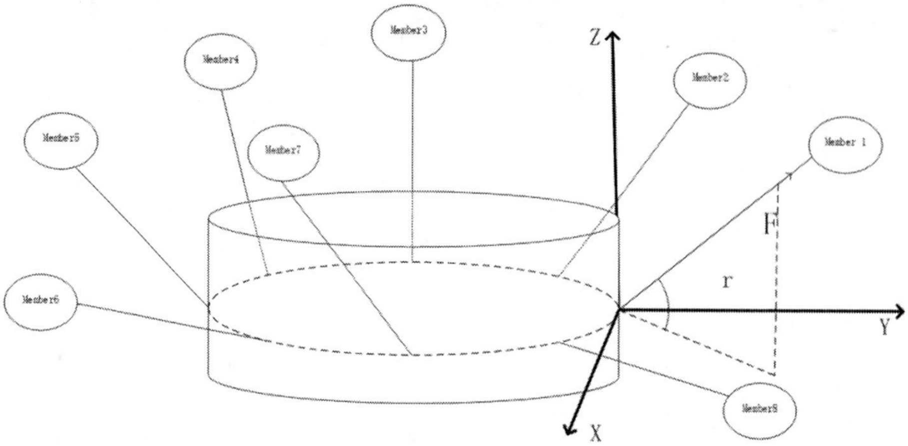


**Fig. 4** Schematic diagram of force on drum surface of category I

Based on the space rectangular coordinate system I, the pull force $f_n$of each rope is decomposed to the z-axis direction, and the component $f_n \sin\theta_n$of each rope pull force along the z-axis direction is obtained. Since the force is vector, there are size and direction, following the principle of vector addition, the pull force components in the z-direction at the moment when each member sends the force are synthesized. At the beginning, 8 members are symmetrically standing, i.e. No.1, No.5 and No.2. And No. 6, No. 3, No. 7, No. 4 and No. 8 stand in the same diameter direction respectively, and the angle between the adjacent two members and the line connecting the drum center is 45° . As shown in Figure 5.

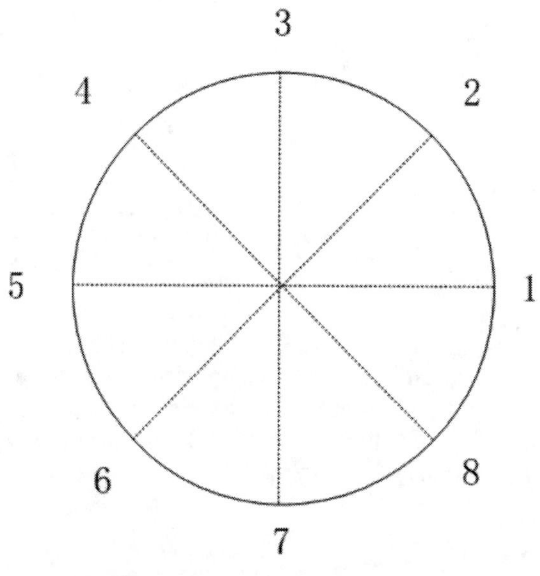


**Fig. 5** Symmetrical arrangement of players

The synthesis result is

$$f_a=f_1 \sin\theta_1-f_5 \sin\theta_5;$$
$$f_b=f_2 \sin\theta_2-f_6 \sin\theta_6;$$
$$f_c=f_3 \sin\theta_3-f_7 \sin\theta_7 ;$$
$$f_d=f_4 \sin\theta_4-f_8 \sin\theta_8.$$

IOP Conf. Series: Materials Science and Engineering **772** (2020) 012119 doi:10.1088/1757-899X/772/1/012119

For the moment expressions of columns $f_a, f_b, f_c, f_d$ , the moment composition is as follows:

$$\sum M_1 = \frac{d}{2} f_1 \sin\theta_1 - f_5 \sin\theta_5 \tag{1}$$

$$\sum M_2 = \frac{d}{2} f_2 \sin\theta_2 - f_6 \sin\theta_6 \tag{2}$$

$$\sum M_3 = \frac{d}{2} f_3 \sin\theta_3 - f_7 \sin\theta_7 \tag{3}$$

$$\sum M_4 = \frac{d}{2} f_4 \sin\theta_4 - f_8 \sin\theta_8 \tag{4}$$

The moment $\sum M_n$ and the moment direction of each group of forces are shown in Figure 6. When the moment value is negative, the moment direction is opposite to that in the figure.

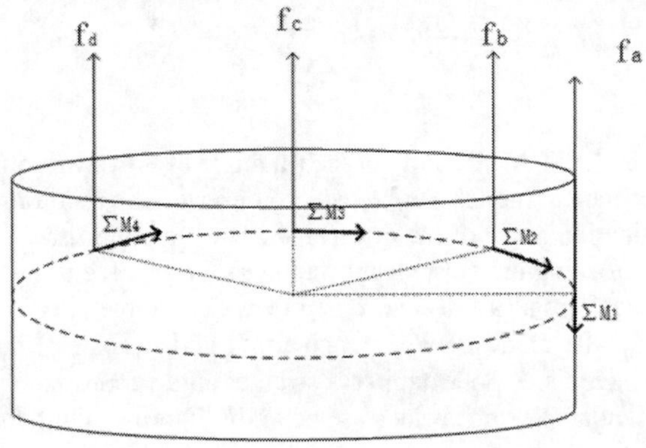


**Fig. 6** Force and moment diagram of each group of forces

Establish the rectangular coordinate system II in the horizontal direction, and decompose the four groups of moments into the X and Y axes of coordinate system II. As shown in Figure 7, the moment components on each coordinate axis are synthesized into $\sum M_X$ and $\sum M_Y$ to obtain the total resultant moment $\sum M_{total}$.

$$\sum M_{total} = \sqrt{\sum M_X{}^2 + \sum M_Y{}^2} \tag{5}$$

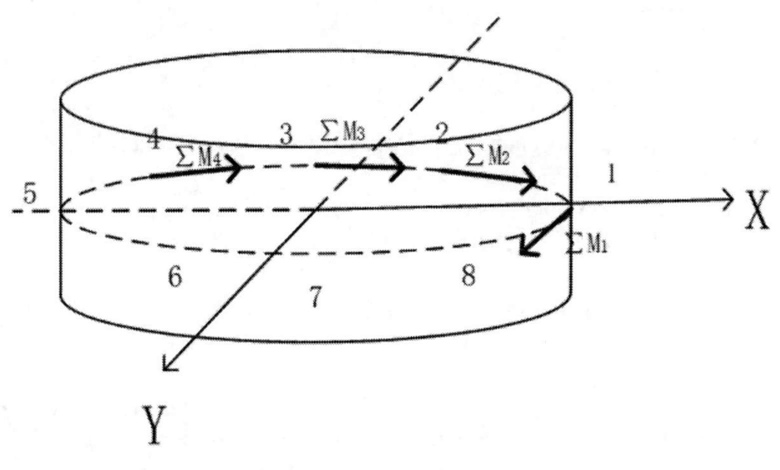


**Fig.7** Diagram of drum moment decomposition

The total moment $\sum M_{total}$ always causes the drum to rotate. The drum is simplified as a thin-walled drum. The expressions of the angular acceleration α, the angular velocity ω and the rotation angle δ of the drum are listed.

$$\sum M_{total} = J_z\alpha \tag{6}$$

$$J_z = \frac{m_1}{12}[3(r_1{}^2 + r_2{}^2) + l^2] \tag{7}$$

α is the angular acceleration, $J_z$ is the moment of inertia around any axis parallel to the drum surface, $m_1$ is the mass of the drum, $r_1$ is the outer diameter of the drum, $r_2$ is the inner diameter of the drum, and l is the height of the drum.

$$\frac{dw}{dt} = \alpha \tag{8}$$

$$\frac{d\delta}{dt} = \omega \tag{9}$$

Write a program with MATLAB to find out the tilt angle of the drum caused by the different strength of the players. This mechanical model can be used to describe the relationship between the strength of the players and the inclination angle of the drum surface at a certain time.

For the second kind of situation, some team members exert force ahead of time. Different from the first kind of situation, there are n (n < 8) team members exert force ahead of time t s, the force of each person pulling rope is $f_n$, the angle between each force and horizontal direction is $\theta_n$, the angle of inclination of drum advance t s is $\beta_1$, establish a straight angle coordinate system, decompose n forces to the direction perpendicular to drum surface as shown in Figure 8, along the vertical direction of drum. The component is $f_n \sin\theta_n \cos\beta_1$.

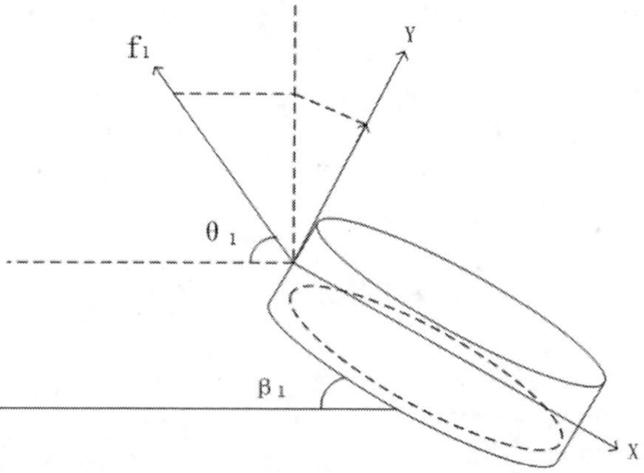


**Fig. 8** Force analysis diagram of drum after its advance force ahead of time t s

Take n forces as the resultant force in the vertical direction $f_a = f_1 \sin\theta_1 - f_5 \sin\theta_5$; $f_b = f_2 \sin\theta_2 - f_6 \sin\theta_6$; $f_c = f_3 \sin\theta_3 - f_7 \sin\theta_7$; $f_d = f_4 \sin\theta_4 - f_8 \sin\theta_8$. For those who do not exert force in advance, $f = 0$ can be taken to calculate the moment $\sum M_1, \sum M_2$ $\sum M_3, \sum M_4$ and the total moment $\sum M_{total}$, According to equation (3) (4) (5), the angle $\beta_1 = \int_0^t w\,dt$ of N team members' inclining force drum in advance time of t s is obtained, and $\beta_1$ can be obtained by MATLAB programming.

After ts, the other 8-n team members exert force. At this time, the direction of the force of n individuals who exert force in advance is determined. The angle between each pulling force and the horizontal direction is $\theta_n$,the 8-n team members exert force, the component force of each member perpendicular to plane 2 can be calculated by the formula $f_n \sin\theta_n \cos\beta_1$, use the first type of model again to calculate the inclination angle $\beta_2$of drum surface relative to plane 2, as shown in Figure 9. Since the two inclination angles are not in the same plane, it is not easy to take the result of adding and subtracting the two angles as the final inclination angle. To solve this problem, the coordinate system is established with plane 2 as the middle plane, and the angle between plane 3 and plane 1 (that is, the horizontal plane) is calculated by using the conclusion that the angle between plane normal vectors is the angle between planes, that is, the total tilt angle. The normal vector of plane 2 can be easily known as (0.0.1); the normal vector of plane 1 can be calculated by $(-\cos(\frac{\pi}{2}-\beta_1), 0, \sin(\frac{\pi}{2}-\beta_1))$; for the normal vector of plane 3, the included angle $\theta$ of the inclined direction of plane 2 and plane 3 should be introduced, and the angle can be calculated by the moment coordinate of the drum surface at the second inclination. In conclusion, the normal vector of plane 3 is$(-\cos(\theta), \sin(\theta), tan\left(\frac{\pi}{2}-\beta_2\right))$, and finally use the normal vector to find the dihedral angle formula.

$$cos(\vec{n_1}, \vec{n}_2)$$

The tilt angle of the last drum can be calculated by MATLAB.

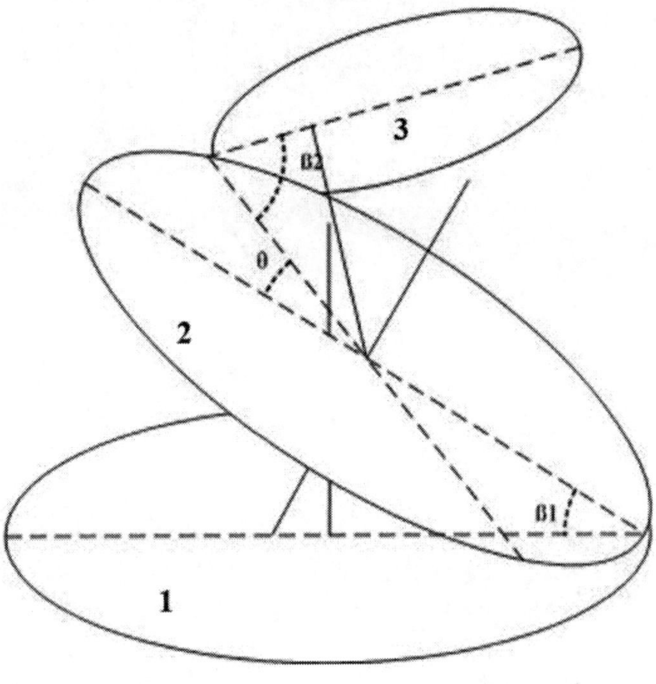


**Fig. 9** Rotation of drum surface when all members apply force after ts advance

## 3. Conclusion

After the above analysis and modeling, by calling the model, it can be found that if the number of players is 8 people, the 2nd and 5th players will advance 0.1 second in advance, and the 1st and 4th players will have a force of 90N, while the strength of other players is 80N,so the angle of the drum surface is 0.0589281 degrees. Similarly, by giving different initial values, the inclination angle of the drum surface in other cases can also be obtained.

## References

[1] Yong Longquan. A class of high order Newton iterative method and its application in linear complementarity problems [J]. Practice and understanding of mathematics, 2019,49 (14): 160-167.

[2] Yang Haitao, Li Kehua, Hu Hangyu, Wu Junyi. Advanced mathematics [M], volume I. Tongji

University Press, 2018
[3] Liu Junqing. Theoretical mechanics [M]. Metallurgical Industry Press, 2008
[4] Zhang Zhiyong, Yang Zuying. Matlab course [M]. Beijing University of Aeronautics and Astronautics Press, 2015
[5] Si Shoukui, sun Zhaoliang. Mathematical modeling algorithm and application [M]. National Defense Industry Press, 2016
[6] Jiang Qiyuan, Xie Jinxing, ye Jun. mathematical model (Fourth Edition) [M]. Beijing: Higher Education Press, 2011.

# AUTHOR INDEX